The 7th International Electronic Conference on Water Sciences

The 7th International Electronic Conference on Water Sciences

Editor

Athanasios Loukas

MDPI • Basel • Beijing • Wuhan • Barcelona • Belgrade • Manchester • Tokyo • Cluj • Tianjin

Editor
Athanasios Loukas
Aristotle University of
Thessaloniki
Greece

Editorial Office
MDPI
St. Alban-Anlage 66
4052 Basel, Switzerland

This is a reprint of Proceedings published online in the open access journal *Environmental Sciences Proceedings* (ISSN 2673-4931) (available at: https://www.mdpi.com/2673-4931/25/1).

For citation purposes, cite each article independently as indicated on the article page online and as indicated below:

LastName, A.A.; LastName, B.B.; LastName, C.C. Article Title. *Journal Name* **Year**, *Volume Number*, Page Range.

ISBN 978-3-0365-7950-4 (Hbk)
ISBN 978-3-0365-7951-1 (PDF)

Contents

About the Editor

Athanasios Loukas

Athanasios Loukas (Prof. Dr.) is a Full Professor of Engineering Hydrology–Water Resources Management and Development in the School of Rural and Surveying Engineering, Aristotle University of Thessaloniki, Greece. He is the Director of the Master Program "Water Resources" (2018–) and the Director of Hydraulic Works and Environmental Management Lab (2023–). He served as the Director of the Transportation and Hydraulic Engineering Dept. (2019–2021). He is a visiting professor at Colorado State University (U.S.A.) and the University of Grenoble Alpes. He has received many academic awards, such as the Fulbright Research Scholarship.

He is Section Editor-in-Chief of the MDPI journal Water (Section of Water Resources Management, Policy and Governance) and Editor-in-Chief of the EWRA European Water journal and an Editorial Board Member of Water Resources Management and Scientific Review Engineering and Environmental Sciences journals. He has been an active reviewer for many international journals.

His research interests include: hydrological modelling, study and forecasting of extreme hydrological events (floods and droughts); climate change impacts on hydrology and water resources; water resources' management and modelling; and GIS and remote sensing in hydrology and water resources. He has co-ordinated and participated in many national and European research projects. He is a reviewer and evaluator of National, International and European research proposals. He has supervised and supervises undergraduate and Master's theses and Ph.D. dissertations. He has delivered many keynote and invited talks. Prof. Loukas is the author and co-author of more than 400 international journal publications, conference proceedings publications and technical reports. His published work has received over 2500 citations, as indicated in the Scopus database (h-index=29), and over 4700 citations, as indicated in the Google Scholar database (h-index = 37, i10-index = 74).

Preface to "The 7th International Electronic Conference on Water Sciences"

This book is the proceedings of the 7th International Electronic Conference on Water Sciences (ECWS-7), held online 15–30 March 2003, promoted by the open access MDPI journal *Water* (ISSN 2073-4441; IF 3.530). The title of ECWS-7 was *"Adaptive Water Resources Management in an Era of Changing Climatic, Environmental and Social Conditions"*. This book of proceedings includes 102 papers, which were internationally and independently reviewed and presented in the conference.

The ECWS-7 conference was focused on adaptive water resources management, policies and governance to tackle the challenges brought about by the ongoing and dramatic climatic, environmental and societal changes. The ECWS-7 conference was a forum to explore and discuss the following key issues: adaptive water resources management; enhancement of water security for humans, their economic activities and the environment; and the formulation of new governance structures for water resources management and development. The book of proceedings intends to assist in the formulation of innovative methodologies and produce recommendations for best practices and the building of institutions that can meet these recommendations.

The ECWS-7 proceedings include papers from researchers from academia, as well as water practitioners from around the globe, contributing original findings, novel ideas, scientific concepts and new technologies and experiences to deal with water resources management, policies and governance, with a special focus on sustainability, resilience, uncertainty, risk and adaptation, making reference to the following topics: integrated water resources management, policies and governance; systems modelling of global change to inform future development paths; hydrological response under climate variability and change; hydro-economic modeling; informing water policy through quantitative models; finding a balance between water availability and environmental protection; policies to enable climate change adaptation and the contribution of stakeholders to adaptive water resources management; instruments to assess environmental impacts and climate change effects (footprints, nexus approach, impact and risk assessment, life cycle assessment, etc.); sustainable management of agricultural water and land resources; the urban water cycle and its adaptation to climate change; advanced technologies to facilitate industrial and municipal wastewater reuse; hydrological extremes under a changing climate; and water disasters as well as approaches to understanding vulnerable landscapes.

I want to express my personal thanks, as ECWS-7 Conference Chair, to the International Conference Committee members for their cooperation in the successful organization of the conference. I also want to thank and highly acknowledge the help and cooperation of the people of the Conference Secretariat (Ms. Shanika Zhu, Ms. Fionna Fu, Ms. Veakie Jia, Mr. Russell Wang) and the technical personnel of MDPI. Finally, I thank the large number of international independent reviewers who reviewed the papers included in the book of proceedings and helped to improve the quality of the papers with their comments.

Athanasios Loukas
Editor

environmental sciences proceedings

MDPI

Editorial

Statement of Peer Review †

Athanasios Loukas

Department of Rural and Surveying Engineering, Aristotle University of Thessaloniki, 54124 Thessaloniki, Greece; agloukas@topo.auth.gr

† Presented at the 7th International Electronic Conference on Water Sciences, 15–30 March 2023; Available online: https://ecws-7.sciforum.net/.

In submitting conference proceedings to *Environment Sciences Proceedings*, the volume editors of the proceedings certify to the publisher that all papers published in this volume have been subjected to peer review administered by the volume editors. Reviews were conducted by expert referees to the professional and scientific standards expected of a proceedings journal.

- Type of peer review: single-blind
- Conference submission management system: Sciforum
- Number of submissions sent for review: 188
- Number of submissions accepted: 110
- Number of submissions published: 102
- Acceptance rate (number of submissions accepted/number of submissions received): 58.5%
- Average number of reviews per paper: 1
- Total number of reviewers involved: 79
- Any additional information on the review process: N/A

Disclaimer/Publisher's Note: The statements, opinions and data contained in all publications are solely those of the individual author(s) and contributor(s) and not of MDPI and/or the editor(s). MDPI and/or the editor(s) disclaim responsibility for any injury to people or property resulting from any ideas, methods, instructions or products referred to in the content.

check for updates

Citation: Loukas, A. Statement of Peer Review. *Environ. Sci. Proc.* **2023**, *25*, 103. https://doi.org/10.3390/environsciproc2023025103

Published: 1 June 2023

environmental sciences proceedings

MDPI

Proceeding Paper

Rainfall-Runoff Modeling Using Artificial Neural Network—A Case Study of Purna Sub-Catchment of Upper Tapi Basin, India †

Usman Mohseni * and Sai Bargav Muskula

Department of Civil Engineering, IIT Roorkee, Roorkee 247667, India; muskula_sbr@ce.iitr.ac.in
* Correspondence: mohseni_ua@ce.iitr.ac.in

† Presented at the 7th International Electronic Conference on Water Sciences, 15–30 March 2023; Available online: https://ecws-7.sciforum.net/.

Abstract: The present study examines the rainfall-runoff-based model development by using artificial neural networks (ANNs) models in the Yerli sub-catchment of the upper Tapi basin for a period of 36 years, i.e., from 1981 to 2016. The created ANN models were capable of establishing the correlation between input and output data sets. The rainfall and runoff models that were built have been calibrated and validated. For predicting runoff, Feed-Forward Back Propagation Neural Network (FFBPNN) and Cascade Forward Back Propagation Neural Network (CFBPNN) models are used. To evaluate the efficacy of the model, various measures such as mean square error (MSE), root mean square error (RMSE), and coefficient of correlation (R) are employed. With MSE, RMSE, and R values of 0.4982, 0.7056, and 0.96213, respectively, FFBPNN outperforms two networks with model architectures of 6-4-1 and Transig transfer function. Additionally, in this study, the Levenberg–Marquardt (LM), Bayesian Regularization (BR), and Conjugate Gradient Scaled (CGS) algorithms are used to train the ANN rainfall-runoff models. The results show that LM creates the most accurate model. It performs better than BR and CGS. The best model is the LM-trained method with 30 neurons, which has MSE values of 0.7279, RMSE values of 0.8531, and R values of 0.95057. It is concluded that the constructed neural network model was capable of quite accurately predicting runoff for the Yerli sub-catchment.

Keywords: artificial neural network; rainfall-runoff modeling; feed-forward back propagation; cascade forward back propagation

Citation: Mohseni, U.; Muskula, S.B. Rainfall-Runoff Modeling Using Artificial Neural Network—A Case Study of Purna Sub-Catchment of Upper Tapi Basin, India. *Environ. Sci. Proc.* **2023**, *25*, 1. https://doi.org/10.3390/ECWS-7-14232

Academic Editor: Athanasios Loukas

Published: 16 March 2023

1. Introduction

Hydrologists have been attempting to understand the translation of rainfall to runoff for many years to estimate streamflow for objectives including water supply, flood control, irrigation, drainage, water quality, power production, recreation, and fish and wildlife propagation [1]. Rainfall-runoff modeling is one of the most prominent hydrological models used to examine the relationship between rainfall and runoff generated by various watershed physical factors [2]. In the real world, all physical catchment features influence rainfall-runoff; hence, generalizing all physical catchment characteristics is a difficult process. It is difficult to depict such a large range of values in a lumped hydrological model since the parameter values must be averaged for each watershed [3].

In the past, academics and hydrologists have presented different ways for effectively forecasting runoff by building several models of rainfall-runoff (RR) [4]. The process of rainfall-runoff is highly nonlinear and incredibly complex and is still poorly understood [5]. Furthermore, several rainfall-runoff models require a substantial amount of data, which are used for calibration and validation time scale, making them computationally intensive and, thus, unpopular [6]. Machine learning techniques are becoming more prevalent due to their ease of use, simplicity, and efficiency [7]. Machine learning techniques are a good option when there are minimal data and the process is complex [8]. In the context of estimating issues, artificial neural networks (ANNs) are a subclass of machine learning that have received significant attention [9]. ANNs are data-processing systems that mimic the capabilities of the human brain [10]. ANNs were first developed in the 1940s and come in a wide variety [11]. ANN models are also known as black-box models [12]. The application of ANNs in the creation of models results in trustworthy and versatile learning ability, which makes ANNs promising for forecasting [13]. ANN models have been extremely prevalent in the domains of hydrology, water resources, and watershed management in recent decades [14]. The ANN contains three layers, an input layer, a hidden layer, and an output layer [15]. The weight of communication is the relationship between the neurons in the consecutive layers [16]. In the given study, the input layer consists of six types of data, namely (rainfall, minimum temperature, maximum temperature, surface pressure, specific humidity, and wind speed). The hidden layer consists of layers with two different sets of number of neurons 10 and 20, respectively. The output layer comprises predicted runoff.

The objectives of the present study are as follows: (i) to develop a rainfall-runoff model for Upper Tapi using an Artificial Neural Networks Technique, (ii) compare ANN rainfall-runoff models developed using NNTOOL with different neural network types, i.e., FFBPNN and CFBPNN, and (iii) to compare ANN rainfall-runoff models trained using LM, BR, and SCG algorithms.

2. Materials and Methods

2.1. Study Area and Data Collection

The current study area comprises a portion of the Upper Tapi Basin known as the Purna sub-catchment (Figure 1). The area lies between Maharashtra and Madhya Pradesh, between latitudes of 20°09′ N to 22°03′ N and longitudes of 75°56′ E to 78°17′ E. It has subtropical to temperate climatic conditions. The mean annual precipitation in the chosen area varies from 833 to 990 mm. Table 1 reveals the sources of data for this study.

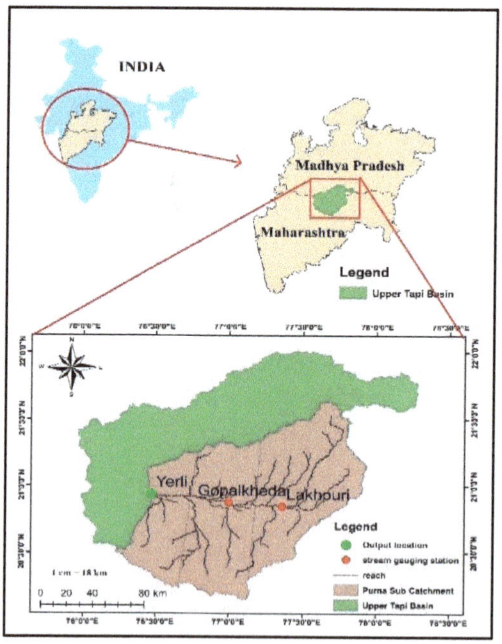

Figure 1. Index Map of the study area.

Table 1. Source of data.

Data Type	Data Source
Digital Elevation Model	USGS Earth Explorer
Rainfall	Central Water Commission
Meteorological data	India Meteorological Department
Discharge	Central Water Commission

2.2. Methodology

The theoretical aspects and research methodology used in the current study to identify the best neural network model to perform the rainfall and runoff modeling for the Yerli sub-catchment have been discussed in this section. Figures 2 and 3 depict the methodological flowchart of NNTOOL and NNSTART respectively.

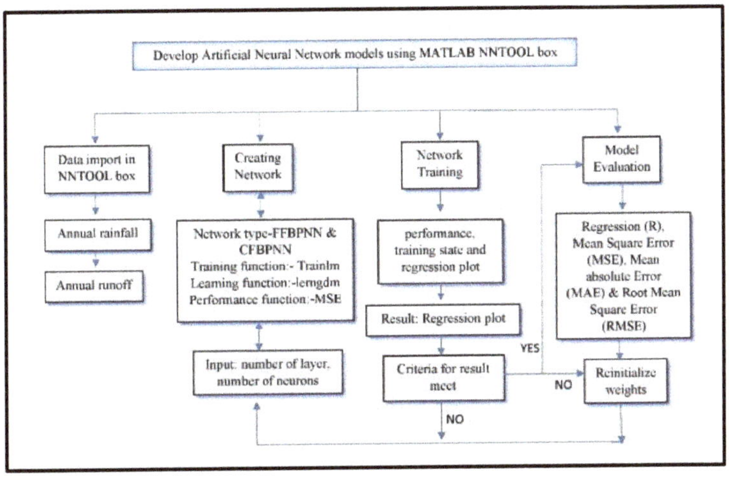

Figure 2. Flow chart of NNTOOL.

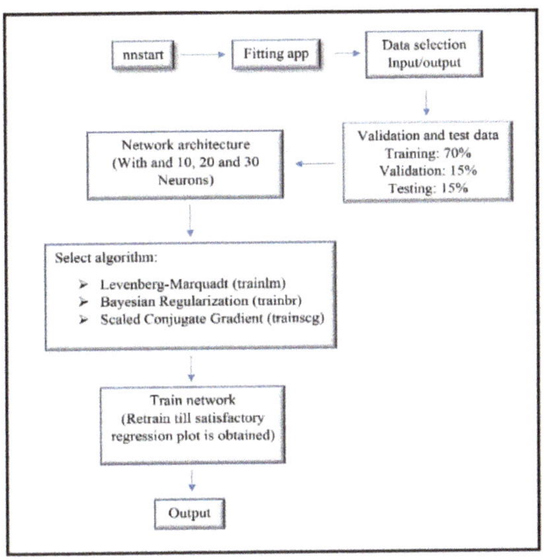

Figure 3. Flow chart of NNSTART.

2.2.1. Following Steps Should Be Performed for Developing an ANN Model Using NNTOOL

- Data Collection: The required observed data (rainfall, runoff, temperature, specific humidity, surface pressure, wind speed) at the prerequisite station are to be collected.
- Import Data: The collected data are imported into the NNTOOL box as input and target data.
- Creating Network: The network is created by selecting a suitable network type, i.e., FFBPNN or CFBPNN. The network architecture is formed (6-2-1, 6-3-1, 6-4-1).
- Number of Neurons: For the given network, the number of neurons is taken as 10 or 20.
- Network Training: The developed network is trained based on performance function.
- Result: Once the network is trained, the result is checked by plotting the regression plot, and the predicted output is obtained.
- Retraining: If the obtained regression plot is not satisfactory, then reinitialization of weights has to be conducted by changing the number of neurons.
- Model Evaluation: Based on statistical parameters such as (MSE), (RMSE), (R^2), and (R).

2.2.2. Following Steps Should Be Performed for Developing an ANN Model Using NNSTART

- Neural Fitting App: The Neural Fitting app will help to select data, create and train a network, and evaluate its performance using mean square error and regression analysis.
- Data Selection: The collected data will be used as both the input and output data. The input data are in a 6 × 36 matrix. On the other hand, the target data are in a 1 × 36 matrix.
- Validation and Test: The data are split as follows, 70% (training), 15% (validation), and 15% (testing).
- Network Architecture: For the given network, the number of neurons is taken as 10, 20, and 30.
- Select Algorithm: For training, the algorithms, namely Levenberg–Marquardt (trainlm), Bayesian Regularization (trainbr), and Scaled Conjugate Gradient (trainscg), were used.
- Train Network: To fit the input and goal data, train the network.
- Retrain: The network is retrained if a satisfactory regression plot is not obtained.
- Output: Desired predicted output is obtained after fixing the regression plot.

2.3. *Model Evaluation Criteria*

The findings of the ANN model applied in this study were evaluated by means of:

- Mean Square Error (MSE):

$$\mathrm{MSE} = \frac{1}{n}\sum_{i=1}^{n}(Q_p - Q_o)^2 \tag{1}$$

- Root Mean Square Error (RMSE):

$$\mathrm{RMSE} = \left[\frac{\sum_{i=1}^{n}\left(Q(i) - \hat{Q}(i)\right)^2}{n}\right]^{0.5} \tag{2}$$

- Regression Coefficient (R): Using Regression Plot between predicted and observed runoff.

where Q_p is the value of predicted runoff; Q_o is the value of observed runoff; $\hat{Q}(i)$ is the n estimated runoff value; and $Q(i)$ is the n observed runoff value.

3. Results and Discussion

3.1. NNTOOL

The multilayer FFBPNN and CFBPNN algorithms with Levenberg–Marquardt (LM) are utilized to optimize the learning approach in this study. Two different models were developed, i.e., (FFBPNN) and (CFBPNN) with three different architectures (6-2-1, 6-3-1 and 6-4-1) using several combinations of transfer functions, i.e., (transig, logsig, and purelin) along with two sets of neurons, 10 and 20, and then compared for their capability to estimate the flow for the period 1981–2016.

3.1.1. Feed Forward Back Propagation Neural Network (FFBPNN)

FFBPN, while considering 6-2-1, 6-3-1, and 6-4-1 architectures, the transig function provides the best value for performance. The most effective model architecture for the Transig function is 6-4-1, which has a value of MSE 0.4982, the value of RMSE 0.7056, and the value of R 0.96213. Table S1 contains the inclusive outcomes. However, in comparison to other transfer functions, the transig transfer function with architecture 6-4-1 yielded better results in the current study. Figure 4 depicts the best regression plot.

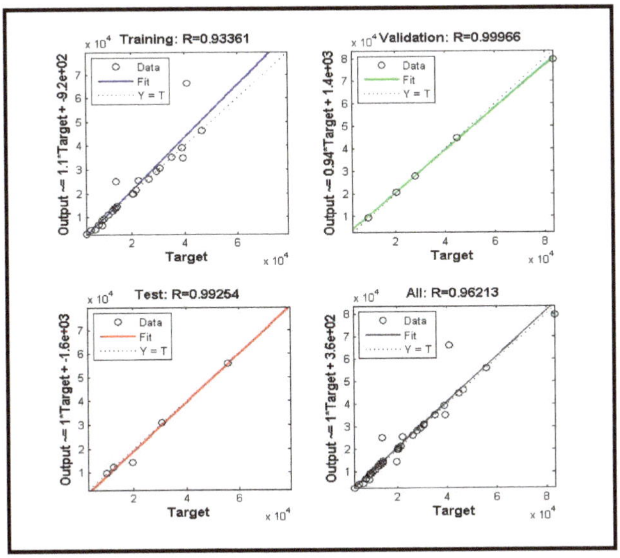

Figure 4. Regression plot for FFBPNN 6-4-1 model.

3.1.2. Cascade Forward Back Propagation Neural Network (CFBPNN)

Similarly, for CFBPNN, while considering 6-2-1, 6-3-1, and 6-4-1 architectures, the transig function provides the best value for performance. The most effective model architecture for the transig function is 6-4-1, which has MSE values of 0.8813, the value of RMSE 0.9387, and the value of R 0.96096. Table S2 contains the inclusive outcomes [17]. However, in comparison to other transfer functions comparison to other, the transig transfer function with architecture 6-4-1 yields better results. Figure 5 depicts the best regression plot.

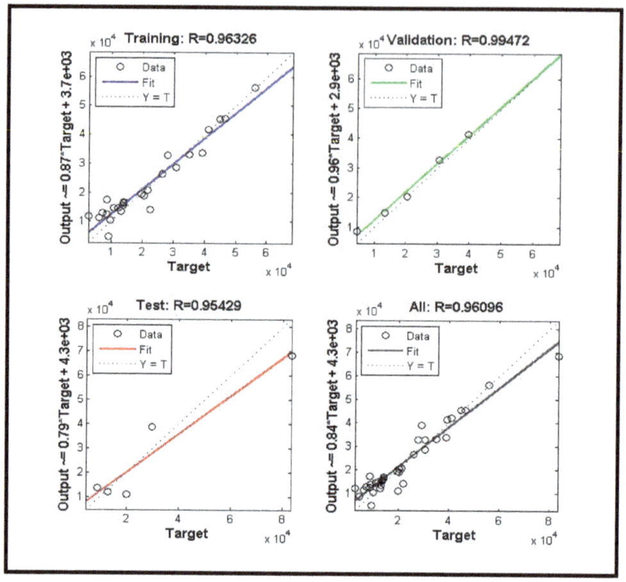

Figure 5. Regression plot for CFBPNN 6-4-1 model.

3.2. NNSTART

In this study, three different algorithms, namely Levenberg–Marquardt (trainlm), Bayesian Regularization (trainbr), and Scaled Conjugate Gradient (trainscg), were used for model development. Table S3 shows the Yerli station results for the ANN trained by LM, BR, and CGS. The study compares ANN models that were trained with LM, BR, and CGS. LM-trained algorithm with 30 neurons is the best model with an MSE value of 0.7279, an RMSE value of 0.8531, and an R value of 0.95057. Figure 6 shows the best regression plot for the LM algorithm with 30 neurons.

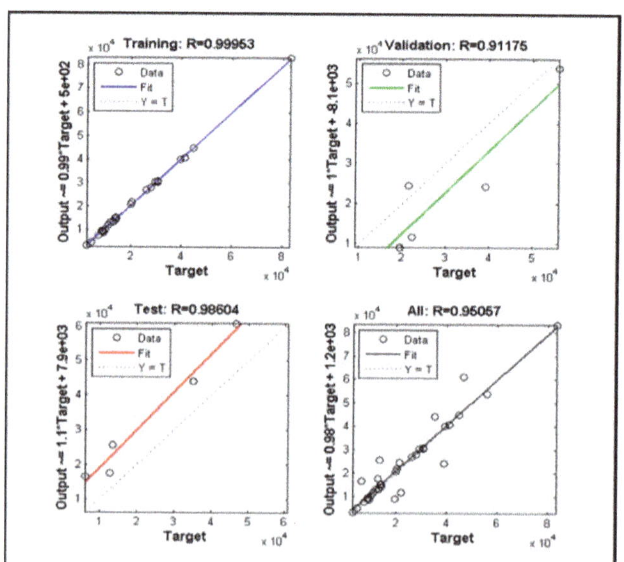

Figure 6. Regression plot for LM algorithm with 30 neurons.

4. Conclusions

This study described how ANN models are used to estimate yearly runoff for the Yerli sub-catchment of the upper Tapi basin. Runoff estimation was undertaken using NNTOOL and NNSTART. Adopting NNTOOL, two different models were developed, i.e., FFBPNN and CFBPNN networks, using several combinations of input data and then comparing their capability of flow estimation for the period 1981–2016. For estimating runoff using NNTOOL, two NNs are used, with the values of MSE, RMSE, and R calculated. For the transig function in FFBPNN, the most prominent model architecture is 6-4-1, which has an MSE value of 0.4982, an RMSE value of 0.7056, and a value of R of 0.96108. The 6-4-1 model architecture for the transig function is the most effective for CFBPNN, with MSE values of 0.8813, RMSE values of 0.9387, and R values of 0.96096. Using three different algorithms, LM, BR, and CGS, were used to predict runoff. Among the three, the LM-trained algorithm with 30 neurons is the best model, with MSE values of 0.7279, RMSE values of 0.8531, and R values of 0.95057. According to the findings, FFBPNN predicts better results than CFBPNN, and the LM algorithm stands out among the other algorithms.

Supplementary Materials: The following supporting information can be downloaded at: https://www.mdpi.com/article/10.3390/ECWS-7-14232/s1, Table S1: Results of FFBPNN for Yerli station, Table S2: Results of CFBPNN for Yerli station, Table S3: Results of NNSTART for Yerli station.

Author Contributions: Conceptualization, U.M.; methodology, U.M.; software, U.M.; validation, U.M.; formal analysis, U.M. and S.B.M.; data curation, U.M.; writing—original draft preparation, U.M.; writing—review and editing, U.M. and S.B.M. All authors have read and agreed to the published version of the manuscript.

Funding: This research received no external funding.

Institutional Review Board Statement: Not applicable.

Informed Consent Statement: Not applicable.

Data Availability Statement: Unavailability of data due to ethical restrictions.

Conflicts of Interest: The authors declare no conflict of interest.

References

1. Tokar, A.S.; Johnson, P.A. Rainfall-runoff modeling using artificial neural networks. *J. Hydrol. Eng.* **1999**, *4*, 232–239. [CrossRef]
2. Namara, W.G.; Damise, T.A.; Tufa, F.G. Rainfall Runoff Modeling Using HEC-HMS: The Case of Awash Bello Sub-Catchment, Upper Awash Basin, Ethiopia. *Int. J. Environ.* **2020**, *9*, 68–86. [CrossRef]
3. Rathod, P.; Borse, K.; Manekar, V.L. Simulation of rainfall-runoff process using HEC-HMS (case study: Tapi river, India). In Proceedings of the 20th International Conference on Hydraulics, Water Resources and River Engineering, Roorkee, India, 17–19 December 2015.
4. Chen, J.; Adams, B.J. Integration of artificial neural networks with conceptual models in rainfall-runoff modeling. *J. Hydrol.* **2006**, *318*, 232–249. [CrossRef]
5. Zhang, B.; Govindaraju, R.S. Prediction of watershed runoff using Bayesian concepts and modular neural networks. *Water Resour. Res.* **2000**, *36*, 753–762. [CrossRef]
6. Lu, P.; Chen, S.; Zheng, Y. Artificial intelligence in civil engineering. *Math. Probl. Eng.* **2012**, *2012*, 145974. [CrossRef]
7. Tayyab, M.; Zhou, J.; Adnan, R.; Zeng, X. Application of Artificial Intelligence Method Coupled with Discrete Wavelet Transform Method. *Procedia Comput. Sci.* **2017**, *107*, 212–217. [CrossRef]
8. Chae, Y.T.; Horesh, R.; Hwang, Y.; Lee, Y.M. Artificial neural network model for forecasting sub-hourly electricity usage in commercial buildings. *Energy Build.* **2016**, *111*, 184–194. [CrossRef]
9. Khayatian, F.; Sarto, L.; Dall'O', G. Application of neural networks for evaluating energy performance certificates of residential buildings. *Energy Build.* **2016**, *125*, 45–54. [CrossRef]
10. Kisi, O.; Shiri, J.; Tombul, M. Modeling rainfall-runoff process using soft computing techniques. *Comput. Geosci.* **2013**, *51*, 108–117. [CrossRef]
11. Rogers, L.L.; Dowla, F.U. Optimization of groundwater remediation using artificial neural networks with parallel solute transport modeling has been successfully applied to a variety of optimization. *Water Resour. Res.* **1994**, *30*, 457–481. [CrossRef]
12. Vandana, M.; John, S.E.; Maya, K.; Sunny, S.; Padmalal, D. Environmental impact assessment (EIA) of hard rock quarrying in a tropical river basin—Study from the SW India. *Environ. Monit. Assess.* **2020**, *192*, 580. [CrossRef] [PubMed]

13. Sahour, H.; Gholami, V.; Vazifedan, M. A comparative analysis of statistical and machine learning techniques for mapping the spatial distribution of groundwater salinity in a coastal aquifer. *J. Hydrol.* **2020**, *591*, 125321. [CrossRef]
14. Orimi, M.G.; Farid, A.; Amiri, R.; Imani, K. Cprecip parameter for checking snow entry for forecasting weekly discharge of the Haraz River flow by artificial neural network. *Water Resour.* **2015**, *42*, 607–615. [CrossRef]
15. Chandwani, V.; Vyas, S.K.; Agrawal, V.; Sharma, G. Soft Computing Approach for Rainfall-runoff Modelling: A Review. *Aquat. Procedia* **2015**, *4*, 1054–1061. [CrossRef]
16. Chang, T.K.; Talei, A.; Quek, C.; Pauwels, V.R.N. Rainfall-runoff modelling using a self-reliant fuzzy inference network with flexible structure. *J. Hydrol.* **2018**, *564*, 1179–1193. [CrossRef]
17. Samantaray, S.; Sahoo, A. Prediction of runoff using BPNN, FFBPNN, CFBPNN algorithm in arid watershed: A case study. *Int. J. Knowl.-Based Intell. Eng. Syst.* **2020**, *24*, 243–251. [CrossRef]

environmental sciences proceedings

MDPI

Proceeding Paper

Bivariate Analysis with Synthetic Hydrograph Shapes for Hydrological Dam Safety Assessment †

Daniel Carril-Rojas * and Luis Mediero

Department of Civil Engineering: Hydraulics, Energy and Environment, Universidad Politécnica de Madrid, 28040 Madrid, Spain; luis.mediero@upm.es
* Correspondence: daniel.carril.rojas@upm.es
† Presented at the 7th International Electronic Conference on Water Sciences, 15–30 March 2023; Available online: https://ecws-7.sciforum.net/.

Abstract: Hydrological dam safety analyses should assess the frequency curve of maximum reservoir water levels in flood events by routing a large set of inflow hydrographs. Therefore, stochastic bivariate analyses are used. Hydrograph shapes obtained by using hydrometeorological simulations are required for each flood peak-hydrograph volume pair. Hydrograph shapes depend on hyetograph shapes. A sensitivity analysis is required to select the appropriate hyetograph shape, focusing on the influence of the hyetograph time step on the hydrograph shape. In this study, the Cuerda del Pozo Dam in central Spain is selected as a case study.

Keywords: hydrological dam safety; bivariate analysis; floods; hyetographs; rainfall-runoff modelling

1. Introduction

Hydrological frequency analyses are usually focused on univariate approaches of flood peak time series. For example, univariate analysis can be used to either design or assess the safety of most bridges and culverts, as they can be designed only with the study of the flow peaks for a given return period. In such cases, maximum water levels usually depend on flood peaks. However, it is necessary to account for the entire flood hydrograph to assess hydrological dam safety correctly [1]. In a dam, inflow hydrographs are routed in the reservoir; therefore, outflow peak flows will be equal or smaller than inflow peak flows, depending on reservoir storage volumes, spillway length and characteristics, the operation strategy in the spillway gates, and inflow hydrograph characteristics, etc. Consequently, a univariate analysis cannot fully assess the exceedance probability of a given hydrological event and multivariate analyses are required. In addition, it is necessary to use stochastic approaches to generate a large set of hydrograph volume–peak flow pairs that keeps the statistical properties of both the marginal univariate frequency distributions and the dependence structure between variables [2].

The hydrological safety in a dam is related to the maximum water level that the reservoir reaches in a flood event associated with a given return period. Therefore, it should be analysed with the frequency curve of maximum reservoir water levels in order to better assess hydrological dam safety. For this purpose, a large set of inflow hydrographs generated using a stochastic approach should be routed through the reservoir. In such an approach, the result will depend on several variables, such as flood peak and hydrograph volume univariate frequency curves, and the dependence structure between these variables [3]. Therefore, it is essential to consider several variables by using a multivariate analysis to assess hydrological dam safety.

In addition, a given hydrograph shape has to be assigned to each flood peak-hydrograph volume pair in the stochastic procedure. Therefore, hydrometeorological simulations are required to characterise the catchment response in flood events and obtain feasible hydrograph shapes that can be generated. Hydrometeorological simulations are based on

Citation: Carril-Rojas, D.; Mediero, L. Bivariate Analysis with Synthetic Hydrograph Shapes for Hydrological Dam Safety Assessment. *Environ. Sci. Proc.* **2023**, *25*, 2. https://doi.org/10.3390/ECWS-7-14175

Academic Editor: Lampros Vasiliades

Published: 14 March 2023

hyetographs that characterise precipitation depths and the temporal variability of rainfall in storms. Several design hyetograph shapes are available in the literature [4]. Hydrograph shapes will depend on the design hyetograph selected [5]; therefore, a design hyetograph that characterises rainfall characteristics in the catchment and simulates hydrograph shapes similar to those observed should be selected.

This paper aims to analyse a set of hyetograph shapes to assess its influence on hydrograph shapes in hydrometeorological simulations. Moreover, a hyetograph shape that will allow for calibrating flood peaks and hydrograph volumes with reasonable hydrological model parameter values will be identified. An approach based on a bivariate analysis using hydrometeorological simulations to assess hydrological dam safety will be used.

2. Case Study

The analysis is carried out in the Cuerda del Pozo Dam located in central Spain (Figure 1). The Dam belongs to the Douro River Basin Authority.

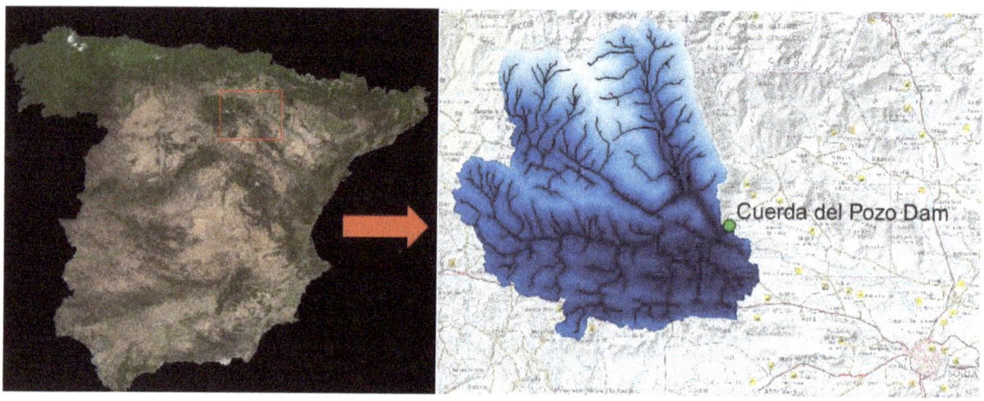

Figure 1. Location of the study and detailed map of the catchment area.

The catchment area that drains into the Cuerda del Pozo Dam is 547 km^2. The dam is located at an altitude of 1080 m above sea level. In the catchment, the maximum altitude is 2146 m, and the mean altitude is 1315 m. The Cuerda del Pozo Dam is located on the River Douro with a height of 40.25 m. The conservation volume is 250 hm^3.

A digital terrain model (DTM) with a cell size of 25 m was used to delineate the extent of the catchment and its properties. It was supplied by the Spanish *Instituto Geográfico Nacional* (IGN). Daily precipitation data at ten rainfall gauging stations were supplied by the Spanish National Meteorological Agency (AEMET, *Agencia Estatal de Meteorología*, in Spanish). Daily streamflow data at two gauging stations and daily reservoir storage volume and outflow releases at the Cuerda del Pozo Dam were supplied by the Centre for Hydrographic Studies of CEDEX and the *Sistema de Información del Agua* of the Spanish Ministry of Environment. Hourly rainfall and streamflow data were supplied by the real-time SAIH system of the Douro River Basin Authority. Figure 2 shows the location of the ten rainfall gauging stations (red dots), the two streamflow gauging stations (yellow dots), and the Cuerda del Pozo Dam (green dot).

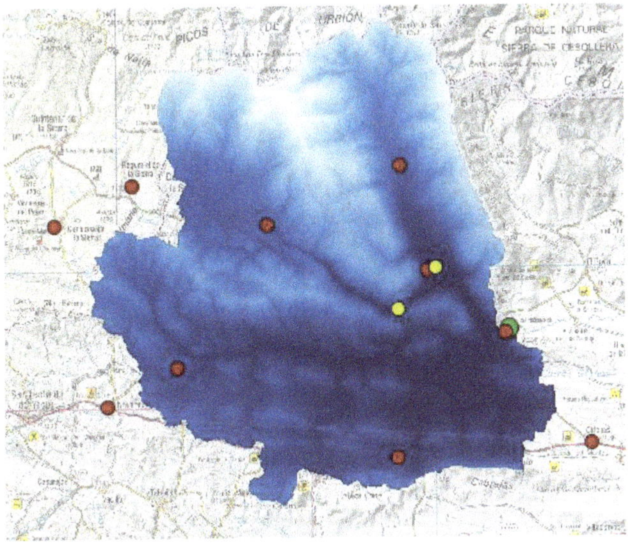

Figure 2. Location of the rainfall (red dots), streamflow (yellow dots) and dam (green dot) gauging stations considered in the study.

3. Materials and Methods

This study is based on a stochastic procedure to generate a large set of inflow hydrographs that is routed through the Cuerda del Pozo Reservoir to assess the hydrological safety of the Cuerda del Pozo Dam. Hydrometeorological simulations are used to generate feasible hydrograph shapes that can be generated in the catchment in flood events. This study simulates the Cuerda del Pozo catchment in natural regime. The following return periods are considered in the analysis: 2, 2.5, 5, 10, 25, 50, 100, 200, 500, 1000, 2000, 5000 and 10,000 years.

The hydrometeorological simulation is based on the curve number method and the Clark unit hydrograph. The SCS curve number is used as the loss method to obtain the net rainfall [6]. The Clark unit hydrograph is used as a transform method for characterising the runoff concentration and diffusion processes in the catchment [7].

This study aims to find an adequate hyetograph shape to generate both the runoff volumes given by the univariate frequency curve of the hydrograph volumes and the flood peaks given by the univariate frequency curve of the flood peaks. This process presents different difficulties; therefore, a sensitive analysis is used to obtain an adequate shape of the hyetograph that leads to feasible values of the hydrological model parameters, calibrating both hydrograph volume and flood peaks.

Flood hydrographs in the Cuerda del Pozo Dam are usually generated by storms with a mean duration of ten days. Therefore, 10-day hyetographs are considered in order to obtain the runoff volumes given by the univariate frequency curve of the hydrograph volumes. In this study, the Chicago hyetograph based on the intensity-duration-frequency curve is considered [8].

In long-duration hyetographs, the time step selection is crucial, as the hydrograph shape will depend on the hyetograph time step. If a small time step is considered in a design hyetograph of several days, sharp hydrographs will be generated with peak flows that can be greater than required. Consequently, unreasonable model parameter values, that would otherwise be necessary, could be needed to smooth the concentration and diffusion processes, decreasing the simulated flood peaks. However, if a large time step is considered, smooth hydrographs could be obtained with flood peaks smaller than required. In this case, model parameter values could not be found to calibrate flood peaks.

In this study, a set of time steps will be considered in the design hyetograph to analyse its influence on the flood hydrograph shape. Time steps of 1, 2, 3, 6, 12 and 24 h are considered.

4. Results and Discussion

A return period of 1000 years has been considered for the sensitivity analysis. Figure 3 shows the hyetographs obtained for the time steps considered in the study.

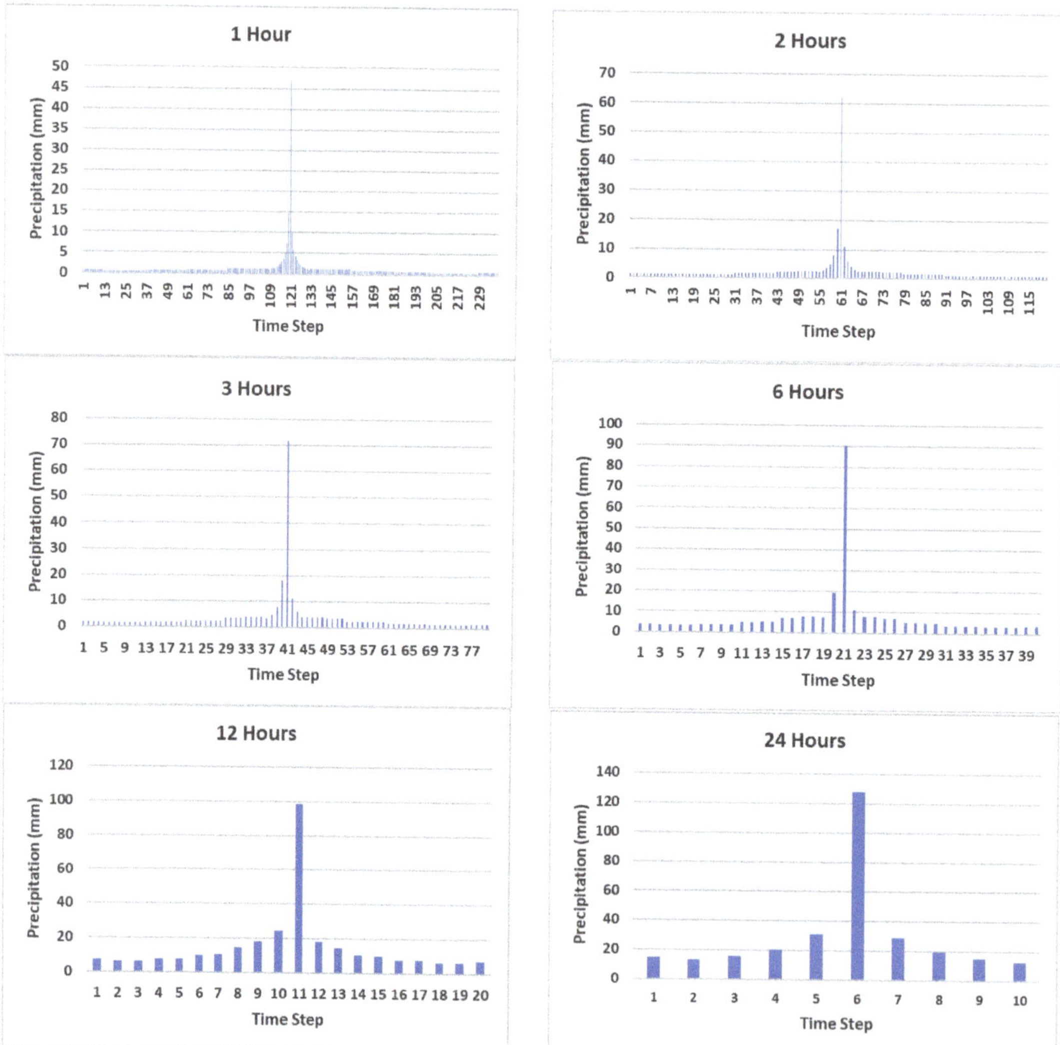

Figure 3. 10-day Chicago hyetographs considered in the Cuerda del Pozo catchment.

Before calibrating the model, flood hydrographs are obtained for each time step to assess the influence of the hyetograph time step on the hydrograph shape (Figure 4). The results show that, for small time steps, sharp hydrograph shapes are obtained with high peak flows. In contrast, for large time steps, smooth hydrographs are generated with low peak flows.

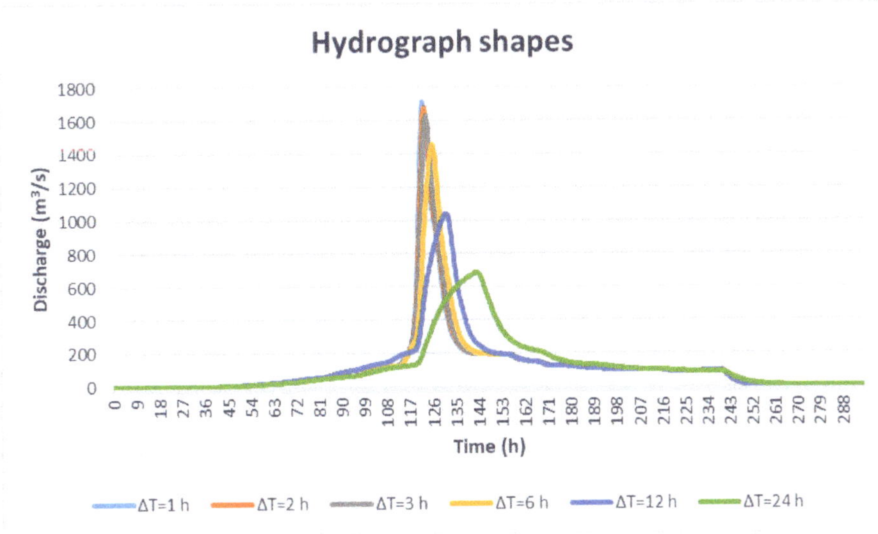

Figure 4. Flood hydrographs obtained before calibrating the hydrological model parameters.

The hydrological model is calibrated to obtain the 1000-year flood peak and the hydrograph volume given by the univariate frequency curves. The peak flow is calibrated with the storage coefficient parameter (K) of the Clark unit hydrograph used as a transform method. The results are shown in Figure 5 and Table 1. The figure shows that, with the calibrated K parameter, similar hydrograph shapes are obtained for the time steps of 1, 2, 3, 6 and 12 h. This is not the case for the 24 h time step because the desired peak flow cannot be reached even with the lowest values of the parameter K.

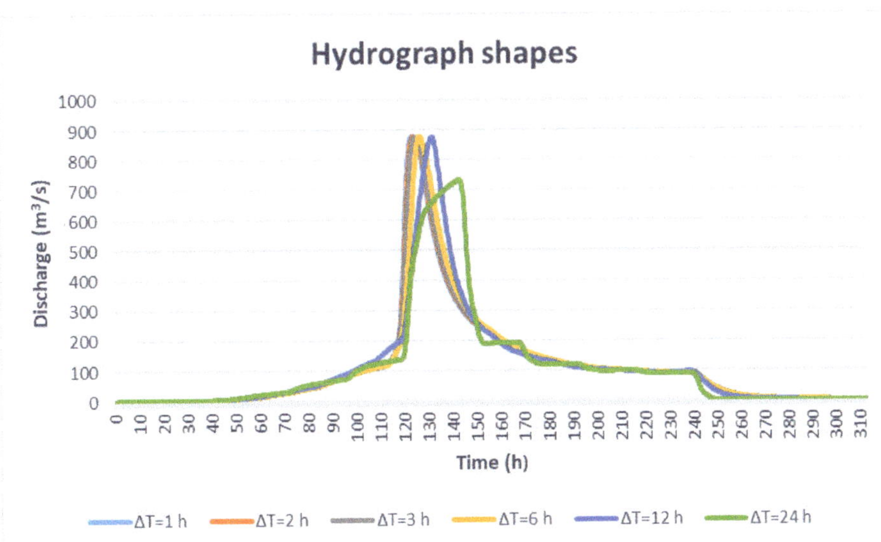

Figure 5. Flood hydrographs obtained after calibrating the hydrological model parameters.

Table 1. Results of the calibration of the storage coefficient parameter K of the Clark unit hydrograph method for each subcatchment.

K Values		Subcatchment		
		SC1	SC2	SC3
Initial		2.833	1.471	1.024
Calibrated	Δt = 1 h	9.631	5.001	14.50
	Δt = 2 h	9.716	5.045	14.60
	Δt = 3 h	9.773	5.075	14.70
	Δt = 6 h	9.632	5.001	14.50
	Δt = 12 h	5.722	2.971	11.15
	Δt = 24 h	0.283	0.147	1.0

In addition, in order to evaluate the results correctly, the values of the coefficient K obtained for each time step after calibrating the hydrological model are analysed. The table shows that, to obtain the desired peak flow value, high K values are required for small time steps and low K values are required for large time steps.

5. Conclusions

In flood events, hydrological dam safety should be assessed using the frequency curve of maximum reservoir water levels. For this purpose, a large set of inflow hydrographs is routed through the reservoir. Therefore, in this study in Cuerda del Pozo, a stochastic bivariate analysis based on hydrometeorological simulations is performed to assess hydrological dam safety. With this aim, a set of hyetograph shapes and their influence on the shape of the hydrographs are analysed.

The results of the study show that the flood hydrographs obtained before calibrating the hydrological model have lower peak flows and smoother shapes with each hyetograph time step increase. In addition, as the hyetograph time interval increases, the calibrated K parameter becomes smaller. However, for a hyetograph time step of 24 h, even with a very low K parameter value, the peak flow required by the univariate flood frequency curve cannot be achieved.

Author Contributions: Conceptualization, L.M.; methodology, L.M.; software, D.C.-R.; validation, D.C.-R.; formal analysis, D.C.-R.; investigation, D.C.-R. and L.M.; resources, D.C-.R. and L.M.; data curation, D.C.-R. and L.M.; writing—original draft preparation, D.C.-R.; writing—review and editing, D.C.-R. and L.M.; visualization, D.C.-R.; supervision, L.M.; project administration, L.M.; funding acquisition, L.M. All authors have read and agreed to the published version of the manuscript.

Funding: This research was supported by the project (PID2019-107027RB-I0/AEI/10.13039/501100011033) "SAFERDAMS: Assessment of the impact of climate change on hydrological dam safety" funded by the *Agencia Estatal de Investigación* of the Spanish Ministry of Science and Innovation.

Institutional Review Board Statement: Not applicable.

Informed Consent Statement: Not applicable.

Data Availability Statement: Data used in the study can be found on https://www.aemet.es/ (accessed on 25 April 2022) and https://www.miteco.gob.es/es/agua/temas/evaluacion-de-los-recursos-hidricos/sistema-informacion-anuario-aforos/ (accessed on 20 June 2022).

Acknowledgments: The authors would like to acknowledge the Spanish *Instituto Geográfico Nacional* (IGN) for supplying the DTM; the Spanish *Agencia Estatal de Meteorología* (AEMET) for supplying the daily precipitation data; the Spanish Centre for Hydrographic Studies of CEDEX and the *Sistema de Información del Agua* of the Spanish Ministry of Environment for supplying the daily streamflow data; and the SAIH real-time system of the River Douro Basin Authority for supplying sub-daily precipitation data.

Conflicts of Interest: The authors declare no conflict of interest.

References

1. Mediero, L.; Jiménez-Álvarez, A.; Garrote, L. Design flood hydrographs from the relationship between flood peak and volume. *Hydrol. Earth Syst. Sci.* **2010**, *14*, 2495–2505. [CrossRef]
2. Klein, B.; Schumann, A.H.; Pahlow, M. Copulas—New Risk Assessment Methodology for Dam Safety. In *Flood Risk Assessment and Management*; Schumann, A., Ed.; Springer: Dordrecht, The Netherlands, 2011. [CrossRef]
3. Requena, A.I.; Chebana, F.; Mediero, L. A complete procedure for multivariate index-flood model application. *J. Hydrol.* **2016**, *535*, 559–580. [CrossRef]
4. Wartalska, K.; Kaźmierczak, B.; Nowakowska, M.; Kotowski, A. Analysis of Hyetographs for Drainage System Modeling. *Water* **2020**, *12*, 149. [CrossRef]
5. Alfieri, L.; Laio, F.; Claps, P. A simulation experiment for optimal design hyetograph selection. *Hydrol. Process.* **2008**, *22*, 813–820. [CrossRef]
6. Mishra, S.K.; Singh, V. *Soil Conservation Service Curve Number (SCS-CN) Methodology*; Water Science and Technology Library (WSTL, Volume 42); Springer: Dordrecht, The Netherlands, 2003. [CrossRef]
7. Ahmad, M.M.; Ghumman, A.R.; Ahmad, S. Estimation of Clark's Instantaneous Unit Hydrograph Parameters and Development of Direct Surface Runoff Hydrograph. *Water Resour. Manag.* **2009**, *23*, 2417–2435. [CrossRef]
8. Keifer, C.J.; Chu, H.H. Synthetic storm pattern for drainage design. *J. Hydraul. Eng.* **1957**, *83*, 1–25. [CrossRef]

environmental sciences proceedings

MDPI

Proceeding Paper

Intervention Time of Porous Asphalt Mixture Evaluation to Prevent Clogging †

Emmanuelle Stefânia Holdefer Garcia and Liseane Padilha Thives *

Department of Civil Engineering, Federal University of Santa Catarina—Campus Florianópolis, Florianópolis 88037-000, SC, Brazil

* Correspondence: liseane.thives@ufsc.br; Tel.: +55-4837212114

† Presented at the 7th International Electronic Conference on Water Sciences, 15–30 March 2023; Available online: https://ecws-7.sciforum.net.

Abstract: Porous pavements are considered an alternative to increasing the city's permeable surfaces and have great potential for rainwater harvesting. Over time, the surface layer clogs, limiting the application and use of porous pavement. This study aims to evaluate the maximum time for an intervention on a porous surface to recover its permeable properties before clogging. Through a constant load permeameter, rainwater runoff simulations were performed to measure the permeability coefficient of porous asphalt mixtures. Void content and interconnected voids were also evaluated. The parameter permeability showed a good indicator for determining when a surface layer will clog, and by taking regular measurements, the necessary intervention time to prevent it can be established. The designed porous mixture surface intervention should occur every year.

Keywords: porous asphalt mixture; clogging; intervention time

check for updates

Citation: Garcia, E.S.H.; Thives, L.P. Intervention Time of Porous Asphalt Mixture Evaluation to Prevent Clogging. *Environ. Sci. Proc.* **2023**, *25*, 3. https://doi.org/10.3390/ECWS-7-14196

Academic Editor: Athanasios Loukas

Published: 14 March 2023

1. Introduction

The growing urbanization process in cities leads to an increase in impermeable areas, generating changes in the natural characteristics of watersheds. Consequently, the recharge of springs and groundwater is compromised, and as the volume of surface runoff increases, the risks of floods and inundations occur. An alternative for integrated rainwater management is using porous pavements and surfaces.

The pavement layer surface mainly comprises porous asphalt mixtures, produced with a high amount of air voids and a percentage of coarse aggregates to form an open-graded granulometry and allow water to flow through it. The open gradation results in higher voids (18–25%) compared to densely-graded mixtures [1,2]. On the other hand, fine aggregates contribute to mixture strength and durability, but increasing their proportion compromises the permeability, which is negligible if the void content is less than 14% [3].

Zhang et al. asserted that studies had shown the benefits of porous mixtures compared to dense ones. Due to the interconnected porosity, the water flows out of the pavement during rainfall, avoiding accumulation on the surface. The road safety improvements include increased skid resistance and reduced hydroplaning [4].

The infrastructure can introduce low-impact development models through increasing permeable areas such as green spaces, porous pavements, and water systems. These models allow rainwater to be absorbed, retained, and released to reduce runoff volume and peak flow, slow down runoff speed, replenish groundwater, and filter rainwater pollutants [5,6]. However, the water that precipitates on the pavement surface carries contaminants and powdery materials (runoff), which, over time, lead to the progressive clogging of the voids.

Due to clogging, functionality is affected, and the porous pavement (or surface) cannot facilitate drainage. Additionally, the porous surface suffers damage due to water stagnation inside. The rate of clogging depends on factors such as the pavement surface, void content, traffic load, pavement slope, maintenance measures adopted, and environmental and local conditions [7]. An evaluation of the progression of clogging can be performed in the field or laboratory using permeameters.

Porous surface pavement layers are beneficial as they contribute to stormwater management by reducing surface runoff. On the other hand, the voids in porous layers clog over time, and the permeable characteristics decrease. Because of this, road agencies avoided applying the porous surface layer to the pavement.

In order to contribute to advancing the use of porous mixtures as a pavement surface layer, the primary motivation of this study was to evaluate the clogging by simple permeability tests. Thus, permeability reduction enabled establishing the needed intervention time to restore porous mixtures permeable characteristics. This study aims to evaluate the maximum time for an intervention in a porous mixture to recover its permeable properties before clogging.

2. Methodology

In this study, the clogging progression of a porous asphalt mixture was measured in the laboratory to evaluate the maximum intervention time for recovery of the permeable properties. The porous asphalt mixtures were produced with asphalt rubber fabricated at a Brazilian refinery (terminal blend) with 15% incorporated rubber. Aggregates of granitic origin from southern Brazil were used, whose gradation curve is shown in Figure 1.

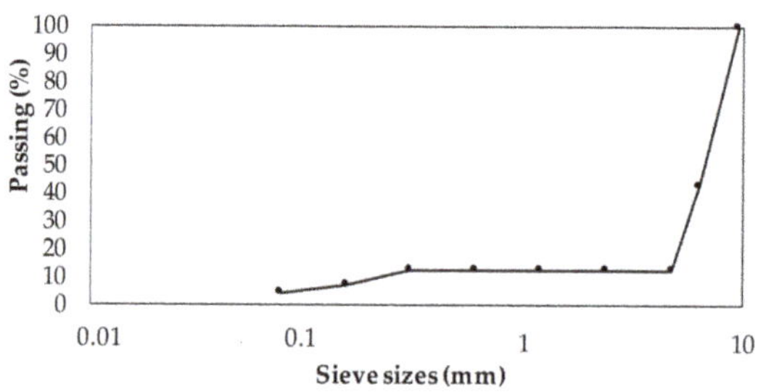

Figure 1. Mixture gradation curve.

The mixture design was performed according to the ASTM D 7064 standard [8] in the CGS (660 kPa, angle 1.25°, 50 gyrations) and resulted in an asphalt content of 5.0% and a void volume of 24.2%. Figure 2 presents the experimental methodology procedure.

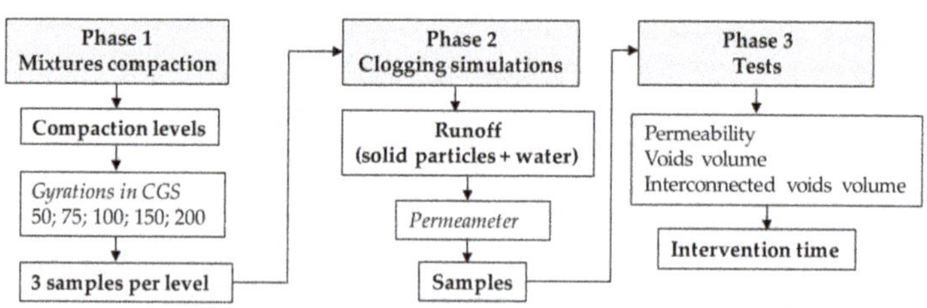

Figure 2. Experimental procedure.

To simulate the traffic influence on the mixture's air void reduction, in Phase 1, the samples (100 mm in diameter; 63.5 in height on average) were moulded at different compaction levels in the Superior Performing Asphalt Pavements (SUPERPAVE) gyratory compactor (CGS) through increasing gyration numbers (50—reference, 75, 100, 150, and 200). Three samples per level were produced to perform the tests.

The solid material that accumulated under an overpass on a medium traffic road was collected to compose the solid particles for runoff experimentation. After drying in an oven, the solid material was selected by sieving, and the fraction retained on sieve no. 100 (0.149 mm) was used, based on the study by Fwa et al. [9]. Adjustments were made to the amount of material for the sample dimensions of this study, resulting in 3.53 grams. To promote sample clogging in the runoff solution, 3.53 grams of the solid material were added to one liter of water. The runoff solution was introduced into the reservoir (Figure 3), and the samples were submitted to simulation cycles in a constant head permeameter developed by the authors of this study (Figure 3). The permeability coefficient was calculated through Equation (1). For porous materials, Fwa et al. [9] defined a constant (n of 0.70) to minimize the significant error between readings in tests using laboratory permeameters, which were also adopted in this study.

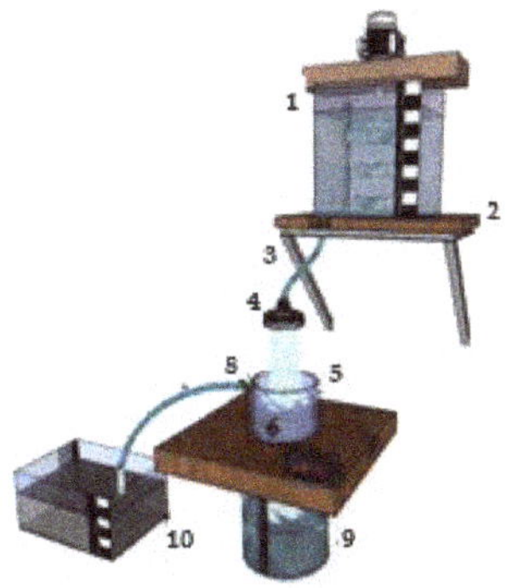

Figure 3. The constant-head permeameter developed in the study. Legend: (1.) Reservoir (volume 5 liters); (2.) reservoir support elevated 50 cm from the sample; (3.) hose (diameter 10 mm); (4.) plastic shower for water flow control; (5.) plastic mould in which the sample is inserted; (6.) sample with lateral waterproofing by silicone; (7.) support to fix the sample; (8.) a spillway (3/8″ gauge) to keep the water level above the sample; and (9.) a water collector from the spillway.

The initial permeability coefficient (k_i) was determined using water, and after five simulations with runoff solution, the final permeability coefficient (k_f) was measured. For each sample, the void volume and the interconnected void volume were evaluated before and after the simulations. It was used as a criterion, which means the sample was considered clogged when the interconnected void volume reached 12%.

For each sample group, the permeability coefficient reduction (kr) was calculated as a function of the mean value distribution (Equation (4)) and the uncertainty range (Equation (5)). The maximum intervention time in months was established from the number of simulations required to reach the criteria, considering the annual precipitation average for Florianópolis (1570 mm). In this way, for this precipitation, each simulation cycle represents 261.67 mm, and one cycle corresponds to two months.

$$k = V/i^n \quad (1)$$

$$V = Q/t \times A \quad (2)$$

$$i = H/L \quad (3)$$

$$kr = a \times e^{-bN} \quad (4)$$

$$u = \mu + u \quad (5)$$

where k is the permeability coefficient (mm/s); V is the runoff speed (mm/s), given by Equation (2); i is the hydraulic gradient (dimensionless), given by Equation (3); n is the constant [9] equal to 0.7; Q is the drained volume (mm^3); t is each cycle simulation time (s); A is the top surface area of the sample (mm^2); H is the height of the water layer (mm); L is the sample height (mm); kr is the permeability coefficient reduction (mm/s); a and b are regression coefficients; N is the number of simulations; u is the uncertainty range; μ is the mean permeability coefficient (mm/s); and u is the standard deviation (mm/s).

3. Results and Discussion

3.1. Influence of Compaction Effort in the Void Reduction

As a function of the number of gyrations, Figure 4a presents the reduction in void volume, and Figure 4b the interconnected void volume.

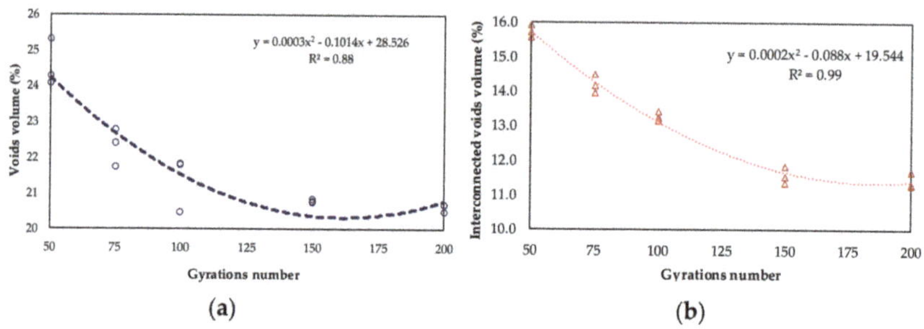

Figure 4. Void reduction as a function of the number of gyrations. (**a**) Void volume; (**b**) interconnected voids.

In Figure 4, it can be observed that as the level of compaction effort increased, the void volume decreased. After 100 gyrations, the void volume tended to stabilize with average values of 21.34%, 20.76%, and 20.62% for 100, 150, and 200 gyrations, respectively (Figure 4a). The interconnected void volume reduction and the increased compaction effort indicated a decrease in permeability (Figure 4b). Similarly, there was a tendency for values to stabilize after 100 gyrations, being, on average, 13.25%, 12.63%, and 12.23% for 100, 150, and 200 gyrations, respectively.

3.2. Influence of Compaction Effort in the Permeability Coefficient Reduction

Table 1 shows the initial permeability coefficients (ki) measured before and after the final (kf) evaluated simulations. The values were normalized for the permeability coefficient reduction (kr) analysis through the obtained average (Figure 5). The uncertainty range presented a reliability of 68.3%. In Figure 5, the increase in compaction effort, which represents the traffic load over time, is reflected in the reduction in the permeability coefficient.

Table 1. Permeability coefficients.

Gyrations Number	ki (mm/s) [1]	kf (mm/s) [2]
50	2.95	1.53
75	2.26	0.84
100	1.98	0.64
150	1.01	0.15
200	1.15	0.10

[1] Initial permeability coefficient; [2] final permeability coefficient.

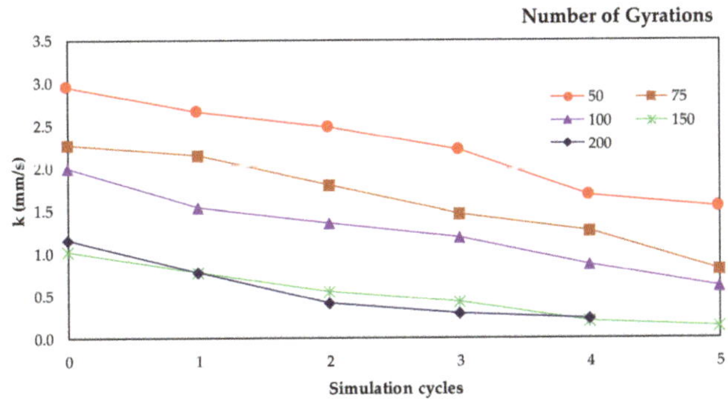

Figure 5. Void reduction as a function of the number of gyrations.

In the literature, there is a lack of studies that evaluated clogging in samples submitted to different compaction efforts. Samples submitted in simulation cycles using a permeameter with more than 20% voids were less affected by clogging [9]. Lin et al. [10] asserted that the initial permeability coefficient of samples was 0.56 mm/s, and after simulations, when it reached 0.49 mm/s, the surfaces were clogged. These studies corroborated the fact that permeability tests help estimate void reduction due to clogging.

3.3. Intervention Time

Table 2 presents the regression coefficients obtained (Equation (2)) for each group of samples (number of gyrations). Table 3 shows the obtained average values of void volume, interconnected void volume, and permeability coefficients for the sample group as a function of the number of gyrations. The permeability coefficient reduction (kr) was calculated by Equation (2) when the interconnected void volume reached a value of 12%, and then the intervention time can be estimated.

Table 2. Regression coefficients (a and b).

Number of Gyrations	a	b
50	3.0786	0.137
75	2.5136	0.199
100	2.0386	0.223
150	1.1594	0.425
200	1.1013	0.429

Table 3. Parameters for the permeability reduction calculation.

Number of Gyrations	Vv [1] (%)	IVv [2] (%)	k_{mean} [3] (mm/s)	k_{min} [4] (mm/s)	IT [5] (months)	Precipitation (mm)
50	24.54	15.73	1.53	1.16	14.16	1853
75	22.28	14.19	0.80	0.68	13.15	1720
100	21.34	13.25	0.60	0.49	12.64	1654
150	20.79	12.63	1.01	0.12	10.66	1395
200	20.68	12.23	0.22	0.21	7.59	994

[1] Void volume; [2] interconnected void volume; [3] mean permeability coefficient; [4] minimum permeability coefficient; and [5] intervention time.

The intervention time for the porous asphalt mixture to recover its permeable properties was proportional to the number of gyrations, reflecting that the traffic effort contributes to the reduction of permeability. For example, in Table 3, the intervention time was fourteen months for the compaction after 50 gyrations, while for 200 gyrations, it was more than seven months. To prevent clogging progression, the intervention by surface cleaning for the porous mixture studied should occur at least every year.

4. Conclusions

A porous asphalt mixture was produced to estimate the necessary intervention time for the maintenance of permeable characteristics. Samples of different compaction efforts in the CGS were submitted based on the increase in the number of gyrations (50, 75, 100, 150, and 200 gyrations) to evaluate the traffic effect over time.

The samples were tested in the laboratory by cycling rainwater (runoff solution) through a constant head permeameter developed in the study. Before and after simulations, the void volume, interconnected voids, and permeability coefficient were measured. A stabilization of void volume and interconnected void volume values was observed after 100 gyrations, indicating that higher compaction levels present slight variations. The permeability reduction of each set of samples was evaluated statistically through the uncertainty range and the analysis of experimental data.

A minimum value of 12% of the interconnected void volume was established, considering that the porous mixture was clogged. The intervention time was estimated after the pavement surface layer was open to traffic. The permeability reduction was proportional to the increase in the number of gyrations. For samples subjected to 50 gyrations, the maximum intervention time was 14 months. On the other hand, for 200 gyrations, considering the effect of traffic, the intervention time was shorter, approximately 7.6 months.

Considering the porous asphalt mixture of this study and the local annual rainfall (1570 mm), based on the evaluated parameters, it is recommended that the intervention time for surface cleaning (with water and air jets) be carried out at least each year. Moreover, it is essential for future studies that the mixture's internal structure be analyzed to estimate the intervention time and permeable parameter measurements. Additionally, experimental stretches must be constructed and monitored for data validation. The study limitations were using the same amount of solid material in the laboratory tests and just one type of porous mixture. The following research steps should involve the development of a stormwater simulator and analysis by tomography.

Author Contributions: Conceptualization, E.S.H.G.; methodology, E.S.H.G.; formal analysis, L.P.T.; investigation, E.S.H.G.; writing—original draft preparation, E.S.H.G.; writing—review and editing, L.P.T.; supervision, L.P.T. All authors have read and agreed to the published version of the manuscript.

Funding: This research was funded by The Brazilian National Council for Scientific and Technological Development (CNPq) grant number 6813497078212769, and The APC was funded by CAPES Foundation.

Institutional Review Board Statement: Not applicable.

Informed Consent Statement: Not applicable.

Data Availability Statement: https://repositorio.ufsc.br/handle/123456789/185442 (in Portuguese) (accessed on 13 March 2023).

Acknowledgments: The first author is thankful to the CAPES Foundation (Financing Code 001) for support with a scholarship.

Conflicts of Interest: The authors declare no conflict of interest.

References

1. Gupta, A.; Rodriguez-Hernandez, J.; Castro-Fresno, D. Incorporation of additives and fibers in porous asphalt mixtures: A review. *Materials* **2019**, *12*, 3156. [CrossRef] [PubMed]
2. Alvarez, A.E.; Martin, A.E.; Estakhri, C. A review of mix design and evaluation research for permeable friction course mixtures. *Constr. Build. Mater.* **2011**, *25*, 1159–1166. [CrossRef]
3. Taylor, P.; Ranieri, V.; Sansalone, J.J.; Shuler, S. Relationships among gradation curve, clogging resistance, and pore-based indices of porous asphalt mixes. *Road Mater. Pavement Des.* **2010**, *11*, 37–41. [CrossRef]

4. Zhang, Z.; Sha, A.; Liu, X.; Luan, B.; Gao, J.; Jiang, W.; Ma, F. State-of-the-art of porous asphalt pavement: Experience and considerations of mixture design. *Constr. Build. Mater.* **2020**, *262*, 119998. [CrossRef]
5. Wei, J.; He, J. Numerical simulation for analyzing the thermal improving effect of evaporative cooling urban surfaces on the urban built environment. *Appl. Therm. Eng.* **2013**, *51*, 144–154. [CrossRef]
6. Huang, W.; Liu, X.; Zhang, S.; Zheng, S.; Ding, Y.; Tong, B. Performance-guided design of permeable asphalt concrete with modified asphalt binder using crumb rubber and SBS modifier for sponge cities. *Materials* **2021**, *14*, 1266. [CrossRef] [PubMed]
7. Kumar, K.; Kozak, J.; Hundal, L.; Cox, A.; Zhang, H.; Granato, T. In-situ infiltration performance of diferent permeable pavements in an employee used parking lot—A four-year study. *J. Environ. Manag.* **2016**, *167*, 8–14. [CrossRef] [PubMed]
8. *ASTM D7064*; Standard Practice for Open-Graded Friction Course (OGFC) Mix Design. American Society for Testing and Materials (ASTM): West Conshohocken, PA, USA, 2004.
9. Fwa, T.F.; Lim, E.; Tan, K.H. Comparison of permeability and clogging characteristics of porous asphalt and pervious concrete pavement materials. *Transp. Res. Rec.* **2015**, *2511*, 72–80. [CrossRef]
10. Lin, W.; Park, D.G.; Ryu, S.W.; Lee, B.T.; Cho, Y.H. Optimum mix design of enhanced permeable concrete–an experimental investigation. *Constr. Build. Mater.* **2016**, *24*, 2664–2671. [CrossRef]

environmental sciences proceedings

MDPI

Proceeding Paper

Ecosystem Services Evaluation from Sustainable Water Management in Agriculture: An Example from An Intensely Irrigated Area in Central Greece †

Yiannis Panagopoulos *, Dimitrios Karpouzos, Pantazis Georgiou and Dimitrios Papamichail

Department of Hydraulics, Soil Science and Agricultural Engineering, School of Agriculture, Aristotle University of Thessaloniki, 54124 Thessaloniki, Greece; dimkarp@agro.auth.gr (D.K.); pantaz@agro.auth.gr (P.G.); papamich@agro.auth.gr (D.P.)

* Correspondence: ypanag@agro.auth.gr; Tel.: +30-2310-998-708

† Presented at the 7th International Electronic Conference on Water Sciences, 15–30 March 2023; Available online: https://ecws-7.sciforum.net.

Abstract: This study presents the provisional, regulating and cultural ecosystem services that can be delivered by the newly constructed multi-purpose reservoir of Lake Karla located in a water-scarce agricultural area in central Greece. The present short paper takes advantage of literature data and outputs produced from a dynamic GIS hydrologic and management model of the study area with SWAT that simulated hydrology, reservoir operation, irrigation practices and crop production. The paper highlights the net provisional services that the local agricultural society can gain from the full operation of Karla and the additional benefits arising, such as flood control, biodiversity maintenance, aesthetic improvement and touristic opportunities.

Keywords: ecosystem services; hydrologic modelling; irrigated agriculture; Karla reservoir; SWAT; water management

check for updates

Citation: Panagopoulos, Y.; Karpouzos, D.; Georgiou, P.; Papamichail, D. Ecosystem Services Evaluation from Sustainable Water Management in Agriculture: An Example from An Intensely Irrigated Area in Central Greece. *Environ. Sci. Proc.* **2023**, *25*, 4. https://doi.org/10.3390/ECWS-7-14250

Academic Editor: Athanasios Loukas

Published: 16 March 2023

1. Introduction

Irrigated agriculture is essential to crop production in the Mediterranean region, which faces water scarcity. Hence, its sustainability requires the efficient management of the available but limited regional water resources [1]. Investments for sustainable water management in the agricultural sector include a large variety of interventions from small ponds to large reservoirs that are able to collect, store and distribute water with the purpose of maximising the total farm income [2]. Particularly in water-scarce agricultural regions, reservoirs increase the availability of water for irrigated agriculture, and at the same time they can preserve the integrity of the local ecosystem while providing environmental benefits [3]. This study presents the newly created Karla reservoir, a large work constructed for the above purposes at the southeastern edge of the agricultural catchment of Pinios (~10,800 km^2) in Central Greece. For decades, the wider area has been a significant producer of agricultural crops, but water shortages during summers and over-abstractions of groundwater have caused significant losses of crop production [4], mainly cotton, which still drives the local agricultural economy [5].

Karla was a natural lake until 1962, when it was drained in favor of increasing agricultural land, but water scarcity and loss of wetland functions and values prompted the need for its reconstruction [6], which was completed almost one decade ago. The new project is today a Natura and Ramsar aquatic ecosystem, and a functional reservoir [7], having as its main water supply purpose the irrigation of crops. Its development was possible through natural winter runoffs from the surrounding areas and water diversions from Pinios river during the wet period of the year [8].

Irrigated crops have high water demands as they need almost 5000 m^3/ha per growing season, water that had been abstracted mostly from groundwater sources in past years. This has led to the overexploitation of aquifers and to a significant increase in the energy cost of pumping [5,8]. The 38 km^2 reconstructed Lake Karla is designed to store water between the lower water depth of 2–2.5 m and the upper depth of 4.5–5 m. These depths correspond to ~100 $\times$ 10^6 m^3 and 180–200 $\times$ 10^6 m^3 of water, respectively. The depth of 2–2.5 m corresponds to the minimum water level of the reservoir, necessary to fulfill the ecological criteria of the wetland. So, the maximum allowable water volume is nearly 200 $\times$ 10^6 m^3, but only half is allowed for extraction. For security reasons, an artificial tunnel can remove excess water to the sea [8].

The purpose of this short paper is to briefly present and evaluate the significance of the expected delivered benefits from the construction of Karla, both in terms of the available water for irrigation and respective crop productivity, and the environmental and socio-economic improvement of the surrounding area. To this end, we use literature information and already published environmental, hydrologic and crop growth modeling data [7–9].

2. Materials and Methods

A modeling system of Pinios basin with SWAT (Soil and Water Assessment Tool) model [10] represented the hydrology, water management and crop productivity under two scenarios: (i) with irrigation water supply from Lake Karla and (ii) with the inexistence of the lake and irrigation water abstracted from groundwater [8]. Figure 1 shows the area of interest in GIS.

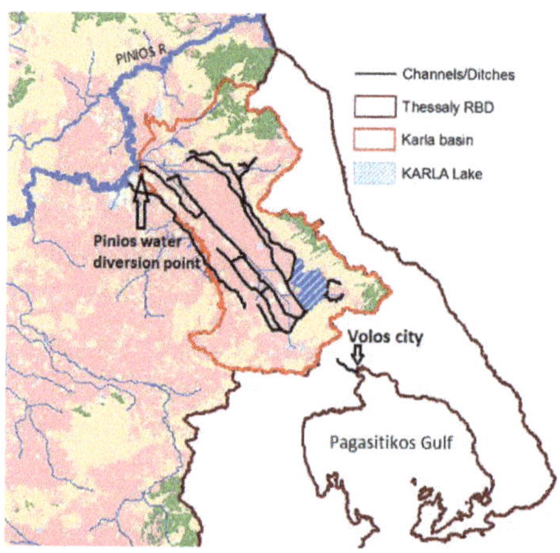

Figure 1. The Karla reservoir (lake) within the Karla basin of the Thessaly River Basin District (RBD) along with the channels/ditches diverting water from Pinios river and the surrounding areas into the lake.

The elaboration of the model was based on recent land use, water management and climate data of the historic period 2001–2010. The second scenario represents a situation without the existence of the lake, thus without water diversions from the river of Pinios, with surface waters of the Karla basin routed naturally to the small nearby rivers, the area of the lake (Figure 1) being fallow and the limited groundwater storage being the only source of irrigation water for the 8000 ha of cropland around the lake's area. In the first scenario, the Karla reservoir was simulated in SWAT based on a water volume–surface area relationship and the calculation of its water balance on a monthly basis by considering precipitation, evaporation, inflows from Pinios diversions and natural runoff, abstractions in the dry period for irrigating 8000 ha of crops, seepage and overflows. SWAT incorporates a crop growth component, which simulates the potential growth and yield when crops do not experience temperature, water and nutrient stress [11]. The model was calibrated and validated based on reported hydrologic data and crop yields [8]. Regarding other data, necessary for discussing non-provisional services, we used the recent studies of Sidiropoulos et al. [7] and Dodouras et al. [9], which describe and evaluate, based on real data, the entire new Lake Karla project.

3. Results and Discussion

The SWAT model was executed for a 10 y period with historic climate (2001–2010) data for the two scenarios, including or not Lake Karla in the simulation, and provided monthly and annual outputs. The outputs of the calibrated model give estimates of provisional services, namely actual irrigation water applied to the 8000 ha crop area and the crop yield on a mean annual basis, while published studies give information about non-provisional services, such as flood control, aesthetic improvement, enhancement of biodiversity and touristic opportunities. All information arising both from the simulation results and the reported data is summarized in Figure 1 and is associated with types of ecosystem services provided by the lake [12] (Figure 2).

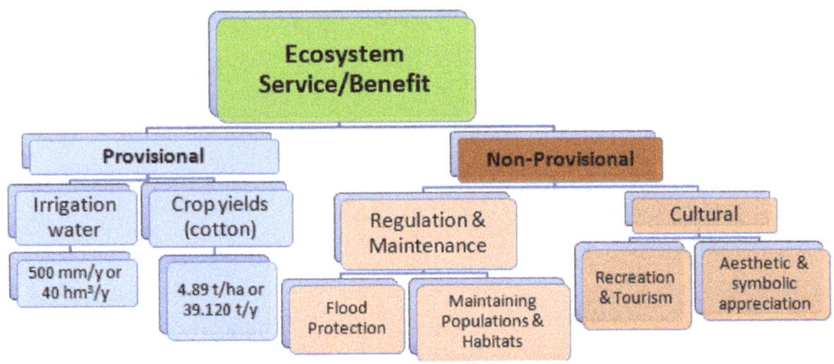

Figure 2. Ecosystem services relevant for Lake Karla (based on [12]).

With no Karla reservoir simulated, the model was forced to apply 500 mm (5000 m^3/ha) of water from May to late September by abstracting groundwater; however, the availability of water in the aquifer could not fully cover the irrigation water doses. Hence, less than 50% of the optimum water could be annually abstracted (215 mm or 2150 m^3/ha). On the other hand, with the Karla reservoir simulated, there was full availability of water for application to cotton fields, thus the model simulated optimum irrigation doses. This had very positive effects on crop production as the crop growth routine of SWAT simulated cotton yield as 4.89 t/ha/y for the 8000 of cotton land, almost double that without the Lake Karla in operation. This is attributed to the absence of water stress days in the accumulation of crop biomass development. Thus, with Lake Karla, the irrigation requirements are fully met (100%), with the effectiveness of covering irrigation needs without Karla being only 43%. For the 8000 ha of irrigated land, total cotton production with Karla included is twice as much as without it; the annual production is 39,120 t and 21,520 t, respectively. The additional income rises up to 880 EUR/ha, which is equivalent to 7,040,000 EUR for the entire cropland area of the 8000 ha based on a guaranteed unit cotton price of 0.40 EUR/kg. For a typical individual land property of 4–5 ha, the annual income increase for a single farmer is estimated as ~4000 EUR.

On the other hand, water saving in the aquifers is considerable as well. Lake Karla results in no pumping, thus saving 17.2 hm^3 annually in the aquifer (2150 m^3/ha for 8000 ha). With the 38 km^2 Karla Lake, continuous seepage (0.1 mm/h) from the lake's bottom can also result in an annual aquifer recharge of 33.2 hm^3/y, almost 30 hm^3/y more than the precipitation that could become net groundwater recharge through percolation on an annual basis from the same area. In combination, increased percolation and zero groundwater abstractions result in significant benefits for the aquifer and the soil stability.

Water availability within Lake Karla relies mainly on Pinios diversions during the wet period of the year with abstracted water quantities that never cause hydrologic pressures on the river. A minimum water level of 2.5 m and a volume of 100 hm^3 in the lake has to be maintained to support biodiversity. SWAT simulated the actual monthly water storage as shown in Figure 3 where the minimum of 100 hm^3 is always respected within the 10-year simulation period.

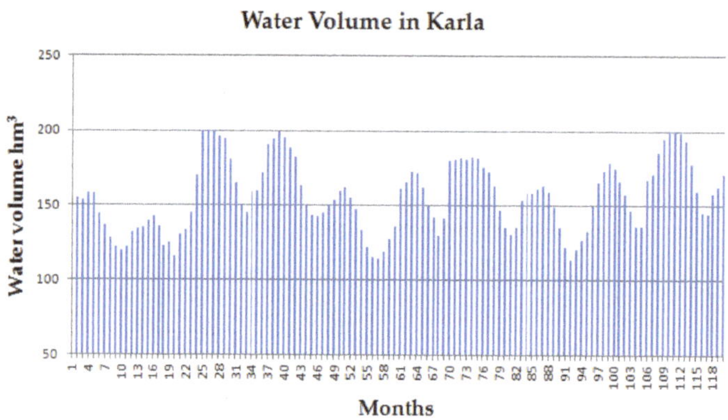

Figure 3. Water stored in Karla reservoir on each month within a 10-year (120 months) simulated historic period starting in 2001 (1 = first month of the simulation).

Thus, apart from the provisioning services of irrigation water and crop productivity (Figure 4a), several other benefits can be obtained through ecosystem services from the restored Lake Karla. The lake offers shelter and food for many species of migratory birds, being a biodiversity hot-spot. Birds favor nesting in the marshes and artificial islands (Figure 4b). The fish community of Lake Karla is composed of six families and thirteen species, with the family of Cyprinidae being the most dominant in terms of abundance and biomass [7].

Figure 4. Ecosystem services provided by Lake Karla: (**a**) crop areas irrigated by the lake's water, (**b**) part of the lake with bird nesting areas, (**c**) the museum, a touristic attraction. Photos taken by the first author of the manuscript in October 2019.

The agricultural floodplain in the perimeter of Karla will be highly protected from inundation, even during seasons with high precipitation and runoff depths. The Karla reservoir will have the capability to trap flooded waters through the operation of the already constructed works. As a result, it will be possible to prevent farmland inundation, enhance groundwater restoration over the years and mitigate desertification. Increased groundwater availability may also support other water uses.

The ecological benefits associated with biodiversity enhancement can also offer great cultural services. Info kiosks, bird-watching sites, the tourist information center and the local museum of natural history (Figure 4c) already exist [13]. The area can offer activities, such as visiting environmental education sites, hiking and cycling, while encouraging the development of additional services/facilities such as accommodation and water sports.

4. Conclusions

This short study summarized the main provisional and non-provisional ecosystem services delivered by the new Lake Karla reservoir located in the southeastern part of Pinios river basin, Thessaly region, central Greece. Based on a previous modelling study and literature information, water supply in irrigated agriculture, the respective crop production and farmers' income in the surrounding cropland areas of Lake Karla can be maximized with the presence of the reservoir, and at the same time, water scarcity and overexploitation of groundwater resources can be mitigated. Maintaining a minimum water level in the reservoir can also support biodiversity and other ecosystem functions uninterruptedly. Cultural (aesthetic and recreational) services also have a great potential to be further developed.

However, for the project's successful operation in the long term, there is a need to raise awareness among all of the stakeholders, not only for the benefits but also for the maintenance and to further develop the actions required. Of high priority is the construction of the remaining parts of the collective irrigation network around the reservoir. For the economic viability of the project, a volumetric pricing of irrigation water that is counterbalanced by crop productivity should be established. Finally, it has to be ensured that the minimum water level in the reservoir is always maintained so that the lake can act as an important biodiversity site and healthy ecosystem.

Author Contributions: Conceptualization, Y.P.; methodology, Y.P.; D.K.; P.G. and D.P.; software, Y.P.; validation, Y.P.; formal analysis, Y.P.; D.K.; P.G. and D.P.; investigation, Y.P.; D.K.; P.G. and D.P.; resources, Y.P.; data curation, Y.P.; D.K.; P.G. and D.P.; writing—original draft preparation, Y.P. and D.K.; writing—review and editing, Y.P.; D.K.; P.G. and D.P.; supervision, D.P. All authors have read and agreed to the published version of the manuscript.

Funding: This research received no external funding.

Institutional Review Board Statement: Not applicable.

Informed Consent Statement: Not applicable.

Data Availability Statement: The data presented in this study are available on request from the corresponding author. The data are not publicly available due to their large size and the fact that they are modelled data.

Conflicts of Interest: The authors declare no conflict of interest.

References

1. Mademli, V.; Karpouzos, D.; Georgiou, P.; Pantelakis, D. Development of a scalable computational tool for irrigation scheduling. *Eur. Water* **2020**, *71*, 15–26.
2. Georgiou, P.E.; Papamichail, D.M. Optimization model of an irrigation reservoir for water allocation and crop planning under various weather conditions. *Irrig. Sci.* **2008**, *26*, 487–504. [CrossRef]
3. Pistocchi, A. *Nature-Based Solutions for Agricultural Water Management*; EUR 31351 EN; Publications Office of the European Union: Luxembourg, 2022; ISBN 978-92-76-60487-7. [CrossRef]
4. Loukas, A.; Vasiliades, L. Probabilistic analysis of drought spatiotemporal characteristics in Thessaly region, Greece. *Nat. Hazards Earth Syst. Sci.* **2004**, *4*, 719–731. [CrossRef]
5. Panagopoulos, Y.; Makropoulos, C.; Kossida, M.; Mimikou, M. Optimal implementation of irrigation practices: Cost-effective desertification action plan for the Pinios basin. *J. Water Resour. Plan. Manag.* **2013**, *140*, 1943–5452. [CrossRef]
6. Zalidis, G.; Takavakoglou, V.; Panoras, A.; Bilas, G.; Katsavouni, S. Reestablishing a sustainable wetland at former Lake Karla, Greece, using Ramsar restoration guidelines. *Environ. Manag.* **2004**, *34*, 875–876. [CrossRef] [PubMed]

7. Sidiropoulos, P.; Chamoglou, M.; Kagalou, I. Combining conflicting, economic, and environmental pressures: Evaluation of the restored Lake Karla (Thessaly-Greece). *Ecohydrol. Hydrobiol.* **2017**, *17*, 177–189. [CrossRef]
8. Panagopoulos, Y.; Dimitriou, E. A Large-Scale Nature-Based Solution in Agriculture for Sustainable Water Management: The Lake Karla Case. *Sustainability* **2020**, *12*, 6761. [CrossRef]
9. Dodouras, S.; Lyratzaki, I.; Papayannis, T. *Lake Karla Walking Guide*; Mediterranean Institute for Nature and Anthropos (med-INA): Athens, Greece, 2014; 228p.
10. Gassman, P.W.; Sadeghi, A.M.; Srinivasan, R. Applications of the SWAT Model Special Section: Overview and Insights. *J. Environ. Qual.* **2014**, *43*, 1–8. [CrossRef] [PubMed]
11. Neitsch, S.L.; Arnold, J.G.; Kiniry, J.R.; Williams, J.R. *Soil and Water Assessment Tool—Theoretical Documentation Version 2009*; Texas Water Resources Institute Technical Report 406; Texas A&M University System College Station: College Station, TX, USA, 2011; Available online: http://swat.tamu.edu/media/99192/swat2009-theory.pdf (accessed on 10 January 2023).
12. Grizzetti, B.; Lanzanova, D.; Liquete, C.; Reynaud, A. *Cook-Book for Water Ecosystem Service Assessment and Valuation*; EUR 27141; Publications Office of the European Union: Luxembourg, 2015. [CrossRef]
13. Management Authority of Lake Karla. Available online: http://www.fdkarlas.gr/Default.aspx (accessed on 10 January 2023).

environmental sciences proceedings

MDPI

Proceeding Paper

Hydrodiplomacy and Climate Change: An Assessment of the Transboundary River Basins of Greece †

Charalampos Skoulikaris

UNESCO Chair INWEB, Department of Civil Engineering, Aristotle University of Thessaloniki, 54124 Thessaloniki, Greece; hskoulik@civil.auth.gr

† Presented at the 7th International Electronic Conference on Water Sciences, 15–30 March 2023; Available online: https://ecws-7.sciforum.net.

Abstract: Hydrodiplomacy is the emerging framework where legal acts, based on technical data and information, aim to support commonly accepted solutions to water-related tensions among states with transboundary waters. In this research, hydrodiplomacy components in relation to (a) policy, (b) preventive, (c) cooperative, and (d) technical aspects are considered together with climate change, which is bound to destabilize the core element of hydrodiplomacy, i.e., water. The study area is composed of the five transboundary river basins of Greece. The coupling of all these different nature elements is conducted with the use of the AHP multicriteria method, and the results of a normalized output that quantifies water transboundary cooperation in the climate crisis era are given.

Keywords: hydrodiplomacy; transboundary cooperation; shared waters; climate change; Maritsa/Meric/Evros; Mesta/Nestos; Struma/Strymonas; Vardar/Axios; Vjosa/Aoos

Citation: Skoulikaris, C. Hydrodiplomacy and Climate Change: An Assessment of the Transboundary River Basins of Greece. *Environ. Sci. Proc.* **2023**, *25*, 5. https://doi.org/10.3390/ECWS-7-14182

Academic Editor: Luis Garrote

Published: 14 March 2023

1. Introduction

Fragile issues that are based on the current status of systems, such as the management of transboundary waters, are highly probable to be destabilized by the induced changes of climate change to water resources. The latest IPCC's report on climate change (AR6) [1] outlines a high confidence that the observed increase in extreme precipitation is associated with an increase in the frequency and magnitude of river floods. High confidence is also expressed on the linkage of the increased frequency and the severity of agricultural/ecological drought with the anthropogenic warming over the last decades. Regarding the future, the report outlines with great confidence mean streamflow decreases in the Mediterranean, as well as flood increases in the same area in terms of magnitude, frequency, and seasonality. Additionally, the upscaling of water management issues from a national level to an international one, i.e., the cases transboundary water resources, will jeopardize existing balances and cooperation agreements. The latter are based on historic water records, with their creditability and accuracy to be doubted under climate change conditions due to the foreseen spatiotemporal variations in relation to the hydrological cycle components.

The mitigation of past hydro-political tensions over shared water systems is confronted by hydrodiplomacy, i.e., the emerging framework that fosters diplomatic processes for resolving or restricting current or imminent disagreements or conflicts between countries that share common water resources [2]. It has been established that in general, nations with active water cooperation share peace, while nations at risk of going to war over issues other than water, tend to not have active water cooperation agreements with the riparian states they have conflicts with [3]. Keskinen et al. [4] introduce a stepwise Water Diplomacy Paths approach for assessing water diplomacy actions, and the authors recognize five key aspects for water diplomacy, and these the political, preventive, integrative, cooperative, and technical aspects.

The research's objective is to assess the impact of climate change on hydrodiplomacy. To do so, and based on the approach proposed by Keskinen et al. [4], the hydrodiplomacy components laid on (a) policy aspects, such as cooperation agreements and common legal frameworks on water management, (b) preventive aspects, such as socio-political tensions, human pressures on hydrosystems, and historical disputes among riparians, (c) cooperative aspects, such as joint development and research programmes and projects, and (d) technical aspects, such as the quality and quantity status of the transboundary waters, are assessed together with climate change features. The latter are expressed as (e) climate change-related studies on transboundary waters and as (f) climate change-related ratified agreements and protocols. The coupling of the various components is communicated through an index, which is applied in the transboundary river basins of Greece and reflects water-related transboundary cooperation.

2. Materials and Methods

2.1. Case Study Area

The case study area consists of the five transboundary river basins that Greece shares with its neighboring countries. Particularly from east to west, Greece shares with Bulgaria and Turkey the Maritsa/Meric/Evros river basin, with Bulgaria the Mesta/Nestos river basin, with Bulgaria and North Macedonia the Struma/Strymonas river basin, with North Macedonia the Vardar/Axios river basin, and with Albania the Vjosa/Aoos river basin [5], Figure 1.

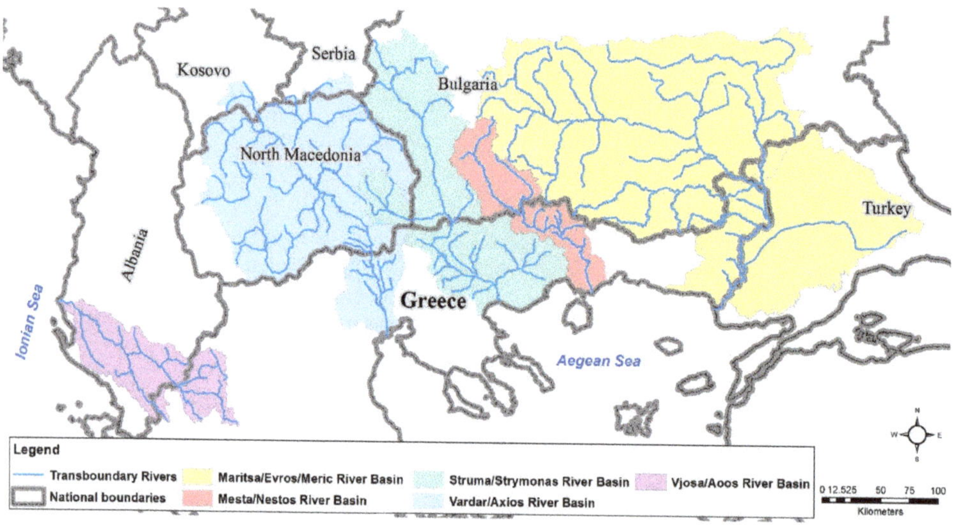

Figure 1. Illustration of the transboundary river basins of Greece that are shared with the neighboring states.

In terms of water policy, Greece and Bulgaria are the only two states that have common environmental policies since both belong to the European Union (EU) and the water governance follows the EU's Water Framework and Floods Directives. As a derivative of the Directives' implementation process, both countries have developed River Basin Management Plans (RBMPs) and Flood Risk Management Plans (FRMPs); thus, the chemical and ecological status of the river water bodies, as well as the pressures on the hydrosystems, have been identified [6]. On the other hand, for the parts of the basins that do not belong in EU Member States, limited information about the water quality status at finer scales is available. However, information relative to population densities, irrigated agriculture demands, and hydropower production can be exploited from the literature [7].

The water-related transboundary disputes between Greece and the other riparian states can be classified into two main categories: one that involves water quantity issues, such as floods, and one connected with water quality issues. Induced floods from the transboundary waters appear in the Maritsa/Meric/Evros and Struma/Strymonas river basins, while degraded water quality inflows from the upstream countries are mentioned in the Mesta/Nestos and Vardar/Axios river basins [6,8–10]. On the other hand, no water-related transboundary pressures appear in the case of the Vjosa/Aoos river basin. The latter basin, together with the Mesta/Nestos one, are basins where a bilateral agreement exists between the riparian states. Toward this direction, since 2010 Bulgaria and Greece have formulated an active joint expert working group that focuses on the implementation of the EU Directives within their shared waters.

Finally, in terms of climate change, all countries of the case study area have ratified the Kyoto protocol and the Paris Agreement for climate change mitigation and adaptation. Greece, Bulgaria, and Turkey have developed national and regional adaptation policies [11]. On the other hand, North Macedonia and Albania have not proceed with implementing the derived obligations from their international commitments. As for the research on climate change impacts on the transboundary waters, which is expressed through the number of scientific publications, the literature shows that for the Maritsa/Meric/Evros and Mesta/Nestos river basins there have been 6 and 9 published scientific articles, respectively, while 5 publications have been conducted in relation to the Struma/Strymonas river basin, e.g., [12–20]. In the case of the Vardar/Axios and Vjoosa/Aoos river basins, 4 publications have been identified for each basin, e.g., [21–23]. To sum up, in the research, all the aforementioned factors, together with climate change, are considered fundamental assets in hydrodiplomacy, with the importance score of each factor/criterion to be given in Table 1.

Table 1. Proposed criteria and scoring scale of each criterion.

No	Criteria	Scoring Scale	No	Criteria	Ranking Scale
1	Rivers' chemical status	Good status > 75% = 5 Good status ~ 50% = 3 Good status < 20% = 0	7	Historical disputes between the riparian	No disputes = 5 Few disputes = 3 Multiple disputes = 0
2	Rivers' ecological status	Good status > 70% = 5 Good status ~ 45% = 3 Good status < 15% = 0	8	Active cooperation agreements	>3 agreements = 5 1–2 agreements = 3 No agreement = 0
3	Vulnerability to floods	No vulnerability = 5 Moderate vulnerability = 3 Vulnerable system = 0	9	Common legislative framework	Yes = 5 Partially = 3 No = 0
4	Population density	Inhabitants/km^2 < 20 = 5 Inhabitants/km^2 ~ 70 = 3 Inhabitants/km^2 > 100 = 0	10	Indicator 6.5.2 of SDG2030	>90% = 5 ~50% = 3 <20% = 0
5	Agriculture activities	Minimum < 10% = 5 Moderate ~ 40% = 3 Extensive > 70% =0	11	International CC policies in national legislation	Yes = 5 Yes, but not in force = 3 No = 0
6	Hydropower generation	Production < 0.1GW = 5 Production < 0.3GW = 3 Production < 0.6GW = 0	12	Publications on CC and transboundary waters	Publications >10 = 5 3–7 publications = 3 Publications <2 = 1

2.2. Hydrodiplomay Criteria, Analytical Hierarchical Process, and Weighting Factors

The mixing of these heteroclites factors was accomplished with the Analytical Hierarchical Process (AHP) multicriteria method [24]. The AHP is grounded on pairwise comparisons of the proposed factors, commonly known as criteria, based on the relevant information about the criteria and on the decision maker's knowledge and experience of the thematic. The method provides decisions when multiple factors/actors are involved in the solution by ranking the proposed criteria, with its applicability in various sectors to be acknowledged in the literature [25,26]. In the research, the AHP was used for estimating the weights of the criteria and thus their contribution to the solution by following the standardized scoring method—from 1 to 9—for each pairwise comparison [27]. It should be mentioned that the scoring is the author's subjective judgment, which nevertheless is based on the collected information that is presented in the previous section. By adopting the following considerations, the final weights and the ranking of the 12 criteria are depicted in Table 2.

Table 2. Criteria weighing and ranking based on the AHP pairwise comparisons.

Name of Criterion	Ranking	Weight
Rivers' chemical status	7	4.30%
Rivers' ecological status	12	1.60%
Vulnerability to floods	8	4.20%
Population density	9	3.40%
Agriculture activities	10	2.10%
Hydropower generation	11	2.00%
Historical disputes between the riparian	12	1.60%
Active cooperation agreements	4	10.10%
Common legislative framework	2	15.30%
Indicator 6.5.2 of SDG2030	5	9.50%
International CC policies in national legislation	1	25.20%
Publications on CC and transboundary waters	3	14.60%

- Climate change is conceived as an important emerging hazard for water resources, thus the climate change-related criteria receive the higher score during the pairwise comparison.
- Criteria related to cooperation agreements and common water policies are very significant and get high scores.
- Water dependencies, e.g., hydropower and irrigation, have a critical role in transboundary waters as they show the dependence of the regional economy on the waters.
- Hydrodiplomacy mechanisms are negatively affected by degraded rivers' water quality, since the water quality affects various human activities.

3. Results and Discussion

The impact of climate change on hydrodiplomacy is expressed by an index score (Table 3) that comes from the normalization of the importance of each criterion on a scale of 1 to 5 (1 = less important, 5 = more important) by multiplying the rank of reach criterion (Table 1) with the weights coming from the AHP method (Table 2). The outputs demonstrate that the more secure cooperation bonds on the transboundary river basin scale under climate change conditions are found in the Mesta/Nestos river basin, which receives a score of 4.11 out of 5 (4.11/5). Particularly, both countries that share the specific basin have common water management policies, implement climate change adaptation plans, and have put in force a joint working group for developing common RBMPs. The lack of extensive knowledge about the water quality status of the Bulgarian river water bodies is one of the criteria that receives a low score. Similarly, the Struma/Strymonas river basin has a high score of 3.70 out of 5, with the flooding problems in the transboundary area impacting the final score. The flooding problems in the transboundary zone as well as the lack of knowledge about the waters originating in Turkey results in classifying the Maritsa/Meric/Evros basin with the third higher score (2.86/5), since climate change will probably exaggerate the existing problems.

On the contrary, the lack of common water management policies together with the lack of development of climate change adaptation plans from North Macedonia and Albania are attributed through the relative low scores for the Vardar/Axios and Vjoosa/Aoos river basins, i.e., 2.41/5 and 2.48/5, respectively. However, both counties are candidate for joining the EU, which is a very promising perspective in terms of common environmental policies with the downstream country.

Table 3. Hydrodiplomacy and climate change index score for the transboundary river basins of Greece.

Basins		Criteria												Score
		1	2	3	4	5	6	7	8	9	10	11	12	
Maritsa/ Meric/Evros	I.S [1]	2.00	3.00	1.60	3.80	1.80	2.40	1.50	4.00	3.00	1.90	2.80	3.00	
	W [2]	0.31	0.30	0.12	0.08	0.04	0.04	0.06	1.01	0.29	0.08	0.10	0.44	2.86
Mesta/Nestos	I.S	5.00	2.00	1.00	3.50	2.20	3.20	4.00	5.00	5.00	2.90	4.00	5.00	
	W	0.77	0.20	0.08	0.07	0.05	0.05	0.17	1.26	0.48	0.12	0.14	0.73	4.11
Struma/ Strymonas	I.S	5.00	1.00	2.00	4.00	1.30	3.20	2.50	5.00	5.00	3.80	2.50	3.00	
	W	0.77	0.10	0.15	0.08	0.03	0.05	0.11	1.26	0.48	0.16	0.09	0.44	3.70
Vardar/Axios	I.S	2.00	1.00	3.50	3.80	1.50	2.00	4.00	2.50	3.00	3.50	2.00	2.00	
	W	0.31	0.10	0.27	0.08	0.03	0.03	0.17	0.63	0.29	0.15	0.07	0.29	2.41
Vjosa/Aoos	I.S	1.00	2.00	5.00	3.90	5.00	4.20	4.50	2.50	1.00	4.00	3.30	2.00	
	W	0.15	0.20	0.39	0.08	0.11	0.07	0.19	0.63	0.10	0.17	0.11	0.29	2.48

[1] I.S: initial score; [2] W: weight.

4. Conclusions

The research proposes a methodology for assessing the hydrodiplomacy mechanism's balance in climate change conditions at the scale of transboundary river basins. The proposed approach is expressed using an index that depicts the dependency of water diplomacy on climate change. The index couples traditional water management factors, such as water quality, floods, and cooperation agreements and protocols with climate change factors with the use of the AHP method. Although the proposed methodology is subjective to the ranking of the criteria and the pairwise comparison scores, the final outputs constitute an important roadmap for the evaluation of the water cooperation status at shared river basins under climatic stress situation.

Funding: This research received no external funding.

Institutional Review Board Statement: Not applicable.

Informed Consent Statement: Not applicable.

Data Availability Statement: Data are fully available upon official request to the author.

Conflicts of Interest: The authors declare no conflict of interest.

References

1. Caretta, M.A.; Mukherji, A.; Renwick, J.; Betts, R.A.; Gelfan, A.; Hirabayashi, Y.; Lissner, T.K.; Cherchi, A.; Gunn, E.L.; Liu, J.; et al. Chapter 4: Water. In *Climate Change 2022: Impacts, Adaptation and Vulnerability. Contribution of Working Group II to the Sixth Assessment Report of the Intergovernmental Panel on Climate Change*; Cambridge University Press: Cambridge, UK, 2022.
2. Schmeier, S.; Shubber, Z. Anchoring Water Diplomacy-The Legal Nature of 209 International River Basin Organizations. *J. Hydrol.* **2018**, *567*, 114–120. [CrossRef]
3. UNECE. Policy Guidance Note on the Benefits of Transboundary Water Cooperation: Identification, Assessment and Communication. In *Convention on the Protection and Use of Transboundary Watercourses and International Lakes United Nations*; Economic Commission for Europe: New York, NY, USA; Geneva, Switzerland, 2015.
4. Keskinen, M.; Salminen, E.; Haapala, J. Water diplomacy paths–an approach to recognise water diplomacy actions in shared waters. *J. Hydrol.* **2021**, *602*, 126737. [CrossRef]
5. Kolokyhta, E.; Skoulikaris, C. Dependencies in transboundary water management in Greece in the face of climate change. In Proceedings of the E-Proc 38 IAHR World Congress, Panama city, Panama, 14–18 August 2019.
6. Skoulikaris, C. Transboundary Cooperation through Water Related EU Directives' Implementation Process. The Case of Shared Waters between Bulgaria and Greece. *Water Resour. Manag.* **2021**, *35*, 4977–4993. [CrossRef]
7. Skoulikaris, C.; Zafirakou, A. River Basin Management Plans as a tool for sustainable transboundary river basins' management. *Environ. Sci. Pollut. Res.* **2019**, *26*, 14835–14848. [CrossRef] [PubMed]
8. Mylopoulos, Y.A.; Kolokytha, E.G. Integrated water management in shared water resources: The EU Water Framework Directive implementation in Greece. *Phys. Chem. Earth Parts A/B/C* **2008**, *33*, 347–353. [CrossRef]
9. UNECE (United Nations Economic Commission for Europe). *Second Assessment of Transboundary Rivers, Lakes and Groundwaters*; United Nations Publications: New York, NY, USA, 2011.

10. Skoulikaris, C. Toponyms: A neglected asset within the water framework and flood directives implementation process; the case study of Greece. *Acta Geophys.* **2022**. [CrossRef]
11. Kolokytha, E.; Skoulikaris, C. WRM and EU policies to adapt to climate change-Experience from Greece. In *Climate Change-Sensitive Water Resources Management*, 1st ed.; Teegavarapu, R.S.V., Kolokyhta, E., Galvao, C.O., Eds.; CRC Press: London, UK, 2020; Chapter 1; pp. 1–23.
12. Chang, H.; Knight, C.G.; Staneva, M.P.; Kostov, D. Water resource impacts of climate change in southwestern Bulgaria. *Geojournal* **2002**, *57*, 159–168. [CrossRef]
13. Alexandrov, V.; Genev, M. Climate variability and change impact on water resources in Bulgaria. *Eur. Water* **2003**, *1*, 25–30.
14. Aksoy, H.; Unal, N.E.; Alexandrov, V.; Dakova, S.; Yoon, J. Hydrometeorological analysis of northwestern Turkey with links to climate change. *Int. J. Climatol.* **2008**, *28*, 1047–1060. [CrossRef]
15. Skoulikaris, C.; Ganoulis, J. Climate change impacts on river catchment hydrology using dynamic downscaling of global climate models. In *National Security and Human Health Implications of Climate Change*; Fernando, H.J.S., Klaic, Z.B., McCulley, J.L., Eds.; Springer: Dordrecht, The Netherlands, 2011.
16. Ganoulis, J.; Skoulikaris, C. Impact of Climate Change on Hydropower Generation and Irrigation: A Case Study from Greece. In *NATO Science for Peace and Security*; Series C: Environmental Security; Springer: Dordrecht, The Netherlands, 2011; Volume 3, pp. 87–95.
17. Skoulikaris, C.; Ganoulis, J. Assessing Climate Change Impacts at River Basin Scale by Integrating Global Circulation Models with Regional Hydrological Simulations. *Eur. Water* **2011**, *34*, 53–60.
18. Skoulikaris, C.; Ganoulis, J. Multipurpose hydropower projects economic assessment under climate change conditions. *Fresenious Environ. Bull.* **2017**, *26*, 5599–5607.
19. Sordo-Ward, A.; Granados, A.; Iglesias, A.; Garrote, L.; Bejarano, M.D. Adaptation effort and performance of water management strategies to face climate change. Impacts in six representative basins of Southern Europe. *Water* **2019**, *11*, 1078. [CrossRef]
20. Stefanopoulou, D.K.; Skoulikaris, C. Assessment of hydrodiplomacy effectiveness under climate change: The case study of the transboundary river basins of Greece. *IOP Conf. Ser. Earth Environ. Sci.* **2022**, *1123*, 012089. [CrossRef]
21. Frasheri, A.; Pano, N. Impact of the climate change on Adriatic Sea hydrology. *Elsevier Oceanogr. Ser.* **2003**, *69*, 92–96.
22. Peters, R.; Berlekamp, J.; Lucía, A.; Stefani, V.; Tockner, K.; Zarfl, C. Integrated Impact Assessment for Sustainable Hydropower Planning in the Vjosa Catchment (Greece, Albania). *Sustainability* **2021**, *13*, 1514. [CrossRef]
23. Skoulikaris, C. Run-Of-River Small Hydropower Plants as Hydro-Resilience Assets against Climate Change. *Sustainability* **2021**, *13*, 14001. [CrossRef]
24. Saaty, T.L. How to Make a Decision: The Analytic Hierarchy Process. *Europ. J. Opera. Res.* **1990**, *48*, 9–26. [CrossRef]
25. Spiliotis, M.; Skoulikaris, C. A fuzzy AHP-outranking framework for selecting measures of river basin management plans. *Desalination Water Treat.* **2019**, *167*, 398–411. [CrossRef]
26. Ogato, G.S.; Bantider, A.; Abebe, K.; Geneletti, D. Geographic information system (GIS)-Based multicriteria analysis of flooding hazard and risk in Ambo Town and its watershed, West shoa zone, oromia regional State, Ethiopia. *J. Hydrol. Reg. Stud.* **2020**, *27*, 100659. [CrossRef]
27. Goepel, K.D. Implementation of an online software tool for the analytic hierarchy process (AHP-OS). *Int. J. Anal. Hierarchy Process* **2018**, *10*, 469–487.

environmental sciences proceedings

MDPI

Proceeding Paper

Myanmar's Planned Resettlement and Social Impact: An Empirical Case Study †

Aung Tun Lin * and Kaiwen Yao

School of Water Resources and Hydropower Engineering, North China Electric Power University, Beijing 102206, China; kwyao@ncepu.edu.cn

* Correspondence: aungtunlin.iitr@outlook.com

† Presented at the 7th International Electronic Conference on Water Sciences, 15–30 March 2023; Available online: https://ecws-7.sciforum.net.

Abstract: This paper explores the planned resettlement process and the social impact induced by the Tha Htay Hydropower Project in Myanmar. Through the empirical case study, the status of direct impact on the resettled people, changes in their livelihood, policy orientation towards practical limitations, practical consequences of resettlement, and inadequate preparation for resettlement implementation were studied as verifiable evidence. In this paper, the survey data were compiled from the responses of resettled people from three resettled villages: Maewa, Payit and Yegauk, which were interpreted as the real condition of the project-affected people because a higher degree of their dissatisfaction negatively affected the project, whereas a higher degree of satisfaction led to positive effects. In this paper, a quantitative method was used to analyze the planned resettlement and its social impact. As a survey descriptive design, a simple random sampling method was utilized to collect samples from the target population with a contribution of structured questionnaires. The collected data were presented by mean and standard deviation to decide the real situation of the resettlement project. The study found that details of social impact were considered to carry out when the construction of the hydropower project began. Then, the policy constraint in the resettlement process was verified, which led to the insufficient preparation and implementation of the resettlement. To meet the future development requirements of the planned resettlement process with fewer social impacts on hydropower development projects, practical contributions and policy recommendations are made for the compensation of farmland, where the people's livelihoods are land-based, and additional livelihood packages.

Keywords: planned resettlement; social impact; resettled people; practical limitation; policy constraint

Citation: Lin, A.T.; Yao, K. Myanmar's Planned Resettlement and Social Impact: An Empirical Case Study. *Environ. Sci. Proc.* **2023**, 25, 6. https://doi.org/10.3390/ECWS-7-14205

Academic Editor: Slobodan Simonovic

Published: 14 March 2023

1. Introduction

Certain hydropower development projects, even in developed countries, have had a long-term effect on the entire human race. It is estimated that infrastructure development projects relocate approximately 15 million people across the world. About four million people are displaced each year worldwide by reservoir projects alone [1]. In Myanmar, because of its abundant land and water resources, hydropower is one of the most significant sectors and supplies the majority of the electricity distribution to locals and exports, with the help of foreign investment [2] and the country's budget allocation. In running hydropower development projects, resettlement often has the most difficult and controversial impact on infrastructure, production development, and social, cultural and environmental issues [3] when aiming to fulfill the satisfaction of all affected people. Additionally, land acquisition is an extremely complex issue, and the previous experiences of resettlement projects have caused a significant conflict and lack of trust [4]. Myanmar's land acquisition and resettlement practice for hydropower development was a combination of the methods of previous similar hydropower development projects and the general methods of national guidelines. Because of the socio-economic loss of the impacted villages and the associated people's livelihoods, the IFC recommends in-kind compensation when the displaced people's livelihoods are land-based, advising replacement land to be in the same condition or superior in a productive land for the people displaced [5].

There are 27 operational hydropower development projects in Myanmar: 9 under construction and 55 additional projects planned [1]. Only the remaining resettlement work of the villages of Payit and Yegauk, due to the Tha Htay Hydropower Project, is anticipated to be completed at the end of 2023 regardless of adding rehabilitation work, although Maewa Village was already resettled in March 2009. This work (because of which 500 households with 1591 persons were displaced to new resettlement locations in March 2009 and March 2021, respectively) is the second project running under the Ministry of Electric Power after the Upper Paunglaung Hydropower Project. All of the resettled people in the Tha Htay Hydropower Project received only cash compensation for each casual household in one million in MMK, for their loss of farmlands, perennial plants, income and other assets. Schmitt-Degenhardt asserted that a landholding of 3 acres was seen as an indicator of the minimal landholding required to reduce poverty, based on agriculture alone in Myanmar [6]. The current resettlement site was selected by the project proponent according to the resettled people's willingness to move near the Thandwe-Taungup road, but it was not the optimal site selection. Every piece of infrastructure of resettlement and the Tha Htay Hydropower Project were mainly based on the union budget, without support from any international financial institutions. During the dam construction and resettlement and severe COVID-19 outbreak, the deficit of bank transfers and the limitations from the union budget to the local bank, as well as an increase in the price of local goods, caused the life of the resettled communities to be more difficult, including a long-term improvement in children's education because parents' job opportunities were uncertain.

In this study, the resettlement and social impacts based on the responses of respondents were examined, highlighting the difference between local and international resettlement practices [7], with the aim of investigating the displaced people's options towards resettlement practice, the social impacts and their satisfaction status with the fundamental infrastructures, transportation, education level [8], healthcare services and job opportunities. Based on an empirical case study including the experiences or perspectives of resettled persons, a judgement on the real situation of resettlement and the constraints of the government policy applied for resettlement was made, since the Tha Htay Hydropower Project is the second resettlement project with national priority proposed by the government.

2. Tha Htay Hydropower Project and Its Regional Context

Six power plants are being constructed, with an installed capacity of 1564 MW, across Myanmar. Among them, the Tha Htay Hydropower Project is under construction, with an installed capacity of 111 MW. In Rakhine state, the most western state of Myanmar, the planned hydropower projects are Lemro 1 (600 MW) and Lemro 2 (90 MW) on the Laymro River, Kyein Ta Li (28 MW) on the Kyein Ta Li River, Mi Chaung (200 MW) on the Kaladan River, Saing Din (77 MW) on the Saing Din River, Than Dwe (39 MW) on the Thandwe River [9], and Ann (10 MW) on Ann Creek, totaling 1044 MW including Tha Htay (111 MW). Despite the eight rivers being identified as having hydropower development potential in Rakhine state, the Tha Htay River is one of the most important rivers after the Lemro River. The Thahtay River flows from northwest to west in Rakhine for around 120 km, before discharging into the Bay of Bengal at Shwe Hlay Town. At the river mouth of the Tha Htay River, the catchment area is 1293 km2 and the discharge is approximately 100–120 m3/s [9]. The 111 MW Tha Htay Hydropower Project in the Thandwe District began construction in 2008; it was developed by MOEP and financed by the government and it has reached over 73% progress as of November 2022. The geographical location of the Tha Htay Hydropower project and its construction site are illustrated in Figures 1 and 2 respectively.

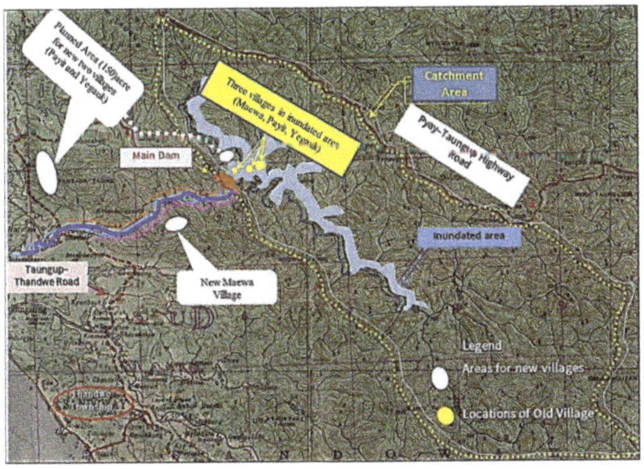

Figure 1. Geographical location of the case study area including two resettlement sites, dam construction site and the inundated area.

Figure 2. Construction site of the Tha Htay Hydropower Project from downstream view. (Source: survey photo).

It is a 92.5 m rock-filled dam despite having a 90.83 m design height. At frist, the Norwegian Government provided a technical assistance for handling environmental and social issues, with the Tha Htay as a pilot project. Depending on an annual budget allocation from the Ministry of Electric Power, all compensations, logistics and required machinery for relocation, land clearing and every activity of the resettlement processes were the responsibility of the project proponent (DHPI). In spite of considering the completion of the project to be 2021, the construction has been ongoing due to the impact of COVID-19, the union budget deficit and the country's political situation. Now, the whole hydropower project is expected to be commissioned in the 2025–2026 fiscal year, depending on the union budget. The Tha Htay power station would be connected to the Oakshitpin-Taungup transmission line of the national power grid. For long-term improvement, it is expected to increase revenue, provide job opportunities to local people during the operation of the hydropower station, and establish small-, medium- and large-scale electricity-based industries, leading to economic development in the local region. The progress of the Tha Htay Hydropower Project and resettlement process so far is shown in Table 1.

The water level of the old Maewa Village was only 16 m above sea level. To escape from the rising water level by constructing a dam wall, Maewa Village was relocated first in March 2009, followed by a plan to relocate the remaining two villages: Payit and Yegauk, to near Mewa Village. However, those two villages had to be left because of the low water level at that time. There was the first democratically elected government of Myanmar in March 2011. The statistics of Maewa, Payit and Yegauk collected in 2008 were compiled again in March 2014. In January 2015, Maewa was compensated for the rebuilding cost of their houses and their lost properties in 2009, with it taking nearly six years for Maewa to receive compensation. In March 2016, the project-affected people from the villages of Payit and Yegauk received compensation for their losses, such as farmlands and perennial plants, but not household compensation. The project proponent first supported the housing compensation batch by batch from 18 March 2022. As per DHPI's 2016 resettlement action plan, it was found that the housing compensation rate of Payit and Yegauk was higher than that of Maewa Village. Moreover, support for the casual household in Payit and Yegauk was about 1,000,000 kyats, while those in Maewa received about 300,000 kyats for damage.

Table 1. Progress of Tha Htay Hydropower Project and resettlement.

Time	Project Progressing Condition
April 2004	Preliminary survey measurements were conducted for the development of the 111 MW Tha Htay Hydropower Project.
December 2004	Myanmar Government approved the Tha Htay Hydropower Project to be implemented.
June 2005	The detailed measurements from the selected dam locations were conducted. After that, location of the dam was confirmed.
April 2008	Construction work of the project started. The inventory of the villages of Maewa, Payit and Yegauk was collected.
March 2009	Maewa Village was relocated to a new location.
March 2014	The statistics of the three villages collected in 2008 were compiled again.
January 2015	Maewa households were first compensated for the rebuilding cost of their houses and the properties they lost.
March 2016	The project-affected people from Payit and Yegauk were compensated for their losses, such as farmlands, garden land, perennial plants and so forth, except household compensation.
March 2020	The resettlement area for Payit and Yegauk was approved by the regional government.
October 2020	The inventory of the remaining two villages to be relocated was made again. Clearing the resettlement area for housing plot and road building.
March 2021	The first batch of housing compensation was given to some households from the villages of Payit and Yegauk, and 75 households from Yegauk and 153 households from Payit were relocated to the proposed resettlement area.
April 2021	The remaining 233 households from Yegauk Village moved to the resettlement area.
November 2022	73% of the whole project work has been completed but resettlement work will be complete in late 2023, except rehabilitation work.
2025-2026 fiscal year	Project commissioning is expected, but it depends on the union budget allocation

Source: DHPI (2022) and interview.

There were immense differences in the customs and cultures among the resettled communities in the relocation of the impacted people [10]. However, the resettlement area of the villages of Payit and Yegauk is 19.31 km away from the dam of the Tha Htay Hydropower Project and 25.75 km away from the original old villages, according to the field survey on the project-affected people. It is in the same district of Thandwe, closer to Shwe Hlay in Thandwe Township, which has the same culture and customs. Due to the heavy rain in the Rakhine state, the agricultural and alluvial lands located upstream of the reservoir are flooded. Now, the resettled people can go to the previously impacted villages to cultivate land for rice and vegetable farming during the implementation of the hydropower project. After impounding the reservoir, they may lose their income if they are not allowed to go to the previous villages to make their earnings.

3. Research Methodology

In this quantitative-approach-based research, a survey descriptive design was used to analyze the planned resettlement and its social impact. Statistical descriptive statistics are used to describe the basic features of the responses of the resettled people and present them by mean and standard deviation, based on the statistical outcomes. A simple random sampling method was used to collect samples from the target population, with a contribution of structured questionnaires from 127 samples: 78 samples from Yegauk Village, 43 samples from Payit Village and 8 samples from Maewa Village (out of 500 households in 3 villages). The target population of the resettlement due to land acquisition for Tha Htay Hydropower Project is 1591, including 166 persons in 39 households in Maewa Village, 433 persons in 153 households in Payit Village, and 922 persons in 308 households in Yegauk Village. Both primary and secondary data were collected. For the primary data collection, a survey with a close-ended questionnaire was conducted. Then, the collected data were tested and analyzed in SPSS software, mainly focusing on the statistical data. The secondary data included those from the resettlement office of hydropower construction unit No.4, related to the inventory of displaced people affected by the project; the baseline data report; the environmental and social management report; the Resettlement Action Plan (RAP) report, including the socio-economy regarding the Tha Htay Hydropower Project; an analysis of the guidelines of IFC standards and national land acquisition, compensation and resettlement; and the rehabilitation policy regarding hydropower development.

4. Findings and Discussion

According to the land acquisition findings, the impact of the Tha Htay Hydropower Project on three submerged villagers was found and is presented in Table 2. About 677.9 acres of farmland, upland, garden land and garden land in the protected forest from three submerged villages were occupied by the project proponent.

Table 2. Impact of the Tha Htay Hydropower Project on three submerged villages.

Items	Maewa Village	Payit Village	Yegauk Village	Total
Household	39	153	308	500
Population	166	433	992	1591
Monastery	1	1	1	3
Basic Education School	-	1	1	2
Rural Health Care Centre	1	1	1	3
Farmland (ac)	49.62	52.67	74.33	176.62
Upland (ac)	64.09	39.92	130.07	234.08
Garden Land (ac)	48.6	55.2	76.1	197.9
Garden Land in the protected Forest (ac)	26.3	26	17	69.3

Source: 2016, Statistics of DHPI's Resettlement Action Plan.

The study firstly found that the basic property infrastructures, including housing, monasteries, schools, rural healthcare centers, playgrounds, water supply systems, roads and housing for educational employees, in the villages of Payit and Yegauk, respectively, have been provided in the resettled villages by the project proponent. As per in Figure 3, there is a library but no housing for educational employees in Maewa. Additionally, there have been no healthcare employees in Maewa Village since 2016 nor in both Payit and Yegauk since 2021. The villages of Payit and Yegauk are separated in the same resettlement area shown in Figure 4 but they are in a close relationship. The people are Buddhist Rakhine people, sharing community facilities such as the community building, rural healthcare center and cemetery. Their way of life has changed differently compared to their previous lifestyle; however, the main source of their livelihood is still casual working such as bamboo cutting, livestock rising and forest product extraction.

Compared to villagers in Yegauk, people from Payit seem to be poorer in their economic situation, because most of them were affected by the flood in 2016 more severely than Yegauk villagers and they received less housing compensation when compared with their previous housing condition.

Figure 3. Resettlement site of Maewa Village (Source: Survey photo).

Figure 4. Resettlement site of Payit Village with green roof and Yegauk with blue roof.

4.1. Background Information of Respondents

The characteristics of 127 respondents were analyzed by gender, age, education and current job condition level according to Table 3, which shows that among 127 respondents, 46% were male and 53.5% were female. Their age was divided into three groups: 16–36 years, 37–56 years and 57–86 years. The study found that the age range between 16 and 36 was largely engaged in the survey, with 52 (41%) respondents; the second-largest group was the age range of 37–56 years, with 48 (37.7%) respondents; and in the remaining age range, between 57 and 86 years, there were only 27 (21.3%) respondents. Concerning the education factor, 56.7% of respondents were illiterate. Moreover, 33.1% and 6.3% of respondents used to go to primary and middle schools, respectively. Two (1.6%) respondents each attended high school and university education, yet a graduated person has been serving as an assistant primary teacher in Yegauk Village since before the displacement of Yegauk Village.

Table 3. Demographic profile of respondents.

Description		Frequency	Percent (%)
Gender	Male	59	46.5
	Female	68	53.5
	Total	127	100
Age (years)	16–36	52	41
	37–56	48	37.7
	57–86	27	21.3
	Total	127	100
Education	Illiterate	72	56.7
	Primary school level	42	33.1
	Middle school level	8	6.3
	High school level	2	1.6
	University level	2	1.6
	Graduated	1	0.8
	Total	127	100

Figure 5 describes the current job conditions of the respondents in three villages: 26.0% of respondents were casual workers and 31.5% relied on cutting bamboo trees for a living. Few respondents were engaged in rice and vegetable farming, driving transport vehicles, running small businesses, logging and serving in the government sector.

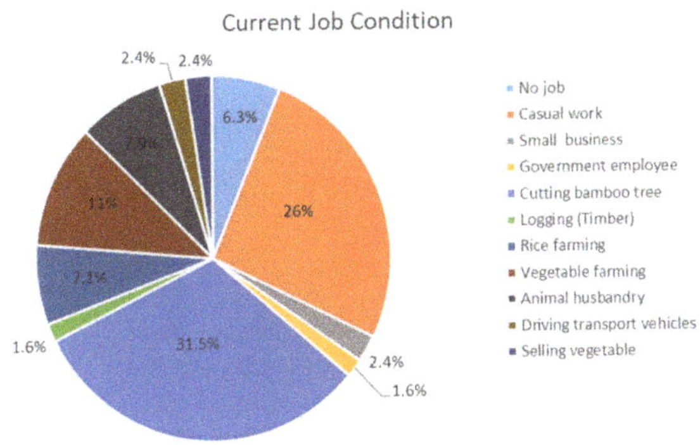

Figure 5. Current jobs of the respondents.

Most villagers who were working odd jobs were illiterate, and a few were peasants. Therefore, they were content with having enough food for daily consumption.

4.2. Social Impacts on Resettled People from Resettlement

Regarding the satisfaction responses of residents for the infrastructure condition, out of 127 respondents, 68 (53.5%) respondents and 4 (3.1%) respondents, respectively, were satisfied with their current resettlement area located beside the Taungup-Thandwe road, which was close to the Shwe Hlay hospital and also good for children's educational development. However, 50 (39.4%) respondents were not satisfied with it because of the lack of agricultural land, cultivation land, or pastureland for their farming adjacent to the resettled area. Only five (3.9%) respondents were neither satisfied nor dissatisfied with it due to the impossibility of living alone in the old village, so they had to move to the resettled area. Concerning the housing provided, there was significant satisfaction for 94 (74%) and 3 (2.4%) respondents regarding receiving the housing compensation for building houses before relocating to the resettlement area. However, 30 (23.6%) respondents made complaints more than six times about the inadequate compensation for housing and for a difficult withdrawal compensation amount from the local bank. According to 122 respondents (96.1%), the roads were reliable; for 1 respondent (0.8%), the interconnection roads were very reliable for all seasons; and few respondents (3.1%) needed to transform the crushed-rock streets in front of their houses into reinforced concrete ones. Then, 95 (74.8%) respondents were satisfied and 2 (1.6%) were very satisfied with electricity conditions, but another 30 (23.6%) respondents from the villages of Payit and Yegauk were not happy with this as the electricity meters in their homes were without internal wiring, causing a lack of electricity access (even for domestic lighting), whereas all respondents from Maewa Village were satisfied with their access to electricity. Next, for 41 (32.3%) and 58 (45.7%) respondents, the situation of fetching water for daily household use was better than the previous conditions, while for another 28 (22.0%) respondents, it was not as good as their previous conditions since they also used to acquire spring water round the clock. However, they all complained about the difficulty of obtaining water during two months only, such as March and April, before the raining season, due to the site being 300 feet above sea level and there only being a little spring waterway in summer. On school facilities and children's educational development, 122 (96.1%) respondents and 4 (3.1%) respondents were satisfied and very satisfied, respectively. Unlike the others, one respondent could not afford to send her daughter to the high school of Shwe Hlay Town due to her daughter's academic tuition fee and the transportation charge per year. According to the field survey, some families from Payit Village took their children to the old village with them to earn money from rice and vegetable farming, neglecting their children's education development. Regarding the healthcare situation, 120 (94.5%) respondents said that their resettled area was close to the Shwe Hlay Hospital, and they could go to any clinic for their health problems. However, according to six respondents, there had been no healthcare personnel at the rural healthcare center in Maewa for seven years, as well as for nearly two years in Payit and Yegauk. Additionally, they explained that they could not go to the hospital or any clinic without any money due to the lack of job opportunities in resettled areas.

Regarding the investigation of their relationship with resettled communities, 18 (14.2%) respondents reported their own problems between the 75 household groups and 233 household groups in Yegauk Village after the resettlement. That is, 75 household groups first moved to a new resettled area in the last week of March 2021, while the remaining 233 household groups demanded 10 million kyats of compensation for their living in the long run. This was in vain, and they had to move late to the resettled area in late April 2021, causing severe tension between the two groups, with the resignation of the latter group from Yegauk Buddhist monastery. However, such an issue was not found in Payit. Yet, for one of the village leaders from Payit Village, he was blamed by the people, who thought he favored the project proponent. Thirteen (10.2%) respondents and 95 (74.8%) respondents had a fairly good or good relationship with the same group. Compared to the relationship with their previous neighbors, for 122 (96.1%) respondents and 5 (3.9%) respondents, respectively, their neighborhood was friendly or very friendly to them. In communication with the host villages, 126 (99.2%) respondents were found to be comfortable, whereas the 1 respondent from Payit Village felt uncomfortable when seeing host villagers due to the despair for her in Payit Village. According to 92 (72.4%) and 2 (1.6%) respondents, they felt socially secure in their respective resettled village. Contrarily, 33 (26.9%) respondents, with the knowledge of theft cases, felt socially insecure in comparison with the old village experience. Concerning people's participation in religious and cultural festivals, over 40% always went to religious festivals, but they sometimes went to cultural festivals. Over 27% often participated in these festivals, while there was over 10% occasionally going to those festivals because of a lack of money.

Regarding people's participation in the resettlement process, 98 (77.2%) respondents were satisfied with it while 27 (21.3%) respondents were unhappy with the village leader's own decision in the site selection, without asking for their agreement, and 2 (1.6%) respondents did not show their satisfaction or dissatisfaction about it. To the question of compensation rate allocation, 102 (80.3%) and 2 (1.6%) respondents said that they had received the highest cash compensation compared to the previous hydropower projects, whereas 22 (17.3%) respondents were dissatisfied with it because the project proponent supported only cash compensation for all they lost, such as houses and farmlands, and as damage, provided one million in MMK for the households living on odd jobs. Such a cash compensation was insufficient for new resettled households to earn a living in the long term. One respondent (0.8%) said that she followed others because other people agreed, and thus so did she.

Then, on satisfaction with their present working opportunities, 64 (50.4%) respondents liked their current jobs for enabling farming or bamboo cutting and trading, while 62 (48.8%) respondents were fed up with it due to the high commuting cost (4500 kyats). In addition, most women were dissatisfied with their unemployment situation once being displaced to a new resettled location. This unsatisfied percentage (48.8%) was close to the percentage of respondents satisfied with their current working condition. Moreover, when finding out changes in their jobs or careers compared to their previous conditions, there was a complete change for 46 (36.2%) respondents, while a slight change was found for 52 (40.9%) respondents. Therefore, the findings pointed out that most of the villagers had changed their jobs. Among them, most women were dependent on only their husbands. When their husbands went to the old villages, they lived alone or with their children at home.

This study sought an in-depth understanding of the responses of resettled people and of policy constraints in the process of resettlement for affected households. It also attempted to identify the development gaps and opportunities in the resettlement program. The findings on impacts from resettlement inform that on average, people were satisfied with the resettlement area selection and housing, with consistently lower responses to them compared to other indicators, according to Table 4. By comparing people's satisfaction, the study found the highest mean score on educational facilities provided, with children's educational development being 4.02, while the lowest mean score on present working opportunities is 3.02. The results reveal that the respondents had more satisfaction in electricity conditions, educational facilities and children's educational development, health conditions, public participation, and compensation rate allocation, and had more consistent responses to them. Overall, the resettlement project was better than the moderate condition. However, the study reveals that the project proponents need to create more relevant job opportunities for resettled people in the long run, as a livelihood restoration plan.

Table 4. Satisfaction and agreement level on basic infrastructures, social protection, people's participation, compensation rate allocation and working condition for resettled people.

Questionnaires	Mean	Std. Deviation
Are you satisfied with the current resettled area selection?	3.2	1.011
Are you satisfied with housing provided?	3.55	0.879
Are you satisfied with the electricity condition at your home?	3.78	0.453
Are you satisfied with educational facilities provided and children's education development?	4.02	0.251
Do you agree that your health care condition is better, compared to previous condition?	3.9	0.433
Do you agree that you feel socially secure here?	3.5	0.899
Are you satisfied with people's participation in the resettlement process?	3.56	0.823
Are you satisfied with compensation rate allocation?	3.66	0.779
Are you satisfied with present working opportunities?	3.02	1.0

Source: Survey Data (2022).

All findings in this study are expected to contribute to practical contribution and policy recommendation. The previous studies really support our findings. These findings based on on-ground data are useful for further studies of resettlement in Myanmar's hydropower development projects.

5. Recommendation and Conclusions

While implementing the Tha Htay Hydropower construction project including a resettlement plan and relying mainly on the union budget allocation, there has been an increase in the focus of the domestic society of Myanmar on the environmental and social impacts of hydropower project construction, resettlement, and employment. Consequently, not only the project proponent but also financial institutions need to plan and implement an appropriate resettlement process by providing both full cash and in-kind compensations to the displaced people, who lost their land-based livelihoods. In fact, following the insufficient land acquisition and resettlement and rehabilitation policy in Myanmar, improper resettlement and severe social impact has led to resettlement conflicts between the project proponent and the displaced people. To solve these problems, this study has explored various entities of emphasis on resettlement, with fewer social impact mechanisms and policies with respect to hydropower projects. For project proponents, they should optimally identify important features of a resettlement site considering the slope, elevation, soil [11], water availability, and even agricultural resources adjacent to the resettled area, even if the site is selected by the project proponent based on the willingness of the affected villagers. A detailed appraisal of agricultural and cultivation land acquisition should be conducted prior to resettlement by coordinating with other regional agencies such as the regional state authority, Forest Department, Agricultural Department and Department of Land Records. It is recommended that "Land for land" policy should be the preferred compensation option when livelihoods are totally land-based. In addition, other livelihood packages should be considered for resettled households so that the resettled areas have sustainable economic and social development. Without agricultural or cultivation lands and livelihood packages, resettled households may face an unsustainable livelihood in the long run. Therefore, it should be arranged to restore the socio-economic life of resettled people through livelihood activities, job opportunities and any social support. After completing the resettlement project in 2023, there should be post-resettlement monitoring and evaluation. Policymakers ought to develop resettlement guidelines to meet international standards and ensure an optimally suitable resettlement site selection prior to issuing resettlement project approval. This is needed to issue specific policies, as well as technical specifications related to land acquisition and the evaluation of social impacts and resettlement.

In this paper, based on the findings of the satisfaction or dissatisfaction of the affected people, the case study presents some essential gaps in applying the current resettlement. First, it is found that Myanmar's resettlement policy allowed cash compensation for all losses of the displaced people from three villages. Instead, an appropriate compensation practice should be thoroughly considered by providing both a cash compensation and in-kind compensation to the displaced people, to cover their previous baseline properties when their livelihoods are land-based. However, the previous and current resettlement projects have not supported the replacement of farmlands to the resettled people. In addition, other livelihood support should be offered to the resettled community's sustainable development because it was found that until the late implementation stage, there were no livelihood activities. Second, the project proponent should participate in any resettlement site selection to ensure it is optimal (considering the slope, its elevation, water resource, available agricultural land adjacent, transportation, job opportunities, etc.). Third, the impacted villages should not be relocated and resettled until the completion of the housing infrastructures in the resettled site. Accordingly, all fundamental infrastructures and other basic requirements need to be ready. Perennial crops should be planted in advance so that they are ready to bear fruit prior to the resettlement project. When all those resettlement issues are completed, their skilled careers may be started.

Considering all the findings and recommendations, the local authorities, the union government, policymakers, technical experts, the displaced community and the local host community need to coordinate to reach an agreement prior to a project, in order to note the potential impacts and consequences of the displacement and resettlement and develop an effective and comprehensive resettlement policy while implementing hydropower projects successfully and sustainably.

Author Contributions: The individual contributions are: Conceptualization, A.T.L.; methodology, A.T.L.; writing—original draft preparation, A.T.L. and K.Y.; writing—review and editing, A.T.L. and K.Y.; supervision, K.Y.; project administration, A.T.L. All authors have read and agreed to the published version of the manuscript.

Funding: This research received no external funding.

Institutional Review Board Statement: Not applicable.

Informed Consent Statement: This article does not contain any studies involving humans or animals performed by any of the authors.

Data Availability Statement: Not applicable.

Acknowledgments: We would like to express a special acknowledgement to the China Scholarship Council and North China Electric Power University. Our thanks also go to all the respondents from the resettled villages for their help with a contribution to our study. Then, we thank some reviewers for their valuable and helpful comments and suggestions to substantially improve this article.

Conflicts of Interest: The authors declare no conflict of interest.

References

1. Gillian Cornish. Women & Resettlement: A Case Study on Gender Aspects at the Upper Paunglaung Hydropower Dam. A Briefer of Spectrum Sustainable Development Knowledge Network. 2018. Available online: https://www.spectrumsdkn.org/en/home/other-sectors/gender/womens-empowerment/303-t-the-upper-paunglaung-hydropower-dam (accessed on 10 June 2022).
2. Kattelus, M.; Rahaman, M.M.; Varis, O. Myanmar under reform: Emerging pressures on water, energy and food security. *Nat. Resour. Forum* **2013**, *38*, 85–98. [CrossRef]
3. Hao, C. Study on evaluation method for sustainable development of reservoir resettlement based on analytic hierarchy process. *Adv. Mater. Res.* **2014**, *962–965*, 2195–2200.
4. International Finance Corporation. *Environmental and Social Impact Assessment Guidelines for Hydropower Projects in Myanmar*; IFC: Washington, DC, USA, 2019.
5. International Finance Corporation. *Handbook for Preparing a Resettlement Action Plan*; Environmental and Social Development Department; IFC: Washington, DC, USA, 2002.
6. Schmitt-Degenhardt, S. A Regional Perspective on Poverty in Myanmar, UNDP, 2013. Available online: https://www.undp.org/sites/g/files/zskgke326/files/migration/mm/UNDP_MM_PvR_RegionalPovertyReport.pdf (accessed on 1 November 2022).
7. Wachenfeld, M.; Yee, H.W.; Oo, M.B.; Bowman, V.; Guest, D. Myanmar's Land Acquisition, Resettlement and Rehabilitation Law 2019—One Step Forward, Two Steps Back? 2020. Available online: https://www.semanticscholar.org/paper/Myanmar%E2%80%99s-Land-Acquisition%2C-Resettlement-and-Law-%E2%80%93-Wachenfeld-Yee/6846e443f7f60d35d25a95f3897299f7a3a1b7c0 (accessed on 14 November 2022).
8. Liu, X.-F.; Song, T.-T. The sustainable livelihoods research of reservoir resettlement in the upper reaches of the Yellow River based on the development of human capital. In Proceedings of the 3rd International Conference on Education, Kuala Lumpur, Malaysia, 20–22 April 2017.
9. International Finance Corporation. *Strategic Environmental Assessment of Hydropower Sector in Myan mar*; Baseline Assessment Report Hydropower; IFC: Washington, DC, USA, 2017.
10. Evrard, O.; Guodineau, Y. Planned resettlement, unexpected migrations and cultural trauma in Laos. *Dev. Chang.* **2004**, *35*, 937–962. [CrossRef]
11. Gautam, K.; Dangol, S. Developing a Resettlement Plan for the People Displaced by Budi Gandaki Hydroelectric Project. In Proceedings of the International Workshop on Role of Lnad Professionals and SDI in Disaster Risk Reduction: In the context of Post 2015 Nepal Earthquake, Kathmandu, Nepal, 25–27 November 2015.

environmental sciences proceedings

MDPI

Proceeding Paper

Anthropization, Salinity and Oxidative Stress in Animals in the Coastal Zone †

Abhipsa Bal [1,2] and Biswaranjan Paital [1,*]

1 Redox Regulation Laboratory, Department of Zoology, College of Basic Science and Humanities, Odisha University of Agriculture and Technology, Bhubaneswar 751003, India; abhipsa.hope@gmail.com
2 Zoology Section, Department of Education in Science and Mathematics, Regional Institute of Education, Bhubaneswar 751022, India
* Correspondence: biswaranjanpaital@gmail.com; Tel.: +91-674-2397029
† Presented at the 7th International Electronic Conference on Water Sciences, 15–30 March 2023; Available online: https://ecws-7.sciforum.net.

Abstract: In coastal and abiding zones, ground water continuously faces a very slight but alarmingly increasing trend in salinity due to several reasons such as the excess loss or use of it, with the constant dissolution of salts from the Earth's surface and heat-trapping pollution from human activities, rising sea levels and finally, high flooding. Many recent studies have indicated that even a slight elevation in ground water salinity may affect freshwater inhabitants, highlighting the importance of research on the effects of low salinity stress on coastal zone freshwater inhabitants. Along with abiotic factors such as salinity, dissolved oxygen, pH, and alkalinity, anthropogenic factors also cause a lot of stress on the inhabitants in coastal zones. Climatic factors also play an important role in influencing the life of coastal water inhabitants. For example, statistics such as those obtained by correlation and discriminant function analysis indicate that sublethal salinity acts as a strong modulator in the physiology of inhabiting fish in fresh as well as coastal water. Parameters such as increase in body weight, feed intake and irregularities in morphometry increase under higher salinities, which are confirmed by a decline in the growth of fishes. Similarly, blood physiology aspects, such as a significant loss in hemoglobin content, the RBC count and eosinophils, are coupled with amelioration in neutrophil count at the higher salinities of 6 and 9 ppt in few freshwater organisms. Normal histoarchitecture is also lost in most fish under high salinity conditions and higher anthropogenic loads. The generation of tissue damage in terms of oxidative stress is prominent under high fluctuations in abiotic factors including higher salinity or under high anthropogenic loads. Hence, a loss in compromised normal physiology due to the toxic effects of low- or high-salinity saline water or in fresh inhabitants including hardy fishes under changing climatic conditions are evident. This raises concerns about maintaining water quality in coastal and allied zones globally in the coming decades.

Keywords: coastal water; environmental effect; global climate change; water salinity; hardy fish; multiple biological factors

check for updates

Citation: Bal, A.; Paital, B. Anthropization, Salinity and Oxidative Stress in Animals in the Coastal Zone. *Environ. Sci. Proc.* **2023**, *25*, 7. https://doi.org/10.3390/ECWS-7-14228

Academic Editor: Athanasios Loukas

Published: 16 March 2023

1. Introduction

Water is an essential component of the ecosystem which anchors life on the Earth and integrates people from varied cultural, societal and economic backgrounds. However, in the context of a sustainable environment, the frequent anthropogenic activities that have been remarkably burdening the ecosystem are emerging as a major concern worldwide [1]. These issues of global concern have currently received prior attention among researchers worldwide as they affect the sustainability of the planet and its inhabitants. Anthropogenic activities have put global water security at risk by negatively affecting the aquatic as well as the terrestrial ecosystem in various ways [1].

Climatic changes and human activities have induced shifts in salinity levels in marine and coastal freshwater ecosystems and have immeasurably affected species diversity in the respective ecotones. Anthropogenic factors have subsequently inundated the geomorphology of coastal wetlands, estuaries, etc. The amount of dissolved salts in water can have a significant impact on the survival of aquatic organisms in the context of varied salinity preferences which are species-specific [2]. The major factor in the decline in species richness in the areas is the oxidative stress caused due to the overproduction of reactive oxygen species (ROS) in response to the shifts in the optimal salinity level [1,2].

The recurring anthropogenically induced climatic disturbances have caused havoc in the biodiversity of the aquatic ecosystem in a specific manner by releasing toxic substances/xenobiotic compounds and further altering hydrological parameters [1]. The source of anthropogenic activities largely includes the use of plastics ranging from nano- to macro-sized plastics in day-to-day life, rapid urbanization and industrialization, residual wastes generated from waste water treatment plants, construction and works, wastes generated from discarded electronic devices, transportation, biocides and fertilizer-based agriculture, etc. The proliferation of injudicious human activities can lead to the depletion of natural resources, habitat loss and a rippling effect throughout the ecosystem [2]. The optimal values of different hydrological parameters such as pH, alkalinity, dissolved oxygen, salinity, turbidity, conductivity, total dissolved solids, biological oxygen demand (BOD), chemical oxygen demand (COD), etc., help in sustaining life under water [3]. Anthropogenic activities induce a paradigm shift in biotic and abiotic parameters and thus lead to disturbances in the dynamics of aquatic biota. Among the abiotic factors, salinity is one of the significant factors which help in maintaining a differential osmotic gradient concentration for the survival of organisms belonging to different aquatic habitats which include freshwater, groundwater and marine ecotones. A swift change in the salinity level can affect the availability of nutrients and physiological processes. Altered salinity induces transformations in biotopes and biocenosis, and causes cytotoxicity, abnormal growth, osmotic imbalance, infertility and/or reproductive disorders, and neurodegenerative disorders along with various morphological and cellular anomalies [4,5]. The increase/decrease in salinity levels of a biota may induce immediate effects on the oxidative stress (OS) physiology of its organisms or might exhibit the dysfunction of mitochondrial complex enzymes upon prolonged exposure (Figure 1).

Further, salinity not only interrupts the osmotic balance but also leads to a disruption of the organelle membranes of aquatic invertebrates as well as those of vertebrates [6,7]. This review aims to assemble and analyze the ecotoxicological impact of human activities driving the rise in salinity and its effect on the oxidative regulation pathway. The assessment of the information in the present review will be helpful for understanding and mitigating the harmful effects of anthropogenic factors and climatic changes on aquatic life in a very precise manner.

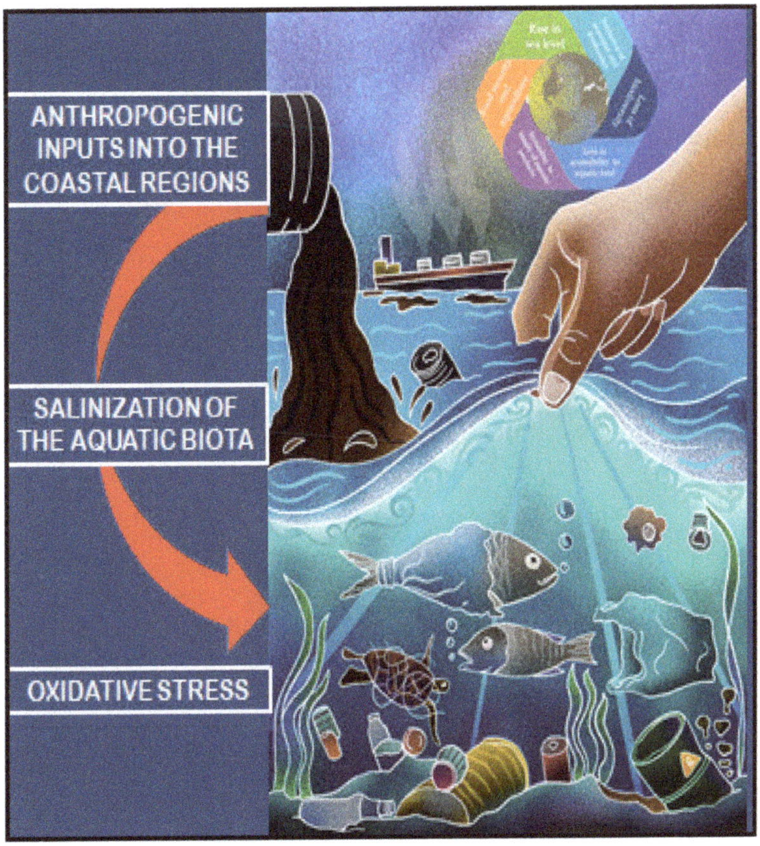

Figure 1. The anthropogenic inputs have intensified the physicochemical parameters as well as the disturbances in climatic changes. Climatic change- and anthropogenic factor-mediated salinity fluctuations have threatened the oxidative health status of aquatic organisms in coastal areas.

2. Anthropogenic Activities, Levels of Salinity and Ecological Threat

The rapid development of industries to meet the daily demands of a rising population has resulted in the production of more human-generated wastes which include CO_2 emissions, undigested pharmaceutical compounds [8], nanoparticles from wastewater treatment plants, return flow from irrigated fields [9], organic and inorganic biodegradable nutrients, heavy metal accumulation, human pathogens [10], etc. Anthropogenic inputs have led to a transformation in the dynamics of rivers, lakes, estuaries, coastal wetlands, etc., through the deposition of sediments, a change in the velocity of water flow [4], a rise in sea levels due to climatic changes, an increase in risks of flood, changes in tidal patterns [11,12], shifts in physicochemical parameters of the aquatic ecosystem (e.g., pH, turbidity, salinity, etc.) [13], an increase in the toxicity and salinity levels of groundwater sources [14], etc.

The accelerated increase in sea levels due to climatic factors is linked to anthropogenic factors [12]. The rise in sea levels has threatened coastal estuaries, lagoons, wetlands and groundwater resources in the respective areas. From the above-cited literature, it is evident that, with the increase in the abundance of microplastics, heavy metals, xenobiotic compounds and other toxic substances in the sediment, the risk of those compounds entering the food chain is apparent [15]. The species richness, species diversity and habitat spatial structure of aquatic inhabitants have become vulnerable due to salinization. In addition, ocean acidification, heavy rainfall and fluctuations in temperature modulate the salinity levels of the aquatic community [16]. Thus, irrespective of the cause behind the salinization of the aquatic community, which may be due to natural or anthropogenic factors, the physiology of coastal water inhabitants is disturbed in general and oxidative stress physiology is disturbed in particular (Figure 2).

Figure 2. Human activities have led to a rise in sea levels and the intrusion of seawater into freshwater biota. The salinization of these ecotones has resulted in a loss of biodiversity and accessibility to aquatic food, and instability in the supply of aquatic products. Thus, this has indirectly threatened the sectors of food security and aquaculture.

Anthropogenic factors are also found to influence many other physiological events in both fresh water and saline water species. An increase in salinity is found to have an influence on the DNA and RNA content of cells [1]. A decrease in hemoglobin content, and alleviated RBC and eosinophil counts are found to be associated with the elevated count of neutrophils at higher salinities of 6 and 9 ppt in few freshwater organisms such as *Heteropneustes fossilis* [1,5,7]. Changing salinity in water is also related to the altered histoarchitecture in many aquatic organisms because the usual histoarchitecture in most freshwater or marine fishes is lost under saline or fresh water intrusion into their habitat, respectively. This fact is also found to be higher under the exposure of such organisms to altered salinity and anthropogenic loads. Higher anthropogenic loads are always conducive to the generation of tissue damage, which is due to the generation of higher reactive oxygen species-accumulated OS in aquatic animals. The generation of tissue damage in terms of OS is prominent under high fluctuations in abiotic factors including higher salinity or under high anthropogenic loads [1,5,7]; see Figures 1 and 2.

3. Salinization Induced Oxidative Stress Physiology in Aquatic Inhabitants

Higher/lower levels of salinity and the associated mediations in physiological disturbances in the inhabitants of coastal water ecosystems are significant. Freshwater species face hypersalinity-mediated OS (whereas marine species are more frequently exposed to hyposalinity-mediated OS (e.g., *Paralichthys olivaceus*). Osmoregulatory physiology plays a major role in the adaptation of an organism to salinity changes in the aquatic community. The energy incurred by the osmoconformers to adapt to salinity changes is lower than that incurred by osmoregulators [17]. However, only a handful of investigations have reported the regulation of OS in the aquatic biota of coastal water environments in response to salinity. The OS incurred by those species and their respective study of antioxidant profiles in invertebrates and vertebrates is precisely described.

3.1. Oxidative Stress Physiology in Aquatic Invertebrates

Salinity induction critically affects sperm quality in free spawning mussels, which perform external fertilization. The sperm is exposed to various stressors like UV radiation and salinity alterations once they are released into the aquatic environment. Studies in *Mytilus galloprovincialis* have reported OS, impaired DNA, and limited mitochondrial activity along with restricted sperm motility, causing a decline in the rate of fertilization [18]. An increase in salinity activates the oxidative stress physiology pathway by accelerating the antioxidant activities of superoxide dismutase (SOD) and catalase (CAT) in monogonot rotifer *Brachionus plicatilis* [19].

3.2. Oxidative Stress Physiology in Aquatic Vertebrates

Hyposalinity can cause oxidative damage and distress to the antioxidant profile of organisms. A decline in the physiological performance of *Paralichthys olivaceu* in coping with the hyposaline conditions has been revealed in the context of decreasing values of hematocrit and hematological parameters [20]. Furthermore, levels of antioxidant enzymes such as catalase increased in response to the oxidative damage caused in the liver cells. With the further intensification of a stressor, apoptotic factors were also reported in the species [21]. The maintenance of osmoregulatory physiology is also associated with antioxidant capacity as well as the oxidative health of species (*Xenopus laevis*) during salinity shifts. An accelerated amount of electron leakage during oxidative phosphorylation caused by the physiological adaptation and osmoregulatory responses that occur to maintain the survival of species often contributes to oxidative stress. Therefore, total antioxidant capacity is inversely proportional to plasma osmolality in hypersaline conditions [22].

Various investigations have recognized the source of anthropogenic factors in establishing a transformed hydrological community in terms of physicochemical parameters, species diversity indexes, etc. Conversely, very few investigations have reported the status of the oxidative health of aquatic inhabitants in response to the anthropogenically mediated salinization of water bodies in coastal areas.

4. Conclusions

The aim of this review was to accentuate the understanding of the oxidative stress physiology of aquatic populations induced by the anthropogenically mediated salinization of aquatic biota in the coastal regions. Anthropogenic factors have either led to an increase or decrease in salinity in different aquatic communities. These slight fluctuations in the optimal level of salinity affect the oxidative health of the inhabitants in these communities by accelerating the production of ROS, and causing an insufficient amount of antioxidant enzymes, limited mitochondrial respiration, etc. Moreover, OS also affects growth, development, physiological adaptations, performance, reproduction, etc. The OS incurred by the species is also associated with the apoptotic factors responsible for cell death. Thus, climate change, anthropogenic loads, abiotic factor stressors, alone or in combination, induce (oxidative) stress in aquatic organisms; therefore, their economic value is influenced.

Author Contributions: Conceptualization, B.P.; writing—original draft preparation, A.B.; writing—review and editing, B.P. and A.B. All authors have read and agreed to the published version of the manuscript.

Funding: Schemes (number ECR/2016/001984 by Science Engineering Research Board, DST, Govt. of India and research award number 1188/ST, Bhubaneswar, dated 1 March 2017, ST-(Bio)-02/2017 by DST, Govt. of Odisha, India) to BRP are acknowledged. Fellowship (1264/ST/BT-MISC-0034-2018) to Abhipsa Bal under Biju Patnaik Research Fellowship to pursue Ph.D. course in Biotechnology, by Department of Science and Technology, Government of Odisha, India is acknowledged.

Institutional Review Board Statement: The study does not require any approval from the institutional review board/ethics committee.

Informed Consent Statement: Not applicable.

Data Availability Statement: Not applicable.

Conflicts of Interest: The authors declare no conflict of interest. The funders had no role in the design of the study; in the collection, analyses, or interpretation of data; in the writing of the manuscript; or in the decision to publish the results.

References

1. Bal, A.; Panda, F.; Pati, S.G.; Anwar, T.N.; Das, K.; Paital, B. Influence of Anthropogenic Activities on Redox Regulation and Oxidative Stress Responses in Different Phyla of Animals in Coastal Water via Changing in Salinity. *Water* **2022**, *14*, 4026. [CrossRef]
2. Banerjee, A.; Shelver, W.L. Micro- and Nanoplastic Induced Cellular Toxicity in Mammals: A Review. *Sci. Total Environ.* **2021**, *755*, 142518. [CrossRef] [PubMed]
3. Doaemo, W.; Betasolo, M.; Montenegro, J.F.; Pizzigoni, S.; Kvashuk, A.; Femeena, P.V.; Mohan, M. Evaluating the Impacts of Environmental and Anthropogenic Factors on Water Quality in the Bumbu River Watershed, Papua New Guinea. *Water* **2023**, *15*, 489. [CrossRef]
4. Stryjecki, R.; Zawal, A.; Krepski, T.; Stępien, E.; Buczynska, E.; Buczynski, P.; Czachorowski, S.; Jankowiak, Ł.; Pakulnicka, J.; Sulikowska-Drozd, A.; et al. Anthropogenic Transformations of River Ecosystems Are Not Always Bad for the Environment: Multi-Taxa Analyses of Changes in Aquatic and Terrestrial Environments after Dredging of a Small Lowland River. *PeerJ* **2021**, *9*, e12224. [CrossRef]
5. Bal, A.; Panda, F.; Pati, S.G.; Das, K.; Agrawal, P.K.; Paital, B. Modulation of Physiological Oxidative Stress and Antioxidant Status by Abiotic Factors Especially Salinity in Aquatic Organisms. *Comp. Biochem. Physiol. C Toxicol. Pharmacol.* **2021**, *241*, 108971. [CrossRef]
6. Paital, B. Antioxidant and Oxidative Stress Parameters in Brain of Heteropneustes Fossilis under Air Exposure Condition; Role of Mitochondrial Electron Transport Chain. *Ecotoxicol. Environ. Saf.* **2013**, *95*, 69–77. [CrossRef]
7. Paital, B.; Guru, D.; Mohapatra, P.; Panda, B.; Parida, N.; Rath, S.; Kumar, V.; Saxena, P.S.; Srivastava, A. Ecotoxic Impact Assessment of Graphene Oxide on Lipid Peroxidation at Mitochondrial Level and Redox Modulation in Fresh Water Fish Anabas Testudineus. *Chemosphere* **2019**, *224*, 796–804. [CrossRef]
8. Hejna, M.; Kapuścińska, D.; Aksmann, A. Pharmaceuticals in the Aquatic Environment: A Review on Eco-Toxicology and the Remediation Potential of Algae. *Int. J. Environ. Res. Public Health* **2022**, *19*, 7717. [CrossRef]
9. Li, C.; Gao, X.; Li, S.; Bundschuh, J. A Review of the Distribution, Sources, Genesis, and Environmental Concerns of Salinity in Groundwater. *Environ. Sci. Pollut. Res. Int.* **2020**, *27*, 41157–41174. [CrossRef]
10. Thai-Hoang, L.; Thong, T.; Loc, H.T.; Van, P.T.T.; Thuy, P.T.P.; Thuoc, T.L. Influences of Anthropogenic Activities on Water Quality in the Saigon River, Ho Chi Minh City. *J. Water Health* **2022**, *20*, 491–504. [CrossRef]
11. De Freitas Souza, C.; Baldissera, M.D.; Verdi, C.M.; Santos, R.C.V.; Da Rocha, M.I.U.M.; da Veiga, M.L.; da Silva, A.S.; Baldisserotto, B. Oxidative Stress and Antioxidant Responses in Nile Tilapia Oreochromis Niloticus Experimentally Infected by Providencia Rettgeri. *Microb. Pathog.* **2019**, *131*, 164–169. [CrossRef]
12. Talke, S.A.; Jay, D.A. Changing Tides: The Role of Natural and Anthropogenic Factors. *Ann. Rev. Mar. Sci.* **2020**, *12*, 121–151. [CrossRef]
13. Paun, I.; Pirvu, F.; Iancu, V.I.; Chiriac, F.L. Occurrence and Transport of Isothiazolinone-Type Biocides from Commercial Products to Aquatic Environment and Environmental Risk Assessment. *Int. J. Environ. Res. Public Health* **2022**, *19*, 7777. [CrossRef] [PubMed]
14. Lin, Y.C.; Lai, W.W.-P.; Tung, H.-h.; Lin, A.Y.C. Occurrence of Pharmaceuticals, Hormones, and Perfluorinated Compounds in Groundwater in Taiwan. *Environ. Monit. Assess.* **2015**, *187*, 256. [CrossRef] [PubMed]
15. Rodrigues, M.; Rosa, A.; Cravo, A.; Jacob, J.; Fortunato, A.B. Effects of Climate Change and Anthropogenic Pressures in the Water Quality of a Coastal Lagoon (Ria Formosa, Portugal). *Sci. Total Environ.* **2021**, *780*, 146311. [CrossRef] [PubMed]
16. Lushchak, V.I.; Bagnyukova, T.V.; Lushchak, O.V.; Storey, J.M.; Storey, K.B. Hypoxia and Recovery Perturb Free Radical Processes and Antioxidant Potential in Common Carp (*Cyprinus carpio*) Tissues. *Int. J. Biochem. Cell Biol.* **2005**, *37*, 1319–1330. [CrossRef] [PubMed]
17. Rivera-Ingraham, G.A.; Barri, K.; Boël, M.; Farcy, E.; Charles, A.-L.; Geny, B.; Lignot, J.-H. Osmoregulation and Salinity-Induced Oxidative Stress: Is Oxidative Adaptation Determined by Gill Function? *J. Exp. Biol.* **2016**, *219 Pt 1*, 80–89. [CrossRef] [PubMed]

18. Cuccaro, A.; De Marchi, L.; Oliva, M.; Monni, G.; Miragliotta, V.; Fumagalli, G.; Freitas, R.; Pretti, C. The Influence of Salinity on the Toxicity of Chemical UV-Filters to Sperms of the Free-Spawning Mussel *Mytilus galloprovincialis* (Lamark, 1819). *Aquat. Toxicol.* **2022**, *250*, 106263. [CrossRef]
19. Han, J., Lee, K.-W Influence of Salinity on Population Growth, Oxidative Stress and Antioxidant Defense System in the Marine Monogonont Rotifer *Brachionus plicatilis*. *Comp. Biochem. Physiol. B Biochem. Mol. Biol.* **2020**, *250*, 110487. [CrossRef]
20. Kim, J.H.; Jeong, E.H.; Jeon, Y.H.; Kim, S.K.; Hur, Y.B. Salinity-Mediated Changes in Hematological Parameters, Stress, Antioxidant Responses, and Acetylcholinesterase of Juvenile Olive Flounders (*Paralichthys olivaceus*). *Environ. Toxicol. Pharmacol.* **2021**, *83*, 103597. [CrossRef]
21. Lee, D.W.; Choi, Y.U.; Park, H.S.; Park, Y.S.; Choi, C.Y. Effect of Low PH and Salinity Conditions on the Antioxidant Response and Hepatocyte Damage in Juvenile Olive Flounder *Paralichthys olivaceus*. *Mar. Environ. Res.* **2022**, *175*, 105562. [CrossRef] [PubMed]
22. Hidalgo, J.; Álvarez-Vergara, F.; Peña-Villalobos, I.; Contreras-Ramos, C.; Sanchez-Hernandez, J.C.; Sabat, P. Effect of Salinity Acclimation on Osmoregulation, Oxidative Stress, and Metabolic Enzymes in the Invasive *Xenopus laevis*. *J. Exp. Zool. Part A, Ecol. Integr. Physiol.* **2020**, *333*, 333–340. [CrossRef] [PubMed]

environmental sciences proceedings

MDPI

Proceeding Paper

Drought Propagation under Combined Influences of Reservoir Regulation and Irrigation over a Mediterranean Catchment †

Omar Cenobio-Cruz [1,*], Pere Quintana-Seguí [1] and Luis Garrote [2]

1 Observatori de l'Ebre, Universitat Ramon Llull-CSIC, 43520 Roquetes, Spain; pquintana@obsebre.es
2 Department of Civil Engineering: Hydraulics, Energy and Environment, Universidad Politécnica de Madrid, 28040 Madrid, Spain; l.garrote@upm.es
* Correspondence: ocenobio@obsebre.es; Tel.: +34-977-500-511

† Presented at the 7th International Electronic Conference on Water Sciences, 15–30 March 2023; Available online: https://ecws-7.sciforum.net/.

Abstract: Drought is a natural phenomenon that is controlled by different factors such as natural climate and catchment controls, and in many worldwide regions, it is now driven by human activities (i.e., reservoirs, irrigation, groundwater abstractions). Reservoirs initially ensure water availability and help cope with drought, especially in semi-arid regions; however, this human modification of the environment may lead to both positive and negative effects on the hydrological cycle, which need to be understood. This involves a better understanding of hydrological processes and incorporating human interactions within coupled human–natural systems to improve drought management. We focused on a strongly irrigated area located in the northeast of the Iberian Peninsula, the northern part of the Canal of Aragon and Catalonia district supplied by the Barasona reservoir. We implemented a simple water management model to simulate the reservoir operation (human-influenced scenario) and examined the contribution of human activities, associated with irrigation, on the water budget and drought propagation. For this purpose, we used simulations performed by the hydrometeorological model SASER (SAFRAN-SURFEX-EauDyssée-RAPID), which provided a natural scenario (without human influence) to contrast with the human-influenced scenario. The model performance was evaluated through the Kling Gupta Efficiency (KGE) metric. The first results demonstrated satisfactory performance to simulate reservoir storage and outflows against observed data, with KGE values of 0.4 and 0.82, respectively. Then we explored the linkages between agricultural drought, associated with evapotranspiration, and hydrological drought. We applied standardized indices to identify different kinds of drought, compared them, and assessed changes induced by human activities. Human modifications modulate the hydrological response of the catchment, and alter the intensity of hydrological drought, while human activities reduce the intensity of agricultural droughts.

Keywords: drought propagation; water management; anthropogenic drought

Citation: Cenobio-Cruz, O.; Quintana-Seguí, P.; Garrote, L. Drought Propagation under Combined Influences of Reservoir Regulation and Irrigation over a Mediterranean Catchment. *Environ. Sci. Proc.* **2023**, *25*, 8. https://doi.org/10.3390/ECWS-7-14239

Academic Editor: Athanasios Loukas

Published: 16 March 2023

1. Introduction

Droughts are usually defined as prolonged periods of below-average precipitation conditions [1,2]. Droughts can be classified into several categories, including meteorological, agricultural, and hydrological [3]. Meteorological droughts are based on precipitation deficits, agricultural droughts refer to soil moisture deficits, and hydrological droughts are mainly based on streamflow. Each of them is characterized by different indices [3,4].

Recent studies have shown that in some regions of the world, such as southern Europe and West Africa, droughts have become more intense and prolonged in recent decades [5,6]. This trend is likely due to a combination of factors, such as climate change and the growing water demand from human activities, such as irrigation and urbanization, which can put additional stress on water resource systems, exacerbating the impacts of droughts [7–9]. Therefore, it is important to consider human influences in the assessment and management of droughts.

Land surface models (LSMs) have been widely recognized as a powerful tool for understanding and simulating the hydrological cycle, including droughts [10–13]. Particularly in Spain, LSMs have been used to evaluate and provide information on water availability and potential drought hotspots [14,15]. Nevertheless, for more realistic modeling of droughts, it is crucial to incorporate the representation of human factors in current-generation LSMs [16].

In this study, we investigate how human activities (irrigation and reservoir operation) impact drought propagation in a coupled human–water system. The twofold objective of this research is (i) implementing a prototype reservoir operation scheme that could be integrated into the SASER (SAFRAN-SURFEX-Eaudyssée-RAPID) modeling chain and which exploits the new SURFEX irrigation scheme [17] and (ii) quantifying the impact of human activities on drought propagation. To address this objective, we evaluate the ability of the new module that simulates reservoir operation and compare it against the simulation performed by the SASER model, which provided a natural scenario (without human influence) to contrast with the human-influenced scenario.

2. Data and Methods

2.1. Study Area and Data

We selected a strongly irrigated area located in the northeast of the Iberian Peninsula, the northern part of the Canal of Aragon and Catalonia (CAyC), with a size of 54,000 ha, which is supplied by the Barasona reservoir (Figure 1). This reservoir has a maximum volume capacity of 84 Hm^3.

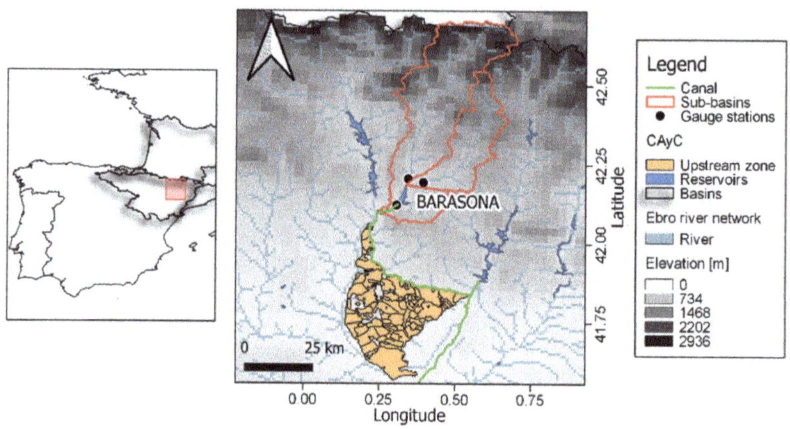

Figure 1. Location of the study area.

The main data used in this study, observed streamflow and reservoir volume, were obtained from the Automatic Hydrologic Information System, SAIH in Spanish. Irrigation demands were collected from the Ebro Hydrographic Confederation (CHE, in Spanish).

In addition, SURFEX-LSM [18], which simulates natural surfaces in the vertical soil column, provided runoff and evapotranspiration (ET) data. Precipitation data were obtained from the gridded meteorological dataset SAFRAN, as depicted in Figure 2a. The version of SASER used in this study incorporates a conceptual reservoir to postprocess the drainage with regionalized parameters, named SASER-reg [19]. In Figure 2a shows the general framework with the main steps used.

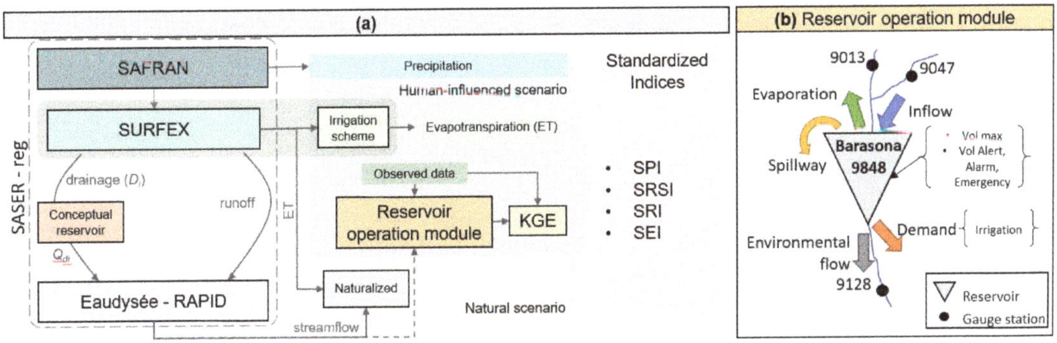

Figure 2. General framework used in our analysis (**a**), and (**b**) schematic representation of the reservoir operation model.

2.2. Reservoir Model Scheme

We implemented a simple reservoir operation scheme as depicted in Figure 2b, based on the Water Availability and Adaptation Policy Assessment (WAAPA) model [20]. This model simulates reservoir operation considering the environmental flows and evaporation losses. The model requires the following input: streamflow, demands, and environmental flows. In this research, before connecting SASER outputs, we assessed the ability of the module to reproduce the dam behavior. Therefore, we first used observed streamflow data as input to the module and then compared the simulated volume and reservoir's outflow against observed data, as Figure 2a indicates.

To evaluate the model performance of the reservoir operation module, we used the Kling–Gupta efficiency, KGE, [21]:

$$KGE = 1 - \sqrt{(1 - r^2) + (1 - \alpha) + (1 - \beta)}; \tag{1}$$

$$\alpha = \frac{\mu_s}{\mu_o} \text{ and } \beta = \frac{\sigma_s}{\sigma_o} \tag{2}$$

where r is the Pearson's correlation coefficient, α represents the bias component, and β is the ratio of variance; μ and σ represent the mean and standard deviation, respectively. Similarly, subscripts s and o represent simulated and observed variables, respectively.

To simulate the natural scenario (without human influence), we used a simulation performed by the SASER model, Figure 2a, and we compared it against the human-influenced scenario, which incorporates a new irrigation scheme developed within the SURFEX model [17], allowing us to estimate a realistic amount of irrigation water, and therefore the evapotranspiration associated with it.

To represent the different types of droughts, we used standardized indices. The Standardized Precipitation Index (SPI) [22] was utilized to characterize meteorological drought. To hydrological drought, we applied the Standardized Runoff Index (SRI) [23], and the reservoir storage was also standardized. For agricultural drought, we used a procedure similar to SPI and calculated the Standardized Evapotranspiration Index (SEI) using the evapotranspiration data associated with irrigation.

3. Results and Discussion

3.1. Reservoir Operation

The results of the reservoir operation module shown here indicate a satisfactory performance to simulate storage and outflows, with KGE values of 0.4 and 0.82, respectively (Figure 3). Creating a complete water management simulation that optimizes water resources by feeding the reservoir module with SASER results is the next step, which is beyond the scope of this study.

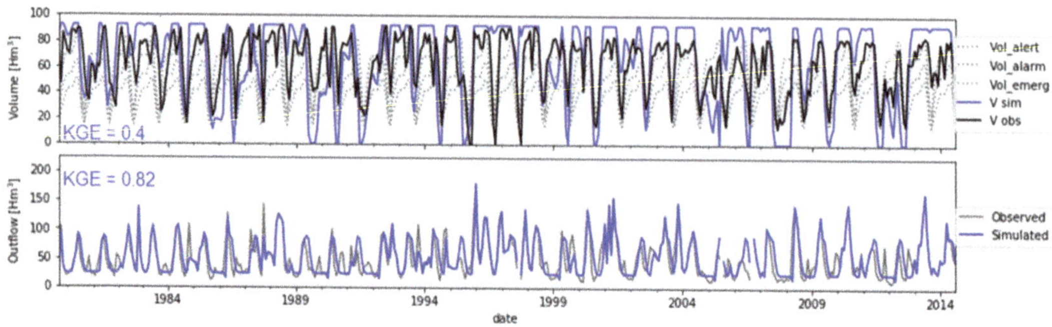

Figure 3. Observed (black line) and simulated (in blue) reservoir storage and outflow for the Barasona reservoir.

It is worth highlighting that for the reservoir simulation, the same irrigation demand was assumed for every year, which does not accurately reflect realistic conditions. Nevertheless, this approach has yielded reasonably good results.

The simulated volume storage follows the same dynamics as observed data (upper panel in Figure 3), except for the events from 1995 to 1997, which correspond to other factors and not to irrigation demands. The simulated and observed outflows show a very good agreement, with a high value of KGE.

3.2. Drought Analysis

Standardized indices at a 12 m time scale were considered to evaluate how the meteorological drought signal propagates through other variables. To understand drought processes, we calculated the frequency of drought events.

Meteorological drought is represented in Figure 4a, and anomalies in the reservoir storage are depicted in Figure 4b. The hydrological drought depicted in Figure 4c shows a similar pattern in both the natural and human scenarios, the solid blue line and red line, respectively. However, the blue shaded area shows the opposite behavior of this index under the human scenario, which is attributable to the reservoir operation.

Anomaly analysis also allows for quantification of the impact of human activities. The frequency, total number of drought events (index < -1), is shown in the bottom right of each panel in Figure 4. For meteorological drought, 10 events are reported. For hydrological drought, in the observed situation, the number of events is higher than for the naturalized scenario (blue and red lines in Figure 4c, respectively). The opposite occurs in the anomalies associated with ET: the number of events is higher in the naturalized scenario than in the scenario where irrigation is active (nine and eight events, respectively). This was expected, as streamflow decreases while ET increases due to irrigation.

We also calculated the total number of months in drought (duration of drought), and we found that in the human scenario, hydrological drought increased from 158 to 176 months, representing an increase of 10%, which suggests that reservoir operation increases the duration of drought events. Whereas for drought associated with evapotranspiration, a similar total duration was obtained in both scenarios (167 months for the human scenario and 163 for the natural scenario).

We observed changes in hydrological drought intensities; for instance, the event between 2005 and 2008 shows lower values in the natural scenario, which suggests that reservoir operation mitigates the effect of drought. Conversely, in 2012, the lower values occurred in the human scenario, suggesting that the reservoir could be aggravating the drought.

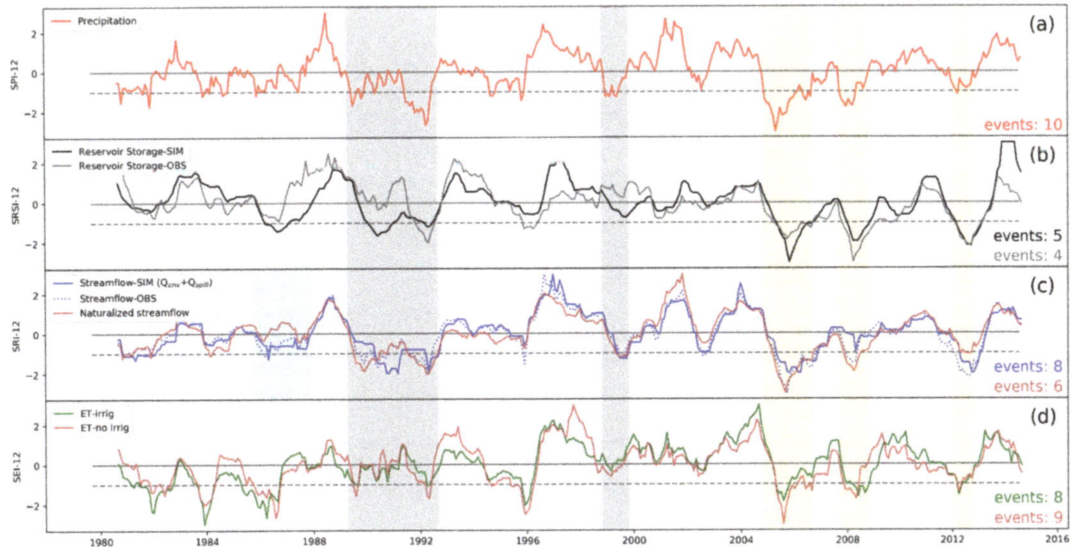

Figure 4. Droughts indices, all to 12 months. (**a**) Standardized Precipitation Index, SPI; (**b**) Standardized Reservoir Storage Index, SRSI; (**c**) Standardized Runoff Index, SRI; (**d**) Standardized Evapotranspiration Index, SEI. Number of drought events is reported in each corresponding panel.

In Figure 3, the gray shaded areas show differences in drought propagation and correspond with the meteorological drought event of maximum duration (40 months). The hydrological drought (a single long event) responds directly to meteorological drought; this response is not reflected in agricultural drought (two short events are reported).

Additionally, we selected three severe droughts (2004–2005, 2008, and 2012, indicated in the yellow shaded areas in Figure 4) to exhibit differences in drought propagation under the human scenario. We found that drought directly propagates from meteorological to hydrological, but not with agricultural (evapotranspiration associated with irrigation) drought. If we focus on the linkage between SRI and SEI, a pattern was found, whereby the first decreases and the other increases, and vice versa. These results show how human interventions contribute to modulating the evapotranspiration and runoff due to extensive irrigation practices.

4. Conclusions

We are currently developing a framework considering the evapotranspiration processes associated with irrigation to evaluate the role of human activities on agricultural and hydrological droughts in a Mediterranean catchment. Thus, we implemented a reservoir simulation scheme as an external module, which allows for a flexible approach (rapid iteration process). The reservoir model shows good performance, considering the model's simplifications. The KGE is good, especially for outflows. Through the SURFEX irrigation scheme, dynamic irrigation demands can be used, which provides a more realistic analysis.

To investigate the impact of human activities on the water cycle and droughts, we used different standardized indices. This allows for analyzing and characterizing drought events (e.g., intensity, duration) and investigating the drought dynamic under a coupled natural–human system. Finally, the impact of irrigation and reservoir operation on the catchment modifies the hydrological response, leading to variations in the severity of hydrological droughts, while at the same time, these human activities have the opposite effect on the intensity of droughts associated with evapotranspiration.

Author Contributions: Conceptualization, O.C.-C. and P.Q.-S.; methodology, O.C.-C., P.Q.-S. and L.G.; software, P.Q.-S. and L.G.; formal analysis, O.C.-C., P.Q.-S. and L.G.; investigation, O.C.-C.; resources, L.G.; data curation, O.C.-C.; writing—original draft preparation, O.C.-C.; writing—review and editing, O.C.-C., P.Q.-S. and L.G.; visualization, O.C.-C.; supervision, P.Q.-S. and L.G. All authors have read and agreed to the published version of the manuscript.

Funding: This research was funded by the HUMID project (CGL2017-85687-R, AEI/FEDER, UE), the Predoctoral grant PRE2018-085027 (AEI/FSE), and the IDEWA project (PRIMA PCI2020-112043/AEI/10.13039/501100011033).

Institutional Review Board Statement: Not applicable.

Informed Consent Statement: Not applicable.

Data Availability Statement: Data will be made available on request.

Conflicts of Interest: The authors declare no conflict of interest.

References

1. Stagge, J.H.; Tallaksen, L.M.; Gudmundsson, L.; Van Loon, A.F.; Stahl, K. Candidate Distributions for Climatological Drought Indices (SPI and SPEI). *Int. J. Climatol.* **2015**, *36*, 2132–2138. [CrossRef]
2. Wilhite, D.A.; Glantz, M.H. Understanding: The Drought Phenomenon: The Role of Definitions. *Water Int.* **1985**, *10*, 111–120. [CrossRef]
3. Mishra, A.K.; Singh, V.P. A review of drought concepts. *J. Hydrol.* **2010**, *391*, 202–216. [CrossRef]
4. Keyantash, J.; Dracup, J.A. The quantification of drought: An evaluation of drought indices. *Bull. Am. Meteorol. Soc.* **2002**, *83*, 1167–1180. [CrossRef]
5. Seneviratne, S.; Nicholls, N.; Easterling, D.; Goodess, C.M.; Kanae, S.; Kossin, J.; Luo, Y.; Marengo, J.; McInnes, K.; Rahimi, M.; et al. Changes in climate extremes and their impacts on the natural physical environment. In *Managing the Risks of Extreme Events and Disasters to Advance Climate Change Adaptation: Special Report of the Intergovernmental Panel on Climate Change*; Cambridge University Press: Cambridge, UK, 2012; pp. 109–230. [CrossRef]
6. Prudhomme, C.; Giuntoli, I.; Robinson, E.L.; Clark, D.B.; Arnell, N.W.; Dankers, R.; Fekete, B.M.; Franssen, W.; Gerten, D.; Gosling, S.N.; et al. Hydrological droughts in the 21st century, hotspots and uncertainties from a global multimodel ensemble experiment. *Proc. Natl. Acad. Sci. USA* **2014**, *111*, 3262–3267. [CrossRef] [PubMed]
7. Bates, B.C.; Kundzewicz, Z.W.; Wu, S.; Palutikof, J.P. *Climate Change and Water. Technical Paper of the Intergovernmental Panel on Climate Change*; IPCC Secretariat: Geneva, Switzerland, 2008; p. 210. ISBN 978-92-9169-123-4.
8. Wada, Y.; Van Beek, L.P.H.; Wanders, N.; Bierkens, M.F.P. Human water consumption intensifies hydrological drought worldwide. *Environ. Res. Lett.* **2013**, *8*, 034036. [CrossRef]
9. Wanders, N.; Wada, Y. Human and climate impacts on the 21st century hydrological drought. *J. Hydrol.* **2015**, *526*, 208–220. [CrossRef]
10. Lehner, B.; Döll, P.; Alcamo, J.; Henrichs, T.; Kaspar, F. Estimating the Impact of Global Change on Flood and Drought Risks in Europe: A Continental, Integrated Analysis. *Clim. Chang.* **2006**, *75*, 273–299. [CrossRef]
11. Vidal, J.-P.; Martin, E.; Franchistéguy, L.; Habets, F.; Soubeyroux, J.-M.; Blanchard, M.; Baillon, M. Multilevel and multiscale drought reanalysis over France with the Safran-Isba-Modcou hydrometeorological suite. *Hydrol. Earth Syst. Sci.* **2010**, *14*, 459–478. [CrossRef]
12. Prudhomme, C.; Parry, S.; Hannaford, J.; Clark, D.B.; Hagemann, S.; Voss, F. How Well Do Large-Scale Models Reproduce Regional Hydrological Extremes in Europe? *J. Hydrometeorol.* **2011**, *12*, 1181–1204. [CrossRef]
13. Van Loon, A.F.; Van Huijgevoort, M.H.J.; Van Lanen, H.A.J. Evaluation of drought propagation in an ensemble mean of large-scale hydrological models. *Hydrol. Earth Syst. Sci.* **2012**, *16*, 4057–4078. [CrossRef]
14. Gaona, J.; Quintana-Seguí, P.; Escorihuela, M.J.; Boone, A.; Llasat, M.C. Interactions between precipitation, evapotranspiration and soil-moisture-based indices to characterize drought with high-resolution remote sensing and land-surface model data. *Nat. Hazards Earth Syst. Sci.* **2022**, *22*, 3461–3485. [CrossRef]
15. Barella-Ortiz, A.; Quintana-Seguí, P. Evaluation of drought representation and propagation in regional climate model simulations across Spain. *Hydrol. Earth Syst. Sci.* **2019**, *23*, 5111–5131. [CrossRef]
16. Pokhrel, Y.N.; Hanasaki, N.; Wada, Y.; Kim, H. Recent progresses in incorporating human land–water management into global land surface models toward their integration into Earth system models. *Wiley Interdiscip. Rev. Water* **2016**, *3*, 548–574. [CrossRef]
17. Druel, A.; Munier, S.; Mucia, A.; Albergel, C.; Calvet, J.-C. Implementation of a new crop phenology and irrigation scheme in the ISBA land surface model using SURFEX_v8.1. *Geosci. Model Dev.* **2022**, *15*, 8453–8471. [CrossRef]
18. Masson, V.; Le Moigne, P.; Martin, E.; Faroux, S.; Alias, A.; Alkama, R.; Belamari, S.; Barbu, A.; Boone, A.; Bouyssel, F.; et al. The SURFEXv7.2 land and ocean surface platform for coupled or offline simulation of earth surface variables and fluxes. *Geosci. Model Dev.* **2013**, *6*, 929–960. [CrossRef]

19. Cenobio-Cruz, O.; Quintana-Seguí, P.; Barella-Ortiz, A.; Zabaleta, A.; Garrote, L.; Clavera-Gispert, R.; Habets, F.; Beguería, S. Improvement of low flows simulation in the SASER hydrological modeling chain. *J. Hydrol. X* **2023**, *18*, 100147. [CrossRef]
20. Sordo-Ward, A.; Granados, A.; Iglesias, A.; Garrote, L.; Bejarano, M.D. Adaptation effort and performance of water management strategies to face climate change impacts in six representative basins of Southern Europe. *Water* **2019**, *11*, 1078. [CrossRef]
21. Gupta, H.V.; Kling, H.; Yilmaz, K.K.; Martinez, G.F.; Kling, H. Decomposition of the mean squared error and NSE performance criteria: Implications for improving hydrological modelling. *J. Hydrol.* **2009**, *377*, 80–91. [CrossRef]
22. Mckee, T.B.; Doesken, N.J.; Kleist, J. The relationship of drought frequency and duration to time scales. In Proceedings of the 8th Conference on Applied Climatology, Anaheim, CA, USA, 17–22 January 1993; pp. 179–184.
23. Shukla, S.; Wood, A.W. Use of a standardized runoff index for characterizing hydrologic drought. *Geophys. Res. Lett.* **2008**, *35*. [CrossRef]

environmental sciences proceedings

MDPI

Proceeding Paper

Estimation of Remotely Sensed Actual Evapotranspiration in Water-Limited Mediterranean Agroecosystems for Monitoring Crop (cotton) Water Requirements †

Marios Spiliotopoulos [1,*], Nicolas Alpanakis [1], Georgios A. Tziatzios [1], Ioannis Faraslis [2], Pantelis Sidiropoulos [1], Stavros Sakellariou [2], George Karoutsos [3], Nicolas R. Dalezios [1] and Nicholas Dercas [4]

1 Department of Civil Engineering, School of Engineering, University of Thessaly, 38221 Volos, Greece; alpanakis@topo.auth.gr (N.A.); getziatz@uth.gr (G.A.T.); psidirop@uth.gr (P.S.); dalezios@uth.gr (N.R.D.)

2 Department of Environmental Sciences, University of Thessaly, Gaiopolis Campus, 41500 Larissa, Greece; faraslis@uth.gr (I.F.); stasakel@uth.gr (S.S.)

3 General Aviation Applications "3D" S.A., 2 Skiathou str, 54646 Thessaloniki, Greece; info@3dsa.gr

4 Department of Natural Resources Development and Agricultural Engineering, Agricultural University of Athens; Iera Odos 75, 11855 Athens, Greece; ndercas1@aua.gr

* Correspondence: spilioto@uth.gr

† Presented at the 7th International Electronic Conference on Water Sciences, 15–30 March 2023; Available online: https://ecws-7.sciforum.net.

Abstract: The role of effective irrigation management for optimal food production is well recognized and can be partially solved through the improvement of water use efficiency (WUE). To control the quantity of the water applied to actual crops, net irrigation water requirements (NIWR) are needed. The computation of NIWR is based on the estimation of crop water requirements (CWR) and soil water balance, where crop evapotranspiration (ET_c) is the main component. Earth observation (EO) using remote sensing (RS) has already become an important tool for the quantification and the detection of spatial and temporal distributions and variability in several environmental variables at different scales. Remotely sensed models are currently considered to be suitable for crop water use estimation in the field and at regional scales.

Keywords: ET; crop water requirements; sentinel imagery; cotton

Citation: Spiliotopoulos, M.; Alpanakis, N.; Tziatzios, G.A.; Faraslis, I.; Sidiropoulos, P.; Sakellariou, S.; Karoutsos, G.; Dalezios, N.R.; Dercas, N. Estimation of Remotely Sensed Actual Evapotranspiration in Water-Limited Mediterranean Agroecosystems for Monitoring Crop (cotton) Water Requirements. *Environ. Sci. Proc.* **2023**, *25*, 9. https://doi.org/10.3390/ECWS-7-14200

Academic Editor: Athanasios Loukas

Published: 14 March 2023

1. Introduction

Currently, the most prevailing group of EO methodologies for the estimation of ET_c is the Energy Balance (EB) algorithms, and more specifically, residual methods [1]. Remotely sensed EB algorithms convert satellite-sensed radiances into land surface characteristics, such as albedo, leaf area index, vegetation indices, surface roughness, surface emissivity and surface temperature, to estimate ET as a "residual" of the land surface Energy Balance equation [2,3]. The most recent EB models differ mainly in how Sensible Heat (H) is estimated [1]. These models include the Two-Source Model (TSM), where the Energy Balances of soil and vegetation are modeled separately, and then combined to estimate total LE, the Surface Energy Balance Algorithm for Land (SEBAL) and the Mapping Evapotranspiration with Internalized Calibration (METRIC), which use hot and cold pixels within the satellite images to develop an empirical temperature difference equation, and the Surface Energy Balance Index (SEBI) based on the contrast between wet and dry areas [4–6]. Other variations in SEBI include the Simplified Surface Energy Balance Index (S-SEBI) and the Surface Energy Balance System (SEBS) [7].

In this paper, an actual daily evapotranspiration (ET_a) computation methodology is presented as a contribution to a European-funded research project named "HubIS". Obtaining useful spatial information and describing difficult physical processes through remote sensing is important for developing better agricultural practices. The applied approach was first operated by the European Space Agency (ESA) using the Sen-ET plug-in, and the proposed methodology is an improvement of ESA's method. The proposed methodology framework consists of seventeen separate steps, with the outcome being the actual daily evapotranspiration flows estimation at a 20 × 20 m spatial resolution. The proposed methodology is applied to cotton in Thessaly, Greece, for the 2021 growing season. The results are very satisfactory and indicate the suitability of Sen-ET SNAP software to estimate daily actual evapotranspiration, as well as spatial variability throughout the crop. The methodology can be applied for effective irrigation management in data-scarce rural regions.

2. Materials and Methods—Study Area

In this study, a combination of Sentinel-2 and Sentinel-3 images for daily crop evapotranspiration estimation is presented and applied to cotton fields in Thessaly, Greece, which is considered to be a water-limited, Mediterranean, agricultural area. The simulation program used is the Sen-ET SNAP software. The plug-in uses satellite images from Sentinel-2 and Sentinel-3 and meteorological data from the Weather Research and Forecast (WRF) model. The purpose of the Sen-ET SNAP plug-in is to enable the estimation of daily actual evapotranspiration rates (and other land–surface energy fluxes) at the field scale.

The initial version of Sen-ET used in SNAP, developed by ESA, uses the two-source Energy Balance (TSEB) model [8] with Sentinel-2, Sentinel-3 and meteorological data from ECMWF (ESA, 2020) [9,10]. The adopted methodology in the present approach follows the initial 17 steps based on the TSEB model (Table 1). However, a modification of the initial Sen-ET SNAP used in the present study takes advantage of meteorological data taken from the WRF model, instead of ECMWF. This approach stands as a modification and improvement from the original plug-in. Real-time data produced by the WRF are inserted into the proposed methodology, finally giving ET_a values in mm/day with a spatial resolution of 20 m × 20 m.

Table 1. The 17 steps created for the operation of Sen-ET.

Steps	Procedures
1	Download Sentinel images
2	Preprocessing of Sentinel-2
3	Creation of a Digital Elevation Model (DEM)
4	Creation of a land use map of the study area
5	Leaf reflection and transmission assessment based on the chlorophyll and water content of the plants
6	Estimation of fraction of green vegetation
7	Production of land use/land cover maps
8	Aerodynamic roughness assessment
9	Pre-processing and downscaling of Sentinel-3 data
10	From warp to template—Resampling into a standard image using GDAL Warp
11	Sharpen land surface temperature
12	Download WRF meteorological data by 3D S.A partner
13	WRF meteorological data preparation based on the High-Resolution Digital Elevation Model (DEM) of the study area
14	Long-wave irradiance estimation based on WRF meteorological data
15	Estimation of net shortwave radiation based on WRF meteorological data and biophysical parameters
16	Estimation of energy fluxes of the land surface
17	Estimation of daily evapotranspiration

The study area is the region of Thessaly, which is the most cultivated region in Greece, (Figure 1). The annual quantity of water consumption is about 1.422 hm^3, with 92%, i.e., approximately 1305.5 hm^3, being used to meet the irrigation needs. Almost 500,000 ha are cultivated in Thessaly, while almost 250,000 are irrigated. Seventy-six (76) % of irrigation amounts comes from groundwater systems through legal or illegal drilling, while only twenty-four (24) % comes from surface water systems [9]. The physical condition of groundwater systems in Thessaly is not satisfactory both in quantitative and quality terms [9]. In the present study, seven experimental plots were used in the research program "HubIS" and were analyzed and processed.

Figure 1. Study area.

3. Results

Figure 2 shows the variations in actual daily ET_a values for cotton during the irrigation period when images from Sentinel-2 and Sentinel-3 were available.

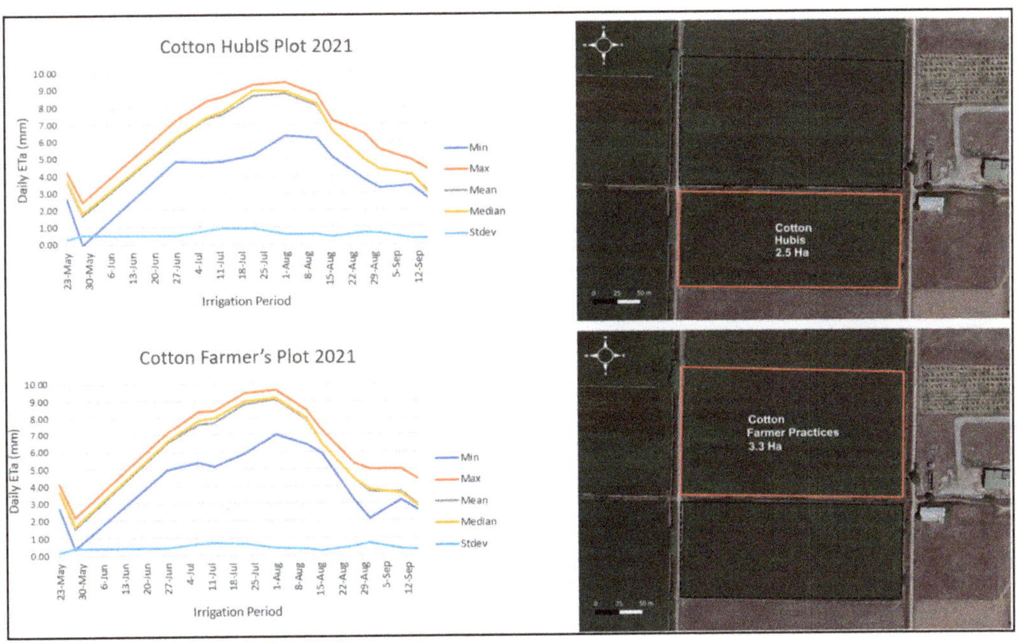

Figure 2. Variation of the actual daily ET_a for cotton plots. Comparison between Hubis methodology and the farmer practices.

Figure 3 shows the spatial distribution of the actual daily ET_a values for the cultivation of cotton in the growing season when images from Sentinel-2 and Sentinel-3 were available. The plots are divided into "traditional" (farmer practices) and "experimental" (Hubis practices) areas. No other explanations are needed for the current study because more results between these plots are expected in future implementation of the program. The daily ET_a values (farmer practices) range from 0 mm (recorded in February) to a maximum of 10 mm (recorded in early August). Figure depicts only a part of the available image dates due to width issues. Additionally, Figure 3 presents the spatial variation in the daily ET_a values for the cultivation of cotton (Hubis practices) for the entire period that satellite images were available. The daily ET_a values range from min 0 mm (recorded in February 2021) to a max of 9.5 mm at the beginning of August 2021.

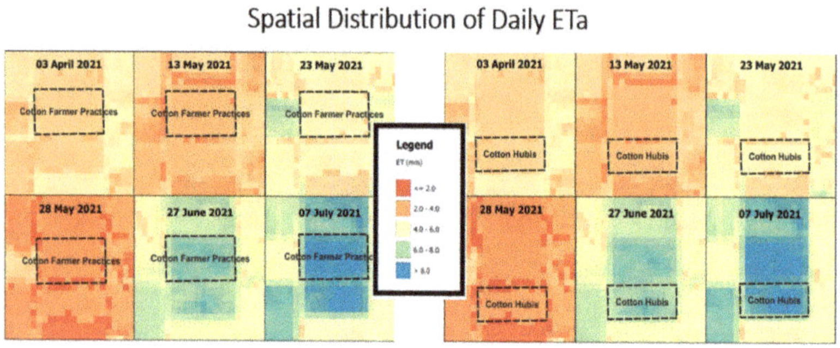

Figure 3. Spatial distribution examples of the actual daily ET_a for the cultivation of cotton (Farmer practices) and cotton (Hubis).

4. Summary

In this study, a methodology developed by the University of Thessaly, Greece, is used for the calculation of actual evapotranspiration values using Sentinel imagery. This methodology suggests a modification and improvement of the initial ESA's SEN-ET methodology. The SEN_ET methodology is supported by SNAP software utilizing Sentinel-2 and Sentinel-3 satellite imagery, as well as meteorological data from the Weather Forecast and Research (WRF) model, to estimate daily ET_a values. The results of the experiment showed that there is a clear overestimation of ET_a in some periods, especially during July and August. At the next step of the method's improvement, it is important to use in-situ measurements.

Author Contributions: Conceptualization, N.A., G.A.T. and I.F.; methodology, I.F.; software, N.A. and G.A.T.; validation, M.S., S.S. and P.S.; formal analysis, P.S.; investigation, G.K.; resources, G.K.; data curation, G.K.; writing—original draft preparation, M.S. and N.A.; writing—review and editing, M.S.; visualization, N.R.D. and N.D.; supervision, N.R.D. and N.D.; project administration, S.S. and N.R.D.; funding acquisition, N.R.D. and N.D. All authors have read and agreed to the published version of the manuscript.

Funding: This research was funded by the project HubIS under the PRIMA 2019 program of the European Commission (HORIZON EUROPE).

Institutional Review Board Statement: Not applicable.

Informed Consent Statement: Not applicable.

Data Availability Statement: Not applicable.

Acknowledgments: The authors acknowledge the European Space Agency for the supporting data and materials.

Conflicts of Interest: The authors declare no conflict of interest.

References

1. Gowda, P.H.; Chávez, J.L.; Howell, T.A.; Marek, T.H.; New, L.L. Surface Energy Balance Based Evapotranspiration Mapping in the Texas High Plains. *Sensors* **2008**, *8*, 5186–5201. [CrossRef]
2. Bastiaanssen, G.M.W. SEBAL-based sensible and latent heat fluxes in the irrigated GedizBasin, Turkey. *J. Hydrol.* **2000**, *229*, 87–100. [CrossRef]

3. Calera, A.; Campos, I.; Osann, A.; D'Urso, G.; Menenti, M. Remote Sensing for Crop Water Management: From ET Modelling to Services for the End Users. *Sensors* **2017**, *17*, 1104. [CrossRef]
4. Bastiaanssen, W.G.M.; Menenti, M.; Feddes, R.A.; Holtslag, A.A.M. A remote sensing surface energy balance algorithm for land (SEBAL) 1. Formulation. *J. Hydrol.* **1998**, *212*, 198–212. [CrossRef]
5. Bastiaanssen, W.G.M.; Pelgrum, H.; Wang, J.; Ma, Y.; Moreno, J.; Roerink, G.J.; van der Wal, T. The Surface Energy Balance Algorithm for Land (SEBAL): Part 2 validation. *J. Hydrol.* **1998**, *212–213*, 213–229. [CrossRef]
6. Allen, R.; Tasumi, M.; Trezza, R. Satellite-Based Energy Balance for Mapping Evapotranspiration with Internalized Calibration (METRIC)-Model. *J. Irrig. Drain. Eng.* **2007**, *133*, 380–394. [CrossRef]
7. Roerink, G.J.; Su, Z.; Menenti, M. S-SEBI: A simple remote sensing algorithm to estimate the surface energy balance. *Phys. Chem. Earth Part B Hydrol. Ocean. Atmos.* **2000**, *25*, 147–157. [CrossRef]
8. Norman, J.M.; Kustas, W.P.; Humes, K.S. Source approach for estimating soil and vegetation energy fluxes in observations of directional radiometric surface temperature. *Agric. For. Meteorol.* **1995**, *77*, 263–293. [CrossRef]
9. Ruescas, A.B.; Peters, M. S3TBX Collocation Tutorial, V1/rev6 1.0; ESA & Brockmann Consult GMBH. Available online: https://step.esa.int/docs/tutorials/Collocation_S3TBX_Tutorial_v1_rev6.pdf (accessed on 7 October 2022).
10. Alpanakis, N.; Tziatzios, G.; Faraslis, I.; Spiliotopoulos, M.; Sidiropoulos, P.; Sakellariou, S.; Blanta, A.; Brisimis, V.; Dalezios, N.; Dercas, N. Modeling and Estimation of Actual Evapotranspiration in 3 Mediterranean agricultural areas in France, Greece, and Portugal by the aspect of the Sentinel 2 and 3 observations. In Proceedings of the 2nd AgroEcoInfo, Volos, Greece, 30 June–1 July 2022; ISBN 978-618-84403-8-8.

environmental sciences proceedings

MDPI

Proceeding Paper

Water Pollutants Removal by Coated Quartz Sand †

Vesna S. Cvetković *, Nataša M. Petrović and Jovan N. Jovićević

Department of Electrochemistry, Institute of Chemistry, Technology and Metallurgy, National Institute, University of Belgrade, Njegoševa 12, 11000 Belgrade, Serbia; vukicevic@ihtm.bg.ac.rs (N.M.P.); jovicevic@ihtm.bg.ac.rs (J.N.J.)

* Correspondence: v.cvetkovic@ihtm.bg.ac.rs

† Presented at the 7th International Electronic Conference on Water Sciences, 15–30 March 2023; Available online: https://ecws-7.sciforum.net/.

Abstract: Presence of major environmental pollutants in water can cause a variety of harmful effects on living organisms in ecosystems. Very often, apart from the ubiquitous arsenic (As), organic compounds, iron, manganese, hydrogen sulfide, ammonia, and similar pollutants may reach dramatically high levels in originally unpolluted waters due to various anthropogenic activities. These elevated levels have major negative effects on the environment and require action. The aim of this research was to propose a new approach to the development and production of cost-effective water pollutant removal media based on natural minerals. The synthesis process is based on chemical and subsequent thermal treatment of quartz sand with a magnesium carbonate water solution. The MgO-coated sand in a 5 min. interaction in column operation with 250 mL of synthetic water with 200 μgL^{-1} of As(V), 0.51 mgL^{-1} of Fe(III), and 1.41 mgL^{-1} of Mn(II) removed more than 90% of the As(V). The final concentrations of arsenic, iron, and manganese in treated waters were very close to zero after 10 min of contact with active materials. The removal process is based on redox and sorption processes on the surface and inside the pores of the active material.

Keywords: arsenic; manganese; iron; removal; MgO-coated quartz sand

check for updates

Citation: Cvetković, V.S.; Petrović, N.M.; Jovićević, J.N. Water Pollutants Removal by Coated Quartz Sand. *Environ. Sci. Proc.* **2023**, *25*, 10. https://doi.org/10.3390/ECWS-7-14188

Academic Editor: Mohamed ElAassar

Published: 14 March 2023

1. Introduction

Since water contamination caused by organic and inorganic pollutants has become a serious worldwide environmental issue, in 2015, the United Nations announced a new Agenda with 17 Sustainable Development Goals (SDG). SDG sixth call should ensure clean and accessible water for present and future generations. Arsenic (As), a toxic metalloid whose reservoir pool is in the earth's crust, is one of the contaminants that have originally occurred in low quantities in unpolluted surface water, but may reach dramatically increased levels due to anthropogenic activities [1]. Very often, apart from ubiquitous As, organic compounds, iron, manganese, hydrogen sulfide, ammonia, and similar pollutants may reach much higher concentrations than permissible levels, which causes water quality deterioration. Ravenscroft et al. announced that around 140 million people in 50 countries have already been drinking waters containing arsenic at levels above the World Health Organization provisional guideline value of 10 $\mu g/L$ [2,3]. Despite the strong global advocacy of prevention and control of high water contaminant exposure, especially exposure to arsenic, this is still a large-scale problem that is recognized as a major public health concern. It is crucial to develop viable and economical technologies for waste water treatments in order to overcome the harmful impact of contaminated waters for human and animal consumption. In the meantime, techniques based on adsorption processes have attracted considerable attention in the attempts to remove pollutants from various aqueous systems [1,4,5]. A number of adsorbents, including activated alumina, coated sand, kaolinite-bentonite ceramics, zerovalent iron, metal oxides, and among them, highly porous $Mg(OH)_2/MgO$, etc., have been used to remove targeted pollutants [1,4,6–8]. It has been shown that a well-designed and carefully operated sand filter can be very effective in removing pollutants and especially arsenic from water [7]. In most cases, the efficiency of adsorbent materials in the removal of pollutants from waters relies on the recyclability of materials and reduction of waste volume by using solid adsorbents [1]. In this study, a new approach to the development and production of water pollutant removal media was adopted. As a result, an inexpensive and effective sorbent based on natural minerals, such as quartz sand, has been synthesized for the treatment of contaminated waters. Moreover, the fabricated active material exhibits a high sorption capacity towards, not only arsenic, but also manganese and iron species' removal from treated waters.

2. Materials and Methods

The mineral chosen in this work was a low cost and naturally abundant quartz sand (98.50 wt.% of SiO_2, grain average size 0.278 mm), with the assumption that there would be no requirement for additional regeneration after polluted water treatment. In order to remove impurities, the quartz sand was washed with deionized water and dried at 105 °C. A freshly prepared concentrated solution of $Mg(CO_3)_4\ Mg(OH)_2 \times 5H_2O$ was then added to the sand in the amount that ensures that Mg in the synthetized material represents the desired mass percentage, for example, 2 wt.% of magnesium. Suspension was than vigorously stirred until it was well homogenized and was dried in an oven at 170 °C for 30 min. After drying, the material was heated at 800 °C for one hour, left to cool down to room temperature, and was powdered in ahate mortar. In this study, the experiments were performed at room temperature, by bringing into contact 5 g of composite active material with 250 mL of a model water solution. For that purpose, the model water solution was made as follows: appropriate amount of arsenic standard solution (H_3AsO_4 in HNO_3 0.5 mol/L 1000 mg/L As Certipur®, Sigma Aldrich, St. Louis, MO, USA), $MnCl_2 \times 4H_2O$, and $FeCl_3 \times 6H_2O$ (Merck, p.a, Darmstadt, Germany) were dissolved in deionized water in the following concentrations: 200 µg As(V) L^{-1}, 1.41 mg Mn(II) L^{-1}, and 0.51 mg of Fe(III) L^{-1}. The model water solution (250 mL) was placed in a glass column and 5 g of composite materials was added. The contact time of the solid and liquid phases were predetermined for each experiment and separation was conducted by filtration. Concentrations of arsenic, manganese, iron, and other pollutant cations were determined by Atomic Absorption Spectrometer (AAS, Varian spectra A-20, Palo Alto, CA, USA). Scanning electron microscopy (SEM-JEOL JSM-5300, Tokyo, Japan), in conjunction with Energy dispersive spectroscopy (EDS) capabilities (LINK QX-2000 Oxford Instruments-XR400, Abingdon, UK), was used to assess the resulting appearance and perform compositional and semi-quantitative chemical analysis of the active material in the composite before and after contact with model water solutions.

3. Results and Discussion

Figure 1 shows the image of the modified quartz sand surface and corresponding EDS analysis. As observed, the quartz sand grains are completely covered with microcrystalline and amorphous stoichiometric and non-stoichiometric oxides with an excess of magnesium in our case. The grains are randomly grouped into more or less separate agglomerates, which are manifested by a highly developed surface area. According to EDS analysis, Figure 1b) apart from magnesium (42.67 at.%) and oxygen (55.96 at.%), a small quantity of Ca (0.21 at.%) and almost no detectible Si were found at the surface of the composite. After contact with the model water solution, the structure and the surface appearance of the MgO-coated quartz sand changed greatly, as shown in Figure 2. It is evident from SEM results Figure 2a) that the unique porous microstructure of the coated quartz sand disappeared, while a more uniform and flatter surface was created. This indicated that, when placed in contact with the model water solution, MgO from the composite dissolved.

This was confirmed by EDS and AAS analyses of the quartz sand and model water solutions, respectively, taken after active contact. EDS analysis of the newly formed surface revealed the presence of Si in a substantial quantity, 23.41 at.%, suggesting that that amount of active material (mainly MgO) on the surface of the coated sand was reduced.

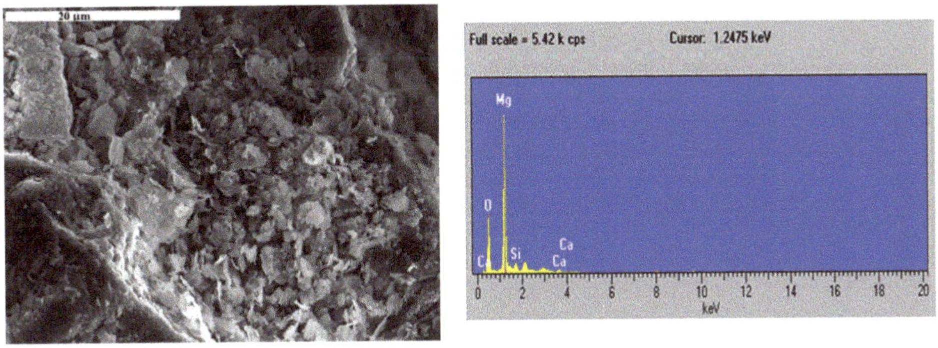

(a) (b)

Figure 1. (**a**) SEM image (magnification 2.500×); (**b**) EDS analysis of the modified quartz sand before treatment with the model water solution.

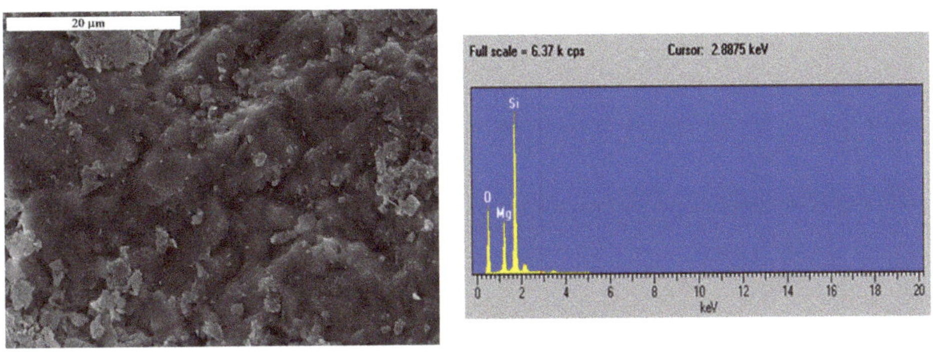

(a) (b)

Figure 2. (**a**) SEM image (magnification 2.500×); (**b**) EDS analysis of the modified quartz sand after contact with the model water solution.

Removal of As(V), Fe(III), and Mn(II) by MgO-coated quartz sand as a function of contact time is shown in Table 1. During contact, magnesium oxides from the surface of the active material are dissolved in model water solutions, providing the electrons in the solution that can participate in a number of reduction processes. The electrochemical reduction processes of As(V), Mn(II), and Fe(III)/Fe(II) are driven by the difference in the electrochemical potential of Mg atoms and pollutants whose standard reduction potentials in this solution are more positive than the standard potential of Mg. Such reduced species are most likely adsorbed/deposited on the surface of the composite material. The reduction process takes place until Mg^{2+} ions continue to solvate themselves in the model water solution. Most likely, part of Mg^{2+} ions react with OH^- ions to form $Mg(OH)_2$, resulting in a rapidly increased pH value, immediately above 10 at the beginning of the process, as seen in Table 1. On the other hand, it can be presumed that the total pH value of the model water solution after treatment is a result of the manganese and iron hydroxide being formed, also with hydroxyl ions present.

According to Wu et al., removal of As(V) could be the result of the As(V) reaction with $Mg(OH)_2$ and other magnesium arsenate formations [9]. This was also revealed in our study when the modified quartz sand after treatment with model water solution containing a very high As(V) concentration (200 mg/L As(V)) was subjected to XRD analysis. Apart from $Mg(OH)_2$ and SiO_2, the analysis showed the presence of Mg_3As_2. It is interesting to note that the As(V) removal efficiency did not decrease with an increase in pH values, as shown in Table 1, which is contrary to the observations made by others [10]. This may be attributed to the precipitation of magnesium, manganese, and even Fe hydroxide on the surface of the solid material, leading to a more effective process [7,11]. In addition, the presence of sludge at the bottom of the column after a prolonged contact time of more than 30 min was observed. Finally, used composite material (coated quartz sand) does not have to be recycled. Instead, it can be used as a component of the materials for road building or similar construction applications, where the adsorbed contaminants can remain captured.

Table 1. Change of As(V), Fe(III), and Mn(II) concentration in model water solution after treatment with MgO-coated quartz sand.

Species and Concentration	Model Water Solution	Contact Time (min)					
		5	10	15	20	25	30
Na (mg/L)	0.025	6.02	3.6	5.12	4.58	4.6	4.52
K (mg/L)	<0.02	0.83	0.86	0.75	0.78	0.69	0.71
Ca (mg/L)	<0.062	9.45	10.45	9.3	9.95	9.9	9.85
Fe (mg/L)	0.51	<0.05	<0.05	<0.05	<0.05	<0.05	<0.05
Mn (mg/L)	1.41	<0.018	<0.018	<0.018	<0.018	<0.018	<0.018
Mg (mg/L)	0.027	9.9	8.1	9.2	8.7	12.75	12.35
As (μg/L)	200	9	7.7	2.6	1.75	1.35	<0.5
pH	4.27	10.49	10.60	10.55	10.59	10.61	10.64

4. Conclusions

Removal of contaminants As(V), Fe(III), and Mn(II) from water by MgO-coated quartz sand was investigated. When placed in contact with model water solutions, MgO from the composite material dissolves, providing electrons for the reduction of present pollutant species. After 5 min of interaction of the MgO-coated quartz sand with 250 mL of model water solution with 200 μgL^{-1} of As(V), 0.51 mgL^{-1} of Fe(III), and 1.41 mgL^{-1} of Mn(II), nearly 95% of the As(V) was removed. Final concentrations of arsenic, iron, and manganese in treated waters after 10 min of contact with active materials were very close to zero. The MgO-coated sand exhibited a strong ability to remove both ionic and colloidal forms of pollutants from treated waters. The removal process is based on redox and sorption processes on the surface and inside the pores of the active composite material.

Author Contributions: Conceptualization and methodology, V.S.C.; investigation, V.S.C. and N.M.P.; data curation, V.S.C. and N.M.P.; writing—original draft preparation, V.S.C.; review and editing, V.S.C. and J.N.J.; All authors have read and agreed to the published version of the manuscript.

Funding: This research received no external funding.

Institutional Review Board Statement: Not applicable.

Informed Consent Statement: Not applicable.

Data Availability Statement: Not applicable.

Acknowledgments: Vesna S. Cvetković and Nataša M. Petrović acknowledge the financial support for the investigation received from the Ministry of Education, Science, and Technological Development of the Republic of Serbia (Grant No. 451-03-68/2022-14/200026).

Conflicts of Interest: The authors declare no conflict of interest.

References

1. Gomaa, H.; Khalifa, H.; Selim, M.M.; Shenashen, M.A.; Kawada, S.; Alamoudi, A.S.; Azzam, A.M.; Alhamid, A.A.; El-Safty, S.A. Selective, Photoenhanced Trapping/Detrapping of Arsenate Anions Using Mesoporous Blobfish Head TiO_2 Monoliths. *ACS Sustain. Chem. Eng.* **2017**, *5*, 10826–10839. [CrossRef]
2. Brammer, H.; Ravenscroft, P. Arsenic in groundwater: A threat to sustainable agriculture in South and South-east Asia. *Environ. Int.* **2009**, *35*, 647–654. [CrossRef] [PubMed]
3. Teoh, Y.P.; Khan, M.A.; Choong, T.S.Y. Kinetic and isotherm studies for lead adsorption from aqueous phase on carbon coated monolith. *Chem. Eng. J.* **2013**, *217*, 248–255. [CrossRef]
4. Cvetković, V.S.; Purenović, J.M.; Purenović, M.M.; Jovićević, J.N. Interaction of Mg-enriched kaolinite–bentonite ceramics with arsenic aqueous solutions. *Desalination* **2009**, *249*, 582–590. [CrossRef]
5. Cvetković, V.S.; Purenović, M.M.; Jovićević, J.N. Manganese removal from water by magnesium enriched kaolinite-bentonite ceramics. *Desalin. Water Treat.* **2010**, *24*, 202–209. [CrossRef]
6. Mohan, D.; Pittman, C.U. Arsenic removal from water/wastewater using adsorbents—A critical review. *J. Hazard. Mater.* **2007**, *142*, 1–53. [CrossRef] [PubMed]
7. Ruiping, L.; Lihua, S.; Jiuhui, Q.; Guibai, L. Arsenic removal through adsorption, sand filtration and ultrafiltration: In situ precipitated ferric and manganese binary oxides as adsorbents. *Desalination* **2009**, *249*, 1233–1237. [CrossRef]
8. Cvetković, V.S.; Purenović, J.M.; Jovićević, J.N. Change of Water Redox Potential, PH and RH in Contact with Magnesium Enriched Kaolinite–Bentonite Ceramics. 2008, 38, 268–278. *Appl. Clay Sci.* **2008**, *38*, 268–278. [CrossRef]
9. Wu, P.-Y.; Jiang, Y.-P.; Zhang, Q.-Y.; Jia, Y.; Peng, D.-Y.; Xu, W. Comparative study on arsenate removal mechanism of MgO and MgO/TiO_2 composites: FTIR and XPS analysis. *New J. Chem.* **2016**, *40*, 2878–2885. [CrossRef]
10. Wu, K.; Wang, H.; Liu, R.; Zhao, X.; Liu, H.; Qu, J. Arsenic removal from a high-arsenic wastewater using in situ formed Fe–Mn binary oxide combined with coagulation by poly-aluminum chloride. *J. Hazard. Mater.* **2011**, *185*, 990–995. [CrossRef] [PubMed]
11. Shafaei, A.; Rezayee, M.; Arami, M.; Nikazar, M. Removal of Mn^{2+} ions from synthetic wastewater by electrocoagulation process. *Desalination* **2010**, *260*, 23–28. [CrossRef]

environmental sciences proceedings

MDPI

Proceeding Paper

Use of Hydrogels for Seawater Desalination Processes: A Patent Landscape Report †

Ahmed Fatimi

Chemical Science and Engineering Research Team (ERSIC), Department of Chemistry, Polydisciplinary Faculty of Beni Mellal (FPBM), Sultan Moulay Slimane University (USMS), Beni Mellal 23000, Morocco; a.fatimi@usms.ma

† Presented at the 7th International Electronic Conference on Water Sciences, 15–30 March 2023; Available online: https://ecws-7.sciforum.net.

Abstract: This patent landscape report encapsulates the state by introducing what has been invented concerning hydrogels for seawater desalination processes. As a result, 81 patent documents were found, and 62% of all patent documents have been published during the last three years. According to the findings, China was ranked as the first jurisdiction, and the most prolific patenting companies are from China and the United States, while William Marsh Rice University (United States) and Tianjin Polytechnic University (China) are the academic institutions leading the way. The patent classifications indicate that most inventions are intended for the treatment of seawater (e.g., desalination); processes specially adapted for manufacturing semi-permeable membranes for separation processes; and the treatment of water using ion-exchange sorbent compositions, dialysis, osmosis, or reverse osmosis. According to the knowledge clusters and expert driving factors in this patent landscape analysis, research and development are centered on methods and apparatus for desalination, which are the focus of the majority of patents, as well as macromolecular gels and synthetic macromolecular substances.

Keywords: seawater; desalination; hydrogels; patent landscape; innovation

check for updates

Citation: Fatimi, A. Use of Hydrogels for Seawater Desalination Processes: A Patent Landscape Report. *Environ. Sci. Proc.* **2023**, *25*, 11. https://doi.org/10.3390/ECWS-7-14184

Academic Editor: Lampros Vasiliades

Published: 14 March 2023

1. Introduction

For humans, animals, agriculture, and many different sectors, including agrifood and pharmaceutics, pure water is a necessary raw material. Owing to climate change and the increase in population, the demand and needs of people are also growing rapidly, which has resulted in problems such as water availability. With the current shortage of such material, other sources or alternatives to improve the current resources are needed [1].

As an alternative, seawater desalination could help to overcome this issue in water-stressed countries. The process implies the use of membranes to remove salt compounds and obtain fresh water, which is necessary for human health and the better functioning of society [2]. In sea basins, for example, the continuous exposure of agriculture to saline water causes soil salinity, which directly affects the morphology, physiology, and metabolism of plants and/or fruits (i.e., poor or no yields, reduced root growth, reduced nutritional status, and so on) [3].

Despite significant advances in desalination technology, seawater desalination continues to consume more energy than traditional fresh water treatment technologies [2]. A novel approach for seawater desalination using hydrogels is currently being explored by researchers around the world [4]. The fundamental benefit of the hydrogel-based process is that it does not require a membrane because the hydrogel surface itself serves as the interface, as well as the high water uptake that such materials can offer. In practical applications, a dry hydrogel is swollen in the seawater, and the hydrogel with absorbed seawater is squeezed to obtain fresh water [5].

Hydrogels are polymer-based materials that can be created from a large number of synthetic polymers, polysaccharides, proteins, or hybrid polymers [6]. The three-dimensional structure of these hydrogels remains in equilibrium in an aqueous medium owing to the balance between the elastic forces of the crosslinked macromolecular network and the osmotic forces of the liquid [7]. Furthermore, the chemical composition and the crosslink density of such materials determine the swelling and permeability properties of the structure [8].

Through the innovation and improvement of raw materials (such as polymers and biopolymers), chemical synthesis, formulation, and chemical engineering related to the desalination process, hydrogels for seawater desalination processes are developing quickly. This is also supported by the rise in the number of patent applications submitted in this field of study and development each year.

This study presents a patent landscape report on innovative and emerging seawater desalination technologies based on hydrogels. A landscape report is a family of techniques for studying the information present in and attached to patents in a specific area [9–12]. It is established as a tool for research planning in accordance with the criteria for patent analysis [13–16]. The results are then examined by responding to specific queries, such as who, what, and where queries regarding patenting patterns.

2. Methods

Patent data for this study were extracted from the Lens online database using different keyword searches in patent titles, abstracts, and claims [17]. Title, abstract, and claim keyword searches are common tools for patent extraction [9–12]. The search was then filtered to include only patent documents (i.e., patent applications, granted patents, and three limited patents) until 2022, and the information was extracted according to patent document type, patent family, publication date, patent applicant, patent owner, patent jurisdiction, legal status of patents, and patent classification [13–16].

3. Results and Discussion

3.1. Documents Type and Patent Families

A total of 81 patent documents were collected between 1985 and 2022 that concern the use of hydrogels for seawater desalination processes. The selection includes 66 patent applications, 12 granted patents, and 3 limited patents. With 57 simple families and 57 extended families, it was confirmed that 57 patent documents covering the same technical subject were published at different times in the same country or in different countries or regions.

3.2. Publication Date of Patent Documents

Based on the earliest priority date, it is possible to precisely assume that 1985 is considered the starting year of patenting of hydrogels for seawater desalination [18]. Thirty-one patent documents were published between 1985 and 2019; however, 62% of all patent documents have been published during the last three years. Furthermore, 2022 was the year with the maximum recorded number of patent documents and patent applications, with 16 and 15, respectively, while 2011 was the year with the maximum number of granted patents, with a total of five (Figure 1).

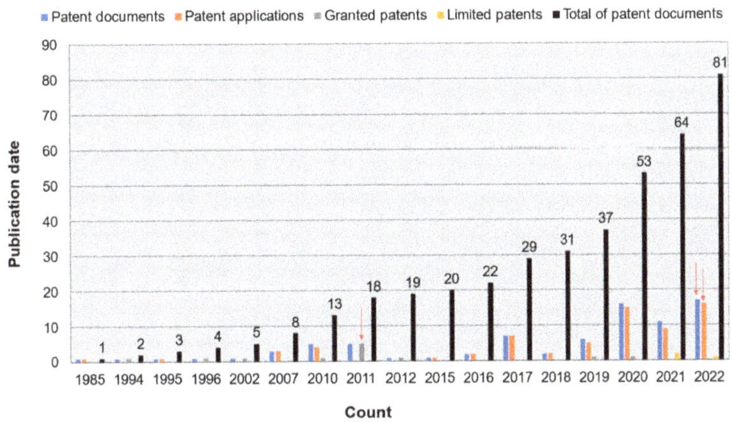

Figure 1. Publication dates of patent documents related to the use of hydrogels for seawater desalination processes.

3.3. *Patent Applicants and Owners*

The patent applicant is the person or organization that has filed a patent application. On the other hand, the patent owner is the organization with rights to the property of the granted patent [15].

The top 10 applicants for patent documents related to the use of hydrogels for seawater desalination processes until 2022 are displayed in Figure 2. The list includes natural persons and organizations (i.e., companies and academic institutions). The companies "Kraton Polymers US LLC" (Houston, TX, United States) and "Baoshan Iron & Steel CO LTD" (Shanghai, China) are the most prolific patenting companies, with nine and four patent documents, respectively, while William Marsh Rice University (Houston, TX, United States) and Tianjin Polytechnic University (Tianjin, China) are the academic institutions leading the way, with four patent documents each. On the other hand, different natural people are the lead applicants, with four or three patent documents.

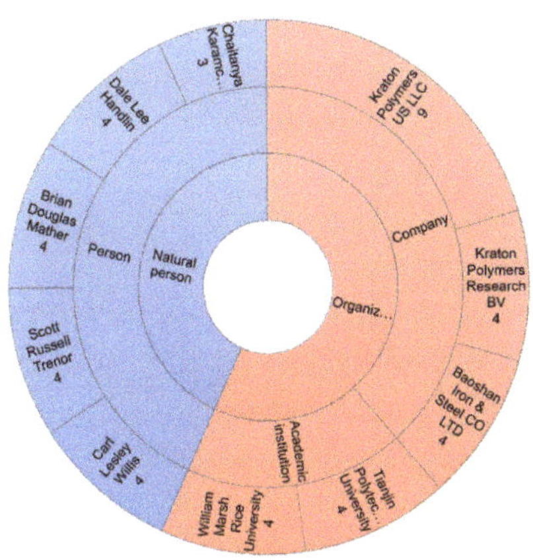

Figure 2. Applicants (top 10) of patent documents related to the use of hydrogels for seawater desalination processes.

Table 1 presents the owners of patent documents related to the use of hydrogels for seawater desalination processes until 2022. The ownership of the rights to the property of the patent concerns eight owners: four companies and four academic institutions. Three of them are, at the same time, already applicants according to the top 10 applicants.

Table 1. Owners of patent documents related to the use of hydrogels for seawater desalination processes.

Owners	Type (Headsquare)	Count
Baoshan Iron & Steel CO LTD	Company (Shanghai, China)	2
Kraton Polymers US LLC	Company (Houston, TX, USA)	2
City University of Hong Kong	Academic institution (Kowloon Tong, Hong Kong)	1
Lockheed Martin Corporation	Company (Bethesda, MD, USA)	1
Saudi Basic Industries Corporation	Company (Riyadh, Saudi Arabia)	1
South China University of Technology	Academic institution (Guangzhou, China)	1
William Marsh Rice University	Academic institution (Houston, TX, USA)	1
Zhejiang University	Academic institution (Hangzhou, China)	1

3.4. Jurisdictions and Legal Status of Patents

Patent jurisdictions correspond to the countries in which the patent is protected through the appropriate patent office at which the patent application is filed. When a patent application is filed, it is examined by the appropriate patent office (e.g., the China National Intellectual Property Administration (CNIPA), the Moroccan Office of Industrial and Commercial Property (OMPIC), the United States Patent and Trademark Office (USPTO), and so on), and a decision is delivered according to the specific patentability criteria that an invention must meet. Finally, a claimed invention through a patent application becomes a granted patent in one or more designated countries if it is qualified for patent protection [13].

For patent documents related to the use of hydrogels for seawater desalination processes, the jurisdictions of patent documents until 2022 are presented in Figure 3. China, through the CNIPA, encompasses 45 patent documents, with a patent contribution per document total of 55.56%. The United States, through the USPTO, encompasses 23 patent documents, with a patent contribution per document total of 28.4%. On the other hand, the global system for filing patent applications, known as the Patent Cooperation Treaty (PCT) and administered by the World Intellectual Property Organization (WIPO), encompasses seven patent documents, with a patent contribution per document total of 8.64%.

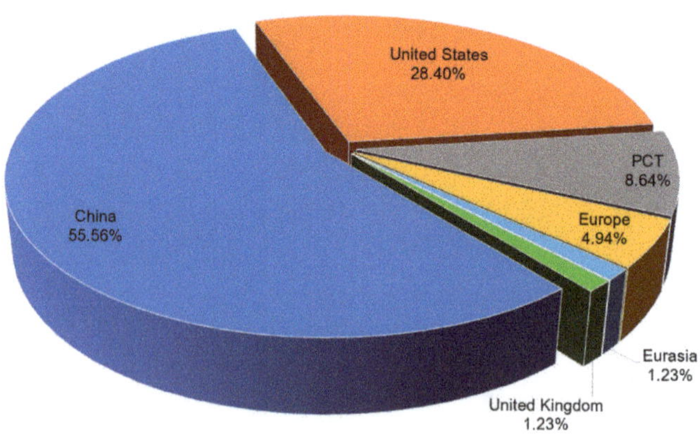

Figure 3. Jurisdictions of patent documents related to the use of hydrogels for seawater desalination processes.

Moreover, these jurisdictions of patent documents related to the use of hydrogels for seawater desalination processes ensure the following five legal statuses:

- 40 patent documents are in force (i.e., active status);
- 30 pending applications (i.e., pending status);
- 6 withdrawn or rejected applications (i.e., discontinued status);

- 4 patents that have reached the term date and are no longer in force (i.e., expired status);
- 1 PCT application that has been granted in one or more designated countries (i.e., patented status).

3.5. Patent Classifications

As a technology indicator in a specific area, the patent classifications help inventors and researchers to categorize claimed inventions and retrieve different patent documents. The International Patent Classification (IPC) is the most widely used hierarchical system in the form of codes. It divides all technology areas into a range of sections, classes, subclasses, groups, and subgroups [13].

For patent documents related to the use of hydrogels for seawater desalination processes, the top 10 IPC codes until 2022 are presented in Table 2. The patent classifications reveal that most IPC code corresponds to C02F103/08, with 29 patent documents. It concerns the treatment of seawater using the desalination process. The IPC codes related to hydrogels and presented into these top 10 correspond to B01J13/00 and C08J3/075, which are categorized for colloid chemistry and macromolecular gels, respectively.

Table 2. IPC codes (top 10) of patent documents related to the use of hydrogels for seawater desalination processes.

IPC	Description	Count
C02F103/08	Treatment of seawater (e.g., desalination)	29
C02F1/14	Treatment of water by distillation or evaporation using solar energy	21
B01D67/00	Processes specially adapted for manufacturing semi-permeable membranes for separation processes or apparatus	14
C02F1/44	Treatment of water by dialysis, osmosis, or reverse osmosis	12
B01J13/00	Colloid chemistry (e.g., the production of colloidal materials or their solutions)	10
C08J3/075	Treatment or compounding of macromolecular substances, such as macromolecular gels	10
B01J20/26	Organic material, such as synthetic macromolecular compounds to make solid sorbent compositions or filter aid compositions sorbents for chromatography processes	9
C02F1/04	Treatment of water by distillation or evaporation using solar energy	8
C02F1/00	Treatment of water, waste water, or sewage	7
C02F1/28	Treatment of water by sorption using ion-exchange sorbent compositions	6

4. Conclusions

The novel approach for seawater desalination using hydrogels was successfully explored by researchers around the world. Inventors proposed various claimed inventions through this innovation. In this study, a patent landscape report was proposed regarding patent document types, patent families, publication dates, patent applicants, patent owners, patent jurisdictions, the legal status of patents, and patent classifications. Therefore, this research reveals that China and the United States are leading the way in innovation, and the most prolific patenting organizations are companies and universities. On the other hand, most inventions are intended for the treatment of seawater using desalination processes specially adapted for manufacturing semi-permeable membranes for separation processes, and the research and development are based on macromolecular gels and synthetic macromolecular compounds.

Funding: This research received no external funding.

Institutional Review Board Statement: Not applicable.

Informed Consent Statement: Not applicable.

Data Availability Statement: The data presented in this study are available within this article content.

Acknowledgments: The author acknowledges the Cambia Institute for the Lens patent dataset used in this study.

Conflicts of Interest: The author declares no conflict of interest.

References

1. Arfin, T. Current innovative chitosan-based water treatment of heavy metals: A sustainable approach. In *Handbook of Biopolymers. In Handbook of Biopolymers: Advances and Multifaceted Applications*; Shakeel, A., Suvardhan, K., Gopalakrishnan, K., Eds.; Jenny Stanford Publishing: New York, NY, USA, 2018; pp. 167–182.
2. Elimelech, M.; Phillip, W.A. The Future of Seawater Desalination: Energy, Technology, and the Environment. *Science* **2011**, *333*, 712–717. [CrossRef] [PubMed]
3. Hafiane, F.Z.; Fatimi, A. An emulsion-based formulation for increasing the resistance of plants to salinity stress: US20160302416A1 patent evaluation. *Environ. Sci. Proc.* **2022**, *16*, 4. [CrossRef]
4. Ali, W.; Gebert, B.; Altinpinar, S.; Mayer-Gall, T.; Ulbricht, M.; Gutmann, J.S.; Graf, K. On the Potential of Using Dual-Function Hydrogels for Brackish Water Desalination. *Polymers* **2018**, *10*, 567. [CrossRef] [PubMed]
5. Höpfner, J.; Richter, T.; Košovan, P.; Holm, C.; Wilhelm, M. Seawater Desalination via Hydrogels: Practical Realisation and First Coarse Grained Simulations. In *Intelligent Hydrogels*; Sadowski, G., Richtering, W., Eds.; Springer International Publishing: Cham, Switzerland, 2013; pp. 247–263.
6. Fatimi, A. Chitosan-based embolizing hydrogel for the treatment of endoleaks after endovascular aneurysm repair. *Int. J. Polym. Mater. Polym. Biomater.* **2019**, *68*, 107–114. [CrossRef]
7. Fatimi, A.; Okoro, O.V.; Podstawczyk, D.; Siminska-Stanny, J.; Shavandi, A. Natural hydrogel-based bio-Inks for 3D bioprinting in tissue engineering: A review. *Gels* **2022**, *8*, 179. [CrossRef] [PubMed]
8. Akhramez, S.; Fatimi, A.; Okoro, O.V.; Hajiabbas, M.; Boussetta, A.; Moubarik, A.; Hafid, A.; Khouili, M.; Simińska-Stanny, J.; Brigode, C.; et al. The circular economy paradigm: Modification of bagasse-derived lignin as a precursor to sustainable hydrogel production. *Sustainability* **2022**, *14*, 8791. [CrossRef]
9. Fatimi, A. Chitosan-based hydrogels: Patent analysis. *Mater. Proc.* **2022**, *9*, 1. [CrossRef]
10. Fatimi, A. Trends and recent patents on cellulose-based biosensors. *Eng. Proc.* **2022**, *16*, 12. [CrossRef]
11. Fatimi, A. Patentability of biopolymer-based hydrogels. *Chem. Proc.* **2022**, *8*, 39. [CrossRef]
12. Fatimi, A. Hydrogel-based bioinks for three-dimensional bioprinting: Patent analysis. *Mater. Proc.* **2021**, *7*, 3. [CrossRef]
13. Fatimi, A. A patent data analysis of the innovation trends in biological control agent formulations. *Recent Adv. Food Nutr. Agric.* **2022**, *13*, 59–69. [CrossRef] [PubMed]
14. Fatimi, A. Cellulose-based hydrogels: Patent analysis. *J. Res. Updates Polym. Sci.* **2022**, *11*, 16–24. [CrossRef]
15. Fatimi, A. Exploring the patent landscape and innovation of hydrogel-based bioinks used for 3D bioprinting. *Recent Adv. Drug Deliv. Formul.* **2022**, *16*, 145–163. [CrossRef] [PubMed]
16. Fatimi, A. Seaweed-based biofertilizers: A patent analysis. *Recent Pat. Biotechnol.* **2022**, *16*, 144–154. [CrossRef] [PubMed]
17. Cambia Institute. The Lens Patent Data Set. *Version 8.7.0.* Available online: www.lens.org (accessed on 4 December 2022).
18. Gould, F.E.; Johnston, C.W. Hydrophilic Polyurethane Acrylate Compositions. UK Patent GB2150938A, 10 July 1985.

environmental sciences proceedings

MDPI

Proceeding Paper

Bridging the Gap between Science and Policy: A Prerequisite for Effective Water Governance †

Maria Karasani, Dionissis Latinopoulos, Nena Ioannidou, Mike Spiliotis and Ifigenia Kagalou *

Department of Civil Engineering, Democritus University of Thrace, 67100 Xanthi, Greece; mkarasan@civil.duth.gr (M.K.); dlatinop@civil.duth.gr (D.L.); partioan@civil.duth.gr (N.I.); mspiliot@civil.duth.gr (M.S.)

* Correspondence: ikagkalo@civil.duth.gr

† Presented at the 7th International Electronic Conference on Water Sciences, 15–30 March 2023; Available online: http://ecws-7.sciforum.net.

Abstract: Water governance in the EU is enshrined in the Water Framework Directive (WFD), with the engagement of stakeholders being one of the governance cornerstones. The inclusion of the interests of scientific and non-scientific groups in decision-making is crucial. Our objective is to examine the contribution of the participatory approach to the effectiveness of local water resource management. Within the Eye4water project, a participatory assessment was applied for the Lissos river basin, through joint identification and evaluation of the main water-related issues. Firstly, we identified the social system engaged to the basin through stakeholders' mapping. Secondly, based on criteria selection, three stakeholders' groups were invited to a workshop. Our preliminary results show that mutual learning should be encouraged at multiple levels. Well-recognized threats such as water pollution, flood risk, and groundwater lowering are present, while biodiversity issues are quite underrepresented.

Keywords: stakeholder analysis; participatory management; local knowledge; Lissos; interactive workshop; basin management

check for updates

Citation: Karasani, M.; Latinopoulos, D.; Ioannidou, N.; Spiliotis, M.; Kagalou, I. Bridging the Gap between Science and Policy: A Prerequisite for Effective Water Governance. *Environ. Sci. Proc.* **2023**, *25*, 12. https://doi.org/10.3390/ECWS-7-14241

Academic Editor: Athanasios Loukas

Published: 16 March 2023

1. Introduction

Water governance describes the legislation, policies, regulation, and institutional frameworks related to the management of water resources, which affect human activities and nature's sustainability. Water governance is a complex process that requires the participation of not only technical experts and the scientific community, but also of the different stakeholder groups in water decision-making [1]. More specifically, the development and implementation of water policies are characterized by challenges which concern the integration of legal requirements, technical issues, scientific knowledge, socio-economic aspects, and the competitive uses of the resource [2] in all stages of the process. For the minimization of conflicts and the measures' success assurance, all voices should be heard, making intensive multi-stakeholder consultations be required for effective, equitable, and sustainable water governance [3]. Lately, the participation of stakeholders in water governance is considered a key element for improving water resources management and is strongly supported (suggested or mandatory) in the majority of water-related EU directives [1,4]. In particular, the Water Framework Directive (WFD) (2000/60/EC) establishes a legal framework to protect and improve the status of aquatic ecosystems, including—among other factors—public participation. It is now documented that stakeholder consultations based on communication and group interaction depend on trust-building science-policy collaborations [5,6]. Common understanding and interpretation of local water issues and solutions, and collaborative production of scientific, local, and bureaucratic knowledge are essential for legitimate decision-making processes and the effective co-creation and implementation of measures [5–7]. The contribution of stakeholders to the design of a good governance scenario together with the gaps in data are considered the most important dependencies in the management of Greek river basins [8]. Having the above in mind when considering the peculiarities of the Lissos basin as concerns stakeholders' identity, this paper aims to outline the participatory bottom-up approach for this basin's management, having the bridging between science and policy as its supreme goal.

2. Materials and Methods

A participatory workshop was conducted under the framework Eye4water, which aims to strengthen the local water management practices in the Lissos river basin by developing supporting technological tools as a result of systematic monitoring of the quality of natural water bodies. To our knowledge, this is the first participatory assessment of a local river basin being applied to the Lissos river basin. Our methodology was conducted in four stages: (1) Process design; (2) Workshop process; (3) Results analysis; and (4) Follow-up.

2.1. Study Area

The Lissos river basin (Rhodope, Thrace, Greece) covers an area of 1486 km^2 and is partly protected by Ramsar. It is a Heavily Modified Water Body (HMWB), which suffers from several anthropogenic pressures such as landfill, Wastewater Treatment Plant, intensive agricultural and industrial activities, livestock, sand extractions, and flow-intercepting construction [9]. It is considered a lesser-researched river of primary importance for the local community [9], of a higher trophic state, receiving important pollution loads in a segmented hydrological network.

2.2. Process Design

2.2.1. Stakeholder Mapping

Firstly, we identified the social system engaged to the Lissos river basin through stakeholders' mapping. We organized a list of stakeholders and we complemented it through Internet searches (Google maps and business lists), on-site contacts, and the use of our own network (NGOs, academic community, farmers, entrepreneurs, administrative authorities). The identified stakeholders were categorized into three groups: (1) farmers, (2) practitioners, and (3) experts. Before the invitations, equal representation and gender equality among the participants were taken into account. The invitations were sent via email, phone calls, posted on the website, and on the social media accountes (Twitter, Facebook, Linkedin, Instagram) of the Eye4water project. The workshop was also announced through a press release (to about 140 media members). A reminder was also sent.

2.2.2. Questionnaires Development

Based on a SWOT-PEST analysis combined with monitoring results, a number of questions were developed. The main aim of the questions was to gather local knowledge and to further understand how the stakeholders value the resource, prioritize pressures, and jointly identify solutions. For each group, a different set of 12 questions was developed, considering each one's relation to the water sector. The context of the questionnaires covered the water uses, the river pressures, the water management and governance, and the possible solutions. The set of questions included open-close, multiple selection, and importance-grading questions where the participants had the liberty to answer as many questions as they wanted from all groups' questionnaires.

2.2.3. Workshop Process

The process was divided into two sessions. In the first session, a formal briefing of the monitoring findings were communicated to the participants, and then the stakeholders were encouraged to participate in two exercises in a free and open manner with the aid of nine facilitators, where a different color of Post-it was attributed to each group. Exercise 1: In this exercise, the stakeholders were invited to answer the questions anonymously by placing a Post-it somewhere on one of three big panels, allowing for the ability to further comment on each question, thereby promoting meaningful discussions. Exercise 2: A follow-up round after the first exercise took place. In this exercise, the stakeholders were asked willingly to answer the questions of the other two groups. The objective was to identify any conflicts and to evaluate their impact on water management.

2.2.4. Workshop Material Analysis

All produced material from the workshop was photo-documented, and processed as follows: The Post-its from the panels were transferred to a database with the qualitative details from the related discussion notes. A categorization followed, aiming to identify thematic convergences and divergences.

3. Results Discussion

We consider the resulting representation marginally sufficient, since, from more than 100 invitations, at least six representatives from each group attended the workshop. The representatives of each group were urged to reply to the questions posed to other groups. Figure 1 is indicative of the participation and interaction among stakeholders. Using this method, each question gathered about five replies. The exercises were complemented by a continuous discussion and a short evaluation feedback of the entire process. The sub-aim of developing simple and understandable queries for linking science to the tools used by both stakeholders and practitioners and further encouraging action and innovation among all stakeholders [10] was achieved, since none of the moderators noticed any misinterpretations or conceptual errors.

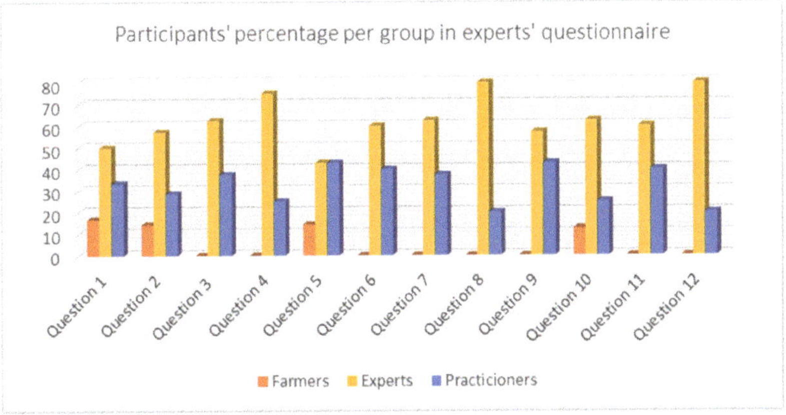

Figure 1. Given answers per group for the experts' questionnaire, indicative of participation and interaction among stakeholder groups.

The results deriving from common queries dealing with the main water uses, pressures, and main issues are presented in Figure 2. It can be clearly seen that irrigation is the major need according to farmers, acknowledging at the same time that agriculture and livestock are among the main activities posing pressure on the watershed. On the contrary, experts and practitioners are more "afraid" of agriculture, livestock, and waste disposal effects, and less of industrial effluents. Different opinions are expressed by the three groups regarding water resource uses or, more simply, their beliefs regarding the needs for water resources allocation, downgrading the importance of biodiversity, industry, and recreation. Despite the large number of low-water crossings and the frequent announcements of flooded areas, floodings were not one of the locals' primary issues.

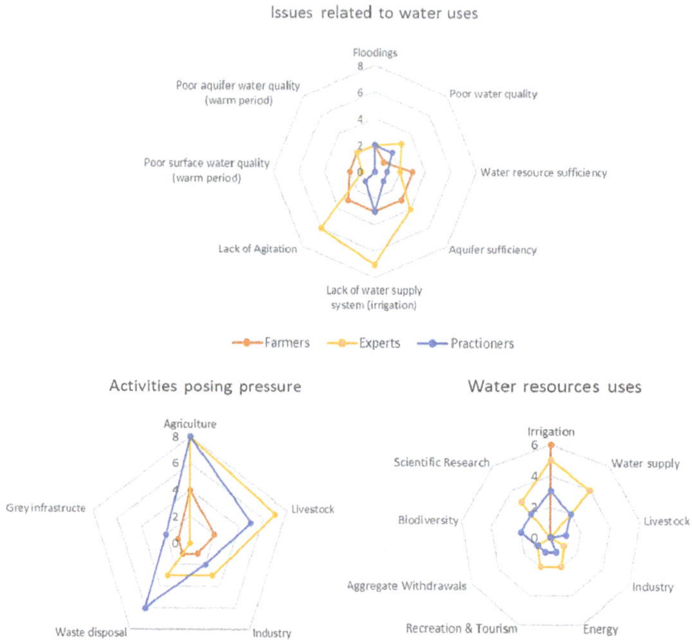

Figure 2. Comparative results regarding common queries from the three stakeholder groups dealing with pressures and uses.

An interesting finding is that experts propose agitation as a measure for land use/land cover alteration in favor of the river system, along with the need for a better irrigation system. Farmers are more anxious of the resource sufficiency, either surface or groundwater. No group deems that there is a seasonal/warm period problem related with water quality. It should be noted that we intentionally omitted queries using a rating scale as a type of answer from the results. The last ones will be used to generate weights for more in-depth analyses through advanced mathematics to suggest some optimal solutions for management of the Lissos basin. It can be stated, though, as a general direction, that better awareness on water issues from the part of higher administrative authorities' and targeted small infrastructure interventions are major components of the solution.

This workshop aimed to bridge the gap between science and policy, and successfully managed to take a "snapshot" of stakeholders' perspective on the management of the Lissos basin. The findings seem to be applicable to decision-making for strategic design and measures implementation, incorporating information of great value that is also based on local knowledge, which could not be gained otherwise. Similar to other research findings [11], Lissos stakeholders seem to be able to implement some management measures (i.e., pollution prevention, channel creation, methodological approach) without official governmental support.

The participation was affected by stakeholders' financial constraints (transport, agricultural duties) while the COVID-19 pandemic situation prevailed in some remote villages. Conflicts between stakeholder groups (farmers vs. practitioners) affected the participatory process. The expressed perception of the different stakeholders' groups did not coincide, except for the activities posing pressure on the watershed. Some points supported by our research and the literature (i.e., seasonal quality and quantity variation, touristic growth potential, and biodiversity) were not supported by public opinion. A major finding can be concluded that "primary production" should not be altered, but rather, should be eased as a measure of water stewardship.

Author Contributions: Conceptualization, I.K. and D.L.; methodology, M.K.; formal analysis, M.K. and N.I.; investigation, D.L.; resources, D.L., M.K. and N.I.; data curation, N.I.; writing—original draft preparation, M.K.; writing—review and editing, D.L. and M.S.; visualization, D.L.; supervision, I.K. and M.S.; project administration, I.K.; funding acquisition, I.K. All authors have read and agreed to the published version of the manuscript.

Funding: Produced for the Eye4Water project, MIS 5047246, implemented under the action: "Support for Research Infrastructure and Innovation" by the Operational Program "Competitiveness, Entrepreneurship and Innovation" in the framework of the Co-financed by Greece and the European Union-European Regional Development Fund.

Informed Consent Statement: Informed consent was obtained from all subjects involved in the study.

Data Availability Statement: Supporting data can be given upon demand in eye4water.com.

Acknowledgments: The authors want to express their gratitude to Ch. Akratos, F. Maris for their contribution to the workshop organization and their presence, to Ntougias S., Adamidis A. for their participation in the first session of the workshop and to C. Papadopoulos, T. Bakas, C. Bakalakou, Ch. Ntislidou, V. Papaevangelou for their working moderation of the workshop and K. Basdeki for the social media and promotion campaign.

Conflicts of Interest: The authors declare no conflict of interest.

References

1. Srdjevic, B.; Medeiros, Y.D.P.; Srdjevic, Z. Empowering small stakeholders' groups in selecting a long-term water management plan. *Water Policy* **2022**, *24*, 1208–1222. [CrossRef]
2. Skoulikaris, C.; Zafirakou, A. River Basin Management Plans as a tool for sustainable transboundary river basins' management. *Environ. Sci. Pollut. Res.* **2019**, *26*, 14835–14848. [CrossRef] [PubMed]
3. Quevauviller, P.; Balabanis, P.; Fragakis, C.; Weydert, M.; Oliver, M.; Kaschl, A.; Arnold, G.; Kroll, A.; Galbiati, L.; Zaldivar, J.M.; et al. Science-policy integration needs in support of the implementation of the EU Water Framework Directive, Environmental. *Sci. Policy* **2005**, *8*, 203–211. [CrossRef]
4. Teegavarapu, R.S.V.; Kolokytha, E.; Galvão, C.D.O. (Eds.) *Climate Change-Sensitive Water Resources Management*, 1st ed.; CRC Press: Boca Raton, FL, USA, 2020. [CrossRef]
5. Romano, O.; Akhmouch, A. Water Governance in Cities: Current Trends and Future Challenges. *Water* **2019**, *11*, 500. [CrossRef]
6. Armitage, D.; de Loë, R.C.; Morris, M.; Edwards, W.; Gerlak, A.K.; Hall, R.I.; Huitema, D.; Ison, R.; Livingstone, D.; MacDonald, G.; et al. Science-policy processes for transboundary water governance. *Ambio* **2015**, *44*, 353–366. [CrossRef] [PubMed]
7. Ricart, S.; Rico, A.; Kirk, N.; Bülow, F.; Ribas-Palom, A.; David, P. How to improve water governance in multifunctional irrigation systems? Balancing stakeholder engagement in hydro social territories. *Int. J. Water Resour. Dev.* **2019**, *35*, 491–524. [CrossRef]
8. Kolokytha, E.; Skoulikaris, C. Dependencies in transboundary water management in Greece in the face of climate change. In Proceedings of the 38th IAHR World Congress, Panama City, Panama, 1–6 September 2019. [CrossRef]
9. Ioannidou, N.; Latinopoulos, D.; Mirli, A.; Bakalakou, C.A.; Karasani, M.; Ntislidou, C.; Kagalou, I.; Akratos, S.C. Is missing knowledge hampering the effectiveness of the sustainable water management? The cases of Laspias and Lissos rivers, Thrace, Greece. In Proceedings of the 7th IAHR Europe Congress, Athens, Greece, 7–9 September 2022.
10. Santillán, D.; Garrote, L.; Iglesias, A.; Sotes, V. Climate change risks and adaptation: New indicators for Mediterranean viticulture. *Mitig. Adapt. Strateg. Glob. Change* **2020**, *25*, 881–899. [CrossRef]
11. Iglesias, A.; Garrote, L. Adaptation strategies for agricultural water management under climate change in Europe. *Agric. Water Manag.* **2015**, *155*, 113–124. [CrossRef]

environmental sciences proceedings

MDPI

Proceeding Paper

Hydrological 2D Modelling of Lithaios River Flows (Greece) Using GIS and Geostatistics for Environmental and Agricultural Water Resources Administration †

Agathos Filintas [1,2,*], George Panoras [3] and George Stamatis [2]

1 Department of Agricultural Technology, University of Thessaly—Campus Gaiopolis, 41500 Larisa, Greece
2 Institute of Mineralogy-Geology, Agricultural University of Athens, Iera Odos 75, 11855 Athens, Greece; stamatis@aua.gr
3 Hellenic Agricultural Organization-Demeter, Water and Soil Resources Institute, 57400 Sindos, Greece; gpanoras@hotmail.com
* Correspondence: filintas@uth.gr

† Presented at the 7th International Electronic Conference on Water Sciences, 15–30 March 2023; Available online: https://ecws-7.sciforum.net.

Abstract: The goal of our investigation is the hydrological 2D modelling of Lithaios River's (Central Greece) streamflow, using GIS and geostatistics for studying water velocity and discharge, stage elevation, and hydraulic features (streamflow depth, water flow area, wettable circumference, hydraulic radius and depth, n Manning's coefficient, Chow's composite n, Froude number, etc.). Moreover, compilations and validations of rating curves (RC) were performed from a series of stage h(t)–discharge Q(t) couples metrics, aiming to use these as a river toolkit to aid environmental and agriculture surface water resources management and help environmental flows calculation, streamflow tracking, and irrigation programming in the regional basin range. The statistical results showed that the Froude number during the study period was $Fr < 1$ showing that Lithaios River's streamflow is classified as subcritical. The models' validation outcomes by using various statistics and geostatistical alternative methods, model simulations and statistics errors criterions, were correlated with the retrieved power models' streamflow data matching for the RC curves and 2D GIS modelling and mapping of river velocity and discharge relationships and were highly satisfying since the stabilities of the deployed relationships were solid. The outcomes of the study results are recommended to provide a hydrological serving toolkit for environmental water resources administration and irrigation programming. This toolkit could assist water supply principalities to rapidly and precisely calculate streamflow volumes and features with a minimal cost rate and workload, and it could be engaged in water supply and agricultural watering administration, the calculation of environmental flows, flood protection, groundwater recharge, and other objectives.

Keywords: hydrological 2D streamflow modelling using GIS and geostatistics; flow velocity; discharge rate; n Manning's coefficient and Chow's composite n; hydraulic properties; rating curve

check for updates

Citation: Filintas, A.; Panoras, G.; Stamatis, G. Hydrological 2D Modelling of Lithaios River Flows (Greece) Using GIS and Geostatistics for Environmental and Agricultural Water Resources Administration. *Environ. Sci. Proc.* **2023**, *25*, 13. https://doi.org/10.3390/ECWS-7-14201

Academic Editor: Lampros Vasiliades

Published: 14 March 2023

1. Introduction

Streamflow velocity and discharge rate, water elevation, hydraulic deep, and flow form are the main themes in hydrology and are closely linked to water supply, quality and administration, flood protection, dewatering, irrigation, dam construction, and other related themes [1–3]. The streamflow velocity and discharge rate have a significant effect on water's retention period and quality [2,4,5]. Thus, these variables are typically needed for hydro-systems modelling. Unfortunately, the ongoing streamflow monitoring on a river's cross-section is commonly unfeasible or very costly [2–6]. Fast and accurate discharges calculation is of high importance for a great number of environmental engineering projects (real-time flood forecasting, water resources administration, etc.) [2,5–7]. The goal of our research is hydrological 2D streamflow modelling of Lithaios River (Central Greece), using GIS, and geostatistics for studying water velocity and discharge, stage elevation, and hydraulic features (streamflow depth, water flow area, wettable circumference, hydraulic radius and depth, n Manning's coefficient, Chow's composite n, Froude number, etc.).

2. Materials and Methods

2.1. Lithaios River Measurements, Instruments Used, and Specifications

The study was conducted in Lithaios River (top width = 15 m) at Trikala monitoring station (M-S), in the region of Thessaly in Central Greece. A propeller current flow [7,8] meter (OTT) was employed together with a modern electronic metering system including a flow computer, data logger, and a real-time display monitor, all calibrated by the manufacturer. River flow data were computed by averaging over a 60 + 60 s measured couple. Vertical measurements of water depths and velocities were performed for temporal monitoring of the cross-section's velocity and discharge variation [8].

2.2. Hydrological Methodology

The river's streamflow velocity and the depths and widths of the defined segments were measured and engaged for the estimation of the cross-section's mean discharge of every segment [2–8]. The overall discharge [2–8] was estimated by the mid-section methodology [5–8]. The features of the cross-section, the water flow velocity of the defined segments and the overall mean flow velocity, were metered, computed, modeled, and depicted in diagrams and GIS maps, respectively, building up a hydrological toolkit for Lithaios River. Water stage elevation and flow measurements were taken monthly for a period of 1 year (January to December). In addition, more measurable variables (streamflow depth, the defined segments' width, overall river width, and water stage elevation) were measured, and more hydraulic features (streamflow depth, water flow area, wettable circumference, hydraulic radius and depth, n Manning's coefficient, Chow's composite n, Froude number, etc.) were computed and depicted in diagrams and saved in the hydrological toolkit. Equation (1) was applied in order to calculate the river flow velocities.

$$V_{i=1}^{n} = a + \left(b_{eq} \times N_{eq}\right) \tag{1}$$

where $V_{i=1}^{n}$ = stream flow velocity (m s^{-1}), n is the number of cross-section segments, a = the initial speed to overcome mechanical resistance, b_{eq} = the system's calibration constant, and N_{eq} = the equipment's rotations per second.

Equation (2) was used for the river's cross-section total discharge.

$$Q_T = \sum_{g=1}^{n} V_{i=1}^{n} A_{J=1}^{n} \tag{2}$$

where Q_T = total discharge (m^3 s^{-1}) of the river's cross-section, g = 1 ... n is the number of cross-section segments, $V_{i=1}^{n}$ = the mean flow velocity of each cross-section segment (m s^{-1}), and $A_{J=1}^{n}$ = the wet flow area of each cross-section segment (m^2).

Couples of stage water elevation *h(t)* and discharge *Q(t)* measurements were utilized to develop mathematical relationships between them. Lithaios River rating curves (*h(t)* − *Q(t)*) [2–8] and changes in the riverbed were computed on the basis of the measured variables using various model equations for regression, ANOVA statistical analysis, and the model's fit F test by utilizing the IBM SPSS v.26 statistics software [2,3,9–25].

2.3. Statistical and Geostatistical Data Analysis, Flow Velocity and Discharge Modelling, and 2D Mapping Methodology

The data were analyzed through the use of IBM SPSS v.26 [2,9,11,12] statistics software. The results are the observations' averages. ANOVA (analysis of variance) [2,3,9,11–25] was used to assess velocity ($V_{i=1}^{n}$), discharge (Q_T), and hydraulic depth effects. In the present study, we used geostatistics (the Kriging method with power model) [2,11–25] for modelling and GIS (geographical information system) hydrological 2D mapping of Lithaios River's water velocity and discharge. Furthermore, the validation of Q_T and $V_{i=1}^{n}$ involves analysis of residual errors, which is the gap between predicted and observed data values and the bias forecast between over- and underestimates. For this purpose, we applied the statistical criteria described by other studies [2,11–16,18–20,22–28], such as the equations for residual sum squares (RSS), standard error (SE), and root mean square error (RMSE).

3. Results and Discussion

The streamflow velocity 2D modelling [2,27] results of the Lithaios River cross-section for the year's maximum (March) and minimum (August) water discharges and the univariate velocity model output statistics are depicted in Figure 1a–d. Lithaios River mean water flow velocity ($V_{i=1}^{n}$) of the cross-section segments for the year's maximum (March) and minimum (August) water discharge results show that $V_{i=1}^{n}$(max) ranged 0.199–0.329 (m s^{-1}) and $V_{i=1}^{n}$(min) ranged 0.098–0.177 (m^3 s^{-1}).

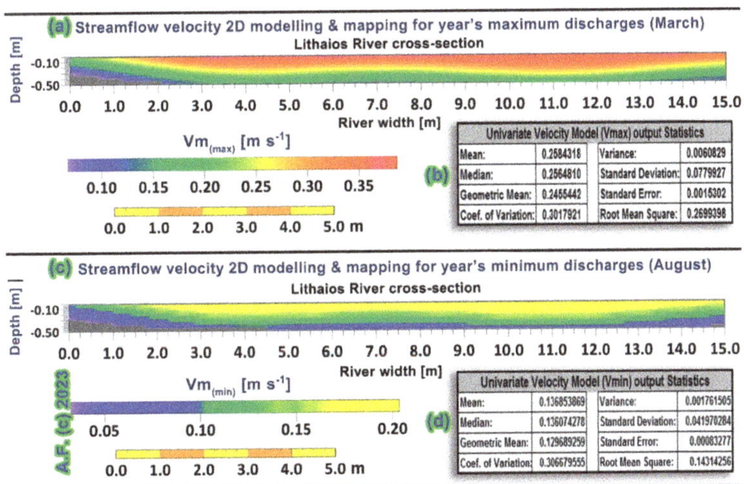

Figure 1. (**a**) Streamflow velocity 2D modelling results on a digital 2D $V_{i=1}^{n}$ map of the Lithaios River cross-section (Trikala M-S) for the year's maximum discharge (March), (**b**) univariate velocity model (Vmax) statistics, (**c**) streamflow velocity 2D modelling results on a digital 2D $V_{i=1}^{n}$ map of the Lithaios River cross-section (Trikala M-S) for the year's minimum discharges (August), and (**d**) univariate velocity model (Vmin) statistics.

The flow velocity ($V_{i=1}^{n}$) statistics [$\bar{x}$ (mean), median, geometric mean, coefficient of variation (CV), s^2 (variance), and s (standard deviation)] for the year's maximum discharges (March) are presented in Figure 1b and the year's minimum discharges (August) are presented in Figure 1d. Velocity fluctuation of a river's cross-section can be specified by means of descriptive statistics [2,3,9,27–29], and of all the descriptive statistics, the coefficient of variation (CV) is the most important measure. [2,9]. The results for both CVs' of the cross-sections velocity variability for the year's maximum (March) (CV = 0.302) and minimum (August) (CV = 0.307) discharges were classified as moderate variability $V_{i=1}^{n}$. The resulting spatial distribution of water flow velocities obtained using river cross-section measurements was best fitted using the Kriging with power model, which resulted in minimum residual sum squares (RSS = 0.0001694), and the RSS used as one of the criteria to choose the greatest model. The other criteria used included the standard error (SE) and root mean square error (RMSE), as in other studies. [2,27]. The best SE for March's velocity modelling was the one using the Kriging with power model (SE = 0.0002236) and for August, it was also the same model (SE = 0.0008328). The RMSE using the Kriging with power model for March's velocity modelling was found to be the best, RMSE = 0.0406329, and for August, it was found to be the best, RMSE = 0.1431426. These results are acceptable since the SE and RMSE scores should be close to zero for accurate prediction and they classified the Kriging with power model as the best model. The abovementioned outcomes, prove the validity and accuracy of the generated 2D digital velocity maps (Figure 1a,c). The relationships between the n Manning's coefficient [2,3,5,7], the Chow's composite n coefficient [5], and the river's water discharges modelling (power model) resulted in high coefficients of determination (R^2) [2,9,12,13,16] for the 12-month measurement study period (Figure 2a,b). The diagrams of the discharges power model (which resulted in being the best model), the Darcy–Weisbach f coefficient multinomial model, and the shear linear model for year's maximum (March) and minimum (August) water discharges are depicted in Figure 2c,d. The R-squared gives a measure of how accurately the observable outputs are reproduced by the model based on the percentage of the total variance that is explained by the model [2,9].

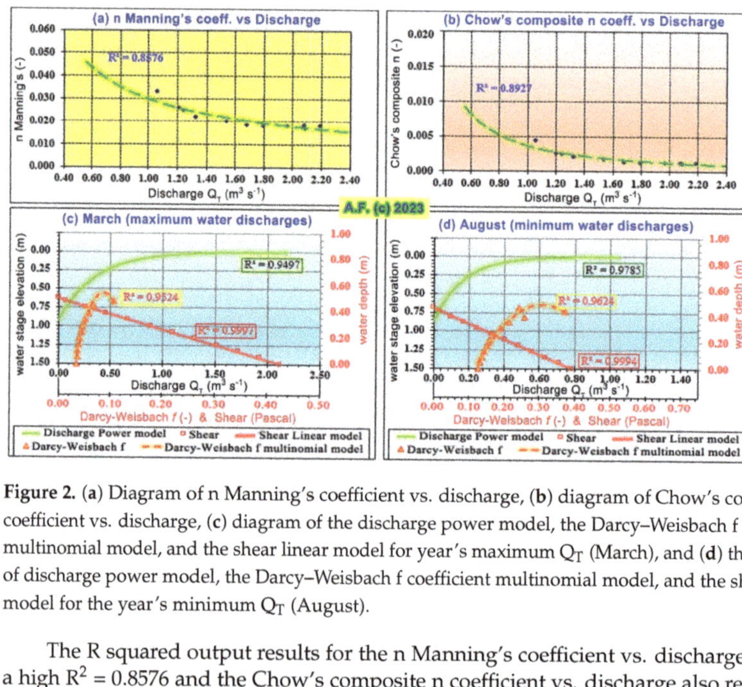

Figure 2. (**a**) Diagram of n Manning's coefficient vs. discharge, (**b**) diagram of Chow's composite n coefficient vs. discharge, (**c**) diagram of the discharge power model, the Darcy–Weisbach f coefficient multinomial model, and the shear linear model for year's maximum Q_T (March), and (**d**) the diagram of discharge power model, the Darcy–Weisbach f coefficient multinomial model, and the shear linear model for the year's minimum Q_T (August).

The R squared output results for the n Manning's coefficient vs. discharge showed a high $R^2 = 0.8576$ and the Chow's composite n coefficient vs. discharge also resulted in a high $R^2 = 0.8927$. The n Manning's coefficient and the Chow's composite n coefficient results show a high degree of correlation with the river's water discharges, with the Chow's composite n coefficient found to have a higher correlation. These results indicate that Chow's composite n [5]—which is built on the hypothesis that the overall force resisting the streamflow in the cross-section is equivalent to the summation of the resisting forces of streamflow in each of the defined segment's regions [2,5]—more accurately approximates the force resisting water flow in Lithaios River. Finally, the statistical results showed that the Froude number during the study period was $Fr < 1$, showing that the Lithaios streamflow is classified as subcritical [2,3,5].

4. Conclusions

The RSS and the prediction error (SE and RMSE) results of spatial and geostatistical 2-dimensional modelling, mapping, and validation of Lithaios River water flows confirmed the validity and accuracy of the generated 2D digital GIS velocity maps of the river's cross-sections. These outcomes have proven that the Kriging with power model had good performance and that it is regarded as very appropriate for 2-D streamflow modelling and digital mapping, as well as being suitable for other hydraulic parameters (n Manning's coefficient, Chow's composite n coefficient, Froude number, shear, Darcy–Weisbach f coefficient, hydraulic radius, etc.). The outcomes of the study are recommended to provide a hydrological toolkit for environmental water resources administration and irrigation programming. This toolkit could assist water supply principalities to rapidly and precisely calculate streamflow volumes and the river's features with a minimal cost rate and workload, and it could be engaged in water supply and agricultural watering administration, the calculation of environmental flows, flood protection, groundwater recharge, and other objectives.

Author Contributions: Conceptualization, A.F.; methodology, A.F., G.S. and G.P.; software, A.F.; validation, A.F., G.S. and G.P.; formal analysis, A.F., G.S. and G.P.; investigation, A.F., G.S. and G.P.; resources, A.F. and G.P.; data curation, A.F., G.S. and G.P.; writing—original draft preparation, A.F.; writing—review and editing, A.F.; visualization, A.F.; supervision, A.F.; project administration, A.F.; funding acquisition, A.F., G.S. and G.P. All authors have read and agreed to the published version of the manuscript.

Funding: This research received no external funding.

Institutional Review Board Statement: Not applicable.

Informed Consent Statement: Not applicable.

Data Availability Statement: All the data of the study are presented in the paper.

Conflicts of Interest: The authors declare no conflict of interest.

References

1. Filintas, A. Land Use Systems with Emphasis on Agricultural Machinery, Irrigation and Nitrates Pollution, with the Use of Satellite Remote Sensing, Geographic Information Systems and Models, in Watershed Level in Central Greece. Master's Thesis, Department of Environment, University of Aegean, Mitilini, Greece, 2005.
2. Filintas, A. Land Use Evaluation and Environmental Management of Biowastes, for Irrigation with Processed Wastewaters and Application of Bio-Sludge with Agricultural Machinery, for Improvement-Fertilization of Soils and Cultures, with the Use of GIS-Remote Sensing, Precision Agriculture and Multicriteria Analysis. Ph.D. Thesis, Department of Environment, University of the Aegean, Mitilini, Greece, 2011.
3. Hatzigiannakis, E.; Filintas, A.; Ilias, A.; Panagopoulos, A.; Arampatzis, G.; Hatzispiroglou, I. Hydrological and rating curve modelling of Pinios River water flows in Central Greece, for environmental and agricultural water resources management. *Desalination Water Treat.* **2016**, *57*, 11639–11659. [CrossRef]
4. Schulze, K.; Hunger, M.; Doll, P. Simulating river flow velocity on global scale. *Adv. Geosci.* **2005**, *5*, 133–136. [CrossRef]
5. Chow, V.T. *Open-Channel Hydraulics*; McGraw-Hill, Inc.: New York, NY, USA, 1959.
6. Munson, B.R.; Young, D.F.; Okiishi, T.H. *Fundamentals of Fluid Mechanics*, 4th ed.; John Wiley and Sons, Inc.: New York, NY, USA, 2002.
7. Herschy, R.W. *Hydrometry: Principles and Practice*, 2nd ed.; John Wiley and Sons: New York, NY, USA, 1999.
8. *ISO 748:2007*; Hydrometry-Measurements of Liquid Flow in Open Channels Using Current-Meters or Floats. International Standard, Fourth Edition 2007-10-15. ISO (International Standards Organizations): Geneva, Switzerland, 2007.
9. Norusis, M.J. *IBM SPSS Statistics 19 Advanced Statistical Procedures Companion*; Pearson: London, UK, 2011.
10. Corato, G.; Moramarco, T.; Tucciarelli, T. Discharge estimation combining flow routing and occasional measurements of velocity. *Hydrol. Earth Syst. Sci.* **2011**, *15*, 2979–2994. [CrossRef]
11. Filintas, A.; Nteskou, A.; Kourgialas, N.; Gougoulias, N.; Hatzichristou, E. A Comparison between Variable Deficit Irrigation and Farmers' Irrigation Practices under Three Fertilization Levels in Cotton Yield (Gossypium hirsutum L.) Using Precision Agriculture, Remote Sensing, Soil Analyses, and Crop Growth Modeling. *Water* **2022**, *14*, 2654. [CrossRef]
12. Stamatis, G.; Parpodis, K.; Filintas, A.; Zagana, E. Groundwater quality, nitrate pollution and irrigation environmental management in the Neogene sediments of an agricultural region in central Thessaly (Greece). *Environ. Earth Sci.* **2011**, *64*, 1081–1105. [CrossRef]
13. Dioudis, P.; Filintas, A.; Koutseris, E. GPS and GIS based N-mapping of agricultural fields' spatial variability as a tool for non-polluting fertilization by drip irrigation. *Int. J. Sus. Dev. Plan.* **2009**, *4*, 210–225. [CrossRef]
14. Filintas, A.; Wogiatzi, E.; Gougoulias, N. Rainfed cultivation with supplemental irrigation modelling on seed yield and oil of *Coriandrum sativum* L. using Precision Agriculture and GIS moisture mapping. *Water Supply* **2021**, *21*, 2569–2582. [CrossRef]

15. Dioudis, P.; Filintas, A.; Papadopoulos, A. Corn yield response to irrigation interval and the resultant savings in water and other overheads. *Irrig. Drain.* **2009**, *58*, 96–104. [CrossRef]
16. Filintas, A.; Dioudis, P.; Prochaska, C. GIS modeling of the impact of drip irrigation, of water quality and of soil's available water capacity on Zea mays L, biomass yield and its biofuel potential. *Desalination Water Treat.* **2010**, *13*, 303–319. [CrossRef]
17. Koutseris, E.; Filintas, A.; Dioudis, P. Antiflooding prevention, protection, strategic environmental planning of aquatic resources and water purification: The case of Thessalian basin, in Greece. *Desalination* **2010**, *250*, 318–322. [CrossRef]
18. Filintas, A. Soil Moisture Depletion Modelling Using a TDR Multi-Sensor System, GIS, Soil Analyzes, Precision Agriculture and Remote Sensing on Maize for Improved Irrigation-Fertilization Decisions. *Eng. Proc.* **2021**, *9*, 36. [CrossRef]
19. Kalavrouziotis, I.K.; Filintas, A.T.; Koukoulakis, P.H.; Hatzopoulos, J.N. Application of multicriteria analysis in the Management and Planning of Treated Municipal Wastewater and Sludge reuse in Agriculture and Land Development: The case of Sparti's Wastewater Treatment Plant, Greece. *Fresenious Environ. Bull.* **2011**, *20*, 287–295.
20. Filintas, A.; Nteskou, A.; Katsoulidi, P.; Paraskebioti, A.; Parasidou, M. Rainfed and Supplemental Irrigation Modelling 2D GIS Moisture Rootzone Mapping on Yield and Seed Oil of Cotton (Gossypium hirsutum) Using Precision Agriculture and Remote Sensing. *Eng. Proc.* **2021**, *9*, 37. [CrossRef]
21. Dioudis, P.; Filintas, A.; Papadopoulos, A.; Sakellariou-Makrantonaki, M. The influence of different drip irrigation layout designs on sugar beet yield and their contribution to environmental sustainability. *Fresenious Environ. Bull.* **2010**, *19*, 818–831.
22. Filintas, A.; Gougoulias, N.; Salonikioti, A.; Prapa, E. Study of soil erodibility by water on tillage and no tillage treatments of a *Helianthus Tuberosus* crop using field measurements, soil laboratory analyses, GIS and deterministic models. *Ann. Univ. Craiova Ser. Biol. Hortic. Food Prod. Process. Technol. Environ. Eng.* **2019**, *29*, 529–536.
23. Koutseris, E.; Filintas, A.; Dioudis, P. Environmental control of torrents environment: One valorisation for prevention of water flood disasters. *WIT Trans. Ecol. Environ.* **2007**, *104*, 249–259. [CrossRef]
24. Filintas, A.; Gougoulias, N.; Papachatzis, A. Soil organic matter modelling and digital mapping of a Triticum turgidum cropfield using as auxiliary variables the plant available water, texture, field measurements, soil laboratory analyses, GIS and geostatistical models. *Ann. Univ. Craiova Ser. Biol. Hortic. Food Prod. Process. Technol. Environ. Eng.* **2019**, *24*, 537–544.
25. Loague, K.; Green, R.E. Statistical and graphical methods for evaluating solute transport models: Overview and application. *J. Contam. Hydrol.* **1991**, *7*, 51–73. [CrossRef]
26. Lu, G.Y.; Wong, D.W. An adaptive inverse-distance weighting spatial interpolation technique. *Comput. Geosci.* **2008**, *34*, 1044–1055. [CrossRef]
27. Filintas, A.; Hatzigiannakis, E.; Arampatzis, G.; Ilias, A.; Panagopoulos, A.; Hatzispiroglou, I. Hydrometry's classical and innovative methods and tools comparison for Stara river flows at Agios Germanos monitoring station in north-west Greece. In Proceedings of the International Conference EGU European Geosciences Union General Assembly 2015, Vienna, Austria, 12–17 April 2015; Volume 17. EGU2015-13601.
28. Filintas, A.; Hatzigiannakis, E.; Panagopoulos, A.; Arampatzis, G.; Ilias, A.; Hatzispiroglou, I. Hydrological modelling of Pinios River (Greece) water flows as assisting tool for environmental and agricultural water resources management using River Analysis Models. In Proceedings of the Fifth International Conference on Environmental Management, Engineering, Planning & Economics (C.E.M.E.P.E. 2015) and SECOTOX, Mykonos Island, Greece, 14–18 June 2015; ISBN 978-960-6865-86-2.
29. Hatzigiannakis, E.; Filintas, A.; Sassalou, M.; Panoras, G.; Zavra, A. Hydro-measurements and water quality sampling monitoring for agricultural use of Pinios River water in Central Greece. In Proceedings of the 13th International Conference on "Protection and Restoration of the Environment", Water Resources Management and Contamination Control, Mykonos Island, Greece, 3–8 July 2016.

environmental sciences proceedings

MDPI

Proceeding Paper

Irrigation Practice Survey for Crops and Urban Greenspaces at Northwest and West Greece and Southeast Italy †

Ioannis L. Tsirogiannis

Department of Agriculture, University of Ioannina, Kostakii Campus, 47100 Arta, Greece; itsirog@uoi.gr

† Presented at the 7th International Electronic Conference on Water Sciences, 15–30 March 2023; Available online: https://ecws-7.sciforum.net.

Abstract: An analytical survey regarding irrigation practice for both crops and greenspaces at the region of Western Greece and the region of Epirus in Northwest and West Greece, respectively, and the region of Apulia in Southeast Italy was performed. On-site interviews (about 500 per region) regarded irrigation systems design, installation and management. Good practices, problems and challenges that concerned irrigation management are pointed out. The need for a frequent and organised survey of irrigation practice coordinated by relevant organisations is underlined, as it could constitute a valuable tool for planning relevant strategies and practical actions.

Keywords: irrigation practice; irrigation water management; agricultural crops; urban landscapes

Citation: Tsirogiannis, I.L. Irrigation Practice Survey for Crops and Urban Greenspaces at Northwest and West Greece and Southeast Italy. *Environ. Sci. Proc.* **2023**, *25*, 14. https://doi.org/10.3390/ECWS-7-14186

Academic Editor: Lampros Vasiliades

Published: 14 March 2023

1. Introduction

Water scarcity is one of the main challenges that both agricultural crops and landscape works face in many countries in the world, particularly in semi-arid and arid regions [1,2]. With the given infrastructure, agricultural (in open field and/or under cover) and landscape irrigation and drainage systems efficiency could be increased promptly if their design, installation and maintenance received regular auditing and if more reasonable water management was applied. Beyond generic statistical registration by international and national organisations and survey authorities, detailed surveys for irrigation practice could provide valuable information for planning relevant strategies and practical actions [3–5]. This study briefly presents the results of a survey regarding irrigation practice that concern applied techniques, usage of new technologies on water irrigation management, etc., for the region of Western Greece, the region of Epirus in Greece and the region of Apulia in Italy.

2. Methods

On-site interviews were performed using special questionnaires (see Supplementary Material) and relevant reports were produced (see Data Availability Statement). The interviews concerned four categories of irrigation system management setups: (a) regional authorities that are responsible for irrigation water and municipalities that are responsible for irrigation of public urban greenspaces (including athletic facilities), (b) public organisations that manage participatory agricultural irrigation systems, (c) agricultural farms and (d) private touristic or leisure greenspaces and gardens. In this framework, four questionnaires were formed depending on the special features of the setups they were addressing, but they all had a common structure and a number of common questions. The questionnaires were divided in six sections as follows:

1. General information regarding organisation/institution/farmer/Land Reclamation Consortium (name, age, address and gender);
2. Information about crops/green areas, irrigation systems and water sources (crops, area and irrigation status);

3. Irrigation management information—Acceptance level of reclaimed water (for irrigation of several crops or for potable use);
4. Economic information—Effect of reclaimed water (plants, soil and productivity);
5. Environmental issues—Concerns regarding reclaimed water (fear factors, quality and treatment);
6. Other information—Pricing of reclaimed water (willingness to pay, price of water and investment in agriculture).

The research was conducted during 2013 and 2014 all around the region of Western Greece, the region of Epirus in Greece, the region of Apulia in Italy (Figure 1) for which 490, 500 and 449 interviews were per region were performed, respectively (Table 1).

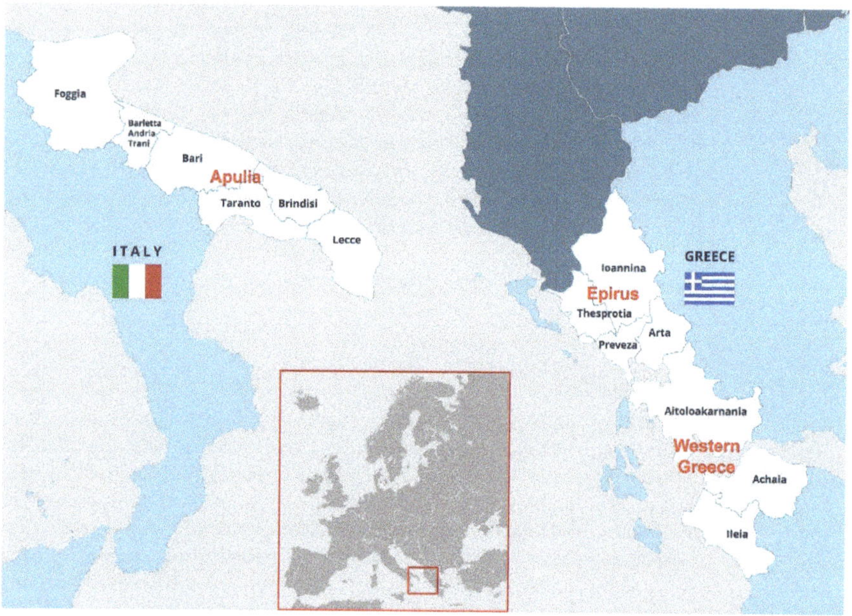

Figure 1. Area of the survey region of Western Greece, the region of Epirus in Greece, the region of Apulia in Italy.

Table 1. Number of interviews that were performed per category of irrigation system management setups and region.

Category of Irrigation System Management Setups	Theme	Total Number of Interviews	Number of Interviews per Region		
			Apulia	Western Greece	Epirus
1. Public administration	Greenspaces	46	30	8	8
2. Local Organisations for Land Reclamation	Agriculture	61	19	24	17
3. Farm level	Agriculture	1037	270	329	438
4. Private landscape/Leisure irrigation system	Greenspaces	296	130	129	37
Total		1440	449	490	500

3. Results and Discussion

Selected topics that concern irrigation management are presented in Table 2 (for regional authorities that are responsible for irrigation water and municipalities that are responsible for irrigation of public urban greenspaces (including athletic facilities) and for public organisations that manage participatory agricultural irrigation systems) and Table 3 (for agricultural farms and for private touristic or leisure greenspaces and gardens).

Table 2. Replies (percentage of total) from (a) regional authorities that are responsible for irrigation water and municipalities that are responsible for irrigation of public urban greenspaces (including athletic facilities) and (b) public organisations that manage participatory agricultural irrigation systems, for selected topics that regard irrigation management (NA: no data available).

	Public Administration			Organisation of Irrigation Water Management		
Reply	**Apulia**	**Western Greece**	**Epirus**	**Apulia**	**Western Greece**	**Epirus**
Awareness of the local water management plan which is applied in the framework of 2000/60/EC	66%	90%	62%	100%	44%	20%
A straightforward strategy regarding irrigation management in the area has been formed	40%	0%	50%	NA	63%	75%
There is efficient irrigation management at basin level	90%	94%	63%	NA	74%	79%
Use of special IT tools and databases for irrigation management	3%	0%	0%	25%	0%	0%
Interest to use special IT tools and databases for irrigation management	75%	13%	9%	NA	7%	5%
Drought is the most significant problem regarding water in the area	50%	46%	50%	NA	57%	50%
Desertification is the most significant problem regarding water in the area	20%	8%	50%	NA	29%	3%
Salinisation of underground aquifers is the most significant problem regarding water in the area	23%	45%	0%	NA	14%	4%
Climate change is the most significant problem regarding water in the area	13%	21%	20%	NA	22%	15%
Provided guidance and rules regarding water distribution are helping to mitigate water shortages in the area	50%	55%	50%	NA	80%	65%
The main cause for water shortages in the area is excessive consumption of water for irrigation	19%	21%	12%	NA	38%	42%
There is inefficient water management at end user level	33%	22%	50%	NA	58%	52%
There is lack of training of end users	60%	58%	50%	16%	84%	76%
The organisation provides end users with advice regarding irrigation, drainage, fertilisation management	100%	53%	87%	50%	0%	53%

Table 3. Replies (percentage of total) from (c) agricultural farms and (d) private touristic or leisure greenspaces and gardens for selected topics that regard irrigation management (NA: no data available).

	Agricultural Farms			Private Greenspaces		
Reply	**Apulia**	**Western Greece**	**Epirus**	**Apulia**	**Western Greece**	**Epirus**
Drought is the most significant problem regarding water in the area	55%	50%	83%	40%	36%	94%
Desertification is the most significant problem regarding water in the area	19%	26%	5%	38%	12%	3%
Salinisation of underground aquifers is the most significant problem regarding water in the area	26%	24%	1%	12%	35%	3%
Climate change is the most significant problem regarding water in the area	35%	11%	15%	45%	9%	1%
The main cause for water shortages in the area is lack of guidance and rules regarding water distribution	28%	16%	14%	24%	0%	22%
Provided guidance and rules regarding water distribution are helping to mitigate water shortages in the area	70%	88%	95%	84%	100%	87%
There is efficient water management at basin level	NA	77%	96%	81%	95%	93%
Support is provided from relevant public authorities	86%	85%	81%	NA	NA	NA
Typically informed regarding agricultural issues, including irrigation by agronomic consultant	95%	85%	95%	NA	NA	NA
Need for training regarding irrigation management / scheduling	32%	28%	6%	NA	NA	NA
Attended at least one educational/training session specialised in irrigation	1%	4%	2%	NA	NA	NA
Provided by the irrigation system's designer/contractor with irrigation scheduling plan	NA	6%	5%	37%	83%	84%
Asked for professional assistance regarding the irrigation schedule	18%	60%	10%	NA	74%	84%
Uses special IT tools and databases for water management	11%	9%	17%	NA	5%	3%
Uses some kind of electronics or IT technology for irrigation management (controllers, sensors, etc.)	NA	7%	6%	49%	71%	57%
Using electronics or IT technologies for irrigation management could lead to water conservation	48%	19%	23%	NA	82%	70%
Willing to use an IT advice service regarding irrigation scheduling	NA	70%	35%	NA	99%	78%
The main drawback when using electronics or IT technologies for irrigation management in complexity	28%	82%	65%	NA	92%	83%

Most of the organisations that regard irrigation management are aware of local water management plans, but they are not convinced that there is a straightforward strategy regarding irrigation water management in the area, although irrigation water management is performed efficiently (Table 2). Probable drought is reported as the most significant concern regarding water availability in the area (Tables 2 and 3), while the setting of rules and the provision of guidance regarding water distribution are regarded helpful as a means to mitigate water shortages in the area (Tables 2 and 3). The organisations do not blame excessive consumption of irrigation water as the main cause for water shortages in the area: they generally do not think that there is adequate irrigation water management at the end user level, although they agree that there is lack of training of end users. In general, they provide end users with advice regarding irrigation, drainage and fertilisation management (Table 2).

End users generally agree that there is efficient water management at the basin level, that the provided rules and guidance regarding water distribution are helping to mitigate water shortages in the area, and that support is provided from relevant public authorities (Table 3). Farmers are typically informed regarding agricultural issues, including irrigation by agronomic consultants, although they do not think that they need special training regarding irrigation management or scheduling and consequently the majority have not attended educational/training sessions specialised in irrigation (Table 3).

For greenspaces, the provision of the irrigation system's designer/contractor with irrigation scheduling plan, and the request for professional assistance regarding the irrigation schedule, is something common but for crops is very rare (Table 3). The use of special IT tools for irrigation management is low to very low for organisations and farmers but high for managers of greenspaces (Tables 2 and 3). While organisations showed low interest to adapt to such tools (Table 2), both categories of end users were quite interested in doing so (Table 3). A significant percentage of both categories of end users believe that using electronics or IT technologies for irrigation management could lead to water conservation: the main drawbacks are the complexity of those systems combined with the lack of training and support (Table 3).

4. Conclusions

An analytical survey revealed good practices and problems regarding design, construction, maintenance and management for the region of Western Greece, the region of Epirus in Greece and the region of Apulia in Italy. The summarised results that concern irrigation management for both agricultural and urban greenspaces irrigation systems were presented. The supervisors of irrigation systems welcomed this initiative. Good practices, problems and challenges were pointed out. The development of a practical surveying and auditing guide, approved by the relevant stakeholders, could be a very useful tool for planning relevant strategies and practical actions with an aim of continuous improvement of irrigation systems efficiency. The role of international irrigation stakeholders such as the European Irrigation Association could be crucial in this effort.

Supplementary Materials: The following are available online at https://www.mdpi.com/article/10.3390/ECWS-7-14186/s1.

Funding: The survey on which this work has been based on was co-financed by EU/ERDF (75%) and from national funds of Greece and Italy (25%) in the framework of the European Territorial Cooperation Programme (ETCP) GREECE-ITALY 2007–2013 (www.greece-italy.eu): IRMA project, subsidy contract no: I3.11.06.

Institutional Review Board Statement: Not applicable.

Informed Consent Statement: Not applicable.

Data Availability Statement: Full reports of the data presented in this study are available at: https://www.irrigation-management.eu/tools/survey (accessed on 15 January 2022).

Acknowledgments: The author, who acted as project manager of the IRMA project, would like to express his gratitude to all the members of the project partners who worked on the survey and on the formation of the relevant project deliverables.

Conflicts of Interest: The author declares no conflict of interest.

References

1. Nikolaou, G.; Neocleous, D.; Christou, A.; Kitta, E.; Katsoulas, N. Implementing Sustainable Irrigation in Water-Scarce Regions under the Impact of Climate Change. *Agronomy* **2020**, *10*, 1120. [CrossRef]
2. Reyes-Paecke, S.; Gironás, J.; Melo, O.; Vicuña, S.; Herrera, J. Irrigation of green spaces and residential gardens in a Mediterranean metropolis: Gaps and opportunities for climate change adaptation. *Landsc. Urban Plan.* **2019**, *182*, 34–43. [CrossRef]
3. Fernández-Cañero, R.; Ordovás, J.; Machuca, M.A.H. Domestic Gardens as Water-wise Landscapes: A Case Study in Southwestern Europe. *HortTechnology* **2011**, *21*, 616. [CrossRef]
4. Tindula, G.N.; Orang, M.N.; Snyder, R.L. Survey of Irrigation Methods in California in 2010. *J. Irrig. Drain. Eng.* **2013**, *139*, 3. [CrossRef]
5. Incrocci, L.; Thompson, R.B.; Fernandez-Fernandez, M.D.; De Pascale, S.; Pardossi, A.; Stanghellini, C.; Rouphael, Y.; Gallardo, M. Irrigation management of European greenhouse vegetable crops. *Agric. Water Manag.* **2020**, *242*, 106393. [CrossRef]

environmental sciences proceedings

MDPI

Proceeding Paper

Seawater Intrusion Vulnerability Assessment Using the GALDIT and the Modified GALDIT–AHP Methods: Application in the Coastal Almyros Aquifer, Thessaly, Greece †

Sibianka Lepuri [1,*], Athanasios Loukas [1] and Aikaterini Lyra [2]

1 Laboratory of Hydraulic Works and Environmental Management, School of Rural and Surveying Engineering, Aristotle University of Thessaloniki, 54124 Thessaloniki, Greece; agloukas@topo.auth.gr
2 Laboratory of Hydrology and Aquatic Systems Analysis, Department of Civil Engineering, School of Engineering, University of Thessaly, 38334 Volos, Greece; lyrakater@gmail.com
* Correspondence: sibiankalepuri@gmail.com
† Presented at the 7th International Electronic Conference on Water Sciences, 15–30 March 2023; Available online: https://ecws-7.sciforum.net/.

Abstract: In the rural and coastal Almyros basin in Magnesia, Greece, the objective of the current study is the assessment of aquifer vulnerability to seawater intrusion using the GALDIT approach. The Almyros aquifer system's quality and quantity have declined as a result of unsustainable groundwater abstraction for irrigation. The Analytical Hierarchy Process (AHP) of Multicriteria Analysis has been used for the modification of the GALDIT index based on the statistics of experts's responses to questionnaires on the influence of hydrological, hydrogeological, and other parameters. For all methodologies and time periods, the aquifer's coastline section had high susceptibility levels, whereas the northeast and southeast had lower values. The most vulnerable area of the aquifer changes over the various time periods of analysis.

Keywords: groundwater; seawater intrusion; GALDIT; vulnerability index; analytical hierarchy process; Almyros

Citation: Lepuri, S.; Loukas, A.; Lyra, A. Seawater Intrusion Vulnerability Assessment Using the GALDIT and the Modified GALDIT–AHP Methods: Application in the Coastal Almyros Aquifer, Thessaly, Greece. *Environ. Sci. Proc.* **2023**, *25*, 15. https://doi.org/10.3390/ECWS-7-14174

Academic Editor: Rodrigo Maia

Published: 3 April 2023

1. Introduction

In the study area, no previous studies have been carried out to assess the vulnerability of the aquifer to seawater intrusion. The over-pumping of water reserves to meet irrigation needs has degraded the quality and quantity of water in the Almyros aquifer. The assessment of vulnerability aims at better management of water resources in the area and protection of the Almyros aquifer system from further degradation [1]. The Almyros basin, which is located at the southernmost edge of the Thessalian plain, is a component of the single Almyros-Pagasitikos basin. The study area's aquifer covers 293 km^2 and has an average elevation of around 108 m and slope of about 5.56%. The Almyros basin experiences a semi-arid Mediterranean environment with 500 mm of annual rainfall on average and an average yearly temperature of 16.5 °C [2]. Five categories have been used to group the most significant geological components of the Almyros aquifer: clay (Neogene), clay–gravel–sand (Neogene), sand (Quaternary), clay–sand (Neogene), and limestone [3]. Following the shift in topographic elevation, the coastal region of the Almyros basin is composed of sandy permeability materials and clay lenses towards the western half of the aquifer. The aquifer's hydrogeological zones are made up of semi-permeable Neogene formations and permeable Quaternary formations [4]. With a geographical average value of 2.3 m per day, hydraulic conductivity values range from 0.1 to 18.7 m per day. In this work, the weights of the parameters are provided using the standard/typical GALDIT method, and the weights are estimated using the Analytical Hierarchy Process (AHP) from the responses of 15 experts, using the GALDIT-AHP method, over three time periods in the

Almyros Basin aquifer. The results on seawater intrusion vulnerability are compared and discussed.

2. Materials and Methods

2.1. Method GALDIT

The GALDIT vulnerability index approach, which was put forth by Lobo-Ferreira and associates [4], determines how vulnerable coastal aquifers are to the salt wedge. The following phrases are abbreviated as the GALDIT method: (a) groundwater occurrence, (b) aquifer hydraulic conductivity, (c) depth to groundwater (level above the sea), (d) distance from the shore to the beach, (e) effects of present seawater intrusion, and (f) aquifer thickness; these are the six factors that must be considered. The three components of the procedure are the calibration of the parameters, the classes of the parameters, and the weights of the parameters. The method was first applied to the Bardez aquifer, Goa (India) [5]. The vulnerability assessment index is calculated from the mathematical type:

$$GALDIT = \frac{\sum_{i=1}^{6}\{(W_i)\,R_i\}}{\sum_{i=1}^{6} W_i} \quad (1)$$

where W_i = the weights of the parameters, R_i = the rating of the parameters. The value range of the index is 2.5–10. Indicators with higher values indicate greater exposure to seawater incursion, whereas those with lower values indicate less exposure.

2.2. AHP Method

Saaty first developed the Analytical Hierarchy Process (AHP) in the 1970s. Since then, it has proven to be a useful tool for creating and modeling scenarios with various, frequently at odds, objectives. The method solves a problem in 6 stages [6] which are: (1) segmentation of the problem, (2) prioritization of objectives, criteria and sub-criteria and alternatives, (3) creation of the table of paired observations, (4) estimation of relevant parameters, (5) estimation of consistency, and finally, (6) general comparison of the method. The reliability of the method is based on the consistency ratio (CR). If CI/RI < 0.10, the degree of consistency is satisfactory, so there is little subjectivity, but if CI/RI is greater than 0.10, there may be major discrepancies and Analytic Hierarchy Method (AHP) conclusions may not be significant. In the context of this work, for each vulnerability method, the corresponding tables of pairwise tables were created and completed by 15 water resources experts including university professors, researchers, and postdocs [1].

2.3. Spearman's Rank Correlation

The Spearman correlation coefficient is named after Charles Spearman and is denoted by the Greek letter ϱ (*rho*) or rs. It is a non-parametric method, which is applied when the parametric conditions are not satisfied (i.e., normality and linearity, the range of observations and the existence of an iso-space scale). The magnitude of agreement is expressed by the sign and magnitude of the Spearman correlation statistic. The equation for calculating the Spearman correlation coefficient is as follows:

$$rho = \frac{6\sum \delta i^2}{\nu(\nu^2-1)} \quad (2)$$

where ν is the number of pairs, ν must be ≥ 4, and δi is the difference in order between the first and second measurements (pairs of measurements). The hypotheses tested when applying the Spearman correlation index are H0: $\varrho = 0$ (lack of correlation between observations) and H1: $\varrho \neq 0$ (existence of correlation between observations). A frequently used significance level is $\alpha = 0.05$. That is, there is a 95% probability that each observed statistical difference is real and not due to chance [7].

3. Results and Discussion

3.1. Calculation of Vulnerability Index GALDIT and Modified GALDIT–AHP

The method was applied for all three study periods 1992–1997, 2004–2009, and 2010–2015. Therefore, parameters that do not remain constant per time period, such as the hydraulic load above sea level, the existing salinity condition, and the aquifer thickness, were calculated for each period using GIS tools. The type of aquifer, based on the geological and hydraulic conditions prevailing in the study area, was considered alluvial/unconfined. In the Almyros basin, as already mentioned, hydraulic conductivity information of the unsaturated zone is provided by the European Soil Data Center (ESDAC) [8]. For the study area, hydraulic conductivity varies between 0.05 and 18.701 m/day or 2.29 m/day on average. For the period 1992–1997, the hydraulic head above sea level was an average of 65.44 m and a maximum value of 217.59 m, in the period 2004–2009 it was 66.85 m and a maximum value of 218. In the period of 2010–2015, it was 66.40 m and a maximum value of 217.39 m. The very low values are located NE–SE of the Almyros basin, near the basin's coastline region, while the high values of the hydraulic load increase towards the center of the aquifer, moving in the direction of the Holorema stream. The highest concentrations of chlorides measured at the measurement sites are shown on the SE side of the basin, in the Xirorema and Platanorema sub-basins. The average thickness for the aquifer media is 29 m but is not constant in all locations in the Almyros basin. The Almyros basin's map and the study area are depicted in Figure 1.

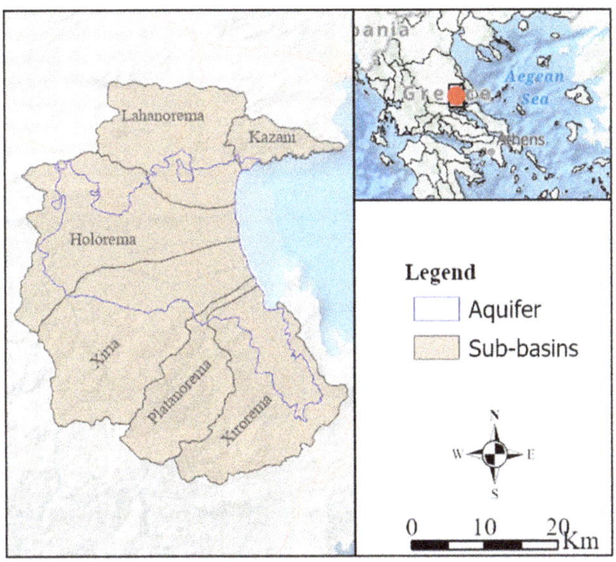

Figure 1. Map of the Almyros basin including the aquifer and the sub-basins.

The weights of the parameters for the statistical indicators median, average, and mode, as determined by the statistical analysis of the 15 experts' responses, were taken from the GALDIT–AHP method's study of the data. The consistency ratio (CR) for each statistical indicator is less than 10%. Specifically, the consistency ratios of the AHP Median, AHP Average, and AHP Mode are 2%, 0.43%, and 8.8%, respectively. The weights of the parameters for each statistical index are presented in Table 1.

Table 1. The weights of each parameter for the GALDIT method and the modified GALDIT–AHP method.

Parameters	Typical	AHP Median	AHP Average	AHP Mode
Groundwater occurrence	0.060	0.276	0.282	0.235
Aquifer hydraulic conductivity	0.200	0.263	0.246	0.235
Level above the sea	0.266	0.190	0.170	0.214
Distance from the shore	0.266	0.118	0.146	0.127
Impact of existing seawater intrusion	0.060	0.081	0.086	0.099
Thickness of the aquifer	0.133	0.072	0.070	0.090

The GALDIT method assigns the greatest weights to the parameters of distance from the coast (D) and hydraulic load above the sea (L). The modified GALDIT–AHP assigns the greatest weights to the parameters of the groundwater occurrence (G) and the aquifer hydraulic conductivity (A). The resulting maps for all the methods for the evaluated time periods are presented in Figure 2.

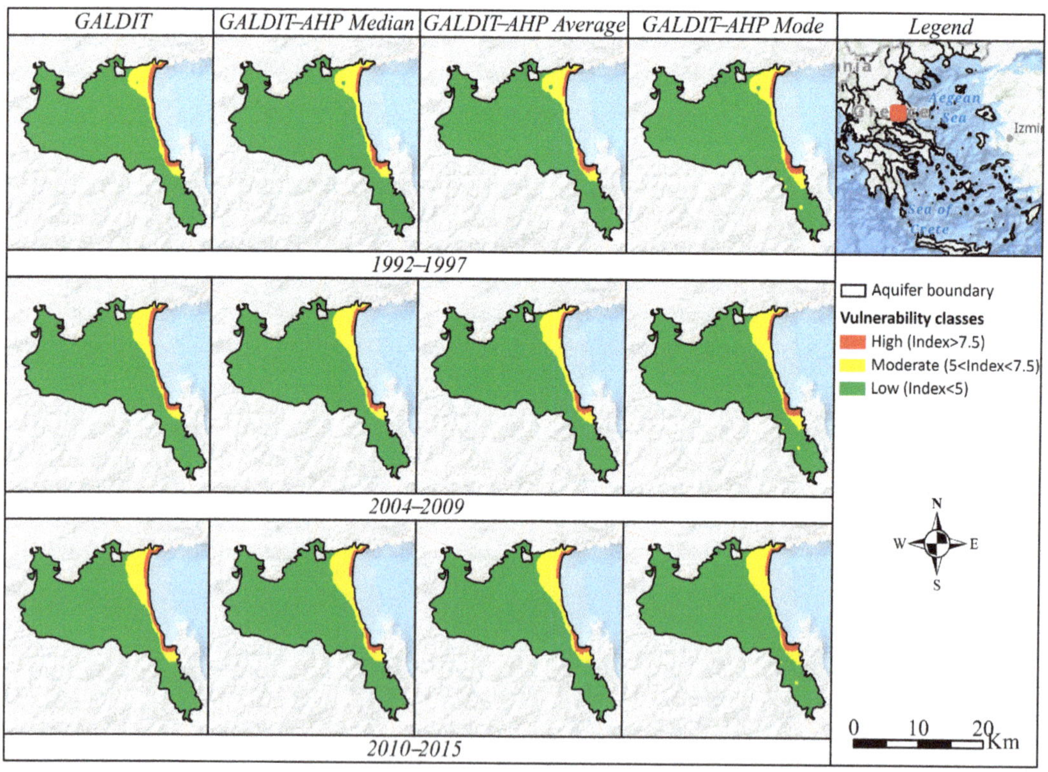

Figure 2. Vulnerability maps of Almyros aquifer with the methods of GALDIT and modified GALDIT–AHP for the periods 1992–1997, 2004–2009, and 2010–2015.

The areas in which a greater extent of high and medium vulnerability is observed are in the Kazani, Lahanorema, and Xirorema sub-basins. In the period 1992–1997, the total percentage of high vulnerability among the indexes covers 1.98% to 3.5% of the aquifer or an area of 5.7 km^2 to 10 km^2. The lowest overall percentage is estimated by the GALDIT–AHP Average and Median indices. Average vulnerability across indices ranges from 7.2% to 8.7%

or an area of 20.7 km^2 to 24.8 km^2. The lowest percentage of average vulnerability was estimated with the weights of the GALDIT index, while the highest percentage of average vulnerability was estimated with the weights of the AHP Mode index. Low vulnerability ranges from 89.17% to 89.92% or an area of 255 km^2 to 258 km^2. In the period 2004–2009, the total percentage of high vulnerability among the indices covers 1.94% to 3.6% of the aquifer or an area of 5.6 km^2 to 10.3 km^2. The lowest overall rate is estimated by the GALDIT–AHP Average and Median indices. Average vulnerability across indices ranges from 8.1% to 9.6% or an area of 23.3 km^2 to 27.6 km^2. The lowest percentage of average vulnerability was estimated with the weights of the GALDIT index, while the highest percentage of average vulnerability was estimated with the weights of the AHP Mode index. Low vulnerability ranges from 88.2% to 88.8% or an area of 253 km^2 to 255 km^2. In the period 2010–2015, the total percentage of high vulnerability among the indices covers 2% to 3.6% of the aquifer or an area of 5.9 km^2 to 10.2 km^2. The lowest overall percentage is estimated by the GALDIT–AHP Average and Median indices. Average vulnerability across indices ranges from 9.1% to 10.4% or an area of 26.2 km^2 to 30 km^2. The lowest percentage of average vulnerability was estimated with the weights of the GALDIT index, while the highest percentage of average vulnerability was estimated with the weights of the AHP Mode index. Low vulnerability ranges from 87.3% to 87.7% or an area of 251 km^2 to 252 km^2. Summarized statistics of the evaluation period (1992–2015) are presented in Table 2.

Table 2. Percentage (%) of the Almyros aquifer under various classes of vulnerability (%) with the typical GALDIT and GALDIT–AHP methods for the period 1992–2015.

Vulnerability Classes	Typical	AHP Median	AHP Average	AHP Mode
High	3.54%	1.99%	1.99%	2.20%
Moderate	8.16%	9.23%	9.21%	9.57%
Low	88.27%	88.70%	88.80%	88.24%

3.2. *Sperman Rank Correlation*

To test the correlation between salinity concentrations (ppt) and GALDIT seawater intrusion index values for all three time periods, the Spearman correlation coefficient was used. Salinity values and the GALDIT vulnerability index were extracted from the sampling sites using the Extract multi values to points tool. The Spearman correlation test was then performed using the SPSS statistical software. Then, using the SPSS statistical package, the Spearman correlation test followed. Summarized statistics of the evaluation period are presented in Table 3.

Table 3. Spearman rank correlation between salinity concentrations and vulnerability indices GALDIT and modified GALDIT–AHP.

Vulnerability Indices	1992–1997	2004–2009	2010–2015
GALDIT	0.44	0.45	0.45
AHP Median	0.43	0.44	0.44
AHP Average	0.44	0.44	0.43
AHP Mode	0.43	0.45	0.46

The significance (p) value of the correlations is less than 0.05, thus the statistical difference is real and not due to chance. Correlation coefficients range from rho = 0.43 to 0.46 per study period.

4. Conclusions

In all the study periods (1992–1997, 2004–2009, and 2010–2015) a gradual increase in high and medium vulnerability values (0.5–2%) was observed, a fact due to changing parameters such as the hydraulic load above sea level, the existing salinity condition, and the aquifer thickness, which change with time. For the GALDIT index, the standard/typical weights and the weights of the AHP Median and AHP Average statistical indicators showed in all study periods a similar overall rate of high vulnerability with a difference of 0.5–1%. Additionally, there are marginal differences in the correlation coefficients between the GALDIT indices and the observed data, with the GALDIT index generated using standard weights displaying the highest connection throughout all research periods. As a result, when compared to the other indices, the standard weights of the GALDIT index slightly better represent the vulnerability assessment both spatially and statistically. The GALDIT method assigns the greatest weights to the parameter of distance from the coast (D) and to the parameter of hydraulic load above the sea (L).

Author Contributions: Conceptualization, methodology, supervision, writing—review and editing, A.L. (Athanasios Loukas); methodology, software, data curation, writing—original draft preparation, writing—review and editing, S.L.; writing—review and editing, A.L. (Aikaterini Lyra). All authors have read and agreed to the published version of the manuscript.

Funding: This research received no external funding.

Institutional Review Board Statement: Not applicable.

Informed Consent Statement: Not applicable.

Data Availability Statement: Not applicable.

Acknowledgments: This research was carried out in the framework of the master's thesis of S. Lepuri, Laboratory of Hydraulic Works and Environmental Management, School of Rural and Surveying Engineering, Aristotle University of Thessaloniki.

Conflicts of Interest: The authors declare no conflict of interest.

References

1. Lepuri, S. Vulnerability of Aquifers of Agricultural and Coastal Basins Due to Nitrate Pollution and Seawater Intrusion. The case of Almyros Basin, Magnesia, Greece. Master Thesis, School of Rural and Surveying Engineering, Aristotle University of Thessaloniki, Thessaloniki, Greece, 2021. [CrossRef]
2. Lepuri, S.; Loukas, A.; Lyra, A.; Voudouris, K. Aquifer vulnerability to nitrate pollution using the canter and drastic methods: The case of almyros in thessaly, greece. Presented at the International Hydrogeological Congress of Greece and Cyprus, Nicosia, Cyprus, 20–22 March 2022.
3. Lyra, A.; Loukas, A.; Sidiropoulos, P.; Tziatzios, G.; Mylopoulos, N. An Integrated Modeling System for the Evaluation of Water Resources in Coastal Agricultural Watersheds: Application in Almyros Basin, Thessaly, Greece. *Water* **2021**, *13*, 268. [CrossRef]
4. Lyra, A.; Loukas, A.; Sidiropoulos, P.; Mylopoulos, N.; Voudouris, K. Seawater intrusion in Almyros aquifer, in Thessaly, Greece. Presented at the International Hydrogeological Congress of Greece and Cyprus, Nicosia, Cyprus, 20–22 March 2022.
5. Chachadi, A.; Lobo-Ferreira, J.P. Assessing aquifer vulnerability to sea-water intrusion using GALDIT method: Part 2–GALDIT Indicators Description. Presented at the 4th Inter Celtic Colloquium on Hydrology and Management of Water Resources, Guimaraes, Portugal, 11–13 July 2005.
6. Saaty, T.L. What is the analytic hierarchy process? In *Mathematical Models for Decision Support*; Mitra, G., Greenberg, H.J., Lootsma, F.A., Rijkaert, M.J., Zimmermann, H.J., Eds.; Springer: Berlin/Heidelberg, Germany, 1988; Volume 48, pp. 109–121.
7. Corder, G.W.; Foreman, D.I. *Nonparametric Statistics for Non-Statisticians*; John Wiley & Sons, Inc: Hoboken, NJ, USA, 2009.
8. ESDAC. European Soil Data Centre, Joint Research Centre, European Commission. Available online: https://esdac.jrc.ec.europa.eu/ (accessed on 21 January 2021).

environmental sciences proceedings

MDPI

Proceeding Paper

Rainwater Harvesting during the COVID Outbreak: A Case Study in Brazil †

Igor Catão Martins Vaz * and Enedir Ghisi

Department of Civil Engineering, Federal University of Santa Catarina, Florianópolis 88040-900, Brazil; enedir.ghisi@ufsc.br

* Correspondence: igor.catao@posgrad.ufsc.br

† Presented at the 7th International Electronic Conference on Water Sciences, 15–30 March 2023; Available online: https://ecws-7.sciforum.net/.

Abstract: This work assessed the potable water savings potential for different scenarios in a flat in Florianópolis, Brazil. An uncertainty analysis was also performed to understand which parameters most influenced the results. First, it was necessary to evaluate the water consumption and calculate the water end-uses during a home-office period due to the coronavirus pandemic. The water end-uses were obtained by monitoring the users' consumptions for sixteen days and comparing them with the water meter on a daily basis. The results were very close to those measured using the water meter, with an average absolute error of 5.6%. The base consumption was, on average, 249.2 litres per capita per day. With a home-office regime and an uninterrupted occupation, the coronavirus pandemic could be postulated to justify the more intense consumption patterns. Regarding the percentage of non-potable end-uses, an average of 25.8% was obtained. Potable water savings were simulated using the computer program Netuno, version 4. Seventy scenarios were evaluated, including different rainwater catchment areas and water and rainwater demands. The main results were that rainwater harvesting through a reduced area, 17.5% of the roof, obtained significant results, compared to the simulation with the whole roof, with a potable water savings potential of 16%.

Keywords: rainwater; water end-uses; potable water savings potential; simulation; rainwater harvesting; coronavirus

Citation: Martins Vaz, I.C.; Ghisi, E. Rainwater Harvesting during the COVID Outbreak: A Case Study in Brazil. *Environ. Sci. Proc.* **2023**, *25*, 16. https://doi.org/10.3390/ECWS-7-14172

Academic Editor: Athanasios Loukas

Published: 14 March 2023

1. Introduction

One of the simplest methods to optimise water consumption is to return to the ancient knowledge of rainwater harvesting systems and use rainwater for non-potable purposes in buildings. Rainwater harvesting is a technique that has been widely known and disseminated in society for thousands of years [1]. According to Gnadlinger [1], there is no single reason why rainwater is no longer the focus of water harvesting techniques. However, the author comments that some factors were climate change, with droughts generating local inefficiency of systems, the desire for a centralised water management system, and the focus on large water supply projects, such as dams and wells.

Studies on the potable water savings potential through the implementation of rainwater harvesting systems are abundant in the literature. Examples of residential buildings [2], industries and agriculture [3,4], schools and universities [5,6], and offices [7] are found in the literature. Local rainwater harvesting can decrease the number of distribution pipes and reservoirs, thus decreasing leakage losses. Lower volumes of water are also withdrawn from rivers and aquifers, which benefits the environment. Finally, using rainwater for non-potable purposes decreases the amount of water treatment chemicals. Studies on life cycle analysis (LCA) have also demonstrated the potential to decrease the environmental impacts of water supply through rainwater harvesting systems [8].

Recently, with the coronavirus pandemic, the discussion has focused on optimised water systems and management. This necessity also includes the reports of the Intergovernmental Panel on Climate Change (IPCC) and the world's focus on water safety projects [9]. Water is a crucial asset toward sustainable and resilient cities and vital toward health and society equity. According to the sixth Sustainable Development Goal (SDG) of the United Nations (UN), one of the global goals is to sustainably ensure water and sanitation.

All of this information, the workplace change, fewer commuting schemes, and less intranational and international travel have caused modifications in many water flows. Not only have the consumption patterns been changing, but the efficiencies of optimisation systems, such as rainwater harvesting and water-saving appliances, have also changed.

Kalbusch et al. [10] evaluated the water consumption changes in Joinville, Brazil, with the outbreak of the coronavirus pandemic. According to statistical tests, changes were observed in the consumption patterns of commerce, industry, and public activities, with significant modifications. An increase of 11% in residential water consumption was also observed but without statistical relevance. Balacco et al. [11] also observed modifications in water consumption in five different Italian cities during the pandemic outbreak. The authors found that users changed to a late wake-up (10:00 a.m.) as a new habit during COVID. This modification, alongside the modification in commuting schemes, influenced the total water demand of the cities.

Thus, this study aims to evaluate one case study with the water end-uses and simulate the potential of potable water savings of a rainwater harvesting system during the pandemic in southern Brazil. An uncertainty analysis was also performed to understand which parameters most influenced the results.

2. Method

The case study consisted of two parts. The first was the water consumption analysis, which measured the use frequencies and flow rates of the appliances while monitoring the water meter of one flat. The measurements were made during the pandemic period, with the social isolation of the users. The second part consisted of simulating the potential for potable water savings through the theoretical implementation of a rainwater harvesting system.

2.1. Object of Study and Water Monitoring

A flat in a multi-family residential building in Florianópolis, Brazil, was chosen to be evaluated for the design of a rainwater harvesting system. The definition considered estimating the water end-uses by monitoring the two residents for sixteen days. Figure 1 shows the location and floor plan of the flat. The green area shows the roof area owned solely by the flat owners, and the red area shows the part of the roof shared with the building and its residents. The flat is located at longitude 48°30′08″ west and latitude 27°36′12″ south.

The monitoring of water consumption was carried out through questionnaires on the uses of the water appliances. Both residents filled out forms for sixteen days. The specific questionnaires for each room presented items regarding the environment, water appliance, frequency of use in the day, and the flow rate (litres/s, litres/cycle, litres/discharge, or litres), which were counted between 00:00 and 23:59 each day.

The volume of water consumed in each water appliance was measured to calculate the water flow. A pre-established volume was measured for showers, taps, and sinks, and the filling time was recorded. For these appliances, an average of three measurements was taken. Regarding bowl-and-tank toilets, the volume of water in each flushing was measured for half and full flush.

For appliances with cycles, such as the washing machine and dishwasher, the consumptions indicated in the appliances' manuals were used. For the drinking water consumption, we used the consumptions indicated by the users at the end of the day, considering the average number of glasses of water drunk and the glass volumes.

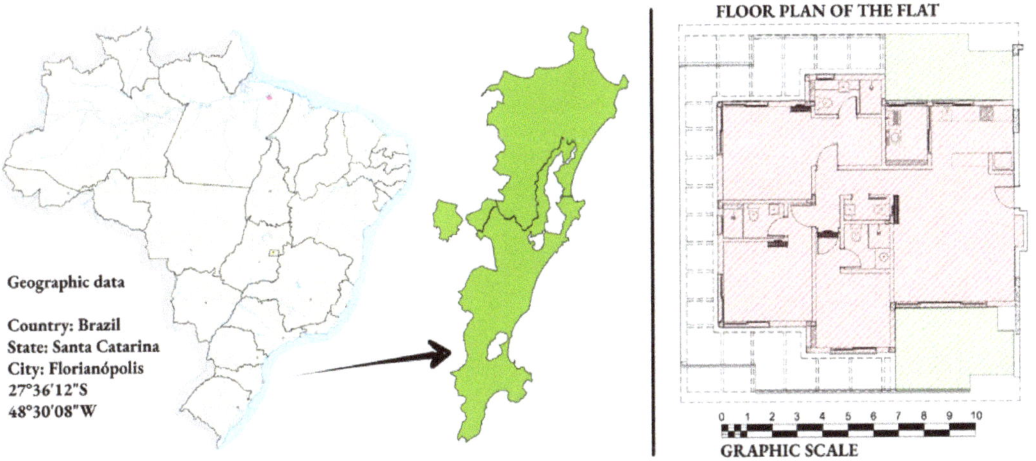

Figure 1. Location of the study and floor plan of the flat [12,13].

2.2. Rainwater Harvesting System

The rainwater harvesting system was modelled based on similar works and using the computer program Netuno, version 4. The program, created by Ghisi and Cordova [14], was based on a deterministic water balance similar to the yield-before-spillage and yield-after-spillage approaches. As for the simulation parameters, the program required local pluviometry data, water consumption characteristics, technical definitions (such as first flush volume and the runoff coefficient of the roof), and upper and lower tank volume definitions.

The goal was to analyse the potential for potable water savings under the consumption found during the pandemic times. However, there is much variability within some of the parameters. To include the uncertainty analysis in the simulation result, 70 different scenarios were modelled based on the range of three parameters. Two values were used for the harvesting area, combined with seven water demands and five rainwater demands. Table 1 shows the data used in the simulations.

Table 1. Parameters used for rainwater harvesting simulation.

Parameter	Value
Pluviometry data	Obtained via INMET [15]
First flush disposal	2 mm
Harvesting area	22 m^2 private roof (PR)/126 m^2 shared and private roof (SPR)
Total water demand	−15%/−10%/−5%/Water demand/+5%/+10%/+15%
Rainwater demand (% of the total water demand)	15%/20%/25%/30%/35% *
Roof runoff coefficient	0.80
Upper tank size	Equal to the average daily rainwater consumption
Lower tank size	Range between 1000 and 6000 litres (step of 250 litres)

* These results were found during the first part of the research and are presented in Section 3.1.

The two harvesting areas were modelled to represent the whole roof and the private part of the roof, including the shared and individual parts. This division occurred because one part was owned solely by the flat owners while the other was shared with other building residents. To simplify the results, the private roof is stated as PR and was 22 m^2. Shared plus the private roof is stated as SPR and was 126 m^2. Figure 1 shows the shared part in red, while the external boundaries in green show the private area.

3. Results and Discussion

3.1. Water Consumption and End-Uses

Water consumption was gathered and analysed in comparison to the water metering. The results were similar, with an absolute mean error of 5.6%. This similarity meant that, on average, daily estimates of water consumption varied by ±5.6% compared to the values registered on the water meter. Daily average water consumption via water meters was 249 litres/capita/day, and the average non-potable water use was estimated to be 25.8%. We considered rainwater could be used only for the washing machine and toilets as a non-potable source.

Table 2 shows the water flow for each of the water appliances. M (millilitres) and T (seconds) stand for the measurement and timing, according to the method shown in Section 2.1.

Table 2. Water flow for the water appliances.

Water Appliance	Room	M1	T1	M2	T2	M3	T3	Average Flow	Unit
Kitchen tap	Kitchen	1225	7.13	1650	8.42	1700	10.27	0.178	litres/s
Dishwasher		According to the manual						8	litres/cycle
Washing machine		According to the manual						9.4	litres/cycle
Drinking water		According to the user's measurements						-	litres/day
Washing tank		1500	5.88	1800	6.97	1600	6.28	0.256	litres/s
Shower	Bathroom 1	2200	11.33	2325	11.05	2340	10.89	0.206	litres/s
	Bathroom 2	5000	24.68	2300	10.17	2400	10.84	0.217	
	Bathroom for guests	Average of other showers						0.212	
Tap	Bathroom 1	470	4.31	450	4.26	430	3.43	0.113	litres/s
	Bathroom 2	275	2.02	390	2.77	450	3.55	0.135	
	Bathroom 3	400	3.55	450	3.84	500	3.63	0.123	
	Bathroom for guests	500	4.32	500	4.12	450	3.87	0.118	
Toilet—One flush	Bathroom 1	Length (34.5)/Width (13.3)/Depth (17.5) [1]						8.030	litres/use
	Bathroom 2	Length (35.2)/Width (14.0)/Depth (19.2) [1]						9.462	
	Bathroom for guests	Average of other one-flush devices						8.746	
Toilet—Half flush	Bathroom 1	Length (34.5)/Width (13.3)/Depth (10.0) [1]						4.589	
	Bathroom 2	Length (35.2)/Width (14.0)/Depth (16.2) [1]						7.983	
	Bathroom for guests	Average of other half-flush devices						6.286	

[1] Dimensions of the bowl-and-tank water volume used in each type of flush in cm.

For the different scenarios of potable water demand (range between −15 and +15%), the minimum and maximum daily water consumption ranged between 250 and 750 litres/day. This range showed how much variability was found within the measurements of daily water demand. Additionally, the daily average water consumption obtained was higher than in previous literature, which states an average figure of 150 litres/capita/day as the Brazilian pattern.

The distributions of the water consumption within rooms and water devices are shown in Figure 2a,b. Figure 2c shows the measured versus metered water consumption. One can see that most of the water consumption occurred in bathroom 1, bathroom 2, and the kitchen, with little demand in the other rooms. Regarding water appliances, consumption was higher for showers, kitchen taps, and toilets.

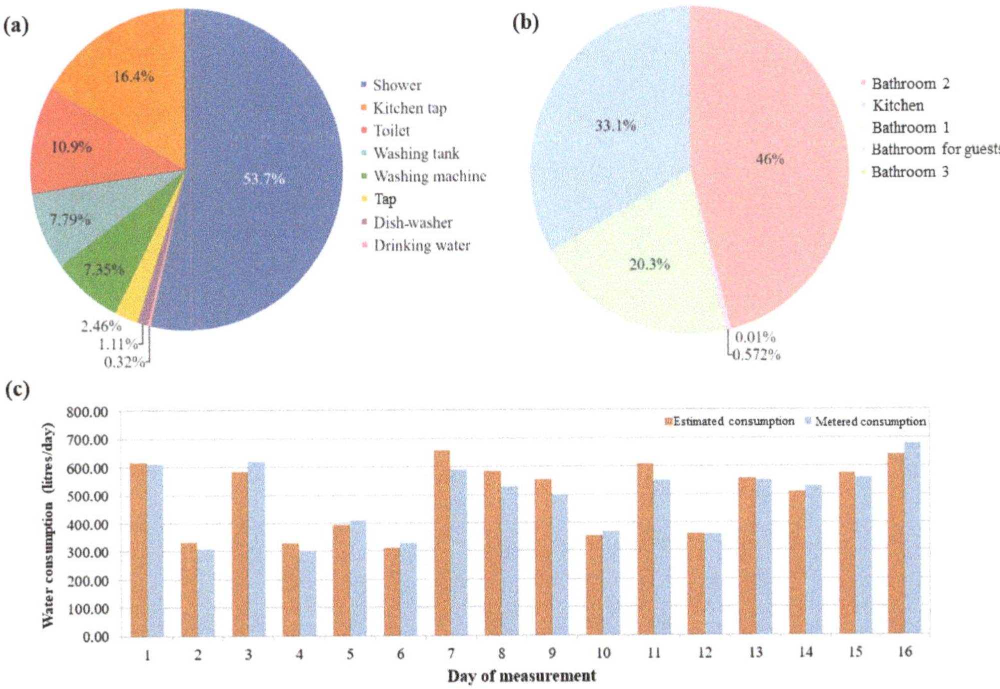

Figure 2. Results of the water monitoring. (**a**) water distribution in water appliances; (**b**) water distribution in rooms; (**c**) measured versus metered water consumption.

Comparing the results to those of Freitas and Ghisi [16], one can see that this case study flat had a higher water consumption per capita than other studies in the same region. Additionally, they obtained a non-potable water consumption of 42.2% of the daily water demand.

3.2. Rainwater Harvesting System

The comparison between the PR and the SPR was the first assessment, resulting in a difference of approximately 8% in potable water savings. This assessment was performed with the baseline consumption and the 25% non-potable water end-uses, presented in Section 3.1. By harvesting rainwater with the 126 m^2 roof, 24% of potable water savings were obtained, while with the 22 m^2 alternative, 16% savings were obtained. Both results were obtained with a lower tank of 3000 litres, which was indicated as the optimal technical solution.

The second assessment was the uncertainty analysis within the results obtained in the water metering. In order to do so, the water demand was ranged, according to Table 1. The results were then checked for PR and SPR. The main conclusion was that the PR, a smaller roof area for harvesting, presented more sensitivity to the total water demand. In this scenario, rainwater was scarcer, and the potable water savings potential dropped when higher water demand was included. For the SPR, almost all demands were met by the rainwater harvesting systems.

The third assessment was the range of the parameter "rainwater demand", which ranged around the figure of 25%. The main result was the opposite of the second assessment, with less impact on PR and more on SPR. Such a result can be explained by the analysis that SPR provided more rainwater. In this scenario, when non-potable water was needed, rainwater would be available in response to a larger harvesting area. The PR area, on the contrary, did not present extra rainwater for the system, being less affected by the parameter.

Both assessments were engaging, showing that PR harvesting proved to be a good alternative. However, if more non-potable (more than 25%) water is needed, the SPR alternative would become more attractive, requiring the approval of the remaining building users. Nevertheless, by dividing the potable water savings potential of the PR by the SPR results, a referential percentage of 65% was obtained. This result means that even with only 17.5% of the total roof area, the users might benefit from more than half of the potable water savings potential. It is easier to install and start a sustainable water practice within the flat. Comparing the results to Freitas and Ghisi's [16], one can see that this study had a much lower potable water savings potential, mainly due to the lower non-potable water demand. One can obtain higher figures in houses with gardens and patio cleanings.

4. Conclusions

The potable water savings potential ranged from 15.80% to 24.43% when considering both roof area possibilities. The results were higher than those obtained in the literature for multi-family buildings and lower than those found for single-family buildings. Additionally, it was found that even the smaller roof area proved to be an exciting approach for the users, starting a sustainable water practice in the flat.

The flat presented higher consumption than the region's average water consumption, and one can postulate that the continuous stay of users due to pandemic isolation might have influenced the results. The non-potable water demand percentage for the flat was similar to previous literature. Further studies can better understand the effects of different user patterns, helping to improve rainwater harvesting dynamics.

Author Contributions: Conceptualization, I.C.M.V. and E.G.; methodology, I.C.M.V.; software, I.C.M.V.; writing—original draft preparation, I.C.M.V.; writing—review and editing, I.C.M.V. and E.G.; supervision, E.G. All authors have read and agreed to the published version of the manuscript.

Funding: This research received no external funding.

Institutional Review Board Statement: Not applicable.

Informed Consent Statement: Informed consent was obtained from all subjects in the study.

Data Availability Statement: The data presented in this study are available on request from the corresponding author.

Acknowledgments: We would like to thank UFSC for their support and Coordenação de Aperfeiçoamento de Pessoal de Nível Superior (CAPES) for the first author's scholarship during this research.

Conflicts of Interest: The authors declare no conflict of interest.

References

1. Gnadlinger, J. Rainwater in the Integrated Management of Water Resources in Semi-Arid Locations: Historical, Biophysical, Technical, Economic, and Socio-Political Aspects. In *Rainwater Harvesting, Manage and Use*; Instituto Nacional do Semiárido: Campina Grande, Brazil, 2015. (In Portuguese)
2. Kuntz Maykot, J.; Ghisi, E. Assessment of A Rainwater Harvesting System in A Multi-Storey Residential Building in Brazil. *Water* **2020**, *12*, 546. [CrossRef]
3. Bertuzzi, G.; Ghisi, E. Potential for Potable Water Savings Due to Rainwater Use in a Precast Concrete Factory. *Water* **2021**, *13*, 448. [CrossRef]
4. Bafdal, N.; Dwiratna, S. Water Harvesting System as an Alternative Appropriate Technology to Supply Irrigation on Red Oval Cherry Tomato Production. *Int. J. Adv. Sci. Eng. Inf. Technol.* **2018**, *8*, 561. [CrossRef]
5. Ghisi, E.; Rupp, R.F.; Triska, Y. Comparing Indicators to Rank Strategies to Save Potable Water in Buildings. *Resour. Conserv. Recycl.* **2014**, *87*, 137–144. [CrossRef]
6. Martins Vaz, I.C.; Ghisi, E.; Thives, L.P. Stormwater Harvested from Permeable Pavements as a Means to Save Potable Water in Buildings. *Water* **2021**, *13*, 1896. [CrossRef]
7. Proença, L.C.; Ghisi, E. Assessment of Potable Water Savings in Office Buildings Considering Embodied Energy. *Water Resour. Manag.* **2013**, *27*, 581–599. [CrossRef]
8. Marinoski, A.K.; Ghisi, E. Environmental Performance of Hybrid Rainwater-Greywater Systems in Residential Buildings. *Resour. Conserv. Recycl.* **2019**, *144*, 100–114. [CrossRef]

9. IPCC. Summary for Policymakers. In *Climate Change 2022: Impacts, Adaptation and Vulnerability*; Cambridge University Press: Cambridge, UK, 2022; pp. 3–33.
10. Kalbusch, A.; Henning, E.; Brikalski, M.P.; de Luca, F.V.; Konrath, A.C. Impact of Coronavirus (COVID-19) Spread-Prevention Actions on Urban Water Consumption. *Resour. Conserv. Recycl.* **2020**, *163*, 105098. [CrossRef] [PubMed]
11. Balacco, G.; Totaro, V.; Iacobellis, V.; Manni, A.; Spagnoletta, M.; Piccinni, A.F. Influence of COVID-19 Spread on Water Drinking Demand: The Case of Puglia Region (Southern Italy). *Sustainability* **2020**, *12*, 5919. [CrossRef]
12. Wikipedia. Map File of Brazil. Author: NordNordWest. Licenced under Creative Commons 4.0. Available online: https://pt.m.wikipedia.org/wiki/Ficheiro:Brazil_location_map.svg (accessed on 15 January 2023).
13. Wikipedia. Map File of Florianópolis. Author: Franzisquin. Licenced under Creative Commons 3.0. Available online: https://pt.wikipedia.org/wiki/Ficheiro:Floripa_2020.svg (accessed on 15 January 2023).
14. Ghisi, E.; Cordova, M.M. Netuno 4. Computer Programme. Universidade Federal de Santa Catarina, Departamento de Engenharia Civil. 2014. Available online: http://www.labeee.ufsc.br/ (accessed on 25 November 2022).
15. INMET. Meteorologic Database for Research and Teaching (Banco de Dados Meteorológicos para Ensino e Pesquisa—BDMEP, in Portuguese). BDMEP. 2021. Available online: https://bdmep.inmet.gov.br/ (accessed on 10 January 2021).
16. Freitas, D.A.; Ghisi, E. Economic feasibility analysis of rainwater harvesting: A case study in Imbituba, Brazil. *Urban Water J.* **2020**, *17*, 905–911. [CrossRef]

environmental sciences proceedings

MDPI

Proceeding Paper

Result-Based Management Tool for the Assessment of Existing Structural Flood Protection and Future Planning; Case Study in the Strymon River Basin, Greece †

Eleni Tzanou [1,*,‡], Charalampos Skoulikaris [2] and Antonios Chatzigiannis [3]

1 School of Surveying and Geoinformatics Engineering, Faculty of Engineering, International Hellenic University, 62124 Serres, Greece
2 Scientific Associate, Civil Engineering Department, Aristotle University of Thessaloniki (AUTH), 54124 Thessaloniki, Greece; hskoulik@civil.auth.gr
3 Consortis Ltd., 57001 Pylaia, Greece; chatzigiannis@consortis.gr
* Correspondence: etzanou@ihu.gr
† Presented at the 7th International Electronic Conference on Water Sciences, 15–30 March 2023; Available online: https://ecws-7.sciforum.net.
‡ This author is responsible for the first authorship in the representative field.

Abstract: Flood occurrence consequences are inextricably linked to the loss of human life and material damage. The latter has a direct economic impact and requires financial resources for repair or reconstruction in order to continue to provide protection. In the study area of the Strymon River in northern Greece, the implemented flood protection and hydraulic structural works combined with failure repeatability, their initial construction costs, and damage/repair costs were thoroughly assessed and correlated. This methodology provides a roadmap to support decision-making procedures to formulate flood protection action plans based on the valuation of current flooding results in established infrastructure.

Keywords: flood protection; management; structural works; economic loss; planning

Citation: Tzanou, E.; Skoulikaris, C.; Chatzigiannis, A. Result-Based Management Tool for the Assessment of Existing Structural Flood Protection and Future Planning; Case Study in the Strymon River Basin, Greece. *Environ. Sci. Proc.* **2023**, *25*, 17. https://doi.org/10.3390/ECWS-7-14204

Academic Editor: Slobodan Simonovic

Published: 14 March 2023

1. Introduction

Flood risk management and protection infrastructure does not comprise an exclusively technical subject. The implementation of flood risk management strategies and their societal integration and acceptance necessitate flood governance. The catchment area of the project, the Strymon River basin in northern Greece, has suffered numerous flood events of varying importance in the past decades with consequences on the natural and socioeconomic sectors. At the European Union (EU) scale, the gradual employment of the Water Framework Directive (WFD) (EC 2000) since the early 2000s and the Directive on the Assessment and Management of Flood Risks (EC 2007), commonly known as the Floods Directive (FD), have provided a critical legislative framework that has led to the development of operational tools, such as the Flood Risk Management Plans (FRMPs) [1].

The FRMPs are the first step in compliance with the European Legislation in the hierarchical pyramid of flood management and protection. Their objectives set for flood risk management focus on reducing the potentially negative consequences that floods have on human health, the environment, cultural heritage, and economic activity, as well as on initiatives to reduce flood occurrence [2]. Although they offer a fundamental plan, their macroscopic perspective in terms of scale does not always give precise results when dealing with large-scale areas.

In order to assess the necessity of flood protection infrastructure on a local scale, further technical and operational data should be coupled together with those derived by the FD. These data include specific information on infrastructure with the site-specific location, structural characteristics, the initial cost of construction, performance, the response to flood events, and the cost of maintenance and reconstruction.

In cases in which structural flood protection measures fail to serve flooding inhibition, the results may affect different sectors, depending on the intensity and extent of the flood event [3]. Material damage is the most common consequence of infrastructure failure, and it has a direct economic impact in that it requires financial resources to be repaired or reconstructed in order to continue to provide the required amount of protection in the future.

The aim of the research is to demonstrate the importance of evaluating the operability of existing flood protection structures in future anti-flood planning at regional and local scale. The operability is evaluated as a synthesis of flood occurrences, cost of structures and cost of structures' maintenance.

2. Materials and Methods

The tools for assessing the financial cost of repairing damage or failure of technical/structural works constructed for flood protection are critical assets in flood project management by the governmental authorities and public organizations engaged. These tools can provide spatial data linked to construction and repair costs for identifying problematic spots and areas that repeatedly show a structural inability to prevent flooding. Particularly, the identification of linkages among spatial data, costs of flood protection works and flood event occurrences could highlight the regions or locations that the current flood protection infrastructures are not effective since flood events continue to occur The outcome of such a process may lead to a flood risk assessment tool that can record information that supports decision-making procedures in order to formulate action and management plans related to flood planning [4].

This study was part of the project "Evaluation of the performance and interoperability of flood protection intervention measures in the area of the Strymon river basin", which was implemented under the INTERREG V-A European Territorial Cooperation Program "Greece-Bulgaria 2014–2020 "Flood Protection—Cross Border Planning and Infrastructure Measures for Flood Protection"; this program aimed to combine the FRMPs with a thorough and detailed record and evaluation of the existing situation in terms of flood protection infrastructure to assess the majority of civil works and previously applied measures in order to evince the areas prone to flooding and provide an Action Plan, in which specific located measures were proposed according to a hierarchical evaluation.

2.1. Study Area

The study was implemented in the catchment area of the Strymon river, with an emphasis on places that show higher flood risk. Strymon is a river of the Balkan Peninsula with a length of 360 km, of which 242 km are in Bulgarian and 118 km are in Greek territory. The total hydrological basin has an area of 16,550 km^2, with only 6344 km^2 located in Greece [5].

The area has suffered numerous flood events, with an increasing intensity of occurrence in the last decade. A results-based approach offers a tool for priority-based planning in which the initial cost of investment in the construction of flood protection work is related to the potential economic failure consequences and the repeated costs of maintenance. Numerous flood events, with an increasing tense of occurrence in the last decade [6] is presented in the case study area, with Figure 1 presenting all the flood events during the period 2012–2018 where official data are available [7].

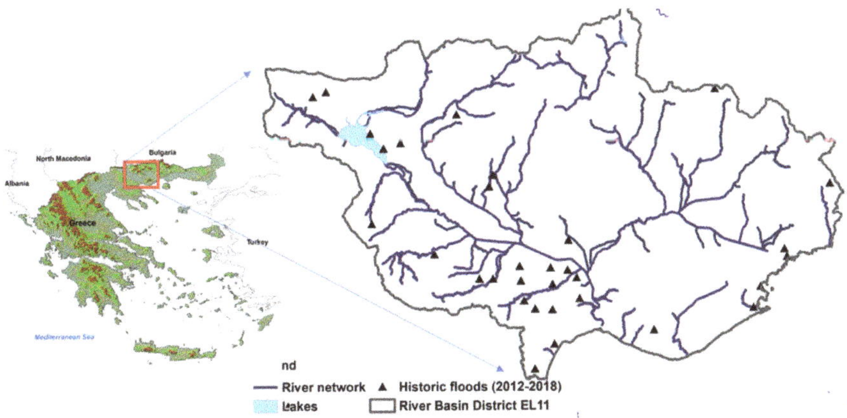

Figure 1. Study area and flood events during 2012–2018.

2.2. Methodology

The methodological approach constituted of the elaboration of the following actions and activities:

1. Development of a geospatial system containing baseline information on all the necessary data [8] (backgrounds, networks, structures, flood events, flood zones, etc.);
2. Recording of the existing legal and institutional framework and its integration into the geospatial system [8] (protected areas, land uses, special zones, special infrastructure, etc.);
3. Introduction of the areas/zones potentially at a high risk of flooding according to the FRMPs;
4. Determination of the TWI (Topographic Wetness Index);
5. Recording of existing measures from the regional and local action and flood protection management plans;
6. Recording of historical flooding phenomena and their damage assessment data;
7. Recording of the results of existing measures, interventions, and infrastructure, depending on their efficiency in flood events;
8. Determination of positive and negative effects of measures on the anthropogenic, natural, and economic environment;
9. Future projection of the operational capacity of the protection and intervention measures;
10. Cost estimation of protection/intervention measures and prioritization.

All data were introduced into a web GIS (Geographic Information System) for visualization and analysis.

3. Results and Discussion

In cases in which past construction data are needed, the main obstacle is the scattered records among the responsible public authorities and the lack of service files. Despite the difficulty of the venture, 652 structural flood protection measures and civil works were recorded in total through field surveys and file searches, constituting the first effort of such a scale in the Greek territory.

These structures were further divided into 21 subcategories according to the type of structure and the frequency at which they occur in the river basin. The occurrence frequency is indicative of the structural type preferred and most constructed in the area. Figure 2 illustrates all categories of structural measures as well as their types and occurrence frequencies in the study area.

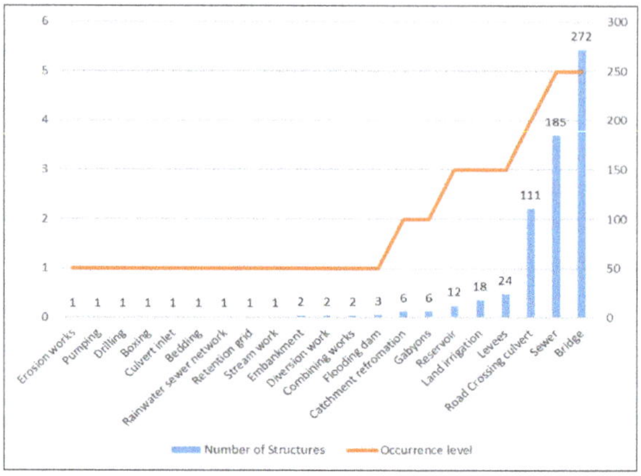

Figure 2. Protection infrastructure according to structure type and occurrence frequency.

All aforementioned data were mapped in QGIS while the geodatabase was populated with information relative to each structure, such as the construction date, cost, maintenance frequency, failure, damage, repair cost, and interoperability with other structures. Figure 3a–c show the mapping outcome of the structure data and their reclassification into six main categories—(i) restoration work, (ii) urgent projects, (iii) dredging work, (iv) cleaning work and work on technical islets (in rivers), and (v) alluvial removal work—according to the type of maintenance necessary over a period of 5 years. Maintenance work included any kind of intervention needed once, while the repeatability of maintenance in a specific structure/location was also recorded, thus providing an index of vulnerability and poor resilience of each structure to flooding events (Table 1).

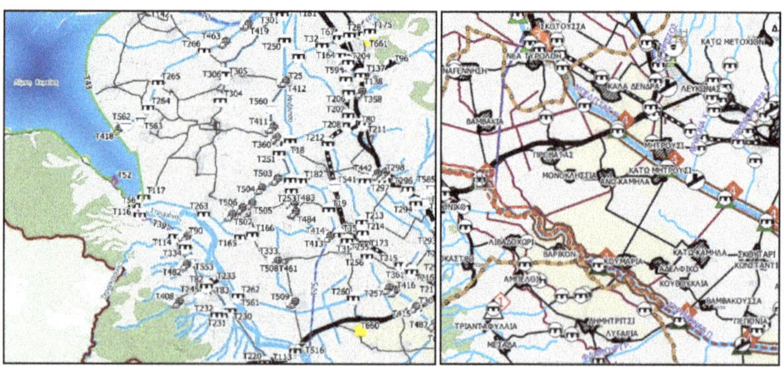

(**a**) (**b**)

Figure 3. Infrastructure mapping (**a**) and maintenance (green mark) /intervention (red mark) repeatability (**b**) in the Strymon river basin.

In total, all 652 recorded flood protection structures showed a repetitive need for intervention actions to be taken due to possible failure or unsuccessful operation at an approximate rate of 1:3, while in most cases, maintenance work needed to be repeated with the reverse rate of 3:1. Two categories proved to show more vulnerability and less of an ability to perform well in potential flooding: (a) restoration work in critical flood protection infrastructure and (b) bridges and stream crossings.

Table 1. Flood protection infrastructure and maintenance interventions.

Type of Structure/Work	Structure Work	Maintenance Work	Maintenance (Including Repeatability)
Stream bottom slope alteration/depth	27	4	10
Restoration work	10	16	17
Islets	4	19	19
Excavations	8	21	44
Removal of alluvium	1	27	75
Technical projects (inlets/culverts)	330	28	96
Bridges	272	138	431
TOTAL	652	253	692

The above were introduced into the GIS model in order to determine the spatial distribution of the data and have a visual outcome that yielded critical locations subject to further study; these locations showed an increased vulnerability to flooding events, as demonstrated by the data.

Figure 3 presents the mapping outcome of infrastructure data visualization (a) and reclassification, with the repeatability of maintenance marked for each structural work (b) related to the number of interventions.

In addition, important information on the initial cost of infrastructure and the maintenance cost was collected. While the initial cost was not always available due to the construction in past years, the maintenance costs for the past decade were easier to obtain. The analysis of the outputs is depicted in Figure 4.

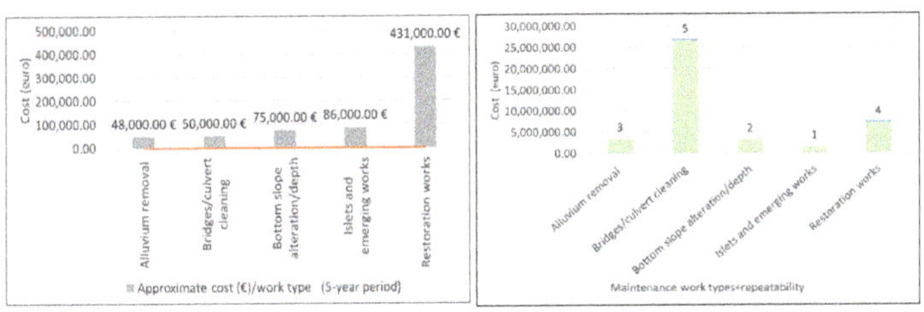

Figure 4. Task repeatability ranking by approximate labor cost (EUR).

The first time a flood protection task is implemented, the restoration work showed a significantly higher cost of EUR 431,000.00, while the repetition of the work/intervention measures placed the interventions in bridges and culvers in the first place with a significant difference. The work on islets showed the lowest cost, implying that it has not yet been necessary to modify them (based on repeatability). The total approximate cost for the maintenance of flood work only rose to EUR 42,711,000 in a 5-year period.

The significance of the structural work and the prioritization of needs in new flood protection infrastructure was also correlated with the FRMPs; all data were overlayed with the food risk zones to spatially determine the most vulnerable infrastructure [9] so as to hierarchize potential interventions.

4. Conclusions

Flood events (major and minor) are inevitable, regardless of the amount of existing protection. Flood protection is critical, yet the financial cost of maintaining high levels of protection can prove rather excessive. In major flood events, failures occur, often at different locations each time depending on the spatial distribution and evolution of the phenomenon. However, it is possible to take into account the parameters that highlight the weaknesses and the problematic or insufficient flood protection design so that rational and reliable alternative proposals and solutions emerge.

From the correlations and analyses of the data in this study, it was derived that an important element for the spatial determination of the new measures is the infrastructure already developed for flood protection. The study showed that, despite the existence of multiple structural measures/work, many of them show poor efficiency, as shown by the high rate of repeated maintenance and its cost.

The methodology followed in this study was based on the results of the performance of existing flood protection measures, their characteristics, and their spatial correlation. The evaluation of structural measures by assessing their vulnerability via their yearly maintenance implemented the economic loss and financial impact that flooding events might have on infrastructure. The new measures should be related to the cost of construction and maintenance each time, applying the rule of cost–benefit [10] and life cycle analysis, and infrastructure that costs more to maintain should be replaced.

Author Contributions: Conceptualization, E.T.; methodology, E.T., C.S. and A.C.; software, E.T., C.S. and A.C.; validation, E.T., C.S. and A.C.; formal analysis, E.T., C.S. and A.C.; investigation, E.T., C.S. and A.C.; resources, E.T., C.S. and A.C.; data curation, E.T., C.S. and A.C.; writing—original draft preparation, E.T.; writing—review and editing, E.T. and C.S.; visualization, E.T. and C.S.; supervision, E.T. and C.S.; project administration, E.T.; funding acquisition, E.T. All authors have read and agreed to the published version of the manuscript.

Funding: This research was funded by the INTERREG V-A European Territorial Cooperation Program "Greece-Bulgaria 2014–2020 Flood Protection—Cross Border Planning and Infrastructure Measures for Flood Protection".

Institutional Review Board Statement: Not applicable.

Informed Consent Statement: Not applicable.

Data Availability Statement: All data, models, and code that support the findings of this study are available from the corresponding author upon reasonable request.

Conflicts of Interest: The authors declare no conflict of interest.

References

1. Directive 2007/60/EC of the European Parliament and of the Council of 23 October 2007 on the Assessment and Management of Flood Risks. Available online: https://eur-lex.europa.eu/eli/dir/2007/60/oj (accessed on 13 December 2022).
2. WHO. *Selecting Measures and Designing Strategies for Integrated Flood Management: A Guidance Document*; Swiss Federal Office for the Environment, Netherlands Ministry of Infrastructure and Environment and Deltares: Delft, The Netherlands, 2017.
3. Raadgever, G.T.; Booister, N.; Steenstra, M.K. Flood risk governance. In *Flood Risk Management Strategies and Governance*; Raadgever, T., Hegger, D., Eds.; Springer: Berlin/Heidelberg, Germany, 2018.
4. Malczewski, J.; Rinner, C. *Multicriteria Decision Analysis in Geographic Information Science*; Springer: Berlin/Heidelberg, Germany, 2015; pp. 3–54.
5. Skoulikaris, C. Transboundary Cooperation Through Water Related EU Directives' Implementation Process. The Case of Shared Waters Between Bulgaria and Greece. *Water Resour Manag.* **2021**, *35*, 4977–4993. [CrossRef]
6. Tolika, K.; Skoulikaris, C. Atmospheric circulation types and floods' occurrence; A thorough analysis over Greece. *Sci. Total Environ.* **2023**, *865*, 161217. [CrossRef] [PubMed]
7. Skoulikaris, C. Toponyms: A neglected asset within the water framework and flood directives implementation process; the case study of Greece. *Acta Geophys.* **2022**, 1–15. [CrossRef]
8. Coutinho-Rodrigues, J.; Simão, A.; Antunes, C. A GIS-based multicriteria spatial decision support system for planning urban infrastructures. *Decis. Support Syst.* **2011**, *51*, 720–726. [CrossRef]
9. Peterson, S.J. *Construction Accounting and Financial Management*; Pearson Prentice Hall: Hoboken, NJ, USA, 2005.
10. Hanley, N.; Splash, C.L. *Cost of Benefit Analysis and the Environment*; Edward Elgar Publishing Ltd.: Cheltenham, UK, 1993.

environmental sciences proceedings

Proceeding Paper

Prediction and Classification of Flood Susceptibility Based on Historic Record in a Large, Diverse, and Data Sparse Country †

Heather McGrath * and Piper Nora Gohl

Natural Resources Canada, Ottawa, ON K1A 0E4, Canada
* Correspondence: heather.mcgrath@nrcan-rncan.gc.ca
† Presented at the 7th International Electronic Conference on Water Sciences, 15–30 March 2023; Available online: https://ecws-7.sciforum.net/.

Abstract: The emergence of Machine learning (ML) algorithms has shown competency in a variety of fields and are growing in popularity in their application to geospatial science issues. Most recently, and notably, ML algorithms have been applied to flood susceptibility (FS) mapping. Leveraging high-power computing systems and existing ML algorithms with national datasets of Canada, this project has explored methods to create a national FS layer across a geographically large and diverse country with limited training data. First, approaches were considered on how to generate a map of FS for Canada at two different levels, (i) national, which combined all training data into one model, and (ii) regional, where multiple models were created, based on regional similarities, and the results were mosaicked to generate a FS map. The second experiment explored the predictive capability of several ML algorithms across the geographically large and diverse landscape. Results indicate that the national approach provides a better prediction of FS, with 95.7% of the test points, 91.5% of the pixels in the training sites, and 89.6% of the pixels across the country correctly predicted as flooded, compared to 65.5%, 80.6% and 75.6%, respectively, in the regional approach. ML models applied across the country found that support vector machine (svmRadial) and Neural Network (nnet) performed poorly in areas away from the training sites, while random forest (parRF) and Multivariate Adaptive Regression Spline (earth) performed better. A national ensemble model was ultimately selected as this blend of models compensated for the biases found in the individual models in geographic areas far removed from training sites.

Keywords: flood susceptibility; Canada; machine learning; flood priority setting

Citation: McGrath, H.; Gohl, P.N. Prediction and Classification of Flood Susceptibility Based on Historic Record in a Large, Diverse, and Data Sparse Country. *Environ. Sci. Proc.* **2023**, *25*, 18. https://doi.org/10.3390/ECWS-7-14235

Academic Editor: Athanasios Loukas

Published: 16 March 2023

1. Introduction

Flooding may occur any time of the year in Canada; however, the risk is generally higher in the spring due to heavy rainfall coupled with a rapidly melting snowpack or ice jamming [1]. The development of flood hazard maps is a provincial responsibility. Thus, to date, the areas selected for mapping are determined provincially, based on their selection criteria [2]. Hazard maps generally cover a geographically small region using high-resolution datasets, which provides high-quality maps for these targeted areas. The limitation to this approach is that only those selected areas have flood maps generated, leaving many communities and large geographic areas without flood maps or indicators of their flood risk.

The emergence of Machine learning (ML) algorithms has shown competency in a variety of fields and are growing in popularity in their application to geospatial science issues. Most recently, and notably, they have been applied to FS mapping [3–6]. There are hundreds of ML algorithms available, which can be generally categorized into groups of classification or regression and may belong to different families, such as neural networks, boosting, random forests, generalized linear models, etc. [7]. These models come from different backgrounds, e.g., statistics (generalized linear models), artificial intelligence and data mining (decision trees), and clustering (K-means), while others represent a connectionist approach (neural networks) [7]. Tools, such as Classification And Regression (caret) library in R, provide an easy mechanism to execute and compare results from a large number of classifiers [8]. The pre-eminence of Random Forest (RF) in solving problems, such as FS, are well supported [3,7,9–12].

In the literature, there is not a consensus on whether a single model or an ensemble model is more appropriate, and different disciplines report varying outcomes. Ensemble learning and hybrid ML methods have been shown to significantly improve performance, especially in cases of high dimensional data [13–15]. Both ensembles and hybrid approaches employ a type of information fusion through differing approaches [14].

Most of the existing literature which has explored ML as applied to the FS problem have covered relatively small geographic regions, and often, single watersheds. Few have explored how these models perform over vast geographic and meteorologically diverse environments. In this research, two questions are explored to create a national FS map for Canada: (i) can a single ML model outperform a collection of regional models across a geographically large and diverse region, and (ii) how model performance is affected with increasing distance from the training sites.

2. Materials and Methods

Following common practices in ML workflows, datasets that may contribute to flooding were first identified, and historic flood events were used as labeled data and split into two classes: flooded/not flooded. The data was pre-processed to scale and centre or set as factors for the nominal datasets. It was split into training and test sets (70/30) using the Classification And Regression Training (caret) create data partition function to create a balanced split [11]. Next, Variable Selection using Random Forest (VSURF) was run to identify the relevant (important) datasets (factors) by finding the model and associated datasets with the smallest out-of-bag (OOB) error. This sub-set of factors was then refined by testing for multicollinearity among the independent variables by assessing the Variance Inflation Factor (VIF) and Pearson correlation coefficient to come up with a final list of important factors. Then, the selected factors were run through a variety of ML models to determine performance. Selection of models include those commonly cited in the literature, found to perform well in FS problems, and aligned with findings from [8] in their comprehensive study. K-fold cross-validation of the training points (K = 3) was repeated for five resampling iterations and the average score was recorded. All ML models were accessed from the caret library in R. The train function in caret was used, which sets up a grid of tuning parameters for a large number of classification and regression algorithms [7]. These control parameters prescribe the computational nuances of the train function. The test data was then run through the trained model and the classification and FS prediction were saved. These results were then analyzed in a confusion matrix, summary statistics were calculated, and the results compared to a historic event database, Figure 1.

2.1. Single Model vs. Multi-Model

As Canada spans a geographically large area and contains a diverse set of climate characteristics, a regional approach was first considered. This approach would see several models developed. National datasets in Canada have been developed based on a variety of frameworks, such as regional, physiological, and ecozone approaches. Commonly accepted in Canada is adopting the National Ecological Framework of Canada, which provides standard ecological units [16]. This classification system "incorporates all major components of ecosystems: air, water, land, and biota" [16]. A second framework considers physiographic regions and is comprised of the shield and borderlands, each sub-divided into additional groupings [17]. As undertaken by [3], in their US work of flood risk, drainage areas were considered as natural region boundaries. In Canada, there are 11 major drainage areas.

In the national approach, all training data were processed together to generate a single model, which was then applied across the whole of the Country.

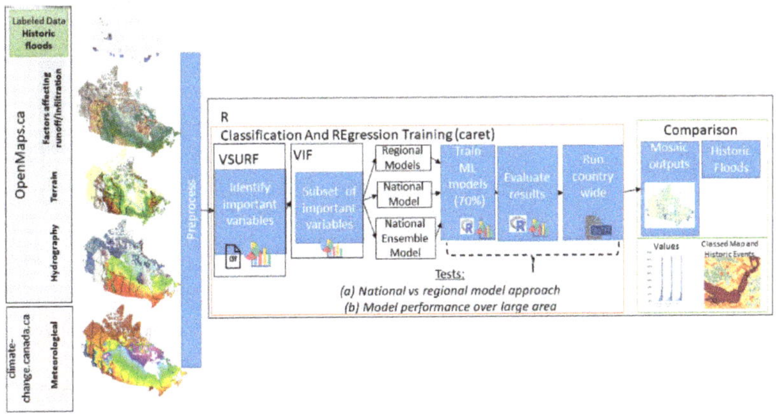

Figure 1. Methodology and tests run to create a flood susceptibility dataset in Canada.

2.2. *Single Model vs. Ensemble over Large Distance*

Most of the existing literature applying ML to the FS problem have tested and applied the results over a relatively small geographic areas, which are not dissimilar from the data found in the training/test set. Given the vast region of interest and the variety in land use, geography, climate, etc., several single ML models, which encompass a variety of classification and regression algorithms, were tested to evaluate their performance, Table 1. In addition, an ensemble model was developed and evaluated. Qualitative and quantitative analysis were performed to evaluate how the different ML models perform over Canada as a whole.

Table 1. Machine learning algorithms tested, C = classification, R = Regression).

Category	Algorithm	Acronym	Type
Decision Trees (DT)	C5.0	C5.0	C
Random Forest (RF)	Random Forest	RF	C, R
	Parallel Random Forest	parRF	C, R
	Regularized Random Forest	RRF	C, R
	Random Forest	ranger	C, R
Boosting (B)	eXtreme Gradient Boosting	xgbDART	C, R
	eXtreme Gradient Boosting	xgbTree	C, R
Support Vector Machines (SVM)	Support Vector Machines with Radial Basis Function Kernel	svmRadial	C, R
	Support Vector Machines with Polynomial Kernel	svmPoly	C, R
	Support Vector Machines with Radial Basis Function Kernel	svmRadialCost	C, R
Neural Networks (NN)	Model Averaged Neural Network	avNNet	C, R
	Neural Networks with Feature Extraction	pcaNNet	C, R
	Neural Network	nnet	C, R
	Multi-Layer Perceptron	mlp	C, R
Multivariate Adaptive Regression Splines	Multivariate Adaptive Regression Spline (MARS)	earth	C, R
Ensemble	Ensemble of Random Forest, Support Vector Machine, Boosting, and Neural Networks	parRF, nnet, svmPoly, xgbTree, pcaNNet	C, R

2.3. ML Models Tested

The single model approach focused on Random Forest, due to its superior performance as found in the literature, [3,7,9–12]. Additional models were considered for the ensemble method and synthesized via a generalized linear model (GLM).

2.4. Validation

Analysis of the results are per confusion matrix, including several measures: Accuracy, Kappa, Sensitivity, Specificity, and F1 score.

$$\text{Accuracy } Acc = \frac{TP + TN}{TP + FN + FP + TN} \tag{1}$$

$$\text{Sensitivity } Sn = \frac{TP}{TP + FN} \tag{2}$$

$$\text{Specificity } Sp = \frac{TN}{TN + FP} \tag{3}$$

$$\text{F1 score } F = 2 \times \frac{P \times Sn}{P + Sn} \tag{4}$$

$$\text{Precision (P)} = \frac{TP}{TP + FP} \tag{5}$$

where *TP* is a true positive prediction, *TN* is true negative, *FP* is a false positive, and *TN* is true negative prediction. The receiver operating characteristics (ROC) area under the curve (AUC), which offers an overall assessment of the model at all classification levels, was also evaluated.

2.5. Data and Study Area

Data used in this project are from nationally available datasets available from a variety of federal government organizations. All data used in this project is publicly accessible and can be found at open.canada.ca (accessed on 10 September 2021) or climate.weather.gc.ca/ (accessed on 20 August 2021). Five sites were selected across Canada, all with a history of flooding and which experienced flooding in the past ten years. The sites include southern British Columbia (BC), which experienced flooding due to atmospheric river in 2021, and flooding south of Lake Athabasca in northern Alberta (AB) due to heavy rainfall. Southern Manitoba (MB), Ontario (ON), and New Brunswick (NB) all have flood events due to the spring freshet along the Red River, Ottawa River, and Saint John River, respectively.

3. Results

In this section, results are presented for (i) single and multi-region model approaches, (ii) single model results and ensemble model across Canada, and (iii) comparison of FS prediction to historic flood events database.

3.1. Single and Multi-Region Model

Results from parallel Random Forest model (parRF) are shown in Table 2 for the multi-region and national approaches using local, regionally important variables and a compiled national set of important variables. For most of the measures in each of the study areas, the results are nearly identical between the local, self-selected factors and the national list of factors. The ON region is the exception, where an increase in model performance was found in all measures using the national list of important factors. Notably, in ON, there was an increase in accuracy from 0.89 to 0.92 in overall accuracy and increase by 0.07 in kappa and specificity when the national list of factors was used. In BC, there is an increase of 0.01 in specificity and decrease of 0.02 in both sensitivity and F1 score between the national and local list of factors. The average of the regional models found a slight increase in accuracy, kappa, and specificity when the national factor list was used.

Table 2. Metrics of the individual models and national model, parallel random forest results shown.

	Regional Models (parRF) Local Variables/National Variables						Single Model
	BC	**AB**	**MB**	**ON**	**NB**	**Average**	**National Model**
Accuracy	0.96/0.96	0.94/0.94	0.82/0.82	0.89/0.92	0.99/0.99	0.91/0.93	0.92
Kappa	0.93/0.93	0.88/0.88	0.64/0.64	0.77/0.84	0.97/0.98	0.84/0.85	0.83
Sensitivity	0.95/0.93	0.91/0.91	0.79/0.79	0.91/0.92	0.98/0.98	0.91/0.91	0.91
Specificity	0.98/0.99	0.97/0.97	0.85/0.85	0.85/0.92	0.99/1.0	0.93/0.95	0.9
Precision	0.98/0.98	0.98/0.97	0.81/0.81	0.89/0.92	0.99/1.0	0.91/0.90	0.9
F1	0.95/0.93	0.91/0.91	0.79/0.79	0.91/0.92	0.98/0.98	0.91/0.91	0.91
AUC-ROC	0.97/0.97	0.96/0.96	0.86/0.86	0.92/0.93	0.99/0.99	0.94/0.94	0.97

3.2. Single Model vs. Ensemble over Large Distance

The National model approach was run on all the models listed in Table 1. The four RF models all perform well, with accuracy of 0.91 or 0.92, and ROC-AUC of 0.96–0.97, respectively. Similar results are found with the two boosting models, xgbDART and xgbTree, and the C5.0 decision tree model. The poorest results come from the mlp model, with an accuracy of 0.76, kappa of 0.52, and ROC-AUC 0.85. The MARS earth model has an accuracy of 0.82 and kappa of 0.64, putting it at the lower end of the performance scale. The results of the models produce results in the range of 0 to 1, where 0 is no flooding and 1 is flooding/wet pixels, and are multiplied by 100 to avoid storing float values in the final dataset.

To compare how the different ML models performed across the country, in areas distinct from the training sites, the national approach was applied to several ML models to evaluate the resultant map, Figure 2. As a reference, the extent of historic flooding for which there is a digital record, is shown in Figure 2a. The RF and earth models present, at a national scale, are somewhat similar susceptibility maps, though the earth model has higher predictions, especially in Nunavut (NU), Northwest Territories (NT), into Saskatchewan (SK) and Manitoba (MB). The earth model computes very high susceptibility values along the western shore of Hudson Bay. This is not found in the other models, nor is it present in the historic record. The RF model, in northern NU, and NT computes a relatively stable prediction without much variation between the islands nor along the shorelines in NU. There are a few spots in western Yukon (YT), which have higher predictions, and this corresponds to the location of meteorological stations. All models capture higher susceptibility to flooding along the southern borders of SK and MB, which aligns with the historic record. The nnet model appears to do well with predictions where the training data exists and provides relatively low susceptibility values in training-sparse regions: northern Quebec (QC), NU, and around the Canadian Rockies have very low susceptibility values. The svmRadial model shows peculiar 'rings' with higher predictions at the outer edges, covering most of NU and eastern NT. These areas may be more diverse from the labelled data than any others, and it is clear the svmRadial model had challenges.

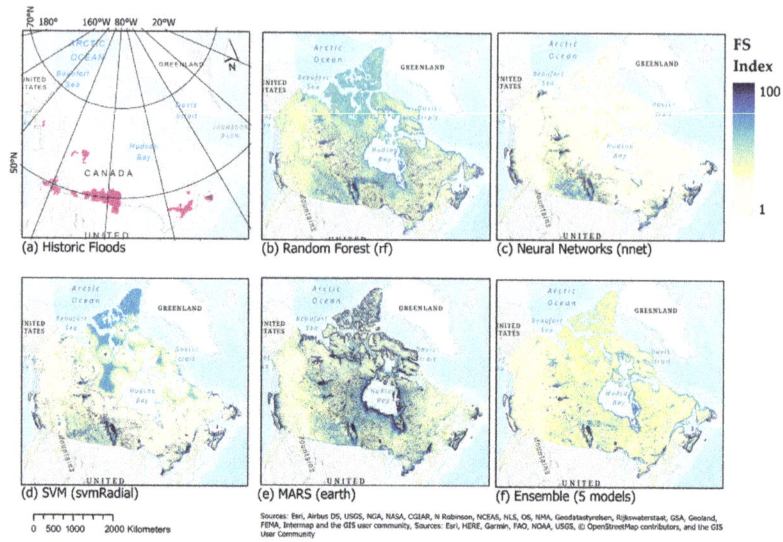

Figure 2. (**a**) Historic flooding in Canada (from EGS Flood Archive and national flood hazard data layer), and results of different single ML models using national approach, (**b–f**) ensemble result.

4. Conclusions

Developing a FS map across a nation as geographically large and diverse as Canada, presents several challenges. In this work, ML algorithms and publicly available national datasets were included to map FS and identify regions more prone to flooding. Testing if a single national model outperformed a regional multi-region model mosaic found that the single national model produced better predictions. However, when a single ML model was extrapolated across the whole of Canada, there were limitations found in several models, including SVM, NN, MLP, and RF. An ensemble approach ultimately produced the best FS map, in comparison to historic flood maps, even though the statistics from the confusion matrix found the ensemble was not the best performer with accuracy and ROC-AUC of 0.89 and kappa 0.78. The resultant dataset provides the first continuous, national picture of flood susceptibility in Canada, with the intended use to support identification and priority setting of flood hazard mapping project and for flood awareness communication.

Author Contributions: H.M.: conceptualization; data curation; formal analysis; investigation; methodology; project administration; resources; software; supervision; validation; visualization; roles/writing—original draft; writing—review and editing. P.N.G.: data curation; writing—review. All authors have read and agreed to the published version of the manuscript.

Funding: This research received no external funding.

Institutional Review Board Statement: Not applicable.

Informed Consent Statement: Not applicable.

Data Availability Statement: Data is available: https://open.canada.ca/data/en/dataset/df106e11-4cee-425d-bd38-7e51ac674128 (accessed on 15 June 2022).

Acknowledgments: Natural Resources Canada Contribution Number: 20220100.

Conflicts of Interest: The authors declare no conflict of interest.

References

1. Canada, P.S. Floods. Available online: https://www.publicsafety.gc.ca/cnt/mrgnc-mngmnt/ntrl-hzrds/fld-en.aspx (accessed on 7 November 2022).
2. Government of Canada, N.R.C. GEOSCAN Search Results: Fastlink. Available online: https://geoscan.nrcan.gc.ca/starweb/geoscan/servlet.starweb?path=geoscan/fulle.web&search1=R=308128 (accessed on 4 May 2022).
3. Collins, E.L.; Sanchez, G.M.; Terando, A.; Stillwell, C.C.; Mitasova, H.; Sebastian, A.; Meentemeyer, R.K. Predicting Flood Damage Probability across the Conterminous United States. *Environ. Res. Lett.* **2022**, *17*, 034006. [CrossRef]
4. Dodangeh, E.; Choubin, B.; Eigdir, A.N.; Panahi, M.; Shamshirband, S.; Mosavi, A. Integrated Machine Learning Methods with Resampling Algorithms for Flood Susceptibility Prediction-ScienceDirect. Available online: https://www.sciencedirect.com/science/article/pii/S0048969719359789 (accessed on 13 May 2022).
5. Shafizadeh-Moghadam, H.; Valavi, R.; Shahabi, H.; Chapi, K.; Shirzadi, A. Novel Forecasting Approaches Using Combination of Machine Learning and Statistical Models for Flood Susceptibility Mapping. *J. Environ. Manag.* **2018**, *217*, 1–11. [CrossRef] [PubMed]
6. Zhao, G.; Pang, B.; Xu, Z.; Tu, T. Mapping Flood Susceptibility in Mountainous Areas on a National Scale in China-ScienceDirect. Available online: https://www.sciencedirect.com/science/article/pii/S0048969717327419 (accessed on 13 May 2022).
7. Fernández-Delgado, M.; Cernadas, E.; Barro, S.; Amorim, D. Do We Need Hundreds of Classifiers to Solve Real World Classification Problems? *J. Mach. Learn. Res.* **2014**, *15*, 3133–3181.
8. Kuhn, M. Building Predictive Models in R Using the Caret Package. *J. Stat. Softw.* **2008**, *28*, 1–26. [CrossRef]
9. Chen, W.; Li, Y.; Xue, W.; Shahabi, H.; Li, S.; Hong, H.; Wang, X.; Bian, H.; Zhang, S.; Pradhan, B.; et al. Modeling Flood Susceptibility Using Data-Driven Approaches of Naïve Bayes Tree, Alternating Decision Tree, and Random Forest Methods. *Sci. Total Environ.* **2020**, *701*, 134979. [CrossRef] [PubMed]
10. Lee, S.; Kim, J.-C.; Jung, H.-S.; Lee, M.J.; Lee, S. Spatial Prediction of Flood Susceptibility Using Random-Forest and Boosted-Tree Models in Seoul Metropolitan City, Korea. *Null* **2017**, *8*, 1185–1203. [CrossRef]
11. Vafakhah, M.; Mohammad Hasani Loor, S.; Pourghasemi, H.; Katebikord, A. Comparing Performance of Random Forest and Adaptive Neuro-Fuzzy Inference System Data Mining Models for Flood Susceptibility Mapping. *Arab. J. Geosci.* **2020**, *13*, 417. [CrossRef]
12. Youssef, A.M.; Pourghasemi, H.R.; El-Haddad, B.A. *Advanced Machine Learning Algorithms for Flood Susceptibility Modeling-Comparison of Their Performance*; Safaga-Ras Gharib Area: Red Sea, Egypt, 2022.
13. Ardabili, S.; Mosavi, A.; Várkonyi-Kóczy, A.R. Advances in Machine Learning Modeling Reviewing Hybrid and Ensemble Methods. In Proceedings of the Engineering for Sustainable Future; Várkonyi-Kóczy, A.R., Ed.; Springer International Publishing: Cham, Switzerland, 2020; pp. 215–227.
14. Kazienko, P.; Lughofer, E.; Trawiński, B. Hybrid and Ensemble Methods in Machine Learning J. UCS Special Issue. *J. Univers. Comput. Sci.* **2013**, *19*, 457–461.
15. Sikora, R.; Al-Laymoun, O. A Modified Stacking Ensemble Machine Learning Algorithm Using Genetic Algorithms. Available online: https://www.igi-global.com/chapter/a-modified-stacking-ensemble-machine-learning-algorithm-using-genetic-algorithms/www.igi-global.com/chapter/a-modified-stacking-ensemble-machine-learning-algorithm-using-genetic-algorithms/122748 (accessed on 21 April 2022).
16. Agriculture and Agri-Food Canada. Terrestrial Ecoregions of Canada-Open Government Portal. Available online: https://open.canada.ca/data/en/dataset/ade80d26-61f5-439e-8966-73b352811fe6 (accessed on 9 May 2022).
17. Natural Resources Canada. Physiographic Regions-Open Government Portal. Available online: https://open.canada.ca/data/en/dataset/028dd58d-320c-53fb-b5bc-8188fd5d5edf (accessed on 9 May 2022).

environmental sciences proceedings

MDPI

Proceeding Paper

An Insight into the Next-Generation Smart Membranes †

Mert Yildirim [1,2,3,*] and Zeki Candan [1,3]

1 Department of Forest Industrial Engineering, Istanbul University-Cerrahpasa, 34473 Istanbul, Turkey; zekic@istanbul.edu.tr
2 Scientific Research Projects Coordinatorship, Istanbul Gelisim University, 34320 Istanbul, Turkey
3 Biomaterials and Nanotechnology Research Group, Istanbul University-Cerrahpasa, 34473 Istanbul, Turkey
* Correspondence: yildirimmert1993@gmail.com

† Presented at the 7th International Electronic Conference on Water Sciences, 15–30 March 2023; Available online: https://ecws-7.sciforum.net/.

Abstract: Membranes are used in desalination or water treatment to separate pollutants from water based on characteristics such as size or charge. Nanofiltration (NF), ultrafiltration (UF), microfiltration (MF), and reverse osmosis (RO) are typical membrane techniques. However, traditional membranes have a number of disadvantages, including fouling both on surfaces and in internal structures, uncontrollable pore size, and membrane features. Smart membranes, also known as stimuli-responsive membranes, have recently attracted attention due to their selectivity, tunable permeability, and tunable and/or reversible attributes. This new generation of smart membranes is created by integrating various stimuli-responsive materials into membrane substrates. These multi-functional smart membranes can self-adjust their physical and chemical features in response to environmental signals such as temperature, pH, light, and other stimuli. Thermo-responsive membranes, pH-responsive membranes, ion-responsive membranes, molecule-responsive membranes, UV-light-responsive membranes, glucose-responsive membranes, magnetic-responsive membranes, and redox-responsive membranes are the current kinds of smart membranes. Because of their smart structures, they have the potential to improve performance by providing high selectivity without reducing permeability, high mechanical stability, and high resistance against fouling, and can meet requirements such as molecular weight cut-off (MWCO), removal efficiencies, and wastewater quality. Smart membranes can show tunable features based on the condition of the stimulus or stimuli present internally or externally, resulting in improved and desirable controllability over the process of pollutant removal from water. Because of their physicochemical stability, repeatability, and long life, stimuli-responsive smart materials (mainly adsorbents and filtration membranes) have the potential to be key materials for membrane production, particularly in the field of water treatment. Smart membranes have a bright future, and it is important to investigate and encourage their use and advancement. This review provides a comprehensive overview of smart membranes.

Keywords: membranes; smart materials; smart membranes; stimuli-responsive membranes; water treatment

Citation: Yildirim, M.; Candan, Z. An Insight into the Next-Generation Smart Membranes. *Environ. Sci. Proc.* **2023**, *25*, 19. https://doi.org/10.3390/ECWS-7-14256

Academic Editor: Athanasios Loukas

Published: 16 March 2023

1. Introduction

Membranes are used in desalination or water treatment to separate pollutants from water based on characteristics such as size or charge. Since the late 1950s, reverse osmosis, nanofiltration, ultrafiltration, and microfiltration techniques have been used in water and wastewater treatment and in different application areas. Rapid developments in membrane technologies in the last 50 years have made these technologies the preferred technologies in water and wastewater treatment. However, traditional membranes have a number of disadvantages, including fouling both on surfaces and in internal structures, uncontrollable pore size, and membrane features.

Developments in membrane production technology have also increased the use of membrane processes in many industries, such as chemistry, petrochemistry, mineral processing, food biotechnology, pharmacy, electronics, paper, etc. Although membrane filters are used as an alternative water recovery process in many areas, clogging is still one of the biggest problems. Clogging in membranes limits the membrane's permeability [1]. In other words, it causes a decrease in the flux passing through the membrane per unit membrane pressure and, therefore, in the production of treated clean water per unit membrane area. In membranes designed for particulate matter or microbial removal, clogging occurs as a result of the accumulation of materials on the membrane surface or in the membrane pores.

Smart membranes, also known as stimuli-responsive membranes, have recently attracted attention due to their selectivity, tunable permeability, and tunable and/or reversible attributes [2]. This new generation of smart membranes is created by integrating various stimuli-responsive materials into membrane substrates. These multi-functional smart membranes can self-adjust their physical and chemical features in response to environmental signals such as temperature, pH, light, and other stimuli [3].

Because of their smart structures, they have the potential to improve performance by providing high selectivity without reducing the permeability, high mechanical stability, and high resistance against fouling, and can meet requirements such as molecular weight cut-off (MWCO), removal efficiencies, and wastewater quality.

This review of smart membranes is briefly summarized.

2. Kinds of Smart Membranes

Positively and negatively responsive smart membranes can self-adjust their physical and chemical properties in response to environmental signals such as temperature, pH, light, and other stimuli.

The responsive gating function is divided into two models: positively and negatively responsive smart membranes [4].

Figure 1 represents positively responsive smart membranes.

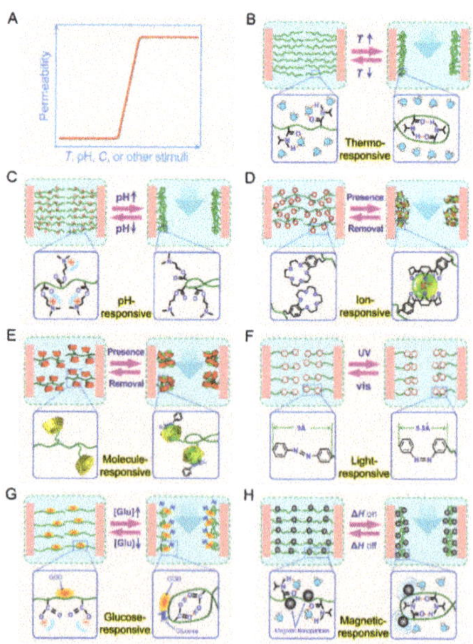

Figure 1. Positively responsive smart membranes (**A**). The permeability of the membrane increases in response to the presence or increase of a stimulus. (**B**) Temperature-responsive, (**C**) pH-responsive, (**D**) specific ion-responsive, (**E**) molecule-responsive, (**F**) UV light-responsive, (**G**) glucose-responsive, and (**H**) magnetic-responsive [4].

Figure 2 represents negatively responsive smart membranes.

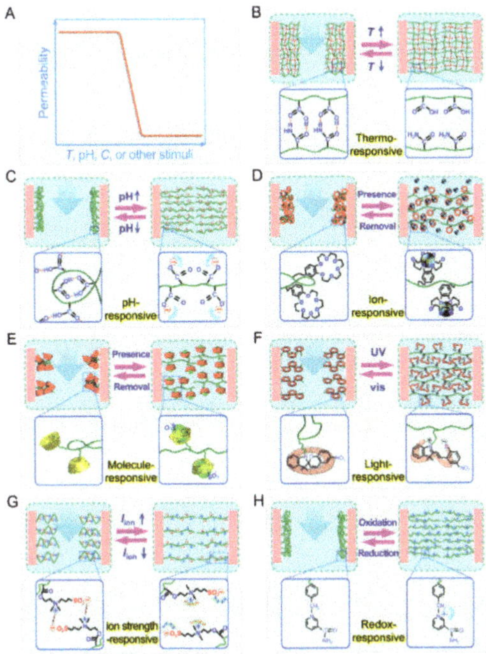

Figure 2. Negatively responsive smart membranes. (**A**) The permeability of the membrane decreases in response to the presence or increase of a stimulus. (**B**) Thermo-responsive, (**C**) pH-responsive, (**D**) ion-responsive, (**E**) molecule-responsive, (**F**) UV light-responsive, (**G**) ion strength-responsive, and (**H**) redox-responsive [4].

3. Conclusions

Membranes play significant roles in sustainable development, especially Goal 6: clean water and sanitation. It is necessary to consider the developments in membrane technologies in terms of new generation membrane production, module development, and related application areas of technology. Smart membranes can show tunable features based on the condition of the stimulus or stimuli present internally or externally, resulting in improved and desirable controllability over the process of pollutant removal from water. Because of their physicochemical stability, repeatability, and long life, stimuli-responsive smart materials (mainly adsorbents and filtration membranes) have the potential to be key materials for membrane production, particularly in the field of water treatment. Although they have advantages, current smart membranes suffer from complicated and difficult-to-scale-up production processes, low flux, and weak mechanical properties. Smart membranes have a bright future, and it is important to investigate and encourage their development, use, and advancement.

Author Contributions: M.Y. and Z.C. contributed equally to this work. All authors have read and agreed to the published version of the manuscript.

Funding: This research received no external funding.

Institutional Review Board Statement: Not applicable.

Informed Consent Statement: Not applicable.

Data Availability Statement: Data and intellectual property belong to the Istanbul University-Cerrahpasa and Istanbul Gelisim University; any sharing needs to be evaluated and approved by the universities.

Acknowledgments: The authors would like to thank the Turkish Academy of Sciences for its support.

Conflicts of Interest: The authors declare no conflict of interest.

References

1. Bandehali, S.; Parvizian, F.; Hosseini, S.M.; Matsuura, T.; Drioli, E.; Shen, J.; Moghadassi, A.; Adeleye, A.S. Planning of smart gating membranes for water treatment. *Chemosphere* **2021**, *283*, 131207. [CrossRef] [PubMed]
2. Tufani, A.; Ozaydin Ince, G. Smart membranes with pH-responsive control of macromolecule permeability. *J. Membr. Sci.* **2017**, *537*, 255–262. [CrossRef]
3. Zou, L.B.; Gong, J.Y.; Ju, X.C.; Liu, Z.; Wang, W.; Xie, R.; Chu, L.Y. Smart membranes for biomedical applications. *Chin. J. Chem. Eng.* **2022**, *49*, 34–45. [CrossRef]
4. Liu, Z.; Wang, W.; Xie, R.; Ju, X.J.; Chu, L.Y. Stimuli responsive smart gating membranes. *Chem. Soc. Rev.* **2016**, *45*, 460–475. [CrossRef] [PubMed]

environmental sciences proceedings

MDPI

Proceeding Paper

Crop Water Stress Detection Using Remote Sensing Techniques †

Muhammad Safdar [1,2,*], Muhammad Adnan Shahid [1,2,*], Abid Sarwar [1], Fahd Rasul [3], Muhammad Danish Majeed [1,2] and Rehan Mehmood Sabir [1,2]

1 Department of Irrigation & Drainage, University of Agriculture, Faisalabad 38000, Punjab, Pakistan; abidsarwar@uaf.edu.pk (A.S.); kambohdanish@gmail.com (M.D.M.); r.m.sabir222@gmail.com (R.M.S.)
2 Agricultural Remote Sensing Lab (ARSL), University of Agriculture, Faisalabad 38000, Punjab, Pakistan
3 Department of Agronomy, University of Agriculture, Faisalabad 38000, Punjab, Pakistan; drfahdrasul@uaf.edu.pk
* Correspondence: safdarsani4340@gmail.com (M.S.); adnan.wmrc@gmail.com (M.A.S.)
† Presented at the 7th International Electronic Conference on Water Sciences, 15–30 March 2023; Available online: https://ecws-7.sciforum.net.

Abstract: To meet the demand for increasing global food production while using limited water resources, crop water stress must be improved in agriculture. Remote-sensing-based plant stress indicators have the benefits of high spatial resolutions, a cheap cost, and short turnaround times. This study discusses the current advancements in agricultural water stress monitoring and irrigation scheduling, some of the challenges that have been met, and the upcoming research needs. Remote sensing systems are prepared to handle the intricate and technical evaluations of agricultural productivity, security, and crop water stress quickly and effectively. We explore the use of remote-sensing systems in the evaluation of crop water stress by looking at the existing research, technologies, and data. This study examines the connection between relative water content (RWC), equivalent water thickness (EWT), and agricultural water stress. Using remote sensing, evapotranspiration, and sun-induced chlorophyll content are examined in connection to crop drought. Spectral indices, remote sensing satellites, and multi-spectral sensing systems, as well as systems that measure land surface temperature, are examined. This critical study focuses on cutting-edge techniques for assessing crop water stress.

Keywords: crop water stress; spectral indices; multi-spectral; remote sensing satellites; thermometric sensing

check for updates

Citation: Safdar, M.; Shahid, M.A.; Sarwar, A.; Rasul, F.; Majeed, M.D.; Sabir, R.M. Crop Water Stress Detection Using Remote Sensing Techniques. *Environ. Sci. Proc.* **2023**, *25*, 20. https://doi.org/10.3390/ECWS-7-14198

Academic Editor: Athanasios Loukas

Published: 14 March 2023

1. Introduction

Arid regions have discovered creative solutions to meet their crop needs, based on their growth phases, kinds, and environmental circumstances, which has led to appreciable yield improvements. A deficiency of irrigation water will cause agricultural water stress at various times throughout the crop cycle, under various environmental conditions. Its primary impact is felt in the rate of photosynthesis, which further causes a disturbance in the rates of transpiration [1,2].

Remote sensing collects information from crops, soil, and ambient elements without direct physical contact [3]. Through the quick identification of crop growth changes that are frequently missed by conventional approaches, it has improved and optimized agricultural production [4]. A highly accurate determination of the crop temperature is made possible by remote sensing systems, which also provide particular information that is important in the study of irrigation scheduling, quantity, and duration [5]. Remote sensing systems can be divided into sensor-based and platform-based systems. In total, two types of sensors may record the reflectivity inside the electromagnetic (EM) spectrum: active sensors and passive sensors. The sensor is mounted onto a variety of remote sensing platforms, including ground vehicles, aircraft, satellites, and handheld devices [6].

Precision irrigation scheduling requires an assessment of crop water stress, which is one of the elements that characterize how a crop interacts with its environment [7]. The

CWS came to be recognized as a common indicator for evaluating this stress on the leaf and canopy scales. This was a more accurate technique to examine the water stress at the plot, regional, and global stages, as well as evapotranspiration. Implementing effective irrigation scheduling techniques is crucial to increase water savings and improve agricultural sustainability [8]. Remote sensing data can reveal information on the geographical and temporal variations of crops [9,10]. Precision agriculture uses spectral reflectance indices from high-resolution hyperspectral sensors on small, unmanned aircraft systems to monitor the crop water status and plan irrigation [11].

An assessment of the crop water deficit using remote sensing devices is the subject of this review. The paper supplies an overview of the many remote sensing systems that can be used to find crop water stress. Optical, thermometric, land-surface temperature, multispectral (spaceborne and airborne), hyperspectral, and LiDAR sensing systems are examined. A consensus about the use of vegetation indices (VIs) as pre-visual indicators of water stress has not yet been reached, due to several confounding factors that affect these VIs on the canopy and landscape scales. This research discusses the current developments in crop water stress monitoring that may be applied to enhance vegetable crop irrigation scheduling and seeks to figure out the most promising method for widespread implementation. To forecast the production conditions and schedule irrigation, the crop water stress needs to be detected during the various growing seasons. Distinguishing this agricultural water stress has been researched using several methodologies. These techniques rely on remote sensing, measurements of the soil water content, and plant responses. The study also considers the fact that different approaches are effectively used for different crops.

2. Comparison of Crop Water Stress Detection Methods

Table 1 provides an overview of various methods used for soil moisture measurement, including the gravimetric method, time domain reflectometer (TDR), neutron probe method, tensiometer method, vegetation indices method by remote sensing (VIs), water indices by remote sensing, water balance indices, remote sensing-based ET estimation by energy balance, CWSI by the infrared thermometer, and LST based CWSI. The table includes a brief description of each method, its advantages, disadvantages, and references. The methods vary in their precision, ease of use, cost, and sensitivity to different soil types and environmental conditions. Some methods require direct contact with soil, while others utilize remote sensing techniques. The choice of method depends on the specific research or application requirements.

Table 1. Comparison of crop water stress detection methods.

Methods	Description	Advantages	Disadvantages	References
Gravimetric Method	A straightforward technique that involves weighing a wet sample, drying it in an oven, reweighing it, and then estimating the amount of water loss as a percentage of the dry soil quantity	Highly precise and reliable technique with hardly any room for instrumental error, not affected by salinity or soil type	Time-consuming, dependent on mass measurements, destructive, and labor-intensive	[12,13]
Time domain reflectometer (TDR)	An electromagnetic method based on the idea that water and other materials, such as soil, have different dielectric constants	Less time-consuming and damaging than gravimetric techniques, reduced labor expenses	Environmentally sensitive, expensive equipment, and calibration dependent on soil texture	[12,13]
Neutron Probe method	Evaluates the soil's volumetric water content	High accuracy, permits observations at various depths, rather simple	Time-consuming monitoring and expensive equipment licensing are required.	[14]
Tensiometer method	Soil-water-potential-based	Cheap, affordable, easy to install, accurate, and for irrigation scheduling	Requires contact with soil and destructive	[14]
Vegetation indices method by remote sensing (VIs)	Indicators of vegetation are used to illustrate its properties	The high temporal and spectral resolution, non-destructive	Precision decreases from leaf scale to canopy scale and image analysis is a difficult task	[15,16]

Table 1. *Cont.*

Methods	Description	Advantages	Disadvantages	References
Water Indices by remote sensing	Determines the reflectance in the SWIR and near-infrared range, which is used to indicate the water content of the canopy. Typical indices include WI, SRWI, NDWI, and MSI	Leaf water content may be measured without causing damage. Excellent direct signs of water stress	The difficulty of ascending to the canopy level	[17]
Water balance indices	Monitors change in the chlorophyll fluorescence and water content of the leaves using the green and SWIR spectral bands. The calculated indices are WABI, WABI-1, and WABI-2	Exhibited excellent performance at the leaf and canopy levels	It is necessary to use an expensive single-spectrum instrument. The penetrability of the SWIR band through heavy atmospheric layers is a problem	[17]
RS-based ET estimation by Energy balance	The surface energy balance equation LE = Rn-G-H Latent Energy includes ET as a residual (LE), Rn = Net Sky Radiation, G = Ground to Air, H = Heat to Air	A single thermal band with the excellent resolution is sufficient and needed METRIC and SEBAL have good consistency and accuracy	It's challenging to determine whether ET is possible. As ET cannot be directly measured, high-resolution thermal imaging is crucial.	[18]
CWSI by infrared thermometer	The canopy temperature and its decrease with the ambient air temperature are used to calculate CWSI	Depends on the direct technique and VPD	Different baselines must be calculated for various crops; this takes time. To evaluate CWSI, many factors must be considered	[19]
LST based CWSI	Utilizing LST and the hot-and-cold pixels approach to calculate CWSI	Using only remote sensing methods Work and time are non-intensive	Depending on this method to calculate LST, LST computation is laborious and varies	[20]

3. Satellite-Based Crop Water Stress Detection

Table 2 provides information on various satellite applications and their advantages and limitations. It includes information on the type of satellite, its applications, the advantages of using it, and any limitations associated with it. Some examples of the satellites included in the table are AMSR-E, AMSR-2, NISAR, Tandem-L, Sentinel-1, and SMAP. The applications of these satellites range from analyzing soil moisture to vegetation status and dynamics observation. The advantages of using these satellites include high precision, excellent resolution, and data collection in all weather conditions, among others. However, some of the limitations include limited frequency ranges, high cost, and limited precision in field determination.

Table 2. Satellite-based crop water stress detection.

Satellite	Applications	Advantages	Limitations	References
AMSR-E	High-efficiency passive microwave soil moisture analysis with drought	Data collection for daily soil moisture measurement with a 12.5 km precision	Just two files every day, one for the day and one for the night	[21]
AMSR-2	Analysis of soil-water-related parameters and global observation of soil moisture (from the soil surface to a few centimeters depth)	More than 99% correct in capturing data both during the day and at night/good resolution and accuracy of data collecting	Only functions in certain frequency ranges, including 6.925, 7.3, 10.65, 18.7, 23.8, 36.5, and 89.0 GHz	[22]
NISAR	Global soil moisture maps with a time horizon of 6 to 12 days	Acquires soil moisture data in all weather conditions and with a precise resolution of 3–10 m	Product assessment in 12–24 h	[23]
Tandem-L	Worldwide soil moisture	Provides extremely accurate measured data with millimeter-level accuracy and excellent resolution between 20 m and 4 km	A significant premium over conventional satellite systems	[24]
Sentinel-1	Dynamics observation	With a precision resolution of 5 to 20 m, field determination is less precise	Easy to create new systems, incorporating sensor structures and application development models	[25]
SMAP	Analyzes the vegetation status and soil surface	High likelihood of mission failure with a 9 km precise resolution	SSM is captured by passive sensors for roughly 36 km	[26]

4. Crop Water Stress Detection Using Spectral Indices

Table 3 below lists several reflectance indices used to indicate plant stress and their respective formulas. Reflectance indices are measures of the amount of light reflected from vegetation at specific wavelengths and can be used to estimate plant health and stress. The plant stress indicators listed in the table are associated with different physiological processes related to plant water statuses, such as stomatal conductance, chlorophyll fluorescence, leaf water potential, and water content. The references listed provide additional information on the use and interpretation of each index.

Table 3. Crop water stress detection using spectral indices.

Reflectance Indices	Formula	Plant Stress Indicators	References
Photochemical Reflectance Index (PRI)	$\frac{R_{570}-R_{531}}{R_{570}+R_{531}}$	Stomatal conductance and chlorophyll fluorescence	[27]
Normalized Photochemical Reflectance Index (NPRI)	$\frac{PRI}{RDI*\frac{R_{700}}{R_{670}}}$	Stomatal conductance and chlorophyll fluorescence	[28]
Normalized Difference Vegetation Index (NDVI)	$\frac{R_{800}-R_{670}}{R_{800}+R_{670}}$	Leaf water potential and stomatal conductance	[29]
Renormalized Difference Vegetation Index (RDVI)	$\frac{R_{800}-R_{670}}{\sqrt{R_{800}+R_{670}}}$	Leaf water potential and stomatal conductance	[30]
Transformed Chlorophyll Absorption in Reflectance Index (TCARI)	$3[(R_{700}-R_{670})-0.2(R_{700}-R_{550})*\left(\frac{R_{700}}{R_{670}}\right)]$	Leaf water potential and stomatal conductance	[31]
Optimized Soil Adjusted Vegetation Index (OSAVI)	$\frac{(1+0.16)(R_{700}-R_{550})}{(R_{800}+R_{670})+0.16}$	Leaf water potential and stomatal conductance	[31]
Normalized Difference Water Index (NDWI)	$\frac{R_{860}-R_{1240}}{R_{860}+R_{1240}}$	Leaf water potential	[32]
Simple Ratio Water Index (SRWI)	$\frac{R_{860}}{R_{1240}}$	Leaf water potential	[33]
Water Index (WI)	$\frac{R_{860}}{R_{1240}}$	Leaf water potential	[33]

5. Crop Water Stress Detection Using Multispectral Sensing Systems

Table 4 provides information on different multispectral sensing systems, their descriptions, advantages, and references. The first system listed is a UAV remote multispectral sensing system called AIRPHEN Multispectral Camera, which has a high-resolution camera and precise CWS (crop water stress) detection. It is also low-cost, cheap, effective, and available with RGB color bands. The second system listed is a spaceborne multispectral sensing system that includes Landsat, Orb view, World view, IKONOS, and Quick bird. These systems are used to figure out agricultural water stress by collecting multispectral high-resolution data, which provides entire crop water stress temporal features.

Table 4. Crop water stress detection using multispectral sensing systems.

Multispectral Sensing Systems	Description	Advantages	References
UAV remote MS sensing system	AIRPHEN Multispectral Camera with a lens of 8 mm focal length, 1280 × 960 pixels, and spectral resolution 10 nm	High-resolution camera, precise CWS detection, low cost, cheap, effective, and available with RGB color bands	[34,35]
Spaceborne MS sensing system	Landsat, Orb view, World view, IKONOS, Quick bird SPOT-5	To figure out agricultural water stress, multispectral high-resolution data should be collected. This will give us entire crop water stress temporal features	[36,37]

6. Future Directions

The target water stress can be located using remote sensing technology. For applications including agricultural growth assessment and irrigation, as well as leaf and canopy phenotypic categorizations that detect crop losses, digital imaging technologies are used. Using information from digital photography, the water stress can be measured. The most recent methods for crop water stress assessment that used digital pictures from remote sensing have shown notable results. Most of the studies showed three degrees of agricultural water stress: minimal stress (optimum moisture), medium stress (mild drought stress), and severe stress (drought stress). With an accuracy that ranged from 83 to 99%, these methods produced encouraging findings for the estimation of agricultural water stress. Machine learning is crucial to raising the calibers and effectiveness of these systems. For an accurate evaluation of the crop water stress, a microcontroller-based signal processor (MSP430) integrated soil and ambient sensors. A dependable resource for examining these crop water levels, and soil water stress factors is an independent wireless sensor system that is made up of a gateway plus a wireless sensory node.

7. Conclusions

Traditional methods, such as measuring the soil moisture, have drawbacks in terms of their sensor costs and installations, and difficulty in obtaining estimates. Plant-based estimates are more dependable and accurate. There are significant relationships between the PRI and NDVI, and attributes such as the LWP, stomatal conductance, crop efficiency, and stem water potential. A crop water stress evaluation is a technical and intricate process in and of itself. Our study suggests new techniques that bring together farmers, researchers, and tech developers. Narrow-band optical indices could be used to plan the irrigation for high-value vegetable crops in water-stressed countries. Conventional irrigation scheduling methods use measurements of the soil moisture and weather, and physiological assessments of the plant response. These methods are ineffective because it is difficult to obtain these measurements, especially for varied soil and crop canopies. This assessment makes recommendations for remote sensing systems and sets the path for creating new facilities that assess a system's effectiveness in diverse environmental scenarios, such as multispectral/hyperspectral and thermal sensing systems that are based on remote sensing features.

Author Contributions: Conceptualization, M.S.; Comparison of Crop Water Stress Detection Methods, M.S. and M.A.S.; Satellite-Based Crop Water Stress Detection, A.S.; Crop Water Stress Detection Using Spectral Indices, F.R.; Crop Water Stress Detection Using Multispectral Sensing Systems, M.D.M., M.A.S., and R.M.S.; Future Directions & Conclusions, writing—original draft preparation, M.S.; writing a review and editing, M.S. All authors have read and agreed to the published version of the manuscript.

Funding: This research received no external funding.

Institutional Review Board Statement: Not applicable.

Informed Consent Statement: Not applicable.

Data Availability Statement: Free sources of satellite data are available for the study, but for better accuracy unavailability of data due to non-free availability.

Conflicts of Interest: The authors declare no conflict of interest.

References

1. Chang, Y.N.; Zhu, C.; Jiang, J.; Zhang, H.; Zhu, J.K.; Duan, C.G. Epigenetic regulation in plant abiotic stress responses. *J. Integr. Plant Biol.* **2020**, *62*, 563–580. [CrossRef] [PubMed]
2. Goldstein, A.; Fink, L.; Meitin, A.; Bohadana, S.; Lutenberg, O.; Ravid, G. Applying machine learning on sensor data for irrigation recommendations: Revealing the agronomist's tacit knowledge. *Precis. Agric.* **2017**, *19*, 421–444. [CrossRef]
3. Tian, H.; Wang, T.; Liu, Y.; Qiao, X.; Li, Y. Computer vision technology in agricultural automation—A review. *Inf. Process. Agric.* **2020**, *7*, 1–19. [CrossRef]
4. Aasen, H.; Honkavaara, E.; Lucieer, A.; Zarco-Tejada, P. Quantitative Remote Sensing at Ultra-High Resolution with UAV Spectroscopy: A Review of Sensor Technology, Measurement Procedures, and Data Correction Workflflows. *Remote Sens.* **2018**, *10*, 1091. [CrossRef]
5. Long, D.S.; Engel, R.E.; Siemens, M.C. Measuring Grain Protein Concentration with In-line Near Infrared Reflectance Spectroscopy. *Agron. J.* **2008**, *100*, 247. [CrossRef]

6. Mulyono, S. Nadirah Identifying Sugarcane Plantation using LANDSAT-8 Images with Support Vector Machines. *IOP Conf. Ser. Earth Environ. Sci.* **2016**, *47*, 12008. [CrossRef]
7. Zhou, Z.; Majeed, Y.; Diverres Naranjo, G.; Gambacorta, E.M.T. Assessment for crop water stress with infrared thermal imagery in precision agriculture: A review and prospects for deep learning applications. *Comput. Electron. Agric.* **2021**, *182*, 106019. [CrossRef]
8. Osroosh, Y.; Peters, R.T.; Campbell, C.S.; Zhang, Q. Automatic irrigation scheduling of apple trees using theoretical crop water stress index with an innovative dynamic threshold. *Comput. Electron. Agric.* **2015**, *118*, 193–203. [CrossRef]
9. Dangwal, N.; Patel, N.R.; Kumari, M.; Saha, S. Monitoring of water stress in wheat using multispectral indices derived from Landsat-TM. *Geocarto Int.* **2015**, *31*, 1–26. [CrossRef]
10. Leroux, L.; Baron, C.; Zoungrana, B.; Traoré, S.B.; Seen, D.L.; Bégué, A. Crop monitoring using vegetation and thermal indices for yield estimates: A case study of a rainfed cereal in semi-arid west Africa. *IEEE J. Sel. Top. Appl. Earth Obs. Remote Sens.* **2016**, *9*, 347–362. [CrossRef]
11. Gago, J.; Douthe, C.; Coopman, R.E.; Gallego, P.P.; Ribas-Carbo, M.; Flexas, J.; Escalona, J.; Medrano, H. UAVs challenge to assess water stress for sustainable agriculture. *Agric. Water Manag.* **2015**, *153*, 9–19. [CrossRef]
12. Sharma, P.K.; Kumar, D.; Srivastava, H.S.; Patel, P. Assessment of different methods for soil moisture estimation: A review. *J. Remote Sens. GIS* **2018**, *9*, 57–73.
13. Tanriverdi, C.; Degirmenci, H.; Gonen, E.; Boyaci, S. A comparison of the gravimetric and TDR methods in terms of determining the soil water content of the corn plant. *Sci. Pap. Ser. A-Agron.* **2016**, *59*, 153–158.
14. Enciso, J.; Porter, D.; Peries, X. *Irrigation Monitoring with Soil Water Sensors (Spanish)*; Texas FARMER Collection; Texas A&M University: College Station, TX, USA, 2007.
15. Romero, M.; Luo, Y.; Su, B.; Fuentes, S. Vineyard water status estimation using multispectral imagery from a UAV platform and machine learning algorithms for irrigation scheduling management. *Comput. Electron. Agric.* **2018**, *147*, 109–117. [CrossRef]
16. Poblete, T.; Ortega-Farías, S.; Moreno, M.A.; Bardeen, M. Artificial Neural Network to Predict Vine Water Status Spatial Variability Using Multispectral Information Obtained from an Unmanned Aerial Vehicle (UAV). *Sensors* **2017**, *17*, 2488. [CrossRef]
17. Rapaport, T.; Hochberg, U.; Shoshany, M.; Karnieli, A.; Rachmilevitch, S. Combining leaf physiology, hyperspectral imaging and partial least squares-regression (PLS-R) for grapevine water status assessment. *ISPRS J. Photogramm. Remote. Sens.* **2015**, *109*, 88–97. [CrossRef]
18. Allen, R.G.; Tasumi, M.; Trezza, R. Satellite-Based Energy Balance for Mapping Evapotranspiration with Internalized Calibration (METRIC)—Model. *J. Irrig. Drain. Eng.* **2007**, *133*, 380–394. [CrossRef]
19. Jackson, R.D.; Idso, S.B.; Reginato, R.J.; Pinter, P.J., Jr. Canopy Temperature as a Crop Water Stress Indicator. *Water Resour. Res.* **1981**, *17*, 1133–1138. [CrossRef]
20. Veysi, S.; Naseri, A.A.; Hamzeh, S.; Bartholomeus, H. A satellite based crop water stress index for irrigation scheduling in sugarcane fields. *Agric. Water Manag.* **2017**, *189*, 70–86. [CrossRef]
21. Kolassa, J.; Gentine, P.; Prigent, C.; Aires, F. Soil moisture retrieval from AMSR-E and ASCAT microwave observation synergy. Part 1: Satellite data analysis. *Remote Sens. Environ.* **2016**, *173*, 1–14. [CrossRef]
22. Kaihotsu, I.; Asanuma, J.; Aida, K. Evaluation of the AMSR2 L2 soil moisture product of JAXA on the Mongolian Plateau over seven years (2012–2018). *SN Appl. Sci.* **2019**, *1*, 1477. [CrossRef]
23. NISAR: The NASA-ISRO SAR Mission. Water: Vital for Life and Civilization. © 2019 California Institute of Technology. Government Sponsorship Acknowledged. Available online: https://nisar.jpl.nasa.gov/system/documents/fifiles/15_NISARApplications_SoilMoisture1.pdf (accessed on 16 July 2021).
24. Tandem-L: A Satellite Mission for Monitoring Dynamic Processes on the Earth's Surface. Available online: https://www.researchgate.net/publication/225007272_Tandem-L_A_Satellite_Mission_for_Monitoring_Dynamic_Processes_on_the_Earth\T1\textquoterights_Surface (accessed on 30 April 2014).
25. Harm-Jan, F.B.; van der Velde, R.; Su, Z. Sentinel-1 soil moisture content and its uncertainty over sparsely vegetated fields. *J. Hydrol. X* **2020**, *9*, 100066.
26. Abbaszadeh, P.; Moradkhani, H.; Gavahi, K.; Kumar, S.; Hain, C.; Zhan, X.; Duan, Q.; Peters-Lidard, C.; Karimiziarani, S. High-Resolution SMAP Satellite Soil Moisture Product: Exploring the Opportunities. *Bull. Am. Meteorol. Soc.* **2021**, *102*, 4–309. [CrossRef]
27. Gamon, J.; Penuelas, J.; Field, C. A narrow-waveband spectral index that tracks diurnal changes in photosynthetic efficiency. *Remote Sens. Environ.* **1992**, *41*, 35–44. [CrossRef]
28. Berni, J.A.J.; Zarco-Tejada, P.J.; Sepulcre-Cantó, G.; Fereres, E.; Villalobos, F. Mapping canopy conductance and CWSI in olive orchards using high resolution thermal remote sensing imagery. *Remote Sens. Environ.* **2009**, *113*, 2380–2388. [CrossRef]
29. Rouse, J.W.; Haas, R.H.; Schell, J.A.; Deering, D.W. Monitoring vegetation systems in the Great Plains with, E.R.T.S. In Proceedings of the Third ERTS-1 Symposium, Washington, DC, USA, 10–14 December 1973; pp. 309–317.
30. Roujean, J.-L.; Breon, F.-M. Estimating PAR absorbed by vegetation from bidirectional reflectance measurements. *Remote Sens. Environ.* **1995**, *51*, 375–384. [CrossRef]
31. Haboudane, D.; Miller, J.R.; Tremblay, N.; Zarco-Tejada, P.J.; Dextraze, L. Integrated narrow-band vegetation indices for prediction of crop chlorophyll content for application to precision agriculture. *Remote Sens. Environ.* **2002**, *81*, 416–426. [CrossRef]

32. Gao, B. NDWI—A normalized difference water index for remote sensing of vegetation liquid water from space. *Remote Sens. Environ.* **1996**, *58*, 257–266. [CrossRef]
33. Zarco-Tejada, P.J.; González-Dugo, V.; Berni, J.A. Fluorescence, temperature, and narrow-band indices were acquired from a UAV platform for water stress detection using a micro-hyperspectral imager and a thermal camera. *Remote Sens. Environ.* **2012**, *117*, 322–337. [CrossRef]
34. Jay, S.; Comar, A.; Benicio, R.; Beauvois, J.; Dutartre, D.; Daubige, G.; Li, W.; Labrosse, J.; Thomas, S.; Henry, N.; et al. Scoring Cercospora Leaf Spot on Sugar Beet: Comparison of UGV and UAV Phenotyping Systems. *Plant Phenomics* **2020**, *2020*, 9452123. [CrossRef]
35. Okujeni, A.; Jänicke, C.; Cooper, S.; Frantz, D.; Hostert, P.; Clark, M.; Segl, K.; van der Linden, S. Multi-season unmixing of vegetation class fractions across diverse Californian ecoregions using simulated spaceborne imaging spectroscopy data. *Remote Sens. Environ.* **2021**, *2021*, 112558. [CrossRef]
36. Ibrahim, E.; Monbaliu, J. Suitability of spaceborne multispectral data for inter-tidal sediment characterization: A case study. *Estuarine Coast. Shelf Sci.* **2011**, *92*, 437–445. [CrossRef]
37. Navarro, A.; Rolim, J.; Miguel, I.; Catalão, J.; Silva, J.; Painho, M.; Vekerdy, Z. Crop Monitoring Based on SPOT-5 Take-5 and Sentinel-1A Data for the Estimation of Crop Water Requirements. *Remote Sens.* **2016**, *8*, 525. [CrossRef]

environmental sciences proceedings

MDPI

Proceeding Paper

Water Quality for Human Consumption from the Public Water Supply System †

Anita Ptiček Siročić *, Kristina Ojdanić, Dragana Dogančić and Lucija Plantak

Faculty of Geotechnical Engineering, University of Zagreb, Hallerova aleja 7, 42000 Varaždin, Croatia; tinovt1234@gmail.com (K.O.); ddogan@gfv.hr (D.D.); lucija.plantak@gfv.unizg.hr (L.P.)

* Correspondence: anitaps@gfv.hr; Tel.: +385-42408957

† Presented at the 7th International Electronic Conference on Water Sciences, 15–30 March 2023; Available online: https://ecws-7.sciforum.net.

Abstract: Adequate, safe, and accessible water constitutes an essential resource for life and is an indispensable factor for the survival of humanity. In order to ensure uncontaminated water for water supplies, industry and agriculture, water quality is defined by microbiological, biological, chemical, and physical indicators. Water monitoring is regulated with the aim to protect human health from the adverse effects of polluted water by monitoring indicator parameters. For the purpose of proving drinking water safety, water sampling was performed at ten different locations, which are part of the water supply network of the town of Virovitica, in Croatia, and its surroundings. The results showed that all the analyzed parameters were in accordance with the ordinance, i.e., that the quality of water for human consumption was satisfactory.

Keywords: water quality; microbiological; biological; chemical and physical indicators; water supply network of the city of Virovitica

check for updates

Citation: Ptiček Siročić, A.; Ojdanić, K.; Dogančić, D.; Plantak, L. Water Quality for Human Consumption from the Public Water Supply System. *Environ. Sci. Proc.* **2023**, *25*, 21. https://doi.org/10.3390/ECWS-7-14230

Academic Editor: Lampros Vasiliades

Published: 16 March 2023

1. Introduction

The problem of water supply affects millions of people worldwide and more than one hundred million people in Europe. Approximately 80% of fresh water in Europe (for drinking and other needs) originates from underground water and rivers, which is why these sources are particularly threatened by climate change, pollution, and overexploitation of water resources [1]. According to the report on water resources in the world, which was issued by United Nations Educational, Scientific and Cultural Organization, UNESCO (Paris, France) in 2003, among 188 countries, Croatia ranked highly, 5th in Europe and 42nd in the world. Croatia is one of the few countries that mostly provides healthy water through the public water supply system, to which 80% of the population is connected. The data show that in the last ten years, less than 10% of the tested samples were unhealthy in Croatia [2]. Water for human consumption must meet the parameters of conformity for water for human consumption as prescribed by the "Regulation on parameters of conformity, methods of analysis, monitoring and safety plans of water for human consumption and the way of keeping a register of legal entities that perform the activity of public water supply". Healthy water for human consumption is water that does not contain microorganisms, parasites and their developmental forms in numbers that represent a potential danger to human health; water that does not contain harmful substances in concentrations that themselves or together with other substances represent a potential danger to human health and water in which the health parameters do not exceed the values prescribed by the ordinance (Official Gazette 125/17) [2]. Water sampling was conducted in the city of Virovitica and surrounding settlements (Podgorje, Đurađ, Lozan and Špišić Bukovica) at different locations over a period of 2 days. The aim of this work was to examine the drinking water quality of the Virovitica water supply system, that is, to determine certain indicators of water quality and to determine whether they meet the standards prescribed by law.

2. Materials and Methods

2.1. Location

The town of Virovitica is located in a lowland area, surrounded by Bilogora, wide open to the north and the Drava valley and is the center of the Virovitica Podravina County. The city area consists of two relief units: the Bilogora part in the south with a pronounced step-type relief where the altitude reaches up to 250 m and the plain part in the north which is no higher than 115 m above sea level and represents a typical plain region. The drainage system of Virovitica was built as a mixed sewage system into which rainwater and wastewater are discharged. In the county, 70% of inhabitants are supplied from organized water supply systems, while the average supply in the Republic of Croatia is around 65% [3]. On the water supply network of the city of Virovitica and its surroundings, sampling of drinking water was carried out at 10 different locations (1 public water supply and 9 local water supply systems).

2.2. Methods

Determination of the water temperature was carried out with the thermometry method in accordance with the SM 2550 B standard [4]. The temperature of water for human consumption was measured during the sampling with an immersion thermometer or with a Hach HQ40d oximeter. To determine water turbidity, a turbidity-testing method in accordance with the HRN EN ISO 70271:2016 standard [5] was used, using a Hach 2100P turbidimeter. Potentiometry is a method used to determine the pH value, in accordance with the HRN ISO 10523:2009 standard [6]. It is based on the measurement of the potential difference of the electrochemical cell using a Mettler Toledo MP220 pH meter. To determine the concentration of iron in water samples, a spectrophotometric method with phenanthroline in accordance with the HRN ISO 6332:2001 standard was used with a HACH DR5000 spectrophotometer [7]. Ammonia concentration in water samples was determined spectrophotometrically with salicylate in accordance with HRN ISO 71501:1998 [8]. The method with 1 N HCl acid was used to determine the nitrate concentration in the samples in accordance with the standard SM 4500NO3B [9]. The nitrite concentration in the samples was determined with sulfanilic acid, in accordance with norm M 221/E [9]. The spectrophotometric method with 1-(2-pyridylazo)2-naphthol was used to determine the concentration of manganese in water samples. Chloride concentration in water samples was determined spectrometrically by the ferrocyanide method, in accordance with the SM 4500CL E standard [9]. Determination of potassium permanganate consumption was carried out using the permanganate index in accordance with the HRN EN ISO 8467:2001 standard [10]. To determine the electrical conductivity in water, a conductometric method was used, in accordance with the HRN EN 27888:2008 standard, using a Mettler Toledo MC226 conductometer [11]. Measurement of free (residual) chlorine was performed directly at the point of consumption in the water sample using the Pocket Colorimeter II colorimeter [2]. The total number of coliform bacteria and Escherichia coli in drinking water samples was determined by membrane filtration methods for water with low background bacterial flora according to HRN EN ISO 93081:2014 [12]. Determination of the number of enterococci in drinking water samples was conducted using a method of membrane filtration according to HRN EN ISO 78992:2000 [12]. The number of Pseudomonas aeruginosa in drinking water samples was determined by the method of membrane filtration according to HRN EN ISO 16 266:2008 [12].

3. Results and Discussion

According to the ordinance [12], the health parameters of water for human consumption (microbiological, chemical), indicators (chemical, microbiological) and parameters of radioactive substances are precisely prescribed. The ordinance specifies the parameters of groups A and B in the monitoring of water for human consumption. The purpose of monitoring water for human consumption on parameters of group A is to obtain basic data on the physical, chemical, and microbiological parameters of conformity of water for human consumption and data on the efficiency of processing water for human consumption, especially disinfection. In contrast to the monitoring of water for human consumption on the parameters of group A, the purpose of this is to obtain all the data on the parameters of the conformity check of water for human consumption, additionally with the parameters of radioactive substances. In the case of a deviation of the indicator parameter from the maximum-allowed concentration, it is necessary for an assessment of the impact of the parameter on human health to be carried out by a competent institute for public health and to take all necessary measures. In Tables 1 and 2, the physical and chemical indicators of the water samples are presented.

Table 1. Physical indicators of samples.

Samples	Temperature (°C)	pH Value	Turbidity (NTU)	Free Chlorine (mg Cl_2/L)	Conductivity (µS/cm)
VT-1	9.60	7.61	0.59	0.06	544
VT-2	11.30	7.65	0.36	0.01	518
VT-3	12.20	7.69	0.53	0.04	525
VT-4	9.10	7.74	0.53	0.06	513
VT-5	11.10	7.62	0.31	0.18	506
VT-6	7.90	7.54	0.52	0.09	519
VT-7	8.00	7.52	0.52	0.01	497
VT-8	9.60	7.44	0.76	0.08	503
VT-9	13.60	7.40	0.50	0.02	490
VT-10	9.40	7.55	0.36	0.11	514

Table 2. Chemical indicators of samples.

Samples	Ammonia (mg NH_4^2/L)	Iron (µg/L)	Manganese (µg/L)	Chlorides (mg/L)	Nitrites (mg NO_2_/L)	Nitrates (mgNO_3_/L)
VT-1	0.01	20.0	5.98	14.04	0.0	1.64
VT-2	0.01	1.0	8.64	14.35	0.0	1.73
VT-3	0.01	19.0	8.28	14.95	0.0	1.64
VT-4	0.01	9.0	9.37	14.86	0.0	1.73
VT-5	0.01	8.0	6.49	14.05	0.0	1.68
VT-6	0.01	14.0	5.17	13.19	0.0	1.64
VT-7	0.01	76.0	11.63	14.14	0.0	1.73
VT-8	0.01	38.0	9.06	14.86	0.0	1.64
VT-9	0.01	0.0	4.68	14.21	0.0	1.59
VT-10	0.01	6.0	6.27	16.01	0.0	1.51

The pH values of the studied samples were in the interval from 7.40 to 7.74. It can be concluded that all the pH values of the analyzed samples are within the limits prescribed by the ordinance, i.e., within the interval of pH values from 6.5 to 9.5 [12]. Water temperature is closely related to the temperature of the environment and depends on the temperature of the soil that surrounds it, the inflow of hot and cold underground water, and the season. The limit value for the drinking water temperature prescribed by the ordinance is 25 °C, while the most optimal is 15 °C. As can be seen from the results, the temperature values of all analyzed samples were below the limited value of 25 °C. If the temperatures are higher, some microorganisms may develop, while low temperatures slow down the flocculation and coagulation processes. Furthermore, water with lower temperature values contains little dissolved gas and has a bad taste. Free residual chlorine is excess chlorine after disinfection, and the free chlorine concentration was measured at the sampling site. The limit value for free chlorine concentration in drinking water is prescribed by the ordinance and is 0.5 mg Cl2/L. As can be seen from Table 1, all values were below the limit value. Turbidity can be determined in all water that does not contain large particles and coarse sediment that settles quickly. The highest turbidity value was obtained in sample VT-8 (0.76 NTU), while the lowest value was in sample VT-5 (0.31 NTU). Meteorological conditions can affect water turbidity, i.e., turbidity increases with the amount of precipitation [13]. Electrical conductivity is an indicator of the conduction of an electric current in a liquid, and it depends on the temperature of the water, the concentration of ions, the type of ions presents in the water, and the mobility of the ions. It is evident from the results that the values were in accordance with the ordinance, i.e., they were below the limit value (2500 μS/cm) [2]. In Table 2, the results of the chemical indicators of water quality are presented. Ammonia is an indicator of the possible presence of microorganisms, animal, and fecal waste, has a characteristic smell and is corrosive to certain materials. Characterization of the studied groundwater samples was carried out in the reductive conditions of the aquifer, where the concentration of free natural ammonia is relatively high [13]. The results of ammonia concentrations in the samples were below the value of the calibration curve, i.e., a negative value, and it can be concluded that the concentration of ammonia in the samples was negligible or that ammonia was not present in the samples at all.

Characterization of the studied samples was carried out to find any elevated contents of dissolved iron, manganese, fossil ammonia and accompanying constituents [13]. An increase in the concentration of iron in water can occur when the pH value and water temperature increase causing changes in the color and smell of the water [14]. The values of iron concentration in the analyzed samples were below the limit value determined by the ordinance (200 μg/L) [2]. Manganese (Mn) is an essential element for the human body in small quantities, and its deficiency manifests itself in slowed growth and development and shortens life span because it participates in the reproductive processes. Comparing the results of the Mn concentration with the limit value of 50 μg/L, it was evident that all values were in accordance with the ordinance [2]. An increased concentration of Mn, as well as iron, occurs due to an increase in the temperature and pH value of the water [15]. Natural water contains a certain concentration of chlorides, and the most common forms are NaCl, $CaCl_2$ and $MgCl_2$. Most drinking water contains 10 to 30 mg of chloride per liter of water, while water with more than 250 mg/L of chlorine in the form of chloride has a salty taste and is not recommended for drinking. High concentrations of chloride in water can occur due to changes in the pH value of water and can affect the development of corrosion in distribution systems [16]. From the results shown in Table 2, it is evident that all the values of the analyzed samples were below the limit value (250 mg/L). Nitrites occur naturally as part of the nitrogen cycle, and in water can be evidence of fresh fecal pollution. They can be found in wastewater, storm sewers of cities and agricultural areas. Table 2 shows that the measured nitrite concentrations in the analyzed samples were 0.0 mg NO_{2-}/L, that is, nitrites were not present in the samples. Increased concentrations of nitrates in water can occur in places where fertilizers are used, during the rotting of plant and animal remains, in industrial waste and wastewater from sludge landfills. The measured values of nitrate concentrations in the analyzed samples are shown in Table 2. The highest values were obtained in samples VT-2, VT-4 and VT-7 (1.73 mg NO_{3-}/L), while the lowest value was obtained in sample VT-10 (1.51 mg NO_{3-}/L), but all values were below the limit value which is 50 mg NO_{3-}/L. The most important representatives of fecal coliform bacteria are Escherichia coli and fecal streptococci (enterococci). The key factors for the growth and reproduction of Escherichia coli in water are the water temperature and the increased amount of nutrients and organic substances in the water [17]. In all studied samples, Escherichia coli as well as Enterococcus were not present, that is, all values were 0 cfu/100 mL. Enterococci are the most suitable group of bacteria for assessing the hygienic quality of water because their numbers correlate with the presence of numerous pathogenic bacteria, total coliform bacteria, and enteroviruses. The presence of enterococci is an indicator of fecal contamination. Increased water temperature and nutrients and organic substances promote the development and reproduction of enterococci in water. The presence of Pseudomonas aeruginosa is related to openings in the water supply system (e.g., taps). Increased water temperature, changes in pH value, nutrients and organic substances in the water and a failure to maintain the distribution system encourage the development and reproduction of Pseudomonas aeruginosa. If the analysis results are in the range of 110 cfu/100 mL, it is necessary to assess the risk of using such water. Pseudomonas aeruginosa was not detected in the analyzed samples.

4. Conclusions

Water quality indicators are determined by different methods, and the determination of indicators is of significant importance for the prevention of water pollution that is used for water supply, agriculture, industrial processes, and leisure. The analyzed samples of water for human consumption of the water supply system of Virovitica and its surroundings, based on microbiological, physical and chemical indicators of the quality of drinking water, comply with the ordinance on conformity parameters, methods of analysis, monitoring and safety plans for water for human consumption and the way of keeping a register of legal entities that perform the activity of public water supply. According to estimates by the World Health Organization, approximately 1.2 billion people fall ill due to contaminated water. Today, according to estimates by the international organization of the United Nations, 1.5 billion people do not have safe drinking water, and in the next 25 years that number will grow to about 4.5 billion. Water is not a commercial product, but a heritage; therefore, it is necessary to make individuals aware of controlled waste disposal, rational water consumption, wastewater treatment, controlled application of agrotechnical substances (pesticides), monitoring the transport of dangerous substances and, ultimately, water conservation and protection.

Author Contributions: Conceptualization, A.P.S. and K.O.; methodology, A.P.S. and K.O.; validation, A.P.S. and K.O.; formal analysis, A.P.S. and K.O.; investigation, A.P.S. and K.O.; resources A.P.S. and K.O.; data curation, A.P.S. and K.O.; writing—original draft preparation, A.P.S., K.O., L.P. and D.D.; writing—review and editing, A.P.S. and L.P. All authors have read and agreed to the published version of the manuscript.

Funding: This research received no external funding.

Institutional Review Board Statement: Not applicable.

Informed Consent Statement: Not applicable.

Data Availability Statement: Not applicable.

Acknowledgments: Virkom d.o.o., Virovitica, Croatia.

Conflicts of Interest: The authors declare no conflict of interest.

References

1. European Environment Agency. The Use of Water in Europe, Quantity and Quality, Face Major Challenges. Available online: https://www.eea.europa.eu/hr/signals/eeasignali2018vodajezi-ot/clanci/uporabavodeueuropi2013 (accessed on 15 December 2022).
2. *Official Gazette 125/2017, Rulebook on Compliance Parameters, Analysis Methods, Monitoring and Safety Plans for Water for Human Consumption and the Way of Keeping the Register of Legal Entities that Perform the Activity of Public Water Supply*; Ministry of Health: Zagreb, Croatia, 2017.
3. Program of Overall Development of the City of Virovitica 2014–2020. ACH. Pocket Pro Temperature Tester. 2021. Available online: https://www.virovitica.hr/wpcontent/uploads/2014/07/PUR_2014.2020._17.09._.pdf (accessed on 15 December 2022).
4. *HRN ISO 78992:2000—Water Quality—Determination of Turbidity—Part 1: Quantitative Methods*, 1st ed.; ISO: London, UK, 2016.
5. *HRN ISO 10523:2009—Water Quality—Determination of pH Value*, 1st ed.; ISO: London, UK, 2009.
6. *HRN ISO 6332:2001—Water Quality—Determination of Iron—Part 1: Spectrophotometric Method*, 1st ed.; ISO: London, UK, 2001.

7. *HRN ISO 71501:1998—Water Quality—Determination of Ammonium—Part 1: Spectrophotometric Method*, 1st ed.; ISO: London, UK, 1998.
8. Kepec, S. *Test Report, Water for Human Consumption Treated Water;* Institute of Public Health "Sveti Rok" of Virovitica-Podravine County, Activity for Health Ecology: Virovitica, Croatia, 2020.
9. *HRN EN ISO 8467:2001—Water Quality—Determination of the Permanganate Index*, 1st ed.; ISO: London, UK, 2001.
10. *HRN EN 27888:2008—Water Quality Determination of Electrical Conductivity*, 1st ed.; ISO: London, UK, 2008.
11. *Pocket colorimeter POCKET Colorimeter II Colorimeter Test Kit for Chlorine Analysis;* HACH: Loveland, CO, USA, 2021.
12. Zhou, Q.; Wang, J.; Tian, L.; Feng, L.; Li, J.; Xing, Q. Remotely sensed water turbidity dynamics and its potential driving factors in Wuhan, an urbanizing city of China. *J. Hydrol.* **2020**, *593*, 125893. [CrossRef]
13. Urumović, K.; Hlevnjak, B. *Hydrogeological Study of the Development of the Virovitica Regional Pumping Station;* Croatian Geological Survey: Zagreb, Croatia, 1997.
14. Zhang, S.; Tian, Y.; Guo, Y.; Shan, J.; Liu, R. Manganese release from corrosion products of cast iron pipe sin drinking water distribution systems: Effect of water temperature, pH, alkalinity, SO42 concentration and disinfectants. *Chemosphere* **2021**, *262*, 127904. [CrossRef]
15. Stets, E.G.; Lee, C.J.; Lytle, D.; Schock, M. Increasing chloride in rivers of the conterminous U.S. and linkages to potential corrosivity and lead action level exceedances in drinking water. *Sci. Total Environ.* **2018**, *613–614*, 1498–1509. [CrossRef] [PubMed]
16. El Zanfaly, H.T.; Hosny, I.; Fayez, M.; Shaban, A. Sanitary significance of fecal streptococci in underground water in Egypt. *Zent. Fur Mikrobiol.* **1989**, *144*, 299–304. [CrossRef]
17. Ofori, I.; Maddila, S.; Lin, J.; Jonnalagadda, S.B. Chlorine dioxide inactivation of Pseudomonas aeruginosa and Staphylococcus aureus in water: The kinetics and mechanism. *J. Water Process Eng.* **2018**, *26*, 46–54. [CrossRef]

environmental sciences proceedings

MDPI

Proceeding Paper

In IWRM, Should Scientific Modeller Perspectives Receive Priority over the Benefit Recipients? †

Rathnayake Mudiyanselage Manjula Pradeep [1,*] and Nallaperuma Thanthirige Sohan Wijesekera [2]

1 Transdisciplinary Research Lab, Faculty of Computing, General Sir John Kotelawala Defence University, Kandawala Road, Ratmalana 10390, Sri Lanka
2 Construction Industry Development Authority, Colombo 00700, Sri Lanka; sohanw2@gmail.com
* Correspondence: pradeep@kdu.ac.lk
† Presented at the 7th International Electronic Conference on Water Sciences, 15–30 March 2023; Available online: https://ecws-7.sciforum.net.

Abstract: Throughout history, discussions on IWRM have established the idea that the recipient stakeholders' poor participation is obstructing sustainable decision making in urban flood management. However, we found that no in-depth study has been carried out to explore the status of stakeholder integrations in such modelling. The present work explored the stakeholder integrations in the modelling through critical literature analysis and expert discussions. We found there are five main components in the modelling, and the recipient stakeholder requirements are satisfactorily integrated with the modelling approach. Nevertheless, we found that integrations of scientific modelling perspectives remain unsatisfactory. This paper urges water resource decision makers to prioritise scientific modellers' perspectives when developing flood management models.

Keywords: IWRM; flood stakeholders; urban flood management; hydrological modelling; GIS modelling; Hydro-GIS modelling

check for updates

Citation: Pradeep, R.M.M.; Wijesekera, N.T.S. In IWRM, Should Scientific Modeller Perspectives Receive Priority over the Benefit Recipients? *Environ. Sci. Proc.* **2023**, *25*, 22. https://doi.org/10.3390/ECWS-7-14167

Academic Editor: Athanasios Loukas

Published: 14 March 2023

1. Introduction

The ultimate aim of the United Nations Sustainable Development Goals (SDG) is the sustainable development of humans in harmony with the environment [1]. Therefore, one of the key undertakings in water resource management is to maintain a satisfactory relationship between natural water cycle needs and social/economical needs. Scientists and water governors have already been working independently to achieve these goals for decades. Nevertheless, with different influences on research, such as the need to incorporate the general public's opinion in decision making [2], as well as stakeholder theory [3], researchers are used to integrating the two agendas. In the meantime, the 1997 UN water conference and 1992 "Dublin Principles" established an international norm for water governance, namely Integrated Water Resource Management (IWRM) [4].

Since IWRM is based on water governance for achieving sustainable goals, it is always better to maintain a close relationship between decision makers and key stakeholders while developing management tools to optimise the requirements [5]. In parallel, various other initiatives, such as Green Infrastructure (GI), Low Impact Development (LID), and the water framework of Economic Co-operation and Development (OECD) [6,7], are also in practice; however, an analysis of the available data found that the effects of flood damage on national economies is increasing [8].

The common excuses for this situation are the recipient stakeholders' poor participation, that governance discourses are limited to stakeholders, and that decisions are mainly theoretical [9,10]. Apart from those, there are dozens of negative reasons for the practical incorporation of recipients' perspectives in administrative decision-making modelling or processes. Furthermore, we found that no study had been carried out to inductively explore the integrations of the requirements of the total stakeholder profile of the process in the current setting. Therefore, our initial work was carried out to evaluate the integration of stakeholder requirements for a specific area of urban flood management [11]. As such, the aim of the present work is to analyse and discuss the results for practical stakeholder integrations in IWRM.

2. Methods

2.1. Evaluation of Levels of Stakeholder Requirements Integration

As there is no established method to carry out this type of transdisciplinary research, which needs to evaluate different components in integrated water management decision making, we required an acceptable research methodology. As such, an in-depth study, including a literature review and expert discussion, was carried out to develop a research methodology for gap identification [12–14].

Then, accordingly, we carried out abductive research using a sequential multi-phase approach of the mixed method. We employed modified constructivist grounded theory, documentary research, and survey strategies to find and verify the main components and their integration depths in the scientific and management model of urban flood management. For the component identification, we studied the GIS2MUSCLE urban flood management tool, 4 hydro-GIS integration models, and 247 works of research; furthermore, for calibration, we utilised 21 experts. The average integration depths among the components were calculated using 32 studies employing Multi-Attribute Utility Theory (MAUT) and Weighted Average Programming (WAP). Finally, we evaluated the results with 70 experts and analysed the result by employing thematic analysis and Multi-Criteria Group Decision Making (MCGDM) methods.

2.2. Data

Through the above steps, we identified five components (main stakeholder categories) as shown in Figure 1. However, in practice, the hydro specialists and GIS specialists were also integrated into the HydroGIS model.

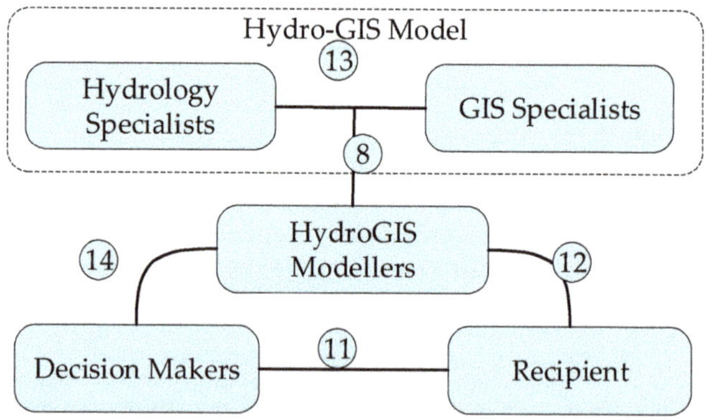

Figure 1. Main components (stakeholder categories) of flood management model. (The lines between components indicating existing integrations. The numbers in the circles indicate how many research works have considered such integration. Source: Author.

Then, rationale was developed to weigh the depth of scientific investigation carried out by the researchers on each integration shown in Figure 1. Furthermore, we analysed the depth of investigation level (scale of Very low to Very high) on each integration carried out by each study by utilising the modified MAUT. However, we observed that studies analysed the integrations in either Very high, High, Medium, or Low depths only (Figure 2).

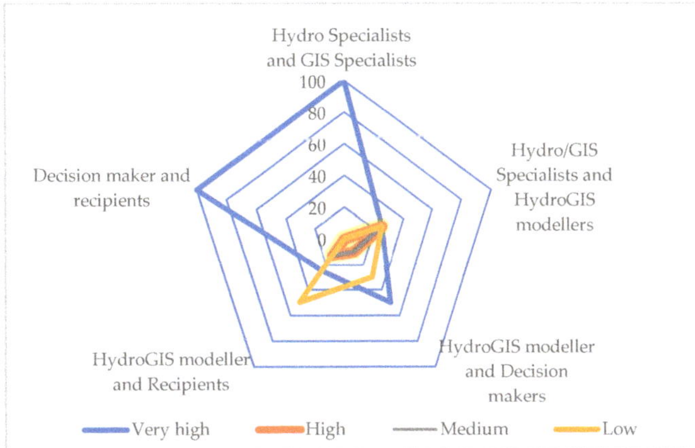

Figure 2. Distribution of investigation depth classes among the integration types. Source: Author.

By generalizing the individual study's investigation depths to develop a final decision, we developed a rationale for weighting the scientific value of the publication [15]. Thereafter, the depth of investigation for each integration was calculated using WAP, and the values are on a 1–5 scale, where 5 is Very high and 1 is Very low. The comparative level of the investigation depths among the integrations was also calculated. Another understanding made during the step is that there are two groups in which the components can be accumulated considering the main undertaking of the flood management model. The present work called them "scientific components" and "management components" (Figure 3).

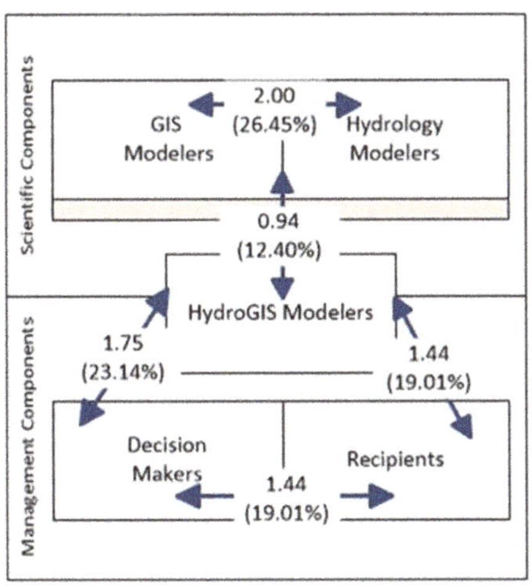

Figure 3. The present investigation depth status of the integration of main components in flood management modelling. The average depth of investigation in each integration is shown as a fraction. The comparative level of the investigation depths is shown as percentages (computed%) (Adapted from Ref. [11]).

Then, we reclassified the investigation depths according to a 1–5 scale and computed the deviation from the mean comparative value of 20% (if equal attention is being paid to all 5 integrations, the 20% is the mean value) using Equation (1). The positive values exhibited exaggerations of attention, while negatives showed an understatement of attention (Table 1).

$$\text{Deviation from mean comparative value} = ((\text{computed}\% \div 20) - 1) \times 100 \quad (1)$$

Table 1. Computation results.

Integration	Identified Depth of Investigation through the Study			Deviation from the Mean Comparative Value (Equation (1) Result)
	Computed	**Classified**	**Computed%**	
Hydro Specialists and GIS Specialists	2.00	Low	26.45%	32% (+ve)
Hydro and GIS Specialists and HydroGIS Modellers	0.94	Very low	12.40%	38%(−ve)
HydroGIS Modellers and Decision Makers	1.75	Low	23.14%	16% (+ve)
HydroGIS Modellers and Recipients	1.44	Very low	19.01%	5%(−ve)
Decision Makers and Recipients	1.44	Very low	19.01%	5%(−ve)

3. Results and Discussion

The resulting flood model development framework, which demonstrates all the roles involved in the flood management modelling with the levels of present attention on integrations, is shown in Figure 4. This work found two definitions for the present level of researchers' interest distributions: (1) The individual interest: the investigation depth of each integration, which is independent of other integrations; (2) the comparative interest: the comparative level of investigation, which demonstrates how the total attention of the researchers is distributed over all possible integrations.

As per the scale of investigation depth defined in the present work, all assessment values that were received were less than 2. This means the present interest in all the integrations is below the "Low" level. Furthermore, we observed that the researchers' attention levels regarding incorporating the perspectives of scientific component modellers with management modellers (hydro/GIS specialists and HydroGIS modellers), HydroGIS modellers with recipients, and recipients with decision makers are in the "Very low" level. Therefore, our findings prove the importance of one of the concepts behind IWRM, namely integrating the recipients into water decision making.

According to the analysis, we found that the researchers understate 37% of the optimum when integrating the scientific modellers' (hydro and GIS modellers) concerns into the management model via the HydroGIS modellers. However, we found that the most challenging requirements being discussed in public at the present include integrating the general public's (recipient) perspective into flood management, which must be satisfactorily attended to, as it received a 5% understatement value. In the meantime, the results show that there is a level of 32% exaggerated attention paid to integrating the hydro modellers' and GIS modellers' perspectives.

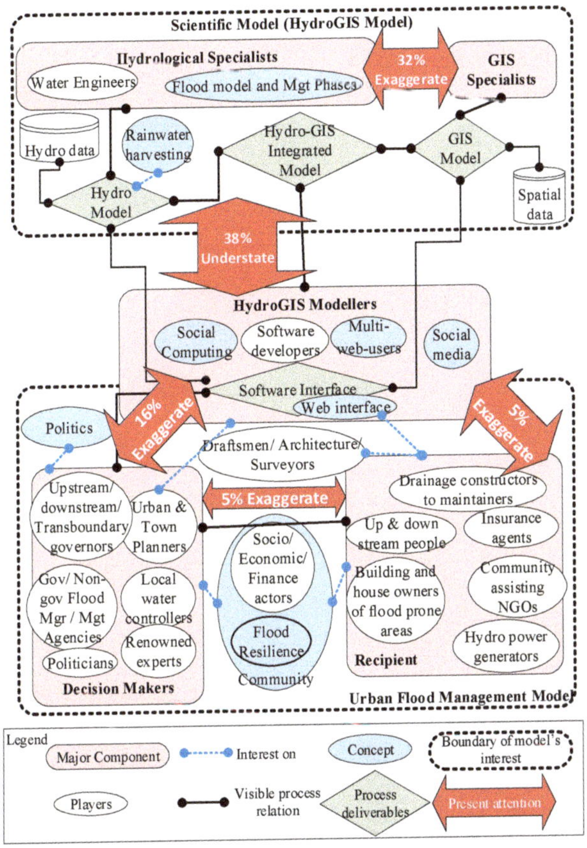

Figure 4. Descriptive view of flood management model development framework. Source: Author.

4. Conclusions

IWRM governs flood management and it requires an understanding to identify all the major components (stakeholder categories) for sustainable flood management modelling.

This study found that there are five main stakeholder categories which need explicit integrations or sustainable flood management modelling. They are hydro modellers, GIS modellers, HydroGIS modellers, decision makers, and recipients.

These five stakeholder categories were grouped into two components: (1) scientific components and (2) management components. The integration of these groups is being carried out by the HydroGIS modeller who develops the flood management model.

The present study shows that perspectives of the components in the scientific model are well integrated with model development, while components within the management model are also satisfactorily integrated. The most discussed recipient stakeholders are also in the management group; hence, we can argue that, at present, recipients' perspectives are satisfactorily incorporated into the flood management initiatives.

Nevertheless, the poorest attention (38% less than optimum) is being paid to integrating the scientific model perspectives into the management model. Therefore, this work concludes by stating that, at the present, there is a need for better attention to all five interactions and therefore, IWRM initiatives should focus more attention to integrating scientific modellers' perspectives thereby fully satisfying the recipient stakeholders' requirements.

Author Contributions: Conceptualization, methodology, software, validation, formal analysis, investigation, resources, data curation, writing—original draft preparation, project administration, funding acquisition, and visualization, R.M.M.P. Writing—review and editing, and supervision, N.T.S.W. All authors have read and agreed to the published version of the manuscript.

Funding: This research received no external funding.

Institutional Review Board Statement: Not applicable.

Informed Consent Statement: Informed consent was obtained from all subjects involved in the study.

Data Availability Statement: No new data were created or analysed in this study. Data sharing is not applicable to this article.

Conflicts of Interest: The authors declare no conflict of interest.

References

1. Open SDG Engagement Platform. Open SDG Engagement Platform—United Nations Partnerships for SDGs Platform. Available online: https://sustainabledevelopment.un.org/partnership/?p=11997 (accessed on 8 January 2021).
2. Arnstein, S.R. A Ladder Of Citizen Participation. *J. Am. Inst. Plan.* **1969**, *35*, 216–224. [CrossRef]
3. Freeman, R.E. *Strategic Management: A Stakeholder Approach*; Pitman Publishing Inc.: Lanham, MA, USA, 1984; ISBN 0273019139.
4. Woodhouse, P.; Muller, M. Water Governance—An Historical Perspective on Current Debates. *World Dev.* **2017**, *92*, 225–241. [CrossRef]
5. Hung, P.; Van Trung, L.; Vo, P. Le Decision Support Tool for Integrated Water Resources Management Based on GIS, Remote Sensing and SWAT Model: A Case Study in the Upper Part of Dong Nai River Basin, Vietnam. In Proceedings of the Advances in Research on Water Resources and Environmental Systems, Matsue, Japan, 21–24 November 2023; Vo, P.L., Tran, D.A., Pham, T.L., Thu, H.L.T., Nguyen Viet, N., Eds.; Springer International Publishing: Cham, Germany, 2023; pp. 361–388.
6. Piyumi, M.; Abenayake, C.; Wijayawardana, N. Low Impact Development (LID) Solutions for Flood-Resilient Urbanization. *J. Inst. T. Plan. Sri Lanka* **2020**, *2*, 15–29.
7. Thoradeniya, B.; Maheshwari, B. Engaging Stakeholders for Water Diplomacy: Lessons for Integrated Water Resource Managment. In *Water Diplomacy in Action: Contingent Approaches to Managing Complex Water Problems*; Islam, S., Madani, K., Eds.; Anthem Press: London, UK, 2017.
8. Pradeep, R.M.M.; Wijesekera, N.T.S. Role of Hydro GIS Tools in Hydrological Modelling and Urban Flood Management: A Literature Review. In Proceedings of the 15th International Research Conference, Ratmalana, Sri Lanka, 29–30 September 2022; KDU Press: Ratmalana, Sri Lanka, 2022.
9. Herath, H.M.M.; Wijesekera, N.T.S. A State-of-the-Art Review of Flood Risk Assessment in Urban Area. *IOP Conf. Ser. Earth Environ. Sci.* **2019**, *281*, 012029. [CrossRef]
10. Ayala-Orozco, B.; Rosell, J.A.; Merçon, J.; Bueno, I.; Alatorre-Frenk, G.; Langle-Flores, A.; Lobato, A. Challenges and strategies in place-based multi-stakeholder collaboration for sustainability: Learning from experiences in the Global South. *Sustainability* **2018**, *10*, 3217. [CrossRef]
11. Pradeep, R.M.M.; Wijesekera, N.T.S. Gaps in the accounting of stakeholder integrations in HydroGIS tools to face the challenge of sustainable urban flood management. *Eng. J. Inst. Eng. Sri Lanka* **2021**, *54*, 1–14. [CrossRef]
12. Pradeep, R.M.M.; Wijesekera, N.T.S. Development of HydroGIS Model Development Framework: Research Methodological Perspectives. In Proceedings of the 15th International Research Conference, Ratmalana, Sri Lanka, 29–30 September 2022; KDU Press: Ratmalana, Sri Lanka, 2022.
13. Pradeep, R.M.M.; Morris, M. The Complexity of Research, and Researching Complexity: A Review of the Options. In Proceedings of the International Symposium of Rajarata University 2021 (ISymRU 2021), Faculty of Technology, Rajarata University of Sri Lanka, Mihinthale, Sri Lanka, 21–22 December 2021; Volume 2021, p. 82.
14. Pradeep, R.M.M.; Morris, M. A philosophical axiom review on "THE METHODOLOGY" of Computing Research. In Proceedings of the 15th International Research Conference, Ratmalana, Sri Lanka, 29–30 September 2022; KDU Press: Ratmalana, Sri Lanka, 2022.
15. Pradeep, R.M.M.; Wijesekera, N.T.S. Journal or Book?: The HydroGIS Perspective on Engineering and Computing Debate. In Proceedings of the 13th International Research Conference, Ratmalana, Sri Lanka, General Sir John Kotelawala Defence University, Ratmalana, Sri Lanka, 15–16 October 2020; pp. 377–386.

environmental sciences proceedings

MDPI

Proceeding Paper

Water and Nitrogen Use and Agricultural Production Efficiency under Climate Change in a Mediterranean Coastal Watershed †

Aikaterini Lyra [1,*] and Athanasios Loukas [2]

1 Laboratory of Hydrology and Aquatic Systems Analysis, Department of Civil Engineering, School of Engineering, University of Thessaly, 38334 Volos, Greece
2 Laboratory of Hydraulic Works and Environmental Management, School of Rural and Surveying Engineering, Aristotle University of Thessaloniki, 54124 Thessaloniki, Greece; agloukas@topo.auth.gr
* Correspondence: klyra@uth.gr; Tel.: +30-(24)-21074153
† Presented at the 7th International Electronic Conference on Water Sciences, 15–30 March 2023; Available online: https://ecws-7.sciforum.net/.

Abstract: This study estimates the effect of climate change on water resource efficiency and crop/agronomic productivity at the Almyros basin in Greece. Groundwater resources are intensively used for irrigation, whereas their quantity and quality are highly downgraded. Climate change impacts have been assessed using Med-CORDEX bias-corrected climatic projections for RCP4.5 and RCP8.5 scenarios. Simulation of coastal water resources was carried out with an Integrated Modelling System (IMS) consisting of the modules of surface hydrology (UTHBAL), reservoir operation (UTHRL), groundwater hydrology (MODFLOW), nitrate leaching/crop growth (REPIC), nitrate pollution (MT3DMS), and salt wedge/salinization (SEAWAT). The indices of Standardized Chloride Hazard (SCHI), Crop Water (CWP) and Economic Water Productivity (EWP), Nitrogen Use Efficiency (NUE) have been employed to analyze water resource management and agronomic scenarios. The findings indicate the water resources' capacity for adaptability and agronomic effectiveness under the influence of salinization and climate change.

Keywords: groundwater; seawater intrusion; coastal agricultural basin; climate change

check for updates

Citation: Lyra, A.; Loukas, A. Water and Nitrogen Use and Agricultural Production Efficiency under Climate Change in a Mediterranean Coastal Watershed. *Environ. Sci. Proc.* **2023**, *25*, 23. http://doi.org/10.3390/ECWS-7-14180

Academic Editor: Rodrigo Maia

Published: 14 March 2023

1. Introduction

Coastal aquifer systems nowadays present an ever-declining quantity and quality degradation as they are intensively used for irrigation purposes. This is especially the case for the aquifer systems located in arid and semi-areas like the Mediterranean basin, where fertile soils and favorable climatic conditions host the productivity of the agricultural and food sectors. It has been predicted that climate change will have a greater impact on the water resources of the Mediterranean areas, and it will alter the water cycle's temporal and geographical distribution. Water scarcity will be advanced in intensity and magnitude while crop yields are expected to decline [1]. Whenever seawater intrusion is an area of concern for water resources in irrigated agriculture as well, it is quite crucial to examine the potential effects of climatic change in coastal arable watersheds [2]. In coastal watersheds where agriculture is the main economic activity, the management of water resource hazards stem from the absence of water storage works, the large amounts of groundwater abstractions that also provoke seawater intrusion, and the large amounts of fertilizers to maximize yields. Irrigation with salinized water causes physiological drought to crops, with effects similar to climatic drought events with regard to crop productivity [3,4]. In general, estimating water resource productivity at the watershed/basin scale is not an extensively studied task either under climate change and/or salinity effects on productivity. The aim of this work is to examine the climate change impacts on water resources efficiency and crop productivity under agronomic and irrigation scenarios, as well as the adaptation potential under water resource development projects at the Mediterranean Almyros basin, a coastal region in Thessaly, Greece, considering, also, the salinity effects on crop yields. The water resources have been simulated based on two climatic Representative Concentration Pathways (RCPs), namely, RCP4.5 and RCP8.5, using an Integrated Modeling System (IMS) formed by Lyra and colleagues for implementation on coastal agricultural watersheds [2,5]. The indices of Standardized Chloride Hazard (SCHI), Crop Water (CWP) and Economic Water Productivity (EWP), and Nitrogen Use Efficiency (NUE) have been employed to analyze water resource adaptation and agronomic alternatives. The findings indicate the water resources' potential for adaptation as well as their agronomic productivity under climate change.

2. Materials and Methods

2.1. Study Area

Almyros basin is located in central Greece, and it is an agricultural coastal basin where wheat, alfalfa, cereals are the main cultivars, while cotton, olives trees, maize, vegetables, orchards, and vineyards are also cultivated. Given the absence of existing freshwater reserves, groundwater has been solely used for irrigation. The hydrology of the region is described by streams with intermittent flows and semiarid climatic conditions [2]. The groundwater in the study basin has been seriously affected and polluted by contaminants from nitrogen leachates and chloride ions from saltwater intrusion. Recently, an urban water supply reservoir, the Mavromati reservoir, has been built, and an irrigation water supply reservoir, the Xirias reservoir, is under construction (Figure 1). Furthermore, a greater irrigation water reservoir has been studied, the Klinovos reservoir (Figure 1).

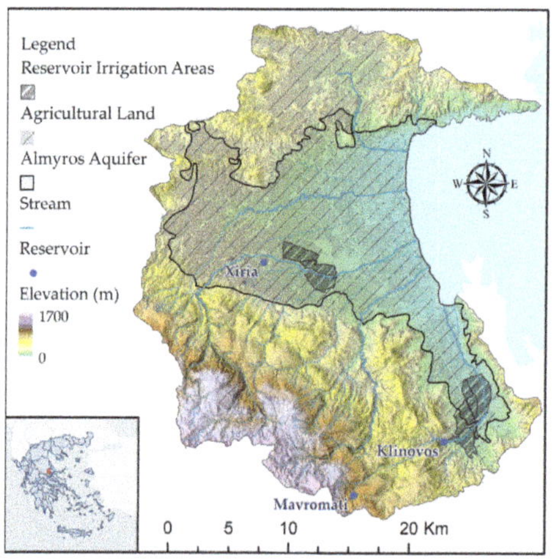

Figure 1. Elevation of the Almyros basin, principal streams, Almyros aquifer system, reservoir locations, and irrigated regions.

2.2. Climate Change and Integrated Modelling System

The simulation is performed for climatic model ensembles based on Med–CORDEX models. Precipitation and temperature ensemble timeseries for the RCPs 4.5 and 8.5 have been bias-corrected using Quantile Mapping. Bias–correction calibration took place during the period 1971–2000, and validation during 2001–2018, counting on observed precipitation and temperature for the studied area, as in a recent earlier study [2]. The simulation of water resources of the study basin was performed using the calibrated and high efficiency Integrated Modelling System consisting of interlinked/coupled models for surface hydrology (UTHBAL), reservoir operation (UTHRL), agronomic schedules/crop growth/nitrate leaching processes (REPIC), groundwater flow (MODFLOW), nitrate transport (MT3DMS), and salt wedge/seawater intrusion (SEAWAT) composed by Lyra and associates [5]. Groundwater simulations for the water table and the nitrate and chloride concentrations started from 1991 because of the availability of historical hydrogeological data.

2.3. Water Resources and Agronomic/Crop Scenarios and Strategies

Strategy A, in which only groundwater is used for irrigation/urban water supply (baseline historical strategy), and Strategy B, in which surface water reservoirs have been developed and used along with groundwater abstractions for irrigation/urban water supply, have been developed. Several agronomic and irrigation scenarios have been developed and simulated with the Strategies A and B, namely historical irrigation and nutrient practices (A0/B0), deficit irrigation and historical nutrient practices (A1/B1), rainfed agriculture and historical nutrient practices (A2/B2), deficit irrigation and reduced nutrient practices (A3/B3), and deficit irrigation and rainfed agriculture and reduced fertilization (A4/B4).

2.4. Salinity, Chlorides Concentration, and Crop Yield

The crop yield is steady within a given range of soil salinity, but, after reaching a maximum tolerance level, the crop output decreases in an idealized, simple linear trend. The electrical conductivity of the water (EC_W) can be approached with a concentration coefficient (X) that depends on the leaching physiology of the cultivated crops, and underlying soil, and the conductivity of soil extract (EC_e) [4]. In order to take into account, the saline implications of the seawater intrusion on agricultural output, the crop yields are adjusted using a relative percentage of yield performance [3]. Electrical conductivity and chloride concentration observations performed by various public and private organizations and former studies, as described in [5], span from 1991 to 2015. Chlorides range from 4 to 1432 mg/L, and EC_W ranges from 0 to 5 dS/m.

2.5. Agronomic Indices and Standardized Chloride Hazard Index (SCHI)

The Standardized Chloride Hazard Index (SCHI) has been used for detecting and characterizing the adaptation of using coastal groundwater for irrigation and the possible impacts of its use on the crop yield, as follows:

$$SCHI = \left(Cl_i - \overline{Cl}\right)/\sigma_{Cl} \quad (1)$$

where Cl_i is the chloride concentrations as simulated by SEAWAT in a monthly timestep, $\overline{Cl}$ is the monthly average, and σ_{Cl} is the standard deviation of the chloride concentrations. The agronomic indices are based on the simulated crop yields by the REPIC model regarding the various alternatives. The index (CWP) quantifies the yield produced for every cubic meter of water applied. The Economical Water Productivity (EWP) indicator determines the performance at each cubic meter of water supplied, and, based on published commodity producer prices for agriculture by OECD–FAO until the 2030s [6], the index scores for the future periods have been projected and estimated. The Nitrogen Use Efficiency index (NUE) quantifies the yield produced for every kg of nitrogen applied.

3. Results-Discussion

3.1. Salinity Impacts on Crop Yield

Based on relevant measurements, it was possible to configure useful information for the relationship of chloride concentrations and electrical conductivity in the Almyros aquifer system. The linear regression equations that connect the two variables are:

$$EC_w(\mathrm{dS/m}) = 0.0032 \times Cl_w\,(\mathrm{mg/L}) + 0.6212 \quad (R^2 = 0.9) \quad (2)$$

$$Cl_w\,(\mathrm{mg/L}) = 272.32 \times EC_w(\mathrm{dS/m}) - 159.17 \quad (R^2 = 0.9) \quad (3)$$

Additionally, the weighted mean leaching fraction and concentration coefficient of the Almyros aquifer are 0.1 and 2.1, respectively. Table 1 presents the rating of seawater intrusion and salinity hazards primarily for the Almyros groundwater body based on the classifications by [3]. According to the simulated chloride concentrations by the SEAWAT model and the crop yields by the REPIC model, Table 2 shows the mean Relative Yield changes caused by salinity for the two RCPs (4.5 and 8.5). Crop production would be reduced under historical irrigation practices regardless of the development of surface water reserves by −0.3% in RCPs 4.5 and 8.5 in 2019–2050 due irrigation groundwater salinity. The implementation of deficit and rainfed agriculture has a stronger positive influence on the decrease of salinization, and, thus, the contribution of the alteration of agronomic practices is evident with a 0.3% gain in productivity in 2019–2050. However, rainfed agriculture is more efficient to maintain and increase the crop production in 2051–2100 since it is not affected by the operation of reservoirs and the cessation of groundwater abstractions and groundwater salinity.

Table 1. Classification of the electrical conductivity of saturated extract (EC_e), the pumped groundwater for irrigation (EC_w), the chloride concentrations (Cl_w) of water, and the Standardized Chloride Hazard Index (SCHI) with regard to the salinity hazards.

EC_e (dS/m)	EC_w (dS/m) (CF = 2.1)	Cl_w (mg/L)	SCH_W Index	Hazard
0–2	0–0.95	0–100	<–1.2	Low with normal yields
2–4	0.95–1.90	100–360	−1.2:−0.9	Low-moderate with yield decrease of sensitive crops
4–8	1.90–3.81	360–878	−0.9:−0.1	Moderate with yield decrease of crops
8–12	3.81–5.71	878–1397	−0.1:0.7	High with yield for tolerant crops
>16	>7.62	>1916	>1.5	Extremely high with yield for tolerant crops

Table 2. Relative Yield changes of the future periods from the historical period for the water resources and agronomic scenarios for both RCPs (4.5 and 8.5).

Time	A–0	A–1/A–3	A–2/A–4	B–0	B–1/B–3	B–2/B–4
2019–2050	−0.3%	0.3%	0.3%	−0.3%	0.3%	0.3%
2051–2100	−14.8%	−0.4%	0.4%	−10.9%	−0.7%	0.4%

3.2. *Agronomic Efficiency Indices and Water Resource Adaptation for Seawater Intrusion*

The agronomic efficiency indices and the water resources adaptation index for seawater intrusion have been estimated and summarized for the time periods of 1991–2018, 2019–2050, and 2051–2100. The SCHI index has been calculated based on the results of the Integrated Modelling System (IMS) and especially the SEAWAT model. SCHI index scores range for all scenarios and time periods from low salinity hazard, −1.14 in the historical period, to almost extremely high salinity hazard, 1.42 in the future period, proving the downgrading of the groundwater from non-saline to very saline. In 2019–2050 under historical and deficit irrigation practices, groundwater use for irrigation will pose low-moderate salinity hazards, whereas rainfed agriculture/deficit irrigation will pose moderate rates in both Strategies A and B and in both RCPs. However, in 2051–2100, the situation will be reversed. Low-moderate salinity hazards will appear during rainfed agriculture/deficit irrigation and moderate salinity hazards for the remaining scenarios for both RCPs.

The agronomic indices have been calculated based on simulations by the Integrated Modelling System (IMS). Crop yields are calculated by the REPIC model and the groundwater abstractions and quality by the MODFLOW and SEAWAT models. Crop yields are multiplied to their relative change, due to salinity impacts of the groundwater as irrigation water, to obtain the simulated production considering the salinity effects. Commodity prices, for use in the EWP index of the crop pattern, were estimated on a weighted spatial average of 0.33 €/kg crop yield in 1991–2018, 0.49 €/kg in 2019–2050, and 0.65 €/kg in

2051–2100. Figure 2 depicts the Crop Water Productivity (CWP), Nitrogen Use Efficiency (NUE), and Economical Water Productivity (EWP) indices' values. CWP and EWP index scores thrive in Strategy B than in A, proving the benefits of surface water reserves on water efficiency. CWP weighted averaged values range from 6.4 tn/m^3 to 19.4 tn/m^3 in RCP8.5, and from 6.5 tn/m^3 to 17.1 tn/m^3 in RCP4.5. CWP variations are approximately 3 to 4 tn/m^3 between deficit irrigation and rainfed agriculture/deficit irrigation practices.

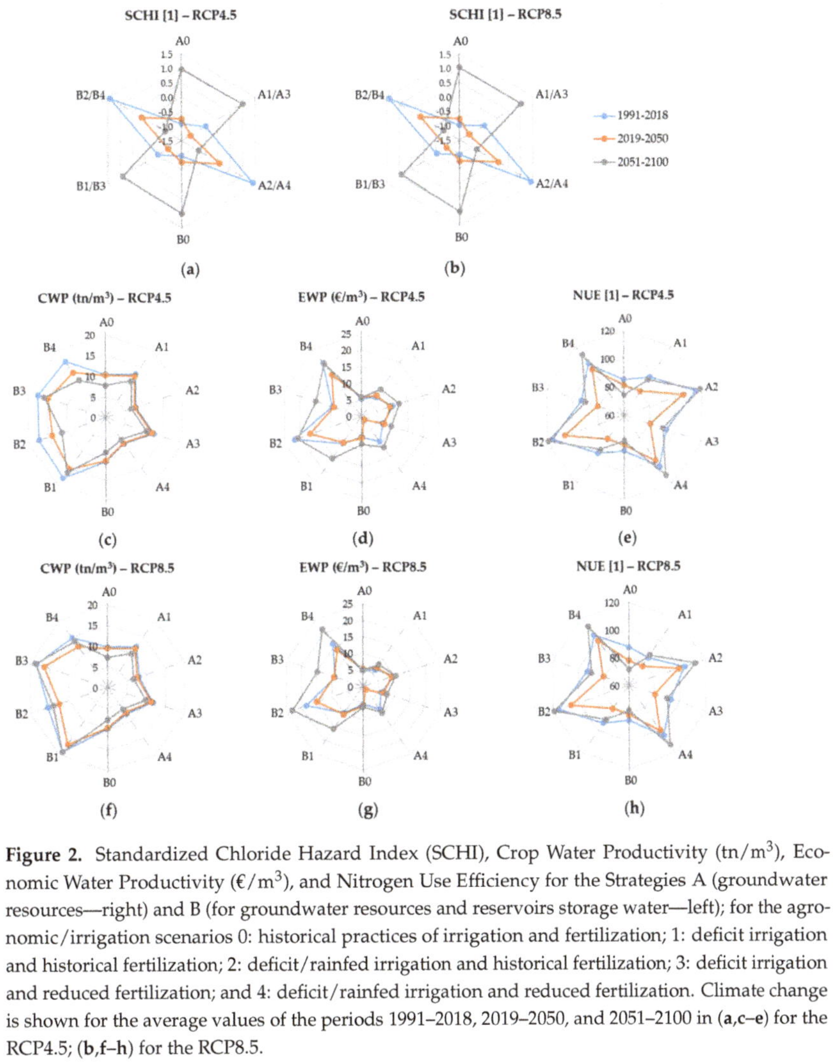

Figure 2. Standardized Chloride Hazard Index (SCHI), Crop Water Productivity (tn/m^3), Economic Water Productivity (€/m^3), and Nitrogen Use Efficiency for the Strategies A (groundwater resources—right) and B (for groundwater resources and reservoirs storage water—left); for the agronomic/irrigation scenarios 0: historical practices of irrigation and fertilization; 1: deficit irrigation and historical fertilization; 2: deficit/rainfed irrigation and historical fertilization; 3: deficit irrigation and reduced fertilization; and 4: deficit/rainfed irrigation and reduced fertilization. Climate change is shown for the average values of the periods 1991–2018, 2019–2050, and 2051–2100 in (**a**,**c**–**e**) for the RCP4.5; (**b**,**f**–**h**) for the RCP8.5.

The groundwater abstractions and salinity effects cause the productivity of water to decline mostly during 2019–2050. EWP scores follow the distribution of CWP. EWP ranges from 1.2 €/m^3 to 10.5 €/m^3 in RCP8.5 and to 12.2 €/m^3 in RCP4.5 for Strategy A. For Strategy B, EWP ranges from 5.4 €/m^3 to 22.9 €/m^3 in RCP8.5 and from 6.1 €/m^3 to 21.8 €/m^3 in RCP4.5. EWP will be increased after 2050, but this would occur possibly due to the increase of future commodity prices. NUE index scores show a declining trend in ongoing fertilizer practices until 2100 in both Strategies and RCPs. NUE values in deficit irrigation and rainfed agriculture/deficit irrigation form a V–shaped evolution with the lowest points to show in 2019–2050 in both Strategies and RCPs. NUE maximizes under the alternative practice of reduced fertilization in both strategies and RCPs. In reduced fertilization, NUE gets scores higher than 100 kg of crop yield per kg of nitrogen applied, while in other alternatives the values range from 71.2 kg yield/kg nitrogen to 93.2 kg yield/kg nitrogen.

4. Conclusions

Water resources adaptation and agronomic efficiency scenarios have been applied under two ensemble RCPs (4.5 and 8.5) in the Almyros basin in Thessaly, Greece, using an Integrated Modeling System (IMS). The results according to SCHI, indicate that the groundwater used for irrigation will pose moderate salinity hazards in 2019–2050 in all strategies. Crop yields are expected to decline for most of the cultivars due to irrigation with salinized water, except for the rainfed agriculture/deficit irrigation alternative. CWP could be improved in the future in Strategy B, and EWP scores might be, also, greatly increased. NUE ratings are equivalent to crop yields, which are affected by the salinity status of groundwater used for irrigation. Hence, the future course of seawater intrusion in the Almyros aquifer is a crucial problem that should be addressed in the near future with drastic measures of water resources adaptation to climatic changes and water needs in order to ensure the success of adaptation measures to climate change and the sustainability of local agriculture.

Author Contributions: Supervision, review and writing, conceptualization, and methodology by A.L. (Athanasios Loukas); writing review and editing, investigation, data curation, software, methodology, formal analysis by A.L. (Aikaterini Lyra). All authors have read and agreed to the published version of the manuscript.

Funding: This research is co-financed by Greece and the European Union (European Social Fund-ESF) through the Operational Programme «Human Resources Development, Education and Lifelong Learning» in the context of the project "Strengthening Human Resources Research Potential via Doctorate Research" (MIS-5000432), implemented by the State Scholarships Foundation (IKY).

Institutional Review Board Statement: Not applicable.

Informed Consent Statement: Not applicable.

Data Availability Statement: The results of this study are freely available.

Acknowledgments: This work was based on the Med-CORDEX initiative (www.medcordex.eu) (accessed on 1 February 2023).

Conflicts of Interest: The authors declare no conflict of interest.

References

1. Cheng, M.; Wang, H.; Fan, J.; Wang, X.; Sun, X.; Yang, L.; Zhang, S.; Xiang, Y.; Zhang, F. Crop yield and water productivity under salty water irrigation: A global meta-analysis. *Agr. Water Manag.* **2021**, *256*, 107105. [CrossRef]
2. Lyra, A.; Loukas, A. Simulation and Evaluation of Water Resources Management Scenarios Under Climate Change for Adaptive Management of Coastal Agricultural Watersheds. *Water Resour. Manag.* **2022**. [CrossRef]
3. Zaman, M.; Shahid, S.A.; Heng, L. Irrigation Water Quality. In *Guideline for Salinity Assessment, Mitigation and Adaptation Using Nuclear and Related Techniques*; Springer International Publishing: Cham, Switzerland, 2018; pp. 113–131.
4. Ayers, R.S.; Westcot, D.W. *Water Quality for Agriculture*; Food and Agriculture Organization of the United Nations: Rome, Italy, 1985; Volume 29.
5. Lyra, A.; Loukas, A.; Sidiropoulos, P.; Voudouris, K.; Mylopoulos, N. Integrated Modeling of Agronomic and Water Resources Management Scenarios in a Degraded Coastal Watershed (Almyros Basin, Magnesia, Greece). *Water* **2022**, *14*, 1086. [CrossRef]
6. Organisation for Economic Co-Operation and Development; Food and Agriculture Organization of the United Nations. *OECD-FAO Agricultural Outlook 2022–2031*; FAO: Rome, Italy, 2022.

environmental sciences proceedings

MDPI

Proceeding Paper

Eco-Hydrological Modelling of a Highly Managed Mediterranean Basin Using the SWAT+ Model: A Preliminary Approach †

Giuseppe Pulighe * and Flavio Lupia

CREA—Research Centre for Agricultural Policies and Bioeconomy, Via Barberini 36, 00187 Rome, Italy; flavio.lupia@crea.gov.it

* Correspondence: giuseppe.pulighe@crea.gov.it; Tel.: +39-0647856814

† Presented at the 7th International Electronic Conference on Water Sciences, 15–30 March 2023; Available online: https://ecws-7.sciforum.net/.

Abstract: Highly managed Mediterranean river basins are facing considerable environmental management challenges. Water resource managers are increasingly concerned about the allocation of limited water resources, environmental quality concerns, and planning under present and future climatic change and uncertainty. We implemented a watershed-scale eco-hydrological model on the Cervaro river basin (southern Italy) using the Soil and Water Assessment Tool (SWAT+) model at daily timesteps from 1990 to 2019. A high-precision land use map derived from the Integrated Administration and Control System (IACS)/Land Parcel Identification System (LPIS) was used to grasp detailed information regarding landscape patterns. This research is a preliminary approach for implementing and running the model with these highly detailed datasets. Future efforts should be oriented to fine-tune the baseline scenario considering agricultural management practices and to evaluate model performances for calibration and validation by assessing the goodness-of-fit objective function values.

Keywords: agriculture; watershed modelling; nutrients; simulation; land use/cover; LPIS

Citation: Pulighe, G.; Lupia, F. Eco-Hydrological Modelling of a Highly Managed Mediterranean Basin Using the SWAT+ Model: A Preliminary Approach. *Environ. Sci. Proc.* **2023**, *25*, 24. https://doi.org/10.3390/ECWS-7-14179

Academic Editor: Silvia Kohnova

Published: 14 March 2023

1. Introduction

Actual agricultural systems are based on intensive farming techniques that require deep plowing, and large usage of nitrogen (N), phosphorus (P), herbicides, pesticides, energy inputs, and land [1]. Leaching losses of sediments, nutrients, and chemicals from agricultural fields can lead to eutrophication of water and soil, deterioration of agroecological systems, and loss of ecosystem services. Agricultural production potential and the capacity to provide other ecosystem services in the future are both reduced as a result of pressures on agricultural systems and natural resources [2], endangering the main soil functions (such as filtering and serving as biological habitats) and their preservation [3]. Agricultural systems face several pressing challenges in resolving the apparent conflict between human needs and resource sustainability [4].

The management of the environment is a significant problem in highly managed Mediterranean river basins. Many solutions have been put out for addressing management issues and local environmental effects on agroecological systems. The use of eco-hydrological modelling for ex-ante quantification of nonpoint source pollution and soil erosion in agroecosystems is a useful approach for implementing sustainable agricultural systems [5]. Many studies have found that eco-hydrological models are effective in simulating the intricate interactions of water, nutrients (N and P), contaminants, and vegetation systems in both natural and agricultural ecosystems.

These models can help researchers and practitioners to better understand and manage these ecosystems, leading to improved crop yields and reduced environmental impact.

Additionally, eco-hydrological models can be used to predict the effects of climate change and other disturbances on agricultural ecosystems, allowing for more effective planning and adaptation. One of the most frequently utilized models is the Soil Water Assessment Tool (SWAT) [6]. Examples include streamflow and nutrient loadings [7], modelling of soil erosion [8], best management practices for reducing fertilizer application [9], and climate change studies [10].

This paper focuses on the application of the SWAT+ model in a Mediterranean river basin to evaluate the influence of anthropogenic management on water balance components, nutrients, and sediment loads. The model employs a high-precision land use map produced from the administrative geodata of the Italian Integrated Administration and Control System (IACS)/Land Parcel Identification System (LPIS) used for monitoring the Common Agricultural Policy (CAP) subsidies. High-resolution land use data provide precise information on landscape patterns and allow us to better represent the hydrological cycle. The model performances were assessed by comparing simulated and observed streamflow data and calculating a goodness-of-fit objective function. The preliminary findings show that the model reproduced the water balance in the watershed well. Future work will be oriented to analyze the effects of alternative management practices on water quality, sediment loads, and soil erosion.

2. Materials and Methods

2.1. Study Area

This study was carried out in the Cervaro river basin, which is located in the Apulia Region (southern Italy) between 41°07′–41°32′ N latitude and 15°06′–15°54′ W longitude (Figure 1). The Cervaro basin has an area of 841 km^2 and ranges in elevation from 0 to 1100 m on the southern side of the Daunia Mountains. The river system consists of the main river course and various major and secondary-order tributaries. The climatic regime is Mediterranean with a bimodal pattern of precipitation distribution with rainy winters and hot summers.

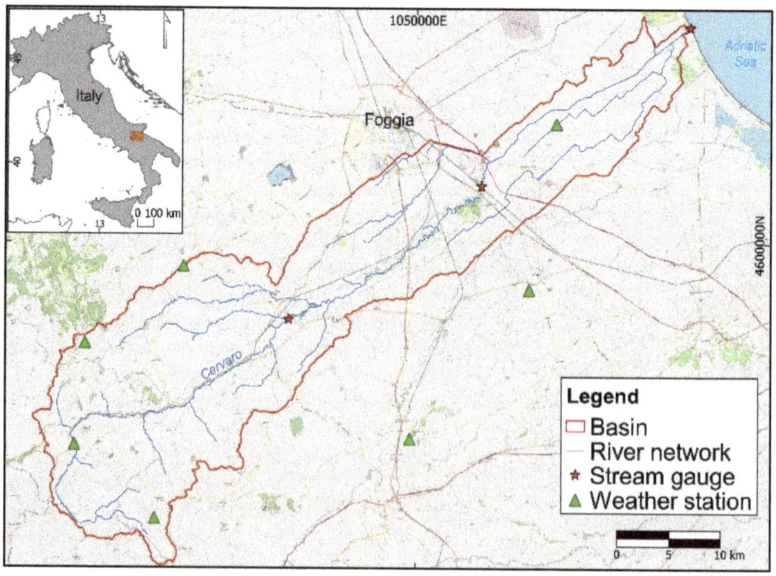

Figure 1. Location of the study area.

2.2. SWAT+ Model Setup

SWAT is a semi-distributed eco-hydrological model that segments a watershed and its sub-watersheds into homogenous geographical units known as Hydrologic Response Units (HRUs), each of which have a distinct combination of land use, soil, and slope [6]. The HRUs are used to determine the soil–water balance. HRUs represent the minimum computational unit within which the soil–water balance is determined. In this study, we used the new restructured SWAT+ model that is more efficient and flexible in terms of model construction and configuration.

2.3. Dataset and Model Setup

The SWAT+ model requires the following geospatial data for implementation (Table 1): land use/cover data, soil data, Digital Elevation Model (DEM), streamflow data, and climate data. In the first step, the DEM was used to delineate the basin and create sub-basin boundaries, while in the second step HRUs were created in conjunction with land use and soil data. In the third step, climate data for the period 1990–2019 were used to set up and run the model as the baseline scenario.

For land use data, we specifically developed a detailed land use map for the study area with 57 thematic classes from the IACS/Land Parcel Identification System (LPIS) conflation [11,12]. More than 70% of the area is devoted to agricultural uses. The SWAT+ model (rev. 60.5.4 – SWAT+ Editor 2.1) was implemented using a QGIS interface and run at daily timesteps from 1 January 1990 to 31 December 2019 considering a 4-year warm-up. The study area was discretized into 7345 HRUs. The Penman–Monteith method was used to calculate potential evapotranspiration.

Table 1. Data sources used to implement the SWAT+ model.

Data	Resolution—Scale	Description	Source
Land use	10 m	57 land use types	[12]
Soil data	10 m	16 soil units	[13]
DEM	10 m	elevation data	[13]
Streamflow	Daily (m^3/s); 1990–2012	runoff at gauge stations	[13]
Climate data	Daily; 1990–2019	temperature, rainfall	[13]

3. Results

The average annual values of the water balance for the whole simulation period are depicted in Figure 2, while monthly water balance components are reported in Table 2. Interannual variability showed that precipitations were concentrated in winter months, while in summer months the average precipitation remained quite stable due to the influence of the gauge stations in the Daunia mountains which have a maximum altitude of 1000 m.

Table 2. Water balance average monthly values [1].

Month	Precipitation	Runoff	Lateral Flow	Water Yield	ET	PET
January	74.1	9.5	9.3	9.5	8.9	9.3
February	59.3	7	25.4	7.0	23.7	25.4
March	67.4	7.2	62.9	7.3	58.3	62.9
April	63.9	2.8	104.6	2.8	95.3	104.6
May	51.7	1.6	151.2	1.6	118.2	151.2
June	43.8	1.5	172.9	1.5	92.7	172.9
July	35.9	1.1	185.2	1.2	58.8	185.2
August	34.3	0.9	158.5	1.0	38.8	158.5
September	58.0	1.7	102.7	1.8	49.0	102.7
October	66.5	2.4	56.0	2.4	36.2	56.0
November	75.5	2.7	17.8	2.8	15.6	17.8
December	71.8	4.5	5.2	4.6	5.1	5.2

[1] All values in mm.

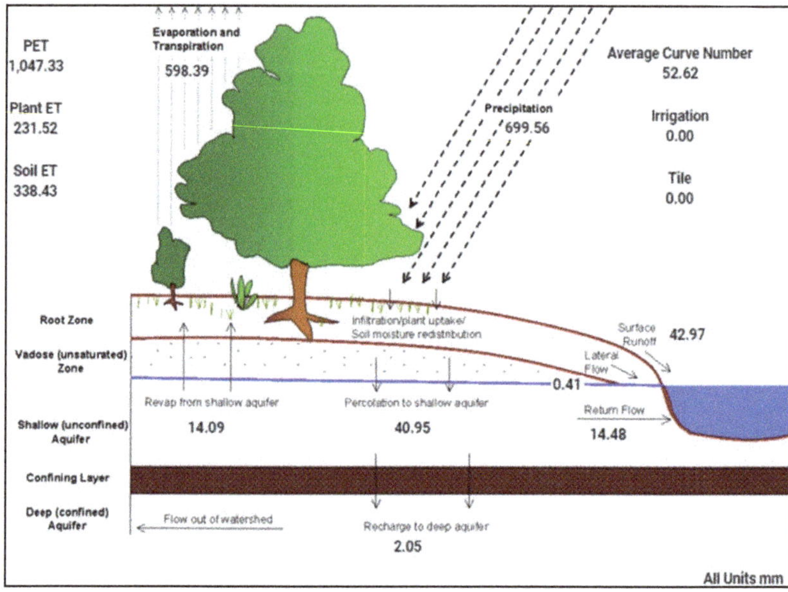

Figure 2. Schematic representation of average annual values of water balance. All units in mm.

The results showed that evapotranspiration was the predominant water balance component, with 598.39 mm year^{-1}, accounting for approximately 85.5% of the annual precipitation (699.5 mm), followed by surface runoff (42.97 mm year^{-1}, representing about 6.1% of the annual precipitation) and water yield (water discharged in the channels, 43.4 mm year^{-1}). The potential evapotranspiration component was equal to 1047.33 mm year^{-1}, while the average precipitation was equal to 699.5 mm year^{-1}. The average annual outflow for the whole simulation period was 0.44 m^3 s^{-1}, in line with the values registered for conterminous river basins in the Apulia region [14], confirming the overall reliability of the model.

4. Conclusions

This study presented a preliminary quantitative assessment of the eco-hydrological processes using the SWAT+ model in a highly managed Mediterranean watershed. The results of the preliminary model set-up of a baseline scenario in the Cervaro river basin (southern Italy) can be summarized as follows:

- The river basin has a vast and heterogeneous agricultural plan with 57 land use types that can strongly influence the water balance with management practices such as planting, irrigation, and fertilization;
- Overall, the SWAT+ model performed well and was able to adequately and reasonably represent the main elements of the hydrological budget; future work should be oriented toward calibrating and validating the model to adjust model parameters toward more realistic model results;
- The use of detailed land use/cover maps as was used in this study can improve the representation of hydrological models in complex agricultural systems, enabling the implementation of best management practices in the future.

Author Contributions: Conceptualization, G.P. and F.L.; methodology, G.P. and F.L.; software, G.P. and F.L.; formal analysis, G.P. and F.L.; data curation, G.P. and F.L.; writing—original draft preparation, G.P. and F.L.; visualization, G.P. and F.L. All authors have read and agreed to the published version of the manuscript.

Funding: The work was carried out in the framework of the project "Open LOD platform based on HPC capabilities for Integrated Administration of Common Agriculture Policy (Open IACS)" co-financed by the Connecting Europe Facility of the European Union by the Innovation and Networks Executive Agency, Agreement No. INEA/CEF/ICT/A2018/1815914.

Institutional Review Board Statement: Not applicable.

Informed Consent Statement: Not applicable.

Data Availability Statement: Regione Puglia Annali e Dati Idrologici Elaborati—Regione Puglia. Available online: https://protezionecivile.puglia.it/annali-e-dati-idrologici-elaborati (accessed on 10 October 2022).

Acknowledgments: We would like to thank Diego Gallinelli and Pietro Macedoni for their contribution to the creation of the land use map.

Conflicts of Interest: The authors declare no conflict of interest.

References

1. United Nations. *Department of Economics and Social Affairs, Population Division. World Population Prospects 2019: Highlights. ST/ESA/SER.A/423*; United Nations: New York, NY, USA, 2019.
2. Tóth, G.; Montanarella, L.; Rusco, E. *Threats to Soil Quality in Europe*; European Commission, Joint Research Centre: Ispra, Italy, 2008; ISBN 9789279095290.
3. Sonderegger, T.; Pfister, S. Global Assessment of Agricultural Productivity Losses from Soil Compaction and Water Erosion. *Environ. Sci. Technol.* **2021**, *55*, 12162–12171. [CrossRef] [PubMed]
4. Springmann, M.; Clark, M.; Mason-D'Croz, D.; Wiebe, K.; Bodirsky, B.L.; Lassaletta, L.; de Vries, W.; Vermeulen, S.J.; Herrero, M.; Carlson, K.M.; et al. Options for Keeping the Food System within Environmental Limits. *Nature* **2018**, *562*, 519–525. [CrossRef] [PubMed]
5. Saiz-Rubio, V.; Rovira-Más, F. From Smart Farming towards Agriculture 5.0: A Review on Crop Data Management. *Agronomy* **2020**, *10*, 207. [CrossRef]
6. Arnold, J.R.; Kiniry, R.; Srinivasan, R.; Williams, J.R.; Haney, E.B.; Neitsch, S.L. *Soil & Water Assessment Tool—Input/Output Documentation Version 2012*; TR-439; Texas Water Resources Institute: College Station, TX, USA, 2012.
7. Pulighe, G.; Bonati, G.; Colangeli, M.; Traverso, L.; Lupia, F.; Altobelli, F.; Dalla Marta, A.; Napoli, M. Predicting Streamflow and Nutrient Loadings in a Semi-Arid Mediterranean Watershed with Ephemeral Streams Using the SWAT Model. *Agronomy* **2019**, *10*, 2. [CrossRef]
8. Himanshu, S.K.; Pandey, A.; Yadav, B.; Gupta, A. Evaluation of Best Management Practices for Sediment and Nutrient Loss Control Using SWAT Model. *Soil Tillage Res.* **2019**, *192*, 42–58. [CrossRef]
9. Wang, W.; Xie, Y.; Bi, M.; Wang, X.; Lu, Y.; Fan, Z. Effects of Best Management Practices on Nitrogen Load Reduction in Tea Fields with Different Slope Gradients Using the SWAT Model. *Appl. Geogr.* **2018**, *90*, 200–213. [CrossRef]
10. Pulighe, G.; Lupia, F.; Chen, H.; Yin, H. Modeling Climate Change Impacts on Water Balance of a Mediterranean Watershed Using SWAT+. *Hydrology* **2021**, *8*, 157. [CrossRef]
11. Lupia, F.; Rizzi, D.; Gallinelli, D.; Macedoni, P.; Pierangeli, F.; Carfì, S.; Pulighe, G. High Resolution Land Use Map for Eco-Hydrological Modelling from IACS/LPIS Geodata Conflation. *Abstr. ICA* **2021**, *3*, 1–2. [CrossRef]
12. Open IACS Project—Open LOD platform based on HPC capabilities for Integrated Administration and Control System of Common Agrarian Policy (Open IACS). Available online: https://researchportal.uc3m.es/display/act544682 (accessed on 10 January 2023).
13. Regione Puglia Sistema Informativo Territoriale Regione Puglia. Available online: http://www.sit.puglia.it/ (accessed on 10 December 2022).
14. D'Ambrosio, E.; De Girolamo, A.M.; Barca, E.; Ielpo, P.; Rulli, M.C. Characterising the Hydrological Regime of an Ungauged Temporary River System: A Case Study. *Environ. Sci. Pollut. Res.* **2017**, *24*, 13950–13966. [CrossRef] [PubMed]

environmental sciences proceedings

MDPI

Proceeding Paper

Effects of Weed Removal Practices on Soil Organic Carbon in Apple Orchards Fields †

Dimitrios Malamataris *, Vassilios Pisinaras, Konstantinos Babakos, Anna Chatzi, Evangelos Hatzigiannakis, Vasiliki Kinigopoulou, Ioannis Hatzispiroglou and Andreas Panagopoulos

Soil and Water Resources Institute, Hellenic Agricultural Organization "DEMETER", 57400 Sindos, Greece; v.pisinaras@swri.gr (V.P.); k.babakos@swri.gr (K.B.); a.chatzi@swri.gr (A.C.); e.hatzigiannakis@swri.gr (E.H.); v.kinigopoulou@swri.gr (V.K.); i.hatzispiroglou@swri.gr (I.H.); a.panagopoulos@swri.gr (A.P.)

* Correspondence: d.malamataris@swri.gr; Tel.: +30-2310798790 (ext. 216)

† Presented at the 7th International Electronic Conference on Water Sciences, 15–30 March 2023; Available online: https://ecws-7.sciforum.net.

Abstract: The accelerated climate crisis has exacerbated the existing water and soil management challenges in the Mediterranean region, which are usually attributed to the combination of both irrational irrigation and unsustainable farming practices. The current conditions and future projections indicate that water-related risks are expected to intensify during the coming decades. Moreover, farmers often do not possess high environmental awareness; they adopt non-sustainable farming practices such as the extensive use of herbicides instead of mowing/mulching for the weeds, thus affecting soil hydraulic characteristics and fertility. To investigate the effects of different weed-management practices on soil organic carbon and thus on soil water holding capacity and infiltrability, an extensive soil-sampling campaign was performed in the semi-arid Mediterranean agricultural pilot basin of Agia—Greece. The pilot is located in the Pinios river basin, which constitutes the most highly productive agricultural plain in the country. The Agia basin was selected since it presents the uneven spatiotemporal distribution of groundwater resources and the wide application of herbicides, while an urgent need exists to sustain and improve agricultural production, with the main crops being apples and cherries. Moreover, the Agia basin constitutes a highly instrumented area where the Pinios Hydrologic Observatory belonging to the International Long Term Ecological Research network has been developed, and thus additional field measurements could contribute to the overall data-collection framework. Soil sampling was conducted in apple orchards in April 2022, just before the beginning of the growing season. Ninety six soil samples in total were collected from eight different fields; half of them applied systematic herbicides treatment, and others mulching. For the upper soil profile (0–10 cm depth), the results indicate that soil organic carbon in the fields applying mowing was found to be higher by more than 30% compared to the fields applying herbicides. The corresponding difference for soil depth of 10–30 cm was 7%, thus demonstrating the effectiveness of mulching in increasing soil organic carbon. The results of the current study could be upscaled at a larger scale in the context of adapting agricultural water-stressed regions to climate change, whilst contributing significantly to the production cost and the preservation of the ecosystemic values of the regional nexus.

Keywords: pinios basin; Mediterranean area; nature-based solutions; soil organic carbon; soil sampling; mulching; sustainable agriculture; environmental protection; soil protection; vegetation

Citation: Malamataris, D.; Pisinaras, V.; Babakos, K.; Chatzi, A.; Hatzigiannakis, E.; Kinigopoulou, V.; Hatzispiroglou, I.; Panagopoulos, A. Effects of Weed Removal Practices on Soil Organic Carbon in Apple Orchards Fields. *Environ. Sci. Proc.* **2023**, *25*, 25. https://doi.org/10.3390/ECWS-7-14185

Academic Editor: Luis Garrote

Published: 14 March 2023

1. Introduction

Soil is considered the largest carbon pool, accounting for about 2500 Gt, followed by the atmospheric (760 Gt) and biotic (560 Gt) pools [1]. Soil organic carbon (SOC) includes plants, roots, fungi, and microbial residues and accounts for more than 60% of global total soil carbon [1]. The understanding of weed-removal practices, including herbicides application and mulching, which affect soil carbon balance dynamics, is crucial for yield-potential preservation, terrestrial ecosystem conservation, and food security [2,3]. Sustainable weed management constitutes a challenge for ensuring agricultural productivity since in the case of improper management, infestations of grass and broadleaf weeds could reduce crops yield and quality. The level of weed and crop competition depends on soil and environmental conditions, weed density, and agricultural practices, including crop row spacing, planting density, and the presence of competitive cultivars [4].

Herbicides constitute the primary tactic to address weed control and are widely used, mainly because they are easily applied and improve the cost/benefit ratio of the agronomic business. However, their repeated use may evolve weed resistance to multiple mechanisms of herbicide action [5] and result in herbicide-resistant weed growth in the case of some cultivations [6]. Moreover, their systematic use may adversely affect soil hydraulic properties. Nowadays, attempts are being made to eliminate the use of herbicides in favor of environmental protection and the appearance of herbicide-resistant weed population avoidance.

Mulching is widely recognized as one of the most efficient weed and soil management techniques in terms of soil structure improvement, soil moisture evapotranspiration reduction, soil water holding, and soil erosion mitigation, all of which support plant growth [7,8]. The effectiveness of mulching depends on the selected technique, the soil characteristics, and the type of crops residue, while the benefits are usually developing slowly and observed after a long period, which may exceed six or more growing seasons [9]. Mowing and mulching techniques seem to increase SOC content and carbon (C) storage, thus reducing greenhouse gas emissions [10], although relevant studies regarding mulching effects on SOC concentration present contradictory results [8,11,12]. This uncertainty relies on complex and labile soil carbon fractions, including light organic carbon, dissolved organic carbon, particulate organic carbon, and easily oxidizable organic carbon, which present a high spatiotemporal distribution based on local climate, soil texture, and soil-management conditions, as well as a large effect on soil biochemical processes, nutrients, and carbon cycling [7].

The purpose of this study is to evaluate the effects of herbicides and mulching technique application on soil organic carbon in low sloping apple orchard fields.

2. Materials and Methods

2.1. Study Area and Soil Analysis

The Agia basin in central Greece, where the Pinios Hydrologic Observatory was established in 2015, covers an area of 52.5 km^2 [13]. Based on the Digital Elevation Model produced by the Hellenic Cadastre S.A. at 5 m resolution, the Agia basin spans altitudes varying from 94 to 1520 m a.s.l. (Figure 1). Land use includes forested, agricultural, and urbanized areas. Agriculture, which is characterized by land fragmentation leading to small land parcels in the order of 1–1.5 ha on average, is the dominant socio-economic activity. Apple and cherries orchards form the key cultivations, the irrigation needs of which are covered by groundwater. Considerable spatio-temporal differentiation of groundwater distribution results in seasonal availability deficiency at specific high productivity zones. During such periods, groundwater abstraction depths are considerable, which, along with the increased energy costs, make irrigation management and water saving essential.

2.2. Field Experiment

Soil texture in the study area is classified as loam (68.38%), sandy loam (24.86%), silt loam (5.00%), silty clay loam (1.15%), sandy clay loam (0.42%), and clay loam (0.18%). Soil also presents a pH of 6.94, 7.05, and 7.06; 42.66%, 42.97%, and 41.09% sand; 38.42%, 38.14%, and 37.16% silt; and 18.88%, 18.85%, and 21.70% clay, at 0–5 cm, 5–15 cm, and 15–30 cm depth, respectively, based on field measurements conducted in July 2020 along with data derived from the SoilGrids system [14].

Field studies were conducted at the beginning of the previous growing season, i.e., April 2022 in the Agia water basin. In particular, soil samples were collected at 12 points in 8 different orchard parcels; half of them by applying herbicides and the rest by implementing the mulching technique (Figure 1). The area of parcels ranges from 1.8 to 19.1 ha with an average value of 7.2 ha. Samples were collected at 2 different depths, i.e., 0–10 cm and 10–30 cm, and thus 96 soil samples in total were collected and analyzed.

Organic carbon content was determined using a wet combustion technique, according to the Wakley–Black method [15]. Soil samples were air-dried, and the fraction of fine earth (<2 mm) was used for the analysis. In total, 1 g of soil was transferred into a 500-mL wide-mouth Erlenmeyer flask; 10 mL of 1 N $K_2Cr_2O_7$ and 20 mL of concentrated H_2SO_4 were added, and the flask was swirled until soil and reagents were mixed. After 30′ standing, 200 mL of deionized water was added to the flask, while 3–4 drops of diphenylamine were used as an indicator for the titration with 0.5 N $FeSO_4$.

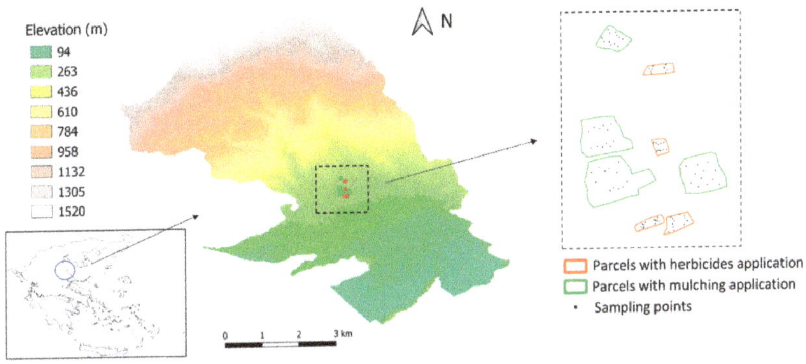

Figure 1. Study area and location of field parcels under study.

3. Results and Discussion

Figure 2 presents high in-field variability of soil organic carbon in terms of both weed removal practice and soil depth. The difference between the mean SOC in fields applying herbicides and mulching was also found to be statistically significant, as indicated using the student's *t*-test method. SOC was found to be 31% and 7% higher for the fields applying mulching compared to those applying herbicides for depths 0–10 cm and 10–30 cm, respectively. The results confirm the importance of mulching application in SOC fractions, particularly in the upper soil layers [7,16].

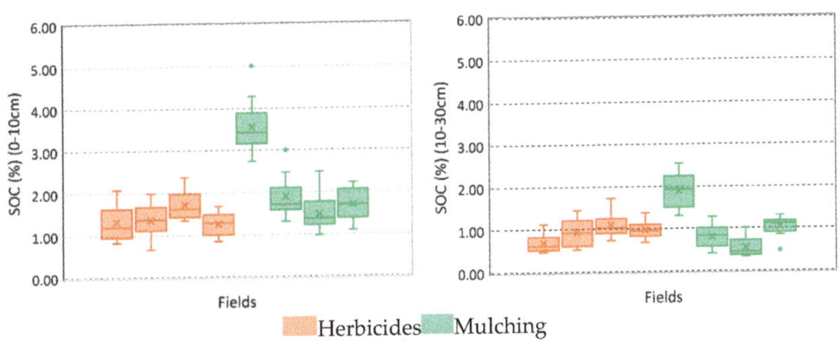

Figure 2. Soil organic carbon (%) for the two different weed removal practices at two different soil depths.

Increased SOC documented in the fields where mulching was applied provides multiple assets to the farmers that have adopted it. Soil fertility significantly improves, especially when SOC increase relates to crop residue retention [17]. Soil water holding capacity is also positively affected, despite the fact that the latest literature suggests that actual improvement is rather limited [18]. Precision agriculture and environmentally friendly pest and weed management are adopted techniques in the framework of the Green Deal and the Farm to Fork policies that call for a 50% reduction in agrochemicals used in the European Union Member States by the year 2030. Hence, compliance with these regulations and the environmental targets set to call for the systematic adoption of techniques such as mulching is necessary. Additionally, the progressive increase in agrochemicals as a result of the energy crisis and the declining number of licensed drastic compounds for use in the member states is a strong incentive to convert to nature-based solutions such as mulching.

4. Conclusions

This research provides a scientific basis and confirms the value of mulching as a sustainable although time-consuming agricultural method to improve soil organic carbon sequestration. Long-term effects of mulching have to be tracked to identify whether SOC improvement reaches a steady state related to the saturation of SOC pools. Future studies are also needed to quantify associated economic benefits and agricultural sustainability related to carbon-transformation processes.

Author Contributions: Conceptualization, V.P., A.P., E.H. and D.M.; methodology, V.P., A.P. and E.H.; investigation, E.H., D.M., I.H. and A.C.; resources, V.P., D.M., V.K. and K.B.; data curation, V.P., D.M. and V.K.; writing—original draft preparation, D.M.; writing—review and editing, A.P. and V.P.; visualization, D.M. and V.P.; supervision, A.P., V.P. and E.H.; project administration, A.P. and V.P.; and funding acquisition, A.P. and V.P. All authors have read and agreed to the published version of the manuscript.

Funding: This research was funded by PRIMA program supported by the European Union, grant number 2041 (LENSES—leaning and action alliances for nexus environments in an uncertain future) (Call 2020 Section 1 Nexus IA).

Institutional Review Board Statement: Not applicable.

Informed Consent Statement: Not applicable.

Data Availability Statement: Not applicable.

Conflicts of Interest: The authors declare no conflict of interest.

References

1. Lal, R. Soil carbon sequestration impacts on global climate change and food security. *Science* **2004**, *304*, 1623–1627. [CrossRef] [PubMed]
2. Beiermann, C.W.; Miranda, J.W.; Creech, C.F.; Knezevic, S.Z.; Jhala, A.J.; Harveson, R.; Lawrence, N.C. Critical timing of weed removal in dry bean as influenced by the use of preemergence herbicides. *Weed Technol.* **2022**, *36*, 168–176. [CrossRef]
3. Gu, Y.J.; Han, C.L.; Kong, M.; Siddique, K.H.; Li, F.M. Film Mulching with Low Phosphorus Application Improves Soil Organic Carbon and Its Decomposability in a Semiarid Agroecosystem. *Agriculture* **2022**, *12*, 816. [CrossRef]
4. Nedeljković, D.; Knežević, S.; Božić, D.; Vrbničanin, S. Critical Time for Weed Removal in corn as influenced by planting pattern and PRE herbicides. *Agriculture* **2021**, *11*, 587. [CrossRef]
5. Owen, M.D. Diverse approaches to herbicide-resistant weed management. *Weed Sci.* **2016**, *64*, 570–584. [CrossRef]
6. Kumar, V.; Jha, P. Influence of nitrogen rate, seeding rate, and weed removal timing on weed interference in barley and effect of nitrogen on weed response to herbicides. *Weed Sci.* **2017**, *65*, 189–201. [CrossRef]
7. Gu, C.; Liu, Y.; Mohamed, I.; Zhang, R.; Wang, X.; Nie, X.; Jiang, M.; Brooks, M.; Chen, F.; Li, Z. Dynamic changes of soil surface organic carbon under different mulching practices in citrus orchards on sloping land. *PLoS ONE* **2016**, *11*, e0168384. [CrossRef] [PubMed]
8. Chen, J.; Heiling, M.; Resch, C.; Mbaye, M.; Gruber, R.; Dercon, G. Does maize and legume crop residue mulch matter in soil organic carbon sequestration? *Agric. Ecosyst. Environ.* **2018**, *265*, 123–131. [CrossRef]
9. Affholder, F.; Jourdain, D.; Quang, D.D.; Tuong, T.P.; Morize, M.; Ricome, A. Constraints to farmers' adoption of direct-seeding mulch-based cropping systems: A farm scale modeling approach applied to the mountainous slopes of Vietnam. *Agric. Syst.* **2010**, *103*, 51–62. [CrossRef]
10. Smith, P.; Martino, D.; Cai, Z.; Gwary, D.; Janzen, H.; Kumar, P.; McCarl, B.; Ogle, S.; O' Mara, F.; Rice, C.; et al. Greenhouse gas mitigation in agriculture. *Philos. Trans. R. Soc. B Biol. Sci.* **2008**, *363*, 789–813. [CrossRef]

11. Qiu, Y.; Xie, Z.; Wang, Y.; Malhi, S.S.; Ren, J. Long-term effects of gravel—Sand mulch on soil organic carbon and nitrogen in the Loess Plateau of northwestern China. *J. Arid. Land* **2015**, *7*, 46–53. [CrossRef]
12. Luo, S.; Zhu, L.; Liu, J.; Bu, L.; Yue, S.; Shen, Y.; Li, S. Sensitivity of soil organic carbon stocks and fractions to soil surface mulching in semiarid farmland. *Eur. J. Soil Biol.* **2015**, *67*, 35–42. [CrossRef]
13. Pisinaras, V.; Panagopoulos, A.; Herrmann, F.; Bogena, H.R.; Doulgeris, C.; Ilias, A.; Tziritis, E.; Wendland, F. Hydrologic and geochemical research at Pinios Hydrologic Observatory: Initial results. *Vadose Zone J.* **2018**, *17*, 1–16. [CrossRef]
14. SoilGrids. Available online: https://soilgrids.org/ (accessed on 24 January 2023).
15. Nelson, D.W.; Sommers, L.E. Total Carbon, Organic Carbon, and Organic Matter. In *Methods of Soil Analysis*; Bigham, J.M., Bartels, J.M., Eds.; Soil Science Society of America, Inc.: Madison, WI, USA; American Society of Agronomy, Inc.: Fitchburg, WI, USA, 1996; pp. 961–1010.
16. Xiang, Y.; Li, Y.; Liu, Y.; Zhang, S.; Yue, X.; Yao, B.; Xue, J.; Lv, W.; Zhang, L.; Xu, X.; et al. Factors shaping soil organic carbon stocks in grass covered orchards across China: A meta-analysis. *Sci. Total Environ.* **2022**, *807*, 150632. [CrossRef] [PubMed]
17. Githongo, M.; Kiboi, M.; Muriuki, A.; Fliessbach, A.; Musafiri, C.; Ngetich, F.K. Organic Carbon Content in Fractions of Soils Managed for Soil Fertility Improvement in Sub-Humid Agroecosystems of Kenya. *Sustainability* **2022**, *15*, 683. [CrossRef]
18. Lal, R. Soil organic matter and water retention. *Agron. J.* **2020**, *112*, 3265–3277. [CrossRef]

environmental sciences proceedings

MDPI

Proceeding Paper

Variation in Water Quality in an Impacted Coastal Lagoon over the Last Decade (Küçükçekmece Lagoon, Turkey) †

Latife Köker *, Emine Gozde Ozbayram, Ayça Oğuz Çam, Reyhan Akçaalan and Meriç Albay

Faculty of Aquatic Sciences, Istanbul University, Istanbul 34134, Turkey; gozde.ozbayram@istanbul.edu.tr (E.G.O.); ayca.oguzcam@istanbul.edu.tr (A.O.Ç.); akcaalan@istanbul.edu.tr (R.A.); merbay@istanbul.edu.tr (M.A.)

* Correspondence: latife.koker@istanbul.edu.tr

† Presented at the 7th International Electronic Conference on Water Sciences, 15–30 March 2023; Available online: https://ecws-7.sciforum.net/.

Abstract: Küçükçekmece Lagoon, located inside the Istanbul metropolitan area, is connected to the Marmara Sea by a small canal. Due to the construction of a dam on the Sazlıdere stream, which is the most important feeding source for the lagoon, there has been a decrease in freshwater inflow, so the amount of salinity in the lagoon has started to increase. While salinity was around 11 ppt in the surface water of the lagoon in the 2010s, the level of salinity concentration exceeds 17 ppt today which also has an impact on the ecosystem. The aim of this study is to evaluate the water quality changes in Kucukcekmece Lagoon in a decade. The water quality revealed a high spatial and temporal variation in the lagoon and the bottom (ca 18 m) waters were rich in H_2S, of which the highest concentration was measured as 215 mg/L. Overall, from the large dataset of water quality for more than ten years, there is an obvious effect of anthropogenic activities and the closure of freshwater inlets on the trophic conditions of the lagoon.

Keywords: pollution; water quality; Küçükçekmece Lagoon; H_2S

Citation: Köker, L.; Ozbayram, E.G.; Oğuz Çam, A.; Akçaalan, R.; Albay, M. Variation in Water Quality in an Impacted Coastal Lagoon over the Last Decade (Küçükçekmece Lagoon, Turkey). *Environ. Sci. Proc.* **2023**, *25*, 26. https://doi.org/10.3390/ECWS-7-14246

Academic Editor: Athanasios Loukas

Published: 16 March 2023

1. Introduction

Coastal lagoons are complex and dynamic ecosystems that show sharp distributional gradients and short-term and seasonal variability in their physical, chemical, and biological properties [1]. As a result of long hydrologic retention times, lagoons are ecosystems which are very vulnerable to chemical contamination [2]; pollutants are consequently retained longer [3].

Küçükçekmece Lagoon, located in Istanbul, Turkey, suffers from many problems, which might lead to environment degradation, differences in salinity levels, and substantial changes in its ecosystem. It has been affected by population pressure around the lagoon, wastewater discharges, algal blooms, and salinity increases due to the cutoff of freshwater inputs [4].

In this study, the spatial and seasonal variability of physicochemical parameters and the chlorophyll-*a* were assessed in a coastal lagoon connected to the sea. Our main goal was to understand the nutrient dynamics, including the importance of water exchanges between the lagoon and the adjoining area, and determination of the effect of anthropogenic activities and the closure of freshwater inlets on the trophic conditions of the lagoon for more than ten years.

2. Materials and Methods

Küçükçekmece Lagoon is located in the south-west of Istanbul (Figure 1). It has a direct connection to the Marmara Sea with a canal. It has a 15.22 km^2 surface area and a maximum depth of 18.5 m. Because of rapid urbanization and industrialization, a significant deterioration in water quality was observed due to intense wastewater discharge.

A gradual increase in eutrophication has resulted in cyanobacterial blooms from the 1990s to the 2010s [5,6].

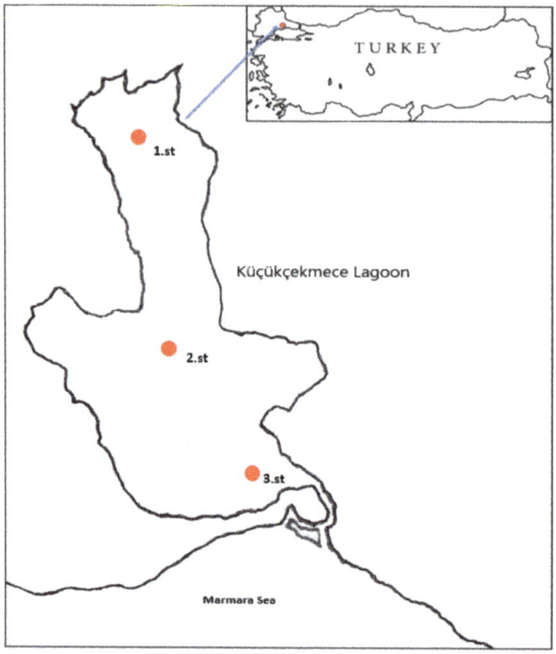

Figure 1. Location of sampling sites in Küçükçekmece Lagoon.

To compare water quality variations, samples were collected from June to December 2012 and 2022 at regular monthly intervals. Water samples were collected from the surface in two stations (first and third station), and to observe the depth profile, the samples were collected throughout the water column (surface, 9 m, and 18 m) from the second station. Dissolved oxygen (DO), salinity, and temperature were measured in situ using the YSI multiparameter instrument (YSI 650 MDS). Chlorophyll-*a* (chl-*a*) was determined using the method of ISO 10260 (1992) [7]. Nutrient analyses for nitrite ($N0_2$), nitrate ($N0_3$), total phosphorus (TP), and silicate were performed according to APHA-AWWA WPCF (1989) [8].

3. Results and Discussion

The physicochemical characteristics of the surface waters of Küçükçekmece Lagoon are given in Figure 2. The water temperature ranged between 8.1 and 28.2 °C. The salinity diagrams (Figures 2 and 3) showed that seawater inflow influences the lagoon's characteristics. It was determined that the salinity level doubled from 2012 (mean: 7.44 ppt) to 2022 (mean: 14.1 ppt). Due to the increase in salinity, the detected cyanobacteria members between 1990 and 2015 were left to marine species instead. This revealed that chl-*a* decreased in 2022. While the mean value of total phosphorus concentration was 712.1 μg/L in 2012, it was measured as 370 μg/L in 2022. The mean value of nitrate and nitrite concentrations increased from 0.72 mg/L to 2.9 mg/L in 2022. All these measurements showed that the trophic status of Küçükçekmece Lagoon was hypertrophic since the 1990s and did not show any sign of enhancement.

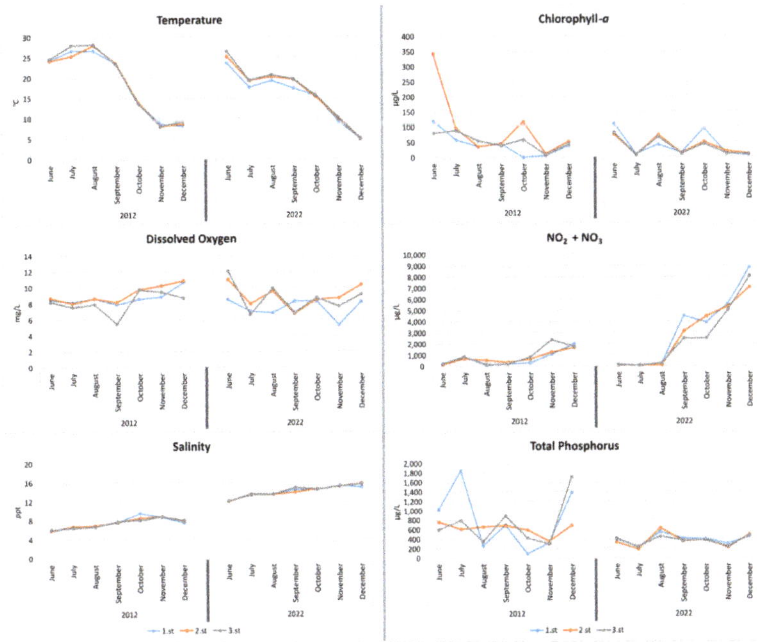

Figure 2. Physicochemical characteristics of surface waters of stations.

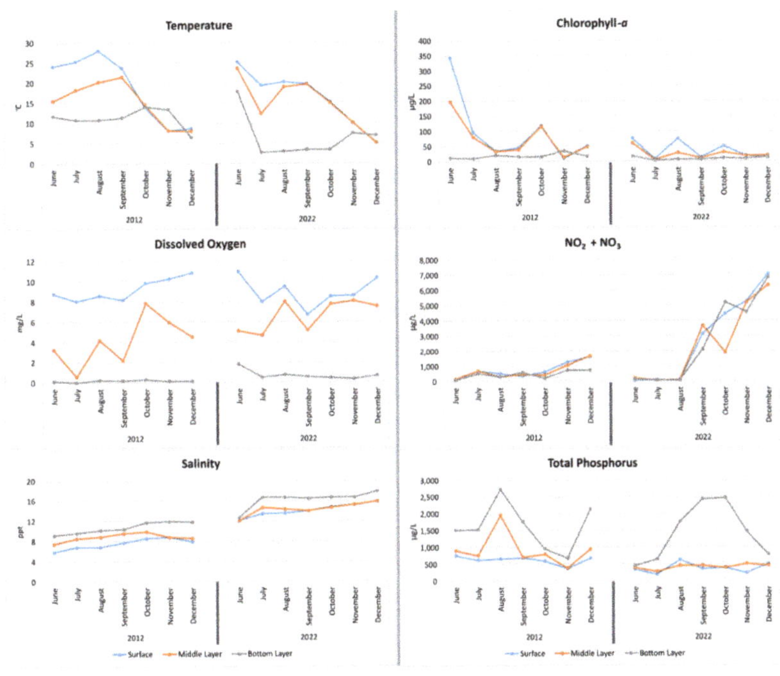

Figure 3. Depth profile of physicochemical characteristics of station 2.

In the water column, data showed that as a result of thermal stratification, the water quality deteriorated at the bottom of the lake because of slow water exchange between layers [9]. As an important indicator of trophic status, oxygen concentration or depletion rates in hypolimnetic water have long been studied [10]. In all sampling periods, the bottom layer was characterized by anoxic conditions (Figure 3). Gürevin et al. reported that the significant release of nutrients from sediment caused long-term eutrophication in the lagoon [11].

4. Conclusions

In this study, due to rapid industrialization and population growth, a significant deterioration in water quality has been observed since the 1990s. From the large dataset of water quality for more than ten years, exposure to wastewater discharges and the closure of freshwater inlets were shown to be related to adverse effects on the water quality in the lagoon. These changes are having a serious effect on the ecosystem dynamics, revealing a shift in local species.

Author Contributions: Conceptualization, R.A. and M.A.; methodology, L.K., E.G.O., and A.O.Ç.; investigation, L.K., E.G.O., and A.O.Ç.; writing—original draft preparation, R.A., M.A., L.K., E.G.O., and A.O.Ç.; writing—review and editing, R.A., M.A., E.G.O., L.K., and A.O.Ç.; visualization, E.G.O.; project administration, E.G.O. All authors have read and agreed to the published version of the manuscript.

Funding: This research received no external funding.

Institutional Review Board Statement: Not applicable.

Informed Consent Statement: Not applicable.

Data Availability Statement: Not applicable.

Acknowledgments: The authors kindly acknowledge Anıl Algedik and Alev Cansu Atar for their kind help in sampling and laboratory work.

Conflicts of Interest: The authors declare no conflict of interest.

References

1. Falcão, M.; Vale, C. Nutrient dynamics in a coastal lagoon (Ria Formosa, Portugal): The importance of lagoon-sea water exchanges on the biological productivity. *Cienc. Mar.* **2003**, *29*, 425–433. [CrossRef]
2. Kjerfve, B. Coastal lagoons. In *Elsevier Oceanography Series*; Elsevier: Amsterdam, The Netherlands, 1994; Volume 60, pp. 1–8.
3. Mee, L.D. Coastal lagoons. In *Chemical Oceanography*; Riley, J.P., Chester, R., Eds.; Academic Press: London, UK, 1978; Volume 7, pp. 441–490.
4. Sönmez, V.Z.; Sivri, N. Temporal changes in water quality index of polluted lagoon ecosystems: A case study on the Küçükçekmece Lagoon. *Environ. Monit. Assess.* **2022**, *194*, 16. [CrossRef]
5. Albay, M.; Matthiensen, A.; Codd, G.A. Occurrence of toxic blue-green algae in the Kucukcekmece lagoon (Istanbul, Turkey). *Environ. Toxicol. Int. J.* **2005**, *20*, 277–284. [CrossRef] [PubMed]
6. Köker, L.; Akçaalan, R.; Dittmann, E.; Albay, M. Depth profiles of protein-bound microcystin in Küçükçekmece Lagoon. *Toxicon* **2021**, *198*, 156–163. [CrossRef]
7. *ISO 10260*; Water Quality—Measurement of Biochemical Parameters—Spectrometric Determination of the Chlorophyll-a Concentration. International Organization for Standardization: Geneva, Switzerland, 1992.
8. APHA. *Standard Methods for the Examination of Water and Wastewater*, 17th ed.; American Public Health Association: Washington, DC, USA, 1989; p. 113.
9. Chimney, M.J.; Wenkert, L.; Pietro, K.C. Patterns of Vertical Stratification in a Subtropical Constructed Wetland in South Florida (USA). *Ecol. Eng.* **2006**, *27*, 322–330. [CrossRef]

10. Nürnberg, G.K. Trophic state of clear and colored, soft-and hardwater lakes with special consideration of nutrients, anoxia, phytoplankton and fish. *Lake Reserv. Manag.* **1996**, *12*, 432–447. [CrossRef]
11. Gürevin, C.; Erturk, A.; Albay, M. Predicting the effects of sediment based internal nutrient loads on eutrophication in Küçükçekmece Lagoon for rehabilitation planning. *Int. J. Sediment Res.* **2017**, *32*, 527–554. [CrossRef]

environmental sciences proceedings

MDPI

Proceeding Paper

Optimal Water Quality Simulation of the Proposed Water Distribution System for the University of Kashmir Using EPANET 2.2 and Leakage Modelling of the Network Using EPANET Extension—WaterNetGen †

Mominah Ajaz [1,*] and Danish Ahmad [2]

1 M-TECH (Water Resources Engineering), National Institute of Technology, Srinagar 190006, India
2 Department of Civil Engineering, National Institute of Technology, Srinagar 190006, India; profdanish27@gmail.com
* Correspondence: mominah@kgpsrinagar.edu.in

† Presented at the 7th International Electronic Conference on Water Sciences, 15–30 March 2023; Available online: https://ecws-7.sciforum.net.

Abstract: Water quality is the most important parameter of portable water. Therefore, water quality simulation is of the utmost importance, along with carrying out the hydraulic analysis of a water distribution network. In the current study, it has been attempted to carry out the water quality simulation of the proposed distribution network for the University of Kashmir using EPANET 2.2 software. The study also aims to obtain the optimal performance of the designed network in terms of water quality parameters. Furthermore, the leakage modelling for the network has been carried out using the EPANET extension—WaterNetGen. It was found that important water quality parameters, like residual chlorine at nodes and water age, were within the standard ranges throughout the simulation period. The minimum concentration of chlorine up to the 11th hour of the simulation was 0.2 mg/L, and the maximum age of water in the storage tank was 12.5 h throughout the simulation period. The total leakage discharge obtained was negligible, equal to 0.1% and 0.15% of the design discharge for WDS part I and part II, respectively. The objective function of maximum efficiency of performance, with respect to water quality of the proposed network, was achieved.

Keywords: water quality simulation; EPANET 2.2; leakage modelling; EPANET extension—WaterNetGen

check for updates

Citation: Ajaz, M.; Ahmad, D. Optimal Water Quality Simulation of the Proposed Water Distribution System for the University of Kashmir Using EPANET 2.2 and Leakage Modelling of the Network Using EPANET Extension—WaterNetGen. *Environ. Sci. Proc.* **2023**, *25*, 27. https://doi.org/10.3390/ECWS-7-14251

Academic Editor: Athanasios Loukas

Published: 16 March 2023

1. Introduction

The quality of water is representative of its suitability for domestic and institutional use. Water quality analysis and modelling is an important aspect of the water distribution system (WDS) design, along with the efficient hydraulic performance of the network [1,2]. A water quality model has to be an optimal solution, like that of the hydraulic model, to achieve the maximum efficiency of the performance of a WDS [3]. Important water quality parameters like concentration of chlorine, decay of chlorine in the system [4,5] and water age [6] have to be modelled, so as to ascertain that the quality of the water is as per the standards [7]. Standard values of these parameters are vital for the optimality of the water quality model, indicating that these parameters are the decision variables for the optimal model, with the standard ranges of these variables as constraints. Leakage modelling is another important requisite of an optimal WDS model. Estimation of the amount of leakage discharge is vital for the efficient performance, with respect to hydraulics, as well as water quality of a WDS [8]. EPANET extension—WaterNetGen is an effective tool for modelling the leakage with a fair degree of accuracy and ease of use [9].

An optimal solution of the hydraulic design of WDS for the University of Kashmir (UOK) was proposed by using EPANET 2.0 in the earlier study. The WDS consists of two separate networks for two different divisions of the study area. Current work is an extension of the work done earlier, such that an optimal water quality model for the proposed WDS is formulated using the pressure driven analysis (PDA) approach of EPANET 2.2. The leakage modelling of the proposed network has been done by WaterNetGen.

2. Methodological Approach and Analysis

A quantitative pressure driven analysis approach (PDA) was used to produce an optimal water quality model of the proposed water distribution system (WDS) for the University of Kashmir (UOK) by using EPANET 2.2. Study of the literature was conducted and the most important water quality parameters, like chlorine concentration, decay of chlorine and water age, were taken as the decision variables for optimal modelling. Standard codes and books were consulted to set out the constraints for the decision variables. Finally, the leakage modelling of the network was carried out by using EPANET extension—WaterNetGen—to access the amount of leakage discharge at the nodes.

2.1. PDA of Water Quality of the Network Using EPANET 2.2

A more realistic PDA approach was used to carry out the water quality modelling, such that the variables were a function of the available pressure head at the nodes. Water quality parameters like chlorine concentration, decay of chlorine and water age were modelled using a PDA approach of the EPANET 2.2 [10]. Various input parameters, like reaction order, reaction coefficient for the bulk and wall reactions of chlorine and limiting concentration of chlorine equal to 0.2 mg/L [7], were provided to run the software successfully. The initial concentration of chlorine added to the supply tank was equal to 2 mg/L (optimum dosage of chlorine, as per the ground water quality test data provided by UOK).

2.2. Leakage Modeling by EPANET Extension—WaterNetGen

The background leakage discharge $Q_k^{\ leak}$ in any pipe (k) of length (L_k) was estimated after entering the values of background leakage coefficient per unit pipe length (β_k) and background leakage exponent (α_k) for each pipe, as per the following equation [11,12]: $Q_k^{\ leak} = \beta_k\ L_k\ (P_k)^{\alpha k}$; $\beta_k = 10^{-7}$, $\alpha_k = 1.18$. The nodal leakage flow at any node 'i' due to the background leakage of the pipes connected at the node was estimated after running the software, as per the following equation [11]: $Q_i^{\ leak} = \frac{1}{2}\ \Sigma\ Q_k^{\ leak}$, where 'k' iterates over all the pipes connected at 'i'

Finally, the emitter discharge at the nodes was obtained after providing the value of emitter coefficient 'β_i' to each node, which was calculated from the following equation [11–13]: $Q_i^{\ leak} = \beta_i\ (P_i)^{0.5}$, where ($P_i$) is the node pressure.

2.3. Optimization of the Water Quality Model

An optimal solution of the water quality model was obtained by selecting the following objective function subject to the decision variables and constraints, as given below:

Objective function: maximization of efficiency of performance, with respect to the water quality of the proposed WDS, without affecting the hydraulic performance.

Decision variables: the following water quality parameters were taken as the decision variables; chlorine concentration, water age.

Constraints: chlorine concentration $\geq$ 0.2 mg/L [7], average water age $\leq$ 1.3 days and maximum water age $\leq$ 3 days [6].

3. Results and Discussion

3.1. Chlorine Concentration at the Nodes

The minimum required concentration of residual chlorine at any point in a WDS is 0.2 mg/L. Figure 1a,b indicates that the chlorine concentration at all the nodes of WDS, part I and part II, at the hour of peak demand is above 0.2 mg/L. From the analysis, 0% of the nodes have a chlorine concentration below 0.58 mg/L at the hour of peak demand in WDS part I, 0% nodes have a chlorine concentration below 0.735 mg/L in WDS part II. Figure 2a,b indicates that there is a drop in the concentration of chlorine, below 0.2mg/L, at the peak demand nodes and the storage tank at 12 pm and onwards. Thus, there is a need to re-add the chlorine at the source node (storage tank) at 12 pm.

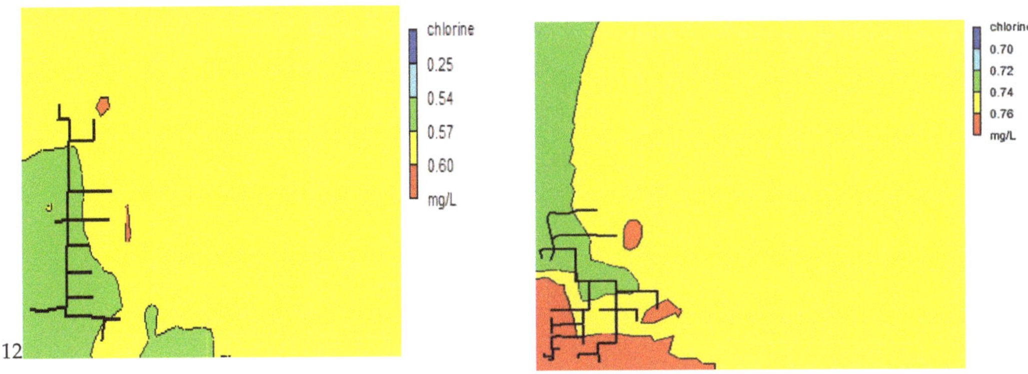

(a) (b)

Figure 1. (**a**) Contour plot of chlorine concentration at nodes at 9:00 am for WDS part I; (**b**) contour plot of chlorine concentration at nodes at 9:00 am for WDS part II.

3.2. Decay of Chlorine in the System

As indicated in Figure 3a,b, the maximum percentage decay of chlorine is taking place in the storage reservoir in both parts of the WDS, due to the reaction within the bulk of the fluid in the storage tank. The decay, due to wall reactions, is lower due to the assumption of the use of lined G.I pipes. The decay percentage is due to the reaction of chlorine in the bulk of the water in pipes.

3.3. Time Series Graph for Age of Water in the Storage Tank

The increased age of water in a WDS is related to the growth of disinfection by products like trihalomethanes, microbial growth, etc. The maximum age of water in a WDS is limited to about 3 days [6]. In both the WDS, part I and part II, the maximum age of water in the storage tank is 12.5 h (Figure 4a,b).

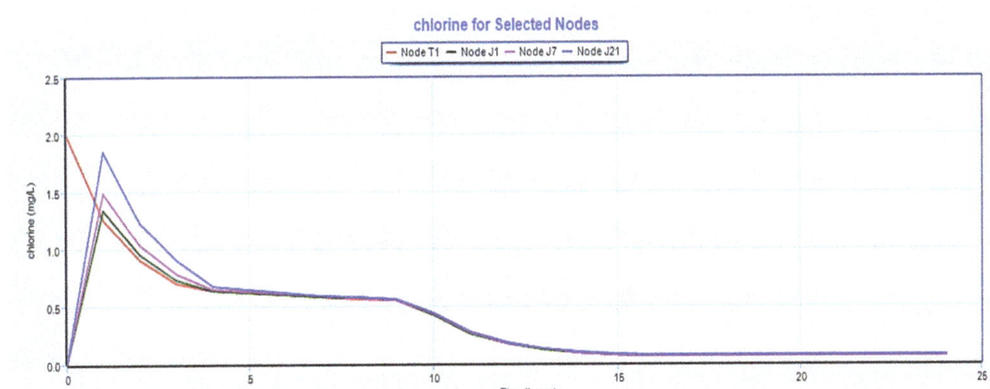

(a)

Figure 2. *Cont.*

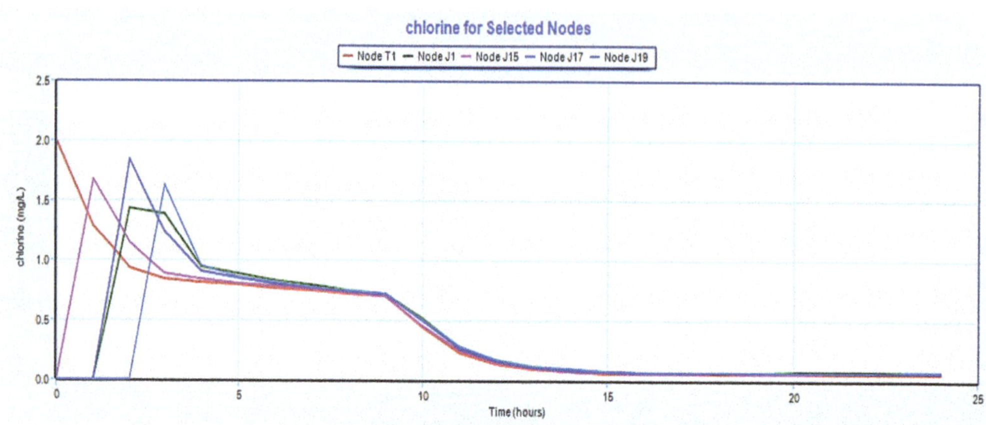

(**b**)

Figure 2. (**a**) Time series plot of chlorine (mg/L) at the storage tank and peak demand nodes for WDS I; (**b**) time series plot of chlorine (mg/L) at the storage tank and peak demand nodes for WDS II.

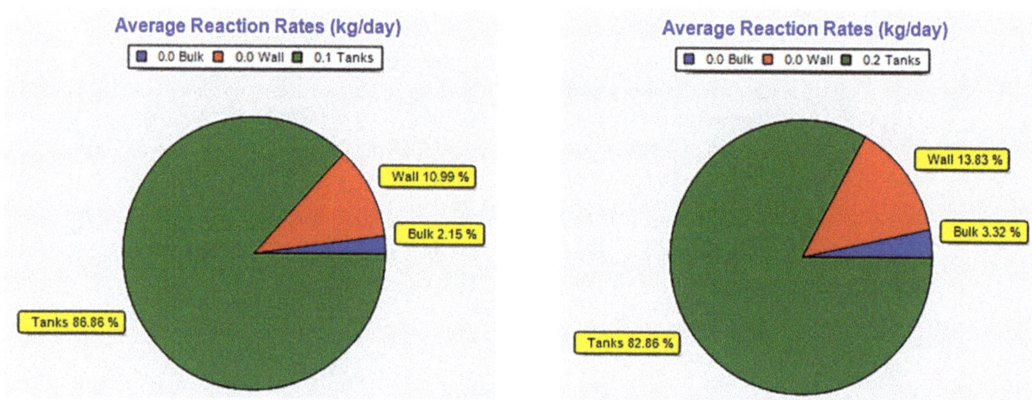

(**a**) (**b**)

Figure 3. (**a**) Pie chart for chlorine decay, WDS part I; (**b**) pie chart for chlorine decay, WDS part II.

3.4. *Leakage Modelling of the Network by EPANET Extension—WaterNetGen*

The emitter discharge at the nodes, which is contributed to the background leakage of the pipes connected at a node, was modelled. The emitter coefficient for each node was evaluated, as explained in Section 2.3. The emitter coefficient corresponding to the time of occurrence of the maximum background leakage and pressure head at the node was taken as the design value. For WDS part I, the emitter coefficient corresponding to 4 h, and for WDS part II, that corresponding to 3 h, was entered for each node. The values of emitter discharge at the nodes at the hour of peak demand were obtained, as in Tables 1 and 2, indicating negligible leakage discharge in the WDS—0.1% for WDS part I and 0.15% for WDS part II.

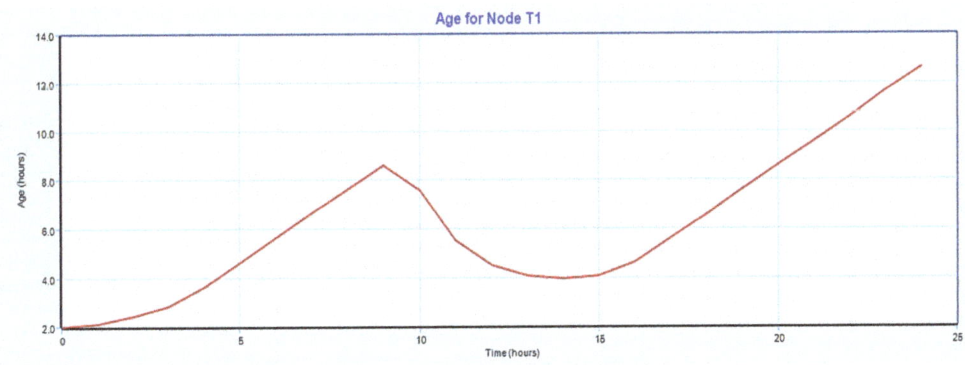

(**a**)

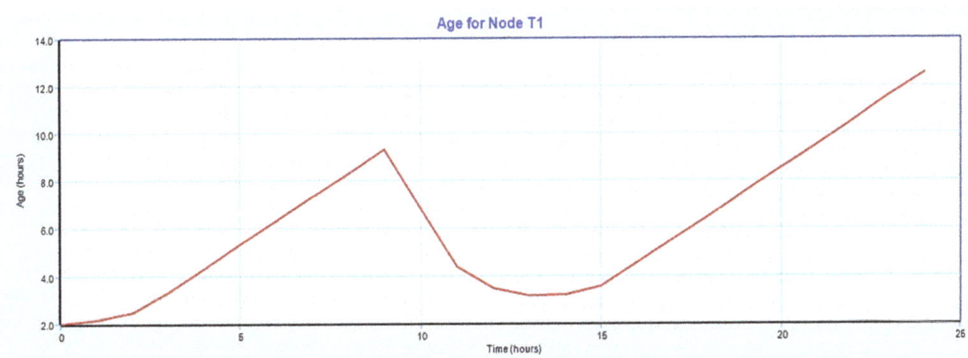

(**b**)

Figure 4. (**a**) Age of water in the storage tank, WDS part I; (**b**) age of water in the storage tank, WDS part II.

Table 1. Emitter flow at nodes for WDS part I.

Node	Pressure (m)	Emitter Flow (lps)	Node	Pressure (m)	Emitter Flow (lps)	Node	Pressure (m)	Emitter Flow (lps)
Junc J1	34.39	0.00038	Junc J11	24.79	0.00069	Junc J21	28.64	0.00055
Junc J2	29.66	0.00033	Junc J12	24.58	0.0002	Junc J22	27.17	0.00018
Junc J3	27.41	0.00016	Junc J13	25.36	0.00036	Junc J23	27.78	0.00019
Junc J4	31.62	0.00041	Junc J14	25.43	0.0005	Junc J24	25.07	0.00015
Junc J5	29.43	0.0004	Junc J15	25.02	0.00039	Junc J25	26.27	0.00049
Junc J6	32.83	0.00036	Junc J16	25.87	0.00011	Junc J26	24.14	0.00014
Junc J7	31.28	0.00026	Junc J17	25.55	0.00042	Junc J27	23.93	0.00036
Junc J8	28.75	0.00073	Junc J18	25.81	0.00041	Junc J28	34.09	0.00041
Junc J9	25.86	0.00024	Junc J19	25.44	0.00037	Junc J29	27.4	0.00073
Junc J10	26.5	0.00113	Junc J20	25.36	0.00005	Junc J30	30.19	0.00031
							total	0.013

Table 2. Emitter flow at nodes for WDS part II.

Node	Pressure (m)	Emitter Flow (lps)	Node	Pressure (m)	Emitter Flow (lps)	Node	Pressure (m)	Emitter Flow (lps)
Junc J1	27.73	0.00012	Junc J19	23.57	0.00043	Junc J37	24.61	0.00036
Junc J2	27.99	0.00039	Junc J20	23.65	0.00084	Junc J38	24.05	0.00014
Junc J3	28.66	0.00021	Junc J21	24.65	0.00023	Junc J39	24.77	0.0001
Junc J4	29.07	0.00017	Junc J22	25.4	0.00021	Junc J40	23.9	0.00031
Junc J5	34.61	0.00025	Junc J23	25.01	0.0002	Junc J41	23.69	0.00031
Junc J6	34.84	0.00067	Junc J24	24.76	0.00011	Junc J42	25.04	0.00097
Junc J7	34.61	0.00088	Junc J25	24.29	0.00019	Junc J43	24.78	0.0001
Junc J8	33.36	0.00072	Junc J26	24.04	0.00019	Junc J44	31.08	0.00037
Junc J9	31.97	0.00021	Junc J27	23.86	0.0002	Junc J45	30.87	0.00029
Junc J10	31.53	0.00027	Junc J28	23.53	0.00031	Junc J46	25.26	0.00014
Junc J11	30.75	0.0002	Junc J29	23.49	0.0001	Junc J47	25.39	0.00008
Junc J12	31.57	0.00065	Junc J30	25.15	0.00058	Junc J48	24.63	0.00022
Junc J13	30.98	0.00057	Junc J31	24.81	0.00035	Junc J49	27.28	0.00052
Junc J14	31.62	0.00028	Junc J32	24.74	0.00017	Junc J50	29.22	0.00131
Junc J15	29.13	0.00089	Junc J33	24.65	0.00024	Junc J51	27.77	0.0009
Junc J16	23.89	0.00036	Junc J34	23.48	0.00011	Junc J52	26.07	0.00094
Junc J17	23.64	0.00048	Junc J35	24.53	0.00035	Junc J53	25.78	0.00111
Junc J18	23.52	0.0004	Junc J36	25.43	0.00009	Junc J54	25.08	0.0001
							total	0.029

4. Conclusions and Future Scope

In this work, an optimal solution of water quality modelling of the proposed WDS for the University of Kashmir has been provided. Chlorine concentration and water age were taken as the decision variables for optimal design. Water quality modelling was carried out by the PDA approach of the EPANET 2.2, and the leakage modelling of the network was done by EPANET extension—WaterNetGen. The objective function of maximum efficiency of water quality performance was achieved, subject to the standard values of the decision variables and minimum percentage of leakage discharge, which was verified without affecting the optimality of the hydraulic design of the network. The main highlights of the work include the following:

The standard minimum chlorine concentration of 0.2 mg/L was maintained at each node up to 11 h of the simulation. However, a re-addition of chlorine to the water in the storage reservoir at 12 h was required to maintain the standard residual chlorine at every point in the WDS. The maximum percentage decay of chlorine took place in the storage reservoir in both parts of the WDS, and a negligible decay was observed in the bulk and at the boundary of the pipes, indicating negligible reaction between pipe material and the water and hence, longer life of the pipes of the network. The age of the water in the storage tank was limited to 12.5 h, indicating prevention of the growth of disinfection by-products and microbial growth. From the hydraulic analysis of the network, it was seen that the water age in the storage tank is inversely related to the pressure head of the tank. The leakage modelling for the network has been completed using WaterNetGen and leakage discharge obtained at the peak demand hour. The total leakage discharge obtained for WDS part I is 0.013 L/s and is 0.029 L/s for WDS part II, respectively, which is 0.1% and 0.15% of the design discharge, respectively, and thus negligible. The very small magnitude of leakage discharge indicates the optimality of the overall design of the network.

The extensions available to the EPANET can be used for water security modelling, real time modelling and fire flow analysis of the designed WDS. EPANET-MSX (multi-species extension), the interaction of multiple chemical agents between each other, with the material of walls of the pipes and the bulk of the fluid, can be modelled. Additionally, the auto decomposition of chloramines to ammonia, formation of disinfection by products and biological regrowth can be modelled. EPANET-RTX (real-time extension) allows for the connection of the operational data with a network model, and the resultant model can be calibrated, verified and tested for precision using the operational data. WaterNetGen can be used for the fire flow analysis of the network model.

Author Contributions: Both authors contributed equally to this work. All authors have read and agreed to the published version of the manuscript.

Funding: This research received no external funding.

Institutional Review Board Statement: Not applicable.

Informed Consent Statement: The study did not involve any humans.

Data Availability Statement: All the data required was obtained from the Engineering wing of the University of Kashmir.

Conflicts of Interest: The authors declare no conflict of interest.

References

1. Grayman, W.M. History of water quality modeling in distribution systems. In Proceedings of the 1st International WDSA/CCWI Joint Conference, Kingston, ON, Canada, 23–25 July 2018.
2. Lipiwattanakarn, S.; Kaewsang, S.; Makpiboon, C.; Changklom, J.; Pornprommin, A. Water quality audit in drinking water distribution networks. *J. Water Resour. Plan. Manag.* **2021**, *147*, 04020113. [CrossRef]
3. Siew, C.; Tanyimboh, T.T.; Seyoum, A.G. Penalty-Free Multi-Objective Evolutionary Approach to Optimization of Anytown Water Distribution Network. *Water Resour. Manag.* **2016**, *30*, 3671–3688. [CrossRef]
4. Nono, D.; Odirile, P.T.; Basupi, I.; Parida, B.P. Assessment of probable causes of chlorine decay in water distribution systems of Gaborone city, Botswana. *Water SA* **2019**, *45*, 190–198. [CrossRef]
5. Mostafa, N.G.; Matta, M.E.; Halim, H.A. Simulation of chlorine decay in water distribution networks using watercad—Case study. *J. Eng. Appl. Sci.* **2013**, *60*, 25–42.
6. Kourbasis, N.; Patelis, M.; Tsitsifli, S.; Kanakoudis, V. Optimizing Water Age and Pressure in Drinking Water Distribution Networks. *Environ. Sci. Proc.* **2020**, *2*, 51. [CrossRef]
7. IS 10500. Indian Standard Drinking Water Specification. Bureau of Indian Standards: New Delhi, India, 2012.
8. Cobacho, R.; Arregui, F.; Soriano, J.; Cabrera, E. Including leakage in network models: An application to calibrate leak valves in EPANET. *J. Water Supply Res. Technol.—AQUA* **2015**, *64*, 130–138. [CrossRef]
9. Muranho, J.; Ferreira, A.; Sousa, J.; Gomes, A.; Sá Marques, A. WaterNetGen: An EPANET extension for automatic water distribution network models generation and pipe sizing João Muranho, Ana Ferreira, Joaquim Sousa, Abel Gomes. *Water Sci. Technol. Water Supply* **2012**, *12*, 117–123. [CrossRef]
10. Rossman, L. *EPANET 2.2 User Manual*; United States Environmental Protection Agency: Washington, DC, USA, 2020.
11. Muranho, J.; Ferreira, A.; Sousa, J.; Gomes, A.; Marques, A.S. Pressure dependent demand and leakage modelling with an EPANET Extension—WaterNetGen. *Procedia Eng.* **2014**, *89*, 632–639. [CrossRef]
12. Avile, A.; Garcı, F. Pressure management for leakage reduction using pressure reducing valves. Case study in an Andean city. *Alexendria Eng. J.* **2019**, *58*, 1313–1326.
13. Walski, T.; Researcher, I.; Weir, M.H. Modeling leakage reduction. *Am. Water Work. Assoc.* **2006**, *98*, 147–155. [CrossRef]

environmental sciences proceedings

MDPI

Proceeding Paper

Water Quality and Risk Assessment in Rainwater Harvesting Ponds †

Emine Gozde Ozbayram *, Latife Köker, Ayça Oğuz Çam, Reyhan Akçaalan and Meriç Albay

Faculty of Aquatic Sciences, Istanbul University, Istanbul 34134, Turkey; latife.koker@istanbul.edu.tr (L.K.); ayca.oguzcam@istanbul.edu.tr (A.O.Ç.); akcaalan@istanbul.edu.tr (R.A.); merbay@istanbul.edu.tr (M.A.)

* Correspondence: gozde.ozbayram@istanbul.edu.tr

† Presented at the 7th International Electronic Conference on Water Sciences, 15–30 March 2023; Available online: https://ecws-7.sciforum.net/.

Abstract: The aim of this study was to investigate the water quality of rainwater harvesting ponds in Istanbul which are used for irrigation. For this purpose, samples were collected from 17 rainwater harvesting ponds during the summer of 2022 and selected physicochemical and biological characterization of these samples was carried out. Cyanobacterial bloom was observed in 2 ponds out of 17 and the dominant species were potentially cyanotoxin producers (*Microcystis*, *Aphanizomenon*, *Dolichospermum*, *Planktothrix*, and *Cuspidothrix*). It is found that one of these ponds was not proper for irrigation purposes due to microcystin presence. To increase the water quality in these reservoirs, onsite management strategies should be taken into consideration.

Keywords: irrigation; rainwater harvesting; rainwater quality; water security; cyanotoxins; climate change

check for updates

Citation: Ozbayram, E.G.; Köker, L.; Oğuz Çam, A.; Akçaalan, R.; Albay, M. Water Quality and Risk Assessment in Rainwater Harvesting Ponds. *Environ. Sci. Proc.* **2023**, *25*, 28. https://doi.org/10.3390/ECWS-7-14245

Academic Editor: Athanasios Loukas

Published: 16 March 2023

1. Introduction

Environmental, economic, and climate changes in many parts of the world put serious pressure on water resources, making a reliable alternative water supply a critical global concern [1]. Rainwater harvesting is a sustainable water management practice that involves collecting and storing rainwater for later use. The collected rainwater can then be used for a variety of purposes, including irrigation, landscaping, and even household uses [1]. Since rainwater harvesting is an off-grid water supply [2], it contributes to reducing the pressure on the central systems, decreasing the reliance on freshwater abstraction [3], particularly in areas where water is in short supply or where the demand for water exceeds the available supply.

When determining the suitability of the water resource, water quality becomes a prominent issue. The harvested rainwater is generally of sufficient quality for non-drinking purposes [4]. However, due to various reasons (e.g., nutrient inputs, temperature increase, draught, etc.), cyanobacteria proliferation can occur in rainwater harvesting ponds and damage the water quality. Furthermore, some species can excrete cyanotoxins that may enter the agricultural fields by irrigation and cause environmental and public health problems. Thus, it is important to evaluate the potential contaminants of harvested rainwater.

The goal of this research was to evaluate the water quality of rainwater harvesting ponds in Istanbul used for irrigation and assess their potential risks for non-potable usage.

2. Materials and Methods

The samples were collected from 17 rainwater harvesting ponds (S1–S17) located in Istanbul (Türkiye) during the summer of 2022 (Figure 1). These ponds were built between 1965–1989 in Istanbul and used for irrigation purposes in the neighboring agricultural fields. The active water volumes of the ponds varied between 11,000–7,000,000 m^3 in which S11 had the highest volume followed by S16 (1,406,405 m^3) and S15 (1,103,386 m^3).

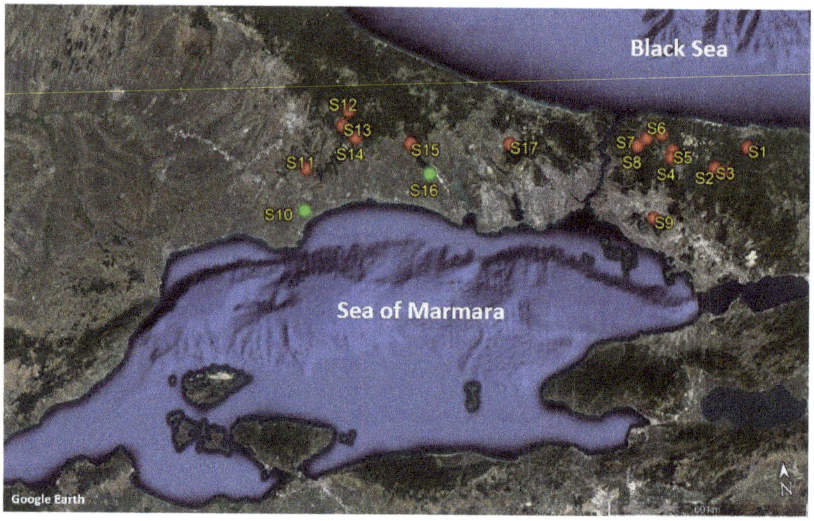

Figure 1. Sampling locations (S): Green dots indicate the rainwater harvesting ponds with cyanobacterial blooms.

The water samples collected from the surface were kept under dark and cold conditions and brought to the laboratory immediately. Water temperature, pH, dissolved oxygen (DO), and conductivity were measured in situ via a portable multiparameter (650 MDS, YSI, Yellow Springs, OH, USA) at each sampling site. Total nitrogen (TN) and total phosphorus (TP) were analyzed according to the methods outlined by the American Public Health Association [5]. The chlorophyll-*a* concentration was determined using the method described in ISO 10260 [6]. Phytoplankton samples were fixed by Lugol's iodine solution, and the phytoplankton enumeration was performed according to Utermöhl (1958) [7]. The microcystin concentrations were measured using liquid chromatography–high-resolution mass spectrometry (LC-HRMS) [8].

3. Results

The physicochemical characteristics of the rainwater harvesting ponds are given in Figure 2. The water temperature ranged between 18–30 °C. The pH was measured between 6.5 and 9.2. A high variation was observed in the conductivity values, in which the highest conductivity was measured in S10 (890 µS/cm) and the lowest was recorded as 112 µS/cm in S14. While TP was measured below 100 µg/L in most of the ponds, the highest value was measured in S11 as 237.5 µg/L. Total nitrogen was detected in the range of 0.7–1.9 mg/L in which the maximum value was detected in S16 followed by S10 (1.6 mg/L). Furthermore, a high variation was observed in the chlorophyll-*a* concentration in which the highest level was detected in S10 as 82 µg/L. Microcystin was found in S16 as 13.80 µg/L.

Cyanobacteria were detected in three samples (S10, S11, and S16) out of 17 ponds (Figure 3A). In two of these ponds, cyanobacterial blooms were observed, with Cyanobacteria comprising 82% of the community in S16 and 72% in S10. In detail, *Microcystis* sp. was responsible for the bloom in S16 while *Aphanizomenon* sp. and *Cuspidothrix issatschenkoi* were in S10 (Figure 3B).

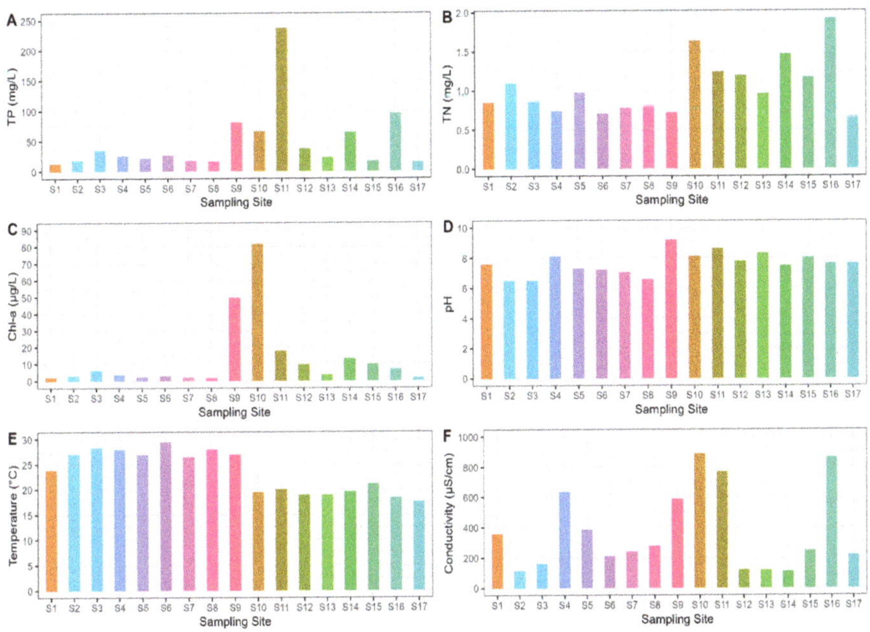

Figure 2. Physicochemical characteristics of the rainwater harvesting ponds during the summer 2022 (**A**) Total Phosphorous, (**B**) Total Nitrogen, (**C**) Chlorophyll-*a*, (**D**) pH, (**E**) Temperature, (**F**) Conductivity.

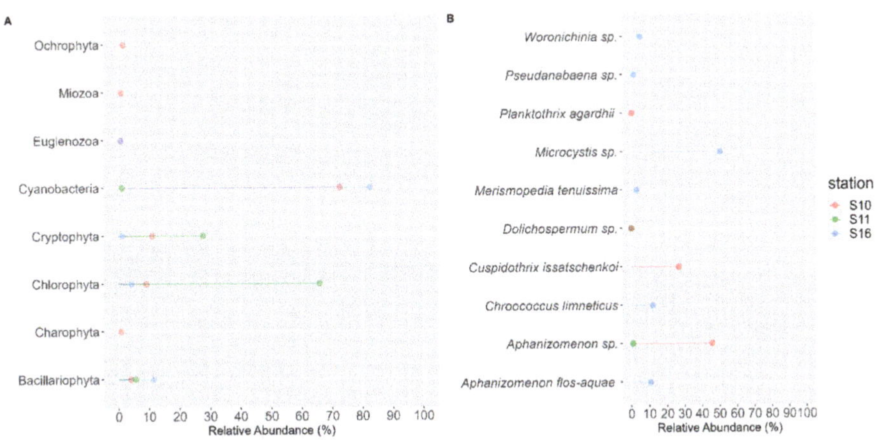

Figure 3. Phytoplankton community composition (**A**) phylum level, (**B**) genus level.

4. Discussion

Rainwater harvesting is a green method of providing water for agricultural practices, since having access to adequate and safe irrigation water is crucial for the success of farming operations. Harvesting of rainwater is also known for a being straightforward, economical, and innovative solution contributing to sustainability and resilience of water sources. Rainwater is considered a high-quality source of irrigation because it is nearly sodium-free and has a low sodium adsorption ratio, which helps to maintain the physical structure of the soil [2]. On the other hand, there could be some other constituents such as emerging untraditional substances which pose a significant challenge to the use of rainwater harvesting for irrigation. Cyanotoxins are one of these components that threaten the ecosystem health.

Certain types of cyanobacteria can excrete toxins that damage the liver (hepatotoxins), harm the nervous system (neurotoxins), and damage cell integrity (cytotoxins) [9]. *Microcystis, Aphanizomenon, Dolichospermum, Planktothrix,* and *Cuspidothrix* are known genera-producing cyanotoxins [10–12]. Irrigation with water that contains cyanotoxins can have negative impacts on the quality and yield of agricultural plants. Since there is not any treatment for these components, these toxins can bioaccumulate in plant tissues. Subsequently, they may enter the food chain with the edible plants and pose environmental and human health risks [9]. Accumulation can vary depending on the type of plant and irrigation method used [13]. Overall, it was found that two ponds were not proper for irrigation purposes. To increase the water quality in these reservoirs, onsite management strategies should be taken into consideration.

Author Contributions: Conceptualization, R.A. and M.A.; methodology, E.G.O., L.K. and A.O.Ç.; investigation, E.G.O., L.K. and A.O.Ç.; writing—original draft preparation, R.A., M.A., E.G.O., L.K. and A.O.Ç.; writing—review and editing, R.A., M.A., E.G.O., L.K. and A.O.Ç.; visualization, E.G.O.; project administration, E.G.O. All authors have read and agreed to the published version of the manuscript.

Funding: This research was funded by the Scientific Research Projects Coordination Unit of Istanbul University, grant number FBG-2022-38851.

Institutional Review Board Statement: Not applicable.

Informed Consent Statement: Not applicable.

Data Availability Statement: Not applicable.

Acknowledgments: The authors kindly acknowledge Hakan Korkusuz for his kind help in sampling.

Conflicts of Interest: The authors declare no conflict of interest.

References

1. Nachson, U.; Silva, C.M.; Sousa, V.; Ben-Hur, M.; Kurtzman, D.; Netzer, L.; Livshitz, Y. New Modelling Approach to Optimize Rainwater Harvesting System for Non-Potable Uses and Groundwater Recharge: A Case Study from Israel. *Sustain. Cities Soc.* **2022**, *85*, 104097. [CrossRef]
2. Deng, Y. Pollution in Rainwater Harvesting: A Challenge for Sustainability and Resilience of Urban Agriculture. *J. Hazard. Mater. Lett.* **2021**, *2*, 100037. [CrossRef]
3. de Sá Silva, A.C.R.; Bimbato, A.M.; Balestieri, J.A.P.; Vilanova, M.R.N. Exploring Environmental, Economic and Social Aspects of Rainwater Harvesting Systems: A Review. *Sustain. Cities Soc.* **2022**, *76*, 103475. [CrossRef]
4. Andualem, T.G.; Hagos, Y.G.; Teka, A.H. Rainwater Harvesting Potential Assessment for Non-Potable Use in Urban Areas. *Sustain. Water Resour. Manag.* **2020**, *6*, 104. [CrossRef]
5. APHA, AWWA, WPFC. *Standard Methods for the Examination of Water and Waste Water*, 17th ed.; APHA, AWAA, WPFC: Washington, DC, USA, 1989; p. 113.
6. *ISO 10260*; Water Quality—Measurement of Biochemical Parameters—Spectrometric Determination of the Chlorophyll-a Concentration. International Organization for Standardization: Geneva, Switzerland, 1992.
7. Utermöhl, H. Zur Vervollkommnung Der Quantitativen Phytoplankton-Methodik: Mit 1 Tabelle Und 15 Abbildungen Im Text Und Auf 1 Tafel. *Int. Ver. Theor. Angew. Limnol. Mitt.* **1958**, *9*, 1–38. [CrossRef]
8. Caixach, J.; Flores, C.; Spoof, L.; Meriluoto, J.; Schmidt, W.; Mazur-Marzec, H.; Hiskia, A.; Kaloudis, T.; Furey, A. Liquid Chromatography–Mass Spectrometry. In *Handbook of Cyanobacterial Monitoring and Cyanotoxin Analysis*; Wiley: Chichester, UK, 2016.
9. Saqrane, S.; Oudra, B. CyanoHAB Occurrence and Water Irrigation Cyanotoxin Contamination: Ecological Impacts and Potential Health Risks. *Toxins* **2009**, *1*, 113–122. [CrossRef] [PubMed]
10. Akcaalan, R.; Köker, L.; Oğuz, A.; Spoof, L.; Meriluoto, J.; Albay, M. First Report of Cylindrospermopsin Production by Two Cyanobacteria (Dolichospermum Mendotae and Chrysosporum Ovalisporum) in Lake Iznik, Turkey. *Toxins* **2014**, *6*, 3173–3186. [CrossRef] [PubMed]
11. Akçaalan, R.; Köker, L.; Gürevin, C.; Albay, M. Planktothrix Rubescens: A Perennial Presence and Toxicity in Lake Sapanca. *Turk. J. Bot.* **2014**, *38*, 782–789. [CrossRef]

12. Stüken, A.; Campbell, R.J.; Quesada, A.; Sukenik, A.; Dadheech, P.K.; Wiedner, C. Genetic and Morphologic Characterization of Four Putative Cylindrospermopsin Producing Species of the Cyanobacterial Genera Anabaena and Aphanizomenon. *J. Plankton Res.* **2009**, *31*, 465–480. [CrossRef]
13. Lee, S.; Jiang, X.; Manubolu, M.; Riedl, K.; Ludsin, S.A.; Martin, J.F.; Lee, J. Fresh Produce and Their Soils Accumulate Cyanotoxins from Irrigation Water: Implications for Public Health and Food Security. *Food Res. Int.* **2017**, *102*, 234–245. [CrossRef]

environmental sciences proceedings

MDPI

Proceeding Paper

Estimating the Potential Evapotranspiration of Egypt Using a Regional Climate Model and a High-Resolution Reanalysis Dataset †

Samy Ashraf Anwar [1,*] and Irida Lazić [2]

1 Egyptian Meteorological Authority, Qobry EL-Kobba, Cairo P.O. Box 11784, Egypt
2 Faculty of Physics, Institute for Meteorology, University of Belgrade, Dobračina 16, 11000 Belgrade, Serbia; irida.lazic@ff.bg.ac.rs
* Correspondence: ratebsamy@yahoo.com

† Presented at the 7th International Electronic Conference on Water Sciences, 15–30 March 2023; Available online: https://ecws-7.sciforum.net/.

Abstract: Station observation is a good data source to monitor the potential evapotranspiration (PET) changes of a specific site particularly for the purpose of crop irrigation activities; however it represents only the site geographic characteristics and provides real-time/historical records. Hence, there was an urgent need to find a promising tool and a simple empirical to predict/project the PET in locations where station observation is not feasible. The Hargreaves–Samani method (HS) is recommended after the Penman–Monteith equation. To address this issue, the Regional Climate Modeling version 4 (RegCM4) with spatial resolution 25 km was used to compute the PET using the HS for the period 1979–2017. Era-Interim reanalysis of 1.5 degrees (EIN15) and NCEP/NCAR reanalysis version 2 of 2.5 degrees (NNRP2) were used to examine the influence of the lateral boundary condition on the simulated PET. The two simulations were designated as EIN15-RegCM4 and NNRP2-RegCM4, respectively. To examine the possible influences on the simulated PET, a comparison was conducted between EIN15-RegCM4 and NNRP2-RegCM4. After that, a comparison was conducted between the original HS formula (HS) and its calibrated version (HSnew) with respect to the 0.1−degree ERA5-land derived reanalysis product (hereafter ERA5) using EIN15-RegCM4 (as an example). Results showed that switching between EIN15 and NNRP2 did not show a notable influence on the simulated PET. Further, calibrating the HS coefficients indicates a considerable improvement in estimating the PET (relative to the original equation) when it is compared with ERA5. Such improvement is confirmed by a significant low mean bias. Over the majority of locations, the RegCM4 shows a good performance using the calibrated HS equation. In conclusion, the RegCM4 can be used to estimate the PET using the calibrated HS either for making a daily forecast or for projecting the future PET under different global warming scenarios.

Keywords: Egypt; ERA-Interim; regional climate model; potential evapotranspiration; ERA5

Citation: Anwar, S.A.; Lazić, I. Estimating the Potential Evapotranspiration of Egypt Using a Regional Climate Model and a High-Resolution Reanalysis Dataset. *Environ. Sci. Proc.* **2023**, *25*, 29. https://doi.org/10.3390/ECWS-7-14253

Academic Editor: Athanasios Loukas

Published: 16 March 2023

1. Introduction

Potential evapotranspiration (PET) plays an important role in the global terrestrial hydrology cycle. Additionally, PET is used for calculating the water needs of different crops and assessing hydrological and meteorological droughts, water balance analysis, and designing and operating irrigation projects [1]. The authors of [2] reported that, the Penman−Monteith (PM) equation is the standard model to compute the PET on various time scales. However, the authors of [3] reported that computing PET (using PM) is not recommended for arid/hyper-arid regions because it requires a surplus of soil moisture and it requires a large number of meteorological variables, which leads to greater uncertainty of the estimated PET.

PET can be computed using simple empirical models, such as temperature-based methods [4,5], radiation-based methods [6] and physically processed-based models (e.g., [2]). Additionally, the authors of [2] reported that the Hargreaves–Samani (HS) method can be recommended directly after PM and it can operate on daily/monthly time scales. Estimating the PET (using HS equation) showed a reliable performance in computing the PET with respect to observations as reported by [7,8]. Further, calibrating the HS showed a reliable performance in computing the PET with respect to PM observations [9–11]. Recently, the authors of [3] calibrated the regional climate model (RegCM4) output using the Climate Research Unit (CRU) and a linear regression model (LRM) at specific locations. However, calibrating the coefficients of the HS equation over Egypt and the role of the lateral boundary condition (used to downscale the RegCM4) in simulating the PET were not considered until the present day. Therefore, the present study aims to:

1. Examine the influence of the lateral boundary condition (EIN15 and NNRP2) on the simulated PET with respect to ERA5–land–derived product (hereafter ERA5; [12]).
2. Address the added value of the calibrated HS equation (relative to the original version) in comparison with the ERA5.
3. Validate the calibrated HS equation (versus the original version) by examining the climatological annual cycle of the simulated PET with respect to ERA5 at locations defined by [3].

Section 2 describes the study area and experiment design; Section 3 shows the results of the study. Section 4 provides the discussion and conclusion.

2. Materials and Methods

2.1. Study Area

A brief description of the study area is available in [3]; model domain dimension is covered in Section 2.2.

2.2. Model Description and Experiment Design

This study used the Abdus Salam International Centre for Theoretical Physics (ICTP) regional climate model version 4.7 (hereafter RegCM-4.7.0; [13]). The RegCM is a broad model used for conducting long-term simulations and future regional climate projections in Intercomparison projects [14]. To address the influence of the lateral boundary condition on the simulated PET, two experiments were conducted over the period 1979–2017. The first two years were considered as spin-up to properly initialize the RegCM4 model following [15], so the actual analysis starts at 1981 and ends at 2017. The two experiments adopted both ERA-Interim reanalysis of 1.5 degrees (EIN15; [16]) and NCEP/NCAR reanalysis version 2 of 2.5 degrees (NNRP2; [17]) to downscale the RegCM4 model. The RegCM4 model domain (Figure 1) was customized with grid spacing of 25 km with 60 grid points in both zonal and meridional directions centered at 27° latitude and 30° longitude. Additionally, the following physical schemes were used in the present study: Emanuel convection scheme over land and ocean [18], radiation scheme of [19] and Holtslag boundary layer scheme [20]. The simulated PET was calculated using the default HS equation as:

$$PET_{HS} = 0.0135 \times SW \times (T2m + 17.8) \quad (1)$$

The calibrated version is written as

$$PET_{HS} = 0.0105 \times SW \times (T2m + 17.8) \quad (2)$$

Several attempts have been made to obtain a reasonable bias of the calibrated HS equation (using ERA5 as the observational dataset). It was found that swtiching the radiation coefficient from 0.0135 to 0.0105 gave more promising results than calibrating the temperature coefficient (17.8). Note that SW (global incident solar radiation) is expressed in units of mm day^{-1} to show how much energy is used to evaporate water [2] and T2m is the 2-m mean air temperature (in °C).

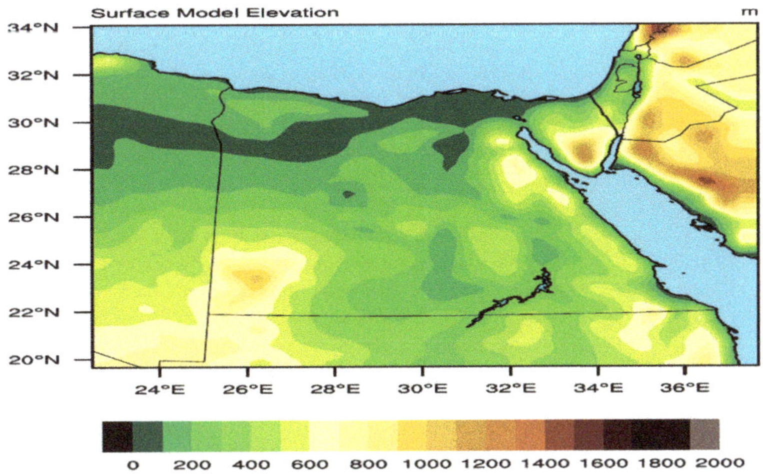

Figure 1. The figure shows the domain dimension and surface elevation (in meters). Please note that the RegCM4 model only supports elevation above mean sea level.

2.3. Validation Data

Various reanalysis products were used to evaluate the RegCM4 performance:

1. ERA5 ([21]): It provides hourly estimates of a large number of atmospheric, land, and oceanic climate variables with 0.25° horizontal grid spacing and 137 vertical levels (up to a height of 80 km). For the purpose of the present study, monthly means were aggregated to the seasonal time scale.
2. ERA5land ([22]): This product provides the surface meteorological variables at high resolution (0.1 degrees) using the land surface model of the ERA5 (titled ECMWF Scheme for Surface Exchanges over Land incorporating land surface hydrology; H-TESSEL).

   Please note that both ERA5 and ERA5land were used to evaluate the simulated SW and T2m of the RegCM4 (since these fields were used as inputs of the HS equation) to take into account the influence of the horizontal grid spacing of the reanalysis product.
3. Station observation is a major source to monitor the PET changes both spatially and temporally. However, availability of long-term records was not sufficient to evaluate the RegCM4 performance (before and after calibrating the HS equation) in this study. Recently, a new high-resolution global gridded PET (hPET) product was developed [12]. This product uses the hourly meteorological variables provided by the offline land model of the ERA5 reanalysis product [21]. Additionally, it adopts the PM equation to compute the PET and it is integrated over the period 1981–2021 in 0.1-degree grid spacing over the global land area. In the present study, monthly mean PET data were used to evaluate the RegCM4 performance both spatially and for locations defined by [3] in Section 1.

For the purpose of the present study, all products were bilinearly interpolated on the RegCM4 curvilinear grid following [10,15].

3. Results

Before assessing the performance of the RegCM4 (in simulating the PET), it is important to quantify the RegCM4 model bias concerning the simulated SW and T2m (as inputs of the HS equation). Figure S1 shows the simulated SW with respect to ERA5 and

ERA5land as well as the difference between ERA5 and ERA5land themselves for the seasons: March−April−May (MAM), June−July−August (JJA), September−October−November (SON) and December−January−February (DJF). From Figure S1, it can be noticed that the RegCM4 is able to reproduce the spatial pattern of the SW with respect to ERA5 and ERA5land in all seasons (Figure S1a–c,g–i,m–o,s–u). Additionally, the RegCM4 overestimates the SW in the MAM season by 10–30 W m^{-2} (Figure S1d,e). In the JJA and SON, the RegCM4 bias ranges from 20 to 40 W m^{-2} (Figure S1j,k,p,q). Lastly in the DJF, the RegCM4 bias becomes 10–20 W m^{-2} with respect to both products (Figure S1v,w). Further, there is no noted difference between ERA5 and ERA5land in all seasons (Figure S1f,l,r,x). It can be noted that the RegCM4 bias is maximized in the JJA and SON and it is minimized in the DJF.

Like SW, the RegCM4 shows a good ability to capture the spatial pattern of the simulated T2m with respect to reanalysis products in all seasons (Figure S2a–c,g—i,m–o,s–u). Additionally, there is no observable difference between ERA5 and ERA5land (Figure S2f,l,r,x), which means that the resolution of the observational dataset does not affect the evaluation of the RegCM4. In addition, an obvious warm-bias is noted in all seasons ranging from 3 to 7 °C (Figure S2d,e,j,k,p,q,v,w) with mostly pronounced warm bias during the summer. Such noted bias can be attributed to the fact that land cover of Egypt is mostly represented by desert leading to a low specific/relative humidity. As a result, the convective activity is affected, producing low total cloud cover (not shown), high SW approaching the earth surface and eventually warming the earth surface and the adjacent air layer close to the earth surface. Another possible reason is that the Holtslag scheme is characterized by high turbulent activity, leading to an enhancement of the warming effect produced by the SW.

3.1. *Influence of Lateral Boundary Condition*

To examine the influence of the lateral boundary condition on the simulated PET, Equation 1 was used to compute the simulated PET. Figure 2 shows the simulated PET (by the EIN15-RegCM4 and NNRP2-RegCM4, respectively) with respect to the ERA5. From Figure 2, it can be noted that the RegCM4 shows good consistency in reproducing the spatial pattern of the simulated PET in comparison with the ERA5 product (see Figure 2a−c,g−i,m−o,s−u). Additionally, it can be observed that there is no significant difference between EIN15 and NNRP2 in all seasons (Figure 2f,l,r,x). Such behavior can be attributed to two reasons: 1—RegCM4 has a similar performance when it is driven either by EIN15 or NNRP2 [23] and 2—RegCM4's physical parameterization dominates over the lateral boundary condition [24]. In addition, it can be observed that both simulations have a bias of 1–2.5 mm day^{-1} in the March−April−May season (MAM; Figure 2d,e).

The bias approaches its maximum in the June−July−August season because the RegCM4 shows a bias of 1–4.5 mm day^{-1} overall Egypt (JJA; Figure 2j,k). In the September-October-November (SON) season, the bias ranges between 1 and 3 mm day^{-1} over coastal regions and middle Egypt and 1–1.5 mm day^{-1} over Upper Egypt (see Figure 2p,q). Lastly, in the December-January-February (DJF) season, the bias is around 0.5–2 mm day^{-1} over majority of Egypt (Figure 2v,w). From a simple check between Figures S1 and S2 and Figure 2; it can be noted that the PET spatial pattern is more consistent with the SW than T2m. Therefore, calibrating the SW coefficient is more effective than T2m (see Equation (2)). This point will be discussed briefly in Section 3.2.

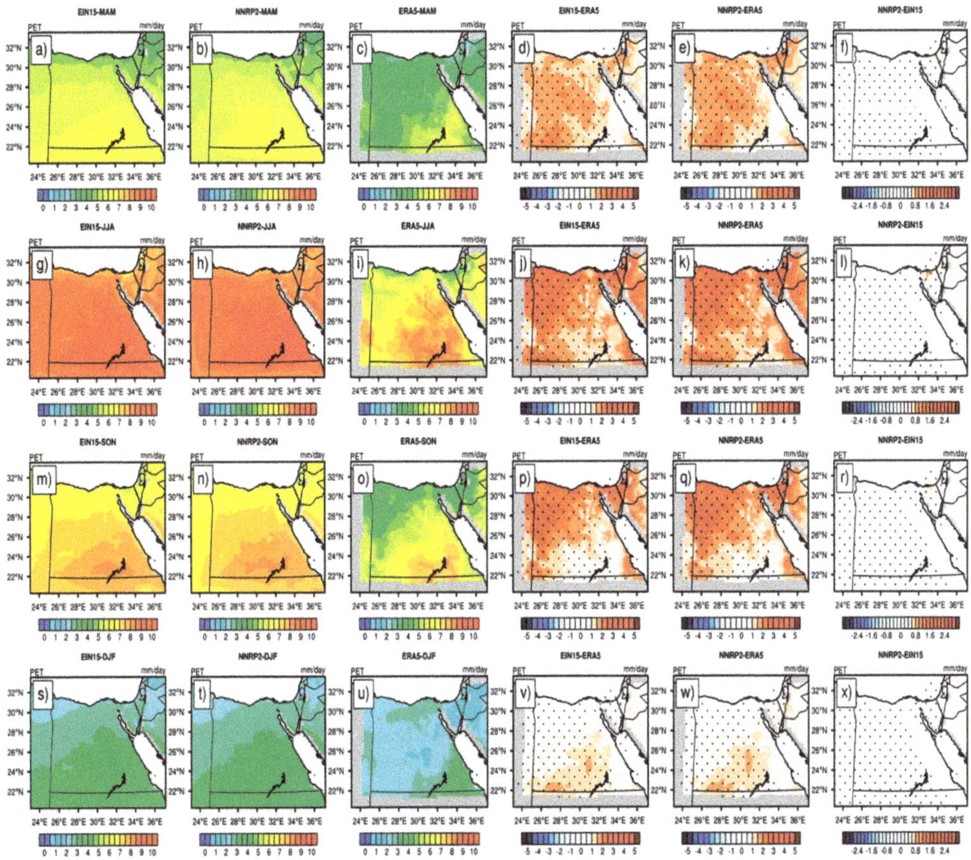

Figure 2. The figure shows the potential evapotranspiration over the period 1981–2017 (PET; in mm day^{-1}) for: MAM season in the first row (**a**–**f**); JJA in the second (**g**–**l**); SON in the third (**m**–**r**); and DJF in the fourth (**s**–**x**). For each row, EIN15 is on the left, followed by NNRP2; ERA5 is the third from left, EIN15 minus ERA5, NNRP2 minus ERA5 and the difference between NNRP2 and EIN15. Significant difference/bias is indicated in black dots using student *t*-test with alpha equals to 5%.

3.2. Added Value of the Calibrated HS Equation

As noted in Section 3.1, there is no significant difference between the two simulations. Therefore, the RegCM4-EIN15 simulation was taken (as an example) to examine the added value of the calibrated HS equation compared to ERA5. Figure 3 shows the simulated PET before calibration (HS), after calibration (HSnew) in comparison with the ERA5 and the difference between HSnew and HS. In general, both simulations are able to capture the spatial pattern of the simulated PET against the ERA5 (Figure 3a–c,g–i,m–o,s–u). However, HSnew shows added value over the HS in all seasons particularly in the JJA. Such value is indicated in two points: 1—better ability to reproduce the PET spatial pattern relative to HS and 2—the RegCM4 bias is significantly reduced in all seasons (particularly in the JJA) compared to the HS. For instance, in the MAM season, the HS shows a bias of 1–2.5 mm day^{-1} over the entire domain (Figure 3d). On the other hand, the HSnew shows a bias of 0.5 mm day^{-1} over the majority of Egypt, with some regions approaching 0.5–1 mm day^{-1} and 0.5 to −1.5 mm day^{-1} around Lake Nasser (Figure 3e). Qualitatively, the HSnew reduces the PET by 0.6–1.2 mm day^{-1} relative to the HS (Figure 3f).

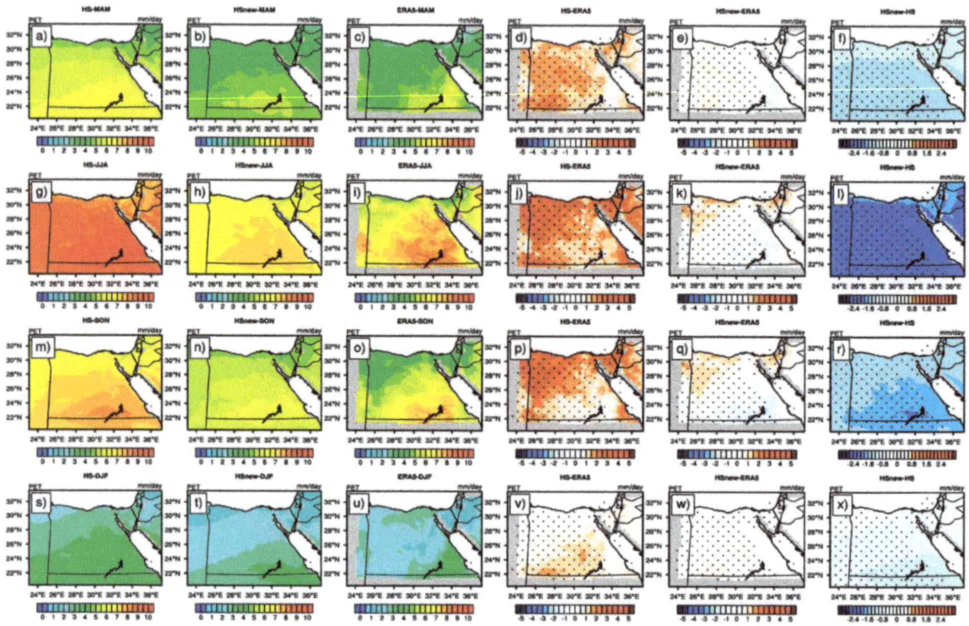

Figure 3. The figure shows the potential evapotranspiration over the period 1981–2017 (PET; in mm day^{-1}) for: MAM season in the first row (**a**–**f**); JJA in the second (**g**–**l**); SON in the third (**m**–**r**); DJF in the fourth (**s**–**x**). For each row, HS is on the left, followed by HSnew; ERA5 is the third from left, HS minus ERA5, HSnew minus ERA5 and the difference between HSnew and HS. Significant difference/bias is indicated in black dots using student *t*-test with alpha equals to 5%.

In the JJA and SON seasons, the HS overestimates the PET over all of Egypt by 1–4.5 mm day^{-1} (see Figure 3j,p). After calibration, the HSnew reduces the PET bias to 0.5–1.5 mm day^{-1} over the north coast of Egypt and the western desert and −0.5 mm day^{-1} around Lake Nasser and middle Egypt (see Figure 3k,q). From a qualitative point of view, the HSnew approximately reduces the simulated PET by 1.6–2.2 mm day^{-1} in the JJA (Figure 3l) and by 0.8–1.6 mm day^{-1} in the SON (Figure 3r). Lastly, in the DJF, it can be observed that HSnew shows its added value over the HS in middle and upper Egypt where the bias was 0.5–2 mm day^{-1} prior to calibration (Figure 3v) and became 0.5–1 mm day^{-1} post calibration (Figure 3w). Further, the HSnew approximately reduces the PET by 0.4–1.2 mm day^{-1} relative to the HS (see Figure 3x). Overall, it can be noted that the added value of the HSnew (over the HS) can be arranged according to season in the following order: 1—JJA; 2—SON; 3—MAM; and 4—DJF. These findings are in agreement with the results reported in Figure S1.

To further explore the added value of the calibrated HS, the climatological annual cycle (Figure 4) of the simulated PET of the HS and HSnew (compared to ERA5) was plotted for locations reported by [3]. Only Port-Said was not mentioned because it shows missing values. From Figure 4, it can be observed that the performance of HS/HSnew varies with location and month. For instance, the HS is close to ERA5 in the months of January, February, November and December, while HSnew is close to ERA5 for the rest of the months in Alexandria. For Arish, Marsa-Matruh and Ismailia; HSnew is closer to ERA5 than HS. In Giza and Asswan, the situation is quite different because HS performs better than HSnew in all months. Further, HSnew shows an improved performance over HS in Assyut. Additionally, the situation in Luxor is similar to the one observed in Alexandria. Finally, in Siwa, Dakhla and Kharga, HSnew shows an improved performance (relative to the HS) in comparison with the ERA5.

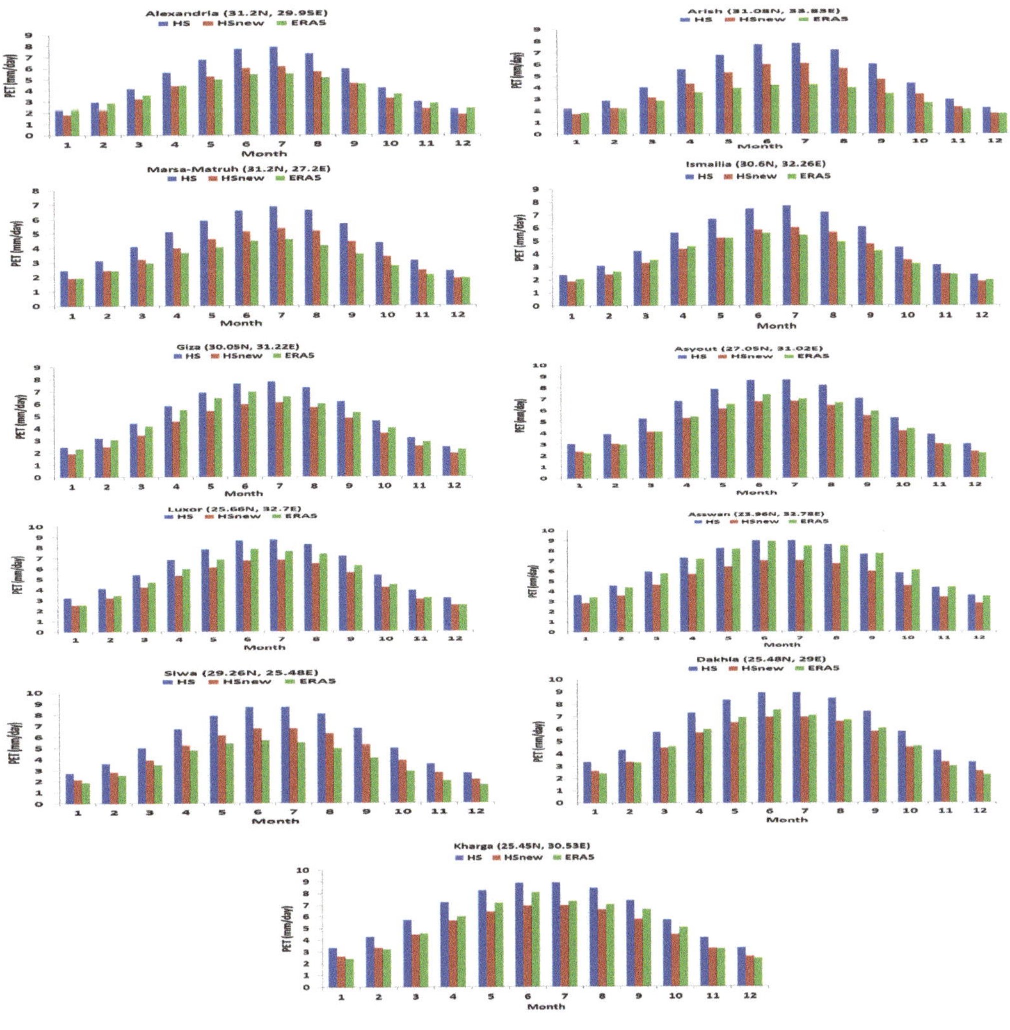

Figure 4. The figure shows the climatological annual cycle of the simulated PET for HS and HSnew compared to ERA5 for locations reported by [3].

4. Discussion and Conclusions

Potential evapotranspiration (PET) is important for monitoring hydrological and meteorological droughts as well as assessing the crop irrigation needs. Additionally, it is a major component in the global terrestrial hydrology cycle. Therefore, the availability of long-term records of PET on a hierarchy of time scales (ranging from hourly to seasonal) is important. The authors of [2] recommend the PM model to compute the PET because it is based on the physical exchange of water and energy between vegetation and atmosphere; however, it requires a large number of meteorological variables (which may not be available for a long time for a variety of locations). Further, uncertainty of the involved meteorological variables may induce a source of uncertainty in the computed PET (and in particular if they are derived from reanalysis products/regional climate models). In addition, it requires a surplus of soil moisture (which is not suitable for the domain of the present study). Hence, there was an urgent need to compute the PET with a simple empirical method (only needs a few meteorological inputs).

Among various empirical methods, the HS model was chosen in this study because it gives a good performance with observational datasets of the PM [7–11]. However, the HS has not been calibrated in Egypt until today. In the present study, the regional climate model (RegCM4) was used to compute the PET comparing between the non-calibrated/calibrated HS with respect to ERA5. The influence of the lateral boundary condition on the simulated PET was also examined. The results showed that switching between EIN15 and NNRP2 did not show a considerable impact on the simulated PET (Figure 2). Spatially, the calibrated HS showed its added value (relative to the original HS model) particularly in the JJA season; such value can be seen by a reduction in the PET bias with respect to the ERA5 (Figure 3). On a point scale, the HS/HSnew performance varies with location and month (Figure 4). Nevertheless, the calibrated HS model can be recommended to construct a regional map of PET of Egypt, predict the daily PET for locations (where station observations are not available) and project the future PET under different global warming scenarios [10,15]. To ensure more robust results of the simulated PET (using the calibrated HS model), a future work will consider the following points:

1. Revising the short/longwave radiation scheme, tuning the parameters of the boundary layer scheme to possibly reduce the uncertainty of the simulated SW, T2m and, eventually PET.
2. Adapting a bias-correction technique (e.g., [3]) to correct the simulated PET over a location of interest.
3. Studying the influence of climate change on the PET of Egypt using the calibrated HS equation [25] and CMIP5/6 simulations [10,15,26].

Supplementary Materials: The following are available online at https://www.mdpi.com/article/10.3390/ECWS-7-14253/s1, Figure S1: global incident solar radiation, Figure S2: 2-m mean air temperature.

Author Contributions: Conceptualization, S.A.A.; methodology, S.A.A.; software, S.A.A.; validation, S.A.A.; formal analysis, S.A.A. and I.L.; investigation, S.A.A. and I.L.; resources, S.A.A.; data curation, S.A.A.; writing—original draft preparation, S.A.A. and I.L.; writing—review and editing, S.A.A. and I.L.; visualization, S.A.A.; supervision, S.A.A. and I.L.; project administration, S.A.A. All authors have read and agreed to the published version of the manuscript.

Funding: This research received no external funding.

Institutional Review Board Statement: Not applicable.

Informed Consent Statement: Not applicable.

Data Availability Statement: Not applicable.

Acknowledgments: Egyptian Meteorological Authority (EMA) is acknowledged for providing the computational power to conduct the model simulations. Hourly potential evapotranspiration (hPET) was retrieved from the web link https://data.bris.ac.uk/data/dataset/qb8ujazzda0s2aykkv0oq0ctp (accessed on 18 October 2022). However, the monthly mean can be acquired from the authors upon request.

Conflicts of Interest: The authors declare no conflict of interest.

References

1. Abdullah, S.S.; Malek, M.A.; Abdullah, N.S.; Kisi, O.; Yap, K.S. Extreme learning machines: A new approach for prediction of reference evapotranspiration. *J. Hydrol.* **2015**, *527*, 184–195. [CrossRef]
2. Allen, G.R.; Pereira, S.L.; Raes, D.; Smith, M. *Crop Evapotranspiration: Guidelines for Computing Crop Water Requirements*; Report 56; Food and Agricultural Organization of the United Nations (FAO): Rome, Italy, 1998; 300p.
3. Anwar, S.A.; Salah, Z.; Khaled, W.; Zakey, A.S. Projecting the Potential Evapotranspiration in Egypt Using a High-Resolution Regional Climate Model (RegCM4). *Environ. Sci. Proc.* **2022**, *19*, 43. [CrossRef]
4. Hargreaves, G.L.; Samani, Z.A. Reference crop evapotranspiration from temperature. *Appl. Eng. Agric.* **1985**, *1*, 96–99. [CrossRef]
5. Hargreaves, G.L.; Allen, R.G. History and evaluation of Hargreaves evapotranspiration equation. *J. Irrigat. Drain. Eng.* **2003**, *129*, 53–63. [CrossRef]

6. Irmak, S.; Irmak, A.; Allen, R.G.; Jones, J.W. Solar and Net Radiation-Based Equations to Estimate Reference Evapotranspiration in Humid Climates. *J. Irrig. Drain. Eng.* **2003**, *129*, 5. [CrossRef]
7. Er-Raki, S.; Chehbouni, A.; Khabba, S.; Simonneaux, V.; Jarlan, L.; Ouldbba, A.; Rodriguez, J.C.; Allen, R. Assessment of reference evapotranspiration methods in semi-arid regions: Can weather forecast data be used as alternate of ground meteorological parameters? *J. Arid. Environ.* **2010**, *74*, 1587–1596. [CrossRef]
8. Potop, V.; Boroneant, C. *Assessment of Potential Evapotranspiration at Chisinau Station*; Mendel a Bioklimatologie: Brno, Czech Republic, 2014; pp. 3–5.
9. Sperna Weiland, F.C.; Tisseuil, C.; Dürr, H.H.; Vrac, M.; Van Beek, L.P.H. Selecting the optimal method to calculate daily global reference potential evaporation from CFSR reanalysis data for application in a hydrological model study. *Hydrol. Earth Syst. Sci.* **2012**, *16*, 983–1000. [CrossRef]
10. Anwar, S.A.; Mamadou, O.; Diallo, I.; Sylla, M.B. On the influence of vegetation cover changes and vegetation-runoff systems on the simulated summer potential evapotranspiration of tropical Africa using RegCM4. *Earth Syst. Environ.* **2021**, *5*, 883–897. [CrossRef]
11. Cobaner, M.; Citakoğlu, H.; Haktanir, T.; Kisi, O. Modifying Hargreaves–Samani equation with meteorological variables for estimation of reference evapotranspiration in Turkey. *Hydrol. Res.* **2017**, *48*, 480–497. [CrossRef]
12. Singer, M.; Asfaw, D.; Rosolem, R.; Cuthbert, M.O.; Miralles, D.G.; MacLeod, D.; Michaelides, K. Hourly potential evapotranspiration (hPET) at 0.1degs grid resolution for the global land surface from 1981-present. *Sci. Data* **2021**, *8*, 224. [CrossRef]
13. Giorgi, F.; Coppola, E.; Solmon, F.; Mariotti, L.; Sylla, M.B.; Bi, X.; Elguindi, N.; Diro, G.T.; Nair, V.; Giuliani, G.; et al. BrankovicRegCM4: Model description and preliminary tests over multiple CORDEX domains. *Clim. Res.* **2012**, *52*, 7–29. [CrossRef]
14. Giorgi, F.; Pal, J.S.; Bi, X.; Sloan, L.; Elguindi, N.; Solmon, F. Introduction to the TAC special issue: The RegCNET network. *Theor. Appl. Clim.* **2006**, *86*, 1–4. [CrossRef]
15. Anwar, S.A.; Diallo, I. Modelling the Tropical African Climate using a state-of-the-art coupled regional climate-vegetation model. *Clim. Dyn.* **2022**, *58*, 97–113. [CrossRef]
16. Dee, D.P.; Uppala, S.M.; Simmons, A.J.; Berrisford, P.; Poli, P.; Kobayashi, S.; Andrae, U.; Balmaseda, M.A.; Balsamo, G.; Bauer, P.; et al. The ERA-Interim reanalysis: Configuration and performance of the data assimilation system. *Q. J. R. Meteorol. Soc.* **2011**, *137*, 553–597. [CrossRef]
17. Kanamitsu, M.; Ebisuzaki, W.; Woollen, J.; Yang, S.K.; Hnilo, J.J.; Fiorino, M.; Potter, G.L. NCEP-DOE AMIP-II Reanalysis (R-2). *Bull. Am. Meteorol. Soc.* **2002**, *83*, 1631–1643. [CrossRef]
18. Emanuel, K.A. A scheme for representing cumulus convection in large-scale models. *J. Atmos. Sci.* **1991**, *48*, 2313–2335. [CrossRef]
19. Clough, S.A.; Shephard, M.W.; Mlawer, E.J.; Delamere, J.S. Atmospheric radiative transfer modeling: A summary of the AER codes, Short Communication. *J. Quant. Spectrosc. Radiat. Transf.* **2005**, *91*, 233–244. [CrossRef]
20. Holtslag, A.A.M.; Boville, B.A. Local versus nonlocal boundary layer diffusion in a global model. *J. Clim.* **1993**, *6*, 1825–1842. [CrossRef]
21. Hersbach, H.; Bell, B.; Berrisford, P.; Hirahara, S.; Horányi, A.; Muñoz-Sabater, J.; Nicolas, J.; Peubey, C.; Radu, R.; Schepers, D.; et al. The ERA5 global reanalysis. *Q. J. R. Meteorol. Soc.* **1993**, *146*, 1999–2049. [CrossRef]
22. Muñoz-Sabater, J.; Dutra, E.; Agustí-Panareda, A.; Albergel, C.; Arduini, G.; Balsamo, G.; Boussetta, S.; Choulga, M.; Harrigan, S.; Hersbach, H.; et al. ERA5-Land: A state-of-the-art global reanalysis dataset for land applications. *Earth Syst. Sci. Data* **2021**, *13*, 4349–4383. [CrossRef]
23. Wang, G.; Yu, M.; Pal, J.S.; Mei, R.; Bonan, G.B.; Levis, S.; Thornton, P.E. On the development of a coupled regional climate–vegetation model RCM–CLM–CN–DV and its validation in Tropical Africa. *Clim. Dyn.* **2016**, *46*, 515–539. [CrossRef]
24. Erfanian, A.; Wang, G.; Yu, M.; Anyah, R. Multi-Model Ensemble Simulations of Present and Future Climates over West Africa: Impacts of Vegetation Dynamics. *J. Adv. Model. Earth Syst.* **2016**, *8*, 1411–1431. [CrossRef]
25. Awal, R.; Rahman, A.; Fares, A.; Habibi, H. Calibration and Evaluation of Empirical Methods to Estimate Reference Crop Evapotranspiration in West Texas. *Water* **2022**, *14*, 3032. [CrossRef]
26. Anwar, S.A.; Diallo, I. A RCM investigation of the influence of vegetation status and runoff scheme on the summer Gross Primary Production of Tropical Africa. *Theor. Appl. Climatol.* **2021**, *145*, 1407–1420. [CrossRef]

environmental sciences proceedings

MDPI

Proceeding Paper

Daily Streamflow Modelling Using ML Based on Discharge and Rainfall Time Series in the Besós River Basin, Spain †

Mohamed Hamitouche [1,*] and Marc Ribalta [2]

1 Sustainable Water Management and Governance in Natural and Agricultural Environments, Mediterranean Agronomic Institute of Zaragoza (IAMZ), International Centre for Advanced Mediterranean Agronomic Studies (CIHEAM), Av. Montañana 1005, 50059 Zaragoza, Spain
2 Eurecat, Technology Centre of Catalonia, Unit of Applied Artificial Intelligence, 08005 Barcelona, Spain; marc.ribalta@eurecat.org
* Correspondence: armoh94@gmail.com
† Presented at the 7th International Electronic Conference on Water Sciences, 15–30 March 2023; Available online: https://ecws-7.sciforum.net.

Abstract: Machine learning (ML)-based data-driven modelling is an efficient approach for good estimates of flow and maximum discharge at certain points within a basin. This paper is mainly aimed at evaluating the predictive capability of ML algorithms for daily streamflow modelling in the Besós River Basin (Spain), based on open source flow discharge and rainfall historical time series. In this sense, two modelling scenarios, without and with consideration of the antecedent hydrologic conditions, were evaluated, and three ML algorithms—support vector machines, random forest (RF) and gradient boosting (GB)—were compared to multiple linear regression (MLR), and were implemented. The prediction results revealed that the SVR model outperformed the other suggested models. Additionally, it was deduced that taking into account preceding hydrologic conditions clearly improves prediction results.

Keywords: streamflow modelling; machine learning; data-driven; preceding hydrologic conditions; virtual sensor

check for updates

Citation: Hamitouche, M.; Ribalta, M. Daily Streamflow Modelling Using ML Based on Discharge and Rainfall Time Series in the Besós River Basin, Spain. *Environ. Sci. Proc.* **2023**, *25*, 30. https://doi.org/10.3390/ECWS-7-14168

Academic Editor: Lampros Vasiliades

Published: 14 March 2023

1. Introduction

Rainfall–runoff modelling is believed to be complex, nonlinear, and time-varying because the basin response depends not only on hydrometeorological parameters but also on spatiotemporal irregularity in basin characteristics and rainfall patterns [1]. Usually, hydrologists develop and use different types of models to simulate hydrological processes. Regardless of their structural variations, these models generally fall into three main types, including physical, conceptual, and data-driven models (DDMs) [2].

Over the past two decades, data-driven approaches based on artificial intelligence (AI), have gained drastically increasing interest from hydrologists [3], due to their significant contribution to improving the accuracy, versus the failure stories, of the classical and conventional methods in terms of spatial scale, time scale, the amount of data needed, facilities, the inability to handle nonlinear and nonstationary hydrological processes, and even in terms of accuracy, viewing the complexity of the equations governing the hydrological cycle's mechanisms which often require simplifications and theoretical assumptions leading to considerable errors and uncertainties [4].

Many different statistical methods, such as multiple linear regression (MLR), and various types of AI and machine learning (ML) algorithms such as support vector machines (SVM) and artificial neural networks (ANN), have been widely applied in recent years [5]. Especially, ANN and SVM have the advantage of handling complex relationships between input and output variables and have been used successfully in various water resources problems [6]. However, despite the application of several ML techniques available in the literature, the gradient boosting (GB) approach has not been widely applied to predict daily flows [7]. Also, for hydrological extremes, GB and random forest (RF) are more often explored for qualitative predictions rather than quantitative predictions [4].

In this sense, this paper presents the application of three regression ML algorithms—support vector regression, random forest regression and gradient boosting regression—were compared to MLR and used to model the daily flow discharge at the outlet of the Besós River basin (Spain) under two scenarios: without and with consideration of the antecedent hydrologic conditions. The objective is to discuss and evaluate the performance of the aforementioned DDMs in daily streamflow prediction based on open-source flow discharge and rainfall historical time series by comparing them with each other and with the MLR model based on several statistical evaluation measures, and to evaluate the impact of using the preceding hydrologic conditions.

2. Materials and Methods

2.1. Study Location and Data Collection

The Besós is a Mediterranean river characterised by a very irregular hydrological regime with highly variable flows related to climatic conditions. It is the collector of various tributaries that originate in the Catalan pre-coastal range, which includes the Mogent, Congost, Tenes, Caldes and Ripoll rivers, as shown in Figure 1. Its hydrographic basin of approximately 1020 km^2 supports a population of almost one million inhabitants with high water consumption, mainly destined for industrial and urban use, since agriculture, especially in its lower section, has been losing importance, and is close to disappearing. The Besós basin shows a pronounced relief formed by the coastal and precoastal ranges with elevations of up to 1350 m a.s.l. and steep slopes. As indicated by the topography, the geology of the area differs between the mountain ranges, composed mainly of granites, slate, limestone and sandstone, and the central valley where clay, sand, and conglomerate deposits dominate.

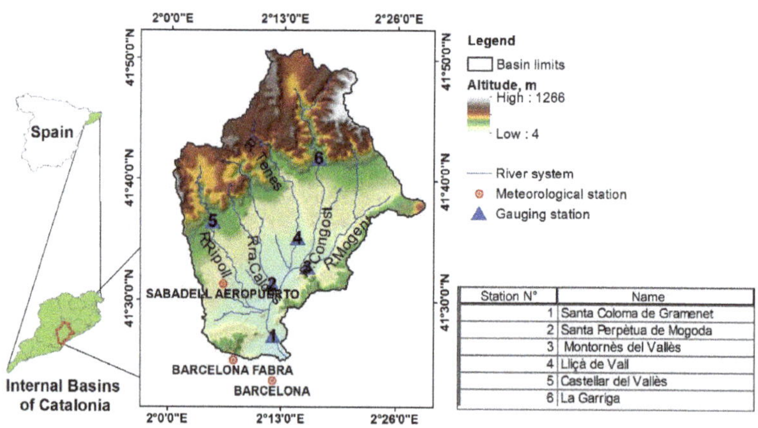

Station N°	Name
1	Santa Coloma de Gramenet
2	Santa Perpètua de Mogoda
3	Montornès del Vallès
4	Lliçà de Vall
5	Castellar del Vallès
6	La Garriga

Figure 1. Study area and location of the meteorological and gauging stations.

In this study, open-source historical series of daily flow and rainfall data between 2003 and 2010 from the Catalan Water Agency and the State Meteorological Agency were gathered. Information was collected on a total of five gauging stations, "La Garriga", "Castellar Valles", "Lliçà de Vall", "Montornès Valles", and "Santa Perpetua de Mogoda", and three meteorological stations, "Barcelona", "Barcelona Fabra" and "Sabadell Aeropuerto", were used to explore the applicability of ML models for daily flow discharge prediction at the target station "Santa Coloma de Gramenet". Figure 1 shows the geographical distribution of the stations.

2.2. Exploratory Analysis and Data Preparation

2.2.1. Data Exploration

The available collected data are distributed over a period from 1 January 2003 to 31 December 2010, as some stations no longer have records after this date. The longest common period with the fewest missing values is from 1 January 2003 to 6 May 2008. Therefore, this observation period was considered in creating the prediction models.

For this common period, the "Castellar del Vallès" gauging station has a considerable number of missing values (about 56.18%), so it was decided that it would not be considered in the modelling process, and the lost information could be therefore obtained from the "Sabadell Aeropuerto" meteorological station. Also, the meteorological stations and the "La Garriga" gauging station presented some missing values in their data series (<16%). For that reason, because data imputation may lead to additional uncertainties to those series due to measurement errors, it was decided that all data rows for which at least one station had a missing value were deleted, and that the models would learn from the rest of the data that were supposed to be sufficient for their training and validation. The resulting data time series contained a total of 1404 daily rainfall and flow records.

The basic information of the observatories and the dataset statistical analysis conducted after dealing with the missing values showed that the Besós basin receives, in its lower part, as represented by the "Sabadell Aeropuerto" station, an average annual rainfall of 415 mm varying from 0 to 122 mm per day. The average flow discharge at its outlet ("Santa Coloma del Vallès") is 4.4 m^3/s, and most flow values are below 20 m^3/s. Only two values exceeded 80 m^3/s during this observation period. Also, the flow discharge has never been zero.

The parameters "skewness" and "kurtosis" were used to examine the data distribution. Practically, all the data time series have very large values of these coefficients and therefore do not have a normal distribution.

A correlation heat map was created to explore the relationship between the different data inputs, showing a good correlation between meteorological stations (rainfall) as well as between gauging stations (flow), while no correlation was detected between flow and rainfall, reflecting the nonlinear rainfall–flow relationship.

2.2.2. Data Model

The selection of input variables is determined by a combination of prior knowledge of causality, examination of time series plots, data availability, and study objectives. In this study, we used rainfall and flow discharge data, with rainfall being the primary driving force of runoff. Since only one meteorological station was located within the basin, we utilized data from two neighbouring stations outside the basin (as shown in Figure 1) to define the rainfall in the differential basin between upstream gauging stations and the prediction point (target station).

Given that flow is effectively made up of contributions from different subareas whose travel time covers a range of values, the next step was the determination of the appropriate lag time concerning the prediction output. This was carried out through a cross-correlation analysis between the flow at the outlet and the upstream and downstream rainfall and flow. The cross-correlation showed the considerable influence of the previous day's (t−1) rainfall on the current value of the outlet flow for the three meteorological stations. From time t−2, this correlation decreased to less than 0.3. Also, regarding the flow at the input stations, it was seen that there was a decreasing correlation from the same day of recording. The antecedent flow subsequent to time instant t−2 did not contribute significantly to the outflow generation. Therefore, the antecedent values of flow and rainfall corresponding to time instants t and t−1 were considered.

When looking at the time step, several features are immediately obvious, such as the day of the month and the month of the year, which may be helpful in understanding the flow periodicity or seasonality. It is then necessary to use variables that extract and preserve hidden information within cyclical data, such as the distance between two events: day 30 or 31 and day 1, or month 12 and month 1. This seems important as the missing values were removed. To do this, "sin" and "cos" were used to assign each cyclic variable (day and month) to a circle so that the smallest value for that variable appeared right next to the largest value. Four cyclic features (day_{sin}, day_{cos}, $month_{sin}$, and $month_{cos}$) with respect to the day and the month of the year were created to obtain a total of 18 input features.

The general representative DDM for the first scenario can be defined as:

$$\hat{Q}_t = f(Q_{it}, Q_{it-1}, P_{kt}, P_{kt-1}, day_{sin}, day_{cos}, month_{sin}, month_{cos}) \quad (1)$$

where $\hat{Q}_t$ represents the predicted flow at time t at the "Santa Coloma de Gramenet" station; Q_{it} is the flow recorded at time t at each of the predictor gauging stations; Q_{it-1} is the flow recorded at these gauging stations at time $t - 1$ ('i' ranges from 1 to 4); P_{kt} represents the rainfall at time t observed at each of the meteorological stations; P_{kt-1} is the rainfall at time $t - 1$ observed at each of the meteorological stations ('k' ranges from 1 to 3); and day_{sin}, day_{cos}, $month_{sin}$, and $month_{cos}$ represent the features or cyclical variables of the day and month.

These variables were used to predict the flow at the "Santa Coloma de Gramenet" station as a first scenario, which was intended to be extrapolated to basins without a gauging station or flow measurement, depending on their physiographic characteristics and climatic conditions. In the second scenario, in addition to these input variables, the antecedent flow discharge of the "Santa Coloma de Gramenet" station was also used as an input variable, since it indirectly describes the soil moisture status. Also, since the rainfall data contain zero values, such flows add more information, meaning that the longer the zero-input interval, the more the output decreases. The above was conducted by performing an autocorrelation for the target variable dataset, between the previous values and the current one up to 5 lags. In this second scenario, DDMs could be used in poorly gauged or previously gauged catchments where flow measurements were interrupted or as a virtual sensor for the Besós river basin itself. The autocorrelation showed the considerable influence of the two previous values (correlation ≥ 0.4). Therefore, the antecedent flows at the target station at times $t - 1$ and $t - 2$ were considered. This was also verified by trial and error experiments.

The general representative DDM, in this second scenario, can be defined as:

$$\hat{Q}_t = f(Q_{it}, Q_{it-1}, P_{kt}, P_{kt-1}, Q_{ot-1}, Q_{ot-2}, day_{sin}, day_{cos}, month_{sin}, month_{cos}) \quad (2)$$

where Q_{ot-1} and Q_{ot-2} are the antecedent flows at times $t - 1$ and $t - 2$ observed at the "Santa Coloma de Gramenet" outlet gauging station.

To prevent input data in larger numerical ranges from dominating those in smaller numerical ranges and to avoid numerical difficulties during computation, all input variables were scaled to the range [0, 1] using the Min-Max Scaler method. The complete chronologically organized dataset was then divided into training and test datasets to obtain an approximate training/test split ratio of 80%/20%.

2.2.3. Hyperparameter Optimisation

Models were trained with a range of hyperparameter values that was determined by examining various hydrological studies. Afterwards, the hyperparameter optimal values were determined through a trial and error process using the Grid Search technique due to its simplicity and robustness, in the training and test datasets. Then, the optimal hyperparameters were maintained and the best models were used to predict the flow rate.

2.2.4. Validation Metrics

Four quantitative validation metrics, including: the coefficient of determination (R^2), the Nash–Sutcliffe efficiency coefficient (NSE), the Mean Absolute Error (MAE) and the Root Mean Squared Error (RMSE), were used to assess the prediction accuracy and to compare the different data-driven models based on their relative performance.

3. Results and Discussion

The values of the DDM-statistical performance metrics for the training and test periods are presented in Table 1. Hydrographs were also plotted to visualize the DDM behaviour, particularly for extreme values.

Table 1. Model performance metrics for training and test periods.

Scenario	Model	Training				Test			
		RMSE	MAE	R^2	NSE	RMSE	MAE	R^2	NSE
1	MLR	1.783	0.780	0.819	0.779	0.806	0.502	0.819	0.808
	SVR	1.685	0.600	0.838	0.774	0.604	0.369	0.898	0.877
	GBR	1.062	0.558	0.936	0.928	0.720	0.480	0.856	0.844
	RFR	0.983	0.259	0.945	0.922	0.758	0.545	0.840	0.834
2	MLR	1.477	0.583	0.876	0.858	0.776	0.388	0.833	0.850
	SVR	1.563	0.492	0.861	0.805	0.578	0.307	0.907	0.890
	GBR	0.171	0.131	0.998	0.998	0.685	0.381	0.869	0.862
	RFR	0.921	0.207	0.952	0.933	0.624	0.368	0.892	0.890

It is clear from Table 1 that the RFR and GBR models were more efficient in predicting the flow for the first scenario, in the training period, with a slight victory for the RFR model, reaching R^2 values of 0.945 and 0.936, respectively; RMSEs of 0.983 m^3/s and 1.062 m^3/s; MAEs of 0.259 m^3/s and 0.558 m^3/s, and approximately equal NSEs. However, the SVR outperformed all models in the test period for all metrics.

MLR performance was not as good as that of the three other DDMs, but it was very acceptable concerning performance metrics. It outperformed the RF model in the test period with respect to the MAE and the SVR model in the training period regarding the NSE. The fact that the MLR prediction values have a good correlation with the observed values is related to the fact that the number of input gauging stations was slightly higher than that of meteorological stations, that the flow discharge values had more weight in the MLR equation than those for rainfall, and that the flow inputs showed a good linear correlation with each other, unlike the rainfall inputs.

Interestingly, the prediction results are satisfactory, and there are some improvements in the performance of the models in the test period compared to the training period, except for GBR regarding the R^2 and NSE and RFR regarding the MAE, R^2 and NSE. This may have been the result of overfitting these models.

The hydrographs in Figures 2 and 3 of the observed and predicted values for each model in the training period, as well as the test period at the target station, indicate that, in general, the predicted flow fits well with the observed flow.

However, for the training period, it can be seen that there were more peaks than there were in the test period. The maximum peak observed in the training period was underestimated by all the models, but it can be seen that MLR, SVR and GBR managed to approximate it well, while RFR presented a high relative error at this point. In general, the RFR had the best performance in the training period.

Possible reasons for this result could be the better generalisation ability of SVR due to the structural risk minimisation approach which led to an optimal global solution. Regarding the overfitting of the GBR and the RFR models, it should first be remembered that these two models are based on building trees from a random Bootstrap sample, which makes both models stochastic, with an uncertainty associated with the predicted value. The high number of trees that was found to have trained the two models may have been behind this overfitting. Also, GBR is a nondeterministic algorithm; i.e., even for the same input, it can present different outputs in different executions.

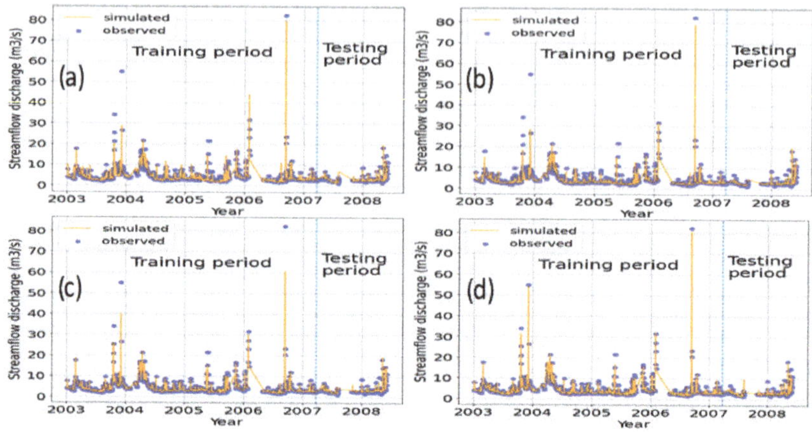

Figure 2. First scenario's hydrographs of the observed and simulated flow discharge in the training and test periods. (**a**) MLR, (**b**) SVR, (**c**) RFR, and (**d**) GBR.

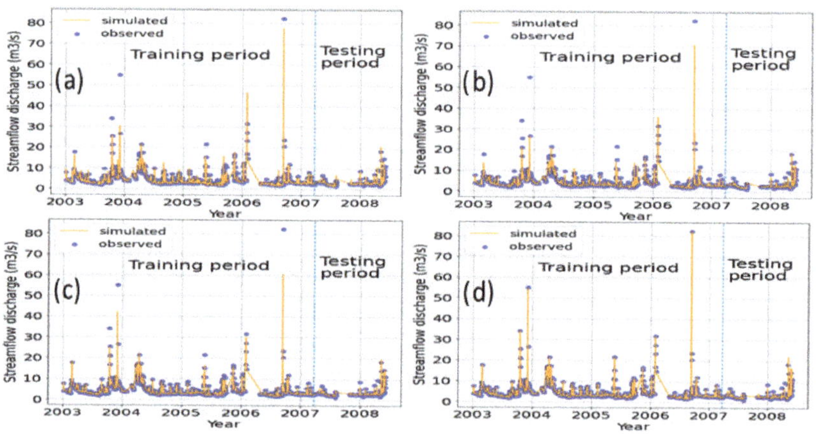

Figure 3. Second scenario's hydrographs of the observed and simulated flow discharge in the training and test periods. (**a**) MLR, (**b**) SVR, (**c**) RFR, and (**d**) GBR.

Regarding the second scenario, the GBR outperformed all models in the training period with near-perfect performance. In the test period, the SVR in this case was also the best regarding all metrics.

Some DDM performance improvements were shown in the test period compared to the training period. Regarding SVR, all metrics improved. For MLR, only RMSE and MAE improved. In contrast, the GBR and the RFR showed a decrease in their performance, except for the RMSE of the RFR, which improved. This overfitting, in addition to the aforementioned reasons, may have been due to the insufficient data size, or that the data's split ratio for training and testing the models was not adequate. Generally, the simulated hydrographs fit well with the observed hydrographs. It is seen that GBR has a great ability to predict flow peaks. All models predicted the flow at the basin outlet well.

When comparing the DDMs, it can be seen that the use of the antecedent flow discharges had a great impact in improving their performance, with a reduction in the MAEs of 23%, 17%, 21% and 33% for the MLR, SVR, GBR and RFR models. The RFR showed the greatest improvement in performance during the test period with a decrease in the RMSE and MAE of 18% and 33%, and an increase in the R^2 and NSE of 6% and 7%, respectively.

4. Conclusions

To predict the daily flow at the outlet of the Besós river basin, the MLR and three data-driven ML models, SVR, RFR and GBR, were used. The obtained results show that the SVR model outperformed the other models whether or not the preceding hydrologic conditions were considered. MLR, as well as the decision tree ensemble models (RFR and GBR), has also shown a good flow prediction capacity. It is worth noting that the proposed DDMs have demonstrated high efficiency in capturing the real trend and the underlying phenomena of rising and falling flow curves. The use of the antecedent flows in the target gauging station had a positive impact on improving the performance of all models.

To improve the prediction capabilities of ML models, in future work, it is recommended to use other variables to build a strong relationship with the streamflow; to perform a sensitivity analysis of input features to bring out those that contribute the most to the flow prediction; to pay close attention to the data length and split ratio; to ensure that the training phase experiences most of the streamflow patterns to allow the models in the test period to simulate the flow discharge with an acceptable level of accuracy.

Author Contributions: M.H. and M.R. conceived, designed, and led the research and contributed to the paper editing; M.H. was behind the research conceptualisation and analytical development. M.R. supervised all actions. All authors have read and agreed to the published version of the manuscript.

Funding: This research was funded by the International Centre for Advanced Mediterranean Agronomic Studies (CIHEAM).

Institutional Review Board Statement: Not applicable.

Informed Consent Statement: Not applicable.

Data Availability Statement: The data, programming codes and additional material were uploaded to: https://www.kaggle.com/datasets/mohamedhamitouche/rainfall-and-runoff-data-for-the-bess-river-basin (accessed on 10 March 2023).

Conflicts of Interest: The authors declare no conflict of interest.

References

1. Cheng, K.; Wei, S.; Fu, Q.; Pei, W.; Li, T. Adaptive management of water resources based on an advanced entropy method to quantify agent information. *J. Hydroinform.* **2019**, *21*, 381–396. [CrossRef]
2. Zhang, L.; Yang, X. Applying a multi-model ensemble method for long-term runoff prediction under climate change scenarios for the Yellow River Basin, China. *Water* **2018**, *10*, 301. [CrossRef]
3. Terzi, Ö.; Ergin, G. Forecasting of monthly river flow with autoregressive modeling and data-driven techniques. *Neural Comput. Appl.* **2014**, *25*, 179–188. [CrossRef]
4. Hamitouche, M.; Molina, J.L. A Review of AI Methods for the Prediction of High-Flow Extremal Hydrology. *Water Resour. Manag.* **2022**, *36*, 3859–3876. [CrossRef]
5. Isunju, J.B.; Kemp, J. Spatiotemporal analysis of encroachment on wetlands: A case of Nakivubo wetland in Kampala, Uganda. *Environ. Monit. Assess.* **2016**, *188*, 203. [CrossRef] [PubMed]
6. Hosseini, S.M.; Mahjouri, N. Developing a fuzzy neural network-based support vector regression (FNN-SVR) for regionalizing nitrate concentration in groundwater. *Environ. Monit. Assess.* **2014**, *186*, 3685–3699. [CrossRef] [PubMed]
7. Fonseca, T.L.; Gorodetskaya, Y.; Tavares, G.G.; de Melo Ribeiro, C.B.; da Fonseca, L.G. A Gradient Boosting Model Optimized by a Genetic Algorithm for Short-term Riverflow Forecast. *Rev. Mundi Eng. Tecnol. Gestão* **2019**, *4*, 3845. [CrossRef]

environmental sciences proceedings

MDPI

Proceeding Paper

Global Change Explorer—A Web-Based Tool for Investigating the Complexities of Global Change †

Slobodan P. Simonovic

Department of Civil and Environmental Engineering, University of Western Ontario, London, ON N6G 4N4, Canada; simonovic@uwo.ca; Tel.: +1-519-200-2692

† Presented at the 7th International Electronic Conference on Water Sciences, 15–30 March 2023; Available online: https://ecws-7.sciforum.net.

Abstract: Global Change Explorer (GCE) is an interactive web-based tool for investigating the complexities of global change. GCE uses the ANEMI simulation model developed at the University of Western Ontario, Canada. ANEMI simulates system dynamics to offer information on Earth's dynamic processes and the behaviours that instigate change. The ANEMI model is an integrated assessment model of global change that emphasizes the role of water resources. The model sectors that comprise ANEMI3 (the current version of the model) are that of the climate system; carbon, nutrient, and hydrologic cycles; population dynamics; land use; food production; sea level rise; energy production; the global economy; persistent pollution; water demand; and water supply development. GCE is designed to allow the use of ANEMI to simulate various future scenarios related to five main themes: climate change; population dynamics; food production; water quality; and water quantity. The users are presented with the opportunity to ask different questions, select simulation runs, and evaluate model outputs.

Keywords: global change; system dynamics; water; decision support

check for updates

Citation: Simonovic, S.P. Global Change Explorer—A Web-Based Tool for Investigating the Complexities of Global Change. *Environ. Sci. Proc.* **2023**, *25*, 31. https://doi.org/10.3390/ECWS-7-14170

Academic Editor: Luis Garrote

Published: 14 March 2023

1. Introduction

In the early 1950s, the concept of global change entered the international stage with a clear indication that Earth is a closed system in which natural resources and the environment determine the boundaries of population growth and economic development.

This paper starts with a simple assumption that global change problems require a systems approach. Global change research mostly uses integrated assessment, defined as an interdisciplinary process of bringing together knowledge from different disciplines.

There are a limited number of tools that allow users to assess, analyze, and adapt to global change. Global Change Explorer (GCE, https://www.globalchange-uwo.ca/, accessed on 26 January 2023) is an attempt to bring global change and its consequences closer to decision makers, scientists, and the general public [1]. GCE allows for the interactive investigation of global change complexities. It uses ANEMI, a system dynamics simulation model developed at the University of Western Ontario [2–4]. ANEMI is named after the Greek Anemoi gods of the four winds: Boreas the north wind (bringing the cold breath of winter), Zephryos the west wind (the god of spring breezes), Notos the south wind (the god of summer rain-storms), and Euros the east wind (associated with the season of autumn).

ANEMI simulates system dynamics to offer information on Earth's dynamic processes and the behaviours that instigate change. It is an integrated assessment model of global change that emphasizes the role of water resources. The main job of ANEMI is to explicitly integrate various sectors (natural, physical, and socioeconomic) into a single model, providing effective consideration of the high-level feedback relationships between the physical environment and social adaptation. At lower levels, this relationship ends in thousands of feedbacks between various model sectors and variables.

The ANEMI model was developed using the Vensim system dynamics simulation environment [5]. The entire model code is archived using Zenodo (https://doi.org/10.5281/zenodo.4025424, accessed on 26 January 2023) and is available to everyone. Details on how to run the model, modify inputs, and view the outputs in graphical or tabular formats are provided in the repository.

GCE is designed to allow the use of ANEMI to simulate various future scenarios related to five main themes: climate change; population dynamics; food production; water quality; and water quantity. The users are presented with the opportunity to ask different questions, select simulation runs, and evaluate model outputs.

The main objective of this paper is to give: (i) a brief description of the ANEMI model and (ii) an illustrative presentation of GCE. The next section of the paper provides a brief description of the ANEMI model. It is followed by the graphical description of GCE's use. The paper ends with a summary statement.

2. The ANEMI Model

The development of the ANEMI model was carried out using the system dynamics simulation (SD) approach. SD is one of the systems analysis tools specifically designed to analyze the dynamic nature of systems that are composed of feedback loops [6,7]. A structure is essential for the interpretation of complex interactions occurring in models of global change. Learning the structure of a system means learning how things are related [6]. The concept of feedback systems emerged as the basis for structuring the observations of global change. The feedback is defined as a closed path connecting in sequence a decision that controls the action, the state of the system, and information about the state of the system [7]. The feedback provides a direct link between the system structure and its dynamic behaviour. The language of system dynamics includes a powerful tool of causal diagramming to capture the system structure (express what we know about a system) and mathematical simulation to generate dynamic system behaviour.

2.1. The ANEMI Model Description

The model sectors that comprise the ANEMI model are climate, carbon, nutrient, and hydrologic cycles; population dynamics; land use; food production; sea level rise; energy production; the global economy; persistent pollution; water demand; and water supply development (detailed descriptions of each model sector are available in [4]). The high-level model structure is shown in Figure 1. The selection of model sectors was made to capture the dynamics of global change with an emphasis on water resources. The main strength of the ANEMI model is its highly endogenous structure and tight coupling of all twelve sectors. This allows for studying global change from a feedback-based perspective to understand the dynamic behavior of global issues (e.g., climate change, food–water–energy resource limitations, and the development of the water supply to support a growing population). Figure 1 depicts the main feedback relationships between the sectors.

The arrows represent causal relationships between the sectors and the main connecting variables. Arrows with positive signs denote positive causality (change in connected sectors/variables in the same direction) and those with a negative sign denote negative causality (change in connected sectors/variables in the opposite direction). The causal links closed in a loop form the feedback relationships. The positive reinforcing feedbacks are formed by an even number of negative causal links. The negative balancing feedbacks are formed by an odd number of negative causal links.

The colored arrows in Figure 1 are used to illustrate thematic feedbacks between major model sectors. For example, the red arrows depict a feedback relationship between the food production, population, energy–economy, carbon, and climate sectors. If food production decreases relative to the global population, this slows population growth and available labour to drive global economies. With less economic productivity, fewer emissions are anticipated, lowering greenhouse gas concentrations and slowing climate change. This has a positive effect on land yield, as global temperatures would not increase as much, having a positive influence on food production. This is an example of a negative or self-regulating feedback loop. The overall behaviour of the global system is governed by how strongly each feedback loop interacts with the system over time.

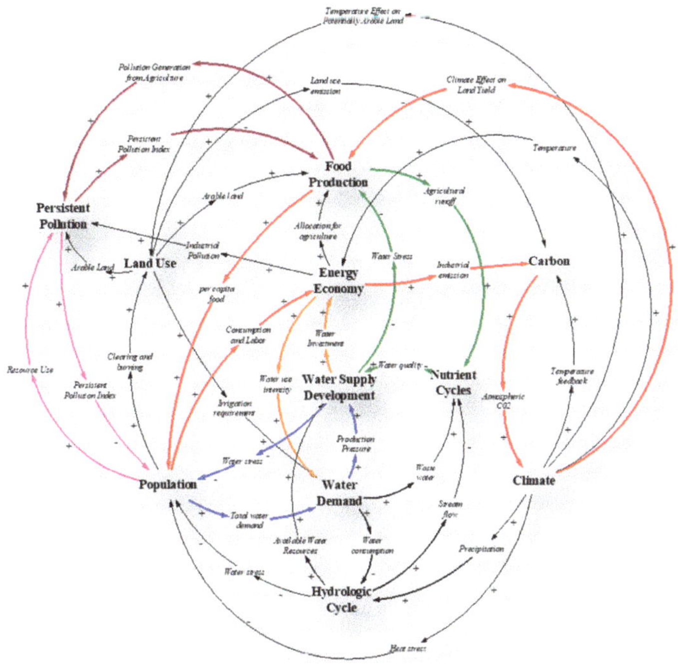

Figure 1. ANEMI model intersectoral feedback diagram (Adapted from Ref. [4]).

Feedback loops between sectors, or intersectoral feedback loops, are responsible for a global change in the Earth system. ANEMI focuses on representing global-scale feedbacks that are driving overall change and assessing their importance and influence within the Earth system. The highly endogenous structure of the model creates a very large number of feedbacks that are driving the dynamics of the Earth system.

The ANEMI model was developed using Venism systems dynamics simulation software [5]. The entire model code is available from the Zenodo archive (https://doi.org/10.5281/zenodo.4483736, accessed on 26 January 2023), with details on running the model, modifying inputs, and viewing the outputs.

2.2. Use of the ANEMI Model

The ANEMI model is parameterized to simulate the Earth system over the period from 1980 to 2100, with an integration time step of 1/128th of a year. Due to the rigid nature of the system equations, the simple Euler integration method is used. The model has been parameterized to capture the initial conditions in 1980 and reproduce the behaviour of key system variables in each sector.

The generalized, feedback-based structure of the ANEMI model allows the user to test policy scenarios related to global change. By altering one of the model parameters at a given point in time, all model sectors will respond through the driving feedback processes which govern the model structure. The output of the model using the baseline parameter set provides a reference scenario to which alternative policy scenarios may be compared. Different parameter sets have the potential to cause a shift in model behaviour compared to the reference scenario, as the dominance of feedback processes in the reference scenario may change. The large number of variables and sectors in the ANEMI model allows for the development of a wide variety of scenarios, each with a different focus. Up to now, the main investigations conducted by the ANEMI model focused on simulating various future scenarios related to five main themes: (i) climate change; (ii) population dynamics; (iii) food production; (iv) water quality; and (v) water quantity. For a detailed description of the tested scenarios, please refer to [4].

3. Global Change Explorer (GCE)

The web version of the GCE (Global Change Explorer) program [1] was designed as a smart interface that allows for the ANEMI model to be simulated through a series of interactive, user-defined scenarios. GCE is available at: https://globalchange-uwo.ca/ (accessed on 26 January 2023).

The communication with GCE starts by activating the *Scenarios* menu, as shown in Figure 2. The user is presented with five scenarios and questions related to each of them. The selection of a question and scenario begins the ANEMI simulations, and changing some input variables opens up possibilities for experimenting with the model.

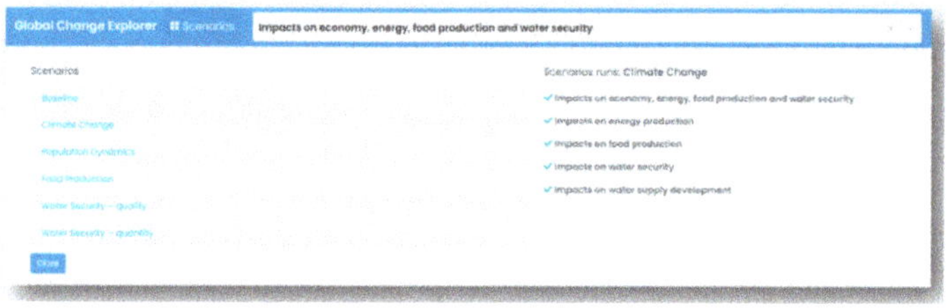

Figure 2. The GCE main menu.

The smart interface automatically provides the *Variables* that can be experimented with based on the selected question and scenario. A variable selection screen for one scenario and one question is shown in Figure 3.

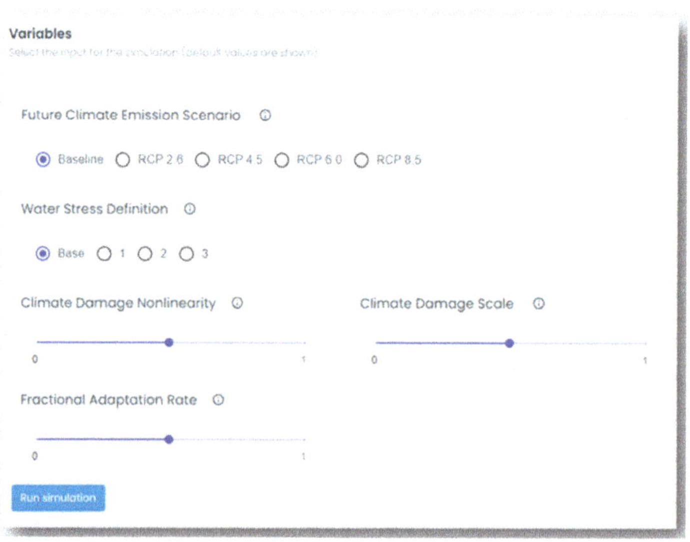

Figure 3. GCE variable selection.

GCE offers assistance by providing a *description of the scenario and variables* that the user can experiment with, as shown in Figure 4.

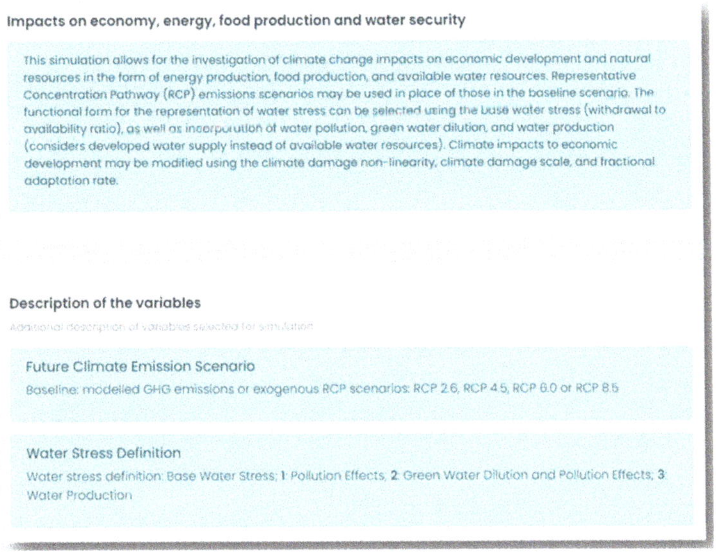

Figure 4. GCE variable description.

Upon selection of the variables, the ANEMI model is executed in the background and simulation results are stored for presentation in GCE. The left side of the GCE screen offers a graphical presentation of the results. Multiple options for downloading the results (graphs, tables, various image formats, etc.) can be selected by choosing the desired options under the menu button (≡).

The upper part of the results screen shows the summary of the model performance, and the combined graph shows the dynamics of the key variables associated with the selected scenario. An example output screen is represented in Figure 5.

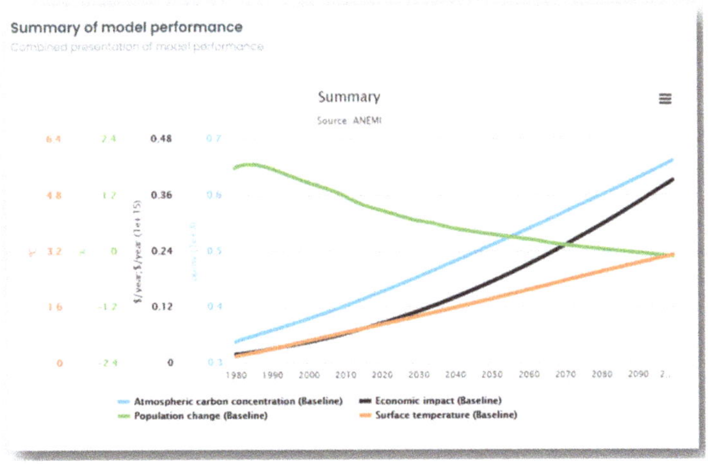

Figure 5. GCE summary of model performance.

The bottom part of the screen shows the *Individual performance metric*, which uses individual graphs to present the dynamic performance of the most significant variables for the selected scenario. As shown in Figure 6, GCE provides these graphs in a comparative form. Each user selection of input variables is memorized and presented as one line, and graphs show all experiments in the same place, enabling the immediate communication of important answers to the user's questions.

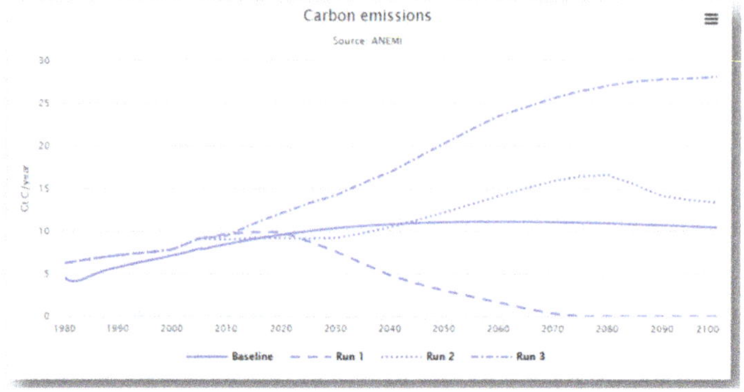

Figure 6. GCE individual performance metric.

4. Summary

This paper has briefly documented the ANEMI model as a new tool for global change analysis. The feedback-based structure is designed to promote understanding of the feedbacks that drive Earth system behaviour and the process of global change occurring within it. The aggregated spatial scale of the model allows for examining global scale feedbacks through the development of scenarios that focus on individual or multiple model sectors.

An intelligent interface (GCE) is also presented as a mechanism to allow for the global change experimentation of users with different skill levels necessary for the use of complex models such as NAEMI.

The main limitation of GCE is that there is a limited number of preselected scenarios focusing on: (i) climate change; (ii) population dynamics; (iii) food production; (iv) water quality; and (v) water quantity. It is my hope that the future use of the model will eliminate this limitation. The interest of the community and various questions by potential users will guide future activities in expanding the ability of GCE.

Funding: This research was funded by the Natural Sciences and Engineering Council (NSERC) of Canada's discovery grant CRDPJ/472152-2014.

Institutional Review Board Statement: Not applicable.

Informed Consent Statement: Not applicable.

Data Availability Statement: The text of the article includes the archive location of the ANEMI model and all the data and scenarios that were investigated up to now, as well as the URL of GCE.

Acknowledgments: I would like to acknowledge the programming support of P. Breach and A. Schardong.

Conflicts of Interest: The author declares no conflict of interest.

References

1. Simonovic, S.P.; Davis, E.; Akhtar, K.; Breach, P.; Schardong, A. *Global Change Explorer (GCE): An Interactive Tool for Investigating Complexities of Global Change—Ver 1.0*; Western University, Facility for Intelligent Decision Support, 2020; Available online: https://globalchange-uwo.ca (accessed on 26 January 2023).
2. Davies, E.G.R.; Simonovic, S.P. ANEMI: A New Model for Integrated Assessment of Global Change. *PLoS ONE* **2010**, *11*, 127–161. [CrossRef]
3. Akhtar, M.K.; Wibe, J.; Simonovic, S.P.; MacGee, J. Integrated Assessment Model of Society-Biosphere-Climate-Economy-Energy System. *Environ. Model. Softw.* **2013**, *49*, 1–21. [CrossRef]
4. Breach, P.A.; Simonovic, S.P. ANEMI: A Tool for Global Change Analysis. *PLoS ONE* **2021**, *16*, e0251489. [CrossRef] [PubMed]

5. *Vensim*; Ventana Systems: Harvard, MA, USA, 2020; Available online: https://vensim.com/ (accessed on 26 January 2023).
6. Forrester, J.W. *Principles of Systems*; Productivity Press: Portland, OR, USA, 1990; ISBN 978-1883823412.
7. Simonovic, S.P. *Managing Water Resources: Methods and Tools for a Systems Approach*; UNESCO, Paris and Earthscan James & James: London, UK, 2009; p. 576. ISBN 978-1-84407-554-6.

environmental sciences proceedings

MDPI

Proceeding Paper

A Multi-Objective Optimization Framework for Water Resources Allocation Considering Stakeholder Input †

Jorge Andres Garcia [1] and Angelos Alamanos [2,*]

[1] The Water Institute, University of Waterloo, Waterloo, ON N2L 3G1, Canada; ja4garciahernandez@uwaterloo.ca
[2] Department of Civil Engineering, University of Thessaly, 382 21 Volos, Greece
* Correspondence: alamanos@civ.uth.gr

† Presented at the 7th International Electronic Conference on Water Sciences, 15–30 March 2023. Available online: https://ecws-7.sciforum.net.

Abstract: Water resources and water-related sectors are increasingly affected by multiple challenges such as climate change and extreme events, issues of ageing infrastructure, natural and qualitative water scarcity, recession, wars, population movements, increased energy and resource demand, etc. In an attempt to balance different goals of water allocation under different constraints, we present a multi-objective optimization model. The model considers various water supply sources (groundwater, surface water, desalinated water, treated wastewater) and water uses (domestic, agricultural, industrial). Water demand, availability, quality parameters, costs, and stakeholder input for the prioritization of the different goals set are synthesized through Goal Programming.

Keywords: water resources management; multi-objective optimization; Goal Programming; conceptual model; stakeholder input; water supply; water scarcity; multi-sectoral water demand

1. Introduction

Water resources and several water-related sectors such as energy, fuels, industry, agriculture, and the economy are increasingly affected by the evident impacts of climate change on environmental resources and extreme events, issues caused by the ageing and mismanagement of existing infrastructure, natural and qualitative water scarcity, and recent changes such as recession, wars, population movements, increased energy and resource demand, and COVID-19. Such a compound of factors affects water allocation, as increasing usage must be met with limited and deteriorating resources and with the maximum efficiency to cope with the increased costs [1,2]. This often creates competition and conflicts among the different users, enhancing mismanagement in terms of water allocation [3]. This problem has been considered through the lens of optimization, maximizing or minimizing predefined goals such as water production, costs, deficits, profits from water-related activities, etc. [4]. Multi-objective optimization techniques have proved useful in assessing the trade-offs among different goals, coupling surface and groundwater sources for various usages [5]. Several studies thus far have accounted for the costs and/or water quality requirements [6]; however, fewer examples exist of applications considering all these parameters together, particularly studies making use of an open source code, thus making the models replicable [7], and studies which allow a direct input from the relevant stakeholders [8]. This study aims to provide a holistic and replicable model accounting for all the above parameters for optimal water allocation. We combine different water supply sources, various water usages, the respective supply costs and water quality requirements, and exploit the capabilities of Goal Programming (GP) to incorporate the input of stakeholders regarding the prioritization of the different goals set. The significance of this work lies in the detailed modeling description that allows its replication and application in different cases and study areas facing similar problems.

Citation: Garcia, J.A.; Alamanos, A. A Multi-Objective Optimization Framework for Water Resources Allocation Considering Stakeholder Input. *Environ. Sci. Proc.* **2023**, *25*, 32. https://doi.org/10.3390/ECWS-7-14227

Academic Editor: Athanasios Loukas

Published: 16 March 2023

2. Conceptual and Mathematical Description of the Model

A multi-objective optimization model was developed which is applicable to any timespan, from a monthly to annual time step and more. GP was used to build this model, as it is a powerful and flexible technique allowing consideration of multiple objectives as well as the possibility to involve stakeholders [9]. The general GP structure is based on linear programing where we set the decision variable(s), specify our desirable goals, define the potential deviations from these goals, and the parameters involved. Each goal can have its own constraints; alternatively, a set of common constraints can be used, depending on the problem at hand.

In this case, the decision variable $Q_{s,u}$ represents the volume of water [m^3/year] from source s allocated to user u. The index s refers to the different water supply sources (groundwater, surface water, desalinated water, treated wastewater = TWW), and u refers to the different water uses (domestic, agricultural, industrial, and hydropower generation). Two deviation variables are introduced for the goals:

- DWD_u: deficit in water demand for user u [m^3/year].
- EWA_s: exceedance (above renewable level) of water extraction of source s [m^3/year].

The parameters of the model are the following:

- WD_u: volume of water demanded by user u [m^3/year].
- WA_s: volume of water availability (renewable resources) of source s [m^3/year].
- $\delta_{s,u}$: binary parameter equal to one if it is feasible to allocate water from source s to user u, and zero otherwise.
- $WQ_{s,q}$: concentration of substance q in water from source s [g/m^3].
- $AQ_{u,q}$: threshold of maximum allowable concentration of substance q to meet quality requirements for user u [g/m^3]. Each user u can have its own mix of quality parameters (e.g., dissolved solids, phosphorous, nitrogen, etc.).
- $cost_s$: unitary water extraction cost of source s [$\$/m^3$].
- Budget: budget allocated for water provision [\$/year].

The Objective Function (Equation (1)) minimizes the deficits in water demand for users and the exceedances of water extraction from the sources:

$$\min z = \sum_u \alpha_u DWD_u + \sum_s \beta_s EWA_s \quad (1)$$

The parameters α_u, β_s penalize the deviation from the water demand and water extraction goals, respectively. Goal 1, water demand (Equation (2)): water supply must be at least sufficient to satisfy the water demand of the users:

$$\sum_s Q_{s,u} \geq WD_u - DWD_u \ \ \forall u \quad (2)$$

Goal 2, water supply (Equation (3)): water supply must not exceed renewable water volumes for each type of source:

$$\sum_u Q_{s,u} \leq WA_s + EWA_s \ \ \forall s \quad (3)$$

Water quality constraint (Equation (4)): the water volume mix supplied to each user must have the concentration of the harmful substances below their maximum allowable thresholds for that user:

$$\sum_s WQ_{s,q} Q_{s,u} \leq AQ_{u,q} \left(\sum_s Q_{s,u} \right) \ \ \forall u, q \quad (4)$$

Budget constraint (Equation (5)): the budget for water extraction must not be exceeded:

$$\sum_{s}\left[cost_s\left(\sum_{u} Q_{s,u}\right)\right] \leq \text{Budget} \tag{5}$$

Feasibility constraint (Equation (6)): if certain variables $Q_{s,u}$ are unfeasible due to practical reasons, the following restriction controls which variables are available for the model:

$$Q_{s,u} \leq \delta_{s,u} M \ \forall s,u \tag{6}$$

where the value of M would be a very large constant, as, for example, shown in Equation (7).

$$M = \sum_{s}\sum_{u} Q_{s,u} \tag{7}$$

The model presented in Equations (1)–(6) finds an optimal balance between the two goals, i.e., having deficits on the demand of water by the different users and incurring in over extractions of the sources. Additionally, the model ensures that the water quality thresholds by substance, as needed by each user, are met and the cost of water extraction is within the allocated budget. The parameters α_u, β_s express the 'cost' for the decision-makers of having deficit on each user and overexploitation of each water source. The conceptual model is also described in Figure 1.

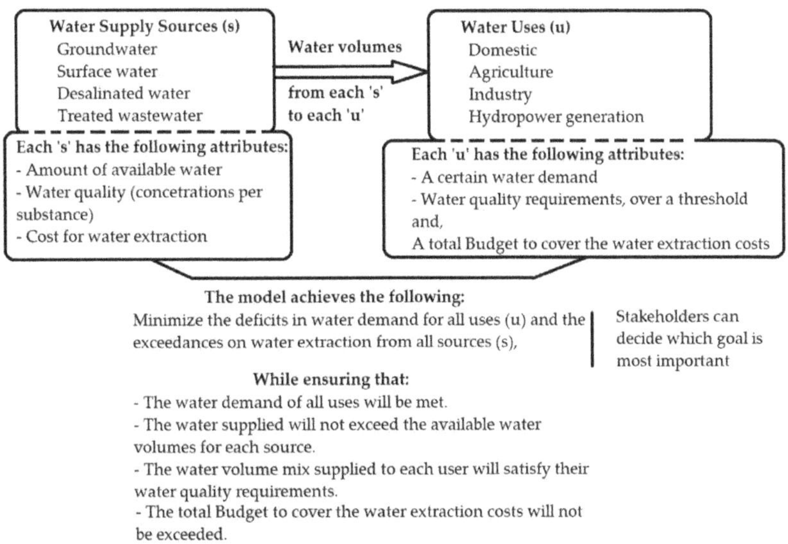

Figure 1. The conceptual diagram representing the proposed model.

The coverage of the various water demands from the supply sources need to be done in a way that will also enhance the water allocation efficiency. While the model ensures that the deficits for users and supply sources will be minimized, it is up to the decision-makers to increase the efficiency of the coverage of the water demand. For example, they could promote more water re-use, or usage of renewable surface water, while using less groundwater and reducing the costly operation of the desalination plants to produce drinking water or reducing the hydropower production to reduce its environmental impact in terms of carbon emissions. The preferences among different supply sources could be inserted in the Objective Function, with different coefficients per source promoting or penalizing its extraction; this, however, would make the model result more complex to interpret.

The model presented here was coded in Python, as it is an open-source programming language that can handle complex and computationally demanding optimization problems. The code is available (see Supplementary Materials) to enhance replicability and allow any necessary modifications to the model.

3. Stakeholder Input

In the previous section where the model was described, it was mentioned that the objectives of the model can be prioritized based on the weights assigned (α and β). GP attempts to minimize this set of deviations from multiple pre-specified (desirable) goals which are introduced simultaneously in the Objective Function. These weights can be assigned on a custom scale, usually a 0–1 scale, and the rationale is to assign higher weights to those goals that are considered more important. Thus, the model will 'penalize' the deviations from these goals, so that lower order goals are considered only after higher order goals.

The weights can be assigned by the analyst (modeler) to test the model and the sensitivity of the various decisions, and, ultimately, a group of stakeholders will define them. This is particularly important and it is the necessary condition to integrate the modeling technology into the social and political components of the planning and management process. Table 1 includes some stakeholder groups that would have a direct or indirect interest in participating in such a process of weighing the different goals.

Table 1. Potential stakeholders that could be involved in the proposed modeling process, with a general description of their role.

Stakeholder Group A: Representatives from the central Government. This group refers to representatives from the Ministry of Environment, the Environmental Protection Agencies (EPA), General Water Directorate, Agency of Land Reclamation Works, or relevant bodies of climate, energy, agriculture, etc., depending on the management structure of each country. These stakeholders operate at a higher-level, providing more general guidelines (e.g., River Basin Management Plans), so they can be key for connecting their more general guidelines to the actual decisions at a smaller scale.
Stakeholder Group B: Representatives from the regional-scale authorities, such as regional governments, the Prefecture, State or Municipal Division level agencies depending on the country. They are often responsible for implementation of the higher-level guidelines at the regional scale and for tracking the progress, so it will be useful to stay connected with all other stakeholder groups.
Stakeholder Group C: Local authorities, industry stakeholders, agricultural co-operations relating to water and agricultural management, organizations of land reclamation, urban regulators, and representatives of municipal institutions. Continuous dialogue with stakeholder groups A and B will help seek the proper expertise and skills and consider the broader picture of the goals discussed, in order to apply any measure with the maximum efficiency. Non-Governmental Organizations (NGOs) can be a part of this group, or a separate one, depending on the connection with the other stakeholders and, often, on the alignment of their environmental policies.
Stakeholder Group D: Experts and experienced professionals; start-ups and technology experts; researchers and academics. These will play the role of the solution holders in theoretical and practical terms, and will also provide feasibility considerations with respect to the application of the different decisions discussed.

It might be challenging to sit together with relevant stakeholders and explain, test, and finalize such models, as stakeholders usually do not accept implementation of modeling within their planning process; however, the ability to appreciate the trade-offs among different objectives is often appealing [3,10]. This is expected to be an element that will draw the attention of stakeholders and decision-makers in the future, as the management of the water sector becomes more challenging: the discovery of the effect of alternative assumptions and goal prioritization through collective workshops and discussions. In many cases, such exercises have helped create a common or shared understanding among stakeholders of the systems they are managing. Involving stakeholders in the model-building process provides them with a sense of ownership, a much better understanding of what the models can do, what answers they can and cannot provide, the assumptions used, the reasoning behind them, and their possible impacts; therefore, it could clarify ways to reduce any uncertainties [11].

Moreover, stakeholder participation in modeling exercises, where they can see models as a tool that they will be able to benefit from, creates also discussions that lead toward a better understanding of everyone's interests and concerns.

4. Concluding Remarks

In this study, a model for optimal water allocation was developed, considering multiple goals regarding water demand, water availability, water quality requirements, and costs. Among the advantages of the model presented here are (i) the parsimony of its formulation, which captures the relevant features of water resource allocation all the while remaining clear, simple, and easy to interpret; (ii) the low data requirements, which makes it easy to implement; and (iii) its versatility to be extended or enriched with study-specific requirements. For example, the model can be coupled with hydrological models to estimate water supply available per source along with its quality and water demand per use, while including additional economic modeling to account for the relevant costs.

This model can be tested under different management strategies or future scenarios (e.g., climate change) by altering certain parameters. For example, various interventions to make water use more efficient, and thus reduce water demand, can be considered for each use u. Water storage infrastructure can be considered to increase water supply; consideration of other supply sources and/or uses are also possible.

Sensitivity and uncertainty analyses are included in our future plans, for example considering: water demand for agricultural water use (the others are more inelastic), depending on management scenarios; water availability from surface water (SW) and groundwater (GW), depending on temperature and precipitation variations (considering also climate change scenarios); costs depending on monetary considerations, or accounting for the full cost of water; and finally, different weights of importance (α, β) for the different goals.

Given the current and future complex challenges of the water sector, solutions and approaches need to be supported by science and integrated. The model presented here, with the opportunities that it offers, can be a good example for such future applications: it is replicable, it can be tailored to similar problems, it allows the input of stakeholders in the model-building process, and it can assist the relevant stakeholders in reaching a common or 'shared' vision of at least how the systems they manage (as represented by the model) work. Finally, such exercises can also be useful for education purposes, for building an understanding of the functions and the interconnectedness of water systems.

Supplementary Materials: The following supporting information can be downloaded at: https://github.com/jorge-antares/water_allocation_model, including the Python script (accessed on 5 February 2023).

Author Contributions: Conceptualization, J.A.G. and A.A.; methodology, J.A.G. and A.A.; writing J.A.G. and A.A. All authors have read and agreed to the published version of the manuscript.

Funding: This research received no external funding.

Institutional Review Board Statement: Not applicable.

Informed Consent Statement: Not applicable.

Data Availability Statement: Not applicable.

Conflicts of Interest: The authors declare no conflict of interest.

References

1. Greve, P.; Kahil, T.; Mochizuki, J.; Schinko, T.; Satoh, Y.; Burek, P.; Fischer, G.; Tramberend, S.; Burtscher, R.; Langan, S.; et al. Global Assessment of Water Challenges under Uncertainty in Water Scarcity Projections. *Nat. Sustain.* **2018**, *1*, 486–494. [CrossRef]
2. Souza da Silva, G.N.; de Moraes, M.M.G.A. Economic Water Management Decisions: Trade-Offs between Conflicting Objectives in the Sub-Middle Region of the São Francisco Watershed. *Reg. Environ. Chang.* **2018**, *18*, 1957–1967. [CrossRef]
3. Alamanos, A.; Koundouri, P.; Papadaki, L.; Pliakou, T.; Toli, E. Water for Tomorrow: A Living Lab on the Creation of the Science-Policy-Stakeholder Interface. *Water* **2022**, *14*, 2879. [CrossRef]
4. Alamanos, A.; Xenarios, S.; Mylopoulos, N.; Stålnacke, P. Integrated Water Resources Management in Agro-Economy Using Linear Programming: The Case of Lake Karla Basin, Greece. *Eur. Water* **2017**, *60*, 41–47.
5. Naghdi, S.; Bozorg-Haddad, O.; Khorsandi, M.; Chu, X. Multi-Objective Optimization for Allocation of Surface Water and Groundwater Resources. *Sci. Total Environ.* **2021**, *776*, 146026. [CrossRef]
6. Majedi, H.; Fathian, H.; Nikbakht-Shahbazi, A.; Zohrabi, N.; Hassani, F. Multi-Objective Optimization of Integrated Surface and Groundwater Resources Under the Clean Development Mechanism. *Water Resour. Manag.* **2021**, *35*, 2685–2704. [CrossRef]
7. Garcia, J.A.; Alamanos, A. Integrated Modeling Approaches for Sustainable Agri-Economic Growth and Environmental Improvement: Examples from Greece, Canada and Ireland. *Land* **2022**, *11*, 1548. [CrossRef]
8. Tarebari, H.; Javid, A.H.; Mirbagheri, S.A.; Fahmi, H. Multi-Objective Surface Water Resource Management Considering Conflict Resolution and Utility Function Optimization. *Water Resour. Manag.* **2018**, *32*, 4487–4509. [CrossRef]
9. Charnes, A.; Cooper, W.W. *Management Models and Industrial Applications of Linear Programming*; John Wiley: New York, NY, USA, 1961; Volumes 1–2.
10. Rojas, R.; Castilla-Rho, J.; Bennison, G.; Bridgart, R.; Prats, C.; Claro, E. Participatory and Integrated Modeling under Contentious Water Use in Semiarid Basins. *Hydrology* **2022**, *9*, 49. [CrossRef]
11. Loucks, D.P.; van Beek, E. *Water Resource Systems Planning and Management: An Introduction to Methods, Models, and Applications*; Springer Nature: Berlin, Germany, 2017; ISBN 978-3-319-44234-1.

environmental sciences proceedings

MDPI

Proceeding Paper

A Systems Approach to Identifying Hazards in the Management of Vegetative Buffers for the Protection of Drinking Water Quality †

Hew Cameron Merrett * and Jao Jia Horng

Centre for Emergency Response Information, National Yunlin University of Science and Technology, Douliu 640, Taiwan; horngjj@gmail.com
* Correspondence: hew.merrett@gmail.com

† Presented at the 7th International Electronic Conference on Water Sciences, 15–30 March 2023; Available online: https://ecws-7.sciforum.net/.

Abstract: Using ecosystem services provided by stream buffers has the potential to complement conventional engineering solutions, such as water treatment, and reduce public health risks to consumers. These buffers interrupt the movement of contaminants and sediments from non-point sources such as agricultural land to surface waters. This study uses System Theoretic Process Analysis (STPA) and Early Warning Signal Analysis based on STPA (EWaSAP) methodologies to systematically examine the sociotechnical structures involved in managing vegetated buffers in surface water catchments using a theoretical scenario representative of typical surface water supplies.

Keywords: drinking water; catchment management; System Theoretic Process Analysis (STPA)

check for updates

Citation: Merrett, H.C.; Horng, J.J. A Systems Approach to Identifying Hazards in the Management of Vegetative Buffers for the Protection of Drinking Water Quality. *Environ. Sci. Proc.* **2023**, *25*, 33. https://doi.org/10.3390/ECWS-7-14173

Academic Editor: Athanasios Loukas

Published: 14 March 2023

1. Introduction

For Water Service Providers (WSPs), the safety and quality of drinking water is paramount to protecting public health and providing a valued product. To ensure safety, a WSP must have a good understanding of the potential hazards of the supply system all the way from catchment to tap [1]. In drinking water source protection, ecosystem services in the catchment area play a vital role in the cost-effective delivery of water quality outcomes through controlling the movement of sediment, nutrients, and contaminants as well as stabilizing banks and slopes. Leveraging these services and viewing catchment areas as water treatment assets has the potential to complement conventional engineering solutions, such as water treatment, and reduce public health risks to consumers [2]. Riparian buffers in catchments are often critical zones for targeted mitigation measures for interrupting the movement of contaminants and sediments from non-point sources such as agricultural land to surface waters [3].

This study uses a comprehensive hazard analysis to identify requirements for the good management of stream buffers for drinking water outcomes. Riparian buffers supply many services as natural water treatment infrastructure by protecting and enhancing natural ecosystem services. Thus, as a drinking water source protection strategy, vegetative buffers can be a practical risk management approach [4]. The services provided by vegetation buffers include soil erosion control through slope stability and water purification by reducing sediments, nutrients, pollutants, and pathogens entering waterways [5]. Replacing such services with constructed assets entails complex water treatment, dam construction, and slope engineering. Furthermore, through the natural purification processes, buffers can reduce operational costs for existing water treatment infrastructure and processes. A review of catchment land cover and chemical costs for water treatment by [6] showed that vegetation buffers between diffuse pollution sources and streams has a negative correlation between stream buffers area and treatment costs. The resultant reduction in treatment costs provides a compelling case for using vegetation buffers as a public health risk management strategy and reducing operational costs associated with water treatment.

The water quality benefits of stream buffers appear to be well understood in the water industry. Many examples exist where stream buffers have been used to improve water quality outcomes [7,8]. In addition, studies have shown that the public is willing to support catchment interventions to provide water quality improvement outcomes. For example, in the water industry in England and Wales, customers have shown an acceptance of beneficiary pays solutions in catchment management for drinking water source protection [9]. However, a review of the Capital Expenditure (CAPEX) bias in the water and sewerage sectors in England and Wales found that for many water service providers, there is a belief that CAPEX solutions are favored over solutions that rely on operational expenditure (OPEX) [10]. Much of this bias arises from a perceived lower certainty of outcomes of operational interventions on natural assets.

Typical hazard analysis methods for drinking water assets include FMEA, HAZOP, Fault Trees amongst many others [11]. This study uses System Theoretic Process Analysis (STPA) to systematically examine the hazards inherent to sociotechnical structures involved in using vegetated buffers for drinking water quality management in surface water catchments. STPA is a hazard analysis methodology based on System Theoretic Accident Modelling Processes (STAMPs), which, being founded on systems theory, views safety as the emergent property of the system [12]. This method has been used in a wide range of applications from aerospace design through to regulation and legal systems. In the study of drinking water source protection programs by Ref. [13], STPA was used to analyze the hazards associated with catchment-level ecosystem services provided by stream buffers. However, the study only considered ecosystem services in general without inspecting the different processes individually. In this study, the sociotechnical system of interest includes technical, social, economic and agency interaction factors which control the water quality outcomes from vegetative buffers. The use of conventional hazard analysis techniques in this type of system or process is limited as they focus on the reliability of individual components and miss the interaction of the various components in the plans. Furthermore, such approaches have limited ability to identify leading indicators of safety or early warning signs.

Building on the hazard analysis using STPA, this study investigates the leading indicators of safety throughout the system based on the Early Warning Signal Analysis based on STPA (EWaSAP) methodology proposed by Ref. [14]. The EWaSAP methodology presented in Ref. [14] builds on the STPA methodology for hazard analysis to identify possible early warning signals that control actions may be failing to enforce the safety objectives of the system.

2. Methods

To test the applicability of STPA and EWaSAP for the assessment of process risks in managing vegetation buffers, a theoretical example was created based on typical real-world conditions. The hypothetical scenario constructed is a surface water catchment used for the public supply of drinking water with a broad mix of land uses controlled by different private and public entities.

2.1. STPA Method

The STPA methodology consists of four key steps: define the purpose of analysis, model of control structure, identify unsafe control actions, and identify loss scenarios. In this study, the four steps of STPA are adapted from Ref. [15] with the parallel EWaSAP steps [13] which are described in the following sections.

2.1.1. Step 1—Define the Purpose of the Analysis

In the definition of the purpose of the analysis, the unacceptable losses are identified, as well as the associated system-level hazards and the corresponding safety constraints. In this STPA step, the EWaSAP tasks include:

1. Identify the agents outside of the system scope that need to be informed about the potential realization of a system-level hazard.
2. Establish a synergy with sensory services within and outside the system scope and identify the system-level safety constraints that have been violated.

2.1.2. Step 2—Model of the Control Structure

The model is not a physical model of the system, rather a model of the hierarchal control of the system components included in the scope of the analysis. The control model is created using a series of feedback and control loops [15].

2.1.3. Step 3—Identify Unsafe Control Actions

The next step of the analysis is to identify the ways the control actions can be unsafe. Ref. [14] provides four prompts for identifying potential unsafe control actions (UCAs): not providing the control action when required; providing the control action that causes a hazard; providing the control action too soon or too late; or is stopped too soon or is applied too long. The concurrent EWaSAP tasks relate to the enforcement of internal awareness actions.

2.1.4. Step 4—Identify Loss Scenarios

The loss scenarios combine causal factors that can lead to the identified UCAs being realized. The EWaSAP methodology focuses on using the available pool of data to indicate the existence of factors that could result in the UCA and, ultimately, the violation of the high-level safety constraints.

3. Results

3.1. Definition of Purpose

In the scenario presented, the key stakeholders are the consumers of the water produced by the drinking water catchment. Therefore, the key losses or accidents considered are the illness or death resulting from pathogens or contaminants introduced from the catchment area. The causes that could lead to such losses are contaminants or pathogens in concentrations too great to be removed effectively through downstream water quality control barriers or water quality that reduces the effectiveness of downstream water quality control processes. For drinking water supplies, these downstream barriers typically include water treatment and disinfection processes.

For step one of ESWaP, the focus is on identifying the agents outside of the system which must be informed of a system-level hazard. In this case, the main agent would be the position in direct control of the drinking water system, which for a typical WSP may be a position such as a Water Quality Operations Manager. In this situation, the indication of degraded water quality would come from a violation of the water quality limits that reflect the verified capability of downstream water treatment and disinfection processes. The high-level hazards, corresponding safety constraints identified, and the associated warning signals are provided in Table 1.

Table 1. The system-level accidents, hazards safety constraints and indicators.

Accident	Hazard	Safety Constraint	Indicators of Safety Constraint Violation
Supply of unsafe drinking water	H1 Stream buffer is not effective in removing pathogens	Buffer must ensure the removal of pathogens to specified levels	Indicated by: levels of pathogens exceed the capability of downstream treatments Sensors: water sampling data
Supply of unsafe drinking water	H2 Stream buffer is not correctly managed and maintained	Stream buffers must be maintained to ensure they function as intended	Indicated by: condition assessment Sensors: visual inspections/condition assessments, land manager feedback
Degradation of water quality	H3 Stream buffer is not effective in removing contaminants to specified levels	Stream buffer must be able to remove contaminants as per the intended purpose	Indicated by: runoff meeting set specifications Sensors: water sampling data, in situ monitoring

3.2. *Safety Control Structure*

In a drinking water supply system like the scenario created for this study, the quality of the water supplied is under the control of the WSP who are accountable for the final supply to the customer. When it comes to catchment management, the management structures and accountabilities for actions to protect water quality outcomes involves multiple landholders and government agencies. The WSP often has limited direct influence over the landholders and government agencies responsible for natural resource management and pollution regulation. As such, included in the safety structure is a role for the government agencies accountable for the management of water resources. Additionally, considered in the safety structure is the role of the public health authority with statutory responsibility for regulating drinking water supplies. While there is no direct responsibility for managing stream buffers as a regulator, there is indirect influence through regulatory actions. Including enabling actors in the safety control structure provides a detailed view of the broader sociotechnical structure which influences the successful management of ecosystem services in drinking water catchments.

For the WSP, several key internal functions are included in the safety control structure as the control of these functions has considerable influence on drinking water quality outcomes. The WSP functions relate to the maintenance, operations, and planning actions related to water quality control processes. For the study scenario, the description of all the key actors involved in managing vegetative buffers in drinking water catchments, and the associated control actions and information are listed in Figure 1.

3.3. *Identification of Unsafe Control Actions*

In this step of the analysis, each of the 18 control actions included in the high-level control structure was reviewed to establish the scenarios in which the control actions can be unsafe and potentially violate the system safety constraints. As a theoretical example, the identification of UCAs was based on the authors' knowledge in conjunction with industry guidance and the WHO guidance document on protecting surface water for public health [4]. The actions considered multiple aspects, from typical planning and operations to strategic management and policy. At this stage of the study, a total of 46 UCAs were identified for the high-level control actions related to the management of stream buffers. A sample of the UCAs for operational and strategic control action is provided in Table 2.

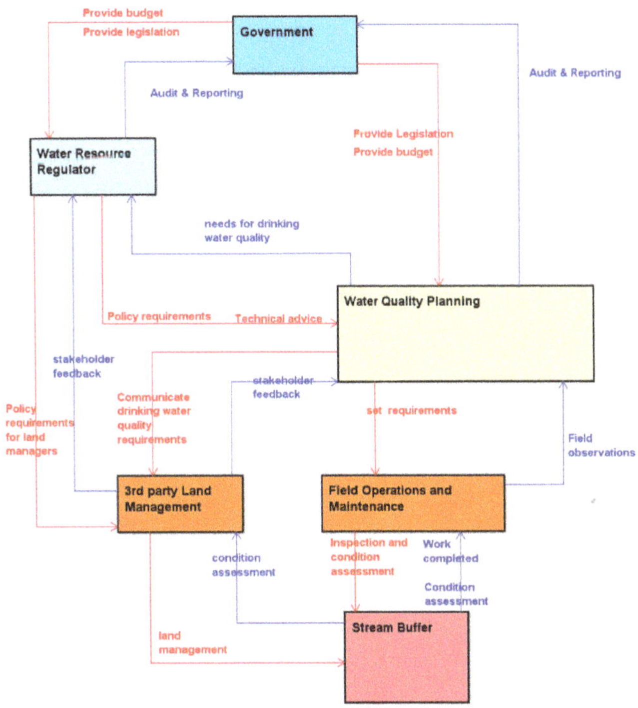

Figure 1. High-level safety control structure with control actions and feedback.

Table 2. Sample of system-level unsafe control actions (UCAs).

Control Action	From	To	Not Providing Control Action	Providing Causes Hazard	Too Early Too Late
Land management	3rd Party Land Management	Stream Buffer	Land manager does not manage the condition of the stream buffer	Land management actions conflict with the drinking water quality objectives	The land manager is delayed in responding to conditions that degrade the performance of buffers
Set performance requirements	Water Quality Planning	Field Operations and Maintenance	There are no performance requirements set during the planning process	The performance requirements set do not meet the needs of the application	The performance requirements are changed and no longer meet the buffer's performance

3.4. Causal Factors, Countermeasures, and Early Warning Signs

The causal factors are the scenarios that result in potentially unsafe control actions and the eventual potential resulting in the violation of the high-level safety constraints previously identified in Table 1. The STPA Handbook [15] includes guidance for the identification of loss scenarios as the fourth step in the STPA method. For EWaSAP, the third step is to enforce internal awareness actions to indicate the occurrence of a flaw and the violation of assumptions made in the design of the system. This step is a proposed add-on to step 3 in the STPA method. For this study, when completing step 3 of EWaSAP in conjunction with the STPA method consideration was given to the potential causal factors when identifying the signs of the flaws occurring. The next step was to consider what countermeasures could be put in place to prevent the identified scenarios leading to unsafe control actions. A total of 73 causal factors were identified from the UCAs, and each causal factor then had a corresponding countermeasure assigned. As some of the causal factors had similar failure mechanisms and therefore had a similar countermeasure assigned, resulting in a total of 61 countermeasures. A sample of the countermeasures and early warning signs is provided in Table 3.

Table 3. A sample of the countermeasures and early warning signs identified for stream buffers.

Causal Factor	Countermeasure	Unsafe Control Action (UCA)	Early Warning Signs	Sensor Requirement
Land manager operations are not compatible with buffer management	Engage and educate 3rd party landowners to ensure their management actions are compatible with drinking water quality objectives	Land management actions conflict with the drinking water quality objectives	Land management activities/actions are impacting buffer condition	Should indicate compliance with accepted land management practices
The challenge of the raw water is greater than the ability of the treatment process	Set service level requirements for the performance of the buffer	The water quality supplied does not ensure public health	Noncompliance of raw water quality with service level agreement	Should measure water quality against the relevant parameters

The sensor element is derived from the control feedback in the safety control structure and supplies the controller with information to control the actions of the actuator. The actual sensor will depend on the specifics of a given situation and may include visual observations, water quality data, etc. The timing of information from the sensor will depend on the rate at which conditions can change. The timing of sensor reading is essential for informing the early warning signs which confirm if the countermeasure is effective and enforcing the required safety constraints in the management structure. This process was completed for all 61 countermeasures identified.

Of all the early warning signs identified, the majority were related to the risk assessment and planning process accounting for 39% of all indicators. The risk assessment and planning processes set the foundations for the overall system, where issues are identified and rectified, and this stage can prevent possible degradation due to management actions. The next highest number of early warning signs can be found with maintenance and operation functions (13%) and government policy and regulation (12%). Like any other asset in the water supply system, stream buffers require ongoing maintenance and operations to ensure the expected level of performance is maintained. In this instance, monitoring the early warning signals related to operations and maintenance functions provides greater certainty in meeting the water quality objectives. As for government policy and regulation, while not directly influenced by a WSP, there is importance in being able to navigate the aspects of policy and regulations which influence stream buffer management. The smallest group of early signs related to water quality sampling accounted for only 3% of all indicators identified. Water quality sampling is often used as the principal indicator for the effectiveness of water quality interventions. While effective for characterizing water quality, monitoring is a lag indicator in this instance as stream buffers may become seriously degraded before any change in water quality results is observed.

4. Conclusions

While the value of ecosystem services has been widely recognized in the water industry, there has been limited investigation into the warning signals in the management systems of such assets. Using a structured hazard analysis provides insight into the management needs to ensure that stream buffers continue to perform as expected in the supply of safe drinking water.

This example provides a valuable test of how taking a highly systematic approach to identifying hazards using STPA can help to better understand management requirements. Furthermore, developing a set of early warning signals and sensors using EWASP can help us to understand the warning signals and sensors for the effective management of natural assets. The uncertainty of outcomes can be a limiting factor when deciding between investing in catchment interventions and conventional water treatment infrastructure. However, the approach of using STPA and EWASP in combination provides a basis for being able to systematically design management regimes for greater assurance of meeting the requirements for the safe and reliable supply of drinking water. Due to the systematic approach, this approach can be expanded as needed to encompass a range of different operational risks, such as the interlinkages with technical aspects (e.g., water treatment processes), and sociotechnical aspects, such as policy development.

The substantial number of early warning signs identified in this case would require significant resources to implement and measure. The methods used, are very helpful in identifying the hazards involved; however, they do not provide a means to assess the relative importance of the warning signs in the overall system. The validation, ranking, and selection of the final early warning signs warrant further research and investigation to assist WSPs in safely managing ecosystem services to protect drinking water quality.

Author Contributions: Conceptualization, H.C.M. and J.J.H.; methodology, H.C.M. and J.J.H.; formal analysis, H.C.M.; investigation, H.C.M.; resources, J.J.H.; data curation, H.C.M.; writing—original draft preparation, H.C.M. and J.J.H.; writing—review and editing, H.C.M. All authors have read and agreed to the published version of the manuscript.

Funding: This research received no external funding.

Institutional Review Board Statement: Not applicable.

Informed Consent Statement: Not applicable.

Data Availability Statement: Data can be provided upon request.

Conflicts of Interest: The authors declare no conflict of interest.

References

1. World Health Organisation. *Guidelines for Drinking-Water Quality: Fourth Edition Incorporating the First Addendum*; WHO: Geneva, Switzerland, 2017.
2. Vincent, J.R.; Ahmand, I.; Adnan, N.; Burwell IIIW, B.; Pattanayak, S.K.; Tan-Soo, J.-S.; Thomas, K. Valuing Water Purification by Forests: An Analysis of Malaysian Panel Data. *Environ. Resour. Econ.* **2016**, *64*, 59–80. [CrossRef]
3. Stutter, M.L.; Chardon, W.J.; Kronvang, B. Riparian buffer strips as a multifunctional management tool in agricultural landscapes. *J. Environ. Qual.* **2012**, *41*, 297–303. [CrossRef]
4. Rickert, B.; Chorus, I.; Schmoll, O. *Protecting Surface Water for Health: Identifying, Assessing and Managing Drinking-Water Quality Risks in Surface-Water Catchments*; WHO: Geneva, Switzerland, 2016.
5. WSAA. *Source Catchments as Water Quality Treatment Assets: Industry Best Practices and Triple Bottom Line Cost Evaluation of Catchment Management Practices*; Water Services Association of Australia Limited: Docklands, Australia, 2016.
6. Freeman, J.; Madsen, R.; Hart, K. *Statistical Analysis of Drinking Water Treatment Plant Costs, Source Water Quality and Land Cover Characteristics*; United States Trust for Public Land: San Francisco, CA, USA, 2007.
7. Yeboah, F.K.; Lupi, F.; Kaplowitz, M.D. Agricultural landowners' willingness to participate in a filter strip program for watershed protection. *Land Use Policy* **2015**, *49*, 75–85. [CrossRef]
8. Buckley, C.; Hynes, S.; Mechan, S. Supply of an ecosystem service—Farmers' willingness to adopt riparian buffer zones in agricultural catchments. *Environ. Sci. Policy* **2012**, *24*, 101–109. [CrossRef]
9. Mathieu, L.; Tinch, R.; Provins, A. Catchment management in England and Wales: The role of arguments for ecosystems and their services. *Biodivers. Conserv.* **2018**, *27*, 1639–1658. [CrossRef]
10. OFWAT. *Capex Bias in the Water and Sewerage Sectors in England and Wales-Substance, Perception or Myth. A Discussion Paper*; OFWAT: Birmingham, UK, 2012.
11. Hokstad, P.; Røstum, J.; Sklet, S.; Rosén, L.; Pettersson, T.J.; Linde, A.; Sturm, S.; Beuken, R.; Kirchner, D.; Niewersch, C. Methods for Risk Analysis of Drinking Water Systems from Source to Tap-Guidance Report on Risk Analysis. Techneau Report. 2009. Available online: https://citeseerx.ist.psu.edu/document?repid=rep1&type=pdf&doi=c32c5de468d1a357dff807261f724749299d6b23 (accessed on 25 September 2022).
12. Leveson, N. A new accident model for engineering safer systems. *Saf. Sci.* **2004**, *42*, 237–270. [CrossRef]
13. Merrett, H.C.; Chen, W.T.; Horng, J.J. A Systems analysis approach to identify critical success factors in drinking water source protection programs. *Sustainability* **2019**, *11*, 2606. [CrossRef]

14. Dokas, I.M.; Freehan, J.; Imran, S. EWaSAP: An early warning sign identification approach based on a systemic hazard analysis. *Saf. Sci.* **2013**, *58*, 11–26. [CrossRef]
15. Leveson, N.; Thomas, J. STPA Handbook. Available online: http://psas.scripts.mit.edu/home/get_file4.php?name=CAST_handbook.pdf (accessed on 18 April 2018).

environmental sciences proceedings

MDPI

Proceeding Paper

Clay-Based Sorbents for Environmental Protection from Inorganic Pollutants †

Iryna Kovalchuk

Institute for Sorption and Problems of Endoecology, National Academy of Science of Ukraine, 03164 Kyiv, Ukraine; kowalchukiryna@gmail.com

† Presented at the 7th International Electronic Conference on Water Sciences, 15–30 March 2023; Available online: https://ecws-7.sciforum.net/.

Abstract: Natural clay minerals are widely used in various environmental protection technologies as cheap, accessible, and effective sorbents. Considering the tasks and conditions for purification of polluted waters, the different methods of surface modification of clay minerals were used. The surface modification of the montmorillonite by cationic surfactants (hexadecyltrimethylammonium bromide); the method of surface modification of montmorillonite by large-sized inorganic cations of Fe, Al, Zr, Ti; applying nanoscale zero-valent iron powder to the montmorillonite surface were used for increasing of efficiency of water purification from chromium (VI) and uranium (VI).

Keywords: contaminated water; heavy metals; radionuclides; clay minerals; surface modification

check for updates

Citation: Kovalchuk, I. Clay-Based Sorbents for Environmental Protection from Inorganic Pollutants. *Environ. Sci. Proc.* **2023**, *25*, 34. https://doi.org/10.3390/ECWS-7-14247

Academic Editor: Athanasios Loukas

Published: 16 March 2023

1. Introduction

Heavy metals and radionuclide compounds are among the ones of all the standard pollutants of surface and underground waters. They have increased toxicity for living organisms and humans, even at very low concentrations in water [1,2]. The main sources of environmental pollution with uranium compounds are nuclear fuel cycle enterprises. In Ukraine, nuclear fuel cycle enterprises are located in the Kryvyi Rih Basin [3]. To a much lesser extent, environmental pollution is associated with the use of depleted uranium munitions, which take place in places of armed conflicts [4]. Chromium occupies a special place among heavy metals due to its high toxicity. Chromium compounds are found in the effluents of electrochemical industries and wastes of the hydrometallurgical industry. The maximum allowable limits in water, according to World Health Organization, are low: 0.03 mg/L for uranium compounds and 0.05 mg/L for chromium compounds [5]. Therefore, problems of environmental protection from inorganic pollutants are becoming more and more relevant.

To remove trace amounts of toxicants, the most effective are sorption methods that combine high selectivity to the target compounds with sufficient productivity. Spesial attention is paid to the development of sufficiently cheap sorbents. Natural clay minerals are widely used in various environmental protection technologies. However, their significant drawback is insufficient sorption capacity, especially to anionic forms of pollutants. Thus, the maximum sorption of Cr(VI) ions on kaolinite with a specific surface of 15.54 m^2/g is only $3.33 \cdot 10^{-7}$ mol/g [6]. Therefore, various methods of surface modification to increase the effectiveness of clay minerals are used. There is the use of surfactants, hydroxo complex of inorganic cations, nanoparticles of active metals, etc. These methods make it possible to obtain cheap "low cost" sorbents with improved sorption properties for their application in environmental protection technologies.

Modification of the surface of layered silicates with organic substances quaternary ammonium bases [7] allows for a significant increase in their sorption properties to nonionic organic compounds and anionic forms of dangerous inorganic toxicants (chromium, arsenic, uranium, etc.). Both negatively and positively charged areas are formed on the surface of particles-active centers of sorption processes. Thus, organobentonite with different degrees of surface modification shows high efficiency to U(VI) ions in a wide pH range [8,9]. The maximum values of sorption on organobentonite modified with hexadecyltrimethylammonium bromide (HDTMA, $(C_{16}H_{33})N(CH_3)_3Br$) are 17.39 and 12.44 mg/g at medium pH of 7 and 9, respectively [10]. Significant amounts of sorption of anionic forms of U(VI) in the alkaline pH region are also achieved for the modified HDTMA fibrous silicate sepiolite [11]. For samples modified by hexadecyltrimethylammonium bromide, hexavalent chromium sorption values are 30 mmol/kg for kaolinite and 795 mmol/kg for montmorillonite at pH < 1 [12]. When modifying the surface of kaolinite, illite, and smectite using HDTMA in quantities sufficient to form a monolayer, the removal of chromate, arsenate, and nitrate ions from aqueous solutions was recorded [13,14].

The intercalation of large inorganic cations into the interlayer space of clays of smectite groups is the perspective method of modification. The inorganic cations migrate into the interlayer space of clay minerals, fixed, and create two-dimensional zeolite-like materials. These materials-pillared interlayered clays (PILC)-have adjustable flat micropore sizes and high heat resistance [15,16]. For modification of montmorillonite Fe, Zr, and Fe/Zr ions were used; the maximum sorption capacity for chromium (VI) ions was 16.24, 19.24, and 22.34 mg/g [17].

The use of nano-sized powders of zero-valent iron in nature protection technologies is associated with significant difficulties due to their high tendency to rapid oxidation and aggregation. An effective method of solving this problem is the immobilization of nanoparticles of zero-valent iron on the surface of organic and inorganic materials: activated carbon, silica gel, diatomite, cellulose, graphene, and clay minerals [18,19]. Montmorillonite-supported zero-valent iron allows for significantly higher removal efficiency of U (VI) compounds from water (99.2%) than using zero-valent iron (48.3%) [20]. For an iron-containing sorbent based on fibrous silicate-sepiolite, the maximum value of sorption of Cr (VI) is 43.86 mg/g, which is one of the highest values obtained during the sorption of hexavalent chromium compounds [21,22].

The aim of our article was the comparison of the purification efficiency of the contaminated uranium and chromium compounds water with the sorbents on the base of montmorillonite modified by the different methods.

2. Materials and Methods

The most common mineral of the smectite group of ones—montmorillonite (MMT) was taken from the Cherkasy deposit (Dashukivka, Lysyansky district of Cherkasy region, Ukraine). The 2:1 structure of montmorillonite is composed of one octahedral sheet that is contained between two tetrahedral sheets. Large amounts of water can be accommodated between the layers of MMT [23]. The general structural formula of montmorillonite is $(Ca_{0.12}Na_{0.03}K_{0.03})_{0.18}(Al_{1.39}Mg_{0.13}Fe_{0.44})_{1.96}(Si_{3.88}Al_{0.12})_{4.0}O_{10}(OH)_2{\cdot}nH_2O$, the cation exchange capacity (CEC) is 1.0 mmol/g.

Modification of the surface of MMT was carried out by (i) cationic surfactant HDTMA [9]; (ii) the salts of inorganic cations Al, Zr, Ti, Fe [24,25]; (iii) zero-valent iron by reduction of ions Fe^{3+} with a solution of sodium borohydride in a nitrogen atmosphere [20,26]. The synthesized samples of sorbents were named MMT-HDTMA, MMT-PILC, and MMT-ZVI.

Low-temperature nitrogen adsorption isotherms were determined on a volumetric automatic apparatus (Quantachrome, Nova 2200e) at −196 °C. The surface morphology of MMT was studied by scanning electron microscopy SEM on a Jeol JSM-6060 instrument (Tokyo, Japan).

Sorption experiments on the removal of uranium (VI) and chromium (VI) ions from water were performed from a solution prepared from uranyl trihydrosulfate salt ($UO_2SO_4{\cdot}3H_2O$) and potassium dichromate $K_2Cr_2O_7$. The ionic strength was created with a solution of NaCl. Experimental studies on the effect of pH were performed at initial concentrations of each of the heavy metals 100 µmol/L. Sorption isotherms were obtained at pH 6. The sorption values from the solutions were determined under static conditions in a thermostated cell at 25 °C with continuous shaking of the samples for 1 h (volume of the aqueous phase-50 mL, the mass of sorbent-0.1 g). After establishing the adsorption equilibrium, the liquid phase was separated by centrifugation (6000 rpm). The equilibrium metal concentration was determined spectrophotometrically (UNICO 2100UV) using an Arsenazo III reagent at a wavelength of 665 nm for U (VI) and a diphenylcarbazide at a wavelength of 546 nm for Cr (VI). The sorption U(VI), Cr (VI)) a, μmol/g were calculated by the formula:

$$a = (C_{in} - C_{eq}) \cdot V/m, \quad (1)$$

which shows C_{in} and C_{eq} as the initial and equilibrium concentration of the metal, μmol/L; the V-volume of the solution, L; and the m-mass of the sample of sorbent, g.

3. Results and Discussion

The modification of the surface of MMT by cationic surfactant HDTMA was confirmed (i) the increase in the basal reflex on the MMT-HDTMA diffractograms from 1.23 nm to 2.81 nm, (ii) the change in the ξ-potential of the particles from −65 to 55 mV, and (iii) the appearance of characteristic bands at 2920 cm^{-1} and 2850 cm^{-1} in the IR spectra of surfactant modified samples [9]. An increase in the values of the basal reflex on the diffractograms from 1.261 for MMT to 1.736 nm for A1-PILC, 1.81 for Zr-PILC, and 1.55 for Ti-Fe-PILC samples indicate an increase in the interlayer space of the mineral. This is the main characteristic of the successful intercalation process that is judged [24,25]. The appearance of peaks 0.252 and 0.202 nm of iron compounds (α-Fe, FeO, FeOOH) in the diffractograms of the synthesized samples MMT-ZVI indicates the modification of the MMT surface with nanoscale zero-valent iron [26].

The nitrogen sorption isotherm on the natural montmorillonite (Figure 1), according to the modified de Boer classification, belongs to type II isotherms and is typical for nonporous sorbents with a small macroporous component [27]. The nature of the nitrogen adsorption curves on the samples of MMT-HDTMA, MMT-ZVI, and MMT-PILC are similar to MMT. In the range of values p/p0 > 0.4, the pronounced hysteresis loops can be observed on isotherms (MMT, MMT-ZVI, MMT-PILC), indicating the presence of a well-developed structure of H3 type meso and macropores typical for such samples.

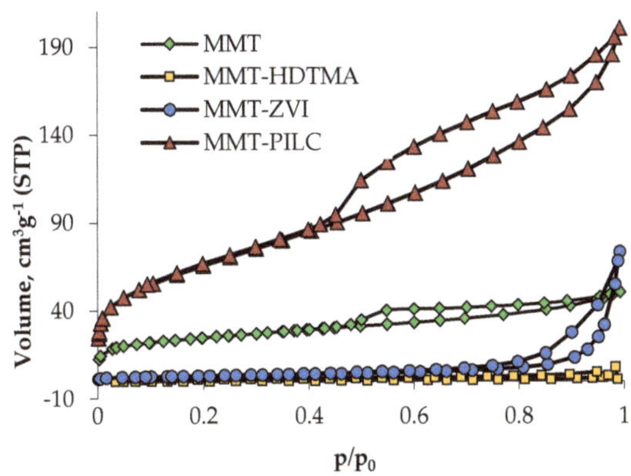

Figure 1. Nitrogen adsorption-desorption isotherms of natural and modified montmorillonites.

The specific surface area of MMT (89.1 m^2/g) sharply decreases after the surface modification and consists of 2.3 m^2/g for MMT-HDTMA and of 24.7 m^2/g for MMT-ZVI. Such reduction is stipulated by practically complete filling of micropores with surfactant molecules or iron nanoparticles, the resultant blocking of the access of nitrogen molecules to these pores. The sample of MMT-PILC has an increased specific surface area of 236.2 m^2/g.

This is due to the formation of stable two-dimensional porous structures in the interlayer space of the mineral.

The layered structure of montmorillonite is shown in Figure 2a. As it is shown in Figure 2b, in the sample MMT-HDTMA the montmorillonite surface is covered by cluster forms of the surfactant mainly. The total organic carbon content of MMT-HDTMA was 9.29% (w/w) which corresponds to 11.14% (w/w) of the HDTMA-ions.

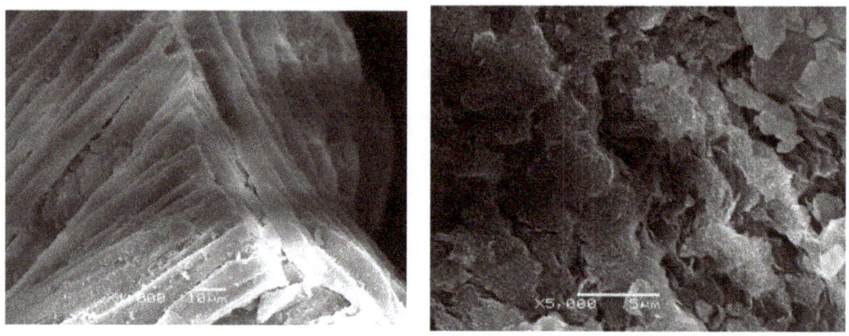

(**a**) (**b**)

Figure 2. SEM microphotographs of the particles of natural (**a**) and HDTMA modified (**b**) montmorillonite.

The dependence of the sorption values of uranium and chromium ions on the surface of natural and modified montmorillonite from pH is presented in Figure 3. The curve MMT for U(VI) (Figure 3a) has a characteristic form with a marked minimum in the acidic and alkaline pH areas. In the neutral pH area, the charge of the MMT surface and the charge of uranium ions have opposite signs. It influences the proceedings of sorption processes and the appearance of the maximum on the curve [26]. Natural MMT practically does not remove chromium anions due to the negative surface charge in a wide pH range (Figure 3b). The forms of heavy metals in the solution affect the sorption values [28]. Uranium in the acidic pH area exists in the uranyl cation UO_2^{2+}, while in the neutral and alkaline area in solution, uranium exists mainly in the form of neutral $UO_2(OH)_2$, UO_2CO_3 and negatively charged species $(UO_2)_2CO_3(OH)_3^-$. In natural waters under aerobic conditions, chromium is in the form of $HCrO_4^-$ and CrO_4^{2-} depending on the pH of the environment.

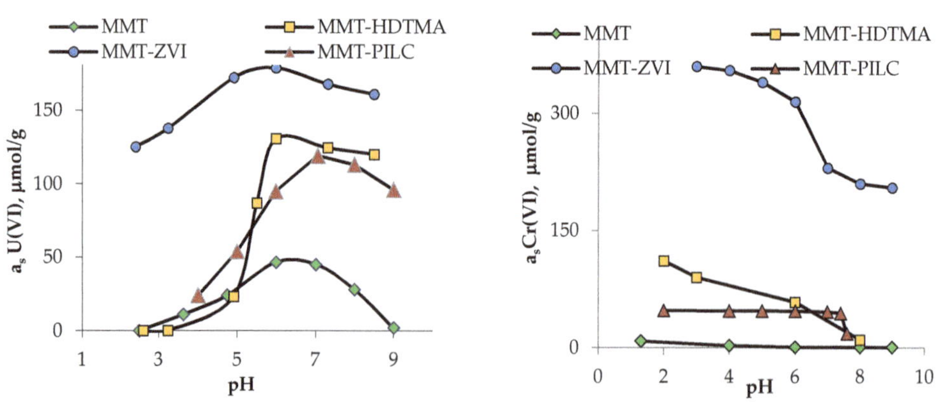

(**a**) (**b**)

Figure 3. Effect pH on U(VI) (**a**) and Cr(VI) (**b**) sorption by the natural and modified MMT.

The surface modification of MMT by organic and inorganic compounds leads to the exchange of its sorption characteristics and expands the pH range of sorbents. The values of uranium sorption on MMT-HDTMA and MMT-PILC samples increase in the alkaline pH region (Figure 3a). The values of chromium sorption on MMT-HDTMA and MMT-PILC samples increase in the acidic and neutral pH region (Figure 3b). MMT-ZVI samples significantly improve the sorption of uranium (VI) and chromium (VI) in all ranges of pH, with a maximum for uranium at pH 4–7 and for chromium at pH 3–5.

Sorption isotherms of uranium and chromium ions show that the sorption characteristics of samples of montmorillonite (Figure 4) modified with surfactant or salts of inorganic cations or zero-valent iron are much higher as compared to those for natural mineral MMT (76 µmol/g U(VI) and 12 µmol/g Cr(VI)). The maximum sorption capacity of modified montmorillonite was for MMT-HDTMA 181 µmol/g U(VI) and 218 µmol/g Cr(VI); for MMT-PILC 145µmol/g U(VI) and 215 µmol/g Cr(VI); for MMT-ZVI 305 µmol/g U(VI) and 345 µmol/g Cr(VI).

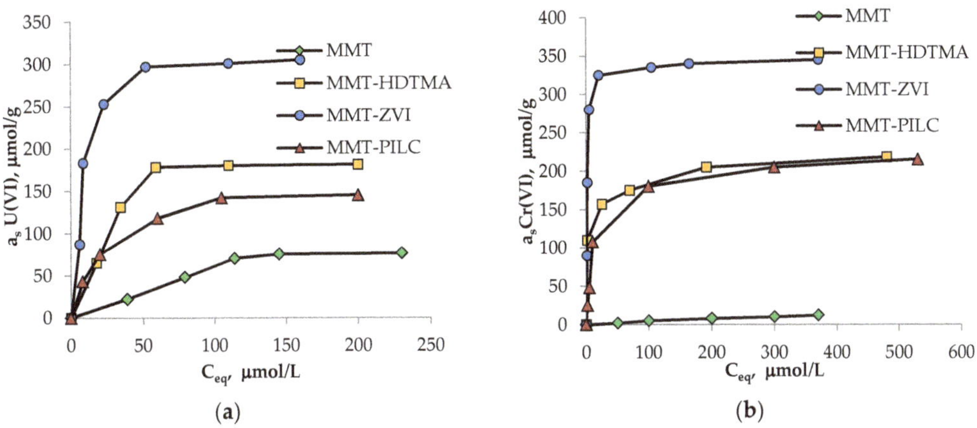

Figure 4. Isotherms of sorption of U(VI) (**a**) and Cr(VI) (**b**) by the natural and modified MMT.

Organoclays MMT-HDTMA can retain harmful substances due to the formation of surface sorption complexes with uranium and chromium compounds. Active sorption centers of MMT-PILC samples are hydroxyl groups of the metals in the interlayer space. Removal of U(VI) and Cr(VI) from water on the MMT-ZVI samples occurs primarily due to the exchange of hydroxyl ions of the hydroxide film on the surface of nanoscale iron particles by uranium or chromium complexes. In addition, for the samples, MMT-ZVI, the reduction of U(VI) to U (IV) and Cr(VI) to Cr(III) can occur with the next immobilization of uranium or chromium compounds on the surface of the sorbent.

4. Conclusions

Sorption materials on the base of montmorillonite (organoclays, pillared clays, clay-supported zero-valent iron) were used for the removal of inorganic pollutants uranium (VI) and chromium (VI) from contaminated water. A significant increase in the sorption characteristics of the synthesized materials was shown. The obtained sorbents can be widely used in water purification technologies and modern environmental protection technologies. It is advisable to carry out organophilization of clays when underground reaction barriers are created in the soil along the path of movement of polluted waters. The use of pillared clays is effective for the deep cleaning of small volumes of water from heavy metals and radionuclides. Sorbents based on zero-valent iron and clay minerals can be widely used both in the mode of water purification on underground permeated reactive barriers and on traditional sorption technological equipment.

Funding: This research received no external funding.

Institutional Review Board Statement: Not applicable.

Informed Consent Statement: Not applicable.

Data Availability Statement: Not applicable.

Conflicts of Interest: The author declares no conflict of interest.

References

1. Bockris, J.M. *Environmental Chemistry*; Springer: Boston, MA, USA, 1977.
2. Shrestha, R.; Ban, S.; Devkota, S.; Sharma, S.; Joshi, R.; Tiwari, A.P.; Kim, H.Y.; Joshi, M.K. Technological Trends in Heavy Metals Removal from Industrial Wastewater: A Review. *J. Environ. Chem. Eng.* **2021**, *9*, 105688. [CrossRef]
3. Babak, M.I.; Koshyk, Y.I.; Avdeev, O.K.; Bezrodnyi, S.A.; Savelyev, Y.Y.; Kucha, P.M. *Mining and Processing of Uranium Ores in Ukraine*; Chernov, A.P., Ed.; ADEF-Ukraine: Kyiv, Ukraine, 2001; p. 238. (In Ukrainian)
4. Sapozhnikov, Y.A.; Aliyev, R.A.; Kalmykov, S.N. *Environmental Radioactivity*; BINOM: Moscow, Russia, 2006; p. 286. (In Russian)
5. World Health Organization. *Guidelines for Drinking-Water Quality: Fourth Edition Incorporating the First and Second Addenda*; World Health Organization: Geneva, Switzerland, 2022.
6. Zachara, J.M.; Cowan, C.E.; Schmidt, R.L.; Ainsworth, C.C. Chromate Adsorption by Kaolinite. *Clays Clay Miner.* **1988**, *36*, 317–326. [CrossRef]
7. Yuan, G.D.; Theng, B.K.G.; Churchman, G.J.; Gates, W.P. Clays and Clay Minerals for Pollution Control. In *Handbook of Clay Sciences*; Elsevier: Amsterdam, The Netherlands, 2013; Volume 5, pp. 587–644.
8. Majdan, M.; Pikus, S.; Gajowiak, A.; Gładysz-Płaska, A.; Krzyżanowska, H.; Żuk, J.; Bujacka, M. Characterization of uranium(VI) sorption by organobentonit. *Appl. Surf. Sci.* **2010**, *256*, 5416–5421. [CrossRef]
9. Kovalchuk, I.A.; Laguta, A.M.; Kornilovych, B.Y.; Tobilko, V.Y. Organophilized layered silicates for removing uranium (VI) compounds from mineralized waters. *Chem. Phys. Technol. Surf.* **2020**, *2*, 215–227.
10. Houhoune, F.; Nibou, D.; Chegrouche, S.; Menacer, S. Behaviour of modified hexadecyltrimethylammonium bromide toward uranium species. *J. Environ. Chem. Eng.* **2016**, *4*, 3459–3467. [CrossRef]
11. Gajowiak, A.; Gładysz-Płaska, A.; Sternik, D.; Pikus, S.; Sabah, E.; Majdan, M. Sorption of uranyl ions on organosepiolite. *Chem. Eng. J.* **2013**, *219*, 459–468. [CrossRef]
12. Krishna, B.S.; Murty, D.S.R.; Prakash, B.S.J. Surfactant-modified clay as adsorbent for chromate. *Appl. Clay Sci.* **2001**, *20*, 65–71. [CrossRef]
13. Li, Z. Oxyanion sorption and surface anion exchange by surfactant-modified clay minerals. *J. Environ. Qual.* **1999**, *28*, 1457–1463. [CrossRef]
14. Li, Z.H.; Bowman, R.S. Retention of inorganic oxyanions by organo-kaolinite. *Water Res.* **2001**, *35*, 3771–3776. [CrossRef] [PubMed]
15. Gil, A.; Korili, S.A.; Trujillano, R.; Vicente, M.A. *Pillared Clays and Related Catalysts*; Springer: New York, NY, USA, 2010; p. 522.
16. Vicente, M.A.; Gil, A.; Bergaya, F. Pillared Clays and Clay Minerals. In *Handbook of Clay Science*; Elsevier: Amsterdam, The Netherlands, 2013; pp. 523–557.
17. Zhou, J.; Wu, P.; Dang, Z.; Zhu, N.; Li, P.; Wu, J.; Wang, X. Polymeric Fe/Zr pillared montmorillonite for the removal of Cr(VI) from aqueous solutions. *Chem. Eng. J.* **2010**, *162*, 1035–1044. [CrossRef]
18. Pasinszki, T.; Krebsz, M. Synthesis and Application of Zero-Valent Iron Nanoparticles in Water Treatment, Environmental Remediation, Catalysis, and Their Biological Effects. *Nanomaterials* **2020**, *10*, 917. [CrossRef] [PubMed]
19. Chen, A.; Shang, C.; Shao, J.; Zhang, J.; Huang, H. The application of iron-based technologies in uranium remediation: A review. *Sci. Total Environ.* **2017**, *575*, 1291–1306. [CrossRef] [PubMed]
20. Sheng, G.; Shao, X.; Li, Y.; Li, J.; Dong, H.; Cheng, W.; Gao, X.; Huang, Y. Enhanced Removal of Uranium(VI) by Nanoscale Zerovalent Iron Supported on Na–Bentonite and an Investigation of Mechanism. *Phys. Chem.* **2014**, *118*, 2952–2958. [CrossRef] [PubMed]
21. Shi, L.N.; Zhang, X.; Chen, Z.L. Removal of Chromium (VI) from wastewater using bentonite-supported nanoscale zero-valent iron. *Water Res.* **2011**, *45*, 886–892. [CrossRef] [PubMed]
22. Fu, R.; Yang, Y.; Xu, Z.; Zhang, X.; Guo, X.; Bi, D. The removal of chromium (VI) and lead (II) from groundwater using sepiolite-supported nanoscale zero-valent iron (S-NZVI). *Chemosphere* **2015**, *138*, 726–727. [CrossRef] [PubMed]
23. Carrado, K.A. Introduction: Clay Structure, Surface Acidity, and Catalysis. In *Handbook of Layered Materials*; CRC Press: Boca Raton, FL, USA, 2004.
24. Guerra, D.L.; Airoldi, C.; Lemos, V.P.; Angelica, R.S. Adsorptive, thermodynamic and kinetic performances of Al/Ti and Al/Zr-pillared clays from the Brazilian Amazon region for zinc cation removal. *J. Hazard. Mater.* **2008**, *155*, 230–242. [CrossRef] [PubMed]
25. Pylypenko, I.V.; Kovalchuk, I.A.; Veremeyenko, V.V.; Spasyonova, L.M. Sorption of cobalt, chromium and uranium ions by Fe/Ti-pillared montmorillonite. *Easter Eur. J. Enterp. Technol.* **2014**, *4*, 57–61.
26. Kornilovych, B.; Kovalchuk, I.; Tobilko, V.; Ubaldini, S. Uranium Removal from Groundwater and Wastewater Using Clay-Supported Nanoscale Zero-Valent Iron. *Metals* **2020**, *10*, 1421. [CrossRef]

27. Sing, K.S.W.; Everett, D.H.; Haul, R.A.W.; Moscou, L.; Pierotti, R.A.; Rouquerol, J.; Siemieniewska, T. Reporting physisorption data for gas/solid systems with special reference to the determination of surface area and porosity. *Pure Appl. Chem. Res.* **1985**, *57*, 603–619. [CrossRef]
28. Langmuir, D. *Aqueous Environmental Geochemistry*; Prentice Hall: Upper Saddle River, NJ, USA, 1997; 600p.

environmental sciences proceedings

MDPI

Proceeding Paper

The Karla Aquifer (Central Greece), an Agricultural Region under Intensive Environmental Pressure Due to Agricultural Activities †

Georgios A. Tziatzios [1,*], Luca Doro [2], John Tzabiras [1], Nikitas Mylopoulos [1], Athanasios Loukas [3] and Nicholas Danalatos [4]

1 Laboratory of Hydrology and Aquatic Systems Analysis, Department of Civil Engineering, University of Thessaly, 38334 Volos, Greece; jjabiras@gmail.com (J.T.); nikitas@civ.uth.gr (N.M.)
2 Texas A&M AgriLife Research-Blackland Research & Extension Center, Temple, TX 76502, USA; luca.doro@brc.tamus.edu
3 Department of Rural and Surveying Engineering, Aristotle University of Thessaloniki, 54124 Thessaloniki, Greece; agloukas@topo.auth.gr
4 Department of Agriculture Crop Production and Agricultural Environment, School of Agricultural Sciences, University of Thessaly, 38446 Volos, Greece; danal@uth.gr
* Correspondence: getziatz@civ.uth.gr; Tel.: +30-2421074153

† Presented at the 7th International Electronic Conference on Water Sciences, 15–30 March 2023; Available online: https://ecws-7.sciforum.net.

Abstract: The Agricultural Policy Environmental eXtender (APEX) model is used to study how different agricultural practices, such as fertilizing, irrigation, and tillage, would affect water quality and runoff in the Lake Karla watershed (Central Greece). The model was calibrated for the potential evapotranspiration with satisfactory results for the period 1980–2008 and for the yields of the main crops grown in the region (cotton, maize, and wheat) from 1980–2015.

Keywords: Agronomic Simulation; APEX; Lake Karla watershed

Citation: Tziatzios, G.A.; Doro, L.; Tzabiras, J.; Mylopoulos, N.; Loukas, A.; Danalatos, N. The Karla Aquifer (Central Greece), an Agricultural Region under Intensive Environmental Pressure Due to Agricultural Activities. *Environ. Sci. Proc.* **2023**, *25*, 35. https://doi.org/10.3390/ECWS-7-14189

Academic Editor: Lampros Vasiliades

Published: 14 March 2023

1. Introduction

The challenges that the agricultural sector must deal with are multidimensional and large. On the one hand, the increase in production is intended to cover the nutritional needs of the rapidly growing population. On the other hand, limiting the use of water, fertilizers, and pesticides to protect the sustainability of agroecosystems while protecting the natural environment from problems, such as nutrient losses with nitrogen losses, is often used as a typical example. In recent years, the difficulties created by these challenges have been aggravated by the projected climate change [1–4]. Scientists apply simulation models to examine all the aforementioned challenges [4]. Simulation models are an approach to represent quantitative knowledge about the system of interest and how the different components of that system interact. Agroecosystem models can help agronomists to understand crop growth, predict crop yields, and assess management for better water and nutrients used. Climate data, soil, and information about the management of the agroecosystem are used to inform these models. Such agroecosystem tools can normally simulate many periods, locations, management styles, and scenarios and can provide useful information to agricultural science and farming, exploring the changing aspects between the atmosphere, plants, soil, and water, assisting in crop agronomy, pest management, plant breeding, natural resources management, and evaluating the effect of climate change [5]. In this article, we present the activities that are currently carried out for an ongoing project where the agricultural policy environmental extender (APEX) model is applied in a rural region in Central Greece to assess crop production and water and nitrogen losses under current and future weather conditions.

2. Methodology

The APEX model has been implemented for the aquifer of the Karla Basin. APEX was developed to help evaluate different land management strategies regarding their environmental impact, erosion, cost, and possible water supplies. APEX simulates the nitrogen and the water process, the crop yield, at the field, farm, or watershed levels, subdividing the simulated area into several units with homogeneous soil, weather, land use, and topography commonly defined subareas [6,7]. The Karla watershed is an area with intense agricultural activity [8]. Figure 1 presents the land uses and crop classification for the Karla aquifer, as displayed within the ArcAPEX interface. After the delineation process, ArcAPEX separated the study area into 34 homogeneous subareas. The model was set to simulate 46 years in total, with the first 10 years used as a spin-up period and not considered in the calibration process. For the calibration, cumulative monthly data from 1961 to 2009 were used for the potential evapotranspiration (PET), while the crop yield of the main crops was calibrated considering the period 1980–2015.

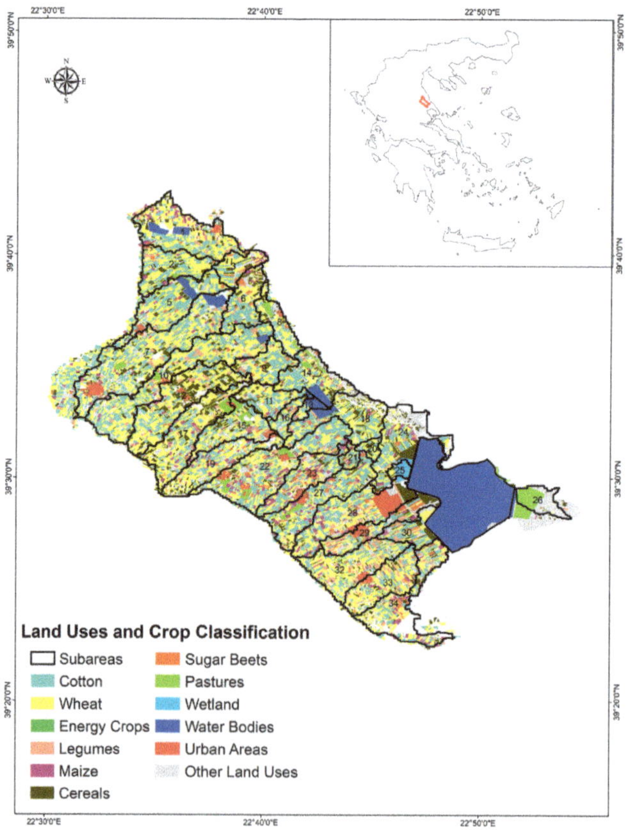

Figure 1. The study area.

Two statistical criteria were used to evaluate the results obtained for PET. The Nash–Sutcliffe model efficiency (Ef) in Equation (1) and the coefficient of determination (R^2) in Equation (2) indicate how well the model describes adaptation in the observed and estimated data:

$$Ef = 1 - \frac{\sum_{t=1}^{T}\left(Y_m^t - Y_o^t\right)^2}{\sum_{t=1}^{T}\left(Y_o^t - \overline{Y_o}\right)^2} \quad (1)$$

where $\overline{Y_O}$ is the mean observed value, Ym is the estimated value by the model, and Yo is the observed at time t. Ef ranges from 1 (best result) to minus infinite.

$$R^2 = \left[\frac{\Sigma(x - \overline{x}) - (y - \overline{y})}{\sqrt{\Sigma(x - \overline{x})^2 \Sigma(y - \overline{y})^2}} \right] \tag{2}$$

where x and y are the observed and the estimated values by the model, $\overline{x}$ and $\overline{y}$ are the mean observed and estimated values by the model, respectively. R^2 ranges from 1 (best result) to 0 (worst result).

3. Results

The APEX model was initially calibrated considering the PET. During the calibration process, four methods for the PET estimation were examined using the Hargreaves approach resulting as the best method. The results for Ef and R^2 are presented in Table 1. Figure 2 shows the scatter plot where the observed and simulated values of PET are compared. As reported in Table 1, the model was able to provide a good estimate of PET, resulting in an Ef value of 0.85 and R^2 of 0.90.

Table 1. Statistical Criteria Results.

Statistical Criteria	Results
Ef	0.85
R^2	0.90

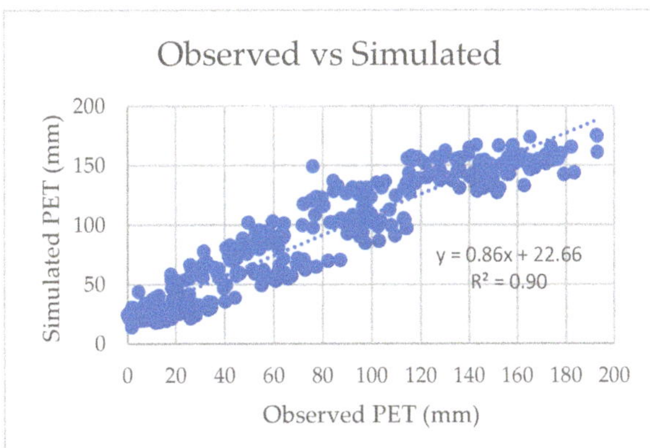

Figure 2. Comparison of the observed and simulated PET.

The work continued with the calibration of yields of the main crops grown in the study area (cotton, maize, and winter wheat). It is worth noting that the calibration of crop yields was based on the average crop yield of winter wheat, cotton, and maize provided by the Greek Ministry of Rural Development and Food [9]. Due to the fact that APEX reports the yield as dry weight, the reported yield data has been adjusted for moisture content. We considered a moisture content between 6.5% to 8% for cotton [10] and 14% for maize and wheat. After adjusting the average observed crop yield for the moisture content, the target crop yield for calibration was 2.6–3.2 Mg ha^{-1} for cotton, 8.6–17.2 Mg ha^{-1} for maize, and 2.0–3.0 Mg ha^{-1} for wheat. Having only one average reported yield available, it was not possible to conduct a statistical assessment of the performance in simulating crop yield. Figures 3–5 show the simulated crop yield for all the simulated years after the calibration process for wheat, cotton, and maize, respectively. The values reported are the average of the yield simulated by the APEX model in all the areas where each crop is cultivated within the watershed.

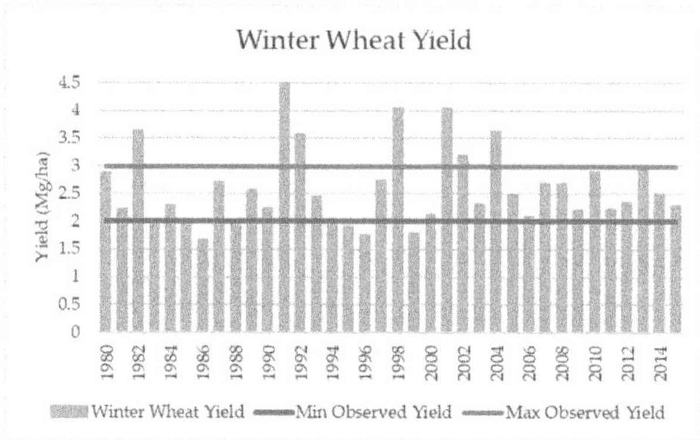

Figure 3. Simulated wheat yield; minimum and maximum average reported yield.

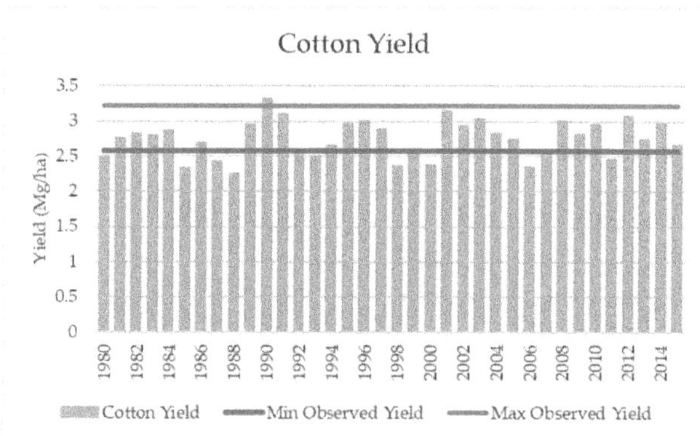

Figure 4. Simulated cotton yield; minimum and maximum average reported yield.

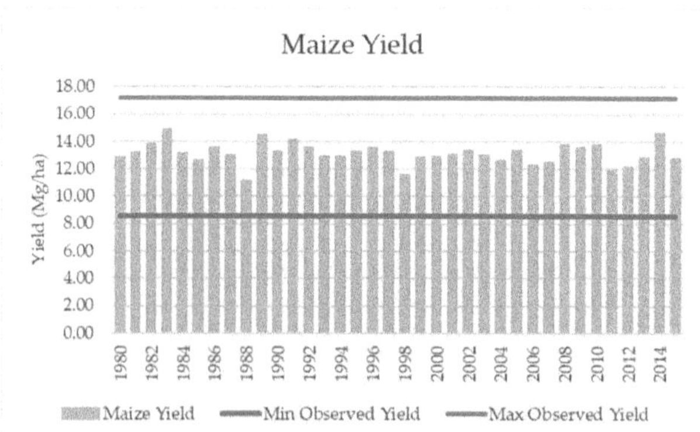

Figure 5. Simulated maize yield; minimum and maximum average reported yield.

After the calibration process, the model was able to provide good results in simulating crop yield. The average simulated yield for wheat was 2.6 Mg ha^{-1}, which was in the range of the average reported yield. In some years, the yield was overestimated, probably due to an overestimation of the crop-available water that, in turn, produced no water stress and a very high crop yield. We will continue to analyze this aspect to improve the quality of the results for this crop. The average simulated yield for cotton was 2.75 Mg ha^{-1}, which is within the range of the average reported yield with some years where the simulated yield is below the minimum or above the maximum average reported yield. Results for maize were better, with an average simulated yield of 13.1 Mg ha^{-1} and yielded within the reported range for all examined years.

To calibrate crop yields, parameters that regulate the simulation of soil water content (soil water lower limits and soil evaporation) and the effect of water stress and high temperature on the harvest index were adjusted. Further, the harvest index for maize was revised to consider the higher harvest index of the new maize hybrids, which were obtained thanks to plant breeding and genetic improvement.

The calibration process will be continued considering the runoff and nitrate leaching. In the final step, the APEX model will be used to study the impacts of climate change scenarios on the agroecosystems of the Karla watershed.

4. Discussion

At the beginning of this research, we were able to design the Karla watershed within the APEX model. The first step was based on the automatic delineation of the watershed using the ArcAPEX interface. After this initial step, the input files generated by the interface required some modification to better represent features of the watershed that were not correctly captured by the ArcAPEX interface, such as the presence of a reservoir in the lower part of the watershed. After setting up all the input data required by the model, we started the calibration procedure. Beginning with the calibration of the PET, which we consider as the starting point to have a good simulation of the water balance, we were able to obtain good results with R^2 of 0.9 and Ef of 0.85. The good results obtained for the PET are followed by satisfactory results in the simulation of the yields of the three main crops cultivated in the studied area. The APEX model was able to produce reasonable results for maize, cotton, and wheat yields with some overestimation that required analysis for the later harvest. The calibration process will be completed by including the analysis of runoff and nitrate leaching, and the model will be used to assess how climate change will affect crop production and water and nutrient losses.

5. Conclusions

This modeling study will allow us to better understand if the APEX model could be considered a useful tool for studying agroecosystems in Mediterranean climates. Obtaining good results in the calibration and validation process will allow us to use the APEX model to assess the impact of land management and climate change in the Karla watershed.

Author Contributions: Conceptualization, G.A.T.; methodology, G.A.T. and L.D.; modeling activity, G.A.T. and L.D.; climate data analysis J.T.; writing—original draft preparation, G.A.T.; writing—review and editing, L.D.; N.M.; A.L., and N.D.; visualization, G.A.T. and L.D.; supervision N.M., A.L., and N.D. All authors have read and agreed to the published version of the manuscript.

Funding: This research received no external funding.

Institutional Review Board Statement: Not applicable.

Informed Consent Statement: Not applicable.

Data Availability Statement: Not applicable.

Acknowledgments: The important contribution of Konstantina's Konovesis and Athanasios Lois from the Organization for Payments and Control of Community Aids, Payments and Guarantees (OPEKEPE), as well as Georgia Pappas and Ioannis Chrysiadou from the Ministry of Rural Development and Food for the provision of the crop data of the study area.

Conflicts of Interest: The authors declare no conflict of interest.

References

1. Velten, S.; Leventon, J.; Jager, N.; Newig, J. What is sustainable agriculture? A systematic review. *Sustainability* **2015**, *7*, 7833–7865. [CrossRef]
2. Prandecki, K.; Wrzaszcz, W.; Zieliński, M. Environmental and climate challenges to agriculture in Poland in the context of objectives adopted in the European green deal strategy. *Sustainability* **2021**, *13*, 10318. [CrossRef]
3. Jordan, N.; Gutknecht, J.; Bybee-Finley, K.A.; Hunter, M.; Krupnik, T.J.; Pittelkow, C.M.; Prasad, P.V.V.; Snapp, S. To meet grand challenges, agricultural scientists must engage in the politics of constructive collective action. *Crop Sci.* **2021**, *61*, 24–31. [CrossRef] [PubMed]
4. Silva, J.V.; Giller, K.E. Grand challenges for the 21st century: What crop models can and can't (yet) do. *J. Agric. Sci.* **2020**, *158*, 794–805. [CrossRef]
5. Asseng, S.; Zhu, Y.; Basso, B.; Wilson, T.; Cammarano, D. Simulation Modeling: Applications in cropping systems. In *Encyclopedia of Agriculture and Food Systems*; Van Alfen, N.K., Ed.; Academic Press: London, UK, 2014; Volume 4, pp. 102–113.
6. Steglich, E.M.; Osorio, J.; Doro, L.; Jeong, J.; Williams, J.W. Agricultural Policy/Environmental eXtender Model User Manual Version 1501. Blackland Research and Extension Center, Temple. 2018. Available online: https://epicapex.tamu.edu/manuals-and-publications/7 (accessed on 17 February 2023).
7. Taylor, R.A.; Jeong, J.; White, M.; Arnold, J.G. Code modernization and modularization of APEX and SWAT watershed simulation models. *Int. J. Agric. Biol. Eng.* **2015**, *8*, 81–94.
8. Sidiropoulos, P.; Tziatzios, G.; Vasiliades, L.; Mylopoulos, N.; Loukas, A. Groundwater nitrate contamination integrated modeling for climate and water resources scenarios: The case of Lake Karla over-exploited aquifer. *Water* **2019**, *11*, 1201. [CrossRef]
9. Tavoularis, K. *Average Crop Yields in Greece*; Ministry of Agricultural Development and Food Agricultural Policy and Documentation Directorate Department of Agricultural Statistics: Athens, Greece, 2012.
10. Available online: https://agresearchmag.ars.usda.gov/ar/archive/2002/may/cotton0502.pdf (accessed on 17 February 2023).

environmental sciences proceedings

Proceeding Paper

Sustainable Water Management in Indus Basin and Vulnerability Due to Climate Change †

Syeda Mishal Zahra [1,2,*], Muhammad Adnan Shahid [1,2,*], Muhammad Aali Misaal [3], Muhammad Zaman [2], Muhammad Imran [2], Sidra Azam [2] and Fazal Hussain [1,2]

1 Agricultural Remote Sensing Lab (ARSL), University of Agriculture, Faisalabad 38040, Pakistan; fazal.civilengineer@gmail.com
2 Department of Irrigation and Drainage, Faculty of Agricultural Engineering and Technology, University of Agriculture, Faisalabad 38040, Pakistan; muhammad.zaman@uaf.edu.pk (M.Z.); m.imrankbr@gmail.com (M.I.); sidraa3679@gmail.com (S.A.)
3 Department of Farm Machinery and Power, Faculty of Agricultural Engineering and Technology, University of Agriculture, Faisalabad 38040, Pakistan; muhammadaalimisaal@gmail.com
* Correspondence: syedamishalzahra212@gmail.com (S.M.Z.); adnan.wmrc@gmail.com (M.A.S.)
† Presented at the 7th International Electronic Conference on Water Sciences, 15–30 March 2023; Available online: https://ecws-7.sciforum.net.

Abstract: Pakistan depends heavily on the Indus River Basin System (IRBS) which is essential for meeting the great majority of Pakistan's agricultural and home consumption requirements. The Indus River is responsible for over 90% of Pakistan's agricultural output and accounts for 25% of the country's GDP. Because of the problems with the water supply, Pakistan may soon face serious food scarcity. By 2025, the water deficit is expected to reach 32%, according to the World Bank's 2020–2021 study, leading to a food deficit of about 70 million tons. Recent predictions suggest that by 2025, the water storage capacity will have reduced by over 30% due to climate change. Extreme events, i.e., temperature and precipitation, occurred in Pakistan, and these affect human beings. Pakistan has a very low per capita water storage capacity, at about 150 m^3. As a result of decreasing surface water supplies and rising groundwater abstraction, the viability of irrigated agriculture may soon be threatened. To maximize the potential for increased storage, Pakistan must enhance its water-use efficiency and implement sustainable strategies for managing its groundwater and surface water resources. The crucial aspects in keeping irrigated agriculture viable in the Indus Basin are developing the infrastructure and eliminating distrust among the provinces.

Keywords: Indus River Basin System; food scarcity; climate change; water resources

Citation: Zahra, S.M.; Shahid, M.A.; Misaal, M.A.; Zaman, M.; Imran, M.; Azam, S.; Hussain, F. Sustainable Water Management in Indus Basin and Vulnerability Due to Climate Change. *Environ. Sci. Proc.* **2023**, *25*, 36. https://doi.org/10.3390/ECWS-7-14203

Academic Editor: Luis Garrote

Published: 14 March 2023

1. Introduction

Pakistan is the fifth most populous and developing economy, with a population of over 22.50 million. The world's best irrigation system lies in Pakistan and 80% of the country's population lives in the Indus Basin. Despite this, Pakistan has become a water-stressed country and, as per the estimates of the United Nations (UN), the availability of per capita water in Pakistan is 1090 m^3 [1,2]. The agricultural sector has been the key contributor towards the economic developments of the country since it achieved independence. The contribution of the agricultural sector to GDP of Pakistan during FY 2021–22 was around 22.7%, comprising 37.4% of the labor force [3]. The agricultural sector and water resources are under immense pressure, as the population of the country is increasing 2.6% [4]. The increasing population demands more food, which requires more areas to be cultivated. The Pakistan Council of Research in Water Resources (PCRWR) have intimated that a water shortage is expected by 2025 if no new water reserves are constructed. Farmers are fulfilling their agricultural water needs by overexploiting ground water. The poor quality of ground water is resulting in salinization, which is an alarming concern for sustainability of agriculture in Pakistan [3]. Pakistan is among the list of countries which are significantly affected by salinity. Conjunctive management practices of ground and surface water can increase the irrigation effectiveness and efficiency [4]. About 70% of irrigated plains are using ground and surface water simultaneously. With the growing economy, the distribution and management of irrigation waterways is a serious concern for the future. The municipal and industrial supplies in urban areas are predicted to reach 14 km^3 from 5.3 km^3 by 2025 [1] and per capita water availability will be reduced by 32%, resulting in a food shortage of 70 million tons [5]. The Indus Basin was designed to supply low-intensity irrigation in large areas of canal command. The increase in cropping intensity is increasing the gap in supply demand and creating incessant unrest.

Climate change is one of the worst developments of that century, and has significantly affected water supply and cropping patterns. Glacial melt and retreat, change in precipitation patterns, and snow melt are the most frequently observed uncertainties resulting from climate change. About 45% of basin flow is generated by snow and glacial melt, which indicates that vulnerability to climate change is very high. The Hindu Kush Himalaya (HKH) range is widespread over 2000 km^2 in the Asian continent and contains rivers such as the Brahmaputra, Indus, and Ganges. Almost 1 billion people in Nepal, India, Bangladesh, Pakistan, China, and Bhutan are dependent on these rivers. Recent studies have predicted that glacial melt will be increased due to global warming, which can increase the river flows by 40% [6], but the average flow rates will be decreased by 60% in Indus Basin [7]. The duration between May and September accounts for 85% of discharge in the Indus Basin in form of snow and glacial melt and monsoon activity. The hydrological cycles are expected to change, as the floods of 2010 and 2022 are examples of a hydrological system [8]. Therefore, there is a need to adopt and implement water management techniques, the building of new reservoirs for water storage, and the allocation of water resources in an optimum way. Additionally, mitigation measures are needed to tackle the effects of climate change to ensure water and food security and safety in Pakistan.

2. Materials and Methods

2.1. Study Area

The area of the Indus Basin is 566,000 km^2 and spread over four provinces of Pakistan, i.e., Punjab, Khyber Pakhtunkhwa, Baluchistan, and Sindh. The basin length of Indus is about 2900 km. It is fed by the eastern (Sutlej and Ravi) and the western (Chenab, Kabul, and Ravi) rivers, as shown in Figure 1. The river flows through an elevation of 18,000 ft from the Himalayas to the plains of Sindh and discharges into the Arabian Sea. The Indus Basin in Pakistan has a mean annual flow of 176 billion m^3 and 90% of the flow is used for irrigation purposes. The irrigation network consists of the main, branch canal, and distributaries with a length of about 57,000 km, with 88,600 outlets.

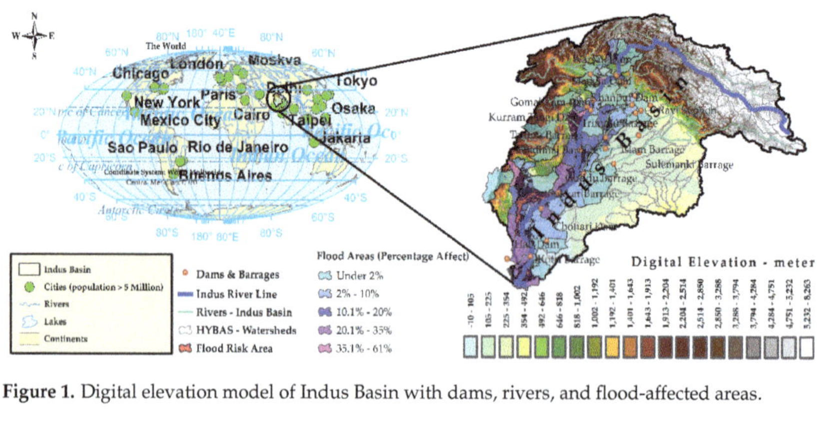

Figure 1. Digital elevation model of Indus Basin with dams, rivers, and flood-affected areas.

2.2. Data Collection and Analysis

Pakistan is one of the most climatically varied countries due to its wide temperature range, which includes extremes such as the Sahara desert's temperature and the arctic cold of Alaska. Data on extreme events of climatic parameters, i.e., temperature and precipitation, were collected for the duration of 1919–2022. The data were collected from the World Meteorological Organization (WMO) [9] and Pakistan Meteorological Department (PMD) [10]. The collected data were processed in GIS to highlight the main areas in which extreme events occurred. As per the intensity of extreme events, the hot spotted areas were graded using Inverse Distance Weighing (IDW) method.

3. Results and Discussion

3.1. Sustainability of Water

Pakistan is the fifth most populous country in the world, with a population growth rate of 2.6%; its population is expected to reach 250 million by 2025 [4]. The urban population of Pakistan comprised 37% of the country's population in 2017, and this is expected to reach 52% by 2025 [5]. The movement of the population towards cities will increase water consumption by 8%. The per capita water availability in Pakistan was 5000 m^3 in 1947, which was reduced to 1100 m^3 in 2005 and is predicted to drop to 800 m^3 by 2025 [5]. According to the estimates of the UN, the water demand in Pakistan is increasing by 10% per annum. The area-wise water withdrawal in Pakistan is calculated as 175 km^3, of which around 71% (124.25 km) is from surface water and 29% (50.75 km) is from groundwater [8]. In Pakistan, surface water has always been a burning topic, but groundwater has also many problems associated with it, i.e., salinity, overdraft, and waterlogging. According to an estimate, there are 0.8 million pumps installed in Pakistan and 50% of agricultural water needs in Punjab are met by them. The over-pumping of groundwater has led to the salinization of 4.5 million hectares; about 50% of the area lies in irrigated plains of the Indus Basin. The inappropriate irrigation practices and waterlogging due to seepage from unaligned canals have affected nearly 1 million hectares of irrigated plain of the Indus Basin, but the problem of salinity is worse in Sindh. The curative measure was taken to counter the problem of salinity, but proved to be futile, and the land productivity badly decreased. The water storage capacity of Pakistan is considerably lower than that of developed countries. The per capita water availability in US and China is 5000 and 2000 m^3, respectively, while it is 150 m^3 in Pakistan [1]. The storage capacity of Pakistan is only 30 days, while our neighboring country India has 120 to 220 days. The maximum storage capacity of water, i.e., 900 days. The storage capacity of any country depends upon its reservoirs. Pakistan has only two main reservoirs, i.e., Tarbela and Mangla, and the capacity of both reservoirs has been reduced by 20 and 32% due to sedimentation [5]. The Lieftinck report of 1968 stated that one Tarbela-sized reservoir every 10 years is necessary to meet the increasing demands of agriculture. It is also recommended to increase the storage capacity to 22 billion m^3 by 2025 to meet the projected requirement of 165 billion m^3. If no new reservoirs are developed, the water availability will be reduced by 12% by the next decade due to sedimentation. Another reason for the development of new reservoirs is the increasing industrialization and urbanization, which demands more energy to fulfill their needs.

3.2. Extreme Events

Extreme high and low temperatures, as well as the heaviest rainfalls, are all features of Pakistan's weather. In Turbat, Balochistan on 28 May 2017, a record-breaking 53.7 °C was measured, making it the hottest day in Pakistan's history [11]. On 26 May 2010, at Moenjo Daro, Sindh, Pakistan, a temperature of 53.5 °C was recorded, making it the second-hottest day in the country's history [12]. This was the fifth-highest temperature ever reported on Earth. On 23 July 2001, Islamabad received 620 mm of rain, the largest amount ever recorded rainfall in just 10 h.

3.2.1. Temperature

Heat waves mostly tend to associate high temperatures in Pakistan with the summer months at any time between April and September. Temperatures exceeding 50 °C were common in parts of southern Pakistan, where they wreak havoc in these areas. The record-breaking heat wave of summer 2010 that occurred in the final ten days of May was the deadliest heat wave in Pakistan's history. The north-central region of Pakistan (Punjab and KPK) was scorching hot. In April 2017, regions of southern Pakistan were struck by a severe heat wave, with temperatures reaching 50 °C. Cities throughout Pakistan set new April high-temperature records due to this heat wave. In Larkana, Sindh, the highest recorded temperature was 50 °C on April 19, breaking the previous record set in April 2000, when the temperature reached 48.5 °C. Mohenjo-Daro, Sindh, saw the hottest temperature ever recorded in Asia at 53.5 °C on May 26, 2010, while Larkana, Sindh, had the second-hottest temperature ever reported in Asia at 53 °C. The intense heatwave of summer 2010 lasted from May 22nd to May 31st, and it caused temperatures to rise beyond 50 °C in 12 different cities throughout Pakistan. At least 18 individuals lost their lives on May 27 because of extreme heat in Pakistan, when temperatures reached above 45 °C in several regions. In addition, 11 cities saw their highest reported temperatures of 50 °C or above during the intense heatwave season, while five locations experienced temperatures of 53 °C or higher. Over and beyond the 45 °C mark, eleven cities also saw highs of less than 50 °C, as illustrated in Figure 2. This breaks the previous record set at Jacobabad, Pakistan, on 12 June 1919, when the temperature reached 52.8 °C, setting a new Asian record.

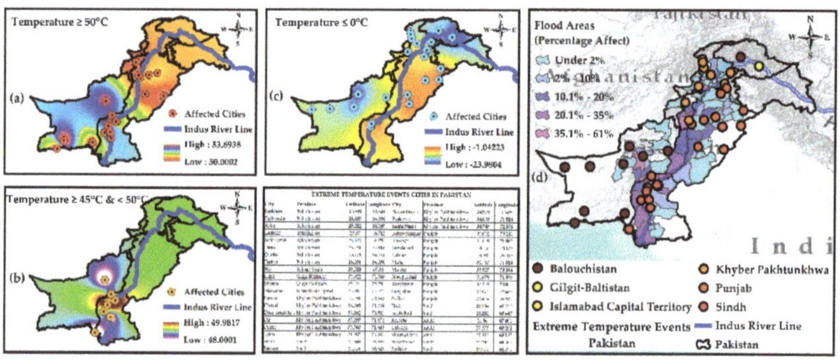

Figure 2. Extreme temperature events in Pakistan with varying ranges of temperature. (**a**) Temperature ≥50 °C; (**b**) temperature ≥40 °C and <50 °C; (**c**) temperature ≤0 °C; (**d**) all the regions in Pakistan in which extreme temperature events occurred.

Most cold waves happen between October and March, when temperatures drop significantly in northern and western Pakistan. The glacial regions of Gilgit Baltistan, where winter temperatures can drop below −20 °C, may be Pakistan's coldest spot. Temperatures of −65 °C have been recorded on K2. Recently, Pakistan saw the worst cold wave on record during the winter of 2020. Cities such as Islamabad, Peshawar, and Lahore on the foothills of the Himalayas may record temperatures below freezing, while even the warmest areas in Pakistan had temperature drops of 2 °C or more.

3.2.2. Precipitation

Monsoon and Western Disturbances were both sources of precipitation in Pakistan. Pakistan experienced widespread flooding from July through September due to the monsoon season. From October–May, precipitation spread over the whole of Pakistan due to western disturbances, with the heaviest rain falling in Northern Pakistan. However, in June, Western Disturbances sometimes affect the northern areas of Pakistan. On rare occasions, the pre-monsoon season might begin in this month as well. In the first week of the month, the country was battered by a Monsoonal downpour reminiscent of August's, when an extremely low air pressure system (29") developed over Kashmir and then swept eastward into Northern Pakistan. Devastating rains fell from 1–5 September, bringing death and destruction to many cities. The rivers Chenab, Jhelum, Ravi, Sutlej, and Indus in Pakistan all overflowed their banks during the latter two days of the period, which were particularly rainy. According to the Pakistan Meteorological Department, the region in northern Pakistan received around 200 mm of rain between 1 and 5 September 2014, as shown in Figure 3. A powerful low-pressure region formed in the Bay of Bengal in the latter days of August, after the extreme drought conditions in Sindh throughout the months of July and August. In the first two weeks of September 2012, a low-pressure region came into Sindh, bringing with it torrential rains in Upper Sindh. Jacobabad received 481 mm of rain in only 7 days, and 441 mm of rain in just 36 h, both world records. Larkana set a record with 239 mm, while Sukkur set a record with 206 mm. The worst of the flooding occurred in the Larkana district. In the last week of July 2010, Pakistan saw unprecedented severe monsoon rainfall, leading to flooding in the provinces of Balochistan and Sindh. Over 200 mm of rain occurred in a 24 h period in several locations in Khyber Pakhtunkhwa and Punjab. The previous 24 h rainfall record in Peshawar was 187 mm in April of 2009, but the recent total of 274 mm smashed that mark. The province of Khyber Pakhtunkhwa had rainfall in the cities of Risalpur, Cherat, Saidu Sharif, Mianwali, and Kohat.

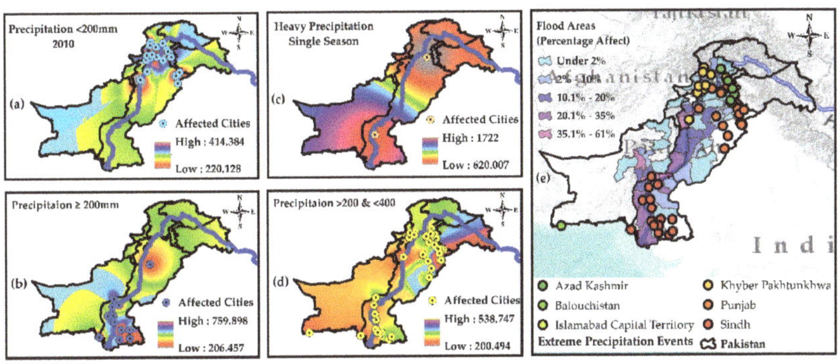

Figure 3. Precipitation extreme events in Pakistan (**a**) Precipitation in 2010 < 200 mm; (**b**) precipitation ≥ 200 mm; (**c**) heavy precipitation in a single season; (**d**) precipitation >200 and < 400 mm; (**e**) all the regions in Pakistan in which extreme precipitation events occurred.

3.2.3. Vulnerability Due to Climate Change

Climate change's impact and the occurrence of extreme events on the Indus Basin's water supply are still impossible to predict. There is a lot of guesswork involved in predicting how melting glaciers, melting snow, glacial retreat, and changes in precipitation will affect specific areas. About 45% of the flow to the basin comes from glacier melt and snow in the Himalayas, making it very susceptible to climate change and glacial melt [6]. Spanning six different countries in Asia—Pakistan, Nepal, India, China, Bhutan, and Bangladesh—is the Hindu Kush Himalaya (HKH) range. There are many major rivers in this area, including the Indus, Brahmaputra, and Ganges. Nearly one billion people rely on these rivers as their primary supply of drinking water. The loss of local glaciers is consistent with global trends, providing more evidence that glacial retreat has increased over the last century. New projections suggest that the pace of glacier melt caused by a rise in global temperature will accelerate in the coming years. This will increase river flow by 40% [3]. In the long term, however, the Indus River's typical flows will decrease by over 60% [2]. In addition, rising temperatures are expected to increase evapotranspiration rates across the irrigated Indus Basin, leading to higher demands for irrigation water and fiercer competition for surface and groundwater supplies among the provinces. Between May and

September, monsoon rainfall, glacier, and snowmelt account for over 85% of the yearly flow in the Indus Basin. The intensity, location, and timing of monsoon activity are all predicted to shift, causing widespread changes to the hydrological system during the next decade or two. One result of this shift in the hydrological system is the devastating flood that hit Pakistan in 2010. The thawing of glaciers had little effect on this.

4. Conclusions

Pakistan is an agriculturally based struggling economy, and water plays an important role in the sustainable development of the nation. The increasing population of the country is creating an unbalance in the demand and supply of food and water, leading to an alarming challenge to food safety and security in the country. Currently, the country is facing severe water and food crises due to bad management and governance over water resources. Technically weak policies, lack of water storage capacity, extreme climatic events due to climate change, and inter-provincial disputes have resulted in crucial conditions. Although the contribution of Pakistan to climate change is less than 1%, it is severely affected by cross-border pollution. The seasons shift with extreme temperature and precipitation events in the country. The summer season is prolonged, while the winter is shortened, leading to the generation of more excessive runoff than usual. The unavailability of water storage structures has resulted in the form of water wastage in the Arabian sea, causing floods. The recent floods of 2022 are an example of vulnerability due to climate change. Appropriate reforms for water and food policies are needed at this time, and will be achieved by developing water reservoirs with efficient distribution and application of water, i.e., drip and sprinkler irrigation, which will ensure sustainable agriculture and development of the country.

Author Contributions: Conceptualization, S.M.Z., M.A.S. and M.A.M.; methodology, S.M.Z. and M.A.M.; software, S.M.Z.; validation, S.M.Z., M.A.M. and M.Z.; formal analysis, S.M.Z.; investigation, M.A.S.; writing—original draft preparation, S.M.Z. and M.A.M.; writing—review and editing, M.I.; visualization, S.A. and F.H., supervision, M.A.S. All authors have read and agreed to the published version of the manuscript.

Funding: This research received no external funding.

Institutional Review Board Statement: Not applicable.

Informed Consent Statement: Not applicable.

Data Availability Statement: The data (extreme climate change events) used in this study are publicly available at [11,12].

Conflicts of Interest: The authors declare no conflict of interest.

References

1. Bhatti, M.T.; Anwar, A.A.; Shah, M.A.A. Revisiting telemetry in Pakistan's Indus basin irrigation system. *Water* **2019**, *11*, 2315. [CrossRef]
2. Dhaubanjar, S.; Lutz, A.F.; Gernaat, D.E.; Nepal, S.; Smolenaars, W.; Pradhananga, S.; Biemans, H.; Ludwig, F.; Shrestha, A.B.; Immerzeel, W.W. A systematic framework for the assessment of sustainable hydropower potential in a river basin—The case of the upper Indus. *Sci. Total Environ.* **2021**, *786*, 147142. [CrossRef] [PubMed]
3. Yaqoob, H.; Teoh, Y.H.; Sher, F.; Ashraf, M.U.; Amjad, S.; Jamil, M.A.; Jamil, M.M.; Mujtaba, M.A. Jatropha curcas biodiesel: A lucrative recipe for Pakistan's energy sector. *Processes* **2021**, *9*, 1129. [CrossRef]
4. Hussain, S.; Malik, S.; Masud Cheema, M.; Ashraf, M.U.; Waqas, M.; Iqbal, M.; Ali, S.; Anjum, L.; Aslam, M.; Afzal, H. An Overview on Emerging Water Scarcity Challange in Pakistan, Its Consumption, Causes, Impacts and Remedial Measures. *Big Data Water Resour. Eng.* **2020**, *1*, 22–31. [CrossRef]
5. Janjua, S.; Hassan, I.; Muhammad, S.; Ahmed, S.; Ahmed, A. Water management in Pakistan's Indus Basin: Challenges and opportunities. *Water Policy* **2021**, *23*, 1329–1343. [CrossRef]
6. Lau, W.K.; Kim, K.M. The 2010 Pakistan flood and Russian heat wave: Teleconnection of hydrometeorological extremes. *J. Hydrometeorol.* **2012**, *13*, 392–403. [CrossRef]
7. Qureshi, A.S.; McCornick, P.G.; Qadir, M.; Aslam, Z. Managing salinity and waterlogging in the Indus Basin of Pakistan. *Agric. Water Manag.* **2008**, *95*, 1–10. [CrossRef]

8. Syed, A.; Sarwar, G.; Shah, S.H.; Muhammad, S. Soil Salinity Research in 21st Century in Pakistan: Its Impact on Availability of Plant Nutrients, Growth and Yield of Crops. *Commun. Soil Sci. Plant Anal.* **2021**, *52*, 183–200. [CrossRef]
9. Watto, M.A.; Mugera, A.W. Groundwater depletion in the Indus Plains of Pakistan: Imperatives, repercussions and management issues. *Int. J. River Basin Manag.* **2016**, *14*, 447–458. [CrossRef]
10. Arshad, S.; Shafqat, A. Food Security Indicators, Distribution and Techniques for Agriculture Sustainability in Pakistan. *Int. J. Appl. Sci. Technol.* **2012**, *2*, 137–147.
11. WMO Verifies 3rd and 4th Hottest Temperature Recorded on Earth | World Meteorological Organization. 2022. Available online: https://public.wmo.int/en/media/press-release/wmo-verifies-3rd-and-4th-hottest-temperature-recorded-earth (accessed on 1 March 2023).
12. Enhanced POWER Data Access Viewer (Beta). Available online: https://power.larc.nasa.gov/data-access-viewer/ (accessed on 1 March 2023).

environmental sciences proceedings

MDPI

Proceeding Paper

A Comparative Analysis of SMAP-Derived Soil Moisture Modeling by Optimized Machine Learning Methods: A Case Study of the Quebec Province †

Mohammad Zeynoddin [1] and Hossein Bonakdari [2,*]

1 Department of Soils and Agri-Food Engineering, Université Laval, Québec City, QC G1V 0A6, Canada; mohammad.zeynoddin.1@ulaval.ca

2 Department of Civil Engineering, University of Ottawa, Ottawa, ON K1N 6N5, Canada

* Correspondence: hossein.bonakdari@uottawa.ca; Tel.: +1-61356-25800 (ext. 6016)

† Presented at the 7th International Electronic Conference on Water Sciences, 15–30 March 2023; Available online: https://ecws-7.sciforum.net.

Abstract: Many hydrological responses rely on the water content of the soil (WCS). Therefore, in this study, the surface WCS products of the Google Earth Engine Soil Moisture Active Passive (GEE SMAP) were modeled by a support vector machine (SVM), and extreme learning machine (ELM) models optimized by the teacher learning (TLBO) algorithm for Quebec, Canada. The results showed that the ELM model is only able to forecast 23 steps with Correlation Coefficient (R) = 0.8313, Root Mean Square Error (RMSE) = 6.1285, and Mean Absolute Error (MAE) = 5.0021. The SVM model could only estimate the future steps, one step ahead, with R = 0.8406, RMSE = 18.022, and MAE = 17.9941. Both models' accuracy dropped significantly while forecasting longer periods.

Keywords: teacher learner; optimization; ELM; SVM; LSTM; forecast

Citation: Zeynoddin, M.; Bonakdari, H. A Comparative Analysis of SMAP-Derived Soil Moisture Modeling by Optimized Machine Learning Methods: A Case Study of the Quebec Province. *Environ. Sci. Proc.* **2023**, *25*, 37. http://doi.org/10.3390/ECWS-7-14183

Academic Editor: Athanasios Loukas

Published: 14 March 2023

1. Introduction

Numerous hydrological reactions depend on the amount of water in the soil. As soil moisture rises, more runoff is created, resulting in increased sediment movement. This environmental element affects the soil's erosion resistance. Runoff, sediment, and erosion are crucial in hydraulic structure design and watershed studies. The variations in the WCS affect the agriculture section. The sustainable management of agricultural water and land resources depends on this factor. Many environmental parameters, such as soil and surface temperature, the amount of precipitation, and groundwater level, influence this parameter. Hydrological extremes and climate variations intensely impact these parameters, which increases the importance of studying WCS under changing climate conditions. The constraints of measuring and expenditure limitation cause this parameter not to be accessible at high spatio-temporal resolutions everywhere, particularly in vast areas like Quebec. Therefore, a strategy should be considered for collecting and modeling this useful parameter in data-scarce locations. This research will use SMAP products to model and forecast the WCS.

Accordingly, Google Earth Engine (GEE) cloud datasets will be used. Using this platform provides the possibility of obtaining curated datasets worldwide. This platform uses high-efficiency computing resources and cloud-based calculations to process planetary-scale data more efficiently. It also allows users to share their products and analysis in the form of an application (app) [1]. One of these valuable apps is SOILPARAM, developed by [2]. This app provides historical records of some soil parameters in the form of a time series.

Using machine learning (ML) methods in modeling and forecasting hydrological data analysis is common. The regression support vector machine (SVM) and extreme learning machine (ELM) models are two of many artificial intelligence (AI) methods that have proven their potential power in modeling natural phenomena. The inherent intense seasonality and stochastic patterns in the WCS make these modeling techniques suitable for forecasting and extracting patterns from the datasets. Both models are considerably fast and structurally simple when compared to other AI methods. They can be used for generating real-time results. ELM is a single-layer feed-forward network model known for its simple structure, fast computational process, and accuracy in forecasting non-linear, highly seasonal datasets [3]. The ELM's accuracy in forecasting rainfall [3], flows in rivers [4], sediment transport [5], etc. has been proven. The authors of [6] used the ELM model and its integration with ensemble empirical mode decomposition to forecast the WCS in the upper layer of soil and compared it with a random forest. The model outcomes showed that ELM outperformed the random forest, and its hybridization increased the accuracy. Likewise, the SVM model has been used widely in modeling datasets because of its simplicity, and derivable equations. For instance, ref. [7] used SVM to forecast the WCS, five steps ahead by feeding the climatic factors as inputs to the model. They reported a good performance for the SVM model as a result of using six meteorological inputs and the first lag of WCS at 0.05 and 0.1 m.

The advantages of these two methods were addressed briefly. However, similar to other AI methods, they suffer from input selection, model parameters tuning, and kernel selection. Since the SVM model is a linear method, it may produce naïve results in intense non-linear data. Optimizing them, using the teacher-learning-based optimization (TLBO) algorithm [8] will reduce the tuning and input selection problems and helps find a better solution. The major advantage of the TLBO is that it has significantly fewer controlling parameters than its equivalents and is readily applied to different models. This study consists of sequence research on the GEE SMAP WCS product completed by [8]. In that study, they used a deep learning long short-term memory (LSTM) model and used the WCS as the sole input of the model with optimization and structural investigation approaches. The outputs of that study showed the potential power of LSTM in forecasting WCS in a dynamic and long-term manner. Therefore, this study investigates whether the introduced models can produce similar results. The TLBO optimization similarly will be used and different lags of WCS as inputs will be checked to obtain the models' capacity. Lastly, the length of their accurate forecast horizon will be determined.

2. Model Descriptions

2.1. Support Vector Machine

This approach is praised for being generalizable, powerful, and precise. Support Vector Machine (SVM) uses statistical theories and risk minimization structural concepts. In this method, a decision function is created to boost model generalization and reduce modeling errors by employing a deep dimensional space called feature space (FS) and therefore optimizing margin border separation [9,10]. This strategy works with datasets containing few samples. The SVM framework is based on the non-linear mapping of input space into a high-dimensional domain for identifying a hyperplane. It minimizes generalization errors [11].

If the target values would be WCS_i $(i = 1{:}l)$ as $\{(L_1,WCS_1), \ldots ,(L_i,WCS_i)\}$ and L_i as the lag inputs, in a training set with i samples, the $F_l(x)$ as a linear function for training the network can be defined as follows:

$$F_l(x) = \sum_{i=1}^{S} (\theta_i - \theta_i^*)(L_i.\, L) + B \quad (1)$$

where θ_i, θ_i^* the slack variables, $\beta_i \in R^N$ is the weights matrix and B equals to bias. The maximum margin size is obtained by calculation of the Euclidean norm of weights. To estimate weights (β), compute the objective function as:

$$Min.:M_P = \frac{1}{2}\|\beta\|^2 + C\sum_{i=1}^{N}(\theta_i+\theta_i^*)\text{Subjected to}:\begin{cases}\forall i: WCS_i - (\beta_i L_i + B) \le \varepsilon + \theta \& \forall i: (\beta_i L_i + B) - WCS_i \le \varepsilon + \theta^* \\ \forall i: \theta_i \ge 0 \& \forall i: \theta_i^* \ge 0\end{cases} \quad (2)$$

C denotes the penalty constant. The F_l function approximates the training points with an ε error and then generalizes it. $L_1.L$ is the input variables' dot products. To avoid performing dot multiplication on transformed data samples, a kernel function is written to replace each occurrence of it.

2.2. Extreme Learning Machine

The extreme learning machine (ELM) is a development of feed-forward neural networks that tries to solve the problem of time-consuming training and local minima trapping. Trapping results in reducing the generalizability and customizability of model parameters [12]. Accordingly, input weights and neuron bias are set stochastically, and output weights are computed by solving a linear equation as follows:

$$\sum_{j=1}^{k} W_j^O\ AF_j\ (x_i) = \sum_{j=1}^{k} W_j^O\ AF(W_j^I \cdot L_i + B_j) = WCS_i, \qquad j = 1,\ldots,z \quad (3)$$

where W^I and W^O are the input and output weights, and AF is the activation functions. L_i is the input variable, and z is the number of samples in each input variable. The iterative technique outlined by [13] is used in the ELM model to regulate the random selection of input weights and bias neurons, and increase generalizability. A total of 1000 iterations are set to find the best weights. Extra iterations did not influence model errors.

3. Evaluation Criteria

This study uses the conventional Coefficient of Determination (R), Root Mean Square Error (RMSE), and Mean Absolute Error (MAE) to evaluate and compare the models.

4. Study Region and Dataset Description

The study point is in the south of Quebec City, Canada, with a latitude of 46.73 N and a longitude of 71.5 W. The region comprises the Jacques-Cartier South, Chaudière, and Sainte-Anne rivers. The WCS data was downloaded from the National Aeronautics and Space Administration (NASA) Enhanced SMAP Global Soil Moisture Dataset uploaded in the GEE environment by NASA [14]. The dataset time range is from 2015 to July 2022, with a 3-day measurement interval. This dataset was averaged weekly to obtain a total of 306 data points. To train and evaluate the model, considering the size of the dataset, it was partitioned by a 70:30 ratio. The first partition, which contains 70% of the time series data points, was used to train the networks and find the optimum weights, while the remaining 30% of the dataset was used to evaluate the model forecasts and estimated weights. The statistical features are presented in Table 1. The dataset's download link is presented in the "Data Availability Statement" section.

Table 1. The dataset's characteristics.

Data	Nbr.	Min.	Max.	1st Q.	Median	3rd Q.	Mean
Train	306	4.725	25.400	20.114	24.122	25.062	21.711
Test	77	4.315	25.387	12.294	21.770	24.717	18.645
Total	383	4.315	25.400	19.285	23.901	25.023	21.095

Nbr., Number of data, Min. and Max., Minimum and Maximum of data, 1st Q. and 3rd Q., first and third Quarters.

5. Data Investigation and Model Tunning

The range for optimization and input definition is considered as [1 lag, 7 lags] based on the ACF results (Figure 1). The range for the ELM hidden neuron size parameter is [1, 34] with 1000 iterations. The ranges for the SVM model are also: C and $\sigma \in [0.01, 2000]$, $\varepsilon \in [0.001, 1]$. The TLBO parameters are population = 20 and maximum iteration = 100.

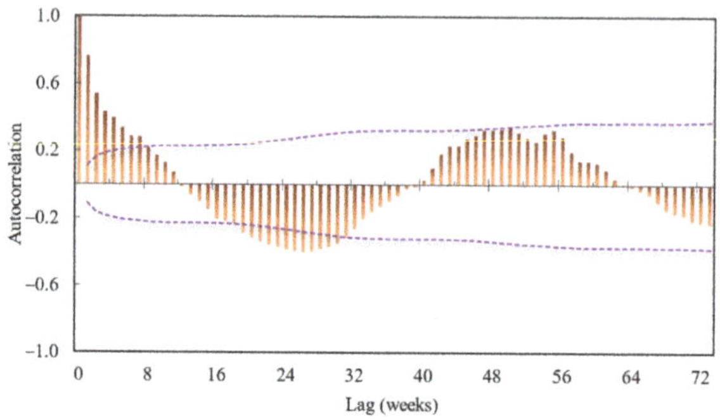

Figure 1. The autocorrelation function of datapoints for 1/4 of train data.

6. Model Results

A core i7 processor, with 16 Gigabytes of Random Access Memory (RAM), performed the modeling and the runtime for the ELM optimization was approximately 8 h. This time for the SVM model was 0.5 h, and, in both models, the optimum values were obtained in early iterations, specifically the SVM model (Figure 2a,b). After modeling, the optimum results were obtained by all seven inputs for both models and the maximum hidden neuron size for ELM. The optimum results of TLBO-ML integrations are presented in Table 2. The overall performance of both SVM and ELM models in the long-term forecast was very poor, and both methods generated very naïve results so that the most accurate outcome was obtained by ELM with R = 0.3654, RMSE = 17.9146, and MAE = 17.8131. The forecast process was performed based on the addition of each estimated step to the historical data, creating input lags and approximating the future step by the previous one. Therefore, both ELM and SVM forecasted the 77-point test period, and the long-term forecast was defined accordingly. This approach to forecasting failed, and it was found that both models' forecasting accuracy is limited to less than 77 steps (Figure 3a,b).

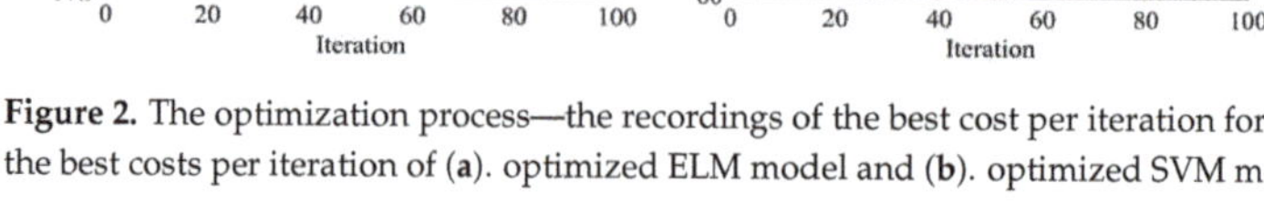

Figure 2. The optimization process—the recordings of the best cost per iteration for ELM and SVM; the best costs per iteration of (**a**). optimized ELM model and (**b**). optimized SVM model.

Table 2. The models' evaluation results for the test period.

Model	R	RMSE (mm)	MAE (mm)
Opt [1]-ELM (Static)	0.3654	17.9146	17.8131
Opt-SVM (Static)	0.2954	60.8881	0.5993
Opt-ELM (23-Steps)	0.8313	6.1285	5.0021
Opt-SVM (Dynamic)	0.8406	18.022	17.9941

[1] Opt: Optimized by TLBO.

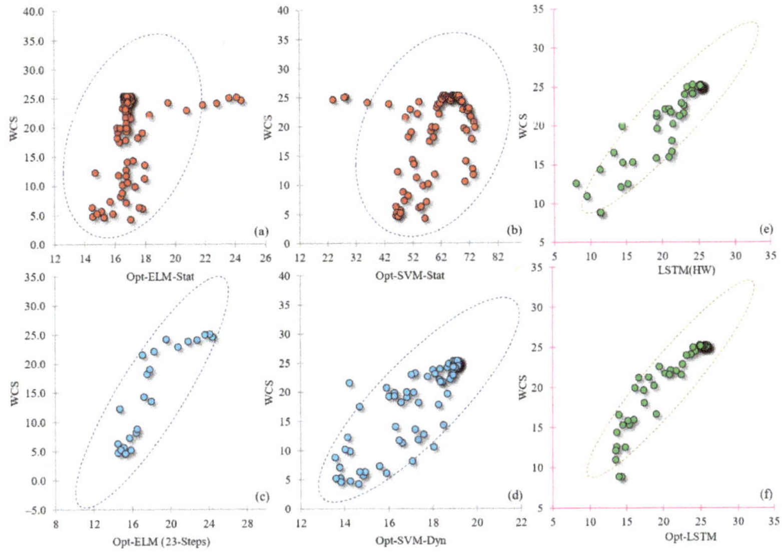

Figure 3. The scatter plots of forecasted data points vs. observed WCS based on duration, stat: long-term-static, 77-step forecast, Dyn: 1-step forecast, Opt: Optimized. (**a**). static forecast of Opt.ELM, (**b**). static forecast of Opt.SVM, (**c**). short-term forecast of Opt.ELM, (**d**). dynamic forecast of Opt.SVM, (**e**). LSTM with HW preprocess, (**f**). forecast of Opt.LSTM, all vs. WCS [8].

By doing more research and defining the different forecasting steps in the modeling process, it was found that the ELM model can predict WCS values up to 23 steps into the future, with the correlation going up by 138%, the RMSE index going down by 65%, and the MAE index going down by 71% (Figure 3c,d). The SVM model's forecasting accuracy is also limited to one step in the future, and considering the severe fluctuation in the dataset, this linear model is not able to forecast more than one step in the future. Nevertheless, the ELM (23-step) model was more successful in short-term forecasting than the SVM. In Figure 3c,d, it can be seen that the majority of the points are located in the 95% confidence intervals and estimations are closer to the linear form than the long-term forecasts.

Ref. [8] undertook a study on the same products of the GEE SMAP by an LSTM model. In that study, they used two approaches for the long-term forecasts of the WCS dataset. The results of both approaches are presented in Figure 3e,f. The LSTM model was more successful in estimating values and patterns than the long-term forecasts of the SVM and ELM. The best results of the LSTM in a 50-step, long-term forecast, were: R = 0.9220, RMSE = 1.9614, MAE = 1.2837 by the Holt–Winters (HW) preprocessing method, and by TLBO optimization it estimated the WCS values by R = 0.9337, RMSE = 1.7809, MAE = 1.1892, which is considerably more accurate than this study's ML methods, even in the 23-step ELM and dynamic SVM forecasts. In conclusion, the ELM model is more capable of estimating the WCS values and fluctuation than the SVM, but it is limited to 23 steps, which is almost half of the dataset's period. In other words, it can forecast up to half of the periodic patterns. However, using sole models without the methodology suggested in [8] cannot produce very accurate results. It is suggested that ELM or SVM integrate preprocessing techniques, such as advanced smoothing methods, or other seasonal methods in seasonal data, such as WCS, to reduce the fluctuations in the dataset's structure, even if the periodic ACF pattern is not significant.

7. Conclusions

In this study, the surface soil moisture products of the GEE SMAP were modeled by SVM and ELM. The TLBO algorithm optimized these models to estimate future steps based on the forecast of each step. The results showed that the ELM model is only able to forecast 23 steps each time with R = 0.8313, RMSE = 6.1285, and MAE = 5.0021. The SVM model was only able to estimate the future steps one step ahead with R = 0.8406, RMSE = 18.022, and MAE = 17.9941. Both models' accuracy dropped significantly while forecasting longer periods than the ones mentioned. Since this study is a sequence to a former study on the same product of SMAP by TLBO-LSTM, a comparison between the results was made. Accordingly, the proposed deep learning LSTM method in the former study is more successful in forecasting longer periods than ELM and SVM, with R = 0.9337, RMSE = 1.7809, and MAE = 1.1892. We suggest that advanced smoothing methods should be integrated, or other seasonal preprocessing techniques, to decrease both fluctuations and correlations in the time series structure.

Author Contributions: Conceptualization, H.B.; methodology, H.B. and M.Z.; software, M.Z.; validation, H.B. and M.Z.; formal analysis, M.Z.; investigation, M.Z.; resources, H.B.; data curation, M.Z.; writing—original draft preparation, M.Z.; writing—review and editing, H.B.; visualization, M.Z.; supervision, H.B.; project administration, H.B.; funding acquisition, H.B. and M.Z. All authors have read and agreed to the published version of the manuscript.

Funding: The authors acknowledge the financial support provided by Fonds de recherche du Québec—Nature et technologies (FRQNT) (#316369) and Natural Sciences and Engineering Research Council of Canada (NCERT) Discovery Grant (#RGPIN-2020-04583) to perform the current research.

Institutional Review Board Statement: Not applicable.

Informed Consent Statement: Not applicable.

Data Availability Statement: The readers can find the dataset by the following GEE app [SOIL-PARAM] developed by [2]: Link to app: https://zemoh.users.earthengine.app/view/soilparam (accessed on 24 December 2022).

Conflicts of Interest: The authors declare no conflict of interest.

References

1. Gorelick, N.; Hancher, M.; Dixon, M.; Ilyushchenko, S.; Thau, D.; Moore, R. Google Earth Engine: Planetary-scale geospatial analysis for everyone. *Remote Sens. Environ.* **2017**, *202*, 18–27. [CrossRef]
2. Zeynoddin, M.; Bonakdari, H.; Gumiere, S.J.; Caron, J.; Rousseau, A.N. SOILPARAM 1.0: A Global-Scaled Enhanced Remote Sensing Application for Soil Characteristics Data Retrieval—Google Engine Environment, An Open-Source Treasure. In Proceedings of the IAHR World Congress From Snow to Sea, Granada, Spain, 18–23 June 2022; Ortega-Sánchez, M., Ed.; International Association for Hydro-Environment Engineering and Research (IAHR): Granada, Spain, 2022; pp. 5309–5319, ISBN 978-90-832612-1-8.
3. Zeynoddin, M.; Bonakdari, H.; Azari, A.; Ebtehaj, I.; Gharabaghi, B.; Madavar, H.R. Novel hybrid linear stochastic with non-linear extreme learning machine methods for forecasting monthly rainfall a tropical climate. *J. Environ. Manag.* **2018**, *222*, 190–206. [CrossRef] [PubMed]
4. Deo, R.C.; Şahin, M. An extreme learning machine model for the simulation of monthly mean streamflow water level in eastern Queensland. *Environ. Monit. Assess.* **2016**, *188*, 90. [CrossRef] [PubMed]
5. Bonakdari, H.; Ebtehaj, I. A comparative study of extreme learning machines and support vector machines in prediction of sediment transport in open channels. *Int. J. Eng.* **2016**, *29*, 1499–1506.
6. Prasad, R.; Deo, R.C.; Li, Y.; Maraseni, T. Soil moisture forecasting by a hybrid machine learning technique: ELM integrated with ensemble empirical mode decomposition. *Geoderma* **2018**, *330*, 136–161. [CrossRef]
7. Khalil, A.; Gill, M.K.; McKee, M. New applications for information fusion and soil moisture forecasting. In Proceedings of the 2005 7th International Conference on Information Fusion, Philadelphia, PA, USA, 24–27 July 2005; IEEE: Piscataway, NJ, USA, 2005; p. 7, ISBN 0-7803-9286-8.
8. Zeynoddin, M.; Bonakdari, H. Structural-optimized sequential deep learning methods for surface soil moisture forecasting, case study Quebec, Canada. *Neural. Comput. Applic.* **2022**, *34*, 19895–19921. [CrossRef]
9. Sharafi, H.; Ebtehaj, I.; Bonakdari, H.; Zaji, A.H. Design of a support vector machine with different kernel functions to predict scour depth around bridge piers. *Nat. Hazards* **2016**, *84*, 2145–2162. [CrossRef]
10. Azimi, H.; Bonakdari, H.; Ebtehaj, I. Design of radial basis function-based support vector regression in predicting the discharge coefficient of a side weir in a trapezoidal channel. *Appl. Water Sci.* **2019**, *9*, 78. [CrossRef]
11. Yapıcı, E.; Akgün, H.; Özkan, K.; Günkaya, Z.; Özkan, A.; Banar, M. Prediction of gas product yield from packaging waste pyrolysis: Support vector and Gaussian process regression models. *Int. J. Environ. Sci. Technol.* **2022**, *20*, 461–476. [CrossRef]

12. Bonakdari, H.; Qasem, S.N.; Ebtehaj, I.; Zaji, A.H.; Gharabaghi, B.; Moazamnia, M. An expert system for predicting the velocity field in narrow open channel flows using self-adaptive extreme learning machines. *Measurement* **2020**, *151*, 107202. [CrossRef]
13. Ebtehaj, I.; Soltani, K.; Amiri, A.; Faramarzi, M.; Madramootoo, C.A.; Bonakdari, H. Prognostication of Shortwave Radiation Using an Improved No-Tuned Fast Machine Learning. *Sustainability* **2021**, *13*, 8009. [CrossRef]
14. Sazib, N.; Mladenova, I.; Bolten, J. Leveraging the Google Earth Engine for Drought Assessment Using Global Soil Moisture Data. *Remote Sens.* **2018**, *10*, 1265. [CrossRef]

environmental sciences proceedings

MDPI

Proceeding Paper

Forecasting of Banana Crop Productivity Using Geospatial Approach: A Case Study of Anand District †

Usha Pandya [1,*], Ashwini Mudaliar [2] and Amol Gaikwad [3]

1 Department of Botany, Government College, Daman 396210, India
2 Department of Botany, Faculty of Science, The Maharaja Sayajirao University of Baroda, Vadodara 390002, India; ashwini144@gmail.com
3 Department of Mathematics, Government College, Daman 396210, India; amolg029@gmail.com
* Correspondence: ushajosh12@gmail.com
† Presented at the 7th International Electronic Conference on Water Sciences, 15–30 March 2023; Available online: https://ecws-7.sciforum.net.

Abstract: The banana is one of the main fruit crops in the world as it has gained importance in the global market for many industries due to its high source of nutrients and its fibre content. Owing to climate change and irregular precipitation, the yield of banana crops is becoming very unpredictable and, thus, there is a need to understand the impact of climatic parameters on the yield. Mathematical models are crucial for strategic and forecasting applications; however, models related to the banana crop are less common, and reviews on previous modelling efforts are scarce, emphasizing the need for evidence-based studies on this topic. This study employs the geospatial approach to establish a relationship between climatic variables and the banana crop productivity of the Anand district of Gujarat, India. Sentinel data was utilized to derive various indices like Normalised Difference Vegetation Index (NDVI), Leaf Area Index (LAI), Enhanced Vegetation Index (EVI), and Normalised Difference Water Index (NDWI). Land Surface Temperature (LST) was also derived using a Landsat dataset. Evapotranspiration (ET) data was also considered while understanding the impact of these parameters on yield. Values were extracted based on the ground control points (GCP) of different agricultural fields of the study area. Derived data was analysed using different statistical tools to understand the relationship between different indices and the productivity of the banana crop. Results indicated that the banana yield is highly dependent on water availability and the ET of the study area, proving that these parameters can be utilized for generating predicting models of banana yield.

Keywords: banana; sentinel data; Landsat; EVI; NDWI; evapotranspiration

Citation: Pandya, U.; Mudaliar, A.; Gaikwad, A. Forecasting of Banana Crop Productivity Using Geospatial Approach: A Case Study of Anand District. *Environ. Sci. Proc.* **2023**, *25*, 38. https://doi.org/10.3390/ECWS-7-14248

Academic Editor: Athanasios Loukas

Published: 16 March 2023

1. Introduction

India is a diverse country with a wide variety of agricultural products owing to its different climatic conditions. Banana ranks second amongst fruit consumption in India after mango due to its low price and high nutritional value [1]. The area under banana cultivation has increased up to 880 thousand hectares, with a production of approximately 35 million metric tons in 2022 [2]. The leading producers of bananas are the Tamil Nadu, Maharashtra, and Gujarat states of India, indicating their important contribution in banana crop production. Anand is one of the important banana producing districts of Gujarat and contributes significantly to the total banana production of the state. However, the latest data of the year 2022 shows that farmers are turning to other fruit crops because the banana is a water-intensive crop and water levels across the state are declining. In addition, the overall yield of the banana is decreasing due to climate change and irregular precipitation, and it is getting every day. In contrast to this, the demand for the banana is increasing gradually due to its high nutritional value and its affordability.

Accurate predictions of crop production are critical for developing effective strategies at the farm level [3]. Prediction of banana crop yield, therefore, has become imperative given the fact that its importance is increasing so incredibly. The availability of different types of spatial data makes it easy to carry out geospatial analysis and understand the impact of different ecological factors on the banana crop. This study utilized various parameters like Normalized Difference Water Index (NDWI), Land Surface Temperature (LST), Evapotranspiration (ET), Normalized Difference Vegetation Index (NDVI), Enhanced Vegetation Index (EVI), and Leaf Area Index (LAI) to forecast the banana yield of the Anand district. In this study, LAI is considered as a proxy for the banana yield given that high LAI indicates healthy crop and high production. Modelling carried out using different techniques aids our understanding of the impact of various parameters on crop yield and helps to generate accurate models for the prediction of the yield. Among the different models, a mathematical model provides a description of the behaviour of real-world systems in mathematical concepts, terms, and languages such as equations, inequalities, functions, variables, and constraints [4]. Moreover, these models allow us to make crop predictions under specific environmental conditions [5].

Therefore, this study attempts to utilize a geospatial approach in order to forecast the banana crop yield of the Anand district of Gujarat using mathematical modeling.

2. Study Area

Anand is primarily an agricultural district of the Gujarat State of India (Figure 1). As per climatological data of the nearest Indian Meteorological Department (IMD) Station at Ahmedabad, the summers are generally hot and the winters are cool. Mean maximum temperature ranges between 28.4 °C during January and about 41.8 °C during May, and the mean minimum temperatures vary between 11.7 °C during January and 27 °C during June. Long-term average annual rainfall for the Ahmedabad IMD station is 799.6 mm. Most of the rainfall is received during the south-west monsoon between June and September [6]. The banana is one of the key crops grown in the district in terms of its value. The planting season of the banana varies between mid-February and the first week of March. More than 10,000 Ha land is under banana cultivation, and since it is an annual crop and needs more investment, even the scale of finance for banana cultivation is reasonably high.

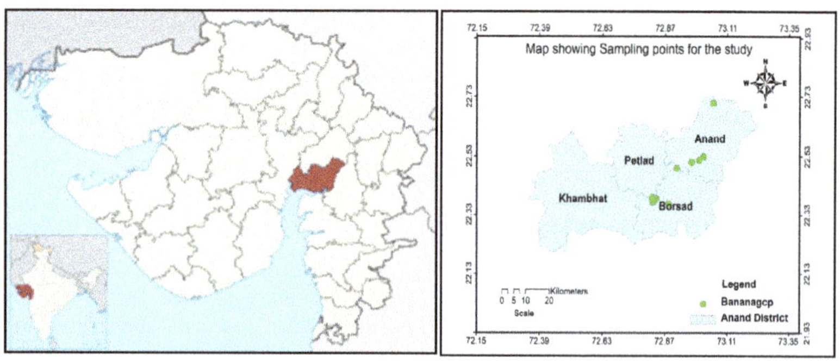

Figure 1. Map of Study area.

3. Methodology and Material

The banana crop yield forecasting needs different parameters to be assessed to understand their impact on yield. A field study was carried out in the Anand district to locate banana fields, and Ground Control Points (GCP) were collected. A total of 11 different fields were identified, which covered the different parts of the district. The duration of the study was from January 2018 to December 2021 for all of the parameters. Sentinel 2 data was utilized to derive various indices like NDVI, EVI, and NDWI. LAI was derived from EVI using the following formula:

$$LAI = (3.618 \times EVI - 0.118)$$

LST was also derived using a Landsat dataset after processing using ERDAS 9.1. ET data was also extracted from MODIS satellite data. Sentinel 2 data has a repetition of every 8 days, whereas MODIS satellite data has the repetition of every 15 days. Data for both the satellites collected from January 2018 to December 2021 were averaged out monthly, and values were extracted corresponding to GCPs using ENVI 5.1 software. Various statistical methods like regression, t-test, etc. were utilized to understand the impact of different parameters on the banana yield and to forecast that yield.

4. Result and Discussion

Agricultural production is significantly affected by environmental factors. Weather influences crop growth and development, causing large intra-seasonal yield variability. The Anand district, which is located in central Gujarat, shows significant variations in various environmental parameters. Application of geospatial approach helps in easy derivation of these parameters for a long timeseries of more than 10 years depending upon the requirement [7]. Various parameters considered for the study showed a significant seasonal variation during the study period. LST varied from 8 °C in winter to 43 °C in summer, indicating the year round variations in temperature (Figure 2). NDWI was found to vary from −1.41 to −0.28, indicating the variation in the water regime of the agricultural fields (Figure 3). Low NDWI values were observed from January to June—months corresponding to the dry season of the year. EVI varied from 0.09 to 0.82 during the study period (Figure 4). Correspondingly, LAI of the banana crop varied from 0.19 to 2.85, indicating the vigour of the crop (Figure 5). EVI and LAI were found to be high during the months from February to July, months of the year representing the plantation period of the banana crop. NDVI values were found to be vary from 0.14 to 0.58, wherein high values indicated the growing season of banana crop (Figure 6). ET varied from 0.1 to 2.41 mm day^{-1}, revealing its seasonal variation (Figure 7). It showed higher values during monsoon season, indicating its role in crop growth.

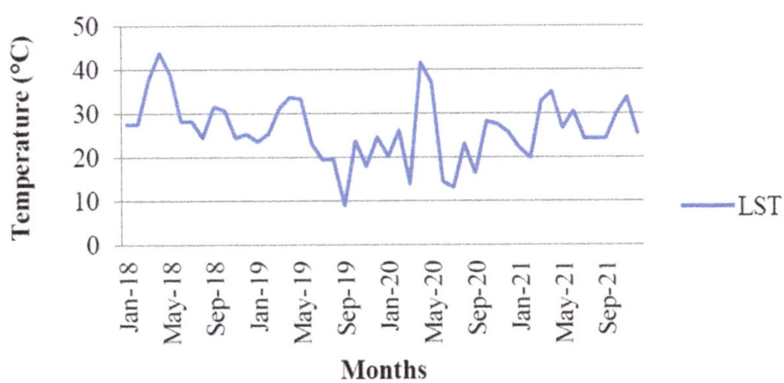

Figure 2. Graph showing variation in LST during the study period.

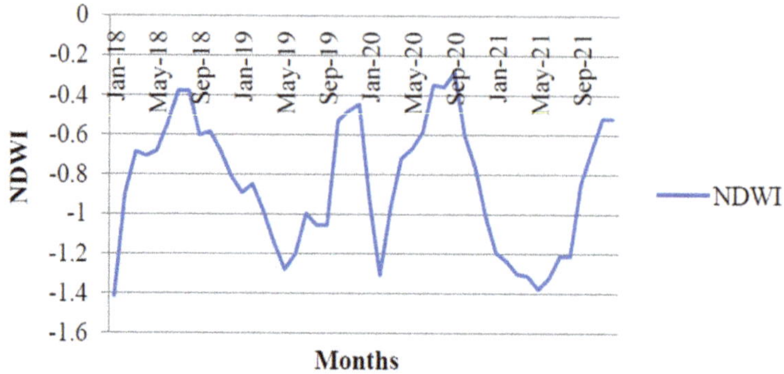

Figure 3. Graph showing variation in NDWI during the study period.

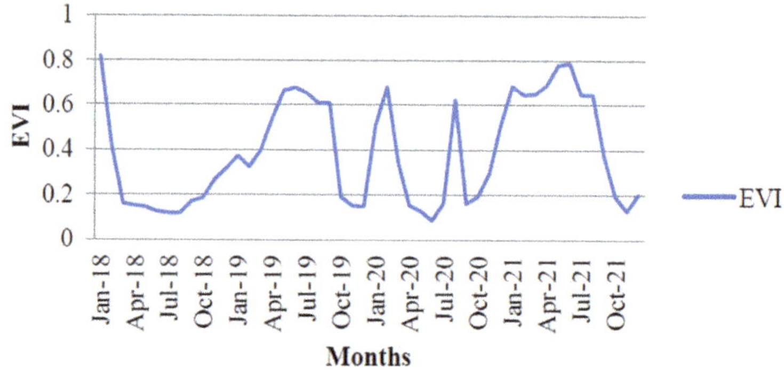

Figure 4. Graph showing variation in EVI during the study period.

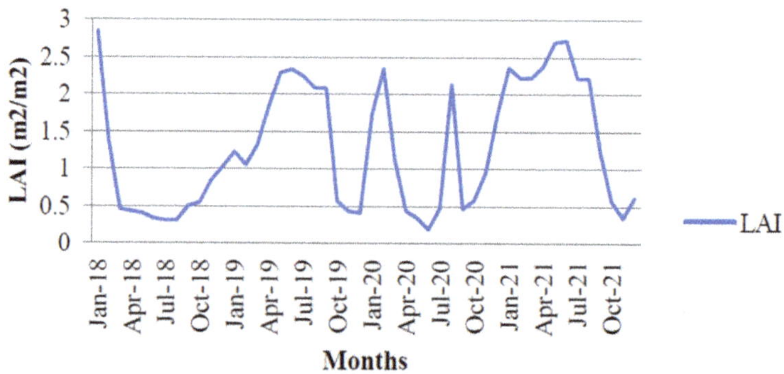

Figure 5. Graph showing variation in LAI during the study period.

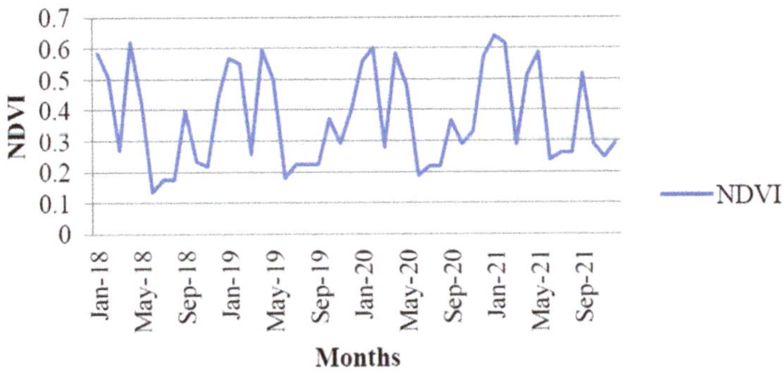

Figure 6. Graph showing variation in NDVI during the study period.

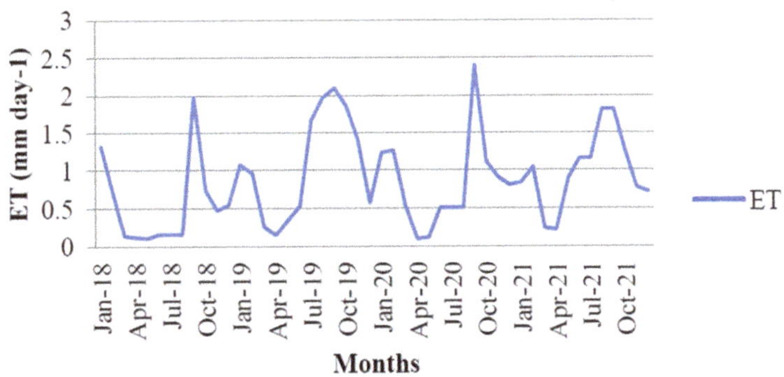

Figure 7. Graph showing variation in ET during the study period.

Values derived from these parameters were analysed statistically to understand the relationship between different parameters. The regression analysis between different parameters revealed an interesting result. It was clear from the analysis that since R^2 was only 0.02, LAI was, therefore, not dependent on LST, indicating that the temperature variations were having less impact on crop growth. The ET and LAI were also poorly related with each other, with an R^2 value as low as 0.01. However, the crop was dependent on NDVI, as revealed by an R^2 value of 0.64 when regressed with LAI. EVI also played a major role in crop growth as it was the main parameter with LAI. Regression analysis between NDWI and LAI showed that these two variables were highly dependent on each other, with an R^2 value as high as 0.76. This indicated that the growth of the banana crop was highly influenced by NDWI, and that the yield of the crop varied with varying values of NDWI. These results were also supported by the T-test.

Therefore, the regression model derived using these two parameters was found to be significant for forecasting the banana yield (Figure 8).

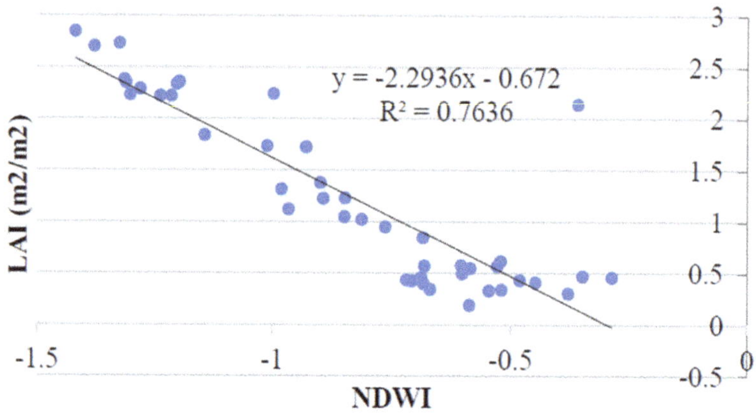

Figure 8. Chart showing regression equation for predicting Banana yield.

5. Conclusions

The banana is one of the most important fruit crops of India and at a global level. Increased water stress due to unpredicted rainfall and competition for water resources is leading to a reduced yield of the crop. This study attempted to generate banana crop productivity based on the various parameters derived using a geospatial approach. Regression analysis revealed that the banana crop is highly dependent on NDWI, i.e., on the ground water condition of the agricultural field, as indicated by high R^2 values. The regression model derived using these parameters can be utilized to predict banana yield.

Author Contributions: Conceptualization, U.P. and A.M.; methodology, U.P.; software, A.M.; validation, U.P, A.M. and A.G.; formal analysis, A.G.; resources, A.M., writing—original draft preparation, U.P.; writing—review and editing, U.P., A.M. and A.G. All authors have read and agreed to the published version of the manuscript.

Funding: This research received no external funding.

Institutional Review Board Statement: Not applicable.

Informed Consent Statement: Not applicable.

Data Availability Statement: This study has been used Sentinel-2 multispectral imagery that is publicly available at https://scihub.copernicus.eu/ (accessed on 11 January 2023) and Landsat-8 multispectral imagery that is publicly available at http://earthexplorer.usgs.gov (accessed on 11 January 2023).

Conflicts of Interest: The authors declare no conflict of interest.

References

1. Dave, A.K.; Zala, Y.C.; Pundir, R.S. Comparative economics of Banana cultivation in Anand district of Gujarat. *Econ. Aff.* **2016**, *61*, 305–312. [CrossRef]
2. Anonymous. 2022. Available online: https://www.statista.com/statistics/1038905/india-production-of-banana/#:~:text=During%20fiscal%20year%202022%2C%20the,during%20the%20measured%20time%20period (accessed on 9 January 2023).
3. Olivares, B.O.; Vega, A.; Rueda Calderón, M.A.; Montenegro-Gracia, E.; Araya-Almán, M.; Marys, E. Prediction of Banana Production Using Epidemiological Parameters of Black Sigatoka: An Application with Random Forest. *Sustainability* **2022**, *14*, 14123. [CrossRef]
4. Chaturvedi, D.K. *Modeling and Simulation of Systems Using MATLAB and Simulink*; CRC Press: Boca Raton, FL, USA, 2017.
5. Medina-Ruíz, C.; Mercado-Luna, I.; Soto-Zarazúa, G.; Torres-Pacheco, I.; Rico-García, E. Mathematical modeling on tomato plants: A review. *Afr. J. Agric. Res.* **2011**, *6*, 6745–6749. [CrossRef]

6. Yadav, S.S. *District Groundwater Brochure Anand District Gujarat*; Government of India Ministry of Water Resources, Central Ground Water Board West Central Region: Ahmedabad, India, 2013.
7. Zhang, H.; Zhang, Y.; Li, X.; Li, M.; Tian, Z. Predicting Banana Yield at the Field Scale by Combining Sentinel-2 Time Series Data And Regression Models. *Appl. Eng. Agric.* **2023**, *39*, 81–94. [CrossRef]

environmental sciences proceedings

MDPI

Proceeding Paper

A Monthly Water Balance Model for Assessing Streamflow Uncertainty in Hydrologic Studies †

Lampros Vasiliades * and Ioannis Mastraftsis

Department of Civil Engineering, School of Engineering, University of Thessaly, Pedion Areos, 38334 Volos, Greece; imastraftsis@uth.gr

* Correspondence: lvassil@civ.uth.gr; Tel.: +30-242-107-4115

† Presented at the 7th International Electronic Conference on Water Sciences, 15–30 March 2023. Available online: https://ecws-7.sciforum.net.

Abstract: The accurate assessment of streamflow is crucial for operational water resource management projects. The aim of this study was to estimate the uncertainties in the surface runoff simulated by a monthly water balance model in a mountainous watershed of the Portaikos river, a tributary of the Pinios river, Thessaly, Greece. The University of Thessaly (UTHBAL) monthly water balance model was developed in the R statistical computing environment language, named 'R-UTHBAL', to estimate surface water balance in data-scarce watersheds. Two sources of uncertainties in hydrological modelling were considered: the uncertainties in input data estimation and in model parameters. The uncertainties were estimated with the use of the R-package '*hydroPSO*', a global Particle Swarm Optimisation (PSO) algorithm for the calibration of environmental models. The R-UTHBAL was integrated with the *hydroPSO* algorithm and advanced sensitivity analyses, and user-friendly evaluation plots were estimated to facilitate the interpretation and assessment of the calibration results. Application of R-UTHBAL with the *hydroPSO* showed that the uncertainty in streamflow estimation should always be accounted for and evaluated in operational water resources management projects.

Keywords: water balance model; UTHBAL; *hydroPSO*; optimisation; sensitivity analysis; uncertainty analysis

check for updates

Citation: Vasiliades, L.; Mastraftsis, I. A Monthly Water Balance Model for Assessing Streamflow Uncertainty in Hydrologic Studies. *Environ. Sci. Proc.* **2023**, *25*, 39. https://doi.org/10.3390/ECWS-7-14192

Academic Editor: Silvia Kohnova

Published: 14 March 2023

1. Introduction

Conceptual rainfall–runoff models are frequently used to estimate the runoff generation mechanisms and the water balance components at various temporal and spatial scales. A satisfactory match between the observed and simulated outputs is often achieved by calibrating the model parameters. However, the results are quite uncertain due to aleatory and epistemic uncertainty. Inaccuracies in the input data (such as precipitation and temperature), the calibration data (such as streamflow), the model parameters, and the mathematical model structure are the four main causes of epistemic uncertainty in hydrological modelling [1]. While the latter two are more model-specific, the first two error sources are influenced by the quality of the data. Hence, assessment of the uncertainties is crucial in hydrological studies, water resource management, climate change assessment, and estimations of water balance in ungauged watersheds [2].

The aim of this study is to estimate the uncertainties in the surface runoff simulated by a monthly water balance model in a mountainous watershed of the Portaikos river, a tributary of the Pinios river, Thessaly, Greece. The University of Thessaly (UTHBAL) monthly water balance model [3,4] is developed in the R statistical computing environment language, named 'R-UTHBAL', to estimate surface water balance in data-scarce watersheds. Two sources of uncertainties in hydrological modelling are considered: the uncertainties in input data estimation and in model parameters. The uncertainties are estimated with the use of the R-package '*hydroPSO*', a global Particle Swarm Optimisation (PSO) algorithm [5] for the calibration of environmental models. The R-UTHBAL is integrated with the *hydroPSO* algorithm and advanced sensitivity analyses, and user-friendly evaluation plots are estimated to facilitate the interpretation and assessment of the calibration results.

2. Materials and Methods

2.1. The R-UTHBAL Model

The monthly water balance model UTHBAL [3] was selected to be redesigned in the R-Environment because it has been applied successfully to simulate surface runoff and groundwater recharge in many studies [6,7]. Monthly time series of precipitation, mean temperature, and potential evapotranspiration are used as inputs by UTHBAL. The snowpack and snowmelt are calculated from total precipitation which is divided into rainfall and snowfall. Using a soil moisture mechanism, the model splits the entire watershed runoff into three components: the surface runoff, the interflow, and the baseflow. The model's first objective is to accurately fulfil actual evapotranspiration. Watershed runoff, actual evapotranspiration, groundwater recharge, and soil moisture are the model's outputs. Mathematical details can be found in a recent study [3].

Figure 1 presents the R-UTHBAL model with the flow diagram and the R-Environment framework using the *hydroPSO* algorithm. Six model parameters should be estimated and are usually determined during calibration based on monthly streamflow data. The model parameters are: the *CN* (Curve Number) of the US Soil Conservation Service, the C_m parameter of monthly melt rate factor, the coefficient α of actual evapotranspiration (*aAET*), the coefficient K of groundwater recharge, the coefficients β and γ of interflow (*CONMR*) and baseflow (*CONGROUND*), respectively (Figure 1a). Previous applications of the UTHBAL model showed that the model parameters are independent, well-defined and simulation streamflow errors are normally distributed [4]. Several command functions were carried out in the R environment for the operation of the models, input and data preparation and transformation, selection of parameter space, calibration method, optimisation function, and statistical analyses (Figure 1b).

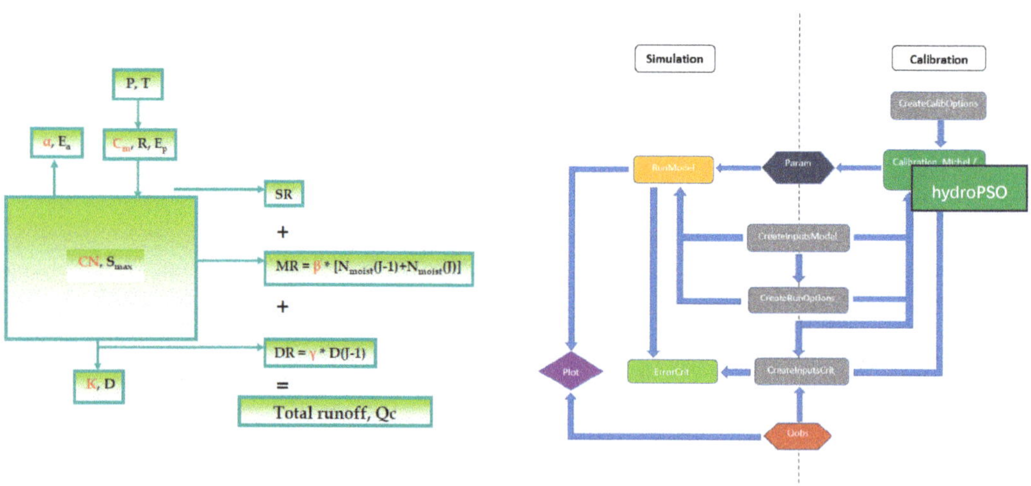

(a) (b)

Figure 1. The R-UTHBAL model: (**a**) Flow diagram and (**b**) the R workflow environment with hydroPSO algorithm.

2.2. Water Balance Modelling Procedure

The study area is the Portaikos River watershed in the outlet at Pyli hydrometric station. It is a forested mountainous watershed and has an area of about 133 km^2. The mountainous watershed is located in Thessaly Region and Portaikos river is one of the main tributaries of Pinios River. Monthly streamflow data were available for the period October 1990–September 1993. Areal input datasets were estimated for the above period using typical engineering methods (i.e., precipitation/temperature gradients, Thiessen polygons) from the available meteorological stations. Potential evapotranspiration was calculated with the Thorthwaite method based on the estimated mean monthly temperature values.

The Latin-Hypercube One-factor-At-a-Time procedure developed by van Griensven et al. [8] for sensitivity analysis of model parameters was firstly applied to identify significant model parameters and the required mathematical model structure. Then, several objective functions, (i.e., Nash–Sutcliffe Efficiency and variations or adaptations, such as Kling–Gupta efficiency (KGE) formulations [9,10] and Root Mean Square Error (RMSE)), addressing different parts of the hydrograph, were used to assess both the skill and the robustness of the R-UTHBAL model to perform consistent streamflow predictions using the temporal split-sample test. Confidence intervals in the simulated runoff due to input data uncertainty, parameter uncertainty, and total uncertainty were calculated using the *hydroPSO* algorithm. *hydroPSO* implements several state-of-the-art enhancements and fine-tuning options to the Particle Swarm Optimisation (PSO) algorithm to meet specific user needs. *hydroPSO* easily interfaces the calibration engine to different model codes through simple ASCII files and/or R wrapper functions for exchanging information on the calibration parameters. Then, it optimises a user-defined goodness-of-fit measure until a maximum number of iterations or a convergence criterion is met. Finally, advanced plotting functionalities facilitate the interpretation and assessment of the calibration results.

3. Results and Discussion

Table 1 presents the sensitivity analysis results for the six (6) model parameters using the modified KGE [10] as the objective function, and 5000 strata for LH sampling with variance fraction 10%. In this table, the parameter ranges are also depicted. Based on Table 1, all model parameters should be included in the mathematical model structure. Using the temporal split-sample test between the first (October 1960–September 1977) and second period (October 1977–September 1993), the models were calibrated for half of the years during both the first and second periods, leaving the remaining half of the years for validation. The KGE2 values of the optimisation process were 0.79 and 0.87 for the first and second period, respectively, for 2000 model realisations. Figure 2 presents the results for the first period and parameter values versus the corresponding goodness-of-fit values (KGE2) obtained during the optimisation procedure.

Table 1. Model parameter range values, sensitivity analysis results, and optimised model parameters using the R-UTHBAL model and *hydroPSO* algorithm.

Parameter	Min Value	Max Value	Ranking Number	Normalised Relative Importance (%)	Optimised Value 1960–1977	Optimised Value 1977–1993
C_m	0	12	6	09.96	6.660	2.149
CN	30	100	5	14.47	45.435	46.546
K	0	1	4	17.16	0.005	0.299
$\alpha = aAET$	0	1	3	18.70	0.002	0.261
$\beta = CONMR$	0	1	2	19.66	0.120	0.091
$\gamma = CONGROUND$	0	1	1	20.02	0.943	0.988

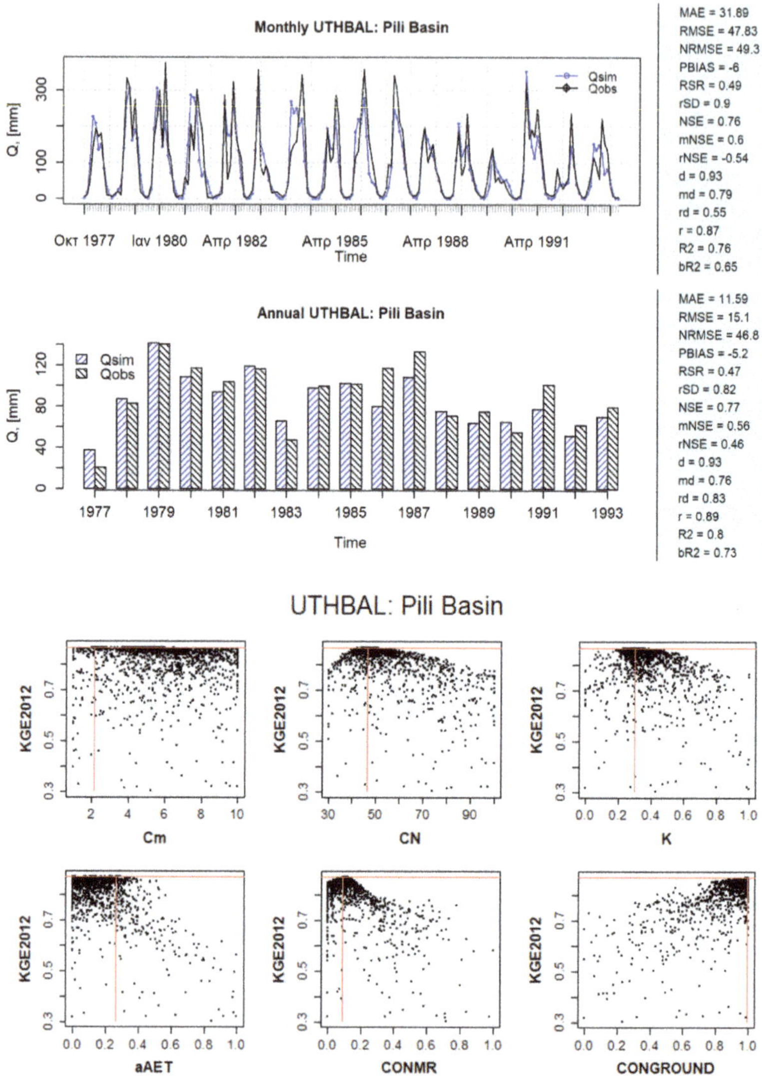

Figure 2. Validation results (monthly and annual graphs using several performance indices) for the first period, October 1960 to September 1993, and parameter values versus the corresponding goodness-of-fit values (KGE2) obtained during optimisation.

Based on a selected threshold of KGE2 > 0.3 as the behavioural threshold, all parameter values above the threshold were selected and weighted quantiles of model parameters were calculated to provide an estimate of the uncertainty in each model parameter. Using the P-factor, which represents the percent of observations that are within the user-defined uncertainty bounds and the R-factor that represents the average width of the user-defined uncertainty bounds divided by the standard deviation of the observations, a quantification of the uncertainty was estimated [5]. Figure 3 presents the results of the uncertainty analysis for the verification period and shows the best simulated streamflows along with the 95 Percent Prediction Uncertainty (95 PPU), and Figure 4 presents the uncertainty in flow duration curve using the one of the best simulated streamflows and the flow duration curves for the 2.5 and 97.5 weighted quantiles of model simulations obtained during the verification period.

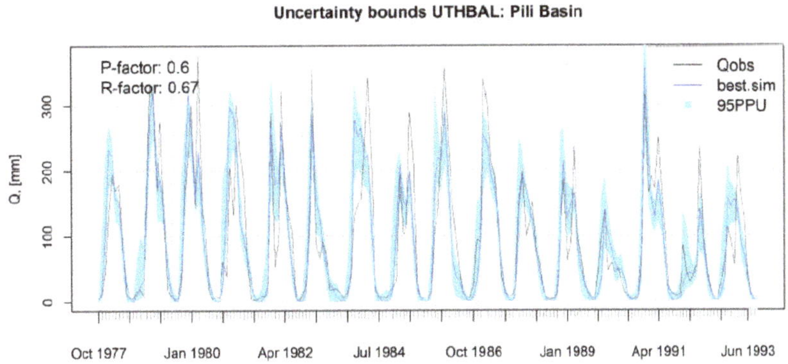

Figure 3. The 95 Percent Prediction Uncertainty (95 PPU) for model simulations during the verification period.

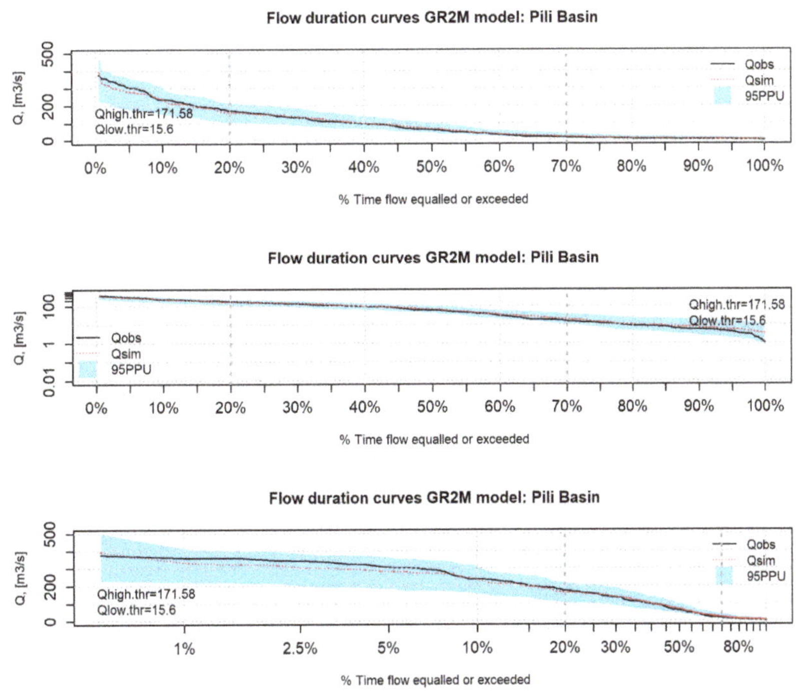

Figure 4. Flow duration curve of the observed (black line) and best simulated (blue line) streamflows. In addition, flow duration curves for the 2.5 and 97.5 weighted quantiles of model simulations obtained during the verification period. The upper panel is the normal flow duration curve, the middle panel focuses on low flows (log = 'y'), and the lower panel focuses on high flows (log = 'x').

4. Concluding Remarks

Accurate assessment of streamflow is crucial for operational water resources management projects. This study estimated the uncertainties in the surface runoff simulated by a monthly water balance model in a mountainous watershed of the Portaikos river, a tributary of the Pinios river, Thessaly, Greece. The R-UTHBAL water balance model was integrated with the hydroPSO algorithm and advanced sensitivity analyses, and user-friendly evaluation plots were estimated to facilitate the interpretation and assessment of the optimisation process. Application of R-UTHBAL with the hydroPSO in Portaikos river basin in Thessaly, Greece, showed that the uncertainty in streamflow estimation should always be accounted for and evaluated in operational water resource management projects.

Author Contributions: Conceptualization, methodology, software, writing—review and editing: L.V.; methodology, software: I.M. All authors have read and agreed to the published version of the manuscript.

Funding: This research received no external funding.

Institutional Review Board Statement: Not applicable.

Informed Consent Statement: Not applicable.

Data Availability Statement: Not applicable.

Conflicts of Interest: The authors declare no conflict of interest.

References

1. Engeland, K.; Xu, C.-Y.; Gottschalk, L. Assessing Uncertainties in a Conceptual Water Balance Model Using Bayesian Methodology/Estimation Bayésienne Des Incertitudes Au Sein d'une Modélisation Conceptuelle de Bilan Hydrologique. *Hydrol. Sci. J.* **2005**, *50*, 9. [CrossRef]
2. Refsgaard, J.C.; van der Sluijs, J.P.; Højberg, A.L.; Vanrolleghem, P.A. Uncertainty in the Environmental Modelling Process—A Framework and Guidance. *Environ. Model. Softw.* **2007**, *22*, 1543–1556. [CrossRef]
3. Loukas, A.; Mylopoulos, N.; Vasiliades, L. A Modeling System for the Evaluation of Water Resources Management Strategies in Thessaly, Greece. *Water Resour. Manag.* **2007**, *21*, 1673–1702. [CrossRef]
4. Vasiliades, L.; Loukas, A. Hydrological Response to Meteorological Drought Using the Palmer Drought Indices in Thessaly, Greece. *Desalination* **2009**, *237*, 3–21. [CrossRef]
5. Zambrano-Bigiarini, M.; Rojas, R. A Model-Independent Particle Swarm Optimisation Software for Model Calibration. *Environ. Model. Softw.* **2013**, *43*, 5–25. [CrossRef]
6. Malamataris, D.; Kolokytha, E.; Loukas, A. Integrated Hydrological Modelling of Surface Water and Groundwater under Climate Change: The Case of the Mygdonia Basin in Greece. *J. Water Clim. Change* **2020**, *11*, 1429–1454. [CrossRef]
7. Lyra, A.; Loukas, A.; Sidiropoulos, P.; Tziatzios, G.; Mylopoulos, N. An Integrated Modeling System for the Evaluation of Water Resources in Coastal Agricultural Watersheds: Application in Almyros Basin, Thessaly, Greece. *Water* **2021**, *13*, 268. [CrossRef]
8. van Griensven, A.; Meixner, T.; Grunwald, S.; Bishop, T.; Diluzio, M.; Srinivasan, R. A Global Sensitivity Analysis Tool for the Parameters of Multi-Variable Catchment Models. *J. Hydrol.* **2006**, *324*, 10–23. [CrossRef]
9. Gupta, H.V.; Kling, H.; Yilmaz, K.K.; Martinez, G.F. Decomposition of the Mean Squared Error and NSE Performance Criteria: Implications for Improving Hydrological Modelling. *J. Hydrol.* **2009**, *377*, 80–91. [CrossRef]
10. Kling, H.; Fuchs, M.; Paulin, M. Runoff Conditions in the Upper Danube Basin under an Ensemble of Climate Change Scenarios. *J. Hydrol.* **2012**, *424–425*, 264–277. [CrossRef]

environmental sciences proceedings

MDPI

Proceeding Paper

A Unified Hydrologic Framework for Flood Design Estimation in Ungauged Basins †

Lampros Vasiliades [1,*], George Papaioannou [2] and Athanasios Loukas [3]

1 Department of Civil Engineering, School of Engineering, University of Thessaly, Pedion Areos, 38334 Volos, Greece
2 Department of Forestry and Management of the Environment and Natural Resources, Democritus University of Thrace, 68200 Orestiada, Greece; gpapaio@fmenr.duth.gr
3 Department of Transportation and Hydraulic Engineering, School of Rural and Surveying Engineering, Aristotle University of Thessaloniki, 54124 Thessaloniki, Greece; gloukas@topo.auth.gr
* Correspondence: lvassil@civ.uth.gr; Tel.: +30-2421074115
† Presented at the 7th International Electronic Conference on Water Sciences, 15–30 March 2023; Available online: https://ecws-7.sciforum.net/.

Abstract: Design flood hydrograph estimation is a key problem in hydrology and is necessary for a variety of applications from the design of hydraulic structures to flood risk mapping processes. Furthermore, in large ungauged basins (>1000 km^2), design flood estimation methods mainly rely on single-event theories using digital elevation models, land use/land cover and soil type data, and relevant meteorological information (temperature and rainfall data). The single event-based deterministic approach was adopted based on three modelling components: (i) a synthetic storm generator; (ii) a hydrological simulation model; and (iii) a hydrological routing model. In this study the 100-year design flood (which is assumed equal to 100-year extreme rainfall) was estimated for the Pinios River Basin, Thessaly, Greece, at Larissa outlet station (upstream of the area by about 6500 km^2). The hydrological approach is based on semi-distributed modelling of the rainfall–run-off process (at the sub-basin scale) using HEC-HMS v.4.10 software and the SCS-CN method for estimating rainfall excess, as well as the unit hydrograph theory and the Muskingum hydrological flow routing method for propagating the surface run-off to the sub-basin outlets.

Keywords: design flood; extreme rainfall; IDF; SCS-CN; unit hydrograph

check for updates

Citation: Vasiliades, L.; Papaioannou, G.; Loukas, A. A Unified Hydrologic Framework for Flood Design Estimation in Ungauged Basins. *Environ. Sci. Proc.* **2023**, *25*, 40. https://doi.org/10.3390/ECWS-7-14194

Academic Editor: Silvia Kohnova

Published: 14 March 2023

1. Introduction

Hydrological extremes such as extreme precipitation and severe floods have consistently posed a risk to human culture. Due to the potential rise in meteorological and hydroclimatological extremes in recent decades, the issue has received significant attention [1]. The primary input for modelling hydrological extremes is rainfall, which has both geographical and temporal characteristics that must be taken into consideration in simulation procedures. In order to design hydraulic structures, design rainfall is often estimated as a univariate variable [2]. Hence, design flood hydrograph estimation is a key problem in hydrology and is necessary for a variety of applications from the design of hydraulic structures to flood risk mapping processes. Furthermore, in large ungauged basins (>1000 km^2), design flood estimation methods mainly rely on single-event theories using digital elevation models, land use/land cover and soil type data, and relevant meteorological information (temperature and rainfall data) [3].

The single event-based deterministic approach was adopted based on three modelling components: (i) a synthetic storm generator; (ii) a hydrological simulation model; and (iii) a hydrological routing model [4]. In this study, the 100-year design flood (which is assumed equal to 100-year extreme rainfall) was estimated for the Pinios River Basin, Thessaly, Greece, at Larissa outlet station (upstream area is about 6500 km^2). The hydrological approach is based on semi-distributed modelling of the rainfall–run-off process (at the sub-basin scale) using HEC-HMS software and the SCS-CN method for estimating rainfall excess, as well as the unit hydrograph theory for propagating the surface run-off to the sub-basin outlets. Design rainfall at the subbasin scale is estimated from 13 intensity–duration–frequency (IDF) rainfall point curves using thiessen polygons and adjusted to the mean elevation of the sub-basin with the developed precipitation gradients. The design flood hydrograph is estimated by combining the IDF approach with standard time profiles for constructing synthetic rainfall events of a certain probability, the SCS-CN method for extracting the excess from the gross rainfall, and the unit hydrograph theory for propagating the surface run-off to the basin outlet.

2. Materials and Methods

2.1. Study Area

The design flood hydrographs were estimated for a larger part of the Pinios River Basin in Thessaly, Greece. Figure 1 shows the hydrological basin upstream of the point of interest near Larissa City with a total area of 6407 km^2. The hydrographic network, depicted also shown in Figure 1, was configured in the application of Water Directive 2000/60/EC and includes the main watercourses of the study area. The most extensive and complex network of watercourses develops in the Pinios Basin, and includes, in addition to the main branch of the river, almost all its important tributaries, namely: (a) in the southern part of the basin, Enipeas, Farsaliotis, Sofaditis, and Kalentzis, (b) in the western and southwestern part, Pamisos, Portaikos, Malakasiotiko, and Murgani, and (c) in the northern part, the Litheos and Neochoritis rivers.

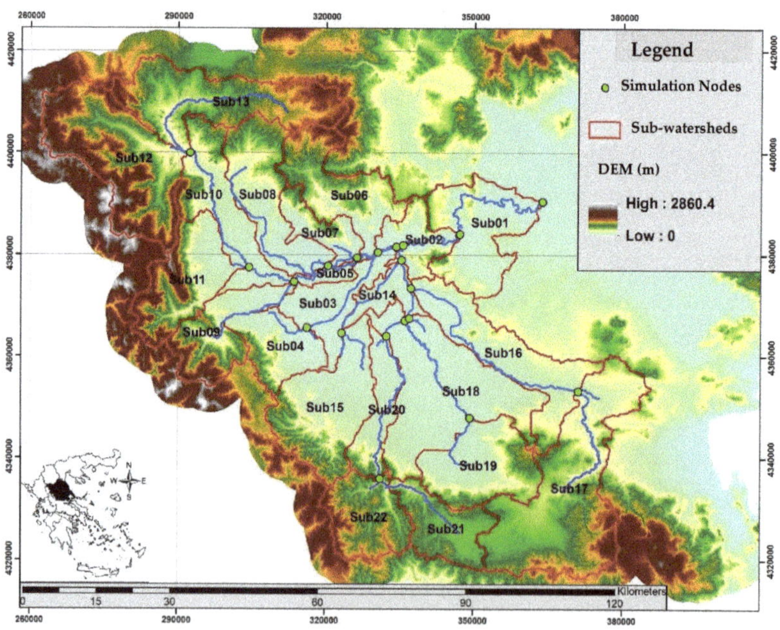

Figure 1. Study area with sub-watersheds, the hydrographic network, and simulation nodes in the Greek Geodetic Reference System 1987 (EPSG: 2100).

Based on the hydrographic network, the study area was divided into 22 watersheds and smaller sub-basins, considering nodes at the entrances of the zones and the confluences of the main watercourses of the study area. The boundaries of the basins are shown in Figure 1 and Table 1 and present the main characteristics of the sub-watersheds. In the mountainous and semi-mountainous parts of the hydrological basin, the hydrological basins were drawn based on a digital elevation model (DEM) with a pixel size of 25 × 25 m, while in the plain areas, a DEM with 5 m resolution was used across the main bed of the Pinios River combined with National Cadastre maps as well as Google Earth satellite images. The mapping based on higher resolution information was necessary in the c riverbed as well as in the lowland sections, because many of the watercourses are diverted and do not fully follow the natural slope of the ground. In addition, in some cases, embankments have been constructed on either side of the bed to prevent lateral runoff, with the result being that these embankments actually define artificial barriers. Artificial interventions in the flow are also created by other projects, such as road embankments, canals, etc. In addition, the information of the hydrographic network was integrated into the DEM (DEM reconditioning) for more accurate mapping of the flow field in the areas of low and very low gradients and for the correct mapping of the sub-basins and the accurate calculation of the geomorphological characteristics of the sub-basins and watercourses.

Table 1. Sub-watershed characteristics.

Code	Sub-Watershed	Area (km^2)	Mean Elevation (m)	Outlet Elevation (m)	Maximum Flow Length (km)	Curve Number (CN_{II})	Time of Concentration (h)
Sub01	Larisa	331.15	165.77	61.7	65.617	53.2	20.980
Sub02	Piniada	138.49	193.46	71.4	34.212	55.8	11.130
Sub03	Karditsa-Keramidi	185.09	94.63	79.8	39.607	53.8	36.917
Sub04	Megas	193.14	244.83	92.0	27.504	57.7	9.793
Sub05	Nomi-Mesdani	37.70	98.23	84.2	28.226	54.8	22.327
Sub06	Neochoritis	293.48	489.99	84.2	45.317	70.1	8.470
Sub07	Litheos_2	114.68	246.94	88.4	29.161	52.1	8.596
Sub08	Litheos	321.53	288.80	93.3	62.304	60.4	14.766
Sub09	Pamisos	234.13	568.03	98.0	38.381	59.2	6.848
Sub10	Sarakina-Mesdani	298.87	298.67	98.0	55.542	57.9	13.455
Sub11	Portaikos	296.42	629.41	104.1	34.830	59.1	6.605
Sub12	Malakasiotikos	518.53	962.30	236.6	42.671	62.1	7.196
Sub13	Mourgani	440.46	698.19	236.6	61.254	68.2	10.229
Sub14	Vlochos	92.68	98.42	83.3	23.084	53.1	23.535
Sub15	Kalentzis	398.20	321.06	92.5	41.734	58.2	11.776
Sub16	Enipeas_1	356.69	201.85	86.3	65.253	56.9	20.166
Sub17	Enipeas_2	682.97	524.27	158.9	69.814	59.9	13.685
Sub18	Farsaliotis_1	565.33	140.91	89.0	56.036	53.9	31.078
Sub19	Farsaliotis_2	310.32	299.73	109.3	39.066	55.7	11.690
Sub20	Sofaditis_1	158.58	210.05	93.2	42.671	58.6	13.228
Sub21	Sofaditis_2	203.92	542.48	249.0	34.009	64.7	7.890
Sub22	Sofaditis_3	234.80	739.57	249.0	27.447	64.1	5.783

The run-off curve number (CN) proposed by the Soil Conservation Service (SCS, 1972) was used to condense the physiographic characteristics of the watershed into a single representative value. In the present study, it is used to estimate the maximum potential retention, which is the input data of the SCS-CN method (the method is applied to estimate hydrological deficits in the context of hydrological modelling with HEC-HMS software). The hydrological simulation model of the Pinios River Basin (just upstream of the confluence of the Pinios River with the Gousbasianiotis stream) includes 22 sub-watersehds, 20 nodes, and 19 river reaches.

2.2. Hydrological Modelling

The surface integration of IDF point measurements at the sub-basin level was carried out using the Thiessen polygon method. The calculated weights resulting from the application of the methodology were used to calculate the intensities (and rainfall heights) for durations of 5 min to 48 h and return periods from 1–1000 years according to the Thiessen method. From the calculated intensities i(d, T) according to the Thiessen method for various durations d and return periods, it is possible to adjust the equation of the rainfall ombrian curves by optimizing the five (5) parameters κ, λ′, ψ′, θ, and η at the sub-basin level. As the objective function of the optimization, the weighted root mean square error (weighted RMSE) was used, resulting from the minimization of the root mean square error of the observed and calculated intensities (RMSEintensity) and rainfall heights (RMSEdepth) for each sub-basin.

In the present study, to estimate the 100-year return period design flood, the flood return period was assumed to be equal to the return period of an extreme typical 100-year return period storm. The duration of the storm was chosen for all sub-basins to be constant and equal to 48 h so that the design storms in the sub-basins had a duration longer than the concentration time of the entire hydrological basin upstream of the point of interest. Finally, the design rainfall was distributed over time using the alternating block method (ABM). Time-distributed precipitation was used as an input hyetograph to the hydrological model to produce the hydrographs for each sub-catchment. Hydrological uncertainty has been expressed in terms of the two typical antecedent moisture conditions (AMC) that are accounted for in the Soil Conservation Service curve number (SCS-CN) approach of moderate (or average—AMC_{II}) represented by CN_{II} and wet (or high—AMC_{III}) represented by CN_{III}. The transformation of the excess rainfall over each sub-basin to the flood hydrograph at the outlet junction (rainfall–run-off model) was achieved using the dimensionless curvilinear unit hydrograph approach of SCS, which is considered the prevailing modelling approach for ungauged basins. A key assumption of the method was the implementation of the concept of varying the (i.e., run-off-dependent) time of concentration, which affects the shape of unit hydrographs, thus introducing further non-linearities to the overall modelling approach. Hence, in order to take account of the dependence of the response time of the basin against run-off, a kinematic-wave-theory-based semi-empirical formula was employed, considering that t_c is inversely proportional to the design rainfall. Further details are given in [3,4]. HEC-HMS software was used, and the design flood hydrographs from the generated storms were estimated for all scenarios.

3. Results and Discussion

Table 2 presents the characteristics of the calculated synthetic unit hydrographs by the SCS-CN method for the 22 sub-watersheds of the study area and for the selected return period T = 100 years. In addition, Figure 2 presents the developed IDF curves and unit hydrographs from the application of the methodology for two sub-basins, the Megas sub-basin with the code name Sub04 and the mountainous sub-basin of Sofaditis (Sofaditis_3) with the code name Sub22, for selected periods return T = 5, 100, and 1000 years.

Table 2. Characteristics of SCS synthetic unit hydrographs for the return period of 100 years in all sub-watersheds.

Code	Sub-Watershed	t_c (h)	t_p (h)	t_b (h)	Q_p (m^3/s)	$P_{D=48h,T=100y}$ (mm)	$P_{D=48h,T=5y}$ (mm)
Sub01	Larisa	14.098	8.584	42.918	80.244	166.2	75.1
Sub02	Piniada	7.754	4.777	23.886	60.300	172.9	83.9
Sub03	Karditsa-Keramidi	25.654	15.517	77.586	24.810	177.3	85.6
Sub04	Megas	6.987	4.317	21.586	93.055	200.0	101.8
Sub05	Nomi-Mesdani	16.171	9.828	49.138	7.979	125.9	66.1
Sub06	Neochoritis	6.483	4.015	20.075	152.038	163.3	95.7
Sub07	Litheos_2	6.439	3.989	19.943	59.804	134.4	75.4
Sub08	Litheos	11.042	6.750	33.751	99.076	135.6	75.8
Sub09	Pamisos	5.158	3.220	16.099	151.245	216.7	122.9
Sub10	Sarakina-Mesdani	10.026	6.141	30.703	101.237	139.3	77.4
Sub11	Portaikos	4.893	3.061	15.304	201.430	226.9	124.5

Table 2. *Cont.*

Code	Sub-Watershed	t_c (h)	t_p (h)	t_b (h)	Q_p (m^3/s)	$P_{D=48h,T=100y}$ (mm)	$P_{D=48h,T=5y}$ (mm)
Sub12	Malakasiotikos	5.575	3.470	17.350	310.816	230.5	138.4
Sub13	Mourgani	7.625	4.700	23.500	194.924	176.7	98.2
Sub14	Vlochos	16.238	9.868	49.338	19.535	195.6	93.1
Sub15	Kalentzis	8.279	5.093	25.463	162.639	211.4	104.5
Sub16	Enipeas_1	13.719	8.356	41.782	88.784	179.6	83.1
Sub17	Enipeas_2	9.696	5.943	29.714	239.040	207.9	104.4
Sub18	Farsaliotis_1	20.668	12.526	62.628	93.879	175.9	77.8
Sub19	Farsaliotis_2	7.810	4.811	24.056	134.160	181.5	81.0
Sub20	Sofaditis_1	8.901	5.465	27.327	60.352	191.6	86.8
Sub21	Sofaditis_2	5.363	3.343	16.714	126.882	210.3	97.2
Sub22	Sofaditis_3	4.016	2.535	12.674	192.678	232.9	112.3

Note: t_c: adjusted time of concentration, t_p: time to peak, t_b: base time, Q_p: peak run-off, $P_{D=48h,T=100y}$: extreme rainfall for the rain duration of 48 h and T = 100 years, $P_{D=48h,T=5y}$: extreme rainfall for the rain duration of 48 h and T = 5 years.

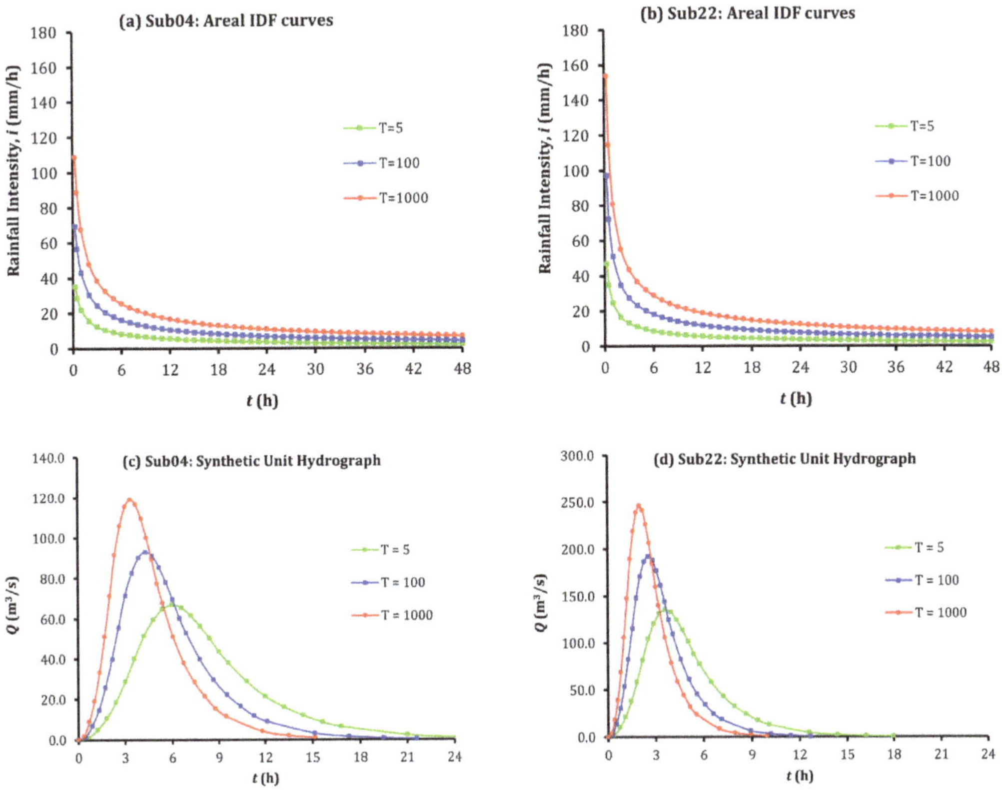

Figure 2. The developed intensity–duration–frequency (IDF) curves for (**a**) the Sub04 and (**b**) Sub22 sub-watersheds. Synthetic unit hydrographs: (**c**) the Sub04 sub-watershed and (**d**) the Sub22 sub-watershed.

Application of the hydrological simulation model of the Pinios River Basin, which includes 22 sub-basins, 20 nodes, and 19 river reaches using (i) the developed IDF curves; (ii) the SCS-CN unit hydrograph model; and (iii) the Muskingum hydrological flow routing model, is shown in Figure 3 for the outlet J1 of the study area for the two employed scenarios and the return period of 100 years.

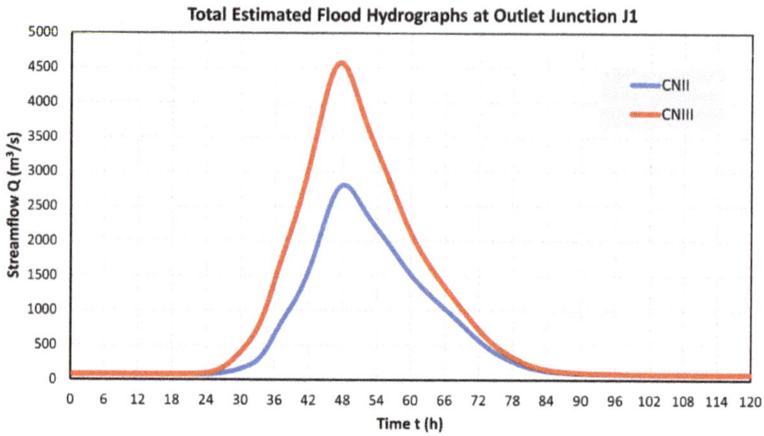

Figure 3. Design flood hydrographs at the outlet of the study area for the employed hydrologic scenarios and the return period of 100 years.

The design flood values for a return period of T = 100 years at the outlet of the study area, which were estimated by the application of semi-distributed methodology, are 2808 m^3/s for average soil moisture conditions (CN_{II}, average scenario) and 4571 m^3/s for wet soil moisture conditions (CN_{III}, adverse scenario). These flood values as well as the flood hydrographs should be used in the design of hydrotechnical projects in the study area.

Author Contributions: Conceptualization, mathematical analysis, and writing—review and editing L.V., G.P. and A.L. All authors have read and agreed to the published version of the manuscript.

Funding: This research received no external funding.

Institutional Review Board Statement: Not applicable.

Informed Consent Statement: Not applicable.

Data Availability Statement: Not applicable.

Conflicts of Interest: The authors declare no conflict of interest.

References

1. Blöschl, G.; Hall, J.; Viglione, A.; Perdigão, R.A.P.; Parajka, J.; Merz, B.; Lun, D.; Arheimer, B.; Aronica, G.T.; Bilibashi, A.; et al. Changing Climate Both Increases and Decreases European River Floods. *Nature* **2019**, *573*, 108–111. [CrossRef] [PubMed]
2. Apel, H.; Thieken, A.H.; Merz, B.; Blöschl, G. Flood Risk Assessment and Associated Uncertainty. *Nat. Hazards Earth Syst. Sci.* **2004**, *4*, 295–308. [CrossRef]
3. Papaioannou, G.; Vasiliades, L.; Loukas, A.; Alamanos, A.; Efstratiadis, A.; Koukouvinos, A.; Tsoukalas, I.; Kossieris, P. A Flood Inundation Modeling Approach for Urban and Rural Areas in Lake and Large-Scale River Basins. *Water* **2021**, *13*, 1264. [CrossRef]
4. Papaioannou, G.; Efstratiadis, A.; Vasiliades, L.; Loukas, A.; Papalexiou, S.; Koukouvinos, A.; Tsoukalas, I.; Kossieris, P. An Operational Method for Flood Directive Implementation in Ungauged Urban Areas. *Hydrology* **2018**, *5*, 24. [CrossRef]

environmental sciences proceedings

MDPI

Proceeding Paper

Utilization of a Telemetry Monitoring System for the Dynamics of Water Quantity and Quality in the Dadahup Swamp Irrigation Area †

Asril Zevri

Directorate General Water Resources, Ministry of Public Work and Housing, Jakarta 12110, Indonesia; asrilzevri19@gmail.com

† Presented at the 7th International Electronic Conference on Water Sciences, 15–30 March 2023; Available online: https://ecws-7.sciforum.net.

Abstract: One of the technological developments supporting irrigation modernization is the installation of telemetry monitoring systems that are built based on three main elements of sensors, internet connections, and data centers. The Dadahup Swamp Irrigation Area is one of the development areas that has been supported by the installation of a telemetry monitoring system at three points located on the upstream and midstream sides of the irrigation area. The water management system of the Dadahup Swamp Irrigation Area has not been working optimally in regulating the water level, which has resulted in the irrigated land not receiving good water quality for rice plant growth. The research objective is to describe the condition of the water quantity and quality profile based on the telemetry monitoring system. The research method was carried out by observing the water level elevation, rainfall, and pH, during the rainy and dry seasons in the upstream, midstream, and downstream parts. The results showed that the dynamics of the water quantity greatly affected the water quality in both the rainy and dry seasons.

Keywords: telemetry monitoring system; water quantity; water quality; Dadahup

check for updates

Citation: Zevri, A. Utilization of a Telemetry Monitoring System for the Dynamics of Water Quantity and Quality in the Dadahup Swamp Irrigation Area. *Environ. Sci. Proc.* **2023**, 25, 41. https://doi.org/10.3390/ECWS-7-14191

Academic Editor: Lampros Vasiliades

Published: 14 March 2023

1. Introduction

The food estate development program is one of the Government's efforts to anticipate a food crisis due to the significant increase in population in recent years [1]. One of the efforts to develop food land into agricultural cultivation is the rehabilitation and improvement of the Dadahup Swamp Irrigation Area network with a potential area of ±21,226 thousand ha in Kapuas Regency, Central Kalimantan Province [2].

Due to the extreme seasonal changes, the Dadahup Swamp Irrigation Area is often flooded, resulting in crop failure [3]. The condition of the channel network system and the irrigation buildings that are not functioning optimally along with the broken and collapsed embankments around the irrigation area has resulted in the conversion of the area into shrubland [4].

The development of technology that supports the polder water system is the installation of the internet of things (IOT) which consists of sensors, internet connections, and data centers [5]. The installation of a telemetry monitoring system in the Dadahup Swamp Irrigation Area aims to observe changes in the water quantity and quality in the upstream and midstream of the main primary channel in order to optimize the operation and maintenance of the irrigation networks and buildings [6].

2. Material and Methods

2.1. Study Area

The research site is located in the Dadahup Swamp Irrigation Area of Kapuas Regency, Central Kalimantan Province, which is traversed by the Barito, Kapuas Murung, and

Mengkatip Rivers. The tertiary blocks of the Dadahup Swamp Irrigation Area are divided into seventeen tertiary blocks dominated by shrubs, shown in Figure 1.

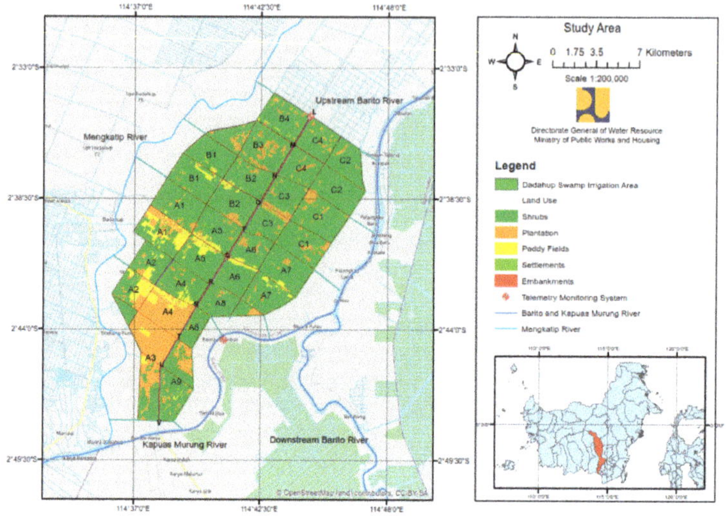

Figure 1. Study Area.

2.2. Methods

The research was conducted at point locations L and Q by collecting data on the water level, rainfall, and pH, which were obtained by direct observation through the telemetry monitoring system shown in Figure 2.

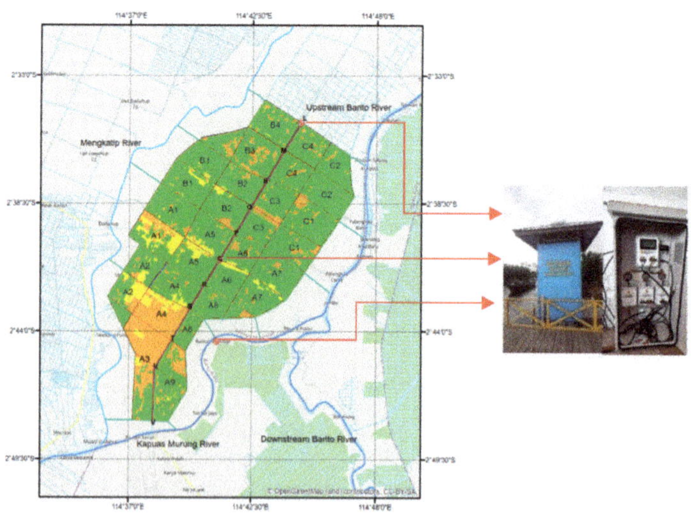

Figure 2. The Dadahup Swamp Irrigation Monitoring Telemetry System.

2.3. Characteristics of the Dadahup Swamp Irrigation Area

Hydro-topography is the relationship between the water table elevation and land elevation, which is a factor in determining the classification type of the swamp irrigation areas, as shown in Figure 3 [7]. The Dadahup Swamp Irrigation Area falls into categories C and D, where the tidal water level never inundates the land elevation; so, the main source of water is rainfall, which is included in the nontidal swamp category shown in Figure 4 [8]. The characteristics of the soil, which is dominated by pyrite peat soil, can reduce the water quality and have a negative impact on rice growth [9]. The polder water system is one of the solutions for the development of the Dadahup Swamp Irrigation Area by utilizing rainfall as a water source by isolating the land, which is regulated through the operation of gate buildings in the north, east, west, and south [10].

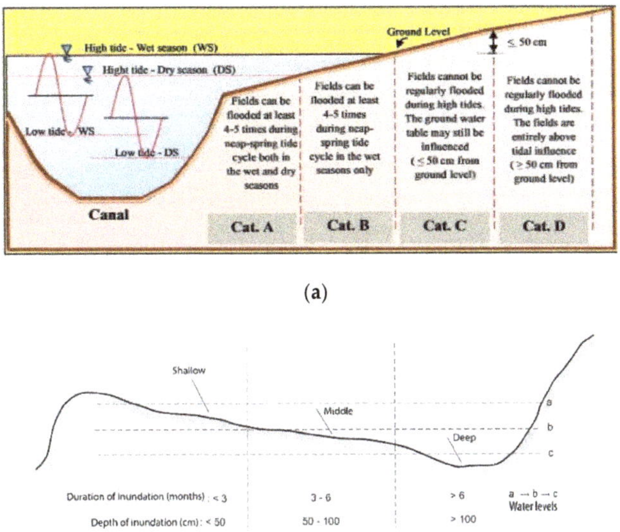

Figure 3. (**a**) Tidal Swamp Hydro-topography and (**b**) Nontidal Swamp Hydro-topography.

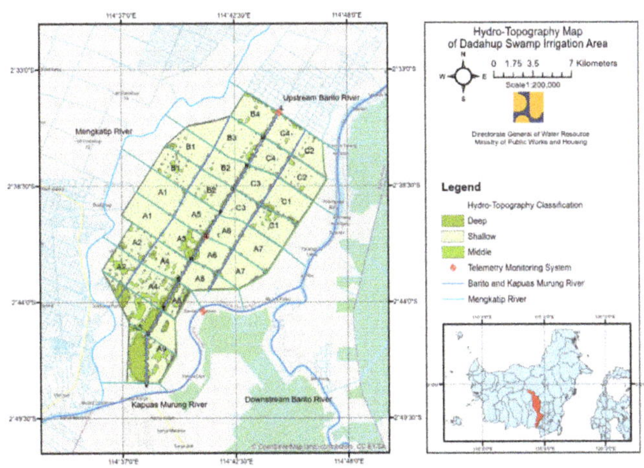

Figure 4. Hydro-topography Map of the Dadahup Swamp Irrigation Area.

2.4. Water Quantity and Quality

The tidal influence of the estuary and the upstream discharge of the Barito River can affect the water level at the intake channel of the Dadahup Swamp Irrigation Area with a distance of up to 158 km to the upstream of the river [11]. The Dadahup Irrigation Area is included in the wet category of climate types B and E, where the rainy season occurs from May to October, and the dry season occurs from November to April. The swamp soil characteristics are influenced by the presence of pyrite formed from marine deposits containing organic matter and sulfate-reducing bacteria. The oxidation of pyrite can cause high acidity, which produces sulfuric acid and jarosite minerals that reduce the pH concentration and interfere with plant growth [12].

3. Results and Discussion

3.1. Water Quantity Dynamics Profile

The dynamics of the water quantity illustrate the profile of the relationship between the rainfall and water level during the dry and wet seasons at points L and Q in the main primary channel shown in Figures 5 and 6.

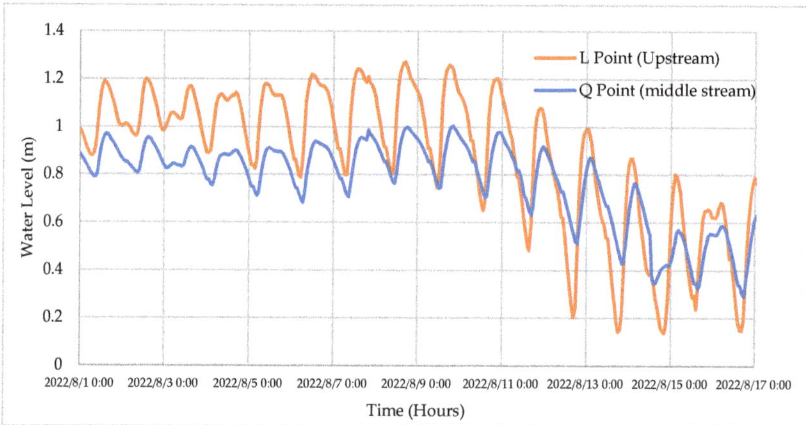

Figure 5. Water Level Dynamics Profile in the Dry Season in the Upper and Middle Sections of the Dadahup Swamp Irrigation Area.

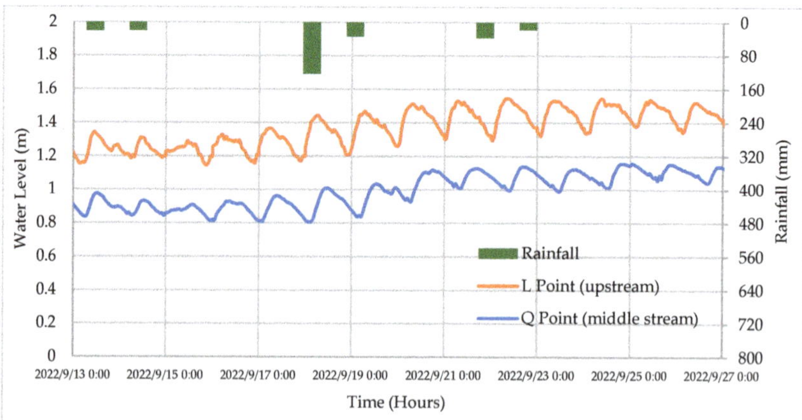

Figure 6. Water Level Dynamics Profile in the Rainy Season in the Upper and Middle Sections of the Dadahup Swamp Irrigation Area.

Based on the results of the description of the dynamic profile of the water level at points L and Q for the dry season and the rainy season, there was a change in the water level elevation influenced by the rainfall with an average increase of 0.3 m, which is shown in Table 1.

Table 1. Changes in the Water Level during the Dry and Rainy Seasons at Points L and Q.

No	Point	Season	Water Level (m)
1	L	Dry	1.2
2	Q		0.9
3	L	Rainy	1.5
4	Q		1.2

3.2. *Water Quality Dynamics Profile*

The water quality dynamics illustrate the relationship profile of the pH during the dry and wet seasons. The results of the water quality dynamics profile at point L upstream and point Q midstream of the Dadahup Swamp Irrigation Area during the dry and wet seasons are shown in Figures 7 and 8.

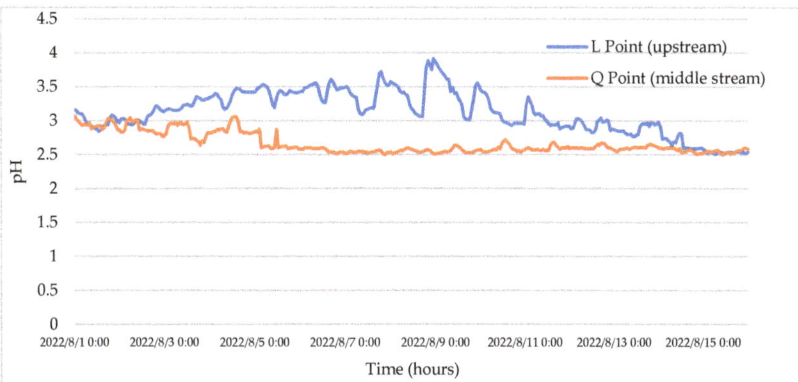

Figure 7. Water Quality Dynamics Profile in the Dry Season in the Upper and Middle Sections of the Dadahup Swamp Irrigation Area.

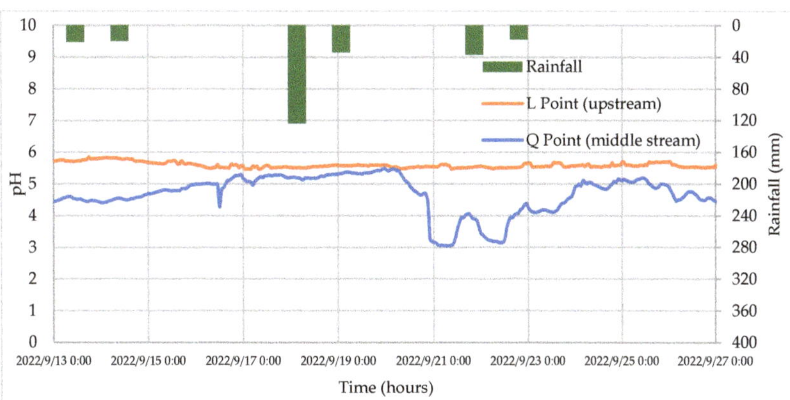

Figure 8. Water Quality Dynamics Profile in the Rainy Season in the Upper and Middle Sections of the Dadahup Swamp Irrigation Area.

Based on the results of the description of the dynamic profile of the water quality, there was a change in the pH value influenced by the rainfall with an average increase reaching level 3, which is shown in Table 2.

Table 2. Changes in the pH Values during the Dry and Rainy Seasons in the Upper and Middle Sections of the Dadahup Irrigation Area.

No	Point	Season	pH
1	L	Dry	3.91
2	Q		3.07
3	L	Rainy	5.84
4	Q		5.48

4. Conclusions

The results explained that the pH value and water level elevation increased as a result of the rainfall and became the basis for optimizing the operation and maintenance of the Dadahup Irrigation network by utilizing a telemetry monitoring system.

Author Contributions: Conceptualization, A.Z.; methodology, A.Z.; formal analysis, A.Z.; investigation, A.Z.; resources, A.Z.; data acuration, A.Z.; writing—original draft preparation, A.Z.; writing—review and editing, A.Z.; visualization, A.Z.; supervision, A.Z.; project administration, A.Z.; funding acquisition, A.Z. All authors have read and agreed to the published version of the manuscript.

Funding: This research was supported by Directorate General of Water Resources Ministry of Public Works and Housing and Department of Civil Engineering and Environmental Engineering, Faculty of Engineering, Universitas Gadjah Mada, Yogyakarta.

Institutional Review Board Statement: Not applicable.

Informed Consent Statement: Not applicable.

Data Availability Statement: Data are available from corresponding author.

Acknowledgments: The author would like to thank all those who have supported the completion of this paper.

Conflicts of Interest: The author declares no conflict of interest.

References

1. Santosa, E. Percepatan Pengembangan Food Estate Untuk Meningkatkan Ketahanan Dan Kemandirian Pangan Nasional. *Risal. Kebijak. Pertan. DAN Lingkung. Rumusan Kaji. Strateg. Bid. Pertan. dan Lingkung.* **2015**, *1*, 80. [CrossRef]
2. Arif, M.I.; Legono, D.; Luknanto, D. Study on the Performance of the Hydraulics System Planning of Swampy Irrigation Area of Dadahup, Kapuas District, Central Kalimantan Province. *IOP Conf. Ser. Earth Environ. Sci.* **2021**, *930*, 1–10. [CrossRef]
3. Arif, M.I.; Legono, D.; Luknanto, D.; Wignyosukarto, B.; Marpaung, M.F. Behavior of Flow of Channel Network of Dadahup Swampy Irrigation Area. In *IOP Conference Series: Earth and Environmental Science*; IOP Publishing: Bristol, UK, 2022; Volume 1091, p. 012035. [CrossRef]
4. PT. Virama Karya Persero (KSO) Cabang Semarang – PT. Wiratman – PT. Archiegama Bangun Cipta Pratama. *Laporan Akhir Survey Investigasi Desain (SID) Rehabilitasi dan Peningkatan Jaringan Irigasi Rawa Wilayah Kerja Blok A*; Kalimantan II River Basin Center Palangka Raya at the Directorate General of Water Resources: Palangka Raya, Indonesia, 2020.
5. Setiadi, D.; Muhaemin, M.N.A. Penerapan Internet of Things (IoT) Pada Sistem Monitoring Irigasi (SMART IRIGASI). *J. Infotronik* **2018**, *2*, 95–98. [CrossRef]
6. Zevri, A.; Rahardjo, A.P.; Legono, D. Swamp Water Parameter Dynamics Induced by Rainfall and Tides in Dadahup Irrigation Area, Kalimantan. In *IOP Conference Series: Earth and Environmental Science*; IOP Publishing: Bristol, UK, 2022; pp. 1–10. [CrossRef]
7. Istianto, H.; Bernard, R.; Suryadi, F.X. Improving the performance of tidal irrigation through the water management, (study case gandus Palembang, South Sumatra). In *Proceedings–International Association for Hydro-Environment Engineering and Research (IAHR)-Asia Pacific Division (APD) Congress: Multi-Perspective Water for Sustainable Development, IAHR-APD 2018*; Curran Associates, Inc.: Red Hook, NY, USA, 2018; Volume 1, pp. 677–685.
8. Wignyosukarto, B.S. Konsep Pengembangan Pola Tata Air Lahan Gambut. *Forum Tek.* **2000**, *24*, 342–357.

9. Purboseno, S.; Suparyanto, T.; Hidayat, A.A.; Pardamean, B. A Hydrodynamic Analysis of Water System in Dadahup Swamp Irrigation Area. In Proceedings of the 2021 1st International Conference on Computer Science and Artificial Intelligence (ICCSAI), Jakarta, Indonesia, 28 October 2021; pp. 400–406. [CrossRef]
10. Noor, M.; Anwar, K.; Kartiwa, B. *Sistem Polder untuk Pengembangan Pertanian Berkelanjutan di Lahan Rawa Lebak*; IAARD PRESS: Kota Bogor, Indonesia, 2017; pp. 22–38.
11. Suriadikarta, D.A. Teknologi Pengelolaan Lahan Rawa Berkelanjutan: Studi Kasus Kawasan Ex PLG Kalimantan Tengah. *J. Sumberd. Lahan* **2012**, *6*, 45–54.
12. Blunden, B.; Indraratna, B. Pyrite Oxidation Model for Assessing Ground-Water Management Strategies in Acid Sulfate Soils. *J. Geotech. Geoenviron. Eng.* **2001**, *127*, 146–157. [CrossRef]

environmental sciences proceedings

MDPI

Proceeding Paper

Comparison of the Performance of CMIP5 and CMIP6 in the Prediction of Rainfall Trends, Case Study Quebec City †

Amirhossein Salimi [1], Tadros Ghobrial [1] and Hossein Bonakdari [2,*]

[1] Department of Civil and Water Engineering, Université Laval, Quebec City, QC G1V0A6, Canada; amirhossein.salimi.1@ulaval.ca (A.S.); tadros.ghobrial@gci.ulaval.ca (T.G.)
[2] Department of Civil Engineering, University of Ottawa, Ottawa, ON K1N6N5, Canada
* Correspondence: hbonakda@uottawa.ca
† Presented at the 7th International Electronic Conference on Water Sciences, 15–30 March 2023; Available online: https://ecws-7.sciforum.net.

Abstract: Climate change affects many meteorological parameters that could result in spatiotemporal variations of the hydrological cycle. These variations can affect local rainfall intensities or design storms; therefore, it is necessary to assess the local effects of climate change in different areas. Therefore, the current research aims at evaluating the accuracy of the precipitation data of the most recent Coupled Model Intercomparison Project phases 5 and 6 (CanESM2 from CMIP5 and CanESM5 from CMIP6 models), over a historical period from 1953 to 2010, as well as the predicted data for the future between 2010 and 2050 for the Quebec City rain gauge Station (Jean Lesage Intl). In this regard, precipitation data were analyzed using a statistical index to find the most accurate model for the study area. The results of this evaluation showed that CanESM5 is more accurate than CanESM2 for most of the evaluation indices. However, both of these models did not perform well since the precipitation prediction for CanESM5 (as the accurate model) R index was 0.48 for the monthly and was 0.75 in the seasonal scale. In addition, the Bias index revealed that both models underestimated rainfall prediction with negative index values for both scales and models. The trend of future precipitation under socio–economic scenarios (4.5 (pessimistic) and 8.5 (optimistic)) shows that the changes in future precipitation are not significant. In addition, for scenario 4.5, the trend of precipitation decreases for almost half of the year, while for scenario 8.5, the magnitude of the decrease and the number of months with a decreasing trend of precipitation are significantly reduced when compared to scenario 4.5.

Keywords: precipitation; CMIP; Quebec; Mann–Kendall test; GCM

Citation: Salimi, A.; Ghobrial, T.; Bonakdari, H. Comparison of the Performance of CMIP5 and CMIP6 in the Prediction of Rainfall Trends, Case Study Quebec City. *Environ. Sci. Proc.* **2023**, *25*, 42. https://doi.org/10.3390/ECWS-7-14243

Academic Editor: Silvia Kohnova

Published: 16 March 2023

1. Introduction

Global warming has caused significant changes in the climate. In recent years, the severity of droughts, floods and extreme events has increased in different parts of the globe. The Intergovernmental Panel on Climate Change (IPCC) was established to identify its effects and especially how human activities affect it. In order to conduct climate change studies, climate variables under the influence of greenhouse gas emissions must first be simulated [1].

One of the important consequences of climate change is the change in the meteorological parameters' trend, especially the precipitation trend [2]. Therefore, a great deal of research has been conducted to evaluate climate change's effect on extreme rainfall events. These studies showed that global warming is affecting and causing climate changes based on the Coupled Model Intercomparison Project Phase 5 (CMIP5) and Coupled Model Intercomparison Project Phase 6 (CMIP6) climate reports in Canada [3]. Compared to other methods such as the multi-model ensemble (MME), the CMIP5 and CMIP6 models showed better performance. In addition, various methods can be used to reduce their uncertainty [4]. With the release of the sixth report (CMIP6), the desire to examine the performance of this report compared to the fifth report (CMIP5) has increased among researchers. One of the major improvements in CMIP6 is the introduction of socio–economic scenarios [5]. Examining the difference between the data from the report 5 and 6 climate models for temperature and precipitation shows that for the fixed time intervals, most of the temperature indices show higher predicted changes in CMIP6 when compared to CMIP5 in Canada. Rainfall changes in CMIP6 mainly occurred in extreme precipitation indices [3]. However, it is clear that the method of General Circulation Models (GCM) ensembles can lead to different estimates of future mean changes and different levels of uncertainty in those estimates [4]. Overall, current research has shown that the CMIP6 ensemble provides a narrower band of the uncertainty of future climate projections specifically for North America and brings more confidence to hydrological impact studies [6]. The importance of such analysis is in assessing risk and future vulnerability, and implementing efficient measures to control the changes made in the flow of rivers and their ability to warn of floods [5,7].

It is necessary to examine the system's response as a general unit for determining the possible effects of climate changes such as the increase in the concentration of greenhouse gases and the impact of socio–economic activities on the climate system [5]. For this purpose, it seems necessary to use climate models. These models include the main stages that occur in the climate system and calculate the corrections of different components when responding to the changes in the forcing factors. Therefore, evaluating the accuracy of the data of these models and choosing the most efficient and adaptable models are important and necessary steps for any forecasting [8]. This assessment is more important for precipitation, which has a more significant behavioral complexity than other meteorological phenomena. Therefore, identifying the mechanism and evaluating the effectiveness of atmospheric general circulation models in estimating precipitation and knowing their temporal and spatial frequency significantly affects the preparedness for such extreme events. Therefore, in this research, the effectiveness of the CMIP5 and CMIP6 for predicting extreme rainfall events is assessed, and the future trend of rainfall reported by the superior model is evaluated.

2. Methods and Materials

2.1. Data and Models

Monthly precipitation records of the Jean Lesage Intl Station (Figure 1) were collected from Canada Gov. historical meteorology data records [9]. The data-set period was from 1953 to 2020 (67 years). In addition, the CMIP5 and CMIP6 models were used to investigate the accuracy and evaluate future climate change under different scenarios. For these purposes, 2 different models (one from each CMIP) were selected according to previous research results [3,5]. These models are reported in Table 1. The historical period for evaluating the accuracy of the selected models was chosen. For CMIP5, this period was between 1953 to 2005 and for CMIP6, the period from 1953 to 2010 was selected.

Table 1. The selected CMIP models.

Model	CMIP	Scenario	Resolution
CanESM2	5	RCP 2.6, 4.5, 8.5	$0.5^\circ \times 0.5^\circ$
CanESM5	6	SSP 2.6, 4.5, 8.5	$0.5^\circ \times 0.5^\circ$

2.2. General Circulation Models

Climatic variables are simulated under the influence of increasing or decreasing greenhouse gases. There are different methods for this task; however, the most reliable is the use of atmospheric general circulation models or GCMs. GCMs can be used to understand the dynamics of the physical components of the atmosphere that are related to climate change phenomena. The purpose of using GCMs is to obtain spatial–temporal patterns of climate changes as well as long-term forecasting of climate variables [5]. Climate modeling is an important tool for understanding past, present and future climate changes [2]. In other words, currently the most reliable tool for investigating the effects of climate change on different systems is the use of a GCM. These models are able to model the trends of atmospheric and oceanic parameters for a long-term period using approved IPCC scenarios [2]. Their main weakness is the low spatial resolution and the simplification they consider for climate processes. To overcome the weakness of low resolution, it is necessary to scale the output of these models before using them in climate-change impact-assessment studies [1].

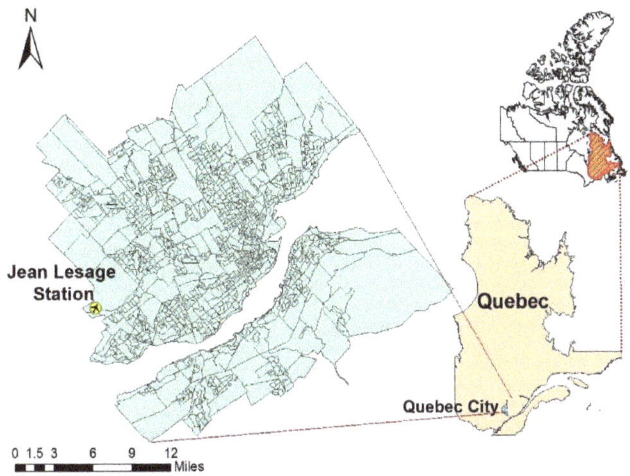

Figure 1. Selected Station position.

2.3. *Mann–Kendall Trend Analysis*

The Mann–Kendall method was first presented by Mann (1945) and then expanded and developed by Kendall (1970). Among the non-parametric tests, the Mann–Kendall test is the best choice for checking the uniform trend in series [10]. This test is used to determine the randomness and trend in the series. First, to determine the non-parametric nature of the statistical series, data are arranged and ranked in ascending order and then based on that, the randomness of the data with no trend is specified. If there is a trend in the data, then it is non-random.

The null hypothesis of the Mann–Kendall test indicates randomness and the absence of a trend in the data series, and the acceptance of the one hypothesis (rejection of the null hypothesis) indicates the existence of a trend in the data series [10].

2.4. *Evaluation of Performance*

Five types of statistical indices were employed to assess the performance of the CMIP data sets. The correlation coefficient (R) (Equation (1)) as a correlation-based index, Normalized Root Mean Square Error (NRMSE) (Equation (4)), Bias (Equation (2)), Root Mean Square Relative Error (RMSRE) (Equation (5)) and Slope (Equation (3)). The mathematical definitions of the mentioned indices are as follows:

$$R = \frac{\sum_{i=1}^{n}(y_i - \overline{y})(x_i - \overline{x})}{\sqrt{\sum_{i=1}^{n}(y_i - \overline{y})^2 \sum_{i=1}^{n}(x_i - \overline{x})^2}} \quad (1)$$

$$Bias = \frac{\sum_{i=1}^{n}(x_i - y_i)}{n} \quad (2)$$

$$SLOPE = \frac{\sum(x_i - \overline{x})(y_i - \overline{y})}{(n-1)S^2x} \quad (3)$$

$$NRMSE = \frac{1}{\overline{y}}\sqrt{\frac{1}{n}\sum_{i=1}^{n}(x_i - y_i)^2} \quad (4)$$

$$RMSRE = \sqrt{\frac{1}{n}\sum_{i=1}^{n}\left(\frac{x_i - y_i}{y_i}\right)^2} \quad (5)$$

where x_i and y_i are the ith samples of the estimated and actual values (respectively), $\overline{x}$ and $\overline{y}$ are the average of the estimated and actual values (respectively), n is the number of samples, and S^2x is the Variance of x.

3. Results

3.1. Evaluation the Performance of the Models

The performance of the CMIPs' models is shown in Table 2, which reported the accuracy of the models to compare to the observation data. Five different indices were used for evaluation at two different time scales. Based on the correlation coefficient (R), CanESM5 had better performance when compared to the CanESM2 model. Nevertheless, the correlation coefficient on a monthly scale was poor since R values were less than 0.5, although if the scale changed to the seasonal, they improved ideally (between 0.65 and 0.75).

Table 2. Performance of CMIP models.

Model	Scale	R	NRMSE	RMSRE	Bias	Slope
CanESM2	Monthly	0.43	0.40	0.36	−85.6	−0.30
CanESM5		0.48	0.54	0.30	−58.7	0.20
CanESM2	Seasonal	0.65	1.51	0.96	−11.8	0.50
CanESM5		0.75	1.08	0.67	−10.3	0.13

In addition, Normalized Root Mean Square Error (NRMSE) results showed that CanESM5 had better performance when compared to CanESM2 on the seasonal scale. These results are promising and show that the model can estimate precipitation within acceptable errors because the NRMSE values are close to one, which means that the deviations in precipitation estimates are small.

The RMSRE is a criterion similar to RMSE; their main difference is that RMSRE is divided by projected values. The best value for these criteria is 0, meaning there is no difference between projected and observed values. Based on Table 2, the CanESM5 model performed better with an RMSRE of 0.3 and 0.67 for the monthly and seasonal scales, respectively.

Mean Bias deviation shows the systematic error in the amount of precipitation. A value of zero indicates that the difference between the observed and predicted precipitation amount is not systematic, while a large Bias indicates that the amount of precipitation deviates greatly from the observed amount of precipitation. The fact that the Bias parameter is close to zero also indicates the model's accuracy in the simulation. A negative Bias indicates underestimation, while a positive Bias indicates overestimation. Based on the results in Table 2, both models underestimated precipitation at this Station.

Finally, the Slope is used to assess the direction of the projection line or the angle coefficient. If the Slope is negative, the relationship between the two variables (X and Y) will be inverse, and the Slope expresses the amount of change in Y relative to each unit of change in X. For this index, a Slope of 1, or regression 1:1 between the two variables, means perfect correlation. The value of the Slope in Table 2 shows that the estimated data were far from the regression line (1:1) and the value of the Slope statistic for both models was equal to or less than 0.5.

Taylor diagrams provide a visual framework for comparing a suite of variables from one or more test data sets to one or more reference data sets. Commonly, the test data sets are model experiments while the reference data set is a control experiment or some reference observations (e.g., Station data sets). Generally, the plotted values are derived from climatological monthly, seasonal or annual means. Because the different variables (e.g., precipitation, temperature) may have widely varying numerical values, the results are normalized by the reference variables. The normalized variances ratio indicates the model's relative amplitude and observed variations [11]. Figures 2 and 3 provide information about monthly and seasonal Taylor diagrams of the CanESM2 and CanESM5. It is clear that CanESM5's performance on the monthly scale is better than CanESM2's, although both models have poor performance on the seasonal scale. All in all, CanESM5 can provide better results compared to CanESM2.

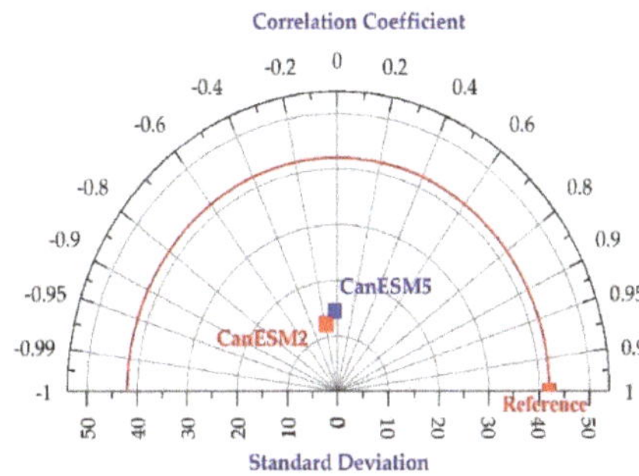

Figure 2. Models' performance in monthly scale.

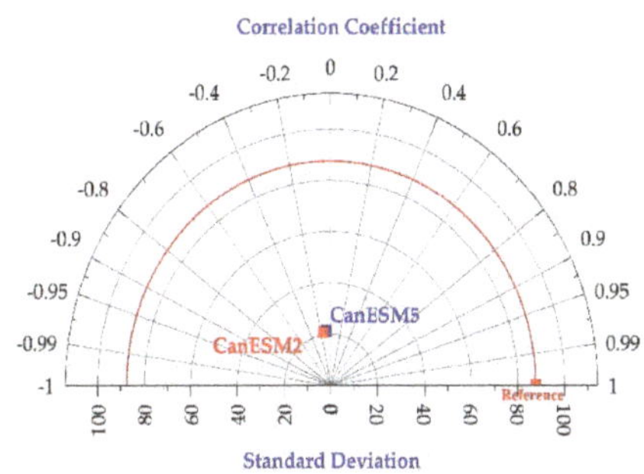

Figure 3. Models' performance in seasonal scale.

In summary, the values obtained by the models show that the efficiency of the CanESM5 model in estimating the amount of precipitation was better than the CanESM2 model. In addition, due to the closeness of the indicators that take into account the number of deviations and compare the estimated and actual time series, the mentioned model can detect fluctuations and precipitation trends in the selected Station.

3.2. Precipitation Future Trend

It is more suitable to use non-parametric methods for series that cannot be fitted with a special statistical distribution and have high skewness or elongation. The Mann–Kendall test is one of the most common and widely used non-parametric trend analysis methods of time series. Data changes are identified using the Mann–Kendall method, and their type and time are determined [12]. According to the essential role of precipitation in providing water resources, it is vital to study the process of its changes in the future. This study will help the authorities with planning and managing water resources. Since international reports have delivered serious warnings about the crisis and even the physical lack of water in the future for different parts of the world, knowing the predicted variability of this important meteorological parameter is essential [13]. Although the evaluation of the models' performance on historical data has shown that the models do not have a high ability to estimate the amount of precipitation, the comparison of the projected and observed time series shows a slight deviation and an acceptable agreement between the two data sets. Therefore, the future forecast of the precipitation by the selected model (CanESM5) is an effective step in understanding the precipitation pattern in the future.

Table 3 presents the Mann–Kendall parameters (Test Z) for the different time-scale rainfall future trends (from 2020 to 2049) using the best-fit model (CanESM5) for the two socio-economic (SSP) scenarios. The selected SSP scenarios for this study were 4.5 (Pessimistic) and 8.5 (Optimistic). The time scales were Monthly (Jan to Dec), Seasonal (Spring to Winter) and Annual for the selected model's scenarios (4.5 and 8.5). For scenario 4.5, rainfall changes were decreasing for February and June to August, while for other months, these changes were increasing. These outputs can be found based on Test Z values in Table 3. Moreover, rainfall changes in the seasonal scale were also decreasing in the Summer and Fall. Trend analysis results for May, for July in the monthly scale and for Summer and Fall in the seasonal scale showed that the downward trend was more intense since the value of Test Z (Kendall's score) reached more than −1 in this period of time.

Table 3. Mann–Kendall test Z values for 4.5 and 8.5 scenarios.

Time Series	4.5 Test Z	8.5 Test Z	Time Series	4.5 Test Z	8.5 Test Z
Jan.	1.48	0.04	Jul.	−1.78	1.89
Feb.	−0.30	0.25	Aug.	−0.36	1.53
Mar.	0.16	−0.71	Sep.	1.30	0.14
Apr.	−0.43	−0.79	Oct.	2.02	2.86
May.	−1.07	1.46	Nov.	1.30	0.18
Jun.	−0.46	−0.09	Dec.	1.48	1.71
Spring	0.71	−0.18			
Summer	−1.25	0.39	Annual	0.00	2.82
Fall	−1.32	2.32			
Winter	1.78	2.21			

On the other hand, for scenario 8.5 (Table 3), rainfall changes were increasing in most of the months, although, like scenario 4.5, the decreasing trend remained. However, the intensity of the downward trend became less prominent when compared to scenario 4.5. In addition, rainfall changes in the spring tended to decrease, unlike the 4.5 scenario.

4. Conclusions

The results showed that the studied models do not have a high ability to estimate precipitation in the Jean Lesage Intl Station. According to the results of the studied statistics such as the correlation coefficient (R) and Slope, the accuracy of the models was poor and the correlation coefficient in all models was less than 0.5 on a monthly scale. However, in the seasonal scale, the correlation value was reached at 0.75 in the best model. The Slope index was also consistent with the correlation coefficient because in the two investigated models, the distribution of precipitation data was rarely very close to the regression line (1:1) and the Slope value was usually less than 0.5. In addition, the results of the two selected models were close to each other, but the CanESM5 model was more accurate than the other model in the studied Station. The deviation of the projected data and the Station data was very small, which can be shown based on the NRMSE index in all the investigated models as less than 2. In addition, in the selected Station, the Bias index indicated both models would underestimate the rainfall trend in both time scales. The comparison of the obtained findings showed that the present research results were largely consistent with some other researchers. For example, Hidalgo and Alfaro (2014) showed that most of the CMIP5 models have a low ability to estimate precipitation in the central regions of the United States [11]. Rupp et al. (2013) showed that although the CMIP5 model rainfall data have less accuracy compared to other grided data such as NCEP and ERA40, they estimate the seasonal cycle of precipitation with the same accuracy as networked data in the northwestern regions of America [12]. Mehran et al. (2014) concluded that the CMIP5 model rainfall data are consistent with GPCP data in most parts of the world but do not perform well in dry areas [13]. Ebtehaj and Bonakdari (2023) concluded that the results of the comparison of the CanESM5 and CanESM2 models strongly depend on the month and season, and that the results of CanESM5 are slightly better compared to the other model [5].

Finally, the precipitation trend analysis results for the CanESM5 model and under the two scenarios 4.5 and 8.5 showed that the trend of precipitation changes at the Jean Lesage Intl Station will not be significant. In addition, in scenario 4.5, the precipitation trend decreased in almost half of the year, while in scenario 8.5, the intensity of the decrease and the number of months with a decreasing trend of precipitation were significantly reduced.

Author Contributions: Conceptualization, H.B. and T.G.; methodology, H.B.; software, A.S.; validation, A.S. and H.B.; formal analysis, A.S.; investigation, A.S.; resources, H.B.; data curation, H.B.; writing—original draft preparation, A.S.; writing—review and editing, T.G. and H.B.; visualization, A.S.; supervision, T.G. and H.B.; project administration, T.G.; funding acquisition, H.B. All authors have read and agreed to the published version of the manuscript.

Funding: This project is partially funded by the ESSOR–MARTHE-ET-ROBERT-MÉNARD scholarship administered by CentrEau (Centre québécois de recherche sur la gestion de l'eau) and the Government of Québec Ministère de Sécurité Publique (MSP) under the project 'Compréhension du comportement des rivières en hiver et mesures de gestion des risques liés aux inondations (FLUTEIS; CPS-18-19-26)'.

Institutional Review Board Statement: Not applicable.

Informed Consent Statement: Not applicable.

Data Availability Statement: GCM datasets are available at Copernicus Climate Change Service, Climate Data Store, (2021): CMIP6 climate projections. Copernicus Climate Change Service (C3S) Climate Data Store (CDS). DOI:10.24381/cds.c866074c (accessed on 25 November 2022).

Conflicts of Interest: The authors declare no conflict of interest.

References

1. Salimi, A.H.; Masoompour Samakosh, J.; Sharifi, E.; Hassanvand, M.R.; Noori, A.; von Rautenkranz, H. Optimized artificial neural networks-based methods for statistical downscaling of gridded precipitation data. *Water* **2019**, *11*, 1653. [CrossRef]
2. IPCC. Summary for Policy Markers Climate Change: The Physical Science Basis. In *Contribution of Working Group I to the Fourth Assessment Report of the Intergovernmental Panel on Climate Change*; Cambridge University Press: Cambridge, UK, 2007; p. 881.
3. Sobie, S.R.; Zwiers, F.W.; Curry, C.L. Climate Model Projections for Canada: A Comparison of CMIP5 and CMIP6. *Atmos. Ocean.* **2021**, *59*, 269–284. [CrossRef]
4. Lovejoy, S. The Future of Climate Modelling: Weather Details, Macro weather Stochastics—Or both? *Meteorology* **2022**, *1*, 414–449. [CrossRef]
5. Ebtehaj, I.; Bonakdari, H. A comprehensive comparison of the fifth and sixth phases of the coupled model intercomparison project based on the Canadian earth system models in spatio-temporal variability of long-term flood susceptibility using remote sensing and flood frequency analysis. *J. Hydrol.* **2023**, *617*, 128851.
6. Martel, J.L.; Brissette, F.; Troin, M.; Arsenault, R.; Chen, J.; Su, T.; Lucas-Picher, P. CMIP5 and CMIP6 model projection comparison for hydrological impacts over North America. *Geophys. Res. Lett.* **2022**, *49*, e2022GL098364. [CrossRef]
7. Miara, A.; Macknick, J.E.; Vörösmarty, C.J.; Tidwell, V.C.; Newmark, R.; Fekete, B. Climate and water resource change impacts and adaptation potential for US power supply. *Nat. Clim. Chang.* **2017**, *7*, 793–798. [CrossRef]
8. Eyring, V.; Gleckler, P.J.; Heinze, C.; Stouffer, R.J.; Taylor, K.E.; Balaji, V.; Guilyardi, E.; Joussaume, S.; Kindermann, S.; Lawrence, B.N.; et al. Towards improved and more routine Earth system model evaluation in CMIP. *Earth Syst. Dyn.* **2016**, *7*, 813–830.
9. Extracted from the Environment and Climate Change Canada Historical Climate Data. Available online: https://climate.weather.gc.ca/index_e.html (accessed on 4 December 2022).

10. Hussain, M.; Mahmud, I. pyMannKendall: A python package for non-parametric Mann Kendall family of trend tests. *J. Open Source Softw.* **2019**, *4*, 1556. [CrossRef]
11. Hidalgo, H.G.; Alfaro, E.J. Skill of CMIP5 Climate Models in Reproducing 20th Century Basic Climate Features in Central America. *Int. J. Climatol.* **2014**, *35*, 3397–3421. [CrossRef]
12. Rupp, D.E.; Abatzoglou, J.T.; Hegewisch, K.; Mote, M. Evaluation of CMIP5 20th century climate simulations for the Pacific Northwest USA. *J. Geophys. Res. Atmos.* **2013**, *118*, 10884–10906. [CrossRef]
13. Mehran, A.; AghaKouchak, A.; Phillips, T.J. Evaluation of CMIP5 continental precipitation simulations relative to satellite-based gauge-adjusted observations. *J. Geophys. Res.* **2014**, *119*, 1695–1707. [CrossRef]

environmental sciences proceedings

MDPI

Proceeding Paper

The Role of Climate Change on Water Resources Management in the Southern Caucasus in the Post-Conflict Period †

Firuz Suleymanov

Department of Geographical and Historical Studies, University of Eastern Finland, 80130 Joensuu, Finland; firuzsul@uef.fi

† Presented at the 7th International Electronic Conference on Water Sciences, 15–30 March 2023; Available online: https://ecws-7.sciforum.net.

Abstract: Climate-change-induced environmental impacts has an especially strong influence on water resources. Declining water availability not only results in droughts, but is also responsible for decreasing the quality of water in water-scarce regions, such as the South Caucasus. Armed conflict between Armenia and Azerbaijan in 2020 occurred during the recent major war in the South Caucasus. In this paper, I ask how will intensifying climate change in the region affect the current political situation? Is there any chance of multilateral cooperation for water management?

Keywords: water management; climate change; southern Caucasus; armed conflict; Azerbaijan; Armenia

check for updates

Citation: Suleymanov, F. The Role of Climate Change on Water Resources Management in the Southern Caucasus in the Post-Conflict Period. *Environ. Sci. Proc.* **2023**, *25*, 43. https://doi.org/10.3390/ECWS-7-14165

Academic Editor: Athanasios Loukas

Published: 14 March 2023

1. Introduction

Climate change has various impacts on different living conditions in societies. Increasing global temperatures are causing huge problems for surface water evaporation, while the warmer atmosphere is maintaining more moisture aloft, which causes both large floods and induces extreme droughts in different parts of the world. Climate change concerning water resources raises temperatures, the frequency and intensity of droughts, flooding, and evaporation in various parts of the world such as the South Caucasus. The South Caucasus mainly faces droughts and low precipitation levels. Frequent instances of armed conflict between and within South Caucasus countries and neighbors work as an accelerator of competition and collaboration regarding water resources.

In the last century, the South Caucasus has faced frequent instances of armed conflict. The long-standing Nagorno-Karabakh conflict (1988–present) between Azerbaijan and Armenia erupted again into war in 2020, resulting in Azerbaijan recapturing much of its territory from local Armenian separatists and Russia negotiating a new truce and installing troops to maintain peace. The conflict between Armenia and Azerbaijan over Karabakh in the fall of 2020 upended the decades-old regional balance established during the 1991–94 war [1]. The long-standing armed conflict damages communication between parties and collaborative bilateral and regional actions. Parties need to build trust with the mediation of neutral parties for efficient water management in post-conflict periods to reduce the harmful impact of climate change. Countries cannot adapt to climate change by themselves and require the regional or international management of water and other natural resources. One of the most difficult development challenges of the twenty-first century is the lack of water. Nearly 3 billion people, or more than 38% of the world's population, live in watercourse regions that are affected by water shortages, water pollution, and geopolitical instability. A lack of water can have distinctly negative effects on welfare and the profitability of agriculture and cause armed conflict, terrorism, business decline, lack of sectoral and social growth, as well as environmental and business deterioration. Without sufficient water resources, the overall productivity of societies is decreasing, influencing demographic change, economic development, urbanization, and globalization, as well as intensifying conflict over water use. This paper illustrates how a lack of enough water resources during climate change involves Azerbaijan and Armenia competing and/or collaborating in water resources management in the South Caucasus in the post-conflict period.

2. Materials and Methods

This article is a single case study of water in post-conflict Azerbaijan and Armenia. The research is largely qualitative in nature, although on a few occasions, some quantitative data were used. A descriptive, analytical approach was used to perform the study. The researcher examined primary and secondary sources and the scholarly literature on the topic. As the theoretical background of the research was rather complex, the approach of its method is multidisciplinary, combining the fields of the environment and international relations. The analysis is based on a one-of-a-kind collection of deliberately chosen interviews with water resources management, climate change, and international relations experts in Azerbaijan throughout the study period. In-depth interviewing was conducted to take into account a broad and proportional representation of various parts of society. Questions were designed and preliminarily discussed by the leading researchers and academicians in order to directly collect information on the research focus. Semi-structured interviews consisted of 12 questions in Azerbaijani. Government records were also reviewed, as well as the literature on the role of climate change in water resources management and the post-conflict situation between Armenia and Azerbaijan to qualify and contextualize the interview data. Studying interviews from a wide range of qualitative perspectives is a technique. Having said that, interviews were conducted largely with an exploratory goal to acquire a deeper understanding of how climate change plays an essential role in water resource management between former war participants. The "snowballing" sampling method of sample selection was centered on locating key informants in the literature and through internet searches in accordance with pertinent knowledge or professional experience. By using "referral" methods—asking the informants to suggest other subject matter experts—more participants were chosen. All interview data were recorded and transcribed word for word and stored digitally. I used the NVivo® program for data systematization and transcription, while there were times when I chose to use manual systematization. In addition to the systematic analysis of documents and the field diary, interpretive analysis rooted in the adopted theoretical framework was used to analyze the data and identify the relationship between climate change, water resources management, and post-conflict periods in this case.

The prisoner's dilemma game model was applied in this study. The prisoner's dilemma is a decision-making conundrum in which two persons acting in their self-interest do not generate the best conclusion. This game simulates a case in which there are benefits to cooperating, although each actor has the incentive to do whatever the other actor does. The standard prisoner's dilemma is presented in such a manner that both sides opt to defend themselves at the expense of the other party. Consequently, both participants are in a worse position than they would be if they had collaborated in the decision-making process. The prisoner's dilemma describes a status in which two sides, separated and incapable to communicate, must choose whether or not to cooperate with each other. In this paper, the players are Armenia and Azerbaijan, respectively, due to recent armed conflicts, and the third party is climate change. Georgia was excluded due to this country having used a peaceful and collaborative approach to Armenia and Azerbaijan.

3. Results and Discussion

First of all, water scarcity issues are currently being experienced by nations all over the world, some of which are minor, and others are severe. A growing population and the effects of climate change will make these issues much more challenging. From the present day to 2050, countries will deal with these issues regularly [2]. Global warming, commonly known as man-made climate change, alters the hydrological patterns that control the water supply on Earth [3]. Climate change affects the quantity, quality, and timeliness of water, which impacts humans more and more [4].

Moreover, climate change will further stress water resources in the South Caucasus region. Precipitation levels are expected to fall, and temperatures are expected to rise, resulting in a drop in the quantity of runoff by 2050 or sooner. At the same time, crop water demands will rise as temperatures rise. The transboundary character of water resources, combined with the overwhelming likelihood that climate change will limit water flow and quantities in general, raises the prospect of disputes over increasingly valuable water resources. These tensions, however, can be mitigated by integrated regional water resource planning. More water storage facilities, for example, are nearly always more effectively built at higher elevations, where the natural terrain may be used to produce reservoirs and where the steeper terrain offers more potential for hydropower generation by the reservoir's outflow. In the World Bank's studies, scientists forecasted grave climate-induced water vulnerabilities in Armenia and Azerbaijan. Climate change has had and will continue to have the greatest influence downstream in the Kura-Aras basin, where the quantity of water flowing from the river is lowest and the agricultural demand is highest. Over the previous century, the volume of glaciers in the South Caucasus has decreased by half, and 94 percent of the glaciers have retreated by 38 m per year [5].

Furthermore, external players, generally international organizations, also played a significant role in reshaping water governance during and after the war [6]. When armed conflict undermines water resource management, local communities also lose a critical climate adaptation option [7]. International organizations or neutral countries can be mediators to improve communication between parties for sustainable water management in post-conflict periods, such as the Indus Waters Treaty [8], and response to climate change more effectively. Neutral international organizations can help to build trust between Azerbaijan and Armenia, at least terms of in bilateral and regional water management. Trust is often seen as a critical enabler of water governance. Water cooperation has the capacity to remove distrust and suspicion between and across countries, introducing prospects for shared profits and establishing a reciprocity paradigm [9]. The governance of water resources has a critical role in both enabling the rebuilding of trust after conflict and avoiding a return to conflict by ending or minimizing existing conflicts [10]. One of the problems that frequently impede improved regional water cooperation is a failure to recognize the benefits of collaboration. Trust might be difficult to establish in areas with a lengthy history of war, such as the cases in the South Caucasus. Many nations are accustomed to considering water resources as an issue of national security, and thus may be reluctant to share information. When countries agree to collaborate, they want institutional structures and mechanisms that will survive over time, even if their relationship is stressed for other reasons.

In addition, armed conflicts in the South Caucasus, in particular between Armenia and Azerbaijan, damage communications and collaborations between parties. Countries cannot adapt to climate change alone and demand collaborative water management. Trust is an integral part of water management, especially in post-conflict periods. On the other hand, building trust between previous conflict parties is a challenge. Table 1 illustrates Armenia and Azerbaijan as actors in the prisoner dilemma and the possible results of their behavior in collaborative water management during climate change. Both sides need to trust each other so that the opposite side will not take advantage of their desire to cooperate in water management. If the first side cooperates and the second side will not cooperate, then the second side will gain more benefits rather than those that will result from the cooperation of both of them, and the first will gain larger harm than those that will result from the non-cooperation of them. As a result, both parties are afraid to cooperate in water management during climate change to avoid the biggest possible biggest negative effect. The lack of trust as a consequence of long-standing armed conflict between parties and the lack of a peace treaty are two essential barriers to collaborative water management in the region.

Table 1. The prisoner's dilemma game model in this case.

		Armenia	Armenia
		Cooperate	**Do not Cooperate**
Azerbaijan	Cooperate	Benefits for the both of them.	More benefits for water management in Armenia; Damage to water management in Azerbaijan.
Azerbaijan	Do not cooperate	Damage to water management in Armenia; More benefits for water management in Azerbaijan.	Medium damage to water management in Armenia; Medium damage to water management in Azerbaijan.

Furthermore, as a result of interviews, most participants agreed that climate change plays a critical role in water resources management between Armenia and Azerbaijan in the post-conflict period. All participants agreed that water resources in the region are scarce, in particular in Azerbaijan, and during climate change, parties need to collaborate. In detail, an academician claimed that the negative impact of climate change on water resources and their management of it is already visible in Azerbaijan [11]. Nagorny Karabakh's conflict began its twentieth year of "no-war-no-peace" in 2014. Since the conflict began 20 years ago, Armenian and Azerbaijani societies have been radicalized to feel mutual hostility [12]. During this period, Armenia and Azerbaijan did not effectively collaborate or perform risk taking for collaboration, and instead, they expressed high-level hostility. A civil society member thought that climate change will force Armenia and Azerbaijan to act immediately to decrease the negative impact of climate change on water resource management [13]. An ecological activist claimed that without collaboration, it is hard to respond to climate change effectively [14], particularly during post-conflict periods. According to an independent researcher, parties must involve Turkey as an upstream country of the major water resources of both countries in water resource management between Armenia and Azerbaijan or in the general context in the South Caucasus [15]. An independent researcher highlights the role of trust and mediation in water resources management. On the other hand, civil society members claimed that collaboration between parties will decrease water resource availability and access in Armenia and Azerbaijan [15].

4. Conclusions

After armed conflict, rebuilding and recovery initiatives should use a constant strategy of utilizing water resources, which will not just be sensitive to national needs and the environment, but will also require the assistance of local communities to encourage collaboration and peace in the long term and reduce the chance of the relapse of conflict between parties. Collaboration among many parties regarding critical survival concerns such as water governance can have a good spillover impact on other, more controversial areas during climate change in post-conflict periods. Establishing a commitment to developing and sharing critical natural resources such as water in a sustainable and equitable manner can support overcoming existing insecurity or distrust between states and create a climate of mutual benefits and the assessment of long-term common interests. Regional water quality management and monitoring in the South Caucasus may help both the economy and the ecology throughout the Kura-Aras basin. The protection of riverine aquatic ecosystems will necessitate collaboration, with the result being improved water quality for all users. According to the interviews, water is already scarce in Armenia and Azerbaijan, and cooperation between parties can support the effective management of water resources and water access, particularly during climate change. The interview participants agree that climate change mainly makes a negative impact on water resources and creates more challenges for its management. The coordination of water resource management, with proper regard for riparian rights and a specific country's needs, has the potential to considerably lessen the consequences of climate change on water supplies, and hence, improve the shared benefits. Future regional water management must take into account non-agricultural water consumers, such as those using local water for hydropower and industrial consumers, as well as the preservation of natural systems. Parties need to build trust between each other via the mediation of neutral actors and repair communication, at least for collaboration in water management to respond to the negative impact of climate change in the South Caucasus.

Funding: This research received no external funding.

Institutional Review Board Statement: Not applicable.

Informed Consent Statement: Not applicable.

Data Availability Statement: Not applicable.

Conflicts of Interest: The authors declare no conflict of interest.

References

1. Saparov, A. Place-name wars in Karabakh: Russian Imperial maps and political legitimacy in the Caucasus. *Cent. Asian Surv.* **2023**, *42*, 61–68. [CrossRef]
2. Salem Hilmi, S.; Pudza Musa, Y.; Yihdego, Y. Water strategies and water–food Nexus: Challenges and opportunities towards sustainable development in various regions of the World. *Sustain. Water Resour. Manag.* **2022**, *8*, 114. [CrossRef] [PubMed]
3. Gehrig, J.; Rogers, M.R. Water and Conflict. In *Water and Conflict, Incorporating Peacebuilding into Water Development*; Warner, D., Seremet, C., Bamat, T., Eds.; Catholic Relief Services: Baltimore, MD, USA, 2009; p. 134.
4. Javeline, D.; Dolšak, N.; Prakash, A. Adapting to water impacts of climate change. *Clim. Change* **2019**, *152*, 209–213. [CrossRef]
5. Boehlert, B.; Neumann, J.; Strzepek, K.; Droogers, P.; Sharrow, S. Chapter 3 Armenia: Risks, Impacts, and Adaptation Menu. In *Building Resilience to Climate Change in South Caucasus Agriculture*; Ahouissoussi, N., Neumann, J.E., Jitendra, P.S., Eds.; The World Bank: Washington, DC, USA, 2014; pp. 41–42.
6. Schillinger, J.; Özerol, G.; Güven-Griemert, Ş.; Heldeweg, M. Water in war: Understanding the impacts of armed conflict on water resources and their management. *WIREs Water* **2020**, *7*, 11–15. [CrossRef]
7. Vivekananda, J.; Schilling, J.; Smith, D. Climate resilience in fragile and conflict-affected societies: Concepts and approaches. *Dev. Pract.* **2014**, *24*, 487–501. [CrossRef]
8. Qureshi, W.A. Water as a Human Right: A Case Study of the Pakistan-India Water Conflict. *Penn State J. Law Int. Aff.* **2017**, *5*, 376–379.
9. Swain, A. Water and post-conflict peacebuilding. *Hydrol. Sci. J.* **2016**, *61*, 1313–1322. [CrossRef]
10. Weinthal, E.; Troell, J.; Nakayama, M. Water and post-conflict peacebuilding: Introduction. Water International. *Post-Confl. Peacebuilding Water Manag.* **2011**, *36*, 143–153. [CrossRef]
11. Interview with Academician, Baku, Azerbaijan. 4 February 2023.
12. Ayunts, A.; Zolyan, M.; Zakaryan, T. Nagorny Karabakh conflict: Prospects for conflict transformation. Nationalities Papers. *J. Natl. Ethn.* **2016**, *44*, 543–559. [CrossRef]
13. Interview with civil society member, Baku, Azerbaijan. 29 December 2022.
14. Interview with ecology activist, Baku, Azerbaijan. 4 February 2023.
15. Interview with an independent researcher, Baku, Azerbaijan. 3 January 2023.

environmental sciences proceedings

MDPI

Proceeding Paper

Vulnerability of Small Rivers Coastal Part Due to Floods: The Case Study of Lesvos North—West Coast †

Stamatia Papasarafianou [1,*], Aliki Gkaifyllia [2], Anna-Eirini Iosifidi [3], Stavros Sahtouris [2], Nathalie Wulf [4], Alexandra Culibrk [2], Maria-Danai Stamataki [2], Theodoros Chatzivasileiou [2], Ilias Siarkos [2], Celia Rouvenaz [5], Eleni-Ioanna Koutsovili [2], Thomas Hasiotis [2] and Ourania Tzoraki [2]

1 Department of Science and Technology, University of Napoli Parthenope, 80143 Napoli, Italy
2 Department of Marine Sciences, University of the Aegean, 80100 Mytilene, Greece
3 Department of Geography, University of the Aegean, 80100 Mytilene, Greece
4 Department of Geography, University of Christian-Albrechts, 24118 Kiel, Germany
5 Department of Civil, Environmental and Geomatic Engineering, University of Environmental Engineering—ETH Zürich, 8092 Zürich, Switzerland
* Correspondence: stamatia.papasarafianou001@studenti.uniparthenope.it; Tel.: +30-6981-718-533

† Presented at the 7th International Electronic Conference on Water Sciences, 15–30 March 2023; Available online: https://ecws-7.sciforum.net.

Abstract: This study presents the development of a vulnerability assessment methodology combining both hydraulic and oceanographic values to evaluate the fragility of the island's coast, subject to floods. The study area covers the coastal part Petra - Molyvos in North-West Lesvos Island, Greece. Petra stream drains a catchment area of 8.06 km^2. The flooded sections of the river's coastal part are analyzed by the HEC-RAS model, while the coastal vulnerability index (CVI) was calculated by the InVEST model. The scenario of habitats' role in beach protection showed 53% of coastal protection and the CVI moderate exposure to sandy beaches. A change in the geomorphology of the estuary was observed during the summer period, due to the river sediment dredging and small delta reclamation processes.

Keywords: HEC-RAS; coastal vulnerability model; erosion; Lesvos

Citation: Papasarafianou, S.; Gkaifyllia, A.; Iosifidi, A.-E.; Sahtouris, S.; Wulf, N.; Culibrk, A.; Stamataki, M.-D.; Chatzivasileiou, T.; Siarkos, I.; Rouvenaz, C.; et al. Vulnerability of Small Rivers Coastal Part Due to Floods: The Case Study of Lesvos North—West Coast. *Environ. Sci. Proc.* **2023**, *25*, 44. https://doi.org/10.3390/ECWS-7-14257

Academic Editor: Athanasios Loukas

Published: 16 March 2023

1. Introduction

Beaches are critical components of the coastal zone; coastal services contribute around 43% of the total benefits provided by human well-being [1,2]. Beaches provide cultural and ecosystem services, such as food, flood protection, natural environmental conservation, and recreation activities [2,3]. With the increases in multiple human activities, the bio-geological resources are being exploited, polluted, and exposed to stressful uses and pressures, increasing the coastal vulnerability. As a result, coastal erosion is threatening tourist zones and services.

Coastal erosion constitutes a contemporary environmental problem and the projections of coastal retreat under climate change and vulnerability highlight the threat to sustainable growth in the Aegean archipelagos [4,5]. The threat of coastal retreat increases as the rivers' dammed sediments are arrested in the reservoirs, which has an indirect impact on water storage due to the transportation of sediments by the surface water flow. For example, flash floods can be perceived negatively as they cause (i) damage to farms, and infrastructures, (ii) bank erosion, (iii) and the transport of sediment and coarse material [6,7]. As a large problem that puts the achievement of multiple sustainable development goals of the United Nations at risk, this should be managed through an integrated approach, considering the possibility of sustainable growth [3].

At the same time, the hydrological regimes of Intermittent Rivers and Ephemeral Streams are characterized by flow cessation and dry events at certain periods of the year [6], and ephemeral streams exhibit certain conditions during the hydro-geomorphological processes of runoff generation. Ephemeral streams are more frequent globally, and more than 50% of the river network is intermittent [8]. The Mediterranean strip is the most affected area in general, but in terms of specific regions or areas within countries, different percentages of the river network are affected. Specifically, in France, only 20% of the river network is affected, while in Sardinia and Sicily, 90% of the river network is affected. Additionally, in the southeast of Spain, more than 70% of the fluvial systems are ephemeral streams [9]. The flowing phase enhances sediment transport, while the non-flowing and dry phases facilitate river access and sand and gravel extraction [5].

This paper combines the main results with an analysis of floods after flushy rain episodes and the coastal erosion vulnerability of the sandy and rocky beaches. The aim is to discuss how ephemeral streams could behave to water and sediment balance in a touristic beach, under human and natural pressures. Vulnerability calculation was done using InVEST Coastal Vulnerability index, combining oceanographic and ecological parameters, while HEC-RAS was used to calculate the critical cross-sections in the Petra River.

2. Materials and Methods

2.1. Study Area

The study area covered the coastal part from Petra to Molyvos in west–north Lesvos Island, Greece (Figure 1). The area is characterized by a typical Mediterranean climate with warm and dry summers and cold and wet winters. The mean annual precipitation is 545 mm, and the mean temperature is 17.39 °C. During the winter, the precipitation records showed January, February, and December as the wet months. The main wind directions are north–northeastern, with a maximum intensity of from 4 to 5 bft and 5 to 6 bft. A Petra stream of 1.75 km in length drains a catchment area of 7.97 km^2, with a mean width of 6.15 m, a mean slope of 1.03%, and a mean depth of 1.63 m, and flows into the Aegean Sea.

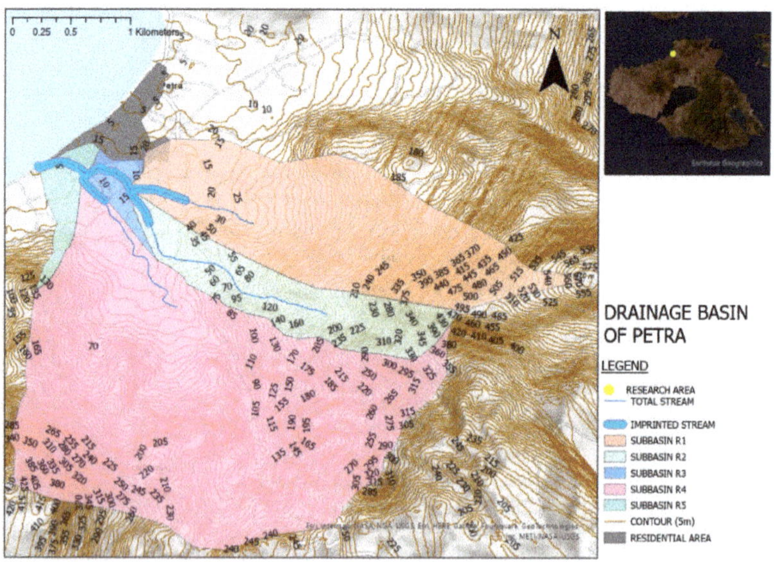

Figure 1. Petra catchment: The basin was divided into 5 subbasins to analyze the mainstream Petra River.

2.2. Coastal Vulnerability Models

The InVEST Coastal Vulnerability model is widely still used to assess coastal erosion, and can provide a framework for regional coastal zone protection and future development planning. To estimate erosion from coastal hazards for people throughout the Petra–Molyvos coastline, we mapped and valued the changes in ecosystems, which could help to convert in the flow, obtaining many different benefits [4]. The model builds on previous, similar indices that account for bio-geophysical components to compare their exposure to erosion and flooding in severe weather [5]. The InVEST CV model produces a qualitative index of coastal exposure to erosion, as well as summaries of human population density in proximity to the coastline. The exposure of erosion in the study area is defined by calculating the coastal vulnerability index (CVI), which ranks sites from lowest exposure (rank = 1) to highest risk of erosion (rank = 5) and inundation using the following variables: habitat type, the local bathymetry and topography, the relative wind and wave force associated with storms, and the population density (Equation (1)).

$$CVI = (R_{Habitats} R_{Shorline\ Type} R_{Relief} R_{Waves} R_{Surge})^{1/5} \quad (1)$$

2.3. HEC-RAS Model

The software (version 6.3.1) Engineers River Analysis System (HEC-RAS) developed by the Hydrologic Engineering Center allows for users to model rivers flowing through open natural channels and is used to compute water surface profiles [10]. The HEC-RAS system can simulate one-dimensional, steady-flow, water-surface profile computations and uses geometric and hydraulic computation routines. The river geometries, such as centerlines, bank lines, flow paths, and cross-sectional lines, are the major parameters processed in HEC-RAS to generate flood-prone areas [10].

3. Results

The Aegean Archipelago climate, water masses' movement, and geomorphology have developed a strong ecological and human resilience to degradation. However, both environmental and anthropogenic changes are threatening the precarious balance between water resources and flood risk [7].

3.1. Coastal Vulnerability

Natural activities play a significant role in coastal vulnerability. The Aegean circulation ends with a northern flow in the eastern basin. The passage between Asia Minor, Chios, and Lesvos has more kinetic energy than the rest of the eastern basin (low kinetic energy) due to the 200 m isobath, which turns to the east. Additionally, the currents near Petra have low kinetic energy [11]. The complex physiography of the Aegean archipelago influences the wind, in addition to the wave climate. In general, wind and waves occur because of the short fetches and relatively mild duration [2]. The wind and waves on the Northern Aegean coastlines are mostly northern in direction, and as a result, they occur frequently on the Petra-Molyvos coastline. Although waves are generally more energetic in the winter season, even in the summer, energetic waves can be found the Aegean Sea and in Petra, which are mainly forced by the strong, dry N–NE 'etesian' winds [12,13].

The results of the InVEST Coastal Vulnerability model showed that the main exposed areas of erosion are sandy beaches. The two vulnerability scenarios—with habitats (Figure 2a) and without habitats (Figure 2b)—showed the importance of habitats in beach protection. A total of 13% of the coastal zone showed a high risk of erosion (rank = 4–5) and 21% of the coastal zone had moderate risk (rank = 3) for the scenario without habitats. The vulnerability scenario with habitats obtained a score of 53% for coastal protection, and the CV index showed moderate risk (rank = 2–3). Additionally, the results from the CV index showed low exposure (rank = 1–2) to rock beaches and high exposure (rank = 4–5) to sandy beaches (Figure 3).

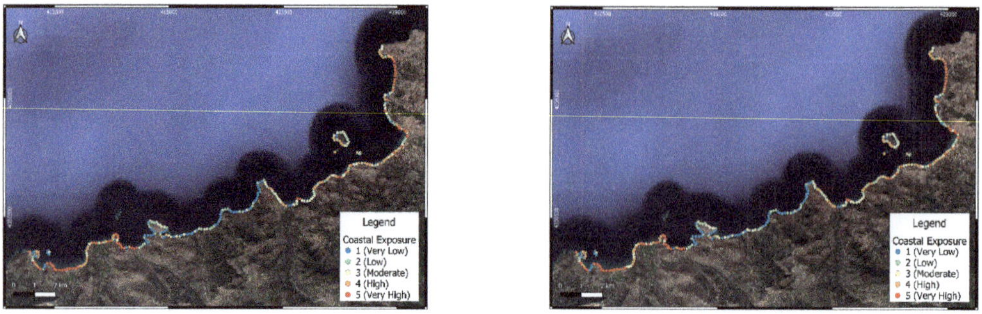

(a) (b)

Figure 2. Map of Petra–Molyvos' exposure to erosion (**a**) with the habitats variable and (**b**) without the habitats variable.

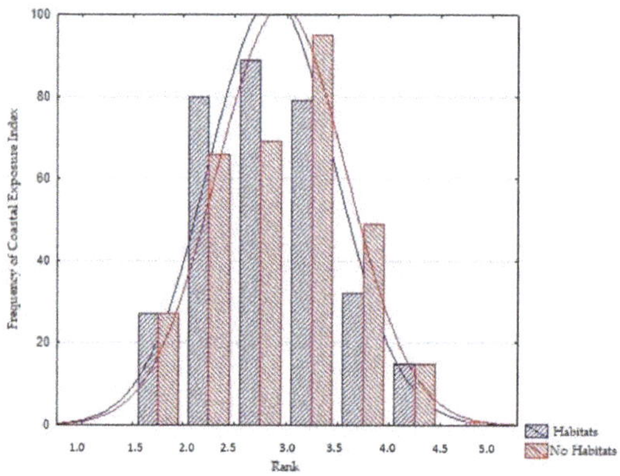

Figure 3. Histogram with a frequency of the occurrence of observed coastal vulnerability index values.

3.2. Hydraulic Study

The Froude value at the location R5 (Table 1) of the cross-section S2 is 0.15 and is characterized as subcritical. This cross-section is located 17 m upstream (Figure 4a,b) the mouth of the stream towards the sea. It presents the lowest velocity and energy line slope values in the branch (v = 0.57 m/s, slope = 0.0005 m/m), and the mean velocity of 1.33 m/s. The Froude value at the cross-section S18 location is 0.14 (Figure 4a,c) and is characterized as subcritical. The S18 has the smallest geometry in the branch. The velocity and slope values of the energy line are also small (v = 0.48 m/s, slope = 0.0005 m/m), and the branch average velocity is 1.10 m/s. The Froude value at the cross-section S26 location is 0.44 (Figure 4a,c) and is characterized as subcritical. The S26 geometry is the second smallest for the branch. The power line velocity and gradient values are low throughout the branch, with the cross-section values of v = 0.87 m/s, slope = 0.0059 m/m, and the branch average velocity is 0.90 m/s. Upstream of the cross-section, there is a hydraulic jump where the flow changes from supercritical to subcritical. The Froude value at the cross-section S32 location is 0.27 (Figure 4a,e), and the velocity and slope of the energy line are significantly lower than all other cross-sections (v = 0.66 m/s, slope = 0.001856 m/m). The wetted surface is the largest present in branch R1, while the branch average velocity is equal to 1.42 m/s. Finally, the Froude value at the cross-section S37 location is 0.43 (Figure 4a,f) and is characterized as subcritical. The velocity and slope of the energy line are the third smallest (v = 0.99 m/s, slope = 0.006421 m/m). The cross-section shows the second smallest geometry, and the branch average velocity is equal to 1.25 m/s. The simulation results show that, under flashy rain events, several cross-sections of the stream are overflooded.

Table 1. Stream of Petra's hydraulic characteristics.

Stream of Petra					
Reach	Subbasin Area (km^2)	Subbasin Δη (m)	Reach Length (m)	Q (m^3/s)	Overflow Section
R1	2.06	560	375	2.15	S32
R2	0.87	485	317	1.40	S37
R3	0.14	15	399	1.55	S26
R4	4.65	345	258	3.90	S18
R5	0.20	145	367	8.50	S2

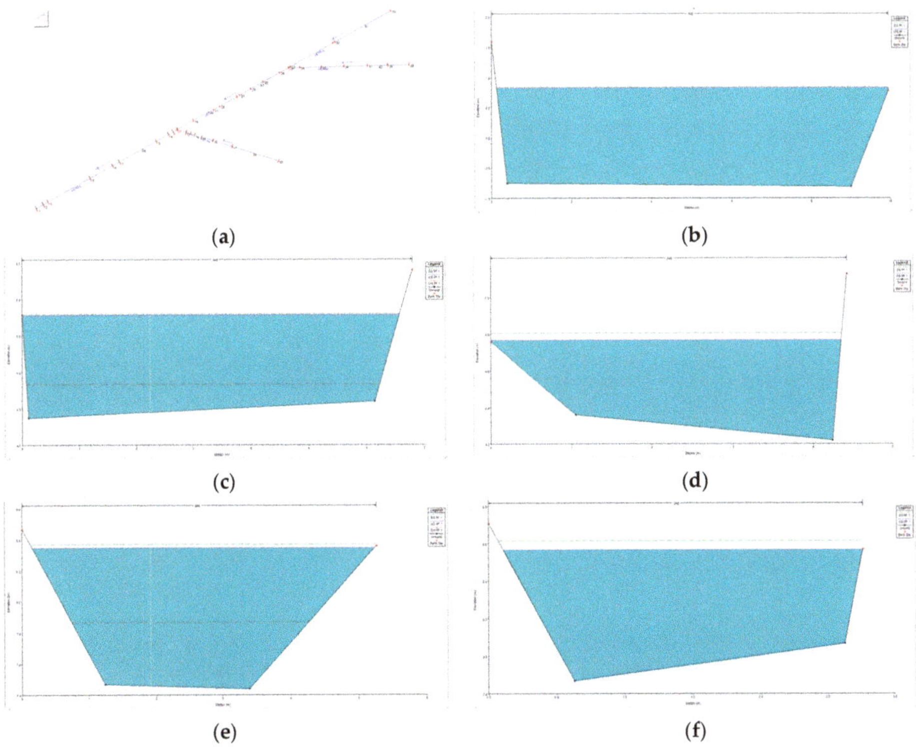

Figure 4. The critical cross-section and the stream plot: (**a**) the Petra stream (**b**) S2, (**c**) S18, (**d**) S26, (**e**) S32, and (**f**) S37.

4. Conclusions

The paper analyses the importance of rivers in the sediment transport balance in the coastal zone of Molyvos-Petra. It discusses the factors affecting coastal streams after flashy rain episodes and the vulnerability of the sandy and rocky beaches. Concluding, how ephemeral streams could behave to sediment balance in a touristic beach, under human and natural activities.

Multiple indicators are used to evaluate the vulnerability of each coastline. According to the results, the habitat is an important part of the coastal zone, and its influence on coastal exposure plays a significant role. All kinds of habitats, whether along the coastline or in the water, have protective effects on the coast [14]. Further, if the coastline is composed of many rocks, low in exposure and high in relief, it can reduce the erosion of waves on the coast. However, the coastline composed of sandy beaches, which are high in exposure and low in relief, has little effect on reducing the impact of wind and waves [14].

Moreover, the hydraulic analysis of the Petra River, during the rainy events showed low velocities in geometrically small cross-sections. People should pay more attention to some impacts, such as storm surges, floods, coastal erosion, and other natural disasters that endanger the economy, life, property safety, and the ecological environment around the coastal zone when people focus on the development of the coastal zone [14].

Author Contributions: Conceptualization, T.H. and O.T.; methodology, S.P.; software, A.-E.I., S.S., T.C., C.R. and I.S.; validation, T.H. and O.T.; formal analysis, A.-E.I., S.S., T.C., A.G. and I.S.; resources, S.P.; data curation, A.-E.I., S.S., A.G., N.W., C.R., A.C., E.-I.K., M.-D.S. and S.P.; writing—original draft preparation, S.P.; writing—review and editing, S.P.; visualization, O.T.; supervision, O.T.; project administration, T.H.; funding acquisition, T.H. and O.T. All authors have read and agreed to the published version of the manuscript.

Funding: This research was funded by Partner, INTERREG Greece–Cyprus, Coastal erosion due to climate change: assessment and ways of effective treatment in tourist areas of the North Aegean and Cyprus.

Institutional Review Board Statement: Not applicable.

Informed Consent Statement: Not applicable.

Data Availability Statement: Not applicable.

Acknowledgments: The authors thank Hasiotis T., Tzoraki O. and Sahtouri S. for their assistance with the operating costs of the present research, which were partly funded by Partner, INTERREG Greece–Cyprus, Coastal erosion due to climate change: assessment and ways of effective treatment in tourist areas of the North Aegean and Cyprus. Koutsovili E.I., Stamataki M.D., Culibrk A., and Chatzivasileiou T. for funded by the Operational Program National Strategic Reference Framework (NSRF) North Aegean 2014-2020, AEGIS+. Papasarafianou S. is funded by the Ph.D. program Phenomena and Environmental Risk by the University of Naples, Parthenope, and Wulf N. This research is part of ERASMUS Plus.

Conflicts of Interest: The authors declare no conflict of interest.

References

1. Levrel, H.; Cabral, P.; Marcone, O.; Mongruel, R. The services provided by marine ecosystems. In *Value and Economy of Marine Resources*; John Wiley & Sons Inc.: Hoboken, NJ, USA, 2014; pp. 1–51.
2. Monioudi, I.N.; Velegrakis, A.F.; Chatzipavlis, A.E.; Rigos, A.; Karambas, T.; Vousdoukas, M.I.; Hasiotis, T.; Koukourouvli, N.; Peduzzi, P.; Manoutsoglou, E.; et al. Assessment of island beach erosion due to sea level rise: The case of the Aegean archipelago (Eastern Mediterranean). *Nat. Hazards Earth Syst. Sci.* **2017**, *17*, 449–466. [CrossRef]
3. Caroa, C.; Marquesa, J.C.; Cunhac, P.; Teixeiraa, Z. Ecosystem services as a resilience descriptor in habitat risk assessment using the InVEST model. *Ecol. Indic.* **2020**, *115*, 106426. [CrossRef]
4. Sharp, R.; Tallis, H.T.; Ricketts, T.; Guerry, A.D.; Wood, S.A.; Chaplin-Kramer, R.; Nelson, E.; Ennaanay, D.; Wolny, S.; Olwero, N.; et al. *InVEST + VERSION + User's Guide*; Stanford University: Stanford, CA, USA, 2018.
5. Hammar-Klose, E.S.; Thieler, E.R. *Coastal Vulnerability to Sea-Level Rise: A Preliminary Database for the U.S. Atlantic, Pacific, and Gulf of Mexico Coasts*; U.S. Geological Survey: Reston, VA, USA, 2001.
6. Pastor, A.V.; Tzoraki, O.; Bruno, D.; Kaletova, T.; Mendoza-Lera, C.; Alamanosi, A.; Brummer, M.; Datry, T.; De Girolamo, A.M.; Jakubínský, J.; et al. Rethinking ecosystem service indicators for their application to intermittent rivers. *Ecol. Indic.* **2022**, *137*, 108693. [CrossRef]
7. Stubbington, R.; Acreman, M.; Acuna, V.; Boon, P.J.; Boulton, A.J.; England, J.; Gilvear, D.; Sykes, T.; Wood, P.J. Ecosystem services of temporary streams differ between wet and dry phases in regions with contrasting climates and economies. *People Nat.* **2020**, *2*, 660–677. [CrossRef]
8. Skoulikidis, N.T.; Sabater, S.; Datry, T.; Morais, M.M.; Buffagni, A.; Dörflinger, G.; Zoogaris, S.; Sánchez Montoya, M.M.; Bonada, N.; Kalogianni, E.; et al. Non-perennial Mediterranean rivers in Europe: Status, pressures, and challenges for research and management. *Sci. Total Environ.* **2017**, *577*, 1–8. [CrossRef]
9. Camarasa-Belmonte, A.M. Flash-Flooding of ephemeral streams in the context of climate change. *Geogr. Res. Lett.* **2021**, *47*, 121–142. [CrossRef]

10. Tamiru, H.; Dinka, M.O. Application of ANN and HEC-RAS model for flood inundation mapping in lower Baro Akobo River Basin, Ethiopia. *J. Hydrol. Reg. Stud.* **2021**, *36*, 100855. [CrossRef]
11. Olson, D.B.; Kourafalou, V.H.; Johns, W.E.; Samuels, G.; Veneziani, M. Aegean Surface Circulation from a Satellite-Tracked Drifter Array. *J. Phys. Oceanogr.* **2007**, *37*, 1898–1917. [CrossRef]
12. Androulidakis, Y.; Kombiadou, K.; Makris, C.; Baltikas, V.; Krestenitis, Y.N. Storm surges in the Mediterranean Sea: Variability and trends under future climatic conditions. *Dyn. Atmos. Ocean.* **2015**, *71*, 56–82. [CrossRef]
13. Soukissian, T.; Hatzinaki, M.; Korres, G.; Papadopoulos, A.; Kallos, G.; Anadranistakis, E. *Wind and wave atlas of the Hellenic seas*; Hellenic Centre for Marine Research Publications: Attiki, Greece, 2007.
14. Ai, B.; Tian, Y.; Wang, P.; Gan, Y.; Luo, F.; Shi, Q. Vulnerability Analysis of Coastal Zone Based on InVEST Model in Jiaozhou Bay, China. *Sustainability* **2022**, *14*, 6913. [CrossRef]

environmental sciences proceedings

MDPI

Proceeding Paper

Groundwater Quality Analysis in Mygdonia Basin, Greece †

Kyriaki Devlioti *, Christos Mattas , Triantafyllos Kaklis and Konstantinos Voudouris

Laboratory of Engineering Geology and Hydrogeology, Aristotle University of Thessaloniki, 57500 Thessaloniki, Greece; cmattas@geo.auth.gr (C.M.); kaklis@geo.auth.gr (T.K.); kvoudour@geo.auth.gr (K.V.)

* Correspondence: kdevliot@geo.auth.gr

† Presented at the 7th International Electronic Conference on Water Sciences, 15–30 March 2023; Available online: https://ecws-7.sciforum.net.

Abstract: The objective of this research is the groundwater quality analysis in the Mygdonia basin (North Greece), where intensive agricultural activities take place. Groundwater sampling was carried out in May (17 samples) and in September (26 samples) for the years from 2013 to 2015. In situ physicochemical parameters (pH, EC, and T) were tested, whereas major ions (Ca^{2+}, Na^{+}, Mg^{2+}, K^{+}, NO_3^{-}, SO_4^{2-}, Cl^{-}, and HCO_3^{-}) were analyzed in the laboratory. Conventional statistical methods were applied for the classification of the groundwater hydrochemical type and the identification of the prevailing hydrogeological processes. Finally, some recommendations are presented for the protection of groundwater quality.

Keywords: aquifers; groundwater quality assessment; sustainable management; Mygdonia basin; North Greece

1. Introduction

Groundwater is critical for supporting human well-being and economic development. During recent decades, groundwater has been under increasing pressure and the threat of pollution due to agricultural and industrial activities, urbanization, and tourism development. In addition, the climate crisis will affect the parameters of the hydrologic cycle (increase in temperature, reduction of rainfall, droughts, floods) and, consequently, the natural recharge of the aquifers, alongside the quantity and quality of groundwater resources. In groundwater, dissolved ions and solids are always present. That defines its physical and physicochemical characteristics. When the concentration of pollutants is greater than threshold values, then it is unsuitable for irrigation and/or other uses.

The Mygdonia hydrologic basin consists of a rare complex of ecosystems, including the lakes, Koronia and Volvi, which constitute an important wetland system of high ecological and international importance (protected by Ramsar Convention), but over the years, a degradation, especially of Lake Koronia, has been observed [1]. The main causes of the water quality deterioration of Lake Koronia are the water withdrawals and the uncontrollable discharge of untreated domestic and industrial effluents [2,3].

This work investigates the groundwater quality of the Mygdonia basin. Conventional statistical methods, hydrogeochemical diagrams, along with data reduction methods, such as factor analysis, were applied in order to describe groundwater quality, classify samples according to their water type, and identify the prevailing hydrogeological processes that take place in the aquifers of the basin. Some proposals to improve the sustainability of groundwater resources in the Mygdonia basin are presented.

2. Study Area-Data Collection

2.1. General Characteristics

Mygdonia basin, covering an area of more than 2000 km^2 (650 km^2 in the lowland), is located in the Thessaloniki Prefecture, central Macedonia, North Greece (Figure 1). It

check for updates

Citation: Devlioti, K.; Mattas, C.; Kaklis, T.; Voudouris, K. Groundwater Quality Analysis in Mygdonia Basin, Greece. *Environ. Sci. Proc.* **2023**, *25*, 45. https://doi.org/10.3390/ECWS-7-14231

Academic Editor: Athanasios Loukas

Published: 16 March 2023

is characterized by the typical Mediterranean and continental climate with wet (October–April) and dry periods (May–September).The mean annual rainfall is 585 mm, and the mean annual temperature is 13.5 °C [4].

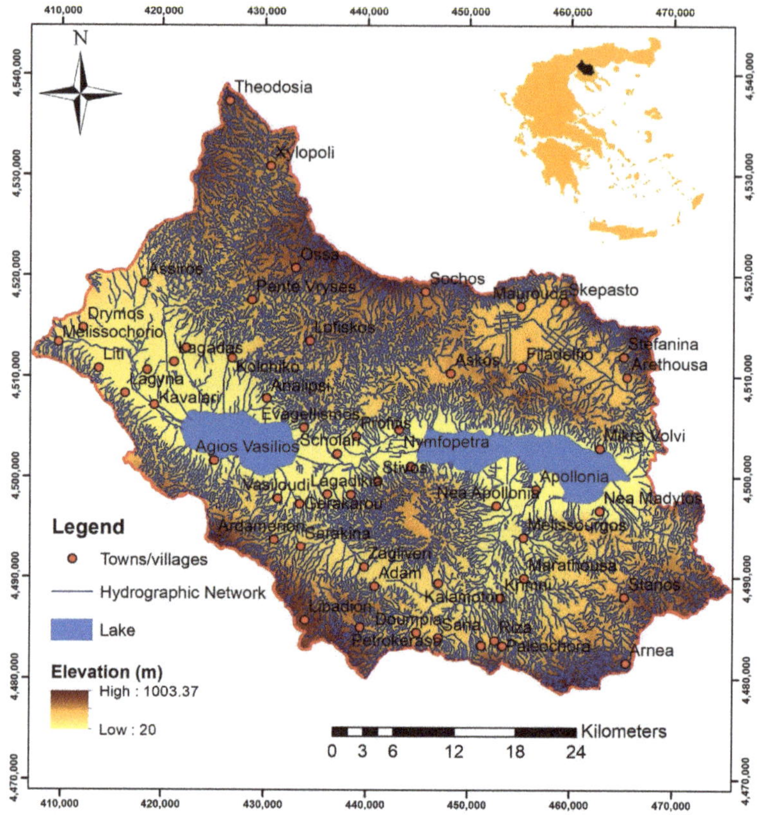

Figure 1. Morphological map of the Mygdonia basin. Map created by the authors based on free data derived by Copernicus (https://www.copernicus.eu/el, accessed on 16 January 2022).

From a geological point of view (Figure 2), as a tectonically active basin, the lowland of the Mygdonia basin is formed from recent unconsolidated materials consisting of sands, clay, and silty sand sediments [5]. The bedrocks are paleozoic gneisses, amphibolites, schists, marbles, and granitic intrusions of the Serbomacedonian massif [6].

Within the Quaternary deposits of lowland, aquifer systems are developed, including an unconfined shallow aquifer and deeper confined aquifers. The groundwater level is 0.7–43 m below the ground's surface. Generally, the highest values are recorded in April (end of wet period) and the lowest in early October (end of dry period) [1].

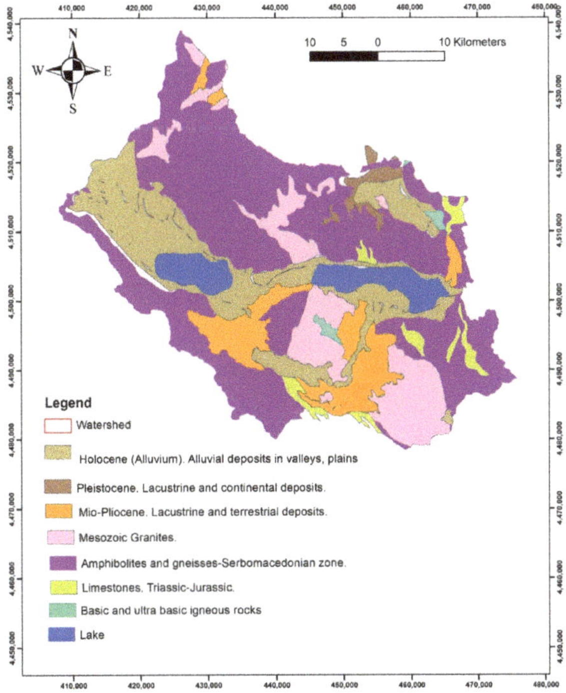

Figure 2. Geological map of the Mygdonia basin. (Adapted from Ref. [7]).

Groundwater recharge occurs through direct infiltration of rainfall, seepage from torrents, lateral subsurface inflows from mountainous areas, and return water flow from irrigation use [8].

2.2. Data Collection

Groundwater samples were collected during the wet (May) and dry (September) periods over the years from 2013 to 2015, showing similar results along with in situ measurements of the physicochemical characteristics. In this work, the results of the year 2014 are presented. The samples were analyzed in the Laboratory of Engineering Geology and Hydrogeology at the Aristotle University of Thessaloniki for eightmajor ions (Ca^{2+}, Na^{+}, Mg^{2+}, K^{+}, NO_3^{-}, SO_4^{2-}, Cl^{-}, and HCO_3^{-}).Sampling point locations aimed to cover the entire basin. Maps showing the spatial distribution of Electrical Conductivity (EC), temperature (T in C°), and pH were created.

The AQUACHEM software was used to create Piper and Durov diagrams, and also the SAR (sodium adsorption ratio) value was calculated. Finally, factor analysis was applied in order to identify the main hydrochemical processes and uncover relationships between ions.

3. Results of Groundwater Quality Analysis

Summary statistics of the results of chemical analyses are presented in Table 1. The electrical conductivity (EC) mean values are equal to 916.5 μS/cm for May and 889 μS/cm for September. High values are attributed to the intensive agricultural activities, the existence of geothermal fields and human activities in the area. It is known that during the previous decades, many textile dyeing units operated in the area. The uncontrolled disposal of untreated waste in the channels resulted in the degradation of soil and groundwater resources.

Table 1. Descriptive statistics of pH, EC, and major ions for groundwater samples.

			μS/cm	mg/L							
		pH	EC	Na	K	Ca	Mg	HCO_3	SO_4	Cl	NO_3
May-14	Mean	7.47	916.47	81.76	3.13	56.78	26.86	307.24	100.47	48.61	33.45
	Max	8.04	1971.00	240.00	5.90	96.00	67.90	580.00	280.00	234.00	110.00
	Min	7.07	444.00	16.00	1.70	20.40	10.00	216.00	26.00	8.00	0.40
Sep-14	Mean	7.49	889.04	66.97	3.03	54.68	33.45	302.31	76.28	38.94	36.00
	Max	8.04	2010.00	233.00	6.10	88.00	72.00	513.00	252.00	217.00	75.20
	Min	6.29	259.00	18.60	0.80	21.20	6.80	158.00	14.00	12.80	4.40

pH values range from 6.29 to 8.04,and groundwater temperatures range from 16.5 °C to 21.8 °C (Figure 3). The spatial variation and especially the high temperatures could be caused by the mixing of groundwater with geothermal fluids.

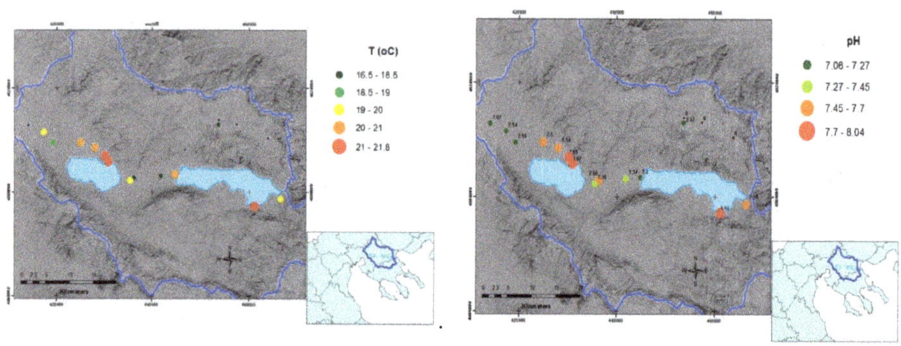

(**a**) (**b**)

Figure 3. Spatial distribution for (**a**) temperature (T) and (**b**) pH for the period of May 2014. (Adapted from Ref. [1]).

A high percentage of the groundwater samples (25%) exceeds the maximum permissible limit for drinking purposes set by Directive 2006/118 EC, the Joint Ministerial Decision 39626/208/E130, and the Ministerial Decision 1811/2011 (Official Government Gazette 3322 B/2011). This is associated with the use of nitrogen fertilizers in agriculture. It is pointed out that the common fertilizers in the area contain nitrogen, phosphate, sulfate, and potassium.

Sodium values exceed the threshold value of 150 mg/L in seven out of forty-four samples. The high sodium values were recorded at boreholes close to the geothermal field and at livestock units. Groundwater quality regarding sodium could be affected by the geothermal fluids, and the use of salt in the livestock units could locally increase the sodium concentration in water samples.

Typically, the distribution of calcium (Ca^{2+}) ions for the two sampling periods is shown in the maps of Figure 4, constituting a typical example given that calcium is one of the most important dissolved ions in water since the content of groundwater is directly linked to the geological environment through which the groundwater moves from the moment of its infiltration up to the entire period of its storage in the aquifer. Its highest concentrations are found in areas where water remains for a long time in the aquifer, as in the area between Koronia and Volvi lakes. High values of sulfates are locally recorded and should be attributed to the use of fertilizers and/or the mixing with geothermal fluids.

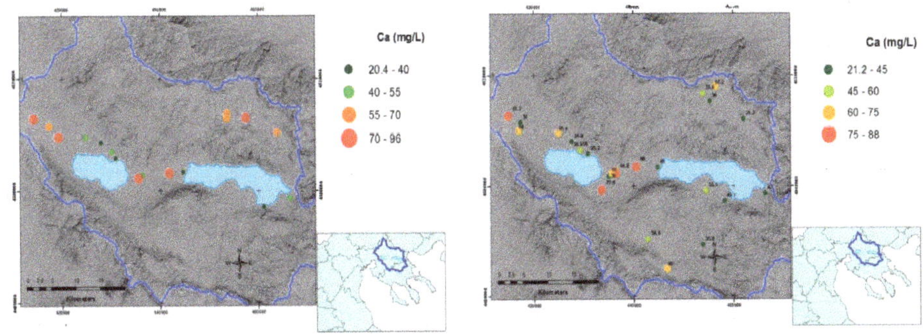

(a) (b)

Figure 4. Distribution of Ca^{2+} ions in the Mygdonia basin in (**a**) May 2014 and (**b**) September 2014. (Adapted from Ref. [1]).

Hydrochemical diagrams were created, providing information about the water type. In Figure 5, Durov diagrams for both sampling periods were created, showing that all samples belong to bicarbonates in terms of anions. Regarding cations, samples belong to calcium and sodium, while based on anions and cations (square), several of them belong to bicarbonate, sulfate, and chloride. It is, therefore, concluded that the groundwater type is calcium-bicarbonate (Ca-HCO_3), while some samples are a mixture of different hydrochemical types of water [9].

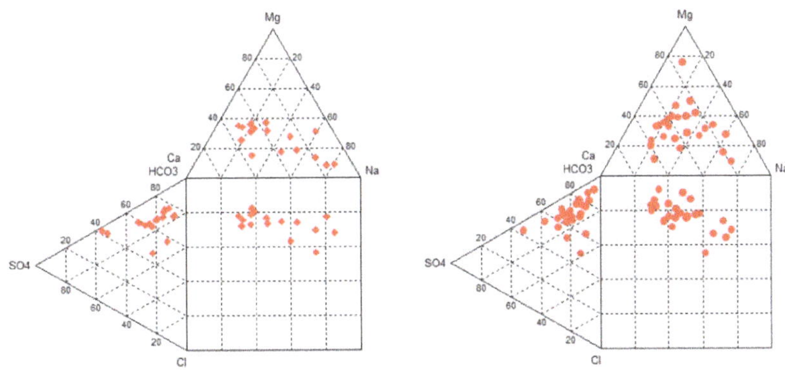

(a) (b)

Figure 5. Durov diagrams including all samples from the Mygdonia basin in (**a**) May2014 and (**b**) September 2014 [1].

Concerning Piper diagrams, all samples regarding anions belong to bicarbonate, while in terms of cations, samples belong to calcium and sodium. Based on anions and cations (diamond), the majority belongs to the category of water coming from mixing processes with increased values of sodium. There are no dominant ions (Figure 6).

The SAR index proposed by the salinity laboratory of the US Department of Agriculture defines the degree of Na^+ adsorption by soil because this ion influences soil permeability [10]. SAR values range between 0.15 and 0.5, indicating that the water is suitable or permissible for irrigation use.

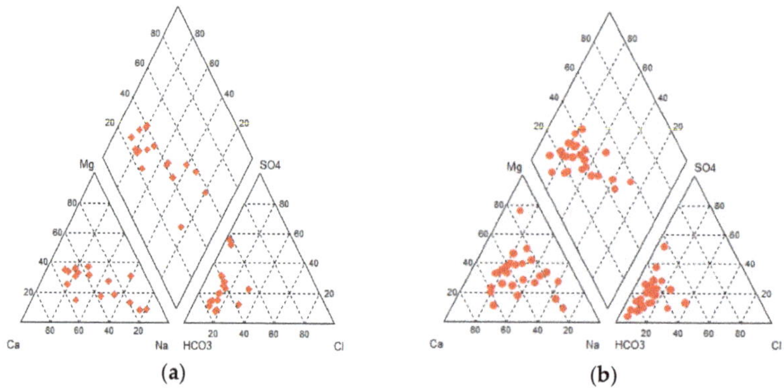

Figure 6. Piper diagrams including all samples from the Mygdonia basin in (**a**) May 2014 and (**b**) September 2014. (Adapted from Ref. [1]).

Statistical Analysis

Application of factor analysis [11–13] in all the chemical analyses of the period 2014–2016 showed that three factors, explaining 71% of the total variance, could describe the main hydrogeological and hydrochemical processes which take place in the Mygdonia basin. The first factor accounts for 32.6% of the total variance of the data and has high loadings on Mg, Na, EC, HCO_3, and Cl. This factor correlates to the mixing activities of water.

The second factor accounts for 21.5% of the total variance and exhibits high loadings with respect to calcium (dissolution of carbonates), and the third factor (16.8% of the total variance) shows high loadings on sulfate indicating an origin from fertilizers and/or geothermal fluids [14].

4. Conclusions and Discussion

Hydrochemical analysis was carried out to draw conclusions about the groundwater quality of the Mygdonia basin. From the analysis, a possible connection with geothermal fluids emerged. More specifically, regarding temperature values in some places, they were slightly increased (values up to 21.8 °C), which is probably connected to the existence of a geothermal field, while SO_4^{2-} appears elevated along the northern coastline of Lake Koronia, which may also be due to their mixing with geothermal fluids. Alongside, nitrate ions show high concentration values in the lowland part of the Mygdonia basin, probably due to the intense agricultural activity and use of fertilizers.

Based on Piper and Durov diagrams, it follows that the groundwater of the area has calcium and sodium as its main cations and bicarbonate as its main anion. That is, the dominant hydrochemical type of groundwater is Ca-HCO_3 and mixing water types. Based on SAR values, the water is suitable for irrigation. Large numbers of samples exceeded the permissible upper limit of 50 mg/L (EU Council) for drinking water.

A set of measures, including monitoring (water quality, groundwater level, level of lakes, climatic data, water abstractions, etc.), should be applied for the protection of water quality and the sustainability of aquifers alongside the socio-economic development of the wider area. The application of rational fertilization waste-water treatment plants, as well as the delineation of protection zones around the boreholes and springs for water supply, will contribute to water quality protection.

Further investigation into the role of the geothermal field on groundwater quality should be implemented in the area of the boreholes in which high-temperature values were recorded. The continuous monitoring of groundwater quality characteristics is essential for the rational and sustainable management of groundwater in the Mygdonia basin.

Author Contributions: Conceptualization, K.V.; methodology, K.D. and K.V.; data curation, K.D., T.K. and C.M.; writing—original draft preparation, K.D. and C.M.; writing—review and editing, K.D., C.M. and T.K.; supervision, K.V. All authors have read and agreed to the published version of the manuscript.

Funding: This research received no external funding.

Institutional Review Board Statement: Not applicable.

Informed Consent Statement: Not applicable.

Data Availability Statement: Not applicable.

Acknowledgments: A part of this research was carried out in the framework of the Thesis of Kyriaki Devlioti, Laboratory of Engineering Geology & Hydrogeology, Department of Geology, Aristotle University of Thessaloniki, Greece (Supervisor: K. Voudouris).

Conflicts of Interest: The authors declare no conflict of interest.

References

1. Devlioti, K. Hydrogeological Research and Groundwater Quality of the Mygdonia Basin. Master Thesis, School of Geology, Aristotle University of Thessaloniki, Thessaloniki, Greece, 2016; p. 128. (In Greek)
2. Ballas, L.G. Use of Fuzzy Rules in Water Resources Management-Application to the Volvi Hydrological Basin. Ph.D. Thesis, Aristotle University Thessaloniki, Thessaloniki, Greece, 2007.
3. Malamataris, D.; Kolokytha, E.; Loukas, A. Integrated hydrological modelling of surface water and groundwater under climate change: The case of the Mygdonia basin in Greece. *J. Water Clim. Chang.* **2020**, *11*, 1429–1454. [CrossRef]
4. Chantzi, P.; Dotsika, E. Mygdonia Basin (N. Greece) in the View of Isotope Geochemistry. In *Energy, Transportation and Global Warming, Green Energy and Technology*; Grammelis, P., Ed.; Springer International Publishing: Cham, Switzerland, 2016. [CrossRef]
5. Psilovikos, A. Paleogeographic Development of the Basin and Lake of Mygdonia (Lagada and Volvi Area, Greece). Ph.D. Thesis, Department of Geology, Aristotle University of Thessaloniki, Thessaloniki, Greece, 1977. (In Greek)
6. Mountrakis, D. *Geology of Greece*; University Studio Press: Thessaloniki, Greece, 1985.
7. Bornovas, J.; Rondoyanni, T. *Geological Map of Greece, Scale 1:500.000*, 2nd ed.; Institute of Geology and Mineral Exploration: Thessaloniki, Greece, 1983.
8. Nimfopoulos, M.K.; Mylopoulos, M.; Katirtzoglou, K. A qualitative-quantitative study of water and environmental pollution at the broader area of the Mygdonia basin, Thessaloniki, n. Greece. In Proceedings of the 6th Pan-Hellenic Geographical Conference of the Hellenic Geographical Society, Thessaloniki, Greece, 3–6 October 2002.
9. Mattas, C.; Kaklis, T.; Devlioti, K.; Vouvalidis, K. Groundwater qualitative characteristics of Mygdonia basin. Investigation of the pollution sources and suitability for human use. *Bull. Geol. Soc. Greece* **2016**, *L*, 854–864.
10. Tizro, T.A.; Voudouris, K. Groundwater quality in the semi-arid region of the Chahardouly basin, West Iran. *Hydrol. Process.* **2008**, *22*, 3066–3078. [CrossRef]
11. Voudouris, K.; Lambrakis, N.; Papatheodorou, G.; Daskalaki, P. An application of factor analysis for the study of the Hydrogeological conditions in Plio-Pleistocene aquifers of NW Achaia (NW Peloponnesus, Greece). *Math. Geol.* **1997**, *29*, 43–59. [CrossRef]
12. Voudouris, K.; Panagopoulos, A.; Koumantakis, J. Multivariate statistical analysis in the assessment of Hydrochemistry of the Northern Korinthia Prefecture alluvial aquifer system (Peloponnese, Greece). *Nat. Resour. Res.* **2000**, *9*, 135–146. [CrossRef]
13. Busico, G.; Cuoco, E.; Kazakis, N.; Colombani, N.; Mastrocicco, M.; Tedesco, D.; Voudouris, K. Multivariate statistical analysis to characterize/discriminate between anthropogenic and geogenic trace element occurrence in the Campania Plain, Southern Italy. *Environ. Pollut.* **2018**, *234*, 260–269. [CrossRef] [PubMed]
14. Poutoukis, D.; Ntotsika, E. Hydrogeochemical and isotopic investigation of the thermal bath water from Lagada and Volvi. In Proceedings of the 2nd Hydrogeological Congress, Patras, Greece, 24–28 November 1993; Greek Committee of Hydrogeology: Athens, Greece, 1994; Volume B, pp. 679–689. (In Greek)

environmental sciences proceedings

MDPI

Proceeding Paper

Towards More Efficient Hydraulic Modeling of Water Distribution Networks Using the EPANET Software Engine †

Athanasios V. Serafeim [1], Anastasios Perdios [1], Nikolaos Th. Fourniotis [2] and Andreas Langousis [1,*]

[1] Department of Civil Engineering, University of Patras, 26504 Patras, Greece; up1056489@upatras.gr (A.V.S.); civ6261@upnet.gr (A.P.)
[2] Department of Civil Engineering, University of the Peloponnese, 26334 Patras, Greece; nfou@uop.gr
* Correspondence: andlag@upatras.gr or andlag@alum.mit.edu
† Presented at the 7th International Electronic Conference on Water Sciences, 15–30 March 2023; Available online: https://ecws-7.sciforum.net.

Abstract: Hydraulic modeling of water distribution networks (WDNs) is a vital step for all water-related professionals towards the development of management practices and strategies that aim for the reduction of water losses and the associated financial cost and environmental footprint. In the current work, we develop an easy-to-implement methodology for the effective modeling of WDNs, which seeks to minimize the computational load without undermining the analysis's accuracy, using the open access EPANET (Environmental Protection Agency Network Evaluation Tool) software package. The effectiveness of the proposed methodology is tested via a large-scale, real-world application for the city of Patras.

Keywords: hydraulic modeling; hydraulic network; EPANET; computational nodes; junctions; sensitivity analysis; leakage allocation

check for updates

Citation: Serafeim, A.V.; Perdios, A.; Fourniotis, N.T.; Langousis, A. Towards More Efficient Hydraulic Modeling of Water Distribution Networks Using the EPANET Software Engine. *Environ. Sci. Proc.* **2023**, *25*, 46. https://doi.org/10.3390/ECWS-7-14166

Academic Editor: Lampros Vasiliades

Published: 14 March 2023

1. Introduction

Today, the problem of the continuing decrease in available freshwater reserves is widely recognized, which is significantly magnified if one considers the effects of climate change on the spatial and temporal distribution of water resources [1–5]. Consequently, there is an urgent environmental and societal need to implement efficient management practices to water distribution networks (WDNs), which constitute the core infrastructure for drinking water supply to users.

The first step towards developing efficient management strategies is the accurate hydraulic modeling of WDNs using a dedicated software package (e.g., EPANET) in order to identify their weaknesses and evaluate their overall operational condition. To do so, one needs to develop a detailed representation (i.e., a model) of the pipeline grid using appropriate hydraulic objects (e.g., pipes, pumps, valves, junctions, reservoirs, tanks, etc.). The effectiveness of the modeling procedure is mainly determined by the modeling accuracy, which is significantly affected by the density of the computational nodes (i.e., junctions). Although a high nodal density model produces more accurate results, it also dramatically increases the computational requirements, leading to time-consuming solutions. Under this setting, the current work focuses on developing a practical methodology for the optimal allocation of computational nodes, in terms of modeling accuracy and computational cost.

2. Area of Application

We have applied the analysis that follows to the four largest and most highly populated pressure management areas (PMAs) of the water distribution network of the city of Patras, in western Greece (namely Boud, Kentro, Panachaiki, and Prosfygika; see Figure 1). The corresponding PMAs, which share similar characteristics regarding their population density as well as land uses and topography, consist of over 200 km of HDPE and PVC pipes, and cover most of Patras's city center (about 4 km^2), providing water to more than 58,000 customers, as reported by the associated public authorities (see Table 1).

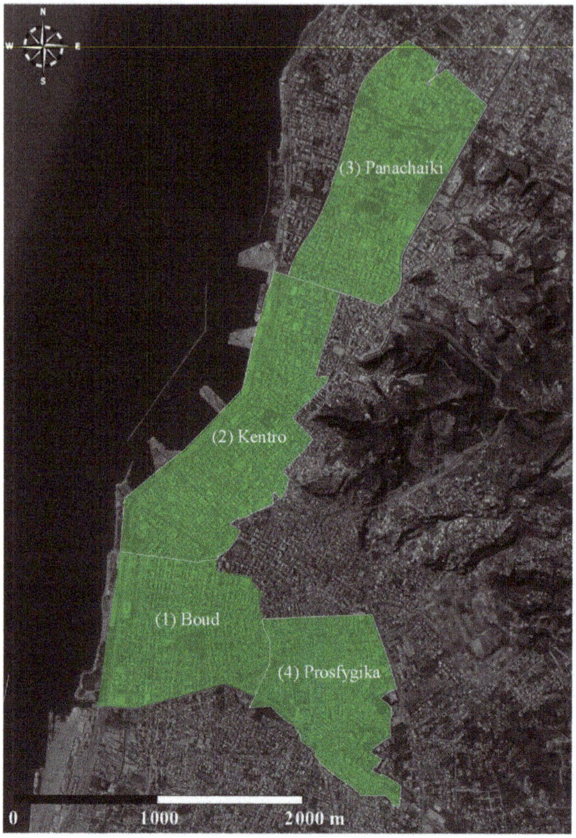

Figure 1. Map showing the position of the four largest pressure management areas located in the central region of Patras. Numbers are linked to the entries listed in Table 1.

Table 1. Name, total area, length of the pipeline grid, and population of the four largest pressure management areas (PMAs) of the city of Patras. Numbers reflect the encompassed areas in Figure 1.

PMA	Area (km^2)	Length of the Pipeline (m)	Population (cap.)
(1) Boud	0.95	44,953	15,361
(2) Kentro	1.21	62,175	13,991
(3) Panachaiki	1.18	51,704	18,002
(4) Prosfygika	0.80	43,246	10,657

3. Methodology

3.1. Hydraulic Model Design

To realistically describe a WDN, the nodal density should be high enough to effectively describe both the area's topographic variability as well as the original connectivity of the network, while considering all necessary hydraulic parameters (e.g., pipe material and diameter). Under this concept, we chose to place computational nodes at the following: (a) intersections between two or more pipes, (b) changes in pipes' diameters and/or material, (c) fire hydrant locations, (d) dead ends, and (e) at the locations of high-water demand consumers [6].

To take into account the area's topographic variability (which highly affects the analysis's pressure outcome; see [7–9]), we used a sensitivity analysis to determine an appropriate nodal density. Figure 2 summarizes the corresponding EPANET simulation's time complexity (i.e., the computational time), in terms of network's nodal density (i.e., number of nodes per km). It can be observed that time complexity increases almost exponentially with increasing nodal density for all four cases, as a result of the heavier computational load. Selection of a proper solution (i.e., nodal density) is achieved through an optimal trade-off between time complexity and the required accuracy of the simulation, tailored to each specific case, as a function of topographic variability. For the purposes of the current study, we opted to incorporate 10 nodes per km (i.e., at least one computational node per 100 m), as for larger nodal densities the computational time increases significantly.

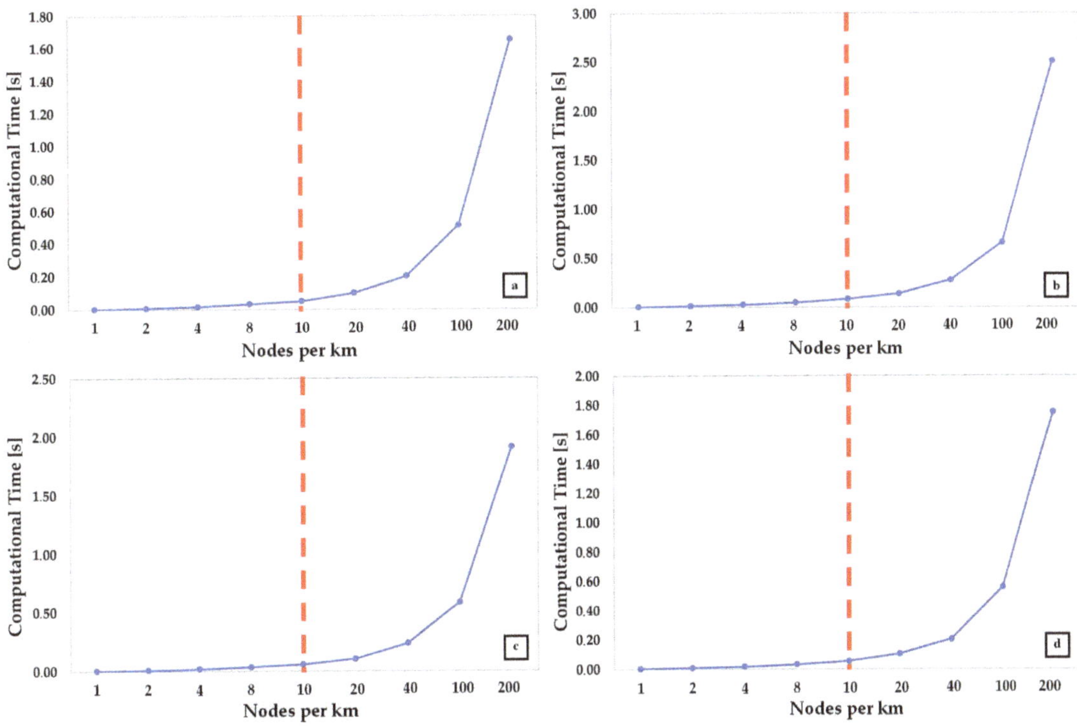

Figure 2. EPANET simulation's time complexity for the four PMAs considered, in terms of network's nodal density: (**a**) Boud, (**b**) Kentro, (**c**) Panachaiki, and (**d**) Prosfygika.

3.2. Real Losses (RL, Leakages) Allocation

To perform the hydraulic simulation, firstly, we determined the total water demand at each network node, and divided it into two parts: a demand-driven component and a pressure-driven component. The former is based on the flow pattern, as consumers' usage varies throughout the day, while the pressure-driven component accounts for network leaks, which increase when the applied pressure increases. The modeling of leaks is done by assuming that the leakage rate is proportional to the square root of the difference between the actual nodal pressure and the minimum pressure necessary to fulfill consumption requirements. To do so, we multiply the initial leakage rates at each computational node by the parameter:

$$c_j = \frac{(s_j - s_j^*)^{0.5}}{\sum_{j=1}^{n} (s_j - s_j^*)^{0.5}}, \text{ for } s_j > s_j^* \quad (1)$$

where s_j is the numerically simulated head at node $j = 1, \ldots, n$ (i.e., the sum of nodal elevation and pressure head), and s_j^* is the minimum threshold head at node j (i.e., the sum of nodal elevation and the minimum required pressure head). The hydraulic simulation is repeated until convergence (see [9]).

4. Results

We implemented the proposed hydraulic modeling methodology (see Section 3) for the four largest pressure management areas of the water distribution network of the city of Patras, based on their geometric characteristics and hydraulic parameters as well as the area's altitudinal variation. In order to estimate the water consumption, we used flow-pressure data at 1 min temporal resolution for the 4-month-long summer period from 1 June 2019–31 August 2019, with the data having been collected from the pressure regulation stations of the water distribution network (WDN) of the City of Patras in Western Greece. Flow and pressure data for each of the four stations were obtained from the Municipal Enterprise of Water Supply and Sewerage of Patras (DEYAP) and were quality checked to identify and eliminate errors resulting from communication issues and other data transmission malfunctions.

Figure 3 illustrates the nodal pressures and water velocity results for PMAs Boud (Figure 3a), Kentro (Figure 3b), Panachaiki (Figure 3c), and Prosfygika (Figure 3d), which were obtained through hydraulic simulations using the EPANET 2.x solver for the design and analysis of water networks. It is noted that in all cases the minimum pressure requirements (21 m in PMA Boud, 24 m in PMAs Panachaiki and Prosfygika, and 28 m in PMA Kentro) and maximum speed requirements were met (based on pipe diameters; for more info, see [6]).

In order to test the accuracy of the proposed methodology, we use on-site pressure data obtained by DEYAP through smart pressure meters located at the most distant nodes of PMAs' pressure-regulating valves (i.e., the points in Figure 3 marked in blue). Table 2 summarizes the calculated pressure values obtained from the smart metering system and the corresponding pressure values obtained by the EPANET solver.

Table 2. Modeled and on-site metered pressure values of the four largest pressure management areas (PMAs) of the city of Patras. Numbers are linked to the positions in Figure 1.

PMA	Model Pressure (m)	On-Site Pressure (m)	Absolute Relative Difference (%)
(1) Boud	48.256	52.234	7.615
(2) Kentro	74.868	66.549	9.219
(3) Panachaiki	99.412	95.834	3.733
(4) Prosfygika	50.226	52.947	5.139

It is observed that the proposed methodology results in almost identical pressure values as the on-site metering, with absolute relative deviations not exceeding 10% for all four cases (7.615%, 9.219%, 3.733%, and 5.139% for PMAs Boud, Kentro, Panachaiki, and Prosfygika, respectively), indicating the robustness of the proposed methodology.

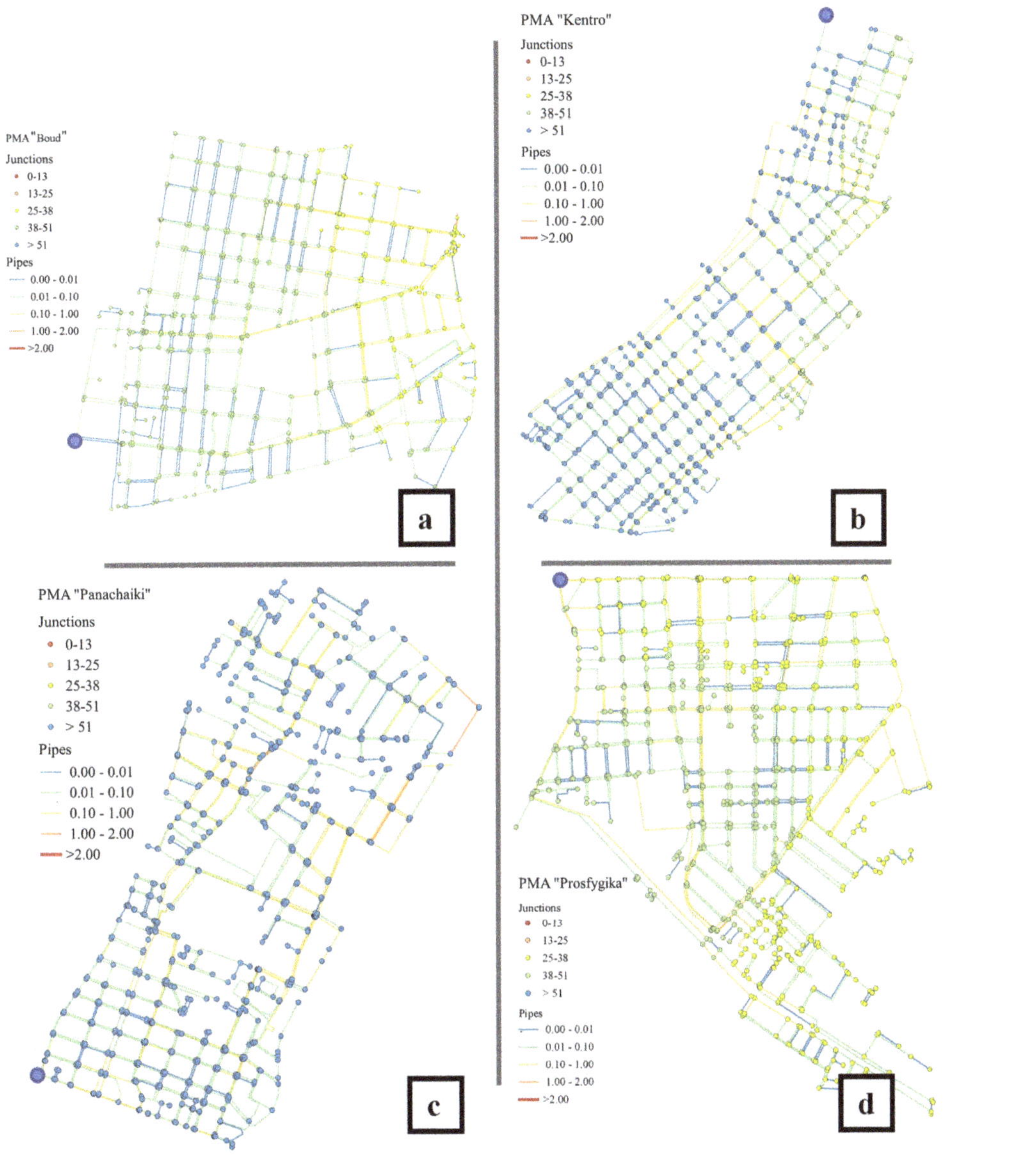

Figure 3. Nodal pressures and water velocity results for PMAs: (**a**) Boud, (**b**) Kentro, (**c**) Panachaiki, and (**d**) Prosfygika, obtained through hydraulic simulations using the EPANET solver. PMA locations are illustrated in Figure 1.

5. Conclusions

Hydraulic modeling of WDNs is an important step towards the development of efficient water management practices and strategies, with the aim of reducing water losses and the associated financial cost and environmental footprint. In the current work, we developed an easily applicable methodology for the effective modeling of WDNs that maintains a sufficient level of estimation accuracy with minimal computational load, using a sensitivity analysis to determine the appropriate nodal density, in order to effectively describe both the topographic variability as well as the original connectivity of the network.

Additionally, the water requirement at each node in the network was established by combining two factors: one being driven by demand and the other by pressure, resulting in more accurate depictions of the operational pressures.

The developed hydraulic models allowed us to implement and test a variety of methodologies regarding water loss estimations (see [7]), the identification of pressure control failures and the release of notifications (see [5]), as well as the optimal partitioning of WDNs into PMAs, without undermining the overall hydraulic resilience of the network (see [9]). The developed approaches can significantly reduce the volume of lost water (30% on average in each PMA), which corresponds to approximately €300,000 in annual savings, based on the balance sheet of the fiscal year 2019 (see [10]).

Author Contributions: Conceptualization, methodological formulation, and interpretation: A.V.S. and A.L.; data preprocessing, formal analysis, verification, visualization, and writing—original draft preparation: A.V.S.; writing—review and editing: A.P., N.T.F. and A.L.; funding acquisition, project administration, and supervision: A.L. All authors have read and agreed to the published version of the manuscript.

Funding: The research project was supported by the Hellenic Foundation for Research and Innovation (H.F.R.I.) under the "1st Call for H.F.R.I. Research Projects to support Faculty Members & Researchers and the procurement of high-cost research equipment grant" (Project Number: 1162).

Institutional Review Board Statement: Not applicable.

Informed Consent Statement: Not applicable.

Data Availability Statement: The data used are protected under a nondisclosure agreement. Acquisition requests should be addressed to DEYAP (https://www.deyap.gr, accessed on 1 November 2022).

Conflicts of Interest: The authors declare no conflict of interest.

References

1. Deng, Y.; Cardin, M.; Babovic, V.; Santhanakrishnan, D.; Schmitter, P.; Meshgi, A. Valuing flexibilities in the design of urban water management systems. *Water Res.* **2013**, *47*, 7162–7174. [CrossRef] [PubMed]
2. Rehan, R.; Knight, M.A.; Unger, A.J.A.; Haas, C.T. Development of a system dynamics model for financially sustainable management of municipal watermain networks. *Water Res.* **2013**, *47*, 7184–7205. [CrossRef] [PubMed]
3. Charalambous, B.; Foufeas, D.; Petroulias, N. Leak detection and water loss management. *Water Util. J.* **2014**, *8*, 25–30.
4. Serafeim, A.V.; Kokosalakis, G.; Deidda, R.; Karathanasi, I.; Langousis, A. Probabilistic estimation of minimum night flow in water distribution networks: Large-scale application to the city of Patras in western Greece. *Stoch. Environ. Res. Risk Assess.* **2021**, *36*, 643–660. [CrossRef]
5. Perdios, A.; Kokosalakis, G.; Fourniotis, N.T.; Karathanasi, I.; Langousis, A. Statistical framework for the detection of pressure regulation malfunctions and issuance of alerts in water distribution networks. *Stoch. Environ. Res. Risk Assess.* **2022**, *36*, 4223–4233. [CrossRef]
6. Langousis, A.S.; Fourniotis, N.T. *Elements of Design of Water Supply and Sewerage Works*; GOTSIS Publications: Patras, Greece, 2020; p. 722, ISBN 978-960-9427-89-0. (In Greek)
7. Serafeim, A.V.; Kokosalakis, G.; Deidda, R.; Karathanasi, I.; Langousis, A. Probabilistic Minimum Night Flow Estimation in Water Distribution Networks and Comparison with the Water Balance Approach: Large-Scale Application to the City Center of Patras in Western Greece. *Water* **2022**, *14*, 98. [CrossRef]
8. Serafeim, A.V.; Kokosalakis, G.; Deidda, R.; Karathanasi, I.; Langousis, A. Probabilistic framework for the parametric modeling of leakages in water distribution networks: Large scale application to the City of Patras in Western Greece. *Stoch. Environ. Res. Risk Assess.* **2022**, *36*, 3617–3637. [CrossRef]
9. Serafeim, A.V.; Kokosalakis, G.; Deidda, R.; Fourniotis, N.T.; Langousis, A. Combining Statistical Clustering with Hydraulic Modeling for Resilient Reduction of Water Losses in Water Distribution Networks: Large Scale Application Study in the City of Patras in Western Greece. *Water* **2022**, *14*, 3493. [CrossRef]
10. Serafeim, A.V.; Kokosalakis, G.; Deidda, R.; Karathanasi, I.; Langousis, A. Water Loss Estimation and Associated Financial Cost in Water Distribution Networks: Large Scale Application to the City of Patras in Western Greece. In Proceedings of the IWA World Water Congress & Exhibition 2022, Copenhagen, Denmark, 11–15 September 2022.

environmental sciences proceedings

MDPI

Proceeding Paper

Detection of Organophosphorus Esters (OPEs) in Groundwater †

Lucija Plantak *, Anita Ptiček Siročić, Ivana Grčić and Ranko Biondić

Faculty of Geotechnical Engineering, University of Zagreb, Hallerova Aleja 7, 42000 Varaždin, Croatia; anitaps@gfv.hr (A.P.S.); ivana.grcic@gfv.unizg.hr (I.G.); ranko.biondic@gfv.unizg.hr (R.B.)

* Correspondence: lucija.plantak@gfv.unizg.hr

† Presented at the 7th International Electronic Conference on Water Sciences, 15–30 March 2023; Available online: https://ecws-7.sciforum.net.

Abstract: Organophosphate esters (OPEs), are used as flame retardants and plasticizers to protect or enhance the properties of plastics, textiles, and many other materials. Sampling was carried out in groundwater from the karst aquifer Bokanjac–Poličnik near the city of Zadar, Croatia. To determine their continuous presence, samples were taken once during each season for one year. In the collected samples, nine OPEs were identified: tris(2-butoxyethyl) phosphate-TBEP, tricresyl phosphate-TCP, triphenyl phosphate-TPPA, tris(1-chloro-2-propyl) phosphate-TCPP, tris(2-chloroethyl) phosphate-TCEP, tris(1,3-dichloroisopropyl) phosphate-TDCPP, diethyl phthalate-DEP, tri-n-butyl phosphate-TBP, and di(2-ethylhexyl) adipate-DEHA.

Keywords: organic pollutants; OPEs; groundwater quality; karst aquifer

1. Introduction and Background

Groundwater is an essential water supply for settlements in many countries across the world, especially in Croatia, where 90% of water withdrawal is from underground aquifers. In Croatia, the quality of groundwater must be in line with the guidelines of the EU Commission Water Framework Directive (WFD) [1], as well as the Groundwater Directive (GWD) [2]. Emerging pollutants, particularly organic pollutants in groundwater, are under-researched due to the absence of monitoring regulations. Nevertheless, monitoring is carried out on a voluntary basis [3].

Organophosphate esters (OPEs), which are triesters, are used as flame retardants and plasticizers to protect or enhance the properties of plastics, textiles, and many other materials. They are high-production-volume chemicals with large variations in their physical–chemical properties that are widely used in many human activities and can be detected in groundwater due to their insolubility in water, especially wastewater. In [4], it was demonstrated that OPEs have been used for decades. Thus, their occurrence in the environment is not new, because since the 1980s, reports have described their detection in surface waters, groundwaters influenced by wastewater, and drinking water. OPEs can be a health hazard for humans due to their toxicity, and some of them have shown carcinogenic properties, e.g., chlorinated OPEs such as TCEP, TCPP, and TDCPP, which can accumulate in the liver and testis, thereby inducing tumors [5].

Various analytical methods for the determination of OPEs in water, i.e., various extraction techniques, have been developed, followed by gas chromatography with mass spectrometry (GC-MS) and liquid chromatography with mass spectrometry (LC-MS) [6].

Sampling was carried out in groundwater from the karst aquifer Bokanjac–Poličnik near the city of Zadar, Croatia, in order to detect OPEs. To determine their continuous presence in the case study aquifer, samples were taken throughout the entire hydrological year, that is, once during each season for one year. The preparation of samples was performed with the Solid Phase Extraction (SPE) method, and they were analyzed with liquid chromatography with mass spectrometry and quadrupole time-of-flight spectrometry (LC-MS/QTOF). In the collected samples, nine OPEs were identified with the Water Screening

Citation: Plantak, L.; Siročić, A.P.; Grčić, I.; Biondić, R. Detection of Organophosphorus Esters (OPEs) in Groundwater. *Environ. Sci. Proc.* **2023**, *25*, 47. https://doi.org/10.3390/ECWS-7-14169

Academic Editor: Carmen Teodosiu

Published: 14 March 2023

Personal Compound Database and Library (PCDL): tris(2-butoxyethyl) phosphate-TBEP, tricresyl phosphate-TCP, triphenyl phosphate-TPPA, tris(1-chloro-2-propyl) phosphate-TCPP, tris(2-chloroethyl) phosphate-TCEP, tris(1,3-dichloroisopropyl) phosphate-TDCPP, diethyl phthalate-DEP, tri-n-butyl phosphate-TBP, and di(2-ethylhexyl) adipate-DEHA. Their properties and usages are listed in Table 1.

Table 1. OPEs properties detected in the case study aquifer and their usages.

OPEs	CAS No.	Formula	Usage	Literature
TBEP	78-51-3	C24H51O4P	Adhesives, sealant chemicals, flame retardants, paint and coating additives, plasticizers, cleaning-furnishing care products	[6,7]
TCP	78-32-0	C21H21O4P	Plasticizers, flame-retardants, solvents for nitrocellulose, as additives to extreme-pressure lubricants, fluids in hydraulic systems, as lead scavengers in gasoline, used to sterilize certain surgical instruments and in flexible PVC.	[8,9]
TPPA	115-86-6	C18H15O4P	Flame retardants, paint and coating additives, plasticizers in automobile upholstery, fireproofing agents, used for impregnating roofing paper, components of lubricating oils and hydraulic fluids.	[6,10]
TCPP	13674-84-5	C9H18Cl3O4P	Adhesives and sealants, building/construction materials—wood products, electrical and electronic products, fabric, textile, and leather products, flame retardants, foam seating and bedding products, insulating foam, plastic and rubber products.	[6]
TCEP	115-96-8	C6H12Cl3O4P	Additive plasticizers and flame retardants in plastics, especially flexible foams used in automobiles and furniture, and in rigid foams used for building insulation.	[11,12]
TDCPP	13674-87-8	C9H15Cl6O4P	Flame retardants, plasticizers, used in polyvinyl chloride, rigid PUF, epoxy resin, and polyester fiber.	[13]
DEP	84-66-2	C12H14O4	Adhesives and sealants, air care products, automotive care products, cleaning and furnishing care products, laundry and dishwashing products, personal care products, toys, playgrounds, sporting equipment, aroma chemicals.	[6]
TBP	126-73-8	C12H27O4P	Flame retardants, functional fluids (closed and open systems), plasticizers, building/construction materials, hydraulic fluids, inks, toners, and colorant products.	[6,12]
DEHA	103-23-1	C22H42O4	Bis(2-ethylhexyl) adipate is a diester. It is used as a plasticizer in the preparation of various polymers. Adhesives and sealants, automotive care products, building/construction materials, electrical and electronic products, and fabric, textile, and leather products. Food packaging, fuels and related products, furniture and furnishings, lubricants and greases, metal products, paints and coatings, personal care products.	[6]

2. Methodology

The sampling of the groundwater was caried out at three piezometer wells in the Bokanjac–Poličnik basin near the city of Zadar in Croatia. Samples was taken four times in one year in order to over all the seasons, i.e., to obtain samples from one hydrological year. For the sampling, sterile glass bottles were used, and they were refrigerated during transport and up to the time when sample preparation began. The three sites where the samples were taken are marked as J, B, and B4. Due to the complexity of the groundwater samples, which were relatively clear and of high quality according to the regulation parameters for human consumption, the small concentrations of the expected organic pollution could not be directly measured, and there was a need to concentrate the samples for the analysis. The analyte prepared for analysis was concentrated using the Solid Phase Extraction (SPE) method to increase the selectivity of the method. The SPE method was used to eluate the concentrated analyte in order to simplify the identification and quantification of the expected compounds in extremely low concentrations. The cartridges used for SPE were Bond elut plexa C18, and the samples were eluted with methanol (MeOH) and filtered through a KX syringe filter (PET 25 mm, 0.22). The filtered samples were analyzed using an Agilent 6530 LC/MS QToF device that detects the ion mass (M/z) and relative abundance of isotopes. With this device, we determined which ions were in the samples based on the masses that were detected [14].

LC/MS QToF analysis was performed with an InfinityLab Poroshell 120 EC–C18 (3.0 × 100 mm, 2.7 μm, Agilent Technologies, Inc.) column. We injected 0,4 mL/min of analyte into the following mobile phases: (A) deionized water with 0.1 % formic acid, (B) MeOH with 0.1 % formic acid, and (C) acetonitrile (ACN) with 0.1 % formic acid. These mobile phases were used in 20 min runs in the positive ionization mode, with a column temperature of 35 °C.

3. Results

The OPEs that were detected with the above-mentioned methodology were identified with the Agilent Water Screening Personal Compound Database and Library (PCDL), and the results are shown in Figures 1–6.

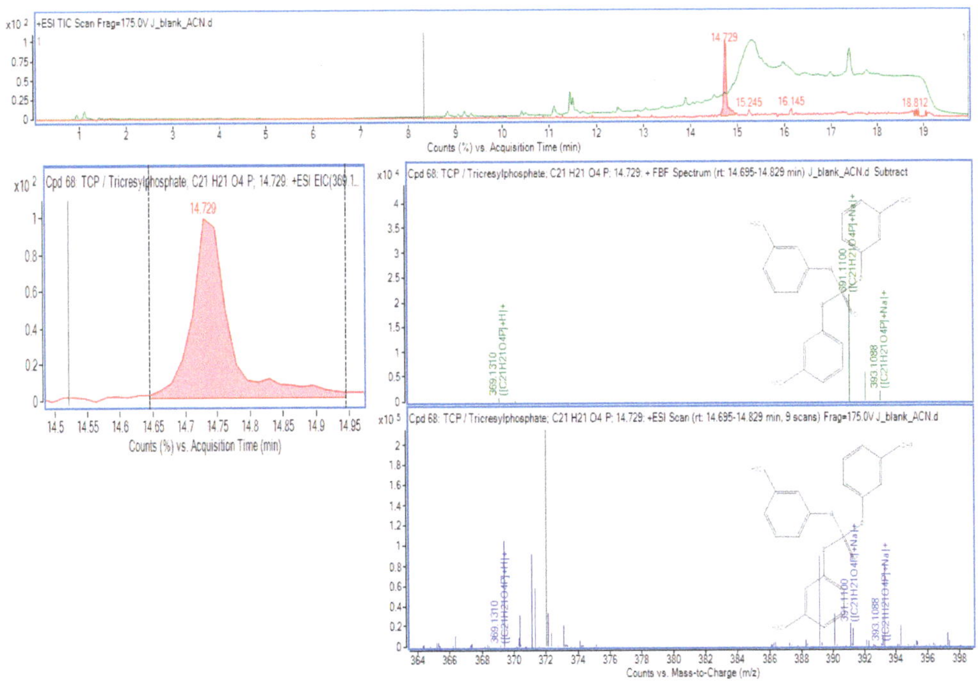

(a)

Figure 1. *Cont.*

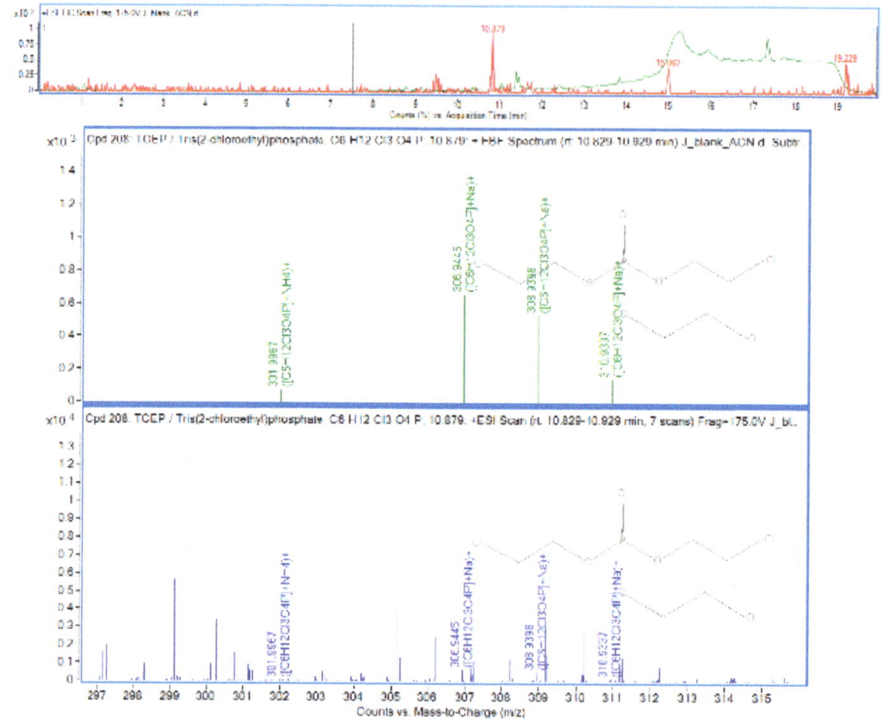

(**b**)

Figure 1. OPEs in sample J: (**a**) TCP; (**b**) TCEP.

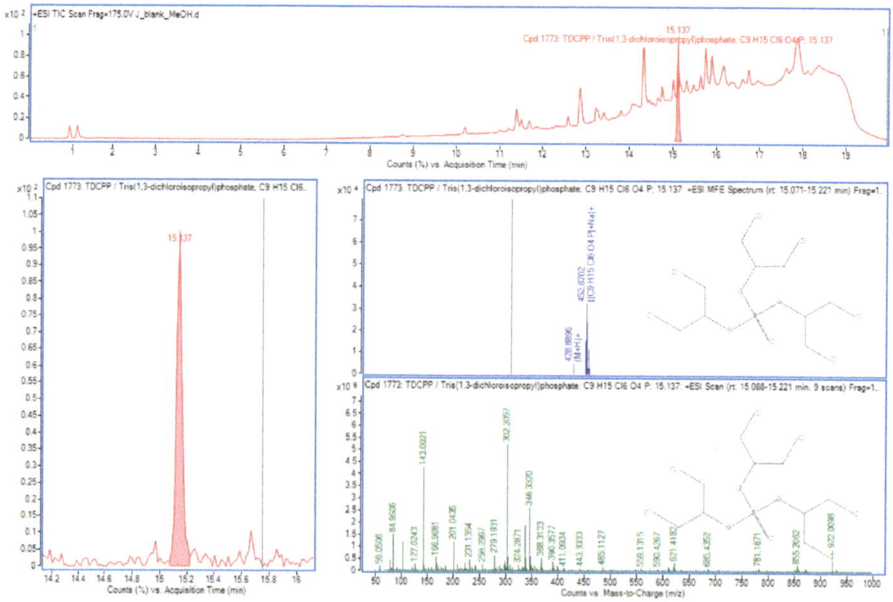

Figure 2. TDCPP in sample J.

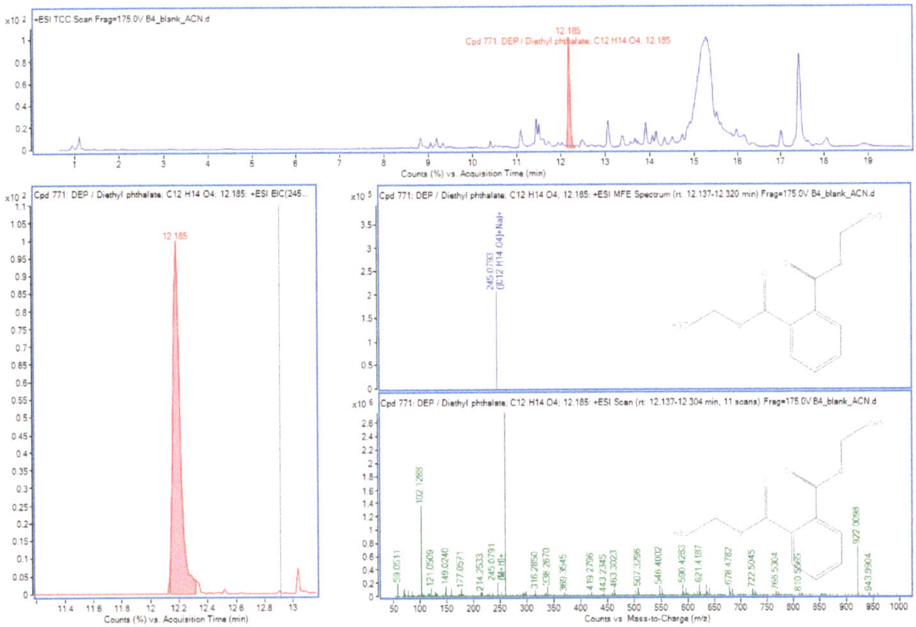

(**a**)

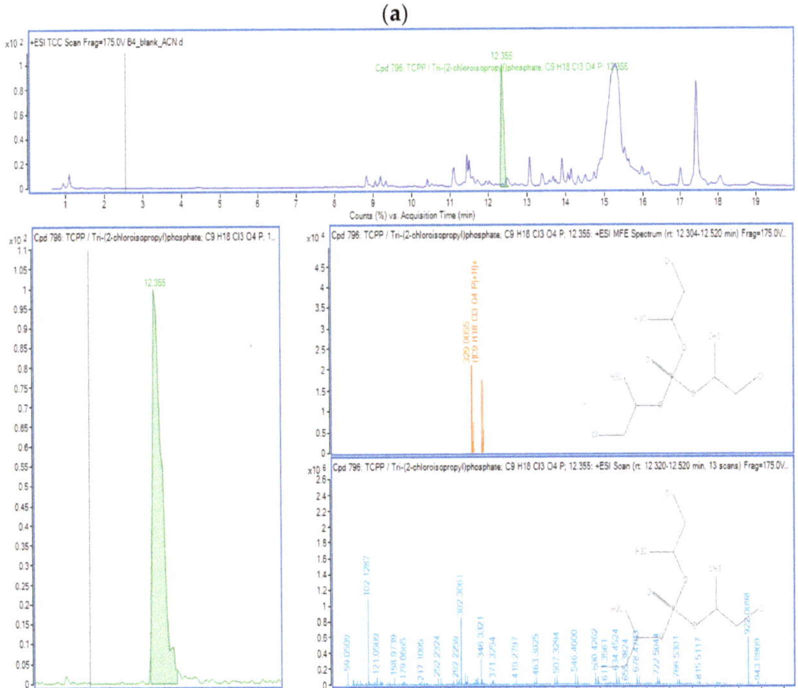

(**b**)

Figure 3. OPEs in sample B4: (**a**) DEP; (**b**) TCPP.

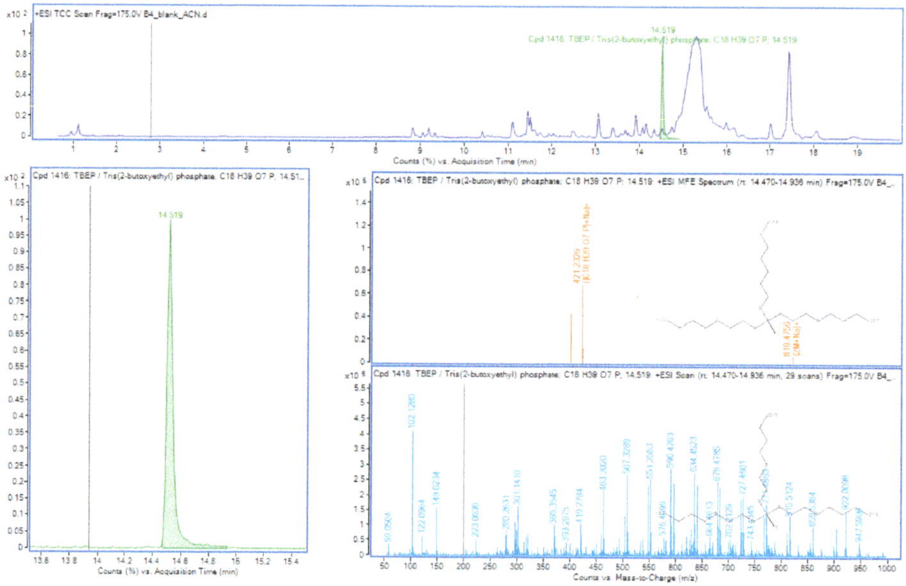

Figure 4. TBEP in sample B4.

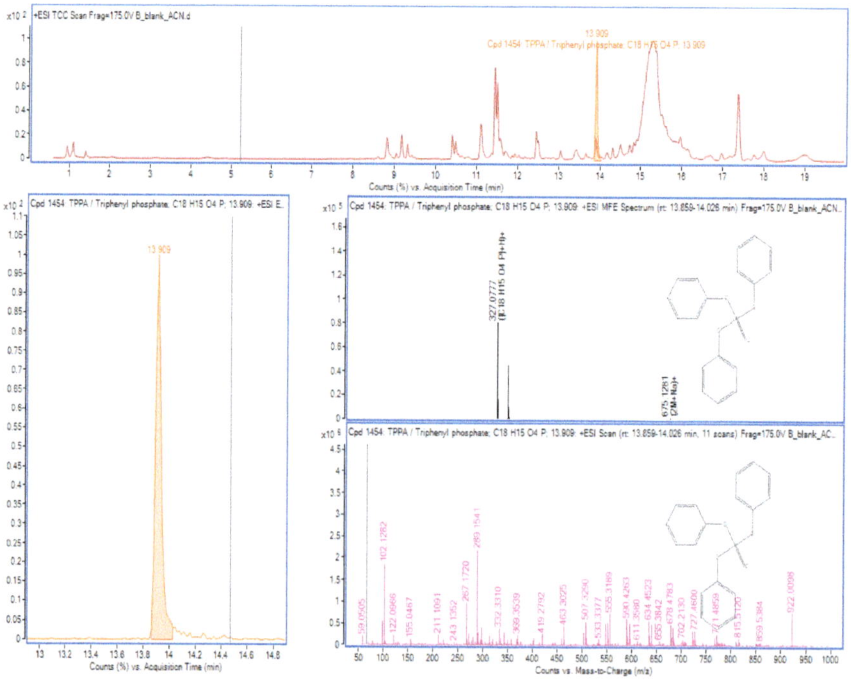

(a)

Figure 5. *Cont.*

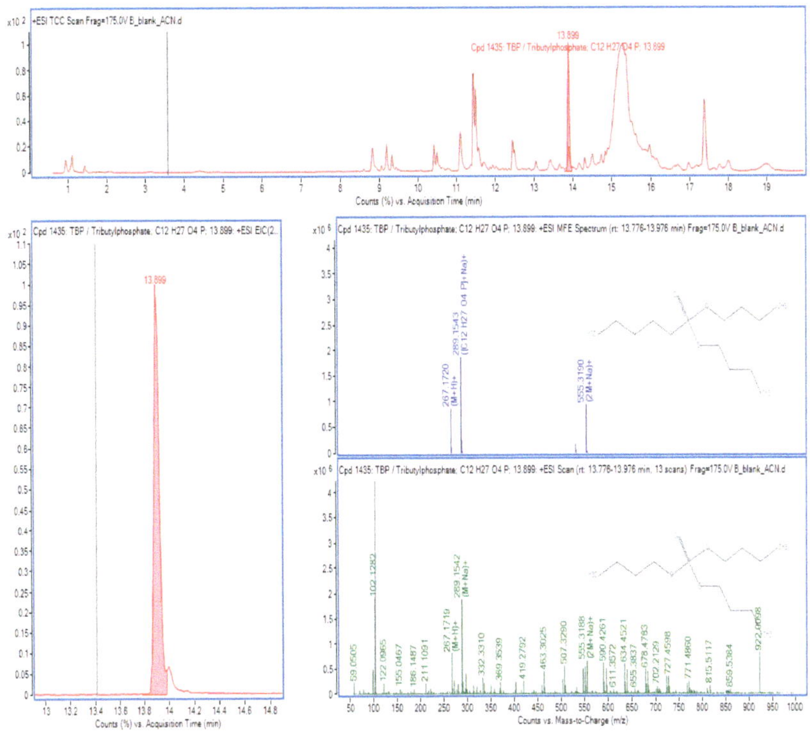

(**b**)

Figure 5. OPEs in sample B: (**a**) TPPA; (**b**) TBP.

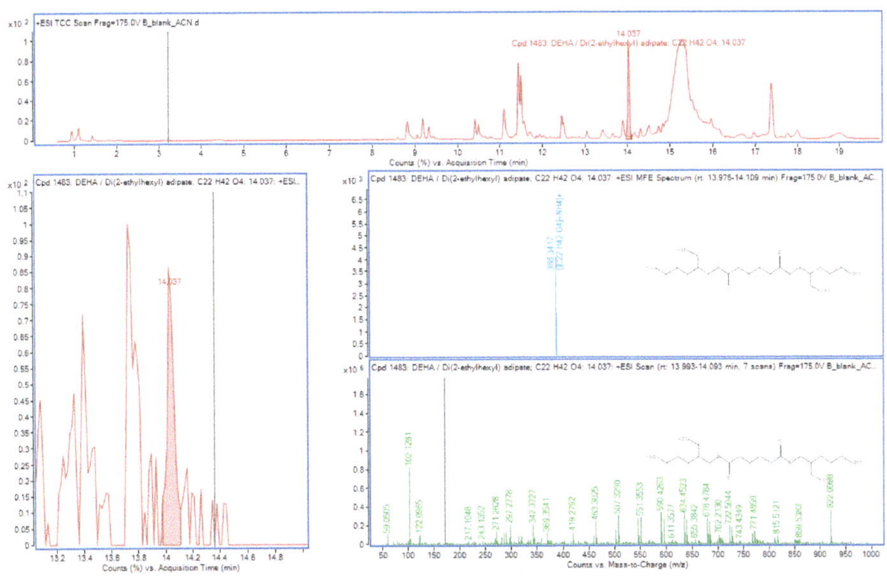

Figure 6. DEHA in sample B.

4. Conclusions and Discussion

OPEs are potential pollutants of groundwaters that are not under regulations, and there are no defined maximum available concentrations for such emerging pollutants. OPEs can be found in nature in relatively small concentrations, and it is not easy to identify their presence in groundwater. In this research, nine OPEs were detected in groundwater samples from a karst basin in Croatia, which were taken during each season for one year. In the areas where the samples were taken, the water used for the water supply system is withdrawn from the karst aquifer, used here as a case study. Consequently, there is a need to raise public awareness of the usage and release of OPEs into the environment. Additionally, some of the OPEs found in the samples are toxic, i.e., dangerous for human health, due to their persistence in the water environment. This is why research on their identification, quantification, and behavior in groundwater is important.

Author Contributions: Conceptualization, L.P. and A.P.S.; methodology, L.P.; validation, I.G., A.P.S. and R.B.; formal analysis, L.P.; investigation, L.P.; writing—original draft preparation, L.P.; writing—review and editing, I.G., R.B. and A.P.S. All authors have read and agreed to the published version of the manuscript.

Funding: This research was funded by EU funded projects, grant numbers KK.01.1.1.04.0006 and KK.05.1.1.02.0022.

Institutional Review Board Statement: Not applicable.

Informed Consent Statement: Not applicable.

Data Availability Statement: Not applicable.

Conflicts of Interest: The authors declare no conflict of interest.

References

1. Directive 2000/60/EC of the European Parliament and of the Council of 23 October 2000 Establishing a Framework for Community Action in the Field of Water Policy. Available online: https://eur-lex.europa.eu/legal-content/EN/TXT/?uri=CELEX:32000L0060 (accessed on 17 October 2022).
2. Directive 2006/118/EC of the European Parliament and of the Council of 12 December 2006 on the Protection of Groundwater against Pollution and Deterioration. Available online: https://eur-lex.europa.eu/legal-content/EN/TXT/?uri=CELEX:02006L0118-20140711 (accessed on 17 October 2022).
3. Common Implementation Strategy for the Water Framework Directive and the Floods Directive, Voluntary Groundwater Watch List. June 2019. Available online: https://circabc.europa.eu/sd/a/e6882891-d4a2-4a64-9cf7-f04e13b0d17e/Voluntary%20Groundwater%20Watch%20List%20(Endorsed%20V3.1%20-%20June%202019).pdf (accessed on 17 October 2022).
4. Reemtsma, T.; García-López, M.; Rodríguez, I.; Quintana, J.B.; Rodil, R. Organophosphorus Flame Retardants and Plasticizers in Water and Air I. Occurrence and Fate. *TrAC-Trends Anal. Chem.* **2008**, 27, 727–737. [CrossRef]
5. Chokwe, T.B.; Abafe, O.A.; Mbelu, S.P.; Okonkwo, J.O.; Sibali, L.L. A Review of Sources, Fate, Levels, Toxicity, Exposure and Transformations of Organophosphorus Flame-Retardants and Plasticizers in the Environment. *Emerg. Contam.* **2020**, 6, 345–366. [CrossRef]
6. Quintana, J.B.; Rodil, R.; Reemtsma, T.; García-López, M.; Rodríguez, I. Organophosphorus Flame Retardants and Plasticizers in Water and Air II. Analytical Methodology. *TrA-Trends Anal. Chem.* **2008**, 27, 904–915. [CrossRef]
7. EPA Chemicals under the TSCA. Available online: https://www.epa.gov/chemical-data-reporting (accessed on 20 November 2022).
8. O'Neil, M.J. *The Merck Index—An Encyclopedia of Chemicals, Drugs, and Biologicals*; Merck and Co., Inc.: Whitehouse Station, NJ, USA, 2006; p. 1676.
9. Weil, E.D. Flame Retardants, Phosphorus. In *Kirk-Othmer Encyclopedia of Chemical Technology (1999–2012)*; John Wiley & Sons: New York, NY, USA, 2001.
10. *Documentation of the TLVs and BEIs with Other World Wide Occupational Exposure Values*, 7th ed.; American Conference of Governmental Industrial Hygienists: Cincinnati, OH, USA, 2013.
11. European Commission/European Chemical Substances Information System (ESIS); European Union Risk Assessment Report, Tris(2-Chloroethyl) Phosphate (115-96-8). July 2009, p. 17. Available online: https://esis.jrc.ec.europa.eu/ (accessed on 3 September 2014).
12. Haz-Map, Information on Hazardous Chemicals and Occupational Diseases. Available online: https://haz-map.com/Agents/1943 (accessed on 20 November 2022).

13. Crump, D.; Chiu, S.; Kennedy, S.W. Effects of Tris(1,3-dichloro-2-propyl) phosphate and Tris(1-chloropropyl) phosphate on Cytotoxicity and mRNA Expression in Primary Cultures of Avian Hepatocytes and Neuronal Cells. *Toxicol. Sci.* **2012**, *126*, 140–148. [CrossRef] [PubMed]
14. Moschet, C.; Lew, B.M.; Hasenbein, S.; Anumol, T.; Young, T.M. LC- and GC-QTOF-MS as Complementary Tools for a Comprehensive Micropollutant Analysis in Aquatic Systems. *Environ. Sci. Technol.* **2017**, *51*, 1553–1561. [CrossRef] [PubMed]

environmental sciences proceedings

MDPI

Proceeding Paper

Quality of Surface and Ground Water in Three States of Nigeria: Assessment of Physicochemical Characteristics and Selected Contamination Patterns †

Francis Olawale Abulude [1,*], Akinyinka Akinnusotu [2], Ebenezer Alaba Adeoya [3], Samson Olatunde Mabayoje [4], Samuel Dare Oluwagbayide [5], Kikelomo Mabinuola Arifalo [6] and Ademola Adamu [6]

1 Environmental and Sustainable Research Group, Science and Education Development Institute, Akure 340001, Ondo State, Nigeria
2 Department of Science Laboratory Technology, Rufus Giwa Polytechnic, Owo 341101, Ondo State, Nigeria; akinnusotuakinyinka@gmail.com
3 National Agency for Science and Engineering Infrastructure (NASENI), Akure 340001, Ondo State, Nigeria; adeoyaeben@yahoo.com
4 Department of Biological Sciences, College of Sciences, Afe Babalola University (ABUAD), Ado Ekiti 360101, Ekiti State, Nigeria; mabayojeso@abuad.edu.ng
5 Department of Agricultural and Bio-Environmental Engineering, Federal Polytechnic, P.M.B 50, Ilaro 112106, Ogun State, Nigeria; samuel.oluwagbayide@federalpolyilaro.edu.ng
6 Department of Chemistry, Bamidele Olomilua University of Education, Science and Technology, Ikere Ekiti 361101, Ekiti State, Nigeria; karifalo@yahoo.co.uk (K.M.A.); adamu.ademola@coeikere.edu.ng (A.A.)
* Correspondence: waleabul@outlook.com; Tel.: +234-803445674

† Presented at the 7th International Electronic Conference on Water Sciences, 15–30 March 2023; Available online: https://ecws-7.sciforum.net.

Abstract: The study of water quality is crucial given the amount of industrial, agricultural, and other human activities at the sampling sites. The aim of the study was to assess the physicochemical characteristics and selected contamination patterns of water samples in Nigeria. This study used conventional analytical techniques to analyze the physicochemical parameters in water samples from 33 sampling sites (dug wells, boreholes, rivers, and rainwater) in three different states (Ekiti, Osun, and Ondo) of Nigeria. These parameters included pH, total dissolved solids (TDS), electrical conductivity (EC), temperature, relative humidity (RH), and four elemental parameters (Ca, Na, Fe, and Cu). The enrichment factor (EF), contamination factor (CF), and metal index (MI) were used to characterize the data. Temperature (28.17 °C), TDS (130.2 mg/L), EC (260.0 μS/cm), pH (6.88), Na (14.47 ppm), Ca (25.74 ppm), Fe (0.49 ppm), and Cu (0.08 ppm) were the average values from the results. Na and Ca had a direct relationship with one another. The levels of heavy metals were below those recommended by the Nigerian Industrial Standard for Drinking Water Quality (NISDQW) and the World Health Organization (WHO). The metal levels in the water samples were over 1.5, which is the threshold value indicated by the EF classifications. In particular, EFs were moderate to significantly enriched. All element CFs were below the Level 1 pollution threshold. The water samples are pure based on the MI's rating of water quality. Human and natural activities may represent a risk to the local public health; hence, it is highly advisable that all stakeholders adopt rapid and long-lasting collective action to limit pollution levels as part of the water quality governance system.

Keywords: anthropogenic activities; water samples; heavy metals; TDS; enrichment factor; WHO; SON

Citation: Abulude, F.O.; Akinnusotu, A.; Adeoya, E.A.; Mabayoje, S.O.; Oluwagbayide, S.D.; Arifalo, K.M.; Adamu, A. Quality of Surface and Ground Water in Three States of Nigeria: Assessment of Physicochemical Characteristics and Selected Contamination Patterns. *Environ. Sci. Proc.* **2023**, *25*, 48. https://doi.org/10.3390/ECWS-7-14258

Academic Editor: Athanasios Loukas

Published: 16 March 2023

1. Introduction

The accessibility and sustainability of water and sanitation for every person worldwide by 2030 are among the key sustainable development objectives (Goal 6) [1,2]. The objectives are to provide everyone with equitable access to clean, inexpensive drinking water; achieve adequate and equitable access to sanitation and hygiene; improve water quality by reducing pollution, eliminating dumping, and minimizing the release of hazardous chemicals and materials; achieve half as much untreated wastewater; implement integrated water resources management at all levels; and achieve protection and restoration [3].

For continued population expansion and development, access to a secure and reliable water supply is a crucial requirement [4]. Both surface and ground water are essential sources of water for the global population. About 90% of the world's readily usable freshwater resources are found in groundwater, with the other 10% being found in lakes, reservoirs, rivers, and wetlands. Moreover, the expansion of an estimated 40% of the world's agricultural production is supported by groundwater irrigation of arable lands [4]. Groundwater is the most dependable source of drinking water in sub-Saharan Africa [5].

Southwest Nigeria faces a number of difficulties, including how to accomplish sustainable development goals and provide drinkable water for its expanding population. Due to the scarcity of surface water supplies in some regions, people are forced to use underground water, which presents a problem in those areas. In the states we examined, groundwater and surface water are significant natural resources that have an impact on both human and animal health and welfare. As a result, primary research and quality control efforts should be directed at the quality of these resources. This premise would not be considered out of place if the qualities of the water samples are determined. The main goal of this study was to assess the quality (physicochemical characteristics and selected contamination patterns) of ground and surface water samples (borehole, dug well, rainwater, and river) from various communities in chosen areas of three states (Osun, Ondo, and Ekiti) in the southwest of Nigeria.

2. Materials and Methods

The Osun, Ekiti, and Ondo States of Nigeria, which make up the study area, are situated in the southwest of the country. The research area's climate generally follows the same pattern. In terms of population, transportation, industry, housing, and agriculture, the states are rapidly expanding. For this experiment in November 2022, 33 water samples were selected from rivers (10), hand-dug wells (14), boreholes (7), and rain (2). The samples were obtained in polyethylene bottles, cleaned properly with distilled water, and then treated with nitric acid before being filtered through membrane filters with pores measuring 0.45 microns. A Garmin global positioning system was used to determine the coordinates of the sampling locations. A portable multi-parameter meter called the Temp/pH/TDS/EC meter (model MI 1399) was used to measure the temperature, electrical conductivity, pH, temperature, and TDS in situ immediately after sample collection in the field. To stop the precipitation of trace elements, nitric acid was used to acidify the water samples. The elements (Na, Ca, Cu, and Fe) were evaluated utilizing conventional methods (AAS, Buck Scientific GVP 210, USA) in the Central Laboratory of Quality Monitoring at Afe Babalola University in Ado-Ekiti, Ekiti State, Nigeria. The metal index, contamination factor, and enrichment factor (EF) were determined. EF was calculated using this factor [6]:

$$\mathrm{EF} = \frac{(\mathrm{C}_i/\mathrm{C}_{\mathrm{ref}})\ \text{sample}}{(\mathrm{B}_i/\mathrm{B}_{\mathrm{ref}})\ \text{Background}} \tag{1}$$

where B_i is the background value of an element of interest and Bref is the background value of the reference element in the study area, C_i is the concentration of trace elements in the sample, C_i is the concentration of the reference element in the sample, and Cref is the background value of the reference analyte in the sample. The reference element used in this study was Fe, which is most widely used for normalization [6]. EF classification: EF < 2 (deficiency to minimal enrichment), $2 \leq 5$ (moderate enrichment), $5 \leq$ EF < 20 (significant enrichment), $20 \leq$ EF < 40 (very high enrichment), and EF $\geq$ (extremely high enrichment).

CF was calculated using this factor [6]:

$$CF = C_i/B_i \tag{2}$$

These values were obtained by calculating the ratio of the element's background concentration to the concentration of the element present in the sample [7]. C_i = concentration of the examined element *i*, and B_i = geochemical background value of the element. The contamination values in increasing order of contaminations are 0 = none, 1 = none to medium, 2 = moderate, 3 = moderate to strong, 4 = strongly polluted, 5 = strong to very strong, 6 = very strong [8]. In terms of metal and metalloid contamination, MI shows an overall trend in water quality [9], where *Hc* is the *i*th parameter's monitored value (in mg/L), and *Hmac* is the *i*th parameter's maximum permissible concentration [10]. According to the MI, the water is either lowly polluted (MI < 10) or moderately polluted (10 < MI < 20) [11]. MI was calculated by Equation (3) [11]:

$$MI = \sum_{i=1}^{n} \frac{Hc}{Hmac} \tag{3}$$

The MI classification for water samples are: <0.3 = very pure (Class I), 0.3–1.0 pure (Class II), 1.0–2.0 slightly affected (Class III), 2.0–4.0 moderately affected (Class IV), 4.0–6.0 strongly affected (Class V), >6.0 seriously affected (Class VI).

3. Results and Discussion

The recorded mean pH levels of the water samples did not differ from one another statistically ($p > 0.05$). The pH in the water samples ranged from 5.66 to 7.89, with a mean value of 6.88 0.60 (Table 1). Dug wells had the lowest pH readings, whereas boreholes had the highest. Except for the lowest pH level, the pH of water was within the range of 6.5–8.5 allowed by WHO [12] and SON [13] for drinking water. The outcome is consistent with the findings of Appiah-Opong et al. [14]. Water with a pH level below 6.5 is considered to be too acidic for human consumption, which could lead to conditions such as acidosis and harm the digestive and lymphatic systems [6]. Statistical analysis revealed that the means were not different ($p > 0.05$) from one another, despite the fact that the electrical conductivity of the water was generally higher in the dug wells than in other samples. The mean EC values in the water samples ranged thus: dug well (84–1003 µS/cm), borehole (136–386 µS/cm), rain (54–59 µS/cm), and rivers (76–297 µS/cm). The overall mean was 260 µS/cm.

Table 1. Basic Description of the Parameters.

Variable	Mean	SE Mean	StDev	CoefVar (%)	Min	Q1	Median	Q3	Max	Skewness	Kurtosis
Temperature (°C)	28.17	0.28	1.67	5.93	24.80	26.90	28.60	29.30	31.20	−0.38	−0.72
TDS (mg/L)	130.20	15.60	89.80	68.99	27.00	71.00	118.00	162.00	511.00	2.49	9.39
EC (µS/cm)	260.00	30.70	176.40	67.86	54.00	144.00	239.00	314.50	1003.00	2.45	9.10
pH	6.88	0.10	0.60	8.00	5.66	6.47	7.02	7.27	7.89	−0.52	−0.46
Na (mg/L)	14.47	0.63	3.62	24.98	9.40	11.05	14.80	17.40	23.50	0.46	−0.36
Ca (mg/L)	25.74	1.74	10.02	38.92	8.20	17.80	22.90	31.15	48.30	0.69	−0.18
Fe (mg/L)	0.49	0.05	0.30	60.14	0.18	0.29	0.40	062	1.35	1.62	2.29
Cu (mg/L)	0.08	0.01	0.04	41.48	0.02	0.06	0.08	0.11	0.16	0.57	−0.27

WHO (2011)—temperature (22–29 °C), TDS (1000 mg/L), EC 1000 (mg/L), pH (6.5–8.5), Na (200), Ca (75), Fe (0.03), Cu (200 mg/L); NISDQW (2015)—temperature (22–29 °C), TDS (1000 mg/L), EC 1000 (mg/L), pH (6.5–8.5), Na (200), Ca (75 mg/L), Fe (0.03), Cu (200 mg/L).

The EC results fell below the 1000 µS/cm drinking water limit set by WHO [12] and SON [13]. Although EC is not a concern for human or aquatic health, it might be a sign of other issues with water quality [14]. The high values of EC in the dug well could be linked to anthropogenic activities, as well as the soil's mineral or salt dissolution [15]. TDS values ranged between 27 and 511 mg/L, with 260 mg/L being the mean. The considerable high variability in the water samples was shown by the coefficient of variation in terms of percentages. No differences in water temperature were found to be statistically significant ($p > 0.05$). The results showed the range, standard error, skewness, and kurtosis of temperature as 24.80–31.20 °C, 0.28, −0.38, and −0.72, respectively. The minimum value was recorded in a stream. The reason could be due to the activities of the sampling which assisted in the aeration of the water. In addition to the time of the sample, other factors that may affect temperature include water depth, season, groundwater influx, and air circulation [16]. These findings concur with those made in the Ivory Coast by Koffi et al. [15].

The concentrations of Na, Ca, Fe, and Cu in the thirty-three sampling points ranged from 9.40 to 23.50 mg/L, 8.20 to 48.30 mg/L, 0.18 to 1.35 mg/L, and 0.02 to 0.16 mg/L, respectively. The elements' average concentrations fell in the following order: Ca > Na > Fe > Cu (Table 1). These results agree with those found by Koffi et al. [15] in the Ivory Coast. The number of components was below what is considered to be acceptable for drinking and irrigation water on a national and international level. Only a few of the water samples had iron contents that were above the WHO limit (0.3 mg/L); over 90% of the samples were regarded as suitable for use in irrigation and human consumption. Undesirable tastes and smells are typically connected to underground water that contains more iron [17]. It is possible that the iron in water samples came from natural sources as well [18]. The concentration of copper was higher in river water samples than in others. This shows that the rivers are either naturally high in copper or absorbed from soils with Cu fertilizers. The calcium concentration was greater than the amount of sodium content for more than 90% of the water samples gathered. The amount of carbonate minerals that make up the water-bearing formations, ion exchange mechanisms, and the precipitation of calcite in the aquifer can all be used to explain this [19].

Enrichment factors of elements followed this order Ca > Cu > Na > Fe. The metal levels in the water samples were over 1.5, which is the threshold value indicated by the EF classifications. In particular, EFs were moderately to significantly enriched. Figure 1 displays the water sample histogram. The samples fell within the very pure, Class I category. This result was compared with the findings of Khosnam et al. [20] for the Silakhor River in Iran.

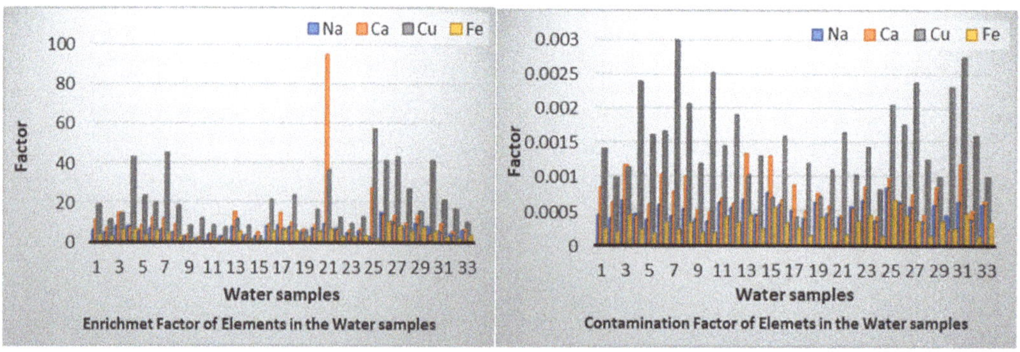

Figure 1. *Cont.*

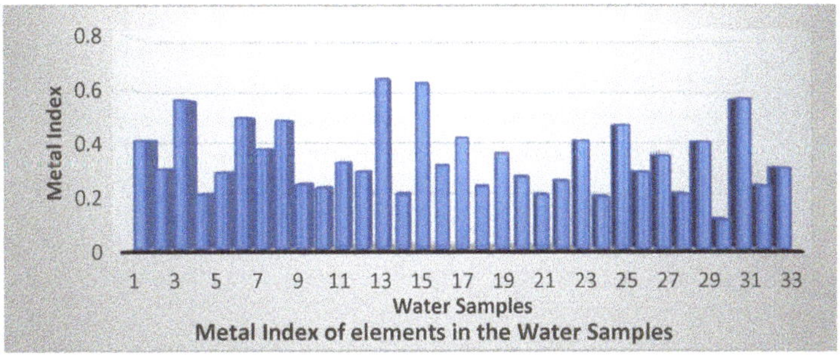

Figure 1. The Metal Index, Contamination Factors, and Enrichment Factor of the Water Samples.

Figure 2 and Table 2 depicted the matrix correlation (Pearson correlation). This showed the correlations between the physicochemical parameters and the elements. There were strong positive relationships between EC and TDS (r = 0.99) and Ca and Na (r = 0.72) depicting that an increase in EC causes an increase in TDS, showing a direct relationship between the variables. Ca and Na had a substantial correlation, which suggested that the components in the water samples might have had identical origins.

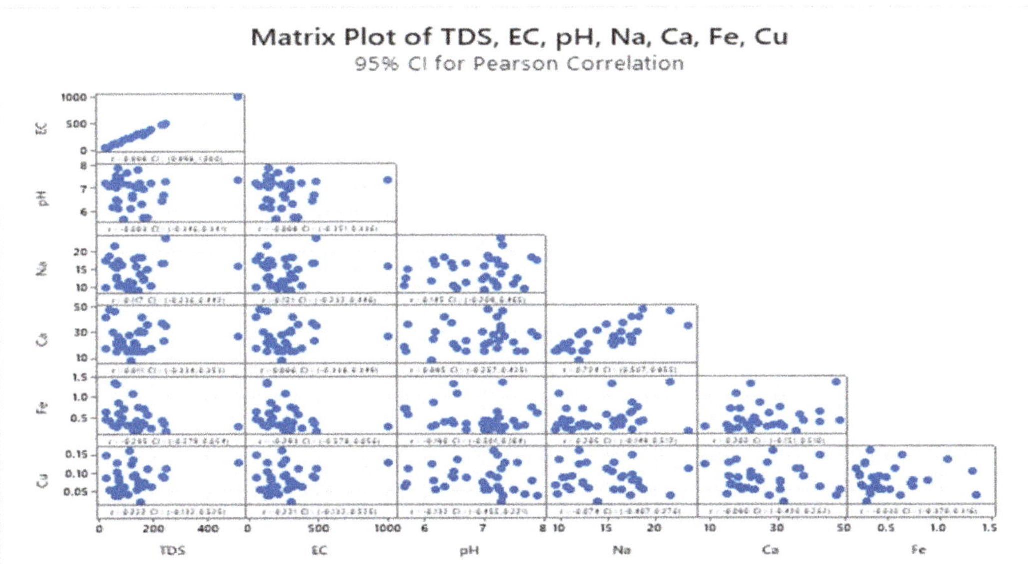

Figure 2. The matrix correlation of the water samples (Pearson correlation).

Table 2. The correlation of the water samples (Pearson correlation).

EC	**0.99**					
pH	−0.00	−0.01				
Na	0.12	0.12	0.15			
Ca	0.01	0.01	0.10	**0.72**		
Fe	−0.30	−0.29	−0.19	0.20	0.20	
Cu	0.22	0.22	−0.13	−0.07	−0.09	−0.03
	TDS	EC	pH	Na	Ca	Fe

4. Conclusions

This study assessed the physicochemical characteristics and selected contamination patterns of thirty-three water samples from three southwest states in Nigeria. Additionally, in the study, the EF, CF, and MI were determined. The physicochemical and elemental levels obtained were within the standard limits of WHO and NISDQW. The EFs of the water samples were moderately to significantly enriched. All elemental CFs were below the Level 1 pollution threshold. The MI rating showed that they are suitable for human and animal consumption and also for irrigation purposes. Although the present study concluded that the water samples were good, efforts should be put in place to prevent them from being polluted.

Author Contributions: Conceptualization, F.O.A.; methodology, A.A. (Akinyinka Akinnusotu) and S.D.O.; formal analysis, E.A.A. and S.D.O.; investigation, S.O.M.; data curation, K.M.A.; writing—original draft preparation, F.O.A.; writing—review and editing, E.A.A., K.M.A., S.D.O. and A.A. (Akinyinka Akinnusotu); supervision, A.A. (Ademola Adamu) and S.O.M. All authors have read and agreed to the published version of the manuscript.

Funding: This research received no external funding.

Institutional Review Board Statement: Not applicable.

Informed Consent Statement: Not applicable.

Data Availability Statement: The raw data are available upon request. Please contact the corresponding authors.

Acknowledgments: The authors are grateful for the permissions granted by the owners of the dug wells and boreholes to obtain samples from their homes and offices.

Conflicts of Interest: The authors declare no conflict of interest.

References

1. Abulude, F.O.; Fagbayide, S.D. Water, sanitation, and poverty in the changing world. Case of Nigeria. *An. Univ. Din Oradea Ser. Geogr.* **2018**, *28*, 91–96.
2. Oluwagbayide, S.D.; Abulude, F.O. Assessment of the quality of dug well water samples in Nigeria and their suitability for drinking and irrigation purposes. *Sustain. Water Resour. Manag.* **2022**, *8*, 149. [CrossRef]
3. United Nations. Department of Economic and Social Afairs Disability. Envision2030 Goal 6: Clean Water and Sanitation. Targets. Available online: https://www.un.org/development/desa/disabilities/envision2030-goal6.html (accessed on 24 November 2020).
4. Saana, S.B.B.M.; Fosu, S.A.; Sebiawu, G.E.; Jackson, N.; Karikari, T. Assessment of the quality of groundwater for drinking purposes in the Upper West and Northern regions of Ghana. *SpringerPlus* **2016**, *5*, 2001. [CrossRef] [PubMed]
5. Iyasele, J.U.; Idiata, D.J. Physico-chemical and microbial analysis of borehole water samples: A case of some boreholes in Edo North, Edo state. *J. Emerg. Trends Eng. Appl. Sci.* **2011**, *2*, 1064–1067.
6. Bhuiyan, M.A.H.; Parvez, L.; Islam, M.A.; Dampare, S.B.; Suzuki, S. Heavy metal pollution of coal mine-affected agricultural soils in the northern part of Bangladesh. *J. Hazard. Mater.* **2010**, *173*, 384–392. [CrossRef] [PubMed]
7. Hakanson, L. An ecological risk index for aquatic pollution control. A sediment logical approach. *Water Res.* **1980**, *14*, 975–1001. [CrossRef]
8. Varol, M. Assessment of heavy metal contamination in sediments of the Tigris River (Turkey) using pollution indices and multivariate statistical techniques. *J. Hazard. Mater.* **2011**, *195*, 355–364. [CrossRef] [PubMed]
9. Edet, A.E.; Offiong, O.E. Evaluation of water quality pollution indices for heavy metal contamination monitoring. A study case from Akpabuyo-Odukpani area, Lower Cross River Basin (southeastern Nigeria). *GeoJournal* **2002**, *57*, 295–304. [CrossRef]
10. Shil, S.; Singh, U.K. Health risk assessment and spatial variations of dissolved heavy metals and metalloids in a tropical river basin system. *Ecol. Indic.* **2019**, *106*, 105455. [CrossRef]
11. Rezaei, A.; Hassani, H.; Hassani, S.; Jabbari, N.; Fard Mousavi, S.B.; Rezaei, S. Evaluation of groundwater quality and heavy metal pollution indices in Bazman basin, southeastern Iran. *Groundw. Sustain. Dev.* **2019**, *9*, 100245. [CrossRef]
12. WHO. *Guidelines for Drinking-Water Quality*, 4th ed.; WHO: Rome, Italy, 2011. Available online: http://www.who.int/about/licensing/copyright_form/en/index.html (accessed on 12 December 2022).
13. *NIS 554: 2015*; Nigerian Standard for Drinking Water Quality (NSDWQ). Standard Organization of Nigeria: Abuja, Nigeria, 2015. Available online: https://africacheck.org/wp-content/uploads/2018/06/Nigerian-Standard-for-Drinking-Water-Quality-NIS-554-2015.pdf (accessed on 12 December 2022).

14. Appiah-Opong, R.; Ofori, A.; Odosuhene, M.; Ofori-Attach, E.; Nunoo, F.K.E.; Tuffour, I.; Gordo, C.; Arhinful, D.K.; Nyarko, A.K.; Fosu-Mensah, B.Y. Heavy metals concentration and pollution index (HPI) in drinking water along the southwest coast of Ghana. *Appl. Water Sci.* **2021**, *11*, 57. [CrossRef]
15. Koffi, K.M.; Coulibaly, S.; Atse, B.C.; Paul, E. Survey of heavy metals concentrations in water and sediments of the estuary Bietri Bay.Ebrie Lagoon, Cote D'Ivoire. *Int. J. Res. Earth Environ. Sci.* **2014**, *1*, 2311.
16. Dan, S.F.; Umoh, U.U.; Osabor, V.N. Seasonal variation of enrichment and contamination of heavy metals in the surface water of Qua Iboe River Estuary and adjoining creeks, South-South Nigeria. *J. Oceanogr. Mar. Sci.* **2014**, *5*, 45–54. [CrossRef]
17. Sethy, S.N.; Syed, T.H.; Kumar, A.; Sinha, D. Hydrogeochemical characterization and quality assessment of groundwater in parts of Southern Gangetic Plain. *Environ. Earth Sci.* **2016**, *75*, 232. [CrossRef]
18. Smedley, P.L.; Edmunds, W.M.; West, J.M.; Gardner, S.J.; Pelig-Ba, K.B. *Vulnerability of Shallow Groundwater Quality Due to Natural Geochemical Environment. 2, Health Problems Related to Groundwater in the Obuasi and Bolgatanga Areas, Ghana*; British Geological Survey: Nottingham, UK, 1995; p. 130.
19. Sebiawu, E.G.; Fosu, S.A.; Saana, S.B.B.M. A physico-chemical and bacteriological analysis of borehole water samples from the Wa Municipality of the Upper West Region, Ghana. *Int. J. Eng. Res.* **2014**, *3*, 1959–1965.
20. Khoshnam, Z.; Sarikhani, R.; Ghassemi Dehnavi, A.; Ahmadnejad, Z. Evaluation of Water Quality Using Heavy Metal Index and Multivariate Statistical Analysis in Lorestan Province. *Iran. J Adv. Environ. Health Res.* **2017**, *5*, 29–37.

environmental sciences proceedings

MDPI

Proceeding Paper

Comparative Analysis of Photodegradation of Ibuprofen and Clotrimazole Water Pollutant Using UVC Rays in Presence and Absence of ZnO Photocatalyst †

Shania Pesik [1], Eric Jobiliong [2] and Eden Steven [1,3,*]

1 SPH Applied Science Academy, Sekolah Pelita Harapan Lippo Village, Tangerang 15810, Indonesia
2 Department of Industrial Engineering, Universitas Pelita Harapan, Tangerang 15810, Indonesia
3 Emmerich Research Center, Jakarta 14450, Indonesia
* Correspondence: eden.steven@gmail.com

† Presented at the 7th International Electronic Conference on Water Sciences, 15–30 March 2023; Available online: https://ecws-7.sciforum.net.

Abstract: The recent surge in pharmaceutical micro-pollutants in water bodies calls for an efficient method to neutralize wastewater to sustain the ecosystem. One of the ways to degrade drug molecules is through photocatalytic degradation using UV rays. ZnO is known to be a common catalyst in the degradation of contaminants found in wastewater. However, due to its toxicity to the environment, there is a need to objectively re-evaluate its necessity and alternatives. In addition, most studies have focused on the utilization of UVA/UVB rays for the photocatalytic degradation process, as such, there are currently limited studies evaluating the efficacy of UVC for such purpose. In this work, we provide a comparative analysis of the photodegradation of drug molecules using UVC rays with and without the ZnO catalyst. Ibuprofen (IBP) and clotrimazole were used for analysis. We found that the use of a ZnO catalyst did not always produce better results. In some cases, we found that IBP was degraded more without ZnO (up to 94.4%) than with the ZnO (1 g/L) (up to 86.6%) in 60 min. However, without ZnO, we observed secondary metabolite by-products of IBP that required a longer treatment period to fully degrade. The inferior degradation strength for treatment with ZnO could be explained by increasing turbidity from adding greater concentrations of ZnO, which decreased the UV transmission to the IBP solution. To support the results, an investigation of the photocatalytic degradation of clotrimazole (an antifungal) with varying concentrations of ZnO as a catalyst was also carried out. The optimum ZnO concentration was determined to be ~1000 ppm, above or under which the efficiency of the degradation suffered. Thus, the use of a ZnO catalyst requires strict dosage control. Such tight regulation is not required for the system using just UVC rays, but it requires a longer treatment time to completely degrade the drug molecules and their by-products.

Keywords: photodegradation; photocatalytic; wastewater treatment

Citation: Pesik, S.; Jobiliong, E.; Steven, E. Comparative Analysis of Photodegradation of Ibuprofen and Clotrimazole Water Pollutant Using UVC Rays in Presence and Absence of ZnO Photocatalyst. *Environ. Sci. Proc.* **2023**, *25*, 49. https://doi.org/10.3390/ECWS-7-14176

Academic Editor: Lampros Vasiliades

Published: 14 March 2023

1. Introduction

The discovery of micro-pollutants in the environment is a major concern that has yet to be addressed clearly [1]. Among the predominant examples of emerging organic contaminants (EDCs) are pharmaceutical and personal care products (PPCPs) [2]. Pharmaceuticals can reach water bodies through industrial, domestic, urban, agricultural, and hospital disposal via sewage systems [3]. Although wastewater treatment plants (WWTPs) are able to neutralize up to approximately 95% of these pharmaceutical components, they still end up in the environment at low concentrations [4]. Furthermore, typically, only large facilities, such as industrial plants, have such wastewater treatment plants.

Studies have shown that the toxicological effects of pharmaceutical micro-pollutants remain even at low concentrations of nanograms or micrograms per liter [5]. The release of micro-pollutants into the environment is highly dangerous for the ecosystem, as it poses health risks to humans and animals and is hazardous to the environment [6]. The rapidly growing population contributes to the increased demand for the production and consumption of such components [7]. This raises the frequency and probability of the improper disposal of pharmaceuticals and the treatment of effluents, which inevitably increases the likelihood of pharmaceutical components being found in the environment as micro-pollutants [8]. Unwanted pharmaceutical pollutants have been found in various water bodies, such as rivers, lakes, groundwater, wastewater treatment plant (WWTP) effluents, and drinking water [9].

In this work, we investigated the photocatalytic degradation effectiveness of UV treatments on drug-contaminated water. UV water treatment is one of the simplest and most readily available methods for various public constituents, from households to large hospital facilities. This photocatalytic degradation study was carried out and analyzed via UV–Vis spectroscopy. Ibuprofen (IBP) and clotrimazole (CTZ) were chosen as the drug molecules of interest.

IBP is the world's third most consumed drug, making it a leading pharmaceutical micro-pollutant, as shown by its high detection rate in water systems around the world [10]. Many studies have explored the use of ZnO and/or TiO2 as catalyst(s) in the photocatalytic degradation of various pharmaceutical micro-pollutants and have shown promising results. The catalysts are activated upon UV irradiation, and highly oxidizing species are generated and aid in the process of degrading ibuprofen to its intermediates [11]. Sabouni et al. demonstrated that 94.5% of ibuprofen was degraded when ZnO was used as a photocatalyst after 120 min of UVA radiation [12]. Similarly, Jallouli et al. reported that ibuprofen was below the detection limit after 30 min of photocatalysis with TiO_2 [13].

CTZ is among the top 10 PPCP compounds when ranked according to its persistence, risk, bioaccumulation, toxicity, and occurrence in various countries [14]. Despite this, the degradation of CTZ is not commonly discussed, and very little information can be found. CTZ has a high degree of persistence, with a half-life of around 60 days and low biodegradability [15]. Naturally, CTZ can be degraded by physicochemical processes in soils and water or by being digested and metabolized by organisms [16]. However, the digestion of clotrimazole in non-target organisms can cause harmful effects. CTZ poses adverse effects on marine environments, especially in algae.

Our work presented herein aimed to provide new insights into the dynamic photocatalytic degradation of IBP and CTZ in UVC using various concentrations of ZnO as a catalyst. The time-dependent degradation of IBP and CTZ was analyzed using an exponential model, from which the maximum degradation efficiency and degradation rate were obtained. A recommendation based on the results is given to those seeking to employ UV waste treatment at their facilities.

2. Materials and Methods

2.1. Sample Preparation

Ibuprofen (IBP) powder was obtained from crushing Proris© tablets. The stock solution of IBP was prepared by adding 1 g of IBP powder into 1 L of methanol and stirring until complete dissolution was achieved. A serial dilution with distilled water was carried out to obtain 0, 10, 20, and 40 ppm IBP solution to construct the standard calibration curve. For example, to make the 40 ppm IBP solution, 4 mL of the IBP stock solution was diluted with 96 mL of distilled water. The sample tested was a 40 ppm IBP solution. For samples containing zinc oxide (ZnO), the solution was dosed with 1 g/L ZnO microparticles (SmartLab, molecular weight 81.39 g/mol) and stirred using a magnetic stirrer at 800 RPM for 30 min prior to irradiation. CTZ solutions (60 ppm) were prepared using similar methods to those used for IBP. CTZ degradation in the presence of ZnO was carried out at ZnO concentrations of 0.5, 1, 1.5, and 2 g/L.

2.2. Experimental Setup

The IBP solution (40 ppm, 100 mL) was put in a 120 mL glass beaker (Figure 1a). A UVC lamp (7-watt) was immersed in the solution. The solution was constantly stirred at room temperature while being irradiated to prevent particles from sedimenting on the bottom of the beaker. The magnetic stirrer was set to 460 RPM throughout the irradiation time. Samples were taken every 15 min during the irradiation time using a clean glass pipette.

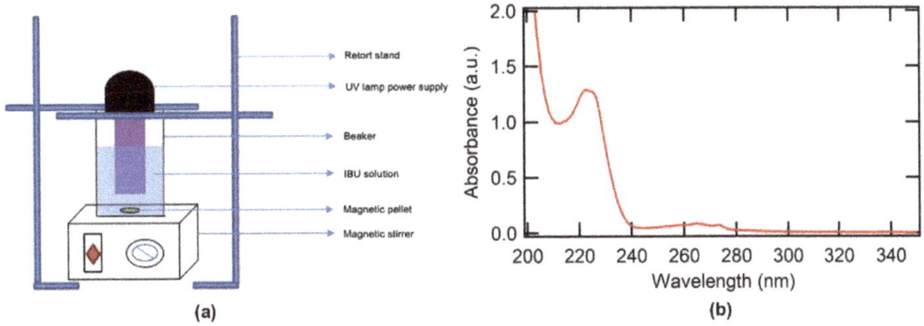

Figure 1. (**a**) Basic experimental setup and (**b**) basic UV–Vis profile of IBP.

2.3. Measurement and Analysis Using UV–Vis Spectrophotometer

UV spectra were obtained using a Jasco V730 UV–Vis Spectrophotometer in the UV region between 200 and 350 nm (Figure 1b). Generally, the main peak at 222.8 nm is considered the main IBP characteristic, whereas the profiles between 243 to 280 nm (represented by the peak at 273 nm) are accepted as characteristics of IBP intermediate degradation by-product molecules [17]. Peak height and background subtraction analysis were carried out using Spectra Manager. The IBP solution sample (40 ppm) was directly irradiated under constant stirring. Samples were taken at every 15-minute interval, and they were transferred into a quartz cuvette and put inside the UV–Vis Spectrophotometer to obtain the UV spectra. For samples containing ZnO, prior to obtaining the UV spectra, samples of IBP solution containing ZnO were first centrifuged for 10 min to separate the ZnO particles from the solution. The same method was applied to CTZ solutions but with peak wavelength observation at 263 nm [18].

3. Results and Discussion

3.1. Time Dependence of Degradation of IBP without and with ZnO Particles

The IBP solution was irradiated for 60 min, and the IBP concentration was analyzed every 15 min to investigate the degradation efficiency in terms of irradiation time (Figure 2). At around ~222 nm, which is the region for the main IBP peak, we observed the expected degradation of IBP upon UVC irradiation. In Figure 2a, we show the IBP concentration (C) decreased during the UV irradiation period with respect to the initial IBP concentration (C_0). A rapid decrease in the first 15 min is observed, and it followed an exponential trend for at least 60 min.

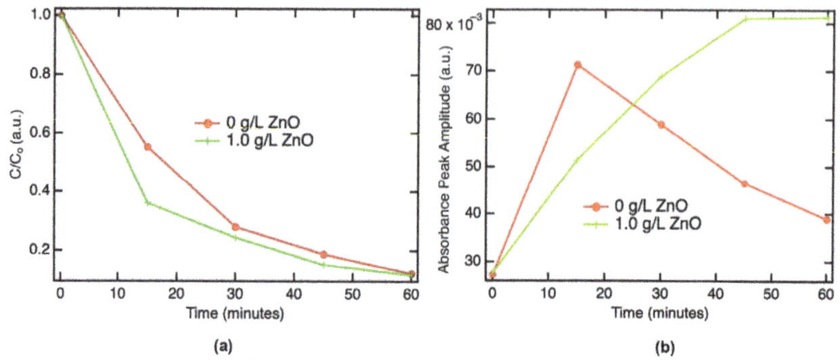

Figure 2. (**a**) IBP concentration (C/C_0) against irradiation time (minutes). The concentration of IBP was obtained by converting the absorbance peak amplitude of the IBP UV spectra at ~222 nm using a standard curve. (**b**) Absorbance (at 273 nm) of IBP by-product against irradiation time (minutes).

By contrast, a completely different situation was observed in the IBP by-product region around ~273 nm (Figure 2b). In the case of treatments without ZnO, we observed a clear formation of by-product molecules during the first interval, in which the most significant IBP degradation took place (Figure 2b). Subsequently, there was a continual decrease in the absorbance of the by-product molecules, which implied that it was also degraded alongside the IBP over the irradiation time. However, when ZnO was added, there was a distinct pattern of by-product formation that showed a gradual increase in absorbance accompanying the degradation of IBP. Unlike in the case of the absence of ZnO, the absorbance of by-products did not decrease over time, which means that the by-product was not degraded alongside the IBP. One possible by-product is 4-isobutylacetophenone, 4-IBAP, formed during the degradation of IBP under UV light [19]. 4-IBAP is known to be more toxic, and it takes longer to degrade than the parent compound, IBP [20].

3.2. Extraction of the Rate of IBP Degradation and Maximum IBP Degradation Potential

To further elucidate the performance characteristics of the UVC irradiation against IBP without and in the presence of ZnO, we applied an exponential model to time-dependent degradation curves shown in Figure 2a. The rate of degradation and maximum degradation potential were calculated with the following equation:

$$\frac{C}{C_0} = y_0 + e^{-kt}, \quad (1)$$

where C is the concentration, C_0 is the initial concentration, y_0 is the maximum degradation potential, k is the rate of degradation in min^{-1}, and t is irradiation time in min.

The extracted rate of degradation (k) and maximum degradation potential (y_0) are shown in Table 1. The rate of IBP degradation was greater in the presence of ZnO, which proves the photocatalytic abilities of ZnO. However, adding ZnO decreased the maximum degradation potential. This could be caused by the increase in turbidity and/or competition for adsorption sites on the surface of ZnO between the IBP and its by-product.

Table 1. k constant and maximum degradation potential of IBP.

ZnO Concentration (g/L)	k (min^{-1})	Maximum Degradation Potential (%)
0	0.045	94.4
1	0.083	86.6

Thus, adding ZnO as a photocatalyst is recommended in situations in which time is limited and the rate is prioritized over achieving the maximum degradation potential.

When time is not a constraint, UVC treatment in the absence of ZnO is recommended. This yielded the greatest maximum degradation potential of 94.4%, although at a slower rate. Another consideration is that UVC treatment in presence of ZnO left a considerably high amount of IBP remaining in the solution, which might still be toxic if released into the environment.

3.3. Clotrimazole Degradation Rate and Maximum Degradation Potential

Similar characterization and analysis were also carried out for the clotrimazole solution. In Table 2, we show the results of the extracted degradation rate (k) and maximum degradation potential (y_0). The value of k increased from 0–1.0 g/L of ZnO and reached a maximum at 1.0 g/L before decreasing when more ZnO was added. At the highest ZnO concentration of 2.0 g/L, k was lower than that of without ZnO, showing again how the excessive addition of ZnO is inefficient and might reverse its photocatalytic abilities. Similar to the degradation of IBP, high concentrations of ZnO increased the solution turbidity, which might limit light transmission and prevent the photolysis of CTZ, hence resulting in lower k values and degradation efficiency.

Table 2. k constant and maximum degradation potential of CTZ.

ZnO Concentration (g/L)	k (min^{-1})	Maximum Degradation Potential (%)
0	0.025	92.0
0.5	0.034	95.2
1.0	0.051	99.9
1.5	0.047	95.9
2.0	0.023	93.0

Overall, the degradation potential of CTZ in UVC showed good results, with over 90% of the maximum degradation potential at all concentrations of ZnO. The lowest potential was reached without ZnO at 92%, whereas the greatest potential was reached when 1.0 g/L of ZnO was added, at 99.9% or close to complete degradation. Again, exceeding this optimum concentration of ZnO decreased the degradation efficiency or the maximum degradation potential, with the lowest at 93% when 2.0 g/L of ZnO was added and an expected continual decrease with the increase of ZnO concentration owing to the increase in turbidity. In the case of CTZ, the greatest k value and maximum degradation potential were achieved when 1.0 g/L of ZnO was added, indicating that this is its optimum ZnO concentration.

It is worth noting that in the case of CTZ, we did not observe the presence of by-products in the UV spectra. Thus, there was likely to be no competition between the adsorption sites of the main drug molecules and their by-products, as was the case for IBP. This could explain why, in general, the addition of ZnO works better for CTZ than IBP.

4. Conclusions

Although it is generally perceived that the addition of ZnO as a photocatalyst increases the effectiveness of UV treatments for degrading pharmaceutical micro-pollutants, our work revealed that there are various outcomes that require careful investigation. We found that IBP was not as efficiently degraded when ZnO was used in the solution during the UVC treatment (the maximum degradation potential was 86.6%) as it was without ZnO (94.4%). Despite the ability of ZnO to degrade the IBP faster, this advantage was superseded by its inability to degrade the IBP by-product. In other cases, such as that of CTZ, a high degradation efficiency with and without ZnO was observed. The use of ZnO for degrading CTZ appears to be better than that without, with maximum degradation potentials of 99.9% and 92%, respectively. The optimal ZnO concentration was determined to be ~1 g/L, above which screening effects due to increased turbidity started to dominate the system, which lowered the photodegradation effectiveness. Overall, it is clear that UVC without ZnO was sufficiently effective in degrading IBP and CTZ, with a maximum degradation potential of more than 90% in both cases. Furthermore, the by-product of IBP was also shown to be more effectively degraded without ZnO. Thus, we hope that our results may further encourage the adoption of a simple UVC batch-stirred treatment system to treat wastewater in households, hospitals, and the pharmaceutical industry.

Author Contributions: Conceptualization, S.P. and E.S.; formal analysis, S.P., E.J. and E.S.; investigation, S.P., E.J. and E.S.; writing, S.P. All authors have read and agreed to the published version of the manuscript.

Funding: This research was funded by SPH Applied Science Academy Program at Sekolah Pelita Harapan Lippo Village, period 2021–2024.

Institutional Review Board Statement: Not applicable.

Informed Consent Statement: Not applicable.

Data Availability Statement: The data presented in this study are available on request from the corresponding author.

Acknowledgments: We thank Sekolah Pelita Harapan Lippo Village and Universitas Pelita Harapan for the generous lab access. Special thanks to the school's and university's leadership in supporting our research through the SPH Applied Science Academy Program.

Conflicts of Interest: The authors declare no conflict of interest.

References

1. Stasinakis, A.; Gatidou, G. *Micropollutants and Aquatic Environment. Treatment of Micropollutants Water and Wastewater*; IWA Publishing: London, UK, 2016; pp. 1–51. [CrossRef]
2. Wang, J.; Wang, S. Removal of Pharmaceuticals and Personal Care Products (PPCPs) from Wastewater: A Review. *J. Environ. Manag.* **2016**, *182*, 620–640. [CrossRef] [PubMed]
3. Bottoni, P.; Caroli, S.; Caracciolo, A.B. Pharmaceuticals as Priority Water Contaminants. *Toxicol. Environ. Chem.* **2010**, *92*, 549–565. [CrossRef]
4. Thiebault, T.; Boussafir, M.; Le Milbeau, C. Occurrence and Removal Efficiency of Pharmaceuticals in an Urban Wastewater Treatment Plant: Mass Balance, Fate and Consumption Assessment. *J. Environ. Chem. Eng.* **2017**, *5*, 2894–2902. [CrossRef]
5. Żur, J.; Piński, A.; Marchlewicz, A.; Hupert-Kocurek, K.; Wojcieszyńska, D.; Guzik, U. Organic Micropollutants Paracetamol and Ibuprofen-Toxicity, Biodegradation, and Genetic Background of Their Utilization by Bacteria. *Environ. Sci. Pollut. Res. Int.* **2018**, *25*, 21498–21524. [CrossRef] [PubMed]
6. Daughton, C. Pharmaceuticals as Environmental Pollutants: The Ramifications for Human Exposure. In *International Encyclopedia of Public Health*; Academic Press: Cambridge, MA, USA, 2008; pp. 66–102. [CrossRef]
7. Rehman, M.S.U.; Rashid, N.; Ashfaq, M.; Saif, A.; Ahmad, N.; Han, J.-I. Global Risk of Pharmaceutical Contamination from Highly Populated Developing Countries. *Chemosphere* **2015**, *138*, 1045–1055. [CrossRef] [PubMed]
8. Rudd, N.D.; Wang, H.; Fuentes-Fernandez, E.M.A.; Teat, S.J.; Chen, F.; Hall, G.; Chabal, Y.J.; Li, J. Highly Efficient Luminescent Metal-Organic Framework for the Simultaneous Detection and Removal of Heavy Metals from Water. *ACS Appl. Mater. Interfaces* **2016**, *8*, 30294–30303. [CrossRef] [PubMed]
9. Rzymski, P.; Drewek, A.; Klimaszyk, P. Pharmaceutical Pollution of Aquatic Environment: An Emerging and Enormous Challenge. *Limnol. Rev.* **2017**, *17*, 97–107. [CrossRef]
10. Chopra, S.; Kumar, D. Ibuprofen as an Emerging Organic Contaminant in Environment, Distribution and Remediation. *Heliyon* **2020**, *6*, e04087. [CrossRef] [PubMed]
11. Tanveer, M.; Guyer, G.T.; Abbas, G. Photocatalytic Degradation of Ibuprofen in Water Using TiO2 and ZnO under Artificial UV and Solar Irradiation. *Water Environ. Res.* **2019**, *91*, 822–829. [CrossRef] [PubMed]
12. Sabouni, R.; Gomaa, H. Photocatalytic Degradation of Pharmaceutical Micro-Pollutants Using ZnO. *Environ. Sci. Pollut. Res. Int.* **2019**, *26*, 5372–5380. [CrossRef] [PubMed]
13. Jallouli, N.; Pastrana-Martínez, L.M.; Ribeiro, A.R.; Moreira, N.F.F.; Faria, J.L.; Hentati, O.; Silva, A.M.T.; Ksibi, M. Heterogeneous Photocatalytic Degradation of Ibuprofen in Ultrapure Water, Municipal and Pharmaceutical Industry Wastewaters Using a TiO2/UV-LED System. *Chem. Eng. J.* **2018**, *334*, 976–984. [CrossRef]
14. Prayitno; Kusuma, Z.; Yanuwiadi, B.; Laksmono, R.W.; Kamahara, H.; Daimon, H. Hospital Wastewater Treatment Using Aerated Fixed Film Biofilter—Ozonation (Af2b/O3). *Adv. Environ. Biol.* **2014**, *8*, 1251–1260.
15. Gonçalves, N.P.F.; del Puerto, O.; Medana, C.; Calza, P.; Roslev, P. Degradation of the Antifungal Pharmaceutical Clotrimazole by UVC and Vacuum-UV Irradiation: Kinetics, Transformation Products and Attenuation of Toxicity. *J. Environ. Chem. Eng.* **2021**, *9*, 106275. [CrossRef]

16. The Risks of Environmental Effects of Pharmaceutical and Medicinal Products. Available online: https://www.greenfacts.org/en/pharmaceuticals-environment/index.htm (accessed on 9 May 2022).
17. Khalaf, S.; Hasan, J.; Lelario, F.; Scrano, L.; Bufo, S.; Karaman, R. TiO_2 and Active Coated Glass Photodegradation of Ibuprofen. *Catalysts* **2020**, *10*, 560. [CrossRef]
18. Mahmood, S.; Ahmad, Z.; Aslam, M.; Naeem, F.; Hussain, A.; Kumar, N. Method Development and Validation for the Estimation and Evaluation of Clotrimazole (an-Antifungal Drug) in Tablet Preparation by UV-VIS Spectroscopy. *Int. J. Pharm. Sci. Rev. Res.* **2015**, *32*, 55–58.
19. Wang, Z.; Srivastava, V.; Ambat, I.; Safaei, Z.; Sillanpää, M. Degradation of Ibuprofen by UV-LED/Catalytic Advanced Oxidation Process. *J. Water Process Eng.* **2019**, *31*, 100808. [CrossRef]
20. Ruggeri, G.; Ghigo, G.; Maurino, V.; Minero, C.; Vione, D. Photochemical Transformation of Ibuprofen into Harmful 4-Isobutylacetophenone: Pathways, Kinetics, and Significance for Surface Waters. *Water Res.* **2013**, *47*, 6109–6121. [CrossRef] [PubMed]

environmental sciences proceedings

MDPI

Proceeding Paper

Short-Term Precipitation Forecasting Based on the Improved Extreme Learning Machine Technique †

Isa Ebtehaj [1], Hossein Bonakdari [2,*], Bahram Gharabaghi [3] and Mohamed Khelifi [1]

1 Department of Soils and Agri-Food Engineering, Université Laval, Quebec City, QC G1V 0A6, Canada; isa.ebtehaj.1@ulaval.ca (I.E.); mohamed.khelifi@fsaa.ulaval.ca (M.K.)
2 Department of Civil Engineering, University of Ottawa, 161 Louis Pasteur Private, Ottawa, ON K1N 6N5, Canada
3 School of Engineering, University of Guelph, Guelph, ON N1G 2W1, Canada; bgharaba@uoguelph.ca
* Correspondence: hbonakda@uottawa.ca; Tel.: +1-613-562-5800

† Presented at the 7th International Electronic Conference on Water Sciences, 15–30 March 2023; Available online: https://ecws-7.sciforum.net/.

Abstract: In this study, an improved version of the Extreme Learning Machine, namely the Improved Weighted Regularization ELM (IWRELM), is proposed for hourly precipitation forecasting that is multi-steps ahead. After finding the optimal values of the proposed method, including the number of hidden neurons, the activation function, the weight function, the regularization parameter, and the effect of orthogonality, the IWRELM model was calibrated and validated. Thereafter, the calibrated IWRELM model was used to estimate precipitation up to ten hours ahead. The results indicated that the proposed IWRELM (R = 0.9996; NSE = 0.9993; RMSE = 0.015; MAE = 0.0005) has acceptable accuracy in short-term hourly precipitation forecasting up to ten hours ahead.

Keywords: extreme learning machine (ELM); hourly precipitation; improved weighted regularization extreme learning machine (IWRELM); machine learning, Quebec; real-time forecasting; water resource management

check for updates

Citation: Ebtehaj, I.; Bonakdari, H.; Gharabaghi, B.; Khelifi, M. Short-Term Precipitation Forecasting Based on the Improved Extreme Learning Machine Technique. *Environ. Sci. Proc.* **2023**, *25*, 50. https://doi.org/10.3390/ECWS-7-14237

Academic Editor: Athanasios Loukas

Published: 16 March 2023

1. Introduction

As a fundamental hydrological variable, precipitation significantly contributes to the land surface and atmospheric processes. Recently, there have been many applications for precipitation forecasting, such as pollutant concentration level monitoring, flood forecasting, and more. Forecasting precipitation is challenging for meteorological scientists due to precipitation timing and quantity variability. As a result of its persistence and complexity, rainfall forecasting has piqued the interest of academics. Moreover, potential flooding as the result of snow melt and heavy precipitation in early Spring in Canada [1,2] may cause significant fatalities and economic damage. Therefore, a quantitative and accurate precipitation forecast can be helpful in formulating appropriate measures and reducing the risk of floods and landslides leading to property damage and loss of life.

It is recognized that existing models use complex statistical models, which are often neither computationally nor economically feasible; downstream applications may also not affected by them. The use of machine learning algorithms in combination with the time series concept is therefore being explored as a possible solution to these shortcomings. The most well-known form of machine learning (ML) is an Artificial Neural Network (ANN) trained with the backpropagation (BP) training algorithm. Although this method has been used by many scholars, it has limitations, such as slow convergence, a time-consuming training process, being stuck in local minima, overfitting, and low generalization proficiency [3–6]. The Extreme Learning Machine (ELM) was introduced by Huang et al. [7] as a single-layer feed-forward neural network designed to overcome the classical ANN.

The main advantages of the ELM are its high generalization capability [7,8], rapid training process, and low number of adjustable parameters by the user.

Along with all these advantages, the random determination of more than 66% of the parameters related to the final model (i.e., two matrices of bias consisting of hidden neurons and input weights) [9] is one of the limitations of this method. Ebtehaj et al. [9] suggested a simple iterative process to reduce the effect of random initialization of these two matrices. Moreover, Deng et al. [9] considered the regularization term in the loss function of the ELM. They defined a weighting process to improve the ELM's generalizability in the presence of outliers, resulting in the Weighted Regularized ELM (WRELM). A comparison of the WRELM with the first version of the ELM introduced by [7] proved the higher ability of the WRELM [8,10]. To the authors' best knowledge, no study has been carried out on the application of the WRELM in hourly precipitation forecasting.

In the current study, a computer program was coded in a MATLAB environment to develop an improved version of the WRELM (i.e., IWRELM) by taking advantage of the iterative process introduced in [9] and the WRELM [10]. The introduced IWRELM is applied for hourly precipitation forecasting. The different parameters of the IWRELM, including the effect of orthogonality, the regularization parameter, the weight function, the activation function, and the number of hidden neurons, have been optimized through different defined models. Moreover, various models have been defined for multi-step-ahead forecasting of the precipitation in Quebec City, Canada.

2. Materials and Methods

2.1. Study Area

The data used in the current study were recorded hourly at the Sainte- Catherine-de-la-Jacques-Cartier station from 12/14/1994 to 10/31/2022 (latitude of 46.8378 and longitude of −71.6217). The collected data were divided into training and testing stages at a 50:50 ratio. The minimum, average, standard deviation, and maximum values of the dataset are 0, 0.0843, 0.5574, and 18.4 mm/h, respectively. All measurements in the dataset were made by the "Ministère de l'Environnement et de la Lutte contre les changements climatiques, de la Faune et des Parcs" [11] of Quebec, Canada.

2.2. Improved Weighted Regularized Extreme Learning Machine (IWRELM)

Suppose we have a dataset with N training samples, x_i (i = 1, 2, ..., S) representing the inputs, and y_i representing the outputs associated with those inputs. Assuming that the f(x) is used as the activation function and that there are h neurons in the hidden layer, the mathematical relationship specified by the IWRELM to map the input variables to the output variables can be stated as follows:

$$\sum_{j=1}^{h} z_j f(\mathbf{o}_j \cdot \mathbf{x}_i + A_j) = y_i, \quad i = 1, 2, .., S \tag{1}$$

where z_j is the vector of the output weight, f(x) denotes the activation function, $\mathbf{o}_j$ is the input weight matrix, A_i is the bias of hidden neurons, $\mathbf{x}_i$ and y_i denote the input and output variables, respectively, S denotes the number of samples, and h is the number of hidden neurons.

Based on Equation (1), which consists of S equations, a matrix representation of the equation can be expressed as:

$$\mathbf{G}z = \mathbf{y} \tag{2}$$

$$z = [z_1, \ldots, z_S]^T \tag{3}$$

$$\mathbf{y} = [y_1, \ldots, y_S]^T \tag{4}$$

$$G = \begin{bmatrix} f(\mathbf{o}_1 \cdot \mathbf{x}_1 + A_1) & \cdots & f(\mathbf{o}_h \cdot \mathbf{x}_1 + A_L) \\ \vdots & \ddots & \vdots \quad f \\ f(\mathbf{o}_1 \cdot \mathbf{x}_S + A_1) & \cdots & f(\mathbf{o}_h \cdot \mathbf{x}_S + A_L) \end{bmatrix}_{S\times h} \quad (5)$$

Due to the random definition of the output weight (i.e., O) and the bias of hidden neurons (i.e., A), the matrix G is known. Therefore, the only unknown variable in Equation (2) is z. Because matrix G is, in most cases, not a square matrix, it is not possible to directly calculate z using Equation (2) [7]. To solve this problem, it is easier to use the minimization of the loss function to calculate the optimal least-squares solution. The loss function in the IWRELM is defined as follows:

$$E = \min_{z} C\|\mathbf{W}(\mathbf{y} - \mathbf{Gz})\|_2^2 + \|z\|_2^2 \quad (6)$$

where C is the regularization parameter and W is the weight of each sample through the weighting process in the IWRELM. In the weighting process, Equation (6) is considered as Equation (7) with different weights assigned to all samples so that the samples with the least error receive the most weight and vice versa:

$$E = \min_{z} C\|(\mathbf{y} - \mathbf{G}z)\|_2^2 + \|z\|_2^2 \quad (7)$$

The solution of the output weight (i.e., w) from Equation (7) is as follows:

$$\hat{z} = \left(\mathbf{G}^T\mathbf{G} + \mathbf{I}/C\right)^{-1}\mathbf{G}^T\mathbf{y} \quad (8)$$

Here, I is the identity matrix. Moreover, C should not be zero. It could be a positive value less than 1 ($0 < C < 1$). The applied functions for the weighting process are defined in Table 1. The process of calculating the output weight using Equation (7), the weighting process using the functions provided in Table 1, and the recalculated loss function based on Equation (6) are repeated according to the iteration number defined by the user.

Table 1. Weight functions applied in the IWRELM.

No.	Function Definition
1	$w_i = \begin{cases} 1 & \lvert 1.349 \times (\mathbf{y} - \mathbf{Gz})/\mathrm{IQR}\rvert \le 2.5 \\ 2 \times (3 - \lvert 1.349(\mathbf{y} - \mathbf{Gz})/\mathrm{IQR}\rvert) & 2.5 \le \lvert 1.349 \times (\mathbf{y} - \mathbf{Gz})/\mathrm{IQR}\rvert \le 3 \\ 10^{-4} & \text{Otherwise} \end{cases}$
2	$w = (\mathrm{abs}(r) < 1 \times (1 - r^2)^2;\ r = (1.349 \times (\mathbf{y} - \mathbf{Gz}))/(4.685 \times \mathrm{IQR})$
3	$w = 1/\max(1, \mathrm{abs}(r));\ r = (1.349 \times (\mathbf{y} - \mathbf{Gz}))/(1.345 \times \mathrm{IQR})$
4	$w = (\mathrm{abs}(r) < \mathrm{pi}) \times \sin(r)/r;\ r = (1.349 \times (\mathbf{y} - \mathbf{Gz}))/(1.339 \times \mathrm{IQR})$
5	$w = 1/(1 + \mathrm{abs}(r));\ r = (1.349 \times (\mathbf{y} - \mathbf{Gz}))/(1.4 \times \mathrm{IQR})$
6	$w = 1/(1 + r^2);\ r = (1.349 \times (\mathbf{y} - \mathbf{Gz}))/(2.385 \times \mathrm{IQR})$
7	$w = \tanh(r)/r;\ r = (1.349 \times (\mathbf{y} - \mathbf{Gz}))/(1.205 \times \mathrm{IQR})$
8	$w = \tanh(r)/r;\ r = (1.349 \times (\mathbf{y} - \mathbf{Gz}))/(2.795 \times \mathrm{IQR})$
9	$w = 1 \times \mathrm{abs}(r) < 1;\ r = (1.349 \times (\mathbf{y} - \mathbf{Gz}))/(2.985 \times \mathrm{IQR})$
10	$w = 1/\max(0.0001, \mathrm{abs}(\mathbf{y} - \mathbf{Gz}))$

IQR is the interquartile range.

3. Results and Discussion

This section details IWRELM-based modeling that was used to find the optimal parameters, including the number of hidden neurons (NHN), the activation function, the weight function, the regularization parameter, and the effect of orthogonality. Finally, the performance of this model in multi-step-ahead precipitation forecasting is investigated using the found optimal values and functions.

Figure 1 shows the statistical indices of the developed IWRELM in hourly precipitation forecasting. Details of the statistical indices provided in Figure 1, including the correlation coefficient (R), Nash–Sutcliffe efficiency (NSE), Mean Absolute Error (MAE), Root Mean Square Error (RMSE), and Corrected Akaike Information Criteria, can be found in recently published studies [12,13].

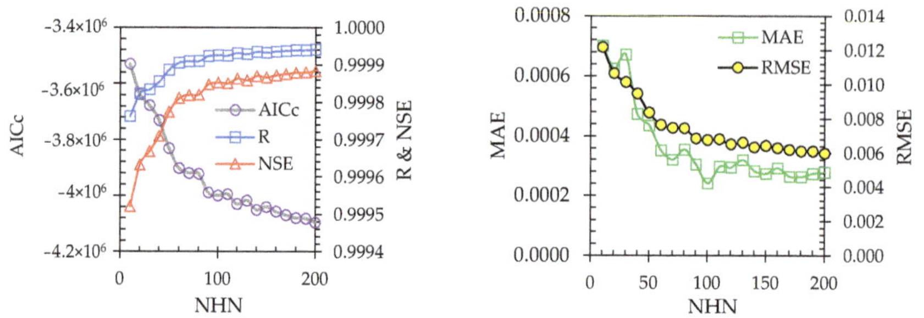

(**a**) Akaike Information Criteria (AICc), correlation coefficient (R) and Nash–Sutcliffe efficiency (NSE) (**b**) Mean Absolute Error (MAE) and Root Mean Square Error (RMSE)

Figure 1. Performance evaluation of the WRELM with different numbers of hidden neurons.

Precipitation at the next time point is estimated as follows using the value of this variable considering one to three delays:

$$Pr(t) = f(Pr(t-1), Pr(t-2), Pr(t-3)) \quad (9)$$

where Pr(t), Pr(t−1), P(t−2), and Pr(t−3) denote the precipitation at time t, t−1, t−2, and t−3, respectively.

The twenty different values in the range of 10 to 200 were checked for the number of hidden neurons. Due to Figure 1b, it can be seen that the R and NSE indices are greater than 0.99, indicating a high level of accuracy in forecasting hourly precipitation with a different number of hidden neurons. Evaluating the effect of NHN in the IWRELM indicates that increasing the number of hidden neurons generally enhances the model's accuracy. Based on Figure 1b, increasing the number of hidden neurons leads to a significant reduction in MAE and RMSE, especially when the number of hidden neurons is more than 100. For less than 100 hidden neurons, an increase of 10 units in NHN significantly changes the values of MAE and RMSE. For example, the MAE and RMSE of NHN = 10 (RMSE = 0.012 and MAE = 0.0007) are more than 10% and 12% lower than the values of these indices for NHN = 20 (RMSE = 0.0106 and MAE = 0.00062), while the difference between the values of these two indices for NHN = 200 and NHN = 190 is less than 2%. However, in some instances, the increase in NHN did not correspond directly to a decrease in MAE and RMSE. For example, the values of these two indices corresponding to NHN = 110, 130, and 150 are higher than those corresponding to NHN = 100, 120, and 140, respectively. The reason for this could be related to the random determination of input weights and the bias of hidden neurons (the two main matrices in ELM-based methods such as the WRELM), which includes more than 66% of the total number of optimized values [8]. Considering that in the models presented in this study the iteration number is equal to 1000, and when facing such conditions its value is increased to 100,000, the best results related to IWRELMs with different numbers of hidden neurons have been presented.

Considering that the values of regression-based indices in different models are not significantly different (similar to Figure 1), R and NSE are not provided to evaluate other parameters of the IWRELM. Figure 2 shows the statistical indices of the IWRELM used to find the optimal activation function, the weight function provided in Table 1, the regularization parameter (defined in Equation (6)), and the effect of orthogonality on modeling performance. For the activation function, six different functions were compared (i.e., sigmoid (Sig), sine (Sin), tangent hyperbolic (Tanh), radial basis function (Radbas), triangular basis function (Tribas), and hard limit (Hardlim)). The RMSE and MAE of the IWRELM with the activation function Hardlim are more than seven and nine times higher than Sig and Radbas, respectively, representing the only functions that performed better than Hardlim in these two indices.

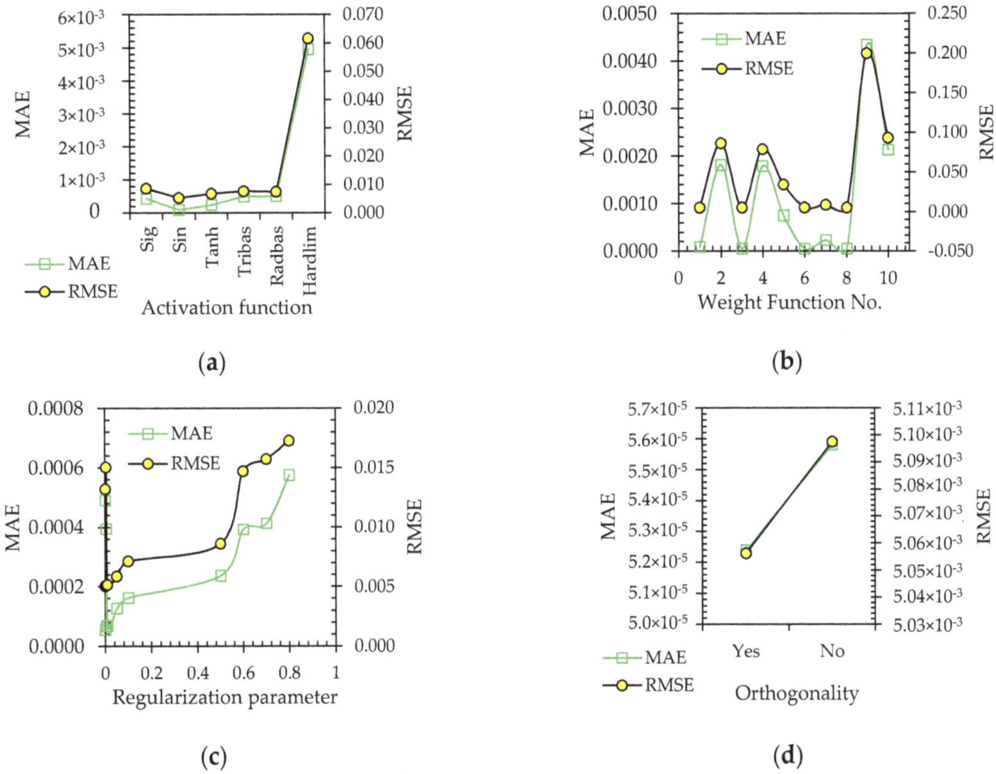

Figure 2. Investigation of the effects of (**a**) the activation function, (**b**) the weight function, (**c**) the regularization parameter, and (**d**) orthogonality on IWRELM performance.

A comparison of other functions indicated insignificant differences between them, with Sin outperforming the others. For the weight functions provided in Table 1, which are the most crucial features of the IWRELM compared to the classical ELM, the difference between all functions is remarkable, with the RMSE and MAE of function 9 more than four times greater than the respective values for functions 1, 3, 6, 7, and 8. The performance levels of functions 1, 3, 6, 7, and 8 are very close, with function 3 minimally outperforming the others (i.e., functions 1, 6, 7, and 8); this function was chosen as the optimal function. The regularization parameter is selected in the range of 0.0001 to 0.8. The results indicated that the best performance for the IWRELM is achieved with the lowest regularization parameter value (i.e., C = 0.0001). Moreover, the statistical indices provided in Figure 2d prove the importance of using the orthogonality function to define random initialized matrices (i.e., the bias of hidden neurons and input weights).

Figure 3 shows the statistical indices of the IWRELM in forecasting hourly precipitation from one to ten hours ahead. To develop these models, three lags were considered as input variables, as in Equation (3) for one hour ahead (1HA). For two to ten hours ahead, the first inputs are Pr(t−2) to Pr(t−10), the second inputs are Pr(t−3) to Pr(t−11), and the third ones are from Pr(t−4) to Pr(t−12). It can be seen that increasing the time ahead for forecasting precipitation (i.e., from 1HA to 10HA) decreases the values of two correlation-based indices (i.e., R and NSE) and increases the values of the RMSE and MAE indices. Additionally, the index values for 10HA precipitation forecasting, where performance is almost the weakest, are acceptable (R = 0.99965; NSE = 0.99926; MAE = 0.00054; RMSE = 0.015). Based on the results of this study, the method developed here (i.e., the IWRELM) has a high ability to forecast precipitation several hours in advance.

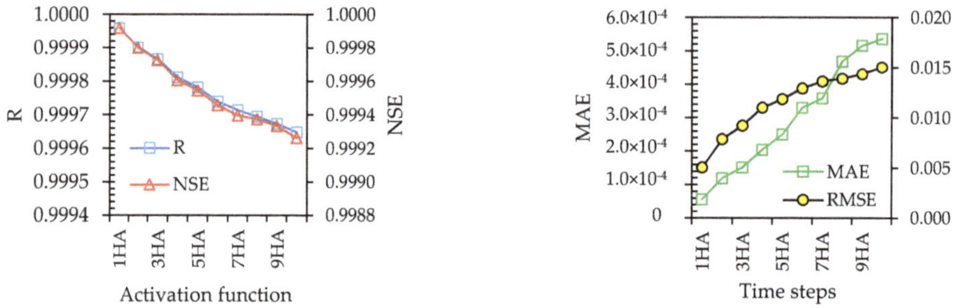

(**a**) Correlation-based indices (**b**) Absolute-based indices

Figure 3. Assessment of the WRELM's ability in multi-step-ahead precipitation forecasting.

4. Conclusions

In this study, the Improved Weighted Regularization Extreme Learning Machine (IWRELM) was proposed for hourly forecasting of precipitation one to ten hours ahead of time. By calibrating this model for one-hour-ahead forecasting, the optimal number of hidden neurons (100), the optimal regularization parameter (C = 0.0001), the most efficient activation function (Sine function), and the most suitable weighting function (e.g., function 3 in Table 1) were found. Using the optimal values of the different parameters, the model's performance was checked in terms of forecasting precipitation up to ten hours ahead of time. The results showed that the model's accuracy decreased as it predicted further into the future, i.e., as time ahead increased from one to ten hours. Nevertheless, the model's accuracy in forecasting precipitation ten hours ahead of time remains acceptable (R = 0.9996; NSE = 0.9993; RMSE = 0.015; MAE = 0.0005). Since there has been no evaluation of input variables, feature selection can be used in future studies and a comparison of the model's performance with optimal inputs can also be performed.

Author Contributions: Conceptualization, I.E. and H.B.; methodology, I.E. and H.B.; software, I.E.; validation, I.E. and H.B.; formal analysis, I.E.; investigation, I.E.; data curation, I.E.; writing—original draft preparation, I.E. and H.B.; writing—review and editing, H.B., M.K., and B.G.; visualization, I.E.; supervision, H.B.; project administration, H.B.; All authors have read and agreed to the published version of the manuscript.

Funding: This research was funded by the Natural Sciences and Engineering Research Council of Canada (NSERC), Discovery Grant (#RGPIN-2020-04583), and the "Fond de Recherche du Québec-Nature et Technologies", Québec government (#B2X–315020).

Institutional Review Board Statement: Not applicable.

Informed Consent Statement: Not applicable.

Data Availability Statement: All data used during the study were provided by a third party. Direct requests for these materials may be made to the provider, as indicated in the Acknowledgements.

Acknowledgments: The authors would like to thank the "Ministère de l'Environnement et de la Lutte contre les changements climatiques, de la Faune et des Parcs" of Quebec, Canada.

Conflicts of Interest: The authors declare no conflict of interest.

References

1. Ebtehaj, I.; Bonakdari, H. A comprehensive comparison of the fifth and sixth phases of the coupled model intercomparison project based on the Canadian earth system models in spatio-temporal variability of long-term flood susceptibility using remote sensing and flood frequency analysis. *J. Hydrol.* **2023**, *617*, 128851.
2. Zaghloul, M.S.; Ghaderpour, E.; Dastour, H.; Farjad, B.; Gupta, A.; Eum, H.; Axhari, G.; Hassan, Q.K. Long Term Trend Analysis of River Flow and Climate in Northern Canada. *Hydrology* **2022**, *9*, 197. [CrossRef]
3. Bonakdari, H.; Ebtehaj, I. A comparative study of extreme learning machines and support vector machines in prediction of sediment transport in open channels. *Int. J. Eng.* **2016**, *29*, 1499–1506.
4. Bonakdari, H.; Ebtehaj, I.; Samui, P.; Gharabaghi, B. Lake Water-Level fluctuations forecasting using Minimax Probability Machine Regression, Relevance Vector Machine, Gaussian Process Regression, and Extreme Learning Machine. *Water Resour. Manag.* **2019**, *33*, 3965–3984. [CrossRef]
5. Bonakdari, H.; Qasem, S.N.; Ebtehaj, I.; Zaji, A.H.; Gharabaghi, B.; Moazamnia, M. An expert system for predicting the velocity field in narrow open channel flows using self-adaptive extreme learning machines. *Measurement* **2020**, *151*, 107202. [CrossRef]
6. Ebtehaj, I.; Bonakdari, H.; Moradi, F.; Gharabaghi, B.; Khozani, Z.S. An Integrated Framework of Extreme Learning Machines for Predicting Scour at Pile Groups in Clear Water Condition. *Coastal Eng.* **2018**, *135*, 1–15. [CrossRef]
7. Huang, G.B.; Zhu, Q.Y.; Siew, C.K. Extreme learning machine: Theory and applications. *Neurocomputing* **2006**, *70*, 489–501. [CrossRef]
8. Ebtehaj, I.; Bonakdari, H. A reliable hybrid outlier robust non-tuned rapid machine learning model for multi-step ahead flood forecasting in Quebec, Canada. *J. Hydrol.* **2022**, *614*, 128592. [CrossRef]
9. Ebtehaj, I.; Soltani, K.; Amiri, A.; Faramarzi, M.; Madramootoo, C.A.; Bonakdari, H. Prognostication of shortwave radiation using an improved No-Tuned fast machine learning. *Sustainability* **2021**, *13*, 8009. [CrossRef]
10. Deng, W.; Zheng, Q.; Chen, L. Regularized extreme learning machine. In Proceedings of the 2009 IEEE Symposium on Computational Intelligence and Data Mining, Nashville, TN, USA, 30 March–2 April 2009; IEEE: Piscataway, NJ, USA, 2009; pp. 389–395.
11. Ministère de l'Environnement et de la Lutte contre les changements climatiques, de la Faune et des Parcs. Données du Réseau de surveillance du climat du Québec, Direction de la qualité de l'air et du climat, Québec. 2022. Available online: https://www.environnement.gouv.qc.ca/ (accessed on 23 November 2022).
12. Walton, R.; Binns, A.; Bonakdari, H.; Ebtehaj, I.; Gharabaghi, B. Estimating 2-year flood flows using the generalized structure of the Group Method of Data Handling. *J. Hydrol.* **2019**, *575*, 671–689. [CrossRef]
13. Zeynoddin, M.; Ebtehaj, I.; Bonakdari, H. Development of a linear based stochastic model for daily soil temperature prediction: One step forward to sustainable agriculture. *Comput. Electron. Agric.* **2020**, *176*, 105636. [CrossRef]

environmental sciences proceedings

MDPI

Proceeding Paper

Time-Series-Based Air Temperature Forecasting Based on the Outlier Robust Extreme Learning Machine †

Isa Ebtehaj [1], Hossein Bonakdari [2,*], Bahram Gharabaghi [3] and Mohamed Khelifi [1]

1 Department of Soils and Agri-Food Engineering, Université Laval, Quebec City, QC G1V 0A6, Canada; isa.ebtehaj.1@ulaval.ca (I.E.); mohamed.khelifi@fsaa.ulaval.ca (M.K.)

2 Department of Civil Engineering, University of Ottawa, 161 Louis Pasteur Private, Ottawa, ON K1N 6N5, Canada

3 School of Engineering, University of Guelph, Guelph, ON N1G 2W1, Canada; bgharaba@uoguelph.ca

* Correspondence: hbonakda@uottawa.ca; Tel.: +1-613-562-5800

† Presented at the 7th International Electronic Conference on Water Sciences, 15–30 March 2023; Available online: https://ecws-7.sciforum.net.

Abstract: In this study, an improved version of the outlier robust extreme learning machine (IORELM) is introduced as a new method for multi-step-ahead hourly air temperature forecasting. The proposed method was calibrated and used to estimate the hourly air temperature for one to ten hours in advance after finding its most optimum values (i.e., orthogonality effect, activation function, regularization parameter, and the number of hidden neurons). The results showed that the proposed IORELM has an acceptable degree of accuracy in predicting hourly temperatures ten hours in advance (R = 0.95; NSE = 0.89; RMSE = 3.74; MAE = 1.92).

Keywords: extreme learning machine (ELM); hourly air temperature forecasting; machine learning; outlier robust extreme learning machine (ORELM); Quebec; real-time forecasting; water resource management

check for updates

Citation: Ebtehaj, I.; Bonakdari, H.; Gharabaghi, B.; Khelifi, M. Time-Series-Based Air Temperature Forecasting Based on the Outlier Robust Extreme Learning Machine. *Environ. Sci. Proc.* **2023**, *25*, 51. https://doi.org/10.3390/ECWS-7-14236

Academic Editor: Athanasios Loukas

Published: 16 March 2023

1. Introduction

One of the most significant challenges facing humanity is climate change mitigation. Despite the existing challenges in forecasting climate change effects on Earth, there is scientific agreement on its detrimental consequences. The effects of climate change have been identified as adversely affecting ecosystems, soil erosion, reducing biodiversity, rising sea levels, extreme temperature changes, and global warming. Additionally, a significant impact is expected on food security, human health, energy consumption, and economy. Forecasting air temperatures, in particular, has become an increasingly crucial climatic aspect in various fields, including tourism, energy, agriculture, industry, and so on [1,2]. There are several applications for air temperature forecasting, including forecasting cooling and energy consumption for residential buildings [3], adaptively controlling greenhouse temperatures [4], and predicting natural hazards [5]. As a result, there is a need to reliably anticipate air temperature, because it would assist in a planning horizon for constructing an energy policy, an insurance policy, and infrastructure upgrades, as well as business development, when combined with the study analysis of additional elements in the topic of interest.

Machine Learning (ML) helps improve several types of systems so that their use is expanded. Although feed-forward neural network (FFNN) is the most well-known and widely-used MA-based technique in modeling complex systems, it has some weaknesses, such as overfitting, trapping in local minima, lengthy training process, and slow convergence [6–9]. Huang et al. [10] proposed a single-layer FFNN (SLFFNN) known as the extreme learning machine (ELM) to overcome the mentioned limitations. Compared with conventional FFNN methods, ELM requires minimal user involvement, provides rapid training, and has high generalizability [10,11]. Although this method has many advantages, it has several limitations, one of which is its low performance in the presence of outliers. Zhang and Luo [12] proposed the outlier robust ELM (ORELM) to improve the model performance by increasing sparsity. Moreover, the second limitation of the ELM is the random determination of the two main matrices (i.e., input weights and bias of hidden neurons), which account for at least 66% of all tuned parameters of the final model [13]. It was suggested by Ebtehaj et al. [13] that, by introducing a simple iterative process, the random initialization of the two matrices would be reduced to a minimum. To the authors' best knowledge, there is no study on applying the ORELM in hourly air temperature forecasting.

This research aimed to develop a novel technique and explore the potential of new data intelligence models by integrating the ORELM and iterative process defined by Ebtehaj et al. [13] to forecast hourly air temperature in Quebec City, Canada. The parameters of the developed improved version of the ORELM (IORELM), including the activation function, regularization parameter, and the number of hidden neurons, were optimized by defining different models. Moreover, the performance of the IORELM was evaluated for multi-step-ahead forecasting of the hourly air temperature.

2. Materials and Methods

2.1. Study Area

A time series of hourly air temperature data was collected at the station of Sainte-Catherine-de-la-Jacques Cartier (latitude: 46.8378 and longitude = −71.6217) from 1 January 2001 to 30 November 2022 for the present study (Figure 1). A 50:50 ratio was used to split the collected data into training and testing stages (i.e., 95,376 samples were collected for each stage). It has a minimum, an average, a standard deviation, and a maximum of −36.37, 1.6, 11.455, and 23.5 °C, respectively. Measurements of the dataset were conducted by the "Ministère de l'Environnement et de la Lutte contre les changements climatiques, de la Faune et des Parcs" [14] of Québec, Canada.

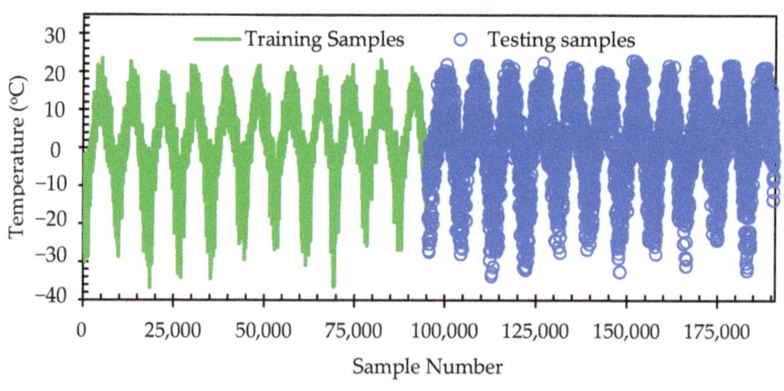

Figure 1. Time series of historical temperature for both the training and testing stages.

2.2. Improved Outlier Robust Extreme Learning Machine (IORELM)

Considering several k random samples as ($\mathbf{x}_i$, t_i) and A(x) as the activation function, the general form of the SLFFNN is defined as follows:

$$\sum_{j=1}^{L} W_j f(\mathbf{InW}_j \cdot \mathbf{x}_i + S_j) = y_i, \qquad i = 1, 2, \ldots, k \tag{1}$$

where k is the number of samples; $\mathbf{x}_i$ and y_i are the input and output variables, respectively; S_i is the bias of hidden neurons; $\mathbf{InW}_j$ is the input weight matrix; A(x) is the activation function; W_j is the vector of the output weight; and L is the number hidden neurons.

A matrix representation of the above-mentioned equation, which is composed of L equations, is given below:

$$\mathbf{HW} = \mathbf{y} \tag{2}$$

$$W = [W_1, \ldots, W_k]^1 \tag{3}$$

$$\mathbf{y} = [y_1, \ldots, y_k]^T \tag{4}$$

$$\mathbf{H} = \begin{bmatrix} A(\mathbf{InW}_1 \cdot \mathbf{x}_1 + SA_1) & \cdots & A(\mathbf{InW}_h \cdot \mathbf{x}_1 + S_L) \\ \vdots & \ddots & \vdots \\ A(\mathbf{InW}_1 \cdot \mathbf{x}_k + S_1) & \cdots & A(\mathbf{InW}_h \cdot \mathbf{x}_k + S_L) \end{bmatrix}_{k \times L} \tag{5}$$

The matrix **H** could be calculated easily owing to the random definition of bias of hidden neurons (i.e., S) and the output weight (i.e., **InW**). Consequently, the only unknown variable in the matrix form of the equations provided in Equation (2) is **W** (i.e., output weight matrix). Ebtehaj and Bonakdari [15] mentioned that the number of tuned parameters through modeling must be less than the number of training samples. Therefore, Equation (2) is not square in most cases and solving it to find the unknown variable is not simple. Thus, the lost function defined for IORELM is as follows:

$$E_{ORELM} = \min_{W} \|\mathbf{e}\|_1 + \frac{1}{C}\|W\|_2^2 \text{ subjected to } \mathbf{y} - \mathbf{HW} = \mathbf{e} \tag{6}$$

where C is the regularization parameter. Because solving this equation results in a constrained convex optimization problem, the augmented Lagrange of Equation (6) is defined as follows:

$$L(W, \mathbf{e}, \lambda) = \|\mathbf{e}\|_1 + \frac{1}{C}\|W\|_2^2 + \lambda^T(\mathbf{y} - \mathbf{HW} - \mathbf{e}) + \frac{\mu}{2}\|\mathbf{y} - \mathbf{HW} - \mathbf{e}\|_2^2 \tag{7}$$

where μ is a penalty parameter, defined as $\mu = 2k/\|\mathbf{y}\|_1$, and λ is the Lagrange multiplier vector. The explicit solution of Equation (6) using the defined augmented Lagrange vector in Equation (7) is as follows:

$$\beta_{k+1} = (\mathbf{H}^T\mathbf{H} + 2/C\mu\mathbf{I})^{-1}\mathbf{H}^T(\mathbf{y} - \mathbf{e_k} + \lambda_k/\mu) \tag{8}$$

where

$$\begin{aligned} \mathbf{e}_{k+1} &= \text{shrink}(\mathbf{y} - \mathbf{HW}\beta_{k+1} + \lambda_k/\mu, 1/\mu) \\ &\cong \max\{|\mathbf{y} - \mathbf{HW}_{k+1} + \lambda_k/\mu| - 1/\mu, 0\} \circ \text{sign}(\mathbf{y} - \mathbf{HW}_{k+1} + \lambda_k/\mu) \end{aligned} \tag{9}$$

where "$\circ$" is the element-wise multiplication and I is the identity matrix.

For all defined iteration numbers by the user, the output weight was calculated and the value of the samples in both training and testing was computed using the three matrices (i.e., input weights, bias of hidden neurons, and output weights) and the activation function. Finally, the optimized values of the mentioned matrices related to the model with the lowest testing error were stored as the final model.

3. Results and Discussion

In this section, the IORELM-based model is discussed in detail in terms of finding the optimal parameters, including the regularization parameter, the activation function, and the number of hidden neurons (NHN). Using the most optimum values of the NHN, the regularization parameter, and the best activation function, the performance of the IORELM in multi-step-ahead forecasting of the hourly air temperature was assessed.

The performance of the IORELM is evaluated in Figure 2 by considering NHN in the range of [2, 20]. The performance of NHN = 1 was also evaluated, but its use in a figure prevented the observation of the differences between other values, so it was not included in this figure. Figure 2 provides information on some statistical indices, including corrected Akaike information criteria, root mean square errors (RMSEs), mean absolute errors (MAEs), Nash–Sutcliffe efficiency (NSE), and correlation coefficients (R). The mathematical definition of these indices can be found in recent publications [16,17]. To forecast hourly air temperatures, the following inputs were taken into consideration:

$$T(t) = f(T(t-1), T(t-2), T(t-3)) \quad (10)$$

where T(t), T(t−1), T(t−2), and T(t−3) are the air temperature at time t, t−1, t−2, and t−3, respectively.

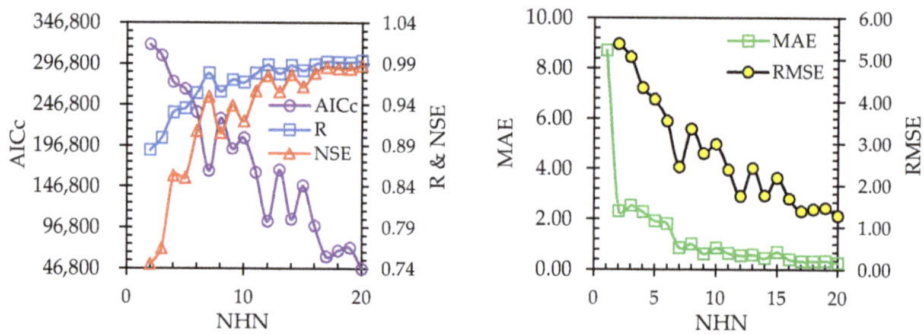

(**a**) AIC and correlation-based indices (**b**) Absolute-based indices

Figure 2. Performance evaluation of the IORELM with the different number of hidden neurons.

The general trend presented in Figure 2 shows that the increase in the number of neurons in the hidden layer has a direct relationship with the modeling accuracy. However, there are some exceptions for some indices. For example, all indices for the models with NHN = 8, 12, 14, 17, and 18 have higher values than those for NHN = 9, 13, 15, 18, and 19, respectively. It should be noted that, as NHN increases, the differences between the presented models decrease. There could be explained by the fact that the input weights and biases of hidden neurons are determined randomly, which accounts for more than 66% of the total number of optimized values [13]. Based on this knowledge, the number of iterations increased to 100,000, and the best solution is presented in Figure 2. It can be seen that increasing NHN by more than 17 does not significantly change the values of R, NSE, RMSE, and MARE indices. Nevertheless, the AICc index, which considers the accuracy and simplicity of the model, should also be evaluated. Considering that the lowest value of AICc was observed in NHN = 20 and that there is an insignificant difference between other indices in the high value of NHN, NHN = 20 was chosen as the optimal number of neurons.

The optimal activation function was found using the optimal number of hidden layer neurons and the functional equation presented in Equation (10). The results presented in Figure 3a show that, among the six functions given in this figure (i.e., Sigmoid (Sig), sine (Sin), tangent hyperbolic (Tanh), radial basis function (Radbas), triangular basis function (Tribas), and hard limit (Hardlim)), the performance of the tribas, radbas, and hardlim functions is much weaker than that of the others. The statistical indices indicated that the Sigmoid function (R = 0.994; NSE = 0.989; RMSE = 1.217; MAE = 0.236) outperformed the others. The RMSEs of the IORELM modeling using Sig are more than 4.9%, 17%, 280%, 236%, and 180% lower than those for Sin, Tanh, Tribas, Radbas, and Hardlim, respectively. For MARE, these ratios are 14%, 41%, 601%, 380%, and 876%, respectively. Therefore, the Sigmoid function was selected as the most optimum activation function.

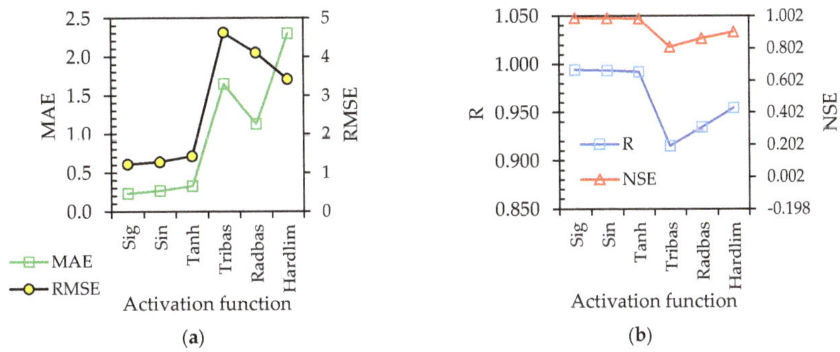

Figure 3. Investigation of the effects of the (**a**) activation function and (**b**) regularization parameter on the IORELM performance.

Finding the optimum regularization parameters (i.e., 0.0001, 0.0005, 0.001, 0.005, 0.01, 0.05, 0.1, and 0.5) is considered in Figure 3b. The trend of changes presented in this figure shows that there is no clear relationship between the regularization parameter and the model's accuracy, so it can be seen that MAE at C = 0.01 is better than the value of this index at C = 0.005 and 0.5. The RMSE and MAE ranges for all regularization parameter values are [1.217, 1.235] and [0.231, 0.275], respectively. Although the difference between the presented values is significant, the best performance of IORELM was obtained at the lowest value of the regularization parameter (i.e., C = 0.0001).

Figure 4 illustrates the statistical indices of IORELM in forecasting hourly air temperatures one to ten hours ahead. Three lags were considered as input variables for developing these models, as described in Equation (10) for one hour ahead (1HA). Two to ten hours ahead, the first inputs are T(t–2) to T(t–10), the second inputs are T(t–3) to T(t–11), and the third inputs are T(t–4) to T(t–12), respectively. The increase in forecast time (i.e., from 1HA to 10HA) decreases the values of two correlation-based indices (R and NSE) while increasing the values of RMSE and MAE. The relative differences of the value of one index for 2HA to 10HA compared with the 1HA indicated that this ratio for R, NSE, MAE, and RMSE is [0.56%, 4.48%], [1.13%, 10.04%], [82.84%, 714.7%], and [40.87%, 207.53%], respectively. However, it can be seen that the results of hourly air temperature for 10HA (R = 0.95; NSE = 0.89; RMSE = 3.74; MAE = 1.92) are acceptable. It should be noted that this model has predicted 95,376 different samples with only 20 neurons in the hidden layer, which confirms the model's simplicity.

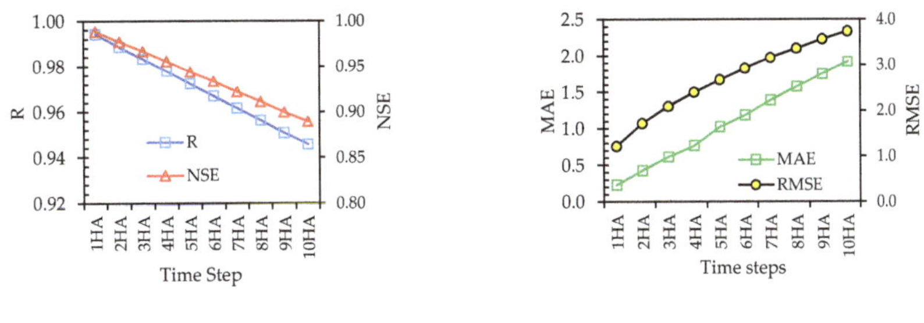

(**a**) Correlation-based indices (**b**) Absolute-based indices

Figure 4. Assessment of the IORELM capacity in multi-step-ahead precipitation forecasting.

4. Conclusions

This study suggests the use of the improved outlier robust extreme learning machine (IORELM) for forecasting air temperature hourly up to ten hours in advance. The most optimal values of hidden neurons (e.g., 20), the most efficient activation function (e.g.,

Sigmoid function), and the regularization parameters (e.g., C = 0.0001) were determined by calibrating this model for one hour in advance (R = 0.994; NSE = 0.989; RMSE = 1.217; MAE = 0.236). Model performance was assessed for one to ten hours in advance using the most optimal values of the different parameters. The more the time ahead increases, the more the accuracy of the model decreases. Nevertheless, the model's performance in terms of ten hours ahead of precipitation forecasting was acceptable (R = 0.95; NSE = 0.89; RMSE = 3.74; MAE = 1.92). As the effect of input variables has not been assessed, it is recommended that future studies use feature selection and compare the model's performance with optimal inputs.

Author Contributions: Conceptualization, I.E. and H.B.; methodology, I.E. and H.B.; software, I.E.; validation, I.E. and H.B.; formal analysis, I.E.; investigation, I.E.; data curation, I.E.; writing—original draft preparation, I.E. and H.B.; writing—review and editing, H.B., M.K. and B.G.; visualization, I.E.; supervision, H.B. and M.K.; project administration, H.B. All authors have read and agreed to the published version of the manuscript.

Funding: This research was funded by the Natural Sciences and Engineering Research Council of Canada (NSERC), Discovery Grant (#RGPIN-2020-04583), and the "Fond de Recherche du Québec-Nature et Technologies", Québec government (#B2X–315020).

Institutional Review Board Statement: Not applicable.

Informed Consent Statement: Not applicable.

Data Availability Statement: All data used during the study were provided by a third party. Direct requests for these materials may be made to the provider, as indicated in the Acknowledgements.

Acknowledgments: The authors would like to thank the "Ministère de l'Environnement et de la Lutte contre les changements climatiques, de la Faune et des Parcs" of Québec, Canada.

Conflicts of Interest: The authors declare no conflict of interest.

References

1. Tol, R.S. Estimates of the damage costs of climate change. Part 1: Benchmark estimates. *Environ. Resour. Econ.* **2002**, *21*, 47–73. [CrossRef]
2. Pachauri, R.K.; Allen, M.R.; Barros, V.R.; Broome, J.; Cramer, W.; Christ, R.; Church, J.A.; Clarke, L.; Dahe, Q.; Dasgupta, P.; et al. *Climate Change 2014: Synthesis Report. Contribution of Working Groups I, II and III to the Fifth Assessment Report of the Intergovernmental Panel on Climate Change;* Intergovernmental Panel on Climate Change: Geneva, Switzerland, 2014.
3. Chen, G.; Hua, J.; Shi, Y.; Ren, C. Constructing air temperature and relative humidity-based hourly thermal comfort dataset for a high-density city using machine learning. *Urban Clim.* **2023**, *47*, 101400. [CrossRef]
4. Dombaycı, Ö.A.; Gölcü, M. Daily means ambient temperature prediction using artificial neural network method: A case study of Turkey. *Renew. Energy* **2009**, *34*, 1158–1161. [CrossRef]
5. Yu, X.; Hu, X.; Wang, G.; Wang, K.; Chen, X. Machine-Learning Estimation of Snow Depth in 2021 Texas Statewide Winter Storm Using SAR Imagery. *Geophys. Res. Lett.* **2022**, *49*, e2022GL099119. [CrossRef]
6. Bonakdari, H.; Ebtehaj, I. A comparative study of extreme learning machines and support vector machines in prediction of sediment transport in open channels. *Int. J. Eng.* **2016**, *29*, 1499–1506.
7. Bonakdari, H.; Ebtehaj, I.; Samui, P.; Gharabaghi, B. Lake Water-Level fluctuations forecasting using Minimax Probability Machine Regression, Relevance Vector Machine, Gaussian Process Regression, and Extreme Learning Machine. *Water Resour. Manag.* **2019**, *33*, 3965–3984. [CrossRef]
8. Bonakdari, H.; Qasem, S.N.; Ebtehaj, I.; Zaji, A.H.; Gharabaghi, B.; Moazamnia, M. An expert system for predicting the velocity field in narrow open channel flows using self-adaptive extreme learning machines. *Measurement* **2020**, *151*, 107202. [CrossRef]
9. Ebtehaj, I.; Bonakdari, H.; Moradi, F.; Gharabaghi, B.; Khozani, Z.S. An Integrated Framework of Extreme Learning Machines for Predicting Scour at Pile Groups in Clear Water Condition. *Coastal Eng.* **2018**, *135*, 1–15. [CrossRef]
10. Huang, G.B.; Zhu, Q.Y.; Siew, C.K. Extreme learning machine: Theory and applications. *Neurocomputing* **2006**, *70*, 489–501. [CrossRef]
11. Ebtehaj, I.; Bonakdari, H. A reliable hybrid outlier robust non-tuned rapid machine learning model for multi-step ahead flood forecasting in Quebec, Canada. *J. Hydrol.* **2022**, *614*, 128592. [CrossRef]
12. Zhang, K.; Luo, M. Outlier-robust extreme learning machine for regression problems. *Neurocomputing* **2015**, *151*, 1519–1527. [CrossRef]
13. Ebtehaj, I.; Soltani, K.; Amiri, A.; Faramarzi, M.; Madramootoo, C.A.; Bonakdari, H. Prognostication of shortwave radiation using an improved No-Tuned fast machine learning. *Sustainability* **2021**, *13*, 8009. [CrossRef]

14. Ministère de l'Environnement et de la Lutte Contre les Changements Climatiques, de la Faune et des Parcs. Données du Réseau de Surveillance du Climat du Québec, Direction de la Qualité de l'air et du Climat, Québec. 2022. Available online: https://www.environnement.gouv.qc.ca/ (accessed on 23 November 2022).
15. Ebtehaj, I.; Bonakdari, H. Discussion of "Comparative Study of Time Series Models, Support Vector Machines, and GMDH in Forecasting Long-Term Evapotranspiration Rates in Northern Iran" by Afshin Ashrafzadeh, Ozgur Kişi, Pouya Aghelpour, Seyed Mostafa Biazar, and Mohammadreza Askarizad Masouleh. *J. Irrig. Drain. Eng.* **2021**, *147*, 07021005.
16. Walton, R.; Binns, A.; Bonakdari, H.; Ebtehaj, I.; Gharabaghi, B. Estimating 2-year flood flows using the generalized structure of the Group Method of Data Handling. *J. Hydrol.* **2019**, *575*, 671–689. [CrossRef]
17. Zeynoddin, M.; Ebtehaj, I.; Bonakdari, H. Development of a linear based stochastic model for daily soil temperature prediction: One step forward to sustainable agriculture. *Comput. Electron. Agric.* **2020**, *176*, 105636. [CrossRef]

environmental sciences proceedings

MDPI

Proceeding Paper

Large-Scale Mapping of Inland Waters with Google Earth Engine Using Remote Sensing †

Mervegul Aykanat Atay [1] and Gordana Kaplan [2,*]

1 Institute of Graduate School, Eskisehir Technical University, Eskisehir 26555, Turkey; maatay@eskisehir.edu.tr
2 Institute of Earth and Space Sciences, Eskisehir Technical University, Eskisehir 26555, Turkey
* Correspondence: kaplangorde@gmail.com; Tel.: +90-536-697-5605

† Presented at the 7th International Electronic Conference on Water Sciences, 15–30 March 2023; Available online: https://ecws-7.sciforum.net.

Abstract: Water resources are becoming scarce due to climate change and anthropogenic activities, necessitating immediate action. The first step in conserving our water supplies is to manage them mindfully and sustainably. To achieve this, water sources must be monitored, mapped, and evaluated regularly. Updating national water maps using conventional methods can be a challenging task. Most of the obstacles have been addressed due to recent breakthroughs in the remote sensing field. In this study, we employed remote sensing data integrated into Google Earth Engine (GEE) to develop an application for mapping Turkey's national inland water bodies. To achieve this aim, we explored the recently developed Multi-Band Water Index (MBWI) in GEE using Sentinel-2 satellite imagery and then applied it throughout the research area. The results showed that GEE is a promising application for handling large amounts of satellite data and can accurately extract water bodies on a national scale. The results of this study could be helpful for various administrative applications that require up-to-date water information. The developed application can be used for different study areas and for spatiotemporal analysis.

Keywords: remote sensing; water; Google Earth Engine; mapping

Citation: Atay, M.A.; Kaplan, G. Large-Scale Mapping of Inland Waters with Google Earth Engine Using Remote Sensing. *Environ. Sci. Proc.* **2023**, *25*, 52. https://doi.org/10.3390/ECWS-7-14171

Academic Editor: Slobodan Simonovic

Published: 14 March 2023

1. Introduction

Water sustainability is critical for the well-being of all organisms on Earth and for the Earth itself. Water resources are becoming scarce due to climate change and anthropogenic activities, necessitating immediate action. The first step in maintaining our water supplies is to practice conscious management and implement long-term solutions. Water sources must be monitored, mapped, and evaluated regularly to achieve this aim. While traditional methods for monitoring water regions are costly and difficult, remote sensing provides an alternative. Remote sensing techniques and data have been employed for more than four decades as an alternative to costly and time-consuming traditional methods for water surface mapping and monitoring. Over the years, many attempts have been made to correctly collect surface water, and researchers are continuously creating alternative models for improved accuracy in diverse study locations. The most widely used water extraction index, the Normalized Difference Water Index (NDWI) [1], is based on the difference between the maximum reflectance of the surface water in the green band and that of non-water surfaces in the near-infrared band, and it has been successfully used in many studies. Several modifications have been made to improve the results [2].

Furthermore, the limitation of the above-mentioned indices has been resolved through the development of multiband water indices [3–5]. The most recently developed water index is the Multi-Band Water Index (MBWI) [6], which outperforms the previously developed indices. In addition to indices, several models have been developed for the minimization of misclassification noise, such as shadows in urban areas [7] or mountainous regions [8]. Remote sensing data and techniques combined with such indices and models have been used for various water-related studies, such as water dynamics monitoring [9], water quality [10], flood mapping [11], etc. It should be noted that most studies are performed across small study areas due to limitations involved in the processing of big data [12]. Following recent developments, these limitations can be easily overcome using the cloud platform Google Earth Engine (GEE). GEE, a cloud computing platform, has been used in the past few years for various water studies, such as dynamics monitoring [13], surface water extraction, and spatio-temporal water changes [14]. In this study, we used GEE for the large-scale surface water mapping of Turkey using Sentinel-2 satellite imagery.

2. Materials and Methods

2.1. Study Area

The Republic of Turkey connects the European and Asian continents (Figure 1). It is a peninsula surrounded by three seas: the Black Sea in the north of Turkey, the Mediterranean Sea in the south, and the Aegean Sea in the west. Turkey has a mountainous and rugged terrain and constitutes approximately 770,760 km^2 of land and 9820 km^2 of water. Among the water areas, Van Lake is the largest natural lake, with 3713 km^2, and Atatürk Dam is the largest artificial lake, with 817 km^2.

Figure 1. Turkey—study area.

2.2. Materials and Methods

The European Commission develops Copernicus satellites in partnership with the European Space Agency (ESA). This includes all-weather radar images from Sentinel-1A and Sentinel-1B, high-resolution optical images from Sentinel 2A and 2B, and ocean and land data from Sentinel 3 that are suitable for environmental and climate monitoring. Sentinel-2 is a wide-field, high-resolution, multi-spectral imaging mission that supports Copernicus Land Monitoring, including the monitoring of vegetation and soil and water cover, as well as the observation of inland waterways and coastal areas. Sentinel-2 consists of 13 bands and outperforms the Landsat program in terms of its spatial and spectral resolutions.

For the purposes of this study, a total of 2806 Sentinel-2 satellite data points were used. The Sentinel-2 data were pre-processed based on region, date, and cloud mask filtering. As a result, the imagery was restricted to Turkey's borders and dates throughout the summer of 2020, with a 10% cloud filter mask added. Using this method, a clean Sentinel-2 picture collection of Turkey was produced. Considering the vast study area, a small number of training and testing samples were selected from the water (90) and non-water classes (190).

The MBWI was chosen for water classification, since it has produced the best results in the literature among the index-based algorithms. The MBWI is based on distinctions between water and other low-reflectance surfaces, restricting the brightness value ranges used to those in the lower or "darker" section of the terrestrial spectral range, being characteristic of water. The MBWI is intended to limit non-water pixels while improving surface water information. Wang et al. provided details of the concept of MBWI [6], and the calculation is given in Equation (1). In addition, to eliminare mountainous shadows that were mistakenly classified as water bodies, we placed a threshold of 5% slope over the study area, and areas with higher slopes were automatically excluded from the water class.

$$MBWI = 2 \times Green - Red - NIR - SWIR1 - SWIR2 \quad (1)$$

In remote sensing analysis, accuracy assessment is a critical evaluator for the results. Thus, in this study, the validation was performed using 100 random sample points from the water class. Two measures of accuracy were tested in this study, namely, the overall accuracy and kappa coefficient. While the overall accuracy provides information about the proportion of correctly mapped reference sites, the kappa coefficient is generated through a statistical test to evaluate the accuracy of the classification. The kappa coefficient essentially evaluates how well the classification performs as compared to the random assignment of values. The kappa coefficient can range from −1 to 1. In remote sensing applications with a mid-spatial resolution, such as Landsat, a kappa value higher than 0.75 is considered acceptable.

3. Results and Discussion

The study area's surface water bodies were extracted with the employed methodology. Thus, we extracted the water bodies in Turkey in the summer of 2020 (Figure 2). The visual inspection showed that the classification yielded good results, considerating the vast study area. In water extraction studies, areas with high slopes and urban areas are the most challenging; however, the developed algorithm also showed good results for these areas.

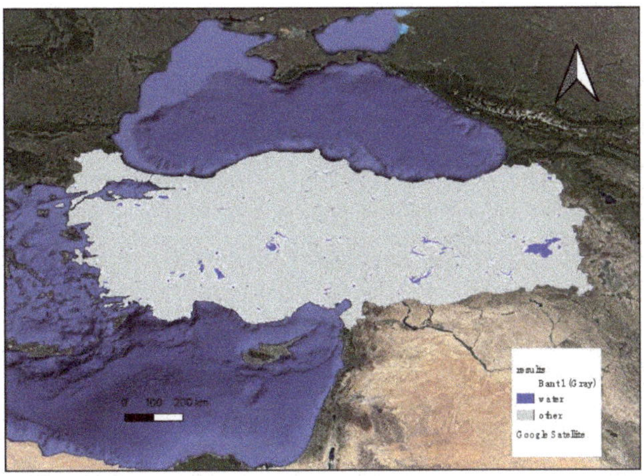

Figure 2. Results.

The accuracy assessment showed an overall accuracy of 0.94 for the water bodies' classification, meaning that 94% of the water areas were classified correctly. The kappa statistics had a significant high value of 0.86. For a vast area, the obtained results are acceptable and highly important from several points of view. As the methodology was developed in GEE, it can be used repeatedly for different dates, smaller study areas, etc., providing fast and reliable information on the water bodies. The water areas can be easily calculated, and spatio-temporal analysis can be performed using the same algorithm. With a small modification, the application can be set to use Landsat data, allowing one to analyze the water bodies for five decades. In this study, we classified the water bodies in the summer of 2020. The same application could be used for near-real-time applications. The greatest disadvantage of the present study is the spatial resolution of the used satellite imagery, which was 10 m in this case. This means that the algorithm is only able to classify water bodies that are larger than 10 m, and very small water bodies will not be extracted. However, the obtained results could be useful in various applications and provide the user with a clear image of the water bodies throughout the study area. The results again showed that GEE is a powerful platform that is able to classify vast areas within a few minutes.

Author Contributions: Conceptualization, M.A.A. and G.K.; methodology, G.K.; software, M.A.A.; validation, G.K.; formal analysis, M.A.A.; investigation, G.K.; resources, G.K.; data curation, G.K.; writing—original draft preparation, M.A.A.; writing—review and editing, G.K.; visualization, M.A.A.; supervision, G.K. All authors have read and agreed to the published version of the manuscript.

Funding: This research received no external funding.

Institutional Review Board Statement: Not applicable.

Informed Consent Statement: Not applicable.

Data Availability Statement: Not applicable.

Acknowledgments: The present study is part of Mervegul Aykanat Atay's master's thesis, supervised by Gordana Kaplan.

Conflicts of Interest: The authors declare no conflict of interest.

References

1. McFeeters, S.K. The use of the Normalized Difference Water Index (NDWI) in the delineation of open water features. *Int. J. Remote Sens.* **1996**, *17*, 1425–1432. [CrossRef]
2. Xu, H. Modification of normalised difference water index (NDWI) to enhance open water features in remotely sensed imagery. *Int. J. Remote Sens.* **2006**, *27*, 3025–3033. [CrossRef]
3. Danaher, T.; Collett, L. Development, optimisation and multi-temporal application of a simple Landsat based water index. In Proceedings of the 13th Australasian Remote Sensing and Photogrammetry Conference, Canberra, Australia, 20–24 November 2006.
4. Fisher, A.; Flood, N.; Danaher, T. Comparing Landsat water index methods for automated water classification in eastern Australia. *Remote Sens. Environ.* **2016**, *175*, 167–182. [CrossRef]
5. Feyisa, G.L.; Meilby, H.; Fensholt, R.; Proud, S.R. Automated Water Extraction Index: A new technique for surface water mapping using Landsat imagery. *Remote Sens. Environ.* **2014**, *140*, 23–35. [CrossRef]
6. Wang, X.; Xie, S.; Zhang, X.; Chen, C.; Guo, H.; Du, J.; Duan, Z. A robust Multi-Band Water Index (MBWI) for automated extraction of surface water from Landsat 8 OLI imagery. *Int. J. Appl. Earth Obs. Geoinf.* **2018**, *68*, 73–91. [CrossRef]
7. Wang, Y.; Li, Z.; Zeng, C.; Xia, G.-S.; Shen, H. An urban water extraction method combining deep learning and Google Earth engine. *IEEE J. Sel. Top. Appl. Earth Obs. Remote Sens.* **2020**, *13*, 768–781. [CrossRef]
8. Kaplan, G.; Avdan, U. Water extraction technique in mountainous areas from satellite images. *J. Appl. Remote Sens.* **2017**, *11*, 046002. [CrossRef]
9. Pickens, A.H.; Hansen, M.C.; Hancher, M.; Stehman, S.V.; Tyukavina, A.; Potapov, P.; Marroquin, B.; Sherani, Z. Mapping and sampling to characterize global inland water dynamics from 1999 to 2018 with full Landsat time-series. *Remote Sens. Environ.* **2020**, *243*, 111792. [CrossRef]
10. Yigit Avdan, Z.; Kaplan, G.; Goncu, S.; Avdan, U. Monitoring the water quality of small water bodies using high-resolution remote sensing data. *ISPRS Int. J. Geo-Inf.* **2019**, *8*, 553. [CrossRef]
11. Soltanian, F.K.; Abbasi, M.; Bakhtyari, H.R. Flood monitoring using Ndwi and Mndwi spectral indices: A case study of Aghqala flood-2019, Golestan Province, Iran. *Int. Arch. Photogramm. Remote Sens. Spat. Inf. Sci.* **2019**, *42*, 605–607. [CrossRef]
12. Nguyen, U.N.; Pham, L.T.; Dang, T.D. An automatic water detection approach using Landsat 8 OLI and Google Earth Engine cloud computing to map lakes and reservoirs in New Zealand. *Environ. Monit. Assess.* **2019**, *191*, 1–12. [CrossRef] [PubMed]
13. Mobariz, M.; Kaplan, G. Monitoring Amu Darya river channel dynamics using remote sensing data in Google Earth Engine. In Proceedings of the 5th International Electronic Conference on Water Sciences, Online, 16–30 November 2020.
14. Albarqouni, M.M.; Yagmur, N.; Bektas Balcik, F.; Sekertekin, A. Assessment of Spatio-Temporal Changes in Water Surface Extents and Lake Surface Temperatures Using Google Earth Engine for Lakes Region, Turkey. *ISPRS Int. J. Geo-Inf.* **2022**, *11*, 407. [CrossRef]

environmental sciences proceedings

MDPI

Proceeding Paper

Artificial Neural Networks and Regression Modeling for Water Resources Management in the Upper Indus Basin †

Muhammad Imran [1], Muhammad Danish Majeed [1,2,*], Muhammad Zaman [1,*], Muhammad Adnan Shahid [1,2], Danrong Zhang [3], Syeda Mishal Zahra [1,2], Rehan Mehmood Sabir [1,2], Muhammad Safdar [1,2] and Zahid Maqbool [2]

1 Department of Irrigation & Drainage, Faculty of Agricultural Engineering & Technology, University of Agriculture, Faisalabad 38000, Pakistan; m.imrankbr@gmail.com (M.I.); adnan.wmrc@gmail.com (M.A.S.); syedamishalzahra212@gmail.com (S.M.Z.); rehan.mahmood@uaf.edu.pk (R.M.S.); safdarsani4340@gmail.com (M.S.)

2 Agricultural Remote Sensing Lab-(ARSL)-NCGSA, University of Agriculture, Faisalabad 38000, Pakistan; zahidgcuf@gmail.com

3 College of Hydrology and Water Resources, Hohai University, Nanjing 210098, China; danrong_zhang@hhu.edu.cn

* Correspondence: kambohdanish@gmail.com (M.D.M.); muhammad.zaman@uaf.edu.pk (M.Z.)

† Presented at the 7th International Electronic Conference on Water Sciences, 15–30 March 2023; Available online: https://ecws-7.sciforum.net.

Abstract: A flood is a natural disaster. Heavy rainfall and overflow frequently cause enclosed land areas to fill with water, resulting in considerable loss of human life and property, including damage to buildings, bridges, electric supply networks, and transportation, and economic concern. This work was carried out in the Upper Indus Basin (UIB). We developed an artificial intelligence model for forecasting the flood events in this study. Long-short term memory (LSTM) and seasonal auto-regressive integrated moving average (SARIMA) were used in this study to forecast flood events. This study used a dataset from 1971–2009 and divided it into training, testing, and forecasting from 1971–2004, 2005–2009, and 2010–2014, respectively. The best statistical analysis result was observed with the LSTM model, which documented the value of root mean squared error (RMSE) at 22.79 and 35.05 for training and testing, respectively. Hence, the results of the study highlight that the LSTM model was the most suitable among the artificial neural networks for flood event forecasts. This current study will help in the forecasting of high storms for effective water resources management.

Keywords: Upper Indus Basin; flood forecasting; LSTM; SARIMA; water resources management

Citation: Imran, M.; Majeed, M.D.; Zaman, M.; Shahid, M.A.; Zhang, D.; Zahra, S.M.; Sabir, R.M.; Safdar, M.; Maqbool, Z. Artificial Neural Networks and Regression Modeling for Water Resources Management in the Upper Indus Basin. *Environ. Sci. Proc.* **2023**, *25*, 53. https://doi.org/10.3390/ECWS-7-14199

Academic Editor: Athanasios Loukas

Published: 14 March 2023

1. Introduction

Most Asian countries depend on agriculture, and Pakistan is the country most dependent on agriculture. A huge part of the gross domestic production (GDP) of Pakistan is based on agriculture, and the agriculture sector of this country is mostly reliant on irrigation water generated by the Upper Indus Basin. The water scarcity report of the International Monetary Fund (IMF) ranked Pakistan in the third position among water scarcity in the year 2018. The International Panel for Climate Change (IPCC) estimated that the total temperature increased by 0.72 °C in the period from 1951 to 2012. The expected temperature will likely increase from 1 °C to 3 °C until 2050 and from 2 °C to 5 °C until 2100, depending on the different gas emission circumstances documented by IPCC-2013. Water reserves are the core of most crises in countries such as Pakistan, where the economy, culture, and textiles are intimately connected to irrigation water. The effects of warming drift on the outer circle are particularly unique and complex [1].

Throughout history, floods have been recorded as one of the most devastating natural disasters capable of causing severe personal damage as well as destroying property. In recent years, the expansion of flood events has been a problem resulting in a large number of deaths every year. Another main reason is that as the human population is increasing, human communities are becoming closer to water resources. Flood-affected people's infrastructure and lives have been severely damaged and disturbed [2].

In this study, data-driven approaches based on the statistical relationship between input and output data were used to forecast flood events in the UIB. A possible alternative to the current approaches for the hydrological forecast of streamflow may be data-driven approaches such as ANNs [3].

2. Materials and Methods

2.1. Study Area

Research study area of Indus basin spans within the following geographical points of 33–75° N and 72–78° E covered by the Swat hillock, along with the Mohamad, Mangla Complex in the north, the district of Charsadda in the southwesterly area, and the Kotli, Mangla, Kalam Complex to the west, as shown in Figure 1. Kalam Station is at a high mean sea level of 5821 km and Munda Dam is at a low mean sea level of 376 km. The research region is characterized by climatic lateral tropical and wet temperate zones, with thunderstorms and snowfall. Summers are hot (41.9 °C) and winters are frigid (0.8 °C) in the chosen location. The effluvium ornament is a cardinal aspect in determining water potential. The flow of a locality is defined by its effluvium compactness, which is the relative amount of rainfall that enters. As a result, the lower the runoff, the higher the chances of recharging.

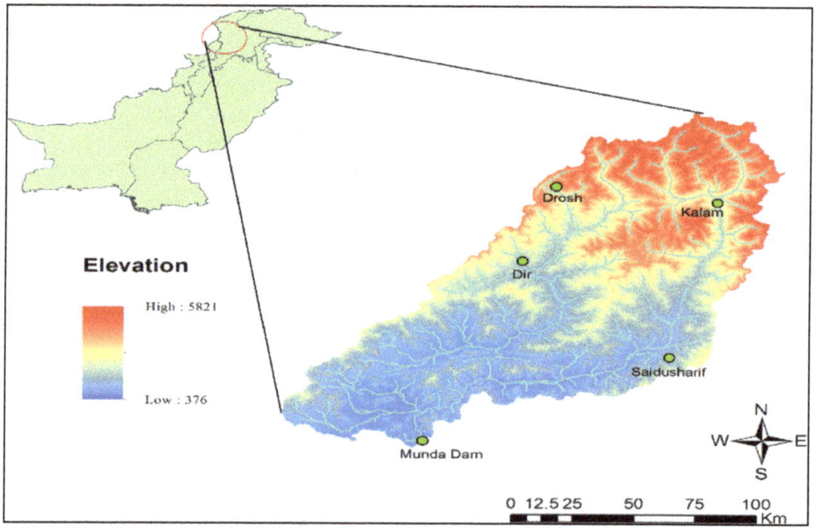

Figure 1. Digital elevation model of the Upper Indus Basin with gauge stations.

2.2. Data Collection and Model Description

Pakistan is one of the most climatically varied countries due to its wide temperature range, and it has a higher occurrence of rainfall in the UIB including extreme flood events. Extreme climatic devastations specifically in the forms of floods have impacted precious lives and financial losses in Pakistan in the last three decades (e.g., 2011, 2020 and 2022). Data on the extreme events of stream flow were collected for the duration from 1971–2009. The data were collected from the Water and Power Development Authority (WAPDA) and the Pakistan Meteorological Department (PMD) [4].

2.2.1. LSTM Model

Deep learning is a type of neural network that uses a larger number of layers and layer types to model complex systems and interactions. Because traditional neural networks cannot retain temporal information, recurrent neural networks were developed using previous time step information. LSTMs are a deep learning version of recurrent neural networks that can remember information for longer. To change the data, LSTM cells use gates, vector addition, and multiplication to remove or add information.

2.2.2. SARIMA Model

SARIMA is a regression model, and all regression models assume that the values in a dataset are independent of one another. When using regression to predict time series, it is critical to ensure that the data are stationary, which means that statistical properties such as variance do not change over time. In ARIMA, "AR" denotes the "autoregressive" component, which is the lag of the stationary series, "MA" denotes the lags of the forecast errors, and "I" denotes the order of differentiation to make the series stationary. The SARIMA (1, 0, 1) × (0, 1, 1) 12 model was used in this study for the forecasting of flow in the UIB. To ensure that the time series data were stationary, the Dickey–Fuller test was used. The resulting p-value was less than 0.03 and the test static was −3.739768, allowing us to reject the null hypothesis and conclude that the data were stationary. The seasonality period(s) was a 12-month moving average, and the minimum AIC score was 138.065 at 29 time steps.

2.2.3. Model Evaluation Criteria

Root mean square error (RMSE) is a statistical method commonly used to compare predicted values with observed values in hydrology fields to evaluate the performance of forecasting models. Based on the relative range of the data, the RMSE is frequently used to evaluate how closely the predicted values match the observed values.

$$\mathrm{RMSE} = \sqrt{\frac{1}{n}\sum_{i=1}^{n}(\mathrm{Y} - \mathrm{Yi})^2} \quad (1)$$

In Equation (1), Y and Yi are actual and predicted discharges at time t, respectively; Y is the mean of actual discharges; and n is the total number of observations.

2.2.4. Model Structure

Our study is related to open-source software libraries. According to the literature, Python [5] is the programming language of choice. The NumPy [6], Pandas [7], and Matplotlib [8] libraries are also imported for data processing, management, and visualization. We created the LSTM model and TensorFlow [9], a Google open-source software library. TensorFlow was originally designed to conduct machine learning, deep learning, and numerical computation research using data flow graphs. However, this framework is comprehensive enough to be applicable to a wide range of domains.

3. Results and Discussions

The forecasting models developed were validated and tested using independent data. When comparing the actual data and forecast values made during the validation and testing process, the RMSE values were used to evaluate the qualitative and quantitative performance of the scenarios. The size of the training datasets varies depending on the scenario. A thirty-four-year dataset (1971–2004) was used for training, while a five-year dataset (2005–2009) was used for testing and a five-year dataset (2010–2014) was used for forecasting.

The RMSE in the two-prediction model for testing and training is shown in Table 1. From the observation of the below RMSE results, the LSTM model performs better for training as well as testing as compared to the SARIMA model.

Table 1. Model Evaluation Results.

RMSE	LSTM	SARIMA
Training	22.79	27.82
Testing	35.05	54.42

Training, Testing, and Forecast Results

Figure 2a–c shows the visual comparison results of the actual and forecasted flow data from Kalam Station. Figure 2a shows the comparison results of the SARIMA model at Kalam Station. The blue line shows the training data, the green line shows the testing data, and the red line shows the forecasted data, which are the output data of our model. Figure 2b,c shows the comparison result of the LSTM model at Kalam Station. In Figure 2b, the green line shows the training data (actual values), and the red line shows the testing data (predicted values), while in Figure 2c, the green line is the observed values from the model and the red line is the forecasted values, which are the output data of our model.

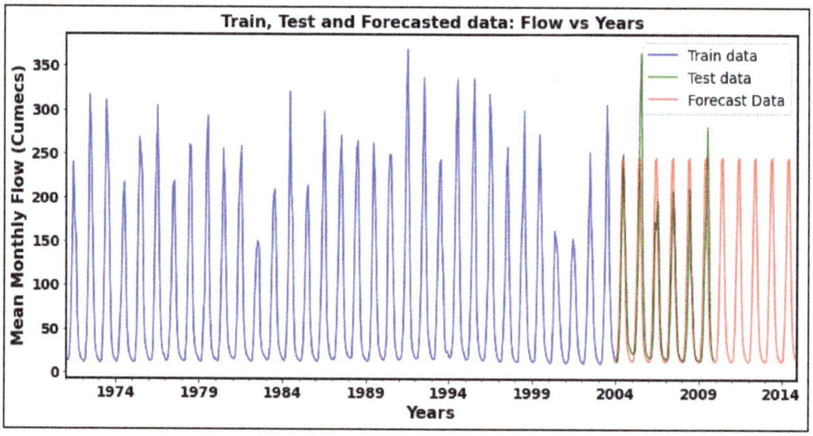

(**a**)

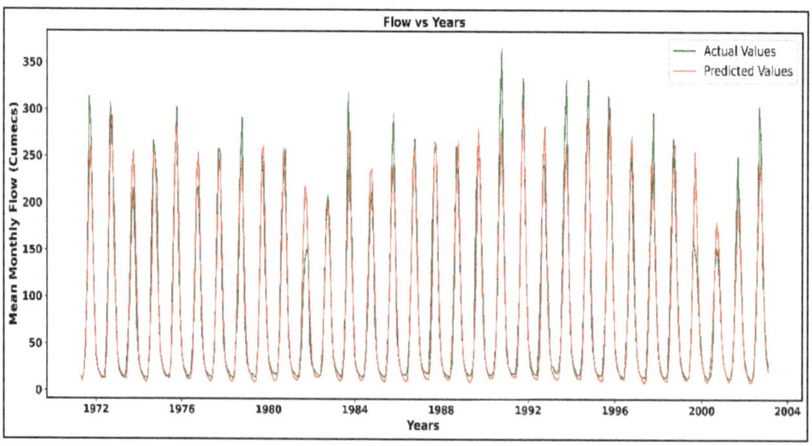

(**b**)

Figure 2. *Cont.*

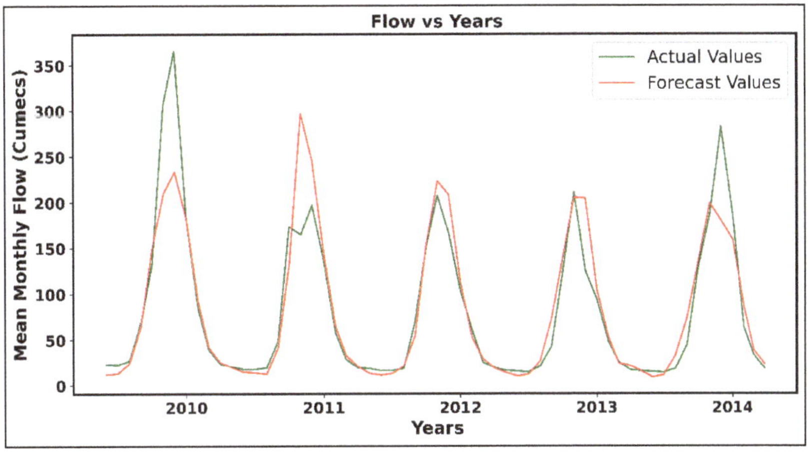

(c)

Figure 2. Visual comparison of actual and forecasted data: (**a**) SARIMA, (**b**) LSTM training and testing, and (**c**) LSTM forecasting.

4. Conclusions

Accurate time series forecasting for upcoming flood events is important, but it is a challenging task such as in Pakistan. The flood forecasting at the Swat River flow gauging station, particularly for downstream stations that lack discharge information, such as Saidu Sharif in this study, plays an important role in early flood warning systems. In this research, we examined the classical statistical models used such as the deep learning model, LSTM, and the regression model, SARIMA. In our comparison of the models for Kalam Station, the LSTM model achieved better results and more accurate forecasting performance than the SARIMA model. Linear data modeling of water flows for SARIMA yield better result as compared to other statistical deep learning models which are good for nonlinear datasets. The RMSE values of the LSTM model fit to the series found for training and testing; for one-year-ahead forecasting, the values are 22.79 and 35.05, respectively. These results indicates that the deep learning algorithm is a dependable ideal solution for flood prediction due to its high precision.

Author Contributions: Conceptualization, M.I., M.Z. and M.D.M.; Models selection for water modeling, M.Z., M.A.S. and S.M.Z.; Data Collection, M.I., M.S. and Z.M. Model Training, Testing, and Forecast, M.I., M.D.M. and R.M.S.; Future Directions & Conclusions, M.Z. and D.Z.; writing—original draft preparation, M.I., M.D.M. and R.M.S.; writing a review and editing, M.I. and M.D.M. All authors have read and agreed to the published version of the manuscript.

Funding: This research received no external funding.

Institutional Review Board Statement: Not applicable.

Informed Consent Statement: Not applicable.

Data Availability Statement: Available on request after due procedure.

Conflicts of Interest: The authors declare no conflict of interest.

References

1. Shahid, F. Climate Change: Impacts on Pakistan and Proposed Solutions. *Pak. Soc. Sci. Rev.* **2021**, *5*, 223–235. [CrossRef]
2. Gude, V.; Corns, S.; Long, S. Flood Prediction and Uncertainty Estimation Using Deep Learning. *Water* **2020**, *12*, 884. [CrossRef]
3. Atashi, V.; Gorji, H.T.; Shahabi, S.M.; Kardan, R.; Lim, Y.H. Water Level Forecasting Using Deep Learning Time-Series Analysis: A Case Study of Red River of the North. *Water* **2022**, *14*, 1971. [CrossRef]

4. Jawad, M.; Nadeem, M.S.A.; Shim, S.-O.; Khan, I.R.; Shaheen, A.; Habib, N.; Hussain, L.; Aziz, W. Machine Learning Based Cost Effective Electricity Load Forecasting Model Using Correlated Meteorological Parameters. *IEEE Access* **2020**, *8*, 146847–146864. [CrossRef]
5. Le, X.H.; Ho, H.V.; Lee, G.; Jung, S. Application of Long Short-Term Memory (LSTM) Neural Network for Flood Forecasting. *Water* **2019**, *11*, 1387. [CrossRef]
6. Van Der Walt, S.; Colbert, S.C.; Varoquaux, G. The NumPy Array: A Structure for Efficient Numerical Computation. *Comput. Sci. Eng.* **2011**, *13*, 22–30. [CrossRef]
7. McKinney, W. Data structures for statistical computing in python. *Proc. 9th Python Sci. Conf.* **2010**, *1*, 56–61. [CrossRef]
8. Hunter, J.D. Matplotlib: A 2D graphics environment. *Comput. Sci. Eng.* **2007**, *9*, 90–95. [CrossRef]
9. Abadi, M.; Agarwal, A.; Barham, P.; Brevdo, E.; Chen, Z.; Citro, C.; Corrado, G.S.; Davis, A.; Dean, J.; Devin, M.; et al. TensorFlow: Large-Scale Machine Learning on Heterogeneous Distributed Systems. *arXiv* **2016**, arXiv:1603.04467.

environmental sciences proceedings

MDPI

Proceeding Paper

Modeling Soil Erodibility by Water (Rainfall/Irrigation) on Tillage and No-Tillage Plots of a *Helianthus* Field Utilizing Soil Analysis, Precision Agriculture, GIS, and Kriging Geostatistics †

Agathos Filintas *, Nikolaos Gougoulias and Eleni Hatzichristou

Department of Agricultural Technology, University of Thessaly—Campus Gaiopolis, 41500 Larisa, Greece; ngougoulias@uth.gr (N.G.); eleni.a.xatz@gmail.com (E.H.)

* Correspondence: filintas@uth.gr

† Presented at the 7th International Electronic Conference on Water Sciences, 15–30 March 2023; Available online: https://ecws-7.sciforum.net.

Abstract: The aim of our study is the modeling at the field level of the soil erodibility (*K* factor) by water (rainfall and irrigation) on traditional tillage (CoTl) and no-tillage (NoTl) plots cultivated with *Helianthus annuus* utilizing plot observations, soil sampling laboratory analyses, GIS, precision agriculture (PA), and Kriging geostatistical modeling. A split-plot layout consisting of four handlings × three replicates of trial blocks (with a southeast facing 7.5% slope) was used. Grid template surface soil core (0.0–5.0 cm) samples were taken to characterize the textures (sandy, silty, clayey, very fine sandy, and gravelly), organic matter concentrations, and the soil's microstructure and water permeability categories. One GPS satellite tracker system was utilized to define the sampled positions, and 40 soil cores were air-dried and sieved with a 2 mm sieve to identify the soil's mechanical micro-texture using the Bouyoucos methodology. The organic matter was extracted by chemical oxidation with 1 mol L^{-1} $K_2Cr_2O_7$ and titration of the remaining reagent with 0.5 mol L^{-1} $FeSO_4$. The soil microstructure and permeability categories were defined following the USDA classification system. The soil erodibility by water modeling of *K* ($Mg \cdot ha \cdot h \cdot ha^{-1} \cdot MJ^{-1} \cdot mm^{-1}$) was derived according to the Wischmeier nomographic method by incorporating it into a developed GIS geospatial model using Kriging geostatistics. The statistical results of the ANOVA test ($p = 0.05$) among the soil erodibility datasets showed significant differences between the two tillage systems, as well as between the four management treatments. Moreover, it was found that the no-tillage (NoTl) plots and the treatment of no tillage plus vegetative coverage were the best tillage and agricultural practices for hillslope farm fields and can be considered environmentally friendly farming methods to curb soil erodibility by water, reduce runoff hazard, and maintain the soil's environment and its beneficial nutrients.

Keywords: soil erodibility by water (rainfall/irrigation); tillage; soil analyses; spatial analysis; precision agriculture and Kriging geostatistical models; *Helianthus annuus* crop; organic matter

Citation: Filintas, A.; Gougoulias, N.; Hatzichristou, E. Modeling Soil Erodibility by Water (Rainfall/Irrigation) on Tillage and No-Tillage Plots of a *Helianthus* Field Utilizing Soil Analysis, Precision Agriculture, GIS, and Kriging Geostatistics. *Environ. Sci. Proc.* **2023**, *25*, 54. https://doi.org/10.3390/ECWS-7-14254

Academic Editor: Athanasios Loukas

Published: 16 March 2023

1. Introduction

The erosion of soil is the phenomenon of soil particles being separated and transported by water or wind [1]. Today, it is a major issue for agricultural growth and food safety at regional, country, and world levels [2,3]. Greece has a developed agricultural sector with a declining farmer population and heavy farming operations that have led to the increased erosion of soils. In an effort to study new ways to decrease soil erosion and preserve precious soil reserves, several erosion models have been successfully deployed and widely tested all over the world. Soil erosion and hazards are considered major problems of the environment in Greece. The soil's erodibility (*K*) is a fundamental parameter in erosion forecasting methods such as the USLE (Universal Soil Loss Equation) [4] and the RUSLE (Revised USLE) [5,6]. The *K* factor is a complicated soil attribute, which is the ease with which the soil is degraded by waterdrop splashing during rain or irrigation (mainly by sprinklers or waterjets), water runoff, or their combination [3]. The capturing of erosion's principal variable (*K* factor) in forecasting modeling has proved to be a difficult task [7]. To overcome this issue, implicit methods are used to assess the *K* factor and allow these studies to be carried out [8]. The aim of our study is the geospatial modeling at field level of the soil erodibility by waterdrops on traditional tillage (CoTl) and no-tillage (NoTl) *Helianthus* plots utilizing observations, soil laboratory analyses, precision agriculture, Kriging geostatistics, and GIS mapping under climate change in Greece.

2. Materials and Methods

2.1. Study Area and Site Description

The trial was carried out in the agricultural hilly erosion-prone area of the Gaiopolis University Campus of the University of Thessaly (Larissa, Central Greece). The region enjoys moderate continental climatic conditions with a hot arid summer and a gentle winter that is characterized as Csa (Koeppen climate classification) [2] and is further classified as a XERIC MOISTURE REGIME [9], with an average annual temperature and precipitation of 17.35 °C and 380.75 mm, respectively. The highest and lowest average monthly precipitation were pr(hi) = 113.40 mm (May) and pr(low) = 12.20 mm (November), respectively. The cumulative precipitation was 652.40 mm year^{-1}. A split-plot layout consisting of 4 handlings (treats) × 3 replicates of trial blocks (with a southeast facing 7.5% slope) was used. *Helianthus annuus* plants were seeded to facilitate plant coverage in a number of treatments: (a) the A-treatment was traditional tillage (CoTl) plus vegetative coverage (VCov), (b) the B-treatment was CoTl with no vegetative coverage (NoVCov), (c) the C-treatment was no tillage (NoTl) plus vegetative coverage (VCov), and (d) the D-treatment was no tillage (NoTl) with no vegetative coverage (NoVCov). The dimensions of the 12 trial plots were 6 m × 22.1 m downslope, with an overall plot area of 1591.2 m^2.

2.2. Soil Sampling, Laboratory Analyses, and Classification

Grid template surface soil core (0.0–5.0 cm) samples were taken to characterize the textures (sandy (*Sa*), silty (*Si*), clayey (*Cl*), very fine sandy (*vfSa*), and gravelly (*Gra*)), organic matter (*OrM*) concentrations, and the soil microstructure plus water permeability categories. One GPS (Global Positioning System) satellite tracker system was utilized to define all the sampled positions, and 40 surface soil cores were air-dried and sieved with a 2 mm sieve to identify the soil's mechanical microtexture using the Bouyoucos methodology [10,11]. The organic matter was extracted by chemical oxidation with 1 mol L^{-1} $K_2Cr_2O_7$ and titration of the remaining reagent with 0.5 mol L^{-1} $FeSO_4$ [11]. The soil microstructure (that is the assemblage of soil particulates and agglomerates via identifiable particles or granules) categories [9] and water permeability categories were defined following the USDA classification system [9,12]. The soil erodibility by water modeling of the *K* factor (Mg·ha·h·ha^{-1}·MJ^{-1}·mm^{-1}) was derived according to the Wischmeier nomographic method [4,12–14], by incorporating it into a developed GIS geospatial model using Kriging geostatistics. The *K* factor Equation (1) [4,12–14] was derived for all soils consisting of less than 70% silt plus vfSa:

$$K = \left[\frac{(2.1 \times 10^{-4}\,(12 - OM) M^{1.14} + 3.25(S-2) + 2.5(P-3))}{100} \right] \times 0.1317 \quad (1)$$

where *K* = soil erodibility of the USLE method (Mg·ha·h·ha^{-1}·MJ^{-1}·mm^{-1}), *M* = product of the percentage of silt plus vfSa and the other soil components except clay (0.002 mm > clay, 0.05 mm > silt > 0.002 mm, and 0.1 mm > sand > 0.05 mm), *OrM* = soil organic material concentration (%), *S* = soil microstructure category, and *P* = water permeability category.

2.3. Statistical and Geostatistical Data Analysis, Soil Erodibility Modeling, and Methodology

Data analysis was performed using the IBM SPSS v.26 [15–21] statistics software package. The outputs showed the means of the soil samples analyses and field metrics.

The ANOVA (analysis of variance) statistic test [14–29] was applied to evaluate the tillage systems and the treatment effectiveness. The Levene statistics test for the homogeneity of variants [14–22] was employed to verify the hypothesis of variance equality for the soil *K* datasets. The LSD test [14–22] was used to separate means when significantly different outputs ($p = 0.05$) among treatments were obtained. In the present study, we used geostatistics (Kriging) and precision agriculture [14,16–19,21–23,27] for modeling and Geographical Information System (GIS) mapping of the soil's textural class, organic matter concentration, and soil structure and permeability categories, respectively, as well as the soil erodibility. A GIS interface (ArcGIS © version 10.2) was employed to treat, store, and model the entry data variables in order to produce a soil erodibility digital map by using geospatial analytics and PA. The evaluation of the *K* factor requires residual errors' analytics among the projected and observable values and the forecast identification range of overestimates and underestimates. For this purpose, we applied the statistic parameters reported earlier in other studies [14–19,21–23,27,30–32], such as the equations for the Mean Prediction Error (MPE), the Root Mean Square Error (RMSE), the Mean Standardized Prediction Error (MSPE), and the Root Mean Square Standardized Error (RMSSE). The soil erodibility modeling outputs of the plots were used in order to extract the *K* data for the validation procedure on the basis of the soil's *K* trained and evaluation datasets.

3. Results and Discussion

Soil erodibility is a function of four parameters: the soil's texture, structure, permeability, and the *OrM* concentration. The soil analysis outputs showed that sand with very fine sand had the ranges of 26.47–46.34% and 21.73–22.08%, respectively. The mean silt and clay contents were 19.91% and 20.22%, respectively. The soil's organic matter [14,17–19,21–23,27] modeling results are depicted in Figure 1a–c.

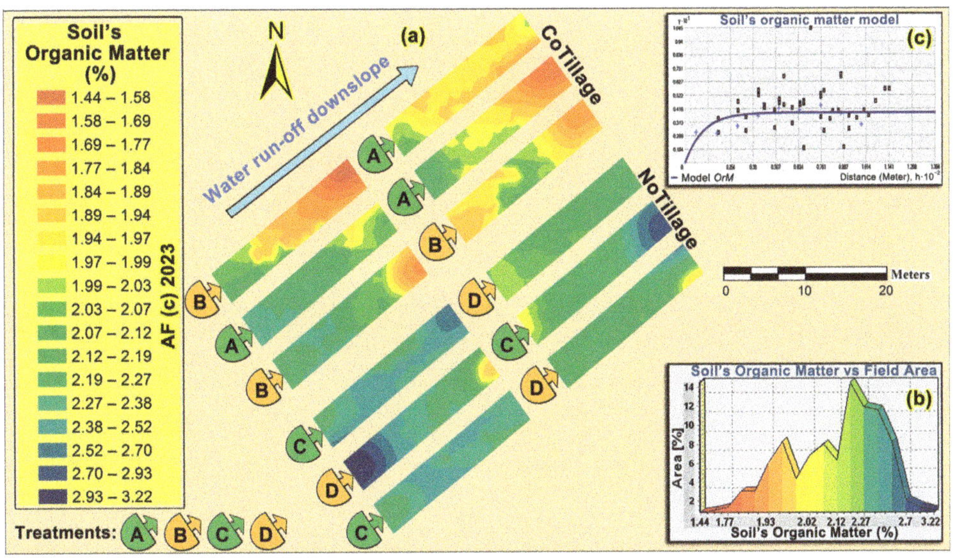

Figure 1. (**a**) Modeling outcomes on a soil organic matter digital GIS map of the *Helianthus* plots. (**b**) Diagram of the *OrM* classes vs. the percentage of the *OrM* area. (**c**) Semivariogram of the model.

Its concentration classes ranged from 1.44% to 3.22% (Figure 1b), indicating the soil's *OrM* had medium to high content. The soil's organic matter geospatial analysis showed that 34.887% of the soil plots' area had medium *OrM* content (1.44–2.00%), while the remaining 65.113% had high *OrM* content (2.00–3.22%). The modeling and statistical outputs revealed that the *K* factor over the measuring time span ranged from a min 0.025 to a max 0.043 Mg·ha·h·ha^{-1}·MJ^{-1}·mm^{-1} (average K = 0.034, standard deviation s = 0.0062). The soil characteristics of the *Helianthus* plots were sampled, analyzed, and digitized in accordance with their GPS-located field positions using the WGS 1984 geographic coordinate system (CS) and stored in a geodatabase. The soil parameters, tillage, and treatment datasets were projected to the UTM Zone 34N CS (Greece's zone). The outputs of the geospatial erodibility modeling are visualized in a digital GIS map of the field in Figure 2a–c. Furthermore, the outcomes of the erodibility categories in relation to the percentage of the K factor area are illustrated in Figure 2b. The validation of the geospatial soil erodibility modeling (Figure 2c) resulted in the following geostatistical outcomes: mean prediction error (MPE) = −0.000000924, root mean square error (RMSE) = 0.00598019, mean standardized prediction error (MSPE) = −0.00518898, and root mean square standard error (RMSSE) = 1.0498154. These results are highly acceptable considering that the MPE, RMSE, and MSPE scores should be close to zero for an optimized forecast, and the RMSSE scores should be close to unity, suggesting an accurate estimate of the forecast variability. The abovementioned results confirmed the reliability and accuracy of the generated soil erodibility digital GIS map for the trial hillslope field of *Helianthus annuus*. Furthermore, these outcomes have proven that the ordinary Kriging exponential model demonstrated a good performance and is regarded as highly appropriate for geospatial modeling and mapping of the K factor as well as other soil parameters (clay, sand, silt, *OrM*, very fine sand, etc.). The output of the ANOVA test (p = 0.05) among the soil erodibility datasets in relation to the tillage method showed that the two tillage systems [traditional (CoTl) and no tillage (NoTl)] differed significantly in certain ways; so, it was necessary to further investigate the pattern of their differences. Therefore, in order to validate the equality hypothesis of variance for the erodibility dataset, the Levene statistical test for homogeneity of variances was conducted.

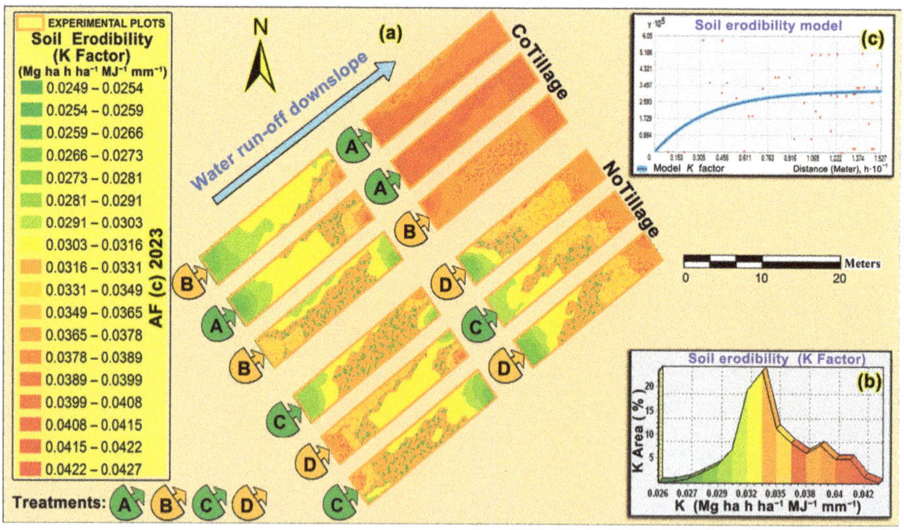

Figure 2. (**a**) The soil erodibility modeling results on a digital GIS map of the Helianthus plots. (**b**) Diagram of erodibility categories vs. percent of K factor area. (**c**) Semivariogram of the model.

The findings of the Levene statistics for the soil erodibility in the tillage systems and treatments showed the variations in homogeneity of the K factor between the tillage systems (CoTl and NoTl) and also between the treatments' (A, B, C, and D) datasets were not significantly different, which means that the hypothesis of equality of variation was confirmed. The Levene hypothesis was found to be true; so, ANOVA and LSD (Least Significant Differences) statistics were performed to evaluate the treatment effects and the mean separation of treatments. The optimum tillage system in Central Greece for hillside plots of high erosion hazard with a downward slope $\geq$ 7.5% was proven to be the NoTl system. The ANOVA results (p = 0.05) revealed that the soil datasets of the erodibility treatments (A, B, C, and D) significantly differed (Sig. = 0.029). The optimum treatment for limiting soil erodibility (K factor) and maintaining a healthy soil environment was judged to be treatment C [(NoTl-VCov) (no tillage plus vegetative coverage)] for hillside plots with a high erosion hazard with a downward slope of $\geq$ 7.5%.

4. Conclusions

The prediction errors' outcome of the validation of the geospatial and geostatistical modeling for the GIS soil erodibility mapping proved the validity and accuracy of the generated K-factor GIS digital map of the *Helianthus annuus* tested plots. All these results have demonstrated that the ordinary Kriging exponential model had good performance and is regarded as very well suited for modeling soil erodibility and many other soil parameters (clay, sand, silt, *OrM*, very fine sand, etc.) and digital mapping. In consideration of the ANOVA test results of the tillage systems and the treatment effects on the soil erodibility, the optimum tillage system found was NoTl (no tillage) and the superior treatment was C [(NoTi-VCov) (no tillage plus vegetative coverage)] for hillside plots of high erosion hazard with a downward slope $\geq$ 7.5%. These can be considered as environmentally friendly farming methods to curb soil erodibility by water, reduce runoff hazard, and maintain the health of the soil's environment and its beneficial nutrients.

Author Contributions: Conceptualization, A.F.; methodology, A.F., N.G. and E.H.; software, A.F.; validation, A.F., N.G. and E.H.; formal analysis, A.F., N.G. and E.H.; investigation, A.F., N.G. and E.H.; resources, A.F. and N.G.; data curation, A.F., N.G. and E.H.; writing—original draft preparation, A.F.; writing—review and editing, A.F.; visualization, A.F. and E.H.; supervision, A.F.; project administration, A.F.; funding acquisition, A.F., N.G. and E.H. All authors have read and agreed to the published version of the manuscript.

Funding: This research received no external funding.

Institutional Review Board Statement: Not applicable.

Informed Consent Statement: Not applicable.

Data Availability Statement: All the data of the study are presented in the paper.

Conflicts of Interest: The authors declare no conflict of interest.

References

1. Foster, G.R.; Meyer, L.D. A Closed-form Soil Erosion Equation for Upland Areas. In *Proceedings of Sedimentation Symposium in Honor Professor*; Einstein, H.A., Sten, H.W., Eds.; Colorado State University: Ft. Collins, CO, USA, 1972; pp. 12:1–12:19.
2. Filintas, A. Land Use Systems with Emphasis on Agricultural Machinery, Irrigation and Nitrates Pollution, with the Use of Satellite Remote Sensing, Geographic Information Systems and Models, in Watershed Level in Central Greece. Master's Thesis, Department of Environment, University of Aegean, Mitilini, Greece, 2005.
3. Filintas, A. Land Use Evaluation and Environmental Management of Biowastes, for Irrigation with Processed Wastewaters and Application of Bio-Sludge with Agricultural Machinery, for Improvement-Fertilization of Soils and Cultures, with the Use of GIS-Remote Sensing, Precision Agriculture and Multicriteria Analysis. Ph.D. Thesis, Department of Environment, University of the Aegean, Mitilini, Greece, 2011.
4. Wischmeier, W.H.; Smith, D.D. Predicting rainfall erosion losses. In *A Guide to Conservation Planning*; Agriculture Handbook 537; USDA-ARS-58: Washington, DC, USA, 1978.
5. Renard, K.; Foster, G.; Weesies, G.; McCool, D.; Yoder, D. Predicting soil erosion by water: A guide to conservation planning with the Revised Universal Soil Loss Equation (RUSLE). In *Agricultural Handbook*; United States Government Printing: Washington, DC, USA, 1997; pp. 65–100.
6. USDA Department of Agriculture—Agricultural Research Service: Revised Universal Soil Loss Equation. 2002. Available online: http://www.sedlab.olemiss.edu/rusle (accessed on 22 April 2022).
7. Panagos, P.; Meusburger, K.; Alewell, C.; Montarella, L. Soil erodibility estimation using LUCAS point survey data of Europe. *Environ. Model. Softw.* **2012**, *30*, 143–145. [CrossRef]

8. Bonilla, C.A.; Johnson, O.I. Soil erodibility mapping and its correlation with soil properties in Central Chile. *Geoderma* **2012**, *189–190*, 116–123. [CrossRef]
9. Soil Survey Staff. *Soil Taxonomy: A Basic System of Soil Classification for Making and Interpreting Soil Surveys*; USDA Natural Resources Conservation Service: Washington, DC, USA, 1975.
10. Bouyoucos, J.G. Hydrometer method improved for making particle size analysis of soils. *Agron. J.* **1962**, *54*, 464–465. [CrossRef]
11. Page, A.L.; Miller, R.H.; Keeney, D.R. *Methods of Soil Analysis Part 2: Chemical and Microbiological Properties*; Agronomy, ASA and SSSA: Madison, WI, USA, 1982; p. 1159.
12. Filintas, A. *Soil Erosion and Environmental Protection*; University of Thessaly: Larisa, Greece, 2022. (In Greek)
13. Wischmeier, W.H.; Johnson, C.B.; Cross, B.W. A soil erodibility nomograph for farmland and construction sites. *J. Soil Water Conserv.* **1971**, *26*, 189–193.
14. Filintas, A.; Gougoulias, N.; Salonikioti, A.; Prapa, E. Study of soil erodibility by water on tillage and no tillage treatments of a *Helianthus Tuberosus* crop using field measurements, soil laboratory analyses, GIS and deterministic models. *Ann. Univ. Craiova Ser. Biol. Hortic. Food Prod. Process. Technol. Environ. Eng.* **2019**, *XXIV*, 529–536.
15. Norusis, M.J. *IBM SPSS Statistics 19 Advanced Statistical Procedures Companion*; Pearson: London, UK, 2011.
16. Stamatis, G.; Parpodis, K.; Filintas, A.; Zagana, E. Groundwater quality, nitrate pollution and irrigation environmental management in the Neogene sediments of an agricultural region in central Thessaly (Greece). *Environ. Earth Sci.* **2011**, *64*, 1081–1105. [CrossRef]
17. Filintas, A.; Wogiatzi, E.; Gougoulias, N. Rainfed cultivation with supplemental irrigation modelling on seed yield and oil of *Coriandrum sativum* L. using Precision Agriculture and GIS moisture mapping. *Water Supply* **2021**, *21*, 2569–2582. [CrossRef]
18. Filintas, A.; Nteskou, A.; Kourgialas, N.; Gougoulias, N.; Hatzichristou, E. A Comparison between Variable Deficit Irrigation and Farmers' Irrigation Practices under Three Fertilization Levels in Cotton Yield (*Gossypium hirsutum* L.) Using Precision Agriculture, Remote Sensing, Soil Analyses, and Crop Growth Modeling. *Water* **2022**, *14*, 12654. [CrossRef]
19. Dioudis, P.; Filintas, A.; Koutseris, E. GPS and GIS based N-mapping of agricultural fields' spatial variability as a tool for non-polluting fertilization by drip irrigation. *Int. J. Sus. Dev. Plann.* **2009**, *4*, 210–225. [CrossRef]
20. Dioudis, P.; Filintas, A.; Papadopoulos, A. Corn yield response to irrigation interval and the resultant savings in water and other overheads. *Irrig. Drain.* **2009**, *58*, 96–104. [CrossRef]
21. Filintas, A.; Dioudis, P.; Prochaska, C. GIS modeling of the impact of drip irrigation, of water quality and of soil's available water capacity on Zea mays L, biomass yield and its biofuel potential. *Desalination Water Treat.* **2010**, *13*, 303–319. [CrossRef]
22. Filintas, A. Soil Moisture Depletion Modelling Using a TDR Multi-Sensor System, GIS, Soil Analyzes, Precision Agriculture and Remote Sensing on Maize for Improved Irrigation-Fertilization Decisions. *Eng. Proc.* **2021**, *9*, 36. [CrossRef]
23. Filintas, A.; Nteskou, A.; Katsoulidi, P.; Paraskebioti, A.; Parasidou, M. Rainfed and Supplemental Irrigation Modelling 2D GIS Moisture Rootzone Mapping on Yield and Seed Oil of Cotton (*Gossypium hirsutum*) Using Precision Agriculture and Remote Sensing. *Eng. Proc.* **2021**, *9*, 37. [CrossRef]
24. Dioudis, P.; Filintas, A.; Papadopoulos, A.; Sakellariou-Makrantonaki, M. The influence of different drip irrigation layout designs on sugar beet yield and their contribution to environmental sustainability. *Fresenious Environ. Bull.* **2010**, *19*, 818–831.
25. Kalavrouziotis, I.K.; Filintas, A.T.; Koukoulakis, P.H.; Hatzopoulos, J.N. Application of multicriteria analysis in the Management and Planning of Treated Municipal Wastewater and Sludge reuse in Agriculture and Land Development: The case of Sparti's Wastewater Treatment Plant, Greece. *Fresenious Environ. Bull.* **2011**, *20*, 287–295.
26. Hatzigiannakis, E.; Filintas, A.; Ilias, A.; Panagopoulos, A.; Arampatzis, G.; Hatzispiroglou, I. Hydrological and rating curve modelling of Pinios River water flows in Central Greece, for environmental and agricultural water resources management. *Desalination Water Treat.* **2016**, *57*, 11639–11659. [CrossRef]
27. Filintas, A.; Gougoulias, N.; Papachatzis, A. Soil organic matter modeling and digital mapping of a Triticum turgidum cropfield using as auxiliary variables the plant available water, texture, field measurements, soil laboratory analyses, GIS and geostatistical models. *Ann. Univ. Craiova Ser. Biol. Hortic. Food Prod. Process. Technol. Environ. Eng.* **2019**, *XXIV*, 537–544.
28. Koutseris, E.; Filintas, A.; Dioudis, P. Antiflooding prevention, protection, strategic environmental planning of aquatic resources and water purification: The case of Thessalian basin, in Greece. *Desalination* **2010**, *250*, 318–322. [CrossRef]
29. Koutseris, E.; Filintas, A.; Dioudis, P. Environmental control of torrents environment: One valorisation for prevention of water flood disasters. *WIT Trans. Ecol. Environ.* **2007**, *104*, 249–259. [CrossRef]
30. Loague, K.; Green, R.E. Statistical and graphical methods for evaluating solute transport models: Overview and application. *J. Contam. Hydrol.* **1991**, *7*, 51–73. [CrossRef]
31. Lu, G.Y.; Wong, D.W. An adaptive inverse-distance weighting spatial interpolation technique. *Comput. Geosci.* **2008**, *34*, 1044–1055. [CrossRef]
32. Filintas, A.; Panoras, G.; Stamatis, G. Hydrological 2D Modelling of Lithaios River Flows (Greece) Using GIS and Geostatistics for Environmental and Agricultural Water Resources Administration. *Environ. Sci. Proc.* **2023**, *25*, 13. [CrossRef]

environmental sciences proceedings

MDPI

Proceeding Paper

Integrated Water Resource Management Using Water Evaluation and Planning Model: A Case Study of Lower Bari Doab Canal, Pakistan †

Hira Khanam [1,*], Sikandar Ali [1,2,*], Muhammad Zaman [1], Muhammad Adnan Shahid [1,2,*], Hafsa Muzammal [1,2], Muhammad Zeeshan Khan [1], Muhammad Safdar [1,2] and Muhammad Danish Majeed [1,2]

1 Department of Irrigation & Drainage, University of Agriculture, Faisalabad 38000, Punjab, Pakistan; muhammad.zaman@uaf.edu.pk (M.Z.); hafzamuzammal990@gmail.com (H.M.); zeeshan896@gmail.com (M.Z.K.); safdarsani4340@gmail.com (M.S.); danish.majeed@uaf.edu.pk (M.D.M.)

2 Agricultural Remote Sensing Lab (ARSL), University of Agriculture, Faisalabad 38000, Punjab, Pakistan

* Correspondence: hirakhanam020@gmail.com (H.K.); sikandar_ali@uaf.edu.pk (S.A.); adnan.wmrc@gmail.com (M.A.S.)

† Presented at the 7th International Electronic Conference on Water Sciences, 15–30 March 2023; Available online: https://ecws-7.sciforum.net.

Abstract: Freshwater scarcity is a global concern and is caused by the overuse of water, exceeding the sustainable level. In Pakistan, overpopulation, climate change alleviation, and economic development influence the asset of water. For water management, development, and conservation, policymakers need to formulate the right plans and projects. For this purpose, they need to evaluate the impact of different projects and plans and evaluate considerable data and information. Different tools are being adopted for integrated water resource management. Among them, WEAP is user-friendly, reliable, and efficient, and is used widely across the globe. It will aid the policymakers to develop suitable projects and plans. This study was carried out by using WEAP, in the Lower Bari Doab Canal command area (LBDC), Punjab, Pakistan. This study sheds light on the current water allocation situation and simulated scenarios of population growth rates and economic growth to evaluate the future water demand situation. The results of the current account year (2015) show that the domestic water demand of Okara (101.51 MCM) and the agricultural water demand (1713 MCM) of Sahiwal was seen to be the maximum among the four districts. Unmet demand was found to be maximum for the districts laying in the tail end (Sahiwal and Khanewal). The results of the scenarios show that, if the current water consumption, population growth, and economic development continue, unmet demand will increase in the future, and if the population grows at a higher rate and economic development increases, it will result in higher unmet water demand in the coming years. WEAP was found to be a user-friendly and efficient model to better understand water demand.

Keywords: demand; WEAP; LBDC; unmet demand

Citation: Khanam, H.; Ali, S.; Zaman, M.; Shahid, M.A.; Muzammal, H.; Khan, M.Z.; Safdar, M.; Majeed, M.D. Integrated Water Resource Management Using Water Evaluation and Planning Model: A Case Study of Lower Bari Doab Canal, Pakistan. *Environ. Sci. Proc.* **2023**, *25*, 55. https://doi.org/10.3390/ECWS-7-14202

Academic Editor: Athanasios Loukas

Published: 14 March 2023

1. Introduction

The sources of water in Pakistan are either natural or artificial. Accessibility to water is declining globally due to environmental and climate changes, industrialization, contamination, and other anthropogenic activities. Around two billion people are living in areas that are under water stress, and they are not able to use freshwater resources due to the uneven distribution of freshwater [1]. The water crisis has been on the list of the top five world crises for the last 6 years and has been placed in fourth position for the year 2019 [2]. Water shortage is a globally escalating phenomenon that endangers the life of whole species of animals and plants.

The system is as follows for the distribution of water in agricultural lands of Pakistan. The water is supplied after 7 days according to the turn of each farmer [3]. This amount of water supplied to each farmer depends on its land holding, i.e., 25 minutes/acre/week. This system is purely supply-based and does not consider the water required by the crops grown within an area. This amount of water is distributed throughout the year, but the water demand varies within a year. Sometimes, more water is supplied than the crops demand and it becomes waste, and sometimes less water is supplied, which results in the inability to fulfill the demand of the crops grown. This ultimately results in increasing the groundwater pumping to fulfill the demand for the crops. Therefore, there is a dire need to develop a system that considers both demand and supply. For the past three decades, many national and international studies were conducted on the water resource management and development of Pakistan. In these studies, water is studied according to different aspects, including water security [4,5], water balance [6], water and food security (GOP, 2017), water sector strategy [7], food, water, and three energy nexuses [8], and these studies use different models, including the Indus Basin Model Revised, to relieve water scarcity.

The inadequate political governance and inappropriate policies also cause the mismanagement of the valuable resource of water. Conducting different studies and implementing different projects and policies requires considerable involvement from various departments; therefore, policymaking in Pakistan is not a facile task. The model used in this study is WEAP. The results obtained from WEAP will help to combat the problems caused by climate change, industrialization, and population growth to water resources and to re-design the water allocation policies by considering both demand and supply. It provides an all-inclusive, adaptable, and easy-to-use structure for planning and policy analysis. WEAP is used for calculating water balance, scenario generation, and policy analysis. It calculates the water balance based on crop water requirements. It also aids in reallocating water resources according to water availability. Different possible future scenarios were built to analyze the water resources. These types of models help to improve water distribution and to achieve efficient water resource management. This work thus aims (1) to investigate the current water allocation scenario in the Lower Bari Doab Canal command area and (2) to evaluate the water allocation plans under different possible future scenarios.

2. Materials and Methods

2.1. Description of Study Area

This study was carried out in the command area of the Lower Bari Doab Canal. This canal originates from Balloki Headworks, which lies on the left bank of the River Ravi beside the Balloki barrage as shown in Figure 1. The gross command area of LBDC is found between the 29°53′0″ to 31°13′6″ N latitudes and the 71°34′49″ to 73°51′56″ E longitudes. Most of the water reaching the LBDC is from River Jehlum and River Chenab via a link canal. The gross command area and culturable command area of LBDC are 0.728 M hac and 0.676 M hac, respectively. The length of the main canal is 124 miles, and it irrigates the districts of Kasur, Okara, Sahiwal, and Khanewal. It provides irrigation supplies to 65 nos. of taking channels and 3500 watercourse outlets. The irrigated area for Rabi is 1,579,003.594 acres, and the irrigated area for Kharif is 1,448,764.597 acres. The average temperature of the study area is 24.77 °C. Its designed head capacity is 9841 cusecs with a water allowance of 33.3 cusecs/1000 acres. The length of the main canal is 201 km.

2.2. Data Collection

The water evaluation and planning model used in this study is found to be very effective for calculating water demand and unmet demand. It is also used for evaluating and analyzing different plans and projects before implementation [9]. Data required to fulfill the objectives of the study and to obtain the results from the WEAP include climatic data (maximum temperature, minimum temperature, wind speed, precipitation, cloudiness fraction, average humidity, wind speed measurement height), crops grown in the area, existing cropping pattern, rotation of crops, the population in the area, domestic demand, and soil types data. Data on the flow in the Lower Bari Doab Canal in all the concerned years are also required. All the mentioned data are collected from various departments. The data along with their sources and format are given in Table 1 below.

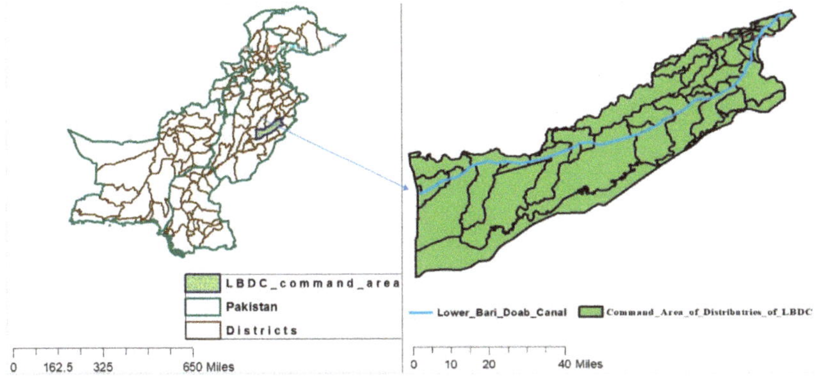

Figure 1. Location of Lower Bari Doab Canal command area.

Table 1. Data used and their sources and format for LBDC command area.

Data	Sources	Time
Climatic data	Pakistan Meteorological Dept., Monthly and global assembled and satellite datasets	Daily, Monthly
Flows of LBDC	Irrigation department, GOP	Daily
Land use data	MODIS	Yearly
Soil data	World Soil map.	Daily Discharge data
Population (District wise), Water consumption, Growth rate	Pakistan Bureau of Statistics, World Bank Data	Yearly record

2.2.1. Soil Data

Soil types within the study area were identified from the world soil map by using different commands in ArcGIS. These commands include the extraction of the study area soil map from the world soil map which was classified and studied to determine the soil types within the area as shown in Table 2. The soil map for the study area is shown in Figure 2 below.

Table 2. Soil type and code.

Sr. no.	SNUM	Soil Types
1	3736	Clay Loam
2	3878	Loam
3	3883	Clay Loam

2.2.2. Agricultural Land Utilization Data

These data were collected from the classification of the land use data of Pakistan in GIS. An arc map has been used to superposition this land use with canal command areas. The land use data for both Rabi and Kharif seasons in each of the districts were extracted from the land use data of Pakistan for both Rabi and Kharif seasons. The extracted land use data were studied in Excel and changed accordingly to be input into the WEAP. The land use data were input in the CSV format into the WEAP. The data holds information about land utilization in all of the four districts laying in the command area of the Lower Bari Doab Canal. These two growing seasons are seen in this study area, i.e., Rabi and Kharif.

The growing season of Rabi is from November to April and the growing season of Kharif is from May to October as shown in Figure 3.

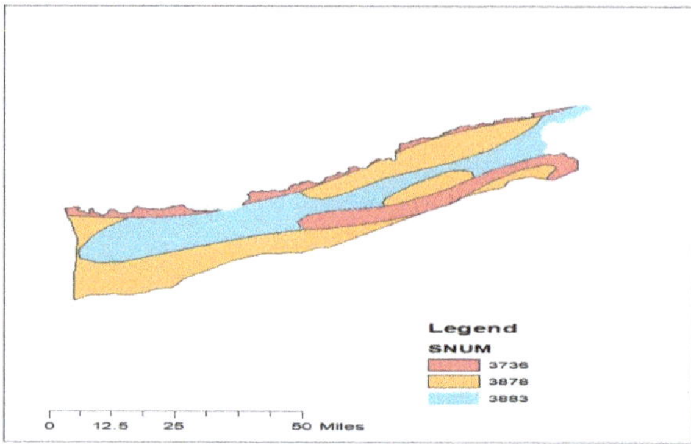

Figure 2. Soil map.

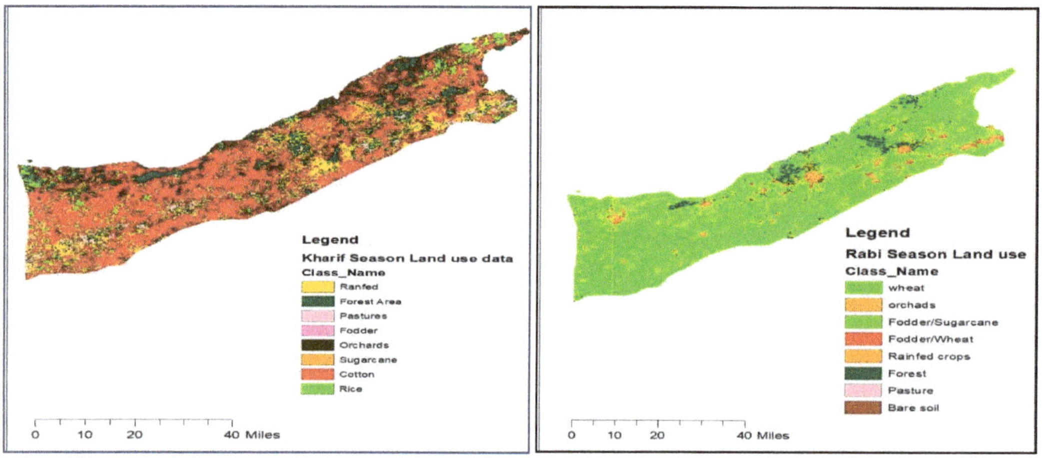

Figure 3. Rabi and Kharif season land use data.

3. Results and Discussion

3.1. Domestic Water Demand of Population within Districts for the Current Account Year (2015)

Domestic water demand includes water requirements for drinking, house chores, and personal cleanliness. The annual water demand of a person is taken as 33.5 m^3. The observed population of Kasur, Okara, Sahiwal, and Khanewal in the year 2015 was 792,045, 2,915,324, 2,414,994, and 730,928, respectively. The corresponding domestic water demand for 2015 for Kasur, Okara, Sahiwal, and Khanewal was found to be 31.21 MCM, 101.51 MCM, 80.6 MCM, and 24.45 MCM, respectively. The demand for Okara was seen to be the maximum demand among the districts as the population of Okara was observed to be 2,915,342, which is the maximum among the four districts. The lowest domestic water demand was found to be for the district of Khanewal due to it comprising the lowest population among these districts. However, the total domestic water demand within the command area was seen to be 238.82 MCM for the year 2015 as shown in Figure 4.

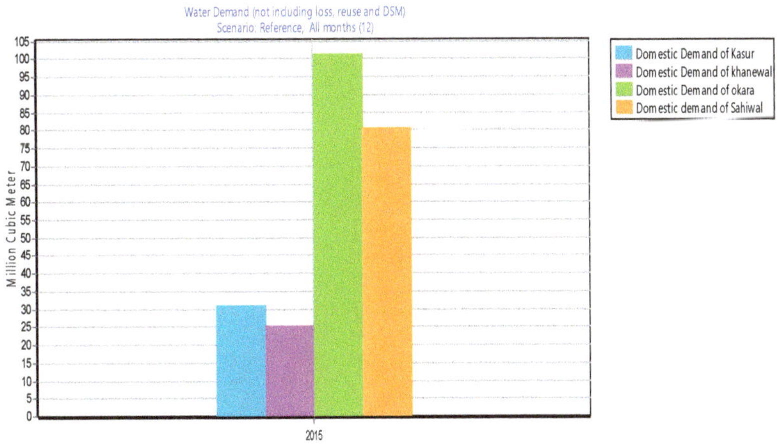

Figure 4. Water demand in MCM.

3.2. Water Demand for Agriculture for the Current Account Year (2015)

Water demand for each of the districts was determined and the water demand of Sahiwal was found to be the most at 1713 MCM as the land cover area in the Sahiwal was almost 742731 hectares, which is the largest land cover of crops among the districts. For ease of understanding, the land cover in the southern and northern buffers was also summed up in the Sahiwal. Moreover, the demand priority of Sahiwal was taken as 3, i.e., its demand will be fulfilled after the demand of Kasur and Okara. Water demand for the district of Kasur was the lowest and calculated as 237.517 MCM as the land use area of crops for Kasur was 17,251 hectares, which is also the lowest land use among the districts. The water demand of Okara was 530.3362, followed by the water demand of Khanewal which was approximated as 445.308 MCM because the land covered by agriculture for Okara was more than that for Khanewal as shown in Figure 5.

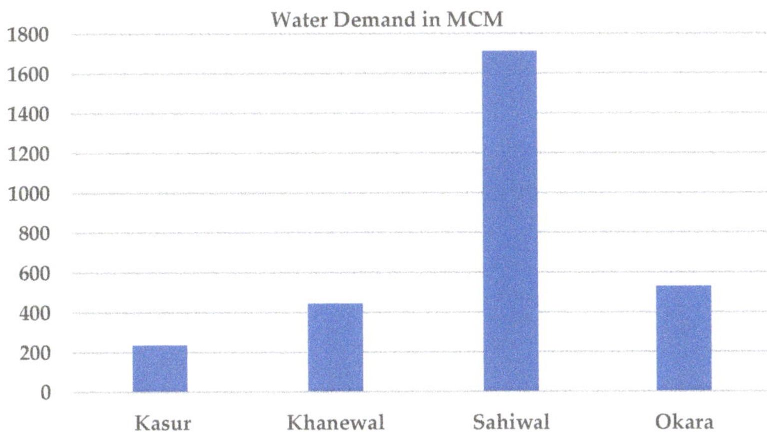

Figure 5. Water demand for Kasur, Okara, Sahiwal, and Khanewal.

3.3. Unmet Demand for Kasur, Okara, Sahiwal, and Khanewal for the Current Account Year (2015)

A demand priority of 1, 2, 3, and 4 were given to the agricultural demand site of Kasur, Okara, Sahiwal, and Khanewal, respectively. Demand priority explains the sequence of fulfilling the water demand of the demand sites. The unmet demand for the districts of

Kasur and Okara was almost zero because Kasur lies at the start of the LBDC and there is more rainfall within the district of Kasur. Similarly, Okara lies after Kasur and its demand were full before Sahiwal and Khanewal. The unmet demand for Sahiwal was 567.69 MCM and the unmet demand for Khanewal was approximated as 344.57 MCM. The unmet demand of Sahiwal was more because its land cover was more than that of Khanewal. This increased cropping pattern results in increased water demand and unmet demand for the district of Sahiwal as shown in Figure 6.

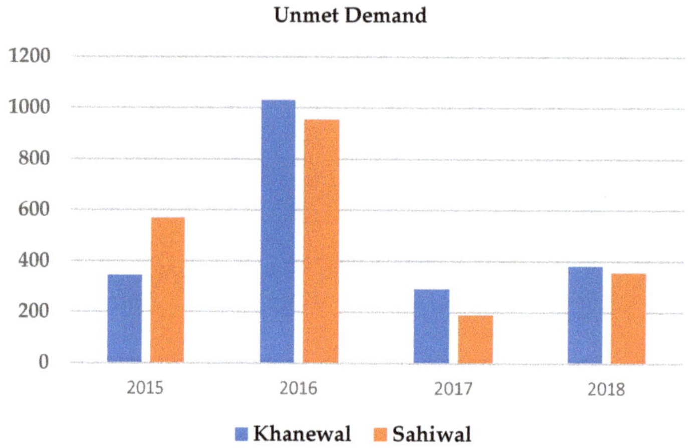

Figure 6. Unmet demand for Sahiwal and Khanewal.

3.3.1. Low Population Growth Rate

This scenario considers that the population is growing at a lower rate of 1.5% after 2020. Following this growth rate, the water demand will be lower for all of the considered years as compared to the higher population growth rate. For Kasur, water demand is approximated as 31.20 MCM in 2015, 34.53 MCM in 2020, 37.20 MCM in 2025, 40.044 MCM in 2030, and 46.511 in 2030. For Khanewal, the domestic water demand was estimated to be 25.449 MCM in 2015, 28.163 MCM in 2020, 32.684 MCM in 2030, and 37.931 MCM in 2040. The highest water demand was observed in Okara for all of the years up to 2040, as compared to all other districts. Okara's water demand sees a rise to 151.291 MCM in 2040. Sahiwal's water demand was observed to be 89.26 MCM in 2020 and 120.22 MCM in 2040 as shown in Figure 7.

3.3.2. Medium Population Growth Rate

The scenario of a medium population growth rate took a 2.03% growth rate per year. Following this growth rate, the water demand in Kasur was observed as 31.206 MCM in 2015, 34.53 MCM in 2020, 42.22 MCM in 2030, and 51.61 MCM in 2040. Water demand increases from 25.44 MCM in 2015 to 42.095 MCM in 2040. Water demand for Sahiwal was observed as 80.66 MCM in 2015, 89.260 in 2020, 109.128 in 2030, and 133.418 in 2040. The highest water demand was observed to be for the district of Okara for all of the years at 101.507 MCM in 2015, increasing to 112.32 MCM in 2020, 137.33 MCM in 2030, and its highest in 2040 for 167.9 MCM in. Similarly, ref [10] took a growth rate of 2.41% for Sindh and 2.13% for Punjab, according to which the water demand will aggravate from 177.8 BCM in 2017 to 192.7 BCM in 2040 as shown in Figure 8.

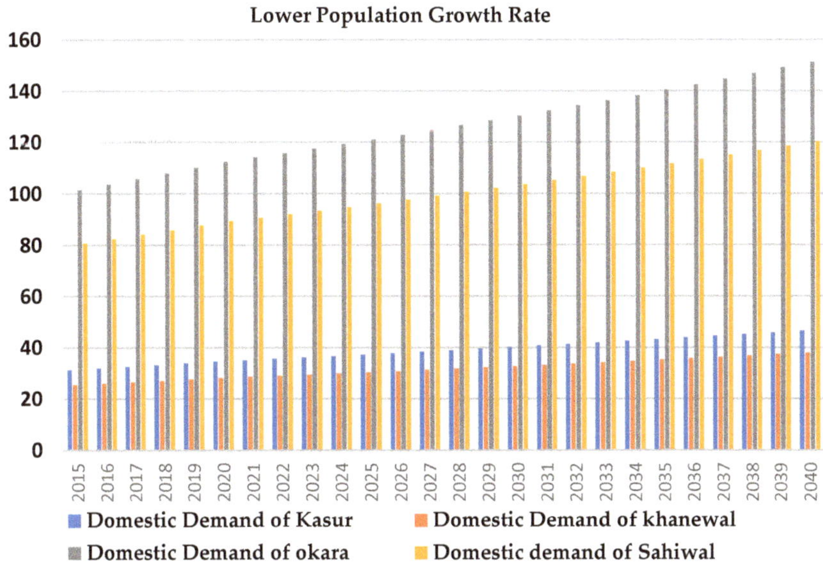

Figure 7. Low population growth.

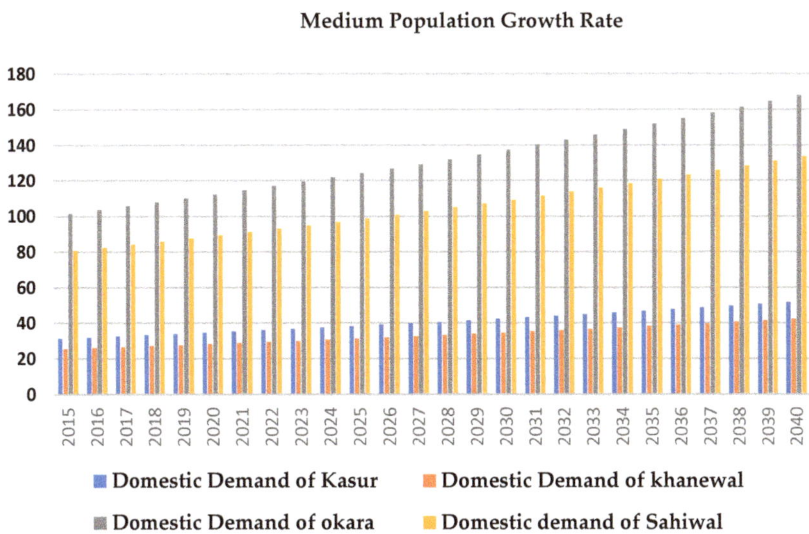

Figure 8. Medium population growth rate.

3.3.3. High Population Growth Rate

This scenario assumes that the high population growth alone results in more development and economic growth. The agricultural land is decreasing as there is more urbanization. Fields are being reduced day by day to make new settlements for a growing population. Economic growth results in an increase in water demand per person and year. This scenario considers the increase in water demand to be of 55 m^3/year. This increase in water use results in a rapid increase in water demand up to the year 2040. In the higher population growth rate scenario, domestic water demand increased to 56.581 MCM, 46.148 MCM, 184.0645 MCM, and 146.263 MCM in Kasur, Okara Sahiwal, and Khanewal, respectively, in 2040. However, this scenario considers the increase in the annual water use rate per person to be 55 m^3/year and the higher population growth rate, i.e., 2.5 %/year. Domestic water demand for Kasur district increased from 31.026 MCM in 2015 to 93.183 MCM in 2040, following the higher population growth rate and increase in annual water use by a person. For the district of Khanewal, the domestic water demand increased from 25.449 MCM to 54.808 in 2020, 62.01 MCM in 2025, 70.15 in 2030, 79.379 in 2035, and 89.810 in 2040 as shown in Figure 9.

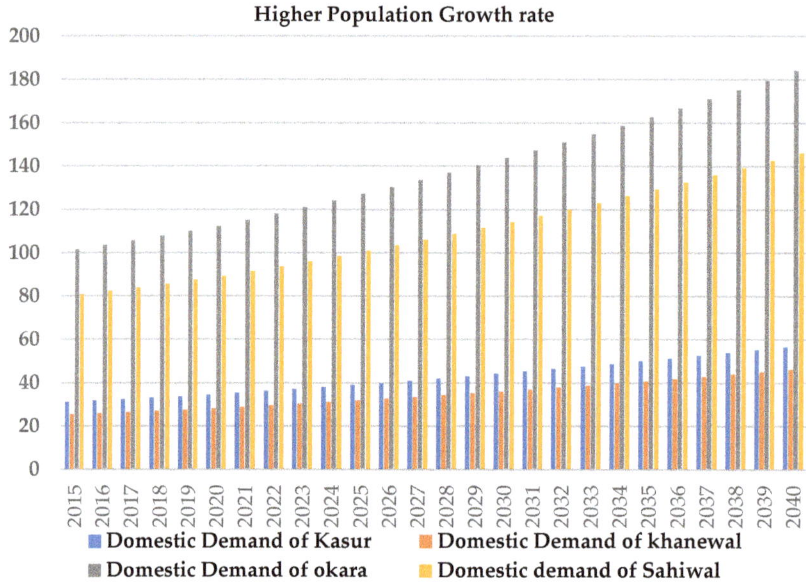

Figure 9. High population growth rate.

3.4. *Higher Living Standard Scenario*

The highest domestic water demand was seen to be for the district of Okara, for which the water demand was recorded as 101.507 MCM in 2015, 218.606 in 2020, 279.83 in 2030, and 358.21 MCM in 2040. For the district of Sahiwal, the water demand was estimated to be 80.66 MCM in 2015, which increased to 284.645 MCM in 2040. Ref. [11] also used WEAP to access the impact of higher population growth rate and urbanization on water demand in the Upper Indus Basin. The water demand increased in both situations. He also concluded that an increase in population and more urbanization results in an increase in water demand and ultimately higher unmet demand. The higher unmet demand will lead to an increase in groundwater abstraction, which causes the rapid deterioration of groundwater quantity and quality as shown in Figure 10.

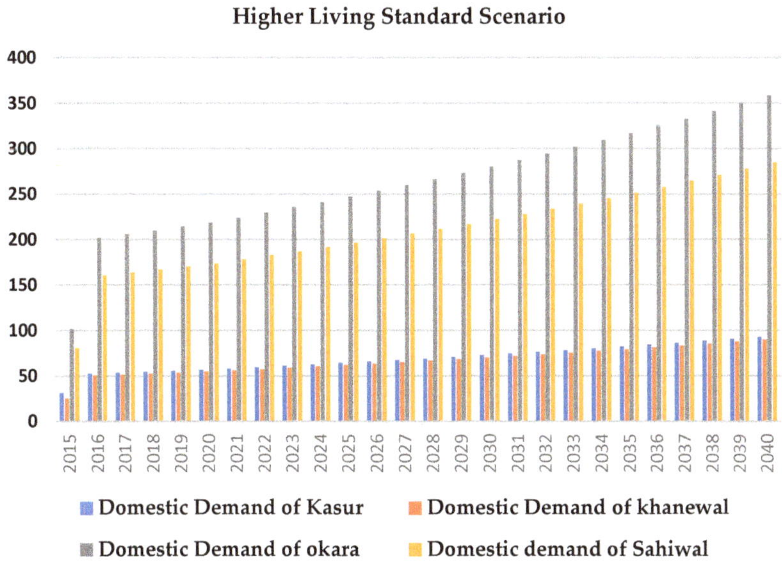

Figure 10. High living standards.

4. Conclusions and Recommendations

Water demand is increasing with increasing population, economic growth, urbanization, industrialization, and climate change. Water demand for the Sahiwal district was found to be at the maximum because it has the largest land use area for crops. Unmet demand was found to be the highest for the districts of Sahiwal and Khanewal. WEAP is user-friendly and can be used easily for the evaluation of different water management strategies and policies before implementation. The constraints faced include the unavailability of institutional support, sufficient and reliable information, and political conflicts.

To meet the unmet demand, protect the groundwater resource from deterioration, and mitigate the effects of climate change, the water should be supplied on a demand basis rather than a supply basis. The wastewater can be treated and then can be used for irrigation purposes. This can help to reduce the irrigation demand, changing the cropping pattern and the use of water-efficient crops in the districts of Sahiwal and Khanewal can reduce the water demand in these areas. Efficient irrigation methods such as sprinklers and drip irrigation can also help in reducing water demand. The domestic demand can be reduced by maintaining a low population growth rate and using water-efficient taps in the houses. Consumer education and awareness level may affect water use and practices, which could result in water demand reduction. WEAP is user-friendly and can be used for identifying the impact of different water management strategies and policies.

Author Contributions: Conceptualization, H.K. and S.A.; Introduction, M.Z. and M.A.S.; Materials and Methods, H.K. and M.Z.K.; Data Collection; H.M.; Results and Discussions, M.S. and M.D.M.; Conclusions and Recommendations, S.A. and M.A.S. writing original draft preparation H.K.; editing, M.S. All authors have read and agreed to the published version of the manuscript.

Funding: This research received no external funding.

Institutional Review Board Statement: Not applicable.

Informed Consent Statement: Not applicable.

Data Availability Statement: Not applicable.

Conflicts of Interest: The authors declare no conflict of interest.

References

1. Oki, T.; Kanae, S. Global hydrological cycles and world water resources. *Science* **2006**, *313*, 1068–1072. [CrossRef] [PubMed]
2. Anonymous. *Global Risks Report 2019*; World Economic Forum: Geneva, Switzerland, 2019; p. 114.
3. Hoff, H.; Bonzi, C.; Joyce, B.; Tielbörger, K. A water resources planning tool for the Jordan River basin. *Water* **2011**, *3*, 718–736. [CrossRef]
4. Ishaque, W.; Shaikh, S. Water and Energy Security for Pakistan; A retrospective analysis. *Grassroots* **2017**, *51*, 1–14.
5. Mahmood, Z.; Liepach, W. *A Productive, and Water-Secure Pakistan: Infrastructure, Institutions, Strategy, the Report of the Water Sector Task Force of the Friends of Democratic Pakistan*; Water Sector Task Force: Islamabad, Pakistan, 2012.
6. Hussain, I.; Hussain, Z.; Sial, M.; Akram, W.; Farhan, M. Water Balance, Supply and Demand and Irrigation Efficiency of Indus Basin. *Water* **2011**, *49*, 13–38.
7. GOP. *Pakistan Vision 2025: One Nationa—One Vision*; Ministry of Planning Development and Reform: Islamabad, Pakistan, 2017.
8. Rasul, G.; Sharma, B. The nexus approach to water–energy–food security: An option for adaptation to climate change. *Clim. Policy* **2016**, *16*, 682–702. [CrossRef]
9. Levite, H.; Sally, H.; Cour, J. Testing water demand management scenarios in a water-stressed basin in South Africa: Application of the WEAP model. *Phys. Chem. Earth* **2003**, *28*, 779–786. [CrossRef]
10. Hassan, D.; Rais, M.; Ahmed, W.; Bano, R.; Burian, S.; Ijaz, M.; Bhatti, F. Future water demand modeling using water evaluation and planning: A case study of the Indus Basin in Pakistan. *Sustain. Water Resour. Manag.* **2019**, *5*, 1903–1915. [CrossRef]
11. Amin, A.; Iqbal, J.; Asghar, A.; Ribbe, L. Analysis of current and future water demands in the Upper Indus Basin under IPCC climate and socio-economic scenarios using a hydroeconomic WEAP Model. *Water* **2018**, *10*, 537. [CrossRef]

environmental sciences proceedings

MDPI

Proceeding Paper

Economic Feasibility of Rainwater Harvesting in Houses in Blumenau, Brazil †

Anthony Midori Fugi, Jéssica Kuntz Maykot *, Enedir Ghisi and Liseane Padilha Thives

Department of Civil Engineering, Federal University of Santa Catarina, Florianópolis 88040-900, SC, Brazil; fugianthony@gmail.com (A.M.F.); enedir.ghisi@ufsc.br (E.G.); liseane.thives@ufsc.br (L.P.T.)

* Correspondence: jessica.maykot@posgrad.ufsc.br

† Presented at the 7th International Electronic Conference on Water Sciences, 15–30 March 2023; Available online: https://ecws-7.sciforum.net.

Abstract: This paper aims to analyse the economic feasibility of using rainwater for non-potable purposes in single-family houses in Blumenau, Brazil. A house was used as a case study to estimate the water end-uses and water consumption. Then, the daily water consumption and water end-uses for non-potable purposes were estimated. Different roof areas, number of residents, daily per capita water consumption and rainwater demand were also considered. The rainwater tank capacities and the potential for potable water savings were estimated using computer simulations. Finally, an economic feasibility analysis was carried out. The potential for potable water savings ranged from 18.76% to 58.06%, and the rainwater harvesting system was found to be economically feasible for most scenarios.

Keywords: rainwater; water end-uses; houses; economic feasibility; potable water savings

check for updates

Citation: Fugi, A.M.; Maykot, J.K.; Ghisi, E.; Thives, L.P. Economic Feasibility of Rainwater Harvesting in Houses in Blumenau, Brazil. *Environ. Sci. Proc.* **2023**, *25*, 56. https://doi.org/10.3390/ECWS-7-14163

Academic Editor: Luis Garrote

Published: 14 March 2023

1. Introduction

The storage and use of rainwater, while providing environmental benefits, can also be an investment to reduce potable water costs. The economic benefit of using rainwater has been addressed in several studies, varying the place of study, building and project type, among other characteristics. Ghisi and Schondermark [1] estimated the potential for potable water savings and performed an economic analysis for single-family homes in five cities in the state of Santa Catarina, Brazil. They obtained variable results depending on the water demand and found that, in most cases, the implementation of the system would be economically feasible.

Morales-Pinzón et al. [2] assessed the economic feasibility of a rainwater harvesting system in Spain. Several types of houses were chosen, covering most of the climates in the country. They observed that rainwater harvesting systems had shorter paybacks. In Italy, Liuzzo et al. [3] analysed the economic feasibility of a rainwater harvesting system in a house in Sicily, with a catchment area of 180 m^2. Rainwater usage was considered only to flush the toilet and for irrigation. The system proved to not always be feasible, with a payback period ranging from 15 to 55 years.

Such studies show that the economic feasibility analysis must be conducted on a case-by-case basis, as it depends especially on water demand, rainfall, water tariff, costs, and catchment area. Blumenau is one of the most populous cities in Santa Catarina; and 80% of the households are single-family houses [4]. These factors, added to urbanisation and, sometimes, heavy rains, make the city prone to flooding [5]. Thus, the main objective of this work is to evaluate the potential for potable water savings and to perform an economic analysis considering rainwater usage in single-family houses in Blumenau.

2. Materials and Methods

The study area was Blumenau, in Santa Catarina state, southern Brazil. A case study was performed in a three-storey single-family house with a roof area of 165 m^2 and four people living in the house. As it is a high-standard building, different scenarios of water end-uses were considered to represent the houses in Blumenau.

2.1. Water Consumption and End-Uses

In order to estimate the water consumption and end-uses, questionnaires were given to the four residents. The questionnaires were left close to each fixture, allowing residents to write the frequency and duration of use of each fixture. For a washing machine, the water level was recorded; for a bowl-and-tank toilet, only the number of flushes per day was recorded. The water consumption measured in the water meter was also registered at the end of each monitoring for comparison purposes. This monitoring was performed over seven days (25–31 August 2019). More details, such as flow rate measurements, are presented by Fugi [6]. Based on the frequency and duration of use of each fixture and the corresponding water flow rate, each water end-use and total water consumption were calculated.

2.2. Computer Simulations

The computer programme *Netuno*, version 4, is capable of performing simulations of rainwater harvesting systems [7]. In this study, the programme was used for sizing the rainwater storage tank, estimating the potential for potable water savings and performing the economic feasibility analysis of the three-storey house and the different scenarios.

Rainfall data for Blumenau were obtained from the Brazilian Water Agency [8]. A first flush equal to 2 mm was adopted as recommended in the Brazilian standard NBR 15527 [9]. Due to losses during rainwater harvesting, a runoff coefficient of 0.8 was adopted. The roof area of the house under study is approximately 165 m^2. For the different scenarios, roof areas equal to 60, 100, 140, and 180 m^2 were adopted. Such values were based on the frequency of areas of Brazilian roofs indicated by Ghisi et al. [10].

The number of residents per household has a major influence on water consumption. For the scenarios considered, 2, 3, 4 and 5 persons were adopted per house; this represents 84.3% of households in Blumenau [11].

The upper tank was sized based on the daily rainwater consumption in each house and scenario, and the sizes were chosen according to availability on the local market. For sizing the lower tank, the minimum and maximum capacities were defined as 500 litres and 20,000 litres, respectively. The programme indicated the capacity to be chosen when the increase in the potential for potable water savings was lower than or equal to 3.5%/m^3.

The total water demand was estimated based on the water consumption and number of residents in the house. Water consumptions equal to 100, 150 and 200 litres/person/day were adopted for the different scenarios. Finally, different rainwater demands were estimated based on the actual house's water end-uses and studies found in the literature: 30%, 40%, 50% and 60% of the total water demand were adopted. In the analysis for the actual house, the water end-use for non-drinking purposes was considered as the rainwater demand.

2.3. Economic Analysis

To perform the economic analysis, the costs of implementing the rainwater harvesting system, water consumption and system operation were obtained. Then, the financial savings regarding the rainwater harvesting system were calculated, i.e., the difference between the water bill with no rainwater harvesting system and that with a rainwater harvesting system. Finally, discounted payback, net present value and internal rate of return were calculated.

The costs of the water tanks and motor pumps were obtained from stores in Blumenau, and the lowest prices found were considered. In order to estimate labour costs, the Brazilian

System of Research on Costs and Indices of Civil Construction was used [12]. The costs of pipes, connections and accessories represented 19% of the total cost related to labour, water tanks and motor pumps [13].

In turn, the energy cost for the pump operation was estimated based on the energy tariff—which was BRL 0.46978 per kWh, according to the local electric utility [14]—and the power and operation of the motor pump. All cash flows from the investment project were brought to day zero, considering the minimum attractive rate of return (MARR). A positive net present value (NPV) indicates that the system is economically feasible. The discounted payback represents the time when savings from using rainwater are equal to the initial investment. The IRR must be higher than the MARR to make the investment feasible.

The minimum attractive rate of return adopted was 0.5% per month, and the analysis period was 20 years. Once it became impossible to predict the future monthly inflation, a constant figure of 0.274% per month was considered.

3. Results

3.1. Water Consumption and End-Uses

The average daily consumption obtained from the water meter was 612.9 litres/day. Based on this consumption and considering that there are 31 days in August, an average monthly consumption of 19.0 m^3 and an average per capita water consumption of 153.2 litres/person/day were estimated. Based on the results from the questionnaires, the consumption was estimated as 3678.9 litres over the period analysed, which is equivalent to 131.4 litres/person/day. Over August, the monthly consumption would be 16.3 m^3.

Water consumption measured by the company responsible for water supply in the city of Blumenau was 21 m^3 in August, when the study was conducted. This volume is 10.5% greater than the estimated consumption. From April to October 2019, consumption varied from 14 to 22 m^3, with a monthly average of 18 m^3 and equivalent to an average daily consumption per capita of 145.2 litres/person/day. Therefore, water consumption when the data were obtained was not atypical.

The difference between measured and estimated consumptions varied from 4.0 to 22.6%, with an average of 11.0%. Such differences may have occurred because of errors in the durations' records and/or because the residents failed to note some uses. One considered that the difference between the estimated and the measured consumptions was evenly distributed among the water fixtures. Thus, the water end-uses were based on questionnaires filled in by the residents. Table 1 shows the water end-uses for an actual house. The water end-uses are similar to those obtained by Freitas [15] and Meincheim [13] studies. The water fixtures with the highest consumptions in single-family homes were the shower, washing machine and toilet.

Table 1. Water end-uses for an actual house.

Water Fixture	Water Consumption	
	Litres	%
Showers	985.4	26.8
Washing machine	925.0	25.1
Toilets	777.7	21.1
Kitchen sink	355.9	9.7
Outdoor taps	253.4	6.9
Washing trough	224.1	6.1
Toilet sinks	103.5	2.8
Dishwasher	30.6	0.8
Drinking fountain	23.3	0.6
Total	3678.9	100.0

3.2. Rainwater Demand

The rainwater harvesting system was designed considering rainwater usage only for non-potable purposes (washing machines, toilets, outdoor taps and washing troughs). Therefore, based on the water end-uses of the actual house, the rainwater demand represents 59.3% of the total water demand. According to this result and the literature review, the different scenarios considered rainwater demands equal to 30%, 40%, 50% and 60% of the total water demand. For each scenario, a lower tank capacity was estimated, while the upper rainwater tank capacity was calculated based on the total rainwater demand.

3.3. Rainfall

In order to carry out the study, daily rainfall data from the Blumenau rain station from February 1989 to January 2019 were considered, and the average annual precipitation in this period was 1770 mm. The maximum, minimum and average monthly rainfall for Blumenau are shown in Figure 1. From September to March, monthly rainfall was higher than the average (147 mm). In November 2008, rainfall was 1001 mm, the year with the most significant flood over the last 17 years.

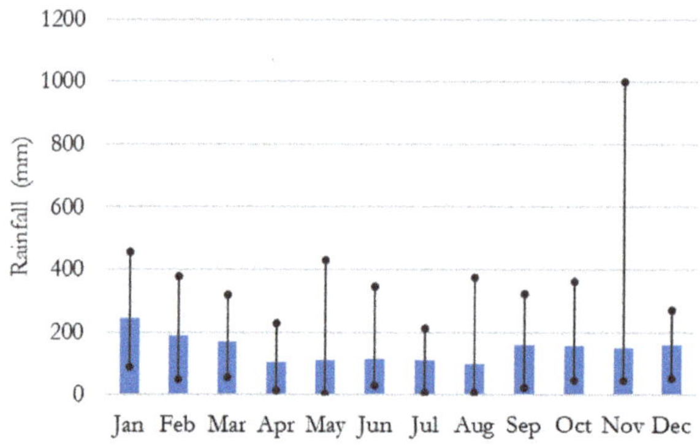

Figure 1. Maximum, minimum and average monthly rainfall for Blumenau over 30 years.

3.4. Potential for Potable Water Savings

Regarding an actual house, the upper rainwater tank capacity was estimated at 363.4 litres. In this way, an upper 500-litre tank was adopted. The ideal capacity for the lower tank was 5000 litres, and the corresponding potential for potable water savings was 50.32%. The house's water consumption was 612.8 litres/day, so that the rainwater system would provide 308.4 litres of rainwater per day. The average monthly consumption of 18 m^3 decreased to 9 m^3. This way, the owners paid only the minimum monthly consumption fee, which is 10 m^3.

Considering the different scenarios simulated, for a potable water demand of 100 litres/person/day or more and equal number of residents, the results were similar. The different roof areas showed little influence on the potential for potable water savings. The rainwater collected from the roof shows that the roof area meets the rainwater demand, so there is no need for a large roof area when the rainwater demand is low. However, as the rainwater demand increases, the roof area significantly influences the potential for potable water savings. The larger the roof area, the smaller the lower rainwater tank capacity. This occurs because the larger the roof area, the more rainwater is harvested, and the replenishment of rainwater in the tank is faster. Similar results were obtained in [1] and [15].

The potential for potable water savings ranged from 18.76% to 58.06%, with an average of 37.90%. As in the study of Lopes et al. [16], it was observed that the larger the rainwater demand and roof area, the greater the potential for potable water savings.

3.5. Economic Analysis

The financial analysis of the implementation and operation of a rainwater harvesting system for the house resulted in the following indices: a net present value of BRL 4814.54, a payback period of 89 months and an internal rate of return of 1.44% per month.

From the 192 different scenarios analysed, 112 scenarios obtained positive net present values, indicating that the rainwater system would be economically feasible for 58.3% of the cases. Payback ranged from 221 to 60 months for economically feasible scenarios. The highest internal rate of return was 2.05% per month.

The scenarios with low water consumption proved to be economically unfeasible. Such infeasibility is due to the flat rate for monthly consumption of up to 10 m^3 of water. Once there is no charge reduction in the water bill but there is still an expenditure of energy for the operation of the pump, the net present value becomes higher than the initial cost. These results were also found by Berwanger and Ghisi [17].

The feasibility analysis showed that the greater the water consumption and the greater the rainwater demand, the more economically feasible the rainwater harvesting system. Figure 2a shows the number of scenarios in which the NPV was positive or negative as a function of the water demand. For water demand equal to 150 and 200 litres/person/day, the NPV was positive for 75% of the cases. For consumptions of 100 litres/person/day, only 25% of the cases had positive NPVs. Figure 2b, in turn, shows the number of scenarios in which the NPV was positive or negative as a function of the roof area. One observes that the scenario number does not vary as a function of the roof. Thus, the roof area showed no influence on the economic feasibility of rainwater harvesting systems.

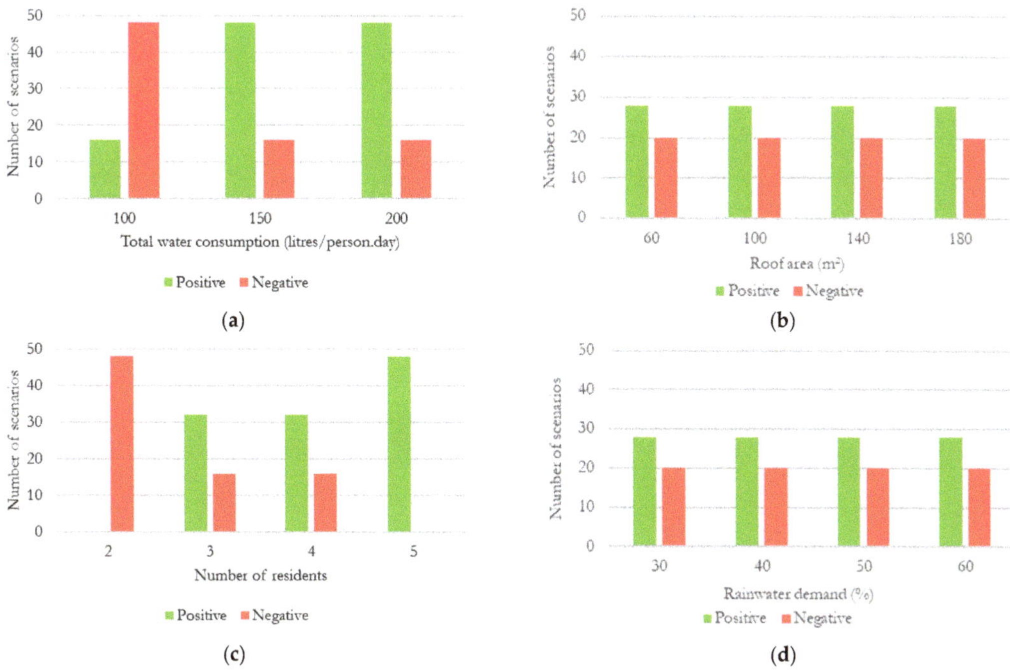

Figure 2. Number of scenarios in which the NPV was positive or negative as a function of the water demand (**a**), roof area (**b**), number of residents (**c**), and rainwater demand (**d**).

The greater the number of residents, the more positive NPVs were obtained. This influence is directly related to water consumption. Figure 2c shows the number of scenarios in which the NPV was positive or negative as a function of the number of residents. For scenarios with two residents, all NPVs were negative; however, for five residents, all NPVs were positive. In 66.6% of scenarios, the NPV was positive for scenarios with either three or four residents. The NPVs were shown to be equally distributed for each rainwater demand. For all demands, the NPV was positive for 58.3% of the scenarios. Figure 2d shows the number of scenarios in which the NPV was positive or negative as a function of the rainwater demand.

As in Ghisi and Schondermark [1]'s study, economic feasibility is directly related to the number of residents and water consumption per capita. Homes with a low number of residents and/or low water consumption should use the rainwater harvesting system only for environmental benefits, not economic ones. Figure 3 shows the NPV as a function of the rainwater demand (in litres/day) for all scenarios. For houses with rainwater demand equal to 60–120 litres/day, all scenarios proved to be economically unfeasible. In cases where the rainwater demand was greater than or equal to 250 litres/day, all scenarios proved to be economically feasible. For cases in which the rainwater demand ranged from 135 to 240 litres/day, it was found that economic feasibility does not have a trend. The absence of a tendency in such cases may occur because high water consumptions and low rainwater demands result in the same rainwater demand as a scenario with low water consumption and high rainwater demand, requiring analysis on a case-by-case basis.

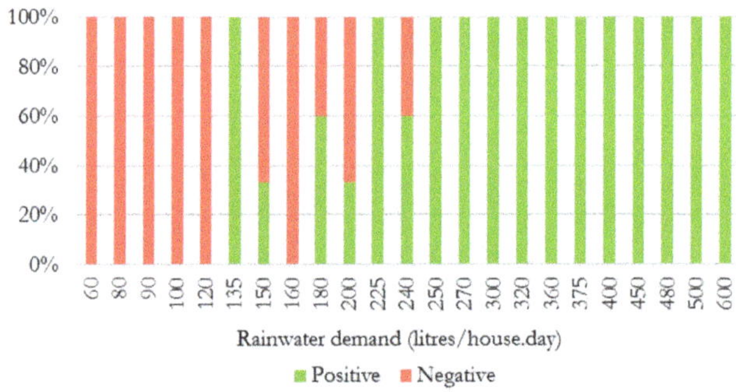

Figure 3. NPV as a function of the rainwater demand for all scenarios.

The actual house obtained better economic rates than the scenarios with 140 and 180 m^2 of roof area and water demand equal to 150 litres/person/day and for the four residents and rainwater demand equal to 60% of the water demand. The comparison was made with these two scenarios, as they have the most similar characteristics to the house. For the scenario with a roof area of 140 m^2, a payback period of 96 months was obtained, and for the 180 m^2, a payback period of 93 months was obtained. The payback period for the actual house was 89 months, indicating better economic feasibility.

4. Conclusions

This study showed that the higher the water consumption and the higher the rainwater demand, the greater the potential for potable water savings. The potential for potable water savings increases as the roof area and the rainwater demand increase. In houses with low water consumption, the roof area had little influence on the sizing of the lower rainwater tank. In contrast, for higher consumptions, the tank capacity increased with increases in the roof area. In houses with high water consumption, the rainwater harvesting system proved to be economically feasible. In cases with high rainwater demand and small roof areas, the potential for potable water savings was low, but they were still economically feasible.

The rainwater harvesting system was not economically feasible for a low number of residents and/or low water consumption cases. Therefore, implementing a rainwater harvesting system for single-family homes in Blumenau is economically feasible for most cases, including the actual house. However, performing the economic feasibility analysis for each case is recommended.

Author Contributions: A.M.F., conceptualization, methodology, formal analysis, data curation, writing—original draft preparation, review and editing; J.K.M., writing—original draft preparation, review and editing; E.G., supervision, conceptualization, methodology, formal analysis, writing—review and editing; L.P.T., supervision; writing—review and editing. All authors have read and agreed to the published version of the manuscript.

Funding: This research received no external funding.

Institutional Review Board Statement: Not applicable.

Informed Consent Statement: Not applicable.

Data Availability Statement: Not applicable.

Conflicts of Interest: The authors declare no conflict of interest.

References

1. Ghisi, E.; Schondermark, P.N. Investment Feasibility Analysis of Rainwater Use in Residences. *Water Resour. Manag.* **2013**, *27*, 2555–2576. [CrossRef]
2. Morales-Pinzón, T.; Lurueña, R.; Rieradevall, J.; Gasol, C.M.; Gabarrell, X. Financial Feasibility and Environmental Analysis of Potential Rainwater Harvesting Systems: A Case Study in Spain. *Resour. Conserv. Recycl.* **2012**, *69*, 130–140. [CrossRef]
3. Liuzzo, L.; Notaro, V.; Freni, G. A Reliability Analysis of a Rainfall Harvesting System in Southern Italy. *Water* **2016**, *8*, 18. [CrossRef]
4. IBGE—Instituto. Brasileiro de Geografia e Estatística CENSO: Blumenau. Available online: https://cidades.ibge.gov.br/brasil/sc/blumenau/pesquisa/23/24304?detalhes=true (accessed on 18 January 2023).
5. ALERTABLU—Sistema de Monitoramento e Alerta de Eventos Extremos de Blumenau Enchentes Registradas. Available online: http://alertablu.cob.sc.gov.br/p/enchentes (accessed on 21 April 2019).
6. Fugi, A. Análise de Viabilidade Econômica de Sistemas de Aproveitamento de Água Pluvial Em Residências Unifamiliares Em Blumenau. Bachelor's Thesis, Universidade Federal de Santa Catarina, Florianópolis, Brazil, 2019.
7. Ghisi, E.; Cordova, M.M. Netuno 4. Universidade Federal de Santa Catarina, Departamento de Engenharia Civil. 2014. Available online: http://www.labeee.ufsc.br/downloads/softwares/netuno (accessed on 5 January 2019).
8. ANA—Agência Nacional de Águas. Quantidade de Água. Available online: http://www3.ana.gov.br/portal/ANA/panorama-das-aguas/quantidade-da-agua (accessed on 8 March 2019).
9. ABNT. *Associação Brasileira de Normas Técnicas NBR 15527—Água de Chuva—Aproveitamento de Coberturas Em Áreas Urbanas Para Fins Não Potáveis*; ABNT: Rio de Janeiro, Brazil, 2019.
10. Ghisi, E.; Gosch, S.; Lamberts, R. Electricity End-Uses in the Residential Sector of Brazil. *Energy Policy* **2007**, *35*, 4107–4120. [CrossRef]
11. IBGE—Instituto, Brasileiro de Geografia e Estatística Panorama: Blumenau. Available online: https://cidades.ibge.gov.br/brasil/sc/blumenau/panorama (accessed on 23 September 2019).
12. *Sistema Nacional de Pesquisa de Custos e Índices da Construção Civil Preço de Insumos*; SINAPI: Brasília, Brazil, 2019.
13. Meincheim, D.L. Potencial de Economia de Água Potável Por Meio Do Aproveitamento de Água Pluvial Em Uma Residência Unifamiliar Localizada Em São José—SC. Bachelor's Thesis, Universidade Federal de Santa Catarina, Florianópolis, Brazil, 2015.
14. CELESC—Centrais. Elétricas de Santa Catarina Tarifas e Taxas de Energia. Available online: https://www.celesc.com.br/tarifas-de-energia#tarifas-vigentes (accessed on 7 October 2019).
15. Freitas, D.A. Análise de Viabilidade Econômica Da Implantação de Sistemas de Aproveitamento de Água Pluvial Em Residências Unifamiliares No Município de Imbituba—SC. Bachelor's Thesis, Universidade Federal de Santa Catarina, Florianópolis, Brazil, 2018.
16. Lopes, A.C.; Rupp, R.F.; Ghisi, E. Assessment of the Potential for Potable Water Savings by Using Rainwater in Houses in Southern Brazil. *Water Sci. Technol. Water Supply* **2016**, *16*, 533–541. [CrossRef]
17. Berwanger, H.; Ghisi, E. Investment Feasibility Analysis of Rainwater Harvesting in the City of Itapiranga, Brazil. *Int. J. Sustain. Hum. Dev.* **2014**, *2*, 104–114.

environmental sciences proceedings

MDPI

Proceeding Paper

Spatial-Temporal Mapping and Delineating of Agulu Lake Using Remote Sensing and Geographic Information Science for Sustainable Development †

Mfoniso Asuquo Enoh [1,*], Chukwudi Andy Okereke [2] and Needam Yiinu Narinua [3]

1 Department of Geoinformatics and Surveying, University of Nigeria—Enugu Campus, Enugu 400102, Nigeria
2 Department of Surveying and Geoinformatics, Imo State University, Owerri 460222, Nigeria
3 Department of Surveying and Geoinformatics, Ken Polytechnic Bori, Bori 502101, Nigeria
* Correspondence: enohmfoniso@yahoo.com
† Presented at the 7th International Electronic Conference on Water Sciences, 15–30 March 2023; Available online: https://ecws-7.sciforum.net.

Abstract: Water is a crucial component of ecosystems and a critical resource that cannot be replaced for social progress or human life. In this study, Agulu Lake, an inland water body located in Anambra, southeast Nigeria, was mapped, classified, and delineated with remotely sensed data so as to monitor the spatial-temporal changes that occurred in the lake's surface water every 15 years, in 1985, 2000, and 2015, in order to achieve sustainable development by 2030. The Sustainable Development Goals (SDGs) of the United Nations emphasize the need to manage the marine environment. Some of the goals of the SDGs have some connection to open surface water, but goal 6a and indicator 6.6.1 are significant to this study. The study adopted Landsat 5 TM (1985), ETM+ (2000), Landsat 8 OLI (2015), ArcGIS 10.5 software, and the maximum likelihood classifier to create various classification maps. The Google Earth image (2015) was also used to show the general overview of Agulu Lake and its environs. The findings demonstrate that during the study period, the land surface class grew while the water surface class (Agulu Lake) shrank.

Keywords: Agulu Lake; GIS; land surface; remote sensing; sustainable development

check for updates

Citation: Enoh, M.A.; Okereke, C.A.; Narinua, N.Y. Spatial-Temporal Mapping and Delineating of Agulu Lake Using Remote Sensing and Geographic Information Science for Sustainable Development. *Environ. Sci. Proc.* **2023**, *25*, 57. https://doi.org/10.3390/ECWS-7-14259

Academic Editor: Athanasios Loukas

Published: 16 March 2023

1. Introduction

Lake water is a crucial component of ecosystems and a critical resource that cannot be replaced for social progress or human life [1]. It supports temperature variation, the cycling of carbon, and other ecological processes. A lake is an important component of infrastructure that promotes the growth of business, trade, agriculture, and transportation, while also offering essential services for human survival [2]. A lake can develop as a widening of water along a river's course, as a network of connected lakes, or as an isolated lake [3]. A headwater lake that receives no input from a single river is maintained by inflow from multiple small tributary streams, direct surface precipitation, and groundwater influx. The study of open surface water, such as lakes, as well as ponds and other freshwater bodies, is known as limnology [4,5].

Lakes are quite promising in all respects, so it is important to monitor their spatial-temporal changes. Remote sensing has been embraced as a tool for monitoring surface water and is preferred over the traditional methods [3,5]. Landsat imagery is a great resource for tracking spatial-temporal changes in surface water as a result of its wide coverage and availability. The lake as an important ecosystem has been recognized by the United Nations. In 2015, the UN's member states put forth the 2030 agenda [6]. These goals have 17 SDGs, whose objective is to solve global problems by 2030. SDG 6 is one of the 17 goals established by the UN. The UN's SDG 6 has six goals with indicators that focus on access to hygienic and clean water [7]. Several studies have been conducted on open surface water. Researchers such as Rokni et al. (2014) [8], Miles et al. (2017) [9], and Sichangi and Makokha (2017) [10] used remotely sensed data to delineate lake water. Their results showed that their areas (the lakes) changed over a period of time. Remote sensing data were used in this study to examine the spatial-temporal changes to Agulu Lake from 1985 to 2015 in order to achieve sustainability by 2030. This study bridges the gap between multiple studies as it relates surface lake water to the UN's SDG. Agulu Lake, an inland body of water located in Agulu, Anambra State, has been experiencing problems of erosion, deforestation, and soil sterility. A visit to the area showed that Agulu Lake was fast depreciating as a result of numerous anthropogenic activities.

2. Description of the Study Area

The study area, "Agulu Lake", is located in Agulu, in the Anaocha Local Government Area of Anambra State, Southeastern, Nigeria. Agulu Lake is a naturally occurring inland body of water with a significant cultural landmark that is slowly being destroyed by flooding, erosion, and landslides [11]. Agulu Lake is the largest lacustrine environment in Anambra State. It is found between latitudes 6 07′ N and 6 09′ N and longitudes 7 01′ E and 7 03′ E of the Greenwich meridian (Figure 1). The study area has a catchment area of 32 km^2 and a depth of 11.2 m at its deepest point, with a mean depth of 5.2 m. The rainfall varies from 1383 mm to 2090 mm per year, with a mean rainfall value of 1851.9 mm. The average temperature is as high as 32.1 °C and as low as 23.5 °C [12].

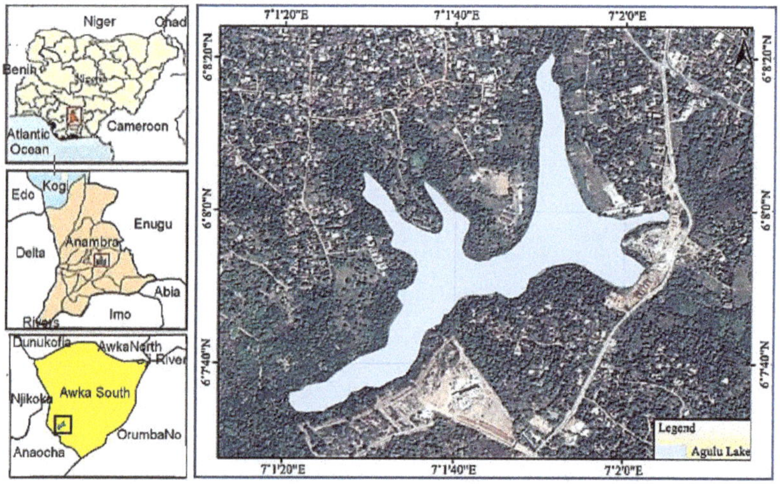

Figure 1. The study area.

3. Materials and Methods

The study used Landsat satellite data for three epochs to monitor the spatial-temporal changes in the lake's water. The three Landsat data sets, Landsat 5 TM (thematic mapper) for 1985, Landsat 7 ETM+ (enhanced thematic mapper plus) for 2000, and Landsat 8 OLI (operational land imager) for 2015, with a spatial resolution of 30 m, were chosen because of their quality and accessibility. These datasets had zero cloud cover and were accessed with a path and row of 188 and 56 from the Global Land Cover Facility (GLCF). To minimize the impact of seasonal variations, the satellite images were collected during the same months. ArcGIS 10.5 and ERDAS Imagine 10.5 were the software used for the classification and change detection processes. False color composite (FCC) analysis, image subsetting, and enhancement were some of the procedures adopted. The Landsat data were aligned with the zone 32N coordinate system of the Universal Transverse Mercator (UTM) using the World Geodetic System (WGS) 1984 ellipsoid. Data layering was carried out via the layer stacking function of the ERDAS Imagine software by combining the image spectral bands in the RGB format. The maximum likelihood classification served as the foundation for the supervised classification approaches carried out with the ERDAS Imagine (2015) program. In this study, land and water bodies were the chosen classification categories. Change detection is an important aspect of the classification process. Here, ArcGIS 10.5 software was used to analyze the change detection process by adopting the postclassification comparison (PCC) method [13]. PCC is widely used by researchers in spatial-temporal change analysis.

4. Results and Discussion

Table 1 provides an overview of the study years' aerial distribution and their corresponding proportions in percentage during the 1985, 2000, and 2015 study years. According to the study, the land class increased from 6,353,100 m^2 (87.56%) in 1985 to 6,358,100 m^2 (87.61%) in 2000 to 6,366,600 m^2 (87.72%) in 2015. The water class sank from 904,500 m^2 (12.46%) in 1985 to 899,500 m^2 (12.46%) in 2000 and to 891,000 m^2 (12.28%) in 2015. Table 2 depicts the spatial-temporal changes (change detection) in the study classes from 1985 to 2000 and from 2000 to 2015. These changes in the study classes may be a result of the various anthropogenic activities taking place in the study area. While Figure 2 demonstrates the study's classification output maps, Figure 3 shows the change detection maps between 1985 and 2000 and 2000 and 2015.

Table 1. Area extent of the study classes.

Class Cover	1985		2000		2015	
	Area (m^2)	**Area (%)**	**Area (m^2)**	**Area (%)**	**Area (m^2)**	**Area (%)**
Land	6,353,100	87.56	6,358,100	87.61	6,366,600	87.72
River	904,500	12.46	899,500	12.39	891,000	12.28
Total	7,257,600	100	7,257,600	100	7,257,600	100

Table 2. Change detection from 1985 to 2000 to 2015.

		1985–2000					2000–2015		
S/N	**1985**	**2000**	**Area (m^2)**	**Area (%)**	**2000**	**2015**	**Area (m^2)**	**Area (%)**	**Change Detection**
1	River	River	544,500	7.50	River	River	448,200	6.18	River constant
2	River	Land	350,000	4.95	River	Land	456,300	6.29	River decreased
3	Land	River	370,000	4.96	Land	River	442,800	6.10	River increased
4	Land	Land	5,993,100	82.58	Land	Land	5,910,300	81.44	Land constant
Total			**7,257,600**	**100.00**			**7,257,600**	**100.00**	

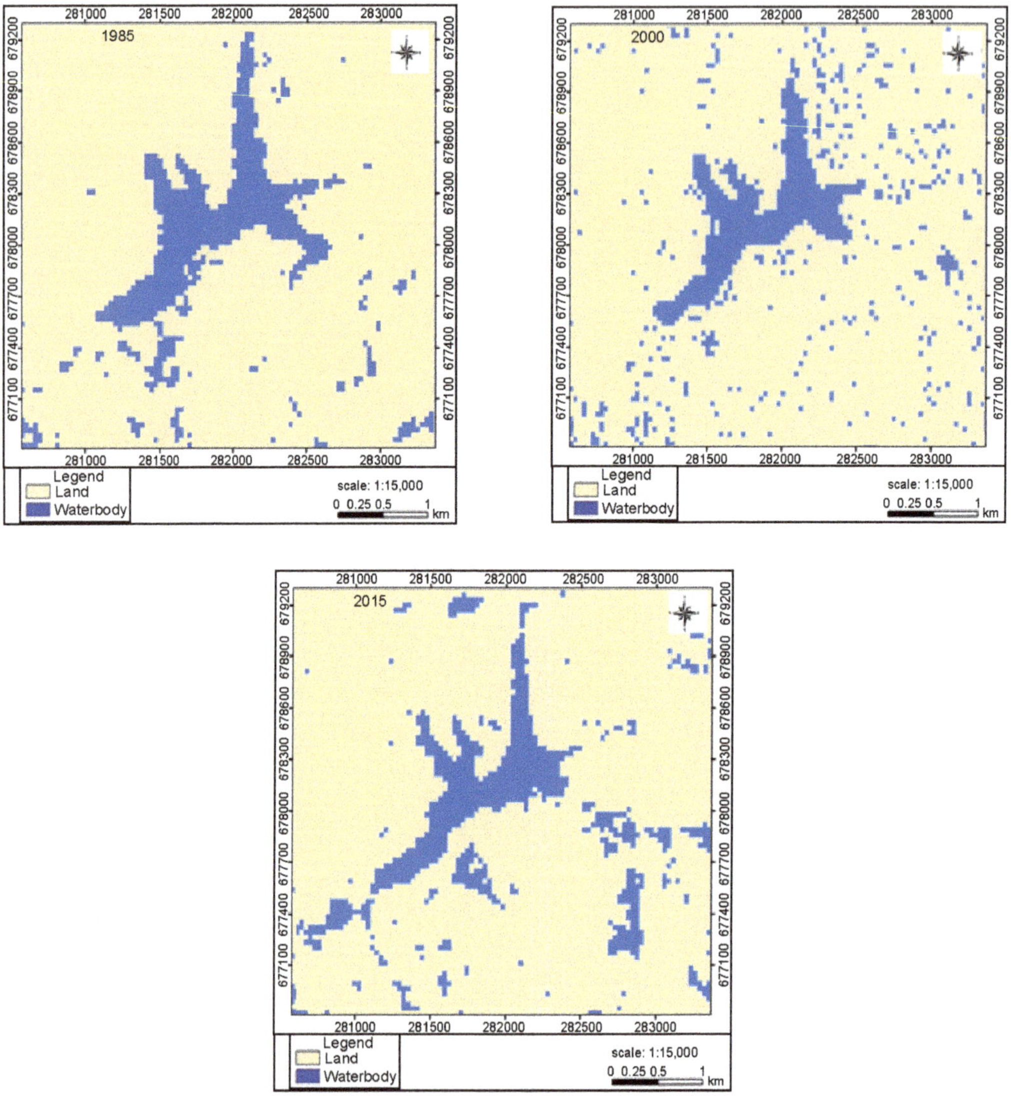

Figure 2. Classification maps.

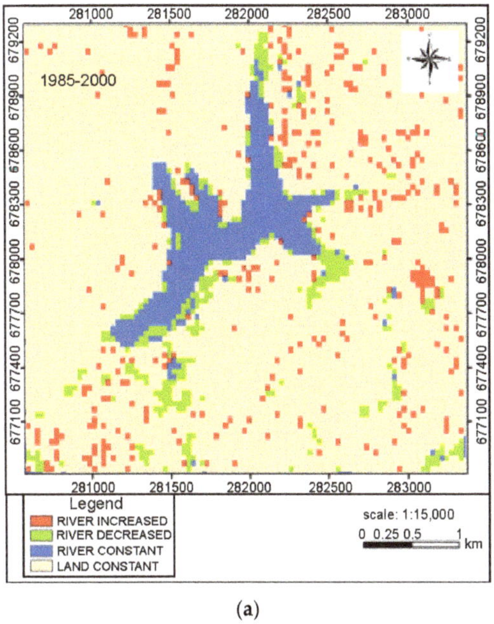

Figure 3. Change detection dynamics maps. (**a**) Change detection between 1985 and 2000. (**b**) Change detection between 2000 and 2015.

5. Discussion

Lakes are essential to achieving the SDGs because they offer solutions to a variety of global problems. The majority of the SDGs are linked to surface water, but this study concentrated on goal 6, target 6a, and indicator 6.6.1. Target 6a emphasizes the involvement of the local community in improving the availability of water and sewage management. Indicator 6.6.1 highlights the extent to which the water-related ecosystem can change over time. In the study, we see that Agulu Lake has been shrinking during the 1985–2000–2015 study years and is anticipated to shrink further by 2030. If we are set to achieve UN SDG 6 by 2030, then there is a need to manage, maintain, and rehabilitate deteriorating lakes and other open water bodies.

6. Conclusions and Recommendations

This study demonstrated the ability to capture spatial-temporal data using GIS and remote sensing every 15 years, from 1985 to 2000 and from 2015 to 2030. Remote sensing technology is widely used for monitoring and mapping water bodies as a result of its availability and wide coverage. The Sustainable Development Goals (SDGs) of the United Nations emphasize the need to manage the marine environment. Altogether, there are 17 SDGs in total, but goals 6a and 6.6.1 are crucial to this study. The study's findings show that Agulu Lake has been shrinking and is anticipated to shrink further by 2030. This shrinkage in the study area (Agulu Lake) may be a result of the numerous anthropogenic activities taking place in the study area's periphery. To accomplish SDG 6 by 2030, it is recommended that sustainable practices be mandated. Due to this, the following suggestions are offered based on the study's findings: An interim master plan should be created to prevent shrinkage, and the local authority should publish a stop notice to all types of development within and around the study area. In conclusion, open surface water should be monitored frequently with remote sensing technology.

Author Contributions: Conceptualization, M.A.E. and C.A.O.; methodology, M.A.E.; software, M.A.E.; validation, M.A.E., C.A.O. and N.Y.N.; formal analysis, M.A.E. and N.Y.N.; investigation, C.A.O.; writing original draft preparation, M.A.E.; writing-review and editing supervision, M.A.E., C.A.O. and N.Y.N.; project administration, M.A.E., funding acquisition, C.A.O. and N.Y.N. All authors have read and agreed to the published version of the manuscript.

Funding: This study had received no external funding.

Institutional Review Board Statement: Not applicable.

Informed Consent Statement: Not applicable.

Data Availability Statement: Not applicable.

Conflicts of Interest: The authors declare that no conflict of interest exists.

References

1. Crétaux, J.-F.; Abarca-Del-Río, R.; Bergé-Nguyen, M.; Arsen, A.; Drolon, V.; Clos, G.; Maisongrande, P. Lake Volume Monitoring from Space. *Surv. Geophys.* **2016**, *37*, 269–305. [CrossRef]
2. Gleeson, T.; Wada, Y.; Bierkens, M.F.P.; Van Beek, L.P.H. Water balance of global aquifers revealed by groundwater footprint. *Nature* **2012**, *488*, 197–200. [CrossRef] [PubMed]
3. Messager, M.L.; Lehner, B.; Grill, G.; Nedeva, I.; Schmitt, O. Estimating the volume and age of water stored in global lakes using a geo-statistical approach. *Nat. Commun.* **2016**, *7*, 13603. [CrossRef] [PubMed]
4. Omondi, R.; Kembenya, E.; Nyamweya, C.; Ouma, H.; Machua, S.; Ogari, Z. Recent limnological changes and their implication on fisheries in Lake Baringo, Kenya. *J. Ecol. Nat. Environ.* **2014**, *6*, 154–163. [CrossRef]
5. Ma, L.; Wu, J.; Liu, W.; Abuduwaili, J. Distinguishing between Anthropogenic and climatic impacts on Lake Size: A modelling approach using data from Ebinur Lake in Arid Northwest China. *J. Limnol.* **2014**, *73*, 350–357. [CrossRef]
6. United Nation. *Sustainable Development Goal 6, Synthesis Report on Water and Sanitation;* United Nations: New York, NY, USA, 2018; ISBN 9789211013702. OCLC 1107804829.
7. United Nations & Nations. Transforming our world: The 2030 agenda for Sustainable Development. In Proceedings of the General Assembly 70 Session, New York, NY, USA, 25–27 September 2015. [CrossRef]
8. Rokni, K.; Ahmad, A.; Selamat, A.; Hazini, S. Water Feature Extraction and Change Detection Using Multitemporal Landsat Imagery. *Remote Sens.* **2014**, *6*, 4173–4189. [CrossRef]
9. Miles, K.E.; Willis, I.C.; Benedek, C.L.; Williamson, A.G.; Tedesco, M. Toward Monitoring Surface and Subsurface Lakes on the Greenland Ice Sheet Using Sentinel-1 SAR and Landsat-8 OLI Imagery. *Front. Earth Sci.* **2017**, *5*, 58. [CrossRef]
10. Sichangi, A.W.; Makokha, G.O. Monitoring water depth, surface area and volume changes in Lake Victoria: Integrating the bathymetry map and remote sensing data during 1993–2016. *Model. Earth Syst. Environ.* **2017**, *3*, 533–538. [CrossRef]
11. Okeke, I.O.C.; Nwokolo, O.C. Conservation and preservation of traditional Agulu Lake in Anambra State, Nigeria. In Proceedings of the Taal: 12th World Lake Conference, Jaipur, India, 15–19 October 2008; Volume 29, pp. 2209–2211.
12. Egboka, B.; Nfor, B.; Banlanjo, E. Water budget analysis of Agulu Lake in Anambra State, Nigeria. *J. Appl. Sci. Environ. Manag.* **2006**, *10*, 27–30. [CrossRef]
13. Singh, S.K.; Mustak, S.; Srivastava, P.K.; Szabó, S.; Islam, T. Predicting Spatial and Decadal LULC Changes Through Cellular Automata Markov Chain Models Using Earth Observation Datasets and Geo-information. *Environ. Process.* **2015**, *2*, 61–78. [CrossRef]

environmental sciences proceedings

MDPI

Proceeding Paper

Evaluating Methods to Enhance the Taste and Health Benefits of Alternative Potable Waters †

D.L. Marrin

Water Sciences & Insights, Santa Barbara, CA 93140, USA; marrin@watersciences.org

† Presented at the 7th International Electronic Conference on Water Sciences, 15–30 March 2023; Available online: https://ecws-7.sciforum.net.

Abstract: As conventional sources of freshwater continue to be impacted, the use of alternatively produced drinking waters, such as desalinated seawater or condensed atmospheric water, are being increasingly consumed. Lacking the minerals and other natural properties of surface and ground waters, alternative waters are often modified or amended to address taste and health issues. This presentation explores some of those treatments in terms of the proposed taste and health benefits, the water quality issues addressed, and the pertinent research on bottled or specialty waters that could assist in identifying how alternative waters might be best amended or modified.

Keywords: water; alternative sources; health; additives; taste

check for updates

Citation: Marrin, D.L. Evaluating Methods to Enhance the Taste and Health Benefits of Alternative Potable Waters. *Environ. Sci. Proc.* **2023**, *25*, 58. https://doi.org/10.3390/ECWS-7-14300

Academic Editor: Athanasios Loukas

Published: 3 April 2023

1. Introduction

The combination of climate change and environmental pollution has impacted the quality and quantity of drinking water available from conventional freshwater sources. In response to these dwindling potable water resources, a growing number of alternative water sources have emerged in the form of technologies that were introduced or markedly improved during the last two decades. These technologies include desalinated ocean or brackish waters, condensed atmospheric water, recycled wastewater, and captured cloud or fog water, all of which produce potable waters that often lack the minerals and other natural properties of ground and surface waters [1]. Hence, alternative waters are increasingly amended with salts or mineral solutions, adjusted for pH or ORP, and treated in other ways to improve taste or enhance human health.

The primary emphasis of water quality has traditionally been identifying and remediating contaminants in potable sources in order to reduce human health risks; however, characterizing quality solely by the absence of toxins overlooks water's potential health benefits in other ways. Public drinking waters have historically been treated to address taste, odor and/or clarity issues [1], but how these treatments may have influenced the nutritional value or health attributes of water (other than providing hydration) has received only limited attention. One exception is fluoride that, while present at detectable levels in some natural waters, remains a controversial additive to potable waters because of the perceived tradeoffs between preventing dental caries and impacting systemic health [2]. Moreover, fluoride is not designated as an essential human nutrient.

A review of the ways in which alternative waters are amended or altered suggests that some are more common, more extensively researched, or potentially more effective than others are. Recent insights into the physics and chemistry of water, combined with an improved understanding of the factors that influence human taste and health, provide a framework for exploring potential enhancements to alternative waters. Many of these enhancements were initially introduced in the form of specialized bottled waters that were marketed as a healthier option than tap water [3].

2. Minerals and Organics

Potential healthful benefits of drinking water have focused predominantly on minerals, also known as electrolytes, that are essential to the optimal functioning of the human body and are sometimes present at sufficient concentrations in source waters to be consequential. Major minerals typically include calcium, magnesium, sodium, chloride, and potassium. At least some trace minerals are normally present in source waters (e.g., selenium, chromium, iodine, phosphorus), whereas others may either be present in source waters or released from storage and conveyance materials comprising water distribution systems (e.g., iron, manganese, copper, zinc). While many of these trace minerals are essential nutrients at low concentrations, elevated levels in drinking water can pose toxicity problems. By contrast, elevated levels of major minerals are usually not a toxicity issue but can adversely affect the water's taste.

Minerals are normally dissolved in water as inorganic ions (positively or negatively charged), although minerals can also be complexed with organic compounds. The nutritional value of minerals in water is highly dependent on a number of variables, including their valence state or electrical charge, the presence of other minerals (especially as ions), a person's particular gut flora, the volume of water normally ingested, and any foods that may be consumed with the water. Phytochemicals or biochemicals present in a variety of different foods can increase or, more commonly, decrease the bioavailability of minerals dissolved in drinking waters. As such, most essential minerals in the human diet are provided by food [4], within which they are typically complexed by organic molecules such as amino acids and carbohydrates.

Drinking water generally supplies less than 5% of the recommended intake of most minerals, with the exception of calcium and magnesium that could account for as much as 20% of the recommended intake under unusual circumstances [4]. A number of epidemiological studies have reported a reduced incidence of cardiovascular disease and hypertension in communities where the tap water is at least moderately hard [5,6], meaning that the combined calcium and magnesium levels are a minimum of 60 to 120 parts-per-million (ppm), reported as calcium carbonate.

As people have a relatively wide range of taste sensitivities for water, assessing subtle changes in the levels of major minerals based on taste alone is difficult. Nonetheless, waters possessing major minerals, which comprise most of the total dissolve solids (TDS), in excess of 500 to 600 ppm are not palatable to many people due to their salty taste. Conversely, waters containing less than 25 ppm of major minerals often have a flat or bitter taste that can be unpleasant. Even at slightly elevated concentrations, many of the trace minerals can impart a metallic taste to water that is considered objectionable.

In addition to the electrolytes (minerals) that may enhance human health, there are usually organic compounds present in natural waters that could do the same. The most notable of the organics are fulvic acids, which are produced from the microbial decomposition of plants and contain various vitamins, amino acids, carbohydrates and lipids that might be beneficial to human health, depending on their purity (i.e., absence of toxins) and bioavailability [7]. Besides potentially harboring contaminants, some organic compounds can also impart an unpleasant taste to water and, consequently, are normally removed using routine water treatment techniques.

3. ORP and pH

In addition to substances dissolved in drinking water, various physical properties of water have been correlated with both taste and human health. Two properties frequently cited are potential hydrogen (pH) and redox potential (measured as an oxidation-reduction potential or ORP), which indicate the water's relative acidity or basicity and oxidative or reductive capacity, respectively. Adjusting the pH of potable water to prevent metals leaching from pipes or to improve taste and disinfection, has long been an accepted water treatment practice. Acidic drinking water can taste bitter or metallic, while basic water often tastes soda-like. Increasing the pH of water specifically to enhance human health, as is the case with various bottled or specialty waters, is a more recent trend that has been applied to alternative waters only to a limited extent.

Artificially adjusted waters with a pH greater than 8.0 are generally referred to as "alkaline", although true alkaline waters have a basic pH resulting from the dissolution of alkali minerals containing sodium or potassium, rather than from modifications via neutralizing filters or electric ionizers. Alkaline water can reportedly act as an antioxidant, hydration enhancer, acid neutralizer, and reducer of blood viscosity [8,9], although some of these health claims are disputed on the basis of their lack of adequate supporting research. Questions surrounding the health claims for alkaline water often relate to the body's efficient systems for tightly regulating the pH of extracellular body fluids [10], regardless of a drinking water's pH unless it is hazardously acidic or basic.

Water with a relatively high pH normally has a correspondingly low ORP, accounting for alkaline water's designation as an antioxidant. Most natural freshwaters have a moderately high or positive ORP value (i.e., oxidizing), rather than a lower or negative ORP value (i.e., reducing), but this depends largely upon the extent to which the waters are oxygenated—either naturally or artificially. Reducing conditions in potable water are often considered to be beneficial inasmuch as they can neutralize free radicals and, thus, minimize the damaging effects. A more recent trend in lowering ORP is the infusion of water with hydrogen gas, produced by electrolysis or dissolving tablets, for purposes of enhancing its antioxidant, anti-inflammatory and metabolic benefits [11,12].

The most common methods for raising the pH and lowering the ORP of alternative drinking water include passing it through a neutralizing media containing calcium carbonate or magnesium oxide and injecting it with sodium hydroxide or sodium carbonate. Sodium-containing additives are sometimes considered less desirable from a health perspective due to excess sodium's role in hypertension. Electrolysis and molecular hydrogen infusion (using elemental magnesium and an organic acid) are more expensive methods of pH/ORP adjustment, and the former technique generates an acidic waste stream.

Reported health benefits for drinking low ORP water, such as its antioxidant effects [13], differ from those posited for drinking highly oxygenated (higher ORP) water, which include improved exercise recovery and liver function [14]. Oxygenated water generally has a fresher taste, particularly when cold, than flatter-tasting waters with less dissolved oxygen or other gases. Oxygenating alternative waters is one of the simplest methods to improve its taste and can be achieved by injecting oxygen or ozone gases. Whereas the reported health benefits of low ORP waters (including hydrogen-infused) and oxygenated waters (including ozone-infused) are plentiful in the popular literature, they too are controversial. Questions focus on the fate of these modified waters following ingestion and on the biochemical mechanisms by which they are presumed to function [15,16].

4. Molecular Nuances

Water's potential health effects may also include nuances of the water molecules themselves. Liquid water is actually a complex and dynamic network of molecules that connect to one another via connections known as hydrogen bonds, which can switch trillions of times per second. This switching among neighboring molecules permits water to flow as a liquid and yet possess much of the molecular structure of a solid (i.e., ice). Besides the bulk liquid network, water molecules can combine to form a variety of geometric assemblages or clusters within which the intermolecular connections persist somewhat longer. Water containing such clusters is sometimes referred to as structured, as it appears to have greater order or more regular clusters than the bulk liquid.

Natural waters are structured as a result of geological, hydrological, fluvial, and other processes. Artificially structured waters are usually produced by introducing various solutes or specific materials into the water or exposing it to an array of fields and energies (e.g., electric, magnetic, vortical, thermal). Drinking structured water has been associated with health benefits such as reducing inflammation and oxidative stresses [17,18]. As a result of the aforementioned hydrogen bond dynamics (even within clusters) and the restructuring of ingested water upon contact with biological molecules or surfaces, questions persist about the mechanisms proposed for structured water's reported health benefits. Unlike pH and ORP that are easily measured in water, molecular structure is often inferred from physical properties such as density, viscosity, and surface tension. A lower surface tension has been associated with a smoother taste and texture of water.

Intracellular water that is structured inside living cells has been hypothesized to behave as a kind of gel in facilitating many cellular functions [19], thus prompting the question of how intracellular water might be influenced by ingesting externally structured water. Theories range from externally structured water's requirement of less metabolic energy for internal restructuring, to its provision of optimum hydration and removal of toxins from the body. Since water molecules enter living cells primarily through small channels (i.e., aquaporins) in a single-file manner [20], any extracellular water structuring is probably lost while transiting the cell membrane. At present, the molecular structuring of alternative waters is performed almost exclusively by consumers. If structuring were to be used for alternative waters on a larger scale, magnetic and vortical methods might be the most likely candidates as they have been applied to various non-potable waters.

Approximately 0.02% of water's hydrogen atoms consist of a heavier isotope (i.e., deuterium), the resulting concentration of which depends on the latitude, elevation, and precipitation temperature of natural waters. So-called deuterium depleted water (DDW) has a concentration that is less than the global average of approximately 150 ppm and is theorized to be healthful because the body's biological processes and structures generally favor hydrogen over deuterium due to the associated water's chemical and physical properties [21]. The mechanisms responsible for DDW's health benefits (e.g., correcting metabolic or genetic processes, inhibiting cancer proliferation), have been more extensively investigated than those for structured waters. DDW is artificially produced via a continuous distillation process that is expensive and, thus, unlikely to be applied routinely to alternative waters.

5. Remineralization

Drinking water produced via reverse osmosis (RO), including that obtained from household systems, water vending machines, and almost all bottled waters, is sometimes remineralized in order to improve taste or nutrition, as well as to achieve specific goals for pH and ORP. Without the remineralization of alternative waters, TDS concentrations are typically less than 20 to 30 ppm. Whereas, the remineralization of low TDS waters is designed to address taste and potential deficiencies of major minerals, there are some interesting questions related to how and why the waters are remineralized.

Perhaps most controversial is whether routinely consuming low TDS water poses a risk to people by osmotically extracting minerals from the body's structures or extracellular fluids that possess higher mineral contents. Whereas mineral leaching has been identified for foods that are cooked or prepared in low TDS water, a similar leaching process has not been demonstrated in human bodies [22,23]. Healthy bodies possess an array of integrated control systems (e.g., sensory, hormonal, excretory) for retaining or expelling specific minerals to maintain the ionic composition of extracellular fluids.

Other questions about water remineralization relate to both the actual salts that are utilized and the ratio of major minerals in added solutions. The most common additives to low TDS waters are calcium carbonate and magnesium oxide, which have limited bioavailability but raise the pH and provide a buffering effect. Other sources of mineralization include various salts, concentrated mineral liquids, and even seawater, most of which are dominated by sodium and chloride. Though seawater has a mineral composition similar to that of blood plasma, it has an ORP higher than most natural freshwaters and often contains organic compounds derived from marine organisms. While some of these compounds have potential health benefits, their addition to alternative waters is unlikely to be implemented.

Some drinking water additives claim to have an ideal ratio of major minerals or a suite of important trace minerals. However, most trace minerals are provided by food and, although drinking waters may emulate the ratio of major minerals in extracellular fluids, there are numerous factors that can result in the differential absorption of those minerals. Consequently, minerals are not necessarily absorbed in proportion to the ratios reflected in liquids, salts, or other additives, requiring the body's homeostatic processes to maintain the precise suite and concentration of minerals comprising those fluids.

When remineralizing alternative waters with a low TDS, an important consideration is taste. Taste is a function of the minerals dissolved within it, as well as its temperature, oxygen content, pH, and organic carbon content. Palatability is also influenced by people's own taste sensitivity and by any materials or chemicals that contact the water during its treatment, storage, or transport [24]. Perhaps as important, from a nutritional perspective, is the presence of calcium and magnesium, the latter of which is a common mineral deficiency due to its relatively low concentration in most foods. The bioavailability of magnesium from mineralized waters is similar to that from foods, and its levels in bodily fluids are stringently controlled by the kidney's excretion and retention processes [25].

6. Conclusions

Much of the evidence (both scientific and anecdotal) presented for the health benefits of amended, modified, or specialty waters is based on one of several types of inquiries. The first includes animal studies and in vitro laboratory investigations that are valuable and represent an essential step in researching health effects, but alone are not definitive. Another is the study of people whose drinking water has one or more of the specified attributes and who appear to be either healthier or longer-lived than most others. These studies frequently represent small or regional populations that are geographically and/or socially isolated and, thus, may possess diets and lifestyles that differ appreciably from those of most postmodern populations.

Another line of evidence for the reported benefits of ingesting modified or specialty waters pertains to the treatment of specific diseases or chronic health problems. Fewer research studies have focused on whether, or how, these waters may improve or sustain wellness among relatively healthy people and, therefore, the routine or prolonged consumption of some modified or specialty waters has been questioned. Additionally, at least some of the hypothesized health benefits are based on product-sponsored research, which may be legitimate but would benefit from corroborating studies by funders who have no financial stake in the findings. Finally, anecdotal observations and customer testimonials can be quite useful, but are not definitive, in evaluating actual health benefits.

Nonetheless, results from all of the aforementioned types of investigations and observations collectively suggest that water's quality is not limited to its simply being free of contaminants. Whereas accepted scientific mechanisms for the efficacy of some of the purported health benefits of water modifications or additives do not currently exist, it does not mean that such benefits are without merit. Research into human nutrition has revealed unexpected, but generally accepted, influences on processes ranging from gene expression to brain plasticity for which the mechanisms are not yet fully understood.

Research studies that investigate both the health effects of modified or amended alternative waters, and mechanisms for their doing so, would be valuable in evaluating whether relatively healthy people might benefit from routinely drinking them. Research into the health benefits of high-quality natural waters [26], which could serve as a model for amending alternative waters, may also be a productive endeavor. Given the likelihood that alternative drinking waters (e.g., recycled, desalinated, atmospheric, cloud captured) will be increasingly consumed as a result of the ongoing depletion and pollution of traditional surface and ground water sources, it would seemingly be worthwhile to better understand how these alternative waters could be optimally modified to improve taste and to potentially promote human health.

Funding: This research received no external funding.

Institutional Review Board Statement: Not applicable.

Informed Consent Statement: Not applicable.

Data Availability Statement: Not applicable.

Conflicts of Interest: The author declares no conflict of interest.

References

1. Nelson, N.; DeLuca, A. Remineralization and stabilization of desalinated water. In *Pathways and Challenges for Efficient Desalination*; Shahzad, M., Dixon, M., Barassi, G., Eds.; IntechOpen: London, UK, 2022; Chapter 5. [CrossRef]
2. Peckham, S.; Awofeso, N. Water fluoridation: A critical review of the physiological effects of ingested fluoride as a public health intervention. *Sci. World J.* **2014**, *2014*, 293019. [CrossRef] [PubMed]
3. Pierce, G.; Lai, L. Toward a comprehensive explanatory model of reliance on alternatives to the tap: Evidence from California's retail water stores. *J. Water Health* **2019**, *17*, 455–462. [CrossRef] [PubMed]
4. Oliveres, M.; Uauy, R. Essential nutrients in drinking water. In *Nutrients in Drinking Water*; World Health Organization: Geneva, Switzerland, 2005; Chapter 4.
5. Gianfredi, V.; Bragazzi, N.L.; Nucci, D.; Villarini, M.; Moretti, M. Cardiovascular diseases and hard drinking waters: Implication from a systematic review with meta-analysis of case-control studies. *J. Water Health* **2017**, *15*, 31–40. [CrossRef] [PubMed]
6. Klevay, L.M.; Combs, G.F. Mineral elements related to cardiovascular health. In *Nutrients in Drinking Water*; World Health Organization: Geneva, Switzerland, 2005; Chapter 7.
7. Carrasco-Gallardo, C.; Guzman, L.; Maccioni, R.B. Shilajit as a natural phytocomplex with potential procognitive activity. *Int. J. Alzheimers Dis.* **2012**, *2012*, 674142. [CrossRef] [PubMed]
8. Koufman, J.A.; Johnston, N. Potential benefits of pH 8.8 alkaline drinking water as an adjunct in the treatment of reflux disease. *Ann. Otol. Rhinol. Laryngol.* **2012**, *121*, 431–434. [CrossRef] [PubMed]
9. Weidman, J.; Holsworth, R.E., Jr.; Brossman, B.; Cho, D.J.; St Cyr, J.; Fridman, G. Effect of electrolyzed high-pH alkaline water on blood viscosity in healthy adults. *J. Int. Soc. Sport. Nutr.* **2016**, *13*, 45. [CrossRef] [PubMed]
10. Brissette, C. Is Alkaline Water Really Better for You? Available online: https://www.washingtonpost.com/lifestyle/wellness/is-alkaline-water-really-better-for-you/2019/08/27/8c646d26-c462-11e9-b72f-b31dfaa77212_story.html (accessed on 2 April 2023).
11. Kang, K.M.; Kang, Y.N.; Choi, I.B.; Gu, Y.; Kawamura, T.; Toyoda, Y.; Nakao, A. Effects of drinking hydrogen-rich water on the quality of life of patients treated with radiotherapy for liver tumors. *Med. Gas Res.* **2011**, *1*, 11. [CrossRef] [PubMed]
12. Song, G.; Li, M.; Sang, H.; Zhang, L.; Li, X.; Yao, S.; Yu, Y.; Zong, C.; Xue, Y.; Qin, S. Hydrogen-rich water decreases serum LDL-cholesterol levels and improves HDL function in patients with potential metabolic syndrome. *J. Lipid Res.* **2013**, *54*, 1884–1893. [CrossRef] [PubMed]
13. Lee, M.Y.; Kim, Y.K.; Ryoo, K.K.; Lee, Y.B.; Park, E.J. Electrolyzed-reduced water protects against oxidative damage to DNA, RNA, and protein. *Appl. Biochem. Biotechnol.* **2006**, *135*, 133–144. [CrossRef] [PubMed]
14. Fleming, N.; Vaughan, J.; Feeback, M. Ingestion of oxygenated water enhances lactate clearance kinetics in trained runners. *J. Int. Soc. Sport. Nutr.* **2017**, *14*, 9. [CrossRef] [PubMed]
15. Piantadosi, C.A. "Oxygenated" water and athletic performance. *Brit. J. Sport. Med.* **2006**, *40*, 740. [CrossRef] [PubMed]
16. LeBaron, T.W.; Sharpe, R. ORP should not be used to estimate or compare concentrations of aqueous H_2: An in-silico analysis and narrative synopsis. *Front. Food. Sci. Technol.* **2022**, *2*, 1007001. [CrossRef]
17. Lindinger, M.I. Structured water effects on animals. *J. Anim. Sci.* **2021**, *99*, skab063. [CrossRef] [PubMed]
18. Lee, H.J.; Kang, M.H. Effect of the magnetized water supplementation on blood glucose, lymphocyte DNA damage, antioxidant status, and lipid profiles in STZ-induced rats. *Nutr. Res. Prac.* **2013**, *7*, 34–42. [CrossRef] [PubMed]
19. Pollack, G.P. *Cells, Gels, and the Engines of Life*; Ebner & Sons: Seattle, WA, USA, 2001.
20. Horner, A.; Zocher, F.; Preiner, J.; Ollinger, N.; Siligan, C.; Akimov, S.A.; Pohl, P. The mobility of single-file water molecules is governed by the number of H-bonds they may form with channel-lining residues. *Sci. Adv.* **2015**, *1*, e1400083. [CrossRef] [PubMed]
21. Basov, A.; Fedulova, L.; Baryshev, M.; Dzhimak, S. Deuterium-depleted water influence on the isotope $^2H/^1H$ regulation in body and individual adaptation. *Nutrients* **2019**, *11*, 1903. [CrossRef] [PubMed]
22. Haring, B.; Van Delft, W. Changes in the mineral composition of food as a result of cooking in hard and soft waters. *Arch. Environ. Health* **1981**, *36*, 33–35. [CrossRef] [PubMed]
23. WQA. *Consumption of Low TDS Water*; Water Quality Association: Lisle, IL, USA, 2013.
24. Dietrich, A.M.; Gallagher, C.D. Consumer ability to detect the taste of total dissolved solids. *J. Am. Water Works Assoc.* **2013**, *105*, E255–E263. [CrossRef]
25. Schneider, I.; Greupner, T.; Hahn, A. Magnesium bioavailability from mineral waters with different mineralization levels in comparison to bread and a supplement. *Food Nutr. Res.* **2017**, *61*, 1384686. [CrossRef] [PubMed]
26. Deutschländer, K. Water and prevention research: The current water study. *Angew. Komplementärmedizin* **2022**, *8*, 1–8.

environmental sciences proceedings

MDPI

Proceeding Paper

Comparison of Geomorphological Parameters Detected Using MERIT and FABDEM Products †

Ankini Borgohain *, Varun Khajuria, Vaibhav Garg, Shiva Reddy Koti and Ashutosh Bhardwaj

Indian Institute of Remote Sensing, Dehradun 248001, India; varunkhajuria10@gmail.com (V.K.); vaibhav@iirs.gov.in (V.G.); shivareddy@iirs.gov.in (S.R.K.); ashutosh@iirs.gov.in (A.B.)

* Correspondence: ankiniborgohain@gmail.com

† Presented at the 7th International Electronic Conference on Water Sciences, 15–30 March 2023; Available online: https://ecws-7.sciforum.net.

Abstract: A morphometric analysis and its comparison was carried out using two multi-resolution DEMs-MERIT and FABDEM for a region in North Eastern Himalayas. The study area includes districts of Kamrup Rural, Kamrup Metropolitan, Dhubri, Bongaigaon, Nalbari, Kokrajhar, and Goalpara, which are located in the state of Assam. The area was selected as it is highly prone to flood every year and was also recently affected by flood in the year 2022. The MERIT DEM developed by Dr. Yamazaki, University of Tokyo by removing multiple error components from existing spaceborne DEMs (SRTM and AW3D) represents the terrain elevations at a 3 sec resolution (~90 m at the equator), whereas the FABDEM is a global elevation map that removes building and tree height biases from Copernicus GLO 30 DEM with 30 m resolution. In this study, watershed delineation and morphometric parameters were computed and analyzed using Archydro tools and HecGeoHMS in Arcmap (v 10.5). The parameters classified as basic, linear, shape and relief aspects were derived and calculated by using standard methods. Some important parameters such as stream length, stream order, bifurcation ratio, drainage density, etc., derived from both the DEMs were compared. From this study, it was observed that MERIT DEM performed better in terms of the drainage delineation and morphometric analysis of the basin for our study area compared to FABDEM, thereby suggesting that MERIT DEM worked well for the study area chosen.

Keywords: DEM; morphometric parameters; watershed

Citation: Borgohain, A.; Khajuria, V.; Garg, V.; Koti, S.R.; Bhardwaj, A. Comparison of Geomorphological Parameters Detected Using MERIT and FABDEM Products. *Environ. Sci. Proc.* **2023**, *25*, 59. https://doi.org/10.3390/ECWS-7-14298

Academic Editor: Athanasios Loukas

Published: 3 April 2023

1. Introduction

Hydrological processes including water movement, generation of surface runoff, etc., rely upon the topography of the surface. A Digital Elevation Model (DEM) represents the bare topographic Earth surface excluding buildings, vegetation and any other objects on the surface. Some of the widely used and notable DEMs are MERIT DEM, FABDEM, etc. DEM is essential digital data for geomorphological study. Studies have been conducted for analysis of accuracy assessment of individual DEM datasets, but very limited work has been conducted wherein comparison of more than one DEM of different resolutions to analyze morphometric parameters derived from them [1]. It is, therefore, important that various openly accessible DEMs must be assessed for their suitability for morphometric analysis. Generally, it is accepted that the higher the resolution of a DEM, the more precise it is, which also indicates better accuracy and finer extraction of components of land surface and the drainage lines present [2].

A morphometric study of a watershed can be used to determine the ways in which different characteristics of an area are related. Morphometric parameter analysis is impacted by the source and resolution of DEMs [3]. Morphometric analysis of watersheds helps us to determine various aspects of linear, areal and relief parameters [4]. These parameters are used as an input for different hydrological applications, flood modelling and management of watersheds. Surface water potential, areas where land degradation occurs, and flood risk can be understood through morphometric parameter analysis [5].

In the present study, we perform comparison between two openly accessible multi-resolution DEMs for their efficiency in terms of geomorphometric parameters for the chosen study area.

2. Materials and Method

The study area, which includes Kamrup and few neighboring districts in Assam, India is shown in Figure 1. To carry out this study, we used two openly accessible multi-resolution DEM—MERIT DEM and FABDEM, for delineation of watershed using ArcHydro tools and HecGeoHMS package in ArcMap (v 10.5). The morphometric parameters computed from both the DEMs were then computed and compared to check their efficiency for the chosen study area.

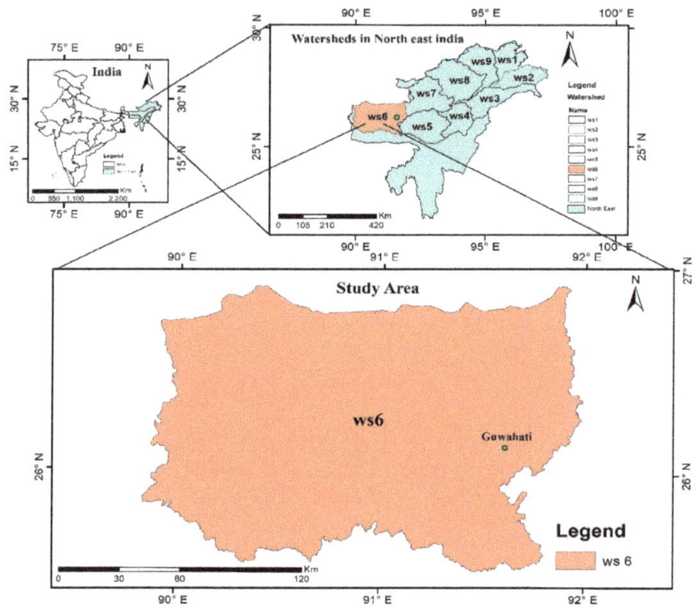

Figure 1. Study area map.

2.1. Pre-Processing of DEM

The specifications of the datasets used in the present study are mentioned in Table 1. The two DEMs used are the MERIT DEM developed by Dr. Yamazaki, University of Tokyo by removing multiple error components from existing spaceborne DEMs (SRTM and AW3D), which represents the terrain elevations at a 3 sec resolution (~90 m at the equator), and the FABDEM, which is the first global DEM (30 m resolution) that used Machine Learning to remove building and tree height biases from Copernicus GLO 30 DEM. The tiles of MERIT DEM and FABDEM were downloaded from the following websites: http://hydro.iis.u-tokyo.ac.jp/~yamadai/MERIT_DEM/ (accessed on 13 August 2022), https://data.bris.ac.uk/data/dataset/25wfy0f9ukoge2gs7a5mqpq2j7 (accessed on 17 August 2022). They were then mosaiced and clipped according to the study area in ArcMap.

Table 1. Specification of datasets used.

Type	Name	Satellite	Resolution (m)	Datum
Fused DEM	ERIT DEM	Fused DEM from SRTM and ASTER	90	WGS84
SAR Interferometry	FABDEM	Derived DEM from TerraSAR-X and TanDEM-X	30	WGS84

2.2. Watershed Delineation

The methodology flowchart used in this study is depicted in Figure 2.

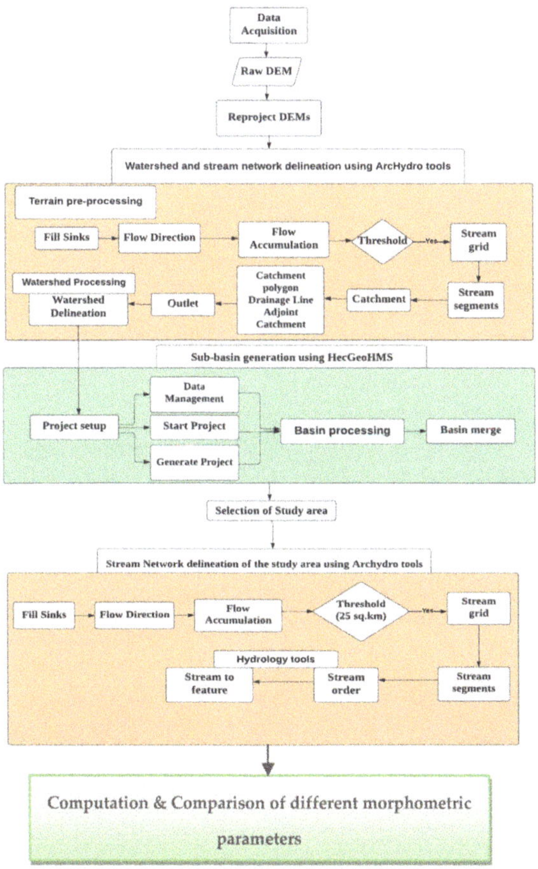

Figure 2. Methodology Flowchart of watershed delineation and computation of morphometric parameters.

For the generation of watershed and sub-basins from both the DEMs, the procedure was carried out using ArcHydro tools and HecGeoHMS in ArcGIS (v 10.5).

The below mentioned steps were followed in sequential order for delineation of watershed and generation of sub-basins:

- Fill Sinks—depressions, if any, present in the DEMs were eliminated using this tool;
- Flow Direction—flow direction was determined using the D8 method;

- Flow accumulation—flow accumulation grid was computed using this tool; for each cell in the input grid, it contains the accumulated number of cells upstream of a cell;
- Stream definition—the total contributing area of both the DEMs was taken as 25 km^2;
- Stream segmentation—the streams created were segmented using this tool;
- Catchment grid delineation;
- All the created raster data were converted to vector format using the following three tools: Catchment polygon processing, Drainage line and Adjoint Catchment tool;
- Drainage point—the cell location within each catchment having the maximum flow accumulation value was denoted by this tool;
- Batch point generation—the outlet points for the sub-basin generated from each of the DEMs were selected at the same location;
- Batch watershed generation—the watershed for the selected area was delineated using this tool.

After completing the above steps, a project was set up in HecGeoHMS using the results obtained from the ArcHydro tools for further computation. As the study area includes Guwahati city and its neighboring districts in Assam, which are highly flood prone and flood occurs in these regions every year; so, out of the 9 major sub-basins extracted using the Basin merging tool, the sub-basin 6 was chosen for further analysis.

2.3. Computation of Different Morphometric Parameters from Both the DEMs

The basic morphometric parameters such as basin area, basin perimeter, etc., were calculated from the attribute table of the sub-basin. Stream order for both the DEMs were computed using Hydrology tools (Spatial Analyst tools). Basin length for the sub-basin was calculated using the measure tool and ArcHydro tool. Other parameters such as stream length, drainage density, relief ratio, ruggedness ratio, form factor, etc., were derived using standard formulas.

3. Results and Discussion

Figures 3 and 4 show the delineated watershed and derived drainage network from MERIT DEM and FABDEM.

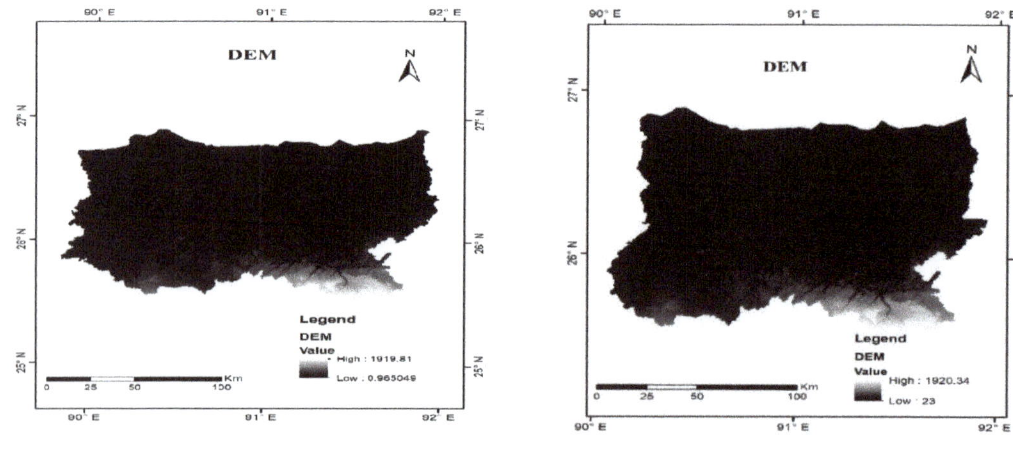

(a) (b)

Figure 3. Openly accessible multi-resolution DEM. (**a**) MERIT DEM and (**b**) FABDEM.

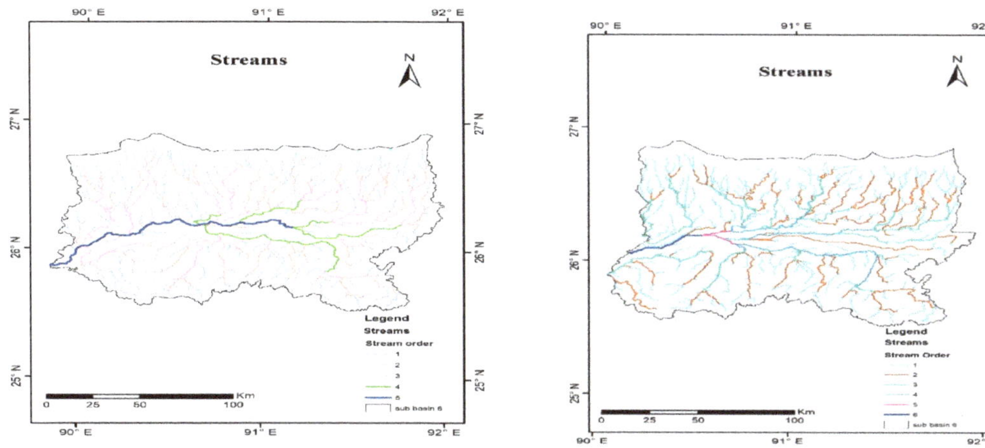

(a) (b)

Figure 4. Drainage network generated from (**a**) MERIT DEM and (**b**) FABDEM.

Different linear, areal and relief aspects including basic parameters such as area and perimeter were computed for the study area. The values of these morphometric parameters derived from both the DEMs are shown in Figure 5. It was observed that the area delineated using MERIT DEM is larger (24,628.2 km^2) compared to FABDEM (21,425.6 km^2). This is likely because in the study area we have chosen, FABDEM was still determined to have random artifacts and pits which were not corrected accurately [6]. Through visual interpretation, it was observed that the streamlines delineated using MERIT DEM were more accurately aligned compared to FABDEM. The sub-basin derived using MERIT DEM showed that it is governed by highest fifth-order stream, while that derived using FABDEM is governed by sixth-order stream, both indicating a dendritic drainage pattern. Although FABDEM has higher resolution compared to MERIT DEM, the total number of streams of FABDEM (493) is smaller than that of MERIT DEM (545) as the sub-basin's area delineated for it is smaller.

The bifurcation ratio values derived indicated that there is no disturbance from any geologic structures for the sub-basin. From the elongation and circularity ratio values, it was seen that the basin is oval to circular. The high basin relief values, low ruggedness number values and relatively low drainage density values obtained from both DEMs implies that less erosion occurs in the study area. It also indicates that the area has a coarse drainage texture and is associated with thick vegetation and high resistance or permeable soil. Low overland flow values (11.9 km from MERIT DEM and 13.1 km from FABDEM) computed from both DEMs also signifies that the study area is indicative of flooding conditions during heavy rainfall.

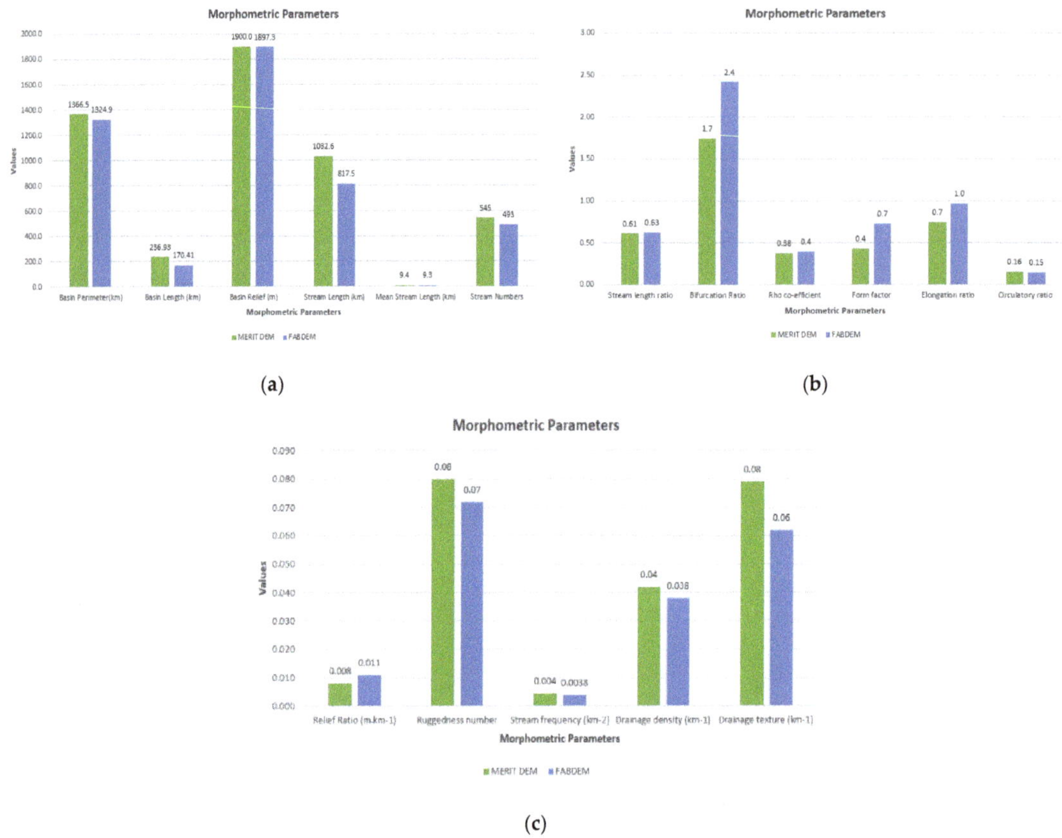

Figure 5. (**a**–**c**) shows different morphometric parameter values derived from MERIT DEM and FABDEM.

4. Conclusions

This study showed the comparison of the morphometric parameters classified as linear, areal and relief aspects from the two openly accessible multi-resolution DEMs, MERIT DEM and FABDEM. From the parameter values derived from two DEMs, it was concluded that the chosen study area is prone to flooding conditions during heavy rainfall with no disturbance from any geologic structures and the area is associated to thick vegetation with less occurrence of soil erosion. Variation is observed in basic parameters such as area and basin length. Among the linear aspects, parameters such as stream order, stream number and stream length showed significant variation between the two DEMs, whereas both the relief and areal aspects did not show much difference, irrespective of the DEM used. From the results, it was observed that variation among some of the parameters mainly depends on the source and resolution of DEM. It is, therefore, important to consider the type of DEM used for watershed delineation and in studies related to morphometric parameter estimation. It was visually observed from the drainage network generated from the two DEMs that drainage lines of the MERIT DEM seem more naturally close compared to FABDEM for our study area. Thus, from hydrological point of view, MERIT DEM can be used in flat regions and its drainage networks are also clearly represented for the study area. On the other hand, from the resolution point of view, FAB DEM with higher resolution (30 m) can be considered for revealing finer details in areas leading to high accuracy.

Author Contributions: Conceptualization, A.B. (Ashutosh Bhardwaj) and S.R.K.; methodology, A.B. (Ashutosh Bhardwaj); software, A.B. (Ankini Borgohain); validation, V.G., V.K. and A.B. (Ashutosh Bhardwaj); formal analysis, A.B. (Ankini Borgohain); investigation, A.B. (Ashutosh Bhardwaj); resources, A.B. (Ankini Borgohain); data curation, A.B. (Ankini Borgohain); writing—original draft preparation, A.B. (Ankini Borgohain); writing—review and editing, A.B. (Ashutosh Bhardwaj); visualization, A.B. (Ankini Borgohain); supervision, A.B. (Ashutosh Bhardwaj); project administration, A.B. (Ashutosh Bhardwaj). All authors have read and agreed to the published version of the manuscript.

Funding: This research received no external funding.

Institutional Review Board Statement: Not applicable.

Informed Consent Statement: Not applicable.

Data Availability Statement: No new data were created or analyzed in this study. Data sharing is not applicable to this article.

Conflicts of Interest: The authors declare no conflict of interest.

References

1. Das, S.; Patel, P.P.; Sengupta, S. Evaluation of different digital elevation models for analyzing drainage morphometric parameters in a mountainous terrain: A case study of the Supin–Upper Tons Basin, Indian Himalayas. *SpringerPlus* **2016**, *5*, 1544. [CrossRef] [PubMed]
2. Anornu, G.K.; Kabo-bah, A.; Kortatsi, B.K. Comparability Studies of High and Low Resolution Digital Elevation Models for Watershed Delineation in the Tropics: Case of Densu River Basin of Ghana. *Int. J. Coop. Stud.* **2012**, *1*, 9–14. [CrossRef]
3. Weydahl, D.J.; Sagstuen, J.; Dick, Ø.B.; Rønning, H. SRTM DEM accuracy assessment over vegetated areas in Norway. *Int. J. Remote Sens.* **2007**, *28*, 3513–3527. [CrossRef]
4. Abboud, I.A.; Nofal, R.A. Morphometric analysis of wadi Khumal basin, western coast of Saudi Arabia, using remote sensing and GIS techniques. *J. Afr. Earth Sci.* **2017**, *126*, 58–74. [CrossRef]
5. Tesema, T.A. Impact of identical digital elevation model resolution and sources on morphometric parameters of Tena watershed, Ethiopia. *Heliyon* **2021**, *7*, e08345. [CrossRef] [PubMed]
6. Hawker, L.; Uhe, P.; Paulo, L.; Sosa, J.; Savage, J.; Sampson, C.; Neal, J. A 30 m global map of elevation with forests and buildings removed. *Environ. Res. Lett.* **2022**, *17*, 024016. [CrossRef]

environmental sciences proceedings

MDPI

Proceeding Paper

A Sustainable Concept for Recovering Industrial Wastewater Using Adjustable Green Resources †

Abdeljalil Adam *, Nabil Saffaj and Rachid Mamouni

Laboratory of Biotechnology, Materials, and Environment, Faculty of Sciences, University IBN ZOHR, Agadir 80000, Morocco; saffaj@gmail.com (N.S.); r.mamouni@uiz.ac.ma (R.M.)

* Correspondence: adam.abdeljalil@gmail.com; Tel.: +212-660-775-429

† Presented at the 7th International Electronic Conference on Water Sciences, 15–30 March 2023; Available online: https://ecws-7.sciforum.net/.

Abstract: In arid environments, specifically in Africa, inadequate water sources have resulted in poor-quality water use in business and agricultural industries. This can also negatively impact the ecosystem, along with the industry's water management. In solar power generation facilities, evaporating basins are usually used to release and store industrial contaminated wastewater. An examination of the environmental implications and concerns of this experience suggests such a form of wastewater discharge to reduce industrial effluents' direct release into the environment. Unfortunately, this strategy could have far-reaching global environmental consequences and issues. In this research, we examined the evolution of the effluent's physicochemical characteristics over a long period using a methodological approach for a power station located in Morocco; the findings of this practical study show a significant increase in the physicochemical characteristics of the wastewater released in the evaporating basins, which might be interpreted by an increase in water pollution. The primary objective of this study is to examine wastewater recycling and the generation of treated water in a solar still, utilizing renewable energy to minimize the environmental and ecological problems associated with wastewater discharged into evaporating basins.

Keywords: evaporating basins; wastewater; ecosystem; environment; physicochemical characteristics

Citation: Adam, A.; Saffaj, N.; Mamouni, R. A Sustainable Concept for Recovering Industrial Wastewater Using Adjustable Green Resources. *Environ. Sci. Proc.* **2023**, *25*, 60. https://doi.org/10.3390/ECWS-7-14302

Academic Editor: Athanasios Loukas

Published: 3 April 2023

1. Introduction

The exploitation of resources is becoming increasingly destructive, and durable resource limitations, particularly in relation to freshwater, continue to pose a significant challenge in the twenty-first century.

Evaporating basins (Figure 1) are ponds made of earth that are lined with a geomembrane and contain a volume of water that evaporates due, mainly, to exposure to full sunlight. The substances that are included in the mixtures begin to crystallize in the saline because of the evaporation of the surface water from the basins, and they are then routinely recovered and disposed of as solid waste [1].

However, there appear to be some disadvantages to employing evaporating basins, such as the need for large amounts of land once the evaporating rate has reduced or if the dumping quantities are very high, and the necessity for such huge tracts when these situations arise [2].

Regardless of their numerous benefits, evaporating basins may cause a variety of environmental problems. It is possible, for illustration, that any industrial effluents from the evaporating basins could have devastating effects on the local ecosystem. Being open waterways, evaporating basins attract a variety of animals, which can lead to an increase in the mortality of some species if the collected effluent exceeds the permissible limits [3].

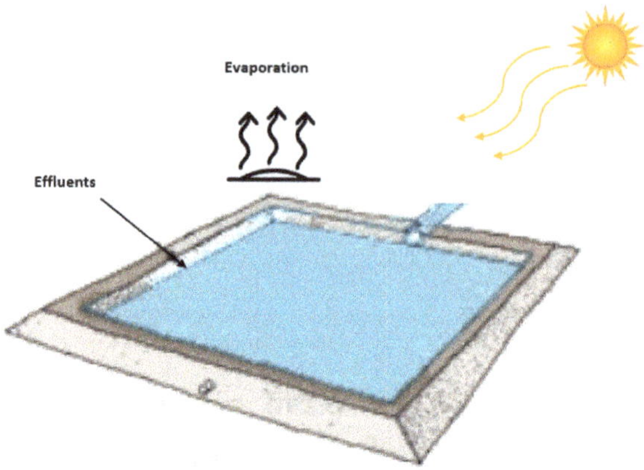

Figure 1. Evaporating Basin.

Wastewater treatment is crucial now because of the need to preserve the planet's precious resources. Many methods have been created to control effluents and reduce water pollution to reach this objective [4].

By analyzing the wastewater from a power plant in Morocco, we expect to demonstrate how serious an issue this has become for all factories, notably for water treatment factories that discharge their effluent into evaporation basins. This study examined the physicochemical factors that contribute to pollution in wastewater.

2. Material and Methods

2.1. Study Area

From November 2021 to October 2022, we obtained effluent samples from the evaporation basin of a Moroccan power plant; the idea was to make sure that we had a representative sample for a whole year.

2.2. Effluent Physicochemical Analysis

The Sigma SD900 Portable Sampler was used to collect the dataset from the power plant between November 2021 and October 2022.

The physicochemical analyses of the wastewater were performed according to accepted and relevant techniques at an independent chemical analysis laboratory in Morocco.

The relevant physicochemical analyses were performed using the methods described below:

Total suspended solids (TSS), electrical conductivity (EC), and pH were measured in accordance with norms ISO 10523:2012, ISO 7888:2001, and EN 872:2013.

The organic pollutants' biochemical oxygen demand (BOD) and chemical oxygen demand (COD) were assessed using the norm 03.7.054 v 2013.

Soluble sulphate (SO4) levels were measured in accordance with NM ISO/TR 896.

2.3. Regression Analysis

The least-squares regression method is used in statistics to make predictions about how control variable could change over time. When employing least-squares analysis, it is essential to focus on the optimum path in conjunction to all other keypoints. It is common practice for users to examine the relationship between the regression model and a set of probable variables by carrying out regression analysis (Equation 1) [5].

$$Y = ax + b \quad (1)$$

where Y = dependent, x = independent, a = intercept, and b = constant.

To better understand the relationships between the various effluent properties, a statistical study has been conducted. The statistical method of regression analysis is used to ascertain the strength and direction of the connection between the variables being investigated [6].

3. Results and Discussion

3.1. Physicochemical Results

The results from a year's length of physicochemical analysis on the power plant's wastewater are summarized in Table 1.

Table 1. Results of physicochemical evaluation.

Parameter	Unit	Limit	11/21	12/21	01/22	02/22	03/22	04/22	05/22	06/22	07/22	08/22	09/22	10/22
pH	pH unit	5.5–9.5	8.1	7.9	7.6	7.9	8.0	7.5	7.6	8.1	8.1	7.1	7.7	8.0
EC	ms/cm	2.7	9	18	19	23	38	18	45	26	33	28	44	36
SO_4	mg/L	600	1800	1840	2320	1985	1300	3500	2900	4256	1987	2692	2569	4200
TSS	mg/L	100	102	79	105	129	153	304	215	116	98	197	203	171
COD	mg O_2/L	500	356	620	560	519	614	652	498	516	563	578	452	512
BOD_5	mg O_2/L	100	95	120	95	115	118	103	142	109	98	102	115	124

Table 1 shows that the main effluent parameters (COD, BOD_5, EC, SO_4 and TSS) exceeded the maximum level imposed by Moroccan regulations for effluent released into the environment [7].

In May of 2022, the highest EC value measured was 45 ms/cm, which is much higher than the threshold value of 2.7 ms/cm. In June of 2022, a maximum SO_4 value of 4256 mg/L was measured, which is over the permissible limit of 600 mg/l. In April 2022, the highest TSS measurement was 304 mg/L, which is above the allowable 100 mg/L. In April 2022, the highest COD value was 652 mg O_2/L, which is over the permissible 500 mg O_2/L. In May of 2022, the highest BOD_5 measurement recorded was 142 mg O_2/L, which is more than the permissible 100 mg O_2/L.

3.2. Regression Results

Descriptive and inferential statistics were employed in the regression study of effluent characteristics.

Table 2 displays the final square of the regression method developed utilizing statistically significant correlations between effluent properties.

Table 2. Regression's least square of the effluent variables.

Y: Dependent	X: Independent	Correlation ®	R^2	a	B (Constant)	Regression Equation (Y = ax + b)
TSS	pH	−0.603	0.364	−127.88	1153.5	TSS = −127.88 pH + 1153.5
TSS	SO_4	0.415	0.172	0.028	81.68	TSS = 0.028 SO_4 + 81.68
TSS	EC	0.314	0.098	1.81	105.13	TSS = 1.81 EC + 105.13
BOD_5	EC	0.649	0.422	0.79	88.91	BOD_5 = 0.79 EC + 88.91
COD	pH	−0.319	0.1023	−84.03	1192.2	COD = −84.03 pH + 1192.2

Figure 2 shows linear connections between the following physicochemical properties of the effluent:

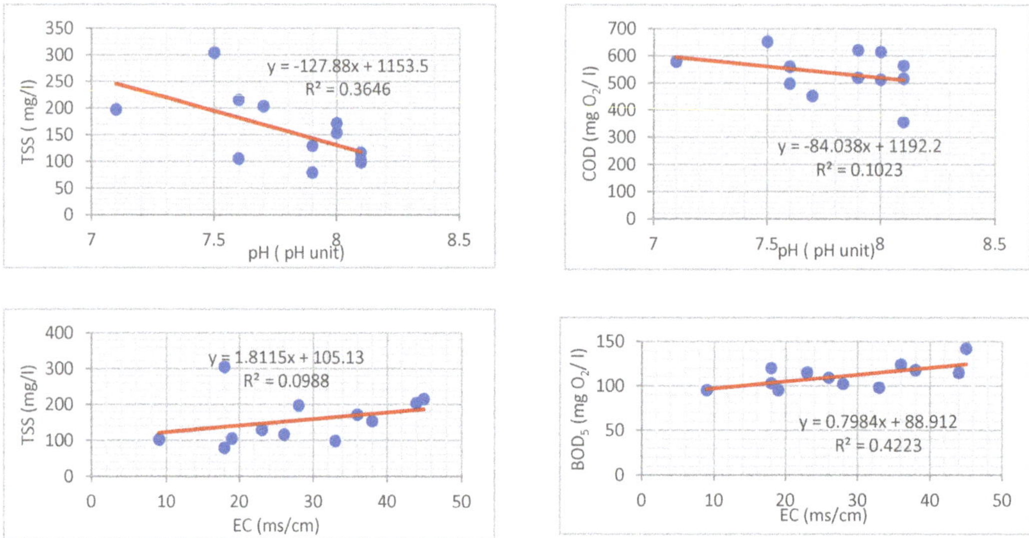

Figure 2. Regression linear plots for effluent variables.

The scatter plots showed positive linear relationships between total suspended solids (TSS) and SO_4, total suspended solids (TSS) and EC, and BOD_5 and EC. Nonetheless, both TSS and COD showed negative linear relationships with pH.

By employing linear regression, we were able to determine the strength of those associations via the obtained R and R^2 values.

The graph demonstrates a positive correlation between SO_4 and EC with TSS, and that increasing TSS likewise increases SO_4 and EC. However, it demonstrates a negative correlation between TSS and COD with pH, and that increasing pH likewise decreases TSS and COD.

The storage of wastewater in the open air has a deleterious impact on the physicochemical and bacteriological properties of industrial effluents, as evidenced by an increase across various effluent properties that exceeds the maximum limits specified by international regulations. Since this phenomenon has the potential to cause ecological and environmental damage, it is imperative that every industry carry out a thorough environmental risk assessment to determine the most appropriate response [8].

According to the results of the physicochemical analysis of the wastewater kept in evaporation ponds, the primary external factors that have altered the physicochemical characteristics of the wastewater over time include biological waste from species that are attracted to the surface of the water, dust, and the concentration of salt in the waste water as a result of evaporation.

Industrial power plants have employed evaporation ponds to retain their effluent based on several criteria for power plant facilities. This approach is intended to prevent the direct discharge of wastewater into the environment. For power plants that still use water vapor as the primary input to power steam turbines, this solution is used to compare the results from this research with those of other current research which suggests evaporation ponds as the ultimate wastewater retention and discharge method.

Wider research might explore the viability of properly recycling effluent instead of storing it in evaporation ponds. It could also look into the possibility of heating the wastewater using the solar collectors already in place for the power plant's process, and then injecting the heated wastewater into a solar sill to create clean water which could be utilized by those solar power plants again, maximizing the surface water utilization to support the required sustainable development.

Considering that evaporation pond techniques are only used in arid, warm climates that are comparable to this study, we may generalize the disadvantages of all worldwide companies that may release their effluents into evaporation basins.

This research and analysis bring us to the conclusion that discharging effluent into evaporating basins is not a viable solution due to the severe negative consequences it has on the environment and its resources.

Consequently, we propose that industries, especially power plants, use environmentally friendly strategies for recycling this effluent. As a result, this might help businesses to justify charging higher rates for water consumption.

Among the most eco-friendly solutions for companies that utilize evaporation ponds is installing huge renewable solar stills; the benefit is conserving the environment as well as reusing wastewater. By doing so, the industry will be able to properly recycle its effluent while spending less on water.

Alternatively, the solar still can be warmed using solar panels that rely solely on sunshine, as shown in the following scheme (Figure 3) for connecting the solar still with evaporating basins. A clear sky all the time reflects a climate that is conducive to this technique.

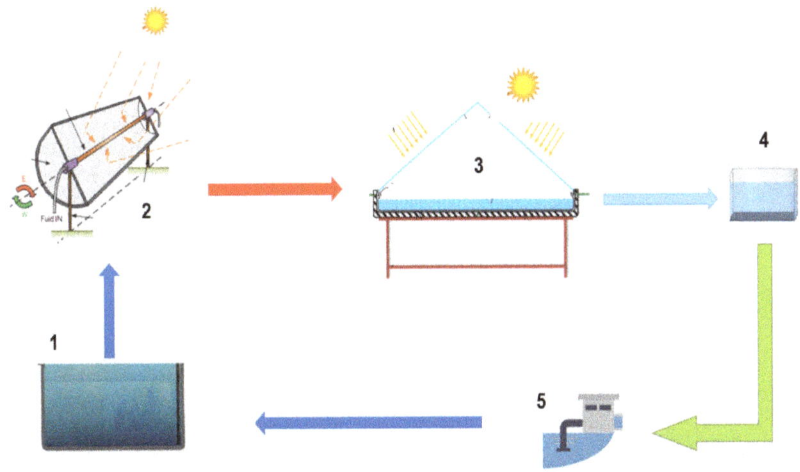

Figure 3. Recycling effluent proposed solution. (1) Effluent; (2) solar collector; (3) solar still; (4) clean water; (5) water processing.

4. Conclusions

This research, which examined the physicochemical features of the effluents released by power plants, revealed that effluent quality worsened when exposed to stagnation or when stored in evaporation basins. Consequently, several physicochemical characteristics, such as electrical conductivity, total suspended particles, sulfates, and chemical and biological oxygen demand, rise substantially as a result of wastewater storage. As a result, there will be a rise in water contamination.

This study used data collected annually from a power plant's wastewater quality assessments to determine the most significant associations between most linked variables.

According to the findings, there is a correlation between the various physical and chemical characteristics of industrial effluent. In comparison, the values obtained for chemical and biological oxygen needs, electrical conductivity, total suspended solids, and sulfates are all greater than the permissible limits for wastewater quality attributes in the research region.

The proposed eco-friendly option for businesses that utilize evaporation ponds is to install huge sustainable solar stills instead. This has the dual benefit of conserving the environment and reusing wastewater. Because of this, businesses will be able to save money on water bills while reusing wastewater.

Author Contributions: The individual contributions are: Conceptualization, A.A.; methodology, N.S.; writing—original draft preparation, A.A.; writing—review and editing, R.M. and N.S.; supervision, N.S.; project administration, A.A. All authors have read and agreed to the published version of the manuscript.

Funding: This research received no external funding.

Institutional Review Board Statement: Not applicable.

Informed Consent Statement: Not applicable.

Data Availability Statement: Not applicable.

Conflicts of Interest: The authors declare no conflict of interest.

References

1. Abdeljalil, A.; Nabil, S.; Rachid, M. Feasibility and Sustainability of Evaporation Ponds as Final Basins for Industrial Wastewater: Statistical Evaluation of Gross Parameters. *Desalination Water Treat.* **2022**, *257*, 41–54. [CrossRef]
2. Ahmed, M.; Shayya, W.H.; Hoey, D.; Mahendran, A.; Morris, R.; Al-Handaly, J. Use of Evaporation Ponds for Brine Disposal in Desalination Plants. *Desalination* **2000**, *130*, 155–168. [CrossRef]
3. Adam, A.; Saffaj, N.; Mamouni, R.; Baih, M. Characterization of industrial Wastewater Physico-Chemical Properties. *Int. J. Tech. Phys. Probl. Eng.* **2022**, *14*, 219–227.
4. Lrhoul, H.; Assaoui, N.E.; Turki, H. Mapping of Water Research in Morocco: A Scientometric Analysis. *Mater. Today Proc.* **2021**, *45*, 7321–7328. [CrossRef]
5. Sow, A.; Traore, I.; Diallo, T.; Traore, M.; Ba, A. Comparison of Gaussian Process Regression, Partial Least Squares, Random Forest and Support Vector Machines for a near Infrared Calibration of Paracetamol Samples. *Results Chem.* **2022**, *4*, 100508. [CrossRef]
6. Halconruy, H.; Marie, N. Kernel Selection in Nonparametric Regression. *Math. Methods Stat.* **2020**, *29*, 32–56. [CrossRef]
7. Moroccan Limit-Values-of-Discharges. Available online: http://www.environnement.gov.ma/fr/78-cat1/1012-valeurs-limites-des-rejets (accessed on 15 July 2017).
8. Abdeljalil, A.; Nabil, S.; Rachid, M. Contribution to Developing an Environmental Emergency Response for Industrial Sites. *J. Environ. Agric. Stud.* **2021**, *2*, 97–102. [CrossRef]

environmental sciences proceedings

MDPI

Proceeding Paper

Numerical Estimation of the Black Sea Circulation near the Continental Slope Using SKIRON and ERA5 Atmospheric Forcing †

Olga Dymova * and Natalia Markova

Wave Theory Department, Marine Hydrophysical Institute, Russian Academy of Sciences, 2 Kapitanskaya St., Sevastopol 299011, Russia

* Correspondence: olgdymova@mhi-ras.ru

† Presented at the 7th International Electronic Conference on Water Sciences, 15–30 March 2023; Available online: https://ecws-7.sciforum.net/.

Abstract: Assessments of the state of sea waters and complex studies of the marine environment in various ocean basins are often based on hydrophysical fields (currents, temperature, salinity, etc.) obtained through the use of numerical modeling. The regular fields of currents are of particular importance for assessing the transport of impurities in sea waters at different depths, including pollutants of various origins. The results of hydrophysical field modeling, in turn, depend on the conditions set at the boundaries of the basin. Therefore, the correct setting of rapidly changing atmospheric conditions is extremely important for the reconstruction of marine dynamics. This paper presents model estimates of the Black Sea circulation obtained using two different datasets, SKIRON and ERA5, as atmospheric forcing. Numerical experiments for 2016 are carried out based on the eddy-resolving MHI-model. ARGO floats and R/V Cruises data are used to validate the simulation results. It was discovered that temperature and salinity RMSE between the model and measurement data are decreased under ERA5 forcing. Near the northeastern continental slope, a change in the direction of the alongshore subpycnocline current, which is detected in the ARGO float trajectory, is modeled using ERA5 rather than SKIRON. Therefore, for a more accurate reconstruction of the Black Sea circulation, ERA5 atmospheric forcing is recommended.

Keywords: Black Sea circulation; modeling; forcing; measurement data

check for updates

Citation: Dymova, O.; Markova, N. Numerical Estimation of the Black Sea Circulation near the Continental Slope Using SKIRON and ERA5 Atmospheric Forcing. *Environ. Sci. Proc.* **2023**, *25*, 61. https://doi.org/10.3390/ECWS-7-14305

Academic Editor: Athanasios Loukas

Published: 3 April 2023

1. Introduction

The circulation in the upper layer of each ocean is in direct contact with the atmosphere and is related to the distribution of meteorological parameters [1–3]. At the same time, the influence of atmospheric forcing on the structure of deepwater circulation is not so clear. For the Black Sea, this problem is complicated by the presence of a strong vertical density gradient (permanent pycnocline) at 50–100 m horizons, which blocks vertical seawater exchange [3].

Regional features of density stratification often arise near the Black Sea continental slope due to the mixing and lowering processes of surface waters along the slope into the deep sea layers [3,4]. The formation of density anomalies here can be caused by external forcing at the boundaries of the basin (including wind, river runoff, etc.), the sinking of denser waters down the continental slope, and the transfer of water with thermohaline characteristics that differ from the ones in eddies [4,5]. These processes are especially important in the northeastern part of the basin due to the narrow and steep continental slope in this region. The seawater density anomalies formed near the slope can lead to the transformation of the velocity field at deep horizons [5]. Thus, the generation of unsteady deepwater undercurrents is found out there [6].

Below the permanent pycnocline, the Black Sea waters become warmer and more saline accumulates towards the bottom [2]. At the same time, anticyclones can form near the shelf edge and then can move along the slope [4,5,7]. In the centers of the anticyclones, subpycnocline waters that are colder and contain less saline deepen, and their movement contributes to the transfer of thermohaline anomalies and the corresponding perturbations of dynamical fields. Such complicated dynamics near the continental slope require a detailed and accurate reconstruction of all hydrophysical characteristics, which is only possible if boundary conditions are correctly specified.

2. Materials and Methods

The Black Sea circulation was reconstructed by an eddy-resolving model from the Marine Hydrophysical Institute (MHI-model) [8]. The model was based on the Navier–Stokes equations in Boussinesq and hydrostatic approximations. Vertical turbulent mixing was described by the Mellor–Yamada closure model 2.5, and horizontal mixing was described using a bilaplacian operator with constant coefficients. The model circulation was driven by atmospheric forcing, including wind stress, heat fluxes, precipitation, and evaporation, on the sea surface. The climatological Black Sea rivers runoff and exchange through the straits were considered. Data assimilation (except for the satellite sea surface temperature data) was not used in the discussed numerical experiments. The MHI-model was implemented on a C-grid with a resolution of $(1/48)^\circ$ longitude, $(1/66)^\circ$ latitude, and 27 z-levels vertically. The detailed model description is presented in [8].

Basin bathymetry was built from EMODnet data [9]. The initial data were obtained from the Black Sea Physical Reanalysis CMEMS [10]. All initial and input fields were linearly interpolated in the MHI-model grid nodes.

In this work, two numerical simulations with identical model setups but different atmospheric forcing were carried out for the year 2016. In the first simulation (SKIRON-experiment), the forcing included 2 h of data on wind velocity, thermal, latent, sensible, and solar heat fluxes, evaporation, and precipitation provided by the SKIRON/Dust modeling system (Greece), with a spatial resolution of 0.1° [11]. In the second simulation (ERA5-experiment), the forcing was based on the freely available hourly data of reanalysis supported by the European Centre for Medium-Range Weather Forecasts for the global climate, with a resolution of 0.25° [12].

Comparative analysis of the SKIRON and ERA5 data showed a significant difference in wind forcing in the Black Sea region. As can be seen in Figure 1, the ERA5 wind stress is stronger than SKIRON one by about 25–30%, and the repeatability of NN-E and N-E wind directions (forming surface cyclonic circulation of the Black Sea) is higher. The remaining fluxes in ERA5 and SKIRON are close to each other, with there being some excess (15–20%) ERA5 data on total heat flux during the year and mass flux (precipitation minus evaporation) in autumn and winter.

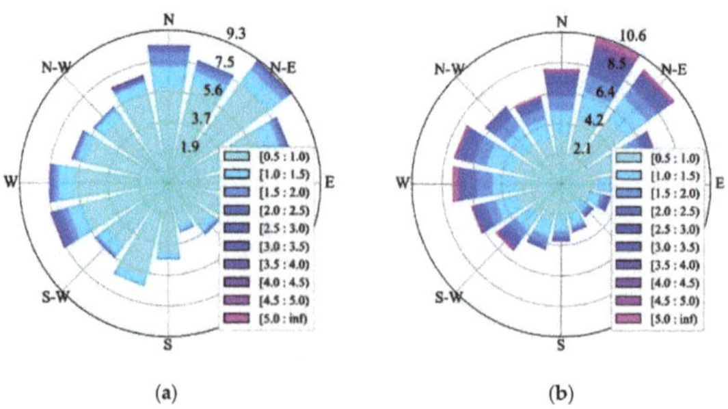

Figure 1. Histograms of the annual mean repeatability of the area-averaged wind directions (digits, %) and wind stress magnitudes (color, 10^{-5} N/cm^2) for the Black Sea area in 2016: (**a**) by SKIRON; (**b**) by ERA5. Data were calculated from the wind velocity at a height of 10 m.

3. Results

Daily data on sea surface height and three-dimensional fields of seawater temperature, salinity, and current velocity for the year 2016 were obtained in the two numerical experiments described above. The next stages of the study compare the simulation results with observational data (validation) and analyze deepwater circulation with a focus on the continental slope region, where the most interesting features of the currents are observed. In the northeastern part of the sea, so-called undercurrents (opposite to the Black Sea surface basin-scale cyclonic gyre—the Rim Current [1–3]) are detected at a depth of 200 m for the period 9 June 2016–14 October 2016, according to ARGO data.

Validation of the model fields was performed based on temperature and salinity measurement data obtained by ARGO profiling floats [13] and R/V «Professor Vodyanitsky» Cruises 87, 89, and 91 [14] in 2016. Our validation methodology is described in [8] (Section 2.2). Root mean square errors RMSE between the model and in situ data for both experiments are presented in Table 1. The temperature RMSE in the upper layer (0–300 m) decrease in the ERA5-experiment compared to the SKIRON-experiment. The highest decrease in temperature error was observed at a depth of 0–30 m. The model salinity in the ERA5-experiment correlated more at a depth of 30–300 m. Therefore, the permanent pycnocline and seasonal thermocline layers in the ERA5-experiment are closer to the measurement data.

Table 1. The temperature and salinity RMSE between simulations and in situ.

Depth, m	Temperature, °C		Salinity, psu	
	SKIRON	ERA5	SKIRON	ERA5
0–5	1.175	0.625	0.224	0.258
5–30	2.390	1.706	0.188	0.212
30–100	0.623	0.489	0.454	0.384
100–300	0.199	0.154	0.423	0.312
300–800	0.036	0.055	0.072	0.084
800–1500	0.030	0.027	0.055	0.075

A difference between the simulation results was primarily found in the velocity fields due to the strong influence of the wind on Black Sea dynamics [1,2]. The increasing wind velocity in ERA5 (Figure 1b) led to a more typical structure of the Rim Current at the end of 2016, when the basin-scale cyclonic gyre was propagated above the continental slope (Figure 2b). The Rim Current was not regenerated in winter, and mesoscale eddies were developed in the central sea part in the SKIRON-experiment (Figure 2a) due to insufficient kinetic energy inflow from the wind [8].

The model circulation in the upper layer was generally cyclonic for both experiments. At the same time, the most significant difference of the current velocity fields was detected below the permanent pycnocline core. Thus, at deepwater horizons, in the ERA5-experiment (Figure 2d), the current field was more intense, and maximal velocity was higher than in the SKIRON one (Figure 2c).

Analysis of ARGO float ID6901833 trajectory data [13] revealed a change in the direction of the alongshore subpycnocline current from the northwestern (cyclonic) to the northeastern (anticyclonic) near the northeastern continental slope. Thus, from 6 September to 14 October, 2016 the float drifted anticyclonicaly at its parking depth of 200 m (Figure 3a, red arrows). Such behavior of the alongshore current was not modeled in the SKIRON-experiment (Figure 3b), but was clearly reconstructed in the ERA5-experiment (Figure 3c). Averaged over the period of anticyclonical movement of the float, the model velocity of the undercurrent riches 0.03–0.05 m/s with instant value up to 0.08 m/s. The undercurrent generation near the Black Sea continental slope is probably associated with the intense mesoscale variability under the permanent pycnocline in the ERA5-experiment (Figure 3c). Here, some eddies were observed along the continental slope. The undercurrents that form near the northeastern slope of the Black Sea seem to be of an anticyclonic nature, similar to the undercurrents formed by anticyclones in the western part of the Bay of Bengal [15].

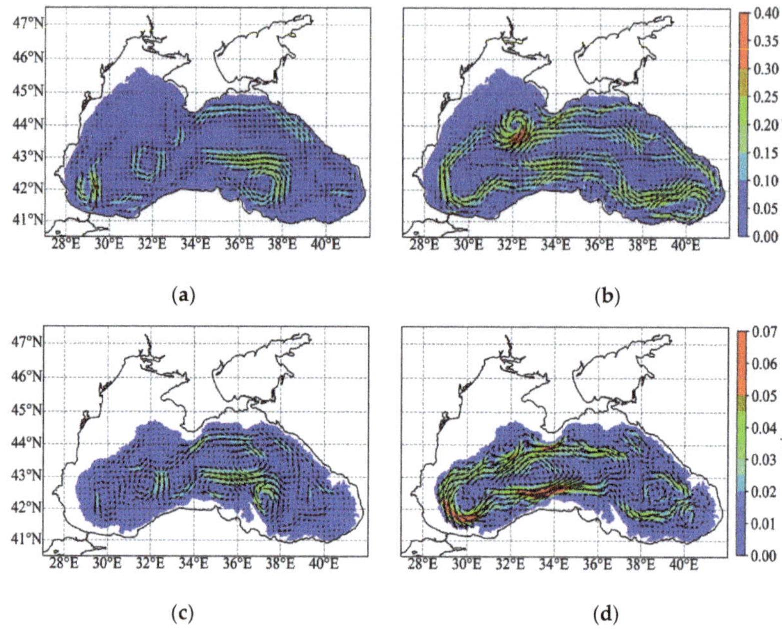

Figure 2. Monthly mean current fields in December 2016 at a depth of 50 m (**a**,**b**) and at a depth of 500 m (**c**,**d**) according to the SKIRON-experiment (**a**,**c**) and the ERA5-experiment (**b**,**d**).

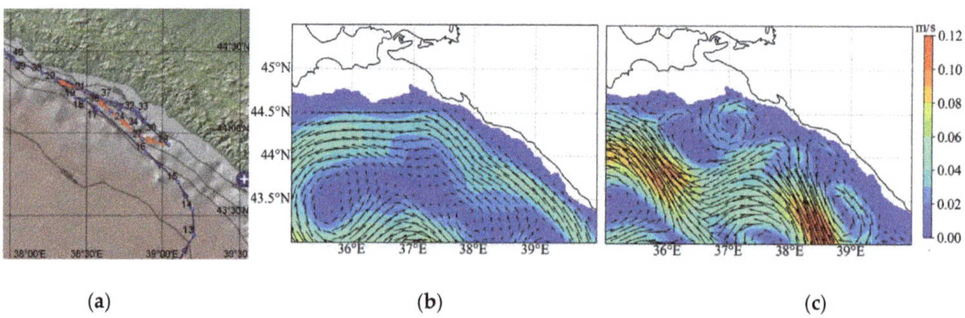

Figure 3. (**a**) ARGO float ID6901833 trajectory at parking depth of 200 m. Model current velocity at 200 m time-averaged for 6 September–14 October 2016 by the SKIRON-experiment (**b**) and the ERA5-experiment (**c**). Blue arrows illustrate the northwestern alongshore current, red arrows correspond to the southeastern current (undercurrent).

The structure of the circulation is inextricably linked with the spatiotemporal variability of seawater thermohaline characteristics [3,7]. The model temperature and salinity fields on the zonal cross-section along 44°N averaged over the period of existing undercurrent are shown in Figure 4. As seen in temperature fields (Figure 4a,b), the upper mixed layer reached a depth of 20–25 m in both experiments, but in the ERA5-experiment its thickness was larger near the eastern coast (up to 25–30 m), and its temperature was higher here as well. The mesoscale anticyclones shown in Figure 3c led to the deepening of isotherms and isohalines near the eastern coast and the formation of an undercurrent along the slope.

There is a downward deflection of the isotherms at zone of 38.3–39.0° N in Figure 4b that corresponds to the anticyclonic current. A similar deflection is also visible in the salinity field (Figure 4d). Thus, the distribution of temperature and salinity in the ERA5-experiment is consistent with the anticyclonic current near the continental slope detected in the ARGO float ID6901833 data [13].

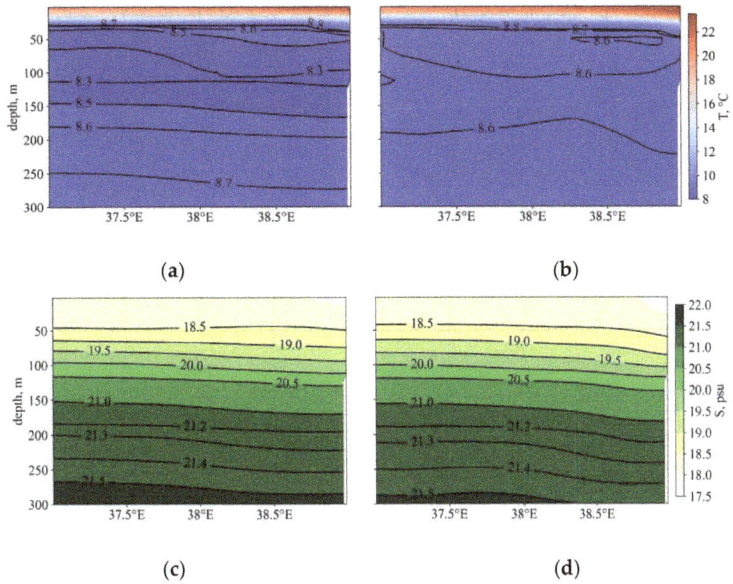

Figure 4. Zonal cross-section along 44° N of the model temperature (**a**,**b**) and salinity (**c**,**d**) fields time-averaged for 6 September–14 October 2016 by the SKIRON-experiment (**a**,**c**) and the ERA5-experiment (**b**,**d**).

4. Conclusions

The important outcome of the study is that atmospheric fluxes can affect the circulation of both the surface and deepwater layers of the Black Sea, and the choice of atmospheric forcing data can be decisive for the correct modeling of hydrophysical fields in the entire basin. As was determined through numerical analysis, with a significant influence of wind forcing, both the upper layer circulation and the deepwater dynamics in the Black Sea depend on the characteristics of the atmosphere. Despite strong density stratification and difficult vertical exchange with deep layers, the atmospheric forcing also affects the circulation at a horizon of 200 m and deeper. Thus, mesoscale features of the model dynamics near the continental slope, such as subpycnocline undercurrents detected from the ARGO observations in the northeastern part of the sea, appear only when using ERA5 forcing. Additionally, the Black Sea thermohaline structure is more accurately reconstructed under ERA5 forcing, and this was confirmed by the TS-measurement data. Atmospheric fluxes in ERA5 were more likely to be intense compared to SKIRON. Thus, for more accurate modeling of Black Sea circulation and its subsequent application for complex studies, from the two widely used meteorological datasets, the use of ERA5 atmospheric forcing data is recommended rather than SKIRON.

Author Contributions: Conceptualization, O.D. and N.M.; forcing preparation, modeling, validation, and visualization O.D.; hydrophysical analysis, N.M.; writing—original draft preparation, O.D. and N.M.; writing—review and editing, O.D. and N.M. All authors have read and agreed to the published version of the manuscript.

Funding: Numerical simulations and validation of the results were funded by the Russian Science Foundation under Grant 22-77-10056 (https://rscf.ru/en/project/22-77-10056/, accessed on 10 March 2023). Observation data processing was carried out within MHI State assignment on theme № FNNN-2021-0004. Estimation of the atmospheric forcing effect on the Black Sea deepwater circulation was supported by the Russian Science Foundation Grant No. 22-17-00150 (https://rscf.ru/en/project/22-17-00150/ (accessed on 10 March 2023)).

Institutional Review Board Statement: Not applicable.

Informed Consent Statement: Not applicable.

Data Availability Statement: The datasets generated and analyzed during the current study are available from the MHI administration upon reasonable and/or special request.

Acknowledgments: The authors are grateful to the developer of the MHI-model, Demyshev S.G., for the opportunity to use the model in the experiments.

Conflicts of Interest: The authors declare no conflict of interest.

References

1. Capet, A.; Barth, A.; Beckers, J.-M.; Marilaure, G. Interannual variability of Black Sea's hydrodynamics and connection to atmospheric patterns. *Deep Sea Res. II* **2012**, *77–80*, 128–142. [CrossRef]
2. Kubryakov, A.A.; Stanichny, S.V.; Zatsepin, A.G.; Kremenetskiy, V.V. Long-term variations of the Black Sea dynamics and their impact on the marine ecosystem. *J. Mar. Syst.* **2016**, *163*, 80–94. [CrossRef]
3. Ivanov, V.A.; Belokopytov, V.N. *Oceanography of the Black Sea*; ECOSY-Gidrofizika: Sevastopol, Ukraine, 2013; pp. 65–71.
4. Zatsepin, A.G.; Korzh, A.O.; Kremenetskii, V.V.; Ostrovskii, A.G.; Poyarkov, S.G.; Solov'ev, D.M. Studies of the hydrophysical processes over the shelf and upper part of the continental slope of the Black Sea with the use of traditional and new observation techniques. *Oceanology* **2008**, *48*, 466–475. [CrossRef]
5. Korotenko, K.; Osadchiev, A.; Melnikov, V. Mesoscale Eddies in the Black Sea and Their Impact on River Plumes: Numerical Modeling and Satellite Observations. *Remote Sens.* **2022**, *14*, 4149. [CrossRef]
6. Demyshev, S.G.; Dymova, O.A.; Markova, N.V.; Korshenko, E.A.; Senderov, M.V.; Turko, N.A.; Ushakov, K.V. Undercurrents in the northeastern Black Sea detected on the basis of multi-model experiments and observations. *J. Mar. Sci. Eng.* **2021**, *9*, 933. [CrossRef]
7. Staneva, J.V.; Dietrich, D.E.; Stanev, E.V.; Bowman, M.J. Rim current and coastal eddy mechanisms in an eddy-resolving Black Sea general circulation model. *J. Mar. Syst.* **2001**, *31*, 137–157. [CrossRef]
8. Demyshev, S.G.; Dymova, O.A. Analysis of the annual mean energy cycle of the Black Sea circulation for the climatic, basin-scale and eddy regimes. *Ocean Dyn.* **2022**, *72*, 259–278. [CrossRef]
9. European Marine Observation and Data Network (EMODnet). EMODnet Digital Bathymetry (DTM 2020)-Tile D3. Available online: https://emodnet.ec.europa.eu/geonetwork/srv/eng/catalog.search#/metadata/19f800a9-f0fd-4055-b4cd-90ed156dc7fc/ (accessed on 12 March 2023).
10. Lima, L.; Masina, S.; Ciliberti, S.A.; Peneva, E.L.; Cretí, S.; Stefanizzi, L.; Lecci, R.; Palermo, F.; Coppini, G.; Pinardi, N.; et al. *Black Sea Physical Reanalysis (CMEMS BS-Currents) (Version 1) Data Set*; Copernicus Monitoring Environment Marine Service (CMEMS), 2020. [CrossRef]
11. Kallos, G.; Nickovic, S.; Papadopoulos, A.; Jovic, D.; Kakaliagou, O.; Misirlis, N.; Boukas, L.; Mimikou, N.; Sakellaridis, G. The regional weather forecasting system SKIRON: An overview. In Proceedings of the International Symposium on Regional Weather Prediction on Parallel Computer Environments, Athens, Greece, 15–17 October 1997; pp. 109–122.
12. Hersbach, H.; Bell, B.; Berrisford, P.; Biavati, G.; Horányi, A.; Muñoz Sabater, J. *ERA5 Hourly Data on Single Levels from 1959 to Present*; Copernicus Climate Change Service (C3S), Climate Data Store (CDS), 2018. [CrossRef]
13. Coriolis. Available online: https://dataselection.coriolis.eu.org/ (accessed on 12 March 2023).
14. Artamonov, Y.V.; Skripaleva, E.A.; Alekseev, D.V.; Fedirko, A.V.; Shutov, S.A.; Kolmak, R.V.; Shapovalov, R.O.; Shcherbachenko, S.V. Hydrological Research in the Northern Part of the Black Sea in 2016 (87th, 89th and 91st Cruises of R/V *Professor Vodyanitsky*). *Phys. Oceanogr.* **2018**, *25*, 229–234. [CrossRef]
15. Francis, P.A.; Jithin, A.K.; Chatterjee, A.; Mukherjee, A.; Shankar, D.; Vinayachandran, P.N.; Ramakrishna, S.S.V.S. Structure and dynamics of undercurrents in the western boundary current of the Bay of Bengal. *Ocean Dyn.* **2020**, *70*, 387–404. [CrossRef]

environmental sciences proceedings

MDPI

Proceeding Paper

A GIS-Based Fuzzy Hierarchical Modeling for Flood Susceptibility Mapping: A Case Study in Ontario, Eastern Canada †

Amir Noori and Hossein Bonakdari *

Department of Civil Engineering, University of Ottawa, 161 Louis Pasteur Private, Ottawa, ON K1N 6N5, Canada; anoor074@uottawa.ca

* Correspondence: hbonakda@uottawa.ca; Tel.: +1-418-656-2131; Fax: +1-418-656-3723

† Presented at the 7th International Electronic Conference on Water Sciences, 15–30 March 2023; Available online: https://ecws-7.sciforum.net.

Abstract: Natural disasters such as floods have severely destroyed the natural environment and infrastructure because of their destructive effects and caused socio-economic losses. In the present study, the authors attempt to present a flood hazard susceptibility map of an eastern region in Ontario, Canada to facilitate flood prevention and mitigation. To this purpose, a combination of Multi-Criteria Decision-Making (MCDM) model and Geographic Information System (GIS) has been considered. Herein, an Analytical Hierarchy Process (AHP) model is applied based on Triangular Fuzzy Numbers (TFNs) in a GIS environment. A total of eight quantitative criteria including elevation, land use/land cover, geology, rainfall, drainage density, slope, soil-type, and distance from river have been used for the flood modeling. Fuzzified pairwise comparison matrices of values have determined the Importance Weights (IWs) of these criteria in Saaty's scale. By calculating IWs, the impact of each effective criterion on flood risk was investigated using the fuzzy AHP method. The consistency Index of each pairwise comparison of criteria has been checked. Based on the calculated IWs result of each criterion, the precipitation, slope, and soil criteria play significant roles as the most eminent flood occurrence criteria. In addition, the obtained results demonstrate percentages of flooded areas and the flood hazard index of the study area.

Keywords: flood susceptibility; GIS; multi-criteria decision-making (MCDM); analytical hierarchy process (AHP); fuzzification; TFNs

Citation: Noori, A.; Bonakdari, H. A GIS-Based Fuzzy Hierarchical Modeling for Flood Susceptibility Mapping: A Case Study in Ontario, Eastern Canada. *Environ. Sci. Proc.* **2023**, *25*, 62. https://doi.org/10.3390/ECWS-7-14242

Academic Editor: Athanasios Loukas

Published: 16 March 2023

1. Introduction

In recent years, natural disasters have been considered the principal problems affecting both developed and developing countries [1]. The frequency of natural disasters has risen considerably over the past decades due to a variety of human and natural parameters such as degradation of the environment, climate change, rapid population growth, and intensified and inappropriate land use [2–4]. Flood is a complicated phenomenon among natural disasters; it causes significant and irreversible damage, leading to considerable human and economic losses [5]. This naturally complex phenomenon happens essentially due to global warming that is responsible for changing the rate and precipitation intensities [6]. It occurs rapidly and negatively impacts city populations and infrastructure [7]. As a result, assessing and regionalizing flood risks are becoming essential and urgent [4]. A crucial step is to map flood susceptibility to predict the probability of a flood. Flood susceptibility mapping makes it possible to determine and predict future flood risks using statistical or deterministic techniques. Mapping areas vulnerable to historic disasters is essential for flood mitigation and management [8].

This paper presents a flood susceptibility map in the region of Ottawa city in Ontario, Canada, to find solutions for the assessment and management of floods by combining

Multi-Criteria Decision-Making (MCDM) and Geographical Information Systems (GIS). The Analytic Hierarchy Process (AHP) is known as one of the most popular techniques [9,10] which has been integrated by GIS to assign a specific weight for each criterion, estimate the Flood Hazard index, and create a special decision-making solution for flood susceptibility mapping [11]. Eight spatial criteria used in this study include land use/land cover, drainage density, precipitation, geology, elevation, slope, soil, and distance from river. Moreover, a fuzzy mathematical set based on Triangular Fuzzy Numbers (TFNs) is utilized in the proposed method in order to reduce the uncertainties and improve the evaluation of flood-susceptible areas [12–16]. Integrating the MCDM method and GIS is considered an appropriate tool and is widely used in various engineering fields. For example, Hammami et al. 2019 applied a GIS-based AHP to evaluate Tunisia's flood susceptibility mapping. They used eight criteria in order to calculate a Flood Hazard Index (FHI) and determine a flood susceptibility map [17]. In a recent study, Msabi and Makonyo 2020 used GIS and multi-criteria decision analysis, considering seven influencing criteria: Elevation, slope, geology, drainage density, flow accumulation, land-use/cover, and soil for mapping the flood susceptibility area in central Tanzania [18]. Souissi et al. 2019 for flood susceptibility mapping in southeastern regions of Tunisia, developed a GIS-based AHP model. They also considered eight criteria in flood modeling: elevation, land use/land cover, lithology, rainfall intensity, drainage density, distance from the river, slope, and groundwater depth. The obtained results showed that the most important flood occurrence criterion was the elevation [19]. Furthermore, Rincón et al. 2018 in another research utilized GIS and an AHP method to define the optimal weight of each criterion related to flood risk to develop the accuracy of flood risk maps of the Don River basin in Toronto, Canada [20]. In the other study related to Flood Susceptibility Mapping, Swain et al. 2020 used an integration of GIS-AHP Technique to investigate flood susceptible areas in India [5]. Some of the limitations of previous studies that use the GIS-based multi-criteria flood susceptibility mapping approach include criteria weighting methods that are not appropriate, or in some cases are not used at all, which can lead to unreliable results. Therefore, this study bridges this gap by applying a hierarchical GIS-based model to assign weights to the criteria, and then determine flood susceptibility map. The rest of this paper is arranged in the following manner: In Section 2, a study area is discussed and describes the framework of the proposed methodology in detail and the third Section of this study presents the results and discussion of this study, and the conclusions are provided in Section 4.

2. Materials and Methods

2.1. Study Area

Ottawa is located at latitude 45°25′29″ N and longitude 75°41′42″ W in the east of southern Ontario, with an elevation of 70 m above sea level (see Figure 1). The climate is semi-continental, with a warm, humid summer and a very cold winter. The area and population of Ottawa district are about 2790 km^2 and 780,000 people, respectively. The temperature typically varies from −14 °C to 27 °C, while the mean precipitation is 920 mm. Rain falls throughout the year in Ottawa. The higher mean monthly rainfall in Ottawa is in July, with an average rainfall of 76 mm, while the least rainfall month is February, with an average rainfall of 12.7 mm. Ottawa experiences extreme seasonal variation in monthly snowfall. The snowy period of the year lasts from October to April with at least 25.4 mm. Snowfall does not occur in July, with an average total accumulation of 0.0 mm. The cold season lasts from December to March, with an average daily high temperature below 1 °C, and the wetter season lasts from April to December. The southern Ontario region has suffered several severe flood events during the last 100 years, resulting in high economic and social impacts.

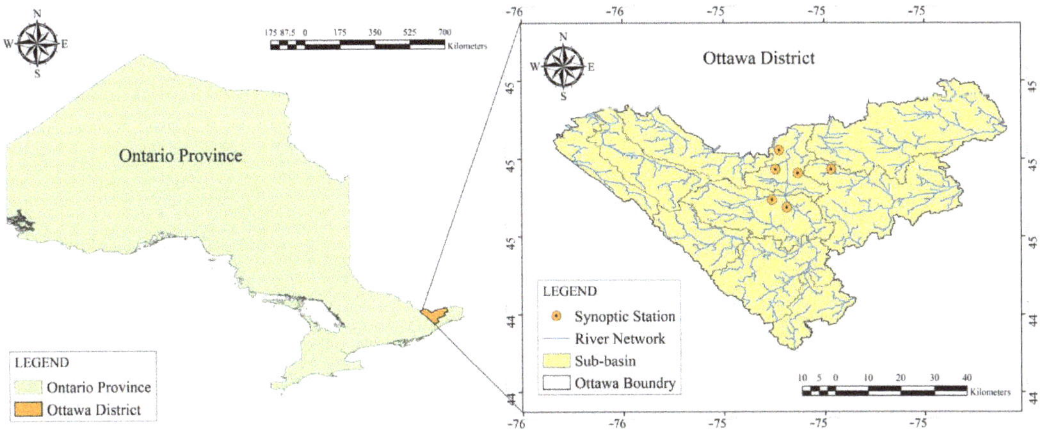

Figure 1. Location of the study area.

2.2. Methodology

The main objective of this paper is to determine flood-susceptible areas based on the combination of MCDM and GIS. Afterward, the Importance Weights (IWs) value of the qualitative criteria and *FHI* are calculated using the AHP method, respectively [21]. The AHP approach is one of the MCDM models that address the complexity involved in the decision-making process using Pairwise comparisons and considering the effects of quantitative criteria [9,22]. The Saaty's scale is proposed in this technique for scoring the importance value of criteria [23] (see Table 1).

Table 1. The Saaty's scale for scoring the importance value of criteria [9].

Numerical Value	Definition
1	Equal important
3	Weakly more important
5	Strongly more important
7	Very Strongly more important
9	Absolutely more important
2, 4, 6, and 8	Intervals of adjacent expressions

Combining the AHP method and a GIS tool into a decision support system holds significant promise for enhancing decision-making in the planning process [15]. The GIS-based hierarchical mode, incorporating qualitative spatial layers and expert opinions, allows the prioritizing of criteria by utilizing the Pairwise Comparison Matrix (PCM) [15]. Moreover, there are different criteria in this study, and there is a lack of agreement among experts within this field due to the presence of uncertainty. In these situations, applying fuzzy memberships to reduce uncertainties is highly recommended. Therefore, in this study, a hierarchical model has been applied based on a combination of AHP methods and GIS in the form of fuzzy sets. First of all, various spatial layers and maps were used and introduced in ArcGIS. However, the present study selected eight criteria, including drainage density, land use/land cover, precipitation, slope, soil-type, elevation, geology, and distance from the river, according to their essential role in flood region selection.

In the next stage, the weight assigned to each criterion has been established through the application of a GIS-based hierarchical integration model [15], respectively. The weighting procedure has been fully implemented in ArcGIS software based on raster layer analysis with 30m cell size. Having selected the criteria, it is imperative to carry out the subsequent steps sequentially. The arrangement of the hierarchical model based on GIS is depicted in Figure 2.

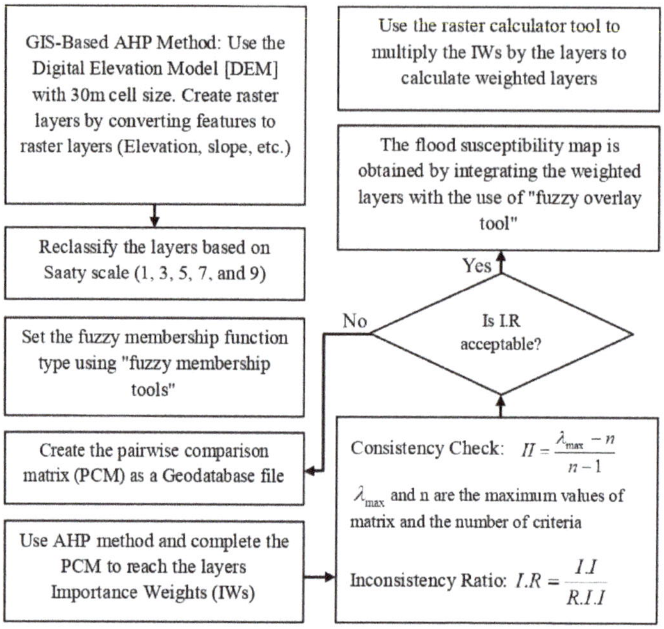

Figure 2. The arrangement of the GIS-based hierarchical model.

In the following, the classification of the layers has been redefined in accordance with common interval scales, which are depicted in Table 1 [24]. The subsequent phase entails computing the criteria weights by applying the AHP based on the PCM [9]. Employing the fuzzy membership tool to establish the nature of the fuzzy membership function is advisable. Lastly, the "fuzzy overlay tool" is utilized to combine the specific weights of each raster layer. Then, the validity of the PCM must be examined. The AHP employs the inconsistency ratio to assess the compatibility of the experts' opinions with the questionnaire. Prior to commencing data processing within a GIS framework, it is crucial to calculate the Inconsistency Index (*I.I*), the Random Inconsistency Index (*R.I.I*), and $\lambda_{\max}$. The calculation of *I.I* can be performed as follows:

$$I.I = \frac{\lambda_{\max} - n}{n - 1} \quad (1)$$

The maximum eigenvalue of the matrix, denoted as $\lambda_{\max}$ and the number of criteria by the variable n, are represented in the Equation (1).

The Ratio of Inconsistency (*I.R*) is obtained by dividing the value of *I.I* by the value of *R.I.I* as stated in Equation (2). The values of *R.I.I* for the matrix are presented in Table 2.

$$I.R = \frac{I.I}{R.I.I} \quad (2)$$

The Inconsistency ratio must be 0.1 or less for the comparisons to be consistent and the respondents to be valid.

Table 2. The Random Inconsistency Index values of matrices.

n	1	2	3	4	5	6	7	8	9	10
R.I.I	0	0	0.58	0.9	1.12	1.24	1.32	1.41	1.45	1.49

The GIS tool was utilized for specifying the weight of criteria in the subsequent step due to its remarkable ability to store and integrate spatial layers. The spatial layers for each criterion were created using ArcGIS software. Each layer was considered the main quantitative criterion for the flood susceptibility mapping process. The primary classification of the critical criteria for flood sustainability mapping in the Ottawa district and their attributed requirements for converting to raster layers are briefly introduced in Table 3. It should be noted that preparing the precipitation data layer involved gathering statistical and synoptic gauge data from the study area, covering the period from 1985 to 2022, focusing on obtaining the long-term annual average. The process of interpolating precipitation data was carried out using the Kriging method, which resulted in the transformation of the data into a raster layer. Furthermore, the drainage density criterion was determined by dividing the total length of the river network by the area of the watershed. The drainage density is calculated using Equation (3) [15].

$$U = \frac{\sum L_i}{A} \tag{3}$$

where L_i stands for the length of the river system measured in kilometers, while A represents the watershed area in square kilometers.

Table 3. Categorization of spatial layers of each criterion with their requirements.

Criteria	References	Attribute Class
Slope	Digital Elevation Model (DEM) of Canada, cell size 30 m	It is of greater significance to have a lower value Slopes 0–2%: the most critical, Slopes exceeding 45%: the least significant [15]
Elevation	DEM of Canada, cell size 30 m	A lower value has higher significance, Altitudes higher than 2500 m have the least importance [15]
Geological class	Canada Geological map with a scale of 1:250,000	The higher value represents more importance of sensitivity level
Precipitation	Annual long-term average precipitation of 6-gauge stations from 1985 to 2022.	More amount of precipitation is more important
Land-Use	Canada Land-use/land cover map with a scale of 1:500,000	Denser landcover has lower importance
Soil-Type	Soil-type classification map with a scale of 1:1,000,000	Soil near the ground level is more significant than the deeper soil below. Additionally, soils that can either shrink or expand are of lesser significance [15]
Distance from River	Shape file rivers of Ontario	Having a shorter distance to the river is considered more important
Drainage Density	shape file rivers and sub-basin boundaries of Ontario	Higher river density of the basin is more important

In the next stage, the reclassification tool is employed to categorize the raster layers according to the available classification system. A fuzzy membership tool based on linear fuzzy sets is utilized to transform the raster layers into fuzzy numbers to accomplish this task. Subsequently, the resulting fuzzified raster layers are subjected to reclassification utilizing the Saaty scale, as presented in Table 1. Then, the reclassified fuzzy raster layers are evaluated using an AHP-based GIS approach to calculate the weight of the fuzzy raster layers. Additionally, the Inconsistency Ratio of PCM is determined using Equations (1) and (2). Furthermore, a demonstration of how to calculate the weighted elevation criterion and form a weighted layer can be seen in Figure 3.

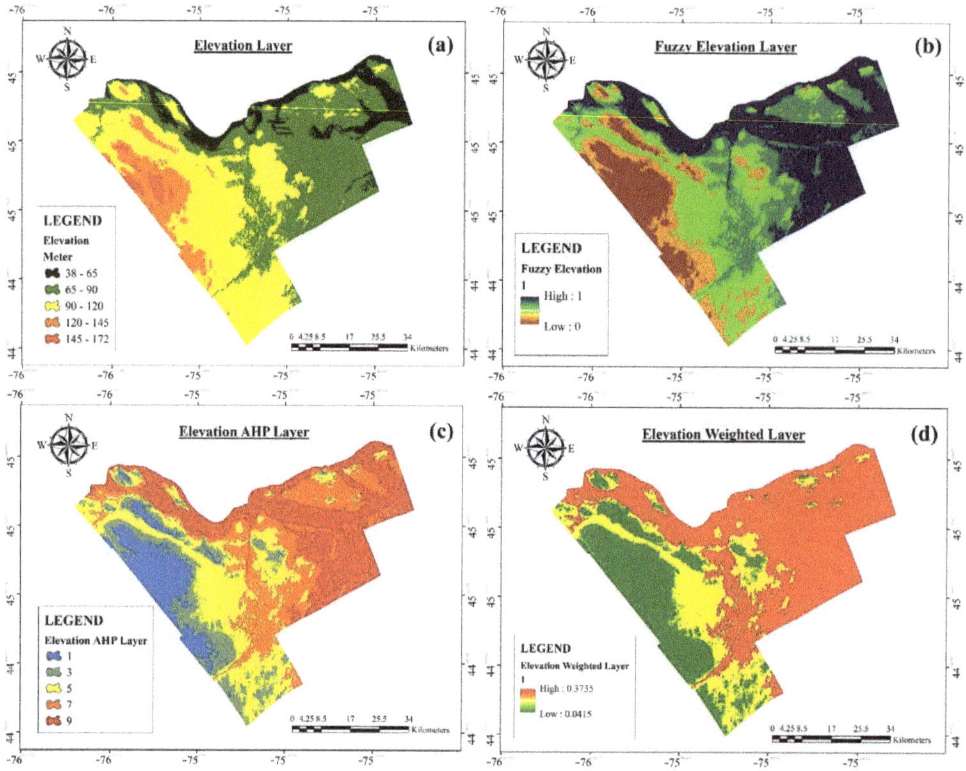

Figure 3. The process of fuzzy weighted Elevation layer within a hierarchical GIS model: (**a**) Elevation raster layer, (**b**) Fuzzy Elevation layer, (**c**) Reclassified AHP layer, (**d**) Weighted Elevation layer.

In the next step, the *FHI* is calculated with the following Equation (4) to evaluate the flooding probability rate:

$$FHI \sum_{i=1}^{n} WiRi \tag{4}$$

where *Ri* is each criterion's raster layer, *Wi* is each criterion's weight, and *n* corresponds to the number of the criteria. Therefore, Equation (5) is written in the following form:

$$\begin{aligned} FH = W_{EL}R_{EL} + W_{LULC}R_{LULC} + W_{DD}R_{DD} + W_{GEO}R_{GEO} \\ +W_{SL}R_{SL} + W_{SP}R_{SP} + W_{PR}R_{PR} + W_{DR}R_{DR} \end{aligned} \tag{5}$$

In this research, to attain *FHI* within a GIS environment, the weights of each criterion are calculated by multiplying them in their respective raster layer through the utilization of the Raster Calculator tool.

Finally, to determine the final flood susceptibility map of the study area, all the fuzzy weighted layers are integrated using the Fuzzy Overlay tool. In the following, the categorization of weighted fuzzy overlay layers of criteria is shown in Figure 4.

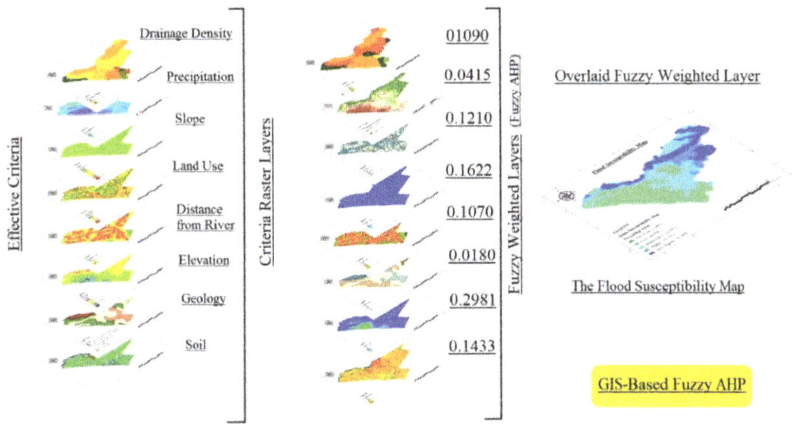

Figure 4. Classification of overlaying weighted fuzzy layers' process based on the criteria specified.

3. Results and Discussion

Table 4 presents the results of the PCM and the average weight per layer. Furthermore, Table 4 showcases the calculated inconsistency ratios of each PCM, which were determined using Equations (1) and (2). In this study, the criteria are eight and the result $R.I.I = 1.41$. Finally, the consistency ratio has been calculated $I.R = 0.041$, since the $I.R$ value was inferior to 0.1 and the consistency of the weight was accepted. Based on the obtained IWs of criteria in Table 4, the precipitation, slope, and soil criteria with the values of 0.298, 0.162, and 0.143 have the highest IW for the flood assessment, respectively. In addition, the geology and elevation criteria with values of 0.018 and 0.041 are the least important in assessing food susceptible areas of the case study.

Table 4. The results of the PCM for the spatial layers and their respective weights.

Layer Name	Soil	Slope	Precipitation	Land Use	Geology	Elevation	Drainage Density	Distance from River	Weight
Soil	1	0.33	0.2	3	6	7	5	3	0.143
Slope	3	1	0.2	5	7	1	0.25	7	0.162
Precipitation	5	5	1	4	8	3	4	7	0.298
Land Use	0.33	0.2	0.25	1	7	7	7	0.2	0.121
Geology	0.167	0.143	0.125	0.143	1	0.5	0.2	0.143	0.018
Elevation	0.143	1	0.33	0.143	2	1	0.167	0.143	0.041
Drainage Density	0.2	4	0.25	0.143	5	6	1	1	0.109
Distance from River	0.33	0.143	0.143	5	7	7	1	1	0.107
I.R	0.041	The matrix is satisfactory in term of consistency							

In addition, the *FHI* was found to evaluate the rate of flooding probability [25], which was calculated as follow:

$$\begin{aligned} FH = {} & 0.041R_{EL} + 0.121R_{LULC} + 0.109R_{DD} + 0.018R_{GEO} \\ & + 0.143R_{SL} + 0.162R_{SP} + 0.298R_{PR} + 0.107R_{DR} = 5.23 \end{aligned} \quad (6)$$

The proposed method, which considers a multitude of quantitative criteria, is utilized to achieve the mapping of flood susceptibility. The final map of the flood susceptibility mapping of the study area was constructed and classified into five major classes with flood potentiality from very low to very high (see Figure 5). Based on Figure 5, 3.18% of the study area represents the very low class, 40.93% of the area represents the low class, 24.37% of the area represents the moderate class, 15.27% of the area represents the high class, and 16.24% of the area represents the very high class.

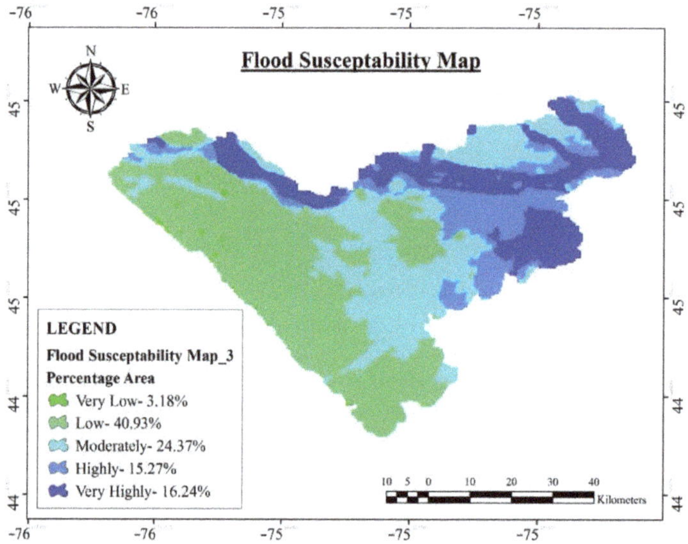

Figure 5. The flood susceptibility map of the Ottawa district.

4. Conclusions

Flood susceptibility mapping is an effective technique that allows for reducing flood hazard dangers in order to assist decision-makers to have proper management over the prone areas, and thereafter ensure appropriate and sustainable socio-economic development. In this paper, to produce the flood susceptibility map of the Ottawa district in southern Ontario, Canada, a MCDM hierarchical approach has been proposed based on the AHP method and GIS with the capability to check the consistency of the obtained model. Eight flood influencing criteria were considered for mapping the flood-susceptible areas, i.e., precipitation, soil, distance from the river, drainage density, elevation, land use/land cover, geology, and slope data. The Importance Weights (IWs) of criteria were evaluated using a hierarchical model based on Geographic Information System (GIS). This approach was employed to establish the IW of each criterion. The evaluation results of the IWs of the precipitation, slope, and soil criteria play a prominent role in investigating flood-prone areas. Moreover, the geology and elevation criteria are far less important than other criteria and have less impact on flood susceptibility mapping. Further, the flood-susceptibility map is classified into five major classes with flood potentiality from very low to very high. We find 3.18% (very low), 40.93% (low), 24.37% (moderate), 15.27% (high), and 16.24% (very high). After that, *FHI* was calculated to evaluate the impact of each criterion on the method, which leads to a better understanding of each criterion on the flood susceptibility map. Finally, the flood susceptibility map presented in this paper will serve as a valuable tool and will play a vital role in assessing flood management not only in the Ottawa region but also in other regions prone to flood events within the country.

Author Contributions: Conceptualization: A.N. and H.B.; methodology: A.N. and H.B.; software: A.N.; validation, A.N.; formal analysis, H.B. and A.N.; data curation, A.N.; writing—original draft preparation, A.N. and H.B.; writing—review and editing, H.B.; visualization, A.N. and H.B.; supervision: H.B. All authors have read and agreed to the published version of the manuscript.

Funding: This research received no external funding.

Institutional Review Board Statement: Not applicable.

Informed Consent Statement: Not applicable.

Data Availability Statement: Data available on request due to restrictions e.g., privacy or ethical.

Conflicts of Interest: The authors declare no conflict of interest.

References

1. Yahaya, S.; Ahmad, N.; Abdalla, R.F. Multicriteria analysis for flood vulnerable areas in Hadejia-Jama'are River basin, Nigeria. *Eur. J. Sci. Res.* **2010**, *42*, 71–83.
2. Rahmati, O.; Haghizadeh, A.; Stefanidis, S. Assessing the Accuracy of GIS-Based Analytical Hierarchy Process for Watershed Prioritization; Gorganrood River Basin, Iran. *Water Resour. Manag.* **2016**, *30*, 1131–1150. [CrossRef]
3. Green, C.H.; Parker, D.J.; Tunstall, S.M. Assessment of flood control and management options. Thematic Review IV.4 prepared as an input to the World Commission on Dams, Cape Town. Available online: https://citeseerx.ist.psu.edu/document?repid=rep1&type=pdfdoi=4dc46bb21cde02a58213f2adba2d1216216a7529 (accessed on 1 November 2000).
4. Fekete, A.; Sandholz, S. Here Comes the Flood, but Not Failure? Lessons to Learn after the Heavy Rain and Pluvial Floods in Germany 2021. *Water* **2021**, *13*, 3016. [CrossRef]
5. Swain, K.; Singha, C.; Nayak, L. Flood Susceptibility Mapping through the GIS-AHP Technique Using the Cloud. *ISPRS Int. J. Geo-Inf.* **2020**, *9*, 720. [CrossRef]
6. Black, R.; Kniveton, D.; Schmidt-Verkerk, K. Migration and Climate Change: Towards an Integrated Assessment of Sensitivity. *Environ. Plan. A Econ. Space* **2011**, *43*, 431–450. [CrossRef]
7. Hapuarachchi, H.A.P.; Wang, Q.J.; Pagano, T.C. A review of advances in flash flood forecasting. *Hydrol. Process.* **2011**, *25*, 2771–2784. [CrossRef]
8. Bout, B.; Jetten, V. The validity of flow approximations when simulating catchment-integrated flash floods. *J. Hydrol.* **2018**, *556*, 674–688. [CrossRef]
9. Saaty, T.L.; Vargas, L.G. The possibility of group choice: Pairwise comparisons and merging functions. *Soc. Choice Welf.* **2012**, *38*, 481–496. [CrossRef]
10. Salimi, A.H.; Noori, A.; Bonakdari, H.; Samakosh, J.M.; Sharifi, E.; Hassanvand, M.; Gharabaghi, B.; Agharazi, M. Exploring the Role of Advertising Types on Improving the Water Consumption Behavior: An Application of Integrated Fuzzy AHP and Fuzzy VIKOR Method. *Sustainability* **2020**, *12*, 1232. [CrossRef]
11. Al-Abadi, A.M.; Shahid, S.; Al-Ali, A.K. A GIS-based integration of catastrophe theory and analytical hierarchy process for mapping flood susceptibility: A case study of Teeb area, Southern Iraq. *Environ. Earth Sci.* **2016**, *75*, 687. [CrossRef]
12. Noori, A.; Bonakdari, H.; Morovati, K.; Gharabaghi, B. The optimal dam site selection using a group decision-making method through fuzzy TOPSIS model. *Environ. Syst. Decis.* **2018**, *38*, 471–488. [CrossRef]
13. Noori, A.; Bonakdari, H.; Salimi, A.H.; Gharabaghi, B. A group Multi-Criteria Decision-Making method for water supply choice optimization. *Socio-Econ. Plan. Sci.* **2021**, *77*, 101006. [CrossRef]
14. Li, Q.; Zhou, J.; Liu, D.; Jiang, X. Research on flood risk analysis and evaluation method based on variable fuzzy sets and information diffusion. *Saf. Sci.* **2012**, *50*, 1275–1283. [CrossRef]
15. Noori, A.; Bonakdari, H.; Hassaninia, M.; Morovati, K.; Khorshidi, I.; Noori, A.; Gharabaghi, B. A reliable GIS-based FAHP-FTOPSIS model to prioritize urban water supply management scenarios: A case study in semi-arid climate. *Sustain. Cities Soc.* **2022**, *81*, 103846. [CrossRef]
16. Zou, Q.; Zhou, J.; Zhou, C.; Song, L.; Guo, J. Comprehensive flood risk assessment based on set pair analysis-variable fuzzy sets model and fuzzy AHP. *Stoch. Environ. Res. Risk Assess.* **2013**, *27*, 525–546. [CrossRef]
17. Hammami, S.; Zouhri, L.; Souissi, D.; Souei, A.; Zghibi, A.; Marzougui, A.; Dlala, M. Application of the GIS based multi-criteria decision analysis and analytical hierarchy process (AHP) in the flood susceptibility mapping (Tunisia). *Arab. J. Geosci.* **2019**, *12*, 653. [CrossRef]
18. Msabi, M.M.; Makonyo, M. Flood susceptibility mapping using GIS and multi-criteria decision analysis: A case of Dodoma region, central Tanzania. *Remote Sens. Appl. Soc. Environ.* **2021**, *21*, 100445. [CrossRef]
19. Souissi, D.; Zouhri, L.; Hammami, S.; Msaddek, M.H.; Zghibi, A.; Dlala, M. GIS-based MCDM—AHP modeling for flood susceptibility mapping of arid areas, southeastern Tunisia. *Geocarto Int.* **2020**, *35*, 991–1017. [CrossRef]
20. Rincón, D.; Khan, U.T.; Armenakis, C. Flood Risk Mapping Using GIS and Multi-Criteria Analysis: A Greater Toronto Area Case Study. *Geosciences* **2018**, *8*, 275. [CrossRef]

21. Drobne, S.; Lisec, A. Multi-attribute decision analysis in GIS: Weighted linear combination and ordered weighted averaging. *Informatica* **2009**, *33*, 459–474.
22. Noori, A.; Bonakdari, H.; Morovati, K.; Gharabaghi, B. Development of optimal water supply plan using integrated fuzzy Delphi and fuzzy ELECTRE III methods—Case study of the Gamasiab basin. *Expert Syst.* **2020**, *37*, e12568. [CrossRef]
23. Saaty, T.L. *Planning, Priority Setting, Resource Allocation*; RWS Publications: Pittsburgh, PA, USA, 1980; p. 281.
24. de Voogd, N.; Rooijen, H.D. Thin-layer response and spectral bandwidth. *Geophysics* **1983**, *48*, 12–18. [CrossRef]
25. Elkhrachy, I. Flash Flood Hazard Mapping Using Satellite Images and GIS Tools: A case study of Najran City, Kingdom of Saudi Arabia (KSA). *Egypt. J. Remote Sens. Space Sci.* **2015**, *18*, 261–278. [CrossRef]

environmental sciences proceedings

MDPI

Proceeding Paper

The Impacts of Climate Change on Monsoon Flood Situations in Pakistan †

Hafsa Muzammil [1,2,*], Muhammad Zaman [1,*], Muhammad Adnan Shahid [1,2,*], Muhammad Safdar [1,2], Muhammad Danish Majeed [1,2] and Rehan Mehmood Sabir [1,2]

1 Department of Irrigation & Drainage, University of Agriculture, Faisalabad 38000, Punjab, Pakistan
2 Agricultural Remote Sensing Lab (ARSL), University of Agriculture, Faisalabad 38000, Punjab, Pakistan
* Correspondence: hafzamuzammal990@gmail.com (H.M.); muhammad.zaman@uaf.edu.pk (M.Z.); adnan.wmrc@gmail.com (M.A.S.)
† Presented at the 7th International Electronic Conference on Water Sciences, 15–30 March 2023; Available online: https://ecws-7.sciforum.net.

Abstract: In this study, we present a comprehensive and detailed analysis to identify and quantify spatial patterns of heavy, very heavy, and extremely heavy rainfall, as well as the trends that have developed over the past seven decades (1951 to 2020) of the monsoon months (June to September) under a warming scenario. We also project these extreme rainfall counts during the near (2036–2060) and late 21st century (2075–2099) for comparison to the historical period (1990–2014). The 5-day maximum rainfall over the provinces of Sindh and Baluchistan is currently about 75% more intense than it would have been without the climate warming by 1.2 °C, and the 60-day rainfall across the basin is currently about 50% more intense. This means that heavy rainfall is now more likely to occur. Due to the high-level of rainfall variability in the area, there is significant uncertainty regarding these estimations, and the causes of the observed changes are not just limited to climate change. However, most of the models and observations we have analyzed for the 5-day rainfall extreme indicate that severe rainfall has been heavier than Pakistan has warmed. According to some of these models, the intensity of the rainfall might have increased by 50% due to climate change under the 5-day event threshold. The model predicts that rainfall intensity will greatly rise in the future for the 5-day event in a climate that is 2 °C warmer than it was in pre-industrial times, while the uncertainty is still quite high for the 60-day monsoon rainfall.

Keywords: climate change; floods; heavy monsoon; national disaster management authority; precipitation variations

Citation: Muzammil, H.; Zaman, M.; Shahid, M.A.; Safdar, M.; Majeed, M.D.; Sabir, R.M. The Impacts of Climate Change on Monsoon Flood Situations in Pakistan. *Environ. Sci. Proc.* **2023**, *25*, 63. https://doi.org/10.3390/ECWS-7-14255

Academic Editor: Athanasios Loukas

Published: 16 March 2023

1. Introduction

The Indus Basin contains 1.10 million square kilometers, with Pakistan contributing 63 percent, India 29 percent, and Afghanistan and China 8 percent. These entire provinces, such as the province of Khyber Pakhtunkhwa, Sindh, and towards the eastern part of Punjab and Baluchistan, cover the Indus Basin River [1]. The Indus River arises in Tibet (China) at Mount Kailash (Mansarovar Lake) and is divided into two sections: Upper Indus and Lower Indus. The water flows from the upper side of the Indus and downstream to the Guddu barrage. The glaciers of Hindu Kush and Karakoram supply water to the Indus River. However, it is the main source of fresh water and it supports 90 percent of agricultural land, and industrial and household requirements [2].

The Indu's basin hydrology is influenced by the interaction of three distinct regimes and their responses to climatic conditions: the glacial, nival (snow melt), and rainy regimes. To assess how climate change may affect the Indus Basin, an understanding of the hydrological processes that determine river flow is essential [3]. About 25 to 35 percent of the water flow in the Indus River is generated by the glacial phase. Due to the region's naturally diversified geography and climate, its flow patterns are influenced by a wide range of responses. Summer temperature is positively connected with runoff rates in the glacial regime (i.e., a higher temperature led to more runoff) and negatively correlated with summer precipitation [4].

Climate Change will have a significant impact on global socio-ecological systems in the future decades, affecting the hydrological cycle, agricultural production, and basic ecosystems [5]. Rising temperatures in basins that rely heavily on glaciers melting will almost certainly result in a rise in stream flow in the short-term, but a decrease in the long-term. This is because the overall quantity of glacier melt is a tradeoff between increasing melt rates on the one hand and decreasing glacier volumes on the other. When the trend of glacier melt shifts from positive to negative [6], due to climate change, this warming effect is predicted to influence rainfall patterns and increase the frequency of extreme weather events, such as flooding, drought, and other natural disasters. Furthermore, rising CO_2 concentrations will have a direct impact on crop growth and development, affecting agricultural land use and crop yields [7].

Planning for climate impacts, creating resistance to such impacts, and enhancing civilization's capabilities to react and recover are all aspects of adjustment [8]. This may help to decrease the damages and disruptions caused by climate change. Governments should play a bigger role in promoting water demand reduction initiatives. Considering the risks of increasing per capita water shortages, provincial governments should take initiative by increasing agricultural, industrial, municipal, and domestic water conservation [9].

This should include more investment for programs that encourage water conservation practices, such as flood and rainwater harvesting. Water pricing rules that properly reflect the true cost of water usage should also be implemented by governments [10].

2. Data and Methods

Observational Data

The National Climate Centre (NCC) of The Pakistan Meteorological Department (PMD) provided the daily gridded rainfall dataset with a spatial resolution of 0.25° × 0.25° during the monsoon period (June—September) in the Pakistan region between 1951 and 2020. A network of 47 stations dispersed throughout the Pakistan landmass was used to create the gridded dataset (Figure 1).

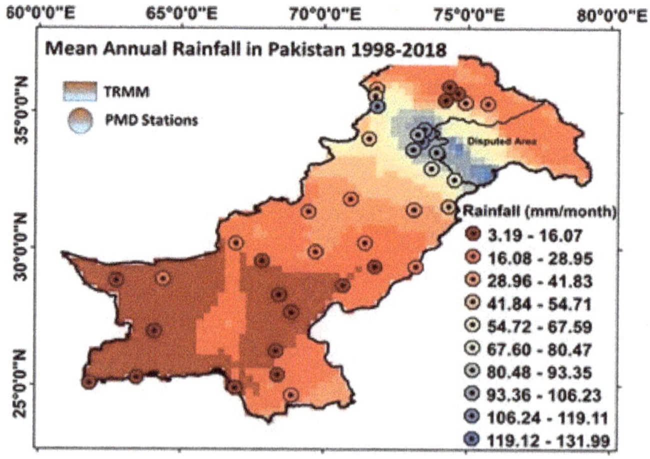

Figure 1. Pakistan Station point [11].

One of the main ways to understand how the climate has changed in the past and how it might change in the future is using climate models. In this analysis, we evaluate predicted changes in the frequency of extreme rainfall events for the near 21st century

(2036–2060) and late 21st century (2075–2099), with a primary focus on the distribution of heavy, very heavy, and extremely heavy rainfall.

With the historical period typically beginning in 1950 (or 1970) and ending in 2005, followed by the Shared Socio-economic Pathways scenarios until the 21st century, we used the model simulations for South Asia that were available from the CMIP 6 (Couple Model Intercomparison Project Phase 6). Global Climate Model (GCM) simulations are downscaled and used to create regional climate model (RCM) simulations.

3. Result and Discussion

The research presented is a comprehensive understanding of how CIMP6 models reflect extreme events at national and global levels. To understand the changes in extreme occurrences, the visualizations offer several viewpoints on past conditions, as well as possibly the future. Future periods and scenarios with a focus on 2025, 2050, 2075, and 2085 can be compared to the baseline climate (Historical Period, 1985–2014, centered on 2000). The extreme indicators do not stand for location-specific (station level) extremes but rather qualitative projection results that directly reflect global model output.

When compared to the mean precipitation, extreme precipitation events often exhibit higher magnitudes of change and distinct indications. As Earth gets warmer, the ability of air to carry more moisture increases exponentially, thus increasing the possibility of more precipitation. This can increase the risk of flooding, since extreme weather events are expected more frequently. Only in regions where precipitation is much less frequent can a trend toward more rainfall be reversed, resulting in an increase in return periods rather than a decrease in the frequency of major occurrences as shown in Figure 2.

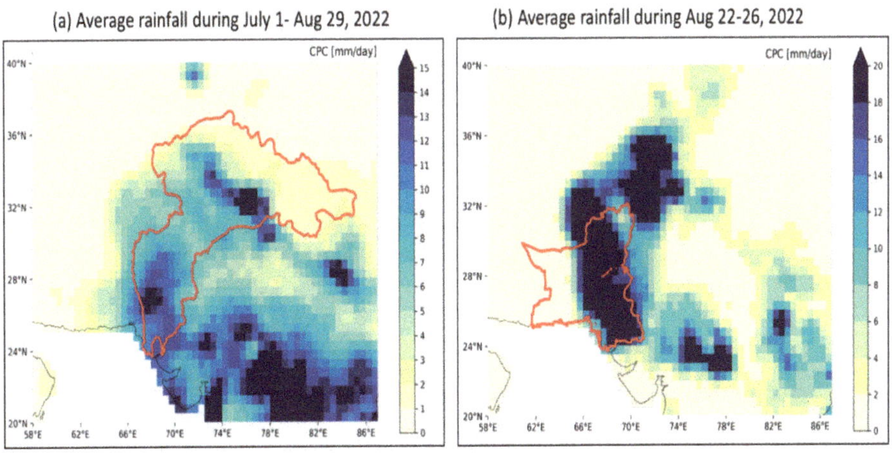

Figure 2. Annual maximum of the mean 60-day precipitation and the annual maximum of the mean 5-day precipitation in monsoon season at the site of Sindh and Baluchistan [12].

The characteristics of the climate as represented by extreme events and distinct from those of the long-term means since they reflect unusual weather events. For a limited number of repetition intervals, return levels (precipitation amounts displayed in mm) are available. In addition, the relationship between the return periods (represented in years) and the spatial distribution of fixed precipitation magnitudes (also known as return levels) can be examined. When the annual exceedance probability is needed, the inverse of the return period is needed. The historical return levels are then used to assess potential adjustments to the related reappearance interval (return period in years). The colors show, for instance, whether an occurrence that has traditionally occurred every 20 years will occur more frequently in the future (green) or less frequently (brown). Projections are therefore displayed as future return times (in years) that refer to the current reference point.

This future change in return period can also be expressed as a change in the chance of future annual extremes. This product, displayed as a factor, supplies a potential increase or decrease in the frequency of an event as shown in Figures 3–6.

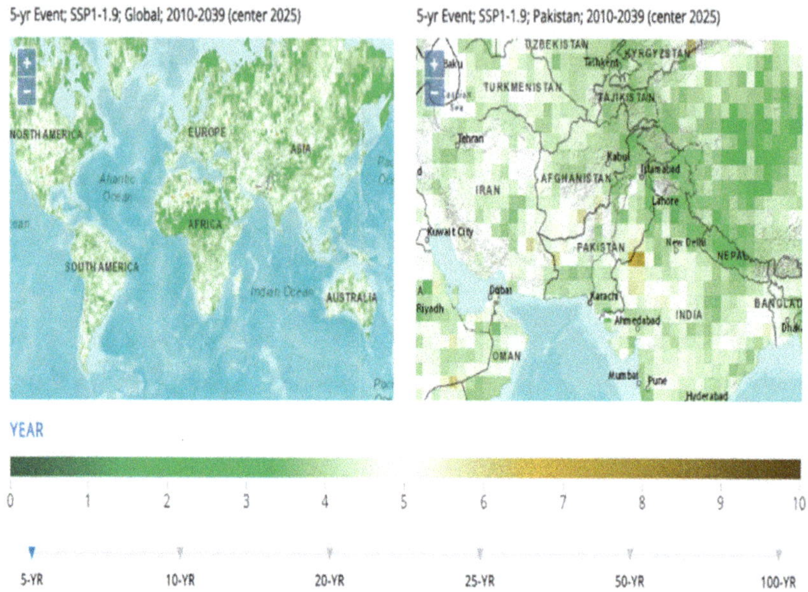

Figure 3. Global precipitation from the year (2010–2039); Source: [13].

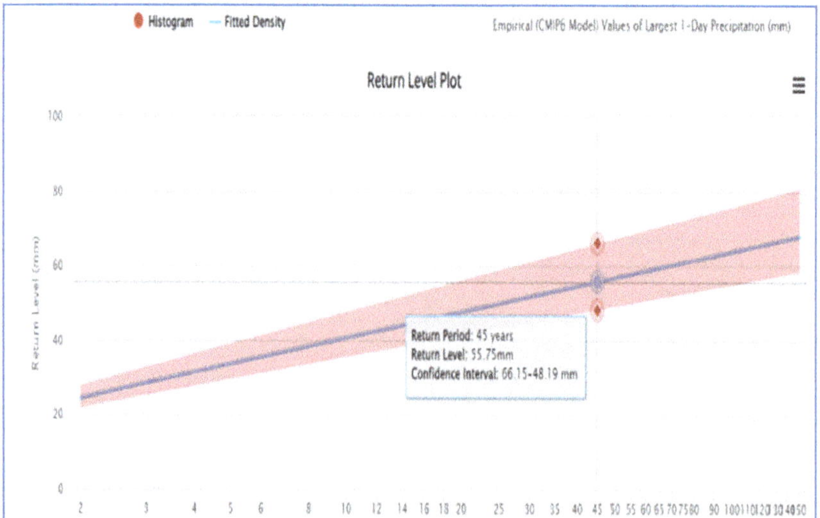

Figure 4. Precipitation return level according to long scale; Source: [13].

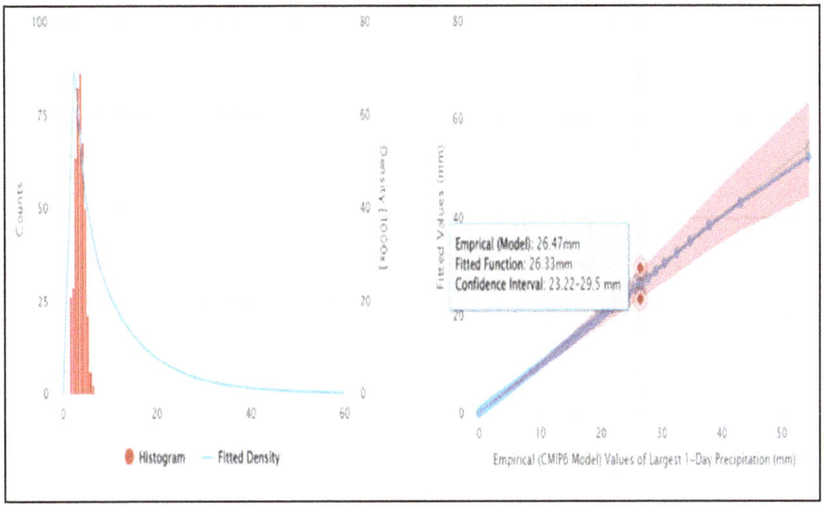

Figure 5. Compare Model and observation value of Precipitation; Source: [13].

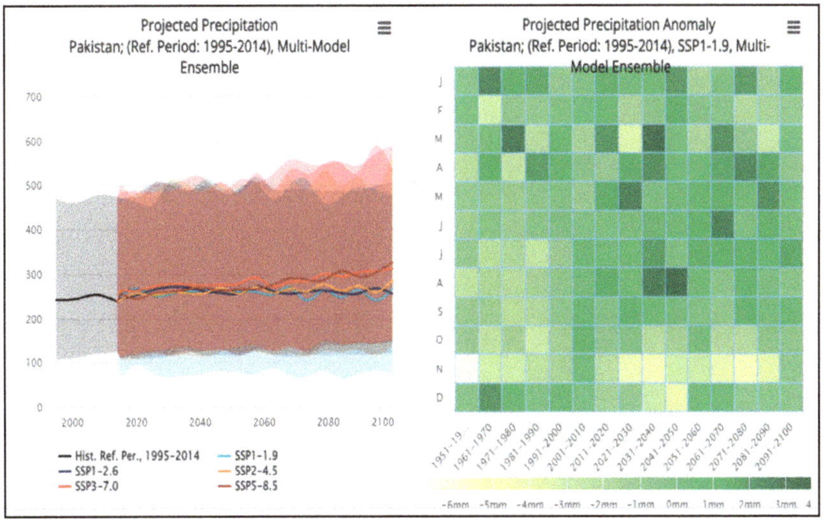

Figure 6. Projected Precipitation in Pakistan; Source: [13].

Table 1 shows the frequency of flood events for different return periods (5, 10, 20, 25, and 50 years) and magnitudes (10th, median, and 90th percentiles) for historical data. The values in each cell represent the flood magnitude in cubic meters per second. Table 1 shows the historical values of precipitation events at different levels, from 25 mm to 200 mm, and their corresponding frequency distributions, represented by the 10th, 50th (median), and 90th percentiles. For example, in historical data, the median precipitation event at 25 mm level occurred 3.16 times, while the 90th percentile precipitation event at 200 mm level occurred 11,719 times.

Table 1. Extreme Precipitation events from the years 1985 to 2014; Source [13].

Return Level, Historical:1985–2014 (Center 2000) (Occurrence Year)															
Event	5 yr			10 yr			20 yr			25 yr			50 yr		
	10	Median	90	10	Median	90	10	Median	90	10	Median	90	10	Median	**90**
Historical	19.9	32.94	53.1	25.12	41.06	58.42	30.55	50.34	85.97	32.24	53.77	91.76	44.58	79.24	140
Event	25 mm			50 mm			100 mm			150 mm			200 mm		
	10	Median	90	10	Median	90	10	Median	90	10	Median	90	10	Median	**90**
Historical	0.97	3.16	11.9	4.27	28.62	377	92.60	852.68	6584	330.3	3160	10,321	792.6	4500	11,719

Table 2 provides information on the intensity of rainfall for different events and return periods. The events are represented by the amount of rainfall in millimeters (mm) for 25 mm, 50 mm, 100 mm, 150 mm, and 200 mm. The return period is expressed in years, ranging from a 10-year to a 90-year return period. The data is presented for two scenarios: historical and a certain location. In the historical scenario, the data shows the median (50th percentile) and 90th percentile rainfall for each event and return period. For instance, the median rainfall amount for the 25 mm event is 0.19 mm, while the 90th percentile rainfall is 0.62 mm. Similarly, for the 200 mm event, the median rainfall amount is 0.02 mm, while the 90th percentile rainfall is 0.01 mm. In the second scenario, the table shows the median and 90th percentile rainfall for the same events and return periods but for a specific location. The location is not mentioned in the table, but it can be assumed that the values represent a location with a certain climate or weather pattern.

Table 2. Annual Exceedance Probability of precipitation.

Historical:1985–2014 (Center 2000) (Occurrence Year)															
Event	25 mm			50 mm			100 mm			150 mm			200 mm		
	10	Median	90	10	Median	90	10	Median	90	10	Median	90	10	Median	90
Historical	0.19	0.62	1.41	0.02	0.08	0.32	0.00	0.01	0.04	0.00	0.00	0.02	0.00	0.00	0.01

Table 3 displays data on different events and the associated return periods in years for each of the five Shared Socioeconomic Pathways (SSPs). For each event and return period combination, the table shows the 10th percentile, median, and 90th percentile of the estimates. The events refer to different climate-related hazards such as floods, drought, and heatwaves, and the return period refers to the average number of years between events of a similar magnitude or severity. The SSPs represent different scenarios of future socio-economic development and associated greenhouse gas emissions. The SSP1-19 scenario assumes a future world with low challenges to mitigation and adaptation, SSP1-26 represents a future world with moderate challenges, SSP2-45 represents a world with intermediate challenges, SSP3-70 represents a world with high challenges, and SSP5-85 represents a world with very high challenges. The numbers in the table represent estimates of the impacts of climate change under each scenario, with higher numbers indicating more severe impacts. For example, the table shows that under SSP3-70, the median estimate for a 50-year flood event is 8.30, while the 10th percentile estimate is 3.20 and the 90th percentile estimate is 14.23. This suggests that there is considerable uncertainty in the estimates, but overall, the expected impact of climate change under SSP3-70 is relatively high compared to the other scenarios.

Table 3. Future Return Period (2010–2039) [13].

Event	5-yr			20-yr			25-yr			50-yr			100-yr		
	10	Median	90	10	Median	90	10	Median	90	10	Median	90	10	Median	90
SSP1-19	2.54	4.36	6.84	4.50	8.74	15.6	9.40	21.76	49.80	15.99	43.49	129.5	26.81	88.10	376
SSP1-26	2.68	4.09	6.57	4.70	7.97	14.81	9.74	19.41	46.40	16.48	37.94	118.4	27.8	74.56	327.0
SSP2-45	2.69	4.14	7.07	4.78	8.08	16.46	9.79	19.66	51.40	16.62	38.91	129.4	27.74	77.05	348.0
SSP3-70	2.71	4.14	6.43	4.75	8.30	14.23	9.70	20.94	44.37	16.41	41.96	118.1	27.74	84.2	336.4
SSP5-85	2.70	4.20	6.64	4.66	8.16	14.78	9.38	19.53	45.02	15.63	38.19	109.5	25.61	74.14	285.2

Table 4 shows the magnitude of different events, expressed as a percentage of a reference level of 10, for different return periods (5-year, 20-year, 25-year, 50-year, and 100-year) under different Shared Socioeconomic Pathways (SSPs). The first column "Event" represents the different types of events being considered, but no further details about the events are provided. The next six columns present the magnitude of the events for each return period, as well as the median, 90th, and 10th percentiles. For example, for the 5-year event, the table shows the median magnitude of the event, as well as the 90th and 10th percentiles, under each SSP. The table presents information for five different SSPs: SSP1-19, SSP1-26, SSP2-45, SSP3-70, and SSP5-85. Each of these SSPs represents a different scenario of future socioeconomic development and greenhouse gas emissions, as defined by the Intergovernmental Panel on Climate Change (IPCC). The information in this table can be used by decision-makers to assess the potential impacts of different events under different SSPs and to inform decision-making regarding adaptation and mitigation measures. It can also be used to assess the relative risks associated with different SSPs and to inform policymaking regarding climate change mitigation and adaptation strategies.

Table 4. Future Return Period of Precipitation (2035–2064) [13].

Event	5-yr			20-yr			25-yr			50-yr			100-yr		
	10	Median	90	10	Median	90	10	Median	90	10	Median	90	10	Median	90
SSP1-19	2.63	4.35	5.45	4.53	8.55	14.36	8.82	21.57	43.64	14.27	43.78	110.3	22.78	89.15	325.5
SSP1-26	2.30	3.73	6.24	3.94	7.14	13.96	6.60	19.41	46.40	16.48	37.94	118.4	27.8	74.56	288.0
SSP2-45	2.35	4..137	7.07	4.78	8.08	16.46	9.79	19.66	51.40	16.62	38.91	129.4	27.74	77.05	299.2
SSP3-70	2.27	3.61	5.96	3.2	8.30	14.23	9.70	20.94	44.37	16.41	41.96	118.1	27.74	84.2	336.4
SSP5-85	2.70	4.20	6.64	4.66	8.16	14.78	9.38	19.53	45.02	15.63	38.19	93.	18.99	59.09	238.3

Table 5 presents information about different events and their corresponding percentiles for different return periods (5-year, 20-year, 25-year, 50-year, and 100-year) based on a reference level of 10. The first column "Event" is the name or type of event being considered, but no further details about the event are given in this table. The next six columns show the percentile values for the magnitude of the event for each return period. For example, the 5-year event has a magnitude at the median of the distribution, as well as at the 90th and 10th percentiles. The magnitude is expressed as a percentage of the reference level of 10. The median represents the middle value of the distribution, with half of the events having a magnitude above the median and half having a magnitude below it. The 90th percentile is the magnitude above which only 10% of the events occur, while the 10th percentile represents the magnitude above which 90% of the events occur. The information in this table could be used by risk assessors and planners to estimate the probability of different events occurring and their corresponding magnitudes. This can help in designing the right measures to mitigate the risks and to protect people and property from potential damages caused by the events.

Table 5. Future Return Period of Precipitation (2060–2089) [13].

Event	5-yr			20-yr			25-yr			50-yr			100-yr		
	10	Median	90	10	Median	90	10	Median	90	10	Median	90	10	Median	90
SSP1-19	2.54	4.50	8.02	4.23	9.03	19.92	8.48	22.59	65.79	13.8	44.92	188.4	22.5	91.74	597.5
SSP1-26	2.27	3.65	6.31	3.78	6.94	14.05	7.34	16.32	43.88	11.5	31.21	113.7	18.0	60.10	331.0
SSP2-45	2.18	3.38	61.2	3.59	6.37	13.21	6.72	14.92	39.34	10.5	28.30	95.80	16.3	53.94	243.8
SSP3-70	1.78	2.95	5.17	2.93	5.48	10.77	5.46	12.47	29.80	8.49	23.56	67.33	13.1	44.08	163.5
SSP5-85	1.57	2.76	5.31	2.55	4.97	10.61	4.55	11.00	28.71	6.88	20.33	69.09	10.1	37.38	174.0

Table 6 presents information on the magnitude of events with different return periods (5-year, 20-year, 25-year, 50-year, and 100-year) under five different Shared Socioeconomic Pathways (SSPs): SSP1-19, SSP1-26, SSP2-45, SSP3-70, and SSP5-85. The SSPs are scenarios that describe possible future developments of the world's social, economic, and environmental systems based on assumptions about factors such as population growth, economic development, and technological change. The values in the table represent the magnitude of the event in a given year for each return period (e.g., the 5-year event has a 2.58 magnitude under SSP1-19, a 2.34 magnitude under SSP1-26, a 2.06 magnitude under SSP2-45, a 1.57 magnitude under SSP3-70, and a 1.43 magnitude under SSP5-85). The magnitude is expressed as a percentage of a reference level (e.g., 10% for the 5-year event) and is shown for three percentiles: the median, the 90th percentile, and the 10th percentile. This information can be used by planners and decision-makers to assess the risks associated with different scenarios and to develop strategies for managing them. For example, they can use this information to identify areas that are more vulnerable to extreme events and to plan infrastructure investments or adaptation measures to reduce their exposure to risks.

Table 6. Future Return Period of Precipitation (2070–2099) [13].

Event	5-yr			20-yr			25-yr			50-yr			100-yr		
	10	Median	90	10	Median	90	10	Median	90	10	Median	90	10	Median	90
SSP1-19	2.58	4.73	7.80	4.47	9.34	18.4	8.99	23.98	64.36	14.97	49.05	186.6	24.72	102.2	592.9
SSP1-26	2.34	3.73	6.47	4.01	7.06	14.3	7.85	16.67	43.76	12.91	32.10	110.7	20.66	61.83	306.6
SSP2-45	2.06	3.33	5.96	3.44	6.27	13.1	6.52	14.56	40.64	10.29	27.71	102.1	15.93	53.08	274.1
SSP3-70	1.57	2.71	4.62	2.72	4.95	9.5	4.96	11.18	27.01	7.62	20.37	61.07	11.72	37.53	154.5
SSP5-85	1.43	2.45	4.50	2.26	4.35	8.9	3.97	9.33	23.77	6.01	16.84	53.14	8.82	30.64	129.3

4. Conclusions

The flooding was a direct result of the excessive monsoon rain that occurred throughout the summer of 2022, which was made even worse by shorter periods of very heavy rain that affected the provinces of Sindh and Baluchistan, especially the month of August. Therefore, for the Indus basin and the two provinces, respectively, we evaluated the maximum rainfall over 60 days and over 5 days during the monsoon season.

In the current environment, the return time for both mentioned events are approximately 1 in 100 years. However, the amount of rain in the Indus basin varies greatly from year to year for a variety of reasons, including its close relationship to the ENSO cycle. Therefore, precise quantification is challenging.

We compared the trends in climate models with and without the human-induced increases in greenhouse gases to find the contribution of human-induced climate change to these observed changes. The affected areas are in the westernmost extremes of the monsoon region. Here, the properties of rainfall in the dry western and moist eastern parts differ significantly.

Many of the climate models that are now available have difficulty simulating these rainfall features. The incidence and intensity of extreme rainfall tend to change significantly less for those that pass our evaluation test than the trends we found in the observations. It is impossible to assess the overall contribution of human-induced climate change because of this gap, which points to the possibility that long-term variability or processes that our study may not have considered can have a significant impact. Our findings are consistent with current IPCC reports.

There is an urgent need to reduce Pakistan's vulnerability to extreme weather, both currently and in case of the possible future increase in high rainfall over Pakistan due to climate change.

Author Contributions: Conceptualization, H.M. and M.Z.; Introduction, M.A.S.; Materials and Methods, H.M., M.Z. and M.D.M.; Results and Discussions, M.S. and R.M.S.; Conclusions, M.Z. and M.A.S. writing original draft preparation, H.M.; editing, M.S. All authors have read and agreed to the published version of the manuscript.

Funding: This research received no external funding.

Institutional Review Board Statement: Not applicable.

Informed Consent Statement: Not applicable.

Data Availability Statement: Not applicable.

Conflicts of Interest: The authors declare no conflict of interest.

References

1. Asadieh, B.; Krakauer, N.Y. Global change in streamflow extremes under climate change over the 21st century. *Hydrol. Earth Syst. Sci.* **2017**, *21*, 5863–5874. [CrossRef]
2. Sharma, B.; Cai, X.; Decondappa, D. *Impact of Climate Change on Water Resources and Agricultural Production in the Indus Basin, South Asia*; IWRA: Madison, WI, USA, 2021.
3. Khan, A.; Koch, M.; Tahir, A. Impacts of Climate Change on the Water Availability, Seasonality and Extremes in the Upper Indus Basin (UIB). *Sustainability* **2020**, *12*, 1283. [CrossRef]
4. Chase, K.J.; Haj, A.E.; Regan, R.S.; Viger, R.J. Potential effects of climate change on streamflow for seven watersheds in eastern and central Montana. *J. Hydrol. Reg. Stud.* **2016**, *7*, 69–81. [CrossRef]
5. Eekhout, J.; de Vente, J. Assessing the effectiveness of Sustainable Land Management for large-scale climate change adaptation. *Sci. Total. Environ.* **2019**, *654*, 85–93. [CrossRef] [PubMed]
6. Lutz, A.F.; Immerzeel, W.; Kraaijenbrink, P.; Shrestha, A.; Bierkens, M. Climate change impacts on the upper indus hydrology: Sources, shifts and extremes. *PLoS ONE* **2016**, *11*, e0165630. [CrossRef] [PubMed]
7. Liu, Z.; Herman, J.D.; Huang, G.; Kadir, T.; Dahlke, H.E. Identifying climate change impacts on surface water supply in the southern Central Valley, California. *Sci. Total. Environ.* **2021**, *759*, 143429. [CrossRef] [PubMed]
8. Dash, J.W. Climate Change Risk Management. In *World Scientific Encyclopedia of Climate Change*; World Scientific Publishing Co. Pte Ltd.: Hackensack, NJ, USA, 2021; pp. 189–202.
9. Parry, J.-E.; Osman, H.; Terton, A.; Asad, S.; Ahmed, T. *The Vulnerability of Pakistan's Water Sector to the Impacts of Climate Change: Identification of Gaps and Recommendations for Action*; United Nations Development Programme: New York, NY, USA, 2017.
10. Li, Z.; Jin, J. Evaluating climate change impacts on streamflow variability based on a multisite multivariate GCM downscaling method in the Jing River of China. *Hydrol. Earth Syst. Sci.* **2017**, *21*, 5531–5546. [CrossRef]
11. Safdar, F.; Khokhar, M.F.; Mahmood, F.; Khan, M.Z.A.; Arshad, M. Observed and predicted precipitation variability across Pakistan with a special focus on winter and pre-monsoon precipitation. *Environ. Sci. Pollut. Res.* **2023**, *30*, 4510–4530. [CrossRef] [PubMed]
12. Blunden, J.; Boyer, T. State of the climate in 2020. *Bull. Am. Meteorol. Soc.* **2021**, *102*, S1–S475. [CrossRef]
13. Harris, I.; Osborn, T.J.; Jones, P.; Lister, D. *Climate Change Knowledge Portal Report. Sci. Data 7*; Climate Change Knowledge Portal: Washington, DC, USA, 2020.

environmental sciences proceedings

MDPI

Proceeding Paper

Groundwater Quality Assessment and Evaluation of Scaling and Corrosiveness Potential of Drinking Water Samples †

Aftab Alam * and Saurabh Kumar

Department of Civil Engineering, National Institute of Technology Patna, Bihar 800005, India; saurabh.ce18@nitp.ac.in

* Correspondence: aftaba.phd19.ce@nitp.ac.in; Tel.: +91-8797527215

† Presented at the 7th International Electronic Conference on Water Sciences, 15–30 March 2023; Available online: https://ecws-7.sciforum.net.

Abstract: This research was to examine water stability and to evaluate the drinking water quality. Groundwater samples from 16 borewells in Aurangabad, Bihar, were taken from the shallow unconfined aquifer and tested for a wide range of physicochemical characteristics. The pH, temperature, TDS, and EC were measured at the sites. Ca^{2+}, Mg^{2+}, F^-, Cl^-, $NO_3{}^-$, $SO_4{}^{2-}$, alkalinity, and hardness concentrations were examined in the laboratory. The groundwater's stability was measured using Corrosiveness Indices including the Langeliar saturation index (LSI), Ryznar stability index (RSI), Puckorius scaling index (PSI), Larson-Skold index (Ls), and Aggressivity index (AI). The data showed that typical values for LSI, RSI, PSI, and LS and AI were −0.92 (±0.47), 9.09 (±0.67), 9.50 (±0.73), 1.73 (±0.78), and 11.05 (±0.48), respectively. Groundwater WQI calculations revealed that 25% of the samples were excellent, 50% were good, 19% were poor, and 6% were extremely poor. All of the water samples tested positively for corrosiveness according to the LSI and PSI indices. Water samples show a strong corrosive potential (87.50% according to RSI) or a low corrosive tendency (12.50% according to RSI). 75% of the water samples have a strong corrosive tendency, 18.50% have a corrosive tendency, and 6.25% have a scaling tendency, according to Ls. According to AI, 93.75% of the water samples had a moderately corrosive tendency, whereas 6.25% were extremely corrosive.

Keywords: groundwater quality; water qk ality index (WQI); statistical analysis; scaling and corrosiveness potential

Citation: Alam, A.; Kumar, S. Groundwater Quality Assessment and Evaluation of Scaling and Corrosiveness Potential of Drinking Water Samples. *Environ. Sci. Proc.* **2023**, *25*, 64. https://doi.org/10.3390/ECWS-7-14316

Academic Editor: Athanasios Loukas

Published: 3 April 2023

1. Introduction

Water is essential for human survival. However, not all the water available in our system may be suitable for human consumption [1]. Population growth has resulted in a massive increase in groundwater demand [2]. Groundwater is an essential source for domestic and industrial purposes [3]. The overuse of groundwater, resulting in the depletion of the groundwater table and the deterioration of groundwater quality as a result of human activities has serious socioeconomic consequences [4]. Among the most effective tools and important parameters for the evaluation and management of groundwater quality, the WQI method is widely used around the world for groundwater quality assessment. WQI is a reflection and ranking of water quality based on the combined effects of different variables [5]. The organic nature of water is affected by corrosion-based physicochemical reactions. Water corrosion and scaling can be predicted with analytical terminology such as Ryznar stability index (RSI), Aggressive index (AI), Langelier saturation index (LSI), Larson-Skold index (LS), and Puckorius scaling index (PSI) [6,7]. The primary goals of this study are to assess the groundwater quality and corrosive potential of drinking water samples. Which help in managing future sustainable groundwater management approaches in the Aurangabad District, Bihar.

2. Study Area

The state of Bihar has a central position in the Gangetic Plain of India. South Bihar Plains (SBP) is a part of the Ganga Basin's periphery alluvial plains, and the region depicted in Figure 1 is its research area, which spans 3389 km^2 between longitudes 84°00′–84°45′ E and latitudes 24°30′–25°15′ N.

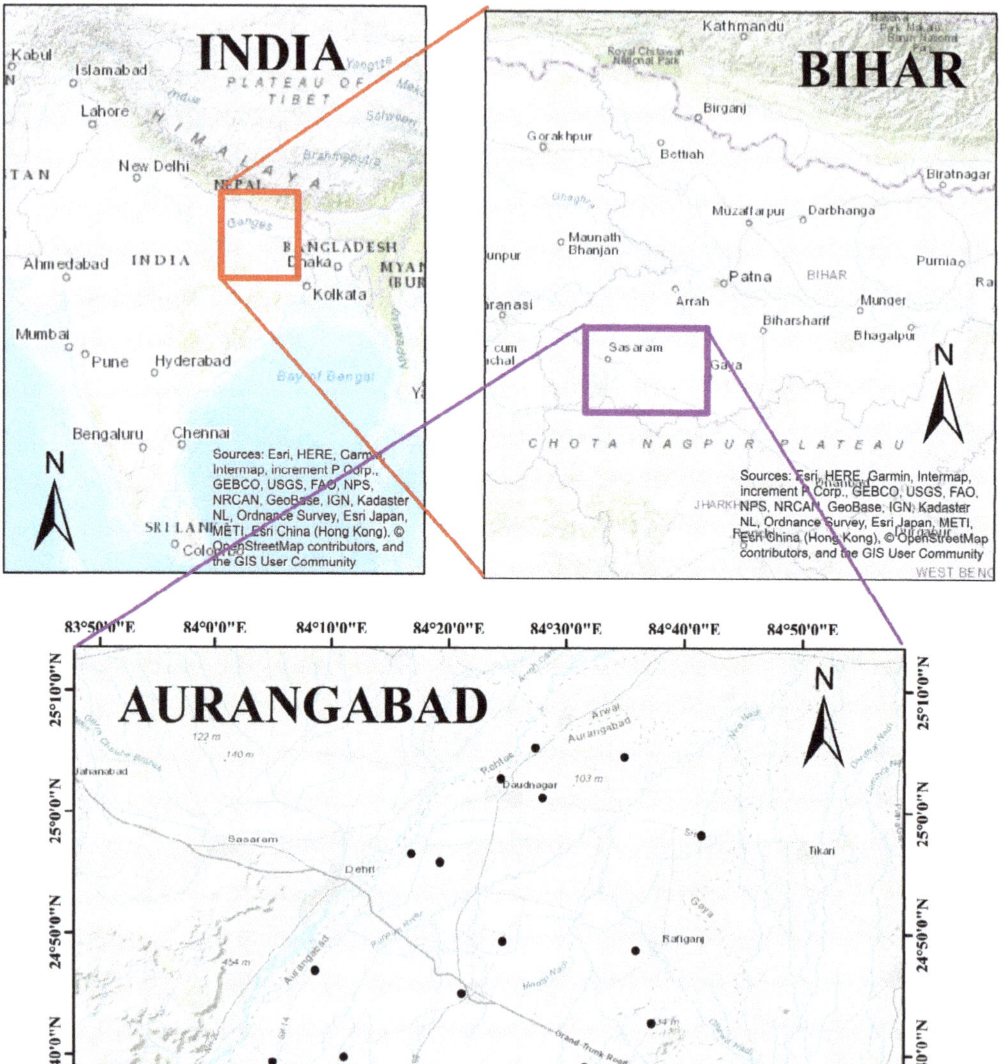

Figure 1. Map of study area.

3. Materials and Methods

3.1. Sampling and Data Analysis

There were 16 samples of groundwater taken from the study area using borewells or hand pumps. The sites were located using GPS. Pre-cleaned 1 L high density PVC bottles were used to collect water samples. Samples were taken for physicochemical analysis using standard methods [8]. A Thermo Scientific Multi-Parameters Kit was used at the site to measure the pH, EC, temperature, and TDS. The concentrations of chloride, sulphate, and phosphate were determined in the laboratory using $AgNO_3$ titration method, turbidimetric method, and colorimetric method, respectively. The standard EDTA method was used to determine total hardness, calcium hard-ness, and magnesium hardness. The dissolved oxygen concentration was determined using Winkler's modified method.

3.2. Determination of Water Quality Index

Water Quality Index was developed by Horton to assess the quality of groundwater [9]. WQI is a rating that assesses the impact of several parameters on water quality. For the calculation of WQI, all stages were performed as described by [10].

3.3. Determination of Corrosiveness Indices

LSI, RSI, PSI, Ls, and AI were used to assess the corrosion potential of water samples. There were seven physicochemical parameters that were used in calculating these indices: pH, temperature, TDS, alkalinity, calcium hardness, chloride, sulphate. As shown in Table 1, corrosiveness indices have been calculated.

Table 1. Equations and classification of Corrosiveness indices.

Index	Equation	Index Value	Tendency of Water
Langelier saturation index (LSI)	LSI = pH − pHs pHs = (9.3 + A + B) − (C + D) A = (Log (TDS) − 1)/10 B = −13.2(Log (°C + 273)) + 34.55 C = Log (Ca^{++} as $CaCO_3$) − 0.4 D = Log (Alkalinity as $CaCO_3$)	LSI < 0 LSI = 0 LSI > 0	Corrosive tendency Neutral tendency Scaling tendency
Ryznar stability index (RSI)	RSI = 2pHs − pH	RSI < 5.5 5.5 < RSI < 6.2 6.2 < RSI < 6.8 6.8 < RSI < 8.5 RSI > 8.5	High Scaling tendency Scaling tendency Neutral tendency Low corrosive tendency High Corrosive tendency
Puckorius scaling Index (PSI)	PSI = 2pHs − pHeq pHeq = 1.465log (Alkalinity) + 4.54 Alkalinity = HCO_3^- + 2(CO_3^-) + OH^-	PSI > 7 PSI < 6	Corrosive tendency Scaling tendency
Larson-Skold Index (Ls)	$Ls = (C_{Cl^-} + C_{SO_4^{2-}})/(C_{HCO_3^-} + C_{CO_3^{2-}})$ C = Concentration in mg/L	Ls > 1.2 0.8 < Ls < 1.2 Ls < 0.8	High corrosive tendency Corrosive tendency Scaling tendency
Aggressive index (AI)	AI = pH + log ((Ca^{++}) × (Alkalinity))	AI < 10 10 < AI < 12 AI > 12	Corrosive tendency Moderately Corrosive Scaling tendency

4. Results and Discussion

4.1. Physicochemical Analysis

Statistical analyses of the groundwater samples are presented in Table 2, along with the consequences of elements above the limit for use in drinking water. The water quality parameters such as pH, Cl^-, NO_3^-, SO_4^{2-} were found to be well below the [11] acceptable limit. The parameters like TDS, F, TH and Ca exceed the acceptable limits of the (WHO, 2006 [11]).

Table 2. Statistical analysis results on the study area groundwater data.

	pH	TDS (mg/L)	EC (μS/cm)	DO (mg/L)	F^- (mg/L)	Cl^- (mg/L)	NO_3^- (mg/L)	SO_4^{2-} (mg/L	Alk. (mg/L)	TH (mg/L)	Ca^{2+} (mg/L)	Mg^{2+} (mg/L)
Min	6.67	139.0	242.5	1.15	0.10	16.81	2.791	3.85	14.66	133.5	82.89	36.84
1st Qu.	7.06	287.0	530.3	1.71	0.52	33.02	11.45	9.38	34.65	201.5	139.30	54.11
Median	7.24	340.2	645.0	2.05	0.71	40.13	18.64	15.78	39.24	276.3	161.18	101.31
Mean	7.26	369.8	721.8	2.02	0.90	51.55	21.63	15.28	39.91	298.7	176.43	122.32
3rd Qu.	7.53	387.9	771.9	2.24	1.11	63.72	31.06	18.48	47.34	351.1	184.20	174.99
Max.	7.84	839.0	1669.0	2.89	2.20	169.28	49.34	38.69	58.81	704.6	492.74	290.12
WHO	7–8	600	-	-	1.5	250	50	250	-	200	100	-

4.2. Water Quality Analysis

In 16 water samples, 25% of the water samples fell in the excellent water quality, 50% of the water sample in good water quality, 19% of water samples are poor and 6% of samples are not suitable for drinking (Figure 2). In the groundwater samples, the WQI values ranged from 33.97 to 201.43 (Figure 3f). A majority of the study area has groundwater of consumable quality and that can be used for drinking as well as domestic purposes.

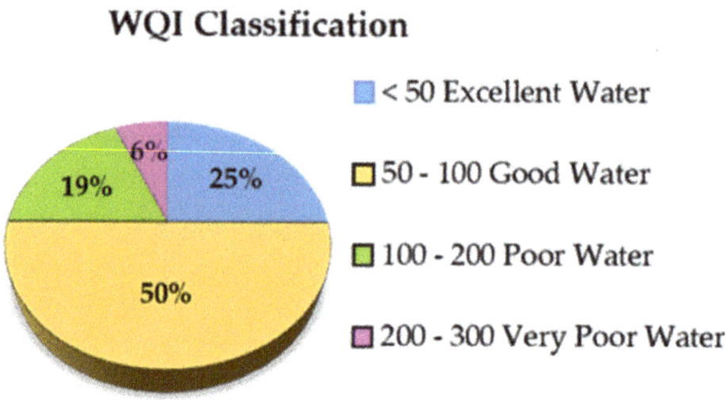

Figure 2. Water quality classification based on WQI value.

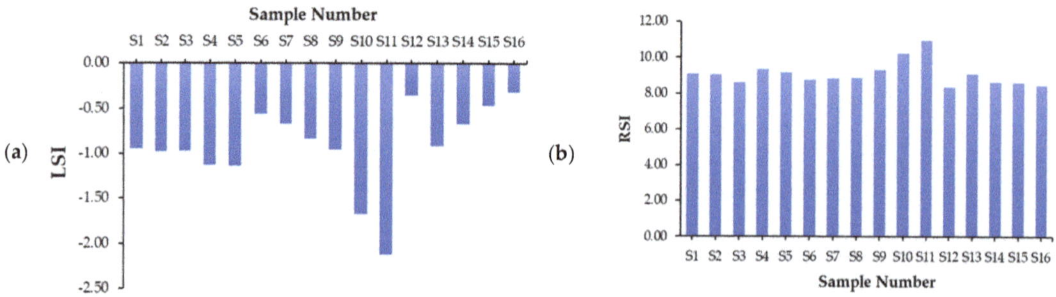

Figure 3. *Cont.*

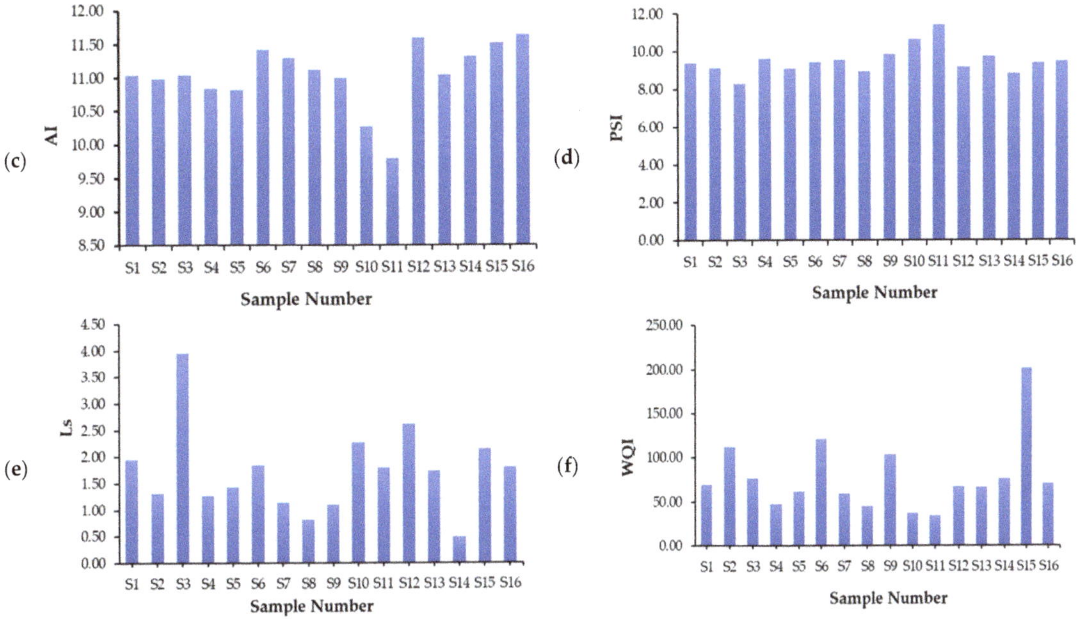

Figure 3. Corrosiveness indices (**a**) LSI, (**b**) RSI, (**c**) AI, (**d**) PSI, (**e**) Ls, and (**f**) WQI of the water samples.

4.3. Corrosiveness Potential of Water Samples

The descriptive study of the corrosiveness indices is shown in Table 3.

Table 3. Descriptive statistics of the Corrosiveness Indices.

Corrosiveness Indices	Minimum	Maximum	Mean ± Standard Deviation
LSI	−2.12	−0.31	−0.92 ± 0.47
RSI	8.36	10.96	9.09 ± 0.67
AI	9.79	11.64	11.05 ± 0.48
PSI	8.29	11.42	9.50 ± 0.73
L_S	0.49	3.95	1.73 ± 0.78

4.3.1. Langelier Saturation Index (LSI)

The LSI includes a valuable indicator for detecting corrosive water. As defined by [12], LSI is the difference between the measured pH and the saturated pH. A negative average LSI value of −0.92 indicates corrosive tendencies in the water samples shown in Table 3. Figure 3a illustrates that all water samples exhibit corrosive characteristics.

4.3.2. Ryznar Stability Index (RSI)

Ref. [13] defines RSI as the difference between the double of saturation pH and the pH of the actual water. The average value of RSI was found to be 9.09, which indicates a high tendency toward corrosion. S12 and S16 had a low corrosive tendency, whereas the rest of the samples had a high corrosive tendency (Figure 3b).

4.3.3. Puckorius Scaling Index (PSI)

The PSI predicts the ultimate amount of sediment by examining the buffering capacity of the water and the amount of precipitation that will occur when the water reaches equilibrium [14]. It fell into the corrosive tendency with an average PSI of 9.50. All samples have a corrosive tendency, as shown in Figure 3c.

4.3.4. Larson-Skold Index (Ls)

Based on the concentration of chloride, sulphate, carbonate and bicarbonate alkalinity the mathematical equation for Ls is derived [15]. There was an average Ls of 1.73, which indicates highly corrosive water. Figure 3d shows that S7, S8, S9 fell into the corrosive tendency, while S14 fell into the scaling tendency, otherwise they fell into the high corrosive tendency.

4.3.5. Aggressive Index (AI)

A formula for AI is calculated by incorporating parameters such as calcium hardness (Ca), pH and total alkalinity [16]. The average AI value was 11.05, indicating a moderate corrosive tendency. Figure 3e depicts that sample S11 has corrosive tendencies, whereas the rest are moderately corrosive.

5. Conclusions

In this study, WQI of groundwater samples revealed that 75% had excellent or good water, and 25% either had poor or very poor water. This study also evaluated the scaling and corrosion potential of drinking water samples. Mainly, iron and steel pipes were used to extract groundwater. LSI and RSI are the main indicators that are utilized for corrosion monitoring in iron and steel pipes. LSI indicates that all water samples had a corrosive tendency, and RSI indicates 87.5% high corrosive tendency. These indices are based on pH, temperature, TDS, Ca^{++}, and alkalinity, all of which have a significant impact on water quality. Chloride and sulfate ions corrode iron and steel pipes and degrade water quality. Ls is also a good tool for monitoring corrosion in iron and steel pipes. In water samples, Ls indicates a 75% high corrosive tendency, an 18.75% corrosive tendency, and a 6.25% scaling tendency. All these indices indicate that water has a corrosive tendency, which can degrade materials and affect water quality. Therefore, the study area regularly monitors groundwater quality for sustainability. Pipe material should be used according to the analysed parameters. So that the lifespan of the pipe may increase while the quality of the water remains unchanged.

Author Contributions: Conceptualization, methodology, software, formal analysis, data curation, writing—original draft preparation, review and editing had done by A.A. and S.K. All authors have read and agreed to the published version of the manuscript.

Funding: This research received no external funding.

Institutional Review Board Statement: Not applicable.

Informed Consent Statement: Not applicable.

Data Availability Statement: Not applicable.

Acknowledgments: Authors are thankful to National Institute of Technology for financial support.

Conflicts of Interest: The authors declare no conflict of interest.

References

1. Wekesa, A.M. Assessment of Groundwater Quality Using Water Quality Index from Selected Springs in Manga Subcounty, Nyamira County, Kenya. *Sci. World J.* **2022**, *2022*, 3498394. [CrossRef] [PubMed]
2. Ram, A.; Tiwari, S.K.; Pandey, H.K.; Chaurasia, A.K.; Singh, S.; Singh, Y.V. Groundwater quality assessment using water quality index (WQI) under GIS framework. *Appl. Water Sci.* **2021**, *2*, 46. [CrossRef]
3. Adimalla, N.; Qian, H. Groundwater quality evaluation using water quality index (WQI) for drinking purposes and human health risk (HHR) assessment in an agricultural region of Nanganur, south India. *Ecotoxicol. Environ. Saf.* **2019**, *126*, 153–161. [CrossRef] [PubMed]
4. Islam, A.R.; Shen, S.; Haque, M.A.; Bodrud-Doza, M.; Maw, K.W.; Habib, M. Assessing groundwater quality and its sustainability in Joypurhat district of Bangladesh using GIS and multivariate statistical approaches. *Environ. Dev. Sustain.* **2018**, *5*, 1935–1959. [CrossRef]
5. Yenugu, S.R.; Vangala, S.; Badri, S. HydroResearch Groundwater quality evaluation using GIS and water quality index in and around inactive mines, Southwestern parts of Cuddapah basin, Andhra. *HydroResearch* **2020**, *3*, 146–157. [CrossRef]

6. Pietrucha-Urbanik, K.; Skowrońska, D.; Papciak, D. Assessment of corrosion properties of selected mineral waters. *Coatings* **2020**, *10*, 571. [CrossRef]
7. Kumar, S.; Singh, R. Qualitative assessment and corrosiveness of the Ganga water: A comparative assessment. *Mater. Today Proc.* **2021**, *45*, 5695–5701. [CrossRef]
8. APHA. *Standard Methods for the Examination of Water and Wastewater*, 23rd ed.; American Public Health Association: Washington, DC, USA, 2017.
9. Horton, R.K. An index number system for rating water quality. *J. Water Pollut. Control Fed.* **1965**, *3*, 300–306.
10. Asadi, E.; Isazadeh, M.; Samadianfard, S.; Ramli, M.F.; Mosavi, A.; Nabipour, N.; Shamshirband, S.; Hajnal, E.; Chau, K.W. Groundwater quality assessment for sustainable drinking and irrigation. *Sustainability* **2020**, *1*, 177. [CrossRef]
11. *WHO Guidelines for Drinking-Water Quality*; WHO: Geneva, Switzerland, 2006; Volume 11, p. 515.
12. Langelier, W.F. The Analytical Control of Anti-Corrosion Water Treatment. *Am. Water Work. Assoc.* **1936**, *10*, 1500–1521. [CrossRef]
13. Ryznar, J.W. A New Index for Determining Amount of Calcium Carbonate Scale Formed by a Water. *Am. Water Work. Assoc.* **1944**, *4*, 472–483. [CrossRef]
14. Taghipour, H.; Shakerkhatibi, M.; Pourakbar, M.; Belvasi, M. Corrosion and Scaling Potential in Drinking Water Distribution System of Tabriz, Northwestern Iran. *Health Promot. Perspect.* **2012**, *1*, 103–111.
15. Kumar, S.; Singh, R.; Maurya, N.S. Water Quality Analysis and Corrosion Potential in the Distribution Network Patna, Bihar. *J. Environ. Eng. Sci.* **2022**, *4*, 164–174. [CrossRef]
16. Kumar, S.; Singh, R.; Maurya, N.S. Assessment of Corrosion Potential Based on Water Quality Index in the Distribution Network of urban Patna, Bihar, India. *J. Nat. Environ. Pollut. Technol.* **2022**, *5*, 2117–2127. [CrossRef]

environmental sciences proceedings

MDPI

Proceeding Paper

Evaluation of GPM IMERG Performance Over the Lake Titicaca Basin at Different Time Scales †

Luis Alberto Quispe *, Eduardo Paxi and Efrain Lujano

Escuela Profesional de Ingeniería Agrícola, Universidad Nacional del Altiplano, Puno 21001, Peru; epaxi@est.unap.edu.pe (E.P.); elujano@unap.edu.pe (E.L.)

* Correspondence: luiquispeco@est.unap.edu.pe

† Presented at the 7th International Electronic Conference on Water Sciences, 15–30 March 2023; Available online: https://ecws-7.sciforum.net.

Abstract: Accurate precipitation measurements are very important as an input for water resources management and various hydrometeorological applications. The Integrated Multi-Satellite Retrievals for Global Precipitation Measurement (GPM) (IMERG) satellite precipitation product (SPP) has been widely used to complement rain gauge measurements. However, it must be evaluated before use and also its application is still lacking in the lake Titicaca basin (LTB). In this research, the evaluation of the performance of GPM IMERG on the LTB at different time scales (daily, monthly and annual) was carried out. The evaluation was performed using rain gauge observations for the period 2003–2016 and three IMERGs, namely early (IMERG-E), late (IMERG-L), and final (IMERG-F). Accordingly, three performance metrics were used that evaluated the accuracy (correlation coefficient, CC), error (root mean square error, RMSE), and bias (percent bias, PBIAS) of the satellite estimates. In general, the monthly IMERG SPP correlated best with the rain gauge measurements. In all the evaluations performed (daily, monthly, and annual), the IMERG-F was in better agreement with the rain gauge measurements at the LTB, with small differences with IMERG-E and IMERG-L. The IMERG SPPs show potential for use in various hydrometeorological applications in the LTB.

Keywords: South American Altiplano; lake Titicaca basin; GPM IMERG; satellite precipitation products

check for updates

Citation: Quispe, L.A.; Paxi, E.; Lujano, E. Evaluation of GPM IMERG Performance Over the Lake Titicaca Basin at Different Time Scales. *Environ. Sci. Proc.* **2023**, *25*, 65. https://doi.org/10.3390/ECWS-7-14324

Academic Editor: Athanasios Loukas

Published: 3 April 2023

1. Introduction

Precipitation is an important variable for hydrological, agricultural, industrial and energy systems [1]. It has a great impact on people's lives and the control of the hydrological cycle, as well as fluctuations that affect water resources management, environmental planning and disaster mitigation [2,3]. Its utility is fundamental as an input to hydrological models, meteorological models and climate models [4,5]. The most accurate precipitation measurements are those taken directly with a rain gauge [6]. However, the availability of such data are limited to the few areas where weather stations have been installed [7].

Climatological and hydrometeorological applications of SPPs have been significantly improved with the appearance of the GPM IMERG [8]. The IMERG combines data from the GPM constellations of satellites to estimate precipitation over most of the earth's surface which lacks terrestrial rain gauges, and offers three runs to meet different users' latency and accuracy requirements, including IMERG Early (IMERG-E), IMERG Late (IMERG-L) and IMERG Final (IMERG-F) [9], which has led many researchers to consider using the IMERG and evaluate its performance.

In recent years, the use of SPPs from IMERG have shown promise in detecting precipitation on different time scales. For example, in mainland China, an evaluation of monthly precipitation products of IMERG and TRMM 3B43 [10] was carried out; in Brazil [11], IMERG grid-level evaluation was conducted at various spatial and temporal scales; in Thailand [12], a hydrological evaluation and application of TRMM and GPM precipitation products in a tropical monsoon basin was conducted; and a comprehensive evaluation of GPM IMERG and MRMS with hourly ground observations was conducted across Canada [13]. Additionally, [14] evaluated GPM IMERG, TMPA 3B42, and ERA-Interim in different topographic and climatic conditions in Iran; in Singapore [15], GPM and TRMM precipitation products were evaluated; [16] compared satellite precipitation products GPM IMERG, TMPA 3B42, and PERSIANN-CDR over Malaysia; [17] focused on a complete comparison of GPM IMERG with nine satellites and reanalysis datasets; while a first validation of IMERG over Spain is presented in [18]. The [19] developed a precipitation dataset through simultaneous use of IMERG, synoptic measurements, and automatic rain gauge measurements in the Philippines; [20] evaluated and compared daily precipitation of GPM and TRMM products over the Mekong River basin; in China [21], an evaluation of the IMERG version 05B precipitation product was conducted and compared with the IMERG version 04A at hourly and daily scales; in Myanmar, TRMM and GPM precipitation products were used for sub-daily scale flood simulations in a sparsely gauged river basin [22]; and a comprehensive evaluation of the latest IMERG and GSMaP precipitation products of the GPM era was conducted in mainland China [23]. Although the GPM IMERG SPP has been used in hydrological modeling in the LTB [24], its performance has not yet been evaluated at different time scales.

Taking the aforementioned studies into account, the objective of this research is to evaluate the performance of the GPM IMERG at different time scales in the lake Titicaca basin, its importance in improving the understanding of climate variability and its impact on flood risk management, hydrological modeling, and hydroclimatic studies. The hypothesis is that the quality and accuracy of GPM IMERG precipitation estimates vary at different time scales in the LTB.

2. Materials and Methods

2.1. Study Area

The LTB is located in southern Peru (Puno department) and west Bolivia (La Paz department) (Figure 1). It is a part of the Titicaca hydrographic region and the Titicaca-Desaguadero-Poopó-Salar of Coipasa (TDPS) endorheic system, bordered by the eastern and western mountain ranges. It covers an approximate area of 53,919.1 km^2. According to the digital elevation model (DEM), its average altitude is 4190.2 m.s.a.l., with a maximum altitude of 6397 m.s.a.l. and a minimum altitude of 3758 m.s.a.l. Most of the LTB has a flat topography, with a mean slope of 13.7%. The mean annual precipitation is 683.3 mm; 59.5% of the annual precipitation occurs in austral summer, 2.3% in winter and 22.1% and 16.1% in the transition periods from wet to dry (autumn) and from dry to wet (spring), respectively.

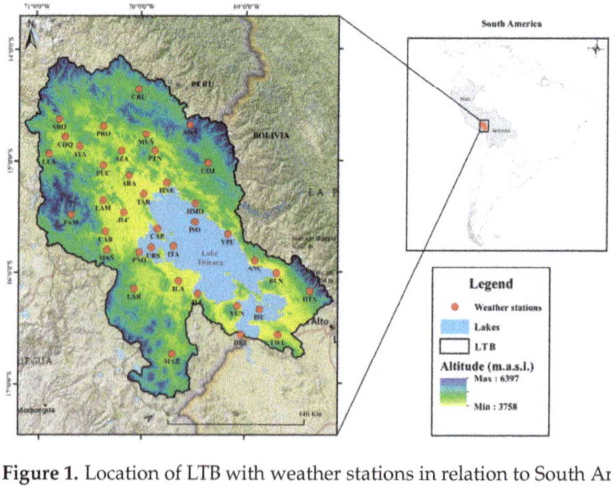

Figure 1. Location of LTB with weather stations in relation to South America.

2.2. Cartographic Information

The DEM was generated by NASA's Shuttle Radar Topography Mission (SRTM) at a spatial resolution of ~90 m, and was obtained from the Google Earth Engine (GEE) platform (https://earthengine.google.com/, accessed on 16 September 2022), Image ID CGIAR/SRTM90_V4 [25].

2.3. Rain Gauge Measurements

Rain gauge measurements were obtained from the Servicio Nacional de Meteorología e Hidrología (SENAMHI) Perú, considering a total of 33 meteorological stations. Moreover, from the Servicio Nacional de Meteorología e Hidrología (SENAMHI) Bolivia, five weather stations within the LTB were considered (Figure 1). The total number of weather stations considered was 38, with a daily recording period from 1 January 2003 to 31 December 2016.

2.4. GPM IMERG Satellite Precipitation Products

In this research, the GPM IMERG SPPs (IMERG-E, L and F) version 6 (V06) were evaluated. GPM produces precipitation data with a temporal resolution of up to 30 min, spatial resolution of $0.1° \times 0.1°$ (latitude 60° N-S) and in three executions (IMERG-E, -L and -F), cohosted by the National Aeronautics and Space Administration (NASA) and Japan Aerospace Exploration Agency (JAXA). In sequence, IMERG-E and L are near real-time data with a delay of 4 hours and 14 hours after observation time respectively, however, IMERG-F has a delay of 3.5 months [9]. The IMERG-E can be used when rapid responses are required, such as possible flood or landslide warnings, while the IMERG-L can be used for agricultural forecasting or drought monitoring [26].

GPM IMERG V06 data were obtained from the National Aeronautics and Space Administration (NASA) GIOVANNI online (Web) server (https://giovanni.gsfc.nasa.gov/giovanni/, accessed on 20 October 2022). The data were collected for the same period as the rain gauge measurements.

2.5. Method

Performance Evaluation of SPPs GPM IMERG

The homogeneity of the rain gauge measurements was verified through the non-parametric CUSUM test using the TREND program (https://toolkit.ewater.org.au/Tools/TREND, accessed on 11 October 2022). TREND is designed to facilitate statistical analysis of trends, changes and randomness in hydrological and time series data [27]. Missing data were filled in using the random forest method incorporated in the MICE (Generates Multivariate Imputations by Chained Equations) package for the R project [28]. Homogeneity was checked with monthly data after filling in the missing data [29,30].

Comparisons between IMERG and rain gauges were performed using a pixel-to-point approach as performed in previous studies [14]. This is based solely on observed precipitation measurements.

In effect, three continuous statistical metrics were used to evaluate performance (Table 1). These metrics aim to quantitatively compare the performance of IMERG measurements with rain gauge measurements. The evaluations were performed with different temporal variations, that is, daily, monthly and annual. The lack of rain gauge measurements in some areas of the LTB could limit the ability to fully evaluate IMERG measurements.

Table 1. Statistical performance metrics.

Metrics	Equation	Range	Optimal Value
Root mean square error (RMSE)	$RMSE = \sqrt{\frac{1}{n}\sum_{i=1}^{n}(O_i - S_i)^2}$	0.0 to ∞	0.0
Correlation coefficient (CC)	$CC = \frac{\sum_{i=1}^{n}(O_i-\overline{O})(S_i-\overline{S})}{\sqrt{\sum_{i=1}^{n}(O_i-\overline{O})^2}\sqrt{\sum_{i=1}^{n}(S_i-\overline{S})^2}}$	−1.0 to 1.0	1.0
Percentage bias (PBIAS)	$PBIAS = \frac{\sum_{i=1}^{n} O_i - S_i}{\sum_{i=1}^{n} O_i} \times 100$	−∞ to ∞	0.0

S is the satellite measurement; *O* the rain gauge measurement; $\overline{S}$ and $\overline{O}$ denote the mean values of *S* and *O* respectively; *n* indicates the number of data pairs.

3. Results

3.1. Daily Evaluation

Figure 2 shows the distribution of continuous statistical quantities compared between rain gauge measurements and the three IMERGs. In summary, the mean CC values in relation to the rain gauge for IMERG-E, IMERG-L and IMERG-F were 0.33, 0.32 and 0.35, respectively. Although low values of CC could be seen, IMERG-F appears to be more consistent with rain gauge observations at the LTB (Figure 2a–c). The mean RMSE value (Figure 2d–f) is between a range of 3.96 mm/day and 7.96 mm/day (mean 5.19 mm/day) for the three IMERGs evaluated. The spatial distribution of PBIAS (Figure 2g–i) showed an underestimation (overestimation) of precipitation at 77% (23%) (mean) of the stations, with overestimates of precipitation in the eastern and northeastern part of the LTB for all the three IMERGs. The mean PBIAS values were −13.52% (IMERG-E), −20.54% (IMERG-L) and 2.68% (IMERG-F).

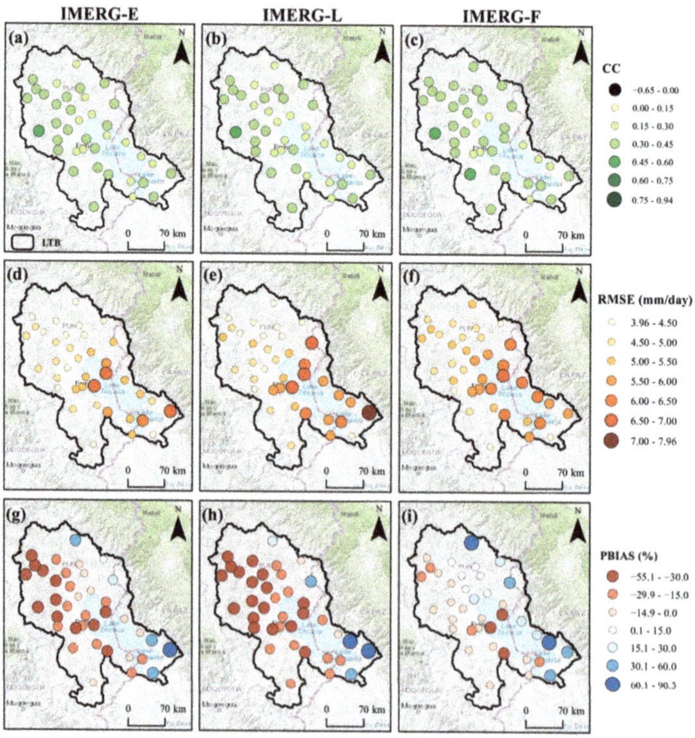

Figure 2. Spatial distribution of CC (**a–c**), RMSE (**d–f**) and PBIAS (**g–i**) of daily rain gauge data in relation to IMERG.

3.2. Monthly Evaluation

The results indicate that IMERG-F was relatively better. The highest correlation (Figure 3a–c) of the monthly evaluation was observed in IMERG-F data in relation to the rain gauges with a mean CC value of 0.90 (the lowest correlation was observed in IMERG-E with a mean CC value of 0.85). IMERG-F showed a correlation greater than 0.79, with a maximum value of 0.94, followed by IMERG-E with a correlation greater than 0.70 and a maximum value of 0.92, while the CC of IMERG-L was between a range of 0.68 and 0.92. The monthly RMSE results (Figure 3d–f) were between a mean range of 32.01 mm/month (IMERG-F) and 42.22 mm/month (IMERG-L) compared to the rain gauge data. IMERG-F compared to IMERG-L and E obtained lower errors at most stations (Figure 3f).

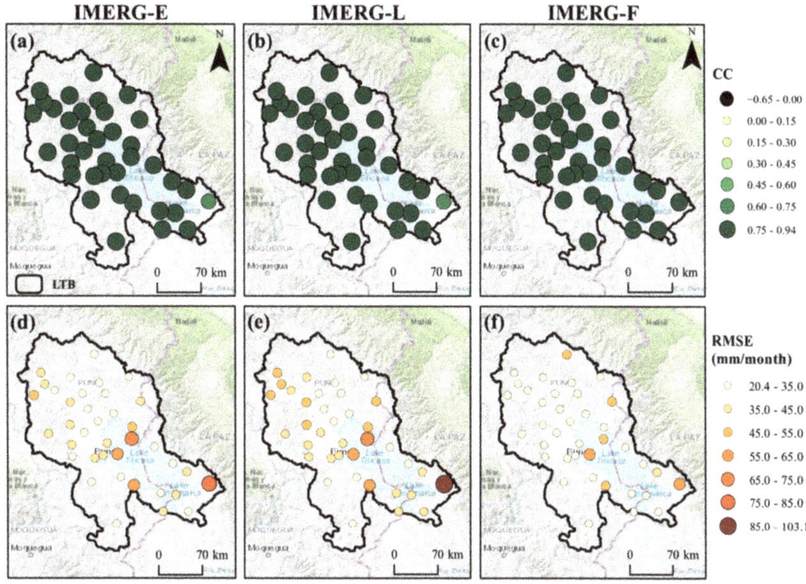

Figure 3. Spatial distribution of CC (**a–c**) and RMSE (**d–f**) of monthly rain gauge data in relation to IMERG.

3.3. Annual Evaluation

In the annual IMERG products, the error increases and the correlation decreases with respect to the monthly evaluation, becoming worse in some stations. Figure 4 shows the spatial distribution of the continuous statistical quantities compared between annual rain gauge measurements and the three IMERGs. The highest correlation (Figure 4a–c) of the annual assessment was observed in the IMERG-F data relative to the rain gauge data with a mean CC value of 0.50 (the lowest correlation was observed in the IMERG-L with a mean CC value of 0.43). IMERG-F showed a CC between −0.55 and 0.85, followed by IMERG-E with a CC between −0.58 and 0.91, while the CC of IMERG-L ranged from −0.65 to 0.92. For IMERG-E, -L and -F, negative correlations were found at 5%, 5% and 3%, while a 3% resulted with a CC < 0.15 (0.11, 0.09 and 0.06) of the total of stations, respectively. Consequently, the CC was greater than 0.15 in 92% of the stations evaluated, with a mean of 0.51 for the three IMERGs. On the other hand, the annual RMSE results (Figure 4d–f) were between a mean range of 175.28 mm/year (IMERG-F) and 262.84 mm/year (IMERG-L) compared to the rain gauge data.

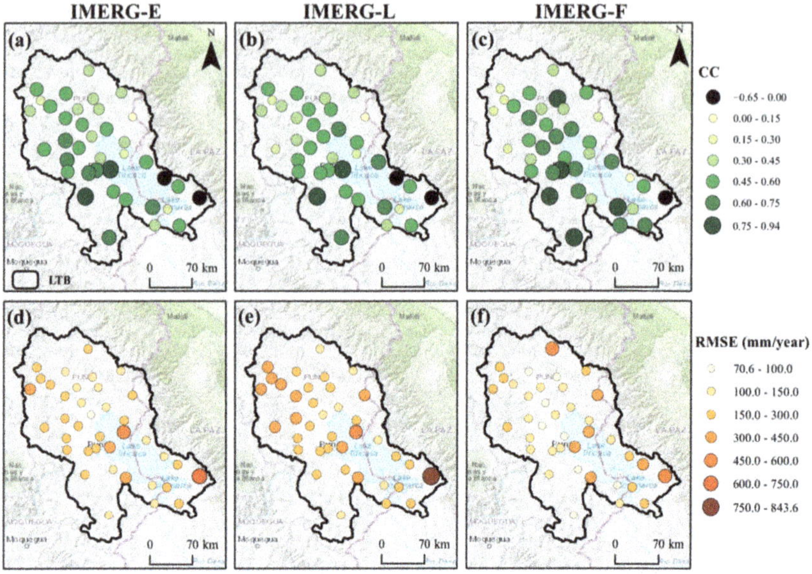

Figure 4. Spatial distribution of CC (**a–c**) and RMSE (**d–f**) of annual rain gauge data in relation to IMERG.

4. Discussion

Validation of precipitation products is very important for climate and hydrological studies [31]. In general, it was possible to find an accuracy of the SPP IMERG on increasing time scales (daily, monthly and annual). Performance was better for monthly data in representing local precipitation in the LTB. The accuracy of the monthly IMERG data relative to the rain gauge data shows variance at some stations, and on average the CC at a monthly scale shows a high acceptance value unlike the other scales (i.e., monthly > annual > daily). This is similar to what was reported at other places [16,17,19]. However, when evaluating the annual IMERG data, negative values of CC and close to zero were found, indicating a deficiency in the measurement of annual precipitation by IMERG. The accuracy of IMERG is good with higher latency and lower with medium latency (i.e., IMERG-F > IMERG-E > IMERG-L), which is why IMERG-F is recommended for use in the LTB. The main reason for the difference in performance is that SPPs are calibrated with terrestrial data [23]. However, the choice of IMERG product will depend, to a greater extent, on the type of application in the LTB. The accuracy of IMERG data may also be affected by the magnitude of precipitation, and there are indeed considerable biases for all the latencies.

5. Conclusions

In this study, the evaluation of the GPM IMERG performance over the lake Titicaca Basin at different time scales was performed by validating an IMERG grid point with rain gauge measurements.

This study concluded that in general, IMERG products provide a valuable opportunity to understand the precipitation characteristics detected by remote sensors. However, the performance could differ on different time scales, with the most promising result, according to the performance metrics, being the monthly time scale, especially IMERG-F, followed by the annual and then the daily scale. The difference between IMERG-E and IMERG-L were minimal due to the fact that they maintain a faster latency. Despite this, considerable biases can be observed in the IMERG data and in future research, bias correction is necessary before using the data for consideration in various hydrometeorological applications.

Author Contributions: Conceptualization, L.A.Q., E.P. and E.L.; methodology, L.A.Q. and E.P.; software, L.A.Q. and E.P.; validation, L.A.Q., E.P. and E.L.; formal analysis, L.A.Q., E.P. and E.L.; investigation, L.A.Q. and E.P.; data curation, L.A.Q.; writing—original draft preparation, L.A.Q. and E.P.; writing—review and editing, L.A.Q., E.P. and E.L.; visualization, L.A.Q. and E.P.; supervision, E.L. All authors have read and agreed to the published version of the manuscript.

Funding: This research received no external funding.

Institutional Review Board Statement: Not applicable.

Informed Consent Statement: Not applicable.

Data Availability Statement: Not applicable.

Acknowledgments: The authors would like to thank the Servicio Nacional de Meteorología e Hidrología of Peru and Bolivia for providing the set of rain gauge measurements and GIOVANNI—NASA for providing the SPPs GPM IMERG V06.

Conflicts of Interest: The authors declare no conflict of interest.

References

1. Carrasco, M.; Yarlequé, C.; Posadas, A. Datos faltantes de precipitación pluvial diaria mediante la Transformada Wavelet. *Rev. Peru. Geo-Atmosférica* **2010**, *88*, 76–88.
2. Hamill, T.; Kiladis, G. Comparison of Global Precipitation Estimates across a Range of Temporal and Spatial Scales. *J. Clim.* **2016**, *29*, 7773–7795. [CrossRef]
3. Lee, J.; Lee, E.; Seol, K. Validation of Integrated MultisatellitE Retrievals for GPM (IMERG) by using gauge-based analysis products of daily precipitation over East Asia. *Theor. Appl. Climatol.* **2019**, *137*, 2497–2512. [CrossRef]
4. Katiraie-boroujerdy, P.; Nasrollahi, N.; Hsu, K.; Sorooshian, S. Evaluation of satellite-based precipitation estimation over Iran. *J. Arid. Environ.* **2013**, *97*, 205–219. [CrossRef]
5. Mahmoud, M.T.; Mohammed, S.A.; Hamouda, M.A.; Mohamed, M.M. Impact of topography and rainfall intensity on the accuracy of imerg precipitation estimates in an arid region. *Remote Sens.* **2021**, *13*, 13. [CrossRef]
6. Sun, Q.; Miao, C.; Duan, Q.; Ashouri, H.; Sorooshian, S.; Hsu, K.L. A Review of Global Precipitation Data Sets: Data Sources, Estimation, and Intercomparisons. *Rev. Geophys.* **2018**, *56*, 79–107. [CrossRef]
7. Ahuja, S.; Dhanya, C. Regionalization of Rainfall Using RCDA Cluster Ensemble Algorithm in India. *J. Softw. Eng. Appl.* **2012**, *5*, 568–573. [CrossRef]
8. Mahmoud, M.T.; Al-Zahrani, M.A.; Sharif, H.O. Assessment of global precipitation measurement satellite products over Saudi Arabia. *J. Hydrol.* **2018**, *559*, 1–12. [CrossRef]
9. Huffman, G.J.; Bolvin, D.T.; Braithwaite, D.; Hsu, K.L.; Joyce, R.J.; Kidd, C.; Nelkin, E.J.; Sorooshian, S.; Stocker, E.F.; Tan, J.; et al. Integrated multi-satellite retrievals for the global precipitation measurement (GPM) mission (IMERG). *Satell. Precip. Meas.* **2020**, *1*, 343–353.
10. Chen, F.; Li, X. Evaluation of IMERG and TRMM 3B43 monthly precipitation products over mainland China. *Remote Sens.* **2016**, *8*, 472. [CrossRef]
11. Gadelha, A.; Coelho, V.; Xavier, A.; Barbosa, L.; Melo, D.C.; Xuan, Y.; Huffman, G.; Petersen, W.; Almeida, C. Grid box-level evaluation of IMERG over Brazil at various space and time scales. *Atmos. Res.* **2019**, *218*, 231–244. [CrossRef]
12. Li, R.; Shi, J.; Ji, D.; Zhao, T.; Plermkamon, V.; Moukomla, S.; Kuntiyawichai, K.; Kruasilp, J. Evaluation and hydrological application of TRMM and GPM precipitation products in a tropical monsoon basin of Thailand. *Water* **2019**, *11*, 818. [CrossRef]
13. Moazami, S.; Najafi, M. A Comprehensive Evaluation of GPM-IMERG V06 and MRMS with Hourly Ground- Based Precipitation Observations across Canada. *J. Hydrol.* **2021**, *594*, 125929. [CrossRef]
14. Sharifi, E.; Steinacker, R.; Saghafian, B. Assessment of GPM-IMERG and other precipitation products against gauge data under different topographic and climatic conditions in Iran: Preliminary results. *Remote Sens.* **2016**, *8*, 135. [CrossRef]
15. Tan, M.; Duan, Z. Assessment of GPM and TRMM precipitation products over Singapore. *Remote Sens.* **2017**, *9*, 720. [CrossRef]
16. Tan, M.; Santo, H. Comparison of GPM IMERG, TMPA 3B42 and PERSIANN-CDR satellite precipitation products over Malaysia. *Atmos. Res.* **2018**, *202*, 63–76. [CrossRef]
17. Tang, G.; Clark, M.; Papalexiou, S.; Ma, Z.; Hong, Y. Have satellite precipitation products improved over last two decades? A comprehensive comparison of GPM IMERG with nine satellite and reanalysis datasets. *Remote Sens. Environ.* **2020**, *240*, 111697. [CrossRef]
18. Tapiador, F.; Navarro, A.; García-Ortega, E.; Merino, A.; Sánchez, J.L.; Marcos, C.; Kummerow, C. The contribution of rain gauges in the calibration of the IMERG product: Results from the first validation over Spain. *J. Hydrometeorol.* **2020**, *21*, 161–182. [CrossRef]
19. Veloria, A.; Perez, G.; Tapang, G.; Comiso, J. Improved rainfall data in the Philippines through concurrent use of GPM IMERG and ground-based measurements. *Remote Sens.* **2021**, *13*, 2859. [CrossRef]

20. Wang, W.; Lu, H.; Zhao, T.; Jiang, L.; Shi, J. Evaluation and comparison of daily rainfall from latest GPM and TRMM products over the Mekong River Basin. *IEEE J. Sel. Top. Appl. Earth Obs. Remote Sens.* **2017**, *10*, 2540–2549. [CrossRef]
21. Xu, S.; Shen, Y.; Niu, Z. Evaluation of the IMERG version 05B precipitation product and comparison with IMERG version 04A over mainland China at hourly and daily scales. *Adv. Space Res.* **2019**, *63*, 2387–2398. [CrossRef]
22. Yuan, F.; Zhang, L.; Soe, K.M.; Ren, L.; Zhao, C.; Zhu, Y.; Jiang, S.; Liu, Y. Applications of TRMM- and GPM-era multiple- satellite precipitation products for flood simulations at sub-daily scales in a sparsely gauged watershed in Myanmar. *Remote Sens.* **2019**, *11*, 140. [CrossRef]
23. Zhou, Z.; Guo, B.; Xing, W.; Zhou, J.; Xu, F.; Xu, Y. Comprehensive evaluation of latest GPM era IMERG and GSMaP precipitation products over mainland China. *Atmos. Res.* **2020**, *246*, 105132. [CrossRef]
24. Asurza, F.; Ramos, C.; Lavado, W. Assessment of Tropical Rainfall Measuring Mission (TRMM) and Global Precipitation Measurement (GPM) products in hydrological modeling of the Huancane river basin, Peru. *Sci. Agropecu.* **2018**, *9*, 53–62. [CrossRef]
25. Jarvis, A.; Reuter, H.; Nelson, A.; Guevara, E. Hole-Filled Seamless SRTM Data V4, International Centre for Tropical Agriculture (CIAT). 2008. Available online: http://srtm.csi.cgiar.org (accessed on 16 September 2022).
26. Sungmin, O.; Foelsche, U.; Kirchengast, G.; Fuchsberger, J.; Tan, J.; Petersen, W.A. Evaluation of GPM IMERG Early, Late, and Final rainfall estimates using WegenerNet gauge data in southeastern Austria. *Hydrol. Earth Syst. Sci.* **2017**, *21*, 6559–6572. [CrossRef]
27. Chiew, F.; Siriwardena, L. *Trend/Change Detection Software. USER GUIDE*; CRC for Catchment Hydrology: Clayton, Australia, 2005.
28. van Buuren, S.; Groothuis-Oudshoorn, K. Mice: Multivariate imputation by chained equations in R. *J. Stat. Softw.* **2011**, *45*, 1–67. [CrossRef]
29. Tomas-Burguera, M.; Vicente-Serrano, S.M.; Beguería, S.; Reig, F.; Latorre, B. Reference 634 crop evapotranspiration database in Spain (1961–2014). *Earth Syst. Sci. Data* **2019**, *11*, 1917–1930. [CrossRef]
30. Woldesenbet, T.A.; Elagib, N.A.; Ribbe, L.; Heinrich, J. Gap filling and homogenization of climatological datasets in the headwater region of the Upper Blue Nile Basin, Ethiopia. *Int. J. Climatol.* **2017**, *37*, 2122–2140. [CrossRef]
31. Wong, J.S.; Razavi, S.; Bonsal, B.R.; Wheater, H.S.; Asong, Z.E. Inter-comparison of daily precipitation products for large-scale hydro-climatic applications over Canada. *Hydrol. Earth Syst. Sci.* **2017**, *21*, 2163–2185. [CrossRef]

environmental sciences proceedings

MDPI

Proceeding Paper

In Situ Biogeochemical Barriers for Contaminated Groundwater Treatment near Uranium Sludge Storages †

Anatoly Boguslavsky [1], Alexey Safonov [2] and Olga Shvartseva [3,*]

[1] V.S. Sobolev Institute of Geology and Mineralogy Siberian Branch Russian Academy of Science, 630090 Novosibirsk, Russia; boguslav@igm.nsc.ru
[2] Frumkin Institute of Physical Chemistry and Electrochemistry, Russian Academy of Sciences, 119071 Moscow, Russia; alexeysafonof@gmail.com
[3] X-BIO Institute of Environmental and Agricultural Biology, University of Tyumen, 625003 Tyumen, Russia
* Correspondence: o.s.shvarceva@utmn.ru
† Presented at the 7th International Electronic Conference on Water Sciences, 15–30 March 2023; Available online: https://ecws-7.sciforum.net/.

Abstract: The contamination of groundwater by uranium, nitrate, ammonium, and sulfate near uranium sludge storage sites due to the degradation of engineering safety barriers is an urgent problem during their long-term operation. The purification of such multicomponent contaminants is a complex task and one of the promising methods for this purpose is in situ bioremediation using the metabolic potential of aborigenic microflora. The work focused on the geochemical, geological, and microbiological parameters of groundwater with multi-component contamination near the uranium sludge storage sites of four chemical plants. In conditions of extreme nitrate contamination (up to 15 g/L), denitrifying bacteria were found to be the dominant group of microorganisms. In conditions of nitrate–ammonium contamination, bacteria responsible for the Anammox process were found. In laboratory, optimal conditions were selected to stimulate microflora to promote nitrate removal. For this purpose, sources of carbon (acetate, whey) were added to the water samples in concentrations necessary for the complete removal of nitrate by microbial denitrification. The experiment was carried out at ambient temperature in hermetically sealed vials. Uranyl nitrate was added to the samples at a concentration of 5 mg/L for uranium. It was found that nitrate removal contributes to the cycle of anaerobic processes of authigenic sedimentation because of sulfate and iron reduction processes, which provide the formation of a mineral geochemical barrier for uranium immobilization. As a result of the experiment, after 3–6 months, depending on the concentration of nitrate in the groundwater sample, the uranium content in the liquid phase decreased by 92–98%.

Keywords: sludge storage; uranium; nitrate ion; sulfate ion; bioremediation

Citation: Boguslavsky, A.; Safonov, A.; Shvartseva, O. In Situ Biogeochemical Barriers for Contaminated Groundwater Treatment near Uranium Sludge Storages. *Environ. Sci. Proc.* **2023**, *25*, 66. https://doi.org/10.3390/ECWS-7-14244

Academic Editor: Athanasios Loukas

Published: 16 March 2023

1. Introduction

At nuclear fuel cycle enterprises, the storage of large volumes of solid radioactive waste after ore processing and uranium enrichment in open sludge storage facilities lead to contamination of the environment with anions of mineral acids (mainly nitrate and sulfate ions) and uranium, primarily of upper aquifers [1]. One of the methods for restoring the quality of water in underground aquifers is in situ bioremediation, since it allows efficient and low-cost immobilization of a pollutant by using the metabolic capabilities of underground microflora by stimulating it by injecting soluble organic compounds into the reservoirs.

The aim of this work was to analyze the behavior of uranium in biogeochemical processes during bioremediation of groundwater in the area of the sludge storage sites of four plants under laboratory conditions and to apply the method in real conditions.

2. Materials and Methods

2.1. Study Objects

The objects of the study were the sludge storage sites of four large nuclear fuel cycle plants in Russia: AECC Angarsk, ECP Zelenogorsk, NCCP Novosibirsk, and CHMZ Glazov. These plants compose different units of the nuclear fuel cycle. However, they all operate on natural uranium, the residual concentrations of which in the solutions are neutralized and discharged into near-surface sludge collectors. Regardless of the stage of the nuclear fuel cycle, at which the liquid uranium-containing wastes are produced, the approved technological approaches to their neutralization cause similar characteristics of phase and mineral composition. Acid uranium-containing solutions are neutralized with $Ca(OH)_2$ lime slurry. During neutralization, the solution is quickly supersaturated with the respect to calcite, gypsum, dolomite, barite, and bassanite. If the original solutions contain high concentrations of fluoride ion, then during neutralization there is an active formation of fluorite, which in some cases can comprise the bulk of the wastes.

Despite the general similarity, the plants have individual differences related both to the operational characteristics of sludge collectors (mineralization and salt composition of discharged slurries, the volume of solutions, and the presence and condition of seals) and the natural conditions of the sludge storage sites (geomorphological position, thickness and protection of aquifers, composition of background groundwater, etc.). For most sites, natural groundwater is hydrocarbonate calcium–magnesium with a mineralization of less than 1 g/L. Because of the interaction of water with the sludge, it becomes nitrate and sulfate–nitrate, and the proportion of sodium and potassium increases and prevails over other cations. The maximum mineralization can reach 15–20 g/L (Table 1).

Table 1. Chemical composition of groundwater in the territory of sludge storage facilities.

		AECC			ECP			NZHK			CHMZ		
		Natural	Middle	Polluted	Natural	Middle	Polluted	Natural	Middle	Polluted	Natural	Middle	Polluted
Eh	mV	115	136	34	166	90	94	−4	−29	57	−70	162	195
pH		7.39	7.10	9.78	7.35	6.70	7.50	7.01	6.51	7.40	7.20	7.80	6.50
CO_3^- + HCO_3^-		96	89	412	267	215	204	393	151	195	54	212	319
SO_4^{2-}	ppm	27	1620	3720	0.55	143	800	24.4	1920	590	21.4	1585	780
Cl^-	ppm	6.45	49	350	1.61	65	17	21	750	2984	10.9	1131	2260
NH_4^+	ppm	0.28	0.20	103	9.1	16.5	68	1.0	1.1	120	7.8	88	292
NO_3^-	ppm	0.98	45.4	2440	0.80	2330	11,500	5.9	900	4740	11.5	3460	7100
Ca^{2+}	ppm	33.9	299.1	114.6	77.8	602	5340	103	556	804	132	1113	2940
Na	ppm	5.25	94.22	1519	11.74	70.73	127	12.40	563	2013	100	1058	1549
Mg	ppm	18	120	5.0	25	60	120	35	160	72	31	29	76
K	ppm	0.72	3.09	139	1.99	5.00	11.0	1.83	8.92	91	16.5	234	47
Si	ppm	7.54	7.11	1.27	5.73	5.47	7.00	6.74	1.19	4.87	3.20	12.4	19.7
Al	ppm	0.04	0.08	0.13	0.07	0.09	0.09	0.11	0.08	0.13	0.14	2.4	3.1
Fe	ppm	13.9	1.66	3.33	1.59	2.03	15.4	5.26	3.46	2.65	1.40	5.10	67
Mn	ppm	0.41	0.07	0.81	1.1	0.73	2.2	1.3	2.1	2.5	0.83	1.0	2.4
P	ppm	0.20	0.31	1.16	0.40	1.29	10.0	0.53	0.47	0.07	0.01	0.01	0.39
U	ppb	0.09	0.24	0.26	0.23	3.0	5.6	2.0	0.86	3422	0.01	1.90	4930

2.2. Analysis Methods

The values of Eh, pH, and electric conductivity were determined using a Hanna pH/Ion Meter (HI 9025 C, Hanna Instruments, Ronchi di Villafranca, Italy) equipped with various combination electrodes.

The trace element composition of water was measured by high-resolution inductively coupled plasma mass spectrometry (ICP-MS) (ELEMENT FINNIGAN MAT) and inductively coupled plasma atomic emission spectrometry (ICP-AES) (IRIS Advantage); anion concentrations were determined with a KAPEL®-105M (LUMEX, Saint Petersburg, Russia) capillary electrophoresis system at the analytical center of the Sobolev Institute of Geology and Mineralogy, SB RAS.

The laboratory modeling of biological processes was conducted by adding electron donors and carbon sources (whey and acetate) to groundwater samples and cultivating at ambient temperature. Groundwater (100 mL) was sealed in hermetic vials with an initial air gas phase. Uranyl nitrate was added to the samples at a concentration of 5 mg/L for uranium. Aliquots were taken for analysis every 5–7 days.

3. Results and Discussion

3.1. Biological Purification of Polluted Water

Aerobic organotrophic, denitrifying, and sulfate- and iron-reducing bacteria were identified based on the results of formation fluid sample inoculations on selective culture media. The maximum number of organotrophic bacteria was in samples taken from points with medium pollution; in the highly polluted site of the AECC and CHMZ, their number was lower than in the uncontaminated areas of the aquifer (Figure 1). Therefore, by example of four plants, it was found that in the nitrate, sulfate, and ammonium pollution zones, microbial community activation processes occurred, followed by an increase in the total number of cells compared with background samples.

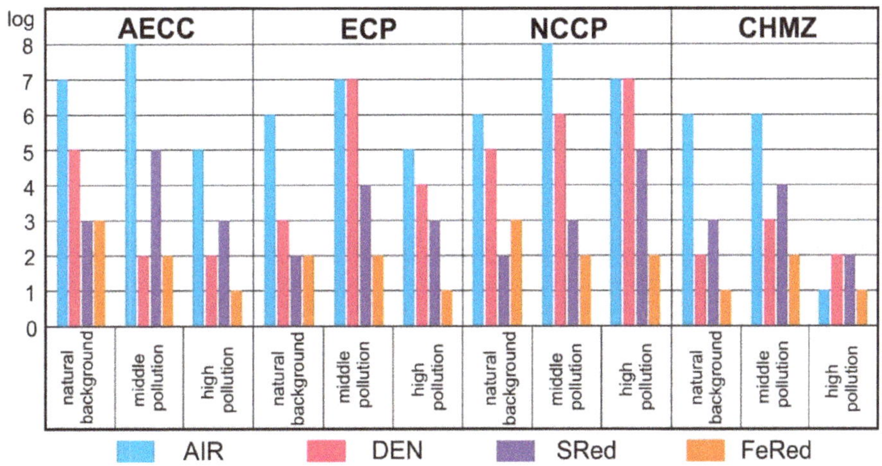

Figure 1. Microbial populations (log cell number/mL) of the main physiological groups in groundwater samples determined by inoculation to selective media.

Table 2 shows data on the rates of removal of nitrate and sulfate by the microbial community of isolated samples found after stimulation with stoichiometric sugar concentrations in laboratory conditions. It was found that in unpolluted samples, the removal of sulfate occurred in 7–14 days; in contaminated samples, the time for removal of sulfate depended on the concentration of nitrate ions. At the point of maximum contamination, the removal time of nitrate reached 180 days and sulfate 210 days. In a zone with a high nitrate ion content, biological in situ stimulation for the nitrate removal may be therefore inefficient.

Thus, the microbial community of groundwater after the stimulation by water-soluble electron donors can lead to the removal of nitrate and the reduction of sulfate with the formation of sulfide mineral phases in reducing conditions. The shift of Eh in the samples to the reduction region led to the intensification of anaerobic processes of denitrification and sulfate reduction and, as a consequence, to the reduction of uranyl ions. Thus, by stimulating the development of the microbial community, the immobilization of uranium significantly increased and the spread of nitrate, sulfate, and uranium contamination decreased.

Table 2. Table of rate of nitrate and sulfate removal in isolated groundwater samples after stimulation of the microbial community with sugar.

The Media		Duration of NO_3^- Decomposition, Days	Rapidity of NO_3^- Decomposition, ppb/Day	Duration of SO_4^{2-} Decomposition, Days	Rapidity of SO_4^{2-} Decomposition, ppb/Day
AECC	natural	0	0.08	12	27 *
	middle	5	1.5	30	324
	polluted	20	32.5	75	186
ECP	natural	0	0.8 *	0	0.55 *
	middle	25	58.2	40	5.7
	polluted	180	54.8	210	4.4
NZHK	natural	0	0.8	7	25 *
	middle	7	20	45	274
	polluted	75	36.4	130	7.9
CHMZ	natural	0	0.8	14	21 *
	middle	45	38.4	90	1585
	polluted	90	54.6	130	780

* Complete dissolution occurred in the first day, the rate is underestimated.

3.2. Future Work on the Formation of Barriers in Situ

The next stage of our research work will include the application of the bioremediation method in situ. For the formation of biogeochemical barriers at the local sites, a number of conditions will be taken into account:

1. Groundwater velocity. In all cases described, this is the water of the first surficial aquifer, which is drained by nearby logs. It is important that the time before the release of this water to the surface is sufficient for the dissolution of nitrate and sulfate (at least 2–3 months), otherwise the water comes to the surface and falls into oxidative conditions in which denitrification and sulfate reduction processes do not occur;

2. The presence of available electron donors and carbon sources (organic matter, hydrocarbonates, and hydrogen);

3. The pH values must be in the range of 6–9. In the case of ammonium contamination, an increase in pH values above 8.5 leads to the formation of ammonia;

4. Biological processes require biogenic elements (P, K, N, and S). If the content of sulfur and nitrogen compounds in contaminated areas is usually high, then in our case the limiting factor will be the content of phosphorus, which must be added when stimulating microbial processes. In addition, the addition of phosphates can lead to the formation of poorly soluble phosphate–uranium phases that contribute to the immobilization of uranium;

5. The concentration of nitrate ions should not be below 5 g/L; in the case of a higher concentration, the rate of removal of nitrate ions and reduction of sulfates may be lower than the rate of water exchange;

6. The composition of the host rocks should be optimal for biofouling (the presence of clay phase and ferriferous minerals apart from quartz). The formation of biofilms on rocks provides the formation of microbial communities resistant to toxic loads. Biofouling of pure quartz is generally not very high [2];

7. The presence of bacteria of a reducing branch of the nitrogen cycle in the microbial community capable of both dissimilative and assimilative reduction of nitrate. Only dissimilative reduction of nitrate at its high concentrations can lead to the accumulation of nitrite, which is more toxic than nitrate and reduces or stops microbial processes [3];

8. To form a stable biogeochemical barrier with reduced redox potential optimal for sulfate reduction and the formation of sulfide-ferriferous mineral phases, conditions of limited oxygen access to the aquifer are required [4].

4. Conclusions

Our studies showed that using the proposed approach results in a partial or full reduction of uranium, ammonium nitrate, and other compounds; distribution can be reached at all sites. While the production facilities are being operated, the injection of reagents for the sustainable development of the biogeochemical barrier should be carried out with a frequency sufficient to change the compounds into solid (U, S) as well as gaseous (N) phases. After decommissioning and preserving the facilities, it is necessary to provide for a margin of stability of barriers, thereby preventing degradation by background water, which will gradually replace contaminated technogenic solutions.

Author Contributions: Conceptualization, A.B., A.S. and O.S.; methodology, A.S.; validation, A.B., A.S. and O.S.; formal analysis, O.S.; investigation, A.B., A.S. and O.S.; writing—original draft preparation, O.S.; writing—review and editing, A.B.; visualization, A.S.; supervision, A.S.; project administration, O.S.; funding acquisition, O.S. All authors have read and agreed to the published version of the manuscript.

Funding: The study was supported by the Russian Science Foundation grant No. 23-27-00362, https://rscf.ru/en/project/23-27-00362/, (accessed on 14 March 2023).

Institutional Review Board Statement: Not applicable.

Informed Consent Statement: Not applicable.

Data Availability Statement: Not applicable.

Conflicts of Interest: The authors declare no conflict of interest.

References

1. Gaskova, O.L.; Boguslavsky, A.E.; Shemelina, O.V. Uranium release from contaminated sludge materials and uptake by subsurface sediments: Experimental study and thermodynamic modeling. *Appl. Geochem.* **2015**, *55*, 152–159. [CrossRef]
2. Vishnyakova, A.; Popova, N.; Artemiev, G.; Botchkova, E.; Litti, Y.; Safonov, A. Effect of Mineral Carriers on Biofilm Formation and Nitrogen Removal Activity by an Indigenous Anammox Community from Cold Groundwater Ecosystem Alone and Bioaugmented with Biomass from a "Warm" Anammox Reactor. *Biology* **2022**, *11*, 1421. [CrossRef] [PubMed]
3. Safonov, A.V.; Babich, T.L.; Sokolova, D.S.; Grousdev, D.S.; Tourova, T.P.; Poltaraus, A.B.; Zakharova, E.V.; Merkel, A.Y.; Novikov, A.P.; Nazina, T.N. Microbial Community and in situ Bioremediation of Groundwater by Nitrate Removal in the Zone of a Radioactive Waste Surface Repository. *Front. Microbiol.* **2018**, *9*, 1985. [PubMed]
4. Safonov, A.; Popova, N.; Boldyrev, K.; Lavrinovich, E.; Boeva, N.; Artemiev, G.; Kuzovkina, E.; Emelyanov, A.; Myasnikov, I.; Zakharova, E.; et al. The microbial impact on U, Pu, Np, and Am immobilization on aquifer sandy rocks, collected at the deep LRW injection site. *J. Geochem. Explor.* **2022**, *240*, 107052. [CrossRef]

environmental sciences proceedings

MDPI

Proceeding Paper

Enhancement of Atmospheric Water Harvesting via Salt-Infused Sponges and Peltier Devices †

Jaewoong Lee [1], Eric Jobiliong [2], Timothy Bastiaan [3], Darren Johanes Manua [1], Ezekhiel Taniara [3] and Eden Steven [1,3,*]

1 SPH Applied Science Academy, Sekolah Pelita Harapan Lippo Village, Tangerang 15810, Indonesia
2 Department of Industrial Engineering, Universitas Pelita Harapan, Tangerang 15810, Indonesia
3 Emmerich Research Center, Jakarta 14450, Indonesia
* Correspondence: eden.steven@gmail.com
† Presented at the 7th International Electronic Conference on Water Sciences, 15–30 March 2023; Available online: https://ecws-7.sciforum.net/.

Abstract: Despite the demand for clean water, it is commonly deficient. In the past two decades, there has been renewed interest in the development of clean water generation processes from atmospheric moisture. Atmospheric water generation is a two-stage process; in the first stage, the moisture is accumulated in an absorber material, and in the second stage, the absorbed moisture is recovered to a vessel by thermal and/or mechanical processes. One of the keys to achieving high efficiency in such processes is the moisture-absorbing agent, which works passively without electricity. Several materials are currently undergoing research, such as metal–organic frameworks (MOF) and hygroscopic salts. However, most approaches would likely be challenging to scale up from technical and economic perspectives. This work aims to develop a commonly accessible, cost-effective, environmentally friendly, and highly effective moisture absorber. Calcium chloride ($CaCl_2$) was chosen as the main salt of interest due to its deliquescence; however, it is known to suffer from agglomeration upon repeated absorption–desorption trials, which decreases efficacy. To overcome this problem, a simple infusion of the salt into the sponges significantly reduced the agglomeration problem of the salt while also improving its absorption rate and maximum water uptake by ~30% at 27 °C and 80% relative humidity (RH) compared to a sample without the cellulose sponge. To elucidate the science behind this synergistic interaction, time-dependent water uptake measurements at controlled conditions were carried out using a microbalance in an environmental chamber. Then, the data were analyzed using a double exponential equation. A physical model of the moisture absorption mechanism in the salt/sponge system was proposed. Finally, a complete atmospheric water generation device prototype was demonstrated by incorporating the salt/sponge absorber into a custom-designed Peltier-based distillation chamber.

Keywords: moisture absorber; clean water generation; Peltier distiller

Citation: Lee, J.; Jobiliong, E.; Bastiaan, T.; Manua, D.J.; Taniara, E.; Steven, E. Enhancement of Atmospheric Water Harvesting via Salt-Infused Sponges and Peltier Devices. *Environ. Sci. Proc.* **2023**, *25*, 67. https://doi.org/10.3390/ECWS-7-14177

Academic Editor: Silvia Kohnova

Published: 14 March 2023

1. Introduction

It is accepted fact that water is the most important substance to living organisms. Nonetheless, some areas lack access to water, which is concerning given how essential it is. Areas that lack natural clear water sources and/or access to water distribution have very few options are available to obtain clean water. One of these may be harvesting rainwater, though it comes with risks of chemical and microbiological contaminations [1]. Furthermore, it is not reliable to expect a consistent amount of rain throughout the year. Thus, more recently, efforts to generate clean water from atmospheric moisture are becoming more prominent [2–7].

Generally, generating clean water from atmospheric moisture consists of a two-stage process. The first stage pertains to the harvesting of atmospheric moistures into a hygroscopic absorbing media [1]. In the second stage, the absorbed moisture is separated from the hygroscopic media into a recovery vessel as clean water. This is achieved using well-established methods such as reverse osmosis [1] or distillation processes [1]. After moisture recovery, the absorbing media can reabsorb moisture and the cycle continues. It should be noted that the efficiency of the atmospheric clean water generation process depends on advancement in both the moisture recovery methods and the moisture-absorbing materials.

Several studies in the past have examined the moisture absorption capabilities of various media, including those that are complex and difficult to produce commercially. Some of the most distinguished media are $CaCl_2$ in an alginate-derived matrix (Alg-$CaCl_2$) [2], multiple versions of metal–organic framework (MOF) materials, including MIL-101(Cr) [3], Cr-soc-MOF-1 [4], Co2Cl2 (BTDD [5], activated carbon such as AC07 [6], and MOF-801 [7], as shown in Table 1.

Table 1. Comparative analysis of water uptake amongst various media containing $CaCl_2$.

Name	Medium	Water Uptake (%)	Conditions (Temperature, Dewpoint [°C], Water Vapor Pressure [mbar], RH [%])	Reference
Alg-$CaCl_2$	Alginate based	288	28, 24.1, 30, 79%	[2]
AC07	Activated Carbon/SiO_2	39	27, 7.9, 10.7, 30%	[6]
MIL-101 (Cr)	MOF	88	30, 10.6, 12.8, 30%	[3]
Cr-soc-MOF-1	MOF	200	25, 19.1, 22.2, 70%	[4]
MOF-801	MOF	30	25, 6,2, 9.5, 30%	[7]
Co_2Cl_2 (BTDD)	MOF	90	25, 6,2, 9.5, 30%	[5]

The most prominent options are Cr-soc-MOF-1, which can absorb 2.0 g of water per gram of salt or 200% of its mass at 25 °C and 22.2 mbar of water vapor pressure, and Alg-$CaCl_2$, which can absorb 2.88 g of water per gram of salt or 288% of its mass at 28 °C, 79% RH, and 30.0 mbar of water vapor pressure [2]. Despite these excellent innovations, scaling up moisture harvester technologies based on the materials may be challenging due to the intricate crystal making requirements—in the case of the MOFs—or the need to extract alginates from their origin; for example, the brown algae.

This work demonstrates the development of a simple and effective moisture harvester alternative material that is scalable and high performing. The moisture harvester is based on a common hygroscopic salt, $CaCl_2$, which is praised as one of the most hygroscopic salts readily available. $CaCl_2$ is also deliquescent, meaning that it absorbs moisture in the air until it dissolves to form a brine. Deliquescence is a property found to be maximized at low temperatures and high humidity [8], as it occurs when the vapor pressure of this brine solution is lower than the partial pressure of the vapor pressure of water in the air [9]. Although calcium chloride itself has excellent water absorbing capabilities, it must be complemented by another material due to its penchant to agglomerate. Agglomeration occurs when fine particles are chemically and physically bonded, "clumping up together in a floc" [10]. As shown in Figure 1, agglomeration occurs in calcium chloride when it liquifies and is dried again to be reused.

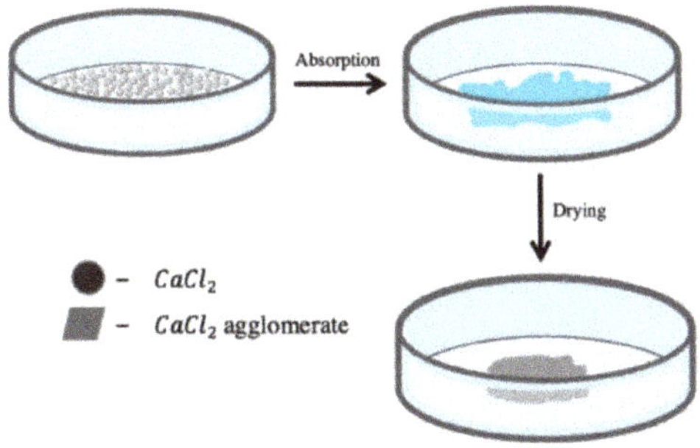

Figure 1. Illustration of agglomeration processes of calcium chloride salts upon moisture absorption and drying. The clumping reduces the amount of surface area in direct contact with the moist environment.

To solve this problem, we investigate the potential of utilizing a simple sponge salt system to overcome the agglomeration problem of deliquescent salts. First, four commercially available types of sponges were screened to determine their best performance. Then, a more thorough investigation was carried out for the best sponge. Moisture absorbance measurement was carried out over time at a constant temperature and humidity. The moisture absorption time dependence was analyzed using a basic exponential model, where the time constant and other absorption parameters were extracted and analyzed. The analysis revealed improvements in the sponge-salt system's water absorption rate and capacity than the salt system alone. A model is then proposed based on two mechanisms that highlight the existence of two distinct absorption mechanisms. Finally, a proof-of-concept water recovery Peltier-based system is proposed.

2. Methods

2.1. Sample Preparation

Each sponge of same size was pre-dried and injected with 0.2 mL of 5 M $CaCl_2$ solution. It was ensured that the solution was evenly distributed. The sponge was then dried in the oven at 60 °C for 12 h on a Petri dish with a lid. The mass was measured to quantify the salt content in the sponge. The sponge salt samples were then ready for water absorption measurement. The salt control was created by evenly spreading out an equal mass of salt as the salt content in the sponge on a Petri dish.

2.2. Moisture Absorption Time Dependence Measurement

The materials were placed on a microbalance inside a temperature- and humidity-regulated chamber. Weight changes of the samples were monitored and logged over time using a custom program written in in LabVIEW. Measurements were carried out at a constant temperature of 27 °C and a humidity level of 80% RH. After the absorption measurement, the sponge was put in the oven for 12 h at 60 °C to run the next trial.

3. Results and Discussion

3.1. Preliminary Experiment

Four types of sponges were tested to decipher the most effective sponge: cellulose, soundproof, magic, and Dacron (Figure 2a). The results were normalized to the salt's mass in the respective sponges. Hence, it is evident that the cellulose sponge was the most effective, as seen in the water uptake vs. time graphs in Figure 2b. The water uptake percentage was defined by $\frac{mass_{moisture\ absorbed}}{mass_{salt}} \times 100\%$. Control measurements of each of the sponges without salts did not show any water absorption effects. Thus, in the next sections, we focus our investigation on the salt–cellulose sponge systems only.

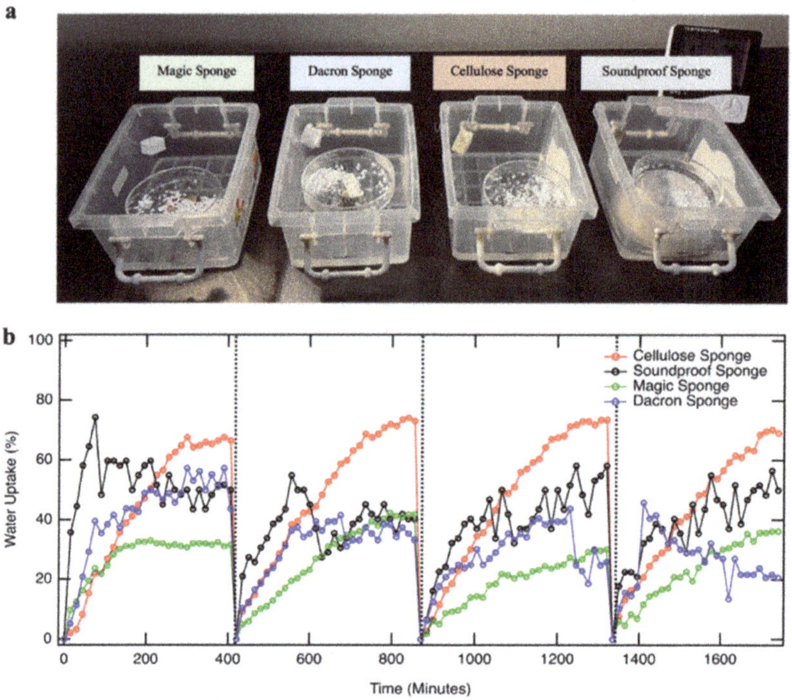

Figure 2. (**a**) Sponge types tested. Each container has a respective sponge and control; change in mass was measured every 15 min for 6–7 h. (**b**) Sponge type curves. Graph shows water uptake by various commercial sponges. They were soaked in 5M solutions, dried, and left to absorb. The vertical dotted lines indicate a new trial. The weight changes during the drying phase were not tracked, except for the beginning and final weight.

3.2. Salt-Cellulose Sponge Water Uptake Time Dependent Measurements

The efficacy of the cellulose sponge was examined by testing its maximum absorption capacity by exposing it to a controlled atmosphere until the salts stopped absorbing moisture, as can be seen in Figure 3. There are several key observations. First, we found that the water uptake dynamics fit very well to a double exponential model (Equation (1)), as listed below:

$$W(t) = W_{saturated} - A_1 e^{\frac{-t}{\tau_1}} - A_2 e^{\frac{-t}{\tau_2}}$$

where $W(t)$ is the water uptake percentage at a given time, $W_{saturated}$ is the maximum water uptake percentage, and A_1, A_2, τ_1, and τ_2 are fitting constants. We confirm that a single exponential model cannot explain such behavior. This suggests that there are two types of water uptake mechanisms in our system. Secondly, it is evident that the sponge salt sample exhibit faster water uptake rate. Finally, the sponge salt sample was found to exhibit a higher maximum water uptake, $W_{saturated}$, of ~305% while for the control salt sample, it was found to be ~272%. The salt sponge sample performed better than most of the previously reported moisture absorbers in the literature (Table 1).

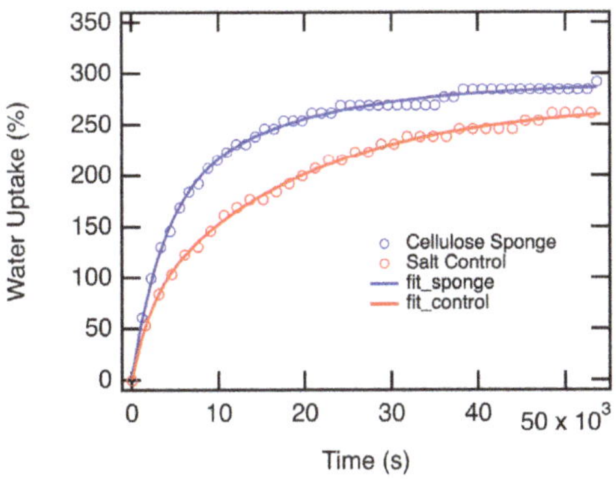

Figure 3. Time-lapsed water uptake of cellulose sponge. Solid line represents the curve fit using the double exponential equation (Equation (1)).

3.3. Proposed Model

Based on the enhancement observed from the water uptake dynamics, we propose a physical model as illustrated in Figure 4. In this model, salts are spread around the pores of the sponge. When the salts absorb moisture, the absorption is homogenous throughout the sponges. This, in effect, causes even salt precipitate distribution upon drying. As a result, the optimal amount of surface area of the salt species is maintained; for example, it does not suffer from an agglomeration problem. The optimal surface area would cause an improved water absorption rate. The infusion of the salts into the cellulose sponge results in an increase in both maximum water uptake and absorption rate. More analysis is needed to explain the improved maximum water uptake and the two absorption mechanisms that were observed in our system. Nevertheless, it is clear that the simple infusion of salts into the cellulose sponge provided significant improvements that are valuable in the context of atmospheric moisture clean water generation.

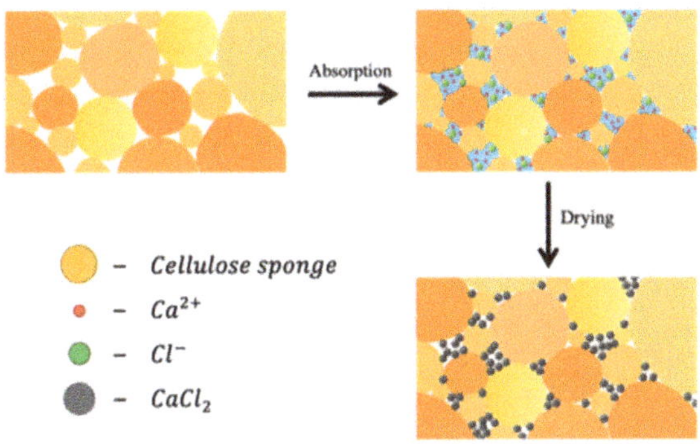

Figure 4. Proposed physical model of the salt-cellulose sponge system. Cellulose sponge with minimal agglomeration.

3.4. Peltier Device Prototype

To recover the absorbed moisture as clean water, a Peltier-based distillation unit was prototyped (Figure 5). The unit was equipped with three water recovery channels to maximize the transfer of water droplet from the evaporation to the condensation chambers. The primary channel was set directly under the acute end of the roof due where most of the vapor was accumulating. The secondary channel directly transferred moist air to the collection chamber using a small DC motor pump. Finally, a tertiary channel was set up across from the acute end of the slanted roof to capture the flow of the larger water droplets accumulating on the top wall. The design also takes into account the temperature gradient generated by the Peltier module, which naturally lowers the temperature of the collection chamber.

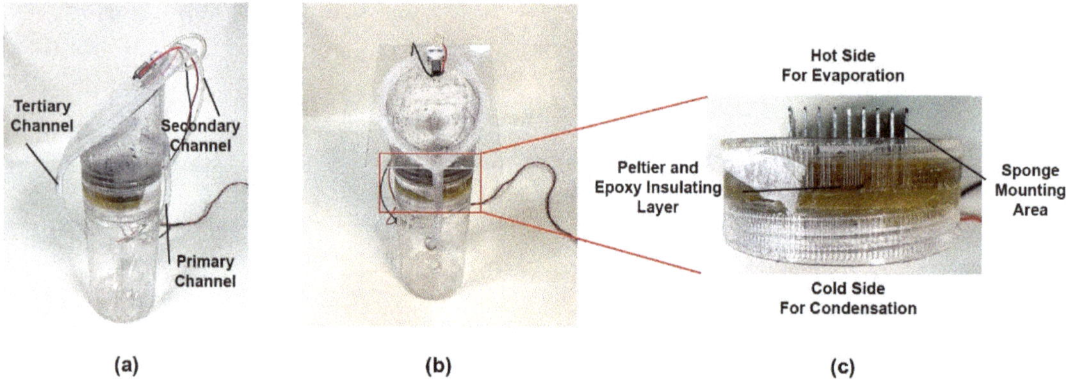

Figure 5. Prototype of the Peltier-based clean water generator. (**a**) Side view showing the three water channels to transfer water from the top evaporation to bottom condensation/collection chambers. (**b**) front view. (**c**) View of the Peltier module.

The device uses a single piece of Peltier module (9 V, 2 A) in a small chamber with a diameter of 8 cm and generates ~5 mL of clean water/hour from atmospheric moisture. Although further optimization is necessary, we note that this system does not waste any sacrificial clean water, as in most reverse osmosis or water distillation systems. Further optimization is also possible, including the incorporation of solar-powered energy sources, development of a multi-module systems, and better material choice to encourage formation and flow of water droplets on the walls of the evaporation chamber.

4. Conclusions

We found that by infusing calcium chloride into a cellulose sponge, faster moisture absorption and an increase in the maximum water uptake capacity (at 27 °C and 80% RH) up to ~305% were achieved compared to that without the sponge at ~272%. The sponge inhibited common problems caused by calcium chloride, such as agglomeration and its deliquescence. The approach is simple, and relatively more cost effective compared to other moisture absorber materials. Cellulose sponge as a medium is simple, cheap, and environmentally friendly; energy efficient; and effective when absorbing moisture, as shown by the salt absorbing moisture up to ~305%. A proof-of-concept device was also demonstrated, utilizing a three-channel water collection pathway using a Peltier device to generate the clean water from atmospheric moisture.

Author Contributions: Conceptualization, J.L., T.B., E.T., D.J.M. and E.S.; formal analysis, J.L., E.J., and E.S.; investigation, J.L., T.B., E.T. and D.J.M.; writing, J.L., E.J. and E.S. All authors have read and agreed to the published version of the manuscript.

Funding: This research was funded by SPH Applied Science Academy Program at Sekolah Pelita Harapan Lippo Village, period 2022–2024.

Institutional Review Board Statement: Not applicable.

Informed Consent Statement: Not applicable.

Data Availability Statement: The data presented in this study are available on request from the corresponding author.

Acknowledgments: We thank Sekolah Pelita Harapan, Universitas Pelita Harapan and Emmerich Research Center for their generous support in providing lab access throughout this research study.

Conflicts of Interest: The authors declare no conflict of interest.

References

1. Hofman-Caris, R.; Bertelkamp, C.; de Waal, L.; van den Brand, T.; Hofman, J.; van der Aa, R.; van der Hoek, J. Rainwater Harvesting for Drinking Water Production: A Sustainable and Cost-Effective Solution in The Netherlands? *Water* **2019**, *11*, 511. [CrossRef]
2. Kallenberger, P.A.; Fröba, M. Water Harvesting from Air with a Hygroscopic Salt in a Hydrogel–Derived Matrix. *Commun. Chem.* **2018**, *1*, 28. [CrossRef]
3. Permyakova, A.; Wang, S.; Courbon, E.; Nouar, F.; Heymans, N.; D'Ans, P.; Barrier, N.; Billemont, P.; Weireld, G.D.; Steunou, N.; et al. Design of Salt–Metal Organic Framework Composites for Seasonal Heat Storage Applications. *J. Mater. Chem. A* **2017**, *5*, 12889–12898. [CrossRef]
4. Towsif Abtab, S.M.; Alezi, D.; Bhatt, P.M.; Shkurenko, A.; Belmabkhout, Y.; Aggarwal, H.; Weselinski, L.J.; Alsadun, N.S.; Samin, U.; Hedhili, M.N.; et al. Reticular Chemistry in Action: A Hydrolytically Stable MOF Capturing Twice Its Weight in Adsorbed. *Water Chem.* **2018**, *4*, 94–105. [CrossRef]
5. Rieth, A.J.; Yang, S.; Wang, E.N.; Dincă, M. Record Atmospheric Fresh Water Capture and Heat Transfer with a Material Operating at the Water Uptake Reversibility Limit. *ACS Cent. Sci.* **2017**, *3*, 668–672. [CrossRef] [PubMed]
6. Tso, C.Y.; Chao, C.Y.H. Activated Carbon, Silica-Gel and Calcium Chloride Composite Adsorbents for Energy Efficient Solar Adsorption Cooling and Dehumidification Systems. *Int. J. Refrig.* **2012**, *35*, 1626–1638. [CrossRef]
7. Kim, H.; Yang, S.; Rao, S.R.; Narayanan, S.; Kapustin, E.A.; Furukawa, H.; Umans, A.S.; Yaghi, O.M.; Wang, E.N. Water harvesting from air with metal-organic frameworks powered by natural sunlight. *Science* **2017**, *356*, 430–434. [CrossRef] [PubMed]
8. Gough, R.V.; Chevrier, V.F.; Tolbert, M.A. Formation of Liquid Water at Low Temperatures via the Deliquescence of Calcium Chloride: Implications for Antarctica and Mars. *Planet. Space Sci.* **2016**, *131*, 79–87. [CrossRef]
9. Britannica, The Editors of Encyclopaedia. "Deliquescence". Available online: https://www.britannica.com/science/deliquescence (accessed on 13 April 2022).
10. Hostomsky, J.; Jones, A.G. Calcium Carbonate Crystallization, Agglomeration and Form during Continuous Precipitation from Solution. *J. Phys. Appl. Phys.* **1991**, *24*, 165–170. [CrossRef]

environmental sciences proceedings

MDPI

Proceeding Paper

Assessment of Cyanobacterial Chlorophyll A as an Indicator of Water Quality in Two Wetlands Using Multi-Temporal Sentinel-2 Images †

Ashwini Mudaliar [1,*,‡] and Usha Pandya [2]

1 Department of Botany, Faculty of Science, The Maharaja Sayajirao University of Baroda, Vadodara 390002, India
2 Department of Botany, Government College, Daman 396210, India
* Correspondence: ashwini144@gmail.com

† Presented at the 7th International Electronic Conference on Water Sciences, 15–30 March 2023; Available online: https://ecws-7.sciforum.net.
‡ Presenter.

Abstract: Cyanobacteria can have a dramatic effect on the quality of water used for human activities, as reported by the World Health Organization. Remote sensing is an appropriate tool for continuous monitoring of the entire water body, given the current state of the lake. In the present study, surface water quality was evaluated using multi-temporal sentinel-2 images based on cyanobacteria's concentration of chlorophyll a (Chl-a) and the water's dissolved oxygen content. Chl-a was used as an indicator of cyanobacterial blooms, and dissolved oxygen was used as an indicator of water quality. Dissolved oxygen was generated using Sentinel 2 dataset. For the present study, two wetlands, Wadhwana and Timbi, in Vadodara City, Gujarat, India, were assessed from 2018 to 2022. Analysis showed that dissolved oxygen is an important environmental factor that influences cyanobacteria abundance. It was seen that the increased concentration of chlorophyll a was associated with a reduction in dissolved oxygen and hence deteriorated the water quality.

Keywords: water quality; Ramsar; wetland; Sentinel-2; chlorophyll a; temp; dissolved oxygen 3

check for updates

Citation: Mudaliar, A.; Pandya, U. Assessment of Cyanobacterial Chlorophyll A as an Indicator of Water Quality in Two Wetlands Using Multi-Temporal Sentinel-2 Images. *Environ. Sci. Proc.* **2023**, *25*, 68. https://doi.org/10.3390/ECWS-7-14252

Academic Editor: Athanasios Loukas

Published: 16 March 2023

1. Introduction

Reservoirs are significant freshwater reserves that have experienced severe, negative effects that have changed their physico-chemical composition both qualitatively and quantitatively. These changes have, in turn, had an impact on the fauna and flora [1]. With the growth of industry and agriculture, there was a significant increase in the nutrient inputs into rivers and lakes, which led to the eutrophication of the water bodies [2]. Due to the readily available nutrients in lakes, algal blooms have become more frequent, intense, and widespread around the world in recent years. Information on the algal composition and dominance in a body of water can provide insight into the dynamics of toxin-producing species such as cyanobacteria. A change in the algal population can have serious environmental consequences, especially when there is an increase in harmful species. The cyanobacteria that produce cyanotoxins endanger both the aquatic ecosystem and human health. To evaluate and reduce the impact of potentially harmful species, it is crucial to monitor and model algal blooms and their composition. A significant group of organisms known as cyanobacteria are in charge of eutrophication-related environmental issues [3]. Although other factors such as water temperature, pH, light, and dissolved oxygen also have an impact on cyanobacteria reproduction, the availability of nutrients plays a major role in this process. Among the factors affecting water quality (such as pH, conductivity, nitrate nitrogen, phosphorus, etc.) dissolved oxygen (DO) levels are one of the most crucial general health indicators when evaluating aquatic ecosystems [4]. For the health of aquatic life, inland waters must maintain high DO levels. Aquatic organisms experience negative effects on metabolic activity, predation risk, and behavior when DO concentrations drop to low thresholds (below 5 mg/L) [5].

The Water Policy Framework can be improved by new technologies and methods that have been developed in recent decades. Blue-green algae or cyanobacteria in particular can be found using spectrometric sensors carried by satellites to monitor phytoplankton growth and composition [6]. Sentinel-2 data, which operates at spatial resolutions of 10 m, 20 m, and 60 m and collects data from 13 different wavelengths, is proving to be very helpful for monitoring blue-green algae blooms in aquatic environments. Based on measurements of their primary pigments Chl-a and phycocyanin, these measurement zones have interesting applications in the evaluation of phytoplankton and cyanobacteria [2,7–10]. Temperature is another crucial indicator of water quality that can be determined using satellite imagery in addition to dissolved oxygen measurement.

An earlier investigation revealed that human activity is to blame for the decline in Wadhwana's water quality [11]. Therefore, the purpose of this study was to estimate chlorophyll a using multi-temporal Sentinel-2 data along with dissolved oxygen and temperature to comprehend the current water quality of two wetlands in the Vadodara District.

2. Study Area

Vadodara is a city in the state of Gujarat. It is 39 meters above mean sea level and can be located at 22°17′59″ North Latitude and 73°15′18″ East Longitude. The Wadhwana Reservoir was constructed in 1910 in what is now Vadodara, formerly known as the Baroda State. The Wadhwana Wetland (Figure 1), in the Dabhoi taluka, is one of the Ramsar sites and is well known among bird watchers in the state. On a global scale, the wetland is significant for its bird life because it provides a wintering habitat for migratory waterbirds, including more than 80 species that migrate along the Central Asian flyway. The middle of winter in 2020 saw the completion of a waterbird census, which counted about 46,000 different species. The wetland serves as an international example of how a wetland originally developed for irrigation has evolved into an important habitat for water birds and a center for ecotourism and nature education. The wetland is used as a global illustration of how a wetland that was initially developed for irrigation has transformed into an essential water bird habitat and a hub for ecotourism and nature education.

Following Wadhwana, Lake Timbi is the largest lake. It attracts many migratory birds. However, both wetlands are under threat from large amounts of irrigation water and other industrial wastewater. Consequently, the aquatic life in these wetlands is in danger.

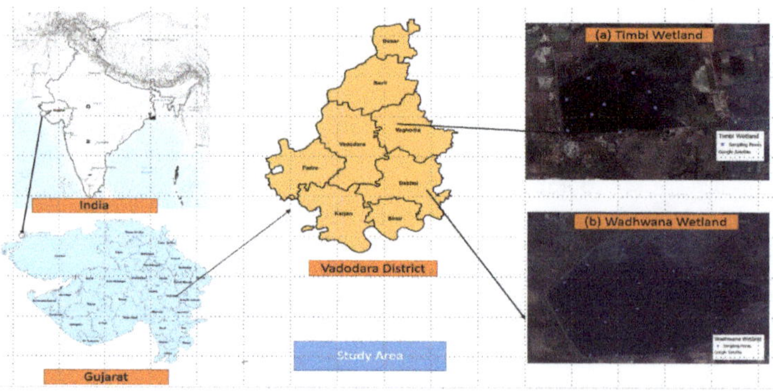

Figure 1. Study area (**a**) Timbi wetland in Waghodia Taluka and (**b**) Ramsar site- Wadhwana Wetland.

3. Materials and Methodology

In this investigation, measurements and water samples were collected in April 2018. Random sampling was used to gather samples from various locations in the wetlands of

Wadhwana and Timbi. Dissolved Oxygen was extracted following the standard procedure by the American Public Health Association [12].

Data

Sentinel-2

Free sentinel data were accessed from Copernicus open access hub (https://scihub.copernicus.eu/, accessed on 22 November 2022) and provided the free Sentinel-2 data (Level 2A product reflectance) used in the current study, which covered the period from January 2018 to December 2022. The data were resampled to a resolution of 10 m. The images were stacked and used for additional processing.

Landsat-8/9

From January 2018 to December 2022, the Landsat 8/9 OLI were downloaded (http://earthexplorer.usgs.gov, accessed on 22 November 2022). Bands 10 and 11 are available in the TIR region on the Landsat 8 TIRS sensor. These thermal bands have a native spatial resolution of 100 m but were resampled and released by USGS at 30m. Digital Numbers (DC) were converted to Top of Atmosphere (TOA) reflectance using radiometric coefficients contained in the metadata. Using Jiménez-Muoz et al. algorithm, the TOA bands were converted into brightness temperatures, and the water surface temperature was generated [13].

4. Results

4.1. Retrieval Method of Chlorophyll A Concentration

By utilizing an empirical chlorophyll model created by Mishra and Mishra (2012), pre-processed Sentinel-2 images were used to derive the normalized differential chlorophyll index (NDCI). Because of backscattering and Chl-a absorption, these bands are most vulnerable to reflection. The spectral difference between bands 665 and 705 is used to calculate NDCI, which is then normalized by the sum of those two bands.

$$NDCI = \frac{[(705) - (665)]}{[(705) + (665)]}$$

Further, Chl-a was quantified using following equation as modelled by Kravitz et al. 2020

$$\mathrm{Chl-a}\left(\frac{\mathrm{mg}}{\mathrm{m3}}\right) = 17.441e^{(4.7038 \times NDCI)}$$

Based on the data coordinates, the values for LST and chlorophyll a from the image bands of the relevant locations were determined (Figure 2). The values were used for further analysis and derivation of dissolved oxygen.

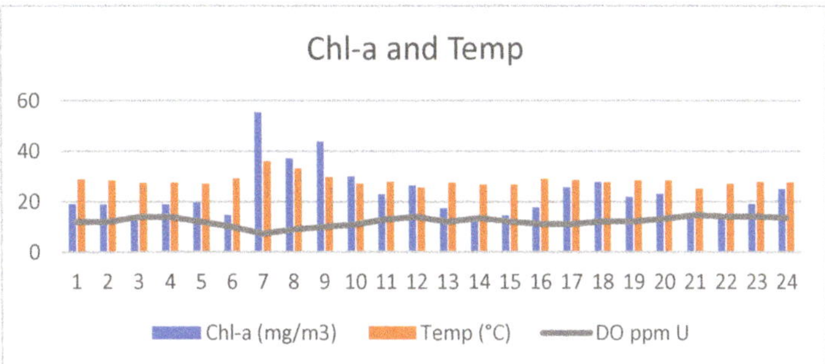

Figure 2. Satellite derived chlorophyll a and temperature along with in situ DO in the Study area.

4.2. Dissolved Oxygen (DO)

The sampling data from Wadhwana and Timbi Wetlands in April 2018 matched the Sentinel-2 image. This resulted in 24 in situ water sample matches on 18th April 2018. The in situ DO values ranged from 7.0 to 13.5 ppm in the Wadhwana wetland, while it ranged from 11 to 14.6 ppm in the Timbi Wetland. Field derived DO was further investigated for its relationship with Chl-a and temperature.

4.3. Analysis for Water Quality of Wadhwana and Timbi Wetland

The Chl-a and temperature data for June, 2018 were utilized with field generated dissolved oxygen were used.

From Table 1, correlation between DO and chlorophyll a (Chl-a) shows −0.7204. This means that there is a strong inverse relationship between Chl-a and DO. R-squared (R^2) equals 0.519. Dissolved oxygen and chlorophyll a are correlated and closely related, since a decrease in dissolved oxygen leads to an increased quantity of algae (chlorophyll a) in wetlands [14]. The correlation between DO and temperature shows −0.8478. This means that there is a very strong inverse relationship between temperature and DO R-Squared (R^2) equals 0.72. According to the results of multiple linear regressions, there is a very strong overall significant relationship between Chl-a, temperature, and DO (F(2,21) = 28.98, p and <0.001, R^2 = 0.73, R^2 adj = 0.71). R-squared (R^2) is 0.734071. The multiple correlation coefficient (R) is 0.85. This indicates that the predicted and observed data are highly correlated. The squared R-adjusted value is 0.71.

Table 1. Relationship between Chla, Dissolved Oxygen and temperature in Wadhwana and Timbi Wetlands.

Regression Model	Empirical Equation	R^2	R
Chl-a and DO	DO = 15.0403 − 0.1271 Chla	0.52	−0.72
Temperature and DO	DO = 31.364 − 0.6834 Temp	0.72	−0.85
MLR with Chl-a, DO and Temperature	DO = 28.902228 − 0.0332848 Chl-a − 0.56885 Temp	0.73	0.71

4.4. Predictive Models for Water Quality of Wadhwana and Timbi Wetland

A comparison of time series data from 2018 to 2022 for the Wadhwana (Figure 3a) and Timbi (Figure 3b) was performed to verify the proposed MLR model's accuracy. The correlation values between Sentinel-2 DO and Chl-a, as well as between the Sentinel-2 DO and temperature, were measured and the test results exceeded 0.6 and 0.89, respectively. Figure 3a,b shows the simulation results in comparison to the measurements. Data prediction by the model was successful. The results of all simulations are good and reasonable when measurement errors and spatiotemporal variations are taken into account. The RMSE for each variable ranges from 0.6 and 1.24.

The Multiple Linear Regression model could effectively simulate interannual dynamics. This model depicts the temporal trend of chlorophyll a with DO and temperature as shown in Figure 3. Increased algal blooms in wetlands and rising temperatures resulted in significant decreases in oxygen, which had far more detrimental effects on the lake's ecosystem than a decrease in algal blooms. Time series data (Figure 3) showing that during the summer months, i.e., as temperatures rose in March–May 2018–2020, dissolved oxygen levels were found to be low and chlorophyll levels high. A significant algal bloom was observed in the Ramsar Wadhwana wetland in 2021/2022, preceded by March to February. Timbi Wetland, which followed a similar pattern, showed a shift in high temperatures from March to February 2021. From these results, it is evident that the concentration of oxygen-consuming algae always increased after the concentration of dissolved oxygen decreased when compared to the concentration of chlorophyll a. In both wetland types, dissolved oxygen was an indicator that could be used to interpret chlorophyll a levels. The chlorophyll a measurement is necessary because it shows the level of eutrophication that could endanger reservoirs around the world.

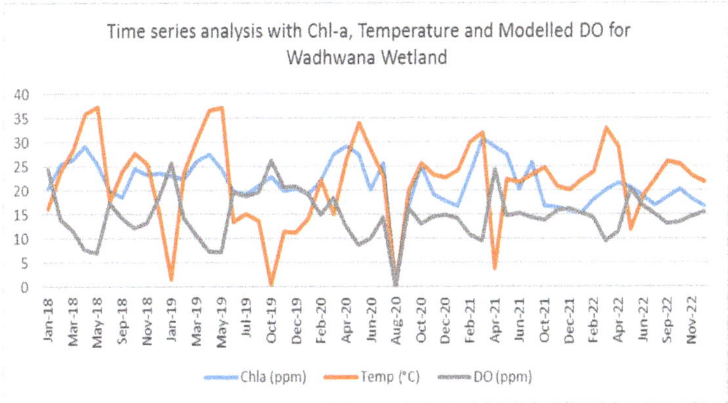

(**a**)

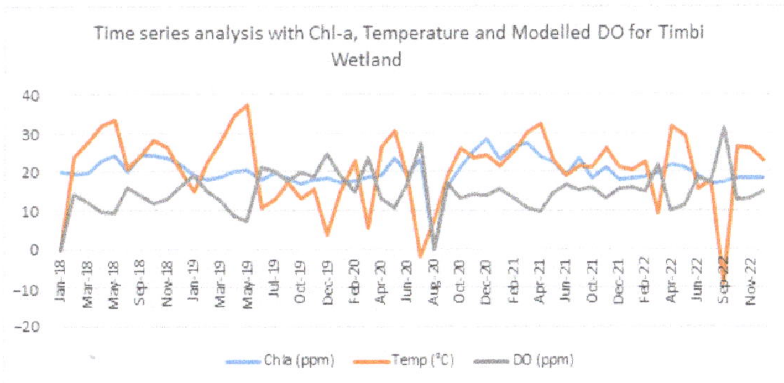

(**b**)

Figure 3. Time series analysis with Chl-a, temperature and modelled DO for (**a**) Wadhwana wetland (**b**) Timbi Wetland.

5. Conclusions

Based on the findings of this research, Sentinel-2 data could be used effectively to accurately map dissolved oxygen levels in wetlands such as the Timbi and Wadhwana Ramsar sites. Accurate mapping of DO, an indicator of water quality parameters, can be used to provide a complete picture of the variability in algal blooms (chlorophyll a) concentrations due to their significant impact on water quality status.

Author Contributions: This research article has been written by two authors: A.M. and U.P.; A.M. involved in data downloading, processing and analysis, draft preparation, review and editing; U.P. in draft paper and review. All authors have read and agreed to the published version of the manuscript.

Funding: This research received no external funding.

Institutional Review Board Statement: Not applicable.

Informed Consent Statement: Not Applicable.

Data Availability Statement: This study used Sentinel-2 multispectral imagery that is publicly available at https://scihub.copernicus.eu/ (accessed on 22 November 2022) and Landsat-8 multispectral imagery that is publicly available at http://earthexplorer.usgs.gov (accessed on 22 November 2022).

Conflicts of Interest: The authors declare no conflict of interest.

References

1. Viso-Vázquez, M.; Acuña-Alonso, C.; Rodríguez, J.L.; Álvarez, X. Remote Detection of Cyanobacterial Blooms and Chlorophyll-a Analysis in a Eutrophic Reservoir Using Sentinel 2. *Sustainability* **2021**, *13*, 8570. [CrossRef]
2. Zhao, C.S.; Shao, N.F.; Yang, S.T.; Ren, H.; Ge, Y.R.; Feng, P.; Dong, B.E.; Zhao, Y. Predicting cyanobacteria bloom occurrence in lakes and reservoirs before blooms occur. *Sci. Total Environ.* **2019**, *670*, 837–848. [CrossRef] [PubMed]
3. Dalu, T.; Wasserman, R.J. Cyanobacteria dynamics in a small tropical reservoir: Understanding spatio-temporal variability and influence of environmental variables. *Sci. Total Environ.* **2018**, *643*, 835–841. [CrossRef] [PubMed]
4. Kauppila, P.; Meeuwig, J.J.; Pitkänen, H. Predicting oxygen in small estuaries of the Baltic Sea: A comparative approach Estuar. *Coast Shelf Sci.* **2003**, *57*, 1115–1126. [CrossRef]
5. Sand-Jensen, K.; Møller, C.L.; Borum, J. High resistance of oligotrophic isoetid plants to oxic and anoxic dark exposure. *Freshw. Biol.* **2015**, *60*, 1044–1051. [CrossRef]
6. Sòria-Perpinyà, X.; Vicente, E.; Urrego, P.; Pereira-Sandoval, M.; Ruíz-Verdú, A.; Delegido, J.; Soria, J.M.; Moreno, J. Remote sensing of cyanobacterial blooms in a hypertrophic lagoon (Albufera of València, Eastern Iberian Peninsula) using multitemporal Sentinel-2 images. *Sci. Total Environ.* **2020**, *698*, 134305. [CrossRef] [PubMed]
7. Kravitz, J.; Matthews, M.; Bernard, S.; Griffith, D. Application of Sentinel 3 OLCI for chl-a retrieval over small inland water targets: Successes and challenges. *Remote Sens. Environ.* **2020**, *237*, 111562. [CrossRef]
8. Kutser, T.; Paavel, B.; Verpoorter, C.; Ligi, M.; Soomets, T.; Toming, K.; Casal, G. Remote Sensing of Black Lakes and Using 810 nm Reflectance Peak for Retrieving Water Quality Parameters of Optically Complex Waters. *Remote Sens.* **2016**, *8*, 497. [CrossRef]
9. Maltese, A.; Capodici, F.; Ciraolo, G.; Corbari, C.; Granata, A.; La Loggia, G. Planktothrix Rubescens in Freshwater Reservoirs: The Sentinel-2 Potentiality for Mapping Phycocyanin Concentration. In Proceedings of the First Sentinel-2 Preparatory Symposium, Frascati, Italy, 23–27 April 2012; Volume 707, p. 37.
10. Mishra, S.; Mishra, D.R. Normalized difference chlorophyll index: A novel model for remote estimation of chlorophyll-a concentration in turbid productive waters. *Remote Sens. Environ.* **2012**, *117*, 394–406. [CrossRef]
11. Kiran, G.S.; Joshi, U.B.; Padate, G.; Joshi, A.G. Preliminary investigation of the water quality of Wadhwana reservoir, Gujarat, India: A case study. *Bull. Environ. Sci. Res.* **2012**, *1*, 9–13.
12. APHA. *Standard Methods for Examination of Water and Waste Water*, 20th ed.; APHA, AWAA, WPCF: Washington, DC, USA, 1998.
13. Jiménez-Muñoz, J.C.; Sobrino, J.A.; Skoković, D.; Mattar, C.; Cristóbal, J. Land Surface Temperature Retrieval Methods from Landsat-8 Thermal Infrared Sensor Data. *IEEE Geosci. Remote Sens. Lett.* **2014**, *11*, 1840–1843. [CrossRef]
14. Ismail, K.; Boudhar, A.; Abdelkrim, A.; Mohammed, H.; Mouatassime, S.; Kamal, A.O.; Driss, E.; Idrissi, E.A.; Nouaim, W. Evaluating the potential of Sentinel-2 satellite images for water quality characterization of artificial reservoirs: The Bin El Ouidane Reservoir case study (Morocco). *Meteorol. Hydrol. Water Manag.* **2019**, *7*, 31–39. [CrossRef]

environmental sciences proceedings

MDPI

Proceeding Paper

Danio rerio: A Sustainable Model for Monitoring Pollutants in Aquatic Environments †

Hakan Çelebi *, Tolga Bahadır, İsmail Şimşek and Şevket Tulun

Department of Environmental Engineering, Aksaray University, Aksaray 68100, Turkey; tolgabahadir61@gmail.com (T.B.); ismailsimsek@aksaray.edu.tr (İ.Ş.); stulun@aksaray.edu.tr (Ş.T.)

* Correspondence: hakancelebi@aksaray.edu.tr; Tel.: +90-382-288-35-98

† Presented at the 7th International Electronic Conference on Water Sciences, online, 15–30 March 2023; Available online: https://ecws-7.sciforum.net.

Abstract: Since the beginning of humanity, many sectors have produced different chemicals. These chemicals are the main causes of environmental pollution and have become a global problem with irreversible effects in terms of health. Because of this, countries have set a target of "a pollution-free planet". We need to determine target-specific strategies to both eliminate pollution and protect health. To date, traditional methods of monitoring in receiving aquatic environments have been used; however, they do not provide information on toxic levels of pollutants. For this reason, researchers have focused on "bio-indicator" or "bio-monitor" organisms. Since these organisms are in equilibrium with the aquatic environment, they can also be considered an integrated sampling tool and may indicate potential contamination. *Danio rerio* (zebrafish) is considered a promising model organism for single health studies in terms of its biological structure. This review aims to present *Danio rerio*'s characteristics, susceptibility to environmental pollutants, and risks associated with pollutants in the aquatic environment.

Keywords: aquatic environment; bio-indicator; *Danio rerio*; environmental pollutants

Citation: Çelebi, H.; Bahadır, T.; Şimşek, İ.; Tulun, Ş. *Danio rerio*: A Sustainable Model for Monitoring Pollutants in Aquatic Environments. *Environ. Sci. Proc.* **2023**, *25*, 69. https://doi.org/10.3390/ECWS-7-14310

Academic Editor: Athanasios Loukas

Published: 3 April 2023

1. Introduction

Environmental change poses a devastating risk to human and environmental health. Environmental pollution is the most difficult challenge for all countries in the world, as it affects all living things and ecosystems under the concept of one health [1–3]. A rapid study of water conditions is necessary for monitoring, assessing, and addressing this global health hazard. Bio-indicators or biological monitors can monitor water quality changes in real-time. Zebrafish (*Danio rerio*) is an ideal bio-indicator for detecting environmental changes due to its biomedical equipment, widespread geographic distribution, and well-characterized specific properties against environmental pollutants [4,5]. *Danio rerio*, which is used to determine the toxicity (teratogenicity, cardiotoxicity, neurotoxicity, hepatotoxicity, and nephrotoxicity, etc.) effect of many chemicals, has also recently been used in the detection and research of diseases [6]. Considering the increasing use of zebrafish as an experimental animal against a wide variety of chemicals and pollutants in toxicity studies, it is also of great importance to improve laboratory conditions in terms of reproductive quality. Despite the zebrafish's numerous advantages, this model organism has several limitations. As a model for monitoring the toxicity of environmental pollutants, zebrafish have some disadvantages, such as low sensitivity and inconvenient statistical experiments [3–6].

In general, a good animal model should have all or most of the following characteristics: (i) structural simplicity that also incorporates the basic cellular processes that more complex organisms have; (ii) easy accessibility for research; (iii) easy and economical to manufacture in the laboratory; (iv) prone to genetic changes; and (v) if possible, have a relatively small and stable genome [7]. The most popular fish species preferred in experimental processes are listed as *Danio rerio*, *Carassius auratus*, *Oryzias latipes*, *Poecia Reticulata*, *Gasterosteus aculeatus*, *Oncorhynchus mykiss*, *Takifugu rubripes*, and *Xiphophorus hellerii*. Research on zebrafish has become more popular in the last decade. *Danio rerio's* high fertility rate (ability to fertilize approximately 200–300 eggs every 5–7 days), economic maintenance, and ease of genetic modification make *Danio rerio* an alternative and valuable vertebrate model compared to other species [8]. The use of zebrafish as a model for toxicity studies is carried out according to standards. These standards are the Zebrafish Toxicity Test, listed as British standard BS/EN/ISO 7346-3-1998, German standard DIN/EN/ISO 7346-3-1998, and Chinese standard GB/T13267-91, OECD n 203-236-473-487-489-490 [9,10]. *Danio rerio* (*Cyprinidae* family: freshwater teleost) is considered a model organism in many research areas, especially in health and pollution detection. In this study, we describe the criteria for the use of this fish species as a model for research in environmental toxicology and argue that *Danio rerio* is an ideal bio-indicator for detecting the toxicity of pollutants.

2. Methodology

A literature search related to zebrafish used as a model in ecotoxicity assessments was conducted in databases such as Web of Science, Science Direct, Scopus, and PubMed. The following keywords were used in the search process: "*Zebrafish*" and "*Ecotoxicity*", "*Danio rerio*" and "*Acute toxicity*", "*Zebrafish development*" and "*Fish assay*", and "*Zebrafish embryo test*" and "*Exposure*", etc. A total of 225 articles were reviewed in all databases. Studies that did not comply with the study's purpose were not included in the evaluation. Of the 225 articles published until January 2023, 92 were thoroughly reviewed. Each article used in the study was compiled according to the following parameters: (i) publication year; (ii) the type of pollutant; (iii) exposure factors; (iv) toxicity; (v) stage of development; (vi) types of tests; and (vii) ecotoxic effects (Figure 1) [1,11,12].

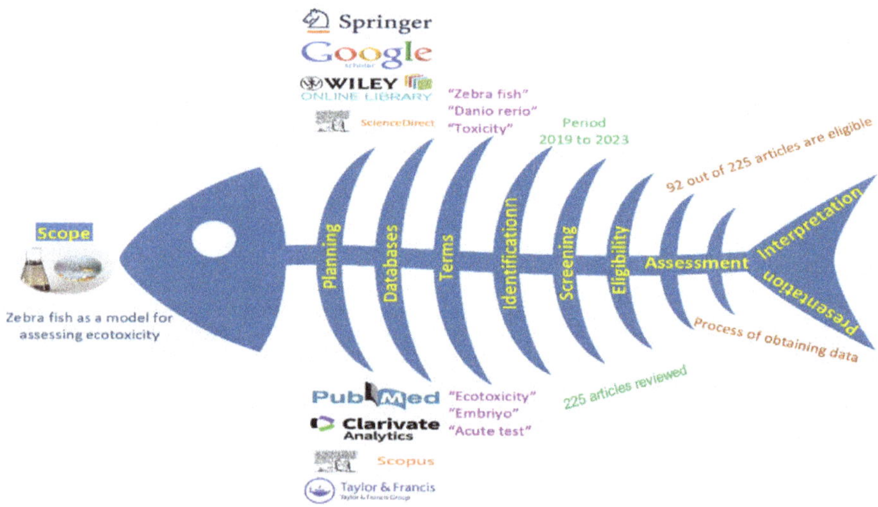

Figure 1. Systematic review methodology.

3. Advantages and Disadvantages of Zebrafish in Ecotoxicity Tests

Zebrafish is a successful monitoring model applied in both medicine (drug production, disease diagnosis, treatment, etc.) and environmental pollution (determination of the effect level of pollutants) [13]. Zebrafish has been widely used in ecotoxicity studies in recent years due to its advantageous features, such as reproduction process and developmental stage [14]. The most important feature that distinguishes zebrafish from other model creatures is that it is used in both sexes, which has brought zebrafish to the forefront in ecotoxicity tests [1]. One of the main advantages of this model is that experimental studies can be carried out in the embryo process [13]. Additionally, embryos (laying from egg: 48–72 h (hpf); organ formation: 120 hpf; adulthood: 3 months) grow rapidly [15,16]. Rapid development is an excellent feature in toxicity studies with animal models as it allows instant monitoring of toxic exposure [17]. All genetic features of zebrafish are stored in a database (www.zfin.org, accessed on 1 March 2023). In addition, 70% of zebrafish genes are similar to human genes [8]. In recent years, zebrafish has been evaluated as an alternative ecotoxicity model to replace mice and other fish species. It has proven to be a bio-indicator that complies with the 3R (Reduce, Reuse, and Recycle) concept, as tests with zebrafish are economical, provide fast results, and the presence of a large number of embryos reduces the chemical requirements [18]. Among the disadvantages of zebrafish, it can be said that different reactions occur according to the sex type, and the embryo development pools are large.

4. Development and Distribution of Zebrafish

The zebrafish was described in 1882 by Francis Hamilton, who found it near the Ganges River in India [19]. After George Streisinger first used the zebrafish to study vertebrate development in the 1980s, it became one of the most important laboratory animals with unprecedented speed [8]. The zebrafish is native to most of the Indian subcontinent, from Pakistan in the west through India, Nepal, and Bangladesh to Myanmar in the east. Zebrafish can be found in a wide variety of habitats. The adult zebrafish is 2 to 4 cm long, and its body is characterized by zebra-like stripes (Figure 2).

Figure 2. Adult zebrafish illustration.

In toxicological studies, especially in order to determine the toxic levels of environmental pollutants, parameters close to the natural environment of zebrafish should be provided in the laboratory environment. In this context, water quality parameters come to the fore (Table 1) [20–23]. The zebrafish is found in rivers, streams, canals, and rice fields in India, Pakistan, Bangladesh, Nepal, Myanmar, and Bhutan [22]. Wild zebrafish distribution includes Brazil, Colombia, Malaysia, Sri Lanka, Thailand, and the United States [21,23]. Native zebrafish is found in more than 3000 institutions in more than 100 countries and is a popular aquarium fish [22,24]. In research studies with zebrafish in the literature, China (26.7%), Brazil (14.8%), France (12.5%), Germany (8.5%), India (6.3%), and Italy (5.1%) take the lead [1]. The University of Oregon in the United States is home to the Zebrafish International Resource Center (ZIRC), the world's first and largest zebrafish resource center located on the Zebrafish Information Network (ZFIN). According to data from the Federation of European Aquaculture Producers (FEAP), this type of fish is produced at a rate of 1.8% (https://feap.info/, accessed on 1 March 2023).

Table 1. Physico-chemical properties of water for lab-raised zebrafish.

Parameters	Unit	Permissible Limits	Optimum Limits
Temperature	°C	6.5–42	26
Hardness	mg/L	72–198	100
pH	-	5.5–9.0	7.7
NaCl	ppt	0.1–14	0.65
Ammonia (NH_4^+)	mg/L	0–0.05	0.02
Nitrite (NO_2^-)	mg/L	0–0.5	0.2
Nitrate (NO_3^-)	mg/L	15–200	50
Phosphate (PO_4^{3-})	mg/L	150–300	250
Dissolved Oxygen	mg/L	5.5–8.0	7.5
Conductivity	µS	200–4000	1250

5. Zebrafish-Based Experimental Applications

In addition to fish species, experiment design is an important factor in ecotoxicity studies. The type of exposure (in vivo or in vitro), exposure route, exposure times, and stage of development are considered experimental parameters in data generation [13]. The majority of tests are performed in vivo (90.3%) compared to in vitro tests (9.7%) (Figure 3). Embryos (7.3%) and larvae (32.1%) are used less frequently than adults (55.8%) in ecotoxicity evaluation according to the development process (Figure 3). In terms of exposure route and duration, water (94.9%) and hour (51.1%) are evaluated at the highest rates, respectively (Figure 3). In Table 2, toxicity studies for which zebrafish are the models are given.

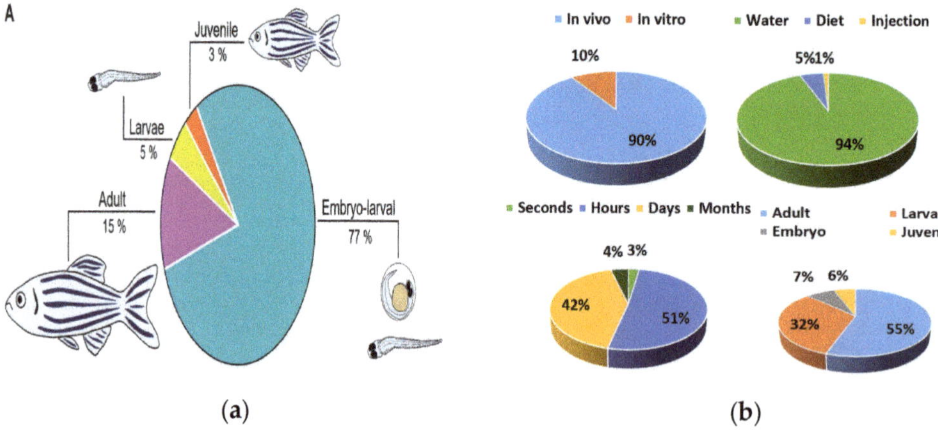

Figure 3. (**a**) *Danio rerio* development; (**b**) type of studies in vivo and in vitro, zebrafish exposure route, exposure time, and developmental stages (Adapted from Refs. [1,11]).

Table 2. Different toxicity studies in the literature.

Test Type	Exposure-Pollutant	References
Zebrafish embryo test	Crude oil component	[25]
Danio rerio long-term exposure	Metformin	[26]
Zebrafish antagonistic test	Deltamethrin	[27]
Adult zebrafish behavioral test	Cannabinoids	[28]
Early life stage zebrafish assay	Sediment	[29]
Acute-Chronic test	Lead	[30]
Early life stage zebrafish assay	Sodium Hypochloride	[31]
Combined toxicity test	Agrochemicals	[32]
Adult zebrafish behavioral test	Triclosan	[33]

6. Results

The zebrafish's small size, resistance to biotic and abiotic conditions, rapid development process, short reproduction period, compatibility with laboratory conditions, easy supply, economy, high fertility, ability to manipulate embryos, and genetic similarity with humans increase the importance of this species for toxicity studies. Thanks to these advantages, zebrafish can be used instead of other living models in scientific research and can support the 3R rule. As a result of studies researched in international databases, it is thought that different uses of this species will be widespread, and it will be a key model organism in the future.

Author Contributions: Conceptualization, methodology, formal analysis, investigation, resources, writing—original draft preparation, writing—review and editing, visualization, H.Ç., T.B., İ.Ş. and Ş.T. All authors have read and agreed to the published version of the manuscript.

Funding: This research received no external funding.

Institutional Review Board Statement: Not applicable.

Informed Consent Statement: Not applicable.

Data Availability Statement: Not applicable.

Acknowledgments: This study was carried out in Aksaray University Central Library and Engineering Faculty Environmental Engineering Department.

Conflicts of Interest: The authors declare no conflict of interest.

References

1. Canedo, A.; Rocha, T.L. Zebrafish (*Danio rerio*) using as model for genotoxicity and DNA repair assessments: Historical review, current status and trends. *Sci. Total Environ.* **2021**, *762*, 144084. [CrossRef] [PubMed]
2. Patel, N.; Khan, M.D.Z.A.; Shahane, S.; Rai, D.; Chauhan, D.; Kant, C.; Chaudhary, V.K. Emerging pollutants in aquatic environment: Source, effect, and challenges in biomonitoring and bioremediation—A review. *Pollution* **2020**, *6*, 99–113.
3. Vareda, J.P.; Valente, A.J.M.; Durães, L. Assessment of heavy metal pollution from anthropogenic activities and remediation strategies: A review. *J. Environ. Manag.* **2019**, *246*, 101–118. [CrossRef]
4. Kelly, J.R.; Shelton, S.G.; Daniel, D.K.; Bhat, A.; Mondal, R.; Nipple, F.; Amro, H.; Bower, M.E.; Isaac, G.; McHaney, G.; et al. Wild zebrafish sentinels: Biological monitoring of site differences using behavior and morphology. *Toxics* **2021**, *9*, 165. [CrossRef] [PubMed]
5. Beffagna, G. Zebrafish as a smart model to understand regeneration after heart injury: How fish could help humans. *Front. Cardiovasc. Med.* **2019**, *6*, 107. [CrossRef] [PubMed]
6. Strungaru, S.A.; Plavan, G.; Ciobica, A.; Nicoara, M.; Robea, M.A.; Solcan, C.; Petrovici, A. Toxicity and chronic effects of deltamethrin exposure on zebrafish (*Danio rerio*) as a reference model for freshwater fish community. *Ecotoxicol. Environ. Saf.* **2019**, *171*, 854–862. [CrossRef]
7. Shen, Q.; Truong, L.; Simonich, M.T.; Huang, C.; Tanguay, R.L.; Dong, Q. Rapid well-plate assays for motor and social behaviors in larval zebrafish. *Behav. Brain Res.* **2020**, *391*, 112625. [CrossRef]

8. Russo, I.; Sartor, E.; Fagotto, L.; Colombo, A.; Tiso, N.; Alaibac, M. The Zebrafish model in dermatology: An update for clinicians. *Discov. Oncol.* **2022**, *13*, 48. [CrossRef]
9. Chatterjee, N.; Lee, H.; Kim, J.; Kim, D.; Lee, S.; Choi, J. Critical window of exposure of CMIT/MIT with respect to developmental effects on zebrafish embryos: Multi-level endpoint and proteomics analysis. *Environ. Pollut.* **2021**, *268*, 115784. [CrossRef]
10. Kelly, J.R.; Benson, S.A. Inconsistent ethical regulation of larval zebrafish in research. *J. Fish Biol.* **2020**, *97*, 324–327. [CrossRef]
11. Saiki, P.; Mello-Andrade, F.; Gomes, T.; Tocha, T.L. Sediment toxicity assessment using zebrafish (*Danio rerio*) as a model system: Historical review, research gaps and trends. *Sci. Total Environ.* **2021**, *793*, 148633. [CrossRef]
12. Canedo, A.; de Jesus, L.W.O.; Bailao, E.F.L.C.; Rocha, T.L. Micronucleus test and nuclear abnormality assay in zebrafish (*Danio rerio*): Past, present, and future trends. *Environ. Pollut.* **2021**, *290*, 118019. [CrossRef]
13. Sieber, S.; Grossen, P.; Bussmann, J.; Campbell, F.; Kros, A.; Witzigmann, D.; Huwyler, J. Zebrafish as a preclinical in vivo screening model for nanomedicines. *Adv. Drug Deliv. Ver.* **2019**, *151–152*, 152–168. [CrossRef] [PubMed]
14. Bambino, K.; Chu, J. Zebrafish in toxicology and environmental health. *Curr. Top. Dev. Biol.* **2017**, *124*, 331–367.
15. Li, X.; Xiong, D.; Ju, Z.; Xiong, Y.; Ding, G.; Liao, G. Phenotypic and transcriptomic consequences in zebrafish early-life stages following exposure to crude oil and chemical dispersant at sublethal concentrations. *Sci. Total Environ.* **2021**, *763*, 143053. [CrossRef] [PubMed]
16. Vaz, R.; Hofmeister, W.; Lindstrand, A. Zebrafish models of neurodevelopmental disorders: Limitations and benefits of current tools and techniques. *Int. J. Mol. Sci.* **2019**, *20*, 1296. [CrossRef] [PubMed]
17. Le Pabic, P.; Dranow, D.B.; Hoyle, D.J.; Schilling, T.F. Zebrafish endochondral growth zones as they relate to human bone size, shape and disease. *Front. Endocrinol.* **2022**, *13*, 1060187. [CrossRef]
18. Chien, L.C.; Wu, Y.H.; Ho, T.N.; Huang, Y.Y.; Hsu, T. Heat stress modulates nucleotide excision repair capacity in zebrafish (*Danio rerio*) early and mid-early embryos via distinct mechanisms. *Chemosphere* **2020**, *238*, 124653. [CrossRef] [PubMed]
19. Zhou, S.; Chen, Q.; Di Paolo, C.; Shao, Y.; Hollert, H.; Seiler, T.B. Behavioral profile alterations in zebrafish larvae exposed to environmentally relevant concentrations of eight priority pharmaceuticals. *Sci. Total Environ.* **2019**, *664*, 89–98. [CrossRef]
20. Aleström, P.; D'Angelo, L.; Midtlyng, P.J.; Schorderet, D.F.; Schulte-Merker, S.; Sohm, F.; Warner, S. Zebrafish: Housing and husbandry recommendations. *Lab. Anim.* **2020**, *54*, 213–224. [CrossRef]
21. Li, S.; Yeo, K.S.; Levee, T.M.; Howe, C.J.; Her, Z.P.; Zhu, S. Zebrafish as a neuroblastoma model: Progress made, promise for the future. *Cells* **2021**, *10*, 580. [CrossRef] [PubMed]
22. D'Angelo, L.; de Girolamo, P. Fish as model systems. In *Laboratory Fish in Biomedical Research Biology, Husbandry and Research Applications for Zebrafish, Medaka, Killifish, Cavefish, Stickleback, Goldfish and Danionella Translucida*, 1st ed.; D'Angelo, L., de Girolamo, P., Eds.; Andre Gerhard Wolff: London, UK, 2022; pp. xix–xxiv.
23. Lee, C.J.; Paull, G.C.; Tyler, C.R. Improving zebrafish laboratory welfare and scientific research through understanding their natural history. *Biol. Rev.* **2022**, *97*, 1038–1056. [CrossRef] [PubMed]
24. Trigueiro, N.; Canedo, A.; Braga, D.; Luchiari, A.C.; Rocha, T.L. Zebrafish as an emerging model system in the global South: Two decades of research in Brazil. *Zebrafish* **2020**, *17*, 412–425. [CrossRef] [PubMed]
25. Gonzalez-Penagos, C.D.; Zamora-Briseno, J.A.; Amendola-Pimenta, M.; Elizalde-Contreras, J.M.; Arcega-Cabrera, F.; Cruz-Quintana, Y.; Santana-Pineros, A.M.; Canizarez-Martinez, M.A.; Perez-Vega, J.A.; Ruiz-May, E.; et al. Pollution and children's health. *Toxicol. Appl. Pharmacol.* **2022**, *445*, 116033.
26. Barros, S.; Ribeiro, M.; Coimbra, A.M.; Pinheiro, M.; Morais, H.; Alves, N.; Montes, R.; Rodil, R.; Quintana, J.B.; Santos, M.M.; et al. Metformin disrupts *Danio rerio* metabolism at environmentally relevant concentrations: A full life-cycle study. *Sci. Total Environ.* **2022**, *846*, 157361. [CrossRef] [PubMed]
27. Jijie, R.; Solcan, G.; Nicoara, M.; Micu, D.; Strungaru, S.A. Antagonistic effects in zebrafish (*Danio rerio*) behavior and oxidative stress induced by toxic metals and deltamethrin acute exposure. *Sci. Total Environ.* **2020**, *698*, 134299. [CrossRef]
28. Licitra, R.; Marchese, M.; Naef, V.; Ogi, A.; Martinelli, M.; Kiferle, C.; Fronte, B.; Santorelli, F.M. A review on the bioactivity of cannabinoids on zebrafish models: Emphasis on neurodevelopment. *Biomedicines* **2022**, *10*, 1820. [CrossRef]
29. Boulanger, E.; Barst, B.D.; Alloy, M.M.; Blais, S.; Houde, M.; Head, J.A. Assessment of environmentally contaminated sediment using a contact assay with early life stage zebrafish (*Danio rerio*). *Sci. Total Environ.* **2019**, *659*, 950–962. [CrossRef]
30. Kataba, A.; Botha, T.L.; Nakayama, S.M.M.; Yohannes, Y.B.; Ikenaka, Y.; Wepener, V.; Ishizuka, M. Environmentally relevant lead (Pb) water concentration induce toxicity in zebrafish (*Danio rerio*) larvae. *Comp. Biochem. Physiol.* **2022**, *252*, 109215. [CrossRef]
31. Ren, Z.; Yu, Y.; Ramesh, M.; Li, B.; Poopal, R.K. Assessment of eco-toxic effects of commonly used water disinfectant on zebrafish (*Danio rerio*) swimming behaviour and recovery responses: An early-warning biomarker approach. *Environ. Sci. Pollut. Res.* **2022**, *29*, 41849–41862. [CrossRef]
32. Hu, G.; Wang, H.; Wan, Y.; Zhou, L.; Wang, Q.; Wang, M. Combined toxicities of cadmium and five agrochemicals to the larval zebrafish (*Danio rerio*). *Sci. Rep.* **2022**, *12*, 16045. [CrossRef] [PubMed]
33. Çelebi, H.; Gök, O. Effect of triclosan exposure on mortality and behavioral changes of *Poecilia reticulata* and *Danio rerio*. *Hum. Ecol. Risk Assess.* **2018**, *24*, 1327–1341. [CrossRef]

environmental sciences proceedings

MDPI

Proceeding Paper

Fuzzy Analytical Solution for the Case of a Semi-Infinite Unconfined Aquifer †

Christos Tzimopoulos [1], Nikiforos Samarinas [1,*], Kyriakos Papadopoulos [2] and Christos Evangelides [1]

1 Department of Rural and Surveying Engineering, Aristotle University of Thessaloniki, 54124 Thessaloniki, Greece; ctzimop@gmail.com (C.T.); evan@topo.auth.gr (C.E.)
2 Department of Mathematics, Kuwait University—Khaldiya Campus, Safat 13060, Kuwait; kyriakos.papadopoulos@ku.edu.kw
* Correspondence: smnikiforos@topo.auth.gr

† Presented at the 7th International Electronic Conference on Water Sciences, 15–30 March 2023; Available online: https://ecws-7.sciforum.net.

Abstract: The solution to the second-order fuzzy unsteady nonlinear partial differential one-dimensional Boussinesq equation is examined. The physical problem concerns unsteady flow in a semi-infinite, unconfined aquifer bordering a lake. There is a sudden rise and subsequent stabilization in the water level of the lake; thus, the aquifer is recharging from the lake. The fuzzy solution is presented by a simple algebraic equation transformed in a fourth-degree polynomial approximation for the head profiles. In order to solve this equation, the initial and boundary conditions, as well as the numerous soil properties, must be known. A fuzzy approach is used to solve the problem since the aforementioned auxiliary conditions are vulnerable to various types of uncertainty resulting from human and machine errors. The physical problem described by a partial differential equation and the generalized Hukuhara derivative and the application of this theory for the partial derivatives were chosen as solving methods. In order to evaluate the accuracy and effectiveness of the suggested fuzzy analytical method, this study compares the findings of fuzzy analysis to those obtained using the Runge–Kutta method. This comparison attests to the accuracy of the former. Additionally, this results in a fuzzy number for water level profiles as well as for the water volume variation, whose α-cuts, provide according to Possibility Theory, the water levels and the water volume confidence intervals with probability $p = 1 - \alpha$.

Keywords: unsteady flow; fuzzy partial derivatives; numerical methods

Citation: Tzimopoulos, C.; Samarinas, N.; Papadopoulos, K.; Evangelides, C. Fuzzy Analytical Solution for the Case of a Semi-Infinite Unconfined Aquifer. *Environ. Sci. Proc.* **2023**, *25*, 70. https://doi.org/10.3390/ECWS-7-14303

Academic Editor: Athanasios Loukas

Published: 3 April 2023

1. Introduction

The horizontal water flow concerning unconfined aquifers without precipitation is described by the one-dimensional second-order unsteady nonlinear partial differential equation, called the Boussinesq equation:

$$\frac{\partial h}{\partial t} = \frac{K}{S}\frac{\partial}{\partial x}\left(h\frac{\partial h}{\partial x}\right), \tag{1}$$

where K = hydraulic conductivity (LT^{-1}), S = effective porosity(L^0T^0), h = piezometric head (L), x = horizontal coordinate (L), t = time (T).

Boussinesq (1904) [1] first proposed the above equation according to the assumption that the horizontal component of velocity ux does not change with depth and is a function of x and t, while the inertial forces are negligible. A unique solution to this nonlinear equation was published by Boussinesq in the "*Journal de Mathématiques Pures et Appliquées*" in 1904. With boundary conditions such as those of soil drained by drains placed in the impermeable substratum, Boussinesq solution dealt with the case of an aquifer atop an impermeable layer. Using the small disturbance method, Polubarinova-Kochina (952, 1962) [2,3] published a solution to Boussinesq's equation. By utilizing polynomial approximation and similarity transformation, Tolikas et al. (1984) [4] found an approximate solution. A weighted residual approach was used by Lockington (1997) [5] to provide an easily applicable analytical solution. Due to an abrupt change in the head at the origin, this approach was used for both the recharging and discharging of an unconfined aquifer. By using Adomian's decomposition method, Moutsopoulos (2010) [6] arrived at a simple series solution with a limited number of terms while he also performed a benchmark test, which demonstrated the benefits of his solution. The problem of flow in a one-dimensional semi-infinite horizontal aquifer with an initially dry head and a power-law function of time at the origin was examined by Lockington et al. in 2000 [7]. In accordance with the numerical outcomes, an approximative quadratic solution was developed. Using the traveling wave method, Basha (2013) [8] obtained a nonlinear solution with an easily applicable logarithmic form. The solution allows for the results to be of practical value in hydrology and is adaptable to any flow situation, whether it be recharge or discharge condition. There are also algebraic formulae for the propagation front velocity, the location of the wetting front, and the linkage between the characteristics of the aquifer. The nonlinear Boussinesq equation was given by a series solution by Chor et al. (2013) [9] in terms of the Boltzmann transform in a semi-infinite domain. An approximate solution was recently obtained by Hayek (2019) [10], who introduced an empirical function with four parameters. Using Microsoft Excel Solver, a numerical fitting approach was used to acquire the parameters. An approximate analytical solution for the recharge and discharge of a homogeneous unconfined aquifer was published by Tzimopoulos et al. in 2022 [11]. Numerous other studies [12–16] offer helpful clarification on the solution, offering a way for testing and accuracy of the numerical methods.

The definition of the initial flow condition, the method of linearizing the Boussinesq equation, the definition of drain spacing and hydraulic conductivity, boundary conditions, etc., are just a few examples of the ambiguities and uncertainties that the physical problem described by the Boussinesq equation presents [17,18]. Without taking into account the ambiguities and uncertainties of the groundwater flow problems, wrong management decisions could be made, leading to a number of significant negative environmental, social, and economic effects. Fuzzy algorithms were used to solve this problem for all the aforementioned reasons.

The fuzzy logic theory is a useful tool for modeling ambiguity, developed by Lofti Zadeh (1965) [19]. Its development has had a significant impact on both theoretical problems [20–23] as well as engineering and hydraulic problems [24]. To solve fuzzy differential equations, some analytical and numerical approaches have recently been put forth. Chang and Zadeh (1972) [25] first introduced the concept of fuzzy derivative, while Dubois and Prade (1982) [26] followed by using the extension principle in their approach. Fuzzy differential functions were studied by Puri and Ralescu (1983) [27], who extended Hukuhara's derivative (H-derivative) [28] of a set of values appearing in fuzzy sets. Kaleva (1987, 1990) [29,30] and Seikkala (1987) [31] developed the fuzzy initial value problem, while Abbasbandy and Allahviranloo (2002) [32] presented a numerical algorithm for solving fuzzy ordinary differential equations based on the second Taylor method. However, this method presented certain drawbacks, and in many cases, this solution was not a good generalization of the classic case. The generalized Hukuhara differentiability (gH—differentiability) was introduced by [33,34], overcoming this drawback. This new derivative is defined for a larger class of fuzzy functions than the Hukuhara derivative. Allahviranloo et al. (2015) [35] introduced the (gH-p) differentiability for partial derivatives as an extension of the above theory. The gH-p differentiability was used by Tzimopoulos et al. (2018, 2020) [17,36], providing a fuzzy linear analytical solution to a parabolic partial differential equation, and also Tzimopoulos et al. (2018) [18] obtained a fuzzy linear analytical solution to the Boussinesq equation in the case of an unconfined aquifer problem.

In this paper, a comparison between the fuzzy analytical nonlinear Boussinesq equation and the Runge–Kutta numerical method is presented for the proposed fuzzy analytical solution accuracy evaluation and effectiveness check. This comparison attests to the accuracy of the former. Additionally, an application of the Possibility Theory [37,38] to the α-cuts of the water level profiles, as well as of the water volume variation, provides the water levels and the water volume confidence intervals with probability $p = 1 - \alpha$. Thus, a combination of fuzzy theory with the Possibility Theory allows managers and engineers to solve practical hydraulic problems, making the right decision.

2. Constructing the Fuzzy Model and Solution

2.1. Crisp Model

As mentioned in the introduction to the case of one-dimensional horizontal flow, the Equation (1) corresponds to Boussinesq equation

The initial and boundaries conditions are as follows:

$$t = 0, h(x,0) = h_0,$$
$$t > 0, h(0,t) = h_1, \ \underset{x\to\infty}{h}(x,t) = h_0$$

The solution [11] of the above Equation (1) is as follows:

$$h = h_0 + (h_0 - h_1)\Omega(\mu,\xi)$$

where

$$\Omega(\mu,\xi) = \ell F(\xi) - \Phi(\xi), \Phi(\xi) = erfc(\xi),$$
$$\xi = \frac{x}{2\sqrt{\frac{Kh_1 t}{S}}}, \ell = (h_0 - h_1)/h_1, \mu = h_1/h_0,$$
$$F(\xi) = -\frac{1}{\pi} + \left(\frac{1}{2} + \frac{1}{\pi}\right)\Phi - \frac{1}{\sqrt{\pi}}(1-\Phi)\xi e^{-\xi^2} + \frac{1}{\pi}(1 - e^{-2\xi^2}) - \frac{1}{2}\Phi^2.$$

2.2. Fuzzy Model

We write Equation (1), in its fuzzy form as follows:

$$\frac{\partial \widetilde{h}}{\partial t} = \frac{K}{S}\frac{\partial}{\partial x}\left(\widetilde{h}\frac{\partial \widetilde{\widetilde{H}}}{\partial x}\right) = \frac{K}{S}\left\{\left(\frac{\partial \widetilde{h}}{\partial x}\right)^2 + \widetilde{h}\frac{\partial^2 \widetilde{h}}{\partial x^2}\right\}, \tag{2}$$

with the new boundary and initial conditions:

$$\begin{array}{ll} \text{Initial conditions} & \text{Boundary conditions} \\ t = 0, \widetilde{h}(x,0) = \widetilde{h}_0, t > 0, & \widetilde{h}(0,t) = \widetilde{h}_1, \ \underset{x\to\infty}{\widetilde{h}}(x,t) = \widetilde{h}_0 \end{array} \tag{3}$$

By converting the aforementioned fuzzy problem into a system of second order crisp boundary value problems—(referred as the *corresponding system* for the fuzzy problem)—and applying the theory of [18,35,39,40], we are able to solve the fuzzy problem (2), subject to the boundary and initial conditions (3). As a result, for the fuzzy problem, eight crisp BVPs systems are feasible under identical initial and boundary conditions.

Note: We shall now turn our attention to the first system's solution as it offers a practical solution to the case of the lake refilling the aquifer.

2.3. Solution of the First System

The nonlinear one-dimensional horizontal flow equations related to the first case (left boundary) and to the second case (right boundary) are provided with the following expressions:

$$\begin{array}{ll} \text{First case} & \text{Second case} \\ \frac{\partial h^-}{\partial t} = \frac{K}{S}\left\{h^-\frac{\partial^2 h^-}{\partial x^2} + \left(\frac{\partial h^-}{\partial x}\right)^2\right\} & \frac{\partial h^+}{\partial t} = \frac{K}{S}\left\{h^+\frac{\partial^2 h^+}{\partial x^2} + \left(\frac{\partial h^+}{\partial x}\right)^2\right\} \end{array} \tag{4}$$

3. Results and Discussion

We have used the data of Lockington (1997) [5], which means K = hydraulic conductivity = 20 m/d, S = effective porosity = 0.27, h_1 = 3m, h_0 = 2m. For the first case, the solution is as follows [11]:

$$h^-(\mu,\xi) = h_0^- + (h_0^- - h_1^-)\Omega(\mu,\xi)$$

Figure 1 illustrates the water level profiles for t = 5d of the new analytical method vs Runge–Kutta method (reference solution) and for α = 0, 1 value. The two methods approach each other closely. In addition, the water table in a period of 5 days approaches 125 m in length. If we consider also the hydraulic conductivity value (K = 20 m/d), the results are absolutely reasonable regarding the physical problem. Figure 2 illustrates the stored volume variation vs. time, and Figure 3 illustrates the membership function of $\widetilde{V}(x,t)$ for t = 5d. According to possibility theory [37,38], every function $[\widetilde{V}]$ estimates the crisp function V, and the α-cut $\{\widetilde{V}\}_a = [V_a^-, V_a^+]$ should be interpreted as the confidence intervals of V with a probability $p \geq 1 - a$. In this regard, in Figure 3, it is seen that for α = 0.05, the value of water volume lies in the interval [5.226, 14.310] with a probability higher than 95%, according to the possibility theory.

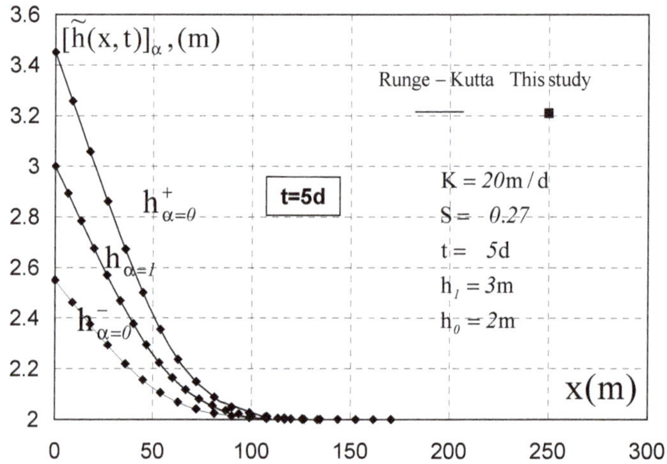

Figure 1. Water level profiles for t = 5d.

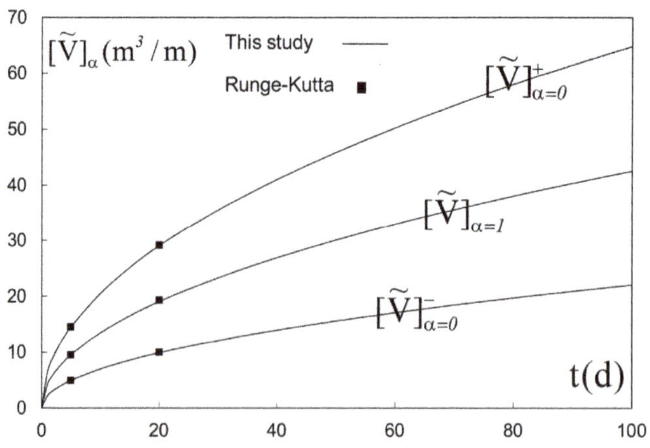

Figure 2. Stored volume variation vs. time.

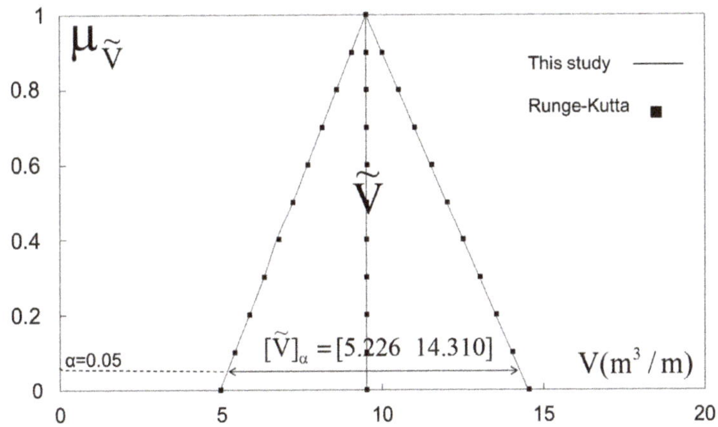

Figure 3. Membership function of $\widetilde{V}(x,t)$.

4. Conclusions

Undoubtedly, groundwater flow problems involve a number of ambiguities and uncertainties, making the use of differential equations even more difficult to solve. However, nowadays, the opportunity is given through the fuzzy partial differential equations theories to include these uncertainties in the final calculations and to provide more accurate results supporting the sustainability of groundwater as well as the researchers and engineers to make better decisions and plannings.

This work presents an innovative analytical solution in the nonlinear Boussinesq equation, which describes the groundwater's unsteady flow. According to the results, the proposed solution completely coincides with the Runge–Kutta method results used as a reference solution for comparison reasons in order to prove the accuracy and reliability of the proposed analytical solution. The volume membership function (Figure 3) could support decision-makers and planners with a higher degree of confidence than the previous years, thanks to the possibility theory.

Author Contributions: Conceptualization, C.T.; methodology, C.T. and N.S.; software, C.T.; validation, K.P. and C.E.; writing—original draft preparation, C.T.; writing—review and editing, N.S. and C.E.; visualization, C.T.; supervision, C.E. All authors have read and agreed to the published version of the manuscript.

Funding: This research received no external funding.

Institutional Review Board Statement: Not applicable.

Informed Consent Statement: Not applicable.

Data Availability Statement: Not applicable.

Conflicts of Interest: The authors declare no conflict of interest.

References

1. Boussinesq, J. Recherches théoriques sur l'écoulement des nappes d'eau infiltrées dans le sol et sur le débit des sources. *J. De Mathématiques Pures Appliquées* **1904**, *10*, 5–78.
2. Polubarinova-Cochina, P.Y. On unsteady motions of groundwater during seepage from water reservoirs. *PMM (Prinkladaya Matematica I Mekhanica)* **1949**, *13*, 2.
3. Polubarinova-Cochina, P.Y. *Theory of Groundwater Movement*; Moscow: Gostekhizdat, Old Soviet Union (Now Russia); de Wiest, R.J.M., Translator; Princeton University Press: Princeton, NJ, USA, 1962. (In Russia)
4. Tolikas, P.K.; Sidiropoulos, E.; Tzimopoulos, C.D. A simple analytical solution for the Boussinesq one-dimensional groundwater flow equation. *Water Resour. Res.* **1984**, *20*, 24–28. [CrossRef]

5. Lockington, D.A. Response of unconfined aquifer to sudden change in boundary head. *J. Irrig. Drain. Eng.* **1997**, *123*, 24–27. [CrossRef]
6. Moutsopoulos, K.N. The analytical solution of the Boussinesq equation for flow induced by a step change of the water table elevation revisited. *Transp. Porous Media* **2010**, *85*, 919–940. [CrossRef]
7. Lockington, D.A.; Parlange, J.Y.; Parlange, M.B.; Selker, J. Similarity solution of the Boussinesq equation. *Adv. Water Resour.* **2000**, *23*, 725–729. [CrossRef]
8. Basha, H.A. Traveling wave solution of the Boussinesq equation for groundwater flow in horizontal aquifers. *Water Resour. Res.* **2013**, *49*, 1668–1679. [CrossRef]
9. Chor, T.; Dias, N.L.; de Zarate, A.R. An exact series and improved numerical and approximate solutions for the Boussinesq equation. *Water Resour. Res.* **2013**, *49*, 7380–7387. [CrossRef]
10. Hayek, M. Accurate approximate semi-analytical solutions to the Boussinesq groundwater flow equation for recharging and discharging of horizontal unconfined aquifers. *J. Hydrol.* **2019**, *570*, 411–422. [CrossRef]
11. Tzimopoulos, C.; Papadopoulos, K.; Evangelides, C.; Spyrides, A. Analytical solution of nonlinear Boussinesq equation. *Desalination Water Treat.* **2022**, *24*, 1127–1147. [CrossRef]
12. Chen, Z.-X.; Bodvarsson, G.S.; Witherspoon, E.A.; Yortsos, Y.C. An integral equation formulation for the unconfined flow of groundwater with variable inlet conditions. *Transp. Porous Media* **1995**, *18*, 15–36. [CrossRef]
13. Parlange, J.-Y.; Hogarth, W.L.; Govindaraju, R.S.; Parlange, M.B.; Lockington, D. On an Exact Analytical Solution of the Boussinesq Equation. *Transp. Porous Media* **2000**, *39*, 339–345. [CrossRef]
14. Pistiner, A. Similarity solution to unconfined flow in an aquifer. *Transp. Porous Media* **2008**, *71*, 265–272. [CrossRef]
15. Olsen, J.S.; Telyakovskiy, A.S. Polynomial approximate solutions of a generalized Boussinesq equation. *Water Resour. Res.* **2013**, *49*, 3049–3053. [CrossRef]
16. Bartlett, M.S.; Porporato, A. A Class of exact solutions of the Boussinesq equation for horizontal and sloping aquifers. *Water Resour. Res.* **2018**, *54*, 767–778. [CrossRef]
17. Tzimopoulos, C.; Papadopoulos, K.; Evangelides, C.; Papadopoulos, B. Fuzzy solution to the unconfined aquifer problem. *Water* **2018**, *11*, 54. [CrossRef]
18. Tzimopoulos, C.; Papadopoulos, K.; Evangelides, C. Fuzzy Solution to the Second Order Unsteady Partial Differential Equation. In Proceedings of the International Conference of Numerical Analysis and Applied Mathematics (ICNAAM 2018), Rhodes, Greece, 13–18 September 2018; Volume 2116, pp. 440004-1–440004-5. [CrossRef]
19. Zadeh, L.A. Fuzzy Sets. *Inf. Control* **1965**. [CrossRef]
20. Bhaskar, V.; Lakshikantham, T.; Devi, V. Revisiting fuzzy differential equations. *Nonlinear Anal.* **2004**, *58*, 351–358.
21. Nieto, J.J.; Rodríguez-López, R. Bounded solutions for fuzzy differential and integral equations. *Chaos Solitons Fractals* **2006**, *27*, 1376–1386. [CrossRef]
22. Aminikhah, H. Approximate analytical solution for the one-dimensional nonlinear Boussinesq equation. *Int. J. Numer. Methods Heat Fluid Flow* **2015**, *25*, 831–840. [CrossRef]
23. Chen, F.; Qiu, X.; Alattas, K.A.; Mohammadzadeh, A.; Ghaderpour, E. A New Fuzzy Robust Control for Linear Parameter-Varying Systems. *Mathematics* **2022**, *10*, 3319. [CrossRef]
24. Guo, M.; Xue, X.; Li, R. The oscillation of delay differential inclusions and fuzzy biodynamics models. *Math. Comput. Model.* **2003**. [CrossRef]
25. Chang, S.S.L.; Zadeh, L.A. On Fuzzy Mapping and Control. *IEEE Trans. Syst. Man Cybern.* 1972. [CrossRef]
26. Dubois, H.; Prade, D. Towards fuzzy differential calculus: Part 3, differentiation. *Fuzzy Sets Syst.* **1982**, *8*, 30–34. [CrossRef]
27. Puri, M.L.; Ralescu, D.A. Differentials of fuzzy functions. *J. Math. Anal. Appl.* **1983**, *91*, 552–558. [CrossRef]
28. Hukuhara, M. Integration des Applications Measurables dont la Valeur est un Compact Convexe. *Funkc. Ekvacioj* **1967**, *10*, 205–233.
29. Kaleva, O. Fuzzy differential equations. *Fuzzy Sets Syst.* **1987**, *24*, 301–307. [CrossRef]
30. Kaleva, O. The cauchy problem for fuzzy differential equations. *Fuzzy Sets Syst.* **1990**, *24*, 389–396. [CrossRef]
31. Seikkala, S. On the fuzzy initial value problem. *Fuzzy Sets Syst.* **1987**, *24*, 319–330. [CrossRef]
32. Abbasbandy, S.; Allahviranloo, T. Numerical solution of fuzzy differential equation. *Math. Comput. Appl.* **2002**, *7*, 41–52. [CrossRef]
33. Bede, B.; Gal, S.G. Generalizations of the differentiability of fuzzy-number-valued functions with applications to fuzzy differential equations. *Fuzzy Sets Syst.* **2005**, *151*, 581–599. [CrossRef]
34. Stefanini, L.; Bede, B. Generalized Hukuhara differentiability of interval-valued functions and interval differential equations. *Nonlinear Anal. Theory Methods Appl.* **2009**, *71*, 1311–1328. [CrossRef]
35. Allahviranloo, T.; Gouyandeh, Z.; Armand, A.; Hasanoglu, A. On fuzzy solutions for heat equation based on generalized Hukuhara differentiability. *Fuzzy Sets Syst.* **2015**, *265*, 1–23. [CrossRef]
36. Tzimopoulos, C.; Papaevangelou, G.; Papadopoulos, K.; Evangelides, C. Fuzzy Analytical Solution to Vertical Infiltration. *J. Softw. Eng. Appl.* **2020**, *13*, 41–46. [CrossRef]
37. Dubois, D.; Prade, H. When upper probabilities are possibility Measures. *Fuzzy Sets Syst.* **1992**, *49*, 65–74. [CrossRef]
38. Mylonas, N. Applications in Fuzzy Statistic and Approximate Reasoning. Ph.D. Thesis, Dimokritos University of Thrace, Komotini, Greece, 2022. (In Greek).

39. Khastan, A.; Nieto, J.J. A boundary value problem for second order fuzzy differential equations. *Nonlinear Anal. Theory Methods Appl.* **2010**, *72*, 3583–3593. [CrossRef]
40. Bede, B.; Stefanini, L. Generalized differentiability of fuzzy-valued functions. *Fuzzy Sets Syst.* **2013**, *230*, 119–141. [CrossRef]

environmental sciences proceedings

MDPI

Proceeding Paper

Water Footprint Score: A Practical Method for Wider Communication and Assessment of Water Footprint Performance †

Konstantina Fotia * and **Ioannis Tsirogiannis**

Department of Agriculture, University of Ioannina, Kostakii Campus, 47100 Arta, Greece; itsirog@uoi.gr
* Correspondence: d_fotia@uoi.gr

† Presented at the 7th International Electronic Conference on Water Sciences, 15–30 March 2023; Available online: https://ecws-7.sciforum.net.

Abstract: In the present study, we propose a simple and practical method for assessing and communicating Water Footprint (WF) performance of a crop. We introduce the concept of "Water Footprint Score" (WFS), a comprehensive and comparable indicator of farmers' water resources management performance, which can be incorporated into agricultural products' labels. WFS as the outcome of the comparison with a water footprint annual reference level, and is a spatiotemporally comparable metric that reflects the convergence to best cultivation practices and can be easily perceived by both farmers and consumers. Examples of water footprint score for two different crops, kiwifruits and table olives, are provided.

Keywords: water footprint annual reference level; water footprint labeling; kiwifruit; table olives

Citation: Fotia, K.; Tsirogiannis, I. Water Footprint Score: A Practical Method for Wider Communication and Assessment of Water Footprint Performance. *Environ. Sci. Proc.* **2023**, *25*, 71. https://doi.org/10.3390/ECWS-7-14311

Academic Editor: Athanasios Loukas

Published: 3 April 2023

1. Introduction

Water Footprint (WF) as a metric of water consumption and pollution along the production chain of a good or a service [1] has evolved during the last two decades into a popular environmental indicator and a valuable tool for water management schemes' development at many levels. Since its first introduction in 2002, numerous WF assessments have been carried out [2,3], providing a large database of product, process and service water footprints. Its application in agriculture, a major water user, provides a clear indication of global crop water consumption and pollution patterns.

Despite its popularity among scientists and policy makers, slower uptake has been noticed at the farmers' and consumers' levels [4], probably due to the significant spatiotemporal variability of WF values that affect its comparability. At the field level, WF computation and further analysis of its components is valuable for farmers, since it provides an explicit insight into the general water management pattern followed, such as the degree of rainwater exploitation as a means of fresh water saving or the degree of water pollution caused by irrational agronomic practices related to fertilization and plant protection. The question that usually arises after computing the WF of a crop is the actual meaning of the WF value as far as further action and response formulation are concerned. Mekonnen and Hoekstra [5] proposed the comparison of a WF of a crop with reference levels of WF and for that reason they have developed global WF benchmark values for crop production. However, although global benchmarks are significant tools for the design and implementation of global policies, at the farmers' level they do not offer practical normalized information since site-specific climatic conditions, soil nutrients and plant health status affect applied agronomic practices depicted in WF computations, and thus the results are not fairly comparable.

Communication of WF to the wider public is a core issue in the growing global market of green products. Water footprint labeling as proposed by Hoekstra et al. [1] can either include the total WF of the product and/or the specification of its components, the degree of environmental sustainability achieved or the volume of water consumption compared to a reference year. The above either offer little or incomplete information regarding applied cultivation practices, require further analysis in order to be comprehensive, or lack comparability. Consumers need to be provided with complete and comprehensive information in order to "reward" farmers' actions towards sustainable water management.

The objective of the present study was to develop a simple and practical method for assessing and communicating in a single score WF performance of an agricultural product at the field level. We introduce the concept of "Water Footprint Score", a comprehensive and comparable indicator of farmers' water resources management performance, which can be easily incorporated into agricultural products' labels.

2. Methodology

2.1. General Concept and Approach

The general approach of the proposed method is to develop a comparable metric that depicts the performance of the applied cropping management practices. Two new terms are introduced:

- Water Footprint Score (WFS), which is proposed as a single score indicator of this performance. It expresses the result of the comparison of the WF of a crop with the WF that could be achieved if the farmer applied the optimal cultivation practices;
- Annual Reference Level of Water Footprint (WF') is introduced as the WF of a certain crop for a certain period that would be achieved if rational cultivation practices were applied at the specific site.

In this way farmers can have a more realistic depiction of how much their agronomic practices distance from the best applicable agronomic practices in the specific spatiotemporal context and not general best agronomic practices, which could be significantly differentiated by site-specific factors. The proposed method aims to provide an effective tool towards adoption of realistic sustainable agronomic practices.

2.2. Standard WF Computation Method by Hoekstra et al.

According to the Water Footprint Assessment Manual [1], the WF of a crop is computed as the sum of three distinctive components that reflect water use and pollution per unit of crop along the entire agricultural process: the green water footprint (WF_{green}), which refers to the amount of rainwater (green water) consumed during the growing period of a crop; the blue water footprint (WF_{blue}), which is linked to the consumption of surface and ground water (blue water); and the grey water footprint (WF_{grey}), which quantifies pollution caused by the application of fertilizers and pesticides and is expressed as the volume of fresh water needed to assimilate this pollution. WF components are computed as Equation (1):

$$WF_{green} = \frac{10 \times \sum_{d=1}^{lgp} ET_{green}}{Y} \quad (1)$$

$$WF_{blue} = \frac{10 \times \sum_{d=1}^{lgp} ET_{blue}}{Y} \quad (2)$$

$$WF_{grey} = \frac{\frac{\alpha \times AR}{c_{max} - c_{nat}}}{Y} \quad (3)$$

where ET_{green} and ET_{blue} (mm) are respectively the green and blue evapotranspiration during the length of the growing period (lgp), which measure the actual water consumption of a crop and if multiplied by ten (10) are expressed in $m^3\ ha^{-1}$; Y is the crop yield (t ha^{-1}); α is the leaching factor of the pollutant; AR is the application rate of the pollutant; and c_{max} and c_{nat} are the maximum allowable and the natural concentration of a pollutant in a certain site, respectively.

2.3. Annual Reference Level of Water Fooptrint

As mentioned above, the annual reference level of a crop's water footrpint is defined as the WF of a certain crop for a certain year at a certain site that would be achieved if optimal cultivation practices were applied.

Optimal cultivation practices regarding irrigation refer to the maximum possible exploitation of rain so that additional irrigation is applied only in the case that effective rainfall cannot meet actual crop water needs as expressed by actual evapotranspiration. Effective rain is defined as the rain that is stored in the soil and remains available to be used by the crop. Exploiting the maximum of the effective rain means that the volume of irrigation water applied to the crop is the minimum possible. This in turn means that maximum level of freshwater saving has been achieved.

Optimal cultivation practices regarding fertilization refer to two possible conditions: either the effective application of rational fertilization based on actual crop needs as determined by soil analysis, which would lead to the minimization of grey water footprint, or the application of organic fertilization, which in this case would lead to zero WF_{grey}.

The annual reference level of water footprint is calculated for green, blue and grey WF.

Hoekstra et al. proposed two methods for ET_{blue} and ET_{green} estimation using the CROPWAT model [1]: the Crop Water Requirements (CWR) option, which assumes optimal conditions and even though it is not very accurate it is employed when no irrigation data are available; and the irrigation schedule option, which is more accurate since it takes into account climate, soil and crop data along with irrigation data in order to calculate actual evapotranspiration (ET_a) using the daily soil water balance approach. While the first option provides the optimal rain water exploitation and thus defines the lowest potential irrigation needs, the second option estimates the real rain water exploitation, defining the actual part of irrigation that covered crop water needs (ET_{blue}).

In the present work's proposed WFS method, actual blue and green WF (WF_{green} and WF_{blue}, respectively) considers actual ET_{green} and ET_{blue} estimation based on the irrigation schedule option specifying the actual irrigation practice, selecting "irrigate at user-defined intervals and application depth" in the CROPWAT model.

The annual reference level of green and blue WF (WF'_{green} and WF'_{blue}, respectively) considers optimal rain water exploitation and thus optimal green and blue ET (ET'_{green} and ET'_{blue}, respectively), which are estimated based on the crop water recuirement option according to [1] as:

$$ET'_{green} = \min (ET_c, P_{eff}) \quad (4)$$

$$ET'_{blue} = \max (0, ET_c - P_{eff}) \quad (5)$$

Alternatively, ET'_{green} and ET'_{blue} can be estimated by applying the irrigation schedule option in the CROPWAT model by selecting "irrigate at critical depletion".

The annual reference level of green and blue WF are calculated according to Equations (1) and (2) subsituting ET_{green} and ET_{blue} with ET'_{green} and ET'_{blue}, respectively (Equations (4) and (5)).

The annual reference level of grey WF (WF'_{grey}) is calculated according to Equation (3) applying for AR the quantity of fertilizer that meets plant needs as determined by soil analysis results. In the case of the application of lower quantities of fertilizers than needed, WF_{grey} is considerd to be equal to WF'_{grey}. In the case of organic cultivation, WF_{grey} is zero.

Regarding yield (Y), for simplification reasons actual crop yield achieved is taken into account so as to avoid considering additional factors that are not entirely related to the agronomic practices applied and thus are not subjected to farmers' control. Additionally, this customized approach normalizes performance and facilitates comparisons.

2.4. Water Footprint Score

The Water Footprint Score (WFS) expresses a crop's WF performance in two major cultivation practices: water management and fertilization, and is defined as the result of the comparison between the actual crop's WF and the annual reference level of the crop's WF

(WF′). Since WF_{blue} only accounts for the volume of irrigation water actually consumed by plants and not the full volume of irrigation water applied, the magnitude of convergence from the best irrigation practice is better described by the ratio WF_{green}/WF_{blue}, which is considered to be a more indicative depiction of irrigation management performance as it encompasses the rainwater exploitation rate. For that reason, instead of comparing the WF_{green} and WF_{blue} with the respective WF'_{green} and WF'_{blue}, the ratio of actual green to blue WF is compared with the ratio of the annual reference level of green and blue WF as:

$$WFS_{green/blue} = 100\% \times \frac{WF_{green}}{WF_{blue}} \Big/ \frac{WF'_{green}}{WF'_{blue}} \tag{6}$$

Actual WF_{grey} is compared to the of annual reference level of WF_{grey} as:

$$WFS_{grey} = 100\% \times \frac{WF'_{grey}}{WF_{grey}} \tag{7}$$

According to this approach, the value of WFS consists of two parts:

- $WFS_{green/blue}$, which reflects water management performance (the ratio of green to blue WF);
- WFS_{grey}, which reflects the fertilization performance (WF_{grey}).

Higher values are associated with better performance. Values equal to 100 for each part indicate excellent WF performance while lower values indicate lower WF performance.

2.5. WFS Labeling

Following the WFS calculation, WFS performance is classified according to three major classes for each component as proposed in Table 1:

Table 1. WFS performance classification.

A (Excellent)	B (Medium)	C (Poor)
100–70	69–30	29–0

WFS classification is proposed as a WFS labeling option, which can explicitly communicate to the consumer a product's WF performance. WFS labeling can offer a quick visualization of the performance in each of the two main agronomic practices, irrigation and fertilization. For this reason, WFS can be expressed with two letters that represent performance in each of the two agronomic practices assessed. For instance, a $WFS_{green/blue}$ of 95% and WFS_{grey} of 22% can be classified as WFS = AC and visualized as (Figure 1):

Figure 1. Visualization of WFS score.

3. Case Studies: Kiwifruit and Table Olive Water Footprint Score

We performed a WFS computation for two major crops in the plain of Arta (Northwestern Greece), kiwifruit and table olive, for the year 2022.

3.1. Materials and Methods

The climate in the area is of a Mediterranean type with moderate rainy winters and dry, hot summers. Average annual temperature is 17.2 °C and annual precipitation is about 1100 mm. The water footprint score was calculated for the growing period of 2022 for a 10 year-old kiwi orchard and a 40 year-old table olive orchard. The kiwifruit orchard covers an area of 10 ha and the plant density is of 650 vines ha^{-1}. The table olive grove covers an area of 0.2 ha and the plant density is of 250 trees ha^{-1}. Meteorological data for the estimation of ET_c were provided by the net of agrometeorological stations established in the plain of Arta by the Department of Agriculture of University of Ioannina [6]. Farmers provided agronomic data such as application rate of fertilizers (AR) and yield (Y). The leaching factor (α) was set at 0.1 [1], c_{max} was set at 50 mg NO_3 L^{-1} (or 11.29 mg N L^{-1}) according to the EU Nitrates Directive, 91/676/EEC and c_{nat} was assumed to be zero [1]. In both orchards, irrigation and fertilization were performed based on farmers' experience. Table 2 summarizes agronomic data regarding the applied net irrigation, fertilization and final crop yield of the two fields.

Table 2. Agronomic data of kiwifruit and table olive orchard.

Field	Applied Net Irrigation (mm)	N-Fertilization (kg ha^{-1})	Yield (t ha^{-1})
Kiwifruit orchard	1114.40	325.10	35
Table olive grove	398	293	17

For both fields, WFS was calculated following the methodology described in the previous section and applying Equations (6) and (7).

3.2. Results and Discussion

According to Equations (6) and (7), in order to calculate WFS we need to calculate the actual WF and the annual reference level of WF components.

Actual WF components (WF_{green}, WF_{blue} and WF_{grey}) were calculated according to the typical method [1] and ET_{green} and ET_{blue} were estimated based on the soil water balance as modeled by CROPWAT [4] when applying the "Irrigation Schedule" option. Table 3 summarizes the actual ET_{green}, ET_{blue} and WF components for the two fields.

Table 3. Actual ET_{green}, ET_{blue} and WF components.

Field	ET_{green} (mm)	ET_{blue} (mm)	WF_{green} (m^3 t^{-1})	WF_{blue} (m^3 t^{-1})	WF_{grey} (m^3 t^{-1})
Kiwifruit orchard	38.78	778.28	11.08	222.37	84.44
Table olive grove	274.6	143.4	161.52	81.76	152.65

The annual reference levels of green and blue WF were computed according to Equations (4) and (5). N-application rate was determined by the soil analysis performed as 182 and 127 kg ha^{-1} for the kiwifruit and table olive, respectively. Table 4 summarizes the annual reference levels of WF components for the two fields.

Table 4. Annual reference levels of ET_{green}, ET_{blue} and WF components.

Field	ET'_{green} (mm)	ET'_{blue} (mm)	WF'_{green} (m^3 t^{-1})	WF'_{blue} (m^3 t^{-1})	WF'_{grey} (m^3 t^{-1})
Kiwifruit orchard	149.78	667.28	42.79	190.65	47.27
Table olive grove	303.2	114.8	178.23	67.06	66.17

The WFS of the two fields was calculated according to Equations (6) and (7):
For the kiwifruit orchard:

$$WFS_{green/blue} = 100\% \times [(11.8/222.37)/(42.79/190.65)] = 22\% \quad (8)$$

$$WFS_{grey} = 100\% \times (47.17/84.44) = 56\% \quad (9)$$

The WFS of the kiwi orchard was classified as (Figure 2):

Figure 2. WFS label for the specific kiwi orchard.

For the table olive grove:

$$WFS_{green/blue} = 100\% \times [(161.52/81.76)/(172.23/67.06)] = 74\% \quad (10)$$

$$WFS_{grey} = 100\% \times (66.17/152.65) = 43\% \quad (11)$$

The WFS of the table olive orchard was classified as (Figure 3):

Figure 3. WFS label for the specific table olive orchard.

For the kiwifruit, performance of both practices (irrigation management and fertilization) was poor to medium. In the case of the table olives, irrigation management proved to be better although fertilization practice was medium. Farmers could achieve a better water management performance through the maximum exploitation of rainfall and applying irrigation based on actual crop's needs. Monitoring soil moisture with sensors and using DSS systems for irrigation are proven to be effective means towards rational irrigation. Regarding fertilization, rational application practice is based on the determination of plants' actual nutrient needs through the performance of soil analysis.

4. Conclusions

The water footprint score is a simple and practical method to assess water footprint performance evaluating the main agronomic practices applied in a crop that have a large impact on the environment: irrigation and fertilization. WFS captures, in a single number, hotspots of agronomic practices and focuses on the specific agronomic practice that needs improvement. Farmers can use the WFS in order to understand how they performed compared to what they could actually do in order to apply sustainable practices at a realistic level. WFS points out which of their agronomic practice is environmentally costly in order to take action towards the specific direction. Additionally, WFS is a practical method to communicate farmers' performance to the wider public. Consumers can have a clear and comprehensive view on a product's "history", facilitating in this way their choice.

Author Contributions: K.F. conceptualized the research and wrote the manuscript, and I.T. reviewed the manuscript. All authors have read and agreed to the published version of the manuscript.

Funding: This research received no external funding.

Institutional Review Board Statement: Not applicable.

Informed Consent Statement: Not applicable.

Data Availability Statement: Full datasets are available upon request.

Conflicts of Interest: The authors declare no conflict of interest.

References

1. Hoekstra, A.Y.; Chapagain, A.K.; Aldaya, M.M.; Mekonnen, M.M. *The Water Footprint Assessment Manual*; Earthscan: London, UK, 2011; ISBN 9781849712798.
2. Feng, B.; Zhuo, L.; Xie, D.; Mao, Y.; Gao, J.; Xie, P.; Wu, P. A quantitative review of water footprint accounting and simulation for crop production based on publications during 2002–2018. *Ecol. Indic.* **2021**, *120*, 106962. [CrossRef]
3. Ma, W.; Opp, C.; Yang, D. Past, present, and future of virtual water and water footprint. *Water* **2020**, *12*, 3068. [CrossRef]
4. Nydrioti, I.; Grigoropoulou, H. Using the water footprint concept for water use efficiency labeling of consumer products: The Greek experience. *Environ. Sci. Pollut. Res.* **2022**, *30*, 19918–19930. [CrossRef] [PubMed]

5. Mekonnen, M.M.; Hoekstra, A.Y. Water footprint benchmarks for crop production: A first global assessment. *Ecol. Indic.* **2014**, *46*, 214–223. [CrossRef]
6. Mamassis, N.; Mazi, K.; Dimitriou, E.; Kalogeras, D.; Malamos, N.; Lykoudis, S.; Koukouvinos, A.; Tsirogiannis, I.; Papageorgaki, I.; Papadopoulos, A.; et al. Openhi.Net: A synergistically built, national-scale infrastructure for monitoring the surface waters of Greece. *Water* **2021**, *13*, 2279. [CrossRef]

environmental sciences proceedings

MDPI

Proceeding Paper

Integration of Microalgae-Microbial Fuel Cell with Microbial Electrolysis Cell for Wastewater Treatment and Energy Production †

Basil Mansoor, Sumreen Ashraf, Umar Rehman, Zia Ullah and Zeshan Sheikh *

Institute of Environmental Sciences and Engineering (IESE), School of Civil and Environmental Engineering (SCEE), National University of Sciences and Technology (NUST), Islamabad 44000, Pakistan; bmansoor.beee18iese@student.nust.edu.pk (B.M.); sashraf.beee18iese@student.nust.edu.pk (S.A.); urehman.beee18iese@student.nust.edu.pk (U.R.); zullah.phdiese@student.nust.edu.pk (Z.U.)

* Correspondence: zeshansheikh@iese.nust.edu.pk

† Presented at the 7th International Electronic Conference on Water Sciences, online, 15–30 March 2023; Available online: https://ecws-7.sciforum.net.

Abstract: The microalgae-microbial fuel cell (mMFC) enables us to perform secondary and tertiary treatment of domestic wastewater while simultaneously producing green electricity. In this work, the treatment of wastewater and producing electricity using a dual-chambered mMFC were demonstrated. Furthermore, the electricity produced by the mMFC was utilized to drive microbial electrolysis cell (MEC) for hydrogen synthesis. Primary domestic wastewater was treated in the anodic compartment of mMFC. Chemical oxygen demand (COD) removal of 67.2 and 54% was achieved in the anodic compartment of batch and continuous mode, respectively. Wetland-treated water was filled in the cathodic compartment, and *Scenedesmus* sp. was used as a catholyte. The overall voltage of 1.85 V was used to run a 1-L MEC. Industrial wastewater was treated with a COD removal of 73% in the MEC and biohydrogen was produced at a rate of 9.8 ± 0.2 mL L^{-1} d^{-1}.

Keywords: microalgae-microbial fuel cell; microbial electrolysis cell; wastewater treatment; energy production; microalgae

check for updates

Citation: Mansoor, B.; Ashraf, S.; Rehman, U.; Ullah, Z.; Sheikh, Z. Integration of Microalgae-Microbial Fuel Cell with Microbial Electrolysis Cell for Wastewater Treatment and Energy Production. *Environ. Sci. Proc.* **2023**, *25*, 72. https://doi.org/10.3390/ECWS-7-14306

Academic Editor: Lampros Vasiliades

Published: 11 April 2023

1. Introduction

Globally, around four hundred billion cubic meters of untreated wastewater is discharged each year [1]. Furthermore, the world's energy need is increasing all the time. Most of our energy requirements are fulfilled by burning fossil fuels. The combustion of fossil fuels produces a large amount of carbon dioxide, which is a greenhouse gas that causes climate change. We need to shift to sustainable green energy sources now more than ever. Microalgae-microbial fuel cell (mMFC) has attracted attention as a long-term wastewater treatment technology as it does not require an external power supply for wastewater treatment (WWT). It uses exoelectrogens to convert the organic matter in wastewater to electricity, and domestic wastewater contains 2–5 kWh/m^3 of energy [2]. In addition, mMFCs can be used in a variety of WWT applications because their biofilm contains not only electroactive microorganisms but also non-electroactive microorganisms such as fermentative, sulfate-reducing, nitrate-reducing, denitrifying, and aerobic microorganisms, which can be used for a variety of functions ranging from the breakdown of complex organics to fermentation [3]. Microbial electrolysis cell (MEC) is a modification of MFC that treats industrial wastewater when operated under anaerobic conditions and produces biohydrogen, which is an excellent green fuel.

In addition to producing bioelectricity and value-added products from microalgal biomass, the microbial fuel cells based on microalgae are effective systems for removing CO_2, nitrogen, and phosphorus from wastewater. These mMFCs have been demonstrated to be effective for the removal of N and P through the symbiotic relationship between microalgae and bacteria in wastewater treatment together with power generation. Since mechanical aeration would otherwise require power, the oxygen produced by microalgae during the day eliminates the need for mechanical aerators [4]. MEC systems need external power at an electrical potential substantially lower than the theoretical one utilized in water electrolysis to create hydrogen. A conventional power source, photovoltaics, wind power, thermoelectric generators, and MFCs are all potential options for external power sources [5].

In this study, mMFCs were used to perform secondary and tertiary wastewater treatment and produce electricity [4]. The secondary wastewater treatment was done in the anodic compartment that contained sludge, whereas the tertiary treatment was performed in the cathodic compartment containing microalgae. A detailed comparison of the batch mode mMFC and continuous flow mMFC was also performed in this study. On the other hand, the main objective of using MEC was to consume the electricity provided by the mMFCs to produce hydrogen using the microbial phenomenon. The MEC also helped in treating industrial wastewater, since high-strength wastewater containing large amounts of COD is required for the synthesis of hydrogen [6].

2. Materials and Methods

2.1. Construction of mMFC and MEC

For mMFC, a two-compartment H-shaped prototype made of acrylic sheets was used. Graphite rods, abraded by sandpaper to increase the surface area, were used as electrodes. Cation exchange membrane (CMI-7000s), which was soaked in a 5% NaCl solution for 12 h prior to its usage, was used as a separator. The working volume of mMFC was 1700 mL and the total volume was 2000 mL. The anodes and cathodes were connected to complete the circuit with low-resistance white copper wires. A stirrer was used to keep the algae in suspension and LED bulbs were used to provide light for the growth of algae. A peristaltic pump was also used to provide a continuous flow to the continuous flow mMFC.

For MEC, a 1000 mL stock bottle was used with similar graphite electrodes as for mMFC. The working volume was 900 mL and the total volume was 1000 mL. Gas syringes and pipes were used for the collection of hydrogen.

2.2. Inoculum and Substrate for mMFC and MEC

For mMFC, 30% activated sludge was obtained from the I-9 wastewater treatment plant, and Islamabad was used to perform secondary treatment of domestic wastewater in the anodic compartment. In the cathodic compartment, microalgae (*Scenedesmus* sp.) was used to perform tertiary treatment of domestic wastewater as a photosynthetic electrolyte. The primary treated domestic wastewater from the membrane bio-reactor (MBR) plant of National University of Sciences and Technology (NUST) was used in the anodic compartment whereas wetland-treated wastewater was used in the cathodic compartment.

For MEC, 20% anaerobic sludge was mixed with 5% cow dung to treat synthetic food industrial wastewater with an initial COD value of 10,000 mg/L.

The obtained sludge was acclimatized to anaerobic conditions before being used in both mMFC and MEC.

2.3. Operation of mMFC and MEC

For mMFC, one mMFC was run in the continuous flow mode (MFC-C) and two in the batch mode (MFC-B). The mMFC was operated at a lab scale under ambient temperature. In the anodic compartment, a pH of 7 was maintained whereas a pH of 7.5 was maintained in the cathodic compartment. The hydraulic retention time (HRT) was 24 h for both MFC-B and MFC-C.

MEC was also operated at a lab scale under ambient temperature in the batch mode. The HRT was 10 days and a pH of 4.5–5.5 was maintained to suppress the growth of methanogens. The MEC was sparged with N_2 gas prior to the start of each batch to ensure there was no oxygen present to be combined with the produced hydrogen.

The mMFCs were linked in series with the copper wires to get the combined voltage supplied to the MEC. A combined voltage of 1.8–1.9 V of all three mMFCs was provided to the MEC.

2.4. Analysis and Calculations

The closed reflux titrimetric method was used to obtain COD. pH was measured by using a Hach multimeter (Model-156). The values of orthophosphate and ammonium nitrogen were obtained using the ascorbic acid method and the salicylate method, respectively [7]. The water samples from both compartments of mMFC were obtained via the valves positioned at the center of the respective compartments. The water was stirred to ensure proper mixing prior to obtaining the samples. A digital multimeter was used to obtain the open circuit voltage whereas a datalogger (Picolog 6) was used to continuously obtain voltages in a closed circuit when a resistor of 470 Ω was used. The polarization curves were obtained by varying the resistances from 1 MΩ to 1 Ω. The voltages corresponding to these resistances were obtained and the current was calculated for each value using Ohm's law. By dividing the current by the surface area of the anode, the current density was obtained to be 0.0024 m^2. Power and power density were calculated using the formulas in Equations (1) and (2), respectively. For algae growth, a UV-visible spectrophotometer (PG Model T6OU, PG Instruments, UK) was used to measure the optical density (OD) at 680 nm [7]. The coulombic efficiency (CE) was found using the formula reported in [8].

$$P = VI \tag{1}$$

$$P = \frac{VI}{A} \tag{2}$$

where V is the voltage drop across each resistor, I is the corresponding current, and A is the surface area of the anode.

3. Results and Discussion

3.1. Sludge and Wastewater Characteristics

The sludge had a pH of 6.5–7, total solid (TS) of 15–25%, volatile solid of 30–35% of TS, moisture content (MC) of 75–85%, oxidation reduction potential (ORP) of −360 to −380 mV, and TKN of 1800–2500 mg/L. The primary treated domestic wastewater used in mMFCs had an initial COD of 240 mg O_2/L (varying for each day), orthophosphate of 18.5 mg/L, turbidity of 136 NTU, conductivity of 1636 µs/cm, pH of 7.3, TSS of 20 mg/L, TDS of 780 mg/L, ammonia nitrogen of 50 mg/L, nitrate nitrogen of 11.4 mg/L, nitrite nitrogen of 1.6 mg/L, and organic nitrogen of 20.16 mg/L.

3.2. Results of mMFC

The maximum open circuit voltage (OCV) was 784 mV in MFC-B whereas the maximum OCV was 723 mV in MFC-C.

As shown in Figure 1a, the COD decreased in the anodic chamber of both MFC-B and MFC-C. The initial COD was 305 mg/L for both MFC-B and MFC-C. The final COD was 100 mg/L for the batch mode, and the removal efficiency was thus 67.2%. In the case of the continuous flow mode, the final COD was 140 mg/L, and the removal efficiency was thus 54%. There was a difference of 13.2% in the removal efficiency. The possible reason for the high COD removal in the batch mode was that there was more time available for the microorganisms to consume the already existing COD, whereas in the continuous flow mode, new COD was pumped in the reactor continuously. Another similar study quoted their COD removal efficiency to be in the similar range of 50 to 65% [8]. Figure 1b shows the COD removal in the cathodic chamber. The removal from the cathodic chamber was 85% in MFC-B and 83% in MFC-C.

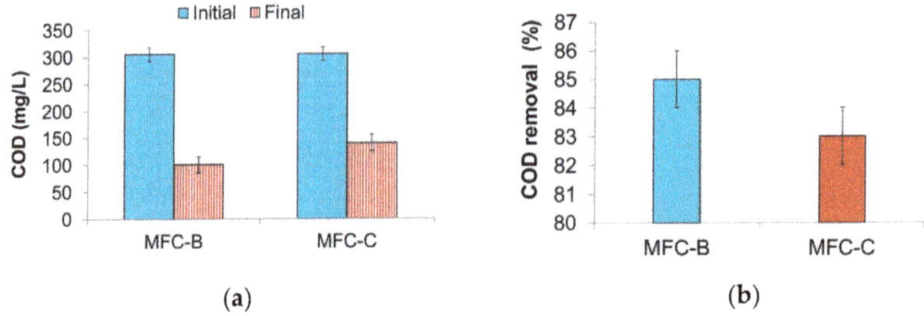

Figure 1. (**a**): COD removal from the anodic chamber. (**b**): COD removal from the cathodic chamber.

The CE indicates the amount of COD removed to be utilized for electricity. As shown in Figure 2, the COD removal was 67.2% in the batch mode while the CE was around 5.8%. In the continuous flow mode, the COD removal was 54% whereas the CE was around 7.7%. A difference of 1.9% was noted and MFC-C had higher CE despite having less removal efficiency. It is worth noting that the values of CE were quite low in comparison to the values of COD removal for both systems, indicating that a greater number of electrons were lost.

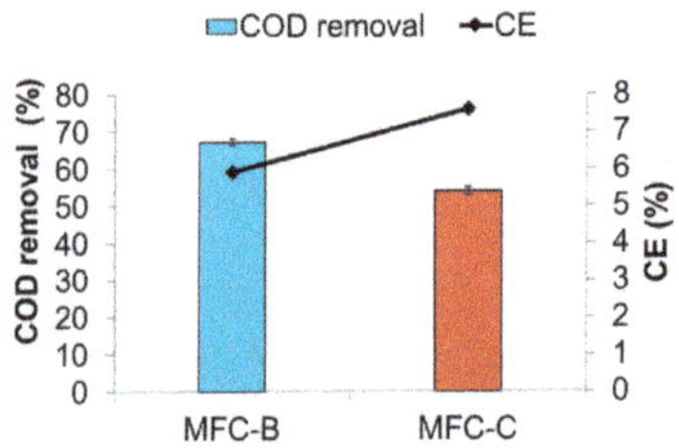

Figure 2. Coulombic efficiency (CE) in MFC-B and MFC-C.

Figure 3a shows that the orthophosphate removal of MFC-B was 80% whereas it was 68% for MFC-C. MFC-B was able to achieve a high orthophosphate removal of 12% as compared to MFC-C. Figure 3b shows that an ammonium nitrogen removal percentage of 60% was achieved in the batch mode and it was around 57% in the continuous flow mode.

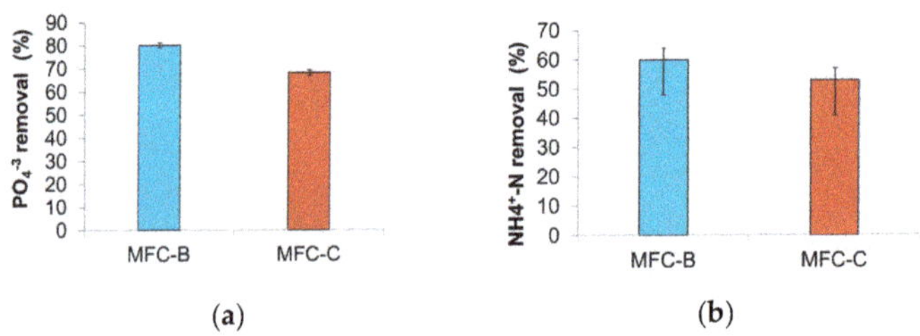

Figure 3. (**a**): Orthophosphate removal in the cathodic chamber. (**b**): Ammonium-N removal in the cathodic chamber.

In Figure 4, an increasing trend can be seen in algal growth. The dotted line shows algal growth in MFC-B and the solid line shows algal growth in MFC-C. There was a similar trend in both mMFCs starting at around 4900 mg/L and growing to about 5700 mg/L in four days.

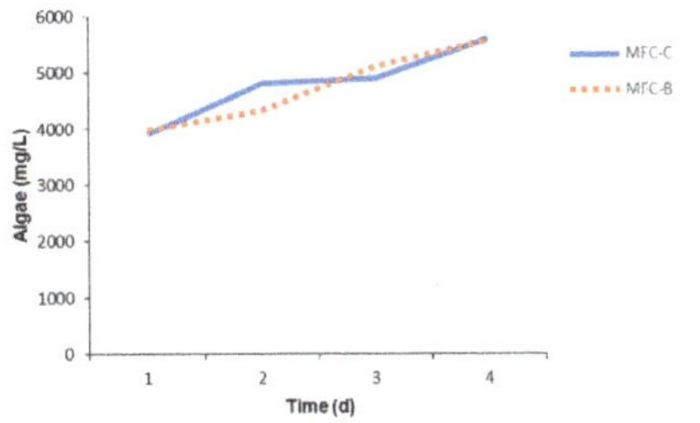

Figure 4. Algal growth in the cathodic chamber.

Figure 5 shows the polarization and power curves for MFC-B and MFC-C. The maximum current density was 375 mA/m^2 in MFC-C and 272.2 mA/m^2 in MFC-B. The maximum power density was 42 mW/m^2 in MFC-C and 25.9 mW/m^2 in MFC-B. By calculating the slopes of the polarization curves, the internal resistance was obtained to be 867.875 Ω in MFC-C and 847.250 Ω in MFC-B. In the continuous flow mode, the internal resistance is more than that in the batch mode, and the power density and current density were also greater. This may be due to the microbial configuration of sludge in the continuous flow mode, in which it was possible that the number of current-producing bacteria was greater in the continuous flow mode than in the batch mode.

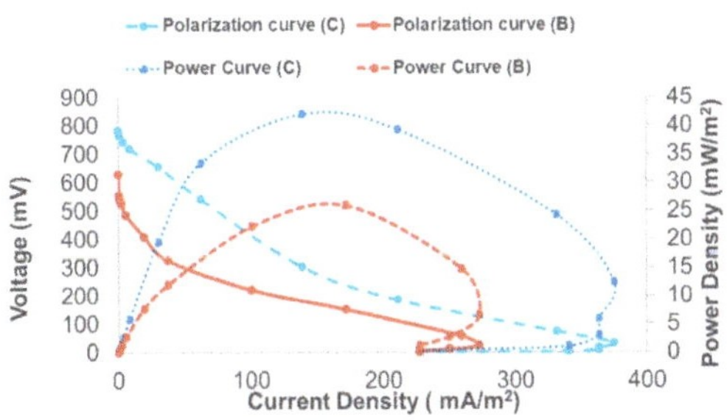

Figure 5. Polarization and power curves.

As shown in Figure 6, the voltage in the continuous flow mode was greater than in the batch mode. In the batch mode, there were dips in the values of voltage after every 24 h, indicating that the batch has ended and most of the COD has been consumed. In addition, the maximum value was achieved in the middle of the batch in the batch mode and the value steadily decreased towards the end of the batch after every 24 h. The curve of the continuous flow mode was relatively steady because of the continuous feed. Thus, COD was always available to the microorganisms, and voltage was constantly produced.

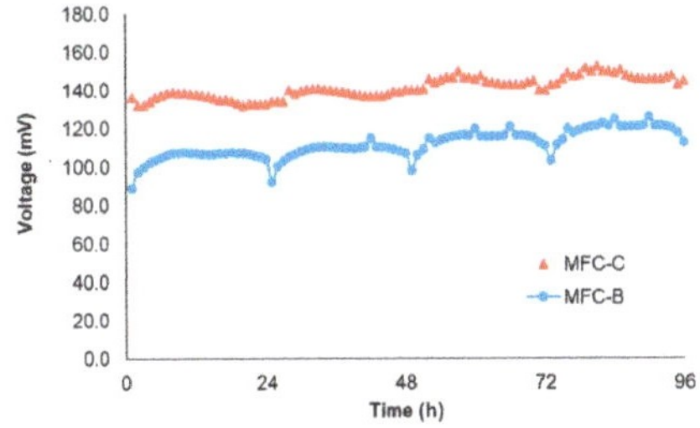

Figure 6. Voltage vs. time graph of MFC-B and MFC-C with a resistor of 470 Ω.

3.3. Results of MEC

Figure 7 shows a decrease in the value of COD of wastewater in MEC. The initial COD of the wastewater was 10,000 mg/L and after running a batch of 10 days, the value had decreased to 2700 mg/L, indicating a removal efficiency of 73%. The voltage provided by the mMFCs initiated a forceful current and was utilized to reduce the protons into hydrogen [6]. The average hydrogen production rate at the same time was 9.8 ± 0.2 mL L^{-1} d^{-1}.

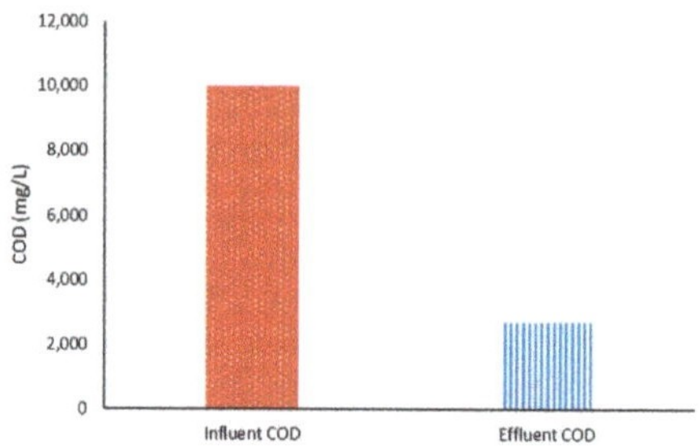

Figure 7. COD Removal in MEC.

4. Conclusions

In this work, three mMFCs (each with a working volume of 1.7 L) were run and compared in the batch mode and continuous flow mode. The system could be applied for both secondary and tertiary treatment as depicted by the results. The secondary treatment was performed as the organics were removed, and the tertiary treatment was performed as nutrient removal was achieved using algae. The COD removal was greater in MFC-B, but MFC-C had a greater CE. OCV was similar in both systems. However, in a closed circuit, MFC-C had a greater voltage output. The combined voltage was enough to power the MEC and the results showed that the supplied voltage overcame its potential barrier, and thus, the hydrogen was produced along with the treatment of wastewater.

Author Contributions: Conceptualization, B.M., S.A., U.R. and Z.S.; methodology, B.M. and Z.U.; software, B.M., Z.U. and S.A.; validation, B.M., S.A., U.R., Z.U. and Z.S.; formal analysis, B.M. and Z.U.; investigation, B.M., S.A., U.R. and Z.U.; resources, S.A. and Z.S.; data curation, B.M., S.A., Z.U. and Z.S.; writing—original draft preparation, B.M.; writing—review and editing, B.M., Z.S. and Z.U.; visualization, B.M., S.A., Z.S. and Z.U.; supervision, Z.U. and Z.S.; project administration, Z.U. and Z.S. All authors have read and agreed to the published version of the manuscript.

Funding: This research received no external funding.

Institutional Review Board Statement: Not applicable.

Informed Consent Statement: Not applicable.

Data Availability Statement: Data is contained within the article.

Conflicts of Interest: The authors declare no conflict of interest.

References

1. Ritchie, H. Max Roser and Pablo Rosado—"Energy". 2022. Available online: https://ourworldindata.org/energy (accessed on 15 December 2022).
2. Yu, J.; Cho, S.; Kim, S.; Cho, H.; Lee, T. Comparison of Exoelectrogenic Bacteria Detected Using Two Different Methods: U-tube Microbial Fuel Cell and Plating Method. *Microbes Environ.* **2012**, *27*, 49–53. [CrossRef] [PubMed]
3. Logan, B.E.; Hamelers, B.; Rozendal, R.; Schröder, U.; Keller, J.; Freguia, S.; Aelterman, P.; Verstraete, W.; Rabaey, K. Microbial Fuel Cells: Methodology and Technology. *Environ. Sci. Technol.* **2006**, *40*, 5181–5192. [CrossRef] [PubMed]
4. Arun, S.; Sinharoy, A.; Pakshirajan, K.; Lens, P.N. Algae based microbial fuel cells for wastewater treatment and recovery of value-added products. *Renew. Sustain. Energy Rev.* **2020**, *132*, 110041. [CrossRef]
5. Aboelela, D.; Soliman, M.A. Hydrogen production from microbial electrolysis cells powered with microbial fuel cells. *J. King Saud Univ.-Eng. Sci.* **2022**. [CrossRef]

6. Sun, M.; Sheng, G.-P.; Zhang, L.; Xia, C.; Mu, Z.-X.; Liu, X.-W.; Wang, H.-L.; Yu, H.-Q.; Qi, R.; Yu, T.; et al. An MEC-MFC-Coupled System for Biohydrogen Production from Acetate. *Environ. Sci. Technol.* **2008**, *42*, 8095–8100. [CrossRef] [PubMed]
7. Saleem, S.; Zeshan; Iftikhar, R.; Zafar, M.I.; Sohail, N.F. Growth kinetics of microalgae cultivated in different dilutions of fresh leachate for sustainable nutrient recovery and carbon fixation. *Biochem. Eng. J.* **2022**, *178*, 108299. [CrossRef]
8. Ullah, Z.; Zeshan, S. Effect of substrate type and concentration on the performance of a double chamber microbial fuel cell. *Water Sci. Technol.* **2019**, *81*, 1336–1344. [CrossRef] [PubMed]

environmental sciences proceedings

MDPI

Proceeding Paper

Flood Vulnerability Mapping Using MaxEnt Machine Learning and Analytical Hierarchy Process (AHP) of Kamrup Metropolitan District, Assam †

Akshayasimha Channarayapatna Harshasimha and Chandra Mohan Bhatt *

Indian Institute of Remote Sensing, Dehradun 248001, India; akshaysimhachhp@gmail.com
* Correspondence: cmbhatt@iirs.gov.in
† Presented at the 7th International Electronic Conference on Water Sciences, 15–30 March 2023; Available online: https://ecws-7.sciforum.net.

Abstract: Addressing a natural hazard's complexity is essential in preventing human fatalities and conserving natural ecosystems as natural hazards are varied and unbalanced in both time and place. Therefore, the main objective of this study is to present a flood vulnerability hazard map and its evaluation for hazard management and land use planning. The flood inventory map is generated for different flood locations using multiple official reports. To generate the vulnerability maps, a total of nine geo-environmental parameters are chosen as predictors form Maximum Entropy (MaxEnt) machine learning and Analytical Hierarchy Process (AHP). Accuracy assessment of the outputs from MaxEnt is performed using the area under the curve. Similarly, for AHP outputs, the accuracy is tested using the generated inventory map and the AUC. It is observed that topographical wetness index, elevation, and slope are significant for the assessment of flooded areas. Finally, flood hazard maps are generated and a comparative analysis is performed for both methods. According to the study's findings, The AUC of the flood map generated by MaxEntis 0.83, whereas the AUC of the flood map generated by AHP is 0.76, which means that the flood map generated by MaxEnt is better. From this study, it can be concluded that hazard maps could be a useful tool for local authorities to identify places that are vulnerable to hazards on a large scale.

Keywords: vulnerability mapping; Maximum Entropy (MaxEnt); Analytical Hierarchy Process (AHP); area under the curve (AUC)

Citation: Harshasimha, A.C.; Bhatt, C.M. Flood Vulnerability Mapping Using MaxEnt Machine Learning and Analytical Hierarchy Process (AHP) of Kamrup Metropolitan District, Assam. *Environ. Sci. Proc.* **2023**, *25*, 73. https://doi.org/10.3390/ECWS-7-14301

Academic Editor: Athanasios Loukas

Published: 3 April 2023

1. Introduction

Around the world, natural catastrophes pose a major threat to property and human lives. Although it is impossible to prevent natural hazards, their negative impact can be reduced by creating effective planning strategies and mitigation techniques. Significant morphological changes in landforms brought on by active tectonics or climatic changes may affect human activity and management. Events such as gully erosion, landslides, and floods are physical phenomena that are active in geological times but uneven in time and space [1–4].

According to (NDMA, 2008), a flood is extra water due to a river being incapable of transferring a large amount of water from the upstream area within its banks after significant rainfall. Floods occur more frequently and are more damaging to local social, economic, and environmental aspects than all other natural catastrophes that occur on a global scale. High intensity precipitation in the watershed, changes in river cross sections caused by sedimentation, sudden dam failure, release of high flow from dams, etc., are just a few causes of floods.

Depending on a variety of criteria that includes velocity, geography, and source, floods can be broadly classified into four categories: fluvial (river) floods, ground water floods, pluvial floods, and surge (coastal) floods. Assam, which is in the monsoon climatic region, has been having an average yearly rainfall from1600 mm to 4300 mm, causing flooding throughout the region (Assam State Disaster Management, n.d.). Overflowing tributaries of the Brahmaputra River also contribute to the volume of flood water in the valley.

Furthermore, this state has unique hydrological, climatic, and unstable geological conditions that intensify the source of numerous geomorphic and geological dangers in the region. Considering all these conditions, the use of remote sensing techniques proves to be a viable solution.

2. Study Area

The Kamrup Metropolitan district, which is located in the state of Assam in the north-eastern part of India, covers an area of 1528 km^2. The study area stretches from 26.07° N latitude to 91.63° E longitude in the lower basin of Brahmaputra, which is prone to rapid flooding nearly every year (Figure 1).

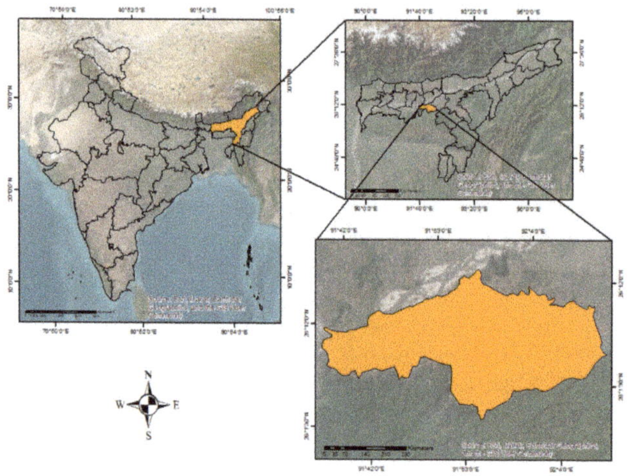

Figure 1. Research study area of the Kamrup Metropolitan district, Assam, India.

In 2021, the districts of Assam had an average annual temperature of 24 °C and an annual rainfall of over 2200 mm. The Kamrup Metropolitan district has major cities and is Assam's administrative center.

3. Materials and Methods

Figure 2 illustrates the methodology that was approached with AHP modeling and MaxEnt modeling.

3.1. Flood Inventory Mapping

A key step for susceptibility mapping is the preparation of an inventory of hazard landforms. The flood inventory for the Kamrup Metropolitan district (Assam, India) is compiled from national and regional documents from various organizations such as Assam State Disaster Management Authority and the North-Eastern Space Applications Centre. About 53 flood areas are listed on the inventory map for floods. For training samples, a random partition approach is used. In the present study, 70% of each hazard is considered for model construction (training) and the remaining 30% of each hazard is used for validation.

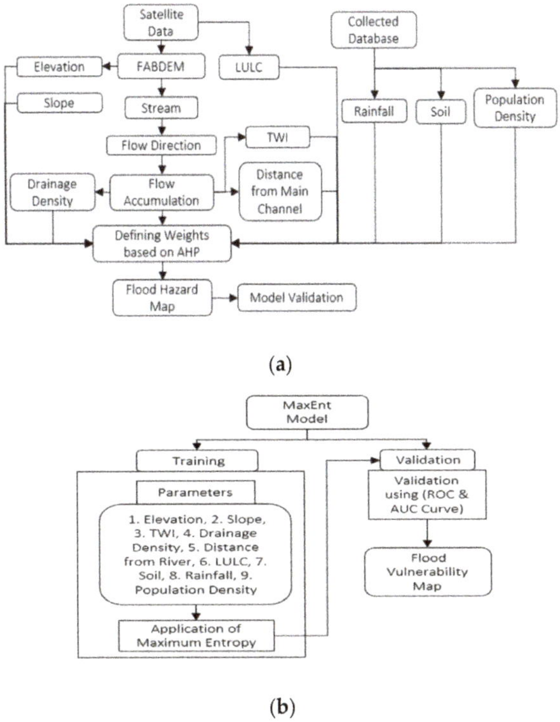

Figure 2. (**a**) AHP flowchart; (**b**) MaxEnt flowchart.

3.2. Flood Conditioning Factors

It is essential to determine the effective factors of different natural hazards and human-made fatalities to perform flood maps [5]. A good understanding of the main hazard-related factors is needed to recognize the susceptible areas.

For this aim, the conditioning factors for the hazard were selected [6–10]. In this study, ArcGIS 10.3 (ESRI, USA) is used to perform the analysis of AHP and to produce and display the data layers. All the factors were processed into a raster grid of 30 × 30 m grid cells. Entire conditioning factors were primarily continuous, and some of them were classified within different categories based on expert knowledge and a literature review [11–14].

3.2.1. Elevation

Elevation is a parameter of great significance for delineating flood hazards and mapping flood zones. During the monsoon, the downstream area generates ideal flood conditions due to sedimentation and surge in river flow. Understanding elevation variation is critical for river basin generation and the propagation of flood waters. In this study, FABDEM (Forest and Building removed Copernicus DEM) data with a spatial resolution of 30 m is used.

3.2.2. Slope

The steepness and length of a region's topography greatly influence its discharge and flooding. The rapid velocity of precipitation runoff is caused by steep or high slopes. Low or flat slopes, on the other hand, are prone to waterlogging, which can lead to high infiltration. The slope map was created using FABDEM (DEM) data.

The geo-environmental parameters used in this study are presented Figure 3 below.

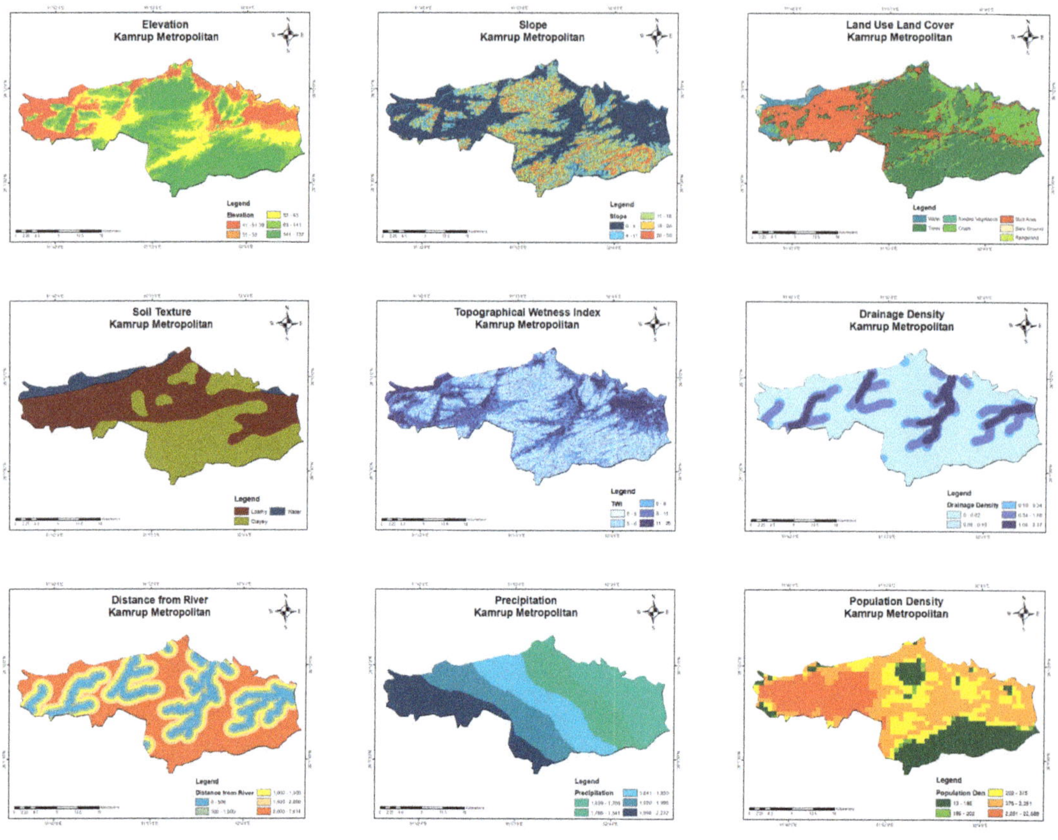

Figure 3. Ninegeo-environmental conditioning factors.

3.2.3. Land Use Land Cover

Land use/land cover plays a significant role in the operation of hydrological and geomorphological processes by directly or indirectly influencing processes such as evapotranspiration, infiltration, runoff generation, and sediment dynamics. The land use/land cover product with a 10 m spatial resolution is obtained from Sentinel-2 using the Google Earth Engine (GEE) platform.

3.2.4. Soil Texture

Soil texture is generally recognized not only as a weighty controlling factor in the mechanism of infiltration and runoff generation but also as being effective forhazard occurrence. This layer was acquired from the NBSSLUP. The soil texture in the study area comprises loam and clay.

3.2.5. Topographic Wetness Index (TWI)

Moore and Grayson [15] and Grabs et al. [16] mention that TWI (Topographic wetness index) represents how the tendency of gravitational forces and the spatial distribution of wetness conditions move water down slope. This layer was generated using DEM. TWI is also important in the regulation of surface runoff since the wetter an area is, the greater its runoff will be.

3.2.6. Distance from River Channel

The distance from the river was estimated in ArcGIS using the Euclidean Distance tool, which displays the distance from the river basin region to the natural drainage. In this context, natural drainage refers to all streams and rivers in the study region, which was categorized into five classes: 500 m, 1000 m, 1500 m, and 2000 m.

3.2.7. Drainage Density

The primary influencing factor that contributes to the occurrence of numerous risks is drainage density. A higher surface runoff ratio results from a higher drainage density. To convert the drainage network pattern to a measurable quantity, the drainage density was determined using an extension of "line density" in ArcGIS 10.3 software.

3.2.8. Rainfall

Rainfall is a key aspect in this study as floods most commonly occur during monsoon season, hence the term "rain-induced floods". The rainfall map of the study area is generated using the Inverse Distance Weighted approach (IDW) from Global Precipitation Measurement datasets. The map is generated considering the annual total rainfall of year 2021 as 2021 was a flood year.

3.2.9. Population Density

One of the critical elements to consider while conducting flood vulnerability research is population density. This component is important for analyzing the social loss and damage suffered by communities in flood-vulnerable areas as a result of floods. The population density map for the study area is obtained from Google Earth Engine databases of population density gridded data with $\approx$1 km of spatial resolution.

4. Results

4.1. Maximum Entropy (MaxEnt)

The MaxEnt software uses the Maximum Entropy model to calculate hazard estimates (version 3.4.4). The MaxEnt model is usually used to estimate species distribution based on the most significant environmental conditions. From a decision-theoretic perspective, we also interpret the maximum entropy estimation as a reliable Bayes estimation. The model relies on a machine learning reaction that generates hypotheses based on skewed data. The result from the model is obtained in ASCII format. The conditioning factors are translated from raster into ASCII format, as required by the software. The most crucial phase in the modeling process is validation. The AUC has been used to assess the built-in hazard model's prediction accuracy, as shown in Figure 4.

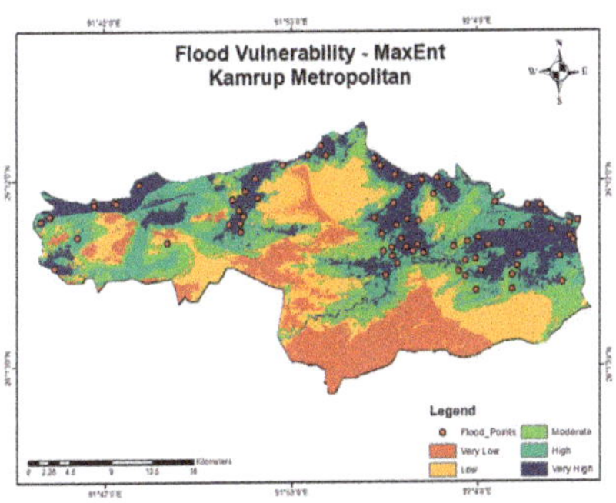

Figure 4. MaxEnt flood mapping.

4.2. Analytical Hierarchy Process (AHP)

Table 1 shows the weights assigned to the nine geo-environmental parameters used to generate the flood hazard map. To obtain the spatial distribution of flood hazards, the parameters evaluated were mapped and normalized into five classes based on a rating scale of 1 to 5, with 1 being the least vulnerable area and 5 being the most vulnerable area, as shown in Figure 5.

Table 1. AHP weights.

Factor	Weight
Slope	0.22
Distance from River channel in meter	0.17
Land use land cover	0.05
Soil Texture	0.10
Elevation	0.07
Rainfall in mm	0.04
Population Density	0.02
TWI	0.21
Drainage Density	0.12

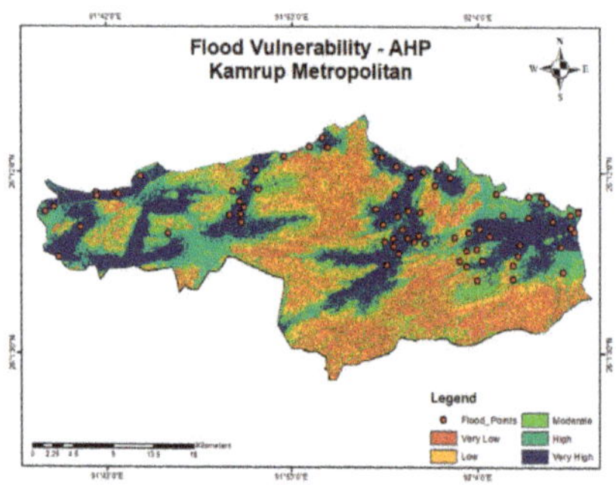

Figure 5. AHP flood Mapping.

4.3. Comparative Analysis of sensitivity and Response Curves

The relative influence of each predictor variable on the outcomes of the predicted maps using the jackknife test was examined using a sensitivity analysis from the AUC. Concerning validating with respect to flood inventory points, we observe that the 0.83 AUC of the MaxEnt model slightly outperformed the 0.763 AUC of the AHP model, as shown in Figures 6 and 7, respectively.

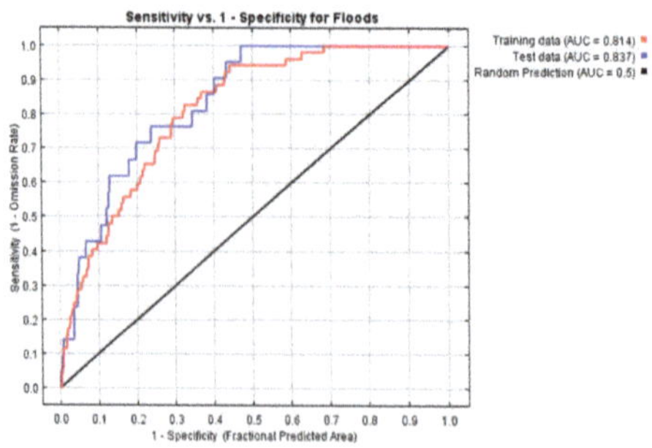

Figure 6. AUC for MaxEnt.

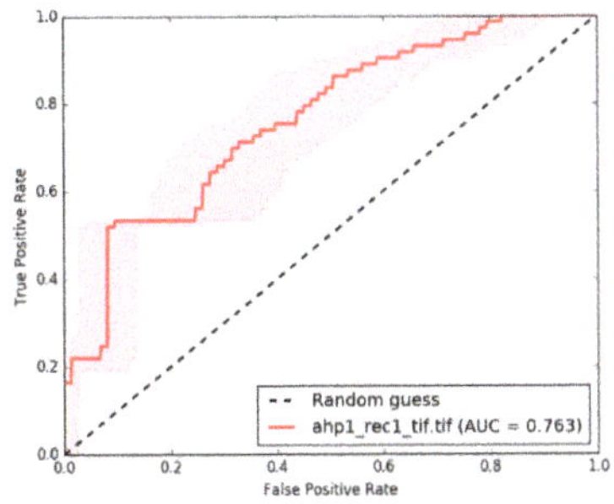

Figure 7. AUC for AHP.

4.4. Spatial Extent of Vulnerability

Flood vulnerability maps generated from the outputs of MaxEnt and AHP show that the areas encircling the river, the surfaces with slopes from 0 degrees to 11 degrees, and the land in the elevation range from 41 m to 52 m showed vulnerability to floods. Subsequently, we observe that the flood map generated by MaxEnt and AHP showed a reasonable resemblance with the historical flood maps of ISRO's Bhuvan.

From the results, we also observe that out of the 1528 km^2 total area of the district, about 650 km^2 was found to be highly vulnerable to floods; moreover, major locations such as Guwahati, Dispur, and Sonapur Gaon in the Kamrup Metropolitan district showed a higher vulnerability to flooding.

5. Conclusions

In this study, flood vulnerability maps are generated for a major district of Assam by utilizing the AHP approach and MaxEnt machine learning. Given its ability to handle huge datasets, a multi-criteria analysis using AHP and MaxEntis identified and proven to be beneficial for flood risk assessment.

Slope, drainage density, TWI, and elevation were the primary flood-causing geo-environmental parameters in the studied area. The AHP method and MaxEnt technique employed in this study are effective and enable the possibility of further research into flood vulnerabilities in various sections of the state or country. The AUC graphs are employed as a validation method in this work, which demonstrates an additional possibility of research validation and applicability in geospatial vulnerability assessment owing to extreme events.

Author Contributions: The initial idea for the work came from A.C.H. with assistance from mentor C.M.B., who is also a corresponding author on this publication. The records, compilation, and choice of the final design of the work were done by A.C.H. All authors contributed to this paper and shared ideas. All authors have read and agreed to the published version of the manuscript.

Funding: This research received no external funding.

Institutional Review Board Statement: Not applicable.

Informed Consent Statement: Not applicable.

Data Availability Statement: Not applicable.

Conflicts of Interest: The authors declare no conflict of interest.

References

1. Achour, Y.; Pourghasemi, H.R. How do machine learning techniques help in increasing accuracy of landslide susceptibility maps. *Geosci. Front.* **2020**, *11*, 871–883. [CrossRef]
2. Arnaud, P.; Bouvier, C.; Cisneros, L.; Dominguez, R. Influence of rainfall spatial variability on food prediction. *J. Hydrol.* **2002**, *260*, 216–230. [CrossRef]
3. Castillo, C.; Gómez, J.A. A century of gully erosion research: Urgency, complexity and studyapproaches. *Earth-Sci. Rev.* **2016**, *160*, 300–319. [CrossRef]

4. Kelarestaghi, A.; Ahmadi, H. Landslide susceptibility analysis with a bivariate approach and GIS in Northern Iran. *Arab. J. Geosci.* **2009**, *2*, 95–101. [CrossRef]
5. Kia, M.B.; Pirasteh, S.; Pradhan, B.; Mahmud, A.R.; Sulaiman WN, A.; Moradi, A. An artificial neural network model for food simulation using GIS: Johor River Basin Malaysia. *Environ. Earth Sci.* **2012**, *67*, 251–264. [CrossRef]
6. Kornejady, A.; Ownegh, M.; Rahmati, O.; Bahremand, A. Landslide susceptibility assessment using three bivariate models considering the new topo-hydrological factor: HAND. *Geocarto Int.* **2018**, *33*, 1155–1185. [CrossRef]
7. Pourghasemi, H.; Moradi, H.; Aghda, S.F.; Gokceoglu, C.; Pradhan, B. GIS-based landslide susceptibility mapping with probabilistic likelihood ratio and spatial multi-criteria evaluation models (North of Tehran, Iran). *Arab. J. Geosci.* **2014**, *7*, 1857–1878. [CrossRef]
8. Rahmati, O.; Pourghasemi, H.R.; Melesse, A.M. Application of GIS-based data driven random forest and maximum entropy models for groundwater potential mapping: A case study at Mehran Region Iran. *CATENA* **2016**, *137*, 360–372. [CrossRef]
9. Tehrany, M.S.; Pradhan, B.; Jebur, M.N. Flood susceptibility mapping using a novel ensemble weights-of-evidence and support vector machine models in GIS. *J. Hydrol.* **2014**, *512*, 332–343. [CrossRef]
10. Conoscenti, C.; Rotigliano, E.; Cama, M.; Caraballo-Arias, N.A.; Lombardo, L.; Agnesi, V. Exploring the effect of absence selection on landslide susceptibility models: A case study in Sicily Italy. *Geomorphology* **2016**, *261*, 222–235. [CrossRef]
11. Das, H.; Sonmez, H.; Gokceoglu, C.; Nefeslioglu, H. Infuence of seismic acceleration on landslide susceptibility maps: A case study from NE Turkey (the Kelkit Valley). *Landslides* **2013**, *10*, 433–454. [CrossRef]
12. Jiménez-Perálvarez, J.; Irigaray, C.; El Hamdouni, R.; Chacón, J. Landslide- susceptibility mapping in a semi-arid mountain environment: An example from the southern slopes of Sierra Nevada (Granada, Spain). *Bull. Eng. Geol. Environ.* **2011**, *70*, 265–277. [CrossRef]
13. Saponaro, A.; Pilz, M.; Wieland, M.; Bindi, D.; Moldobekov, B.; Parolai, S. Landslide susceptibility analysis in data-scarce regions: The case of Kyrgyzstan. *Bull. Eng. Geol. Environ.* **2015**, *74*, 1117–1136. [CrossRef]
14. Jaafari, A.; Najaf, A.; Pourghasemi, H.; Rezaeian, J.; Sattarian, A. GIS-based frequency ratio and index of entropy models for landslide susceptibility assessment in the Caspian forest, northern Iran. *Int. J. Environ. Sci. Technol.* **2014**, *11*, 909–926. [CrossRef]
15. Moore, I.D.; Grayson, R.B. Terrain-based catchment partitioning and runoff prediction using vector elevation data. *Water Resour. Res.* **1991**, *27*, 1177–1191. [CrossRef]
16. Grabs, T.; Seibert, J.; Bishop, K.; Laudon, H. Modeling spatial patterns of saturated areas: A comparison of the topographic wetness index and a dynamic distributed model. *J. Hydrol.* **2009**, *373*, 15–23. [CrossRef]

environmental sciences proceedings

MDPI

Proceeding Paper

Investigation of Groundwater Resources Quality for Drinking Purposes Using GWQI and GIS: A Case Study of Ottawa City, Ontario, Canada †

Amir Noori [1], Farzad Ranjbari [2] and Hossein Bonakdari [1,*]

[1] Department of Civil Engineering, University of Ottawa, 161 Louis Pasteur Private, Ottawa, ON K1N 6N5, Canada

[2] Department of Hydrology and Water Resources, Shahid Chamran University of Ahvaz, Ahvaz 61357-33135, Iran

* Correspondence: hbonakda@uottawa.ca; Tel.: +1-418-656-2131; Fax: +1-418-656-3723

† Presented at the 7th International Electronic Conference on Water Sciences, 15–30 March 2023; Available online: https://ecws-7.sciforum.net.

Abstract: Evaluating groundwater quality for certain purposes requires accurate quantitative and qualitative management, accessibility to the study area, and knowledge of the governing environmental processes. Groundwater resources are used to supply drinking water consumption alongside surface water in most countries. This study aims to investigate the quality of groundwater resources in the city of Ottawa, located in Ontario, Canada, using the Schoeller diagram and the Canadian Groundwater Quality Index (GWQI) in a fuzzy environment. To determine the water quality, the qualitative groundwater parameters including Ca, Mg, Na, Cl, SO_4, HCO_3, NO_3, F, pH, TDS, TH, K, EC, and Alkalinity were considered in the Schoeller diagram and GWQI. Each parameter's interpolated water quality map layer was prepared using the Kriging method in a GIS environment. The results of Schoeller's diagram indicated that the range of drinking water quality was non-potable to inappropriate in more than 22% of the investigated groundwater resources. Moreover, the obtained results of the groundwater quality interpolation map layer based on the GWQI revealed that more than 70% of the groundwater resources were examined in the good and excellent range for drinking purposes. Finally, the obtained interpolated map layers of the Schoeller diagram and GWQI were integrated using GIS. Accordingly, the results indicate that the interpolation values of an integrated layer in the study area are well within the permissible limits, and the quality of the groundwater is suitable for drinking and other consumption purposes.

Keywords: water quality; groundwater; Schoeller diagram; Groundwater Quality Index (GWQI); GIS

Citation: Noori, A.; Ranjbari, F.; Bonakdari, H. Investigation of Groundwater Resources Quality for Drinking Purposes Using GWQI and GIS: A Case Study of Ottawa City, Ontario, Canada. *Environ. Sci. Proc.* **2023**, *25*, 74. https://doi.org/10.3390/ECWS-7-14314

Academic Editor: Athanasios Loukas

Published: 3 April 2023

1. Introduction

Water is one of the substantial requirements in planning, developing, protecting, and controlling water resources. Improper and inefficient assessment and management of surface and groundwater could provide essential risks in the fields of human health and well-being, food security, industrial development, and the life of ecosystems [1–4]. Groundwater is considered one of the most important resources worldwide in the drinking and agriculture sectors. During the last few years, urbanization and population growth have led to an increase in the use of groundwater resources. Therefore, water quality evaluation is one of the significant problems in groundwater studies [5]. Variation in the quality of water resources is a great danger in usage by the agricultural, urban, and industrial sectors [6,7]. Several methods have been developed for water quality determination. Among these key methods to evaluate and manage groundwater resources for drinking purposes, the Schoeller diagram and the Water Quality Index (WQI) methods are the most common. In this study, to assess the water quality of groundwater, the Schoeller diagram, Canadian Groundwater Quality Index (GWQI), and Geographic Information System (GIS) were combined. Schoeller's semi-logarithmic diagram is widely used to compare groundwater quality. This graph shows the concentration differences between water samples. It is classified based on several physio-chemical parameters to evaluate the quality of groundwater [8]. The GWQI method has high capability for groundwater quality assessment across the world. In the GWQI index based on GIS, several chemical parameters affecting the quality of groundwater are integrated. GIS is used for the interpolation and classification of water quality parameters. For this purpose, the kriging method was applied to interpolate each data layer of water quality parameters. Further, to reduce the uncertainties of the obtained results, interpolation map layers were converted to fuzzy set in GIS environment.

During recent years, multiple studies have been presented to evaluate water quality for drinking uses with the Schoeller diagram, water quality indices, and GIS software in different parts of the world [9,10]. In another study, NickPeyman and Mohammadzadeh, 2013, studied groundwater quality in the Mashhad plain aquifer by estimating the GQI index [11]. Soleimani et al., 2013, conducted a study entitled "Investigation of qualitative changes in water resources of east Koohsorkh using the GQI quality index in the GIS environment" [12]. Sadat-Noori et al., 2014, used a combination of the Water Quality Index (WQI) and GIS to determine the groundwater quality of the Saveh-Nobaran aquifer in Arak province, Iran. They used the kriging method in GIS for creating spatial distribution maps of pH, TDS, EC, TH, Cl, HCO, SO_4, Ca, Mg, Na, and K [13]. In another study, Alavi et al., 2016, assessed the water quality of Dez eastern in Iran for drinking and agricultural uses with Schoeller and Wilcox diagrams, and the zoning water quality in a GIS environment considering physical and chemical parameters. They used the kriging interpolation method in GIS [14]. Farhan et al., 2020, investigated the Canadian Water Quality Index (CCME WQI) for drinking and domestic use in Mosul, Iraq. This research examined ten sampling sites along the river to collect samples from 2008 to 2014. The results showed that the water quality of the Tigris River was between 3.66 and 7.93, which is in the good and moderately good category [15]. Pourkhosravani et al., 2021, tried to evaluate groundwater resources' chemical quality for drinking and agricultural purposes using Schoeller and Wilcox diagrams in the Sirjan Plain of Iran. Their classification map of each effective parameter was prepared using IDW-based GIS [16]. Given the fact that a comprehensive study on chemical parameter variations of groundwater quality has not been carried out in the study area, this study aimed to investigate the variation in groundwater quality parameters for drinking purposes in Ottawa city using the interpolation of GWQI and the Schoeller diagram in a GIS environment.

2. Materials and Methods

2.1. Study Area

Ottawa city, with the area of 2790 km^2, is located in in the east of southern Ontario. This city is located at latitude 45°25′29″ N and longitude 75°41′42″ W, with an elevation of 70 m above sea level. The climate is semi-continental, with a warm, humid summer and a very cold winter. The temperature typically varies from −14 °C to 27 °C, while the mean precipitation is 920 mm. Rain falls throughout the year in Ottawa. The highest mean monthly rainfall in Ottawa is in July, with an average rainfall of 76 mm, while the lowest rainfall month is February, with an average rainfall of 12.7 mm. The study area involved different residential and industrial regions which supply the needed water from groundwater resources. Groundwater is one of the main sources of water in this area. There are lots of wells in the study area's aquifers that have high potential as a source of drinking water. The location of the study area is illustrated in Figure 1.

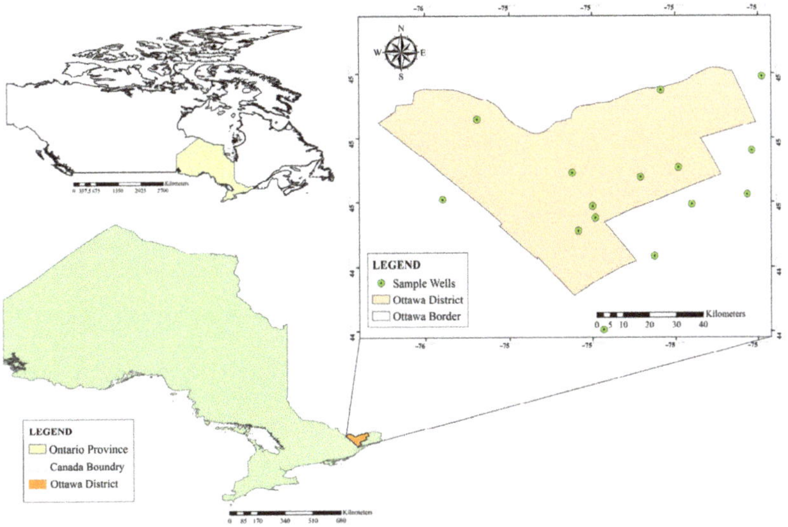

Figure 1. Location of the study area and the sampling wells.

2.2. Methodology

In this study, the available groundwater level and chemistry data of 15 sample wells distributed across Ottawa city were first collected and analyzed from the Provincial Groundwater Monitoring Network (PGMN) Program of Ontario Province on the Ministry of the Environment, Conservation and Parks website [17]. Long-term qualitative data were used during a 17-year statistical period from 2002 to 2019. In this phase, the important parameters were considered to include Ca, Mg, Na, Cl, SO_4, HCO_3, NO_3, F, pH, TDS, TH, K, EC, and alkalinity of groundwater quality.

In order to classify the water quality of groundwater and determine its type and characteristics, the GWQI index and Schoeller diagram method were used with the help of GIS software. Before that, the Schoeller diagram, a highly recommended method, was applied to investigate the quality of drinking water considering eight parameters including TDS, TH, Na, Cl, SO_4, HCO_3, Mg, and Ca. This diagram shows the concentration differences between the sample wells. It is drawn in six classes, including good, acceptable, average, inappropriate, completely inappropriate, and non-potable, based on several physio-chemical parameters to evaluate the quality of groundwater [8]. In the Schoeller diagram, an axis is considered separately for the parameters mentioned above, which determines drinking water quality [18]. Table 1 shows the classification of water quality using the Schoeller diagram method.

Table 1. The classification of water quality using the Schoeller diagram method (mg/L).

Water Classification	TDS	TH	Na	Cl	SO_4
Good	<500	<250	<115	<175	<145
Acceptable	500–1000	250–500	115–230	175–330	145–280
Average	1000–2000	500–1000	230–460	330–700	280–580
Inappropriate	200–4000	1000–2000	460–920	700–1400	580–1150
Completely Inappropriate	4000–8000	2000–4000	920–1840	1400–2800	1150–2240
Non-Potable	>8000	>4000	>1840	>2800	>2240

Next, The GWQI Index was utilized to determine the water quality of groundwater based on Ca, Mg, Na, Cl, SO_4, HCO_3, NO_3, F, pH, TDS, TH, K, EC, and alkalinity parameters.

After the definition of the parameters, three factors to determine the GWQI must be calculated. The values of the three factors were calculated as follow [19]:

F_1 shows the percentage of failed parameters relative to all of the measured parameters:

$$F_1 = \left(\frac{Number\ of\ failed\ parameters}{Total\ number\ of\ parameters}\right) \times 100 \tag{1}$$

F_2 demonstrates the percentage of failed tests:

$$F_2 = \left(\frac{Number\ of\ failed\ tests}{Total\ number\ of\ parameters}\right) \times 100 \tag{2}$$

F_3 indicates the value whereby failed test values did not meet their guidelines. F_3 is calculated in three steps.

Step 1. The number of times an individual's concentration exceeds the guideline is called an "excursion" and is expressed as follows. When the test value should not exceed the guideline:

$$excursion_i = \left(\frac{Failed\ test\ value_i}{Objective_i}\right) - 1 \tag{3}$$

For cases where the test value should not be less than the guidelines:

$$excursion_i = \left(\frac{Objective_i}{Failed\ test\ value_i}\right) - 1 \tag{4}$$

Step 2. The cumulative amount by which individual tests are out of compliance is calculated by summing the excursions of individual tests from their guidelines and dividing by the total number of tests. This parameter, called the normalized sum of excursions, or *nse*, is determined as follows:

$$nse = \frac{\sum_{i=1}^{n} excursion_i}{Number\ of\ tests} \tag{5}$$

Step 3. Then, F_3 is calculated by an asymptotic function that yields the normalized sum of excursions from instructions (*nse*) to a range between 0 and 100.

$$F_3 = \left(\frac{nse}{0.01nse + 0.01}\right) \tag{6}$$

Once the three factors have been obtained, the index itself can be calculated by summing the three factors as a vector and using the Pythagorean theorem. Therefore, the sum of squares of each factor is equal to the square of GWQI.

$$GWQI = 100 - \left(\frac{\sqrt{F_1^2 + F_2^2 + F_3^2}}{1.732}\right) \tag{7}$$

A divisor of 1.732 normalizes the resulting values to a range between 0 and 100, where 0 represents the worst water quality and 100 represent the best water quality.

Computed GWQIs were classified into five categories including excellent, good, fair, marginal, and poor for human consumption in Table 2 [19]. The outcome of the index includes a number between 0 (worst water quality) and 100 (best water quality) [20,21].

Afterward, a spatial classification map for each important parameter of the Schoeller diagram and GWQI Index method was prepared as a raster layer based on the kriging interpolation technique in GIS. It should be noted that, in order to reduce the uncertainty of all the classified raster parameters, each parameter layer was fuzzified based on the linear membership function in GIS [22–24]. Then, the final interpolation layers of the two methods (Schoeller and GWQI) were created by integrating fuzzy Raster layers as effective parameters using the fuzzy overlay tool. Finally, the classification map of the groundwater quality of the study area was generated by integrating the Schoeller and GWQI classified maps based on the overlaying method.

Table 2. The classification of water quality using the GWQI [19].

Rank	Water Quality Ranking System
Poor	0–44.9
Marginal	45–64.9
Fair	65–79.9
Good	80–94.9
Excellent	95–100

3. Results and Discussion

In this study, the water quality parameters of sample wells in the city of Ottawa for the drinking sector were studied using the Schoeller diagram and the GWQI Index. Therefore, based on Schoeller's classification, the amounts of cations such as calcium (Ca), magnesium (Mg), sodium (Na), and anions such as chloride (Cl), sulfate (SO_4), bicarbonate (HCO_3), and two important parameters of total dissolved solids (TDS), and the total hardness (TH), were checked in the diagram, which is shown in Figure 2.

Regarding the discussion of limitations and practical implications in the current research, the groundwater quality in the urban area of Ottawa was explored. Using the approach implemented in this study, water quality could be analyzed concerning the environment, agriculture, and industry if more parameters were obtainable. However, due to the absence of pertinent environmental parameters such as BOD, COD, and DO, and industrial and agricultural parameters such as heavy metals including Fe, Cu, Zn, and As, and the time constraint involved in producing this article, only the drinking perspective was investigated concerning the quality of groundwater.

According to the diagrams (see Figure 2), the water quality of wells W_1, W_2, W_4, W_7, and W_{14} was in the good range; wells W_3, W_5, W_6, W_8, W_9, W_{10}, and W_{11} were in the acceptable range, and the rest of the wells were in the lower quality range. In general, it was concluded that about 71% of the parameter values were in the good range and 29% of the rest of the parameters were in other quality classes.

In the following, the values of each parameter of Schoeller's diagram were interpolated based on the six classes of Schoeller's classification in the ArcGIS software using the kriging method, and finally six classified maps were integrated with the Fuzzy Overlay tool in the form of the water quality classification map of Ottawa city (see Figure 3).

Based on to the classified map of Schoeller, the water quality in the classification for the Ottawa city area was in six categories, from good to non-potable. In the south, west, and north-west areas towards the center, the water quality was good and fair for drinking purposes, and from the central area towards the east and northeast, the water quality decreased. Based on the location of two wells, W_{12} and W_{13}, it can be pointed out that saltwater infiltrated the groundwater resources of these areas, considering the high amount of Na and Cl ions and also the proximity of these two wells to the Ottawa River. According to Schoeller's classification map, some water samples may have good drinking quality, while those ones can contain other harmful and toxic substances; therefore, to solve this problem more parameters were used in the GWQI Index. Hence, those parameters such as electrical conductivity (EC), pH, alkalinity (mg/L $CaCO_3$), Calcium (Ca), Sodium (Na), Magnesium (Mg), Potassium (K), Sulphate (SO_4), Chloride (Cl), Fluoride (F), and Nitrate (NO_3) were evaluated for the GWQI classification map.

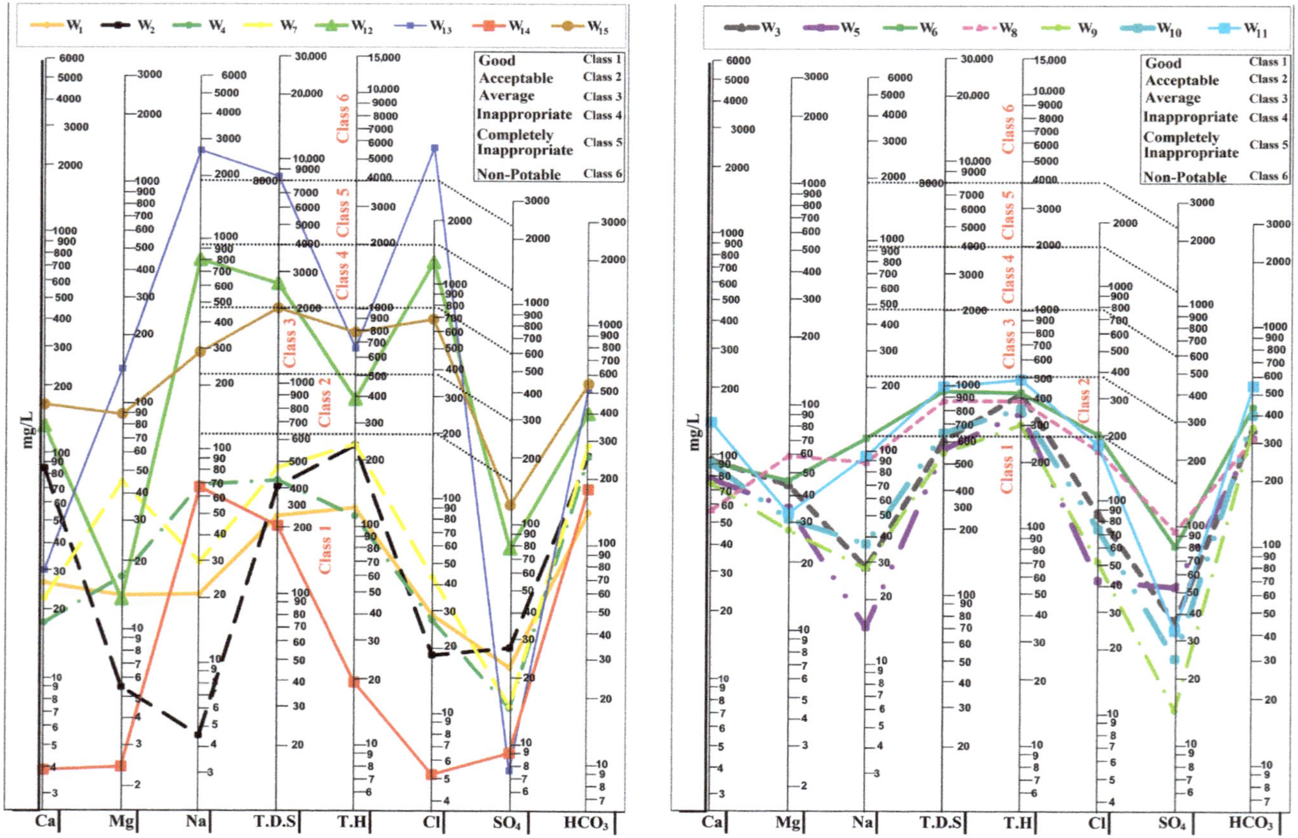

Figure 2. The Schoeller diagram of the studied wells.

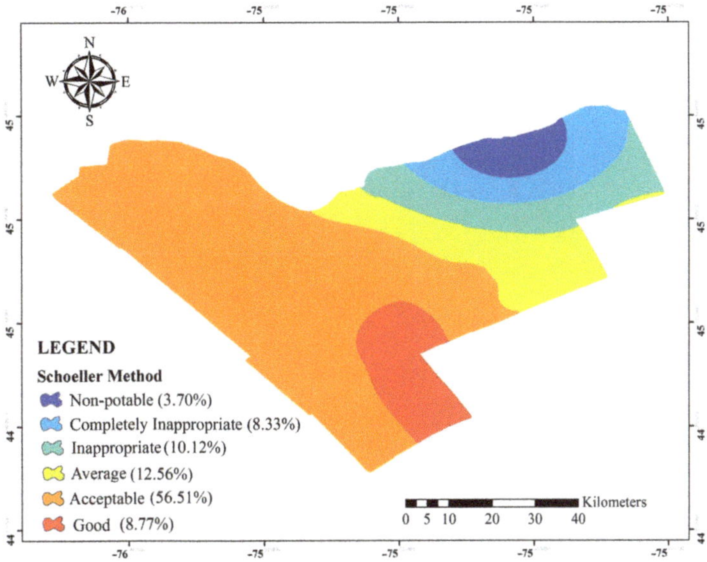

Figure 3. Water quality classification map of Ottawa city using the Schoeller method.

Considering Equations (1)–(7), the annual average values of GWQI, considering Table 2, were calculated in the range of values 33 to 100. Therefore, the water quality in most wells for drinking purposes was rated in the range of excellent to good. The GWQI values were interpolated in the ArcGIS environment using the kriging method, shown in Figure 4. According to Figure 4, the southwest, west, and northwest regions towards the center of the study area have excellent and good drinking water quality, and from the central region toward the southeast, east, and northeast, the water quality is decreased due to the increase of NaCl ions.

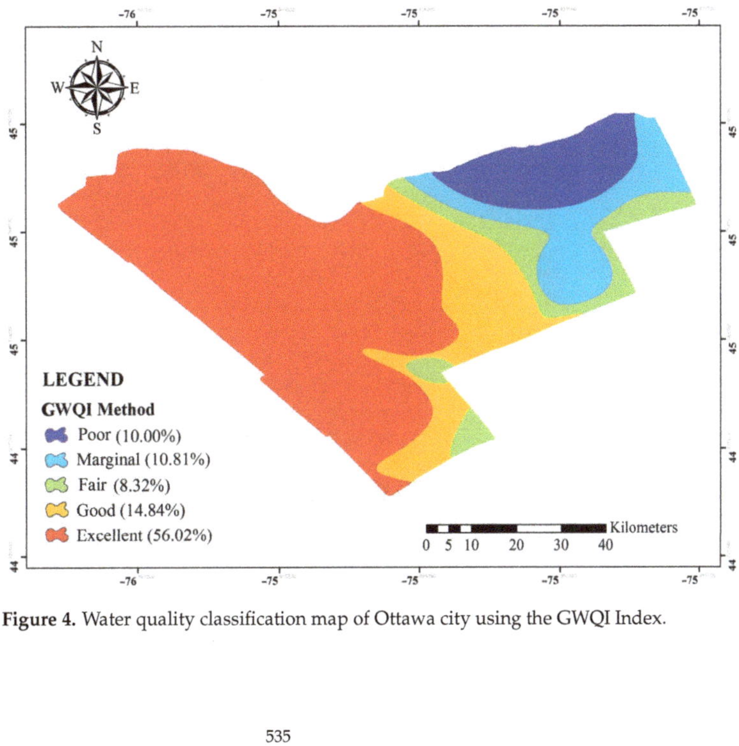

Figure 4. Water quality classification map of Ottawa city using the GWQI Index.

Finally, after preparing Schoeller and GWQI classification maps, these two classified maps overlapped with the Fuzzy Overlay tool in ArcGIS. Additionally, the overlaid map was categorized into five classes from Excellent to Poor (see Figure 5). Furthermore, by investigating the integrated classification map, it was found that the values of groundwater quality were in the excellent and good class range for the area of Ottawa city in the south, west, and the northwest regions towards the center, and also the water quality range decreased in the east and northeast regions.

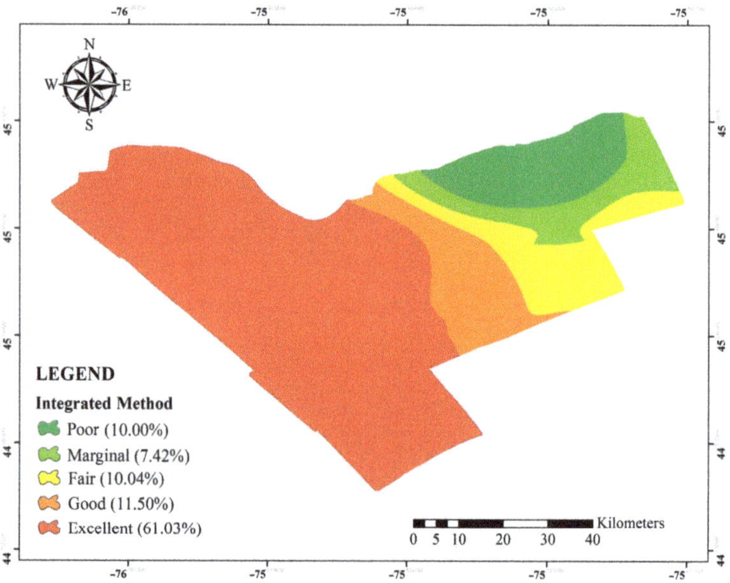

Figure 5. Integrated water quality classification map of Ottawa city.

4. Conclusions

In this study, the main aim was to investigate groundwater quality through the combination of the Schoeller diagram and the Canadian Groundwater Quality Index (GWQI). This study tried to evaluate and analyze the water quality in Ottawa city in Ontario, Canada, based on long-term qualitative data from 15 sample wells between 2002 and 2019. The classified water quality maps of each chemical parameter were prepared using the kriging method based on a fuzzy set in the GIS environment. The obtained results showed that, based on the Schoeller diagram, most of the studied wells were located in good to acceptable quality regions regarding drinking purposes. According to the Schoeller classification map of groundwater resources, the acceptable class, with 56.51% of the aquifer area, and the non-potable class, with 3.70% of the aquifer area, made up the highest and lowest portions of the aquifer, respectively. Moreover, according to the GWQI water quality classification map results, 79.18% of the wells were in the fair to excellent range, and 10% were in the poor range. Finally, the results of assessing the integrated Schoeller and GWQI water quality classification map used for drinking purposes showed that, based on the values of these two methods, the water quality in the central areas and near-west areas were categorized into excellent and good classes, and from the central regions to the east, the water quality had gradually decreased. Moreover, the transferability of the proposed method and results can be discussed in light of the obtained outcomes. Our approach integrated the GWQI index, Schoeller diagram, and GIS to develop a model that can be utilized for agricultural, industrial, and environmental purposes by including several chemical parameters. The GWQI index enabled us to expand the model to encompass other water quality diagrams, such as Wilcox and Piper. Moreover, our study's reliability and transparency provide significant insights for researchers and decision-makers to analyze and make informed decisions about the quality of drinking water. These results show that the proposed method is transferable to more extensive case studies, providing valuable insights for various water quality applications.

Author Contributions: Conceptualization: A.N., F.R. and H.B.; methodology: A.N. and F.R.; software: A.N.; validation, F.R.; formal analysis, H.B. and A.N.; data curation, A.N. and F.R., writing—original draft preparation, A.N. and H.B.; writing—review and editing, H.B.; visualization, A.N., F.R. and H.B.; supervision: H.B. All authors have read and agreed to the published version of the manuscript.

Funding: This research received no external funding.

Institutional Review Board Statement: Not applicable.

Informed Consent Statement: Not applicable.

Data Availability Statement: Data available on request due to restrictions e.g., privacy or ethical.

Conflicts of Interest: The authors declare no conflict of interest.

References

1. Ahmad, S.; Murray, R. *World Environmental and Water Resources Congress 2020: Water Resources Planning and Management and Irrigation and Drainage*; American Society of Civil Engineers (ASCE): Reston, VA, USA, 2020. [CrossRef]
2. Noori, A.; Bonakdari, H.; Morovati, K.; Gharabaghi, B. Development of optimal water supply plan using integrated fuzzy Delphi and fuzzy ELECTRE III methods—Case study of the Gamasiab basin. *Expert Syst.* **2020**, *37*, e12568. [CrossRef]
3. Salimi, A.H.; Noori, A.; Bonakdari, H.; Samakosh, J.M.; Sharifi, E.; Hassanvand, M.; Gharabaghi, B.; Agharazi, M. Exploring the role of advertising types on improving the water consumption behavior: An application of integrated fuzzy AHP and fuzzy VIKOR method. *Sustainability* **2020**, *12*, 1232. [CrossRef]
4. Noori, A.; Bonakdari, H.; Salimi, A.H.; Gharabaghi, B. A group Multi-Criteria Decision-Making method for water supply choice optimization. *Socioecon. Plann. Sci.* **2021**, *77*, 101006. [CrossRef]
5. Umar, R.; Ahmed, I.; Alam, F. Mapping groundwater vulnerable zones using modified DRASTIC approach of an alluvial aquifer in parts of central Ganga plain, western Uttar Pradesh. *J. Geol. Soc. India* **2009**, *73*, 193–201. [CrossRef]
6. Cloern, J.E. Patterns, pace, and processes of water-quality variability in a long-studied estuary. *Limnol. Oceanogr.* **2019**, *64*, S192–S208. [CrossRef]
7. Radfard, M.; Soleimani, H.; Azhdarpoor, A.; Faraji, H.; Mahvi, A.H. Dataset on assessment of physical and chemical quality of groundwater in rural drinking water, west Azerbaijan Province in Iran. *Data Br.* **2018**, *21*, 556–561. [CrossRef] [PubMed]
8. Güler, C.; Thyne, G.D.; McCray, J.E.; Turner, A.K. Evaluation of graphical and multivariate statistical methods for classification of water chemistry data. *Hydrogeol. J.* **2002**, *10*, 455–474. [CrossRef]
9. Pour, H.V.; Sayari, M.; Bayat, N.; Forutan, F. Qualitative and Quantitative Evaluation of Groundwater in Isfahan Najaf Abad Study Area. *J. Middle East Appl. Sci. Technol.(JMEAST)* **2014**, *16*, 523–530.
10. Jamshidzadeh, Z.; Mirbagheri, S.A. Evaluation of groundwater quantity and quality in the Kashan Basin, Central Iran. *Desalination* **2011**, *270*, 23–30. [CrossRef]
11. Nickpeyman, V.; Mohammadzadeh, H. Evaluation of spatial variations of groundwater quality in Mashhad Plain using GQI index. Presented at the National Conference on Exploration of Ground Resources. In *National Conference on Exploration of Ground Resources*; Hamedan University: Hamedan, Iran; 2013. (In Persian)
12. Soleimani, S.; Mahmoudi-Gharaei, M.; Ghasemzadeh, F.; Planet, A. Investigation of quality changes of west water resources of the red mountain using GQI quality index in GIS environment. *J. Earth Sci.* **2013**, *89*, 175–189. (In Persian)
13. Sadat-Noori, S.M.; Ebrahimi, K.; Liaghat, A.M. Groundwater quality assessment using the Water Quality Index and GIS in Saveh-Nobaran aquifer, Iran. *Environ. Earth Sci.* **2014**, *71*, 3827–3843. [CrossRef]
14. Alavi, N.; Zaree, E.; Hassani, M.; Babaei, A.A.; Goudarzi, G.; Yari, A.R.; Mohammadi, M.J. Water quality assessment and zoning analysis of Dez eastern aquifer by Schuler and Wilcox diagrams and GIS. *Desalination Water Treat.* **2016**, *57*, 23686–23697. [CrossRef]
15. Farhan, A.F.; Al-Ahmady, K.K.; Al-Masry, N.A.-A. Assessment of Tigris River Water Quality in Mosul for Drinking and Domestic Use by Applying CCME Water Quality Index. *IOP Conf. Series: Mater. Sci. Eng.* **2020**, *737*, 012204. [CrossRef]
16. Pourkhosravani, M.; Gohari, J.; Sayari, N.; Pourkhosravani, M.; Gohari, F.J.; Sayari, N. Groundwater Quality and Suitability for Different Uses in the Sirjan County. *Desert Ecosyst. Eng. J.* **2021**, *3*, 43–58.
17. Ministry of the Environment, Conservation and Parks. Available online: https://www.ontario.ca/page/ministry-environment-conservation-parks (accessed on 15 January 2023).
18. Melloul, A.; Collin, M. The 'principal components' statistical method as a complementary approach to geochemical methods in water quality factor identification; application to the Coastal Plain aquifer of Israel. *J. Hydrol.* **1992**, *140*, 49–73. [CrossRef]

19. CCME. Canadian Water Quality Guidelines for the Protection of Aquatic Life: CCME Water Quality Index User's Manual 2017 UPDATE, Canadian Environmental Quality Guidelines. *Can. Counc. Minist. Environ.* **2017**, 1–23. Available online: https://ccme.ca/en/res/wqimanualen.pdf (accessed on 10 October 2020).
20. Khan, A.A.; Paterson, R.; Khan, H. Modification and application of the Canadian Council of Ministers of the Environment Water Quality Index (CCME WQI) for the communication of drinking water quality data in Newfoundland and Labrador. *Water Qual. Res. J. Can.* **2004**, *39*, 285–293. [CrossRef]
21. Cash, K.; Wright, R. *Canadian Water Quality Guidelines for the Protection of Aquatic Life*; CCME: Ottawa, ON, Canada, 2001.
22. Ketata, M.; Gueddari, M.; Bouhlila, R. Use of geographical information system and water quality index to assess groundwater quality in el khairat deep aquifer (enfidha, central east tunisia). *Arab. J. Geosci.* **2012**, *5*, 1379–1390. [CrossRef]
23. Lo, C.P.; Yeung, A.K. *Concepts and Techniques of Geographic Information Systems*; Pearson Prentice Hall: London, UK, 2007.
24. Noori, A.; Bonakdari, H.; Hassaninia, M.; Morovati, K.; Khorshidi, I.; Noori, A.; Gharabaghi, B. A reliable GIS-based FAHP-FTOPSIS model to prioritize urban water supply management scenarios: A case study in semi-arid climate. *Sustain. Cities Soc.* **2022**, *81*, 103846. [CrossRef]

environmental sciences proceedings

MDPI

Proceeding Paper

Advances in the Knowledge of Las Loras UNESCO Global Geopark in the Framework of the UNESCO IGCP-730 Project †

África de la Hera-Portillo [1,*], Manuela Chamizo-Borreguero [1], Marwan Ghanem [2], Julio López-Gutiérrez [1], Luis Moreno-Merino [1], Jose Ángel Sánchez-Fabián [3], Karmah Salman [3], Nour-Eddine Laftouhi [4], Rod Fensham [5,6], Alsharifa Hind Mohammad [7], Elena Galindo Rodríguez [1], Mª Mar Corral Lledó [1], Badia Choulli [8] and Fagr Kh. Abdel-Gawad [9]

1 Centro Nacional Instituto Geológico y Minero de España, Consejo Superior de Investigaciones Científicas (IGME-CSIC), 28003 Madrid, Spain; m.chamizo@igme.es (M.C.-B.); j.lopezgu@igme.es (J.L.-G.); l.moreno@igme.es (L.M.-M.); e.galindo@igme.es (E.G.R.); mm.corral@igme.es (M.M.C.L.)
2 Faculty of Geography, Birzeit University, Ramallah P.O. Box 14, Palestine; marwan.ghanem2012@gmail.com
3 Las Loras UNESCO Global Geopark, 34800 Palencia, Spain; geoloras@gmail.com (J.Á.S.-F.); info.geoloras@gmail.com (K.S.)
4 Department of Earth Sciences, Faculty of Sciences Semlalia of Marrakech, Cadi Ayyad University, Marrakech 40000, Morocco; laftouhi@gmail.com
5 Queensland Herbarium (DES), Mt Coot-tha Road, Toowong, QLD 4066, Australia; rod.fensham@des.qld.gov.au
6 Department of Biological Sciences, University of Queensland, St Lucia, QLD 4072, Australia
7 Water, Energy and Environment Center, University of Jordan, Amman 11942, Jordan; alsharifahind@gmail.com
8 Water Research and Technologies Centre, University of Carthage, Tunis 2085, Tunisia; bchoulli@yahoo.fr
9 Centre of Excellence for Research and Applied Studies on Climate Change and Sustainable Development, Water Pollution Research Department, National Research Centre, 33 El Buhouth St., Dokki 12622, Egypt; fagrabdlgawad@gmail.com
* Correspondence: a.delahera@igme.es
† Presented at the 7th International Electronic Conference on Water Sciences, 15–30 March 2023; Available online: https://ecws-7.sciforum.net.

check for updates

Citation: de la Hera-Portillo, Á.; Chamizo-Borreguero, M.; Ghanem, M.; López-Gutiérrez, J.; Moreno-Merino, L.; Sánchez-Fabián, J.Á.; Salman, K.; Laftouhi, N.-E.; Fensham, R.; Mohammad, A.H.; et al. Advances in the Knowledge of Las Loras UNESCO Global Geopark in the Framework of the UNESCO IGCP-730 Project. *Environ. Sci. Proc.* **2023**, *25*, 75. https://doi.org/10.3390/ECWS-7-14313

Academic Editor: Athanasios Loukas

Published: 3 April 2023

Abstract: The Las Loras UNESCO Global Geopark is a recent geopark (2017) with a high wealth of manifestations of interaction between the landscape and the water cycle. It is being studied under the umbrella of the International Geosciences Programme Project-730. In this work we present the main results obtained, related to geology, geomorphology, hydrogeology, structure and relationship between structure and hydrogeology. The components of the geodiversity are identified and each one of them open a new line of research which will be object of study in new projects in an early future.

Keywords: groundwater; Duero-Ebro basin divide; geodiversity; Ubierna Fault System

1. Introduction

The Las Loras UNESCO Global Geopark (UGGp) has been chosen as one of the two Spanish geoparks studied under the umbrella of the IGCP-730 Project [1]. The focus of the study is the interaction between surface water and groundwater.

The different aspects of this research are dissected in the following epigraphs where a brief summary is presented on the geology, structure, hydrogeology and the components of the geological diversity recognized in Las Loras UGGp.

2. Methodology

2.1. Las Loras UGGp Description

The study area has an extension of 950.76 km^2 in the provinces of Palencia and Burgos, in the north of Spain (Figure 1) an area with 13,076 inhabitants. The Agriculture is the economic support. Some relevant industries exits in Aguilar de Campoo, and the tourist companies around the Palentina Mountain [2].

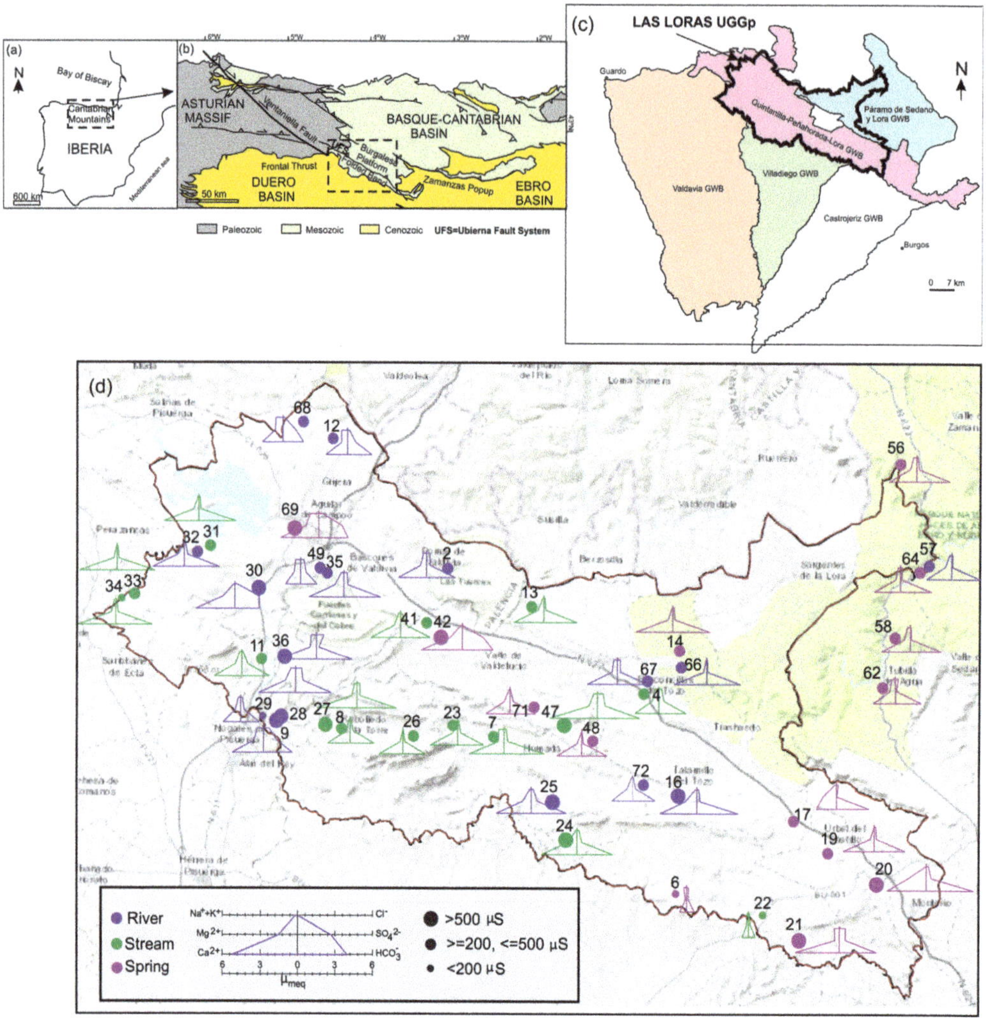

Figure 1. (**a**) Las Loras UGGp in the North of Spain (Cantabrian Mountains). (**b**) Geological context (modified from Ref. [2]) highlighting the trace of the Ubierna Fault System (UFS) and the two areas differentiated as Burgalesa Platform and Folded Band. (**c**) Hydrogeological context: indication of the groundwater bodies recognized in the Ebro-Duero basin divide. (**d**) Stiff diagram of the main water samples obtaining in the field campaigns.

The Ubierna Fault is the main structure which divide Las Loras UGGp into two different domains. These domains correspond to two types of folded relief (Figure 1): (1) The Burgalesa Platform extends from the Ebro river to the Ubierna Fault, characterized by calcareous moors affected by a slight deformation. And (2) the Folded Band covers the area from the Ubierna Fault to the frontal thrust of the Cenozoic materials of the Duero basin, whose folds give name to the area, "loras".

The sedimentary materials of Las Loras UGGp is formed by an almost complete record from the Upper Triassic to the Paleogene [2]. The Keuper facies appears on the main faults acting as a detachment layer with high plasticity.

With regard to climate, Aguilar de Campoo (2009–2021) presents an average annual precipitation of 570 mm, with the temperature ranges between 27.08 °C in July 2013, and −4.66 °C in December 2009, being 9–11 °C the average annual temperature between 2009 and 2021.

Las Loras UGGp includes two rivers flowing into different seas: Duero, towards Atlantic Ocean; and Ebro, towards Mediterranean Sea. Within the second one, the most important rivers are the Lucio River, Rudrón River, Urbel River, and Odra and Brulles Rivers.

Hydrogeologically, the study area includes two adjoining groundwater bodies (GWBs): to the north, the Páramo de Sedano y Lora GWB, (Ebro); and, to the South, Quintanilla-Peñahorada-Lora GWB (Duero) (Figure 1d) [3].

2.2. Field Campaigns

46 water samples were obtained from the field campaigns. Figure 1d show the Stiff diagram obtained for the most relevant samples for this research.

2.3. Review of Data Bases Information

Different resources available at the IGME data bases were used to explore the existence of mineral paragenesis or mining or associated with metallogenetic deposits which could be associated to the Ubierna Fault through hydrotermal activity.

2.4. Water Samples

Different field campaigns were conducted in 2021 and 2022. INAQUAS software [4] was used for the hydrochemical information analysis and graphical representation in hydrochemical diagrams.

2.5. Compilation of Data and Intagration in a GIS

Data were compiled from literature review, and data bases consulted from different sources.

3. Results

3.1. Ubierna Fault System and Its Relationship with Potential Hydrotermalism

The Ubierna fault is the most important geological element in the study area. It presents slightly sinuous NW-SE orientation and important implications for the geomorphology, hydrogeology and groundwater and surface water interactions.

3.2. Components of Geological Diversity

3.2.1. Loras

The loras are a type of synclines subject to compression with sub-sequent erosion by Quaternary fluvial incision.

3.2.2. Moorlands

Moorlands have formed on extensive platforms of carbonate materials [2]. They form the Páramo de la Lora de Valdivia and la Pata del Cid in the northern half of the study area. From the point of view of the extension, the moorlands may have several kilometers meanwhile the loras have a much smaller extension, in some cases, with a few tens of meters.

3.2.3. Caves

The caves appear associated with karstic environment and have been formely studied [5].

3.2.4. Travertine Buildings

A good number of examples exist associated with active springs: Orbaneja del Castillo, Tubilla del Agua, Covalagua, Barriolucio, Hoyos del Tozo, Villaescobedo, Moradillo del Castillo, Revilla de Pomar, Rebolledillo de la Orden and Fuenteodra.

3.2.5. Peat Bogs

They appear restricted to specific examples in La Piedra, Urbel del Castillo, Humada, Fuencaliente de Puerta, Corralejo del Valdelucio.

3.2.6. Canyons and Sickles

Associated to the Ebro, Rudrón and Pisuerga valleys.

3.2.7. Sinkholes

Huron River disappear a few kilometers after its river source to birth as a new river called Rudrón River in the Cueva de Los Moros.

3.2.8. Springs

There are more than two hundred springs in the study area. The most relevant for their cultural significance are: Cueva del Agua (Orbaneja del Castillo), Pozo Azul (Covanera); nevertheless, there are others with hydrogeological relevance as Pisadera and Valdelateja due to both of them present, past or/and current symptoms of hydrothermal activity.

3.3. Main Aquifers Identified and Hydrogeological Model of Functioning

Two types of aquifers are distinguished in the study area: an permeable unconfined aquifer; and a deep confined or semi-confined aquifer [2] made up of permeable materials of Jurassic age.

3.4. Characteristics of Some Representative Springs

The most relevant springs are shown in Figure 1d. They could be classified in two main groups according to their geological pattern and associated aquifer formation: (a) Upper Cretaceous springs; (b) Jurassic springs.

The waters in Las Loras show an undifferentiated hydrochemical facies of bicarbonated waters. Almost all samples present a low mineralization.

4. Conclusions

The Las Loras UGGp present a high concentration of geological diversity components. The most part of them are associated to water circle shown a wide range of manifestations (springs, cave, canyons, travertine buildings, etc.) related to active processes whose action keep transforming the landscape.

The hydrogeological study carried out in this work show three types of sediments according to their behaviour: aquifers themselves, aquitards and aquifuges. An upper aquifer in the Las Loras UGGp includes materials between the Lower Cretaceous and the Quaternary. A deep regional aquifer is represented by marine Jurassic sediments. The first one, is characterized by aquifers with springs located above rivers. The deep regional aquifer is formed by confined aquifers.

Both aquifer systems do not show evidences of hydrogeological connection. The Lower Cretaceous materials acts as an aquitard that separates the upper from the deep aquifer. No signs of hydrothermalism have been recognized in the Ubierna Fault zone.

Author Contributions: Conceptualization, Á.d.l.H.-P.; methodology, Á.d.l.H.-P.; software, Á.d.l.H.-P.; validation, M.G., N.-E.L., M.C.-B., A.H.M., B.C., R.F., E.G.R., M.M.C.L., F.K.A.-G.; formal analysis, L.M.-M.; investigation, Á.d.l.H.-P., J.L.-G.; resources, M.G., J.Á.S.-F., K.S.; data curation, Á.d.l.H.-P.; writing—original draft preparation, Á.d.l.H.-P.; writing—review and editing, Á.d.l.H.-P., R.F.; visualization, Á.d.l.H.-P.; supervision, all co-authors; project administration, M.G.; funding acquisition, M.G., J.Á.S.-F., K.S. All authors have read and agreed to the published version of the manuscript.

Funding: This research was funded by UNESCO International Geosciences Programme (IGCP) and by Severo Ochoa extraordinary grants for excellence IGME-CSIC (AECEX2021).

Institutional Review Board Statement: Not applicable.

Informed Consent Statement: Not applicable.

Data Availability Statement: Data supporting reported results can be found in Ref. [2].

Acknowledgments: The authors particularly thank the anonymous reviewer for the great contribution to the clarity and completeness of this paper.

Conflicts of Interest: The authors declare no conflict of interest.

References

1. IGCP-730 Hydrogeological Significance of Mediterranean Geoparks. Available online: https://en.unesco.org/international-geoscience-programme/projects/730 (accessed on 15 January 2023).
2. De la Hera-Portillo, Á.; López-Gutiérrez, J.; Moreno-Merino, L.; Llorente-Isidro, M.; Fensham, R.; Fernández, M.; Ghanem, M.; Salman, K.; Sánchez-Fabián, J.Á.; Gallego-Rojas, N.; et al. Geodiversity of Las Loras UNESCO Global Geopark: Hydrogeological Significance of Groundwater and Landscape Interaction and Conceptual Model of Functioning. *Resources* **2023**, *12*, 14. [CrossRef]
3. CHD Information System. Duero River Basin Web Page. Available online: https://mirame.chduero.es (accessed on 15 January 2023).
4. Moreno and De la Losa. Interpretación de Análisis Químicos de Aguas Subterráneas Mediante la Hoja INAQUAS. Luis Moreno Merino, Almudena de la Losa Román, Amalia Romero Prados, Sergio Carro Martín. 2016. Available online: https://www.igme.es/Publicaciones/Consulta/Libro/36830 (accessed on 15 January 2023).
5. Rioseras, M.; Acha Alarcía, J.; Grupo Espeleológico Edelweiss. *Cavidades de Orbaneja del Castillo I. El Conejal-Entrecaminos-Pico Resanquillo*; Valle de Sedano: Burgos, Spain, 2016; pp. 12–33. Available online: http://grupoedelweiss.com/pdf/Cubia20.Orbaneja_I.pdf (accessed on 15 January 2023).

environmental sciences proceedings

MDPI

Proceeding Paper

Assessing the Economic Damage of Potential Flooding Zones by Combining Cadaster and Land Use Data in the Larnaca Region, Cyprus †

Alexandra Ćulibrk * **and Ourania Tzoraki**

Department of Marine Sciences, School of Environment, University of the Aegean, 81100 Lesvos, Greece; rania.tzoraki@aegean.gr

* Correspondence: aculibrk@marine.aegean.gr

† Presented at the 7th International Electronic Conference on Water Sciences, 15–30 March 2023; Available online: https://ecws-7.sciforum.net/.

Abstract: Precipitation changes caused by climate change are a phenomenon that may have far more serious consequences in urban areas than previously thought. Preventing the flood extent in rural and urban areas is crucial for the protection of human well-being and private and public land, and for the prevention of human casualties. Modeling potential floods and studying historic ones are key parameters to prevent damages from this natural disaster. European Union Directive 2007/60/EK obligated Member States to compile flood risk maps, showing the potential flood's extent and depth. Using Cyprus river flood risk maps of the region of Larnaca, Corine land cover data, and open-access cadaster data, this study attempts to assess the economic damage. Cadaster data of buildings such as floor number and basement existence are incorporated in the damage functions used for flood recurrence intervals of 20, 100, and 500 years return periods. The study found that in the 500 year return flood period, 562 buildings would be potentially affected. The flood risk areas in the region of Larnaca consist of mainly urban (48%) and cultivation zones (50%), both economically important. The results suggest that economic damage assessment based on the applied data, combined with flood risk maps, can provide valuable information to disaster responders.

Keywords: Cyprus; Larnaca; depth damage curve; flood damage; economic loss assessment

Citation: Ćulibrk, A.; Tzoraki, O. Assessing the Economic Damage of Potential Flooding Zones by Combining Cadaster and Land Use Data in the Larnaca Region, Cyprus. *Environ. Sci. Proc.* **2023**, *25*, 76. https://doi.org/10.3390/ECWS-7-14323

Academic Editor: Athanasios Loukas

Published: 12 April 2023

1. Introduction

The gradual increase in the global temperature brings multiple problems which will significantly affect people's lives, with problems that are directly related to people's everyday life, but also ones that will appear unexpectedly [1]. The increase in precipitation phenomena globally [2] is one of the results of climate change that should particularly concern humanity. In Europe, precipitation trends differ depending on the latitude; the northern part receives a higher percentage of rain by 10–40%, while the southern part of the continent in some areas faces up to a 20% decrease [3]. Despite the fact of decreased rainfall in some areas, a growing number of studies show that there is evidence of an increase in extreme precipitation events [4–6]. Future precipitation extreme trends suggest that the sub-daily extreme rainfall of 50 year and 100 year return periods will triple under the Representative Concentration Pathway (RCP) 8.5 scenario [7]. With southern Europe's extreme hot and dry days, extreme precipitation events will have much worse consequences [8], making flood events a greater risk than expected.

For the management and monitoring of floods in urban areas, the European Commission established the Flood Directive (2007/60/EC) [9]. The main scope of the framework is the reduction of adverse flood effects on the environment, human health, cultural heritage, and economic activities that are directly linked to society. The member states of the European Union were obligated to establish a River Basin Flood Risk Management

Plan, identifying vulnerable areas considering long-term developments and future flood occurrence. The key objective of the flood risk assessment on a country's level is flood hazard maps, which show, based on the flood probability, the flood extent, the depth, and the flow speed. In addition, the flood risk maps assess the potential adverse consequences for the economic activity of the area, as well as for potentially affected residents.

The Republic of Cyprus, as a member state of the European Union, had already incorporated the provisions of Directive 2007/60/EC into Cypriot law in 2010. With the main evaluation parameters being the size of the river basin area (greater than 10 km^2), the proximity to urban developed areas, and the historical overflow of the river, 19 areas of potentially significant flood risk were defined. A key element in the flood risk assessment is the listing of important public elements that will be affected, as well as the economic activities of the area. An integrated coast damage assessment, however, is not analyzed in the flood management plans.

In the study of flood damage assessment for each continent, Huizinga et al. [10] created a methodology based on extensive literature reviews and surveys to assess the economic damage induced by flood depth. For the purpose of overcoming the barrier of comparing flood damage assessments from different countries, normalized damage functions at a national scale were utilized. Adapting the methodology followed by the above-mentioned study at a regional scale, the Cyprus depth damage curve (DDC), at a national scale, was updated based on the Gross Domestic Product of 2021. Furthermore, having one of the biggest cities of Cyprus, Larnaca, as a pilot study area, the methodology was applied within the extent of the potential flooding zones of 20, 100, and 500 year return periods. Adapting the methodology for the data of the residential buildings from Cyprus's cadaster, the study aims to compare the differences in the damage assessment based on national- and local-based economic data.

2. Methods

Larnaca is the third largest city in Cyprus, with a population of 51.468. Located on the northeast coast of Cyprus, the city area covers an area of 32 km^2 with a mean elevation of 26 m. The climate is typically Mediterranean, with rainfall levels reaching 320 mm. Based on assessments by the Water Development Department (WDD), there are 3 potentially high flood-risk zones in the province of Larnaca. In this study, the area of the biggest coverage will be used as an area of interest. The area is located in the central part of the city, Aradippou-Livadia. With a length of 11.2 km, the Archangelos-Kammitsis and Parapotamos rivers pass through the agricultural, residential, and industrial areas before ending up in the sea. Areas located in this potential hazard zone have been exposed to 7 floods in the past, according to the records of the Water Development Department (WDD), of which 5 were of low severity, one was moderate, and one was high.

A key element of this research is the flood risk maps obtained from the WDD of Cyprus, which visualize the potential flood extents of 50, 100 year, and 500 year return periods. The potential flood data contain flow velocity values as well as flood depth. In addition to flood data, building data were used. The buildings' data were digitized by the open-access cadastral platform of Cyprus. The obtained data describe only residential buildings. The location of each building, the estimated value of the property, and their size (sqm) were collected. The gathered data were, furthermore, analyzed in GIS software and verified with field visits. In order to calculate the damage within the urban zone area, Corine's Land Use data were used as a guide to locate the extent of the urban zone while obtaining the data.

To calculate the economic damage of the residential buildings in the potential flood area, the approach of the DDC was used based on the methodology of Huizinga et al. (2017). Their approach calculated the damage curves for each continent, differentiating them based on the maximum damage values per country. These maximum damage values were based on construction cost surveys by multinational construction companies, which provide a coherent set of detailed construction cost data in dozens of countries. A fixed set of maximum flood damage values was calculated for all countries, using statistical regressions with socioeconomic indicators of global development.

To calculate the construction price, the GDP of Cyprus in 2021 was used with residential construction estimation constants. The structure cost index was then used in the calculation of the maximum damage, which is expressed by Equation (1). The components of the equation are defined by a series of fixed values which, according to the study and case, can change the value adjusted to a scale of 0–1.

$$\text{MaxDamage} = \text{StructureCost} \times \text{DepreciatedValue} \times (1\text{-Undamageable part}) \times \text{MaterialUsed} \quad (1)$$

To construct the DDC, the maximum damage cost is related to the normalized scale of the damage function, where the maximum value is matched to the damage scale value 1 and the other values are correspondingly calculated through Equation (2).

$$\text{Damage} = \text{DamageFunction} \times \text{MaxDamageCost}/\text{MaxDamage Function} \quad (2)$$

To calculate the economic damage using the residential building data in the flood area of Larnaca, the flood height of each building, the estimated property value based on 2021, and the size of the building were used. Initially, the flood heights were calculated in relation to the height of each flooded building. The construction cost of the building was calculated based on the value of the property divided by its size, and consecutively the maximum cost was calculated using Equation (1). Beyond the difference in the calculation of the construction price, the Depreciated Value index was overlooked. Since the DDC represents the damage in 2021 and the residential value data from the Cyprus cadaster are estimated for the same year, the Depreciated Value Index was not taken into account. Lastly, to calculate the total damage, the residential buildings were grouped by flood depth, and their damage costs were calculated separately. In order to compare the total damage by flood scenario, the economic data of the Cyprus damage curve were used for the potential flood depths in the Larnaca area. The buildings' point data were grouped based on the area of the urban fabric. The 562 buildings were grouped at 28 polygons of residential areas and the total damage was calculated based on the equation of Cyprus' DDC.

3. Results and Discussion

The potential flood zones in the northern part of Larnaca cover an area with a high density of urban fabric. Based on the data of the Water Development Department, the visualization of the potential flood areas is shown on the map below (Figure 1). Looking at the potential flood areas per scenario, it appears that the flood area increases significantly per flood return period. Specifically, from the 20-year return period to the 500-year, the extent of the flooding area is 85% larger, covering 2.5 square kilometers. The inundation depths were noted to range from 0 to 6.8 m maximum depth, with average depth values per scenario being 0.85 m at the 20-year return and 0.84 m at the 100- and 500-year returns. It is worth noting that values greater than 3 m were not recorded in any of the scenarios outside the river basin extent.

The extent of the urban fabric obtained from Corine (code 112) was used to make a targeted collection of residential data from the cadastral register of Cyprus. It was observed that, in the eastern urban fabric zone, no residential buildings were noted, only auxiliary industrial buildings (warehouses). For the research, only residential data were taken into account; therefore, the eastern area under urban fabric land was not taken into consideration. With the spatial extents of the maximum flood scenario of the 500-year return period (Figure 2), the residential buildings' data were collected from the cadastral register of Cyprus. The locations for each building, the height, and the value of the building were collected for 562 buildings within the flood extent. For research purposes, parks, plots without the presence of buildings, and commercial buildings were not recorded in the database.

Figure 1. Larnaca city area of potential flooding area. The three different areas represent the different potential flood return periods.

Figure 2. Residential building distribution in the spatial extent of the 500-year return period.

Since all buildings belonged to the same urban zoning code, the heights of the buildings had fixed values based on their floors, i.e., buildings with one floor were 5 m high, buildings with two floors were 10 m high and buildings with three floors were 13.5 m high. Of the 562 buildings, prices ranged from EUR 40,000 minimum to EUR 3,000,000 maximum, with an average price of EUR 590,000. Based on the number of floors, 33% of the buildings had one floor, 63% of the buildings had two floors, and 4% of the buildings had three floors.

The damage curves below represent two different assessments: the DDC for Cyprus (Figure 3b), and the DDC for Larnaca's sub-region (Figure 3a). Based on the revised financial data of 2021, on a country scale, Cyprus' flood damage costs ranged from 50 euros/m^2 to a maximum price of 700 euros/m^2. Larnaca's regional results for the potential flood area of a 500-year return period, however, showed that the damage values in the residential zone ranged from 25 euros/m^2 to a maximum value of 200 euros/m^2. Beyond the financial differences, the graphs show some other significant differences. In the horizontal axis of flood depth values, Figure 3b has a maximum value of 6 m flood depth, while Figure 3a has a maximum value of 3 m. In the case study of Larnaca, no buildings were noted in flood areas at water depths higher than 3 m. While the two figures have the same axis, their content is rather different. Figure 3b expresses the approximate damage cost of a mean building in Cyprus based on surveys and bibliography at all possible flood depths. In the case of Larnaca (Figure 3a), the DDC is based on the data of 562 buildings within the potential flood limits. Furthermore, the damage cost does not represent a single building's damage per flood depth, but each one of the buildings' flood depths per flood scenario.

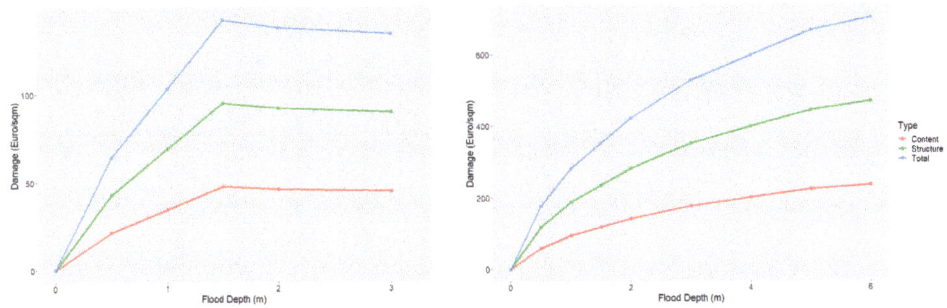

(a) (b)

Figure 3. Damage curve of residential buildings in the region in Larnaca city (**a**), Damage curve of Cyprus based on the GDP of 2021 (**b**).

The economic flood damage results for Larnacas area and Cyprus are presented in Table 1. The costs of total damage for all three potential flood scenarios, based on Larnacas residential data, showed that the damage is calculated at 32.6 EUR M in the 500-year return period, 16.8 EUR M in the 100-year return period, and 13.7 EUR M in the 20-year return period. Additionally, the states' damage prices per flood scenario calculated a maximum cost of 69.2 EUR M in the 500-year return period, 27.7 EUR M for the 100-year return period, and 0.48 EUR M for the 20-year return period. The cost differences for the flood scenarios of 500- and 100-year return periods had a significant difference (37 and 11 EUR M), with the higher costs calculated by the county's damage values; the first scenario seemed to have a negative difference, with the total damage in Larnaca estimated at 13 EUR M more. As mentioned above, this occurred due to the difference in the methodology of assessing the total cost, where the damage cost is based on the number of buildings.

Table 1. Total flood damage cost, based on the Larnaca residential database and Cyprus residential damage costs.

	Flood Damage Based on DDC of Larnaca			**Flood Damage Based on DDC Data of Cyprus**		
Return Period (yrs)	**Structure EUR M**	**Content EUR M**	**Total EUR M**	**Structure EUR M**	**Content EUR M**	**Total EUR M**
500	21.7	10.89	32.6	46.1	23	69.2
100	11.2	5.6	16.8	18.5	9.2	27.7
20	9.1	4.6	13.7	0.32	0.16	0.48

4. Conclusions

Further investments and urban sprawl within the flood risk extent increases the risk of damage [11], and for this reason systematic coast damage evaluation is deemed necessary for management reasons. In the Cyprus flood risk management plans, cost damage is not analyzed in depth. Apart from the references to economic activities in the risk area, the economic damage is assessed at a qualitative level. Huizinga et al. (2017)'s technical report is a great tool for cost damage assessment based on the economy of the national extent. In this approach, a database of the endangered residential buildings and their values were used for the calculation of the flood damage. Even though many of the empirical parameters of the damage indices were used with their default values, the inclusion of damage costs of 562 buildings was an important addition to the local application methodology. Although, with the data available, other models for calculating total flood damage could be used [12], the above application was preferred since the pilot area belonged to an EU country. Due to the time-consuming manual process of gathering the building data, only residential buildings were assessed through the study. With the immediate and collective provision of cadastral data for the entire floodplain area, it would be possible to assess other building categories with the same methodology (e.g., parks and commercial buildings).

Author Contributions: Conceptualization, A.Ć. and O.T.; methodology, A.Ć.; software, A.Ć.; validation, O.T.; formal analysis, O.T.; investigation, A.Ć. and O.T.; resources, A.Ć.; data curation, O.T.; writing—original draft preparation, A.Ć.; writing—review and editing, O.T.; visualization, A.Ć.; supervision, O.T.; project administration, A.Ć. and O.T. All authors have read and agreed to the published version of the manuscript.

Funding: This research received no external funding.

Institutional Review Board Statement: Not applicable.

Informed Consent Statement: Not applicable.

Data Availability Statement: Restrictions apply to the availability of these data. Data was obtained from Cyprus cadastral platform, for policy reasons data can't be shared.

Conflicts of Interest: The authors declare no conflict of interest.

References

1. McMichael, A.J.; Woodruff, R.E.; Hales, S. Climate change and human health: Present and future risks. *Lancet* **2006**, *367*, 859–869. [CrossRef] [PubMed]
2. Mohammed, H.I.D. Climate change and changes in global precipitation patterns: What do we know? *Environ. Int.* **2005**, *31*, 1167–1181.
3. Dai, A.; Fung, I.Y.; Del Genio, A.D. Surface observed global land precipitation variations during 1900-88. *J. Clim.* **1997**, *10*, 2943–2962. [CrossRef]
4. Klein Tank, A.; Wijngaard, J.; Können, G.; Böhm, R.; Demarée, G.; Gocheva, A.; Mileta, M.; Pashiardis, S.; Hejkrlik, L.; Kern Hansen, C.; et al. Daily dataset of 20th-century surface air temperature and precipitation series for the European Climate Assessment. *Int. J. Climatol.* **2002**, *22*, 1441–1453. [CrossRef]
5. Zolina, O.; Simmer, C.; Belyaev, K.; Kapala, A.; Gulev, S. Improving Estimates of Heavy and Extreme Precipitation Using Daily Records from European Rain Gauges. *J. Hydrometeorol.* **2009**, *10*, 701–716. [CrossRef]
6. Madsen, H.; Lawrence, D.; Lang, M.; Martinkova, M.; Kjeldsen, T.R. Review of trend analysis and climate change projections of extreme precipitation and floods in Europe. *J. Hydrol.* **2014**, *519*, 3634–3650. [CrossRef]
7. Hosseinzadehtalaei, P.; Tabari, H.; Willems, P. Climate change impact on short-duration extreme precipitation and intensity–duration–frequency curves over Europe. *J. Hydrol.* **2020**, *590*, 125–249. [CrossRef]
8. Zbigniew, K.W.; Radziejewski, M.; Pińskwar, I. Precipitation Extremes in the Changing Climate of Europe. *Clim. Res.* **2006**, *31*, 51–58. Available online: http://www.jstor.org/stable/24869262 (accessed on 20 February 2023).
9. EUR-Lex, Access to European Union Law. Available online: https://eur-lex.europa.eu/eli/dir/2007/60/oj (accessed on 20 February 2023).
10. Huizinga, J.; De Moel, H.; Szewczyk, W. *Global Flood Depth-Damage Functions: Methodology and the Database with Guidelines*; Publications Office of the European Union: Luxembourg, 2017. [CrossRef]
11. Kundzewicz, Z.W.; Krysanova, V.; Benestad, R.E.; Hov, Ø.; Piniewski, M.; Otto, I.M. Uncertainty in climate change impacts on water resources. *Environ. Sci. Policy* **2018**, *79*, 1–8. [CrossRef]
12. Hammond, M.J.; Chen, A.S.; Djordjevic, S.; Butler, D.; Khan, D.M.; Rahman, S.M.M.; Haque, A.K.E.; Mark, O. The development of a flood damage assessment tool for urban areas. In Proceedings of the 9th International Conference on Urban Drainage Modelling, Belgrade, Serbia, 3–6 September 2012.

environmental sciences proceedings

MDPI

Proceeding Paper

Activated Carbon, CNTs and GO Based Polymeric Nanocomposites Membranes for Textile Wastewater Treatment: Preparation, Performance, and Fouling Control †

Imran A. Khan * and Nasir M. Ahmad *

School of Chemical and Materials Engineering (SCME), National University of Sciences and Technology (NUST), H-12 Sector, Islamabad 44000, Pakistan

* Correspondence: iahmad.phdscme@student.nust.edu.pk (I.A.K.); nasir.ahmad@scme.nust.edu.pk (N.M.A.); Tel.: +92-51-90855213 (N.M.A.)

† Presented at the 7th International Electronic Conference on Water Sciences, 15–30 March 2023; Available online: https://ecws-7.sciforum.net.

Abstract: Textile wastewater is a challenging area for treatment by membrane separation technology due to its complex structure and the presence of reactive components. Therefore, most of the conventional technologies appear incapable of offering satisfactory treatment for the effluents. This paper studies the application of activated carbon, carbon nanotubes, and graphene oxide base polymeric nanofiltration membranes (NF) in the textile industry, which usually produces large volumes of wastewater containing complex contaminants from its daily operation. Hence, it is accepted that NF membranes offer solutions to the problem. The primary performances of NF membranes have been examined in terms of dye rejection, salt rejection, permeate flux, and COD rejection. Some of the NF membranes achieved maximum separation of dye and salts while some attained higher flux. This is due to the large variability of the parameters of textile wastewater and the NF membranes selected. However, for all these attempts, the general issue of (bio)fouling represents a key barrier to full-scale industrial implementation. The low fouling tendency of NF membranes has lately gained substantial attention since they are an exciting addition to conventional technologies (i.e., adsorption, biological degradation, coagulation, and ultrafiltration). Polymers membrane blended with oxidized activated carbon, carbon nanotubes, and graphene oxide increased hydrophilicity, textile dyes, slat rejection, BSA rejection, antibacterial activity, and water flux enhancement from 60% to 100%. We present some nanocomposite membrane developments and demonstrate how they can be used to reduce textile dyes. In addition, the process of membrane fouling and the various approaches for preventing and controlling fouling are discussed.

Keywords: activated carbon; anti-fouling; carbon nanotubes; graphene oxide; membranes; water treatment

Citation: Khan, I.A.; Ahmad, N.M. Activated Carbon, CNTs and GO Based Polymeric Nanocomposites Membranes for Textile Wastewater Treatment: Preparation, Performance, and Fouling Control. *Environ. Sci. Proc.* **2023**, *25*, 77. https://doi.org/10.3390/ECWS-7-14307

Academic Editor: Athanasios Loukas

Published: 3 April 2023

1. Introduction

In all facets of life on Earth, water is a vital resource. Although water covers around 71 percent of the Earth's surface, the majority of the water on the planet is salt water stored in the oceans. Freshwater makes up 2.5 percent of the total water volume, with just 0.3 percent of that in liquid form on the surface [1]. According to a recent UN water study, an estimated 3.6 billion people (almost half of the world's population) live in locations where water is potentially scarce for at least one month every year [2]. According to a World Bank report, the textile sector contributes 20% of the wastewater produced by all industries of the world [3]. The textile sector is among those with the greatest water consumption rates, as it utilizes 230 to 270 tons of water to process one ton of product [4]. Hazardous chemicals, non-degradable organic compounds, surfactants, and salts are all included in the wastewater produced by finishing procedures. Additionally, this wastewater typically has high levels of total dissolved solids (TDS) and chemical oxygen demand (COD). The applicable techniques for treating textile wastewaters are physical (adsorption, ion exchange, and membrane processes), chemical (coagulation, electrocoagulation, and oxidation), and biological (aerobic and anaerobic) processes [5]. To eliminate dyes using membrane technology, several studies have been conducted. Since membranes' small pore diameters can remove dye chemicals and release high-quality effluent, they are one of the best options for treating textile dye wastewater [6].

The main purposes of this review are: (i) to investigate the membrane technology for the treatment of textile wastewater, (ii) to examine the effect of AC, CNTs, and GO for filtration and fouling control process, and (iii) to highlight the needs for future research based on the current studies.

1.1. Current Water Treatment Capabilities

Membrane technology provides up to 53% of the world's clean water production operations and is an effective water treatment method [7]. With the development of polymeric materials, water treatment technology has progressed substantially over the previous century [8]. Micro-, ultra-, and nanofiltration, as well as reverse osmosis, are all powered by hydraulic pressure, which drives water through the pores of the membrane, leaving the other constituents in the retentate behind. The type of membranes used in these procedures is determined by the pore size range and species selectivity of the membranes as shown in Figure 1 [9,10].

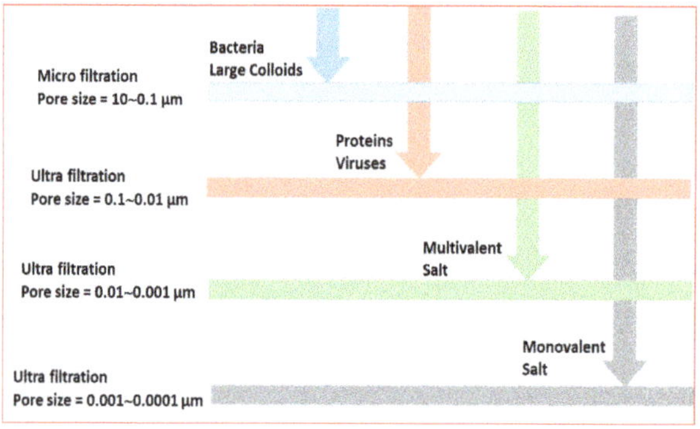

Figure 1. Classification of membranes with regards to pore size.

1.2. Membrane Fabrication Methods

Although cellulose acetate (CA) was initially used as the casting solution when it was first invented by Loeb and Sourirajan in the 1960s [11], it may also be used with other polymers that can be dissolved into homogenous solutions. A polymer, a solvent, and a coagulation medium are the only three components typically needed for the phase inversion process (or non-solvent). Figure 2 illustrates the fabrication of a polymer membrane using the phase inversion method.

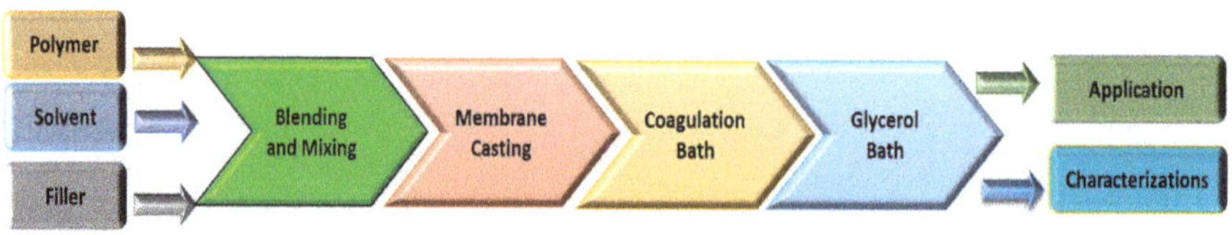

Figure 2. Steps involved in the phase inversion method of membrane fabrication.

Another process, called "electrostatic spinning", is a quick way to make nanofibers from a variety of polymeric materials for use in a variety of applications. First, the required material (such as PVDF) is introduced into the spinneret as a solution or melt (normally a syringe with a metallic needle attached). Next, voltage is applied across the conductive collector and spinneret (typically in the range of 10–40 kV) as shown in Figure 3. Once the solution is immediately drawn into microscopic fiber using a pump, it is quickly injected into the syringe and travels through the electric field from the spinneret to the collector Deposition of fibers on a grounded collector is the last step in the electrospinning process [12].

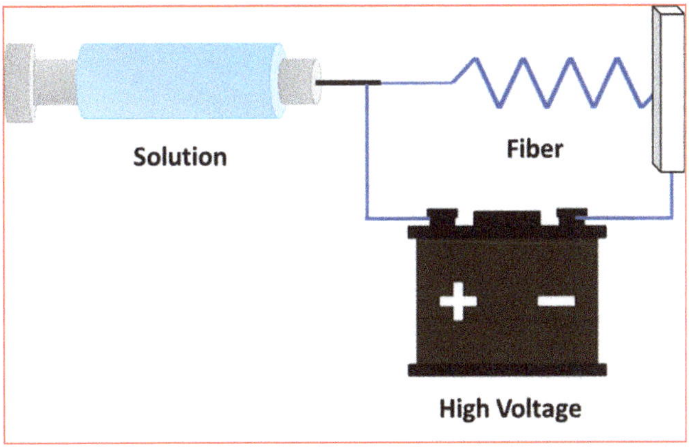

Figure 3. Electrostatic spinning setup for membrane preparation.

1.3. Membrane Fouling

Membrane fouling has been a barrier in membrane water treatment technology since its inception, limiting process efficiency by reducing water permeation flux, degrading product water quality, and increasing energy usage. As per Figure 4, membrane fouling can be brought on by biological fouling, organic adsorption, cake formation, inorganic precipitation, pore-clogging, and other processes that all have the potential to temporarily or permanently reduce flow [13].

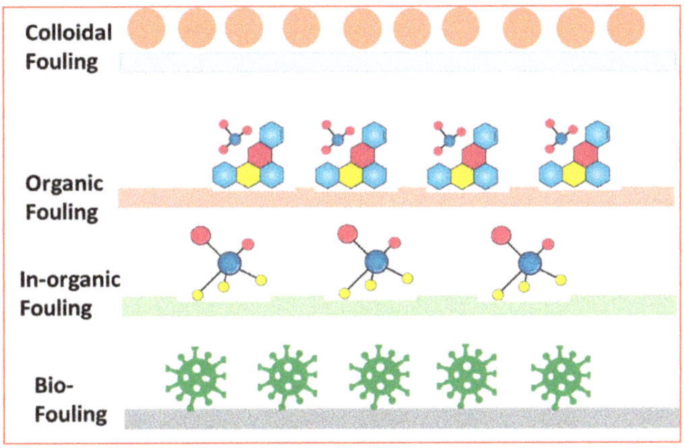

Figure 4. Types of fouling in a water filtration membrane.

2. Carbon-Based Materials for Filtration

A series of carbon-based nanomaterials (CBNs) are cytotoxic to bacteria due to their distinct physicochemical characteristics, such as carbon nanotubes (CNTs) and graphene-based nanomaterials (GO), which in particular exhibit exceptional bacterial cytotoxicity [14]. The most common commercial adsorbent, activated carbon, is a porous substance with a large surface area. Because of the large surface area, the adsorption capacity of chemicals from gases and liquids is very high [15].

Meghmik et al. [16] prepared polysulfone-highly uniform activated carbon spherical mixed-matrix membranes by use of the wet casting of an ultrasonication-assisted dispersion with AC concentrations ranging from 0.5, 1, 3, and 7% with regards to the PSF weight. In comparison to the pristine membrane, the produced membrane (containing 7% AC) had considerably improved antifouling performance, with flux recovery ratios of 90%, dye rejection of 91.13%, and water flux increase of 209.84%.

Yasin et al. found that when mesoporous carbon nano-composite polyethersulfone membranes with 0.20 wt percent MCNs were synthesized, a strong antibacterial ability and protein rejection were reported. Mesoporous carbon PES nano-composite membranes showed higher water flow, hydrophilicity, and antibacterial activity.

Carbon nanotubes were synthesized using a novel approach by M. Mohammed et al. [17], who functionalized them with an OH group, purified them, and employed them as an adsorption medium to remove colors from wastewater. In the elimination of contaminants, the effects of pH, contact duration, CNT dose, and dye concentration were investigated. The removal effectiveness of textile colors was inversely proportional to dye concentration and directly related to contact duration, CNT concentration, and pH.

Tiraferri et al. [18] created a single-walled carbon nanotube composite with polyamide (PA) and found that it had a strong antibacterial action and a 60% reduction in bacterial adhesion. Yun et al. [19] studied the increased hydrophilicity, BSA rejection, and water flow from 60% to 100% in a polysulfone (PS) membrane mixed with oxidized CNTs.

To purify artificial wastewater, S. Leaper et al. [20] examined the efficacy of PVDF mixed-matrix membranes integrating graphene oxide and GO functionalized with Aminopropyl-triethoxysilane (APTS). The surface porosity and mean pore size were found to be greatly enhanced by the addition of very small volumes. The highest-performing membranes overall produced flux improvements of 52% and 86% for GO and GO-APTS membranes, respectively, above the pure PVDF membrane. Due to increased hydrophilicity and greater pore size at higher loading concentrations, the GO membrane's rejection performance somewhat decreased. It was shown that during the phase inversion process, the interactions between the nanofiller and the polymer solution accelerated the rate of inter-diffusion between the solvent and non-solvent. As a result, a structure with high porosity develops, leading to greater flux values.

Later on, S. Leaper et al. [21] investigated the impact of reduced graphene oxide with various degrees of reduction on membrane structure and performance. The observed flux increased from 6.4 to 7 LMH when the carbon to oxygen (C/O) ratio of the GO flakes increased from 2.30 (for GO) to 5.45 (for rGO). However, the flow decreased to 3.9 LMH at a greater level of reduction (C/O = 7.36), showing less favorable pore-forming properties with the further elimination of the oxygen functions. It has been shown in some instances that the pore-forming behavior of GO materials enhances the membranes' porosity and pore structure while simultaneously enhancing their mechanical properties. The high strength of the nanomaterial and its successful interaction with the matrix material are responsible for this.

3. Conclusions

This review paper concentrated on using membrane technology to treat textile effluent and address the issue of water scarcity. Fouling and biofouling are two major issues with membranes that can be mitigated by the creation of polymer-nanocomposite membranes. The technology of phase inversion is frequently used to cast membranes from several new carbon-based nanomaterials. In comparison to bare membranes, nanocomposite membranes perform better overall. The size, shape, surface qualities, and composition of nanoparticles, as well as the kind of polymers used, have a significant influence on the ultimate performance of nanocomposite membranes. According to the literature, there is an ideal concentration of nanoparticles and membrane materials for each combination of nanoparticles and polymeric membrane materials. Higher nanofiller concentrations do not always imply improved performance. To attain the best results, detailed experimental evaluations of particular combinations of nanoparticles and polymeric membranes are always required. Nanocomposite membranes' long-term stability and performance are yet unknown. Furthermore, the scaling-up of nanocomposite membrane production has not been well examined. Nanocomposite membranes will benefit tremendously from advances in mechanistic knowledge of nanoparticle polymer matrix interactions and scalable fabrication. For various applications, future efforts should focus on building more stable, high-performing, and scalable nanocomposite membranes.

Author Contributions: Conceptualization, I.A.K.; methodology, I.A.K.; writing—original draft preparation and writing, I.A.K.; review and supervision; N.M.A. All authors have read and agreed to the published version of the manuscript.

Funding: This research was funded by Higher Education Commission (HEC) of Pakistan, under NRPU Project # 6020.

Institutional Review Board Statement: Not applicable.

Informed Consent Statement: Not applicable.

Data Availability Statement: Not applicable.

Acknowledgments: School of Chemical and Materials Engineering (SCME), National University of Sciences and Technology (NUST), H-12 Sector, Islamabad 44000, Pakistan.

Conflicts of Interest: The authors declare no conflict of interest.

References

1. Mekonnen, M.; Hoekstra, A.Y. Sustainability: Four billion people facing severe water scarcity. *Sci. Adv.* **2016**, *2*, e1500323. [CrossRef] [PubMed]
2. Brusseau, M.L.; Ramirez-Andreotta, M.; Pepper, I.L.; Maximillian, J. *Environmental Impacts on Human Health and Well-Being*, 3rd ed.; Elsevier: Amsterdam, The Netherlands, 2019; ISBN 9780128147191.
3. Tavangar, T.; Jalali, K.; Alaei Shahmirzadi, M.A.; Karimi, M. Toward real textile wastewater treatment: Membrane fouling control and effective fractionation of dyes/inorganic salts using a hybrid electrocoagulation—Nanofiltration process. *Sep. Purif. Technol.* **2019**, *216*, 115–125. [CrossRef]
4. Tahri, N.; Masmoudi, G.; Ellouze, E.; Jrad, A.; Drogui, P.; Ben Amar, R. Coupling microfiltration and nanofiltration processes for the treatment at source of dyeing-containing effluent. *J. Clean. Prod.* **2012**, *33*, 226–235. [CrossRef]
5. Keskin, B.; Ersahin, M.E.; Ozgun, H.; Koyuncu, I. Pilot and full-scale applications of membrane processes for textile wastewater treatment: A critical review. *J. Water Process Eng.* **2021**, *42*, 102172. [CrossRef]
6. Keskin, B.; Ağtaş, M.; Ormancı-Acar, T.; Türken, T.; Imer, D.Y.; Ünal, S.; Menceloğlu, Y.Z.; Uçar-Demir, T.; Koyuncu, I. Halloysite nanotube blended nanocomposite ultrafiltration membranes for reactive dye removal. *Water Sci. Technol.* **2021**, *83*, 271–283. [CrossRef] [PubMed]
7. Obotey Ezugbe, E.; Rathilal, S. Membrane Technologies in Wastewater Treatment: A Review. *Membranes* **2020**, *10*, 89. [CrossRef] [PubMed]
8. Kyzas, G.Z.; Mitropoulos, A.C. Polymeric Materials for Water and Wastewater Management. *Polymers* **2021**, *13*, 168. [CrossRef] [PubMed]
9. Baker, R.W. *Membrane Technology Enhanced Reader*, 3rd ed.; Wiley: Hoboken, NJ, USA, 2012.
10. Kochkodan, V.; Johnson, D.J.; Hilal, N. Polymeric membranes: Surface modification for minimizing (bio)colloidal fouling. *Adv. Colloid Interface Sci.* **2014**, *206*, 116–140. [CrossRef] [PubMed]
11. Strathmann, H. The Formation Mechanism of Phase Inversion Membranes. *Desalination* **1977**, *21*, 241–255. [CrossRef]
12. Yu, F.; Shi, H.; Shi, J.; Teng, K.; Xu, Z.; Qian, X. High-performance forward osmosis membrane with ultra-fast water transport channel and ultra-thin polyamide layer. *J. Memb. Sci.* **2020**, *616*, 118611. [CrossRef]
13. Kochkodan, V.; Hilal, N. A comprehensive review on surface modified polymer membranes for biofouling mitigation. *Desalination* **2015**, *356*, 187–207. [CrossRef]
14. Francois, P.; Andreia, F.; Menachem, E. Antimicrobial Properties of Graphene Oxide Nanosheets Why Size Matters. *ACS Nano* **2015**, *9*, 7226–7236. [CrossRef]

15. Li, Z.; Hanafy, H.; Zhang, L.; Sellaoui, L.; Schadeck Netto, M.; Oliveira, M.L.S.; Seliem, M.K.; Luiz Dotto, G.; Bonilla-Petriciolet, A.; Li, Q. Adsorption of congo red and methylene blue dyes on an ashitaba waste and a walnut shell-based activated carbon from aqueous solutions: Experiments, characterization and physical interpretations. *Chem. Eng. J.* **2020**, *388*, 124263. [CrossRef]
16. Manoukian, M.; Fashandi, H.; Tavakol, H. Polysulfone-highly uniform activated carbon sphere mixed-matrix membrane intended for efficient purification of dye wastewater. *Mater. Res. Express* **2019**, *6*, 055313. [CrossRef]
17. Mohammed, M.I.; Abdul Razak, A.A.; Hussein Al-Timimi, D.A. Modified multiwalled carbon nanotubes for treatment of some organic dyes in wastewater. *Adv. Mater. Sci. Eng.* **2014**, *2014*, 201052. [CrossRef]
18. Tiraferri, A.; Vecitis, C.D.; Elimelech, M. Covalent binding of single-walled carbon nanotubes to polyamide membranes for antimicrobial surface properties. *ACS Appl. Mater. Interfaces* **2011**, *3*, 2869–2877. [CrossRef] [PubMed]
19. Li, Y.X.; Gao, Y.; Yang, C.; Wang, Z.Q.; Xue, G. Facile and controllable assembly of multiwalled carbon nanotubes on polystyrene microspheres. *Chin. J. Polym. Sci. Engl. Ed.* **2014**, *32*, 711–717. [CrossRef]
20. Leaper, S.; Abdel-Karim, A.; Faki, B.; Luque-Alled, J.M.; Alberto, M.; Vijayaraghavan, A.; Holmes, S.M.; Szekely, G.; Badawy, M.I.; Shokri, N.; et al. Flux-enhanced PVDF mixed matrix membranes incorporating APTS-functionalized graphene oxide for membrane distillation. *J. Memb. Sci.* **2018**, *554*, 309–323. [CrossRef]
21. Leaper, S.; Abdel-Karim, A.; Gorgojo, P. The use of carbon nanomaterials in membrane distillation membranes: A review. *Front. Chem. Sci. Eng.* **2021**, *15*, 755–774. [CrossRef]

environmental sciences proceedings

MDPI

Proceeding Paper

Flood Disaster Mapping Using Geospatial Techniques: A Case Study of the 2022 Pakistan Floods †

Asif Sajjad [1,*], Jianzhong Lu [2], Rana Waqar Aslam [2] and Muhammad Ahmad [1]

1 Department of Environmental Sciences, Faculty of Biological Sciences, Quaid-i-Azam University, Islamabad 45320, Pakistan; muhammad380036@gmail.com

2 State Key Laboratory of Information Engineering in Surveying, Mapping and Remote Sensing, Wuhan University, Wuhan 430079, China; lujzhong@whu.edu.cn (J.L.); ranawaqaraslam@whu.edu.cn (R.W.A.)

* Correspondence: asifsajjad@qau.edu.pk

† Presented at the 7th International Electronic Conference on Water Sciences, 15–30 March 2023; Available online: https://ecws-7.sciforum.net.

Abstract: Remote sensing images are an essential tool for mapping the amount of flood inundation after flood events. For early flood estimation, flood mapping is a crucial component. This study used an integration of geospatial techniques to evaluate the flood extent in District Dera Ghazi Khan, Pakistan. The modified normalized difference water index (MNDWI) was utilized to estimate the flood extent using Landsat data. For a thorough flood investigation, pre-flood, during, and post-flood images were obtained. The analysis enabled us to delineate flood extent as well as flood duration. The result showed that the flood continued for nearly 5 weeks in the study area. This proposed geospatial technique provides a framework for the identification of inundated areas, which allows emergency responses to be focused on newly flooded areas. Hence, the current study offers a novel perspective on flood mapping and significantly contributes to flood monitoring.

Keywords: flood mapping; geospatial techniques; Landsat data; MNDWI; emergency response

check for updates

Citation: Sajjad, A.; Lu, J.; Aslam, R.W.; Ahmad, M. Flood Disaster Mapping Using Geospatial Techniques: A Case Study of the 2022 Pakistan Floods. *Environ. Sci. Proc.* **2023**, 25, 78. https://doi.org/10.3390/ECWS-7-14312

Academic Editor: Athanasios Loukas

Published: 3 April 2023

1. Introduction

Flood disasters are frequently occurring with more severity worldwide [1,2], and they account for more than 50% of all natural disasters that happen [3,4]. In the recent decade, flood disasters have caused over 0.25 million causalities and also registered a total of about USD 4.8 billion in economic losses [5,6]. Flooding is the primary global cause of catastrophic loss and destruction, and the number of people exposed to floods is increasing faster than the world's population [7,8].

South Asian countries, such as Bangladesh, India, and Pakistan, claim the highest number of flood-related causalities and economic losses [7,9–11]. In the recent decade, flood disasters have been common in Pakistan and have caused severe damage to all socio-economic sectors [12–14].

Floods in Pakistan generally occur in the monsoon season because of an excessive amount of rainfall in the upstream watershed of the main rivers, the Indus and the Chenab [15–17]. As a result, floods with high magnitudes have occurred in floodplain areas, which ultimately resulted in a huge impact on people's lives, property, and the overall economy of the country [18,19]. Despite the fact that this catastrophic flood menace cannot be entirely prevented, the effects can be lessened by employing effective remote sensing-based flood risk mapping [17,20]. The remote sensing technique is the most affordable way to provide extensive data on all stages of a flood disaster, which may be used for flood mapping [9,13].

Therefore, the main goal of the current research is to delineate flooded areas using suitable satellite-derived MNDWI water indices. Results from the water indices propose a step-wise approach for accurately performing flood mapping. Such an approach gives an insight into the flood situation in this flooded area with minimum inclusion of other non-water classes. In addition, in case of flood disaster occurrence, such insights can assist in proper decision making for emergency flood management in District Dera Ghazi Khan.

Study Area

Tehsil Taunsa, District DG Khan was chosen for this study. Its geographic range is between 30.4078° N and 70.5265° E (Figure 1). It encompasses the south-western portion of Punjab and is bordered by the Punjab districts of Muzaffargarh, Layyah, and Rajanpur. The entire study area covered an area of 11,294 km^2 [21]. The highest and lowest rainfalls ever recorded were 50 mm in July and 2 mm in October, respectively. The average annual rainfall is 22.18 mm [22]. The locals' primary economic activity is agriculture. In the study area, cotton, sugarcane, wheat, rice, and sunflowers are the primary crops grown.

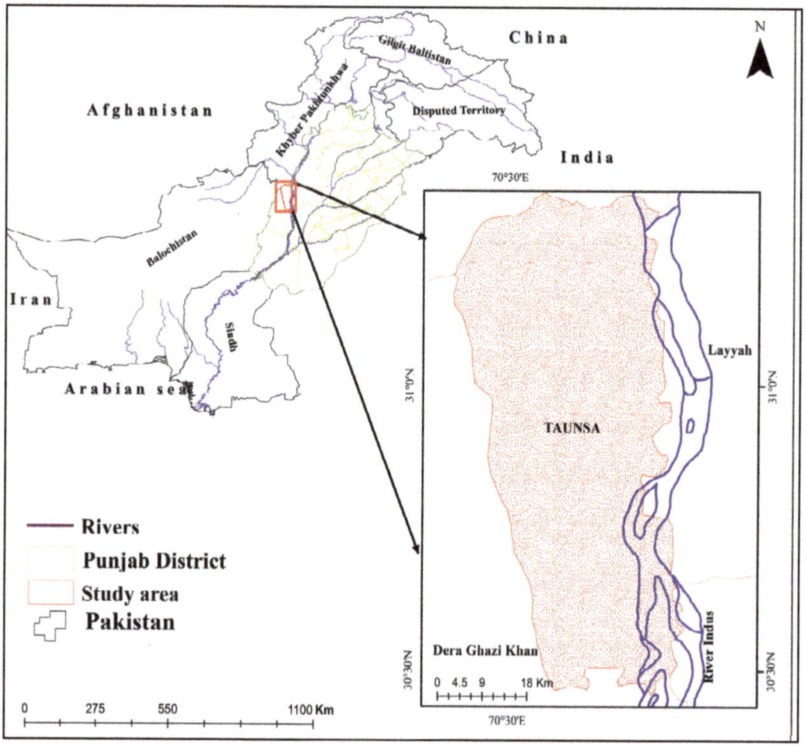

Figure 1. Location of Study Area.

2. Data and Methodology

The Landsat 8–9 OLI/TIRS data were obtained from the United States Geological Survey (USGS) [23–25]. We obtained 6 Landsat images from 21 July and 21 September 2022. Furthermore, these images were utilized to estimate MNDWI water indices for flood mapping using ArcGIS 10.8 [26,27]. This index, first presented by Xu [28], effectively calculates inundated areas. In essence, it employs the green and shortwave infrared bands to extract the flood extent, as illustrated in Equation (1):

$$\text{MNDWI} = \frac{(band3) - (band6)}{(band3) + (band6)} \quad (1)$$

The MNDWI index reduces accumulated noise while highlighting the water surface reflectance using the green band; the resulting values range from 1 to +1. The generated images show negative values for built-up regions and positive values for water areas, respectively, depending on how much water area reflectance is high and how little built-up area reflectance is low in the shortwave infrared band.

Optical Landsat data, which provide up-to-date information with comparatively high temporal resolution have been widely used in the detection of flooded areas [7,17,23]. On the other hand, the presence of clouds lowers the availability of Landsat images during flooding. Certainly, synthetic aperture radar (SAR) and radar satellites can pierce clouds and obtain images in all weather situations and are commonly employed for flood mapping under these circumstances [29,30]. SAR data are normally very expensive and, in our case, the free SAR data were not available and the study area was entirely cloud-free during the 2022 flood.

3. Results and Discussion

Spatio-Temporal Flooded Area Mapping

The 2022 flood extent was identified to delineate the most inundated areas in Taunsa, Dera Ghazi Khan, Pakistan. The results show that the cumulative flood inundation from pre-flood to post-flood instances is as shown in Figures 2 and 3. The results also show that the flood inundation spans an area of approximately 526 km^2 in the 21 July image, whereas the highest flood inundation of 1462.24 km^2 can be seen in the 31 August image. Additionally, the peak flood scenario remained constant until 7 September as rain and hill torrent flow persisted from the Suleiman range and reduced flood inundation to an area of 1196 km^2 at a pace of 38 km^2/day. On 14 September, the inundation receded to about 702 km^2, with a decreasing rate of 72 km^2/day. Lastly, the floodwaters continued to recede and the normal stage was seen as a reducing trend persisting throughout the month of September, to the point that on September 21st, the extent of the inundation covered only 491 km^2 and returned to its pre-flood stage. The trend of the flood extent during three flood instances is presented in Figures 2 and 3.

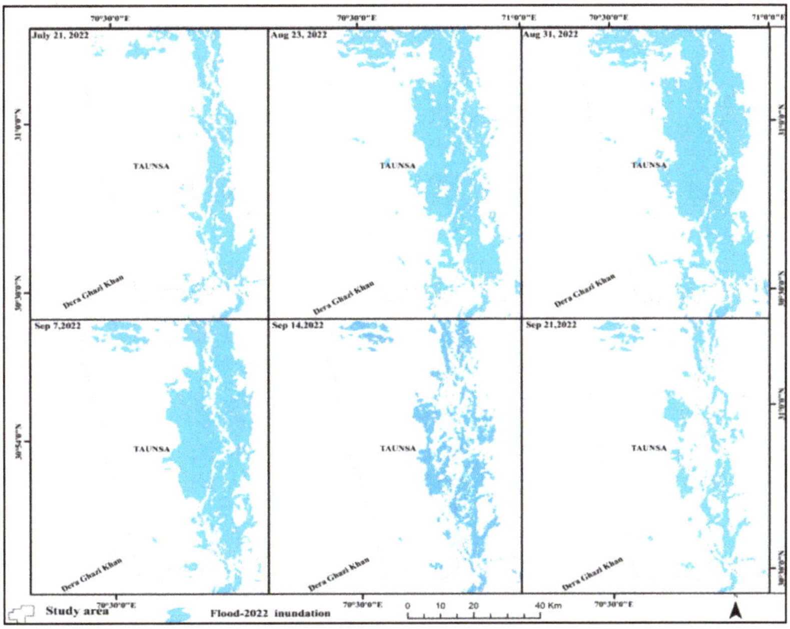

Figure 2. Temporal flood extent, peak flood extent, and duration.

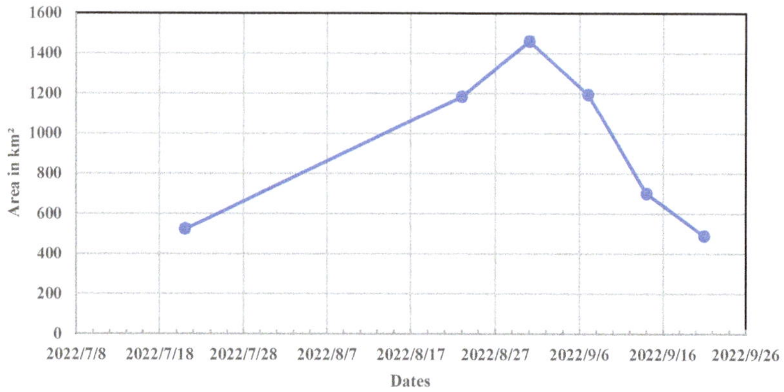

Figure 3. Temporal flood extent from during-flood (21 July to post-flood 21 September 2022).

Areas that have been flooded for several days can be detected using Landsat data [15,17,25]. Landsat data were utilized because they are also capable of accurately detecting flooded areas in built-up and agricultural areas [23,25,30]. However, to properly estimate the extent of flooding, field research and high-resolution data are required. However, due to their high cost, radar and SAR data are not used in the research area. To really observe the flooding scenario for this study, a field survey was also carried out.

As can be seen in Figures 2 and 3, the water remained in the study area for about 5 weeks after the maximum flood peak was recorded on 31 August 2022. The floodwaters subsided in two stages; during the peak stage, the water decreased gradually until the 7th of September at a rate of about 38 km^2 per day. Up to 21 September, the floodwaters in the moderate stage significantly decreased at a rate of 72 km^2 each day. The assessment of the results, however, showed that the agricultural and built-up areas suffered a great deal due to the massive inundation and duration of approximately 5 weeks.

Our results reveal that Landsat images, together with satellite-derived MNDWI indices permit detailed flooded area delineation with reliable accuracy. A recent review paper, similarly, analyzed the same indices used to depict areas under flood water and regarded MNDWI to be the most suitable in terms of its ability to differentiate between turbid water and mixed pixels [31]. Suitable satellite data are critical for flood mapping [6,9]. The temporal relationship between satellite characteristics with flood occurrence is a vital parameter in flood mapping. For example, a low-resolution moderate resolution imaging spectroradiometer (MODIS) satellite (~250 m), with its daily revisit time, has been utilized during many floods to acquire flood mapping, but with questionable accuracy [9]. However, due to the lack of during-flood SAR satellite data, such as Sentinel-1, we employed Landsat data with a resolution of 30 m for flood mapping [2,17,30]. Albertini et al. [31], in their recent review article, also concluded Landsat is the most commonly used satellite for floodwater spatial coverage detection. Despite the better spatial resolution, Landsat satellite data are constrained and cannot obtain geospatial data on time, which usually reduces their applicability to flooded area mapping [27,29]. Furthermore, the MNDWI index used achieved reliable accuracy in terms of flooded area extraction and has been efficiently used in other studies as well for detecting areas under flood water [32].

4. Conclusions

Tehsil Taunsa is particularly vulnerable to frequent riverine and hillside torrent floods. Locals living along nullahs and the Indus River deal with this flood threat almost every year. In the study region in 2022, floods produced by a two-week-long persistent wet spell occurred upstream of the Indus and in the foothills of the Suleiman range, which resulted in flooding and had a significant impact. Standing crops that were ready to be harvested, homes, animals, and all kinds of infrastructures were affected as a result.

These frequent floods are a severe problem that calls for effective preparation and impact mitigation measures, particularly through effective post-flood monitoring procedures, and particularly in Tehsil Taunsa, District D. G. Khan.

This study used remote sensing data integration and jointly employed appropriate methods to demonstrate the flood inundation severity in Taunsa. In order to estimate the inundated areas, Landsat images of flood instances were analyzed using the GIS-based MNDWI index.

The flood mapping also revealed that the flood water persisted in the study region for a month, which increased and exacerbated flood damage in the study areas. This study found that geospatial techniques can be used to carry out advanced flood mapping, which is important for flood management. As a result, the current study offers an alternative perspective on mapping flood inundation utilizing free satellite data and techniques. Flood risk mapping is the first step toward analyzing flood risk. This study can be used as a basis for further evaluating flood risk assessment and management in the study area. The methodology employed can be integrated with other radar datasets and the Analytic hierarchy process (AHP) techniques to further identify the severity of future flood risk while developing flood risk zonation.

Author Contributions: Writing—original draught, A.S.; review and editing, A.S. and J.L.; image analysis, flood inundation maps, A.S.; validation, A.S., M.A. and R.W.A. All authors have read and agreed to the published version of the manuscript.

Funding: This research received no external funding.

Institutional Review Board Statement: Not applicable.

Informed Consent Statement: Not applicable.

Data Availability Statement: The data presented in this study are available on request from the first and corresponding author.

Conflicts of Interest: The authors declare no conflict of interest.

References

1. Syvitski, J.P.; Brakenridge, G.R. Causation and avoidance of catastrophic flooding along the Indus River, Pakistan. *GSA Today* **2013**, *23*, 4–10. [CrossRef]
2. Bhatt, C.M.; Rao, G.S.; Farooq, M.; Manjusree, P.; Shukla, A.; Sharma, S.V.S.P.; Kulkarni, S.S.; Begum, A.; Bhanumurthy, V.; Diwakar, P.G.; et al. Satellite-based assessment of the catastrophic Jhelum floods of September 2014, Jammu & Kashmir, India. *Geomat. Nat. Hazards Risk* **2016**, *8*, 309–327.
3. Zhang, P.; Lu, J.; Feng, L.; Chen, X.; Zhang, L.; Xiao, X.; Liu, H. Hydrodynamic and Inundation Modeling of China's Largest Freshwater Lake Aided by Remote Sensing Data. *Remote Sens.* **2015**, *7*, 4858–4879. [CrossRef]
4. Schumann, G.; Bates, P.D.; Apel, H.; Aronica, G.T. Global Flood Hazard Mapping, Modeling, and Forecasting: Challenges and Perspectives. In *Global Flood Hazard: Applications in Modeling, Mapping, and Forecasting*; John Wiley and Sons: Hoboken, NJ, USA, 2018; pp. 239–244.
5. Mahmood, S.; Sajjad, A.; Rahman, A. Cause and damage analysis of 2010 food disaster in district Muzaffar Garh, Pakistan. *Nat. Hazards* **2021**, *107*, 1681–1692.
6. Khalid, B.; Cholaw, B.; Alvim, D.S.; Javeed, S.; Khan, J.A.; Javed, M.A.; Khan, A.H. Riverine flood assessment in Jhang district in connection with ENSO and summer monsoon rainfall over Upper Indus Basin for 2010. *Nat. Hazards* **2018**, *92*, 971–993. [CrossRef]
7. Halgamuge, M.N.; Nirmalathas, A. Analysis of large flood events: Based on flood data during 1985–2016 in Australia and India. *Int. J. Disaster Risk Reduct.* **2017**, *24*, 1–11. [CrossRef]
8. Milly, P.C.D.; Wetherald, R.T.; Dunne, K.A.; Delworth, T.L. Increasing risk of great floods in a changing climate. *Nature* **2002**, *415*, 514–517. [CrossRef]
9. Haq, M.; Akhtar, M.; Muhammad, S.; Paras, S.; Rahmatullah, J. Techniques of Remote Sensing and GIS for flood monitoring and damage assessment: A case study of Sindh province, Pakistan. *Egypt J. Remote Sens. Space Sci.* **2012**, *15*, 135–141. [CrossRef]
10. Islam, A.S.; Bala, S.K.; Haque, M. Flood inundation map of Bangladesh using MODIS time-series images. *J. Flood Risk Manag.* **2010**, *3*, 210–222. [CrossRef]
11. Gaurav, K.; Sindha, R.; Panda, P.K. The Indus flood of 2010 in Pakistan: A perspective analysis using remote sensing data. *Nat. Hazards* **2011**, *59*, 1815–1826. [CrossRef]
12. Hashmi, H.N.; Siddiqui, Q.T.M.; Ghuman, A.R.; Kamal, M.A.; Mughal, H. A critical analysis of 2010 floods in Pakistan. *Afr. J. Agric. Res.* **2012**, *7*, 1054–1067.

13. Mahmood, S.; Rahman, A.; Sajjad, A. Assessment of 2010 flood disaster causes and damages in district Muzaffargarh, Central Indus Basin, Pakistan. *Environ. Earth Sci.* **2019**, *78*, 63. [CrossRef]
14. Sajjad, A.; Lu, J.; Chen, X.; Chisenga, C.; Mahmood, S. The riverine flood catastrophe in August 2010 in South Punjab, Pakistan: Potential causes, extent and damage assessment. *Appl. Ecol. Environ. Res.* **2019**, *17*, 14121–14142. [CrossRef]
15. Sajjad, A.; Lu, J.; Chen, X.; Chisenga, C.; Saleem, N.; Hassan, H. Operational Monitoring and Damage Assessment of Riverine Flood-2014 in the Lower Chenab Plain, Punjab, Pakistan, Using Remote Sensing and GIS Techniques. *Remote Sens.* **2020**, *12*, 714. [CrossRef]
16. Naeem, B.; Azmat, M.; Tao, H.; Ahmad, S.; Khattak, M.U.; Haider, S.; Ahmad, S.; Khero, Z.; Goodell, C.R. Flood Hazard Assessment for the Tori Levee Breach of the Indus River Basin, Pakistan. *Water* **2021**, *13*, 604. [CrossRef]
17. Sajjad, A.; Lu, J.; Chen, X.; Chisenga, C.; Mazhar, N.; Nadeem, B. Riverine flood mapping and impact assessment using remote sensing technique: A case study of Chenab flood-2014 in Multan district, Punjab, Pakistan. *Nat Hazards* **2022**, *110*, 2207–2226. [CrossRef]
18. National Disaster Management Authority((NDMA). *Annual Flood Report*; Regional Office: Islamabad, Pakistan, 2021.
19. Punjab Provincial Disaster Management Authority (PPDMA). *Annual Flood Report*; Regional Office: Lahore, Pakistan, 2021.
20. Siddiqui, M.; Haider, S.; Gabriel, H.F.; Shahzad, A. Rainfall–run off, flood inundation and sensitivity analysis of the 2014 Pakistan flood in the Jhelum and Chenab River basin. *Hydrol. Sci. J.* **2018**, *63*, 13–14. [CrossRef]
21. Federal Flood Commission Islamabad (FFCI). *Annual Flood Report*; Ministry of Water and Power, Government of Pakistan: Islamabad, Pakistan, 2021.
22. Pakistan Meteorological Department (PMD). *Annual Report*; Regional Meteorological Observatory: Lahore, Pakistan, 2022.
23. Munasinghe, D.; Cohen, S.; Huang, Y.F.; Tsang, Y.P.; Zhang, J.; Fang, Z.F. Intercomparison of Satellite Remote Sensing-Based Flood Inundation Mapping Techniques. *J. Am. Water Resour. Assoc.* **2018**, *54*, 834–846. [CrossRef]
24. Revilla-Romero, B.; Hirpa, F.A.; Pozo, J.T.; Salamon, P.; Brakenridge, R.; Pappenberger, F.; De Groeve, T. On the use of global flood forecasts and satellite-derived inundation maps for flood monitoring in data-sparse regions. *Remote Sens.* **2015**, *7*, 15702–15728. [CrossRef]
25. Shen, L.; Li, C. Water body extraction from Landsat ETM+ imagery using adaboost algorithm. In Proceedings of the IEEE 2010 18th International Conference on Geoinformatics, Beijing, China, 18–20 June 2010; pp. 1–4.
26. Alphan, H.; Doygun, H.; Unlukaplan, Y.I. Post-classification comparison of land cover using multitemporal Landsat and ASTER imagery: The case of Kahramanmaras, Turkey. *Environ. Monit. Assess.* **2009**, *151*, 327–336. [CrossRef] [PubMed]
27. Rokni, K.; Ahmad, A.; Selamat, A.; Hazini, S. Water feature extraction and change detection using multitemporal Landsat imagery. *Remote Sens.* **2014**, *6*, 4173–4189. [CrossRef]
28. Xu, H. Modification of normalized difference water index (NDWI) to enhance open water features in remotely sensed imagery. *Int. J. Remote Sens.* **2006**, *27*, 3025–3033. [CrossRef]
29. Uddin, K.; Matin, M.A.; Meyer, F.J. Operational Flood Mapping Using Multi-Temporal Sentinel-1 SAR Images: A Case Study from Bangladesh. *Remote Sens.* **2019**, *11*, 1581. [CrossRef]
30. Notti, D.; Giordan, D.; Caló, F.; Pepe, A.; Zucca, F.; Pedro Galve, J. Potential and Limitations of Open Satellite Data for Flood Mapping. *Remote Sens.* **2018**, *10*, 1673. [CrossRef]
31. Albertini, C.; Gioia, A.; Iacobellis, V.; Manfreda, S. Detection of Surface Water and Floods with Multispectral Satellites. *Remote Sens.* **2022**, *14*, 6005. [CrossRef]
32. Güvel, Ş.P.; Akgül, M.A.; Aksu, H. Flood inundation maps using Sentinel-2: A case study in Berdan Plain. *Water Supply* **2022**, *22*, 4098–4108. [CrossRef]

environmental sciences proceedings

MDPI

Proceeding Paper

Automatic and Non-Invasive Monitoring of Water Stress in Vineyards †

Pietro Brach del Prever [1,*], Gabriele Balducci [2], Alice Ballestra [2], Carlo Ghiglione [2], Laura Mascheretti [2], Margherita Molinari [2], Giuseppe Nicoletti [2], Valter Carvelli [3], Chiara Corbari [4], Stefano Invernizzi [5] and Stefano Mariani [4]

1 Politecnico di Torino, Corso Duca degli Abruzzi 24, 10129 Torino, Italy
2 Politecnico di Milano, Piazza Leonardo da Vinci 32, 20133 Milano, Italy; gabriele.balducci@asp-poli.it (G.B.); alice.ballestra@asp-poli.it (A.B.); carlo.ghiglione@asp-poli.it (C.G.); laura.mascheretti@asp-poli.it (L.M.); margherita.molinari@asp-poli.it (M.M.); giuseppe.nicoletti@asp-poli.it (G.N.)
3 Department of Architecture, Built Environment and Construction Engineering, Politecnico di Milano, Piazza Leonardo da Vinci 32, 20133 Milano, Italy; valter.carvelli@polimi.it
4 Department of Civil and Environmental Engineering, Politecnico di Milano, Piazza Leonardo da Vinci 32, 20133 Milano, Italy; chiara.corbari@polimi.it (C.C.); stefano.mariani@polimi.it (S.M.)
5 Department of Structural, Geotechnical and Building Engineering, Politecnico di Torino, Corso Duca degli Abruzzi 24, 10129 Torino, Italy; stefano.invernizzi@polito.it
* Correspondence: pietro.brachdelprever@asp-poli.it

† Presented at the 7th International Electronic Conference on Water Sciences, 15–30 March 2023; Available online: https://ecws-7.sciforum.net.

Abstract: In this study, a non-invasive system is proposed for monitoring the health of vine plants by measuring their water stress, with the goal of mitigating frequent extreme meteorological events such as droughts. The envisioned system measures the spatial distribution of the Crop Water Stress Index (CWSI) on the crop field and provides the farmers with precise control over their vine's health and, therefore, on the final quality of their product. To ensure the accurate acquisition of the parameters needed to compute the CWSI, data are collected by field sensors on the ground and by exploiting satellite data. Data fusion then allows us to obtain an associated georeferenced heatmap of the vineyard. The solution has been tested via a prototype, which allowed the collection of information in a vineyard.

Keywords: vineyard; crop; monitoring; water management; CWSI; vegetation index; drones; sensors; satellite

Citation: Brach del Prever, P.; Balducci, G.; Ballestra, A.; Ghiglione, C.; Mascheretti, L.; Molinari, M.; Nicoletti, G.; Carvelli, V.; Corbari, C.; Invernizzi, S.; et al. Automatic and Non-Invasive Monitoring of Water Stress in Vineyards. *Environ. Sci. Proc.* **2023**, *25*, 79. https://doi.org/10.3390/ECWS-7-14164

Academic Editor: Lampros Vasiliades

Published: 14 March 2023

1. Introduction

Climate change is becoming more and more relevant for the agricultural sector, and its most catastrophic events are causing huge losses in crop productivity. A chance to effectively react to such extreme events and adapt to a constantly changing scenario is offered by innovative Agri-Tech technologies. Unfortunately, such solutions are mostly adopted only in large-scale production industries, whereas smaller companies often lack technological expertise, aptitude for innovation, or enough financial resources for the initial investments. The goal of this project is to study and develop a specific solution capable of monitoring the health of a crop field in an automatic and non-invasive way and that does not need a high economic effort nor advanced technical skills to be applied. This study focuses on Italian vineyards since they are usually part of small–medium-sized companies, forming a fragmented economic context based on traditional techniques and with a generally low diffusion of technological innovation. Vineyards are now facing the impact of climate change and, due to their peculiar characteristics, are more prone to its consequences. A further objective is to foster the integration of the traditional know-how of older farmers with the technological expertise of younger ones, thus increasing the resilience towards future environmental scenarios.

One effective parameter to monitor the health status of vineyards is the Crop Water Stress Index (CWSI), a quantitative estimation of a plant's need for water. In detail, water availability is a key parameter influencing the quantity and quality of vineyard production. The knowledge of water stress is an important feature for controlling the optimal water conditions needed to achieve the desired characteristics of production. Hence, the aim of this study is to design and implement an intelligent autonomous system to measure the spatial and temporal distribution of CWSI on a crop field, offering a valuable instrument to have precise insight into the health of vineyards and to schedule targeted interventions to recover optimal values. Furthermore, geospatial and historical analysis of CWSI can be performed to gain insight into the characteristics of the crop field and infer connections with the productivity of the harvest.

Some solutions for the monitoring of vine water stress are available, but their application is still at an early stage and the technology is not yet refined [1]. To the best of our knowledge, no water resource optimization system, via active and adaptive crop monitoring, is available on the market at affordable prices.

The potentialities of the proposed solution have been tested in a data collection session in a vineyard owned by "Azienda Agricola Balladore Pallieri" (Asti, Italy) through a low-cost prototype. For this purpose, miniaturized sensors have been assembled on a Raspberry Pi. This on-field campaign helped us to verify the validity of the concept and its implementation to collect climatic parameters and elaborate the data into real CWSI heatmaps. Moreover, it contributed to highlighting its advantages and the further improvements of the design.

The system provides almost real-time information on the hydration condition of the field at near-real-time with a great level of geographic detail, enabling wise water management and boosting plant care. Thanks to the CWSI maps of the field, it is possible to focus watering only on the areas of the vineyard needing water, thus reducing water use.

The system fits well with the current state of society, responding to one of the most urgent challenges: climate change. The proposed solution increases the resilience of the economic system to climate change as it improves the farmer's ability to control and optimize the water usage in their fields. Moreover, the flexibility of the proposed sensing system and its software suite allows for adaptive and evolving technology. Indeed, the project could successively be adapted to fit the needs of other kinds of crops, thus expanding its market and impact on the agriculture industry.

2. Background, Parameters, and Methods

The Crop Water Stress Index [2,3], or CWSI, is chosen as the key vegetation index in this work as it is strongly related to the water status of the crop. It measures the ability of the plant to exploit the available water in the soil. It is related to the health of the plant because it describes the ability to successfully bring water from the roots to the leaves and fruit. Instead of measuring it directly through the evaluation of stomatal pressure, which would require complex instrumentation, it can be evaluated through climatic and atmospheric data in the area surrounding the plant as it is defined as:

$$\text{CWSI} = 1 - \frac{\text{ET}}{\text{PET}} \tag{1}$$

where ET and PET are measures of evapotranspiration potential, closely linked to the water potential of the plant. A way to calculate this index based on meteorological data was proposed by [3]:

$$\text{CWSI} = \frac{\gamma\left(1 + \frac{r_c}{r_a}\right) - \gamma}{\Delta + \gamma\left(1 + \frac{r_c}{r_a}\right)} \tag{2}$$

$$\frac{r_C}{r_a} = \frac{\frac{\gamma r_a R_n}{\rho c_P} - \frac{T_C - T_A}{\Delta + \gamma} - (p_{v,sat}(T_C) - p_v)}{\gamma\left[(T_C - T_A) - \frac{r_a R_n}{\rho c_P}\right]} \quad (3)$$

where:

- γ is the psychometric constant;
- r_C, r_A are the canopy and aerodynamic resistances;
- Δ is the slope of saturated vapor pressure-temperature relation;
- $p_{v,sat}$ is the saturated vapor pressure;
- p_v is the vapor pressure;
- c_P is the heat capacity of air;
- ρ is the air density;
- R_n is the net solar radiation;
- T_C is the canopy temeprature;
- T_A is the air temeprature.

The value of CWSI ranges from 0 to 1, where 0 stands for a healthy plant with no water stress, while 1 represents a critical situation of water deprivation. CWSI values can be subdivided into four ranges of plant conditions (Table 1), exploiting the correlation between CWSI and Leaf Water Potential available in the literature [4].

Table 1. CWSI values and corresponding water stress of the plant.

CWSI	Plant Water Stress
>0.8	Critical water stress
0.5–0.8	Medium-high water stress
0.2–0.5	Low-medium water stress
<0.5	No water stress

2.1. Satellite Data

The input meteorological data were all taken from the European Database Copernicus, which provides time series with hourly precision over specific regions of the earth, apart from the canopy temperature, which was downloaded directly from Landsat Satellites through Google Earth Engine as the Land Surface Temperature (LST) parameter. Whereas Copernicus offers hourly data every day, Landsat Satellite only detects the data over a specific route once every 15 days, so we filtered the Copernicus data only for the days where we could also have the LST to ensure the best compatibility of the data coming from two different sources.

2.2. Design of the System

The system has two main parts: the sensing part collects information on the vines and the vineyard status, while the analysis part computes the Crop Water Stress Index (CWSI) and generates maps that can be used to monitor the current status of the vineyard in detail, as well as to make predictions about the future development of the grapes and the plants.

The sensing subsystem is a battery-powered module composed of specific sensors interfaced with a Raspberry Pi 4 Model B. To collect climatic parameters and elaborate the data into real CWSI heatmaps, the following sensors were adopted (Figure 1a):

1. Temperature and Humidity sensor (DHT22);
2. Surface Temperature sensor (MLX90614-GY906);
3. Pressure, Temperature and Humidity sensor (BME280);
4. Ambient Light and InfraRed intensity sensor (TSL2591);
5. Ultraviolet intensity sensor (LTR390);
6. Global Positioning System (GY-NEO6MV2 NEO-6M GPS);
7. External anemometer (BT-100-WM), not shown in the figure.

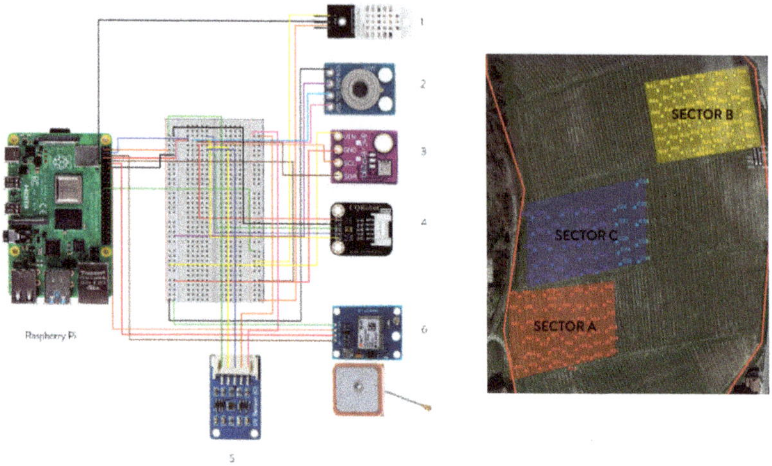

(a) (b)

Figure 1. (**a**) Simplified scheme of the raspberry box; (**b**) Bird view of the selected field of "Azienda Agricola Balladore Pallieri", A, B, C and D (small area south of Sector B) show waypoints.

2.3. Data Collection

In order to verify the validity of the proposed concept, an on-field data collection session in a vineyard owned by "Azienda Agricola Balladore Pallieri" in Calosso (Asti, Italy, see Figure 1b) was carried out on 4 June 2022. The location is characterized by a mild climate, scarcity of droughts, and strong hailstorms. The test was carried out in late spring to investigate the status of mature vines, and the good weather conditions ensured a more reliable measurement of irradiation.

Precise waypoints were chosen in the crop to create a network of measurements roughly every 10 m, to increase the resolution of the CWSI map with respect to the satellite measurements (Figure 1b). The waypoints were divided into three sections in different zones of the crop (A, B, C, D) to investigate the side-by-side areas variation of CWSI linked to changes in slope, which lead to different water drainage.

3. Results

The data collected are presented by thematic maps. From the atmospheric and climatic data collected on the field, the CWSI for each waypoint has been evaluated and associated with its geographical coordinates in GeoTIFF format. This way, the georeferenced CWSI map can be layered on the crop area through a Geographic Information Software (QGIS).

The map from satellite data is plotted to show the different values of the CWSI among the different regions in the field. Figure 2a shows that the CWSI is quite homogeneous and ranges from 0.4 to 0.9 over the whole field. The distribution of CWSI values based on the ground-collected data is more irregular and uneven. Areas A and B show values of CWSI ranging mainly from 0.5 to 1, with few squares having an index value below 0.5, while areas C and D show values ranging from 0 to 0.2.

As pointed out in Table 1, the values of CWSI can be grouped into four classes that differentiate between 0, "no water stress", up to 1, "critical water stress". The map in Figure 2b highlights that grouping. To further inspect the importance of each factor to the CWSI value, we have performed a Global Variance Based Sensitivity Analysis [5], computing the main and total effects of the contribution of each variable to the variance of the output. The results are displayed in Figure 3.

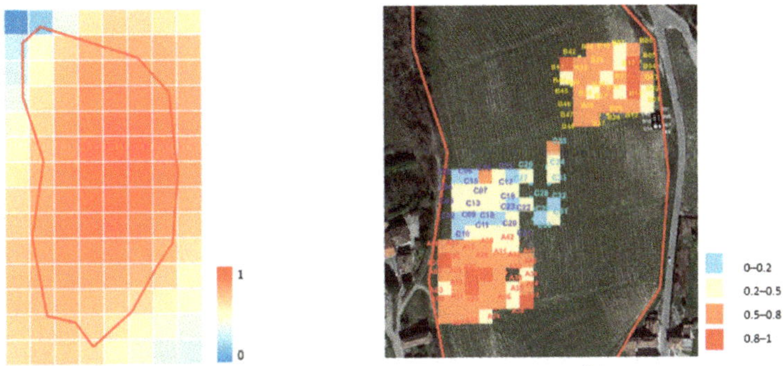

(a) (b)

Figure 2. (**a**) CWSI map evaluated through atmospheric data by Landsat 8 and Copernicus database (14 July 2018); 30 × 30 m resolution. (**b**) CWSI map based on ground-collected data (4 June 2022); 10 × 10 m resolution.

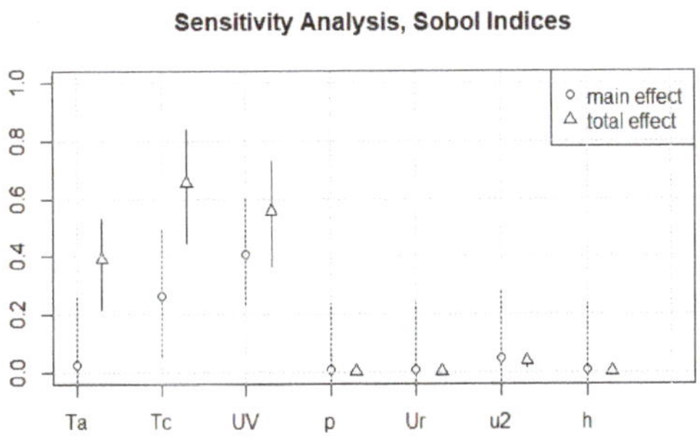

Figure 3. Main and total effects, and 0.95 confidence intervals for variables: Atmospheric Temperature (Ta), Canopy Temperature (Tc), UV Radiation (UV), Atmospheric Pressure (p), Relative Humidity (Ur), Wind Speed (u2) and Ground Height (h).

4. Discussion

The two maps shown in Figure 2 are both informative, but a comparison between the kind of information that they provide represents a valuable outcome. The map based on the satellite-collected data provides information about a larger area, and virtually it can provide information about any portion of ground on the planet. However, there are some issues related to this source. First, the satellites cannot always provide pictures of a specific portion of the ground, and therefore a field cannot be continuously monitored through satellite images and satellite-collected data. Second, the resolution of the satellite images is quite low, meaning that a value of CWSI can be computed for each portion of dimension 30 × 30 m. On the other hand, on-field sensors can hypothetically monitor the field 24/7, and the resolution for CWSI maps based on the on-field collected data is quite high, with portions of dimension 1 × 10 m. Such dimensions can be further reduced if the waypoints are placed closer to each other. Therefore, on-field data collection for CWSI maps computation represent a much more reliable source of information on the water stress and health of the plants. It is worth noting that large fields on a homogenous terrain are unlikely to present considerable differences in CWSI values between neighboring areas, but small fields and fields with non-homogenous terrain require a much finer resolution to properly consider the local variation of CWSI in the management of the resources and of the field.

The sensitivity analysis by total effects (Figure 3) points out that the most influential parameters are the air temperature (Ta), the UV radiation (UV) and the canopy temperature (Tc). It clearly shows that the accuracy of the corresponding sensors needs particular attention to improve their reliability and lower their uncertainty of CWSI measures.

5. Conclusions

Careful resource management is becoming more and more relevant due to frequent extreme meteorological events, such as droughts. Therefore, a non-invasive system of sensors that can inspect the conditions of the plants in near-real time, with high resolution and reliability, represents a valid tool to optimize resource consumption.

The extensive monitoring of the plant's health can be achieved through climatic data detection, which can, in turn, be exploited to evaluate the Crop Water Stress Index, a vegetation index related to the ability of the plant to exploit the available water. The focus on CWSI is dictated by the current desertification trend that leads to hotter and drier summers, especially in the temperate regions where viticulture is extensively diffused (e.g., Italy and France). Due to this change in climate, equipping crops with irrigation systems balancing the lack of rainfall and keeping the crop in the optimal condition of Water Stress will become fundamental, and a close monitoring of the critically stressed areas could lead to better resource management.

Through this non-invasive monitoring system, it is possible to draw maps of georeferenced CWSI on crop areas. The maps offer a resolution, precision, time and operational cost more convenient than a human operator or a satellite without altering the vineyard structure or obstructing the cultivation operations.

With an on-field data collection session, the working principle of the system has been validated through a prototype, and high-resolution CWSI maps have been obtained. Through sensitivity analysis, it has been possible to determine which climatic variables affect the CWSI the most: these variables need to be detected with the most reliable sensors and with the highest possible resolution. The sensitivity analysis has offered an insight into system improvements that would entail both fixed and moving sensors: the former for the variables that can be taken as an average over the whole crop and the latter for the variables that need a higher resolution.

Further validation is needed to better study the correlation between the values of CWSI obtained with the on-field sensors and the ones obtained through satellite data taken by the Copernicus database and Landsat Satellite. Furthermore, the moving sensors could be mounted on autonomous drones to completely automatize the data collection step and create a full system that collects and analyses the data and offers as an output an easy-to-read map of the crop.

Finally, if the system is implemented on a crop for a long enough time span, a new intriguing possibility could be the correlation of the CWSI trend with the seasonal vineyard productivity and final grapes' organoleptic quality. These future developments are beyond the scope of this paper and will be investigated in further studies.

Author Contributions: Conceptualization, all; methodology, P.B.d.P., G.B., A.B., C.G., L.M., M.M. and G.N.; software, C.G., L.M. and G.N.; validation, P.B.d.P., G.B., A.B., C.G., L.M., M.M. and G.N.; data curation, C.G. and G.N.; writing—original draft preparation, P.B.d.P., G.B., A.B., C.G., L.M., M.M. and G.N.; writing—review and editing, all; visualization, L.M.; supervision, V.C., C.C., S.I. and S.M.; project administration, L.M. and S.M.; funding acquisition, V.C., C.C., S.I. and S.M. All authors have read and agreed to the published version of the manuscript.

Funding: The financial support by Alta Scuola Politecnica to the Multidisciplinary Project "AIS4SIA: Artificial Intelligent Systems for a Smart and Innovative Agriculture" is gratefully acknowledged.

Institutional Review Board Statement: Not applicable.

Informed Consent Statement: Not applicable.

Data Availability Statement: The data presented in this study are available on request from the corresponding author.

Conflicts of Interest: The authors declare no conflict of interest.

References

1. Zhou, Z.; Majeed, Y.; Naranjo, G.D.; Gambacorta, E.M. Assessment for crop water stress with infrared thermal imagery in precision agriculture: A review and future prospects for deep learning applications. *Comput. Electron. Agric.* **2021**, *182*, 106019. [CrossRef]
2. Anda, A. Irrigation Timing in Maize by Using the Crop Water Stress Index (CWSI). *Cereal Res. Commun.* **2009**, *37*, 603–610. [CrossRef]
3. Jackson, R.D. Canopy Temperature and Crop Water Stress. In *Advances in Irrigation*; Elsevier: Amsterdam, The Netherlands, 1982; Volume 1, pp. 43–85.
4. Deloire, A.; Pellegrino, A.; Rogiers, S. A Few Words on Grapevine Leaf Water Potential. In *IVES Technical Reviews, Vine and Wine*; International Viticulture and Enology Society: Bordeaux, France, 2020.
5. Sobol, I.M. Sensitivity Analysis for Non-Linear Mathematical Models. *Math. Model. Comput. Exp.* **1993**, *1*, 407–414.

environmental sciences proceedings

MDPI

Proceeding Paper

The Hydrochemical Characteristics of Drinking Water in Central Settlements of Sukhbaatar Province, Eastern Mongolia †

Odsuren Batdelger *, Enkhjargal Togtokh, Oyun-Erdene Boldsaikhan, Renchinbud Badrakh and Gan-Erdene Enkhbold

Institute of Geography and Geo-Ecology, Mongolian Academy of Sciences, Ulaanbaatar 15170, Mongolia; enkhjargalt@mas.ac.mn (E.T.); oyunerdeneb@mas.ac.mn (O.-E.B.); renchinbudb@mas.ac.mn (R.B.); ganerdene_e@mas.ac.mn (G.-E.E.)

* Correspondence: odsurenb@mas.ac.mn

† Presented at the 7th International Electronic Conference on Water Sciences, 15–30 March 2023; Available online: https://ecws-7.sciforum.net.

Abstract: A focus of this study was to investigate water quality in the central settlement area of Sukhbaatar province, which is characterized by semi-arid climate. In total, 47 water samples were collected from 13 soums in 2021, and major ions and trace elements were analyzed to evaluate the suitability of groundwater for drinking purposes. The dominant hydro-chemical facies of groundwater were the Na-HCO_3 type, which represents 46.8% of the total analyzed samples. The water supply wells of Baruun-Urt soum, Asgat, Khalzan, and Erdenetsagaan soums do not meet the requirements of drinking water standards due to the content of magnesium, fluoride, and uranium ions. Additionally, the fluoride ion content in most wells exceeds the drinking water standard, while the fluoride content of Dariganga soum water was less than the drinking water standard; however, the Munkhkhaan soum was suitable.

Keywords: groundwater; hydrochemistry; semi-arid; Sukhbaatar province

check for updates

Citation: Batdelger, O.; Togtokh, E.; Boldsaikhan, O.-E.; Badrakh, R.; Enkhbold, G.-E. The Hydrochemical Characteristics of Drinking Water in Central Settlements of Sukhbaatar Province, Eastern Mongolia. *Environ. Sci. Proc.* **2023**, *25*, 80. https://doi.org/10.3390/ECWS-7-14304

Academic Editor: Athanasios Loukas

Published: 3 April 2023

1. Introduction

One of the main factors affecting the development of any country is clean and fresh water. Groundwater quality issues are crucial for sustainable water resources management in many countries worldwide, especially in arid and semi-arid regions. In the region, drinking water supply mainly depends on the groundwater, with no surface water resources due to natural moisture supply conditions. There is a natural pattern of poor water quality due to poor groundwater recharge in the region. Chemical elements in drinking water could possibly affect human health [1].

The main objective of this work was to study the drinking water quality of the soum centers in the Sukhbaatar province.

2. Materials and Methods

The Sukhbaatar province has a total area of 82.3 thousand km^2 and is included in the steppe zone. In terms of climate, it is dry, with cool summers and cold winters, while the average temperature in July is +21.3 °C, and the average temperature in January is −19.9 °C [2].

We collected 43 groundwater (GW) and 4 spring water samples from soum centers in June (summer) 2021 (Figure 1). All sampling sites were used for drinking water supply. Electrical conductivity (EC), Total Dissolved Solid (TDS), pH, ORP, and water temperature were measured on-site using Multiparameter Hanna HI98195. The sampling site locations were verified using a portable GPS meter (GPSMAP 76S; Garmin Ltd.). All water samples were sealed in 1000 mL (anion and cation) and 100 mL (heavy metal) polyethylene bottles for chemical analysis. We measured Ca^{2+}, Mg^{2+}, Na^+, K^+, HCO_3^-, and Cl^- via titration.

NH_4^+, NO_2^-, NO_3^-, SO_4^{2-}, F, and Fe were analyzed using a T-60 U spectrophotometer. The heavy metals were analyzed via ICP.

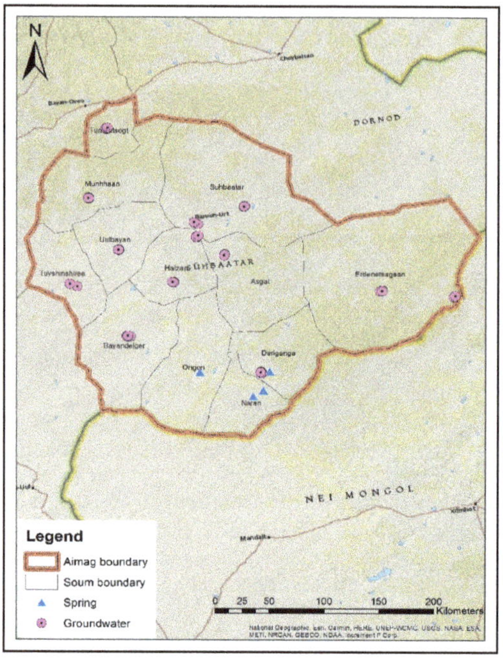

Figure 1. Study area.

As a quality check on the analysis, the charge balance error (CBE $\pm$ 10%) was calculated using the following equation, and the calculated error did not exceed $\pm$10%. The data were compared to the Mongolian standard (MNS0900:2018) [3].

$$\text{CBE} = \frac{\sum \text{cations(eq)} - |\sum \text{anions}|}{\sum \text{cations(eq)} + |\sum \text{anions}|} \times 100\% \quad (1)$$

3. Results and Discussion

Figure 2 shows the classification of the hydro-chemical characteristics of water samples based on the main ion concentrations. The dominant hydro-chemical facies of groundwater were the Na-HCO_3 type, which represents 46.8% of the total analyzed samples, while Ca-HCO_3 and HCO_3-mixed types each represent 17%, the HCO_3-Na-Mg type and HCO_3-Mg-Na type each represent 8.5%, and the mixed-Na-Mg type represents 2.1% of the total samples (Figure 2). The water supply wells of Baruun-Urt soum and Asgat, Khalzan, and Erdenetsagaan soums do not meet the requirements of drinking water standards due to their content of magnesium, fluorine, and uranium ions. Additionally, the fluoride ion content in most wells exceeds the drinking water standard, while the fluoride content of Dariganga soum water was less than the drinking water standard; however, Munkhkhaan soum was suitable. Almost 60% of the water samples exceeds the standard by magnesium.

In total, 21 or 44.7% of all samples do not meet drinking water standards due to uranium content, and 11 or 23.4% of all samples have nitrate pollution. These findings suggest that appropriate actions have to be taken to improve groundwater quality and the protection of public health in the Sukhbaatar province.

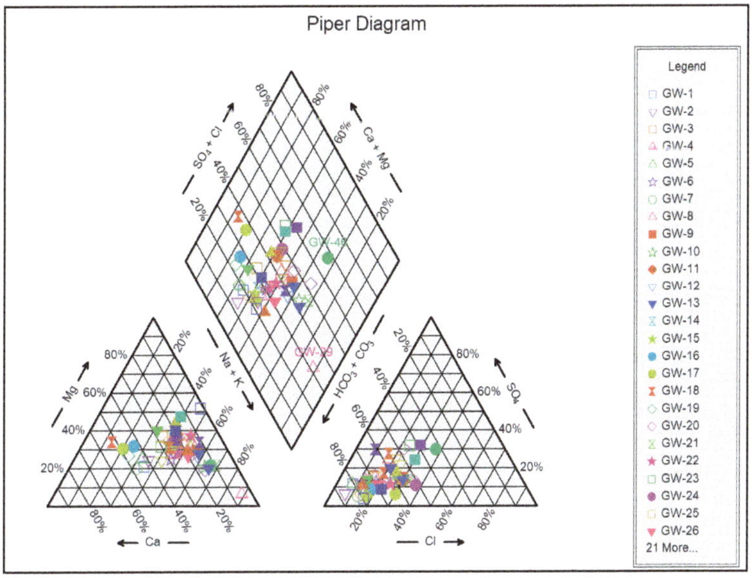

Figure 2. Piper diagram of the collected water samples in the study area.

A Gibbs diagram was used for the geochemical evolution of groundwater, as well as evaluating evaporation, weathering, and precipitation in arid and semi-arid regions [4]. We can see in Figure 3 that most of the samples are plotted in the field of the rock dominance area, indicating that the main geochemical process is rock–water interaction in the study area.

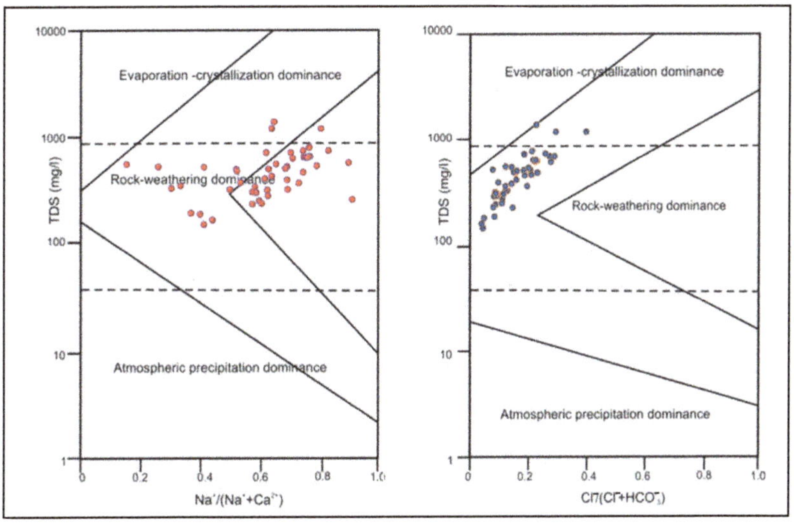

Figure 3. Gibbs diagram of water samples in the study area.

4. Conclusions

In total, 47 water samples were collected from 13 soums to study the suitability of groundwater for drinking purposes. The major conclusions are:

1. The Na-HCO_3 type represents 46.8% of the total samples;
2. In total, the drinking water in 11 soums does not meet drinking water standards regarding fluoride ion out of 13 soums, with Dariganga soum water being less than the standard, and Munkhkhaan soum being suitable;
3. Magnesium, fluoride, uranium, and nitrate were among the elements found higher than the standard, and the worst water soum was Baruun-Urt.

Author Contributions: O.B. wrote the paper and E.T. carried out data preparation. The field study was performed by all (O.B., E.T., O.-E.B., R.B. and G.-E.E.). All authors have read and agreed to the published version of the manuscript.

Funding: This research was funded by Governor of Sukhbaatar province.

Institutional Review Board Statement: Not applicable.

Informed Consent Statement: Not applicable.

Data Availability Statement: Not applicable.

Acknowledgments: The authors acknowledge the financial support received by the Governor of Sukhbaatar province.

Conflicts of Interest: The authors declare no conflict of interest.

References

1. Addisie, M.B. *Assessment of Drinking Water Quality and Determinants of Household Potable Water Consumption in Simada District, Ethiopia*; Cornell University: Ithaca, NY, USA, 2012.
2. Sukhbaatar, C.; Batdelger, B.; Togtokh, E.; Enkhbold, E.; Boldsaikhan, O.; Badrakh, R. *Drinking Water Quality Survey, Conclusions and Recommendations in the Soum Centers of Sukhbaatar Province*; Mongolian State University of Education Publishing: Ulaanbaatar, Mongolia, 2021.
3. *MNS 900:2018*; Environment. Health Protection. Safety. Drinking Water. Hygienically Requirements, Assessment of the Quality and Safety. Mongolian Standard: Ulaanbaatar, Mongolia, 2018.
4. Gibbs, R.J. Mechanisms Controlling World Water Chemistry. *Science* **1970**, *170*, 1088–1090. [CrossRef] [PubMed]

environmental sciences proceedings

MDPI

Proceeding Paper

Open SolWat System with Cooling of the Secondary Wastewater Effluent from a WWTP on the Front Surface of the Photovoltaic Module for efficient Energy Generation and Reclaimed Water Production †

Julia Torres López *, Marta Vivar García, Manuel Fuentes Conde and Ana María Palacios Villa

Grupo IDEA, EPS Linares, Universidad de Jaén, 23700 Linares, Spain; mvivar@ujaen.es (M.V.G.); mfuentes@ujaen.es (M.F.C.); apvilla@ujaen.es (A.M.P.V.)

* Correspondence: juliatoorres@gmail.com

† Presented at the 7th International Electronic Conference on Water Sciences, 15–30 March 2023; Available online: https://ecws-7.sciforum.net/.

Abstract: Energy consumption in wastewater treatment plants (WWTP) is a critical part of their operation and maintenance costs, with tertiary treatments being one of the most energy demanding stages, although as they are not required by law, they are not usually included in the wastewater treatment line. In this research, a photochemical–photovoltaic hybrid system was developed and studied: Open SolWat, which allows solar disinfection of the water while cooling the temperatures on the front surface of the photovoltaic module by means of a thin layer of water flowing from the top, thanks to a pumping system. In comparison to the SolWat technologies studied so far, the improved system allowed better quality reclaimed water (RD1620/2007, R(EU)2020/741) to be obtained from the secondary effluent of a WWTP, with the simultaneous generation of energy. However, this time it also productively improved its energy efficiency (15–21%). The tests were carried out under a 4 h SODIS treatment with real sunlight. Therefore, the possible implementation of the system as a tertiary treatment of a WWTP is considered, as it could improve environmental sustainability and reduce energy consumption.

Keywords: solar energy; photovoltaics; active water cooling; reclaimed water

Citation: Torres López, J.; Vivar García, M.; Fuentes Conde, M.; Palacios Villa, A.M. Open SolWat System with Cooling of the Secondary Wastewater Effluent from a WWTP on the Front Surface of the Photovoltaic Module for efficient Energy Generation and Reclaimed Water Production. *Environ. Sci. Proc.* **2023**, *25*, 81. https://doi.org/10.3390/ECWS-7-14321

Academic Editor: Athanasios Loukas

Published: 3 April 2023

1. Introduction

Solar energy is a renewable energy source capable of addressing environmental concerns and concerns about energy sustainability. The best known and most widely used technology in renewable energy generation is the use of photovoltaic (PV) modules that convert solar energy into electrical energy through photovoltaic cells. Admittedly, much of the solar energy absorbed by PV cells causes an increase in the temperature of the PV module, which causes a reduction in power output and energy efficiency, and negatively affects its performance and lifetime [1]. In addition, the most common climatic factors that influence the conversion efficiency of PV modules are solar radiation [2], ambient temperature and module surface temperature [1], relative humidity, wind speed and accumulated dust or shading problems on the module surface [3].

In recent years, researchers have conducted many reviews and comparative analyses on various proposed cooling techniques [4,5], of which water cooling systems stand out as a good solution to improve electrical efficiency and decrease the degradation rate of PV cells. On the other hand, Vivar et al., 2010 [6], proposed a novel technology called SolWat, which was later proposed [7,8] to improve environmental sustainability (scarcity, water stress and deterioration of water quality) and reduce the energy consumption of a wastewater treatment plant (WWTP) by implementing this technology as a tertiary treatment. This proposal would be very beneficial since energy consumption in the WWTP is a critical part of its operation and maintenance costs, especially for tertiary treatments that require high energy consumption, although as they are not required by law, they are not usually included in the wastewater treatment line. In view of these problems, possible solutions can be found in the use of renewable treatment plants [9,10] (to obtain energy) that also allow water to be reused.

The results obtained by Vivar et al., 2021 [7], and Torres et al., 2022 [8], managed to obtain reclaimed water from the secondary effluent of wastewater from a WWTP, after SODIS disinfection treatment [11] under real sunlight for 4 h, and to produce energy simultaneously. Now, this study proposes a new design in SolWat technology, the Open SolWat system. Following the objectives already proposed in previous research, this work aims to: (a) improve the final quality of reclaimed water for reuse and for final discharge to the environment; and (b) obtain energy, but this time the new prototype will significantly increase the energy efficiency in the SolWat PV module by means of a cooling system based on a thin film of water on the front surface of the module.

2. Materials and Methods

The SolWat technology is based on a solar–photovoltaic hybrid system (water disinfection reactor coupled to a photovoltaic module) that uses only solar energy to purify water and generate electricity. This technology receives solar radiation and uses its broad electromagnetic spectrum for (1) water disinfection thanks to the germicidal effect of ultraviolet (UV) radiation and the thermal effect of far infrared radiation (FIR) that drastically reduce the amount of pathogenic microorganisms present in the water, and (2) electricity production thanks to the visible (VIS) and near infrared (NIR) radiation reaching the photovoltaic module.

The new prototype is called the "Open SolWat system" (Figure 1) and has a dynamic operation mode (with water recirculation). The system consists of a PV module through which a thin water film (1 mm) circulates, allowing cooling of the PV module and solar disinfection of the water, from the top of the module until it is collected in an open water tank, which is exposed to solar radiation and therefore also acts as a solar water disinfection reactor. A reference PV module was also used experimentally as a control system.

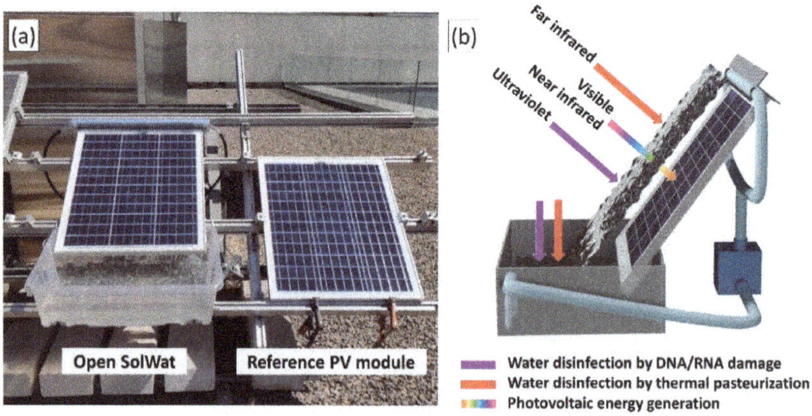

Figure 1. (**a**) Open SolWat system and reference photovoltaic module. (**b**) Use of the electromagnetic spectrum in Open SolWat.

The PV modules were polycrystalline silicon cells (PV LOGIC, Bredon, UK) consisting of 36 solar cells connected in series with dimensions of 634 × 535 × 25 mm (0.251 m^2 cell area) and a nominal power of 45 Wp. This prototype has two water reactors: (a) a reactor formed only by an L-shaped aluminium profile at the top of the photovoltaic module, where the perforated pipe with 53 water outlet microtubes (3 mm water inlet and 2 mm water outlet) with a separation of 1 cm between them is attached; and (b) the open and translucent water tank (62 × 45 × 18 cm). Water recirculation in the SolWat was performed by a pumping system, with a Xylem Flojet magnetic coupling water pump (NDP14/2, supply voltage: 230 V, input power: 10 W) through a set of pipes that propelled the water sample from the open water tank to the SolWat, with an average flow rate of 8.9 L/min. The water pump was primed with 200 mL of the experimental sample. Experimental tests with 4.2 and 6.2 L volumes were carried out with wastewater and purified and ultrapure (Milli-Q) water samples to study the influence of water flow and turbidity on PV performance. The homogenisation of the samples was achieved thanks to the drop of the water sheet into the open tank and the water pipe with 7 orifices distributed inside the tank.

Experimental Set-Up

Six experiments were carried out during the spring and summer of 2022, on the rooftop facilities of the E.P.S. de Linares (Spain), at the University of Jaén. Linares has a temperate climate. The tests were carried out outdoors, under sunny weather conditions.

The Linares WWTP provided the wastewater samples of the wastewater effluent obtained directly after secondary treatment in different seasons of the year and with varying microbiological loads. The experiments always started around 11:00–12:00 p.m. local time (2–3 h before solar noon). The total exposure of the water samples was carried out under sunlight (SODIS treatment) for 4 h, as this provided an adequate treatment that could be carried out during 3 shifts over the course of a day. Microbiological analyses of the initial raw water and after 4 h of SODIS treatment were performed, which included *E. coli*, *E. faecalis* and *C. perfringens* (including spores) as microbiological indicators, to assess the microbiological quality of the water according to Spanish Royal Decree 1620/2007 and Regulation (EU) 2020/741 for water reuse. Similarly, the basic physicochemical parameters of the water samples were analysed. The climatic conditions were monitored and the electrical parameters of this experiment were measured in the same way and under the same methodology previously explained in Vivar et al. (2021) [7] as a first stage of research to obtain reclaimed water and simultaneous energy production.

3. Results and Discussion

Table 1 shows the results obtained from the meteorological conditions in Open SolWat, together with the temperature of the water sample, the initial and final sample volumes obtained, the evaporation percentage and the leakage of the water sheet after 4 h of SODIS treatment.

Water temperature is one of the most important parameters of water quality, as it affects water chemistry and the functions of water microorganisms. The water samples used showed increases between 8.1–10.7 °C, during the experimental tests, in their temperature, influenced by: solar radiation, ambient temperature, heat transfer of the SolWat PV module and the volumes used experimentally (4.2 and 6.2 L). The water losses were mainly due to the evaporation of the water sample (with full exposure to the environment during the experimental treatment), which was related to the heat loss in the PV module cooling, and to the leakage of the water sheet in the PV module, which reached 37.1–52.4%. These water losses varied with the solar intensity, PV module temperature, water sample temperature and water volume used. On the other hand, the turbidity of the water samples (experimental studies carried out with wastewater and Milli-Q) and the thin water film did not show significant evidence with regards to the final energy production.

Under similar experimental conditions, it was observed that, when using a larger sample volume, bacterial inactivation became slightly more difficult. At no time was total bacterial inactivation (*E. coli*, *E. faecalis* or *C. perfringens*) achieved in the treated water samples, but optimal disinfection results were obtained for *E. coli* (99.79–99.98%) and *E. faecalis* (92.65–97.75%), as shown in Figure 2. *C. perfringens* was shown to be the most resistant of the bacteria tested, with low inactivation percentages.

Table 1. Volumes of the water samples, evaporation and leakage volume percentage, climatic conditions and water temperature during the experimental tests, 4 h under sun exposure, in the Open SolWat systems. Avg.: average, Max.: maximum.

Date/Parameter	Water Sample	Initial Volume (L)	Final Volume (L)	Evaporation Volume and Leaks (%)	Solar Global Irradiance (W/m^2)		UV Irradiance (W/m^2)		UV Radiation dose		T $_{Water}$ (°C)		T $_{Ambient}$ (°C)		Wind Speed (m/s)		Relative Humidity (%)	
					Max.	Avg.	Max.	Avg.	(Wh/m^2)	(kJ/m^2)	Max.	Avg.	Max.	Avg.	Max.	Avg.	Max.	Avg.
9 May 2022	Wastewater	4.2	2.4	43.8	1036	944	50.3	45.3	89.81	323.30	32.7	29.7	32.9	29.4	9.4	4.2	38.5	27.5
10 May 2022		6.2	3.8	38.7	1036	930	50.4	44.8	89.95	323.81	38.5	35.7	32.4	29.2	11.0	5.0	36.9	27.7
3 June 2022	Milli-Q	4.2	2.3	45.2	1021	939	51.3	47.0	94.41	339.87	27.1	25.0	28.5	25.8	10.9	5.7	58.3	43.7
7 June 2022		6.2	3.7	40.3	1007	930	50.0	46.1	92.67	333.61	32.7	30.3	37.3	34.3	10.3	4.6	18.7	14.6
7 July 2022	Wastewater	4.2	2.0	52.4	1106	907	50.2	44.1	88.49	318.56	36.9	33.4	40.7	36.4	8.9	4.2	37.3	26.5
8 July 2022		6.2	3.9	37.1	1002	942	48.7	45.2	90.80	326.87	35.7	33.6	40.5	37.0	9.0	4.8	32.7	19.8

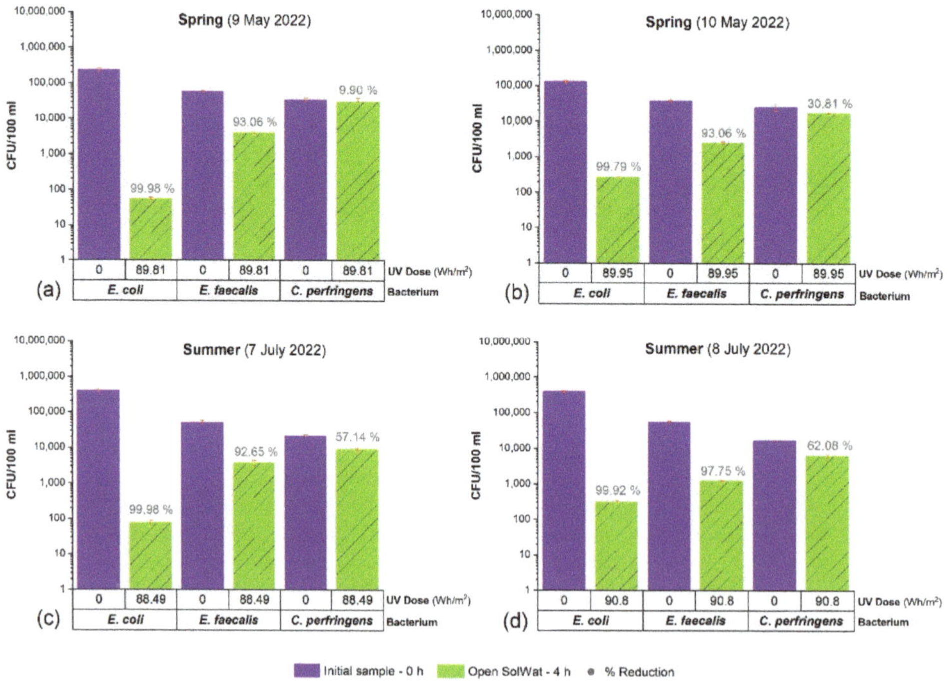

Figure 2. Concentration of *E. coli*, *E. faecalis* and *C. perfringens* in the experimental water samples vs. UV dose for the Open SolWat. Tests carried out in spring, (**a**) 9 May 2022 and (**b**) 10 May 2022, and in summer, (**c**) 7 July 2022 and (**d**) 8 July 2022.

However, due to the exposure of the water sample to the environment there was a tendency for certain physicochemical parameters to increase in concentration as a consequence of climatic conditions and water losses during the SODIS treatment. Moreover, together with the incorporation of other particles in suspension in the water sample, this caused an increase in the most important parameters, turbidity and TSS, which were detrimental to the possible uses for reclaimed water according to Spanish and European regulations. Even so, this problem could be solved with the use of filters in the system, which could also help in the microbiological disinfection of the experimental water. The quality of the reclaimed water obtained during the experimental tests was good and covered the possible uses shown in Table 2.

Table 2. Summary of the possible uses of the reclaimed water obtained experimentally thanks to the Open SolWat, according to the maximum admissible values (bacteriological and physicochemical limits (TSS, turbidity, BOD_5 and other established criteria) established in RD 1620/2007 [12] and in RD (EU) 2020/741 [13].

Use/Experiment			*E. coli* (CFU/100 mL)	TSS (mg/L)	Turbidity (NTU)	Other Criteria	Open SolWat (4 h)			
							9 May 2022	10 May 2022	7 July 2022	8 July 2022
RD 1620/2007	Urban	Quality 1.2: SERVICES (a), (b), (c), (d) Quality type: B	≤200	20	10	***	** Turbidity = 14.4 UNT		** Turbidity = 27.17 NTU	
	Agricultural	Quality 2.1: (a) Quality type: B	≤100	20	10	***	** Turbidity = 14.4 UNT, ** Cr > 0.1 mg/L	** Cr > 0.1 mg/L	** Turbidity= 27.17 NTU	✔
		Quality 2.2: (a), (b), (c) Quality type: C	≤1000	35	*	***	** Cr > 0.1 mg/L	** Cr > 0.1 mg/L	✔	✔
		Quality 2.3: (a), (b), (c) Quality type: D	≤10,000	35	*		** Cr > 0.1 mg/L	** Cr > 0.1 mg/L	✔	✔
	Industrial	Quality 3.1: (a), (b) Quality type: C	≤10,000	35	15	***	✔	✔	** Turbidity = 27.17 NTU	✔
		Quality 3.1: (c) Quality type: C	≤1000	35	*		✔	✔	✔	✔
	Recreational	Quality 4.1: (a) Quality type: B	≤200	20	10		** Turbidity = 14.4 UNT		** Turbidity = 27.17 NTU	
		Quality 4.2: (a) Quality type: D	≤10,000	35	*	*** P_{total}: 2 mg P/L (in standing water)	** P_{total} > 2 mg/L	✔	** P_{total} > 2 mg/L	** P_{total} > 2 mg/L
	Environmental	Quality 5.1: (a) Quality type: C	≤1000	35	*	N_{total}: 10 mg N/L, NO_3^-: 25 mg NO_3/L. Art. 257 to 259 of RD 849/1986.	✔	✔	✔	✔
		Quality 5.3: (a), (b) Quality type: E	*	35	*	***	✔	✔	✔	✔
		Quality 5.4: (a) Quality type: F	The minimum quality required will be studied on a case by case basis.				✔	✔	✔	✔
R(EU)2020/741	Agricultural	Quality type: B	≤100	In accordance with Directive 91/271/EEC. ≤35 mg/L	*	BOD_5 (mg/L): In accordance with Directive 91/271/EEC, ≤25 mg/L O_2	✔		✔	
		Quality type: C	≤1000		*		✔	✔	✔	
		Quality type: D	≤10,000		*		✔	✔	✔	✔

* No limit is set. ** The problem must be solved for the possible use of reclaimed water. *** OTHER POLLUTANTS (Annex II of RD 849/1986 of 11 April) contained in the wastewater discharge authorization → limit entry into the environment. In the case of dangerous substances (Annex IV of RD 907/2007, of 6 July), respect for the NCA must be ensured [12].

Table 3 shows the electrical results from Open SolWat and the reference photovoltaic module, after 4 h of solar exposure. Open SolWat allowed the front surface of the SolWat photovoltaic module to be cooled by a thin sheet of water that constantly flowed from the top of the module, thanks to a pumping system. The influence of the water cooling and the effect of the evaporation of the water sample allowed the high temperatures reached in the PV module during the spring and summer tests to be reduced, with a decrease of 16.2–30.6 °C with respect to the reference photovoltaic module, which resulted in increased electrical performance. As a consequence, energy efficiency was simply and effectively improved by 15–21%, which at the same time guarantees a longer service life. Thus, the system managed to generate an average of 18.6 ± 1.8% more energy compared to the individual photovoltaic module in the experimental studies. In addition, cooling with water allowed cleaning of the module surface.

Table 3. Main electrical results obtained from the photovoltaic module integrated in the Open SolWat system and the reference photovoltaic module, after 4 h of experimental treatment under real sun, on the roof of the E.P.S. of Linares. Avg.: average, Max.: maximum.

Experiment/Date			9 May 2022	10 May 2022	3 June 2022	7 June 2022	7 July 2022	8 July 2022
Global irradiance (W/m^2)		Max.	1036	1036	1021	1007	1106	1002
		Avg.	944	930	939	930	907	942
Ambient temperature (°C)		Max.	32.88	32.44	28.54	34.33	40.68	40.54
		Avg.	29.39	29.22	25.75	37.28	36.41	36.97
Generated energy (Wh)		Single PV module	132.99	132.82	138.62	131.14	125.42	130.55
		Open SolWat	157.75	156.60	160.17	155.76	151.25	157.12
SolWat energy losses vs. the single PV system (%)		Open SolWat Energy/Single PV Energy	−18.62	−17.90	−15.55	−18.77	−20.58	−20.35
PV module temperature (°C)	Single PV module	Max.	68.9	67.9	53.8	71.8	73.8	70.8
		Avg.	61.3	59.8	46.2	64.7	63.3	66.1
	Open SolWat	Max.	39.4	38.5	30.0	38.1	43.0	41.0
		Avg.	36.2	35.7	32.6	35.6	38.6	38.9
Temperature difference between the SolWat PV module and the Single PV module—(T_{SolWat} − T_{Single}) (°C) Average			−25.1	−24.1	−16.2	−30.6	−27.1	−28.2
I_{SC} (A)	Single PV module	Max.	2.604	2.605	2.528	2.529	2.781	2.504
		Avg.	2.350	2.325	2.315	2.327	2.246	2.358
	Open SolWat	Max.	2.650	2.649	2.586	2.570	2.884	2.586
		Avg.	2.389	2.369	2.372	2.368	2.337	2.437
Equivalent to radiation losses (%)		I_{SC} Open SolWat/I_{SC} Single, Avg	−1.65	−1.89	−2.44	−1.73	−4.07	−3.35

4. Conclusions

A hybrid system, Open SolWat, was developed that uses only solar energy to purify water and generate electricity. The system consists of a PV module through which a thin film of water (1 mm) circulates, allowing cooling of the module and solar disinfection of the water, from the top of the module until it is collected in an open water tank, which is exposed to solar radiation and thus also acts as a solar water disinfection reactor. The experiments took place in spring and summer during sunny days, on the rooftop at the E.P.S. of Linares (Jaén, España), with a duration of 4 h under real sunshine. Water samples from the secondary effluent of a WWTP were used. In comparison to the SolWat technologies studied so far, Open SolWat was able to obtain better quality reclaimed water and generate energy simultaneously, but this time it also productively improved its energy efficiency (15–21%). Finally, this technology showed a potential that should be studied in more detail to fulfil the final objective of being implemented as s tertiary treatment in a WWTP (the system would have water availability, and three batches per day of 4 h each could be carried out in sunny conditions), with the use of larger photovoltaic modules and solving the problems mentioned above. With the use of energy-efficient pumps and thanks to the electrical energy generated in the prototype, the energy obtained would be able to be used for the consumption of the pumping system and for other energy costs in a WWTP facility. Furthermore, other parameters stipulated in the Spanish regulations for the use of reclaimed water (e.g., intestinal nematodes, *Salmonella* and *Legionella*) that have not been taken into account in this study should be analysed.

Author Contributions: Conceptualization, J.T.L., M.V.G. and M.F.C.; methodology, J.T.L., M.V.G. and M.F.C.; software, J.T.L., M.V.G. and M.F.C.; validation, J.T.L., M.V.G. and M.F.C.; formal analysis, J.T.L. and A.M.P.V.; investigation, J.T.L., M.V.G., M.F.C. and A.M.P.V.; resources, J.T.L., M.V.G. and M.F.C.; data curation, J.T.L. and M.V.G.; writing—original draft preparation, J.T.L.; writing—review and editing, J.T.L. and M.V.G.; visualization, J.T.L., M.V.G. and M.F.C.; supervision, M.V.G. and M.F.C.; project administration, M.V.G.; funding acquisition, M.V.G. and J.T.L. All authors have read and agreed to the published version of the manuscript.

Funding: This research was funded by the Ministry of Science and Innovation, Program "Research Challenges 2018 R+D+i Projects" (grant number RTI2018-099308-A-I00) and "Predoctoral Training Contracts" (contract number PRE2019-089349). In addition, thanks go to M.J. Rodrigo and his organization (Aqualia Group, Aguas de Linares "Linaqua") for their collaboration with the project. M. Vivar recognizes the funding of the Ministry of Economy and Competitiveness, "Ramón y Cajal" Program (grant number RYC-2015-17306).

Institutional Review Board Statement: Not applicable.

Informed Consent Statement: Not applicable.

Data Availability Statement: Not applicable.

Conflicts of Interest: The authors declare no conflict of interest.

References

1. Dubey, S.; Sarvaiya, J.N.; Seshadri, B. Temperature Dependent Photovoltaic (PV) Efficiency and Its Effect on PV Production in the World – A Review. *Energy Procedia* **2013**, *33*, 311–321. [CrossRef]
2. Yildirim, M.A.; Noeak-Oclón, M. Modified Maximum Power Point Tracking Algorithm under Time-Varying Solar Irradiation. *Energies* **2020**, *13*, 6722. [CrossRef]
3. Mekhilef, S.; Saidur, R.; Kamalisarvestani, M. Effect of dust, humidity and air velocity on efficiency of photovoltaic cells. *Renew. Sustain. Energy Rev.* **2012**, *16*, 2920–2925. [CrossRef]
4. Sato, D.; Yamada, N. Review of photovoltaic module cooling methods and performance evaluation of the radiative cooling method. *Renew. Sustain. Energy Rev.* **2019**, *104*, 151–166. [CrossRef]
5. Dwivedi, P.; Sudhakar, K.; Soni, A.; Solomin, E. Case Studies in Thermal Engineering Advanced cooling techniques of PV modules: A state of art. *Case Stud. Therm. Eng.* **2020**, *21*, 100674. [CrossRef]
6. Vivar, M.; Skryabin, I.; Everett, V.; Blakers, A. A concept for a hybrid solar water purification and photovoltaic system. *Sol. Energy Mater. Sol. Cells* **2010**, *94*, 1772–1782. [CrossRef]
7. Vivar, M.; Fuentes, M.; Torres, J.; Rodrigo, M.J. Solar disinfection as a direct tertiary treatment of a wastewater plant using a photochemical-photovoltaic hybrid system. *J. Water Process Eng.* **2021**, *42*, 102196. [CrossRef]
8. Torres, J.; Vivar, M.; Fuentes, M.; Palacios, A.M.; Rodrigo, M.J. Performance of the SolWat system operating in static mode vs. dynamic for wastewater treatment: Power generation and obtaining reclaimed water. *J. Environ. Manage.* **2022**, *324*, 116373. [CrossRef] [PubMed]
9. Bundschuh, J.; Hoinkis, J. *Renewable Energy Applications for Freshwater Production*; CRC Press and IWA Publishing: London, UK, 2012; Volume 2.
10. Mahian, O.; Wei, J.; Taylor, R.A.; Wongwises, S. *Solar-Driven Water Treatment. Re-Engineering and Accelerating Nature'S Water Cycle*, 1st ed.; Elsevier: Amsterdam, The Netherlands, 2022.
11. McGuigan, K.G.; Conroy, R.M.; Mosler, H.J.; du Preez, M.; Ubomba-Jaswa, E.; Fernandez-Ibañez, P. Solar water disinfection (SODIS): A review from bench-top to roof-top. *J. Hazard. Mater.* **2012**, *235–236*, 29–46. [CrossRef] [PubMed]

12. Guía para la Aplicación del, R.D. 1620/2007 por el que se establece el Régimen Jurídico de la Reutilización de las Aguas Depuradas. C.D.E. Publicaciones. 2007. Available online: https://www.miteco.gob.es/es/agua/publicaciones/GUIA%20RD%201620_2007__tcm30-213764.pdf (accessed on 10 March 2023).
13. The European Parliament and the Council. «Regulation (EU) 2020/741, Minimum requirements for water reuse». *Off. J. Eur. Union* **2020**, *177/33*, 32–55. Available online: https://eur-lex.europa.eu/eli/reg/2020/741/oj (accessed on 10 March 2023).

environmental sciences proceedings

MDPI

Proceeding Paper

Challenges of Estimation Precision of Irrigation Water Management Parameters Based on Data from Reference Agrometeorological Stations †

Chris Koliopanos [1,*], Ioannis Tsirogiannis [1] and Nikolaos Malamos [2]

1 Department of Agriculture, University of Ioannina, Kostakii Campus, 47100 Arta, Greece; itsirog@uoi.gr
2 Department of Agriculture, University of Patras, Nea Ktiria Campus, 30200 Messolonghi, Greece; nmalamos@upatras.gr
* Correspondence: ch.koliopanos@uoi.gr

† Presented at the 7th International Electronic Conference on Water Sciences, 15–30 March 2023; Available online: https://ecws-7.sciforum.net/.

Abstract: In this study, operational decision support systems (DSSs) for irrigation water management that utilize data from weather stations (W/S) or weather data services are presented. The challenges and the ways in which various systems address them are summarized based on a review of the relevant scientific literature and information provided on the websites of the systems under consideration. The selected systems that are presented are categorized into those that utilize W/S data (IRMA_SYS, CIMIS, BlueLeaf, CoAgMet) as well as those that employ remote sensing data (Manna irrigation, Irrisat, Sencrop). Remote sensing DSSs are included in this study because their functionality is closely related to that of W/S-based systems, as it is explained in the study. Additionally, Foreca and OpenET are also examined as they provide data to DSSs for irrigation management. The discussion about the challenges encountered in the use of DSSs based on W/S data aims to stimulate further research and development in this field by the scientific community and system developers.

Keywords: evapotranspiration; open-access; agrometeorological stations; remote sensing; weather stations

check for updates

Citation: Koliopanos, C.; Tsirogiannis, I.; Malamos, N. Challenges of Estimation Precision of Irrigation Water Management Parameters Based on Data from Reference Agrometeorological Stations. *Environ. Sci. Proc.* **2023**, *25*, 82. https://doi.org/10.3390/ECWS-7-14319

Academic Editor: Athanasios Loukas

Published: 3 April 2023

1. Introduction

A Decision Support System (DSS) for irrigation water management is an essential tool for ensuring optimal water usage and crop growth. Many DSSs for irrigation management have adopted the approach of using data from evapotranspiration, precipitation, and irrigation in order to calculate water balance. The rate of evapotranspiration is influenced by several factors such as temperature, humidity, wind speed, solar radiation, soil water availability, and crop type.

The Penman–Monteith [1,2] method is a widely accepted standard for calculating evapotranspiration (ET) from meteorological data. Developed by Allen et al. [3], it is based on the energy balance at the land surface, which includes both the energy used for evaporation and transpiration. The method uses measurements of temperature, wind speed, solar radiation, and atmospheric pressure to calculate ET.

The FAO (Food and Agriculture Organization) Penman–Monteith method [3], also known as FAO-56, is a modified version of the Penman–Monteith method developed by the FAO to calculate crop water requirements. The FAO-56 method is based on the original Penman–Monteith method, but it includes some additional modifications and simplifications that make it more suitable for use in practical applications such as irrigation management. The FAO-56 Penman–Monteith method requires several meteorological measurements to calculate reference crop evapotranspiration (ET_0): Net radiation (R_n) at the crop surface, Soil heat flux (G), Air temperature (T), Actual vapor pressure (v_p), Saturation vapor pressure (e_s), Actual vapor pressure (e_a), Wind speed (u_2) at 2 m above the crop surface, Air pressure (P), Slope of the saturation vapor pressure–temperature curve (Δ), and Psychrometric constant (γ). There are also many other methods to estimate ET_0 using less data that are referred below [4,5].

2. M/S Network Topology and ET_0 Calculation

The topology of a meteorological network for calculating reference crop evapotranspiration (ET_0) refers to the spatial arrangement of weather stations within the network and the method in which data from these stations are used to estimate ET_0. There are several different approaches to design a meteorological network topology, depending on the specific goals and objectives of the network, such as power and communication coverage. Some common approaches include [6]:

Density: Placing weather stations densely across the area of interest to achieve a high spatial resolution of weather data. This approach is useful for studying small-scale variations in ET_0 and microclimates.

Stratified: Dividing the area of interest into different regions or grid cells based on factors such as vegetation type, land use, or topography and placing weather stations in each stratum to represent conditions in that region. This is useful for studying large-scale variations in ET_0 and the effects of different land uses on ET_0.

Random: Placing weather stations randomly across the area of interest to achieve a representative sample of weather conditions. This is useful for studying the average ET_0 for an area.

Hybrid: Combining elements of the above topologies by using a combination of density, stratified, and random arrangements of weather stations.

Examples of use of meteorological networks and their topologies include Blueleaf [7] and Sencrop [8] using a density approach; CIMIS [9], IRMA_SYS [10] and CoAgMet [11] using a hybrid approach. Remote sensing systems such as Manna Irrigation [12] and Irrisat [13] use meteorological weather forecast models to estimate ET_0 and cannot be categorized into a specific type.

The Penman–Monteith FAO-56 method for calculating reference crop evapotranspiration (ET_0) relies on the availability of agrometeorological weather stations that measure various natural parameters. In situations where such stations are not available, ET_0 can be calculated using other models that make simplifying assumptions, such as Hargreaves, Hargreaves–Samani, Kimberly Penman, Makkink, Thornthwaite, Jensen–Haise, Blaney–Criddle, Priestley–Taylor, and Simplified Surface Energy Balance [14]. These models calculate ETo using more easily available quantities such as temperature, radiation balance, and remote sensing data such as NDVI (Normalized Difference Vegetation Index) and LST (Land Surface Temperature). Systems such as Blueleaf, Sencrop, IRMA_SYS, and CIMIS use the FAO-56 method for calculating ET_0, while the CoAgMet system uses the Kimberly Penman method.

W/S-based systems calculate ET_0 using their own data in various time steps and do not forecast ET_0. CoAgMet calculates ET_0 for any point using the data from the nearest W/S. IRMA_SYS and CIMIS use interpolation strategies to estimate ET_0 by taking into account the measurements of various weather stations. Remote sensing systems such as Manna Irrigation and Irrisat use meteorological data to form a virtual W/S at the unknown area of interest. Irrisat uses meteorological data produced by the Cosmo LEPs weather model of ECMWF [15], while Manna Irrigation takes the weather data directly from FORECA [16].

It is important to note that the provided weather information nowadays is a combination of multiband satellite images, ground-based weather stations, and mathematical forecast models. Although it may seem that W/S have been replaced by satellite weather data, ECMWF states that all meteorological models take into account the ground W/S of Europe to evaluate satellite measurements and initialize forecast models. In addition, FORECA [17] states the usage of national weather institutes' observations for their forecasts. Manna Irrigation and Irrisat suggest the implementation of an in situ weather station to improve their services. FORECA also uses any available W/S system to adjust errors in predicted meteorological values using simple or complex mathematical models. The accuracy of virtual weather station data is lower compared to in situ W/S, but Irrisat and Manna Irrigation take advantage of weather forecasts to project future ET_0 values in time. OPEN_ET [18] describes the challenges of calculating ET using satellite images and meteorological models and uses about 800 weather stations from grid-MET, Spatial CIMIS, DAYMET, PRISM, and NLDAS, and six different models to estimate ET. They also describe known issues such as reflection problems from large water masses, shadowed areas, cloud issues, model limitations, and resolution issues.

3. Water Balance—Challenges of Precipitation and Irrigation Water Calculations

Calculating ET_0 and the amount of water loss from crops is only half of the task in setting up an irrigation decision support system (DSS). The other half is calculating the amount of water available for crops, which mainly comes from both precipitation and irrigation water.

All W/S-based systems monitor rainfall straightforwardly by using rain gauges. However, measuring rainfall without a rain gauge near a field is challenging because rain is a highly localized phenomenon and cannot be simply calculated using interpolation methods. Systems such as Blueleaf, Sencrop, CIMIS, CoAgMet, and IRMA_SYS can be accurate, depending on the density of the W/S and the spatial distribution of rain. All weather-based systems retrieve irrigation water measurements, mostly from farmers by hand or automatically, to calculate the water balance and produce irrigation recommendations.

Remote sensing DSSs do not have accurate measurements from rain gauges, but meteorological data can provide rainfall timeseries to them. Both Irrisat and Manna Irrigation take into account precipitation forecasts and suggest the installation of weather stations with rainfall meters as an option. Irrigation water measurements can also be registered manually from farmers, but Irrisat and Manna Irrigation do not require rainfall and irrigation water information to produce recommendations.

Irrisat does not provide information about the implemented water balance estimation method. Manna Irrigation [19] uses a Kc-t (Crop coefficient versus time) plot to determine crop milestones and estimate the Kc progress for the current season. After that, using NDVI satellite measurements, the actual Kc is calculated and compared to the estimated Kc. It also refers to a predicted Kc and an AI method using meteorological data. Applying this methodology, Manna Irrigation calculates the water balance and produces irrigation recommendations.

None of the weather-based systems use remote sensing data or any type of weather forecast in the same way that remote sensing DSS does. It would be interesting to study and adapt remote sensing methods such as Manna Irrigation and Irrisat to improve their efficiency, as NDVI data are often unreliable due to cloud coverage and other parameters, as described analytically by the OPEN_ET system.

4. Resolution

The resolution of all systems refers to the minimum spatial area to which their conclusions can be applied. It is usually expressed in square meters or in hectares. There is no specific way for every system to certify their resolution. The results of a specific interpolation method can produce a different accuracy for different datasets coming from the same W/S network. In W/S-based systems, the resolution is dependent on the interpolation resolution. In remote sensing systems, the resolution is limited by the resolution of satellite images and the resolution of weather forecast models.

The spatial resolution of all systems can vary. For W/S-based systems, IRMA_SYS reports 200 m while CIMIS reports 2 km. For remote sensing systems, the Manna Irrigation resolution is 5 km, Irrisat reports ET_0 with a spatial resolution of 375 m, and that of OPENET is 33 m. Comparing the resolution of all systems, there is no sure conclusion about which type of system is more accurate or not.

5. Results and Conclusions

The study of operational Decision Support Systems (DSSs) for irrigation water management based on weather stations has yielded some useful results. DSSs can be classified into those that use in situ data and those that use remote sensing information. Despite this, weather stations are still necessary for remote sensing systems, as remote sensing data rely on in situ measurements for evaluation and there is often an option/suggestion for the installation of a weather station in the field of interest.

All systems use the evapotranspiration method (ET_0) to calculate crop water needs. To calculate the water balance, all systems require or have the option of irrigation water volumes to be added from farmers. Remote sensing systems face the challenge to accurately estimate precipitation. Manna Irrigation is the only remote sensing system that describes the actual method used to produce irrigation suggestions. It also has a unique approach to estimate water balance using Kc values and the correlation between Kc and the NDVI index.

Remote sensing systems introduced the concept of virtual weather stations by using meteorological data to measure ET_0, incorporating forecasting methods and AI algorithms to produce results. IRMA_SYS has also adopted the method of a virtual weather station based on the interpolation of measurements that can be placed virtually on a field.

The spatial resolution of the systems varies. Theoretically, W/S-based systems are expected to be more accurate due to the use of actual measurements compared to the combination of satellite and meteorological data. A simultaneous comparison between systems on the same field, using several benchmarks against actual data, could provide a means for evaluating their performance.

Author Contributions: C.K., I.T. and N.M. conceptualized the research, C.K. wrote the manuscript, I.T. and N.M. reviewed the manuscript. All authors have read and agreed to the published version of the manuscript.

Funding: This research received no external funding.

Institutional Review Board Statement: Not applicable.

Informed Consent Statement: Not applicable.

Data Availability Statement: Not applicable.

Conflicts of Interest: The authors declare no conflict of interest.

References

1. Allen, R.G.; Walter, I.A.; Elliot, R.L.; Howell, T.A.; Su, D.I.; Jensen, M.E.; Snyder, R.L. *The ASCE Standardized Reference Evapotranspiration Equation*; American Society of Civil Engineers: Reston, VA, USA, 2005.
2. Monteith, J.L. Evaporation and Environment. In *Symposia of the Society for Experimental Biology*; Cambridge University Press: Cambridge, UK, 1965; Volume 19, pp. 205–234.
3. Allen, R.G.; Pereira, L.S.; Raes, D.; Smith, M. *Crop Evapotranspiration-Guidelines for Computing Crop Water Requirements*; FAO Irrigation and Drainage Paper 56; FAO: Rome, Italy, 1998; Volume 300, p. D05109.
4. Ghiat, I.; Mackey, H.R.; Al-Ansari, T. A Review of Evapotranspiration Measurement Models, Techniques and Methods for Open and Closed Agricultural Field Applications. *Water* **2021**, *13*, 2523. [CrossRef]
5. Wanniarachchi, S.; Sarukkalige, R. A Review on Evapotranspiration Estimation in Agricultural Water Management: Past, Present, and Future. *Hydrology* **2022**, *9*, 123. [CrossRef]
6. World Meteorological Organization. Guide to Instruments and Methods of Observation (WMO-No. 8). Available online: https://community.wmo.int/activity-areas/imop/wmo-no_8 (accessed on 20 January 2023).
7. Blueleaf. Available online: https://www.blueleaf.com/ (accessed on 25 December 2022).
8. Sencrop. Available online: https://sencrop.com/eu/ (accessed on 20 January 2023).
9. CIMIS. Available online: https://cimis.water.ca.gov/ (accessed on 17 January 2023).
10. IRMA_Sys. Available online: https://irmasys.com (accessed on 20 January 2023).
11. CoAgMET. Available online: https://coagmet.colostate.edu/ (accessed on 15 January 2023).
12. Manna Irrigation. Available online: https://manna-irrigation.com/ (accessed on 20 January 2023).
13. IrriSat. Available online: https://www.irrisat.com/en/home-2 (accessed on 20 January 2023).
14. Hunduma, S.; Kebede, G. Assessment of Different Models to Estimate Reference Evapotranspiration/ETo: A Review. *World J. Agric. Sci.* **2020**, *16*, 448–462.

15. ECMWF. Available online: https://www.ecmwf.int/en/forecasts/datasets (accessed on 20 January 2023).
16. FORECA. Available online: https://corporate.foreca.com/en/ (accessed on 19 January 2023).
17. FORECA Forecasts. Available online: https://corporate.foreca.com/en/weather-forecasting (accessed on 20 January 2023).
18. OPENET. Available online: https://openetdata.org/known-issues/ (accessed on 20 January 2023).
19. Manna Irrigation. Available online: https://manna-irrigation.com/webinars/estimating-kc-on-a-plot-level-using-remote-sensing-and-artificial-intelligence/ (accessed on 18 January 2023).

environmental sciences proceedings

MDPI

Proceeding Paper

Application of EPANET 2.2 Software and Jal-Tantra Web System for Optimal Hydraulic Design of Water Distribution System for University of Kashmir †

Mominah Ajaz [1,2,*] and Danish Ahmad [3]

1 Demonstrator Civil Engineering, Kashmir Government Polytechnic, Srinagar 190008, India
2 M-TECH (Water Resources Engineering), National Institute of Technology, Srinagar 190006, India
3 Department of Civil Engineering, National Institute of Technology, Srinagar 190006, India; profdanish27@gmail.com
* Correspondence: mominah@kgpsrinagar.edu.in

† Presented at the 7th International Electronic Conference on Water Sciences, 15–30 March 2023; Available online: https://ecws-7.sciforum.net/.

Abstract: The availability of portable water in sufficient quantity and standard quality is one of the basic requirements of a civilization. The study area of this present work is the University of Kashmir, where the existing WDS has served its function for more than 60 years. The UOK is a growing campus, and the existing WDS is unable to meet the demands of a growing population and suffers from excessive leakage losses and an insufficient pressure head at the nodes. In this work, we attempt to replace the existing WDS with an optimally designed WDS for the University of Kashmir so that the objective functions of maximum efficiency of performance and minimum capital cost in terms of pipe diameters are met. From the literature available, EPANET has been found to perform the hydraulic and water quality modelling with a fair degree of accuracy, ease of use, and no cost. Therefore, for our hydraulic analysis, EPANET 2.2 (Environmental Protection Agency, Washington, DC, USA) was used. To augment the assessment of the efficiency of the performance of the WDS, the TPIs (technical performance indices) for pressure and velocity at the hour of peak demand were evaluated. Finally, the cost-optimality of the network in terms of the pipe diameters was validated by using the Jal-Tantra web system developed by IIT Bombay.

Keywords: optimal hydraulic design; (WDS) water distribution system; EPANET 2.0; Jal-Tantra web system; performance analysis; (TPI) technical performance indices

Citation: Ajaz, M.; Ahmad, D. Application of EPANET 2.2 Software and Jal-Tantra Web System for Optimal Hydraulic Design of Water Distribution System for University of Kashmir. *Environ. Sci. Proc.* **2023**, *25*, 83. https://doi.org/10.3390/ECWS-7-14234

Academic Editor: Lampros Vasiliades

Published: 16 March 2023

1. Introduction

A well-designed, efficiently-performing WDS is a prerequisite for both urban and rural civilizations. Various software packages are available for modelling a WDS. The public domain software that is available free of charge include EPANET, Pipes and Loops, Branch, etc., and the paid ones available includes WaterCAD, Watergems, Synergy Water, etc. [1]. The paid versions are found to be more efficient in reference to functionality and compatibility with different computational systems, a graphical user interface, and searching and optimizing algorithms [2]. However, EPANET is freely available software that offers a fair degree of accuracy and ease of use when cost is a constraint [3]. EPANET 2.0 was used by various researchers to design the WDS for different areas [4,5]. The design includes the layout of the WDS, modelling the pressure head at nodes, and the velocity of flow in the links. However, no consideration was given to maximisation of the efficiency of performance or minimisation of the capital cost of the network.

The design of a WDS by EPANET must be such that it serves as an "optimal" solution, which can be obtained by setting out objective functions of maximum efficiency of performance and minimum capital cost of the network [6,7]. The decision variables and the constraints for the optimality of the network design have to be decided after conducting a study of the factors affecting the optimality of the design and consulting various codes and manuals laying down the standard range of values for those parameters [8,9]. The decision variables for optimal hydraulic design include the pressure head at nodes, the velocity of flow in the links, and the staging height of the storage tank. The check for cost optimality of the network formulated by EPANET can be conducted using the Jal-Tantra web system [10,11]. The optimality of the network can be further checked by carrying out a performance analysis using various indices [12–14].

The objective of this current study is to provide an optimally designed WDS for the University of Kashmir using EPANET 2.0 software and the Jal-Tantra web system. The existing distribution system has already served for more than 60 years and is unable to meet the requirements of the growing population of the campus. Moreover, the worn-out links and other components of the network lead to increased pipe and joint losses, reducing the efficiency of the existing distribution system. Therefore, there is an immediate need to redesign the water distribution network such that the design is an optimal solution meeting the objective functions of maximum efficiency of performance and minimum capital cost of the network, rather than just a simulation of the pressure head and velocity of flow.

2. Methodology

2.1. Study Area Description

The study area for our current work is at the University of Kashmir, located in Hazratbal, Srinagar, Jammu and Kashmir, India. It is a residential campus comprising academic blocks, administrative blocks, hostels, a medical unit, a cafeteria, and staff quarters. It has a residential and floating population of 5324 and 1050, respectively, for the Naseem Bagh Campus and 8066 and 1720, respectively, for the Hazratbal Campus.

2.2. Methodological Approach

A quantitative research approach was employed to arrive at the optimal design of the WDS for the University of Kashmir by using various software. A literature review related to the work was carried out, and the related standard books and codes were consulted. A preliminary survey of the study area was followed by a division of the study area into two parts, with separate WDNs for each of them (the Hazratbal campus and the Naseem Bagh campus). The water source for the networks is the groundwater reserve within the campus, which is pumped by intermittently operating pumps into the overhead water tanks.

2.3. Methods of Data Collection

The quantitative data of the existing population at various academic blocks, hostels, family quarters, and administrative blocks was obtained. The maximum depth of water in the bore wells below ground level is 10 m. In addition, the campus plan of the study area was obtained. The data availability was facilitated by the engineering wing of the UOK. Elevation data was obtained using Google Earth Pro.

2.4. Methods of Analysis

2.4.1. Evaluation of the EPANET 2.2 Input Parameters

A dead-end type of distribution system was provided for both parts of the campus. The population forecast for the design period of 30 years, with a constant increase of 5% per decade in both residential as well as the floating population was carried out [15,16]. The values of ADD for both parts of the network were evaluated using water demand as 135 lpcd for the residential population and 45 lpcd for the floating population [17]. The MDD was evaluated using a peak factor of '1.8' [18], and a peak factor of '3' was used to calculate the MHD from the AHD [18]. Provisions to meet the fire demand were kept as per Indian standards [19]. The base demand multiplier equals the coincident draft in litres per person per minute (Table 1). Two different demand patterns were used for the network nodes: one for the hostels and quarters (p1), and the other for academic buildings and administrative blocks (p2). The capacity of the storage tank was evaluated by the mass curve analysis for both parts of the network. The discharge to be provided by the pump was worked out by considering 12 h of pumping per day. The staging height of the tank above ground level was selected after iterations to successfully run the software and achieve standard pressure heads at the nodes. The economical diameter of the rising main was obtained using Dupit's expression. The pressure head developed by the pump has to be equal to the sum of the delivery head, suction head, and frictional head loss in the rising main. The input power required to run the pump equals ($\omega QH/n$), and the efficiency of the pump (n) was taken as 75%. Energy required for a 12-hour per-day operation of the pump was evaluated to arrive at the cost of the electrical energy at Rs 5 per unit per day. The design flow rate of the pump was equal to the MHD in litres per minute (Table 2). After entering the EPANET 2.2 input data [20] and successfully running the software, a hydraulic modeling of the networks was carried out. Simple controls (obtained after iterations to successfully run the software and achieve standard pressure heads at the nodes), as given in Table 3, were used to perform the hydraulic analysis.

Table 1. Demand calculations for WDS Parts I and II.

S NO	PARAMETER	WDS PART I	WDS PART II
1	Forecasted Residential Population	5324	8066
2	Forecasted Floating Population	1030	1720
3	Avg. Daily Demand (ADD)	891,000 L	1,350,270 L
4	Max Daily Demand (MDD)	1,603,800 L	2.43×10^6 L
5	Avg. Hourly Demand (AHD)	37,125 L	56,262 L
6	Max Hourly Demand (MHD)	111,375 L	168,786 L
7	Fire Demand	216,000 L	216,000 L
8	Coincident Draft	1,107,000 L	1,566,270 L
9	Base Demand Multiplier	0.104	0.1

Table 2. Storage tank and pump characteristics.

S NO	PARAMETER	WDS PART I	WDS PART II
1	Total storage capacity of the OHT	6.17×10^5 L/day	8.24×10^5 L/day
2	Volume of the OHT	617 m^3	824 m^3
3	Height of the OHT	10 m	12 m
4	Diameter of the OHT	9 m	9.5 m
5	Discharge to be provided by the pump	0.037 cumec	0.056 cumec
6	Staging Height of the tank	30 m	30 m
7	Economical diameter of the rising main	250 mm	300 mm
8	Pressure head to be developed by the pump	51 m	53 m
9	Input power for the pump	24.7 kW	37.4 kW
10	Cost of electrical energy/day at Rs 5 per unit	1482 rupees	2245 rupees
11	Design flow rate of the pump	1856 lpm	2813 lpm

Table 3. Simple control for WDS Parts (I, II).

WDS Part I
link Pu1 closed if node T1 above 10
link Pu1 open if node T1 below 4
WDS Part II
link Pu1 closed if node T1 above 12
link Pu1 open if node T1 below 6

2.4.2. Evaluation of the Technical Performance Indices (TPIs)

The behavior of the nodes and pipes of the WDS can be evaluated using the TPIs (technical performance indices), which are indicative of the performance of the network with respect to the hydraulic parameters such as pressure head and flow. Two types of TPIs are used: $TPI_{pressure}$ [14] and $TPI_{velocity}$ [14]. The TPI values range from 0 to 1, '0' for poor service, and '1' for efficient service. The TPI values for both parts of the network were evaluated and compared with the standard values as follows:

$$TPI_{pressure} = \Sigma (Q_i \times TPI_i)/\Sigma Q_i,$$

where Q_i = nodal demand and (i) iterates over all the nodes of the network.

$$\begin{aligned} TPI_i &= 0, && P_i < P_{min} \\ &= 1, && P_{min} < P_i < P_{ma} \\ &= 1 - \{(P_i - P_{max})/(P_{max} - P_{min})\}, && P_{max} < P_i < 100 \\ &= 0, && P_i > 100 \end{aligned}$$

Here, P_i = prevailing pressure head at the node, $P_{min =}$ 17 m, P_{max} = 70 m, and TPI_i = TPI. For the node i:

$$TPI_{velocity} = \Sigma(Q_j \times TPI_j)/\Sigma Q_j,$$

where Q_j = flow in pipe (j), and (j) iterates over all the pipes of the network.

$$\begin{aligned} TPI_j &= 0, && V_i < V_{min} \\ &= (V_i - V_{min})/(V_{mean} - V_{min}), && V_{min} < V_i < V_{mean} \\ &= (V_i - V_{max})/(V_{mean} - V_{max}), && V_{mean} < V_i < V_{max} \\ &= 0, && V_i > V_{max} \end{aligned}$$

Here, V_i = prevailing flow velocity in pipe j, $V_{min =}$ 0.2 m/s, V_{max} = 3 m/s, and TPI_j = TPI for the pipe j, $V_{mean} = (V_{min} + V_{max})/2$.

2.4.3. Input Parameters for the Jal-Tantra Web System

The input data for nodes (elevations, base demand, and minimum pressures) and pipes (start nodes, end nodes, length, and roughness) is entered into the Jal-Tantra system. Commercially available pipe diameters and their cost per unit length according to the prevailing market rates are also provided. The input data is given in Tables 4 and 5.

Table 4. General input data.

Network Name	WDS Part I	WDS Part II
Organization Name	UOK	UOK
Minimum Node Pressure	17	17
Default Pipe Roughness 'C'	100	100
Minimum Head loss per KM	0.010	0.01
Maximum Head loss per KM	65.000	65.000
Maximum Water Speed	3.000	3.000
Maximum Pipe Pressure	70.000	70.000
Number of Supply Hours	24	24
Source Node ID	31	55
Source Node Name	Tank (T1)	Tank (T1)
Source Elevation	1624.00	1623.00
Source Head	1628.00	1629.00

Table 5. Commercial pipe data.

Diameter (mm)	Cost (Rs)
15	123
20	155
25	225
32	281
40	320
50	439
65	553
80	712
100	1031
125	1392
150	1626
200	2760

2.4.4. Optimization of the Hydraulic Model

Objective functions:

The first objective function is the maximization of the performance efficiency of the proposed WDS in terms of the hydraulics. The second objective function is the minimization of the capital cost of the network in terms of the pipe diameters and valves.

Decision variables:

The decision variables are as follows:

- P_i = prevailing pressure head at the node.
- V_i = prevailing flow velocity in the pipe.
- H = staging height of the storage reservoir.

Constrains:

- $17 \leq P_i \leq 70$ (m) [18].
- $0.25 \leq V_i \leq 3.0$ (m/s) [18].
- H = 30 m (obtained after iterations to achieve standard values of hydraulic parameters)

3. Results and Discussions

3.1. Distribution Network Layout

A dead-end type of distribution network was drawn for both parts of the study area using EPANET 2.2, as shown in Figure 1a,b. The input data for nodes—elevation; base demand; and demand pattern—was provided for each node, and the input data for pipes—pipe diameter; roughness coefficient; and length—was provided. The pipe network was arranged for both WDS, keeping in mind the geometry of the study area; the dead-end type of distribution system was used for both parts of the study area. Moreover, the basic assumption of the Jal-Tantra web system of having no loops in the network has to be satisfied.

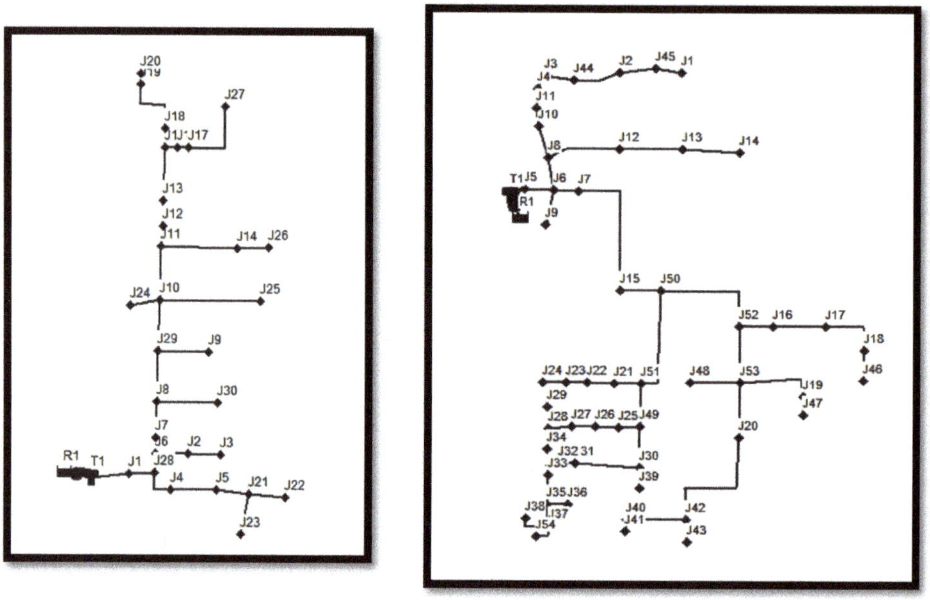

(a) (b)

Figure 1. (**a**,**b**): Distribution Network Layout for WDS parts I and II.

3.2. Contour Plan of Pressure Head at Nodes at 9 a.m. (Peak Demand Hour)

The contour plans given in Figure 2a,b, show the distribution of the pressure head across the layout of the network, parts I and II. Different colours representing the range of the pressure head are depicted by the scale provided in each map. The analysis indicated that the pressure head at all the nodes is greater than 17 m, thus the water at each node can reach up to a minimum of three stories of the building [18]. There is no need to use the PSV_S. Moreover, the pressure head is also less than 70 m at each of the nodes in both parts of the WDS, so there is no need to install the PRVs at any of the nodes.

3.3. Colour-Coded Diagram of Velocity in Links at 9 a.m. (Peak Demand Hour)

The colour-coded velocity diagram given in Figure 3a,b, shows the range of velocities of flow in each pipe of network parts I and II. The scale showing the range of velocity represented by each colour is given for each of the networks. The analysis indicated that the velocity of flow for all the pipes in both networks is in the range of 0.25 m/s to 3 m/s [18]. Thus, there is no danger of erosion and deposition in the pipes during peak flow.

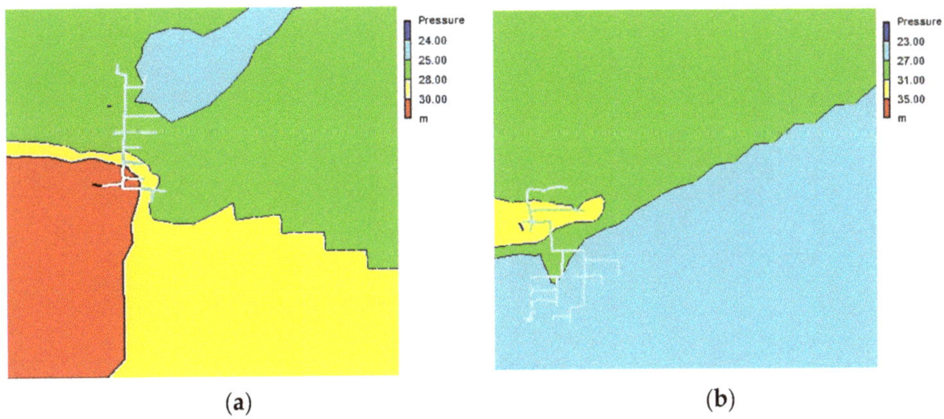

Figure 2. (**a**,**b**): Contour plan of pressure head at nodes at 9 a.m., for WDS parts I and II.

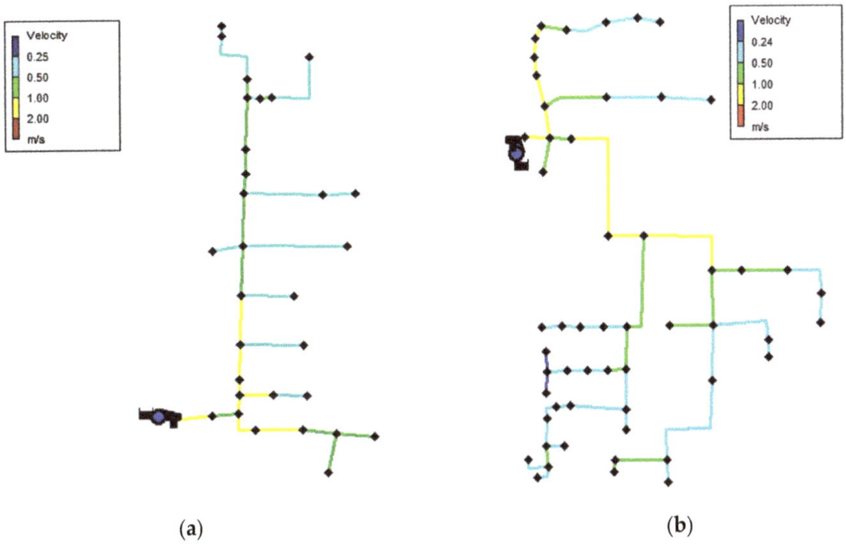

Figure 3. (**a**,**b**): Colour-coded diagram of velocity in links at 9 a.m., WDS parts I and II.

3.4. Time Series Plots

The temporal variation of the pressure head at the peak demand nodes and the water level in the tank are shown in Figure 4a,b, and the velocity in the pipes feeding the peak demand nodes is shown in Figure 4c,d. The pressure head lies within the recommended ranges with respect to all the nodes shown throughout the day. However, the velocity of flow in the pipes at the hours of minimum demand is less than 0.25 m/s, which can be attributed to the demand pattern adopted and the fact that the diameters have been selected to provide sufficient pressure head and cater to the demands at the peak hour. Moreover, the discharge in the pipes at these hours is less, and thus there is no serious hazard of deposition in the pipes. However, a velocity less than 0.25 m/s may be adopted to meet the hydraulic standards only if there is a suitable method of scouring the pipes [18].

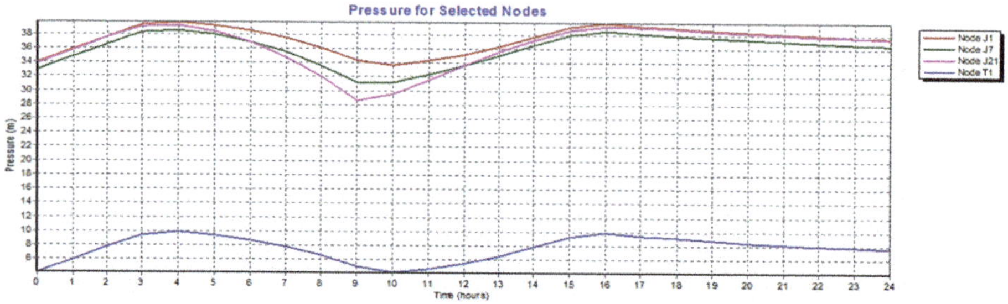

(**a**)

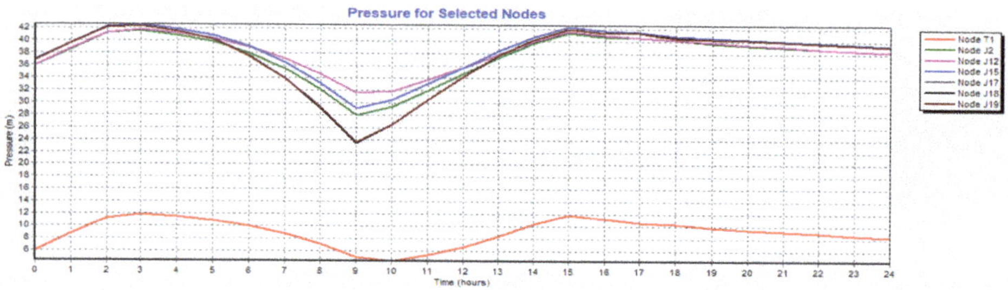

(**b**)

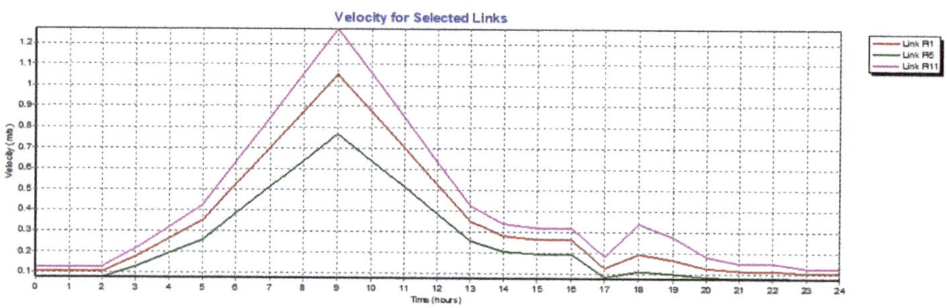

(**c**)

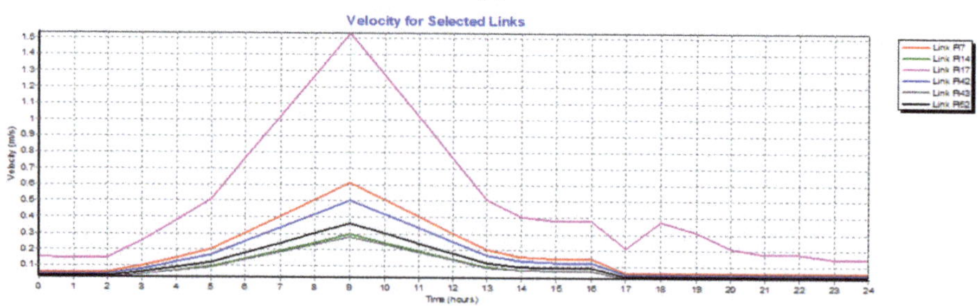

(**d**)

Figure 4. (**a**,**b**): Time series plot of pressure at peak demand nodes and storage tank for WDS parts I and II. (**c**,**d**): Velocity in links feeding peak demand nodes for WDS parts I and II.

3.5. Performance Evaluation by TPI

The TPI $_{\text{pressure}}$ and TPI $_{\text{velocity}}$ for WDS at the peak hour were evaluated for both parts of the study area. The TPI $_{\text{pressure}}$ for WDS part I = 1, indicating 100% efficiency of the network in terms of pressure head at the nodes at the peak hour. The TPI $_{velocity}$ for WDS part I = 0.615, indicating 61.5% efficiency of the network in terms of the velocity of flow in the pipes at the peak hour. The TPI $_{pressure}$ and TPI $_{velocity}$ for WDS part II are 1 and 0.645, respectively. Indicating the efficiency in terms of pressure head at nodes = 100% and in terms of velocity in the links = 64.5%. The lower values of TPI $_{velocity}$ can be attributed to the lower values of velocity during off-peak hours due to the demand pattern adopted.

3.6. Cost Optimization by Jal-Tantra Web System

The optimized value of the cost of the WDS in terms of pipe diameters was achieved by minimizing pipe diameters while maintaining the minimum specified pressure head for the nodes, as given in Table 6. However, the total cost before and after the optimization is approximately the same because the minimization of diameters can be seen in a smaller number of pipes and equals a maximum of 50 mm. Moreover, there is an increase in diameter of about 25 mm in some pipe lengths. The smaller difference in the costs indicates the cost-optimal design of the WDS for the study area by EPANET 2.2.

Table 6. Cost results for pipes, WDS parts I and II.

	WDS Part I		
Diameter	**Length**	**Cost**	**Cumulative Cost**
20.00	93	14,444	14,444
32.00	367	103,197	117,642
40.00	170	54,397	172,038
50.00	117	51,578	223,617
65.00	146	80,596	304,212
80.00	166	118,212	422,424
100.00	198	203,939	626,363
125.00	125	174,683	801,046
150.00	327	531,218	1,332,264
200.00	98	270,342	1,602,606
Total	1807	1,602,606	
	WDS Part II		
Diameter	**Length**	**Cost**	**Cumulative Cost**
15.00	126	15,449	15,449
20.00	72	11,103	26,552
25.00	243	54,690	81,242
32.00	182	51,142	132,384
40.00	191	61,120	193,504
50.00	230	100,827	294,331
65.00	541	299,353	593,684
80.00	609	433,608	1,027,292
100.00	298	307,238	1,334,530
125.00	178	247,931	1,582,461
150.00	312	507,895	2,090,356
200.00	61	167,642	2,257,998
Total	3043	2,257,998	

4. Conclusions

In this work, an "optimal" solution for the design of WDS for the University of Kashmir has been proposed to replace the existing and worn-out WDS, which has already served its function for more than 60 years. The design not only includes the modelling of hydraulic parameters such as pressure head and flow velocity, but the components are also designed in such a way that the objective functions of maximum efficiency of hydraulic performance and minimum capital cost in terms of pipe diameters and valves are satisfied. The main highlights of the work indicating the optimal hydraulic design include the following:

- The WDS is designed to have a minimum pressure head of 17 m at each node throughout the 24 h of a day, so that the water can reach up to a minimum of the third floor level of all the buildings. The pressure head at all the nodes in both parts of the network is such that $17 \leq P_i \leq 70$ (m), indicating the absence of points of deficiency or excess in the pressure head in the network. As such, there is no need to install the pressure sustaining valves (PSV) or the pressure regulating valves (PRV), resulting in a further reduction in the capital cost of the network.
- The velocity of flow in all the pipes during the hours of maximum discharge is within the standard range of $0.25 \leq V_i \leq 3.0$ (m/s), such that there is no danger of erosion or deposition during such hours. However, during the few hours of least discharge, the velocity of flow is less than 0.25 m/s in some pipes. Therefore, a scoring arrangement may be provided to remove the depositions, if any.
- The TPIs were evaluated for both parts of the WDS. The performance at peak hour with respect to pressure at nodes is 100%, while the performance with respect to velocity in pipes is 61.5% for WDS part I and 64.5% for WDS part II. The lower values of TPI velocity can be attributed to the lower values of velocity at off-peak hours due to the demand pattern adopted.
- Finally, the cost-optimality of the network was checked by using the Jal-Tantra web system, where the diameter of a few pipes was reduced by a maximum of 50 mm, while it increased by about 25 mm for a few pipes. The difference in the cost of the WDS as per EPANET modelling and that obtained by the Jal-Tantra is negligible, indicating the cost-optimal design of the WDS by EPANET.

Author Contributions: Introduction, M.A. and D.A.; Methodology, M.A.; Results and Discussions, D.A.; Conclusions, M.A. and D.A. All authors have read and agreed to the published version of the manuscript.

Funding: This research received no external funding.

Institutional Review Board Statement: Not applicable.

Informed Consent Statement: Not applicable.

Data Availability Statement: All the data used was obtained from the engineering wing of the University of Kashmir.

Conflicts of Interest: The authors declare no conflict of interest.

References

1. Awe, O.M.; Okolie, S.T.A.; Fayomi, O.S.I. Review of Water Distribution Systems Modelling and Performance Analysis Softwares Review of Water Distribution Systems Modelling and Performance Analysis Softwares. *J. Phys. Conf. Ser.* **2019**, *1378*, 022067. [CrossRef]
2. Sonaje, N.P.; Joshi, G.M. A review of modeling and application of water distribution networks (wdn). *Int. J. Tech. Res. Appl.* **2015**, *3*, 174–178.
3. Nagvekar, Y.; Pandri, Y.; Shinde, T.; Ukarde, A. A review study on use of epanet for water distribution networks. *Int. J. Innov. Eng. Technol.* **2019**, *6*, 42–44.
4. Anisha, G.; Kumar, A.; Kumar, J.A.; Raju, P.S. Analysis and Design of Water Distribution Network Using EPANET for Chirala Municipality in Prakasam District of Andhra Pradesh. *Int. J. Eng. Appl. Sci.* **2016**, *3*, 53–60.
5. Ramana, G.V.; Sudheer, C.V.S.S.; Rajasekhar, B. Network Analysis of Water Distribution System in Rural Areas using EPANET. *Procedia Eng.* **2015**, *119*, 496–505. [CrossRef]

6. Siew, C.; Tanyimboh, T.T.; Seyoum, A.G. Penalty-Free Multi-Objective Evolutionary Approach to Optimization of Anytown Water Distribution Network. *Water Resour. Manag.* **2016**, *30*, 3671–3688. [CrossRef]
7. Sharma, A.K.; Swamee, P.K. Cost considerations and general principles in the optimal design of water distribution systems. In *Water Distribution Systems Analysis Symposium 2006*; Buchberger, S.G., Clark, R.M., Grayman, W.M., Uber, J.G., Eds.; ASCE: Reston, VA, USA, 2012.
8. Jacobs, H.E.; Strijdom, J.L. Evaluation of minimum residual pressure as design criterion for South African water distribution systems. *Water SA* **2009**, *35*, 183–191. [CrossRef]
9. Kourbasis, N.; Patelis, M.; Tsitsifli, S.; Kanakoudis, V. Optimizing Water Age and Pressure in Drinking Water Distribution Networks. *Environ. Sci. Proc.* **2020**, *2*, 51. [CrossRef]
10. Hooda, N.; Damani, O. JalTantra: A System for the Design and Optimization of Rural Piped Water Networks JalTantra: A System for the Design and Optimization of Rural Piped Water Networks. *INFORMS J. Appl. Anal.* **2019**, *49*, 1–12.
11. Hooda, N.; Damani, O. Inclusion of tank configurations as a variable in the cost optimization of branched piped-water networks. *Drink. Water Eng. Sci.* **2017**, *10*, 39–44. [CrossRef]
12. Bonora, M.A.; Caldarola, F.; Maiolo, M.; Muranho, J.; Sousa, J. The New Set Up of Local Performance Indices into WaterNetGen and Application to Santarém's Network. *Environ. Sci. Proc.* **2020**, *2*, 18. [CrossRef]
13. Muranho, J.; Ferreira, A.; Sousa, J.; Gomes, A.; Marques, J.A.S. Technical Performance Evaluation of Water Distribution Networks based on EPANET. *Procedia Eng.* **2014**, *70*, 1201–1210. [CrossRef]
14. Avile, A.; Garcı, F. Pressure management for leakage reduction using pressure reducing valves. Case study in an Andean city. *Alexendria Eng. J.* **2019**, *58*, 1313–1326.
15. Garg, S.K. *Environmental Engineering (vol I)*; Khanna Publishers: Bangalore, India, 2010.
16. Modi, P.N. *Environmental Engineering (vol I)*; Standard Book House: Delhi, India, 2018.
17. *IS Code 1172: 1993—Code of Basic Requirements for Water Supply, Drainge and Sanitation*; Bureau of Indian Standards: Old Delhi, India, 2002.
18. CPHEEO Manual. Manual on Water Supply and Treatment. 1999. Available online: http://cpheeo.gov.in/cms/manual-on-sewerage-and-sewage-treatment.php (accessed on 22 September 2022).
19. *IS 9668: 1990—Provision and Maintenance of Water Supplies For Fire Fighting*; Bureau of Indian Standards: Old Delhi, India, 2010.
20. Rossman, L.; Woo, H.; Tryby, M.; Shang, F.; Janke, R.; Haxton, T. *EPANET 2.2 User Manual*; U.S. Environmental Protection Agency: Cincinnati, OH, USA, 2020.

environmental sciences proceedings

MDPI

Proceeding Paper

Effect of Different Water Stress on Growth Index and Yield of Semi-Late Rice †

Yixin Zhang [1], Xinyu Zhao [1,*], Fangping Liu [2], Lvdan Zhu [1] and Honglei Yu [2]

1 School of Hydraulic and Ecological Engineering, Nanchang Institute of Technology, Nanchang 330099, China; 2021301026@nit.edu.cn (Y.Z.); zhulv-dan1986@126.com (L.Z.)

2 Jiangxi Provincial Irrigation Experimental Station, Nanchang 330201, China; shong62@163.com (F.L.); 13576015504@163.com (H.Y.)

* Correspondence: spxyz@nit.edu.cn; Tel.: +86-18970091896

† Presented at the 7th International Electronic Conference on Water Sciences, 15–30 March 2023; Available online: https://ecws-7.sciforum.net/.

Abstract: The aim of this research is to find out the effects of different degrees of drought stress on the growth index and yield of semi-late rice. In this experiment, a pot experiment was conducted with hybrid Tian-you-hua-zhan rice as material. Intermittent irrigation was used as control (CK) at the tillering stage, booting stage, heading stage, and milk stage. Three water stress gradients of light drought (70% saturated moisture content), medium drought (60% saturated moisture content), and severe drought (50% saturated moisture content) were set to monitor the soil moisture content. The plant height, number of tillers, leaf area, and chlorophyll content under different water gradients at different growth stages were measured, and the effects of different water stress treatments at different growth stages on rice growth index, yield, and its components were analyzed. The results showed that the inhibition of the plant height at the jointing and booting stages was the most significant; moreover, the number of tillers and leaf area decreased most significantly at the tillering stage, and the heading and flowering stages had the greatest effect on chlorophyll synthesis. As far as yield is concerned, the most sensitive stages are the heading and flowering stages. Different stress treatments reduced the effective panicle number, 1000-grain weight, and seed setting rate of rice, and reduced the yield by 27.57%, 44.23%, and 46.32% respectively, compared with the normal control. The correlation analysis showed that the correlation degree of affecting yield from large to small was 1000-grain weight, seed setting rate, and effective panicle. Therefore, ensuring water supply at the heading and flowering stage can effectively improve the 1000-grain weight, effective panicle, and seed setting rate, and increase the rice yield and water use efficiency.

Keywords: water stress; growth index; 1000-grain weight; yield

Citation: Zhang, Y.; Zhao, X.; Liu, F.; Zhu, L.; Yu, H. Effect of Different Water Stress on Growth Index and Yield of Semi-Late Rice. *Environ. Sci. Proc.* **2023**, *25*, 84. https://doi.org/10.3390/ECWS-7-14318

Academic Editor: Athanasios Loukas

Published: 3 April 2023

1. Introduction

There is a great shortage of water resources in China, and the biological characteristics of rice have great water-saving potential [1]. Therefore, in the process of rice planting, it is of great significance to study the tolerance of water stress at different growth stages and its effects on the physiological index and yield of rice [2–5]. Differences in the effects of water stress and the degree of water stress on rice growth and yield at different growth stages have been reported; compared with rice under severe water shortage, rice recovered better and had a higher compensation effect under mild water stress. Water stress at the booting stage decreases the number of effective panicles, grains per panicle, and seed setting rate, resulting in a decrease in yield [6,7]. Water stress at the flowering stage results in a significant decrease in 1000-grain weight and seed setting rate, resulting in a decrease in yield per plant [8,9]. There have been many studies on the effects of water stress on rice yield, but the conclusions are different due to different planting sites, rice varieties, and experimental methods. There are relatively few studies on the effects of water stress on rice yield and growth index in a reproductive growth stage. In this experiment, according to the actual situation of seasonal drought of rice mid-season in Jiangxi, the effects of water stress on rice yield and growth index were studied, providing theoretical support for the study of rice water requirement and water-saving irrigation in Jiangxi.

2. Materials and Methods

2.1. Experimental Design

The experiment was carried out in the irrigation experiment center of Jiangxi Province, using pot experiment, pot length 38 cm, pot width 26 cm, carried out in a mobile rain shelter to avoid the effect of rainfall on the pot culture. The middle rice variety Yangliangyou was selected, and four middle rice varieties were selected at tillering stage (T1), jointing and booting stage (T2), heading and flowering stage (T3) and milk stage (T4). The degree of drought was controlled by the soil water content after the water layer was dried, and three levels were set, which were 50% of the saturated moisture content (W1, severe drought), 60% (W2, moderate drought) and 70% (W3, light drought) respectively. The two factors were combined into 12 treatments, and another reference treatment (CK) was set up. There were 13 treatments for late rice. The reference treatment (CK) intermittent irrigation always maintains a 2–3 cm thin water layer in the green period, for 7–10 days in the late tillering period, and naturally dries in the yellow mature stage; the rest of the growth period irrigates 2–3 cm each time and dries naturally for 3–5 days, that is, the soil moisture content reaches more than 70% of the saturated moisture content, and then irrigates again and again. Each process sets three repeats, with a total of 39 cells (see Table 1).

Table 1. Experimental design of semi-late rice subjected to drought.

Serial Number	Treatment Mode	Water Condition of Paddy Field at Different Stages					
		Rejuvention Stage	Tillering Stage	Booting Stage	Heading Stage	Milk Stage	Yellow Ripening Stage
1	CK	Normal	Normal	Normal	Normal	Normal	Dorp dry
2	W1T1	Normal	70%	Normal	Normal	Normal	Dorp dry
3	W2T1	Normal	60%	Normal	Normal	Normal	Dorp dry
4	W3T1	Normal	50%	Normal	Normal	Normal	Dorp dry
5	W1T2	Normal	Normal	70%	Normal	Normal	Dorp dry
6	W2T2	Normal	Normal	60%	Normal	Normal	Dorp dry
7	W3T2	Normal	Normal	50%	Normal	Normal	Dorp dry
8	W1T3	Normal	Normal	Normal	70%	Normal	Dorp dry
9	W2T3	Normal	Normal	Normal	60%	Normal	Dorp dry
10	W3T3	Normal	Normal	Normal	50%	Normal	Dorp dry
11	W1T4	Normal	Normal	Normal	Normal	70%	Dorp dry
12	W2T4	Normal	Normal	Normal	Normal	60%	Dorp dry
13	W3T4	Normal	Normal	Normal	Normal	50%	Dorp dry

2.2. Observation Items and Methods

The control methods of the soil moisture are as follows: (1) before rice planting, weigh the test chamber, measure the quality of wet soil and test box, measure the soil moisture content by drying method, and then calculate the dry soil mass in each soil box; (2) in the process of drought test after rice planting, weigh the weight in the test box, calculate the soil water content in the test box, and control the soil water content.

Plant height: measure the distance from the surface of the potted soil layer to the top of the longest leaf before the heading and flowering stages, and the distance from the surface of the potted soil layer to the longest ear from the heading stage to the later stage.

Number of tillers: from the tillering stage, the number of tillers per pot was recorded and measured every five days.

Chlorophyll: using a hand-held SPAD-502 chlorophyll tester, six plants were selected for each plant, one leaf was completely opened for each plant, the SPAD values at the middle part of the leaf and at the upper and lower parts of the leaf were measured, and the average SPAD values were taken.

Leaf area: the first three plants of each pot were measured directly with a ruler, three inverted sword leaves were measured in each plant, and the midvein length and maximum leaf width of each leaf were measured, respectively.

Yield and quality: according to the results of the effective panicle investigation, three plants were selected in each pot area for seed testing; at the same time, the yield of all rice plants in each pot was measured. The content of seeds examination include the effective panicle length, effective panicle number, total grains per panicle, filled grain weight per panicle, 1000-grain weight, and actual yield.

3. Results

3.1. Analysis of Growth Indexes of Rice under Different Water Stress

It can be seen from Figure 1a that drought stress has an inhibitory effect on the plant height of rice during the whole growth stage, and the order of the plant height from high to low under each water stress is as follows: CK > W3T4 > W2T4 > W1T4 > W3T3 > W2T3 > W1T3 > W3T2 > W2T1 > W2T2 > W1T1 > W1T2. It can be seen from the figure that drought stress has the most significant inhibitory effect on the plant height at jointing and booting stages, and the effect of drought stress on the plant height under drought stress is less than that in the previous two stages, which proves that water has a greater effect on the plant height growth at the vegetative growth stage.

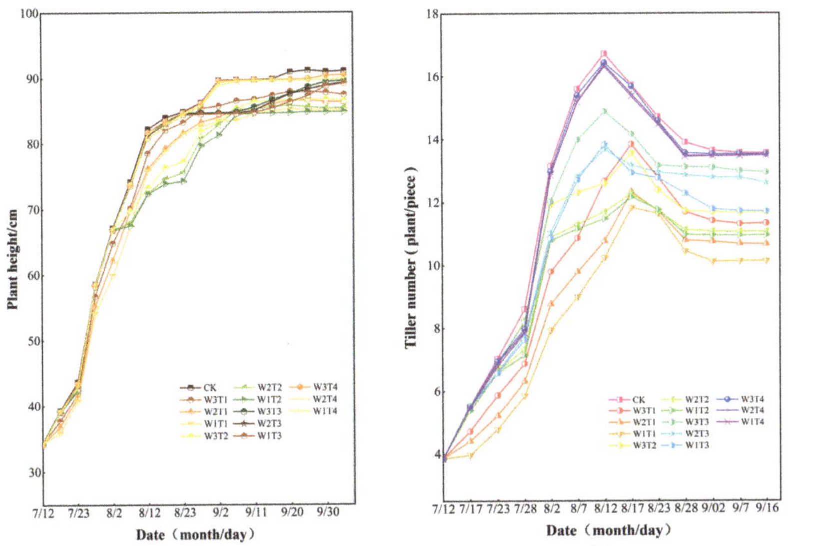

Figure 1. (**a**) Rice plant height under different water stress; (**b**) Rice tillering under different drought treatments.

It can be seen from Figure 1b that water stress inhibited the tillering of rice, which decreased most obviously at the tillering stage (T1) and the booting stage (T2), followed by the heading stage (T3), and the milk stage (T4). It was proved that drought stress had a strong inhibitory effect on rice tillering at the vegetative growth stage.

It can be seen from Figure 2a that the SPAD value of rice during the overall growth period is in a downward trend, the chlorophyll content of rice will increase briefly during the same growth period, and some treatments are higher than the normal control. The treatment at stage T1 is as follows: W2 > W1 > W3 at T2 and W3 > W2 > W1 at T3, but it is basically lower than that in the normal control. During the period of T4, W3 > W2 > W1 and W3 increased briefly during drought and was higher than that of the normal control. The gradient change of the W2 treatment was smaller than that of other treatments, and the decrease of the W1 treatment was the most obvious.

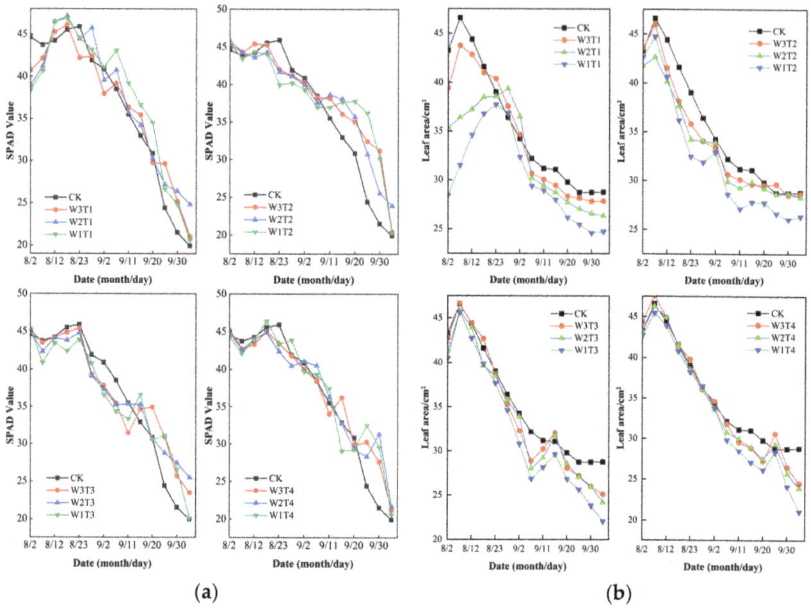

Figure 2. (**a**) Rice SPAD under different drought treatments; (**b**) Leaf area of rice under different drought treatments.

It can be seen from Figure 2b that water stress can inhibit the growth of the rice leaf area, in which the tillering stage and jointing stage are the most obvious, followed by the heading stage. Moreover, the effect on milky rice is less, the overall performance is W3 > W2 > W1, that is, the more serious the water deficit, the smaller the leaf area.

3.2. *Analysis of Yield and Its Components under Drought Treatment*

It can be seen from Table 2 that the effect of drought stress on the rice yield is the greatest at the heading stage, and the highest reduction in rice yield is 46.32%. In the same period, it is W3 < W2 < W1, and the effect at the tillering stage is the least significant.

Table 2. Examination results of yield indexes in growth period of each treatment.

Treatment	Number of Effective Ears (Piece/Pot)	1000-Grain Weight (g)	Spike Length (cm)	Number of Real Grains	Number of Real Grains	Total Grains	Seed Setting Rate (%)	Actual Yield (g/Pot)	Increase or Decrease Compared with CK (%)
CK	62	19.86	19.54	3916	245	4161	94%	77.54	0
W3T1	61	19.7	19.3	3869	341	4210	92%	76.22	−1.7
W2T1	56	19.31	18.9	3795	368	4163	91%	73.28	−5.49
W1T1	52	18.81	17.7	3640	402	4042	90%	68.47	−11.7
W3T2	58	19.2	19.3	3649	450	4099	89%	70.06	−9.64
W2T2	54	18.1	18.6	3377	484	3861	87%	61.12	−21.17
W1T2	49	17.4	18.4	3027	610	3637	83%	56.15	−27.58
W3T3	57	17.7	16.9	3173	733	3906	81%	56.16	−27.57
W2T3	53	16.3	16.8	2453	979	3432	71%	43.24	−44.23
W1T3	47	16.1	16.7	2385	1076	3461	69%	41.62	−46.32

Table 2. *Cont.*

Treatment	Number of Effective Ears (Piece/Pot)	1000-Grain Weight (g)	Spike Length (cm)	Number of Real Grains	Number of Real Grains	Total Grains	Seed Setting Rate (%)	Actual Yield (g/Pot)	Increase or Decrease Compared with CK (%)
W3T4	57	18.58	18.51	2354	610	2964	79%	58.6	−24.42
W2T4	54	17.87	18.27	2337	795	3132	75%	54.27	−30.01
W1T4	49	16.96	17.84	2264	866	3130	72%	50.27	−35.17

The yield components include 1000-grain weight, grain number per spike, seed setting rate, and effective panicle number. From the point of view of the effective panicle, in the same growth period, the order is CK > W3 > W2 > W1. In terms of seed setting rate, severe drought stress at the heading stage had the greatest effect on the rice seed setting rate, which was 25% lower than that of CK. In the same growth period, it was CK > W3 > W2 > W1. In terms of effective panicle length, severe drought stress at the heading stage had the most significant effect on the effective panicle length, with an average decrease of 2.84 cm per plant, and the order was CK > W3 > W2 > W1 in the same period. In terms of grain number per ear, light drought stress at the milk stage had the greatest effect on the grain number per ear, which was 28.77% less than that of CK. In the same growth period, the milk stage was CK > W2 > W1 > W3. In terms of 1000-grain weight, the effect of severe drought at the heading stage on the 1000-grain weight was the most obvious, which was 18.93% lower than that of CK, and the order was CK > W3 > W2 > W1 in the same period.

3.3. Correlation Analysis between Growth Indexes and Stage Water Consumption of Rice

The following conclusions can be drawn from Table 3: Under the normal irrigation level, the correlation between water consumption and plant height at the jointing stage is the largest. The correlation between water consumption at heavy drought levels and plant height at the tillering stage is the strongest, 0.947. The weak correlation between light drought and moderate drought levels at different growth stages shows that severe drought at the tillering stage will significantly impact the rice plant height. Under the conventional irrigation mode, most of the water consumption at the jointing stage is used to increase the height of the rice plant. The water consumption at light drought level is negatively correlated with the rice tillering at all stages, with the strongest correlation at the jointing stage being 0.587. The water consumption at heavy drought levels strongly correlates with the rice tillering at all stages, with the most significant correlation at the tillering stage being 0.926. Normal irrigation and moderate drought levels are weakly correlated with other treatments, indicating that under light drought level, most of the water consumed by the rice at the jointing stage was not used for the increase of tiller number, while under the heavy drought level, most of the water consumed at each growth stage was used for the increase of tiller number of rice, especially for the rice at the tillering stage, which had a significant positive correlation.

The following conclusions can be drawn from Table 4: The water consumption of rice under moderate and severe drought levels has a significant positive correlation with SPAD at the tillering, jointing, and milking stages, which are 0.919, 0.569, and 0.835, respectively. In contrast, the water consumption under normal irrigation and light drought levels have no significant impact on the SPAD value, proving that under severe stress at the tillering stage, most water consumption is used to synthesize chlorophyll. The water consumption is the main factor affecting the SPAD value change under moderate stress at the jointing stage and heading stage. Under normal irrigation, the water consumption and the leaf area at the milky stage are negatively correlated, with a coefficient of 0.765. Therefore, the water consumption does not promote the leaf area growth at the milky stage. In contrast, the water consumption and the leaf area at the jointing stage are positively correlated, with a coefficient of 0.59, indicating that the water consumption promotes the leaf area growth. Under moderate and severe drought levels, the rice leaf area and water consumption are positively correlated, with coefficients of 0.718 and 0.719, respectively. It was proved that water stress had a significant impact on the leaf area.

Table 3. Correlation analysis of water consumption with plant height and tiller number.

Index / Water Gradient	Child-Bearing Period							
	Plant Height/cm				Tiller Number/Tiller			
	T1	T2	T3	T4	T1	T2	T3	T4
CK	−0.217	0.805 **	0.653 *	0.531	−0.023	0.415	0.435	0.400
W3	−0.379	−0.332	−0.523	−0.520	−0.467	−0.587 *	−0.524	−0.576 *
W2	−0.363	−0.242	0.232	0.336	−0.209	−0.014	−0.152	−0.112
W1	0.947 **	0.292	0.291	0.386	0.926 **	0.637 *	0.701 **	0.689 **

T1 represents the tillering stage of rice, T2 represents the booting stage of rice, T3 represents the heading stage of rice, T4 represents the milky stage of rice. At 0.01 level (double tail), the correlation was significant (**). At 0.05 level (double tail), the correlation was significant (*).

Table 4. Correlation analysis of water consumption with SPAD and leaf area.

Index / Water Gradient	Child-Bearing Period							
	SPAD Value				Leaf Area/cm^2			
	T1	T2	T3	T4	T1	T2	T3	T4
CK	−0.418	0.548	0.190	−0.765 **	0.416	0.590 *	0.321	−0.765 **
W3	−0.389	0.427	−0.063	0.425	−0.384	0.043	0.493	0.425
W2	−0.476	0.569 *	0.835 **	0.107	−0.229	0.719 **	0.027	0.107
W1	0.919 **	−0.286	−0.627 *	−0.198	0.718 **	−0.317	0.534	−0.198

At 0.01 level (double tail), the correlation was significant (**). At 0.05 level (double tail), the correlation was significant (*).

3.4. Correlation Analysis of Water Consumption and Yield at Each Growth Stage

Table 5 shows rice's water consumption and actual yield at different growth stages under different water stress levels. Pearson correlation analysis was used to calculate the correlation coefficient between water consumption and rice yield at each growth stage. The calculation results are shown in Appendix A Table A2.

Table 5. Water consumption and actual yield at each growth stage.

Variable Treatment Mode	Water Consumption/mm				Yield/(g/Pot)
	Tillering Stage	Booting Stage	Heading Stage	Milk Stage	
CK	275.86	367.35	121.86	140.37	77.54
W1T1	274.07	366.35	120.44	117.51	76.22
W2T1	262.54	355.00	116.28	110.54	73.28
W3T1	257.15	351.78	115.40	103.32	68.47
W1T2	233.92	350.88	111.68	138.84	70.06
W2T2	212.11	311.93	101.54	131.57	61.12
W3T2	204.14	312.45	97.73	127.59	56.15
W1T3	261.87	331.18	83.70	135.93	56.16
W2T3	259.09	323.00	66.24	137.25	43.24
W3T3	248.60	299.33	58.27	132.61	41.62
W1T4	254.49	243.80	107.36	129.50	58.60
W2T4	262.01	198.53	107.84	134.21	54.27
W3T4	252.82	177.78	101.58	129.50	50.27

From the results of Pearson correlation analysis in Appendix A Table A2, the following conclusions can be drawn: The water consumption at the booting stage and heading stage is significantly correlated with the yield of rice, that is, the water deficit at the booting stage and heading stage will have a severe impact on the yield of rice, among which the correlation at the heading stage is the strongest, which is 0.88. Therefore, it is inappropriate to lack water excessively at this growth stage. In contrast, the water consumption at the tillering and milky stages are weakly correlated and negatively correlated, respectively, indicating that appropriate water stress at the tillering stage has little effect on the final yield of rice. The same is accurate at the milky stage.

4. Discussion

4.1. Effect of Water Stress on the Growth of Semi-late Rice

In this study, it was found that with different water gradients, the overall plant height of rice under heavy drought was lower than that under light and moderate drought, and, under the same water condition, the plant height was CK > T4 > T3 > T1 > T2. It can be inferred that water stress has a greater inhibitory effect on the rice plant height at the jointing stage and tillering stage, which is consistent with the research results of some scholars [10,11].

Tillering shows that under the same water stress, T1 < T2 < T3 < T4 < CK. This is because, when rice is at the reproductive growth stage, mild and moderate water stress have little effect on rice tillering at the reproductive growth stage, and it is beneficial to reduce ineffective tiller. Severe water stress leads to damage of the late filling and dry matter distribution, resulting in the death of the effective tiller, which is similar to the conclusion of Yang X L [12].

The overall photosynthesis showed a downward trend, as the SPAD value of rice at the booting stage increased during the short-term drought period, and most of the rice was higher than CK. Short-term water deficit could promote the photosynthetic potential of chlorophyll. When the drought time is longer than 10 days, the SPAD value of rice decreases continuously, which is basically consistent with the results of previous studies. This is because, at the early stage of the reproductive growth, water can affect the synthesis of chlorophyll in rice, and proper water stress can increase the concentration of chlorophyll, thus promoting the photosynthetic effect of rice.

Water stress can inhibit the growth of the rice leaves, thus inhibit the growth of the leaf area, and the overall performance is W3 > W2 > W1, that is, the more serious the water deficit, the smaller the leaf area.

4.2. Effect of Water Stress on Yield of Semi-Late Rice

In this study, from the yield and its composition, it was found that under different water treatments, the effective panicles were T1 > T2 > T4 > T3. Drought stress inhibited the differentiation of effective panicles, resulting in a significant decrease in the number of effective panicles [13,14]. It was found that the number of effective panicles of rice was very low under the irrigation mode of severe water deficit, which was similar to the results of this experiment [15]. Proper water addition was needed to improve the differentiation of effective panicles at the middle and later stages of reproductive growth.

Under the same water treatment, the panicle length of light drought was T1 $\geq$ T2 > T4 > T3, the moderate drought was T1 > T2 > T4 > T3, and the severe drought was T2 > T4 > T1 > T3. At the late growth stage, the inhibitory effect of water stress on the rice panicle length was very obvious. This is because the rice flag leaf is the main organ for the photosynthesis, therefore water deficiency will lead to the curl of the rice flag leaves and affect the opening and closing of the stomata and the synthesis of chlorophyll, so that the ear cannot grow normally [16,17].

The effect of light drought on the number of grains per panicle was the least at the tillering stage and the greatest at the milking stage. Water played a key role in the millet filling at the later stage of rice reproductive growth, and the water deficit inhibited the grain filling and starch accumulation in rice. Studies have shown that water stress significantly reduces the total amount of dry matter and limits the quantity and rate of stem and sheath carbohydrate transport to grains at the heading and flowering stages. At the same time, it accelerated the senescence of the rice leaves, led to insufficient grain filling time, and reduced the yield [18].

Under the same water treatment, the 1000-grain weight was T1 > T2 > T4 > T3, which proved that water stress had more significant inhibitory effect on the rice grain filling at the heading stage. Some scholars found that drought stress at the heading stage led to a significant decrease in the rice yield. The main factor leading to yield reduction is the decrease of the seed setting rate and 1000-grain weight due to insufficient water supply under drought stress [12], which is similar to the results of this experiment.

5. Conclusions

Based on the above analysis, the effects of different water stress on the rice growth index and yield were as follows:

(1) Water stress at different growth stages had different effects on rice growth indexes. The plant height was most significantly inhibited at the jointing and booting stages, and its water consumption was significantly positively correlated with the plant height, which was 0.805. The tillering stage most obviously affected the tiller number and leaf area. The correlation between water consumption, tiller number, and leaf area under heavy drought levels was very significant, which was 0.926 and 0.718. The heading and flowering stages had the most significant impact on chlorophyll. The correlation between water consumption and chlorophyll under medium drought levels was significant, which was 0.835.

(2) Different stress treatments reduced the adequate panicle number, 1000-grain weight, and seed-setting rate of rice. The most sensitive periods were the heading and flowering stages. Compared with the standard control, the yield of the three stress levels decreased by 27.57%, 44.23%, and 46.32%, respectively. The correlation analysis showed that the correlation degree of the yield was the 1000-grain weight, seed setting rate, and adequate panicle number (see Appendix A Table A1). The 1000-grain weight can have a more significant impact on the yield, which is 0.980.

(3) There was a specific relationship between water consumption and rice yield at different growth stages. The water consumption at the booting and heading stages was significantly correlated with the rice yield. The correlation between water consumption and yield at the heading stage was the strongest, which was 0.880.

Therefore, increasing water input at the tillering and jointing stages is beneficial to the tillering of rice, can effectively promote the stem and leaf growth and the panicle differentiation, and improve the plant height and leaf area of the rice. It is unsuitable for saving water excessively at the heading and flowering stages. Increasing the water supply can improve the 1000-grain weight, effective panicle, and seed setting rate of rice, and improve the rice yield and water use efficiency.

Author Contributions: Y.Z., conceptualization, methodology, formal analysis, data curation, writing—original draft preparation, review and editing; F.L. and L.Z., review and editing; X.Z., supervision, conceptualization, methodology, formal analysis, writing—review and editing; H.Y., data curation, Investigation. All authors have read and agreed to the published version of the manuscript.

Funding: This research was funded by Natural Science Foundation of Jiangxi Provincial, Grant number 20202BABL204068.

Institutional Review Board Statement: Not applicable.

Informed Consent Statement: Not applicable.

Data Availability Statement: Not applicable.

Conflicts of Interest: The authors declare no conflict of interest.

Appendix A

Table A1. Correlation analysis of yield and its components.

Dimension	Effective Panicle Number	1000-Grain Weight	Number of Grains per Spike	Seed Setting Rate	Actual Yield
Effective panicle number	1				
1000-grain weight	0.820 **	1			
Number of grains per spike	0.517	0.643 *	1		
Seed setting rate	0.684 **	0.912 **	0.835 **	1	
Actual yield	0.762 **	0.980 **	0.755 **	0.961 **	1

At 0.01 level (double tail), the correlation was significant (**). At 0.05 level (double tail), the correlation was significant (*).

Table A2. Correlation analysis of water consumption and yield at each growth stage.

Dimension	Tillering Stage	Booting Stage	Heading Stage	Milk Stage	Actual Yield
Tillering stage	1				
Booting stage	0.078	1			
Heading stage	0.175	0.159	1		
Milk stage	−0.125	−0.236	−0.389	1	
Actual yield	0.247	0.585 *	0.880 **	−0.377	1

At 0.01 level (double tail), the correlation was significant (**). At 0.05 level (double tail), the correlation was significant (*).

References

1. Shao, X.W.; Liu, H.D.; Du, Z.Y.; Yang, J.; Meng, F.X.; Ma, J.Y. Effects of water treatments at different stages on growth and yield of rice. *J. Soil Water Conserv.* **2007**, 193–196. (In Chinese)
2. Luo, W.B.; Meng, X.J.; Li, Y.L.; Zou, F.; Zhang, W. Research progress on comprehensive effect of water-saving irrigation on rice in Southern China. *J Water Resour. Water Eng.* **2020**, *31*, 145–151. (In Chinese)
3. Li, S.X.; Guo, H.; Ma, J.; Li, M.; Zhu, P.; Chen, Y. Effects of water stress on partial physiological characteristics and yield compensation in rice at booting stage. *Agr. Sci. Tech.* **2013**, *14*, 1750–1755.
4. Guo, H.; Ma, J.; Li, S.X.; Li, M.; Zhu, P.; Chen, Y. Effects of water stress on partial physiological characteristics and yield compensation in rice at booting stage. *J. South Agric.* **2013**, *44*, 1448–1454. (In Chinese)
5. Feng, X.L.; Zhong, K.Y.; Tang, X.R.; Li, Y.; Liu, Q. Effects of water stress at booting stage on stomatal traits and CAT activity in flag leaves of rice. *J. Irrig. Drain.* **2014**, *33*, 135–137.
6. Hu, M.M.; Gong, J.; Lan, Y.; Peng, L.G.; Wang, J.; Duan, Q.; Wu, C.Y.; Li, T. Effects of water stress at different growth stages on yield and quality of rice. *J. Jiangxi Agric. Univ.* **2021**, *43*, 971–982.
7. Zheng, C.J.; Li, S. Effect of water stress at flowering stage on rice growth and rice quality. *China Rice* **2017**, *23*, 43–45.
8. Sun, Z.Y.; Xiao, M.G.; Zhao, B.P.; Zhang, Q.; Zheng, F.Y.; Wu, B.; Li, X.B.; Leng, C.X.; Wu, L.C.; Wang, Y.J. High yield cultivation techniques of rice at reproductive growth stage in cold regions. *Anhui Agric. Bull.* **2021**, *27*, 42–44.
9. Li, X.Y.; He, Y.X.; Li, S.W.; Wang, C.T. Study on agronomic regulation of rice to drought stress. *J. Southwest Agric.* **2005**, *3*, 244–249.
10. Wang, Y.L.; Wang, J.; Du, J.Z.; Guan, Y.A. Effects of drought stress in different periods on agronomic characters of millet. *North China J. Agric.* **2012**, *27*, 125–129.
11. Duan, S.M.; Yang, A.Z.; Huang, Y.D.; Wu, W.G.; Xu, Y.Z.; Chen, G. Effects of drought stress on growth, physiological characteristics and yield of rice. *J. Nucl. Agric.* **2014**, *28*, 1124–1132.
12. Yang, X.L.; Wang, B.F.; Chen, L.; Cao, C.G.; Li, P. Effects of drought at heading stage on physiological characters and yield of rice. *China Rice* **2015**, *21*, 138–141.
13. Zhu, H.P.; Li, G.Y.; Xia, Q.M.; Long, R.P.; Deng, A.F.; Huang, J.; Xiang, H.Z.; Yang, C.D. Effects of drought stress at different stages on rice yield and growth characteristics. *China Rice* **2017**, *23*, 135–138.
14. Zhang, L.Q. Research on the relationship between leaf area index and its composition at the early growth stage of rice and yield. *J. Huaiyin Inst. Technol.* **2004**, *1*, 71–76.
15. Yang, X.L.; Wang, B.F.; Li, Y.; Zhang, S.Z.; Li, J.L.; Yu, Z.Y.; Cheng, J.P. A review of effects of drought stress on agronomic characteristics and yield of rice. *Hubei Agric. Sci.* **2020**, *59*, 39–43.

16. Xu, W.; Cui, K.; Xu, A.; Nie, L.; Huang, J.; Peng, S. Drought stress condition increases root to shoot ratio via alteration of carbohydrate partitioning and enzymatic activity in rice seedlings. *Acta Physiol. Plant.* **2015**, *37*, 1–11. [CrossRef]
17. Yang, X.; Wang, B.; Chen, L.; Li, P.; Cao, C. The different influences of drought stress at the flowering stage on rice physiological traits, grain yield, and quality. *Sci. Rep.* **2019**, *9*, 3737–3742. [CrossRef] [PubMed]
18. Shao, X.W.; Zhang, R.Z.; Qi, C.Y.; Tong, S.Y.; Yang, M. Effect of water stress at jointing and booting stage on rice growth and yield. *J. Jilin Agric. Univ.* **2004**, *3*, 237–241.

environmental sciences proceedings

MDPI

Proceeding Paper

Applying a Flexible Fuzzy Adaptive Regression to Runoff Estimation †

Mike Spiliotis [1,*] and Luis Garrote [2]

1 Department of Civil Engineering, School of Engineering, Democritus University of Thrace, 67100 Xanthi, Greece
2 Department of Civil Engineering: Hydraulics, Energy and Environment, Universidad Politécnica de Madrid, 28040 Madrid, Spain; l.garrote@upm.es
* Correspondence: m.spiliotis@gmail.com; Tel.: +30-25410-79613
† Presented at the 7th International Electronic Conference on Water Sciences, 15–30 March 2023; Available online: https://ecws-7.sciforum.net.

Abstract: A smart, flexible, fuzzy-based regression is proposed in order to describe non-constant behavior of runoff as a function of precipitation. Hence, for high precipitation, beyond a fuzzy threshold, a conventional linear (precise) relation between precipitation and runoff is established, while for low precipitation, a curve with different behavior is activated. Between these curves and for a runoff range, each curve holds to some degree. Hence, a simplified Sugeno architecture scheme is established on few logical rules. Alternatively, the model can be enhanced by using a combination between the fuzzy linear regression of Tanaka and the aforementioned simplified Sugeno architecture. The training process is achieved based on the Particle Swarm Optimization (PSO) method.

Keywords: fuzzy IF-THEN rules; particle swarm optimization (PSO); least square method; fuzzy regression; runoff estimation

check for updates

Citation: Spiliotis, M.; Garrote, L. Applying a Flexible Fuzzy Adaptive Regression to Runoff Estimation. *Environ. Sci. Proc.* **2023**, *25*, 85. https://doi.org/10.3390/ECWS-7-14308

Academic Editor: Athanasios Loukas

Published: 3 April 2023

1. Introduction

Due to the complexity and the inherent uncertainty of hydrological processes, it is rather impossible to apply white box (or physical-based) models to treat hydrological phenomena. Among the black box models, which can be used, the least square model is widely used. The scope of the article is to enhance the utility of the regression model by using fuzzy sets and logic. Hence, two models are proposed. The first one is based on the combination between the fuzzy reasoning and the least square method, and the other is based on the couple between the fuzzy reasoning and fuzzy regression.

Firstly, a simple relationship between precipitation and runoff at the annual scale could be successfully described by the law presented in Equation (1):

$$R = k(P - P_0) \tag{1}$$

where R is annual runoff in mm, P is annual precipitation in mm, and k (dimensionless) and P_0 (mm) are two parameters that may be estimated through linear regression [1,2]. P_0 may be interpreted as a rainfall depth threshold below which the runoff is zero and k is a runoff coefficient at the annual scale. For humid climates, where annual precipitation is always greater than the runoff threshold, this simple relation is always valid. However, in certain arid or semi-arid climates, annual precipitation may be lower than the runoff threshold, and this would yield negative runoff, meaning that the relationship may not be valid for low precipitation years. In this work, a solution for this problem is proposed. For a high annual cumulative precipitation, beyond a threshold, a conventional (crisp) relation between precipitation and runoff is established, while for low precipitation, a curve with a lower slope must be derived. Between these curves, and for a precipitation range close to the runoff threshold, each curve holds to some degree. Hence, initially, the conventional regression is used for each area. Furthermore, the use of the fuzzy model proposed by Tanaka is later examined.

A simplified Sugeno architecture scheme is proposed based on only two logical rules. The training process is achieved based on an interplay between the Particle Swarm Optimization (PSO) method and either the conventional least square analysis or the widely used fuzzy regression model of Tanaka. Although there are many examples that illustrate the application of Sugeno Systems to water engineering problems based on the MATLAB toolbox, and even if the errors remain within acceptable range, sometimes the rational and logical basis of the model is ambiguous [3,4]. On the other hand, there are also applications of the if-then systems based on a logical explanation, but these lack a proper training process. The method proposed for the application presented in this work was successfully applied by [3] to assess bedload transport in gravel-bed rivers as a function of discharge. The combined use of the simplified Sugeno architecture with the fuzzy regression is proposed for the first time in this work.

2. Proposed Simplified Reasoning System by Using Crisp Regression

The independent variable (here the annual rainfall) takes only two linguistic values (high and low), which correspond to quantitative fuzzy sets, and hence, only two rules exist. Consequently, there are two areas without uncertainty where only one regression equation is activated. However, between two crisp areas (Figure 1), there is a grey area where both rules are activated to some degree. Based on Figure 1, it is obvious that only two rules were structured, whilst in fact, the grey area is between β_1 and β_2.

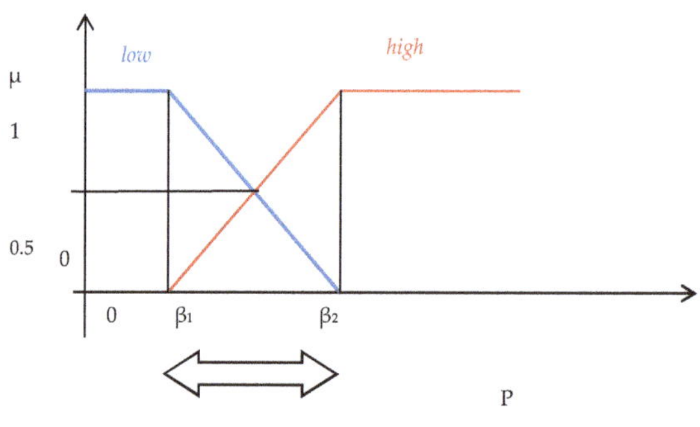

Figure 1. Proposed simplified architecture of the fuzzy rule-based systems.

In this work, only one independent variable (annual precipitation) appears, P. For given variables β_1 and β_2, the following membership functions are modulated (Figure 1). From the picture, it is obvious that the following property holds:

$$\mu_1(P) + \mu_2(P) = 1 \tag{2}$$

As aforementioned, the reasoning consists of two rules:

IF P (annual precipitation) is low, THEN y (annual Runoff) is $(y = a_{10} + a_{11}P)$
IF P (annual precipitation) is high, THEN y (annual Runoff) is $(y = a_{20} + a_{21}P)$ (MODEL 1)

In this model, the coefficients $(a_{10},\ a_{11},\ a_{20},\ a_{21})$ are crisp numbers, so finally a nonlinear crisp curve is produced as follows:

$$\begin{aligned} y &= \frac{\mu_1(P)(a_{10}+a_{11}P)+\mu_2(P)(a_{20}+a_{21}P)}{\mu_1(P)+\mu_2(P)} = \\ &\frac{\mu_1(P)}{\mu_1(P)+\mu_2(P)}a_{10} + \frac{\mu_2(P)}{\mu_1(P)+\mu_2(P)}a_{20} + \frac{\mu_1(P)\cdot P}{\mu_1(P)+\mu_2(P)}a_{11} + \frac{\mu_2(P)\cdot P}{\mu_1(P)+\mu_2(P)}a_{21} = \\ &\mu_1(P)a_{10} + \mu_2(P)a_{20} + \mu_1(P)\cdot P\cdot a_{11} + \mu_2(P)\cdot P\cdot a_{21} \end{aligned} \tag{3}$$

where $\mu_1(P)$, $\mu_2(P)$ are the values of the membership functions of the linguistic terms low and high, respectively (Figure 1). Each rule is activated with respect of the values of the corresponding membership function.

In the case that the thresholds β_1 and β_2 are known, then the next linear system of algebraic equations is produced [3,4] in order to determine the vector θ:

$$\mathbf{\Lambda} \cdot \theta = \mathbf{b}$$

$$\mathbf{\Lambda} = \begin{bmatrix} \mu_1(P^1) & \mu_2(P^1) & \mu_1(P) \cdot P^1 & \mu_2(P) \cdot P^1 \\ & & \cdots\cdots & \\ \mu_1(P^M) & \mu_2(P^M) & \mu_1(P^M) \cdot P^M & \mu_2(P^M) \cdot P^M \end{bmatrix}$$

In which M is the number of data. The superscript for P indicates the examined number of data.

$$\begin{array}{c} \theta = \begin{bmatrix} a_{10} & a_{20} & a_{11} & a_{21} \end{bmatrix}^T \\ \mathbf{b} = \begin{bmatrix} y_1 & \ldots & y_M \end{bmatrix}^T \end{array} \tag{4}$$

In addition, **b** is the matrix that contains the values of the measured runoff (dependent variables)

Hence, according to the usual least squares method, the optimal vector θ^* can be found:

$$\theta^* = \left[\mathbf{\Lambda}^{\mathrm{T}} \cdot \mathbf{\Lambda}\right]^{-1} \mathbf{\Lambda}^{\mathrm{T}} \mathbf{b} \tag{5}$$

The coefficient of determination can be adopted to evaluate the proposed model [3,5]:

$$R^2 = (\mathbf{b} - \mathbf{\Lambda} \cdot \theta^*)^{\mathrm{T}} \cdot (\mathbf{b} - \mathbf{\Lambda} \cdot \theta^*) \tag{6}$$

Hence, if the thresholds β_1 and β_2 are known, the coefficients of the crisp linear equations can be determined with respect to Equations (3) and (5).

3. Proposed Simplified Architecture of the Fuzzy Rule Based System by Using Fuzzy Regression

The main difference with the previous model is that for each rule, a fuzzy regression is activated. Hence, the model will produce a fuzzy band, where all the data must be included within the produced fuzzy band [6,7]. Therefore, the following rule-based system is produced:

$$\left.\begin{array}{l} \text{IF P is low, THEN Annual Runoff, y is } (\tilde{y} = \tilde{a}_{10} + \tilde{a}_{11} P) \\ \text{IF P is high, THEN Annual Runoff, y is } (\tilde{y} = \tilde{a}_{20} + \tilde{a}_{21} P) \end{array}\right\} \text{ (MODEL 2)}$$

In this new model, where the coefficients $(\tilde{a}_{10}, \tilde{a}_{11}, \tilde{a}_{20}, \tilde{a}_{21})$ are fuzzy symmetrical triangular numbers, a nonlinear fuzzy curve is produced. Hence, the problem concludes to the following equation:

$$\begin{array}{c} y = \frac{\mu_1(P)(\tilde{a}_{10} + \tilde{a}_{11} P) + \mu_2(P)(\tilde{a}_{20} + \tilde{a}_{21} P)}{\mu_1(P) + \mu_2(P)} = \\ \mu_1(P)\tilde{a}_{10} + \mu_2(P)\tilde{a}_{20} + \mu_1(P) \cdot P \cdot \tilde{a}_{11} + \mu_2(P) \cdot P \cdot \tilde{a}_{21} \end{array} \tag{7}$$

The tilde means the fuzziness and the characteristic of these numbers are the central value and the semi-widths. The aforementioned numbers are selected to be fuzzy symmetrical triangular numbers for simplicity reasons [6].

Even if the new proposed model produces more complexity, the proposed model has the advantage that all the data are located within the produced fuzzy estimation of the annual runoff. However, if the produced fuzzy band is large, the proposed model has no practical sense. Hence, the minimization of the width regarding the produced fuzzy estimation of the annual runoff is the key question in order to evaluate the applicability of the model. The mathematical background of the Tanaka model can be found in [6,7].

4. The Proposed Learning Process with the Use of PSO

Particle Swarm Optimization (PSO) is a heuristic, stochastic, global optimization method based on the behavior of the swarm [8]. Each possible solution is called a particle, and the set of potential solutions in each iteration creates the *'swarm'*. A *'swarm'* has a dimension N', in which N' is the number of examined solutions. Each examined solution is made up of D variables, in which D is the dimension of the problem [8]. Here, $D = 2$; that is, the couple (β_1, β_2). More details about the method can be found in [8–11].

Firstly, a swarm of candidate solutions is randomly structured. Each of them contains only the (crisp) values of the β_1, β_2, which are the aforementioned parameters of the membership functions. Two cases can be distinguished. The first one is the use of crisp regression, and hence, the least square method is activated, whilst the second choice is the use of the fuzzy model of Tanaka. In the case of the Tanaka fuzzy regression being used, for each candidate solution, a linear programming problem is used.

5. Case Study

The proposed method was successfully applied to the basins of two water resource basins in Spain: the Rio Piedras and Rio de Aguas basins. The Piedras River is a coastal river in the southwest of Spain. It drains a contributing basin of 550 km^2, running from north to south along 40 km in the Huelva province. The mean annual precipitation is 574 mm/year, and mean annual runoff is 106 mm/year. It is regulated by the Piedras and Los Machos reservoirs, which are operated for water supply and irrigation. The Aguas river is a short coastal river in the south of Spain, running along 65 km through the east of the province of Almería. The contributing basin is 547 km^2. The climate is semiarid, with mean annual precipitation of 334 mm/year and mean annual runoff of 41 mm/year.

Based on the proposed adaptive crisp regression between rainfall and runoff, the results are presented in Figure 2 in the case of Rio Piedras.

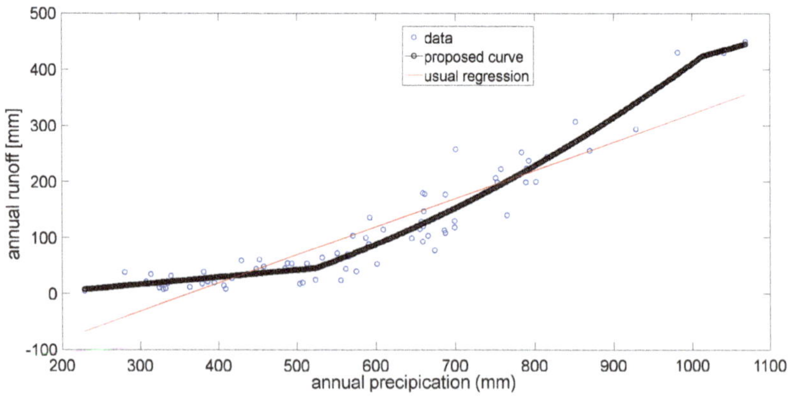

Figure 2. The proposed method and the conventional regression applied in order to assess a relationship between annual precipitation and runoff in the case of Rio Piedras.

The proposed method recognizes the grey region between regarding the precipitation (524.06, 1013.43); that is, a large non-linear behavior. The produced equation is:

$$DICH(P) = \mu_1(P)(-22.1599) + \mu_2(P)(35.8763) + \mu_1(P)0.1287{\cdot}P + \mu_2(P)0.3828{\cdot}P \quad (8)$$

Another important question is the evolution of the swarms. After s significant number of iterations, but not immediately, the convergence of the swarm is obvious (Figure 3).

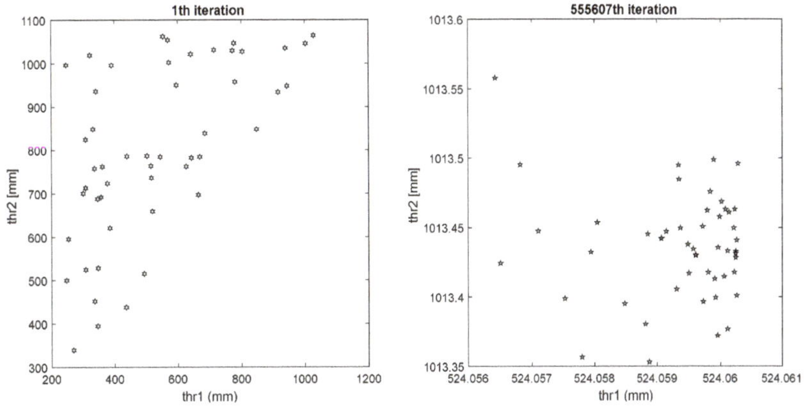

Figure 3. The evolution of the parameter (β_1, β_2) after a sufficient number of iterations in the case of Rio Piedra.

The method is also applied in the case of the Rio de Aguas Basin. The results are shown in Figure 4. In both cases, a significant grey zone exists, as can be shown from Figures 2 and 4. The results for the examined cases indicate that the proposed method better simulates the relationship between the annual precipitation and the annual runoff compared to the conventional crisp regression.

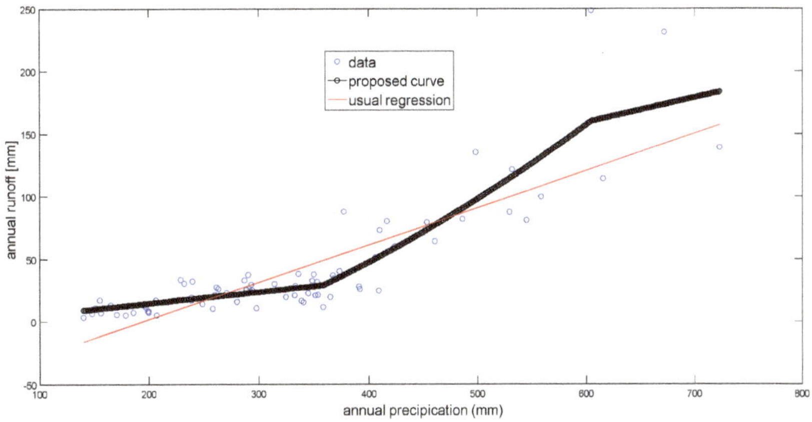

Figure 4. The proposed method and the conventional regression applied in order to assess a relationship between annual precipitation and runoff in the case of Rio de Aguas.

Next, the proposed fuzzy rules with the simultaneous use of the fuzzy regression model of Tanaka is used in the case of Rio Piedras. The following fuzzy curve is produced (Figure 5):

$$
\begin{aligned}
DICH(P) = {} & \mu_1(P)(-6.3182, 3.3334) + \mu_2(P)(250.4427, 76.0990) \\
& + \mu_1(P)(0.0933, 0.0510)\cdot P + \mu_2(P)(0.6130, 0.0029)\cdot P
\end{aligned} \tag{9}
$$

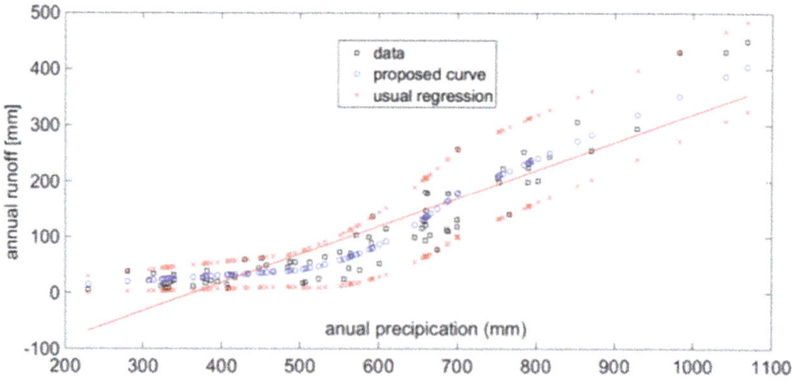

Figure 5. The proposed method with fuzzy regression curves applied in order to assess a fuzzy relationship between annual precipitation and annual runoff in the case of Rio Piedras.

In the bracket, the first term denotes the central value, and the second term expresses the semi-width. For instance, the fuzzy estimation of the annual runoff in the case that the measured annual runoff is y = 98.8538 mm (in the case of Rio Piedras) is depicted in Figure 6.

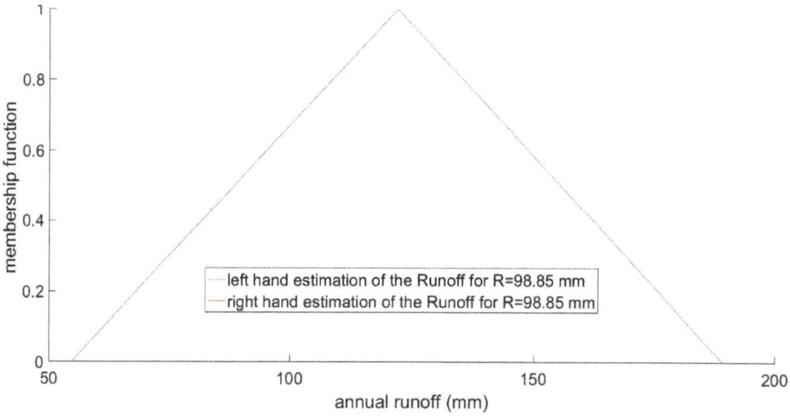

Figure 6. Fuzzy estimation of the runoff in the case that the annual runoff is y = 98.8538 mm in the case of Rio Piedras.

The main difference between model 1 and model 2 is that in the last model, all the data must be included within the produced fuzzy estimation of the annual runoff. The proposed method is suitable to estimate the annual water yield, but it is not intended to assess the peak flow under a significant rainfall. However, the second model requires more computational time.

Author Contributions: Conceptualization, M.S. and L.G.; methodology, M.S. and L.G.; software, M.S. and L.G.; validation, M.S. and L.G.; formal analysis, M.S. and L.G.; investigation M.S. and L.G.; resources, L.G.; data curation, M.S. and L.G.; writing—original draft preparation, M.S. and L.G.; writing—review and editing, M.S. and L.G.; visualization M.S. and L.G.; supervision, M.S. and L.G.; project administration, M.S. and L.G.; funding acquisition, M.S. and L.G. All authors have read and agreed to the published version of the manuscript.

Funding: This research received no external funding.

Institutional Review Board Statement: Not applicable.

Informed Consent Statement: Not applicable.

Data Availability Statement: The data can be found in https://www.miteco.gob.es/es/agua/temas/evaluacion-de-los-recursos-hidricos/evaluacion-recursos-hidricos-regimen-natural/ (accessed on 28 March 2023).

Conflicts of Interest: The authors declare no conflict of interest.

References

1. Ponce, V.M. *Engineering Hydrology: Principles and Practices*; Prentice Hall: Englewood Cliffs, NJ, USA, 1989; pp. 66–67.
2. Spiliotis, M.; Garrote, L. Assessment of a Precipitation-Runoff relation based on a new fuzzy adaptive regression. In Proceedings of the 7th IAHR Europe Congress "Innovative Water Management in a Changing Climate", Athens, Greece, 7–9 September 2022.
3. Spiliotis, M.; Kitsikoudis, V.; Hrissanthou, V. Assessment of bedload transport in gravel-bed rivers with a new fuzzy adaptive regression. *Eur. Water* **2017**, *57*, 237–244.
4. Sen, Z. *Fuzzy Logic and Hydrological Modelling*; Taylor and Francis Group, CRC Press: Boca Raton, FL, USA, 2010; p. 223.
5. Chen, G.; Pham, T.T. *Introduction to Fuzzy Sets, Fuzzy Logic, and Fuzzy Control Systems*; CRC Press: Boca Raton, FL, USA, 2001; pp. 89–103.
6. Tanaka, H. Fuzzy data analysis by possibilistic linear models. *Fuzzy Sets Syst.* **1987**, *24*, 363–375. [CrossRef]
7. Kitsikoudis, V.; Spiliotis, M.; Hrissanthou, V. Fuzzy regression analysis for sediment incipient motion under turbulent flow conditions. *Environ. Process.* **2016**, *3*, 663–679. [CrossRef]
8. Spiliotis, M.; Mediero, L.; Garrote, L. Optimization of Hedging Rules for Reservoir Operation during Droughts Based on Particle Swarm Optimization. *Water Resour. Manag.* **2016**, *30*, 5759–5778. [CrossRef]
9. Parsopoulos, K.E.; Vrahatis, M.N. Recent Approaches to Global Optimization Problems through Particle Swarm Optimization. *Nat. Comput.* **2002**, *1*, 235–306. [CrossRef]
10. Eberhart, R.C.; Simpson, P.K.; Dobbins, R.W. *Computational Intelligence PC Tools*; Academic Press Professional: Boston, MA, USA, 1996.
11. Poli, P.; Keenedy, J.; Blackwell, T. Particle swarm optimization. *Swarm Intell.* **2007**, *1*, 33–57. [CrossRef]

environmental sciences proceedings

MDPI

Proceeding Paper

Determining the Critical Points of a Basin from the Point of View of Water Productivity and Water Consumption Using the WaPOR Database †

Iman Hajirad [1], Sanaz Mohammadi [2] and Hossein Dehghanisanij [3,*]

1 Irrigation and Reclamation Engineering Department, College of Agriculture and Natural Resources, University of Tehran, Karaj 31587-77871, Iran; i.hajirad@ut.ac.ir
2 Water Management and Engineering Department, Collage of Agriculture, Tarbiat Modares University, Tehran 14115-336, Iran; sanaz.mohammadi@modares.ac.ir
3 Agricultural Engineering Research Institute, Agricultural Research Education and Extension Organization, Karaj 3135913533, Iran
* Correspondence: dehghanisanij@yahoo.com
† Presented at the 7th International Electronic Conference on Water Sciences, 15–30 March 2023; Available online: https://ecws-7.sciforum.net/.

Abstract: Actual evapotranspiration is one of the major components of the soil water balance equation. Several methods have been presented for estimating actual evapotranspiration, but the older methods are not practical because of their spatial and temporal dependence. Recently, the Food and Agriculture Organization of the United Nations (FAO) created the WaPOR open-access system on water productivity with the aim of covering countries experiencing water crises in Africa and the Middle East, and the estimation of actual evapotranspiration (ETa) is one of its main products. This portal makes it possible to determine water consumption and water productivity on a large scale with minimal time and cost, so it can be used to manage the agricultural sector. The Google Earth Engine System (GEE), introduced by Google in 2010, is an effective remote sensing instrument for gathering important data from satellite imagery. In this study, the actual evapotranspiration maps of the Maroon-Jarahi basin for 2017 were extracted using the methodology introduced by the FAO and provided in the WaPOR database, and the coding was performed in the GEE system. The results showed that the actual evapotranspiration during this period was highest in July and decreased with the onset of the fall season. Limited water resources are a major obstacle to ensuring food security. Considering that the agricultural sector consumes most of the water in Iran and worldwide, water management in the agricultural sector is of great importance. Water productivity is a key indicator in studying and improving agricultural water management as well as one of the Sustainable Development Goals. In this study, the actual evapotranspiration (AET), net primary production (NPP), and water productivity (WP) for the basin were estimated using the WaPOR portal and Google Earth Engine over a 10-day period with a spatial resolution of 250 m (decadal data). Based on the obtained results, areas with low water productivity were identified. By studying the existing cropping patterns, the type of irrigation system used, and the water and soil conditions in these areas, it was possible to investigate the reason for the low water productivity and propose solutions to improve it.

Keywords: irrigation management; water consumption; WaPOR; evapotranspiration

Citation: Hajirad, I.; Mohammadi, S.; Dehghanisanij, H. Determining the Critical Points of a Basin from the Point of View of Water Productivity and Water Consumption Using the WaPOR Database. *Environ. Sci. Proc.* **2023**, *25*, 86. https://doi.org/10.3390/ECWS-7-14322

Academic Editor: Athanasios Loukas

Published: 3 April 2023

1. Introduction

The agricultural sector will need to increase food production by 60% in developed countries and 110% in developing countries by 2050 to satisfy future food demand because more than 70% of the world's freshwater is used for irrigation. In recent decades, water resources have declined because of population growth and climate change. Long droughts, increased demand for agricultural products, a lack of water, and the availability of agricultural land are posing serious threats to food production. Therefore, increasing crop production per unit of water consumption, known as crop water productivity, is considered a key strategy for ensuring food security [1]. Water productivity is a key factor in studying the production of agricultural products and the efficiency of water resources [2]. Crop yield and evapotranspiration (ET) are typically used to estimate water productivity. Crop yield is influenced by a variety of factors, including soil fertility, disease control, and agricultural practices, whereas ET changes depending on the type of irrigation and drainage system used as well as the climatology and rainfall patterns [3]. From the field scale to the continental scale, remote sensing is increasingly the best option for determining water output and its components at various temporal and spatial resolutions [4]. The proper scheduling and management of a basin's water resources, which necessitates knowledge of the hydrological behavior of the system, including temporal and spatial changes in crucial components such as the actual ET in the basin, have an impact on the economic, agricultural, and social development of a region. Thus, it is necessary to develop techniques that can correctly determine the actual ET [5].

Due to its complexity, estimating ET is challenging. There are several ways to determine this parameter. Because they can be compared with numerical climate models and have a wider coverage than traditional approaches, remote-sensing-based techniques are more widely used. There are several different remote sensing approaches available for estimating actual ET, including surface energy balance [6,7], Penman–Monteith [8], and experimental methods based on plant indices [9,10]. As mentioned, the use of models based on remote sensing to estimate actual evapotranspiration requires validation and recalibration, which may be problematic because of the lack of observational data. WaPOR ET products have recently been presented by the World Food Organization (FAO). The actual ET rate in the FAO WaPOR product is measured using the ETLook algorithm, and the equations used to calculate the actual ET rate in this algorithm were described in detail in a previous study [11]. To introduce this algorithm, the ETLook model is based on the Penman–Monteith equation, which is generally applied to estimate the total potential ET (including the two components of evaporation and transpiration) using common meteorological data (such as solar radiation, air temperature, vapor pressure, and wind speed) [12]. In the ETLook algorithm, with a slight change from this equation, a combination of remote sensing information (including NDVI, surface albedo, soil moisture, solar radiation, land surface cover, and a digital elevation model) and meteorological data (including temperature, humidity, wind speed, and precipitation) is used to estimate the actual ET, and the results are available to the public for free on the FAO website.

On-farm management practices play an important role in improving water productivity. Therefore, in order to spatially identify regions with appropriate and poor farm management practices and to assess the efficacy of crop management strategies, the creation of water productivity maps at the basin scale may be necessary [3]. To know the relationship between water and food, tracking water productivity, and following efficiency objectives, information on water productivity with precise spatial resolution is of vital importance. However, due to the nature of spatial variability at the basin scale and general measurement problems, such data are not readily accessible in the majority of regions, and field measurements are unable to capture spatial trends over large areas. Therefore, preparing water productivity maps at the basin scale can help with solving these problems.

The purpose of this study was to prepare water productivity and actual ET maps to determine critical points from the viewpoints of water consumption and water productivity in the Maroon-Jarahi basin and to provide management solutions to increase water productivity in the basin.

2. Materials and Methods

2.1. Area of Study

The Maroon-Jarahi basin (Figure 1) has an area of 24,307 km^2. Most of this basin is located in Khuzestan Province. The long-term average water flow of the Maroon River in Khuzestan Province is equal to 2105 million m^3. The length of the Maroon-Jarahi River in the whole area is 691 km, with an average slope of 14%. There are more than 20,000 ha of palm gardens in the Maroon-Jarhi basin, 75% of which are in the Shadgan region, downstream of the basin. Intensive rice cultivation is carried out on over 14,000 ha in this basin, 65% of which is located upstream of the basin. At the same time, more than 15,000 ha of palm trees in Shadgan are under water stress, although the cultivation of 4800 ha of paddy downstream of the basin in Shadgan adds to this water stress.

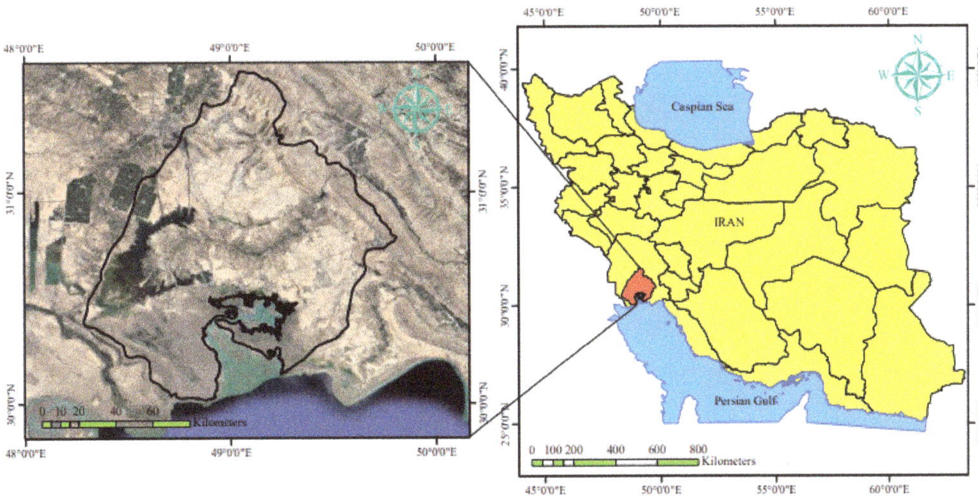

Figure 1. Study area and locations of major agricultural sites in the Maroon-Jarahi basin.

2.2. ETLook Model

The Penman–Monteith equation is solved twice by the ETLook model: once for soil evaporation (E) and once for canopy transpiration (T):

$$\lambda E = \frac{\delta(R_{n,soil} - G) + \rho_{air} C_p \frac{(e_s - e_a)}{r_{a,soil}}}{\delta + \gamma\left(1 + \frac{r_{s,soil}}{r_{a,soil}}\right)} \quad (1)$$

$$\lambda T = \frac{\delta\left(R_{n,canopy}\right) + \rho_a C_p \frac{(e_s - e_a)}{r_{a,canopy}}}{\delta + \gamma\left(1 + \frac{r_{s,canopy}}{r_{a,canopy}}\right)} \quad (2)$$

Regarding the net available radiation ($R_{n,soil}$ and $R_{n,canopy}$), as well as the aerodynamic and surface resistance, the two formulae are different. Additionally, for transpiration, the earth heat flow (G) is not considered. Figure 2 shows a graphic illustration of the key ideas of the ETLook model [8].

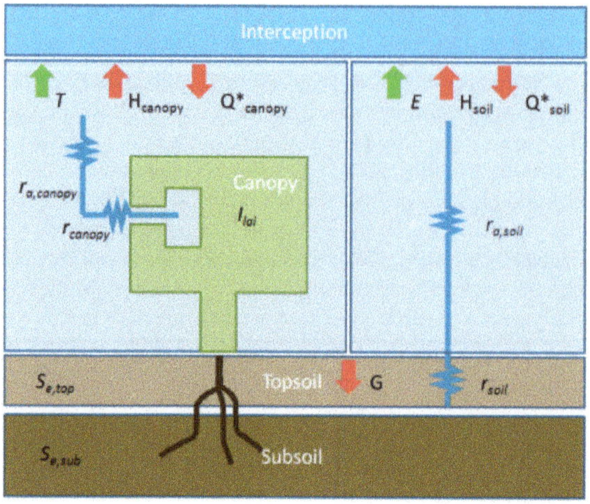

Figure 2. Conceptual diagram of ETLook algorithm (adapted with permission from Ref. [8]. 2018, FAO).

2.3. Net Primary Production

The most important feature of an ecosystem is its net primary production (NPP), which describes how photosynthesis transforms carbon dioxide into biomass. The term NPP refers to a group of definitions that define the exchange of carbon between an ecosystem and the atmosphere. Satellite images and meteorological data are used to derive the NPP. Veroustraete et al. [13] provided a detailed explanation of the methodology's basic concepts, and Eerens et al. [14] provided information on how it is implemented in practice. The Copernicus Global Land Component enhanced these methods, with the inclusion of biome-specific light use efficiency (LUE) being the most significant improvement. All three levels of the NPP were provided on a decadal basis, with pixel values representing the average daily NPP for that particular decade in $gC/m^2/day$. Sometimes it is more appropriate to measure dry matter production (DMP), such as for agricultural objectives (in kg DM/ha/day). The DMP can be calculated by multiplying the NPP by a constant of 0.45 gC/gDM. Therefore, 1 $gC/m^2/day$ (NPP) = 22,222 kg DM/ha/day (DMP). Although greater values (theoretically up to 320 kgDM/ha/day) can occur, typical NPP values range between 0 and 5.4 $g\ C/m^2/day$ (NPP) or 0 and 120 kgDM/ha/day (DMP) [8]. In this study, net primary production (NPP) data were downloaded from the WaPOR and Google Earth Engine (GEE) systems at a pixel resolution of 250 m.

2.4. Crop Water Productivity

Crop water productivity (CWP) is calculated as the yield (Y) divided by actual evapotranspiration and interceptions (ETIa). A low ETIa and a high yield cause high water productivity because the plant is producing a lot while not using much water. A high ETIa and low yield cause low water productivity because the plant is not producing much while using a lot of water. CWP was calculated for the study region using the yield (NPP) map and ETIa from Equation (3):

$$CWP(kg/m^3) = \frac{Yield(kg/ha)}{ETIa(mm)} \tag{3}$$

Actual evapotranspiration and interception (ETIa) data were downloaded in 250 m pixel resolution from the WaPOR and Google Earth Engine systems. The total amount of soil evaporation (E), canopy transpiration (T), and evaporation from raindrops caught by leaves (I) is known as the ETIa [8,15].

2.5. Land Use/Land Cover Map

The basin land use or land cover is one of the factors affecting ET and the hydrological cycle. Knowing the different types of terrain is important for comparing ET. The land-use map of Khwaja Nasir Toosi University [16] with 13 classes was used in this research to assess ET in various locations and to comprehend its various uses. A map of the studied basin's land cover is shown in Figure 3.

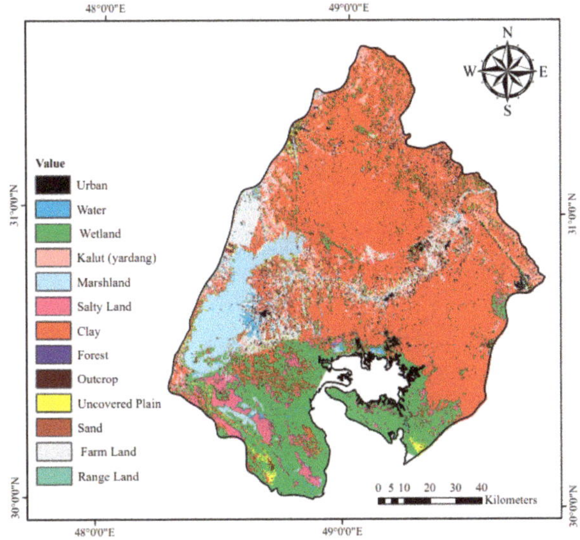

Figure 3. Land use/land cover map of the Maroon-Jarahi basin.

3. Results and Discussion

3.1. Actual Evapotranspiration and Interception

For a better spatial and temporal analysis of the actual evapotranspiration and interception in the Maroon-Jarahi basin, an annual evapotranspiration and interception map for the agricultural class is shown in Figure 4. Irrigated lands are shown in blue, and nonirrigated lands with the least evapotranspiration are shown in red. Compared with nonirrigated lands, irrigated lands occupied a smaller area of the basin.

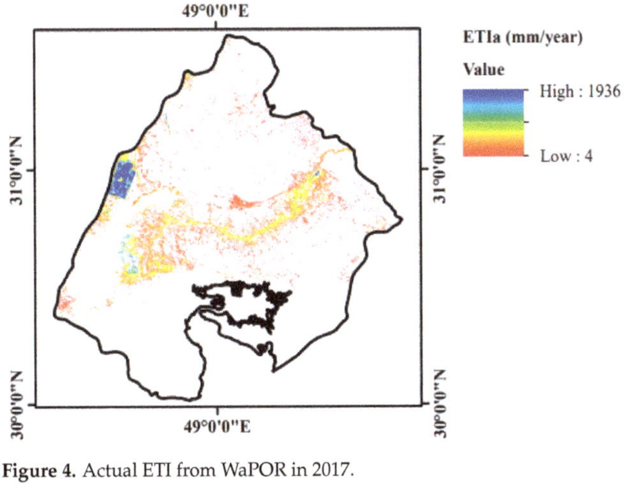

Figure 4. Actual ETI from WaPOR in 2017.

The normalized difference vegetation index (NDVI) for Maroon-Jarahi basin in 2017 (Figure 5) shows good agreement between pixels with high evapotranspiration and pixels with high NDVI values. In other words, areas with higher vegetation cover had higher evapotranspiration.

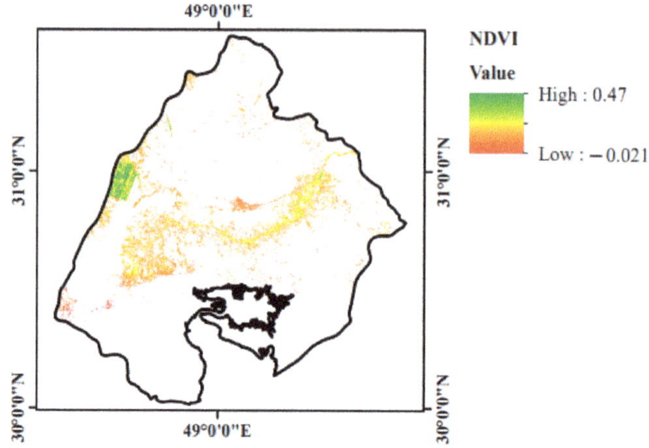

Figure 5. Year-averaged NDVI in 2017 using the MODIS product.

3.2. Net Primary Production

Figure 6 shows the spatial changes in the net primary production at the level of the Maroon-Jarahi basin area, which indicates the conversion of carbon dioxide caused by photosynthesis into biomass. The blue points represent the highest NPP, and the red points represent the lowest NPP at the basin level.

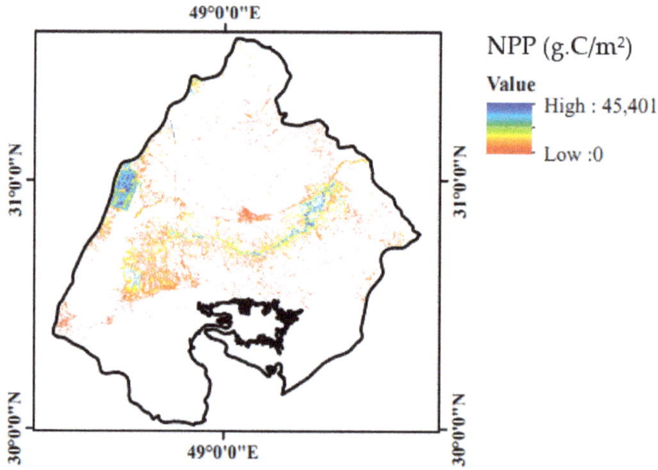

Figure 6. NPP from the WaPOR in 2017.

3.3. Crop Water Productivity Mapping

Figure 7 shows the spatial changes in water productivity in the Maroon-Jarahi basin. In Figure 7, the red dots indicate low WP values, indicating land where agricultural and irrigation management practices have not been performed properly. The blue points indicate fields with high WP values. However, the water productivity map shows that agricultural and irrigation management can significantly improve water productivity. The blue points indicate that farmers properly considered in-farm management to maximize the yield per drop of water.

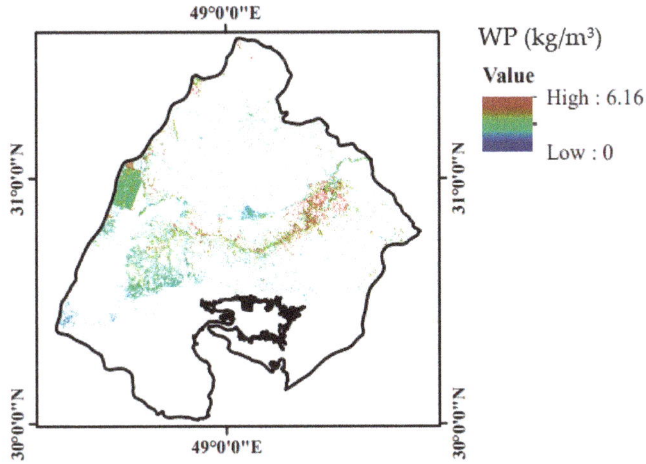

Figure 7. Water productivity map of the Maroon-Jarahi basin.

4. Conclusions

The excessive use of water in the agricultural sector can cause serious problems in arid and semiarid regions. Water consumption is strongly correlated with the amount of ET; therefore, ET estimation can provide valuable information for water consumption planning and management. The Maroon-Jarahi basin's situation in regions with low WP values and high evapotranspiration rates, which are regarded as critical points, was the main focus of this study. The preparation of evapotranspiration and water productivity maps can help local farmers and decision makers identify the best agricultural practices used in high-productivity fields and adopt a similar approach in low-productivity areas to increase productivity in the future. Given the high level of water consumption for agriculture, using remote sensing to monitor agricultural water productivity can help identify gaps in water productivity and evaluate potential solutions to address those gaps.

The irrigation water management system in the Maroon-Jarahi basin agriculture project has not been developed or modified since the beginning of the project. Therefore, one of the management tools of the project can be remote sensing technology used in water allocation, crop water requirement assessment, and water productivity monitoring to identify productivity gaps and develop appropriate solutions to address them.

According to the results obtained after determining the critical points from the point of view of water consumption and water productivity, it is possible to propose management strategies to reduce water consumption for the studied basin, and some of these strategies are as follows:

1. Using different types of mulch to reduce evaporation;
2. Using transplantation;
3. Land leveling;
4. Regulating deficit irrigation;
5. Using subsurface drip irrigation.

These factors can help reduce water consumption and increase water productivity.

Author Contributions: Conceptualization, methodology, supervision, project administration, and funding acquisition: H.D.; software, validation, formal analysis, visualization, and investigation: I.H.; resources, data curation, analysis, and writing—review and editing, S.M. All authors have read and agreed to the published version of the manuscript.

Funding: This research received no external funding.

Institutional Review Board Statement: Not applicable.

Informed Consent Statement: Not applicable.

Data Availability Statement: All important data are available in the paper.

Conflicts of Interest: The authors declare no conflict of interest.

References

1. Ghorbanpour, A.K.; Kisekka, I.; Afshar, A.; Hessels, T.; Taraghi, M.; Hessari, B.; Duan, Z. Crop Water Productivity Mapping and Benchmarking Using Remote Sensing and Google Earth Engine Cloud Computing. *Remote Sens.* **2022**, *14*, 4934. [CrossRef]
2. Bessembinder, J.J.E.; Leffelaar, P.A.; Dhindwal, A.S.; Ponsioen, T.C. Which crop and which drop, and the scope for improvement of water productivity. *Agric. Water Manag.* **2005**, *73*, 113–130. [CrossRef]
3. Zwart, S.J.; Bastiaanssen, W.G.; de Fraiture, C.; Molden, D.J. A global benchmark map of water productivity for rainfed and irrigated wheat. *Agric. Water Manag.* **2010**, *97*, 1617–1627. [CrossRef]
4. Marshall, M.; Aneece, I.; Foley, D.; Xueliang, C.; Biggs, T. Crop Water Productivity Estimation with Hyperspectral Remote Sensing. In *Advanced Applications in Remote Sensing of Agricultural Crops and Natural Vegetation*; CRC Press: Boca Raton, FL, USA, 2018; pp. 79–96.
5. Javadian, M.; Behrangi, A.; Gholizadeh, M.; Tajrishy, M. METRIC and WaPOR estimates of evapotranspiration over the Lake Urmia Basin: Comparative analysis and composite assessment. *Water* **2019**, *11*, 1647. [CrossRef]
6. Allen, R.G.; Tasumi, M.; Trezza, R. Satellite-based energy balance for mapping evapotranspiration with internalized calibration (METRIC)—Model. *J. Irrig. Drain. Eng.* **2007**, *133*, 380–394. [CrossRef]
7. Bastiaanssen, W.G.M.; Menenti, M.; Feddes, R.; Holtslag, A.A. A remote sensing surface energy balance algorithm for land (SEBAL). 1. Formulation. *J. Hydrol.* **1998**, *212–213*, 198–212. [CrossRef]
8. FAO. *WaPOR Database Methodology: Level 1. Remote Sensing for Water Productivity Technical Report: Methodology Series*; FAO: Rome, Italy, 2018; p. 72.
9. Glenn, E.P.; Huete, A.R.; Nagler, P.L.; Nelson, S.G. Relationship between remotely-sensed vegetation indices, canopy attributes and plant physiological processes: What vegetation indices can and cannot tell us about the landscape. *Sensors* **2008**, *8*, 2136–2160. [CrossRef] [PubMed]
10. Nagler, P.L.; Glenn, E.P.; Nguyen, U.; Scott, R.L.; Doody, T. Estimating riparian and agricultural actual evapotranspiration by reference evapotranspiration and MODIS enhanced vegetation index. *Remote Sens.* **2013**, *5*, 3849–3871. [CrossRef]
11. Bastiaanssen, W.G.M.; Cheema, M.J.M.; Immerzeel, W.W.; Miltenburg, I.J.; Pelgrum, H. Surface energy balance and actual evapotranspiration of the transboundary Indus Basin estimated from satellite measurements and the ETLook model. *Water Resour. Res.* **2012**, *48*, W11512. [CrossRef]
12. Allen, R.G.; Pereira, L.S.; Raes, D.; Smith, M. *FAO Irrigation and Drainage Paper No. 56. Crop Evapotranspiration (Guidelines for Computing Crop Water Requirements)*; Food and Agriculture Organisation of the United Nations: Rome, Italy, 1998; p. 300.
13. Veroustraete, F.; Sabbe, H.; Eerens, H. Estimation of carbon mass fluxes over Europe using the C-Fix model and Euroflux data. *Remote Sens. Environ.* **2002**, *83*, 376–399. [CrossRef]
14. Richards, J.A. *Remote Sensing Digital Image Analysis*; Springer: New York, NY, USA, 2022.
15. Poudel, U.; Stephen, H.; Ahmad, S. Evaluating Irrigation Performance and Water Productivity Using EEFlux ET and NDVI. *Sustainability* **2018**, *13*, 7967. [CrossRef]
16. Ghorbanian, A.; Kakooei, M.; Amani, M.; Mahdavi, S.; Mohammadzadeh, A.; Hasanlou, M. Improved land cover map of Iran using Sentinel imagery within Google Earth Engine and a novel automatic workflow for land cover classification using migrated training samples. *ISPRS J. Photogramm. Remote Sens.* **2020**, *167*, 276–288. [CrossRef]

environmental sciences proceedings

MDPI

Proceeding Paper

GIS-Based Multi-Criteria Decision Analysis for Flash Flood Hazard and Risk Assessment: A Case Study of the Eastern Minya Watershed, Egypt †

Kamal Darwish

Department of Geography, Minia University, El Minia 61519, Egypt; kamal.srogy@mu.edu.eg

† Presented at the 7th International Electronic Conference on Water Sciences, 15–30 March 2023; Available online: https://ecws-7.sciforum.net.

Abstract: Flash floods are considered one of the most devastating and frequent extreme climatological natural hazards in the world. El Minya is one of the most vulnerable areas in Egypt for flash flood problems. It was affected by several hazardous historical flash flood events. These events could lead to both catastrophic losses of life and severe damage to the infrastructures of the study area. The study area is located in the middle of Egypt, about 240 km south of Cairo. It is situated along the Limestone Plateau facing El Minia governorate. The main objective of this study is to assess the risk of flash flood hazard on the human activities in the study area. Integration of remote sensing; geographic information systems, analytical hierarchy process, and the multi-criteria decision analysis (MCDA) techniques were applied in this study. Eight different significant effective factors collected from multisource geospatial data, including lithology, hydrology, topography, soil type, land cover, and rainfall data, were selected to evaluate the flood risk map. Remote sensing imagery was used for land use/cover mapping to detect the vulnerable human activities. ArcGIS-based weighted overlay modeling was used to combine the criteria to calculate the final decision map.

Keywords: flash flood; multi-criteria decision-making analysis; Minya; GIS-AHP; weighted overlay modeling; hierarchical modeling; remote sensing; GIS technology

Citation: Darwish, K. GIS-Based Multi-Criteria Decision Analysis for Flash Flood Hazard and Risk Assessment: A Case Study of the Eastern Minya Watershed, Egypt. *Environ. Sci. Proc.* **2023**, *25*, 87. https://doi.org/10.3390/ECWS-7-14315

Academic Editor: Athanasios Loukas

Published: 3 April 2023

1. Introduction

Flash floods are one of the most devastating natural disasters in the world, causing damage to properties, killing more than 5000 people every year, and registering the highest mortality rate among the other riverine and coastal flooding disasters [1].

Egypt is considered one of the countries that suffer from flash flood hazards and risk in mountains and nearby areas; every year, people, human activities, urbanized areas, and infrastructure are threatened by rainwater inundation. In the last few years, Egypt has experienced several flash flood events that caused serious damage and loss of lives, infrastructures, and buildings. The analysis of historical flash flood events over Egyptian land indicates that the eastern desert received highly destructive repetitive flash flood events starting in 1979 along El-Quseir and Marsa Alam, which killed 19 people and destroyed the coastal highway. The flash flood of Marsa Alam city, which happened in 1991 and caused big damage, the Alexandria city flash flood that killed 21 people in 1993, as well as a well-known flash flood that happened in 1994 and caused severe damage to infrastructure and loss of lives in Assiut governorate [2].

In October 2016, a devastating flash flood event struck Ras-Gharib city, killed dozens, and caused damage to infrastructure [3]. The Minya area has been affected by several flash flood events since 1975; heavy rainfall happened in Upper Egypt. It destroyed about 180 houses and displaced 1500 citizens [4]. In November 1997, a flash flood affected the area of study and killed 53 people, destroying 260 houses [4]. Recently, in March 2020, a heavy-rainfall flash flood struck the area, causing land subsidence and huge damage to the El Geish highway, infrastructure, and settlements as shown in Figure 1.

Figure 1. Photographs show the destruction of El Geish highway by the March 2020 flash flood. Source: https://www.youm7.com (Online Newspaper published on 14 March 2020).

The Geographic Information System (GIS) is an important technique that provides the capacity to design geospatial identities and analyze and manipulate spatial information. This information can be managed and organized through attribute tables. The tabulated data is linked to the geographic features, which may contain multiple quantitative and qualitative information.

GIS-based AHP calculations are very important in revealing spatial trends and relationships between geospatial data and retrieving valuable information for decision-making [5]. Multi-criteria decision analysis (MCDA) is a decision-making technique developed to make solutions to complex decision problems [6]. The GIS-MCDA uses the analytical hierarchical process (AHP) method for controlling and arranging the parameters to investigate complex decisions [7]. The GIS-MCDA method has the ability to process and combine different types of geospatial data (rainfall map, land cover, soil types, slope map, drainage density); results can be visualized and presented in maps [5]. It is a very important spatial decision tool for spatial planning and management issues [8,9]. Integration of GIS and MCDA has been globally applied to assess flood hazards and risk assessment in Greece [6,10], Iran [7], Malaysia [11], Saudi Arabia [12], India [13], and Egypt [14].

2. Study Area

The Minya area is located about 240 km south of Cairo city, the capital of Egypt, on the crossing of Longitude 31°30′ N, Latitude 28° E. The drainage basin selected for study has two large valleys (Wadi al-Tarfah and Wadi al-Bustan) crossing the limestone plateau. The mouth of the basin is situated in eastern Beni Mazar city in northeastern Minya city. The area of study extends to cover most of the eastern part of Minya governorate; it covers an area of 10,682.9 sq. km, as shown in Figure 2.

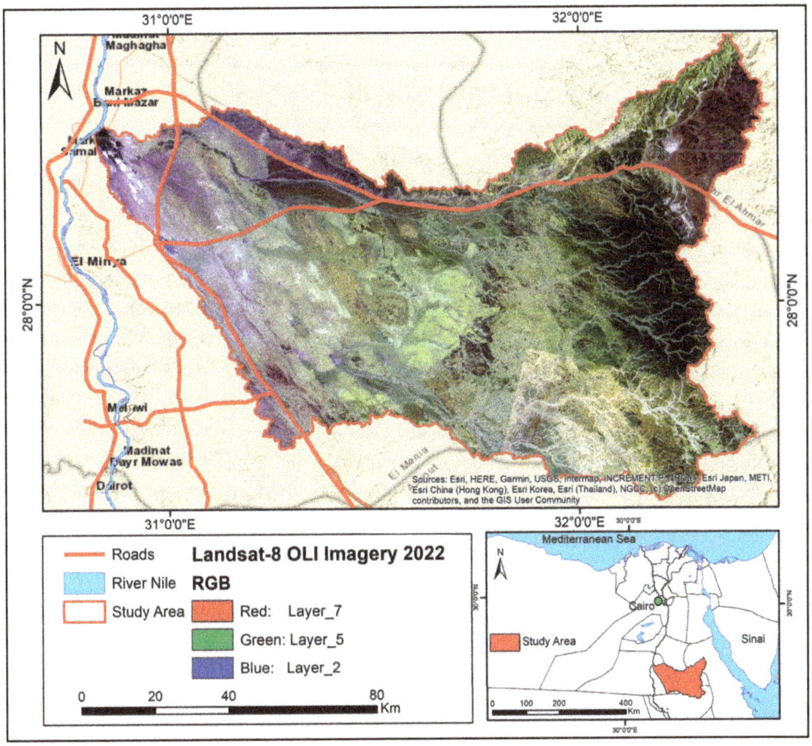

Figure 2. Location map of study area.

3. Materials and Methods

Satellite remote sensing and GIS data were collected for this study to generate the flash flood hazard map based on the AHP technique, as shown in the flowchart in Figure 3. Six significant flood-controlling factors were selected based on the physical and natural properties of the study area, including elevations, slopes, soils, hydrology, land cover, geology, and rainfall. GIS-based MCDA has the ability to examine multi-criteria factors thematic maps using weighted overlaid analysis to assess the flash flood hazard map and risk along human activities [6,7].

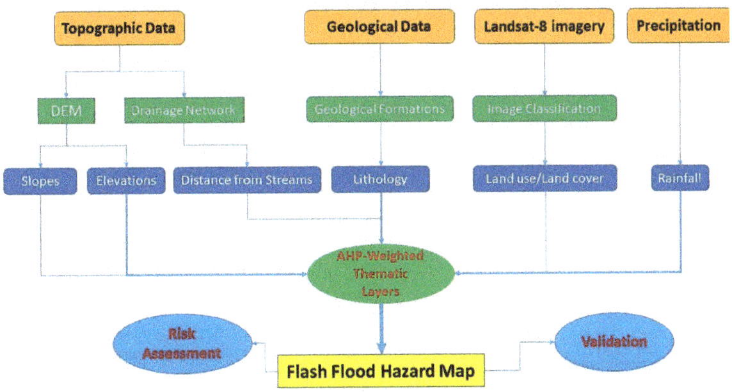

Figure 3. Flowchart shows the MCDA method applied in this study.

The analytical hierarchy process (AHP) has been applied in this study to assess the flash flood hazard and risk in the watershed of the study area; spatial criteria used in this study come from geology, topography, rainfall, hydrology, soil, and land cover maps.

3.1. Lithological Data

The Egyptian geological map of the study area scale 1:500,000 was digitized and analyzed to identify the lithological units, and three major geological formations were detected: (1) Quaternary deposits occupy 13.3% of study area, including sand sheets, wadi deposits, and gravels. (2) The Eocene limestone formation is covering 82.4% of the study area, including the Samalut, Minia, Maghagha, Observatory, Qarara, and Thebos Formations. In addition, (3) the Cretaceous formations cover 4.3% of study area, which include Rakhiyat, Galala, Sudr, Umm Omeiyed, Hawashiya, and Wadi Qena Formations, as shown in Figure 4. Geology is one of the main factors effecting in runoff due to the permeability of rocks and faults may increase the infiltration rate.

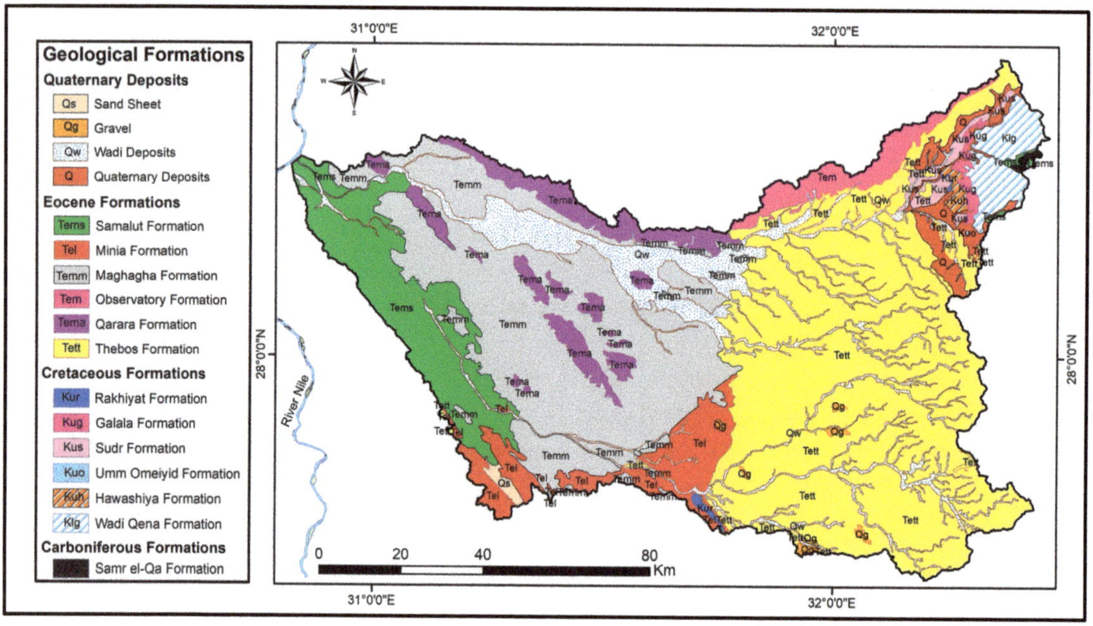

Figure 4. Geological formations of study area.

3.2. Topography

Topographic analysis is a significant factor for flash flood hazard and risk assessment. Satellite-based Digital Elevation Models (DEMs), including SRTM-1arc second (30 m spatial resolution), were downloaded and processed to extract elevations, contours, slopes, aspects, and hydrological analyses shown in Figure 5a–c.

3.3. Watershed Delineation

Hydrology is considered the main effective factor in flash flood intensity and its risk. The ArcGIS-based spatial modeler is used to delineate and map stream orders, flow accumulation, flow direction, and watersheds using the Strahler equation. The drainage basins that have been selected for study contain two large valleys, as shown in Figure 5e.

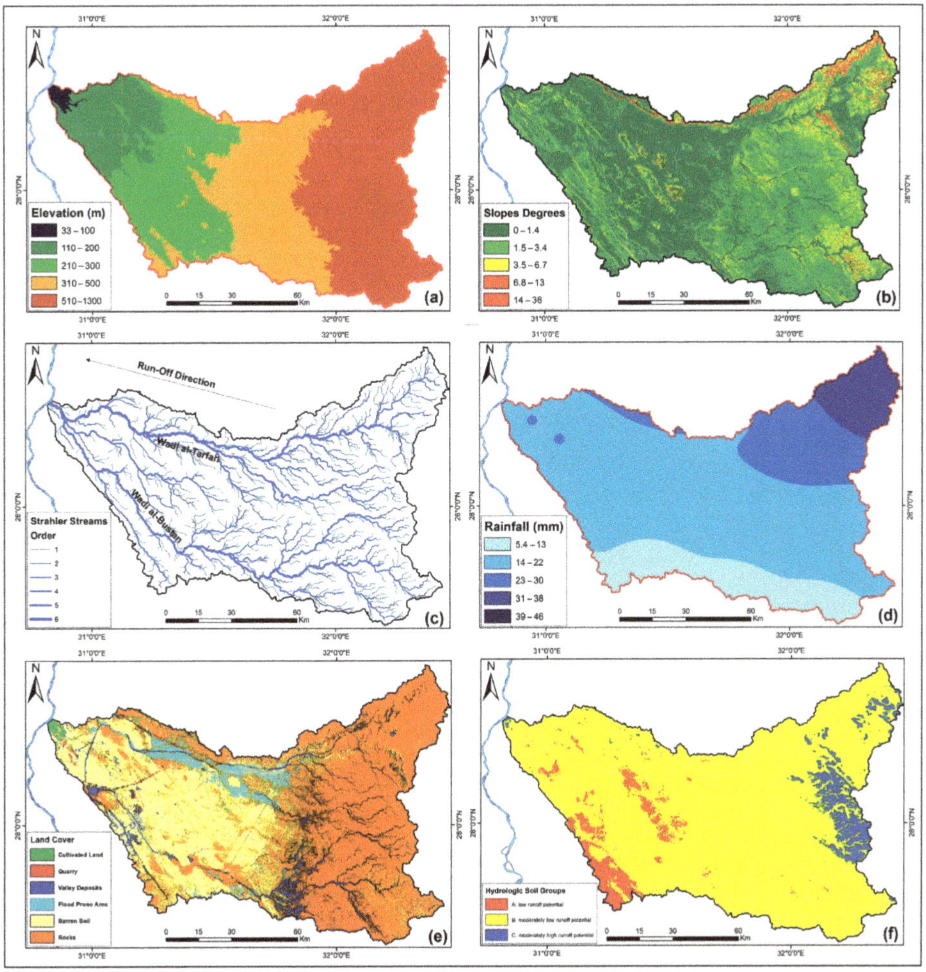

Figure 5. Multi-criteria applied in this study. (**a**) Elevation (m); (**b**) Slopes Degrees; (**c**) Streams; (**d**) Rainfall from 2010 to 2020 in mm; (**e**) Land use/Land cover in 2020; (**f**) Hydrological Soil Group based HYSOGs250m.

3.4. *Rainfall (mm)*

The climate is one of the most important factors which directly affects the intensity of flash floods and run-off processes. Satellite-based rainfall data of this study area were collected and analyzed from 2015 to 2020 (https://power.larc.nasa.gov/data-access-viewer/ (accessed on 5 February 2023)). Figure 5d shows a spatial surface estimation of the rainfall data using the IDW interpolation method.

3.5. *Land Use/Land Cover (LU/LC)*

Landsat-8 OLI imagery taken in 2022 was downloaded from the USGS website: https://earthexplorer.usgs.gov/ (accessed on 28 January 2023). Preprocessing was done, and then unsupervised classification algorithms were used to map the land use/cover classes for the study area. This map is significant for flood hazard and risk assessment.

3.6. Hydrological Soil Group (HYSOGs250m)

The global hydrological soil group data were downloaded from the website: https://daac.ornl.gov/SOILS/guides/Global_Hydrologic_Soil_Group.html (accessed on 7 February 2023) to identify the geographic distribution map of soils inside the study area; three different soil categories were detected in the study area as shown in Figure 5f.

4. Results and Discussion

Eight selected spatial criteria that applied in this study have been reclassified to hazard degrees for each factor separately from (1 to 5 values) (very low to very high), respectively, as shown in Figure 6.

4.1. Topographic Hazard Zonation

Five hazard degrees were assigned for topography; the lowest elevation land considers a higher rate of flash flood than the highest elevated area.

4.2. Slopes Hazard Zonation

The slope degrees are a topographic factor that refers to the flow speed of the rainfall-runoff water. The areas with fewer slope degrees are considered at risk of flood and inundation than the steep slope cliffs.

4.3. Distances from Wadis (Rivers) Zonation

The surrounding land of the channels is much more prone to effect by floodwater. In this study, the 3rd, 4th, 5th, and 6th stream orders are considered the main channels that are filled by water during floods. A 200 m buffer zone is considered a very high-risk zone, while the land far from 1000 m from the main channels is a safe area.

4.4. Drainage Density Hazard Zonation

It is a hydrological factor that refers to the number of streams in the study area. GIS is capable of calculating the line density of streams in the sq. km. Areas with a higher density are considered more at risk than lower-density areas.

4.5. The Permeability Hazard Map

The study area has different geological formations, and the quaternary deposits are considered permeable land (low-risk area), while the Eocene limestone formations were considered a semi-permeable zone (moderate risk). The carboniferous formation is much older and considered an impermeable zone (high risk).

4.6. The Soil Group Hazard Zonation

Three soil groups were selected. Group (A) is considered the lowest hazard zone because it refers to sand-deep sandy soils with very high intrusion rates. Group (B) is a moderate hazard zone because it has relatively fine grains of soil with moderate intrusion rates. Moreover, group (C) is considered the highest hazard degree zone in this study area because it shows fine grains of soil with low intrusion rates.

4.7. Land Use/Land Cover Hazard Map

The classification of satellite imagery of the study area produced several classes of land cover categories. The classes of dry valleys, Wadi paths, and flood-prone areas are considered the highest-risk areas, while the cultivated land is less risky.

4.8. The Precipitation Hazard Map

Rainfall data density was classified into five hazard zones according to the amount of rainfall (mm). The higher rainfall area is assigned as a higher risk area.

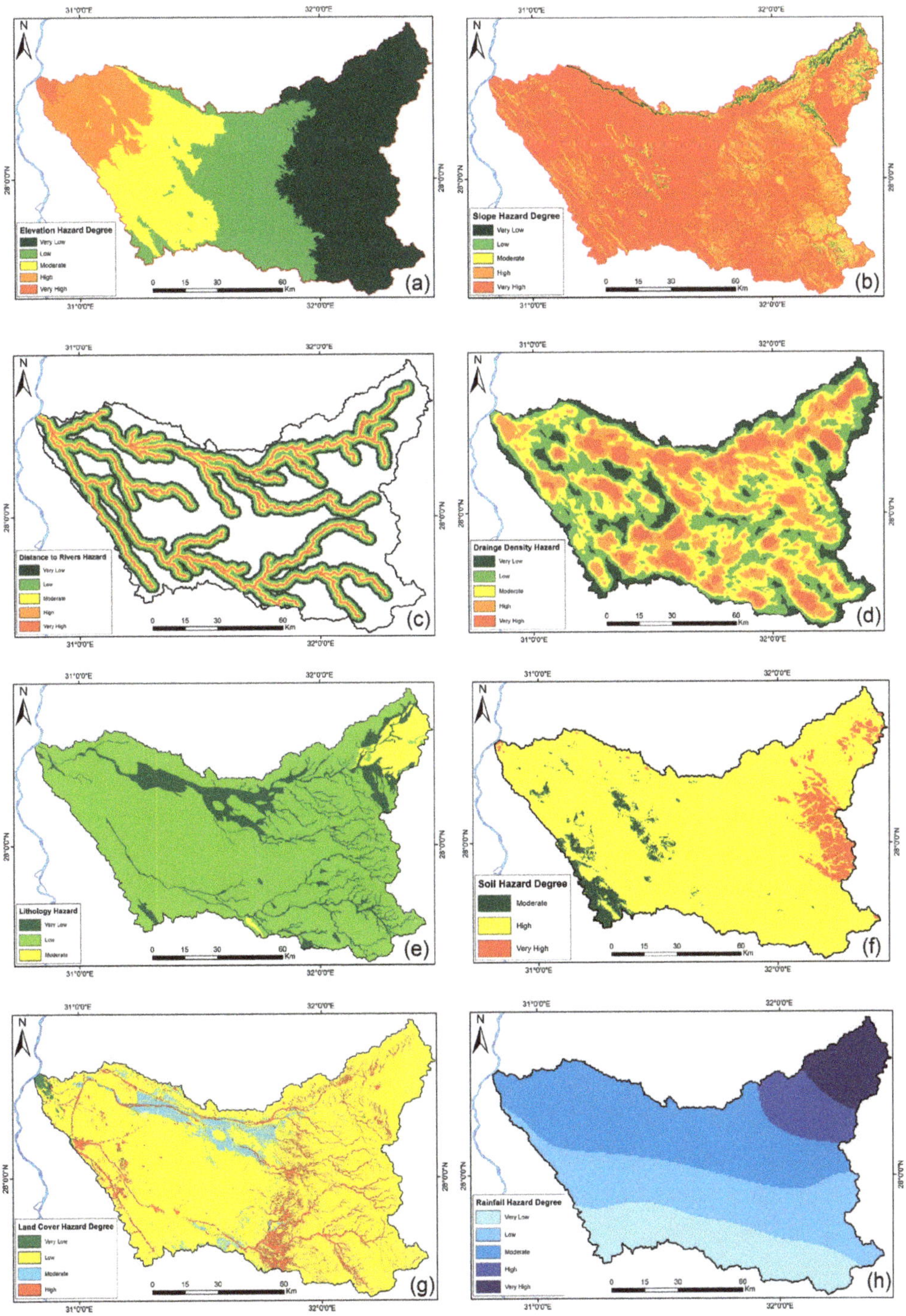

Figure 6. Spatial Multi-Criteria Hazard Zonation. (**a**) Elevation; (**b**) Slopes; (**c**) Distance from Rivers; (**d**) Drainage Density; (**e**) Lithology; (**f**) Soil Hazard; (**g**) Land Cover Hazard; and (**h**) Rainfall.

4.9. AHP Weighted Overlay Calculation

Multi-criteria decision-making analysis techniques include using the analytical hierarchy process analysis, which links multi-spatial data on the same scale. To apply these techniques, weighted overly modeling was used to calculate the weight of each factor and link all factors together using mathematical equations. Table 1. shows a pairwise comparison matrix for factor criteria (selected eight factors). In addition, the percentage of importance criteria values calculated in this study is shown in Table 2.

Table 1. Pairwise comparison matrix for factor criteria.

Factors	El	SL	Li	RF	DoR	DD	LC	Sg
El	1	2	1	1	3	7	1	1
SL	1/2	1	2	8	2	6	4	2
Li	1	1	1	3	3	3	2	3
RF	1	1/8	1/3	1	1	2	2	2
DoR	1/3	1/2	1/3	1	1	1	2	4
DD	1/7	1/6	1/3	1/2	1	1	3	3
LC	1	1/4	1/2	1/2	1/2	1/3	1	3
Sg	1	1	1/3	1/2	1/4	1/3	0	1
Sum	5.98	5.04	5.83	15.50	11.75	20.67	15.33	19.00

El: Elevation; SL: Slopes; Li: Lithology; RF: Rainfall; DoR: Distance from Rivers; DD: Drainage Density; LC: Land Cover; Sg: Soil Groups.

Table 2. Percentage values of importance using the analytic hierarchy process (AHP).

Factors	El	SL	Li	RF	DoR	DD	LC	Sg	Sum	Criteria Weight	Criteria Weight %
El	0.1673	0.3967	0.1714	0.0645	0.2553	0.3387	0.0652	0.0526	1.5118	0.1680	17
SL	0.0837	0.1983	0.3429	0.5161	0.1702	0.2903	0.2609	0.1053	1.9677	0.2186	22
Li	0.1673	0.0992	0.1714	0.1935	0.2553	0.1452	0.1304	0.1579	1.3203	0.1567	16
RF	0.1673	0.0248	0.0571	0.0645	0.0851	0.0968	0.1304	0.1053	0.7314	0.1013	11
DoR	0.0558	0.0992	0.0571	0.0645	0.0851	0.0484	0.1304	0.2105	0.7511	0.0935	10
DD	0.0239	0.0331	0.0571	0.0323	0.0851	0.0484	0.1957	0.1579	0.6334	0.0804	9
LC	0.1673	0.0496	0.0857	0.0323	0.0426	0.0161	0.0652	0.1579	0.6167	0.0785	8
Sg	0.1673	0.0992	0.0571	0.0323	0.0213	0.0161	0.0217	0.0526	0.4677	0.0620	7
Sum										1.00	100

The resulting map from this study is presented in Figure 7. Three different hazard and risk zones were identified using the previous criteria. The higher hazard degree zone was displayed in the red color in the downstream run-off water. The main highway of Elgeish Road and Minia–Ras Ghareb highway, as well as the cultivated land along the fluvial fans, will be seriously affected by the downstream water. The moderate hazard zone covers a larger area surrounded by human activities.

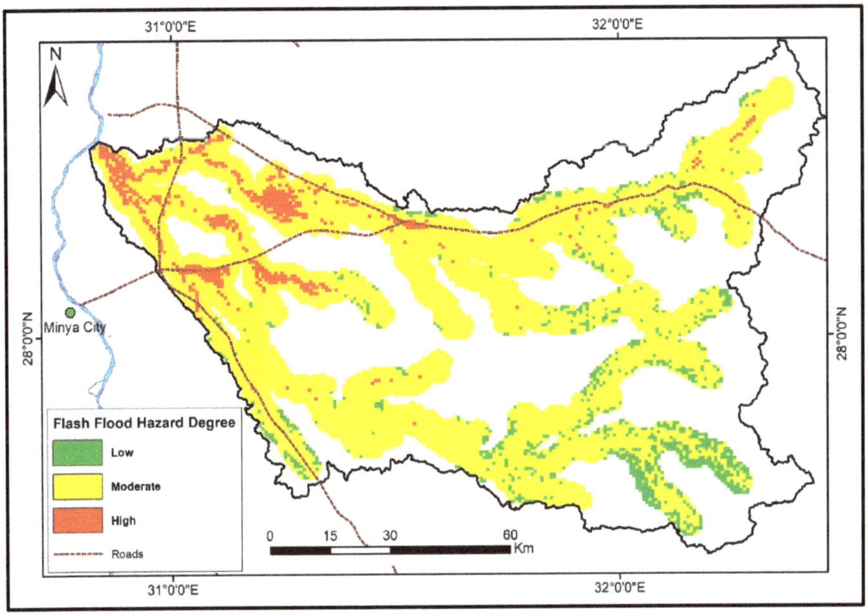

Figure 7. Flash flood hazard map based on MCDA.

5. Conclusions

This study applied the multi-criteria decision-making analysis approach in Geographic Information Systems. It was used in several studies around the world and recommended for flash flood hazard and risk mapping. There are no previous studies that applied this technique for assessing flash floods in this study area. This technique is considered the most effective because it has the ability to link multi-sources spatial data at the same scale to detect the best geospatial solution for decision-makers. In this study, the available criteria were used and combined in the weighted overly modeling in GIS to map the vulnerable areas with a flash flood in a watershed basin in Middle Egypt. The results of this study show a higher flood zone in the downstream area; this area is occupied by highways, cultivated lands, and human settlements. It is highly recommended to use MCDA-based AHP techniques for flash flood hazard and risk mapping in future studies.

Funding: This research received no external funding.

Institutional Review Board Statement: Not applicable.

Informed Consent Statement: Not applicable.

Data Availability Statement: All the data of the study are presented in the paper.

Conflicts of Interest: The author declares no conflict of interest.

References

1. WMO. World Meteorological Organization: Flash Flood Guidance System (FFGS) with Global Coverage. 2016. Available online: https://community.wmo.int/hydrology-and-water-resources/flash-flood-guidance-system-ffgs-global-coverage (accessed on 27 January 2023).
2. El Gohary, R. Environmental Flash Flood Management in Egypt. In *Flash Floods in Egypt*; Negm, E.E., Ed.; Advances in Science, Technology & Innovation; Springer: New York, NY, USA, 2020; pp. 85–105.
3. Elnazer, A.A.; Salman, S.A.; Asmoay, A.S. Flash flood hazard affected Ras Gharib city, Red Sea, Egypt: A proposed flash flood channel. *Nat. Haz.* **2017**, *89*, 1389–1400. [CrossRef]

4. El-Saadawy, O.; Gaber, A.; Othman, A.; Abotalib, A.Z.; El Bastawesy, M.; Attwa, M. Modeling Flash Floods and Induced Recharge into Alluvial Aquifers Using Multi-Temporal Remote Sensing and Electrical Resistivity Imaging. *Sustainability* **2020**, *12*, 10204. [CrossRef]
5. Awange, J.; Kiema, J. Fundamentals of GIS. In *Environmental Geoinformatics. Environmental Science and Engineering*, 2nd ed.; Springer: Cham, Switzerland, 2019; pp. 203–212.
6. Karymbalis, E.; Andreou, M.; Batzakis, D.-V.; Tsanakas, K.; Karalis, S. Integration of GIS-Based Multicriteria Decision Analysis and Analytic Hierarchy Process for Flood-Hazard Assessment in the Megalo Rema River Catchment (East Attica, Greece). *Sustainability* **2021**, *13*, 10232. [CrossRef]
7. Allafta, H.; Opp, C. GIS-based multi-criteria analysis for flood prone areas mapping in the trans-boundary Shatt Al-Arab basin, Iraq-Iran. *Geomat. Nat. Hazards Risk* **2021**, *12*, 2087–2116. [CrossRef]
8. Malczewski, J. GIS-based multicriteria decision analysis: A survey of the literature. *Int. J. Geogr. Inf. Sci.* **2006**, *20*, 703–726. [CrossRef]
9. Drobne, S.; Lisec, A. Multi-attribute Decision Analysis in GIS: Weighted Linear Combination and Ordered Weighted Averaging. *Informatica* **2009**, *33*, 459–474.
10. Kazakis, N.; Kougias, I.; Patsialis, T. Assessment of flood hazard areas at a regional scale using an index-based approach and Analytical Hierarchy Process: Application in Rhodope-Evros region, Greece. *Sci. Total Environ.* **2015**, *538*, 555–563. [CrossRef] [PubMed]
11. Elsheikh, R.F.A.; Ouerghi, S.; Elhag, A.R. Flood Risk Map Based on GIS, and Multi Criteria Techniques (Case Study Terengganu Malaysia). *J. Geogr. Inf. Syst.* **2015**, *7*, 348–357. [CrossRef]
12. Abdelkarim, A.; Al-Alola, S.S.; Alogayell, H.M.; Mohamed, S.A.; Alkadi, I.I.; Ismail, I.Y. Integration of GIS-Based Multicriteria Decision Analysis and Analytic Hierarchy Process to Assess Flood Hazard on the Al-Shamal Train Pathway in Al-Qurayyat Region, Kingdom of Saudi Arabia. *Water* **2020**, *12*, 1702. [CrossRef]
13. Das, S. Flood susceptibility mapping of the Western Ghat coastal belt using multi-source geospatial data and analytical hierarchy process (AHP). *Remote Sens. Appl. Soc. Environ.* **2020**, *20*, 100379. [CrossRef]
14. Abu El-Magd, S.A.; Amer, R.A.; Embaby, A. Multi-criteria decision-making for the analysis of flash floods: A case study of Awlad Toq-Sherq, Southeast Sohag, Egypt. *J. Afr. Earth Sci.* **2020**, *162*, 103709. [CrossRef]

environmental sciences proceedings

MDPI

Proceeding Paper

Correlation between Land Transformation and Climate Change with Flooding Vulnerability: Nature-Based Solutions (NBS) Applied in the Mar Menor Mediterranean Watershed †

Salvador García-Ayllón

Department of Mining and Civil Engineering, Technical University of Cartagena, Paseo Alfonso XIII, 52, 30203 Cartegena, Spain; salvador.ayllon@upct.es

† Presented at the 7th International Electronic Conference on Water Sciences, 15–30 March 2023; Available online: https://ecws-7.sciforum.net.

Abstract: The combination formed by the intense transformation of the territory and climate change in the Spanish Mediterranean basins has configured an explosive cocktail from the point of view of the risk of flooding in these areas. Climate change is making new, intense-type cold-drop rains called DANA a more frequent phenomenon in Mediterranean basins. The vulnerability of the coastal territory in these areas because of the DANA phenomenon now requires the authorities to implement new strategies and policies that reduce the significant economic damage and loss of human life suffered in recent years. However, correlating these two phenomena with the increased risk of flooding is a difficult problem to diagnose, and even more difficult to solve. For this reason, a structured GIS methodology is proposed based on a geostatistical indicators analysis that correlates the transformation of the territory with the increase in vulnerability due to floods. This methodology is evaluated and put into practice in a case study of the basin area of a coastal lagoon in Spain. The impacts originating from the last three cold-drop phenomena developed in recent years are evaluated. The analysis carried out allows us to observe how the phenomenon of diffuse territorial anthropization in the territory has increased the damage caused by floods due to the loss of the natural hydrographic structure of the basin. Based on the results obtained in this analysis, risk mitigation actions will be proposed through the improvement of land management by using nature-based solutions.

Keywords: flooding risk; GIS; DANA; Mar Menor; land transformation; nature-based solutions

Citation: García-Ayllón, S. Correlation between Land Transformation and Climate Change with Flooding Vulnerability: Nature-Based Solutions (NBS) Applied in the Mar Menor Mediterranean Watershed. *Environ. Sci. Proc.* **2023**, *25*, 88. https://doi.org/10.3390/ECWS-7-14240

Academic Editor: Athanasios Loukas

Published: 16 March 2023

1. Introduction

There are numerous effects that the scientific community has managed to associate with climate change, with those related to the problem of floods being some of the most dangerous. Among the areas of the planet where these changes are forcing administrations to rethink the methodologies to address this problem are the Spanish Mediterranean regions. There, the appearance of new cold-drop phenomena called DANA (Spanish acronym for upper-level isolated atmospheric depression) has replaced the traditional flash floods associated with historical fall storm events.

In these regions, a great process of territorial anthropization has developed in recent decades. The Mediterranean area is undoubtedly the region in which tourism, agriculture, infrastructure expansion, etc., has experienced the greatest degree of growth in recent decades. However, it is not easy to assess how this new phenomenon of anthropization directly affects the growing problem of flooding in a territory. Phenomena of a diffuse nature and with a rather indirect impact, such as the sealing effect of the soil due to urbanization processes of the land, orographic alterations produced by changes in land use and dam micro-effects that linear communication infrastructures can generate currently do not have specific methodological approaches in the traditional scientific literature. Several approaches for specific case studies can be found, but comprehensive large-scale evaluation methods from a territorial point of view are still lacking.

This study raises an innovative methodological approach based on the assessment of GIS indicators using geostatistics and nature-based solutions as a mitigation strategy. This new method assesses the spatial correlation between the territorial transformation derived from anthropic procedures with the growing risk of a territory to floods. To evaluate this new framework, the proposal is applied to the Campo de Cartagena area in south-eastern Spain, experiencing devastating floods in the last decade.

2. Area of Study and Methodology

2.1. Area of Study

The area of study is located in the Region of Murcia, a semi-arid area in south-eastern Spain (Figure 1). This region is distinguished by limited rainfall (<300 mm/year), which mainly occurs during storm events in autumn and winter. This watershed has been extensively transformed from a relatively traditional dryland agricultural area to one of intense human activity in the last 50 years despite having significant environmental value [1].

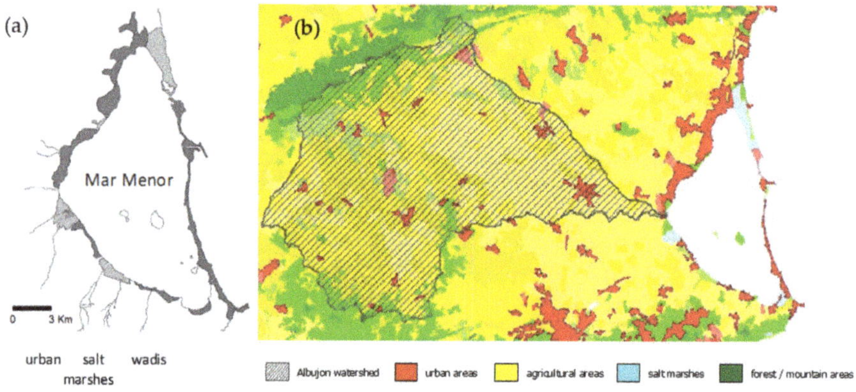

Figure 1. (**a**) Wadis going to the Mar Menor lagoon (**b**) Watershed nourishing Albujon wadi.

Freshwater inputs into Mar Menor come mainly from six main ephemeral watercourses called 'wadis'. These wadis can reach lengths of over 50 km. The watershed area that nourishes some of them, the Albujon, even exceeds that of the Mar Menor Lagoon itself. These wide and shallow channels are not usually active but are able to hold huge amounts of water and sediment during the usual flood periods of autumn and winter in the Mediterranean. The torrential nature of the water supply is now worsened by the impermeable soils and scarce vegetation cover of the watershed areas, going directly to the coastal lagoon. In recent decades, there has been a great transformation of the territory that has coincided with a sharp increase in the damage caused by the rains, being especially relevant those produced in the years 2016, 2018, and 2019 ([2]).

2.2. Methodology

The framework used to analyze the impact of the phenomenon of diffuse territorial anthropization and its relationship with the increase in damage caused by floods is detailed below. A geostatistical assessment was developed to analyze the spatial correlation of the evolution of the selected indicators of territorial anthropization and the increase in damage derived from floods. This spatial statistical analysis was developed using the Local Indicators of Spatial Association (LISA). These statistical indicators evaluate the level of correlation between the behavior patterns of each of the spatial indicators of transformation of the territory generated and the distribution of the levels of damage detected in the last three DANAs. Once the numerical analysis is carried out, the level of two-dimensional correlation is analyzed from the graphic point of view through the Getis-Ord Gi* statistic indicator. We next detail the methodological process of these stages [3].

A couple of GIS datasets were introduced in this model to develop the indices that acted as assessment parameters in the geostatistical evaluation:

- Georeferenced rates of the damage that happened during the three DANAs analyzed.
- Spatiotemporal behavior of the most relevant GIS land anthropization indices used.

To examine the spatial distribution of damage produced by DANA, we generated an index called the Flood Damage Severity index (I_{FDS}). From the information provided by the emergency services from various local and regional authorities which oversee processing the damage files of those affected by the flooding events, we generated a spatial qualitative punctual database of damage (Figure 2). These data were obtained on aggregate units such as tourist resorts, residential buildings, industrial estates, shopping centers, etc., to make georeferenced treatment possible while preserving the legal requirements of anonymity. To achieve a uniform structure of discrete values, the alphanumeric data were classified into three categories based on the level of significance of the damage following criteria, as in Table 1: minor, relevant, and catastrophic damage.

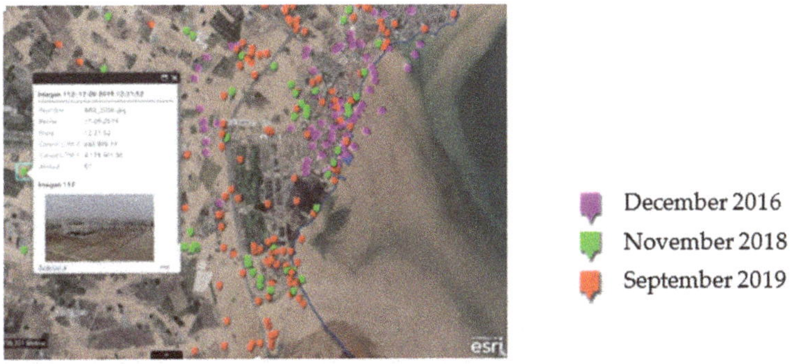

Figure 2. Detail of spatial distribution of the GIS dataset developed for damage. Source: authors using data from several emergency administrations.

Table 1. I_{FDS} criteria for the level of significance of damage.

	Minor Damage	Relevant Damage	Catastrophic Damage
Impact on commu-nications	Temporary pathblocks	Temporary roadblocks	Block on highways or critical road infrastructure
Damage to buildings and material goods	Flooding of garages in buildings	Flooding inside homes and main areas of buildings	Structural destruction or complete loss of the value of material goods
Agriculture and Livestock	Partial crop damage	Complete loss of the crop	Tree uprooting or drowning of ani-mals
Operation of public services	Malfunction of non-essential services (e.g., streetlamps)	Temporary outages of essential supply services (water, elec-tricity)	Permanent or prolonged outages of essential supply services (water, electricity)
Environmental damage	Damage to unprotected areas with no environ- mental, historical, or cultural value.	Recoverable damage to pro-tected areas (e.g., loss of the beach line)	Irreversible damage in protected areas

On the other hand, the land transformation phenomena were assessed from a spatial point of view using GIS indicators of anthropization. This spatial transformation generates sub-phenomena such as the "dam effect" and "soil sealing" or alterations in the orography of the land that, according to the literature checked, may have important effects on the hazard of flooding of a territory. By using historic GIS cartographic datasets, the progress of various dimensionless indices linked with these sub-phenomena was calculated statistically over time. The indices used to evaluate the patterns of land transformation in the area of study can be found below:

Linear infrastructure density index (LID): weighted indicator of the level of territorial density of the linear communication infrastructures (these elements usually generate "dam micro-effects").

$$LID = \frac{\sum h_i L_i^2}{S_{tr}}$$

L_i = length of existing linear infrastructures (m)/h_i = weighting coefficient (high-capacity motorway = 1, traditional road = 0.75, country path = 0.5)/S_{tr} = sector of the area of study (m^2)

Index of soil artificialization (SA): weighted indicator of the level of artificial soil transformation associated with urbanization phenomena by applying the Corine Land Cover criteria (these processes usually generate "soil-sealing" effects) [4].

$$SA = \frac{\sum h_i A_i}{S_{tr}}$$

A_i = Urbanized surface area (m^2)/h_i = weighting coefficient (highly artificial surface = 1, medium artificial surface = 0.75, low-waterproof artificial surface = 0.5)/S_{tr} = sector of the area of study (m^2)

Index of alteration of the terrain orography (ATO): weighted indicator associated with orographic transformations or land-use changes in the territory [5].

$$ATO = \frac{\sum h_i A_i}{S_{tr}}$$

A_i = land-use-altered surface (m^2)/h_i = weighting coefficient (relevant orographic alterations = 1, medium alterations = 0.75, similar or partial land-use change = 0.5)/S_{tr} = sector of the area of study (m^2)

3. Results

Based on the framework proposed in the methodology section, the following results were achieved. First, the existence of geostatistical significance in the distribution of the behavior patterns of the different indicators analyzed was verified. Subsequently, the levels of statistical correlation between each of the indicators of anthropic transformation and the spatial distribution of the damage detected in the last DANA were studied numerically. This evaluation was carried out considering an evolutionary approach in the 1956–2020 period to analyze the incidence of each of the indicators of territorial transformation in the global land transformation process. Then, the level of statistical correlation of each of those anthropic sub-phenomena with the intensity of damage caused by floods was assessed. To implement this analysis, the hot- and cold-spot indicators from LISA two-dimensional statistics were used.

If we parameterize the values reached to the numerical level in a bivariate Getis-Ord Gi* statistic of hot and cold spots from a spatial perspective, we can see clearly differentiated areas in the watershed, as displayed in Figure 3.

As can be verified, diffuse territorial anthropization is a phenomenon whose effects are not easy to appreciate since it gradually accumulates over decades. However, once the impacts corresponding to this phenomenon surface, the consequences, as seen in the latest DANA in the area, can become catastrophic. In this sense, it is interesting to propose large-scale solutions that do not imply the development of aggressive infrastructures with nature to mitigate the effects of these phenomena.

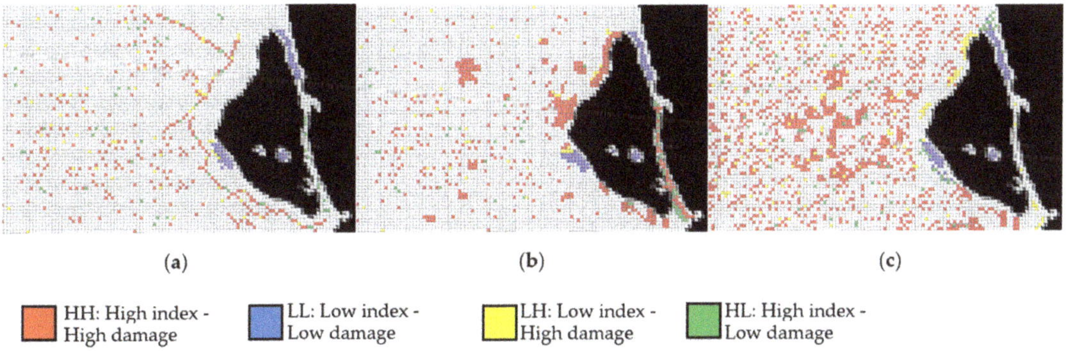

Figure 3. Hot and cold spot Getis-Ord Gi* statistical analysis for spatial statistical correlation between damage and anthropization indices: (**a**) LID - I_{FDS} (**b**) SA - I_{FDS}, and (**c**) ATO - I_{FDS}.

So-called nature-based solutions (NBS) are alternatives that are friendlier to the territory and whose ability to integrate into the natural environment makes them ideal alternatives to solve large-scale problems of this nature. They are tools traditionally used on an urban scale, such as the so-called sustainable urban drainage systems (SUDS). However, their combined approach by aggregating various actions on a large supra-municipal scale could be an innovative solution in this case from the point of view of integrated land management. In this context, it has been proposed, as a conceptual approach the assignment of different alternative solutions based on nature at a subregional scale as mitigation elements of the effects of this problem in the Mar Menor drainage basin. The LISA analysis of hot and cold spots carried out shows how the least-anthropized areas with the highest surface runoff retention and absorption capacity have been those that respond with the best capacity to extreme weather events.

Consequently, strategic planning is proposed at a subregional level, assigning the most appropriate alternatives to the local problems that have been observed based on the spatial statistical analysis carried out. Based on the levels of damage and the anthropic problems detected as their cause, the following NBSs have been proposed based on the results obtained for the three previously studied GIS indicators: vegetated roofs, permeable pavements, and rain gardens in urban areas, infiltration ditches, vegetated ditches, and floodable bioretention beds near linear infrastructures, and, finally, green filters, detention tanks, and retention ponds in large peri-urban areas. An initial allocation of proposals as a first approximation to this action can be seen in greater detail at a supra-municipal scale in Figure 4.

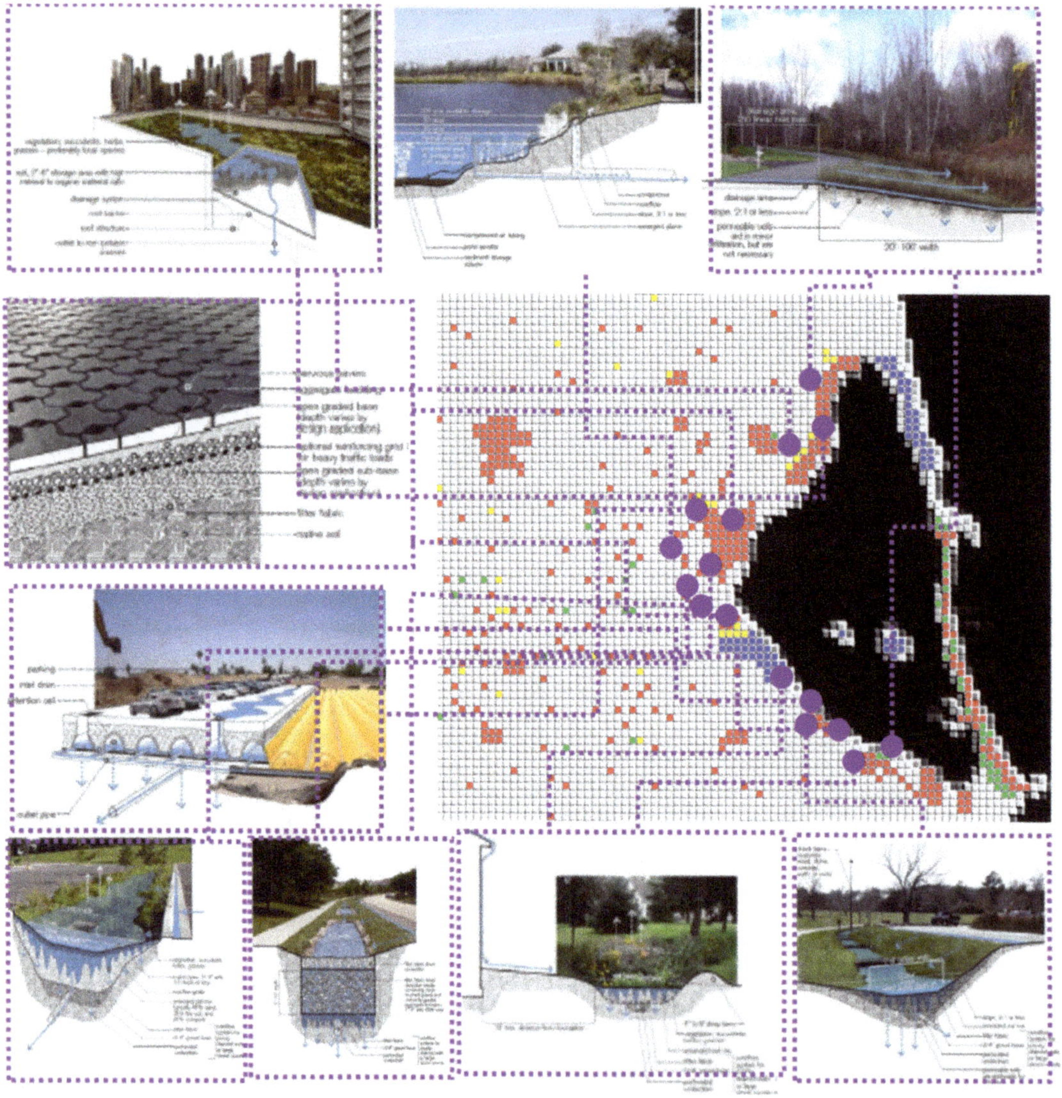

Figure 4. Different alternative solutions based on nature proposed in the context of the Mar Menor watershed as a solution to mitigate the effects of damage caused by floods.

4. Discussion and Conclusions

The work above establishes a new methodological approach which can be presented as a disruptive option in the assessment of the hazard of flooding. The analysis of the relationship between climate change, land transformation, and the increase in vulnerability to flooding is an unusual approach in the study of flood risk, which frequently tends to focus its assessment mostly on hydrological parameters rather than land-use parameters.

At the specific level, it has been possible to verify how the use of land management as a nature-based solution can be a very useful tool for mitigating this phenomenon. We have the atmospheric phenomenon of DANAs in coastal Mediterranean regions which, as highlighted by several studies [6], are increasing in frequency and intensity because of climate change. This issue, although fitting rather to other scientific disciplines, forces us to be vigilant to this phenomenon from the point of view of land-use planning, since its effects on the local environment will also be growing.

Considering the findings achieved through the geostatistical analysis, the anthropization of the territory in the area during recent decades has had a substantial impact on the expanding risk of the Mar Menor watershed to flooding episodes. The loss of the natural hydrographic network of the basin has visibly boosted vulnerability to flooding with a distinct impact in the harm caused to people, economic activities, and highly valued ecological protected areas. Nevertheless, as in other parts of the Spanish Mediterranean façade [7], this relationship of cause and effect does not seem to be homogeneous.

Funding: This research received no external funding.

Institutional Review Board Statement: Not applicable.

Informed Consent Statement: Not applicable.

Data Availability Statement: Data are unavailable due to privacy restrictions.

Conflicts of Interest: The author declares no conflict of interest.

References

1. Garcia-Ayllon, S. Long-Term GIS Analysis of Seaside Impacts Associated to Infrastructures and Urbanization and Spatial Correlation with Coastal Vulnerability in a Mediterranean Area. *Water* **2018**, *10*, 1642. [CrossRef]
2. AEMET. *Informe Meteorologico Región de Murcia Septiembre 2019*; MITECO: Madrid, Spain, 2019.
3. Garcia-Ayllon, S.; Radke, J. Geostatistical Analysis of the Spatial Correlation between Territorial Anthropization and Flooding Vulnerability: Application to the DANA Phenomenon in a Mediterranean Watershed. *Appl. Sci.* **2021**, *11*, 809. [CrossRef]
4. Pistocchi, A.; Calzolari, C.; Malucelli, F.; Ungaro, F. Soil Sealing and Flood Risks in the Plains of Emilia-Romagna, Italy. *J. Hydrol. Reg. Stud.* **2015**, *4*, 398–409. [CrossRef]
5. Jodar-Abellan, A.; Valdes-Abellan, J.; Pla, C.; Gomariz-Castillo, F. Impact of Land Use Changes on Flash Flood Prediction Using a Sub-Daily SWAT Model in Five Mediterranean Ungauged Watersheds (SE Spain). *Sci. Total Environ.* **2019**, *657*, 1578–1591. [CrossRef] [PubMed]
6. Pastor, F.; Valiente, J.A.; Palau, J.L. Sea Surface Temperature in the Mediterranean: Trends and Spatial Patterns (1982–2016). *Pure Appl. Geophys.* **2018**, *175*, 4017–4029. [CrossRef]
7. Martínez Ibarra, E. A Geographical Approach to Post-Flood Analysis: The Extreme Flood Event of 12 October 2007 in Calpe (Spain). *Appl. Geogr.* **2012**, *32*, 490–500. [CrossRef]

environmental sciences proceedings

MDPI

Proceeding Paper

Estimation of Water Turbidity in Drinking Water Treatment Plants Using Machine Learning Based on Water and Meteorological Data †

Vanessa Fernandez Alvarez *, Daniela Granada Salazar, Cristhian Figueroa, Juan Carlos Corrales and Juan Fernando Casanova

Department of Telematics, University of Cauca, Popayan 190002, Colombia;
danioandre@unicauca.edu.co (D.G.S.); cfigmart@unicauca.edu.co (C.F.); jcorral@unicauca.edu.co (J.C.C.); juancasanova@unicauca.edu.co (J.F.C.)
* Correspondence: jaidy@unicauca.edu.co

† Presented at the 7th International Electronic Conference on Water Sciences, 15–30 March 2023; Available online: https://ecws-7.sciforum.net.

Abstract: In rural areas, water treatment plants use rudimentary techniques to evaluate turbidity. However, the incorrect measurement of turbidity can result in poor water quality and, as a result, health issues for its users because it is a crucial indicator to determine the application of adequate treatment to the water. Aquarisc was a project financed with royalties that sought to strengthen the mechanisms and tools for decision-making of the authorities and territorial institutions related to water supply for human consumption. This project installed sensors to assess turbidity in some plants in rural areas of the department of Cauca, Colombia. However, when the project ended, these sensors were removed. Therefore, it became necessary to create machine learning models to predict turbidity values without sensors, considering only pH, temperature, vapor pressure, and precipitation data captured manually by plant operators. In this study, the Linear Regression, Random Forest Regressor, k-Neighbors Regressor, and Extra Trees Regressor algorithms were trained with data provided by the Aquarisc project and the Institute of Hydrology, Meteorology, and Environment Studies of Colombia (IDEAM). As a result, we selected the Random Forest Regressor since it had the best RMSE among all the models and was also the one that best matched the situation of the studied treatment plants. Furthermore, this model did not consider outliers, resulting in an RMSE of 20.98 and 3.49 for the training and test dataset, respectively. Finally, we determined that this algorithm was able to estimate the water's turbidity acceptably and supports the operators in making decisions for the application of adequate treatment to drinking water.

Keywords: turbidity; water treatment; random forest; turbidity estimation

check for updates

Citation: Fernandez Alvarez, V.; Granada Salazar, D.; Figueroa, C.; Corrales, J.C.; Casanova, J.F. Estimation of Water Turbidity in Drinking Water Treatment Plants Using Machine Learning Based on Water and Meteorological Data. *Environ. Sci. Proc.* **2023**, *25*, 89. https://doi.org/10.3390/ECWS-7-14326

Academic Editor: Lampros Vasiliades

Published: 3 April 2023

1. Introduction

The quality of water supplied by potable water treatment plants is crucial for ensuring consumers' health. Turbidity is one of the most critical parameters to determine this quality, as it indicates the presence of colloidal, mineral, and organic substances in the water [1].

The presence of turbidity in water stimulates the proliferation of bacteria and provides coverage and food for pathogens in the water. Moreover, it hinders proper water disinfection and risks consumer health [2]. For this reason, entities supplying water for human consumption in Colombia must ensure compliance with the water quality standards established in Resolution 2115 of 2007, which establishes that turbidity should not exceed 2 Nephelometric Turbidity Units (NTU) [3].

Currently, there are various pieces of equipment and instruments, such as turbidity tubes, Secchi disks, and spectrophotometers, which are used to measure the turbidity of water. However, according to studies carried out by [4,5], and the project "Vulnerability and Risk in Drinking Water Systems of Cauca–Aquarisc", it has been demonstrated that some municipal water treatment plants in Colombia face difficulties in determining the physicochemical conditions of the water. These limitations prevent treatment plants from measuring raw water parameters, including turbidity. It is essential to know the turbidity, as this parameter is crucial for establishing the criteria required at each stage of the purification process and carrying out a planned and precise water purification process [6].

An effective strategy to address this limitation is establishing a model for estimating source water turbidity to support operators in decision-making. Several studies support this assertion, including those conducted in [7–10], where artificial neural networks were developed to predict or estimate turbidity concentrations and other water quality parameters. In addition, studies such as [11,12] performed regression analyses considering satellite images and data acquired by citizen scientists for turbidity prediction. Moreover, more research related to the topic was found, but it was noted that research of this type was scarce in tropical zones.

The lack of exploration in these areas creates a gap in the analysis of the behavior of meteorological variables in the type of climate that occurs and the impact it has on water quality parameters, such as turbidity. Colombia is one of the countries belonging to the tropical zone, so it has a relatively varied climatic classification due to the lack of seasonality. Therefore, creating a local experience based on what has been evaluated by countries with different climatic zones than Colombia would contribute to research in analyzing the behavior of meteorological and hydrological variables about the turbidity parameter.

Therefore, this study evaluates the performance of models that have not been implemented in tropical zones and compares them to determine their applicability in environments with hydrological and meteorological characteristics similar to the local environment, to expand experience in this type of climatic zone for the estimation of turbidity and its relationship with various parameters.

2. Materials and Methods

2.1. Dataset

The Aquarisc project team collected the data using sensors to measure water quality parameters and meteorological variables in some water treatment plants in the department of Cauca (Colombia). The measurement dataset was recorded from January 2020 to January 2021. However, some measurement failures were found in the precipitation parameter when performing an initial analysis. For this reason, we used a meteorological database provided by the Institute of Hydrology, Meteorology and Environment Studies (IDEAM) accessed through the open data portal on its website [13].

The following parameters were used in this study: pH, dissolved oxygen (DO), conductivity, oxidation-reduction potential (ORP), turbidity, temperature (T), relative humidity (RH), vapor pressure (VP), barometric pressure (BP), wind speed, precipitation (P), and radiation.

2.2. Data Modeling

2.2.1. Data Pre-Processing

The dataset preparation process consisted of three phases: data structuring, cleaning, and fusion.

- Structuring phase: The first step in the process of structuring the dataset was to organize the information in a single format since there were two datasets, one of the hydrological parameters and the other of meteorological parameters. The meteorological dataset had shifted columns and needed to be organized to execute the analysis correctly.
- Data cleaning phase: Considering the technical specifications of the different sensors, it was determined which data were outside the specified range for each measurement tool. We continued reviewing which treatment plants had sufficient turbidity data,

which resulted in a dataset consisting of one treatment plant in the urban area and three treatment plants in the rural area.

- Data fusion phase: Once the data were organized and cleaned, a data fusion was performed with the dataset taken by IDEAM. Since both sets did not have the same time scale, a temporal adaptation was made to the IDEAM dataset to match each point in the database obtained from the previous phases.

2.2.2. Exploratory Analysis of Data

- Distribution of water turbidity: When evaluating the distribution of turbidity in the four treatment plants, distributions with long skews to the right are found. To correct the variable's asymmetry and to obtain a better view of its distribution, a logarithmic transformation was applied to make the water turbidity as close as possible to the normal distribution. Given the large number of zero values in the turbidity measurements, an additional unit was added to avoid $-\infty$ values. The following equation was applied for the logarithmic transformation of the water turbidity variable:

$$logTurbidity = log(turbidity + 1) \quad (1)$$

We can better understand the distribution of variables by plotting histograms for each station. For example, Figures 1 and 2 show the histograms for the log transformation of water turbidity at four relevant stations.

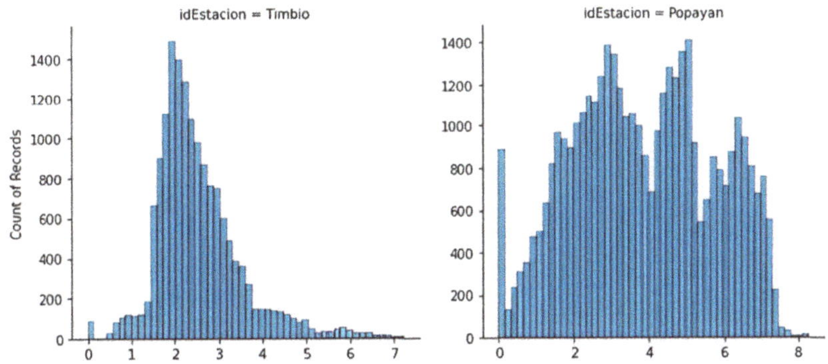

Figure 1. Distribution of the logarithmic transformation of the water turbidity variable in the four stations with available data (Timbio and Popayan) after data cleaning.

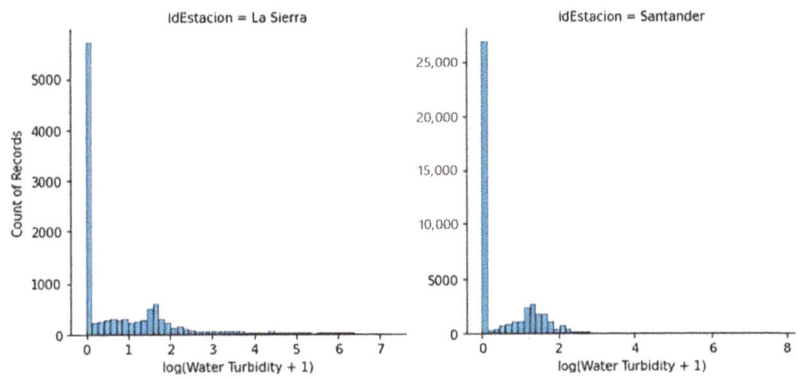

Figure 2. Distribution of the logarithmic transformation of the water turbidity variable in the four stations with available data (La Sierra and Santander) after data cleaning.

It can be observed that there is more significant variability in the Popayan station, which is in the urban zone. However, this indicator cannot be taken as a conclusion since other water treatment stations may have even more significant variability and higher generic water turbidity levels, but the data are simply insufficient.

- Analysis of the relationship of variables with turbidity: A correlation analysis was conducted to investigate the possible relationship between turbidity and various water variables at all stations. In the initial phase of the study, no significant correlation was found between turbidity and variables, such as pH, DO, conductivity, and ORP. Thus, we performed an additional analysis to improve the correlation, consisting of the logarithmic transformation of the turbidity variable and the correlation with the water variables from the first analysis. Although an increase in correlation coefficients was observed, they were still considered low, indicating no linear relationship between turbidity and the examined water variables. This result suggests the need for a non-linear model to capture the complexity of the relationship between these variables. Otherwise, models could generate inaccurate data or significant error metrics, making interpretation and decision-making difficult.

2.2.3. Data Modeling

For data modeling, a process of experimentation was carried out in which the following variations were considered:

- Linear and non-linear models (linear regression, tree-based models).
- With or without outliers.
- With random or temporal division.
- With all stations, with Popayan information only, or with all stations except Popayan.
- With dummy variables that categorize the station.
- With all parameters or only with the most important ones.

The metric used to determine the best model was the RMSE.

3. Results and Discussion

During the linear modeling process, the database of the four stations was considered, and some variations were performed to determine which model offered better performance. The results obtained are presented in Table 1.

Table 1. Summary of linear modeling.

Database	Data Division	Outlier Limit (NTU)	RMSE—In the Sample	RMSE: Sample Output
Stations (dummies), predictors: all parameters, turbidity prediction	Random	-	202.30	206.93
Stations (dummies), predictors: all parameters, box cox transformation and, log-turbidity prediction	Random	-	225.83	223.54
Stations (cat codes), predictors: pH, conductivity, potOxyReduction, PA, T and P, turbidity prediction, and data normalization.	Random	-	202.16	203.39
Stations (cat codes), predictors: pH, conductivity, potOxyReduction, PA, T and P, turbidity prediction, and data normalization.	Random	62.6	8.85	8.85

Upon analyzing the results obtained from linear modeling, it was observed that high RMSE values were obtained, suggesting no clear linear relationship between the parameters and the turbidity variable. Consequently, alternative models that could better fit the data were sought, and a non-linear analysis was conducted using different algorithms.

The k-Neighbors Regressor algorithm was employed as the first non-linear model, and the database composed of the four stations was considered. The second model was based on the extra Trees Regressor algorithm and evaluated solely at the Popayan station. Finally, the best-performing model was based on the Random Forest algorithm and evaluated in the database of the three rural stations.

For all three models, all parameters were used as predictors, and it was found that their performance significantly improved when a temporal data split was performed, using the month of December as the test set. Table 2 summarizes the results of these models and their corresponding performance metrics.

Table 2. Summary of non-linear modeling.

Model	Outlier Limit (NTU)	RMSE—In the Sample	RMSE: Sample Output
Stacking-Estimator K-Neighbors-Regressor	300	0.0	8.0
Stacking-Estimator Extra-Trees-Regressor	-	7.38	9.26
Make-union Extra-Trees-Regressor	300	17.94	83.56
Make-union Extra-Trees-Regressor	-	79.68	118.61
Stacking-Estimator Random-Forest-Regressor	300	1.83	2.37
Stacking-Estimator Random-Forest-Regressor	-	20.92	3.76

Based on the obtained results, it has been verified that the random forest model performs best in predicting water turbidity. On the other hand, the K-Neighbors-based model showed overfitting during data training, while the Extra-Trees-based model yielded a high RMSE value. As a result of these findings, a more detailed evaluation of the best-performing model was carried out to reduce the number of predictors and adapt it to the limited instrumentation used in rural water treatment plants.

As a result of this evaluation, a model that only uses predictors of pH, T, VP, and P has been obtained, which has been the most effective in predicting water turbidity. In addition, the selection of these predictors has reduced the number of parameters considered in the model without compromising the accuracy of the predictions. It is important to note that, during the evaluation, it was observed that meteorological parameters significantly influence the variation of water turbidity, suggesting that the weather conditions in the area significantly impact water turbidity.

In summary, the research results have demonstrated the possibility of obtaining a non-linear predictive model of water turbidity with fewer predictors, which is particularly beneficial in rural areas where instrumentation is limited. Furthermore, the importance of including meteorological parameters as input variables in the model have been emphasized. Their significant impact on water turbidity can be crucial in predicting the parameter and, therefore, helping to make decisions for appropriate water treatment.

4. Conclusions

Implementing machine learning models enables the analysis of water turbidity behavior and facilitates decision-making in treatment plants with limited instrumentation. The results indicate no linear relationship between the database parameters and turbidity, which led to the evaluation of non-linear models that better estimate water turbidity. Although these models show slightly lower performance when reducing the number of predictors, they allow for the appreciation of climatic influence and can be applied more efficiently in treatment plants with limited resources. In addition, by setting a limit on turbidity to exclude outlier values, the performance of the models was improved. The evaluation of models that consider outlier values and achieve better performance could be the subject of future research.

Author Contributions: Conceptualization, J.F.C. and J.C.C.; methodology, C.F., V.F.A. and D.G.S.; software, V.F.A.; formal analysis, V.F.A.; investigation, V.F.A. and D.G.S.; resources, J.F.C.; writing—original draft preparation, V.F.A. and D.G.S.; writing—review and editing, V.F.A., D.G.S. and C.F.; supervision, C.F.; project administration, V.F.A. All authors have read and agreed to the published version of the manuscript.

Funding: This research received no external funding.

Institutional Review Board Statement: Not applicable.

Informed Consent Statement: Not applicable.

Data Availability Statement: Data Availability Statements are available in section "water treatment plant" at https://www.kaggle.com/datasets/jvanessafernandez/water-treatment-plant, accessed on 2 April 2023.

Acknowledgments: We would like to express our gratitude to the Aquarisc project for their valuable collaboration in providing data for the development of this project.

Conflicts of Interest: The authors declare no conflict of interest.

References

1. Miljojkovic, D.; Trepsic, I.; Milovancevic, M. Assessment of physical and chemical indicators on water turbidity. *Phys. A Stat. Mech. Its Appl.* **2019**, *527*, 121171. [CrossRef]
2. World Health Organization. Water quality and health-review of turbidity: Information for regulators and water suppliers. 2017. Available online: https://apps.who.int/iris/handle/10665/254631. (accessed on 10 January 2023).
3. Ministerio de Ambiente, Vivienda y Desarrollo Territorial Y Ministerio de Protección Social. Decreto no 2115 de 2007. (June 2007). Available online: https://minvivienda.gov.co/sites/default/files/normativa/2115%20-%202007.pdf (accessed on 10 January 2023).
4. Marimon, W.; Jimenez, N.; Jiménez, C.; Chavarro, J.; Domínguez, E. Comparative Analysis of Water Quality Indices and Their Relationship with Anthropogenic Activities, Case Study: Bogotá River. *Res. Square* **2021**. [CrossRef]
5. García-Rentería, F.-F.; Nieto, G.A.C.; Cortez, G.H. Evaluation of wastewater discharge reduction scenarios in the Buenaventura Bay. *Water* **2023**, *15*, 1027. [CrossRef]
6. Anexo 1. Available online: http://cinara.univalle.edu.co/images/convocatorias/Convocatoria2/Anexos/ANEXO%201 (accessed on 13 January 2023).
7. Khairi, M.T.M.; Ibrahim, S.; Yunus, M.A.M.; Faramarzi, M.; Yusuf, Z. Artificial neural network approach for predicting the water turbidity level using optical tomography. *Arab. J. Sci. Eng.* **2015**, *41*, 3369–3379. [CrossRef]
8. El Din, E.S.; Zhang, Y.; Suliman, A. Mapping concentrations of surface water quality parameters using a novel remote sensing and artificial intelligence framework. *Int. J. Remote Sens.* **2017**, *38*, 1023–1042. [CrossRef]
9. Kim, S.E.; Seo, I.W. Artificial neural network ensemble modeling with conjunctive data clustering for water quality prediction in rivers. *J. Hydro-Environ. Res.* **2015**, *9*, 325–339. [CrossRef]
10. Delpla, I.; Florea, M.; Rodriguez, M. Drinking water source monitoring using early warning systems based on data mining techniques. *Water Resour. Manag.* **2019**, *33*, 129–140. [CrossRef]
11. Sharaf El Din, E. Enhancing the accuracy of retrieving quantities of turbidity and total suspended solids using Landsat-8-based-principal component analysis technique. *J. Spat. Sci.* **2019**, *66*, 493–512. [CrossRef]
12. Miguel-Chinchilla, L.; Heasley, E.; Loiselle, S.; Thornhill, I. Local and landscape influences on turbidity in urban streams: A global approach using citizen scientists. *Freshw. Sci.* **2019**, *38*, 303–320. [CrossRef]
13. DHIME. Available online: http://dhime.ideam.gov.co/webgis/home (accessed on 25 October 2022).

environmental sciences proceedings

MDPI

Proceeding Paper

Water Security and Environmental Impact Assessment: A Study for Developing Economies †

Muhammad Bilal [1], Muhammad Usman [1], Mehwish Nadeem [2] and Syyed Adnan Raheel Shah [3,*]

[1] Department of Agriculture, Forest and Range Management, Bahauddin Zakariya University, Multan 66000, Pakistan; bilal679.edu@gmail.com (M.B.); usmanisrite@gmail.com (M.U.)

[2] Department of Civil Engineering, Bahauddin Zakariya University, Multan 66000, Pakistan; mehwishnadeem@piet.edu.pk

[3] Department of Civil Engineering, NFC-Institute of Engineering & Technology, Multan 66000, Pakistan

* Correspondence: syyed.adnanraheelshah@uhasselt.be; Tel.: +92-300-791-4248

† Presented at the 7th International Electronic Conference on Water Sciences, 15–30 March 2023; Available online: https://ecws-7.sciforum.net/.

Abstract: Both ecosystems and human societies acknowledge the importance of water. The impact of human activities on both land and water has become more apparent due to various global changes. These include climate change, urbanization, socioeconomic development, and population growth. Although it is widely believed that water security is the key to sustainable development, studies on its evolution and various environmental factors are still in the early stages of development. This study aims to provide an overview of the concept. This concept aims to provide all people with safe water. It goes beyond merely providing adequate supplies to every person in the world, and it also aims to ensure healthy and productive lives. Despite the technological advancements being made in the water resource management industry, the lackadaisical approach to addressing the various challenges associated with water security continues to be a major issue globally. This study will provide an overview of the various facets of the water security concept and its evolution in developing countries due to the environmental changes that have occurred. It also explores the multiple sustainable methods that can be used to address these issues.

Keywords: water security; water scarcity; climate change; socioeconomic changes; sustainable development

check for updates

Citation: Bilal, M.; Usman, M.; Nadeem, M.; Shah, S.A.R. Water Security and Environmental Impact Assessment: A Study for Developing Economies. *Environ. Sci. Proc.* **2023**, *25*, 90. https://doi.org/10.3390/ECWS-7-14325

Academic Editor: Luis Garrote

Published: 3 April 2023

1. Introduction

Water is essential for everyone and the basis of life. Global freshwater is under tremendous anthropogenic strain [1], which is primarily caused by human population expansion and climate change. Freshwater availability and distribution, as well as the effects of water-related disasters such as floods and droughts, are projected to change because of climate change, which will also likely increase the demand for river water [2] and have an effect on groundwater availability [3]. Water security, a perennial human problem, has consequently emerged as a crucial area of policy for the Anthropocene [4,5]. Whether originally seen from a biophysical or sociological standpoint, the concept of "water security" is complicated, contentious, and dynamic. To define and realize it, complex and dynamic thinking or adaptive thinking is required. As part of the transition from water government to water governance, adaptive approaches to water management and governance have been pushed for at least three decades [6].

Assessments of economic development, ecological services [4], and their interaction [7,8] prominently include water challenges. However, assessments of water resources on a global scale [9] frequently use data that are fragmented and expressed as statistics at the country level, which severely restricts the attempts to prioritize their preservation and restoration [10]. Although there is a clear demand, high-resolution spatial analyses have not yet been used in the official assessment process for freshwater resources [9]. This is despite the fact that they have advanced understanding of the human impact on the world's oceans [11,12] and the human imprint on land [13]. Finding a balance between the use of human resources and the preservation of ecosystems is essential to the success of integrated water management strategies [9,14–16]. Systematic accounting is necessary to evaluate how far this target has come in the world and to determine how valuable it might be in the future.

Climate change, population growth, and economic development are all contributing factors to the increasing vulnerability of water resources to variable and extreme flows. These changes can also affect the availability and use of water. These have also had an impact on water-dependent ecosystems' ability to provide ecosystem services. Satisfying human demands is frequently achieved at the expense of the environment [14,17], with long-term consequences for socioecological systems as a whole. Water issues can often—but not always—be traced to governance shortcomings as opposed to the state of the resource base itself. Governmental failures range in scope from the local to the global, are numerous, and have diverse effects on industrialized and developing nations. Additionally, they are impacted by factors that operate concurrently at several levels of governance [18]. Poor governance, which includes a lack of efficiency and effectiveness of current resource-constrained governing systems, allegations of corruption, and the absence of civil society, pose challenges for all forms of development in many developing nations [7,19]. The lack of resources is a problem for emerging nations regardless of their institutional, economic, or physical conditions. In fact, some are regarded as failed states. Most of these nations have failed to provide for fundamental human needs as well as the demands of their own civilizations in terms of health and education and there is paradigm change in water management [20]. Many industrialized nations, however, struggle with the excessive regulation brought on by stiff bureaucracy, industry fragmentation, unsustainable consumption patterns, and a general preference for economic over environmental concerns [21].

The complexity of water systems' interconnections is often accompanied by uncertainties and risks. For instance, with difficult hydrology, which is characterized by the combination of flood vulnerability, natural aridity, and significant intra- and interannual variability, the interconnections become more complex. The complexity of the water systems' interconnections increases the risk that they will not be able to meet their customers' needs. This is especially true in developing regions, where the lack of resources and infrastructure is the most common issue [22].

This study covers the water security issue and environmental impact assessment for developing continents. It explores the utilization of water and the right to secure water with reference to continent-wise populations.

2. Materials and Methods

Water security is a critical issue that affects populations across the world. Different continents face different challenges when it comes to water security, and a significant factor in this regard is the population and water share of each continent. Developing continents, such as Africa and Asia, face significant water security challenges due to their high population growth rates and limited access to freshwater resources. The following analysis provides an overview of the water security issues facing these continents based on their population and water share. So, the analysis included comparative assessment of continents to evaluate the share of each continent. This analysis was based on the population and the freshwater availability in different continents. However, a framework can be utilized to analyze the water security issue as shown in Figure 1. The critical factors can be population, the density of the population, and the available freshwater percentage. Africa has a population of over 1.3 billion people, and the continent is home to some of the fastest-growing economies in the world. However, despite its vast water resources, Africa faces significant water security challenges due to its limited access to freshwater.

Similarly, Asia is the most populous continent in the world, with a population of over 4.6 billion people. The region is home to some of the fastest-growing economies and is facing significant water security challenges due to its high population growth rates and limited access to freshwater resources. The average water availability per capita in Asia is just 3920 cubic meters, well below the global average. So, these continents are critical for analysis and study.

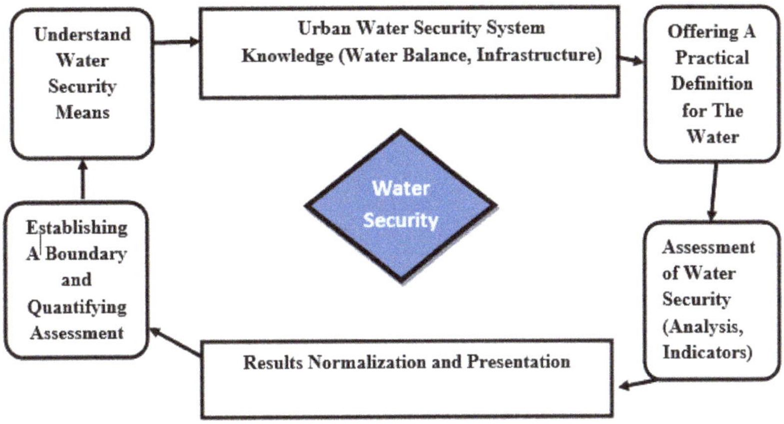

Figure 1. Methodology framework.

3. Results and Analysis

The increasing economic growth and standards for water-secure continents have changed the societal measures of the risk associated with water-related issues. This has resulted in the emergence of competing environmental and social tradeoffs that are becoming more challenging to manage.

Water consumption is different in different continents, based on the freshwater availability according to their population. Asia is a major portion of land consisting of higher freshwater density than other continents. It is more than the entire calculation for Africa, Australia, Europe, and the Americas, respectively, which was 8%, 7%, 13%, and 26% + 11% of the total, as shown in Figure 2.

Developing countries often face challenges in managing their limited water resources and balancing economic development with environmental protection.

In developing countries, access to clean and reliable water can be limited due to a lack of infrastructure and investment. This can lead to water scarcity and increased competition for resources, which can exacerbate existing social and economic inequalities. The statistics shown in Table 1 explain that freshwater availability is a major issue in Africa, which covers the majority of underdeveloped countries.

Table 1. Water consumption with respect to the population of continents.

Continent	World Population%	Density (p/km^2)	Fresh Water%
Asia	59.54%	150	35
Africa	17.2%	45	8
Australia	0.55%	5	7
South America	5.53%	25	26
North America	7.6%	28	11
Europe	9.59%	34	13

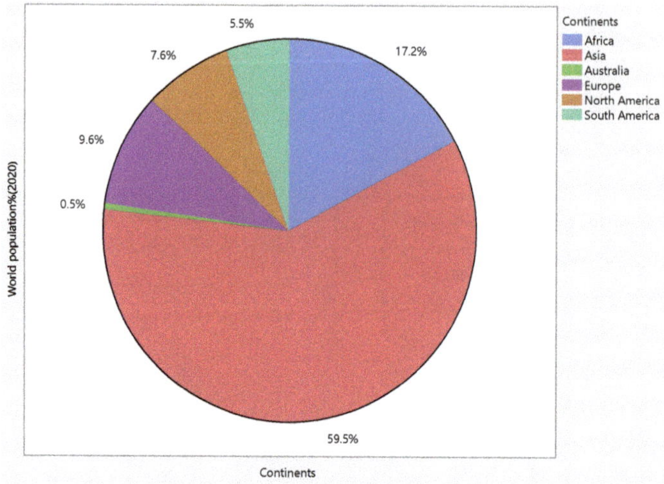

Figure 2. World population in different continents.

Water security is a significant issue in Africa, as many countries on the continent face challenges in managing their limited water resources and providing access to clean and reliable water for human and environmental needs.

One of the main challenges facing Africa is the high level of water scarcity on the continent, as shown in Figure 3. Many countries have limited water resources, and as populations grow and economies develop, competition for water resources is increasing. Climate change is also exacerbating this problem, as it leads to changes in precipitation patterns, increased evaporation, and more extreme weather events, all of which can impact water availability.

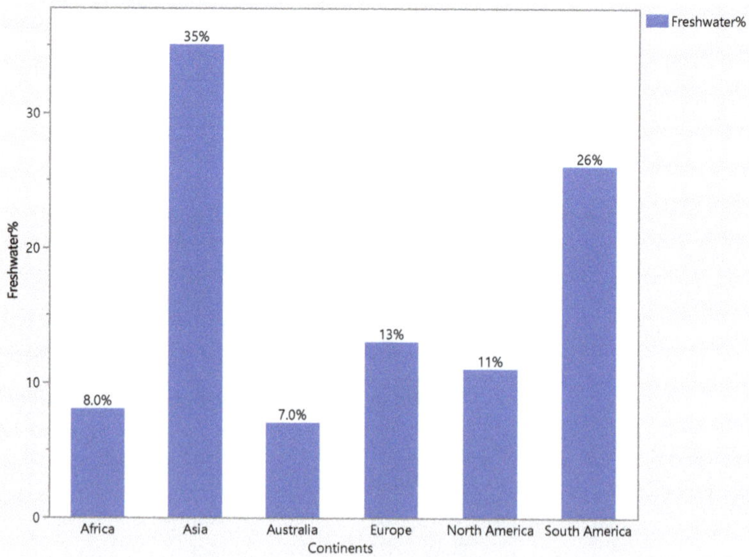

Figure 3. Division of the population and fresh water in the world, by continent.

Further study, the application of fresh science-based approaches, and the adoption of integrated water resources management concepts are all necessary to meet these challenges and sustainably resolve the diverse water-related problems. Understanding and managing water availability and quality is crucial everywhere but especially in developing nations.

4. Discussion

This study demonstrates that freshwater systems are directly endangered by human activities and could be further impacted by anthropogenic climate change; water is often regarded as the most important of all natural resources. Because freshwater is a limited resource, many countries around the world are experiencing water scarcity as water demand rises. Many countries have developed national and regional water management plans to address the growing water crisis. We discovered that nearly 80% of the world's population (4.8 billion) lives in areas where the human water security threat exceeds the 75th percentile (for 2000). Due to the intense agriculture and dense population, the USA, almost all of Europe, and significant sections of central Asia, the Middle East, the Indian subcontinent, and eastern China have a high incident hazard. Central Mexico, Cuba, North Africa, Nigeria, South Africa, Korea, and Japan all have smaller contiguous zones with a high threat of occurrence. The world's growing population is the primary cause of contaminated water. A large portion of fresh water is used in agricultural practices, which also causes a drop in the groundwater level. The 'global' imperative is a recent addition to the discourse of water security. Both the International Water Resources Association (IWRA) and the World Water Congress (WWC) are currently promoting a water strategy for global and continental water security that promotes adaptive approaches and Integrated Water Resources Management.

5. Conclusions

For these problems to be successfully solved, it is necessary to conduct additional research, implement new scientifically based approaches, and support the principles of integrated water resources management. Especially for the African and Asian continents, it is necessary to review the water security policy. Understanding and managing water availability and quality is crucial everywhere but especially in developing nations. Without first achieving a water-secure world, it will not be possible to achieve the United Nations Sustainable Development Goals 2030.

Author Contributions: Conceptualization, M.B.; Data curation: M.U.; Formal analysis: M.B.; Methodology: M.U.; Validation; M.N. and S.A.R.S.; Writing—Original draft: M.B.; Writing—Review and Editing, M.U. All authors have read and agreed to the published version of the manuscript.

Funding: This research received no external funding.

Institutional Review Board Statement: Not applicable.

Informed Consent Statement: Not applicable.

Data Availability Statement: The data are available on suitable demand.

Conflicts of Interest: The authors declare no conflict of interest.

References

1. Rockström, J.; Steffen, W.; Noone, K.; Persson, Å.; Chapin, F.S.; Lambin, E.F.; Lenton, T.M.; Scheffer, M.; Folke, C.; Schellnhuber, H.J.; et al. A safe operating space for humanity. *Nature* **2009**, *461*, 472–475. [CrossRef] [PubMed]
2. Tir, J.; Stinnett, D.M. Weathering climate change: Can institutions mitigate international water conflict? *J. Peace Res.* **2012**, *49*, 211–225. [CrossRef]
3. Mitchell, M.; Curtis, A.; Sharp, E.; Mendham, E. Directions for social research to underpin improved groundwater management. *J. Hydrol.* **2012**, *448*, 223–231. [CrossRef]
4. Vörösmarty, C.J.; McIntyre, P.B.; Gessner, M.O.; Dudgeon, D.; Prusevich, A.; Green, P.; Glidden, S.; Bunn, S.E.; Sullivan, C.A.; Liermann, C.R.; et al. Global threats to human water security and river biodiversity. *Nature* **2010**, *467*, 555–561. [CrossRef] [PubMed]

5. Cook, C.; Bakker, K. Water security: Debating an emerging paradigm. *Glob. Environ. Change* **2012**, *22*, 94–102. [CrossRef]
6. Pahl-Wostl, C. Transitions towards adaptive management of water facing climate and global change. *Water Resour. Manag.* **2007**, *21*, 49–62. [CrossRef]
7. Sullivan, C.; Meigh, J. Targeting attention on local vulnerabilities using an integrated index approach: The example of the climate vulnerability index. *Water Sci. Technol.* **2005**, *51*, 69–78. [CrossRef] [PubMed]
8. Esty, D.C.; Emerson, J.W. From crises and gurus to science and metrics: Yale's Environmental Performance Index and the rise of data-driven policymaking. In *Routledge Handbook of Sustainability Indicators*; Routledge: London, UK, 2018; pp. 93–102.
9. UNESCO; World Water Assessment Programme (United Nations). *UN-Water: Water in a Changing World*; United Nations: New York, NY, USA, 2009.
10. Vörösmarty, C.J.; Green, P.; Salisbury, J.; Lammers, R.B. Global water resources: Vulnerability from climate change and population growth. *Science* **2000**, *289*, 284e288. [CrossRef] [PubMed]
11. Halpern, B.S.; Walbridge, S.; Selkoe, K.A.; Kappel, C.V.; Micheli, F.; d'Agrosa, C.; Bruno, J.F.; Casey, K.S.; Ebert, C.; Fox, H.E.; et al. A global map of human impact on marine ecosystems. *Science* **2008**, *319*, 948–952. [CrossRef] [PubMed]
12. Halpern, B.S.; Ebert, C.M.; Kappel, C.V.; Madin, E.M.; Micheli, F.; Perry, M.; Selkoe, K.A.; Walbridge, S. Global priority areas for incorporating land–sea connections in marine conservation. *Conserv. Lett.* **2009**, *2*, 189–196. [CrossRef]
13. Sanderson, E.W.; Jaiteh, M.; Levy, M.A.; Redford, K.H.; Wannebo, A.V.; Woolmer, G. The human footprint and the last of the wild: The human footprint is a global map of human influence on the land surface, which suggests that human beings are stewards of nature, whether we like it or not. *BioScience* **2002**, *52*, 891–904. [CrossRef]
14. Vidal, F.; Sedan, D.; D'Agostino, D.; Cavalieri, M.L.; Mullen, E.; Parot Varela, M.M.; Flores, C.; Caixach, J.; Andrinolo, D. Recreational exposure during algal bloom in Carrasco Beach, Uruguay: A liver failure case report. *Toxins* **2017**, *9*, 267. [CrossRef] [PubMed]
15. Koetz, T.; Farrell, K.N.; Bridgewater, P. Building better science-policy interfaces for international environmental governance: Assessing potential within the Intergovernmental Platform for Biodiversity and Ecosystem Services. *Int. Environ. Agreem. Politics Law Econ.* **2012**, *12*, 1–21. [CrossRef]
16. Dudgeon, D.; Arthington, A.H.; Gessner, M.O.; Kawabata, Z.I.; Knowler, D.J.; Lévêque, C.; Naiman, R.J.; Prieur-Richard, A.-H.; Soto, D.; Stiassny, M.L.J.; et al. Freshwater biodiversity: Importance, threats, status and conservation challenges. *Biol. Rev.* **2006**, *81*, 163–182. [CrossRef] [PubMed]
17. Zeggini, E.; Scott, L.J.; Saxena, R.; Voight, B.F.; Marchini, J.L.; Hu, T.; de Bakker, P.W.; Abecasis, G.R.; Almgren, P.; Andersen, G.; et al. Meta-analysis of genome-wide association data and large-scale replication identifies additional susceptibility loci for type 2 diabetes. *Nat. Genet.* **2008**, *40*, 638–645. [CrossRef] [PubMed]
18. Gupta, G.; Tarique, K. Prevalence of Musculoskeletal Disorders in Farmers of Kanpur-Rural. *India. J. Community Med. Health Educ.* **2013**, *3*, 2161-0711. [CrossRef]
19. Pahl-Wostl, C.; Lebel, L.; Knieper, C.; Nikitina, E. From applying panaceas to mastering complexity: Toward adaptive water governance in river basins. *Environ. Sci. Policy* **2012**, *23*, 24–34. [CrossRef]
20. Pahl-Wostl, C.; Jeffrey, P.; Isendahl, N.; Brugnach, M. Maturing the new water management paradigm: Progressing from aspiration to practice. *Water Resour. Manag.* **2011**, *25*, 837–856. [CrossRef]
21. Orlowsky, B.; Hoekstra, A.Y.; Gudmundsson, L.; Seneviratne, S.I. Today's virtual water consumption and trade under future water scarcity. *Environ. Res. Lett.* **2014**, *9*, 074007. [CrossRef]
22. Grey, D.; Sadoff, C.W. Sink or swim? Water security for growth and development. *Water Policy* **2007**, *9*, 545–571. [CrossRef]

environmental sciences proceedings

MDPI

Proceeding Paper

Prediction of Flooding Area in Batang Sinamar River Basin Based on Design Return Period Simulation by Using Rainfall Runoff Inundation Model †

Angga Putra Arlis [1], Mohammad Farid [2,3], Ahmad Nur Wahid [2,3,*], Yadi Suryadi [2] and Arno Adi Kuntoro [2,3]

1 Master Program in Water Resources Management, Institut Teknologi Bandung, Jalan Ganesha 10, Bandung 40132, Indonesia; anggaputraarlis@gmail.com

2 Water Resources Engineering Research Group, Institut Teknologi Bandung, Jalan Ganesha 10, Bandung 40132, Indonesia; mfaridrizani@gmail.com (M.F.); suryadi37@yahoo.com (Y.S.); arnoak@hotmail.com (A.A.K.)

3 Center for Water Resources Development, Institut Teknologi Bandung, Jalan Ganesha 10, Bandung 40132, Indonesia

* Correspondence: ahmdnrwhd@gmail.com; Tel.: +62-82271396357

† Presented at the 7th International Electronic Conference on Water Sciences, 15–30 March 2023; Available online: https://ecws-7.sciforum.net/.

Abstract: Lima Puluh Kota Regency in West Sumatera Province is one of the regencies in Indonesia that has had a flood problem every year in the last decade. In the case of such large-scale flooding, it is important to classify the hazard zone for efficiency of the flood mitigation. In this paper, the Rainfall-Runoff-Inundation (RRI) model is applied to the Batang Sinamar River Basin in order to predict the widespread inundation, where both rainfall-runoff from surrounding mountain and rainfall on flood plain contributes to the flood event. Flood simulation was conducted by using a nationally available dataset, high resolution digital surface model, and rainfall ground station data. The simulation was calibrated with discharge observation data in Batang Sinamar, and provides a good result with the Nash–Sutcliffe Efficiency index and correlation value of 0.768 and 0.908, respectively. The result of the simulation using 10-year and 25-year return, showed the increasing discharge by 15.72% from 406.77 m^3/s to 470.74 m^3/s. Furthermore, the average peak of inundation water level increased from less than 1.5 m to more than 1.5 m. Based on these results, it can be concluded that the model can predict the potential inundation area in the Batang Sinamar River Basin in Lima Puluh Kota Regency.

Keywords: rainfall runoff inundation model; Batang Sinamar River Basin; flood hazard

Citation: Arlis, A.P.; Farid, M.; Wahid, A.N.; Suryadi, Y.; Kuntoro, A.A. Prediction of Flooding Area in Batang Sinamar River Basin Based on Design Return Period Simulation by Using Rainfall Runoff Inundation Model. *Environ. Sci. Proc.* **2023**, *25*, 91. https://doi.org/10.3390/ECWS-7-14320

Academic Editor: Athanasios Loukas

Published: 3 April 2023

1. Introduction

Flood has become an annual disaster that is probably the most destructive, boundless, and causes significant losses in many countries [1]. This inundation phenomenon usually inundates areas which are normally dry, and causes significant losses physically, socially, and environmentally [2].

Lima Puluh Kota Regency is a regency in West Sumatera Province Indonesia that is located in sub-catchment Batang Sinamar River Basin, and is part of the upstream of Indragiri catchment area with a total area about 1330.65 km^2. This regency is located on the hillside of the Sago Mountain that has experienced floods from Batang Sinamar River Basin every year in the last decade, from 2010 to 2019. The annual floods cause significant losses in agriculture and public facilities; therefore, a flood prediction model is needed to evaluate this phenomenon.

Hydrodynamic and hydrological models are widely applied to represent flood assessment. Hydrological models such as HEC-HMS, SWAT, and MIKE 11 are usually applied to reproduce the precipitation-runoff process. Hydrodynamic models such as MIKE 21, HECRAS, and DELFT 2D can simulate flow along rivers and floodplains [3,4]. The selection of the model is considered on the purpose of the model and availability of time, funds, and data.

Availability and quality of hydrometeorological data are the main problems in developing countries. The lack of long-term hydro-meteorological observation data and river's topography affects the result of the hydraulic and hydrological model. However, over the last decades, a large number of satellites have been developed by international agencies. Even though the qualities and resolution of these data are not as detailed and accurate as observation data, they provide sufficient results for this phenomenon [5,6].

Common model practitioners run their flood model in two different models, one for hydrological process for the input in the upstream boundary condition, and another for hydrodynamics for the flood inundation. These methods are quite difficult to identify in larger basins if many floods happen. Therefore, the Rainfall-Runoff-Inundation (RRI) model was developed with fully two-dimensional distributed rainfall-runoff inundation [7]. This model is widely applied in flood risk mapping [8,9] and flood damage assessment and management [10–12].

In this paper, the flood inundation model using the Rainfall-Runoff-Inundation model is presented. The model is calibrated with a 2019 flood event and simulated with 10- and 25-years' time return to see the performance and the maximum inundation area.

2. Methods and Materials

2.1. Rainfall-Runoff-Inundation Model

The Rainfall-Runoff-Inundation model is a two-dimensional-based model that can simulate rainfall runoff and inundation processes simultaneously [7,13,14]. This model is able to simulate the water that flows on slopes using 2D diffusion wave equations, and the discharge in rivers using 1D diffusion wave equations. For more realistic results from rainfall-runoff simulation processes, this model also takes consideration of lateral subsurface flow, vertical infiltration, and surface flow. The infiltration process is calculated using the Green-Ampt method [13] with parameters including saturated hydraulic conductivity (K) [mm/hour], soil surface porosity (ϕ), suction head (Sf) [mm], and depth of saturated soil (d) [mm]. The river geometry parameter that is the input of the model can be obtained using Equations (1) and (2) as a function of catchment area A [km^2]. The river geometry is considered as a rectangle with W [m] as the width of the river and D [m] as the depth of the river.

$$W = C_w A^{S_w} \tag{1}$$

$$D = C_d A^{S_d} \tag{2}$$

2.2. Target Flood Event

Batang Sinamar River Basin is a sub-catchment of Indragiri River's catchment area. The area of Batang Sinamar River Basin catchment is 1330.65 km^2. This river is used to irrigate 7766 Ha agriculture areas. It is on average 1246 m at the upstream and 486 m at the downstream above mean sea level.

In December 2019, there uncommon flood events happened at two adjacent times. The rainfall high is within the average, but the volume increased because of the length of the rain event. It affected six districts and damaged residents, public facilities, and agriculture. Besides the flood event in 2019, the model is performed for the flood event in 2013.

2.3. Data Used and Preparation

The general workflow of this study is shown in Figure 1, and the required input data gathered for this study include high daily rainfall in Suliki and Tanjung Pati Station for ten years, and Batang Sinamar Water Surface Elevation Station by Water Resources Agency of West Sumatera, digital elevation model (dem) by DEMNAS Indonesia (you can access it here: DEMNAS (indonesia.go.id)), and land use and soil type by Indonesia Geospatial Agency as shown in Figure 2. The surface raster is processed to obtain the flow characteristic raster data of the area as shown in Figure 3. The river's geometry, which are scarce data, is digitally measured using satellite images and dem shown in Figure 4. Digital measurements are applied in several cross sections along the river to obtain geometry coefficients C_w, S_w, C_d, and S_d with the result values of 1.8931, 0.3772, 0.162, and 0.4772, respectively. Rainfall data are processed using the polygon Thiessen method to obtain high rainfall design with 10- and 25-years as shown in Figure 5.

2.4. Performance Assessment

The RRI model input parameter is sensitive especially for the roughness and infiltration, which need a calibration for approach real events [14]. The model is compared and calibrated with real discharge observation data. Furthermore, the performance of the model is evaluated using Nash–Sutcliffe Efficiency (NSE) [15] and its correlation value is checked with the following equation.

$$NSE = 1 - \frac{\sum_1^n \left(Q_t^{obs} - Q_t^{sim}\right)^2}{\sum_1^n \left(Q_t^{obs} - \overline{Q^{obs}}\right)^2} \tag{3}$$

$$R = \frac{n\sum_1^n Q_t^{obs} Q_t^{sim} - \sum_1^n Q_t^{obs} \sum_1^n Q_t^{sim}}{\sqrt{n\sum_1^n Q_t^{obs\,2} - n\sum_1^n \left(Q_t^{obs}\right)^2}\sqrt{n\sum_1^n Q_t^{sim\,2} - n\sum_1^n \left(Q_t^{sim}\right)^2}} \tag{4}$$

where Q_t^{obs} is observed discharge at time t, Q_t^{sim} is simulated discharge value at time t, and n is number of available discharge value.

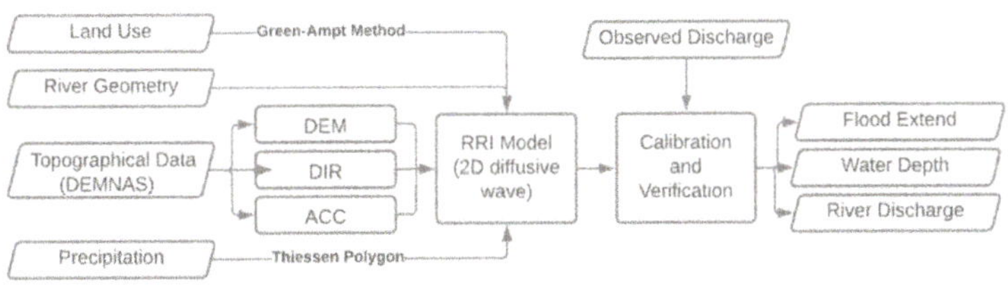

Figure 1. Workframe of the study.

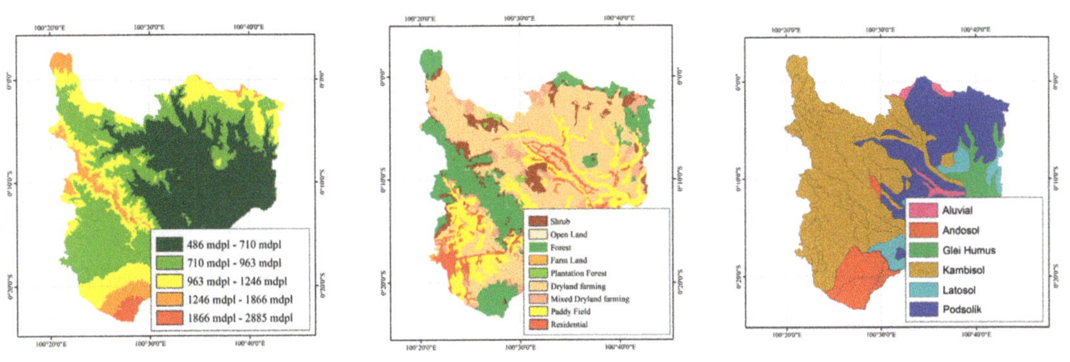

(**a**) (**b**) (**c**)

Figure 2. General information of the study area: (**a**) focus area elevation map; (**b**) land use classification map; (**c**) Soil type map.

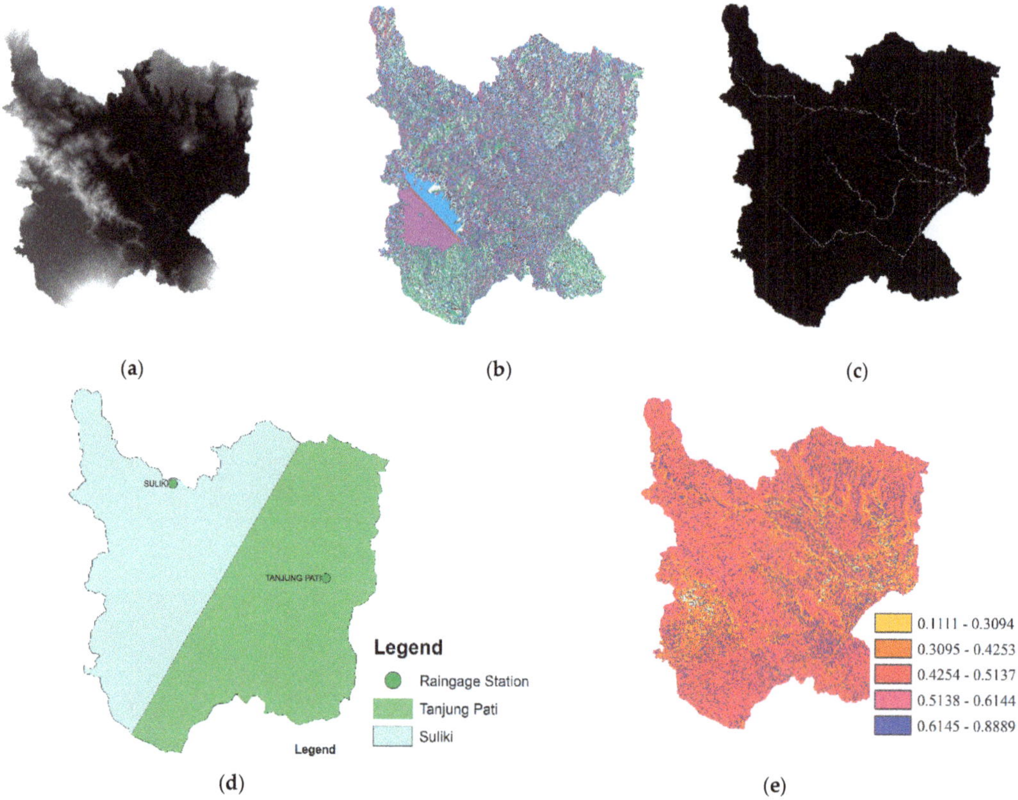

Figure 3. Topographic information as input RRI Model: (**a**) dem; (**b**) flow direction raster; (**c**) flow accumulation raster; (**d**) Thiessen polygon area; (**e**) roughness coefficient.

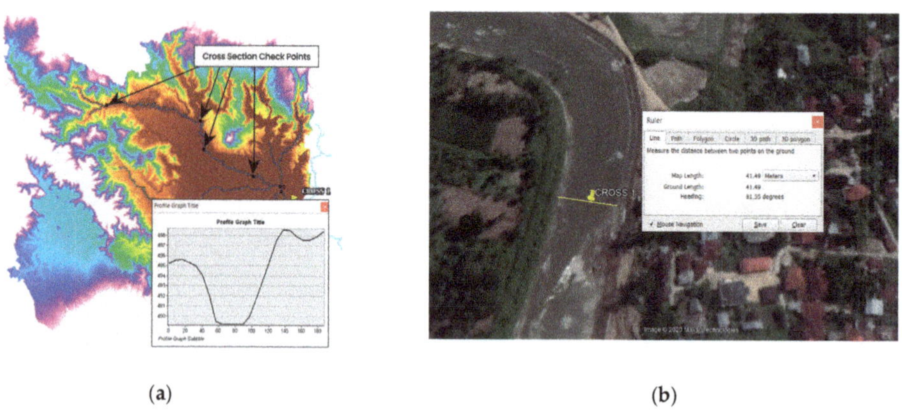

Figure 4. River geometry input for model RRI: (**a**) cross-section checking using dem; (**b**) river width checking using satellite images (source: Google Earth).

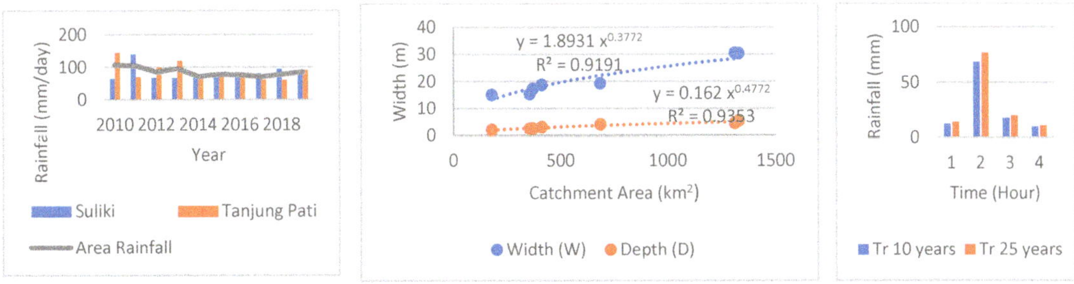

(a) **(b)** **(c)**

Figure 5. (**a**) Annual rainfall data, (**b**) river geometry parameter analysis, (**c**) rainfall distribution in time recurrence.

3. Result and Discussion

The RRI model is applied to simulate two scenarios in the Batang Sinamar River Basin in two scenarios; the first scenario is the flood event on 20 December 2019 and the second is 3 March 2017 as shown in Figure 6. The results of the model were calibrated with flow observation station data and the NSE and correlation value were checked. The simulation for the first scenario used rainfall for 17 days from 9 to 26 December 2019, with the peak on 20 December 2019. Meanwhile the second was simulated for 31 days from 8 February to 10 March 2017. Based on observed and simulated result, the highest discharges of the hydrograph on 20 December are 222.20 m^3/s and 224.65 m^3/s, respectively, and on 3 March are 257.50 m^3/s and 190.48 m^3/s, respectively. The model provides satisfactory results with the NSE and correlation value for the first scenario as 0.768 and 0.908, respectively, and for the second scenario, 0.531 and 0.828, respectively.

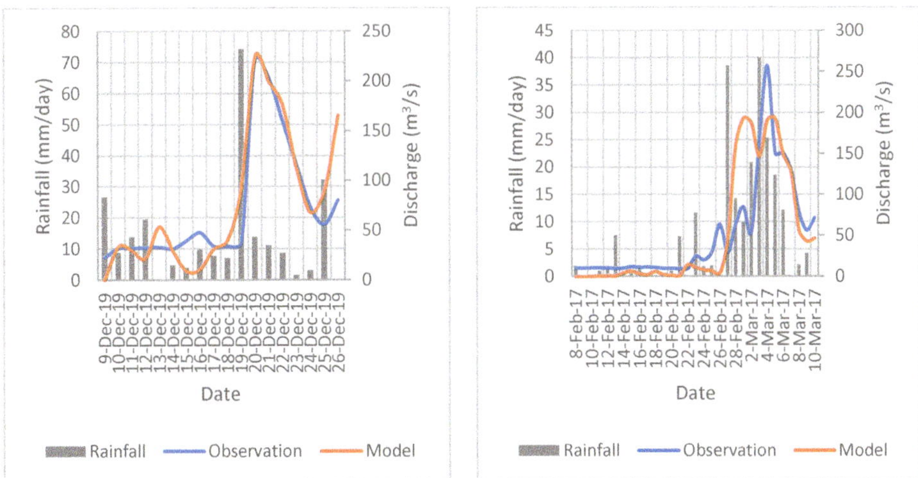

(a) **(b)**

Figure 6. The hydrograph results of Rainfall-Runoff-Inundation modelling compared with daily observation discharge data (**a**) flood scenario 9 to 26 December 2019; (**b**) flood scenario 8 February to 10 March 2017.

The evaluation of flood inundation used 10-year and 25-year design rainfall. These threshold rainfalls are used in the Indonesia standard flood modelling. The result provides a similar pattern between design rainfall as shown in Figure 7. The flooding area increase by about 12.6% from 59.963 km^2 using 10-year design rainfall, and 68.592 km^2 using 25-year design. The highest hydrograph discharge from the simulation is also increased by about 13.6% from 406.72 m^3/s using 10-year design to be 470.74 m^3/s.

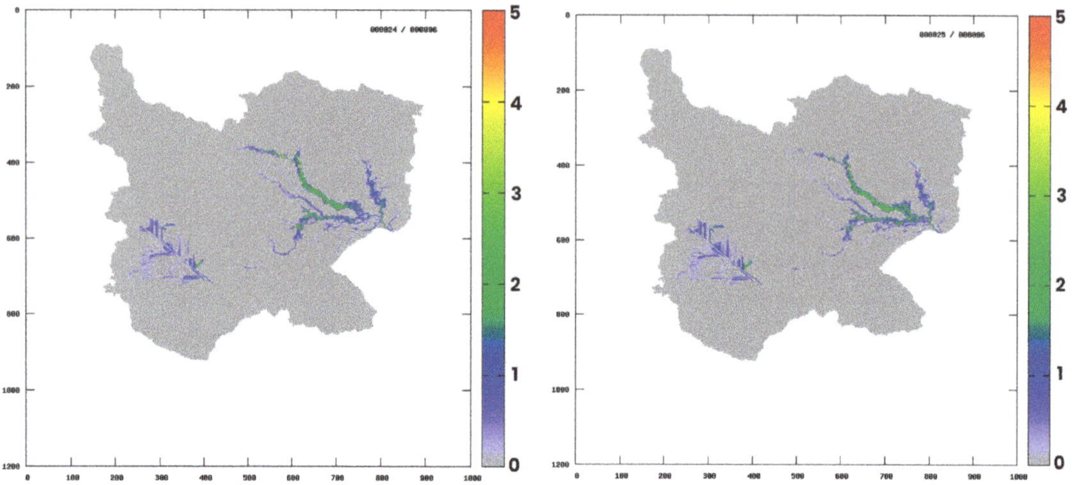

(a) (b)

Figure 7. Simulation result for time return design rainfall: (**a**) 10-year return; (**b**) 25-year return.

4. Conclusions

In this study is observed the application of rainfall-runoff and the inundation model in Batang Sinamar River Basin. The lack of the river's geometry survey data, which are the main input for the model, can be replaced by digital measurement using satellite images. Surprisingly, this provides a good agreement result when it is compared with discharge observation data in two flood events. In the first scenario the model is able to catch the peak of the flow, even though in the second scenario the model is underestimated. It is proven that the NSE values for both scenarios are 0.768 and 0.531, respectively, and the correlation about 0.908 and 0.828, respectively. The result of flooding with 10-year return period has 59.963 km^2 inundated area and 406.72 m^3/s as the highest discharge. Furthermore, in 25-year return period, the inundated area increases by about 13% or becomes 68.592 km^2, and the peak flow grows to 470.74 m^3/s.

This application can be a good alternative when money and time are a limited resource in the process. However, in higher decision making, surveys and field observation must be performed for the exact result.

Author Contributions: Conceptualization, M.F., Y.S. and A.A.K., methodology, M.F., Y.S. and A.A.K., software, A.P.A., validation, A.P.A., formal analysis, A.P.A., investigation, A.P.A., resources, M.F., data curation, A.P.A., writing-original draft preparation, A.P.A. and A.N.W., writing-review and editing, M.F., Y.S. and A.A.K., visualization, A.P.A. and A.N.W., supervision, M.F., Y.S. and A.A.K., project administration, M.F., funding acquisition, M.F. All authors have read and agreed to the published version of the manuscript.

Funding: This research was funded by the Institute for Research and Community Services of Institut Teknologi Bandung (LPPM ITB) through ITB Excellent Research 2023 and by the Indonesia Endowment Funds for Education (LPDP) through grant scheme PRJ-102/LPDP/2021.

Institutional Review Board Statement: Not applicable.

Informed Consent Statement: Not applicable.

Data Availability Statement: Digital elevation model, soil type and land use classification data downloaded freely from Indonesia Geospatial Information Agency site in Indonesia Geospatial Portal. Rainfall and discharge observation data received from West Sumatera Water Resources Agency.

Conflicts of Interest: The authors declare no conflict of interest.

References

1. Teng, J.; Jakeman, A.J.; Vaze, J.; Croke, B.F.W.; Dutta, D.; Kim, S. Flood inundation modelling: A review of methods, recent advances and uncertainty analysis. *Environ. Model. Softw.* **2017**, *90*, 201–216. [CrossRef]
2. Barus, L.; Tambunan, R.; Arif, V. Effect of Changes in Land Use in Flood Disasters in Baleendah District, Bandung Regency. *J. Strateg. Glob. Stud.* **2019**, *2*, 3. [CrossRef]
3. AL-Hussein, A.A.M.; Khan, S.; Ncibi, K.; Hamdi, N.; Hamed, Y. Flood Analysis Using HEC-RAS and HEC-HMS: A Case Study of Khazir River (Middle East—Northern Iraq). *Water* **2022**, *14*, 3779. [CrossRef]
4. Kadam, P.; Sen, D. Flood inundation simulation in Ajoy River using MIKE-FLOOD. *ISH J. Hydraul. Eng.* **2012**, *18*, 129–141. [CrossRef]
5. Cea, L.; Álvarez, M.; Puertas, J. Estimation of flood-exposed population in data-scarce regions combining satellite imagery and high resolution hydrological-hydraulic modelling: A case study in the Licungo basin (Mozambique). *J. Hydrol. Reg. Stud.* **2022**, *44*, 101247. [CrossRef]
6. Hurtado-Pidal, J.; Acero Triana, J.S.; Espitia-Sarmiento, E.; Jarrín-Pérez, F. Flood Hazard Assessment in Data-Scarce Watersheds Using Model Coupling, Event Sampling, and Survey Data. *Water* **2020**, *12*, 2768. [CrossRef]
7. Sayama, T.; Tatebe, Y.; Iwami, Y.; Tanaka, S. Hydrologic sensitivity of flood runoff and inundation: 2011 Thailand floods in the Chao Phraya River basin. *Nat. Hazards Earth Syst. Sci.* **2015**, *15*, 1617–1630. [CrossRef]
8. Zenkoji, S.; Oda, S.; Tebakari, T.; Archevarahuprok, B. Spatial Characteristics of Flooded Areas in the Mun and Chi River Basins in Northeastern Thailand. *J. Disaster Res.* **2019**, *14*, 1337–1345. [CrossRef]
9. Nastiti, K.D.; Kim, Y.; Jung, K.; An, H. The Application of Rainfall-Runoff-inundation (RRI) Model for Inundation Case in Upper Citarum Watershed, West Java-Indonesia. *Procedia Eng.* **2015**, *125*, 166–172. [CrossRef]
10. Sriariyawat, A.; Pakoksung, K.; Sayama, T.; Tanaka, S.; Koontanakulvong, S. Approach to Estimate the Flood Damage in Sukhothai Province Using Flood Simulation. *J. Disaster Res.* **2013**, *8*, 406–414. [CrossRef]
11. Kuribayashi, D.; Ohara, M.; Sayama, T.; Konja, A.; Sawano, H. Utilization of the Flood Simulation Model for Disaster Management of Local Government. *J. Disaster Res.* **2016**, *11*, 1161–1175. [CrossRef]
12. Try, S.; Lee, G.; Yu, W.; Oeurng, C.; Jang, C. Large-Scale Flood-Inundation Modeling in the Mekong River Basin. *J. Hydrol. Eng.* **2018**, *23*, 05018011. [CrossRef]
13. Sayama, T.; Ozawa, G.; Kawakami, T.; Nabesaka, S.; Fukami, K. Rainfall–runoff–inundation analysis of the 2010 Pakistan flood in the Kabul River basin. *Hydrol. Sci. J.* **2012**, *57*, 298–312. [CrossRef]
14. Sayama, T.; Tatebe, Y.; Tanaka, S. An emergency response-type rainfall-runoff-inundation simulation for 2011 Thailand floods. *J. Flood Risk Manag.* **2017**, *10*, 65–78. [CrossRef]
15. Nash, J.E.; Sutcliffe, J.V. River flow forecasting through conceptual models part I—A discussion of principles. *J. Hydrol.* **1970**, *10*, 282–290. [CrossRef]

environmental sciences proceedings

MDPI

Proceeding Paper

Flood Wave Dynamics in the Transboundary Dniester River Floodplain under Reservoirs Impact †

Ana Jeleapov

Institute of Ecology and Geography, Moldova State University, MD-2028 Chisinau, Moldova; anajeleapov@gmail.com

† Presented at the 7th International Electronic Conference on Water Sciences, 15–30 March 2023; Available online: https://ecws-7.sciforum.net.

Abstract: The present research aims to evaluate the impact of stream reservoirs on the Dniester river floods wave dynamics. The main approach consisted in the evaluation of the changes that occur in the flood parameters for three periods: natural flow, flow regulated by Dubasary reservoir, flow regulated by the Dniester hydropower complex and Dubasary reservoir. Using the hydrological time series from the stations situated upstream and downstream of the reservoirs, the following were calculated and compared: flood characteristics, peaks of 10, 5, 1, 0.5, and 0.1% probability, and environment flow components such as high-flow pulses, small floods, large floods. The results show that high flood protection is specific to the Dniester hydropower complex, while through the Dubasary reservoir the flood wave passes mainly in transit. Due to the flow regulating impact, small floods as well as their average peaks and duration were reduced toward the reservoirs' downstream part. High-flow pulses increased in number after the Dniester hydropower complex construction due to hydropeaking effect; however, their reduction is observed downstream of the Dubasary reservoir. Large floods increase in number in the upper part but are transformed into small floods toward the downstream, thus increasing the flood protection capacity.

Keywords: the Dniester river; floods; reservoirs impact; environment flow components

Citation: Jeleapov, A. Flood Wave Dynamics in the Transboundary Dniester River Floodplain under Reservoirs Impact. *Environ. Sci. Proc.* **2023**, *25*, 92. https://doi.org/10.3390/ECWS-7-14193

Academic Editor: Silvia Kohnova

Published: 14 March 2023

1. Introduction

Floods are the most frequent calamities registered in the limits of the Republic of Moldova. From the total number of natural disasters that occurred in the last four decades, floods accounts for 50%; the damage share rising to the same value (45%) [1,2]. The main factor that determines the flood wave generation is represented by extreme climatic conditions expressed by fast snow melt in winter and heavy rains in summer. The largest floods are formed in the floodplains of transboundary big rivers, the Prut and Dniester, in conditions of large water volumes brought from the upper parts of the river basins from Ukraine [1]. Moreover, significant damages are caused by flash floods generated as a result of the heavy rains and excessive slope runoff [1,2]. Certain measures are taken in order to reduce the impact of floods on humans, infrastructure, and economic activity, among which are reservoirs and levee systems. The present research aims to evaluate the impact of stream reservoirs on the Dniester river summer floods dynamics. In order to reflect the tendency of flood change determined by the reservoirs operation, several objectives were designed: identification of certain parameters that can show flood change; collection and analysis of hydrological data; evaluation of flood modifications in space and time from upper to lower course and from natural flow to regulated flow.

2. Study Area

The Dniester is a transboundary river and flows through Ukraine and the Republic of Moldova. The river length is 1362 km and the basin area is 72,100 km^2. Over 70% of the basin is situated in Ukraine, 27% belong to Republic of Moldova, and 0.34% to Poland. The upper part of the basin lies in Carpathians and represents only 30% of the basin area, but due to the high amounts of precipitations, 70% of Dniester runoff is generated in this area. Average annual flow in about 300 m^3/s [2,3]. The flow of the Dniester River is regulated by three reservoirs situated on the stream and one positioned lateral to the river. Three of these reservoirs form the Dniester hydroelectric complex (DHC): the Dnestrovsk reservoir with HPP-1 (water volume—2.6 km^3), the buffer reservoir with HPP-2 (volume of 37 mil. m^3), the artificial reservoir with pumped storage hydroelectric power plant (volume of 41.4 mil. m^3). DHC is situated at the border of Ukraine and the Republic of Moldova. Also, regulation of the Dniester flow is performed by the Dubasary reservoir, positioned in the limits of the Republic of Moldova [4,5].

3. Materials and Methods

The main approach applied in order to investigate the impact of the reservoir cascade on the Dniester river flood dynamics consisted in comparative analysis of the hydrological data collected in natural conditions of flow generation, as well as during the impact of the DHC and Dubasary reservoir operation. This approach is in line with the usually applied main designs: (1) Paired-Before–After Control–Impact, (2) Before–After, (3) Control–Impact, (4) Hydrological Classification, and (5) Predicted Hydrological indices [6]. Moreover, one of the modern approaches is determination of the main Hydrological Alteration Indicators and Environment Flow Components, which is applied for the estimation of impact of reservoirs operation on river flow [7–9]. Thus, hydrological characteristics were comparatively analyzed for three periods: the first corresponds to natural runoff, the second coincides with Dubasary reservoir functioning (1956–1982 years), the third represents the entire flood protection system operation (from 1987 till present). The analyzed times series were considered from the hydrological stations: Zalishchyky (situated upstream of the DHC), Hrushca (situated downstream of the DHC), Bender (situated downstream of the Dubasary reservoir, in the lower part of the basin). Based on the hydrological time series, the following were calculated and compared for three time periods: flood characteristics, peak discharge attenuation coefficient, peaks of 10, 5, 1, 0.5, and 0.1% probability, and environment flow components such as high-flow pulses, small floods, large floods.

The hydrological information used in this study was provided by the responsible data organizations in Moldova and Ukraine: the State Hydrometeorological Service (SHS) [10] and State Water Agency. Data were collected through UNDP in Moldova, Ministry of Environment of the Republic of Moldova, the Commission on Sustainable Use and Protection of the Dniester River Basin (the Dniester Commission).

4. Results and Discussions

4.1. Dynamics in Flood Characteristics

Long time series on flow for the time before dam construction was present only for Zalishchyky and Bender. These stations are also representative for the estimation of reservoirs impact on flood, due to the fact that the first one is situated upstream and the second downstream of hydrotechnical constructions. For three time periods, the average flow peaks are 1781 m^3/s, 1609 m^3/s, 1558 m^3/s at Zalishchyky and 1172 m^3/s, 1024 m^3/s, 882 m^3/s at Bender. Spatially, from the upper to lower part of the river, the maximum flow is reduced by 609 m^3/s, 585 m3/s, and 687 m^3/s, flow change being large in the last period (Figure 1). Floods duration is 13–16 days at Zalischyky and 17–20 days at Bender, with not much differences between the three periods. A certain impact of reservoirs is observed on rising and recession limbs of hydrographs. Thus, rising limb average duration at Zalischyky is 4 days for all periods. In the downstream of DHC, Grushka, the number of days increased from 4 days in the second period to 6–7 days in the third one; however, at Bender the number of days is 7 for all times. Flood wave recession limb is maintained within 11–12 days in the upper part of the DHC, while downstream of the DHC, it has a slight decreasing tendency, at Grushka, from 14 in natural regime to 11 days in regulated regime, but at Bender the number of days is stable as 12.

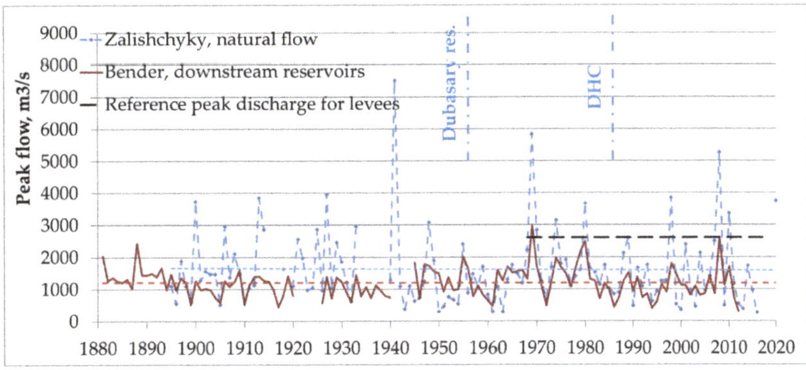

Figure 1. Flood peaks.

4.2. Peak Discharge Attenuation Coefficient Changes

The Dniester river floods propagation processes have changed in time and space due to certain modifications in the river bed and floodplain. During the time period that corresponds to natural flow, flood dynamics were distributed in the natural floodplain and reduction in the peaks was influenced by floodplain width and increasing basin area. In the second and, especially, the third time periods, hydrotechnical structures (reservoirs and levees) constructed in the river floodplain influenced river flood propagation by decreasing the peaks, redistribution of volumes under reservoirs impact, and decreasing flooding area due to levee system. For evaluation of flood dynamics under different river bed conditions, peak discharge attenuation coefficient (K) was estimated. It was calculated as the ratio between peaks from Zalishchyky and Bender station. Thus, this coefficient for natural runoff period is about 0.40, being relatively constant. The peaks of the flood of 1900, 1906, 1913, 1948 at Zalishchyky were at about 3000–4100 m^3/s, at Bender station reduction being to 1300–1700 m^3/s. K ranged from 0.34 to 0.51.

Operation of Dubasary dam influenced the Dniester flood wave dynamics in a certain way. Analyzing the hydrographs of flood events of 1969, 1970, 1974, 1980, a change in the shape was observed. In the upper part, the shape is classical triangular, while in the downstream part it is already trapezoid. The peaks of the mentioned floods were measured to the values of about 3000–6000 m^3/s at Zalishchyky and about 1700–3000 m^3/s at Bender. The K for this time period varied from 0.5 to 0.6, the average being 0.58. Dubasary reservoir is a relatively small one, and intensive siltation processes decrease its flood protection capacity. The flood regulation storage decreased from 630 mln.m^3 (in 1956) to 401 mln.m^3 (in 1982) [2,11], and in these conditions, large floods would flow in transit through it, the reduction capacity being estimated only to 15%.

In conditions of the whole flood protection system functioning, including DHC and levees caused a better flood management. The peaks of larger floods that occurred in 1989, 1998, 2008, 2010 were in the limits of 2700 m^3/s (1989, 2010) and 4000–5600 m^3/s (1998, 2010) at Zalishchyky, while at Bender their values decreased to 1500–2600 m^3/s. Estimated K ranges from 0.44 to 0.62 and average value is about 0.52, which is smaller in comparison to the second period but higher than the one in the first period.

As a result of evaluation of peak discharge attenuation coefficient, it was observed that in natural condition, flood peak attenuation was higher in comparison with actual situation. However, the need of lands for agriculture and settlements development caused construction of flood protection system that led to a decrease in the flood prone areas as well as capacity to reduce the peaks. Furthermore, an effective flood protection is the result of a complex of measures, including DHC, levees, forecast, integrated flood management.

4.3. Changes in Statistical Parameters

Reservoir impact on flood flow was estimated through calculation and comparison of statistical parameters of hydrological time series measured at different river stations. The impact of the DHC was evaluated on the basis of analysis of the time series of the pluvial flood peak discharges from Hrushca (time series presented only for second and third time periods) as well as their probability distribution for the periods before and after construction of DHC (Figure 2). It was estimated that the average peak discharge for the third period is 552 m^3/s lower than for the second one, and the peak discharge of 0.1–20% probability decreases with 905–3586 m^3/s (35–41%). Coefficient of variation (Cv) changes insignificantly from 0.54 to 0.48. The comparative analysis of the discharges of 1–10%, estimated on the basis of the dataset from the Dubasary reservoir, and probable peak discharge from its Operation Rules [11] indicate that the estimated probable peak discharge for the operation period is smaller with approx. 320–700 m^3/s (Figure 3), thus the reservoir having a lower effect in regulating the maximum runoff (12–15%) [1].

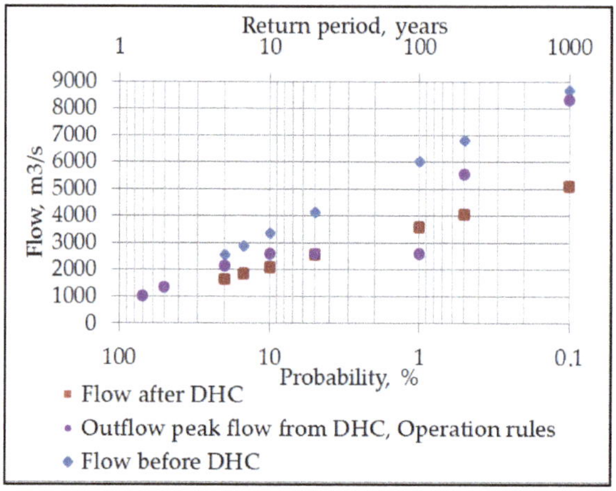

Figure 2. Distribution of peaks of different probabilities, Hrushca st.

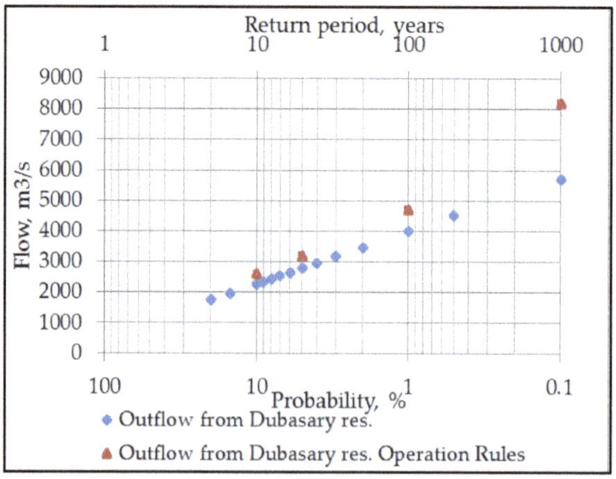

Figure 3. Distribution of peaks of different probabilities, Dubasary st.

As a result of the analysis of the time series from Bender st., it was determined that the average values of peak discharges are lower for the period of the entire flood protection system operation. Cv is 0.34, 0.43, and 0.44 for the three periods. However, the probability distribution shows an increase in the peak discharges of low and medium probability (0.1–10%) with ~22–44% in the second period, and with 1–21% in the third period, in comparison to those of the first period (Figure 4). The increase in extreme values can be explained by the fact that flood wave flowing in conditions of anthropogenic impact propagates through a narrower floodplain, limited by the levees, which determines the increase in both the discharges and levels, and also by the flood control by the Dubasary reservoir, and by occurrence of extreme synoptic situations that favored the generation of more significant floods during the second and third periods compared to the first one.

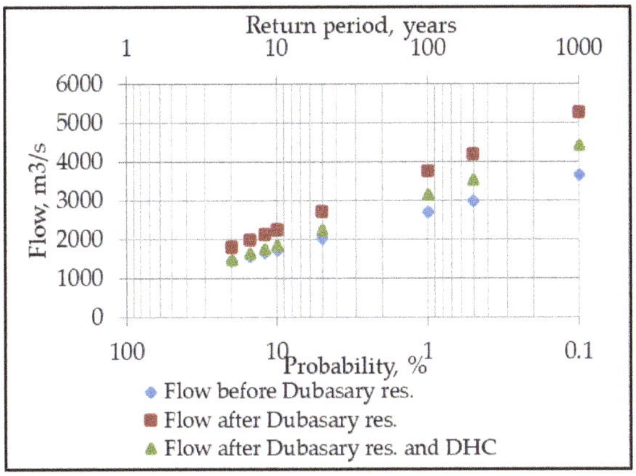

Figure 4. Distribution of peaks of different probabilities, Bender st.

The flood protection system reduces the probable peak discharges of 10–20% by 1–6%. It has a greater effect on regulating the flood runoff with the probability of up to 10%, especially during the period after the DHC construction. Thus, this hydropower complex has a more significant positive influence on the regulation of the flood runoff compared to the Dubasary reservoir [1].

4.4. Environmental Flow Components

Assessment of reservoirs impact on the Dniester floods was also performed by analyzing the characteristics of the environment flow components: high-flow pulses, small floods, large floods. Large floods, with peak discharges exceeding 2600 m^3/s (reference peak flow indicated in DHC rules [12], as well as for levee system) at Hrushca st. were recorded in 1969, 1974, 1980 before the DHC and in 2008 after beginning of its operation (Figure 1). Small floods (with peak discharges exceeding 1200 m^3/s—the multiannual average of peaks for entire monitoring period at Bender) are bigger in number. For the years 1945–1955 (before the Dubasary reservoir construction), at Zalishchyky and Bender on average one case of small floods in 2 years was recorded. In the period after the beginning of reservoir operation, the average number of these events was 1.4 event/year at Zalishchyky, 1.5 event/year at Hrushca, 0.8 event/year at Bender st. in the second period and 1 event/year at Zalishchyky, and 0.4 event/year for both stations in the downstream in the third period. During the Dubasary dam operation, in the upstream (at Hrushca), three small floods were registered in 1968, 1976, 1978, 1980, 1981; in the downstream (at Bender), the same number was recorded only in 1968; and in the other mentioned years their number decreasing to 1–2 events. In the years after the DHC construction, at Zalishchyky, three small floods were registered in 1991, 2001, 2005; five cases in 1998; and in the downstream not more than two small floods/year were recorded, but the number of years without small floods increased (Figure 5). The average value of small floods peak discharges (Figure 6) was 1584 m^3/s at Zalishchyky and 1692 m^3/s at Bender for the years 1945–1955; 1596 m^3/s at Zalishchyky, 1670 m^3/s at Hrushca, and reduced to 1600 m^3/s at Bender st. (downstream from the Dubasari dam) for the period of only the Dubasari res. operation; and 1662 m^3/s at Zalishchyky, 1451 m^3/s at Hrushca, and 1634 m^3/s at Bender for the period after the entire flood protection system construction. On an average, small floods duration was for the Dubasary dam operation period—10 days at Zalishchyky, 25.4 days at Hrushca, and 65 days at Bender; and for the period after DHC construction—15 days at Zalishchyky, 23 days at Hrushca, and 47 days at Bender [1].

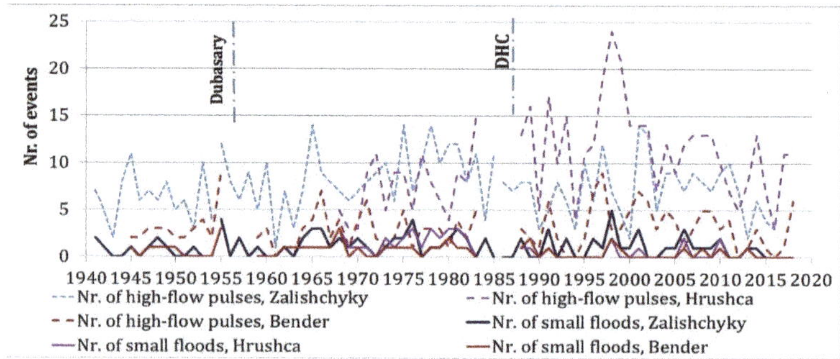

Figure 5. Number of cases of small floods and high flow pulses.

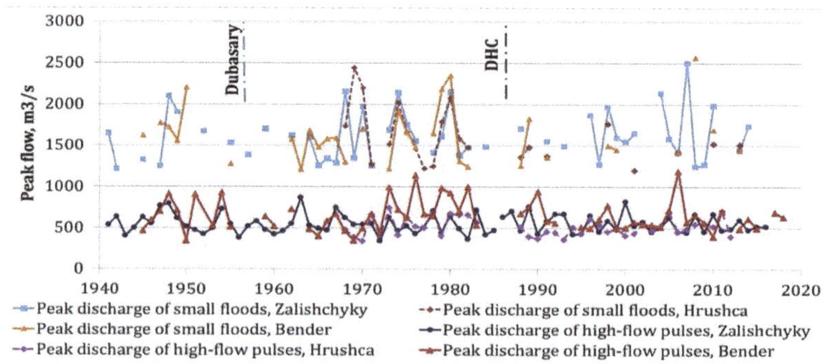

Figure 6. Average peak discharge of small floods and high flow pulses.

Annual frequency of high-flow pulses of ±30% (cases with peak discharges between 310/470 and 1200 m^3/s) for the 1945–1955 is characterized by values between 2 and 10 events/year (on average 6.3 events/year) with an average discharge of 578 m^3/s at Zalishchyky and values between 2 and 9 events/year (on average 3.2 events/year) with an average discharge of 664 m^3/s at Bender. Thus, in natural regime, a decrease in the number of high-flow pulses is observed for the sector from Zalishchyky to Bender; the reduction being 2–5 times. In the years following the Dubasary dam operation, the average frequency of high-flow pulses is 8.5 event/year at Zalishchyky, 7.8 events/year at Hrushca, and only 2.8 events/year at Bender; the decrease in the number of these events under reservoir impact being even with 8–10 events. In this respect, examples can serve the years 1972 and 1977 when at Hruscha 11 and at Bender only 2–3 high-flow pulses were registered. A different situation is specific for the DHC post-construction period. The number of high-flow pulses is 7 events/year at Zalishchyky but at Hrushca it has increased to 11. In 1997–2000, at Zalishchyky, the number of events was 5–14 while at Hrushca, it increased considerably up to 19–24. Thus, in some years, the number of high-flow pulses has doubled (1989, 1996, 2009, 2014) or even tripled (1991, 1998, 1999, 2000, 2013) at the DHC downstream station in comparison to its upstream station. In the same years at Bender the number of these events is already 3–9, i.e., 2–8 times less. The average peak discharges of high-flow pulses are 534 m^3/s at Zalishchyky, 530 m^3/s at Hrushca, and

674 m^3/s at Bender during the Dubasary res. operation and 557 m3/s at Zalishchyky, 496 m^3/s at Hrushca, and 632 m^3/s at Bender after the DHC construction. Their duration is approximately the same: at Zalishchyky—3.8 days for the first and second period and 4.5 days for the third one; at Hrushca—4.5 days for all periods; and at Bender–14.5, 23.8, and 15 days for all periods.

5. Conclusions

Evaluation of the effect of reservoir cascade on flood dynamics of the Dniester river shows that high flood protection is specific to DHC, while through the Dubasary reservoir the flood wave passes mainly in transit. The flood protection system has a greater effect in regulating the floods with medium probability, especially after the DHC construction. The reservoirs caused a slight increase in the coefficient of attenuation of peak discharges from 0.30 to 0.40 (in natural conditions) to 0.50–0.60 (in regulated flow conditions). Due to the flow regulating impact, small floods, their average peaks and duration were reduced in the reservoir downstream part. High-flow pulses increased in number after DHC construction due to hydropeaking effect; however, their reduction is observed in the downstream of Dubasary reservoirs. At present, large floods increase in number in the upper part but are transformed into small floods in the downstream, thus protecting the region from inundation. The increasing frequency and occurrence of floods in the Dniester river basin require improved flood management strategies both in Ukraine and the Republic of Moldova.

Funding: This research received no external funding.

Institutional Review Board Statement: Not applicable.

Informed Consent Statement: Not applicable.

Data Availability Statement: Not applicable.

Conflicts of Interest: The author declares no conflict of interest.

References

1. Jeleapov, A. Assessment of the Anthropogenic Impact on the Pluvial Floods of the Rivers of the Republic of Moldova. Ph.D. Thesis, Institute of Ecology and Geografy of the ASM, Chisinau, Moldova, 2019; p. 241. Available online: https://ieg.asm.md/sites/default/files/TEZAAJ_finENWeb.pdf (accessed on 20 January 2023).
2. Jeleapov, A. *The Study on Pluvial Floods in the Context of Human Impact on Environment*; Tipogr. Impressum: Chișinău, Moldova, 2020; p. 254. (In Romanian)
3. Jeleapov, A.; Melniciuc, O.; Bejan, I. Assessment of flood risk areas in the Dniester River basin (in the limits of the Republic of Moldova). In *Management of Water Quality in Moldova*; Duca, G., Ed.; Springer International Publishing: Cham, Switzerland, 2014; pp. 157–173.
4. Jeleapov, A. Assessment of the impact of the Dniester Hydropower Complex on hydrological state of the Dniester River. *Cent. Eur. J. Geogr. Sustain. Dev.* **2022**, *4*, 24–49. [CrossRef]
5. Regulations to Operate Water Reservoirs of the HPP and PSPP Dniester Cascade with Buffer Storage Reservoir Normal Headwater Level 77.10 m. Draft UKRGIDROENERGO. 2017. Available online: https://bit.ly/3U515GU (accessed on 12 September 2022).
6. Peñas, F.J.; Barquín, J.; Álvarez, C. Assessing hydrologic alteration: Evaluation of different alternatives according to data availability. *Ecol. Indic.* **2016**, *60*, 470–482. [CrossRef]
7. Richter, B.; Baumgartner, J.; Powell, J.; Braun, D. A method for assessing hydrologic alteration within ecosystems. *Conserv. Biol.* **1996**, *10*, 1163–1174. [CrossRef]
8. Richter, B.; Baumgartner, J.; Robert, W.; Braun, D. How much water does a river need? *Freshw. Biol.* **1997**, *37*, 231–249. [CrossRef]
9. The Nature Conservancy. Indicators of Hydrologic Alteration Version 7.1 User's Manual. 2009. Available online: https://www.conservationgateway.org/Documents/IHAV7.pdf (accessed on 12 September 2021).
10. Hydrological Database of the State Hydrometeorological Service of Moldova and Ukraine 2020.
11. *Operation Rules of the Dubasari Reservoir*; Moldovan National Institute of Water Management Design: Chisinau, Moldova, 1983. (In Russian)
12. Operation Rules of the Reservoirs of the DHC. 2022. 38 c. Available online: https://bit.ly/3RUkhFI (accessed on 12 September 2022). (In Russian).

environmental sciences proceedings

MDPI

Proceeding Paper

Hesitant Intuitionistic Fuzzy Approach in Optimal Irrigation Planning in India †

Sangita V. Pawar [1], Premlal Lal Patel [1,*] and Ashwini B. Mirajkar [2]

1 Department of Civil Engineering, Sardar Vallabhbhai National Institute of Technology, Surat 395007, India; sangita.pawar23@gmail.com
2 Department of Civil Engineering, Visvesvaraya National Institute of Technology, Nagpur 440010, India; ashmirajkar@gmail.com
* Correspondence: plpatel@ced.svnit.ac.in; Tel.: +91-9904003857
† Presented at the 7th International Electronic Conference on Water Sciences, 15–30 March 2023; Available online: https://ecws-7.sciforum.net/.

Abstract: The hesitant intuitionistic fuzzy optimization method optimizes multi-objective optimization problems under uncertainty and hesitation, and reflects the practical aspects of better decision-making. Hesitant intuitionistic fuzzy optimization (HIFO), a new optimization technique, has been suggested in the current study to find the best cropping pattern in the Kakrapar Right Bank Main Canal (KRBMC) command area of Ukai-Kakrapar Water Resources Project in India. The HIFO multi-objective fuzzy linear programming (HIFO MOFLP) result includes three objectives: maximization of net irrigation benefits (NIB), maximization of employment generation (EG), and minimization of cost of cultivation (CC), along with the appropriate constraints set. The performance of the aforesaid model is evaluated based on irrigation intensity, degree of acceptance (α^r), and degree of rejection (β^r) for inflows corresponding to 75% exceedance probability. The irrigation intensity from the study HIFO MOFLP model has been found to be 82.05%, while NIB, EG, and CC from the proposed model are 5572.31 million Rs, 14,287.27 thousand-man days, and 3429.99 million Rs, respectively. The proposed HIFO MOFLP model has been compared with the IFO MOFLP approach for the same command area and found to give improved results in the form of the irrigation intensity of the command area and objective function values. The current study demonstrates how hesitant fuzzy membership functions and non-membership functions can be applied to deal with uncertainty and hesitation in a real-world problem.

Keywords: hesitation and uncertainty; hesitant intuitionistic fuzzy optimization; Kakrapar right bank main canal; intuitionistic fuzzy optimization

Citation: Pawar, S.V.; Patel, P.L.; Mirajkar, A.B. Hesitant Intuitionistic Fuzzy Approach in Optimal Irrigation Planning in India. *Environ. Sci. Proc.* **2023**, *25*, 93. https://doi.org/10.3390/ECWS-7-14190

Academic Editor: Athanasios Loukas

Published: 14 March 2023

1. Introduction

Water resource planning and management has grown extremely complex as a result of competing and conflicting user needs. Furthermore, the estimation of irrigation demand, availability of labor, crop prices, and reservoir inflows are imprecise and uncertain in nature in water resource management and planning. In these situations, fuzzy optimization is the best suited to cope with uncertainty and impreciseness in the system. For the best irrigation planning for the Kakrapar right bank main canal (KRBMC) command area in India, Mirajkar and Patel [1] used multi-objective linear programming with a fuzzy decision set. They took into account three objectives and provided a compromise result with a degree of satisfaction (λ) of 0.503. Crisp linear programming was utilized by Mirajkar and Patel [2] to resolve three conflicting objectives: maximization of net irrigation benefit, employment creation, and minimization of cultivation costs. This problem was solved concurrently by finding a multi-objective fuzzy solution with maximum-minimum operators. Two-phase and fuzzy compromised techniques were used to improve the solution using maximum-minimum operator. These algorithms were validated for the KRBMC in the Ukai command area of India. Chen et al. [3] introduced a new fuzzy robust control method for linear parameter varying systems. Results from the proposed model revealed that the methodology has good efficiency and performance. The generalization of fuzzy sets by Atanassov [4] included the hesitation margin, which is equal to one minus the membership degree and the non-membership degree. The use of IFO in multi-objective linear programming was examined by Bharati and Singh [5], who also investigated the effects of linear and non-linear membership functions on optimization. Non-linear membership as well as non-membership functions outscored linear membership and non-membership functions in the IFO technique. Angelov [6] found that a membership degree, which is an objective function's degree of acceptance, is not complimentary to non-membership or rejection. The study demonstrated that intuitionistic fuzzy optimization (IFO) solutions outperformed the traditional fuzzy and crisp optimization methods. Sahoo et al. [7] solved an MOFLP problem for the development and implementation of the land-water-crop system of the Mahanadi-Kathajodi Delta in Eastern India. A unique compromise ratio technique for multiple criterion decision-making in water resources management was introduced by Hashemi et al. [8] based on Atanassov's intuitionistic fuzzy sets idea. Garai and Roy [9] developed the idea of the hesitation index for the optimization of a fictitious mathematical problem where the objective functions included maximization of the degree of acceptance, minimization of the degree of rejection, and hesitation. In order to assist policymakers in reducing water shortages, Li et al. [10] developed an intuitionistic fuzzy (IF) multi-objective non-linear programming (IFMONLP) method for allocating water for irrigation purposes in both dry and rainy circumstances. An ambiguous transportation issue was studied by Ebrahimnejad and Verdegay [11]. Intuitionistic fuzzy multi-objective geometric programming was proposed by Jafarian et al. [12] as a solution to multi-objective nonlinear programming issues. In order to establish the cropping pattern in the KRBMC command area of the Ukai-Kakrapar water resources project in India, Pawar et al. [13] adopted the IFO methodology as a new optimization method. Other uncertainty characteristics, such as the degree of acceptance, rejection, and hesitation, were also presented in addition to the best cropping pattern. To maximize net irrigation benefit, create jobs, and minimize cultivation costs, Pawar et al. [14] applied the IFO MOFLP model to optimize crop allocation in the Ukai-Kakrapar water resources project. For the multi-objective optimization issue, Bharati [15] provided a hesitant intuitionistic fuzzy approach, and an exemplary example demonstratingthe superiority of the suggested model.

2. Description of Study Area

One of the biggest multipurpose projects in India is the Ukai water resources project, which is situated on the Tapi River. The Kakrapar weir is situated 29 kilometers downstream of the Ukai Dam at latitude 21°22′ N and longitude 73°16′ E. The Kakrapar weir consists of two major canals for irrigation of the crops on both sides of the weir, namely, KRBMC and the Kakrapar left bank main canal (KLBMC), with 113,123 hectares and 145,335 hectares of culturable command areas, respectively. Pawar et al. [13] included the index map of the KRBMC inside the Ukai-Kakrapar project.

3. Methodology and Model Development

In this study, the best cropping pattern for the KRBMC command area is determined by maximizing net benefits, creating jobs, and minimizing cultivation costs under a specific set of constraints, namely, available water resources and land. Mirajkar and Patel [16] provided an outline of each objective and constraint, while Pawar et al. [13] described objective functions and constraints briefly.

3.1. Objective Functions and Constraints

In this study, the three objective functions, i.e., maximizing net irrigation benefits (Z_1), employment generation (Z_2), and minimizing cultivation costs (Z_3), are taken into account along with constraints on planting area, affinity, and water allocation. Section 4 describes the formulation of hesitant intuitionistic fuzzy optimization while using optimal solutions of individual objective functions.

3.2. Hesitant Intuitionistic Fuzzy Optimization (HIFO)

KRBMC has been taken into account in this study when using the HIFO approach. In Figure 1, a flow chart describing the details of HIFO implementation for the KRBMC area is shown. Using Equations (1) and (2), the hesitant membership and non-membership functions for the maximization type objective function are briefly described below.

$$\mu_m^{E^r}(x) = \delta^r \begin{cases} 0 & if \quad Z_m(x) \le L_m^{\mu} \\ \frac{Z_m(x) - L_m^{\mu}}{U_m^{\mu} - L_m^{\mu}} & if \quad L_m^{\mu} \le Z_m(x) \le U_m^{\mu} \\ 1 & if \quad f_m(x) \ge U_m^{\mu} \end{cases} \tag{1}$$

$$v_m^{E^r}(x) = \delta^r \begin{cases} 0 & if \quad Z_m(x) \ge U_m^{v} \\ \frac{U_m^{v} - Z_m(x)}{U_m^{v} - L_m^{v}} & if \quad L_m^{v} \le Z_m(x) \le U_m^{v} \\ 1 & if \quad Z_m(x) \le L_m^{\nu} \end{cases} \tag{2}$$

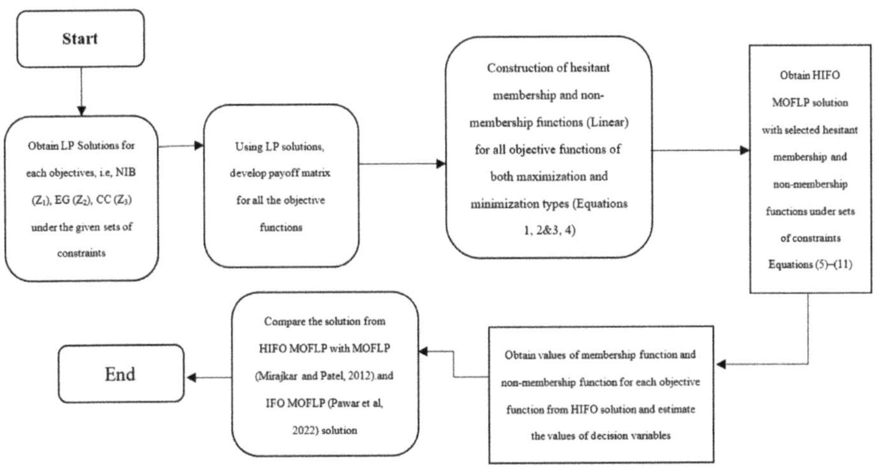

Figure 1. Methodology for development of HIFO MOFLP solution [2,13].

Here, $\mu_m^r(x)$ is hesitant membership, $\nu_m^r(x)$ is hesitant membership and non-membership function for m^{th} objective function, $Z_m(X)$ is the value of the objective functions $U_m^{\mu} = \max(Z_m(X))$, $L_m^{\mu} = \min(Z_m(X))$, and $L_m^{\nu} = L_m^{\mu}$, $U_m^{v} = \times U_m^{\mu}, \in [1,3]$. Here, m = 1, 2, and 3 are the numbers of objective functions, Ø is a hesitant parameter and δ^r is a multiplication factor with $0 \le \delta^r \le 1$wherein r = 1, 2, ... ,9.

(a) The membership and non-membership functions were formed with Equations (3) and (4) in order to minimize the objective function.

$$\mu_m^{E^r}(x) = \delta^r \begin{cases} 0 & if \quad Z_m(x) \ge U_m^{\mu} \\ \frac{U_m^{\mu} - Z_m(x)}{U_m^{\mu} - L_m^{\mu}} & if \quad L_m^{\mu} \le Z_m(x) \le U_m^{\mu} \\ 1 & if \quad Z_m(x) \ge L_m^{\mu} \end{cases} \tag{3}$$

$$v_m^{E^r}(x) = \delta^r \begin{cases} 0 & if \quad Z_m(x) \le L_m^{v} \\ \frac{Z_m(x) - L_m^{v}}{U_m^{v} - L_m^{v}} & if \quad L_m^{v} \le Z_m(x) \le U_m^{v} \\ 1 & if \quad Z_m(x) \ge U_m^{\mu} \end{cases} \tag{4}$$

Here, $U^{\mu}_{m} = \max(Z_m(X))$, $L^{\mu}_{m} = \min(Z_m(X))$, $L^{\nu}_{m} = L^{\mu}_{m}$, and $U^{v}_{m} = \times U^{\mu}_{m}, \in [1,3]$. Here, m = 1,2, and 3 where $0 \leq \delta^{r} \leq 1$, r = 1, 2, ... ,9.

(b) With the use of linear membership and non-membership excluding the hesitation index, the multi-objective fuzzy linear optimization problem (MOFLP) for objectives subject to constraints set can be described using Equations (5)–(11) (Bharti, [15]).

$$\text{Maximize} \sum (\alpha^r - \beta^r) \tag{5}$$

$$\text{Subject to } \alpha^r \leq \mu^{E^r}_{m}(x), \tag{6}$$

$$\beta^r \geq v^{E^r}_{m}(x), \tag{7}$$

$$\alpha^r + \beta^r \leq 1, \tag{8}$$

$$\alpha^r \geq \beta^r, \tag{9}$$

$$\beta^r \geq 0, \tag{10}$$

$$g_j(x) \leq b_j, x \geq 0, \tag{11}$$

$$m = 1,2,\ldots\ldots p; j = 1,2,\ldots\ldots q$$

Here, α^r signifies the degree of acceptance of objective functions, and β^r indicates the degree of rejection of objective functions under sets of constraints.

The objective functions, expressed using Equation (5), under the constraints set, were solved using the modified simplex method.

The application of the HIFO MOFLP approach for the KRBMC command area resulted in effective solutions for each of the objective functions, along with the crop areas for that command area. The three objectives considered are the maximization of NIB, maximization of EG, and minimization of CC. In addition to the suggested optimization technique, it provided better performance parameters, such as the degree of acceptance and the degree of rejection. Table 1 implies the Pay-off Matrix for three objective functions derived from individual LP solutions.

Table 1. Pay-off Matrix from LP solutions for three objective functions (Pawar et al. [13]).

Type of Objective Function	Objective Function	Net Benefit	Employment Generation	Cost of Cultivation
Maximum	Net benefit, in million Rs	5593.77 ^{U}	5559.32	1552.55 ^{L}
Maximum	Employment generation, in 1000 man-days	14,631.25	14,824.02 ^{U}	5499.49 ^{L}
Minimum	Cost of cultivation, in million Rs	3457.54 ^{L}	3454.37	1076.75 ^{U}

Note: ^{U} and ^{L} are the upper and lower values of an objective function.

4. Results and Discussion

For the selected study region, HIFO techniques from Bharati [15] have been applied. The performance of the HIFO MOFLP with MOFLP technique, as well as the ideal values for each of the objective functions and the appropriate crop pattern for the command area (Mirajkar and Patel [2]) and IFO MOFLP approach (Pawar et al. [13]) are described in the following paragraphs.

4.1. Sensitivity of δ^r on Hesitant Membership and Non-Membership Function

The values of δ^r, determining the membership and non-membership function, range from 0 to 1.0. The HIFO MOFLP for the KRBMC area has been optimized using Equations (5)–(11) for different values of δ^r and equivalent values of α^r and β^r were obtained. The values of α^r and β^r have been estimated as follows: $\alpha_1 = 0.97$, $\alpha_2 = 0.98$, $\alpha_3 = 0.99$, $\alpha_4 = 0.05$, $\alpha_5 = 0.03$, $\alpha_6 = 0.02$, $\alpha_7 = 0.01$, $\alpha_8 = 0.01$, $\alpha_9 = 0.01$, $\beta_1 = 0.01$, $\beta_2 = 0.01$, $\beta_3 = 0.01$, $\beta_4 = 0.95$, $\beta_5 = 0.97$, $\beta_6 = 0.98$, $\beta_7 = 0.00$, $\beta_8 = 0.00$, and $\beta_9 = 0.00$. The decision maker must select the value of δ^r for its execution in the command area based on the level of certainty.

4.2. Optimal Cropping Pattern

Here, corresponding to Ø (hesitant parameter) = 1, the optimal values forδ^1, δ^2, δ^3, δ^4, δ^5, δ^6, δ^7, δ^8, and δ^9 are 0.96, 0.98, 1, 0.96, 0.98, 1, 0.96, 0.98, and 1, respectively. The recommended modeling approach, suggested by Bharati [4] resulted in the values of NIB, EG, and CC as 5572.31 million Rs, 14,287.27 thousand-man days, and 3429.99 million Rs, respectively. The consequent values of an individual objective function with MOFLP and IFO MOFLP were 3585.05 million Rs, 10,189.21 thousand-man days, and 2260.13 million Rs, respectively.

It should be emphasized that HIFO MOFLP shows higher values of NIB, EG, and CC vis-à-vis MOFLP and IFO MOFLP approaches. Hence, the HIFO MOFLP solution can be used more reliably with respect to MOFLP and IFO MOFLP approaches.

The optimal values of the cropping pattern obtained from the HIFO MOFLP model, with Ø = 1, are included in Table 2, which can be applied in the KRBMC command area with a greater degree of certainty. Table 2 indicates areas allocated by the various optimization models used in the current study. The irrigation intensity for the same command area corresponding to the aforementioned condition has been found to be 82.05%.

Table 2. Areas allocated to different crops (in ha) by IFO MOFLP, MOFLP, and HIFO MOFLP models.

		Solutions Obtained from Various Models					
		Crisp Linear Programming Individual Solutions *			**IFO MOFLP ***	**MOFLP ***	**Hesitant Intuitionistic Algorithm**
Crop No. (i)	Crops	Net benefit	Employment Generation	Cost of Cultivation	$\alpha = 0.503$, $\beta = 0.282$, $\pi = 0.215$, $S_f = 0.3$	$\lambda = 0.503$	$\alpha_1 = 0.97$, $\alpha_2 = 0.98$, $\alpha_3 = 0.99$, $\alpha_4 = 0.05$, $\alpha_5 = 0.03$, $\alpha_6 = 0.02$, $\alpha_7 = 0.01$, $\alpha_8 = 0.01$, $\alpha_9 = 0.01$, $\beta_1 = 0.01$, $\beta_2 = 0.01$, $\beta_3 = 0.01$, $\beta_4 = 0.95$, $\beta_5 = 0.97$, $\beta_6 = 0.98$, $\beta_7 = 0.00$, $\beta_8 = 0.00$, $\beta_9 = 0.00$
		Cropareas allocated in ha					
1	Paddy (k)	13,100	16,965	13,100	13,386.38	13,386.38	13,100
2	Juwar/Bajra (k)	11,310	11,310	8100	11,310	11,310	11,310
3	Vegetables (k)	1131	1131	690	1131	1131	1131
4	Wheat (r)	3654	3654	3654	16,965	16,965	3654
5	Vegetables (r)	1120	1120	1120	1120	1120	1120
6	Juwar/ Bajra (r)	10,091	10,091	10,091	10,091	10,091	10,091
7	Paddy (hw)	8145	8145	8145	8145	8145	8145
8	Groundnut (hw)	192	192	192	192	192	192
9	Cotton (ts)	860	860	860	860	860	860
10	Vegetables (ts)	5655	5655	1335	5655	5655	3077.996
11	Sugarcane (p)	38,337.21	37,350.34	4998	17,529.92	17,529.92	39,503
12	Banana (p)	633	633	633	633	633	633
	Total	94,228.21	97,106.34	52,918	87,018.3	87,018.3	97,106.34
	Irrigation Intensity %	83.30	85.84	46.78	76.92	76.92	82.05
Maximum	Net benefit, in million Rs	5593.77	5559.32	1552.55	3585.05	3585.05	5572.31
Maximum	Employment generation, in 1000 man-days	14,631.25	14,824.02	5499.49	10,189.21	10,189.21	14,287.27
Minimum	Cost of cultivation, in million Rs	3457.54	3454.37	1076.75	2260.13	2260.13	3429.99

Note: hw—hot weather; k—kharif; p—perennial; r—rabi; ts—two-season; *—taken from Pawar et al. [13].

Table 2 shows the optimal cropping pattern values from the HIFO MOFLP model for Ø = 1, which may be implemented in the KRBMC command region with a higher degree of reliability.

5. Conclusions

The hesitant intuitionistic fuzzy optimization with multi-objective fuzzy linear programming (HIFO MOFLP) has been implemented for the KRBMC of Ukai-Kakrapar irrigation project. The HIFO approach, given by Bharati [15], has been used for the selected study area. Finally, the outcomes of HIFO MOFLP were compared with those of MOFLP and IFO MOFLP for the same research area, same sets of constraints, and objective functions. The primary conclusions from the current research are as follows:

(a) The HIFO MOFLP model, recommended by Bharati [15], has been applied over KRBMC with compromised optimal values of NIB, EG, and CC as 5572.31 million Rs., 14,287.27 thousand man-days, and 3429.99 million Rs, respectively.
(b) The optimal cropping pattern, determined by HIFO MOFLP, is shown in Table 2 with an irrigation intensity of 82.05%.
(c) The results obtained from the proposed model give improved results in terms of irrigation intensity and objective function values as compared with Mirajkar and Patel [2] and Pawar et al. [13] for the same study area and objective functions.
(d) The proposed methodology can be applied to the whole Ukai-Kakrapar command area while giving due consideration to more objectives and corresponding constraints such as inflows, outflows of the reservoir, evaporation losses from the reservoir, etc., which are uncertain in nature. The hesitant intuitionistic fuzzy optimization approach can be discovered further when membership function and non-membership functions are non-linear in nature.

Author Contributions: Conceptualization, S.V.P., P.L.P. and A.B.M.; methodology, S.V.P.; software, S.V.P.; supervision and validation, S.V.P.; formal analysis, S.V.P. and A.B.M.; resources, S.V.P.; data curation, S.V.P. and A.B.M.; project administration, S.V.P., P.L.P. and A.B.M.; writing—original draft, S.V.P.; writing—review and editing, S.V.P., P.L.P. and A.B.M. All authors have read and agreed to the published version of the manuscript.

Funding: This research received no external funding.

Institutional Review Board Statement: Not applicable.

Informed Consent Statement: Informed consent was obtained from all subjects involved in the study.

Data availability Statement: All the data, models, or codes that support the findings of the current study are available with the corresponding authors. The data can be shared on the public portal only after getting approval from the data providing agency.

Conflicts of Interest: The authors declare no conflict of interest.

References

1. Mirajkar, A.B.; Patel, P.L. A fuzzy based optimal irrigation planning for kakrapar right bank canal command area, gujarat, india. *ISH J. Hydraul. Eng.* **2011**, *17*, 43–50. [CrossRef]
2. Mirajkar, A.; Patel, P. Optimal irrigation planning using multi-objective fuzzy linear programming models. *ISH J. Hydraul. Eng.* **2012**, *18*, 232–240. [CrossRef]
3. Chen, F.; Qiu, X.; Alattas, K.A.; Mohammadzadeh, A.; Ghaderpour, E. A New Fuzzy Robust Control for Linear Parameter-Varying Systems. *Mathematics* **2022**, *10*, 3319. [CrossRef]
4. Atanassov, K.T. Intuitionistic fuzzy sets. *Fuzzy Sets Syst.* **1986**, *20*, 87–96. [CrossRef]
5. Bharati, S.K.; Singh, S.R. Solving Multi Objective Linear Programming Problems Using Intuitionistic Fuzzy Optimization Method: A Comparative Study. *Model. Optim.* **2014**, *4*, 10–15. [CrossRef]
6. Angelov, P.P. Optimization in an intuitionistic fuzzy environment. *Fuzzy Sets Syst.* **1997**, *86*, 299–306. [CrossRef]
7. Sahoo, B.; Lohani, A.K.; Sahu, R.K. Fuzzy Multiobjective and Linear Programming Based Management Models for Optimal Land-Water-Crop System Planning. *Water Resour. Manag.* **2006**, *20*, 931–948. [CrossRef]

8. Hashemi, H.; Bazargan, J.; Mousavi, S.M. A Compromise Ratio Method with an Application to Water Resources Management: An Intuitionistic Fuzzy Set. *Water Resour. Manag.* **2013**, *27*, 2029–2051. [CrossRef]
9. Garai, A.; Roy, T.K. Intuitionistic fuzzy optimization: Usage of hesitation index. *NIFS* **2013**, *19*, 60–68. [CrossRef]
10. Li, M.; Fu, Q.; Singh, V.P.; Ma, M.; Liu, X. An intuitionistic fuzzy multi-objective non-linear programming model for sustainable irrigation water allocation under the combination of dry and wet conditions. *J. Hydrol.* **2017**, *555*, 80–94. [CrossRef]
11. Ebrahimnejad, A.; Verdegay, J.L. A new approach for solving fully intuitionistic fuzzy transportation problems. *Fuzzy Optim. Decis. Mak.* **2018**, *17*, 447–474. [CrossRef]
12. Jafarian, E.; Razmi, J.; Baki, M. A flexible programming approach based on intuitionistic fuzzy optimization and geometric programming for solving multi-objective nonlinear programming for solving multi-objective nonlinear programming problems. *Expert Syst. Appl. Int. J.* **2018**, *93*, 245–256. [CrossRef]
13. Pawar, S.V.; Patel, P.L.; Mirajkar, A.B. Intuitionistic fuzzy approach in multi-objective optimization for KRBMC irrigation system, India. *ISH J. Hydraul. Eng.* **2022**, *28*, 463–470. [CrossRef]
14. Pawar, S.; Patel, P.L.; Mirajkar, A. Intuitionistic fuzzy optimization approach in optimal irrigation planning of Ukai-Kakrapar irrigation project, India. *ISH J. Hydraul. Eng.* **2022**, 1–11. [CrossRef]
15. Bharati, S.K. Hesitant intuitionistic fuzzy algorithm for multiobjective optimization problem. *Oper. Res.* **2022**, *22*, 3521–3547. [CrossRef]
16. Mirajkar, A.B.; Patel, P.L. Multi-objective Two-Phase Fuzzy Optimization Approaches in Management of Water Resources. *J. Water Resour. Plan. Manag.* **2016**, *142*, 1–16. [CrossRef]

environmental sciences proceedings

MDPI

Proceeding Paper

Oasification and Desertification under the Framework of Land Degradation Neutrality †

Jaime Martínez-Valderrama [1,2,*], Dongwei Gui [3] and Zeeshan Ahmed [3]

[1] Estación Experimental de Zonas Áridas, Consejo Superior de Investigaciones Científicas, Crtra. Sacramento s/n, La Cañada de San Urbano, 04120 Almería, Spain
[2] Instituto Multidisciplinar para el Estudio del Medio, Universidad de Alicante, San Vicente del Raspeig, 03690 Alicante, Spain
[3] Xinjiang Institute of Ecology and Geography, Chinese Academy of Sciences, Urumqi 830011, China; guidwei@ms.xjb.ac.cn (D.G.); zeeshan@ms.xjb.ac.cn (Z.A.)
* Correspondence: jaimonides@eeza.csic.es
† Presented at the 7th International Electronic Conference on Water Sciences, 15–30 March 2023; Available online: https://ecws-7.sciforum.net.

Abstract: To meet population growth, the excessive abstraction of water resources for irrigating water-intensive crops has become an increasing crisis in arid regions of Northwest China. This contains the perennial contradiction between development and desertification, typical of drylands, which we also find within the United Nations Convention to combat Desertification (UNCCD), where some interpret it as a developmentalist convention and others as a conservationist one. The Land Degradation Neutrality (LDN) concept has been set up as the main tool to combat desertification by the UNCCD and is included in SDG 15.3. LDN refers to a state of zero net land degradation, where "the amount and quality of land resources necessary to support ecosystem functions and services and enhance food security remain stable or increase within specified temporal and spatial scales and ecosystems". Under the LDN framework, we apply two of its main pillars, prevention and land planning. The aim is to understand the underlying biophysical and socio-economic mechanisms of oasis expansion in NW China, a phenomenon known as oasification. The objective is to detect under what conditions oasification tackles desertification and when it triggers land and economic degradation. From this knowledge, it will be possible to propose guidelines of action to balance land use and comply with LDN.

Keywords: oases; cash crops; irrigated agriculture; land-use planning; causal diagrams; system dynamics; NW China

Citation: Martínez-Valderrama, J.; Gui, D.; Ahmed, Z. Oasification and Desertification under the Framework of Land Degradation Neutrality. *Environ. Sci. Proc.* **2023**, *25*, 94. https://doi.org/10.3390/ECWS-7-14238

Academic Editor: Athanasios Loukas

Published: 16 March 2023

1. Introduction

China is one of the countries affected seriously by desertification [1]. China has 6.6 million km^2 of drylands (expanded by 8.3% during 1980–2015 [2]) that support approximately 580 million people [3]. It is in this vast territory where desertification occurs, i.e., the degradation of drylands due to climate variations and inadequate human activity [4]. Desertification affects seven main dryland provinces and autonomous regions (Xinjiang, Inner Mongolia, Tibet, Qinghai, Gansu, Ningxia and Sha'anxi) and 12 main deserts and sandy lands [1]. Although the ambiguity of this complex problem makes it difficult to assess and map [5,6], it is estimated that the direct annual economic loss caused by desertification is RMB 33.1–94.9 $\times$ 10^9 [7]. In addition to these effects, the irreversible nature of desertification narrows the development options of a territory [8], and poverty will be exacerbated [9].

In NW China, the most recognizable expression of desertification is serious water scarcity [10]. The increasing water gap results both from the falling supply of water, and the rising demand for water. The former is related to the effects of climate change. Climate projections in China point to a greater risk of extreme events and aridification in arid and semi-arid regions. In addition, changing snowmelt patterns [11] are a serious threat to the region's oases, which have historically depended on this water flow [12].

The latter is explained by the fast transition from food crops to cash crops. The change in land use was triggered in the early 1980s, when farmers were given more autonomy in land use [13]. In the least developed regions, farmers have few choices for increasing their income (i.e., their opportunity cost is low) besides expanding their agricultural scale [14]. Hence, this oasification process—the natural or artificial expanding the boundaries of oases—has a double edge, as it can tackle or favor desertification. Oasification leads to greater vegetation cover in places that, due to low rainfall, it is very sparse or inexistent. In addition, the quality and quantity of the soil is improved, as shown by various indicators (e.g., soil organic carbon). However, oasification can result in extensive lands being abandoned due to water shortage [15], threatening its very survival. We are, therefore, faced with an archetypal problem in drylands: it is necessary to promote economic development and try to take advantage of water resources to green the landscape, but poor management of these resources can cause the whole system to collapse (i.e., lead to desertification).

This is the ambiguity of the concept of desertification, which basically stems from all that the definition of desertification [16]—recall that it is the degradation of drylands—includes in the word degradation: "the reduction or loss of the biological or economic productivity, complexity of ecosystems, and biodiversity" [4]. This controversy explains why the UNCCD has been so ineffective, and embodied the conflicting expectations of parties: developing countries regard it mainly as a development convention, while developed countries regard it as an environmental convention [17]. This dilemma is not only found in desertification, but also in the Sustainable Development Goals (SDGs), where there are important (and irresolvable) trade-offs between environmental goals (e.g., SDGs 13 or 15) and economic growth goals (e.g., SDGs 1, 2, and 8). The answer to this dilemma can be provided by Land Degradation Neutrality (LDN), with which the UNCCD could re-energize international action on desertification and regain lost protagonism [18].

LDN is defined as "a state whereby the amount and quality of land resources necessary to support ecosystem functions and services and enhance food security remain stable or increase within specified temporal and spatial scales and ecosystems". LDN has been set up as the main tool to combat desertification in SDG 15.3. LDN is pragmatic because it links "development" and its implicit, associated economic benefits to restoration commitments, including offsets. It could be achieved by: (a) managing land more sustainably, which would reduce the rate of degradation; and (b) increasing the rate of restoration of degraded land, so that the two trends converge to give a zero net rate of land degradation.

This research will delve into the socio-economic drivers of oasification, where a research gap has been detected [16]. To this end, causal diagrams—one of the first stages of System Dynamics (SD) modeling [19]—will be used to understand the mechanisms underpinning oasification processes by detecting feedback loops, nonlinearities, and delays between variables. Gaining an in-depth comprehension of the functioning of the socioecosystem paves the way for proposing solutions in the framework of LDN [20].

2. Methods

Modeling can be an important technique in analyzing and describing processes of desertification change in addition to furthering our understanding of the relationship between the socio-economic and ecological factors involved and their effects on desertification [21]. SD is a suitable tool for this challenge. SD is a modeling methodology grounded on the theories of nonlinear dynamical systems and feedback control developed in mathematics, physics, and engineering. SD states that the main, but easily overlooked, cause of the behavior of a complex system lies in its underlying structure of relationships, which includes feedback loops, non-linear relations, delays and decision rules. Formally, an SD model is a set of first-order ordinary differential equations that makes a stock-and-flow representation of the studied system; stock variables show the state of the system over time, and flow variables represent the processes that change the stocks [19,22]. The main advantages of SD are [23,24]: (i) it improves system understanding, and develops system thinking skills, even from the first stage of its development as causal or sketch diagrams; (ii) SD models can incorporate empirical and process-based approaches, and help integrate interdisciplinary knowledge, (iii) the SD literature provides abundant information about related methodologies, and (iv) user-friendly software platforms allow easy access for non-modeler users.

In this paper, we stress the usefulnes of causal diagrams (see Sterman (2000) [22] for details). They are built by establishing the relationships between explanatory and explained variables. The polarity between the independent variables (x) and the dependent variables (y) can be (i) direct (+) when x and y move in the same direction, i.e., as x increases, y increases, or as x decreases, y decreases; or (ii) negative—or inverse—($-$) when x and y move in the opposite direction, i.e., more x and less y or vice versa. The concatenation of the causal relationships between the variables gives rise to a network of feedback loops. The sign of the feedback loops illustrates their behavior: positive feedback is self-reinforcing and behind the explosive or exponential behaviour systems; negative feedback loops are self-correcting and represent the stable performance of the system. An even number of negative arrows or their absence gives rise to a positive feedback loop; an odd number of negative arrows gives rise to a negative feedback loop.

3. Results and Discussion

Causal diagrams are very useful tools for channeling associative thinking that needs to be structured into a tangible hypothesis [25]. In this process of concreteness, operational variables, i.e., those that can be measured and have a real counterpart, coexist with ambiguous or imprecise concepts. This is the case of Figure 1A, where we show some of the relationships established in the exploitation of groundwater resources in NW China. As we can see, 'climate change' is a concept that allows us to reflect on the fact that snowmelt inputs are going to decrease. As the model becomes more concrete, we will have several options to make this idea fully functional. On the one hand, we will be able to develop a whole climate model that explains the dynamics of this snowmelt. On the other hand, we can implement different climate change scenarios that are reflected in different snowmelt volumes.

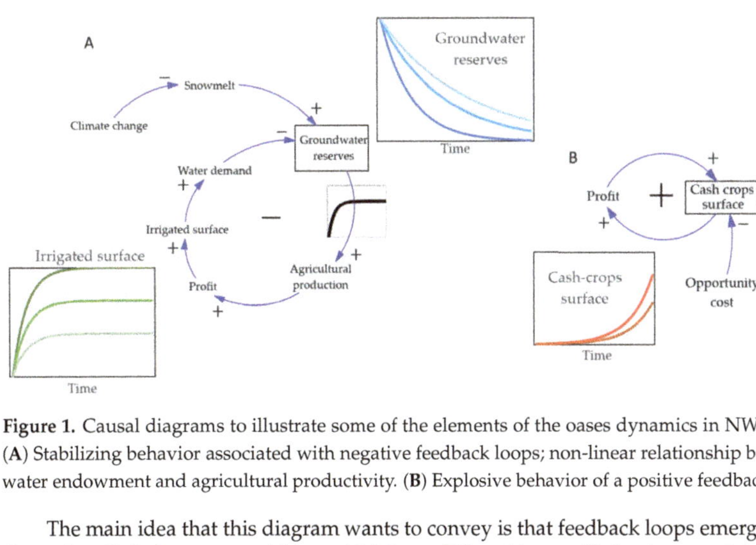

Figure 1. Causal diagrams to illustrate some of the elements of the oases dynamics in NW China. (**A**) Stabilizing behavior associated with negative feedback loops; non-linear relationship between water endowment and agricultural productivity. (**B**) Explosive behavior of a positive feedback loop.

The main idea that this diagram wants to convey is that feedback loops emerge from the various links between variables, which have different types of behavior associated with them. In this case, the loop is negative and has a stabilizing behavior. The water reserve will tend towards an equilibrium (which may be zero) whose value depends on various circumstances. For example, as the water table is deeper, the cost of pumping is higher and therefore discourages activity [26]. To what extent it discourages it will depend on the profitability per cubic meter, the cost of energy or the opportunity cost. It is also relevant to consider that the value of this balance is not irrelevant. When certain thresholds are crossed, processes, such as aquifer subsidence, marine intrusion, or desiccation of surrounding systems, can be triggered [27]. Note that the stabilization of the water reserve is associated with another equilibrium, that of the irrigated area.

In complex systems, such as socioecological systems, different types of loops coexist. In Figure 1B we show a positive one, which results in exponential growth (or decay) of the variables (typical behavior of the Anthropocene [28]). Indeed, as the profitability of a certain type of crop increases, in this case, cash crops, and the other alternatives are very unprofitable (i.e., low opportunity cost), more and more farmers opt for this alternative, which leads to an increase in the irrigated area. This leads to more profits and new farmland, leading to exponential growth in cash crop area. As it is easy to understand, this exponential behavior, towards infinity, cannot be sustained over time, but it allows us to see that, in isolation, there are parts of the system that for a time behave in a threatening way, putting the sustainability of the whole system at risk.

In addition to holistically understanding this structure, SD has specific tools to highlight the delays between cause and effect (e.g., farmers' decision-making time [29]), and the nonlinear nature of many of the system relationships (e.g., water productivity is not linear, but starts from zero, grows first exponentially and then linearly, and then saturates, reaching a maximum). These considerations, which lead us to more elaborate phases of the SD model, are what explain the counter-intuitive nature of the system in the face of certain solutions that try to alleviate water scarcity. Thus, for example, improving the efficiency of irrigation systems often leads to higher water consumption [30,31].

4. Conclusions

The dynamic behavior of a system depends on what kind of loops drive it. When it is a positive one, it tends to behave more abruptly, while if negative ones dominate, it tends to stabilize. Therefore, understanding how these loops arise, how they are activated and how they interact, gives us clues about the sustainability or collapse (i.e., desertification) of a given system. In this sense, SD is an exploratory rather than a predictive tool. That is, the interesting thing is to qualitatively interpret outputs that respond to different 'what if' questions that are processed by differential equations (i.e., the SD model, a quantitative method).

The power of SD lies in the fact that it is easy for people who have no previous experience in conceptualizing complex problems, such as the search for sustainable development. The confrontation between the need to create wealth and well-being and maintain the stock of natural resources that provide a series of fundamental services for nature to continue to function requires holistic tools in which various stakeholders, academics or decision makers participate. In a short time, causal diagrams conceptualizing this debate are put up, highlighting the trade-offs, contradictions, and multiple points of view of the participants. With a little more effort, these diagrams translate the discussion into models that can be simulated so that very simple experiments with counterfactual situations can be implemented to analyze what would have happened under various scenarios (e.g., climate and land-use change).

This approach fits very well with the implementation of LDN, which should be based on land planning and governance. What LDN aims to do is to channel existing plans, so that there are no contradictions or overlaps, and the impact of the various economic activities is counterbalanced by restoring measures. The results should lead us to a neutral human footprint, allowing the coexistence of human societies and nature.

Author Contributions: Conceptualization, J.M.-V. and D.G.; methodology, J.M.-V.; software, J.M.-V.; formal analysis, J.M.-V. and D.G.; investigation, J.M.-V., D.G. and Z.A.; writing—original draft preparation, J.M.-V.; writing—review and editing, J.M.-V., D.G. and Z.A.; visualization, J.M.-V.; supervision, J.M.-V. and D.G.; funding acquisition, J.M.-V. and D.G. All authors have read and agreed to the published version of the manuscript.

Funding: This research was funded by MICINN through European Regional Development Fund [SUMHAL, LIFEWATCH-2019-09- CSIC-4, POPE 2014-2020] and the regional collaborative innovation project for Xinjiang Uygur Autonomous Region (Shanghai Cooperation Organization Science and Technology Partnership project), grant number 2021E01016.

Institutional Review Board Statement: Not applicable.

Informed Consent Statement: Not applicable.

Data Availability Statement: No new data were created or analyzed in this study. Data sharing is not applicable to this article.

Conflicts of Interest: The authors declare no conflict of interest.

References

1. Ci, L.; Yang, X. *Desertification and Its Control in China*; Springer: Beijing, China, 2010; ISBN 9787040257977.
2. Abatzoglou, J.T.; Dobrowski, S.Z.; Parks, S.A.; Hegewisch, K.C. TerraClimate, a High-Resolution Global Dataset of Monthly Climate and Climatic Water Balance from 1958–2015. *Sci. Data* **2018**, *5*, 170191. [CrossRef] [PubMed]
3. Li, C.; Wang, S.; Stringer, L.C.; Wang, Y. Drivers and Impacts of Changes in China's Drylands. *Nat. Rev. Earth Environ.* **2021**, *2*, 858–873. [CrossRef]
4. UN (United Nations). *United Nations Convention to Combat Desertification in Countries Experiencing Serious Drought and/or Desertification, Particularly in Africa. Document A/AC. 241/27, 12. 09. 1994 with Annexe*; United Nations: New York, NY, USA, 1994.
5. Cherlet, M.; Hutchinson, C.; Reynolds, J.; Hill, J.; Sommer, S.; von Maltitz, G. (Eds.) *World Atlas of Desertification*; Publication Office of the European Union: Luxembourg, 2018; ISBN 978-92-79-75350-3.
6. Reynolds, J.F. Desertification Is a Prisoner of History: An Essay on Why Young Scientists Should Care. *Ecosistemas* **2021**, *30*, 2302. [CrossRef]
7. Cheng, L.; Lu, Q.; Wu, B.; Yin, C.; Bao, Y.; Gong, L. Estimation of the Costs of Desertification in China: A Critical Review. *Land Degrad. Dev.* **2018**, *29*, 975–983. [CrossRef]
8. del Barrio, G.; Sanjuán, M.E.; Martínez-Valderrama, J.; Ruiz, A.; Puigdefábregas, J. Land Degradation Means a Loss of Management Options. *J. Arid Environ.* **2021**, *189*, 104502. [CrossRef]
9. Feng, Q.; Ma, H.; Jiang, X.; Wang, X.; Cao, S. What Has Caused Desertification in China? *Sci. Rep.* **2015**, *5*, 15998. [CrossRef]
10. Wang, J.; Huang, J.; Rozelle, S.; Huang, Q.; Zhang, L. Understanding the Water Crisis in Northern China: What the Government and Farmers Are Doing. *Int. J. Water Resour. Dev.* **2009**, *25*, 141–158. [CrossRef]
11. Qin, Y.; Abatzoglou, J.T.; Siebert, S.; Huning, L.S.; AghaKouchak, A.; Mankin, J.S.; Hong, C.; Tong, D.; Davis, S.J.; Mueller, N.D. Agricultural Risks from Changing Snowmelt. *Nat. Clim. Chang.* **2020**, *10*, 459–465. [CrossRef]
12. Gui, D.; Xue, J.; Liu, Y.; Lei, J.; Zeng, F. Should Oasification Be Ignored When Examining Desertification in Northwest China? *Solid Earth Discuss.* **2017**, 1–24. [CrossRef]
13. Zhang, X.; Zhang, L.; He, C.; Li, J.; Jiang, Y. Quantifying the Impacts of Land Use/Land Cover Change on Groundwater Depletion in Northwestern China—A Case Study of the Dunhuang Oasis. *Agric. Water Manag.* **2014**, *146*, 270–279. [CrossRef]
14. Zhou, D.; Wang, X.; Shi, M. Human Driving Forces of Oasis Expansion in Northwestern China Durin the Last Decade—A Case Study of the Heihe River Basin. *Land Degrad. Dev.* **2016**, *28*, 412–420. [CrossRef]
15. Gui, D.; Wu, Y.; Zeng, F.; Lei, J. Study on the Oasification Process and Its Effects on Soil Particle Distribution in the South Rim of the Tarim Basin, China in Recent 30 Years. *Procedia Environ. Sci.* **2011**, *3*, 69–74. [CrossRef]
16. Liu, Y.; Xue, J.; Gui, D.; Lei, J.; Sun, H.; Lv, G.; Zhang, Z. Agricultural Oasis Expansion and Its Impact on Oasis Landscape Patterns in the Southern Margin of Tarim Basin, Northwest China. *Sustainability* **2018**, *10*, 1957. [CrossRef]
17. Ortiz, E.F.; Tang, G. *Review of the Management, Administration and Activities of the Secretariat of the United Nations Convention to Combat Desertification (UNCCD)*; United Nations: Geneva, Switzerland, 2005.
18. Grainger, A. Is Land Degradation Neutrality Feasible in Dry Areas ? *J. Arid Environ.* **2015**, *112*, 14–24. [CrossRef]
19. Forrester, J.W. *Industrial Dynamics*; The MIT Press: Cambridge, MA, USA, 1961.
20. Martínez-Valderrama, J.; Olcina, J.; Delacámara, G.; Guirado, E.; Maestre, F.T. Complex Policy Mixes Are Needed to Cope with Agricultural. *Water Resour. Econ.* **2023**, *6*. [CrossRef]
21. Danfeng, S.; Dawson, R.; Baoguo, L. Agricultural Causes of Desertification Risk in Minqin, China. *J. Environ. Manag.* **2006**, *79*, 348–356. [CrossRef]

22. Sterman, J.D. *Business Dynamics: Systems Thinking and Modeling for a Complex World*; Mc Graw Hill: New York, NY, USA, 2000; ISBN 0-07-231135-5.
23. Davies, E.G.R.; Simonovic, S.P. Global Water Resources Modeling with an Integrated Model of the Social-Economic-Environmental System. *Adv. Water Resour.* **2011**, *34*, 684–700. [CrossRef]
24. Kelly, R.A.; Jakeman, A.J.; Barreteau, O.; Borsuk, M.E.; ElSawah, S.; Hamilton, S.H.; Henriksen, H.J.; Kuikka, S.; Maier, H.R.; Rizzoli, A.E.; et al. Selecting among Five Common Modelling Approaches for Integrated Environmental Assessment and Management. *Environ. Model. Softw.* **2013**, *47*, 159–181. [CrossRef]
25. Martínez-Valderrama, J.; Ibáñez, J. System Dynamics Tools to Study Mediterranean Rangeland's Sustainability. *Land* **2023**, *12*, 206. [CrossRef]
26. Martinez-Valderrama, J.; Ibáñez, J.; Alcalá, F.J.; Dominguez, A.; Yassin, M.; Puigdefábregas, J. The Use of a Hydrological-Economic Model to Assess Sustainability in Groundwater-Dependent Agriculture in Drylands. *J. Hydrol.* **2011**, *402*, 80–91. [CrossRef]
27. Margat, J.; van der Gun, J. *Groundwater around the World. A Geographic Synopsis*; CRC Press (Taylor & Francis): Boca Raton, FL, USA, 2013; ISBN 9780203772140.
28. Steffen, W.; Broadgate, W.; Deutsch, L.; Gaffney, O.; Ludwig, C. The Trajectory of the Anthropocene: The Great Acceleration. *Anthr. Rev.* **2015**, *2*, 81–98. [CrossRef]
29. Ibáñez, J.; Martínez-Valderrama, J. Global Effectiveness of Group Decision-Making Strategies in Coping with Forage and Price Variabilities in Commercial Rangelands: A Modelling Assessment. *J. Environ. Manag.* **2018**, *217*, 531–541. [CrossRef] [PubMed]
30. Di Baldassarre, G.; Wanders, N.; AghaKouchak, A.; Kuil, L.; Rangecroft, S.; Veldkamp, T.I.E.; Garcia, M.; van Oel, P.R.; Breinl, K.; Van Loon, A.F. Water Shortages Worsened by Reservoir Effects. *Nat. Sustain.* **2018**, *1*, 617–622. [CrossRef]
31. Sears, L.; Caparelli, J.; Lee, C.; Pan, D.; Strandberg, G.; Vuu, L.; Lawell, C.Y.C.L. Jevons' Paradox and Efficient Irrigation Technology. *Sustainability* **2018**, *10*, 1590. [CrossRef]

environmental sciences proceedings

MDPI

Proceeding Paper

Analyses of Maximum Precipitation in Brazil and the Variability of Diurnal Cycle †

Aléxia Monteiro Valentim [1,2], Cláudio Moisés Santos e Silva [1,*], Daniele Tôrres Rodrigues [1,3] and Paula Andressa Alves de Araújo [1,2]

1 Center for Exact and Earth Sciences—CCET, Department of Atmospheric and Climate Sciences, Campus of Natal-RN, Federal University of Rio Grande do Norte, Natal 59078-970, Brazil; alexiavalentim_@hotmail.com (A.M.V.); mspdany@gmail.com (D.T.R.)

2 Senai Institute of Renewable Energy Innovation, Natal 59063-400, Brazil

3 Department of Statistics, Federal University of Piauí, Teresina 64049-550, Brazil

* Correspondence: claudio.silva@ufrn.br

† Presented at the 7th International Electronic Conference on Water Sciences, 15–30 March 2023; Available online: https://ecws-7.sciforum.net.

Abstract: According to recent works, the diurnal cycle is more geographically pronounced in places such as South America; and this analysis aims to observe how climate variability is associated with meteorological phenomena at different scales. For this, a set of hourly data from rain gauges throughout Brazil was collected, and through 411 automatic rain gauges, the data were selected between 1 January 2008 and 31 December 2020. Clustered multivariate statistics were performed for regional characterization of the data, with sets of 4, 5, and 6 groups. The identification of the occurrence of different daily cycles on the sub-daily scale demonstrates intense rainfall associated with different meteorological phenomena and spatial variations.

Keywords: sub-daily; cluster analysis; climate change

check for updates

Citation: Valentim, A.M.; Santos e Silva, C.M.; Rodrigues, D.T.; de Araújo, P.A.A. Analyses of Maximum Precipitation in Brazil and the Variability of Diurnal Cycle. *Environ. Sci. Proc.* **2023**, *25*, 95. https://doi.org/10.3390/ECWS-7-14229

Academic Editor: Athanasios Loukas

Published: 16 March 2023

1. Introduction

The diurnal variability is directly related to surface drag such as friction drag, heat flows, mass, and energy [1]; thus, it is the first climatic mode affected by climate change, with possible implications for the occurrence of events in extreme conditions. Studies concerning the diurnal cycle depend on observing data collected in situ, since there are few sets of meteorological data with sub-daily sampling for most continental parts of the globe despite some incipient initiatives (for example, [2]).

Systematic analysis of diurnal cycle can be used for understanding relevant scientific issues in different environmental variables, such as radiation, cloudiness, air and sea surface temperature [3,4], atmospheric activity of rays [5,6], particulate matter, and air pollution in urban areas and forested regions [7,8]. Due to the complexity of the physical processes that modulate the diurnal cycle of precipitation, simulations of this cycle have been used to evaluate the efficiency of climate models [9–11]. In other way, some studies have analyzed the diurnal cycle in a perspective of the difference between present and future climate, identifying changes in the precipitation intensity during 0300 and 1200 UTC worldwide, including South America [12,13].

From January 2008, the National Institute of Meteorology (Instituto Nacional de Meteorologia) (Brasília, Brazil) started to monitor different meteorological variables with one-hour sampling through automatic stations in all continental areas of Brazil. From this perspective, the present research is the broadest analysis of sub-daily precipitation performed with data collected in loco in Brazil. In this sense, the objective is to characterize the space–time aspects of the diurnal cycle of precipitation in Brazil during a continuous period of 13 years. In addition, we demonstrated the regional aspects using multivariate statistical analysis and showing the seasonal variability of the diurnal cycle.

2. Materials and Methods

2.1. Database

The Brazilian territory is the fifth largest country in the world with 8,547,403 km^2 of territorial extension, of which 50% are constituted by the Amazon region. The database used was obtained from the project "Meteorological Data Set for Teaching and Research" of INMET (www.inmet.gov.br, accessed on 30 September 2020) and consists of rainfall collected by 411 automatic weather stations (Figure 1) with 1-hour sampling during the period from 1 January 2008 to 31 December 2020.

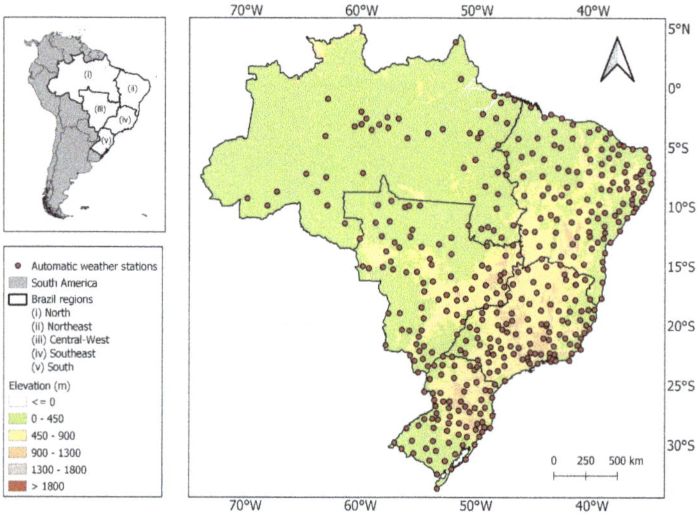

Figure 1. Geographic location of Brazil with spatial distribution of automatic stations used in this study.

2.2. Clustering Method

The diurnal precipitation cycle was analyzed from a perspective of homogeneous regions. In order to separate homogeneous groups, we applied a multivariate statistical technique of cluster analysis. This method joins elements with similar characteristics, thus increasing the similarity between the groups and the difference between the groups [14,15]. The first step of the process is to estimate the degree of similarity or dissimilarity among the data and, in the present study, the Euclidean distance that according to [16] was used, which is frequently used for this purpose in climatic studies in Brazil [17,18]. Additionally, we choose the Ward method, which is a hierarchical clustering [19] based on the minimum variance within a cluster [20]. For the grouping was performed in four, five, and six groups.

3. Results

The annual average of the diurnal precipitation cycles for the three types of groups is shown in Figure 2. Hourly rates are less than 0.35 mm/h when set up for 4 or 5 groups. However, with 6 groups, a maximum of 0.42 mm/h was observed in the new cluster, group 6 in the 1900 UTC range, located in the coastal and inland region of the Amazon.

In the configuration of four clusters, it is made by three dominant types of variability. The first, with the presence of group 1 and 2, has a maximum at 1900 UTC and is typical of continental regimes and associated with a strong convective activity in the middle and late afternoon. Group 1 is more observed in the South and also in the Southeast and

Midwest. Group 2 is geographically located in the Midwest region of the country; the highest rainfall intensity is more pronounced in the South, in relation to the next variability group, group 3, which comprises the northeastern coastal region, with a maximum of 0.30 mm/h at 09:00 UTC.

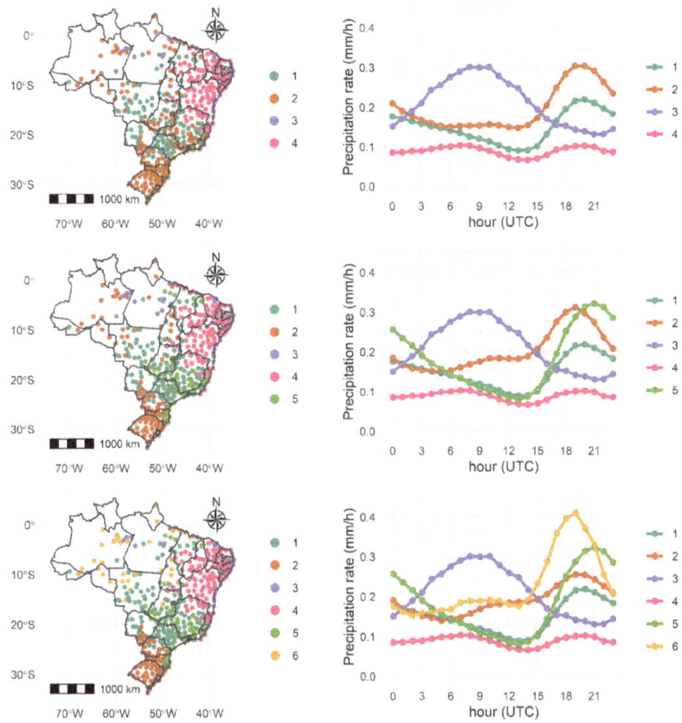

Figure 2. Annual Average of the daytime precipitation cycles for 4, 5, and 6 clusters in the Brazilian territory.

Another mode is group 4 characterized by maximums at 07:00 UTC and 19:00 UTC. And the Amazon, with a mixture of diurnal cycles, in addition to the rain in the middle of the afternoon (more common in group 1) also has precipitation during the night and early morning.

4. Conclusions

From an analysis of precipitation in the Brazilian territory, it was possible to understand the general aspects of diurnal cycle of precipitation in Brazil. Therefore, from a multivariate analysis in data from 1 January 2008 to 31 December 2020, by cluster clustering method, it was possible to conclude that there are different types of diurnal cycle and several meteorological mechanisms that model it. This study demonstrates the need to better understand the cycle of rainfall in Brazil, how they behave in the face of atmospheric phenomena, and how the climate is undergoing changes. The evaluation of the correlation between precipitation and other atmospheric variables, can further improve the accuracy of the forecasts.

Author Contributions: Conceptualization, A.M.V. and C.M.S.e.S.; methodology, A.M.V.; software, D.T.R.; validation, A.M.V., C.M.S.e.S. and D.T.R.; formal analysis, D.T.R.; investigation, P.A.A.d.A.; data curation, D.T.R. and P.A.A.d.A.; writing—original draft preparation, A.M.V.; writing—review and editing, A.M.V.; visualization, C.M.S.e.S.; supervision, C.M.S.e.S.; project administration, C.M.S.e.S.;

funding acquisition, C.M.S.e.S. All authors have read and agreed to the published version of the manuscript.

Funding: This research was funded by National Council for Research and Development (CNPq) grant number PIB14178-2017.

Institutional Review Board Statement: Not applicable.

Informed Consent Statement: Not applicable.

Data Availability Statement: The data can be obtained from the meteorological database of the National Institute of Meteorology. https://portal.inmet.gov.br/ (accessed on 30 September 2020).

Conflicts of Interest: The authors declare no conflict of interest.

References

1. Meredith, E.P.; Ulbrich, U.; Rust, H.W. The diurnal nature of future extreme precipitation intensification. *Geophys. Res. Lett.* **2019**, *46*, 7680–7689. [CrossRef]
2. Lewis, H.W.; Sanchez, J.M.C.; Arnold, A.; Fallmann, J.; Saulter, A.; Graham, J.; Bush, M.; Siddorn, J.; Palmer, T.; Lock, A.; et al. The UKC3 regional coupled environmental prediction system. *Geosci. Model Dev.* **2019**, *12*, 2357–2400. [CrossRef]
3. Webster, P.J.; Clayson, C.A.; Curry, J.A. Clouds, Radiation, and the Diurnal Cycle of Sea Surface Temperature in the Tropical Western Pacific. *J. Clim.* **1996**, *8*, 1712–1730. [CrossRef]
4. Yan, B.; Chan, P.W.; Li, Q.; He, Y.; Shu, Z. Dynamic analysis of meteorological time series in Hong Kong: A nonlinear perspective. *Int. J. Climatol.* **2021**, *41*, 4920–4932. [CrossRef]
5. Boccippio, D.J.; Koshak, W.J.; Blakeslee, R.J. Performance Assessment of the Optical Transient Detector and Lightning Imaging Sensor. Part I: Predicted Diurnal Variability. *J. Atmos. Ocean. Technol.* **2002**, *9*, 1318–1332. [CrossRef]
6. Minobe, S.; Park, J.H.; Virts, K.S. Diurnal cycles of precipitation and lightning in the tropics observed by TRMM3G68, GSMaP, LIS, and WWLLN. *J. Clim.* **2020**, *33*, 4293–4313. [CrossRef]
7. Backman, J.; Rizzo, L.V.; Hakala, J.; Nieminen, T.; Manninen, H.E.; Morais, F.; Aalto, P.P.; Siivola, E.; Carbone, S.; Hillamo, R.; et al. On the diurnal cycle of urban aerosols, black carbon and the occurrence of new particle formation events in springtime São Paulo, Brazil. *Atmos. Chem. Phys.* **2012**, *12*, 11733–11751. [CrossRef]
8. Wimmer, W.; Mazon, S.B.; Manninen, H.E.; Kangasluoma, J.; Franchin, A.; Nieminen, T.; Backman, J.; Wang, J.; Kuang, C.; Krejci, R.; et al. Ground-based observation of clusters and nucleation-mode particles in the Amazon. *Atmos. Chem. Phys.* **2018**, *18*, 13245–13264. [CrossRef]
9. da Rocha, R.P.; Morales, C.A.; Cuadra, S.V.; Ambrizzi, T. Precipitation diurnal cycle and summer climatology assessment over South America: An evaluation of Regional Climate Model version 3 simulations. *J. Geophys. Res.* **2009**, *114*, D10108. [CrossRef]
10. Santos e Silva, C.M.; Lúcio, P.S.; Spyrides, M.H.C. Distribuição espacial da precipitação sobre o Rio Grande do Norte: Estimativas via satélites e medidas por pluviômetros. *Rev. Bras. Meteorol.* **2012**, *27*, 337–346. [CrossRef]
11. Johnson, G.L.; Daly, C.; Taylor, G.H.; Hanson, C.L. Spatial variability and interpolation of stochastic weather simulation model parameters. *J. Appl. Meteorol. Climatol.* **2000**, *39*, 778–796. [CrossRef]
12. Reboita, M.S.; Dutra, L.M.M.; Dias, C.G. Diurnal cycle of precipitation simulated by RegCM4 over South America: Present and future scenarios. *Clim. Res.* **2016**, *70*, 39–55. [CrossRef]
13. Watters, D.; Battaglia, A.; Allan, R.P. The diurnal cycle of precipitation according to multiple decades of global satellite observations, three CMIP6 models, and the ECMWF reanalysis. *J. Clim.* **2021**, *34*, 5063–5080. [CrossRef]
14. Hair, J.F.; Black, W.C.; Babin, B.J.; Anderson, R.E.; Tatham, R.L. *Análise Multivariada de Dados*, 5th ed.; Adonai, T., Sant'Anna, S., Anselmo, C.N., Eds.; Bookman: Porto Alegre, Brazil, 2006; pp. 425–478.
15. Mingoti, S.A. *Análise de Dados Através de Métodos de Estatística—Multivariada—Uma Abordagem Aplicada*; UFMG: Belo Horizonte, Brazil, 2005.
16. Mimmack, G.M.; Mason, S.J.; Galpin, J.S. Choice of Distance Matrices in Cluster Analysis: Defining Regions. *J. Clim.* **2001**, *14*, 2790–2797. [CrossRef]
17. Oliveira, P.T.; Silva, S.; Lima, K.C. Climatology and trend analysis of extreme precipitation in subregions of Northeast Brazil. *Theor. Appl. Climatol.* **2017**, *130*, 77–90. [CrossRef]
18. de Abreu, L.P.; Mutti, P.R.; Lima, K.C. Variabilidade espacial e temporal da precipitação na bacia hidrográfica do Rio Parnaíba, Nordeste do Brasil. *Rev. Bras. Meio Ambiente* **2019**, *7*, 3524759. [CrossRef]
19. Ward, J.H. Hierarchical Grouping to Optimize an Objective Function. *J. Am. Stat. Assoc.* **1963**, *58*, 236–244. [CrossRef]
20. Hervada-Sala, C.; Jarauta-Bragulat, E. A program to perform Ward's clustering method on several regionalized variables. *Comput. Geosci.* **2004**, *30*, 881–886. [CrossRef]

environmental sciences proceedings

MDPI

Proceeding Paper

Advance Ensemble Flood Warning System: A Case Study for Nullah Lai †

Muhammad Aamir Siddiqui [1], Mudasser Muneer Khan [1], Rabia Khan [2] and Syyed Adnan Raheel Shah [3,*]

1 Department of Civil Engineering, Bahauddin Zakariya University, Multan 66000, Pakistan; siddiqui.aamir27@gmail.com (M.A.S.); mudasserkhan@bzu.edu.pk (M.M.K.)
2 Department of Agriculture, Forest and Range Management, Bahauddin Zakariya University, Multan 66000, Pakistan; rabiajabeen789@gmail.com
3 Department of Civil Engineering, NFC-Institute of Engineering & Technology, Multan 66000, Pakistan
* Correspondence: syyed.adnanraheelshah@uhasselt.be; Tel.: +92-3007914248

† Presented at the 7th International Electronic Conference on Water Sciences, 15–30 March 2023; Available online: https://ecws-7.sciforum.net.

Abstract: River flow forecasting is an essential tool to manage floods in the current era, especially for flash flooding scenarios in urban areas. This study focuses the flash flooding scenario in the Nullah Lai basin, which comprises the twin cities Islamabad and Rawalpindi. Steep slopes in the Margalla hills and Islamabad create high numbers of flash floods in the lower reaches of Rawalpindi, which are densely populated. When high-intensity rainfall occurs in the steep slopes of Margalla and Islamabad, high-volume floods with high velocity pour down, which instantaneously reaches the less-sloped Rawalpindi regions, which causes the raising of the water level in the stream, and flooding occurs. The section of the Nullah Lai Rawalpindi starting from the Qatarian bridge to the Gawalmandi bridge has always faced flash flooding over time. In the period of few hours, the water level reaches several fts in the nullah, which is why it is not possible to alert the people living on the banks in a timely manner, a problem that illuminates the need for a forecasting system at Nullah Lai. In the current research, the China Metrological Agency forecast center (CMA)'s ensemble forecast data have been utilized to achieve forecasts in the Nullah Lai. For this purpose, two initial objectives were set to achieve which basic needs are required process the data available in grib format from data centers. A digital model of the Nullah Lai was made using hydrology tools available in ArcGIS 10.3. A digital equation was obtained from gene expression modeling (GEP), which was later used to generate the ensemble stage forecast against the ensemble rainfall forecast. The results obtained show that the flash flooding phenomenon in Nullah Lai can, with some uncertainty, be predicted well in time. Using 3-days-ahead forecast data from CMA, the same floods were predicted 3 days before the event. This research also provides the procedure to use the ensemble forecast data in developing an automated model to generate the ensemble stage forecast for coming events. This study will help the administrative authorities better manage the upcoming floods and save lives and capital costs lost in the flash flooding phenomena which continuously happen in the basin of the Nullah Lai.

Keywords: ensemble; flash flooding; Nullah Lai; flood forecasting; catchment; gene expression modeling; THORPEX; TIGGI; CMA; ArcGIS

check for updates

Citation: Siddiqui, M.A.; Khan, M.M.; Khan, R.; Shah, S.A.R. Advance Ensemble Flood Warning System: A Case Study for Nullah Lai. *Environ. Sci. Proc.* **2023**, *25*, 96. https://doi.org/10.3390/ECWS-7-14197

Academic Editor: Silvia Kohnova

Published: 14 March 2023

1. Introduction

The development and implementation of policies and procedures related to water management and transportation are influenced by climate forecasts. These forecasts can help decision-makers in various areas, such as farming, urban flooding, and transportation. In addition to helping manage the flow of water, they can also help disaster administration and supervisors keep track of the allocation of water to diverse users [1]. The occurrence of flash flooding is a major concern for scientists in normal risks and hydrologic studies due to its top-positioning in various categories of disasters [2]. In the event of a flash flood, people should anticipate the coming surges and prepare for the secure passage of the water flow. This is carried out through the process of de-terminating the stream flow. It is additionally essential for monitoring the water supplies for various purposes, such as the water system, route, mechanical and hydroelectric utilization, and natural watering [3]. The use of climate models is currently scheduled for various global forecasting centers. These models are utilized for the prediction of atmospheric and climate conditions [4]. The NWP system uses the laws of thermodynamics and material science to predict the long-term state of the climate by taking into account the various factors that affect its development. The EPS system is a collection of expectations that are generated by the model. These expectations are then used to generate forecasts [5–9]. A collection of realizations known as individuals' members in an ensemble forecast is composed of various arrangements of similar parameter types. One of the most common realizations is the control estimate, which is based on the best available conditions in the atmosphere. This allows for a more reasonable depiction of the demonstration. An ensemble framework is a representation of the vulnerability of the forecast [10].

The TIGGE (THORPEX Interactive Grand Global Ensemble) network [11,12] was established in 2005 to create strategies that combine different weather-forecasting frameworks. Through its data accumulation centers, the public can access worldwide weather forecasts that are issued by various weather-forecasting organizations. The different frameworks utilized by different forecasting centers vary in their approach to forecasting. For instance, the Nullah Lai center is located in the twin cities of Islamabad and Rawalpindi. This region has been featured in hydrological modeling before [13]. During the monsoon season, heavy rainfalls can cause flash flooding in low-lying areas of the city of Rawalpindi. It is very difficult to remove the people from these areas due to the growing urban population. The penalty for forecasting high-climate events is based on their ability to predict such situations. Despite the increasing number of end users for weather forecasts, the majority of them are not focused on forecasting the exact weather conditions for the upcoming season [14].

This study was conducted to find out what kind of work needs to be conducted to improve the prediction capabilities for the Nullah Lai. There is currently no system that can precisely forecast the likelihood of a certain event happening. This study aims to provide a comprehensive analysis of the various aspects of the river's flow and its development.

2. Procedural Background

The research was based on statistical and digital modeling of the Nullah Lai, which was then utilized in the process of ensemble prediction. The first step for the modeling was to gather data from different data providers. The research was conducted using large datasets to establish a relationship between the rainfall and the stage of the Nullah Lai in the first phase. In the second phase, the data collected from different stations were linked to form a comprehensive analysis.

2.1. Forecast Data

Researchers use different models to predict the amount of rainfall that will occur in a given area. The predictions are then analyzed using data collected by the China Metrological Agency. For the study, the agency's data were used to forecast the flow of water in the Nullah Lai stream. Over 500 forecasts have been made regarding the rainfall that will happen in the watershed of the Nullah Lai. These predictions were made using a format known as grib, which is a representation of the latitude and longitude network. Data on the precipitation were collected from the data center of the China Metrological Agency from 2017 to 2019. The data collected during the study were gathered by researchers using grib files, which were collected monthly. They only used this method for the predictions since they were only focused on extreme weather events that usually occur during the monsoon season. Each file was analyzed separately and used for forecasting the rainfall. Two separate files containing 48 h and 72 h worth of rainfall data were downloaded. These were then split into two grib files, one containing a forecast and the other a plot of the accumulated rainfall. The grib files were then processed using a utility known as wgrib2. The NetCDF file was then processed using the ARC-GIS 10.3 platform. This method generated a total of 48 h worth of forecast data. After that, a table extract tool was used to extract the data into a format that can be used to display the forecast. All 12 events' data were then analyzed and processed in order to produce a graphical representation of the data.

2.2. Rainfall and Stage Data

The data collected by PMD included the daily total rainfall from various gauge stations located in Chaklala, Bokra, Golra, RAMC, Saidpur, and PMD. These measurements were taken from 2010 to 2020. Location shown in Figure 1.

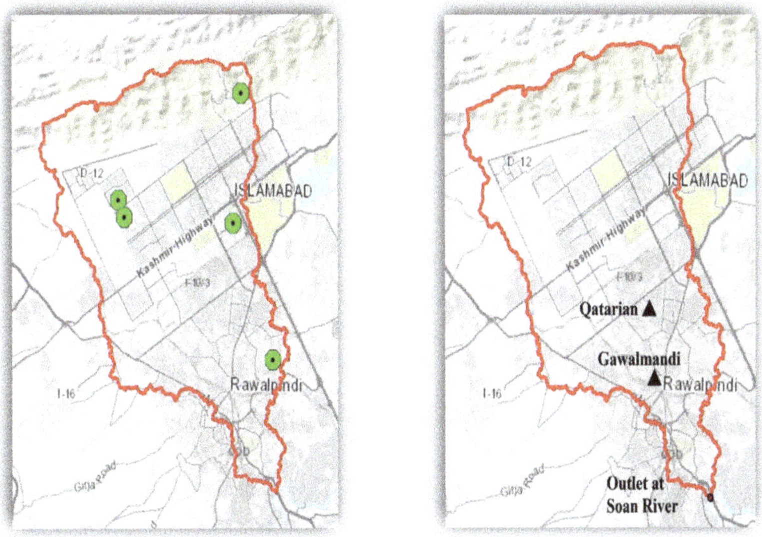

(a) (b)

Figure 1. (**a**) Rain-Gauging Station; (**b**) Stage-Measuring Stations (right-2) Position.

2.3. Digital Modelling of Nullah Lai Catchment

A catchment area is a region of land where rain falls and flows toward a common outlet. Without a digital modeling process, the drawing of a watershed area can be very time-consuming and challenging. This process can be carried out with the help of tools such as the ArcGIS software.

The following describes the steps in the process of digital modeling of Nullah Lai's catchment area. After downloading the data from the USGS website, using the DEM format, the area's digital elevation can be marked with the coordinates of its source. In addition, using the spatial analytic tool, a "fill" tool can be applied to the area.

The "flow direction" tool can be used to obtain a view of the flow directions in the water. The "basin tool" can also be used to delineate the various drainage basins. The conversion tool for the Raster-to-Polygon and Raster-to-Polygons conversions can be used to convert the image into a polygon. Finally, the "clip tool" can be used to clip the watershed from other areas.

The "flow accumulation" tool can be used to create a file that contains the water flow data. The "Raster Calculator" tool to can then be used to make a water stream network.

In the conversion segment, the "Raster" tool can be used to transform the streams into polygons. Finally, the "clip" tool can be used to clip the steam lines from other areas.

The input outlet point coordinates can be used to draw the catchment area to a specific point. The "pour point" tool can also be used to create a watershed area that is respected by each outlet point. Finally, the "Theisen Polygons" tool can be used to create a representation of the rain gauge's polygons.

2.4. Observed Rainfall—Stage Modelling

After collecting data about the rainfall and runoff in the area, a model was developed using the DTREG numerical modeling software. This tool was used to create a Gene Ex-pression Program, which was able to mimic the biological process of developing the phenomenon. The process of gene expression programming involves developing various types of systems, such as decision trees and networks of neural and polynomial concepts. In DTREG, this type of programming is referred to as symbolic regression.

The following relation was obtained by analyzing the 7 years of rainfall and stage data.

$$S = \mathrm{Sqrt}(P - 6.2275699) + ((-26.983058)/P) + 1.8792162 + \mathrm{Sqrt}(\mathrm{Sqrt}P)) \tag{1}$$

Stage is denoted as "S"
Basin mean of rainfall is denoted as "P".

3. Materials and Methods

The research utilized statistical and digital modeling techniques to develop a model of Nullah Lai catchment, shown in Figure 2. The data collected from various sources, such as TIGGI and PMD, were then processed using wgrib2, DTREG, MS Excel, and ArcGIS. The main objective of this study was to create a statistical model of the rainfall–runoff event and to relate this to the data collected by the CMA ensemble.

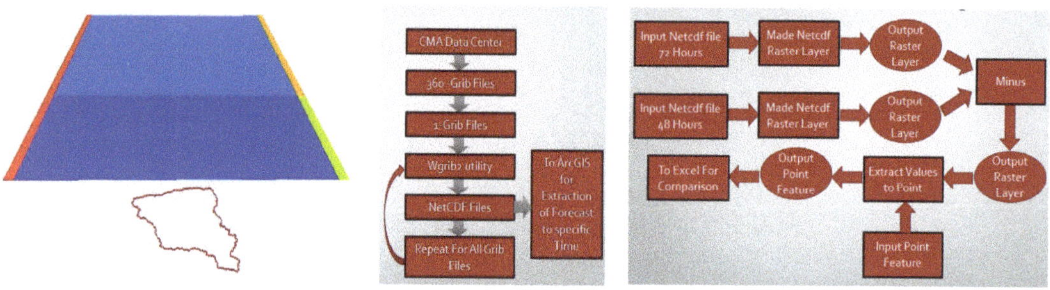

(a) (b) (c)

Figure 2. (**a**) Raster layer over Nullah Lai catchment; (**b**) CMA data processing; (**c**) Extracting forecast using ArcGIS.

4. Results and Analysis

The results of this study are compared with the actual runoff and rainfall produced by the different ensembles. The twelve extreme events that occurred during the 2017 to 2019 monsoon season are discussed here to highlight the significant runoff in the Nullah Lai. Researchers are more likely to focus on the ensemble forecast due to the varying atmospheric conditions. Due to the varying topography and flow conditions of the Nullah Lai catchment area, it is difficult to predict the exact amount of rainfall that will occur. This section aims to provide a comprehensive analysis of the rainfall–runoff and provides a comparison with the ensemble forecast.

Ensemble Predicted Stage vs. Actual Stage

This section compares the ensemble stage, forecasted stage, and actual observed data from the obtained results. It also provides a range of possible outcomes for a given event. Figures 3–5 graphically explain the twelve (12) extreme events from 2017 to 2019.

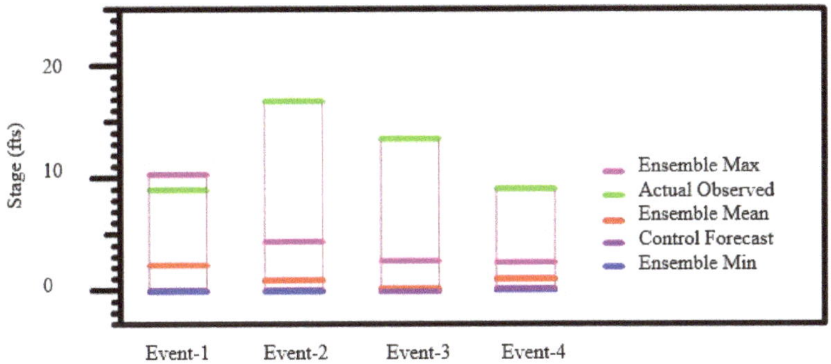

Figure 3. Event 1, 2, 3, and 4: Stage graphical representation.

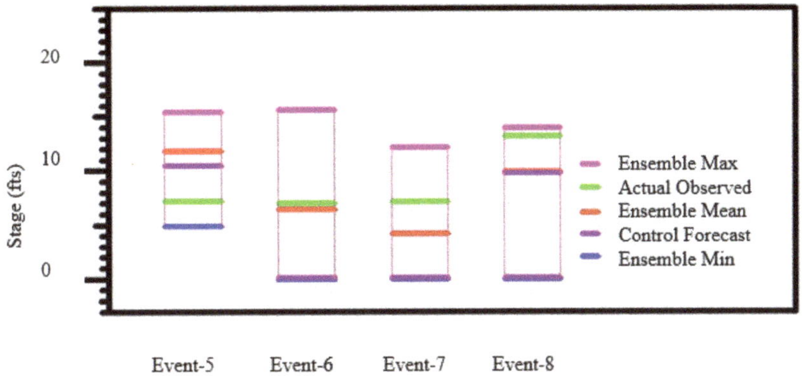

Figure 4. Event 5, 6, 7, and 8: Stage graphical representation.

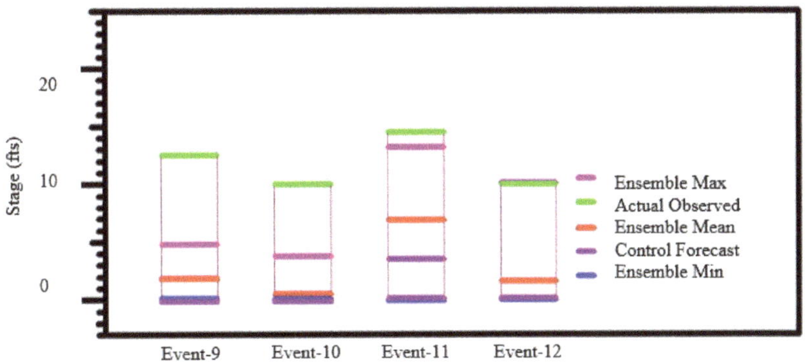

Figure 5. Event 9, 10, 11, and 12: Stage graphical representation.

The predicted outcome of the controlled forecast is 16.67%. This means that only two out of the twelve predictions were carried out correctly. It can be concluded that the forecast is misleading and cannot take into account other factors that can affect the prediction. The ensemble prediction system was able to provide a precise estimate of the flow's expected range.

The accuracy rate of forecasting an event out of 12 events was 7 out of 12. This means that the predicted outcome is 58.33% accurate. The advantage of using an ensemble forecast is that it provides a better idea of the likelihood of an extreme event happening within the next 3 days.

5. Conclusions

The following conclusions can be drawn from the conducted study:

1. The Nullah Lai watershed was digitally modelled using the latest version of ArcGIS 10.3.1. This model was made using the most accurate digital elevation data from the USGS;
2. The model was able to produce a root mean square of 77% and a correlation coefficient of 0.88. The data collected during the period indicated that the model is very accurate;
3. The forecasting system was used to study 12 extreme events that occurred from 2017 to 2019. Out of these, seven events were within the range of the ensemble forecast. The results show that the ensemble forecast is more accurate than the control forecast when it comes to forecasting events.

Author Contributions: Conceptualization, M.M.K.; Data curation: M.A.S.; Formal analysis: M.A.S.; Methodology: M.A.S.; Validation: R.K.; Writing—original draft: M.A.S.; Writing—review and editing. S.A.R.S. All authors have read and agreed to the published version of the manuscript.

Funding: This research received no external funding.

Institutional Review Board Statement: Not applicable.

Informed Consent Statement: Not applicable.

Data Availability Statement: Data will be available upon reasonable demand.

Acknowledgments: This paper is based on thesis submitted by Muhammad Aamir Siddiqui at CED, BZU.

Conflicts of Interest: The authors declare no conflict of interest.

References

1. Khummongkol, R.; Sutivong, D.; Kuntanakulwong, S. Water Resource Management Using Multi-Objective Optimization and Rainfall Forecast. In Proceedings of the 2007 International Conference on Convergence Information Technology (ICCIT 2007), Gwangju, Republic of Korea, 21–23 November 2007.
2. Borga, M.; Anagnostou, E.; Blöschl, G.; Creutin, J.-D. Flash flood forecasting, warning and risk management: The HYDRATE project. *Environ. Sci. Policy* **2011**, *14*, 834–844. [CrossRef]
3. Siccardi, F.; Boni, G.; Ferraris, L.; Rudari, R. A hydrometeorological approach for probabilistic flood forecast. *J. Geophys. Res Atmos.* **2005**, *110*. [CrossRef]
4. Cloke, H.; Pappenberger, F. Ensemble flood forecasting: A review. *J. Hydrol.* **2009**, *375*, 613–626. [CrossRef]
5. Taylor, J.; Buizza, R. Neural network load forecasting with weather ensemble predictions. *IEEE Trans. Power Syst.* **2002**, *17*, 626–632. [CrossRef]
6. Shoaib, M.; Shamseldin, A.Y.; Khan, S.; Khan, M.M.; Khan, Z.M.; Sultan, T.; Melville, B.W. A comparative study of various hybrid wavelet feedforward neural network models for runoff forecasting. *Water Resour. Manag.* **2018**, *32*, 83–103. [CrossRef]
7. Shaukat, R.S.; Khan, M.M.; Shahid, M.; Shoaib, M.; Khan, T.A.; Aslam, M.A. Quantitative Contribution of Climate Change and Land Use Change to Runoff in Tarbela Catchment, Pakistan. *Pol. J. Environ. Stud.* **2020**, *29*, 3295–3304. [CrossRef] [PubMed]
8. Hammad, M.; Shoaib, M.; Salahudin, H.; Baig, M.A.I.; Khan, M.M.; Ullah, M.K. Rainfall forecasting in upper Indus basin using various artificial intelligence techniques. *Stoch. Environ. Res. Risk Assess.* **2021**, *35*, 2213–2235. [CrossRef]
9. Shoaib, M.; Shamseldin, A.Y.; Melville, B.W.; Khan, M.M. Runoff forecasting using hybrid wavelet gene expression programming (WGEP) approach. *J. Hydrol.* **2015**, *527*, 326–344. [CrossRef]
10. Krzysztofowicz, R. The case for probabilistic forecasting in hydrology. *J. Hydrol.* **2001**, *249*, 2–9. [CrossRef]
11. Khan, M.M.; Shamseldin, A.Y.; Melville, B.W. Impact of ensemble size on forecasting occurrence of rainfall using TIGGE precipitation forecasts. *J. Hydrol. Eng.* **2014**, *19*, 732–738. [CrossRef]

12. Khan, M.M.; Shamseldin, A.Y.; Melville, B.W.; Shoaib, M. Impact of ensemble size on TIGGE precipitation forecasts: An end-user perspective. *J. Hydrol. Eng.* **2015**, *20*, 04014046. [CrossRef]
13. Ahmad, B.; Kaleem, M.S.; Butt, M.J.; Dahri, Z.H. Hydrological modelling and flood hazard mapping of Nullah Lai. *Proc. Pak. Acad. Sci.* **2010**, *47*, 215–226.
14. Morss, R.E.; Emanuel, K.A.; Snyder, C. Idealized adaptive observation strategies for improving numerical weather prediction. *J. Atmos. Sci.* **2001**, *58*, 210–232. [CrossRef]

environmental sciences proceedings

MDPI

Proceeding Paper

Analysis of Extreme Rainfall Events on a Sub-Daily Scale in Northeast †

Paula Andressa Alves de Araújo [1,*], Cláudio Moisés Santos e Silva [1,*], Daniele Tôrres Rodrigues [2] and Aléxia Monteiro Valentim [1]

1 Department of Atmospheric and Climate Sciences, Universidade Federal do Rio Grande do Norte (UFRN), Natal 59078-970, Rio Grande do Norte, Brazil; alexia.valentim.127@ufrn.edu.br

2 Department of Statistic, Universidade Federal do Piauí (UFPI), Teresina 64049-550, Piauí, Brazil; mspdany@gmail.com

* Correspondence: paulaand12010@hotmail.com (P.A.A.d.A.); claudio.silva@ufrn.br (C.M.S.e.S.)

† Presented at the 7th International Electronic Conference on Water Sciences, 15–30 March 2023; Available online: https://ecws-7.sciforum.net.

Abstract: Given the complex climate of the Brazilian Northeast, its variable rainfall regime in spatial and temporal scales due to multiple geographic factors and the action of various atmospheric systems, as well as intense precipitation events becoming increasingly present and strong, are incipient the initiatives to analyze extreme rainfall on a sub-daily scale. Thus, characterizing the spatial and temporal distribution of IPESS on the NEB, analyzing the seasonality of the number of events for each location, and the influence of interannual variability in the occurrence of these events became the objective of that work. For this, hourly precipitation observations were collected from 119 automatic surface weather stations spread across the NEB between 2009 and 2018. The night period presented the highest amounts of IPESS. The meteorological stations Ibimirim/PE and Ilhéus/BA had the highest and lowest number of occurrences, respectively. The amplitude of the values of Q presented the thresholds of the stations with the values of 3.4 mm as a minimum, and 28.6 mm as a maximum, referring to the stations of Petrolina/PE and São Luís/MA, respectively. Among the capitals, Teresina/PI had the highest number of EPIES, as well as João Pessoa/PB had the lowest, not only among the capitals but also the second with the lowest number in general. It was observed that during 2009 the events were more frequent, while 2012 presented the smallest amounts.

Keywords: percentile 95; EPEIS; hourly analysis

Citation: de Araújo, P.A.A.; Silva, C.M.S.e.; Rodrigues, D.T.; Valentim, A.M. Analysis of Extreme Rainfall Events on a Sub-Daily Scale in Northeast. *Environ. Sci. Proc.* **2023**, *25*, 97. https://doi.org/10.3390/ECWS-7-14299

Academic Editor: Athanasios Loukas

Published: 3 April 2023

1. Introduction

Since precipitation is one of the most important climate variables, as its variations directly affect numerous human activities, the knowledge about this variable's spatial and temporal variability is crucial to subsidize the decision-making and actions planning to minimize impacts caused by extreme precipitation events.

The North East of Brazil (NEB) is a region with a difficult climatic characterization as a result of geographic factors associated with different atmospheric systems. Then, there are different factors that fluctuate the precipitation regime of the NEB on a wide time and space scale [1–4].

A large part of the climate studies about analyses of precipitation that focus the NEB report mainly the water deficit. Trying to understand the complex social functions of these climatic characteristics is very important since they cause economic losses in addition to deaths. In contrast, heavy rainfall is also reported, which, together with water deficits, are the main types of adverse natural phenomena from the NEB, so those cause significant impacts that can be catastrophes.

Since intense rainfall events have become increasingly more common and strong, according to projections made by the Intergovernmental Panel Climate Change (IPCC) [5], there is vast scientific literature on extreme indices on a daily scale around the world, including in the NEB [6]. However, initiatives to analyze extreme rainfall on a sub-daily scale are still incipient, as well as the hourly analysis, mainly due to the lack of meteorological data with hourly or minute sampling. That said, the objective of the present study was to characterize the spatial and temporal distribution of the Subdaily Scale Intense Precipitation Events (EPIES) over the NEB from hourly data collected in the period between 2009 and 2018, analyzing the seasonality of the number of events for each location and the influence of interannual variability on the occurrence of EPIES.

2. Materials and Methods

2.1. Study Area

The NEB is located latitudinally between 1° S and 18° S and longitudinally between 48° W and 34° W. It comprises a territorial extension of 1,558,000 km^2, according to the Brazilian Institute of Geography and Statistics [7], comprising 9 states.

2.2. Data Collection

The dataset refers to precipitation observations from 119 automatic surface meteorological stations that are distributed in the NEB and are operated by the Brazilian National Institute of Meteorology (INMET). Data were collected hourly for the period between 1 January 2009 and 31 December 2018.

To organize the dataset and apply the methodology, Microsoft Excel software was used to calculate the results. The exploratory analysis of the heavy precipitation indices of each meteorological station was conducted using the R statistical computational system, realized through a box-plot cartogram, which is a graphic technique that shows the behavior of the data in space.

2.3. Methods

The methodology is divided into several stages, so the first consists of selecting the days with accumulated daily precipitation equal to or greater than 1 mm, keeping only the days on which they occurred (RR).

$$\mathrm{RR} \geq 1\ \mathrm{mm} \tag{1}$$

The second step is equivalent to using the percentile 95 [8] in order to find the extreme precipitation measurement values (Q) for each station. Thus, this method was applied to the accumulated precipitation values of all days validated in the previous step.

After completing the first and second stages, the third intended to emphasize the days with precipitation that went beyond the Q value. For this purpose, only those days with daily accumulations greater than the Q value of the station were maintained. That's it:

$$\mathrm{RR} \geq \mathrm{Q} \tag{2}$$

With the limits applied, these conditions were adopted as the definition of an EPIES for these days: it is considered an EPIES if, in a single hour, the accumulated precipitation rate (RI) represents more than 25% of the daily total with extreme rain. Thus, it is necessary to locate the hour with the highest precipitation value (max(RR)) for each day with extreme rain and calculate the percentage value of this record in relation to the daily accumulation:

$$\mathrm{RI} = (\max(\mathrm{RR})/\mathrm{RR}) \times 100,\ \mathrm{EPIES} = \mathrm{RI} \geq 25 \tag{3}$$

Finally, the occurrences of EPIES are counted, and the number of these events is analyzed from a temporal and spatial perspective, presenting them graphically with the support of Microsoft Office Excel.

3. Results

During the exploratory analysis, the Q values varied between 3.4 mm and 28.6 mm, referring to the Petrolina/PE and São Luís/MA stations, respectively. This indicates the great discrepancy of precipitation indices between meteorological stations. The distribution structure of the Q values of the meteorological stations was shown using the Box Plot. Via the interquartile amplitude, it is visualized that 50% of the Q values obtained are between 8.5 mm and 18.55 mm.

Graphs referring to the occurrence of EPIES on the hourly scale were generated using an x-y plane and geo-referenced in terms of latitude and longitude coordinates of the meteorological stations. The results are presented and divided into four-quarters of the day: between 00:00 h and 05:00 h (Figure 1a), between 06:00 h and 11:00 h (Figure 1b), between 12:00 h and 17:00 h (Figure 1c), and between 18:00 h and 23:00 h (Figure 1d). All the graphs were subject to the same categorization about the amount of EPIES occurrences. The quantity of occurrence was separated by color: the color gray belongs to the group without events, and the colors blue, green, yellow, orange, red, and burgundy belong to groups with a number of occurrences varying between 1 and 5, 6 and 10, 11 and 15, 16 and 20, 21 and 25, and between 26 and 30, respectively.

Analyzing Figure 1a, between 00:00 h and 02:00 h, the majority of meteorological stations showed a number of occurrences below or equal to five events. Patos/PB station stood out for presenting quantification of EPIES equal to or higher than 21 events during this time. At 04:00 h, there is an increase in the number of stations, with between 6 and 10 occurrences. Teresina/PI station stood out for presenting from 16 to 20 events in this subgroup.

In looking at Figure 1b, which refers to the range from 06:00 h and 11:00 h, the group composed of two stations that contained 11 to 15 events was the highest number of occurrences during the first hour of this category. At 07:00 h and 08:00 h, a vast quantity of stations composes the group from 1 to 5 events, being the subgroup with the largest quantity of stations, although the majority of the capitals of the states are more concentrated in the subgroup with 6 to 10 EPIES. At 09:00 h, the larger amount of registered occurrences was between 21 and 25 events. While at 10:00 h, the largest quantity was from 16 to 20 EPIES. With respect to 11:00 h, it was found that all Potiguar stations were located between 1 and 5 events. The subgroup from 16 to 20 EPIES was the occurrence limit in this interval.

According to the analyses in Figure 1c, at 12:00 h, the group with 1 to 5 events showed the larger number of stations, so all stations located in Maranhão, Piauí, and Bahia were found in this group. At 13:00 h, 10 stations without events were counted. Most of the capitals are in the subgroup, referring to the number of 1 to 5 events. During 14:00 h, only the Fortaleza/CE station had more than 10 occurrences of EPIES. At 15:00 h, the largest number of occurrences identified was limited to the corresponding subgroup between 11 and 15 events. While at 17:00 h, the station located in Itapipoca/CE presented more than 25 EPIES, reporting the maximum observed value.

In the analysis of Figure 1d, during 18:00 h, a significant part of the capitals presented from 1 to 5 EPIES. The highest number of occurrences of EPIES did not exceed 25 events. At 19:00 h and 20:00 h, the group with from 11 to 15 events, as so as 16 and 20, even though they are not composed of capitals, presented significant numbers of stations. However, at 19:00 h, the largest quantity of EPIES was in the subgroup with values greater than 20 events. With respect to 21:00 h, all stations had at least one occurrence during this time since no stations with no events were observed. Most of the stations in the capitals presented from 1 to 5 EPIES. The subgroups between 11 and 15 events and between 16 and 20 events contain significant numbers of meteorological stations. In analysis at 22:00 h, as well as the previous time, there is a large number of stations in the subgroups between 11 and 15 and between 16 and 20, although both do not present capitals in their compositions. Concluding the analysis of Figure 1d, at 23:00 h, stations composing the subgroup of occurrences between 21 to 25 EPIES were observed.

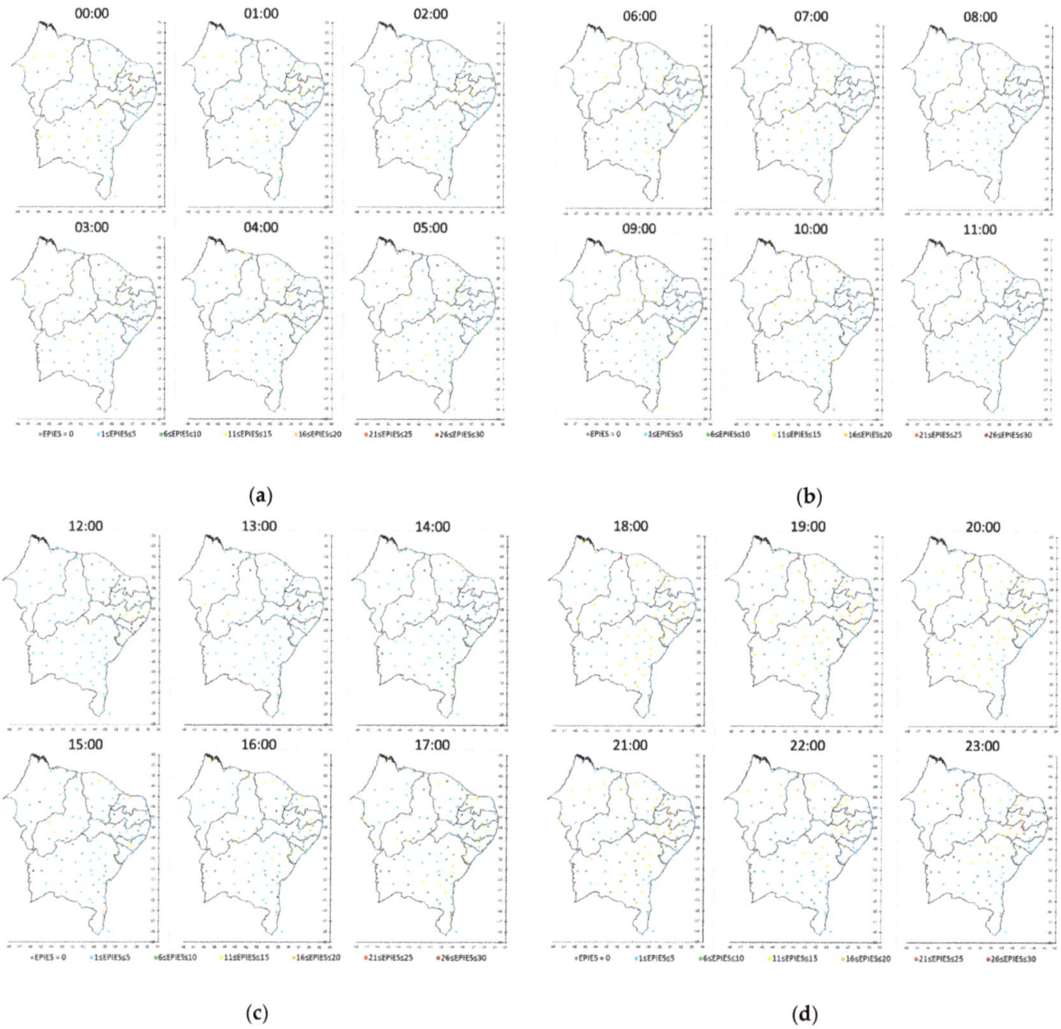

Figure 1. Graphs referring to the amounts of EPIES separated into subgroups: (**a**) between 00:00 and 05:00 h; (**b**) between 06:00 h and 11:00 h; (**c**) between 12:00 h and 17:00 h; (**d**) between 18:00 h and 23:00 h.

Performing the sum of occurrences per hour, we have the quantities of EPIES in the Northeast hourly. Therefore, it was possible to plot such information in a line graph described in Figure 2, to analyze the subsidiary periods that presented the highest and lowest numbers of occurrences of EPIES in the NEB. At night, the highest number of EPIES was presented. While at 13:00 h, the least amount of events was observed.

Still carrying out the sum of the observed values, it is possible to order the five stations that presented the highest and lowest quantity of occurrence. Therefore, the meteorological stations of Ibimirim/PE, Delfino/BA, Cabrobó/PE, Oeiras/PI e Tauá/CE had the highest numbers of EPIES in this exact order. At the same time, Ilhéus/BA, João Pessoa/PB, Palmares/PE, Abrolhos/PE, and Valença/BA were the stations that presented the lowest number of events, respectively.

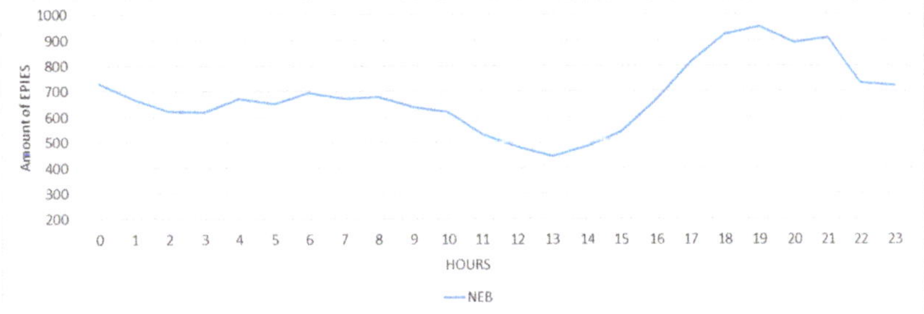

Figure 2. Line graph of the distribution of EPIES occurrences in the Northeast, hourly.

Implementing the same analysis to the states' capitals, we have the quantification of EPIES in descending order: Teresina/PI, São Luis/MA, Aracaju/SE, Salvador/BA, Fortaleza/CE, Natal/RN, Maceió/AL, Recife/PE, and João Pessoa/PB. Aracaju and Salvador had the same number of events.

Considering the monthly occurrences of all years, through the sum of all occurrences in the Northeast, Figure 3 shows a graphic of this information. Therefore, analyzing the line graph, it is notable that the largest amounts of EPIES occurred during the month of April, while the smallest are presented between August and September.

Figure 3. Line graph of the distribution of EPIES occurrences in the Northeast, monthly.

With the same objective, a graphic was made with the annual sum of the number of EPIES (Figure 4). The year with the highest amount of events was reported as 2009, while the one with the lowest amount was 2012.

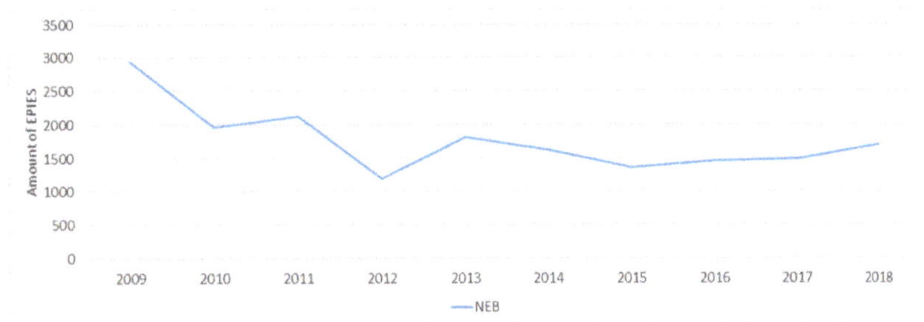

Figure 4. Line graph of the distribution of EPIES occurrences in the Northeast, annually.

4. Conclusions

Aiming to characterize the spatial and temporal distribution of EPIES on the NEB from hourly data in order to analyze the seasonality of events and the influence of interannual variability on the occurrence of EPIES, the present study demonstrated that the night period presented greater amounts of occurrences of this type of event. Ibimirim/PE meteorological station had the biggest number of events, and Ilhéus/BA had the smallest during the interval studied. It was shown that Q values varied between 3.4 mm and 28.6 mm, referring to the Petrolina/PE and São Luís/MA stations, respectively. Among the capitals, Teresina/PI presented the highest number of occurrences, as well as João Pessoa/PB the lowest number of EPIES, not only among the capitals but also proved to be the second meteorological station with the lowest amounts of events among all of the Northeast. It was observed that during the year 2009, the events occurred more frequently, while in 2012, they were less frequent. It is proposed for future work an increase in the period of study and ways to ease the failures present in the databases, increasing the reliability of the results. However, given the conditions of data availability, the present work proved to be competent in terms of the objective presented.

Author Contributions: Conceptualization, P.A.A.d.A. and C.M.S.e.S.; methodology, P.A.A.d.A.; software D.T.R.; validation, P.A.A.d.A., C.M.S.e.S. and D.T.R.; formal analysis, D.T.R.; investigation, A.M.V.; data curation, D.T.R. and A.M.V.; writing-original draft preparation, P.A.A.d.A.; writing-restew and editing, P.A.A.d.A.; visualization, C.M.S.e.S.; supervision, C.M.S.e.S.; project administration, C.M.S.e.S.; funding acquisition, C.M.S.e.S. All authors have read and agreed to the published version of the manuscript.

Funding: This research was funded by National Council for Research and Development (CNPq) grant number PIB14178-2017.

Institutional Review Board Statement: Not applicable.

Informed Consent Statement: Not applicable.

Data Availability Statement: The data can be obtained from the meteorological database of the National Institute of Meteorology. https://portal.inmet.gov.br/ (accessed on 30 September 2020).

Conflicts of Interest: The authors declare no conflict of interest.

References

1. Teixeira, R.F.B. O fenômeno da brisa e sua relação com a chuva sobre Fortaleza-CE. *Rev. Bras. De Meteorol.* **2008**, *23*, 282–291. [CrossRef]
2. Coutinho, M.D.L.; Gan, M.A.; Rao, V.B. Método objetivo de identificação dos vórtices ciclônicos de altos níveis na região Tropical Sul: Validação. *Rev. Bras. De Meteorol.* **2010**, *25*, 311–323. [CrossRef]
3. Torres, R.R.; Ferreira, N.J. Case Studies of Easterly Wave Disturbances over Northeast Brazil Using the Eta Model. *Weather. Forecast.* **2011**, *26*, 225–235. [CrossRef]
4. Reboita, M.S.; Gan, M.A.; Rocha, R.P.; Ambrizzi, T. Regimes de precipitação na América do Sul: Uma revisão bibliográfica. *Rev. Bras. De Meteorol.* **2010**, *25*, 185–204. [CrossRef]
5. Knutti, R. IPCC Working Group I AR5 snapshot: The rcp45 experiment. *World Data Cent. Clim. (WDCC) DKRZ* **2014**. [CrossRef]
6. Bezerra, B.G.; Silva, L.L.; Santos e Silva, C.M.; De Carvalho, G.G. Changes of precipitation extremes indices in São Francisco River Basin, Brazil from 1947 to 2012. *Theor. Appl. Climatol.* **2019**, *135*, 565–576. [CrossRef]
7. Instituto Brasileiro De Geografia E Estatística. *Instituto Brasileiro de Geografia e Estatística*; 2010. Available online: https://www.ibge.gov.br/ (accessed on 10 January 2023).
8. Martins, M.E.G. Percentis. *Rev. De Ciência Elem.* **2014**, *2*, 3.

environmental sciences proceedings

MDPI

Proceeding Paper

Batch Adsorption Studies Incorporating Response Surface Methodology for the Elimination of Acephate †

R. Shiny Raj [1,2] and K. Anoop Krishnan [1,*]

1 Biogeochemistry Group, National Centre for Earth Science Studies (NCESS), Trivandrum 695011, India; shinyraj.raj710@gmail.com
2 Department of Applied Chemistry, Cochin University of Science and Technology (CUSAT), Trivandrum 695019, India
* Correspondence: sreeanoop@rediffmail.com

† Presented at the 7th International Electronic Conference on Water Sciences, 15–30 March 2023; Available online: https://ecws-7.sciforum.net.

Abstract: Banned pesticides are continuously preferred by the planters of the Idukki District irrespective of their toxicity. Among the banned pesticides, acephate is preferred because of its high solubility in water and persistent character. Unfortunately, it detriments the biota, leading to neurogenic, carcinogenic, and physiological disorders in fish. The plantation near the Periyar River basin is contaminated with residues of pesticides, which eventually drain into the river. There is an urgent need for the removal of acephate. Therefore, we have focused on the removal of acephate into the lab scale. Batch adsorption studies were carried out for the removal of acephate. We selected a material Fe-MMT (Fe_3O_4-montmorillonite), which is benign and possesses a high adsorption capacity towards acephate. Adsorbent properties were examined by various analytical tools XRD, SEM, FTIR, and a Surface area analyzer. Adsorption followed Langmuir with first-order kinetic. Kinetic plots exhibited multistage adsorption, indicating film diffusion and pore diffusion during the adsorption or the mechanism of adsorption is chemisorption, physisorption, and Lewis's acid-base interaction. Response surface methodology involving CCD (central composite design) was extracted to maximize the adsorption of acephate onto Fe-MMT. Dosage and concentration seem to be the major parameters that influenced the adsorption. Adsorption achieved a peak (83.18%) at optimum conditions corresponding to pH 6, initial acephate concentration of 2 mg/L, and adsorbent dosage corresponding to 0.5 g/L.

Keywords: acephate; magnetically modified montmorillonite; multistage adsorption; response surface methodology

Citation: Raj, R.S.; Krishnan, K.A. Batch Adsorption Studies Incorporating Response Surface Methodology for the Elimination of Acephate. *Environ. Sci. Proc.* **2023**, *25*, 98. https://doi.org/10.3390/ECWS-7-14309

Academic Editor: Athanasios Loukas

Published: 3 April 2023

1. Introduction

Acephate, (O, S dimethyl acetyl phosphoroamidothioate) an organophosphorous pesticide, and its metabolite have immense potential to harm the biota. Acephate residues are present in the cardamom plantations of the Periyar River basin and are eventually found in the river in ppb ranges [1]. Additionally, residues are present in blood and breast milk, thereby transferring to the newborn [2]. Immediate removal is mandatory, as it affects the biodata and causes death. Among several techniques, adsorption is preferred due to its easy and cost-effective nature. Then, the selection of adsorbent paves another important role in the adsorption process. Fe-MMT was selected as the adsorbent for the removal of acephate, as it is environmentally benign in nature [3]. The paper focuses on the properties of the adsorbent characterized by SEM, FTIR, and the surface area analyzer. Additionally, the adsorbent performance for the removal of acephate was evaluated.

2. Materials and Methods

FTIR spectra of the samples were obtained using a spectrometer (500–4000 cm^{-1}). XRD experiments were carried out using a diffractometer (Make: PAN Analytical Philips: Almelo, Netherland, and Model: XPERT-PRO). Morphology was studied by SEM (Make: TESCAN: Brno, Czech Republic, and Model: VEGA 3 LMU) and chemical composition was evaluated (EDX). Surface area by surface area analyzer (Make: Micrometrics: Norcross, GA, USA, and Model: Tristar II 3020 Version 3.02). The concentration of acephate was measured with the help of a UV-Vis spectrophotometer (Make: Shimadzu, Japan, Model: UV-1800). Montmorillonite was purchased from Sigma-Aldrich Company Response Model and Central Composite Design were employed to investigate the effect of adsorption efficiency of acephate onto Fe-MMT.

3. Results and Discussions

Montmorillonite exhibits typical diffraction peaks at 2θ = 19.5, 26.7, and 40.04, and additional peaks at 2θ = 26.5 and 47 corresponding to silica and aluminum. Diffraction peaks at 2θ of 36 and 42 indicate the presence of magnetite, suggesting that the modified phase is magnetite (Figure 1a), which leads to a dramatic increase in the adsorption efficiency, up to 87%, rather than the bare MMT possessing 60% adsorption capacity. FTIR interpretation revealed that Fe-MMT exhibited strong characteristic peaks at 536 cm^{-1}, 1438 cm^{-1}, and 3377 cm^{-1}. Bands at 520–570 cm^{-1} represent Fe-O stretching vibration (symmetrical). Bands at 3377 cm^{-1} are due to Fe-OH (Figure 1b). SEM images indicate flower-like morphology with EDX spectrum, indicating a rise in the content of Fe (Figure 1c), which improved the material adsorption capacity. Figure 1d pictures the adsorption–desorption isotherms of Fe-MMT. The BET surface area of MMT (256 m^2/g) is higher than that of the modified material (28 m^2/g), highlighting chemical adsorption onto Fe-MMT rather than physisorption.

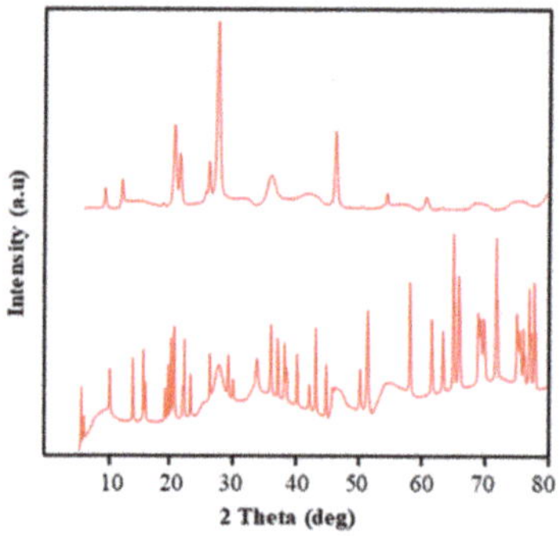

(a)

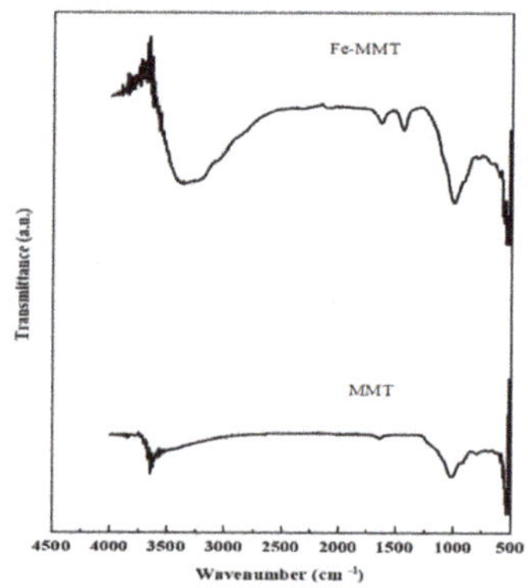

(b)

Figure 1. *Cont.*

(c)

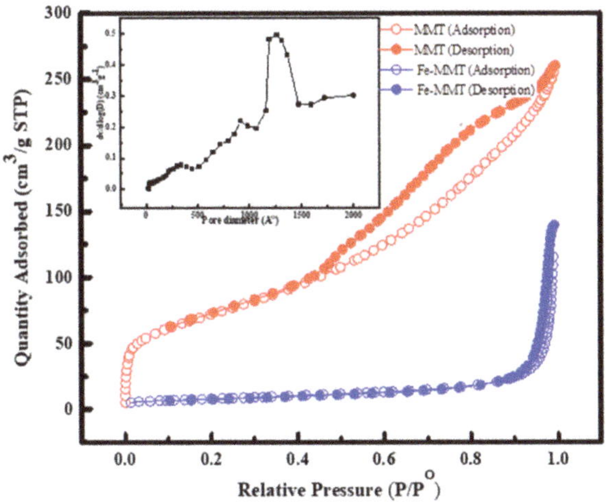

(d)

Figure 1. Images of (**a**) XRD (**b**) FTIR (**c**) SEM-EDAX (**d**) N_2 adsorption-desorption isotherms of Fe-MMT.

4. Batch Adsorption Studies

Optimization of Parameters for Acephate Adsorption

RSM (Response surface methodology), based on the central composite design (CCD), was employed to examine the performance of adsorption. CCD was opted to achieve accurate predictions around extremes of the factors. Concentration, pH, and dosage were the factors that were identified as key parameters contributing to adsorption. Response surface plots were generated using a design expert for the determination of the effect of the input factors on the responses. pH parameter was optimized for the batch adsorption of acephate onto Fe-MMT by testing with pH (1–10) [4]. Figure 2a shows that the acidic medium favored the adsorption and the peak of adsorption was achieved at 5. As the pH increases from 5 to 8, the adsorption of acephate decreases. The reduced adsorption of acephate onto clay may be due to the electrostatic repulsion between the adsorbent and the adsorbate. Adsorption increased with an increase in the initial acephate. The amount of adsorption increased with an increase in the initial acephate concentration and was found to be 0.9, 2.175, 4.2, 5.925, and 7.4 mg/g for the initial acephate concentrations of 2, 5, 10, 15, and 20 mg/L, respectively. Dosage was another vital factor that influenced the adsorption of acephate onto Fe-MMT. As dosage increases (0.1–0.5), adsorption increases, and the peak of adsorption is reached at 0.5g/L, with an efficiency of 90%. Temperature [5] is another parameter that affects the adsorption. As the temperature increases, adsorption efficiency is also enhanced (endothermic nature) [5].

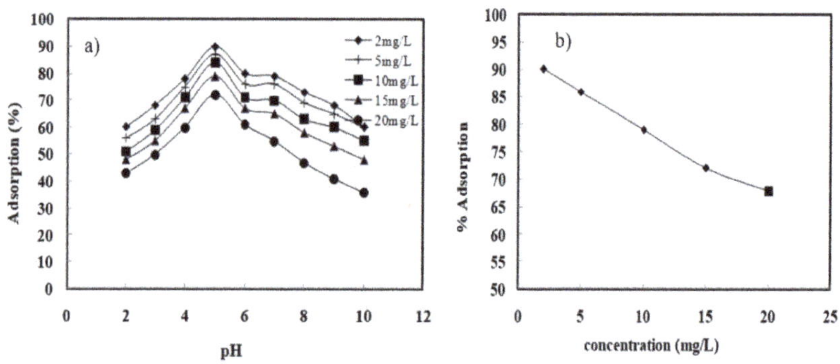

Figure 2. Effect of (**a**) pH and (**b**) initial acephate concentration on the adsorption of acephate onto Fe-MMT.

5. 3D Response Surface Plots

pH, dosage, and concentration significantly influenced the adsorption of acephate by Fe-MMT, summarized in Figure 3. Figure 3 reveals the synergistic effect of two parameters on the adsorption capacity of acephate onto Fe-MMT. The other parameter was retained at the zero level. According to Figure 3, the concentration and adsorbent dose are the most influential factors in the adsorption capacity. Meanwhile, pH partly contributes to the adsorption results. It was found that higher dosage and lower concentration gave the highest adsorption capacity.

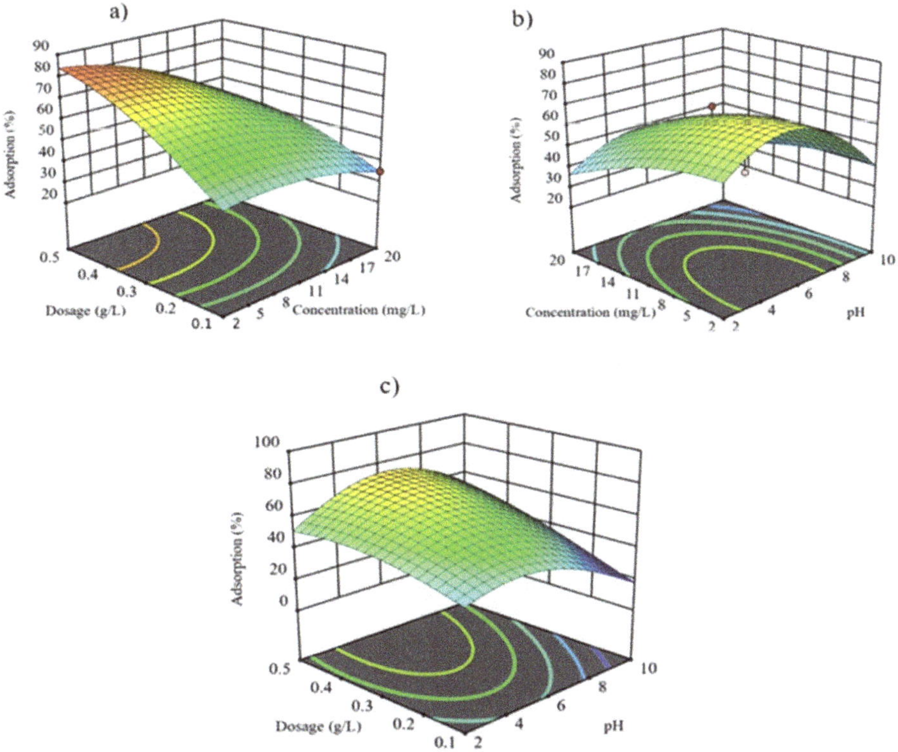

Figure 3. Effect of (**a**) pH and (**b**) initial acephate concentration and (**c**) dosage on the adsorption of acephate onto Fe-MMT.

6. Adsorption Kinetics, Isotherm, and Thermodynamic Parameters

Lagergren's pseudo-first-order, pseudo-second-order, Elovich, and Weber and Morris's Intraparticle diffusion models were scrutinized to study the kinetics of adsorption. Kinetics of adsorption and the model of isotherm obeyed pseudo-first-order ($R^2 > 0.99$) and the Langmuir model (adsorbed homogeneously on a monolayer surface of Fe-MMT). The maximum Langmuir capacity at pH 5 and 303 K is 13.66 mg/g. Thermodynamic parameters revealed that the adsorption is endothermic $\Delta H^0 = 22.855$ kJ/mol, $\Delta S^0 = 105.920$ J/K) and spontaneous in nature ($\Delta G^0 = -7.99$, -8.087, 9.245, and -10.357 kJ/mol).

7. Mechanism of Adsorption

The mechanism of adsorption corresponds to chemisorption (positive group of phosphorous in acephate and deprotonated silanol group), physisorption (hydrogen-bearing amino group of acephate and the oxygen atoms of MMT), and Lewis's acid–base interaction nitrogen atom of the amino group of acephate and Al^{3+} montmorillonite.

Author Contributions: R.S.R.: methodology, investigation, writing–original draft, resources, K.A.K.: conceptualization, writing-review & editing, supervision, funding acquisition. All authors have read and agreed to the published version of the manuscript.

Funding: University Grants Commission, Awardee: R. Shiny Raj, Sr No: 2121410217, Ref No: 21/12/2014(ii) EU-V.

Institutional Review Board Statement: Not applicable.

Informed Consent Statement: Not applicable.

Data Availability Statement: Not applicable.

Conflicts of Interest: The authors declare that they have no known competing financial interests or personal relationships that could have appeared to influence the work reported in this paper.

References

1. Gayathri, S.; Dev, V.V.; Shiny Raj, R.; Krishnakumar, A.; Vishnu Maya, T.M.; Anoop Krishnan, K. Spatiotemporal evaluation of hydrochemical facies and pesticide residues in the cardamom plantations of Southern Western Ghats, India. *Environ. Nanotechnol. Monit. Manag.* **2021**, *16*, 100599. [CrossRef]
2. Lin, Z.; Pang, S.; Zhang, W.; Mishra, S.; Bhatt, P.; Chen, S. Degradation of Acephate and Its Intermediate Methamidophos: Mechanisms and Biochemical Pathways. *Front. Microbiol.* **2020**, *11*, 02045. [CrossRef] [PubMed]
3. Saeidi, M.; Naeimi, A.; Komeili, M. Magnetite nanoparticles coated with methoxy polyethylene glycol as an efficient adsorbent of diazinon pesticide from water. *Adv. Environ. Technol.* **2016**, *2*, 25–31. [CrossRef]
4. Mahadevan, H.; Nayana, A.R.; Viswadas, V.; Antony, S.; Dev, V.V.; Sudhakaran, S.; Priya Pious, H.; Anoop Krishnan, K. A pilot level approach to remove anionic species from industrial effluents using a novel carbonate-steam pyrolysed activated charcoal system. *Adv. Powder Technol.* **2019**, *30*, 98–110. [CrossRef]
5. Dev, V.V.; Baburaj, G.; Antony, S.; Arun, V.; Krishnan, K.A. Zwitterion-chitosan bed for the simultaneous immobilization of Zn(II), Cd(II), Pb(II) and Cu(II) from multi-metal aqueous systems. *J. Clean. Prod.* **2020**, *255*, 120309. [CrossRef]

environmental sciences proceedings

MDPI

Proceeding Paper

Green Synthesis of Magnetite Nanoparticles Using Waste Natural Materials and Its Application for Wastewater Treatment †

Hussein M. Ahmed [1,*], Mohamed A. El-khateeb [2], Neama A. Sobhy [1], Mohamed M. Hefny [3] and Fatehy M. Abdel-Haleem [4]

1 Housing and Building Research Center (HBRC), Sanitary and Environmental Institute, Giza 12613, Egypt; neamaahmedreiad@yahoo.com
2 Water Pollution Research Department, National Research Centre, Giza 12622, Egypt; elkhateebcairo@yahoo.com
3 Chemistry Department, Faculty of Science, Cairo University, Giza 12613, Egypt; mmhefny_cu005@yahoo.com
4 Centre for Hazard Mitigation and Environmental Studies and Research (CHMESR), Cairo University, Giza 12613, Egypt; fatehy@sci.cu.edu.eg
* Correspondence: hussein_fee@yahoo.com

† Presented at the 7th International Electronic Conference on Water Sciences, 15–30 March 2023; Available online: https://ecws-7.sciforum.net/.

Abstract: In this study, a simple, environment-friendly, and cost-effective method is developed to synthesize metallic nanoparticles (NPs) from natural waste residues, such as onion, potato, tea, and moringa, and the effect of extract residues on efficiency, yield, size, shape, and morphology of the magnetite nanoparticle is discussed. The synthesized nanoparticle was characterized by a Fourier transform infrared spectrometer (FT-IR), X-ray diffraction (XRD), X-ray fluorescence (XRF), and energy dispersive spectroscopy (EDX). The promising applications of nanotechnology are their efficiency in wastewater treatment, including the removal of chemical and physical parameters. The study proposes that magnetite NPs can be synthesized using onion, potato, tea, and moringa residues' extract as the reducing agent. The results of the XRD pattern confirmed the synthesized magnetite NPs using onion, potato, tea, and moringa as the crystalline phase of α-Fe_2O_3. EDX spectroscopy showed the presence of elemental iron and oxygen, indicating that the nanoparticles were essentially present in oxide form. UV absorption in the range of 190–340 nm confirmed the formation of Fe/NP, and a Fourier transform infrared spectrometer (FT-IR) indicated the formation of iron oxide crystalline NPs in which reducing and capping agents, such as flavones, and the intensity of the absorption peak in the FT-IR spectrum depends on the type of extract. The synthesized Fe/NPs were tested for treatment of wastewater under different conditions such as contact time (0–60) min and dose (0.1–0.5) g; the results indicate that magnetite NPs of moringa and onion are more effective in degradation and adsorption processes at optimum dose (0.4 g, and time 45 min).

Keywords: green method; iron oxide nanoparticle; extract natural materials; scanning electron microscopic; energy dispersive spectroscopy

Citation: Ahmed, H.M.; El-khateeb, M.A.; Sobhy, N.A.; Hefny, M.M.; Abdel-Haleem, F.M. Green Synthesis of Magnetite Nanoparticles Using Waste Natural Materials and Its Application for Wastewater Treatment. *Environ. Sci. Proc.* **2023**, *25*, 99. https://doi.org/10.3390/ECWS-7-14181

Academic Editor: Carmen Teodosiu

Published: 14 March 2023

1. Introduction

Recently, there has been a great development in the use of nanotechnology in many applications such as medical and environmental fields, which have made many people, believe that this technology can improve their current standard of living [1,2]. The nanoparticles are characterized by many characteristic approaches, such as shape and size, which allow these particles to be used in many life applications including water treatment [2]. The particles are called nanomaterials at particles sizes ranging from 1 to 100 nm [3]. The nanoparticles are characterized by a large surface area, which distinguishes them from the other bulk materials with the same composition, which made this technology improve properties and features such as catalytic activity, electrical conductivity, hardness, and antimicrobial [3]. The nanoparticles are used in water treatment due to these particles having a large functional surface area capable of binding, absorbing, and carrying other compounds [2]. Among the nanoparticles that are widely used in water treatment, such as iron particles with magnetic properties, these particles are characterized by unique properties such as surface area; these properties of Fe/NPs make them applicable in various areas, such as catalysis, magnetic storage media, biosensors, magnetic resonance, and wastewater treatment [2,4].

The preparation of raw materials is carried out by various methods including physical, chemical, enzymatic, and biological. Physical methods are divided into the grinding of large particles, thermal evaporation, plasma arcing, spray pyrolysis, spray deposition, and layer-by-layer growth. Chemical methods are divided into the sol–gel method, electrophoresis, chemical vapor deposition, chemical solution deposition, and hydrolysis. The biological method, which uses a one-step biological extraction method, is environmentally friendly, as it uses environmentally friendly materials such as plant materials, bacteria, fungi, microalgae, and is called the green synthesis method [3,5].

Green synthesis uses plants and microorganism; the synthesis of nanoparticles is an environmentally friendly, economically viable method for large-scale production and a cost-effective method without any harmful and expensive chemicals. The green synthesis of nanoparticles is produced by the biological method in order to overcome the problems with more efficiency than physical and chemical methods due to the length of time and the multiplicity of steps during the preparation process [6]. The green synthesis method depends on the mechanism of the bio-reduction of nanoparticles due to many bio-molecules (vitamins, amino acids, proteins, phenolic acids, and alkaloids) in plant and microorganisms. Phenolic acids are powerful antioxidants, possessing hydroxyl and carboxyl groups that are able to bind metals. The active hydrogen may be responsible for the reduction of metal ions in the formation of nanoparticles [2–4,7].

Iron nanoparticles are prepared from plant extracts such as fruit and vegetable extracts. Iron nanomaterials are considered effective materials in water treatment because of their magnetic properties. Iron particles are found in the forms of Fe_2O_3 and Fe_3O_4 [3,8,9].

Iron nanoparticles have recently gained great research interest in environmental applications since they offer high surface reactivity due to the high surface area. Environmental applications of iron nanoparticles include the detection and elimination of pollutants in wastewater treatment. The application of iron nanoparticles in the environment offers advantages such as improved performance, lower energy consumption, and reduction in residual waste. Iron nanoparticles are one of the most researched and efficient nanoparticles for the removal of pollutants from wastewater [10,11].

The prepared nanoparticles are characterized to know the extent of their formation through many methods such as scanning electron microscopy (SEM), X-ray diffraction (XRD), Fourier transform infrared spectroscopy (FT-IR), and visible and ultraviolet spectroscopy [3,12–14].

In this study, the preparation of iron nanoparticles (Fe/NPs) was based on the green synthesis method, where extracts of different natural materials such as moringa leaves, potato peels, tea waste, and onion peels used for the synthesis of iron nanoparticles (Fe/NPs). The iron nanoparticles were characterized using different techniques such as X-ray diffraction (XRD), Fourier transform infrared spectroscopy (FT-IR), and visible and ultraviolet spectroscopy (UV spectrum). The particle size, magnetic properties, and morphology of Fe/NPs depended on the conditions of the materials used, such as the extract of onion peels, potato peels, tea waste, and moringa leaves; the obtained nanoparticles had different particle sizes, morphologies, yields, and magnetic properties. These Fe/NPs were used in wastewater treatment; the different parameters were applied to determine the efficiency of iron nanoparticles, such as contact time (0–60) min and dose (0.1–0.5) g/L, using this technology after the sedimentation stage of raw wastewater.

2. Materials and Methods

2.1. Preparation of the Extracts of Waste Natural Materials and Iron Nanoparticles

This work aims to prepare iron nanoparticles from extracted waste natural materials (WNMs), as shown in Figure 1. The magnetic iron nanoparticles were synthesized using a mixture of $FeCl_3 \cdot 6H_2O$ and $FeCl_2 \cdot 4H_2O$, aided with onion, potato, tea, and moringa extracts. These materials were obtained from the local markets, Giza, Egypt. They were prepared by washing onion peels, potato peels, tea waste, and moringa leaves several times using tap water to remove any dust, and then they were rinsed and dried at room temperature. The onion peels, potato peels, and moringa leaves were cut into small pieces. Then, approximately 50 g of every waste extract was weighed and boiled in 500 ml of tap water for 30 min. The filtrate of the extract was kept at 4 °C in the refrigerator [4,15]. The magnetite nanoparticles were prepared from the extracts by adding 5 ml of extract to a bottle of iron solution with the simultaneous drop-wise addition of NaOH (I N) solution (this process was carried out at 80 °C, and the solution was mixed at 1000 rpm for 2 h) [16]. The synthesis of nanoparticles was observed by the changing color of the solution from orange to black; the formation of Fe/NP was confirmed by the appearance of a black precipitate [15]. Fe/NPs were separated by centrifugation at 1000 rpm/min, collected, and dried in a dry oven at 50 °C for 48 h [11].

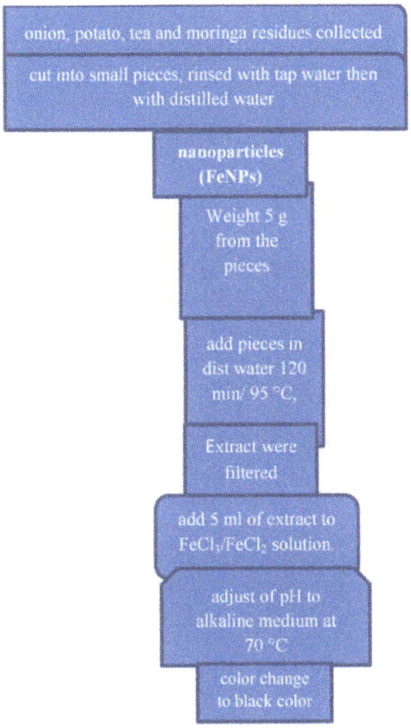

Figure 1. Schematic diagram of onion, potato, tea, and moringa residues in production of iron nanoparticles.

2.2. Characterization of the Synthesized Fe/NPs

The prepared magnetite nanoparticle (Fe/NPs) compounds were characterized using several instruments. For example, a UV–visible spectrophotometer (T-70 spectrophotometer at the Housing and Building Research Center, Chemistry Lab) was used for the analysis of synthesized Fe/NPs periodically as a function of time in the wavelengths ranging from 190–340 nm with a resolution of 0.5 nm. Crystallographic study of Fe/NPs was carried out using X-ray diffraction (XRD 6100, Shimadzu, Tokyo, Japan) with CuK α radiation from

40 kV/30 mA using the 2θ range of 20–70°. Chemical functional group identification on Fe/NPs was determined using FT-IR (FT-IR 8400S, Shimadzu, Tokyo, Japan) in the spectral range of 400–4000 cm^{-1} and elemental analysis was carried out in the Na-U channel using EDX (EDX 720, Shimadzu, Tokyo, Japan).

2.3. Characteristics of Raw Wastewater

2.3.1. Sample Sites and Analysis of Raw Samples

The grey water (collected from the Orasqualia station for wastewater treatment), characteristics indicated that such wastewater was relatively strong, as exhibited by the ammonia, COD, BOD, TDS, EC, pH, PO_4, TP, TN, TKN, NH_3, NO_3, TSS, and phosphate. The characteristics of grey water are shown in Table 1 compared with the Egyptian Environmental Association Affair (EEAA) [17]. The COD and BOD were 560 and 302 mg/L, respectively. The PO_4, TP, TN, TKN, NH_3, and NO_3, were 3.3, 0.66, 33.6, 28.2, 13.2, and 5.4. The pH, EC, TSS, and TDS were 7.2, 1099, 330, and 611 mg/L; turbidity was 89.5 NTU. ORP was −19.7 mV, respectively.

Table 1. Physicochemical characteristics of raw grey water.

Test	Unit	* Average Values of Raw Samples
pH	-	7.2
TDs	mg/L	611
EC	µs/cm	1099
ORP	mV	−19.7
Turbidity	NTU	89.5
COD	mg/L	560
BOD	mg/L	302
TSS	mg/L	330
NH_3	mg/L	13.2
NO_3	mg/L	5.4
TKN	mg/L	28.2
TN	mg/L	33.6
PO_4	mg/L	3.3
TP	mg/L	0.66

Notes: * Average for three samples, TDS is total dissolved solid, TSS is total suspended solid, COD is chemical oxygen demand, BOD is biological oxygen demand, NH_3 is ammonia, NO_3 is nitrate, TKN is total Kjeldahl nitrogen, TN is total nitrogen, TP is total phosphorus, and PO_4 is phosphate. ORP is oxidation reduction potential.

2.3.2. Reagents

All chemicals were of analytical grade. The chemical reagents used included phosphoric acid (H_3PO_4, 85.0%, Fischer scientific, Loughborough, UK), potassium dichromate ($K_2Cr_2O_7$, 99.0%, Merck, Darmstadt, Germany), boric acid (H_3BO_3, 99.5%, LOBA Chemie, Mumbai, India), hydrochloric acid (HCl, 37.0%, Fischer scientific, Loughborough, UK), nitric acid (HNO_3, 70.0%, sodium hydroxide (NaOH, 99.0%, Merck, Darmstadt, Germany), and ammonia solution (NH_3, 35.0%, Fischer scientific, Loughborough, UK).

2.3.3. Instruments and Characterization Techniques

The following instruments were used through this work: a furnace, the drying oven (Fisher Scientific Equipment, American provisioner of scientific instruments, Waltham, MA, USA), digital electronic balance (PCE-BSK 310, PCE Instruments, Southampton, UK), and DIW system (Millipore, Darmstadt, Germany) were used. The pH measurements of the samples were achieved using a pH meter (AD110, ADWA, Szeged, Hungary). Fourier transform infrared spectroscopy (FT-IR) (Thermo Fisher Scientific, Oxford, UK), an orbital shaker instrument (Thermo Fisher Scientific, Waltham, MA, USA), a COD digestor instrument (auto time-controlled, MAC, Udaipur, India), a BOD_5 incubator (Airco, Mumbai, India), UV–Vis spectrophotometers (PG Instruments, Lutterworth, UK), an incubator (Thermo Fisher Scientific, Oxford, UK), a portable multiparameter water-quality measurement (HORIBA company, Irvine, CA, USA), and a Kjeldahl digestion instrument (ESEL, India) were used.

Samples of raw bone waste materials were analyzed by a scanning electron microscope (SEM) model (Quanta 250 FEG—field emission gun—attached with accelerating voltage 30 kV (FEI Company, Eindhoven, Netherlands) [18].

2.4. Batch Experiments Fe/NPs

The treatment wastewater experiments were carried out in 1L treated by Fe/NPs optimized by varying the dose (0.1, 0.3, 0.4, and 0.5 g/L agitated 30 min) and contact time (0, 15, 30, 45, and 60 min agitated with 0.4 g). All experiments were carried out in a jar test at 100 rpm. The same set of the experiment was repeated three times. All experiments were conducted at room temperature. The residual concentration of pollutants in the filtrate was detected using the EPA method. The removal efficiency (R%) was calculated from the Equation (1) [9,19,20]:

$$R\% = \frac{C_0 - C_e}{C_0} \times 100 \tag{1}$$

where R% is the removal efficiency, C_0 is the initial concentration (mg/L), and C_e is the concentration after adsorption (mg/L).

3. Results and Discussion

3.1. Characterization

3.1.1. UV–Vis Spectral Analysis of Fe/NPs

Fe/NPs formations were confirmed by the color change that immediately occurred after the addition of the plant extract iron solution and the adjusted pH. The dark color was a result of surface plasmon excitation vibrations in the Fe/NPs [11]. The absorption peaks were at 215, 210, 257, and 210 nm for onion, potato peels, tea waste, and moringa, respectively. This indicates the presence of Fe/NPs [2]. Figure 2 shows the UV–visible absorption spectrum of Fe/NPs synthesized using each extract waste residue. The formation of Fe/NPs is known to take place through complexation of Fe salts followed by the capping of Fe with phenolic compounds [11].

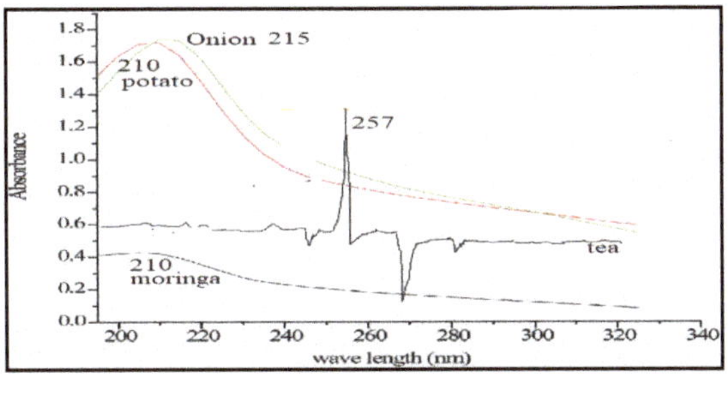

Figure 2. UV–Vis absorption spectrum of onion, potato peels, tea waste, and moringa magnetite nanoparticles.

3.1.2. Appearance of Synthesized Fe/NPs

The appearance of the black color of the Fe/NPs solution indicates the formation of Fe/NPs with the increasing time; this is shown in Figure 3. The color changes arise due to the excitation of the surface plasma resonance phenomenon typically of Fe/NPs [2]. The nanoparticles' formation was confirmed by the color immediately converted from transparent brown to black in a few seconds, demonstrating the synthesis of iron nanoparticles [11].

Figure 3. Onion, potato peels, tea waste, and moringa magnetite nanoparticles.

3.1.3. XRD Pattern Analysis of Fe/NPs

The XRD was obtained to investigate the presence of nanoparticles on moringa, potato, onion, and tea surface. The XRD pattern of the synthesized adsorbent was in the angle range of 2θ, applying Cu kα radiation (λ = 1.5 Å). The XRD technique was used to identify the structure of the prepared iron nanoparticles and is depicted in Figure 4.

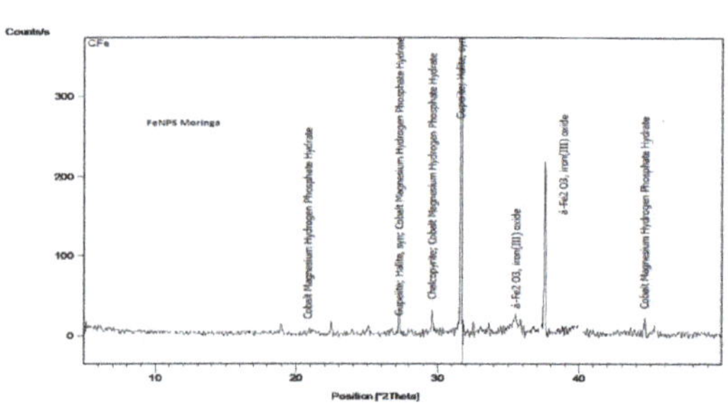

(**a**)

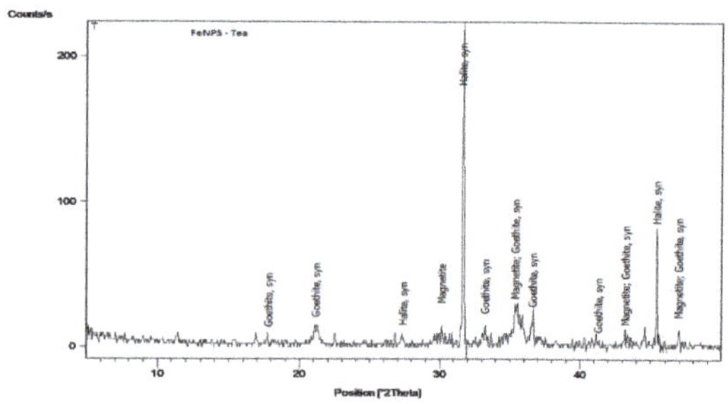

(**b**)

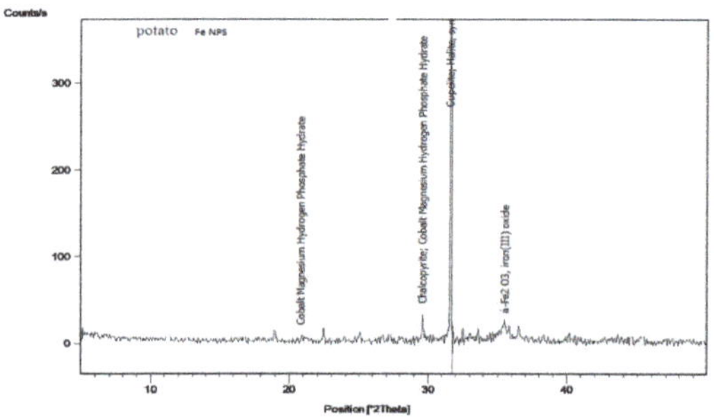

(**c**)

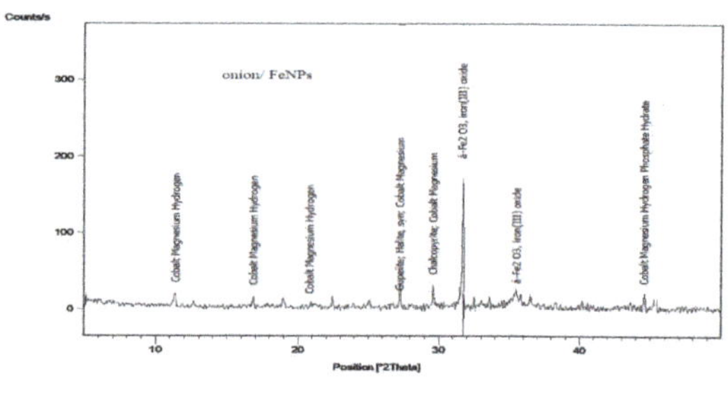

(**d**)

Figure 4. The XRD pattern of (**a**) moringa, (**b**) tea waste, (**c**) potato peels, and (**d**) onion magnetite nanoparticles.

The characteristic nanoparticle peak occurred at approximately 2θ = (10° − 60°). The analysis of the spectrum XRD technique was used for particle-size analysis of Fe/NPs; the XRD pattern shows that the peaks the nanoparticle were 44°, 35°, 35–32°, and 30–35–44° for moringa, potato, onion, and tea, respectively. This resulted in no clear reflection peak in potato and onion due to the other crystalline phase, which might be present as impurity. Thus, the nanoparticles essentially consisted of a binary mixture of the two spinel magnetic iron oxides, magnetite-Fe_2O_3 and solid elements [1,6]. In this pattern, the peak at the angle of 32, 30, 35, and 44° confirmed the presence of Fe_2O_3 particles in the adsorbent structure.

Generally, the XRD analysis confirmed that the Fe_2O_3 particles were successfully coated on the moringa, potato, onion, and tea surface [11].

3.1.4. Energy Dispersive X-ray Analysis

EDX analysis was then performed on the surface of the Fe/NPs, as shown in Figure 5; the EDX results of onion, potato peels, tea waste, and moringa magnetite nanoparticles reveals the elemental composition of the prepared nanoparticles. The EDX profile shows intense peak signals of iron with a Kα peak at 6.5 keV, 6.2 keV, 0.9 keV, and 0.7 keV. Other signals observed include that of oxygen and carbon; the presence of C and O peaks were related to polyphenols or any other C and O containing a compound in the natural materials' extract. The existence of elemental iron and oxygen demonstrate that the nanoparticles were essentially present in oxide form [11]. The percent of detected elements were carbon (C) 12%, iron (Fe) 52%, and oxygen (O) 28%. These results indicate the extract of moringa leaves and potato peels were more highly efficiency than onion and tea for the formation of magnetite iron nanoparticles. Very similar results were reported for Fe nanoparticles prepared with the other leaf [11].

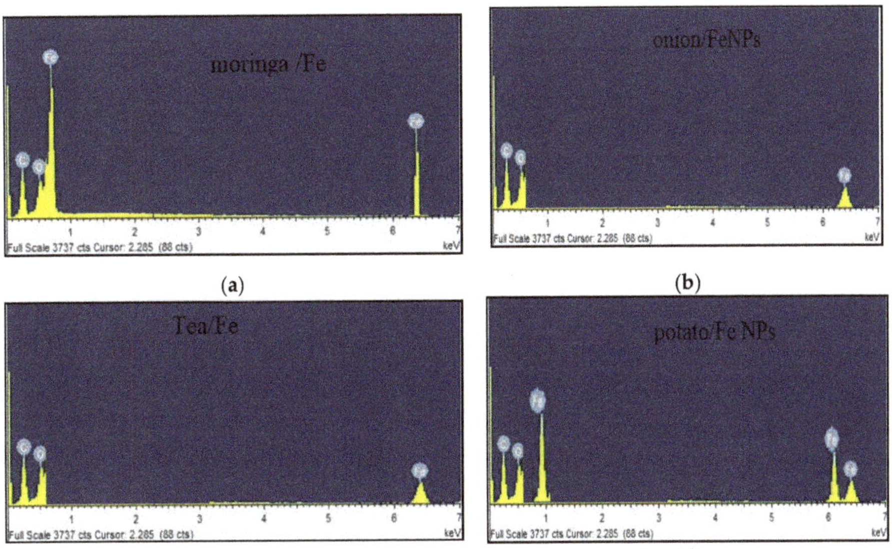

Figure 5. The EDX pattern of (**a**) moringa, (**b**) onion, (**c**) tea waste and (**d**) potato peels magnetite nanoparticles.

3.1.5. The FT-IR Spectra of Fe/NPs

The FT-IR measurements were carried out to identify the possible bio-molecules responsible for the reduction of ferrous chloride and capping of the reduced Fe/NPs. The FT-IR spectra of Fe/NPs after preparation by the green synthesis method from onion, potato peels, tea waste, and moringa extracts are shown in Figure 6. All of the above peaks which can be detected in the spectrum of synthesized Fe/NPs were subjected to FT-IR that showed various bands; the O-H stretching at approximately 3400 cm^{-1} showed the presence of hydroxyl groups from the polyols such as flavones, terpenoids, and polysaccharides present in the various extracts. The decrease in intensity of band O-H stretching in onion and potato might be due to the interaction of nanoparticles. The bands at 1645 cm^{-1} and 1041 cm^{-1} denote the presence of organic material in the sample, majorly contributed by onion, potato peels, tea waste, and moringa iron magnetite particles (Fe/NPs). These bands confirmed the presence of compounds such as flavonoids and terpenoids and, hence, may be held responsible for the efficient capping and stabilization of the obtained magnetite nanoparticles [11].

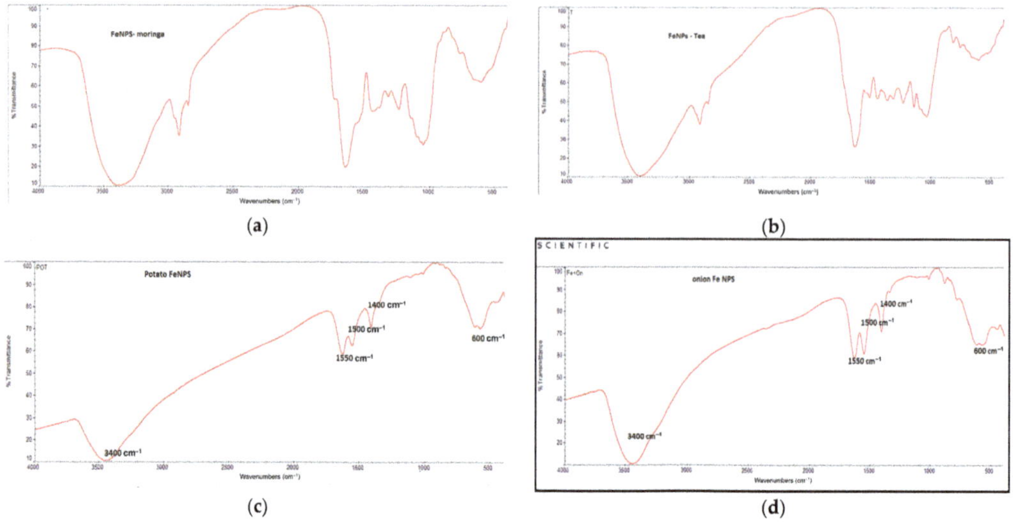

Figure 6. FT-IR spectrum of (**a**) moringa, (**b**) tea waste, (**c**) potato peels and (**d**) onion magnetite nanoparticles.

3.1.6. XRF Analysis of Banana, Orange, and Pomegranate

The XRF pattern of iron oxide nanoparticle prepared from onion, potato peels, tea waste, and moringa are shown in Table 2. The results show the Fe_2O_3 composite in onion, potato peels, tea waste, and moringa Fe/NPs. The percent of magnetite nanoparticles (Fe_2O_3) from onion, potato peels, tea waste, and moringa were 67.3%, 53.92%, 40.86%, and 46.86%, respectively. The Fe_2O_3 percent in magnetite nanoparticles was onion > potato peels > moringa > tea waste.

Table 2. The XRF of onion, potato peels, tea waste, and moringa magnetite nanoparticles.

Sample Name	Fe_2O_3	Na_2O	MnO	CaO	SiO_2	Al_2O_3	K_2O	SO_3^-	Cr_2O_3	MgO	Cl^-	LOI	Total
Fe-Tea	40.86	19.8	0.33	0.47	0.38	0.11	0.03	0.13	0.01	0.07	15.6	22.2	99.99
Fe-Potato	53.92	16.1	0.53	0.3	0.37	0.09	0.05	0.09	0.05	0.04	9.63	18.8	99.98
Fe-Onion	67.3	8.39	0.63	0.47	0.38	0.08	0.14	0.05	0.05	0.04	11.6	9.51	99.94
Fe-Moringa	46.62	22.1	0.46	0.27	0.36	0.07	0.07	-	-	0.04	14.9	15.1	99.99

3.1.7. Yield of Iron Oxide Nanoparticles

The iron oxide nanoparticles were dried at 50 °C until the magnetite nanoparticles were completely dry. The weight of magnetite nanoparticles related to the type of extract used in the synthesis of nanoparticles. Table 3 shows that the weight of the magnetite nanoparticles for moringa, potato, onion, and tea were 38.235, 19.116, 19.116, and 12.899, respectively.

Table 3. Yield of onion, potato peels, tea waste, and moringa magnetite nanoparticles.

Iron Oxide Nanoparticle	Weight (g)
Fe-Metal	7.625
Fe-Moringa	38.235
Fe-Potato	19.116
Fe-Onion	16.114
Fe-Tea	12.899

3.2. Effect of Contact Time

Figure 7 illustrates the effect of contact time on the efficiency of magnetite nanoparticles for the removal of pollutants from grey wastewater, The following condition were applied: 0.1 g/L solution of the adsorbent, optimal pH (pH = 7.5 ± 0.1), and the contact time of (0–60) min, as indicated in Table 4; the removal efficiencies were increased sharply up to equilibrium at 45 min and then slightly increased after 45 min until 60 min. The sharp increase in the removal efficiency may be due to the existence of enormous vacant active sites in the surface. However, by raising the contact time, the availability of pollutants to the active sites on the adsorbent surface was limited [21], which makes the adsorption efficiency reduce. In a similar study, this phenomenon was investigated using different adsorbents [1]. Iron nanoparticles were used for removing pollutants from wastewater, due to the interaction between the compounds and the functional groups at the surface of the absorbent. The functional groups served to define the effectiveness, selectivity, capacity, and reusability of an absorbent. Furthermore, in the case of high iron oxide loading at the surface, the higher the rate of reduction in the nitrate and ammonium ion [22]. The results showed that the optimum time for metal removal by magnetic nanoparticles was obtained in 45 min [1,23].

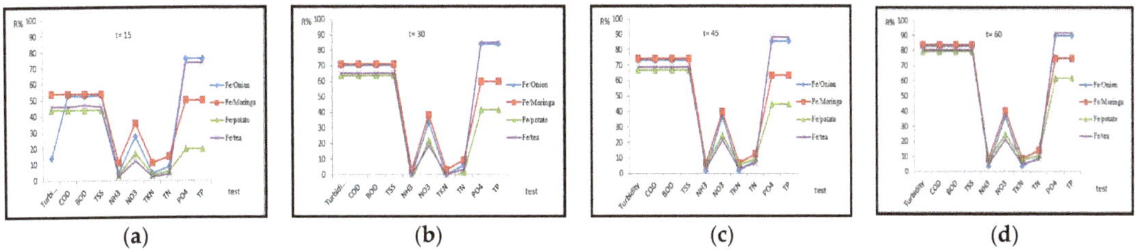

Figure 7. Effect of contact time on removal efficiency of iron magnetite nanoparticles at agitation speed 200 rpm, dose 0.1 g/L, and 20 ± 5 °C. (**a**) 15, (**b**) 30, (**c**) 45, and (**d**) 60 min, iron magnetite nanoparticles such as (FeNPs/moringa, FeNPs/potato, FeNPs/onion and FeNPs/tea).

Table 4. The residual concentration of pollutants and removal efficiency of all adsorbents at optimum time.

Parameter	Unit	Raw	Residual Concentration (mg/L) of Pollutants at Optimum Contact Time 45 min—Turbidity as NTU				Removal Efficiency of Adsorbents (%)			
			Fe/onion	Fe/moringa	Fe/potato	Fe/tea	Fe/onion	Fe/moringa	Fe/potato	Fe/tea
pH	—	7.2	8.3	8.19	8.6	8.4	-	-	-	-
TDS	mg/L	611	688	745	752	978	-	-	-	-
EC	µs/cm	1099.8	1238.4	1341	1353.6	1760.4	-	-	-	-
ORP	mV	−19.7	−80	−72	−95	−83	-	-	-	-
Turbidity	NTU	89.5	24.31	23.19	29.75	28.15	72.85	74.10	66.78	68.57
COD	mg/L	560	152	145	186	176	72.85	74.10	66.78	68.57
BOD	mg/L	302	82.08	78.3	100.44	95.04	72.85	74.10	66.78	68.57
TSS	mg/L	330	89.68	85.55	109.74	103.84	72.85	74.10	66.78	68.57
NH_3	mg/L	13.2	12.98	12.34	12.45	12.79	1.66	6.51	5.68	3.10
NO_3	mg/L	6.4	4.05	3.85	4.82	5.03	36.71	39.84	24.68	21.40
TKN	mg/L	28.24	27.7	26.4	26.6	27.3	1.66	6.51	5.68	3.10
TN	mg/L	34.64	31.8	30.2	31.4	32.4	8.14	12.67	9.19	6.48
PO_4	mg/L	3.3	0.489	1.205	1.8	0.398	85.18	63.48	44.57	87.93
TP	mg/L	0.66	0.097	0.241	0.36	0.079	85.18	63.48	44.57	87.93

Notes: TDS is total dissolved solid, TSS is total suspended solid, COD is chemical oxygen demand, BOD is biological oxygen demand, NH_3 is ammonia, NO_3 is nitrate, TKN is total Kjeldahl nitrogen, TN is total nitrogen, TP is total phosphorus, and PO_4 is phosphate.

3.3. Effect of Adsorbent Dosage

The effect of amount (0.1–0.5 g/L) of magnetic nanoparticles (Fe/NPs) on the removal of pollutants from wastewater such as chemical oxygen demand, biological oxygen demand, total suspended solid, turbidity, ammonia, Kjeldahl nitrogen, total nitrogen, phosphate, and nitrate were studied in samples before and after treatment. The dosage of nanoparticles affected its ability to sorbent contaminants. As shown in Figure 8, when the magnetic particles' dosage increased, the removal efficiency increased, as shown in Table 5. Usually, the reduction of pollutant concentration with the increased dose of iron nanoparticles is a surface function. It may be assumed that the availability of active sites on the nanocomposite increases at higher doses [8,23].

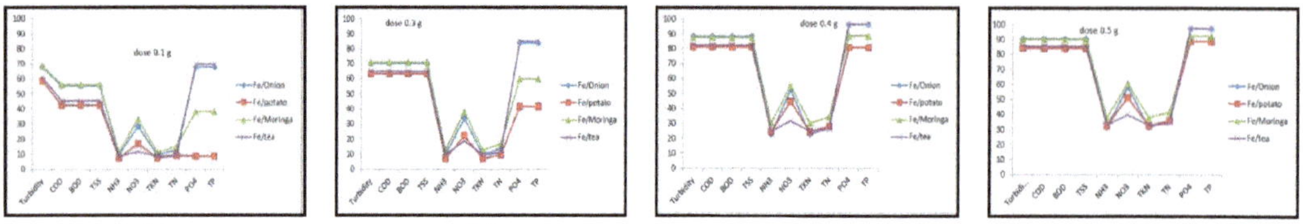

(a) (b) (c) (d)

Figure 8. Effect of different doses on efficiency of iron magnetite nanoparticles (agitation speed 200 rpm, time 45 min, and 20 ± 5°C). (**a**) 0.1, (**b**) 0.3, (**c**) 0.4, and (**d**) 0.5 g.

Table 5. The residual concentration of pollutants, and removal efficiency of all adsorbents at optimum dose 0.4.

Parameter	Raw	Residual Concentration (mg/L) of Pollutants at Optimum Dose 0.4—Turbidity as NTU				Removal Efficiency of Adsorbents (%)			
		Fe/onion	Fe/potato	Fe/moringa	Fe/tea	Fe/onion	Fe/potato	Fe/moringa	Fe/tea
pH	7.2	8.35	8.6	8.25	8.52	-	-	-	-
TDS	611	703	750	752	1005	-	-	-	-
EC	1099	1265	1350	1353	1809	-	-	-	-
ORP	−19.7	−85	−105	−75	−89	-	-	-	-
Turbidity	89.5	10	16	10.	15	88.39	81.25	87.85	82.5
COD	560	65	105	68	98	88.39	81.25	87.85	82.5
BOD	302	35	56	36	52	88.39	81.25	87.85	82.5
TSS	330	38	61	40	57	88.39	81.25	87.85	82.5
NH_3	14.6	11.3	11.05	10.2	11	22.86	24.57	30.17	24.91
NO_3	6.4	3.02	3.54	2.85	4.3	52.81	44.68	55.46	31.87
TKN	31.3	24.1	23.6	21.8	23.5	22.86	24.57	30.17	24.91
TN	37.7	27.2	27.1	24.7	27.9	27.94	27.98	34.45	26.09
PO_4	3.3	0.11	0.62	0.36	0.097	96.66	81.06	88.93	97.06
TP	0.66	0.022	0.125	0.073	0.0194	96.66	81.06	88.93	97.06

Notes: TDS is total dissolved solid, TSS is total suspended solid, COD is chemical oxygen demand, BOD is biological oxygen demand, NH_3 is ammonia, NO_3 is nitrate, TKN is total Kjeldahl nitrogen, TN is total nitrogen, TP is total phosphorus, and PO_4 is phosphate.

The effect of different amounts of Fe/NPs on the adsorption capacity and efficiency under the optimal condition (pH = 7.4, t = 45 min and 200 rpm) is illustrated in Figure 8. It can be observed that, with an increase in the adsorbent dosage from 0.1 to 0.5 g/L, the removal efficiencies at optimum dose (0.4), for onion, potato, moringa, and tea, were from 88.39, 88.39, 88.39, 88.39, 22.86, 52,81, 22.86, 27.94 and 96.66%, 81.25, 81.25, 81.25, 81.25, 24.57, 44.68, 27.98, and 81.06%, 87.85, 87.85, 87.85, 87.85, 30.17, 55.46, 34.45, and 88.93%, 82.5, 82.5, 82.5, 82.5, 24.91, 31.87, 26.09, and 97.06% of COD, BOD, TSS, turbidity, ammonia, TKN, TN, phosphate, and nitrate, respectively. The rise in the adsorption efficiency was related to the increase in the availability of active sites on the adsorbents which can give rise to the adsorption of pollutants. Jung et al. reported that, with an increase in the dosage of various adsorbents, the pollutants' removal was enhanced. However, a decrease in the adsorption capacity with an increase in the adsorbent dosage was probably due to the instauration of the active sites on the adsorbent surface during the adsorption process. This phenomenon can also be due to the aggregation resulting from high adsorbate concentrations, leading to the decrease in the active surface area of the adsorbent [1,23].

4. Conclusions

In the present study, the synthesized iron nanoparticles aided with natural waste material such as onion, moringa, tea waste, and potato were used as reduction and stabilization agents for the synthesis of iron nanoparticles and as an adsorbent for the treatment of wastewater. The synthetization of iron nanoparticles was confirmed by different characterization techniques such as EDX, XRF, FT-IR, XRD, and UV spectrum. The XRF and XRD pattern of the iron nanoparticles revealed a crystalline structure of the nanoparticles. The results illustrated that the synthesized adsorbent showed that iron magnetite aided with moringa and onion had a high efficiency than magnetite aided with potato and tea. The optimum conditions for the adsorption process were obtained with a contact time of 45 min and a dose of 0.4 g. Moreover, due to the favorable performance of onion and moringa in the removal of pollutants and its feasible separation from the aqueous solutions, it can be used as an efficient adsorbent in the treatment of water and wastewater; after conducting the necessary tests for the outlet samples of water after the treatment process, it was found that there were no negative effects on the water samples.

Author Contributions: Conceptualization, H.M.A. and M.A.E.-k.; methodology, H.M.A.; software, N.A.S.; validation, F.M.A.-H.; formal analysis, M.M.H.; investigation, H.M.A.; resources, H.M.A.; data curation, H.M.A.; writing—original draft preparation, H.M.A. and M.A.E.-k.; writing—review and editing, H.M.A. and M.A.E.-k.; visualization, H.M.A. and M.A.E.-k.; supervision, M.A.E.-k. All authors have read and agreed to the published version of the manuscript.

Funding: This research received no external funding.

Institutional Review Board Statement: Not applicable.

Informed Consent Statement: Not applicable.

Data Availability Statement: The data is unavailable due to privacy or ethical restrictions.

Conflicts of Interest: The authors declare no conflict of interest.

References

1. Kakavandi, B.; Kalantary, R.R.; Farzadkia, M.; Mahvi, A.H.; Esrafili, A.; Azari, A.; Yari, A.R.; Javid, A.B. Enhanced chromium (VI) removal using activated carbon modified by zero valent iron and silver bimetallic nanoparticles. *J. Environ. Health Sci. Eng.* **2014**, *12*, 115. [CrossRef] [PubMed]
2. Kale, R.D.; Barwar, S.; Kane, P.; Bhatt, L. Green Synthesis of Magnetite Nanoparticles using Banana Leaves. *Eur. J. Sci.* **2018**, *1*, 26–34. [CrossRef]
3. Kandasamy, S.; Rajendran, S.P. Methods of synthesis of nano particles and its applications. *J. Chem. Pharm. Res.* **2015**, *7*, 278–285.
4. Das, G.; Shin, H.-S.; Kumar, A.; Vishnuprasad, C.N.; Patra, J.K. Photo-mediated optimized synthesis of silver nanoparticles using the extracts of outer shell fibre of cocos nucifera L. fruit and detection of its antioxidant, cytotoxicity and antibacterial potential. *Saudi J. Biol. Sci.* **2021**, *28*, 980–987. [CrossRef] [PubMed]
5. Abdel-Shafy, H.I.; Mansour, M.S.M. Green Synthesis of Metallic Nanoparticles from Natural Resources and Food Waste and Their Environmental Application. In *Green Metal Nanoparticles*; Scrivener Publishing: Beverly, MA, USA, 2018; pp. 321–385.
6. Fahmy, H.M.; Mohamed, F.M.; Marzouq, M.H.; Mustafa, A.B.E.; Alsoudi, A.M.; Ali, O.A.; Mohamed, M.A.; Mahmoud, F.A. Review of Green Methods of Iron Nanoparticles Synthesis and Applications. *Bionanoscience* **2018**, *8*, 491–503. [CrossRef]
7. Carter, N. Physical Properties of Iron Oxide Nanoparticles. Ph.D. Thesis, Maine University, Orono, ME, USA, 2015.
8. Al-Musawi, T.J.; Mengelizadeh, N.; Al Rawi, O.; Balarak, D. Capacity and Modeling of Acid Blue 113 Dye Adsorption onto Chitosan Magnetized by Fe2O3 Nanoparticles. *J. Polym. Environ.* **2022**, *30*, 344–359. [CrossRef]
9. Amany, R.G.; Nagwa, M.S.; Ahmed, S.H.; Abdelmoteleb, Y.A. Biosynthesis, characterization and antimicrobial activity of iron oxide nanoparticles synthesized by fungi. *Az. J. Pharm Sci.* **2020**, *62*, 164–178.
10. Badmus, K.O.; Coetsee-Hugo, E.; Swart, H.; Petrik, L. Synthesis and characterisation of stable and efficient nano zero valent iron. *Environ. Sci. Pollut. Res.* **2018**, *25*, 23667–23684. [CrossRef] [PubMed]
11. Da'na, E.; Taha, A.; Afkar, E. Green Synthesis of Iron Nanoparticles by Acacia nilotica Pods Extract and Its Catalytic, Adsorption, and Antibacterial Activities. *Appl. Sci.* **2018**, *8*, 1922. [CrossRef]
12. Sun, Y.-P.; Li, X.; Cao, J.; Zhang, W.; Wang, H.P. Characterization of zero-valent iron nanoparticles. *Adv. Colloid Interface Sci.* **2006**, *120*, 47–56. [CrossRef] [PubMed]

13. Majlesi, M. Preparation and adsorption properties of chitosan-bound Fe_3O_4 magnetic nanoparticles for phenol removal from aqueous solution. *World Rev. Sci. Technol. Sust. Dev.* **2016**, *12*, 371–380.
14. Tiwari, D.; Behari, J.; Sen, P. Application of nanoparticles in waste water treatment. *World Appl. Sci. J.* **2008**, *3*, 417–433.
15. Abdul Elah Mohammad, D.; Mohammad Taher, E. Antimicrobial activity of silver nanoparticles fabricated from some vegetable plants. *J. Phys. Conf. Ser.* **2019**, *1294*, 062048. [CrossRef]
16. Kheilkordi, Z.; Ziarani, M. Recent advances in the application of magnetic bio-polymers as catalysts in multicomponent. *RSC Adv.* **2022**, *12*, 12672–12701. [CrossRef] [PubMed]
17. Egyptian Environmental Association Affair (EEAA). Book, Law 48, 1982, Permissible Values for Wastes in River Nile and Law 44, Law of the Environmental Protection 1994, Updating 2000. Available online: https://faolex.fao.org/docs/pdf/egy18642.pdf (accessed on 1 December 2022).
18. Ahmed, H.; El-Khateeb, M.; Ahmed, N. Effective Granular Activated Carbon for Greywater Treatment Prepared from Corncobs. *Egypt. J. Chem.* **2022**, *65*, 255–263. [CrossRef]
19. Mahmoud, M.A. Thermodynamics and Kinetics Studies of Mn (II) Removal from Aqueous Solution onto Powder Corn Cobs (PCC). *J. Chromatogr. Sep. Techol.* **2015**, *6*, 301. [CrossRef]
20. Sadon, F.N.; Ibrahem, A.S.; Ismail, K.N. An overview of rice husk applications and modification techniques in wastewater treatment. *J. Purity, Util. React. Environ.* **2012**, *1*, 308–334.
21. Hassan, S.S.M.; Abdel-Shafy, H.I.; Mansour, M.S.M. Removal of pyrene and benzo(a)pyrene micropollutant from water via adsorption by green synthesized iron oxide nanoparticles. *Adv. Nat. Sci. Nanosci. Nanotechnol.* **2018**, *9*, 015006. [CrossRef]
22. Nizamuddin, S.; Siddiqui, M.T.; Mubarak, N.M.; Baloch, H.A.; Abdullah, E.C.; Mazari, S.A.; Griffin, G.J.; Srinivasan, M.P.; Tanksale, A. Iron Oxide Nanomaterials for the Removal of Heavy Metals and Dyes From Wastewater. In *Nanoscale Materials in Water Purification*; Elsevier: Amsterdam, The Netherlands, 2019; pp. 447–472.
23. Khan, S.U.; Zaidi, R.; Hassan, S.Z.; Farooqi, I.H.; Azam, A. Application of Fe-Cu binary oxide nanoparticles for the removal of hexavalent chromium from aqueous solution. *Water Sci. Technol.* **2016**, *74*, 165–175. [CrossRef] [PubMed]

environmental sciences proceedings

MDPI

Proceeding Paper

Jakarta's 2020 New Year Flood Assessment with a Rainfall–Runoff–Inundation (RRI) Model †

Yeremia Immanuel Sihombing [1,*], Akbar Rizaldi [1], Mohammad Farid [1], N. Fajar Januriyadi [2] and Idham Riyando Moe [3]

1 Center for Coastal and Marine Development, Institute for Research and Community Services, Institut Teknologi Bandung, Jalan Ganesha No. 10, Kota Bandung 40132, Indonesia; akbar@ftsl.itb.ac.id (A.R.); mfarid@ftsl.itb.ac.id (M.F.)

2 Department of Civil Engineering, Pertamina University, Jalan Teuku Nyak Arief, RT.7/RW.8, Simprug, Kec. Kby. Lama, Kota Jakarta Selatan 12220, Indonesia; nurul.fj@universitaspertamina.ac.id

3 Directorate General of Water Resources, Ministry of Public Works and Housing, Jalan Pattimura No. 20, Kebayoran Baru, Jakarta Selatan 12110, Indonesia; idham.moe@gmail.com

* Correspondence: yeremia.sihombing97@gmail.com

† Presented at the 7th International Electronic Conference on Water Sciences, 15–30 March 2023; Available online: https://ecws-7.sciforum.net/.

Abstract: Floods hit Jakarta and several areas in the Ciliwung–Cisadane Watershed. The rain that occurred on 31 December 2019 stopped briefly and continued until 1 January 2020. As a result, several areas were flooded for several days. It is said that the rain that occurred was the largest in history. At least, the rainfall station at Halim Perdanakusuma Airport recorded rainfall with an intensity of 377 mm/day. That prompts a question about how much discharge was generated by the rainfall. This study was conducted to assess the flood discharge and the inundated area caused by the rain on 2020 New Year's Eve. The rainfall–runoff–inundation (RRI) model was utilized to simulate the flood discharge and inundation using the 1D–2D hydraulic–hydrology model. This model also calculated infiltration and subsurface flow with the Green–Ampt equation. In addition, the rainfall data used rain data recorded by the ground station, and the topography used SRTM data from the United States Geological Survey (USGS). Then, the flood discharge obtained from the model was compared with the flood return period. The return periods of the flood that were compared were at 2, 5, 10, 25, 50 and 100 years. The results showed that the flood that occurred on 1 January 2020 was larger than the flood with a return period of 100 years. This means that rainfall had the biggest effect on the flood, rather than other factors.

Keywords: flood; Jakarta; 2020; RRI

Citation: Sihombing, Y.I.; Rizaldi, A.; Farid, M.; Januriyadi, N.F.; Moe, I.R. Jakarta's 2020 New Year Flood Assessment with a Rainfall–Runoff–Inundation (RRI) Model. *Environ. Sci. Proc.* **2023**, *25*, 100. https://doi.org/10.3390/ECWS-7-14317

Academic Editor: Athanasios Loukas

Published: 3 April 2023

1. Introduction

Flooding is a natural process of a body of water, which rises to overflow land that is not normally submerged, due to high flow of runoff or sea-surge water [1–3]. There are several types of floods, including fluvial floods, pluvial floods and coastal floods [4–6]. Pluvial floods could be caused by the rain process, which produces an amount of water flow, on the surface, called runoff. Runoff is simply determined by rainfall, area of catchment and catchment response, which is represented by a runoff coefficient [7]. Many factors can cause flooding, including meteorological, geomorphological and anthropogenic factors [8–15]. Due to meandering river shape conditions, especially if the riverbed narrows throughout, a large discharge of water can derive an overflow [16]. Some places that previously did not experience floods have become areas affected by flooding due to changes in land use in the upstream areas [17,18]. In urban areas, flooding occurs due to illegal settlement and sedimentation, which reduce the capacities of rivers and channels [19–21]. In cold-climate regions, early spring snowmelt combined with heavy rainfall can also cause flooding [22].

Seawater levels rising and high rainfall, which are effects of climate change, are also causes of flooding in some areas [23–26]. Social behavior that occurs in the community can also indirectly have an impact on runoff: for example, urbanization. Urbanization will affect the land use of some areas, making changes in land cover that will affect runoff [27,28].

Jakarta was hit by flooding again in early 2020, on January 1st. One hundred and three locations were submerged by the flood around the provinces of DKI Jakarta and West Java, which are located in the Ciliwung–Cisadane Watershed [29]. Flooding occurred due to high rainfall intensity on 31 December 2019, exactly 1 day before. At least two of the eight stations around the DKI Jakarta area recorded rainfall of more than 300 mm; this is the highest rainfall of the last 25 years [30].

Flooding in the Ciliwung–Cisadane Watershed is not a new issue. It has happened since the 1660s [31]. An increase in discharge in the Ciliwung river was allegedly triggered by a change in the land use in the upstream area to become a tea plantation [32]. Some countermeasures to deal with increased runoff in the Ciliwung River have been carried out since the 1970s, and some have been implemented to date [33]. However, over time, as some areas began to develop, the problem of flooding in the Ciliwung–Cisadane Watershed became increasingly complex, and flooding still occurred.

The floods, which have not yet been solved, have caused several areas in the Ciliwung–Cisadane Watershed to still suffer from flooding. This has caused material and nonmaterial losses. These losses were not only caused by the size of the affected area but also due to the unpreparedness of the area in the face of disaster. The more prepared an area is in dealing with disasters, the more it will reduce the losses suffered [34]. This preparedness is not only related to the protection of an area against flooding but all the effort in that area in dealing with the flood disaster itself. This preparedness includes structural and nonstructural efforts [35]. Integration and sustainability of preparation, protection and the ability to respond effectively are the keys to resilience to flood disaster [36].

Many studies about the floods in the Ciliwung–Cisadane Watershed have been carried out, with different considerations. Climate change, land subsidence, land-use change and even social phenomena such as urbanization have been considered in several studies about flooding in the Ciliwung–Cisadane Watershed [37,38]. Future projection of flooding in the Ciliwung basin has been discussed in several studies as well. Moe et al. conducted a study related to the possibility of flooding in Jakarta, with scenarios caused by land subsidence combined with land-use changes [39,40]. Emam et al. showed how climate change and land-use change influence the flood behavior in Jakarta; it increased the peak flow of a 50-year return period in 2030 by 130% [41]. Januriyadi et al. not only conducted a study about future flooding but also analyzed the flood risk, due to climate change and urban development, for 2050 and showed that the risk has multiplied extremely [42].

Although many studies have been conducted about flooding in the Ciliwung–Cisadane Watershed, the occurrence of floods almost every year makes flooding still of interest to be investigated. To predict future changes is of great interest, but projecting current events to past predictions is also beneficial. The news related to Jakarta floods in the early 2020s, and moreover, the information that the rain that occurred was the most in history, startled Indonesia. The question is how large the flood discharge was due to that rain event, and to be more specific, which return period would have been equivalent to the coming flood on 1 January 2020. Today, it is not impossible to assess the flood after it happened. In 2018, Moe et al. conducted a rapid assessment to predict the affected area due to flooding in the Upper Citarum River Basin [43]. The aim of this study is to know the characteristics of the flood on 2020 New Year's Eve using rapid assessment.

2. Materials and Methods

2.1. Study Area

The study area was located at the Ciliwung–Cisadane Watershed, as shown in Figure 1. The watershed consists of 15 river basins, which accumulatively have an area of 5269.84 km^2. It administratively covers 3 provinces and 9 cities/municipalities, as follows: the Province of DKI Jakarta (5 cities), Bogor, Depok, Bekasi and Tangerang. The study area was located between latitude 5°59′28″ S and latitude 6°47′18″ S and between longitude 106°24′45″ E and longitude 107°12′54″ E.

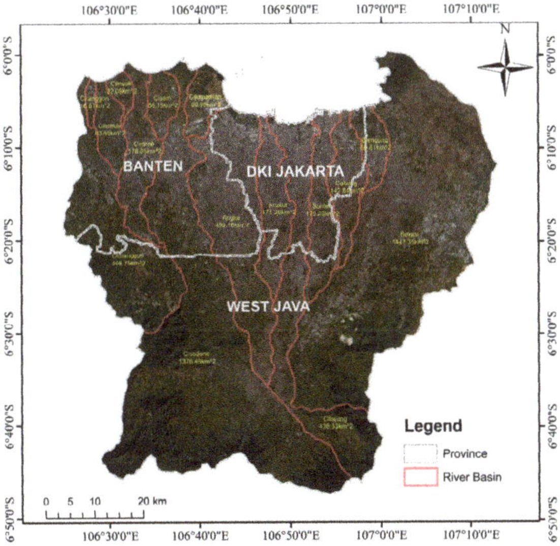

Figure 1. Study area.

2.2. Data

2.2.1. Precipitation Data

Precipitation data were collected from 24 rainfall stations in the Ciliwung–Cisadane Watershed during the flood event (Table 1). All of the rain stations are located around the study area, as shown in Figure 2a,b, which presents the rainfall distribution throughout the river area using the Inverse Distance Weighted Interpolation (IDW) method. The IDW method is the most frequently used deterministic method and can be applied for data whose distribution is characterized by a very large range [44]. From the data, the maximum rainfall during the studied event reached 376 mm. This condition is higher than the predicted rainfall for a 100-year return period in Halim Station, which was 340 mm [45].

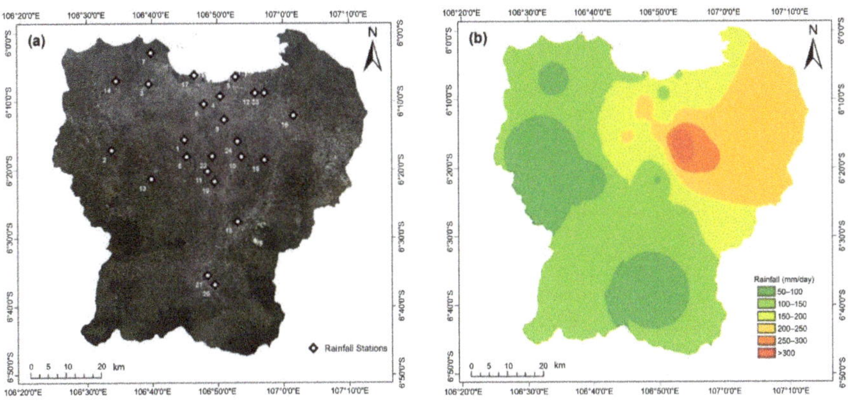

Figure 2. Rainfall datasets in the study area. (**a**) Rainfall station locations. (**b**) Rainfall distribution on 1 January 2020.

Table 1. List of rainfall stations.

No.	Rainfall Station	No.	Rainfall Station
1	Stasiun Klimatologi Tanggerang Selatan	13	AWS Puspitek
2	Stasiun Meteorologi Curug	14	ARG Sepatan
3	Stasiun Meteorologi Cengkareng	15	ARG Jatiasih
4	Stasiun Meteorologi Kemayoran	16	ARG Teluk Pucung
5	Stasiun Maritim Tanjung Priok	17	ARG Muara
6	Pos Hujan Bd Ciputat	18	ARG Jagorawi
7	Pos Hujan Teluk Naga	19	AWS UI
8	ARG Tomang	20	ARG Katulampa
9	ARG Manggarai	21	AWS IPB
10	AWW TMII	22	Pos Hujan Ragunan
11	ARG Ciganjur	23	Pos Hujan Rorotan
12	ARG Sukapura	24	TNI AU Halim

The predicted return-period rainfall was utilized as input in a calibrated model to determine the characteristics of various return periods of flooding. The calculation for return periods of rainfall was developed by Januriyadi et al. [42]. Our forecasting used a BMKG (Meteorological and Geophysical Institution of Indonesia) rainfall dataset for the period of 1986–2010 as a reference. The data were distributed throughout the study area, as presented in Figure 3. Compared with the rainfall that occurred on 1 January, the maximum rainfall of that event exceeded the maximum data in a 100-year period, which corresponds well with the previous study.

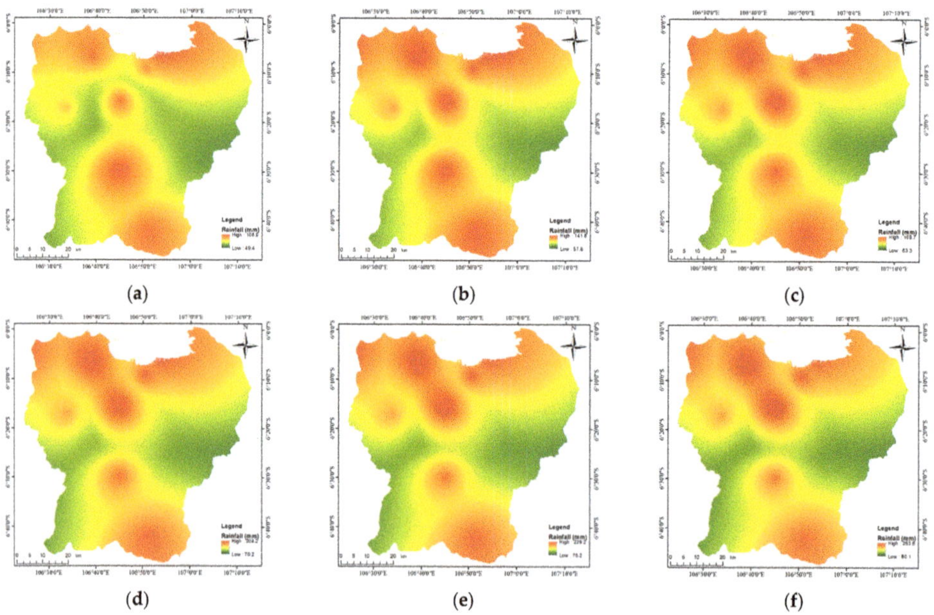

Figure 3. Rainfall return-period datasets. (**a**) Rainfall return-period of 2 years. (**b**) Rainfall return-period of 5 years. (**c**) Rainfall return-period of 10 years. (**d**) Rainfall return-period of 25 years. (**e**) Rainfall return-period of 50 years. (**f**) Rainfall return-period of 100 years.

2.2.2. Topography

The topography data was derived from the Digital Elevation Model (DEM) obtained from the Shuttle Radar Topography Mission (SRTM). The data was open-source data provided by the United States Geological Survey (USGS). For the SRTM, the vertical accuracy was 16 m for a 90% confidence level [46].

Figure 4 presents the topographical conditions of the Ciliwung–Cisadane Watershed. The resolution of the DEM that was used in this calculation was based on 1 arc second, or about 30 m. Nevertheless, the resolution was scaled up to 100 m due to computational issues. The numbers of rows and columns of pixels are 885 and 891. The upscaling of the DEM resolution increased the topography index (TI) of the DEM because of the nesting process of several grids with different TIs into one grid with one TI [47]. The consequences were areas or grids that should have been submerged becoming dry areas, and vice versa.

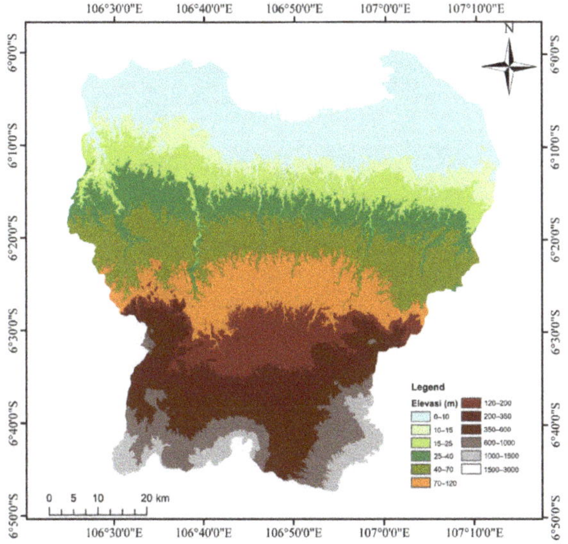

Figure 4. Digital Elevation Model (DEM) of the Ciliwung–Cisadane Watershed.

2.2.3. Land Cover

Land cover data was obtained from the Global Land Cover Characterization Version 2 (GLCC-V2). This database was developed by The U.S. Geological Survey (USGS), the University of Nebraska–Lincoln (UNL) and the European Commission's Joint Research Centre (JRC) in 1992. The land-cover projection had 1 km nominal spatial resolution and unique geographic elements. The land classification for this model has been simplified from the GLCC-V2 for calculation purposes (Figure 5).

Each land-cover classification has different characteristics in the model, based on soil conditions, as presented in Table 2. In this model, the river was distinguished from other water bodies. The river location was autogenerated by the RRI model from the Digital Elevation Model.

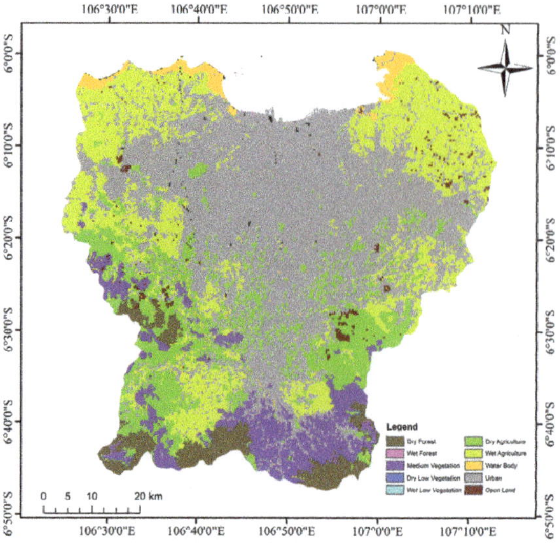

Figure 5. Land cover of the Ciliwung–Cisadane Watershed.

Table 2. The characteristics of the land cover.

Parameter	Land Cover		
	Clay	**Loam**	**Sandy Clay Loam**
Soil Depth (m)	1	1	1
Porosity (-)	0.475	0.463	0.398
k_v (m/s)	0–8.33 × 10^{-8}	0–9.44 × 10^{-7}	0–4.17 × 10^{-7}
S_f	0.361	0.089	0.219
k_a (m/s)	0–0.3	0–0.3	0–0.3
Unsat. Porosity (-)	0	0	0
Beta	8	8	8

2.3. Rainfall–Runoff–Inundation (RRI) Model

The rainfall–runoff–inundation (RRI) model is a two-dimensional model with a simplified equation. This model is capable of simulating rainfall–runoff and flood inundation at the same time; it was also designed to be used immediately after a disaster and it can be useful as a tool to analyze large-scale flooding as well [48]. In addition, this model assumes that the river channel location is in the same grid cells as the slope. A river channel is considered a centerline in a grid cell. It indicates an extra flow path between the grid cells and the actual river course. On the other hand, the slope cells function as the two-dimensional simulation area of the lateral flow. Hence, there are two water depths for slope grid cells in water channels, i.e., of the channel and of the slope (floodplain) itself.

The inflow–outflow interaction between the river and the slope is based on different overflowing formulae. The calculation depends on water-level and levee-height conditions. This model was generated based on mass-balance Equation (1) for governing the equation of flow rate. The momentum equation was derived from the governing equation of the model in the x direction (Equation (2)) and the y direction (Equation (3)):

$$\frac{\partial h}{\partial t} + \frac{\partial q_x}{\partial x} + \frac{\partial q_y}{\partial y} = r - f \tag{1}$$

$$\frac{\partial q_x}{\partial t} + \frac{\partial u q_x}{\partial x} + \frac{\partial v q_x}{\partial y} = -g\,h\,\frac{\partial H}{\partial x} - \frac{\tau_x}{\rho_w} \tag{2}$$

$$\frac{\partial q_y}{\partial t} + \frac{\partial u q_y}{\partial x} + \frac{\partial v q_y}{\partial y} = -g\,h\,\frac{\partial H}{\partial y} - \frac{\tau_y}{\rho_w} \tag{3}$$

where h is the height of the water from the local surface; q_x and q_y are the unit width discharges in the x and y directions, respectively; u and v are the flow velocities in the x and y directions, respectively; r is rainfall intensity; H is the height of water from the datum; ρw is the density of the water, g is gravitational acceleration; and τ_x and τ_y are the shear stresses in the x and y directions, respectively.

The RRI model separates the calculations of the discharge and the hydraulic gradient relationship. Hence, simulations of surface and subsurface flow proceed in the same algorithm. In addition, kinematic-rainfall–runoff-wave and diffusive-wave approximation are also derived in this model. The kinematic wave is calculated with the assumption that the water-surface slope is the hydraulic gradient. On the other hand, diffusion-stream approximation is utilized to form the streamflow equation.

The calibration process was carried out via a simulation of the flood events on 1 January. Then, the inundation area from the simulation was compared with the inundation map obtained from remote sensing by satellites at the same time. The model would have been well-calibrated if the result showed similarities to the inundation obtained from the remote sensing.

3. Results and Discussion

3.1. Model Calibration

The model was calibrated before being used to simulate the return-period flood. It was calibrated with a flood event that occurred on 1 January 2020. The simulation calculated the distribution of flood inundation with the condition of maximum water depth and compared it to the inundated area's satellite data at the same event (Figure 6). Figure 6 shows the comparison of the simulated flood inundation and the flood inundation from Sentinel 1A acquired on 2 January 2020. The flood inundation from Sentinel 1A was generated using an algorithm that was proposed by Chini et al. [49]. The algorithm can detect flood water not only on bare soil but also in urban regions. Even though the Sentinel 1A data was acquired a day after the flood event, some inundation still remained on the land. The results shows similar inundation in the northeast part between the simulation and satellite data, whereas in the middle part, near the ocean, the figure shows that the simulated inundation areas are larger than in the satellite data because the flood waters in the urban area of Jakarta receded on 2 January 2020.

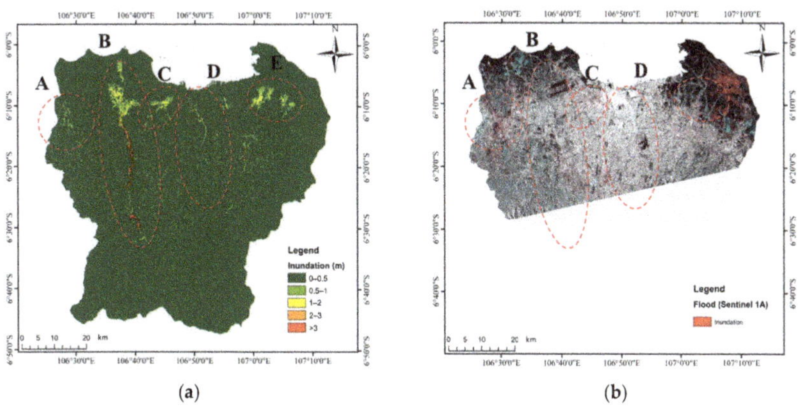

Figure 6. Rapid assessment of Jakarta flood inundation: (**a**) simulated 1 January flood inundation and (**b**) flood inundation data from Sentinel 1A acquired on 2 January 2020.

The inundation in this model consisted of local inundation and large-scale inundation. Local inundation was defined as local water depth that was barely moving. Accordingly, this type of inundation has a limited range of area and shallow water depth. On the other hand, large-scale inundation is water depth that was growing around the river with a wide range of areas.

Circle A, B, C, D and E represent large-scale inundation during the simulation period. The locations of the circles are in the Cimanceuri, Cisadane, Angke, Ciliwung and Bekasi River Basins, respectively. Table 3 shows the approximation of the total affected area due to river inundation. The total area of river inundation was approximately 106.54 km^2. Hence, the percentage of river inundation area during this event was 56.51% of the total inundation area.

Table 3. Large-scale inundation area.

Location	River Basin	Inundation Area (km^2)	Affected City(ies)
A	Cimanceuri	6.3	Tanggerang
B	Cisadane	51.05	Tanggerang
C	Angke	11.9	West Jakarta
D	Ciliwung	5.03	East Jakarta, South Jakarta, Central Jakarta
E	Bekasi	32.26	Bekasi

3.2. Model Application

The model that was calibrated was used to perform the return-period analysis of the flood. The flood return periods were simulated in mostly the same conditions as the main simulation. Nevertheless, the precipitation used as input in these simulations was modified for rainfall return periods that were forecast from historical rainfall data. The periods of rainfall that were used in this calculation were 2 years, 5 years, 10 years, 25 years, 50 years and 100 years. The results of these simulations are shown in Figure 7. Most of the large-scale inundation in each return period was found in the same locations, i.e., the Cimanceuri, Cisadane, Angke, Ciliwung and Bekasi River Basins. Therefore, the location of the large-scale inundation was the same as in the main simulation.

The area and volume of the inundation were calculated in the flood return period simulation. Then, the area and volume of the inundation were compared with the inundation characteristics of the flood that occurred on 1 January 2020. Table 4 shows that the closest return-period flood area to the flood on 1 January was the period with 100 yearly floods. The inundated areas of both flood maps were quite similar (Figure 8). This strengthens the evidence that the flood that occurred on 1 January was a flood with a return period of 100 years, whereas the rainfall on that day was greater than the rainfall for a return period of 100 years. The flood may not have been as large as the rainfall due to the spatial distribution of rainfall. The variability of rainfall spatial distribution could have affected the amount of flood discharge, which generates different flooding [50,51].

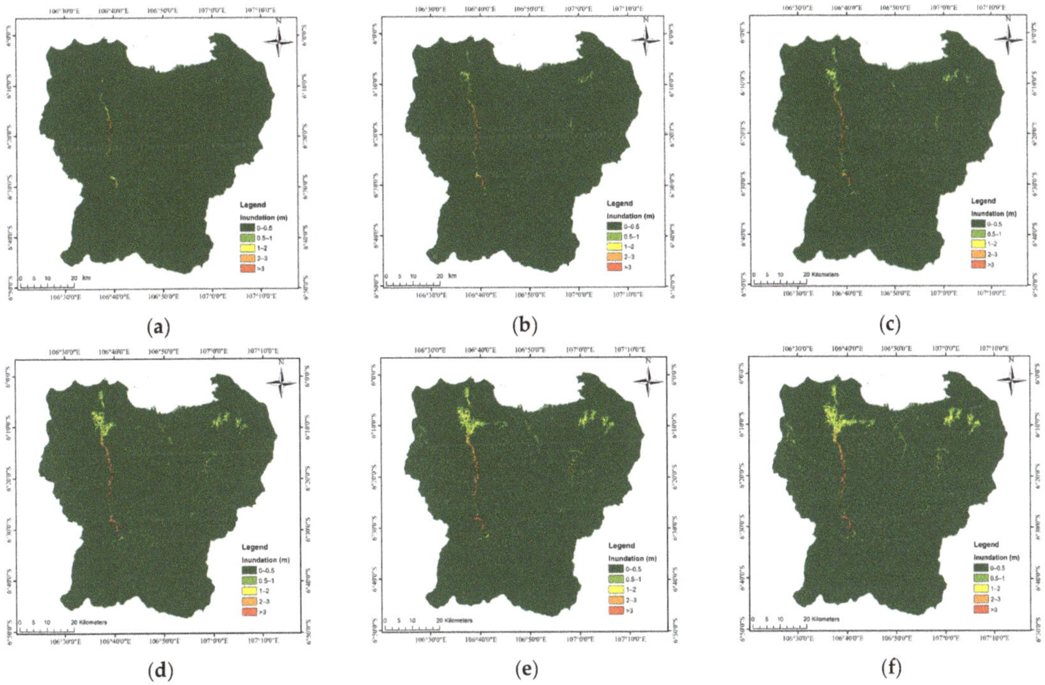

Figure 7. The inundation of flood return periods. (**a**) Flood return period of 2 years. (**b**) Flood return period of 5 years. (**c**) Flood return period of 10 years. (**d**) Flood return period of 25 years. (**e**) Flood return period of 50 years. (**f**) Flood return period of 100 years.

Table 4. The characteristics of flood return periods.

Flood Simulation	Volume (1000 m^3)	Max. Discharge (m^3/s)
Return Period of 2 Years	16,475	199.75
Return Period of 5 Years	22,059	260.88
Return Period of 10 Years	25,797	297.85
Return Period of 25 Years	30,397	340.95
Return Period of 50 Years	33,450	365.67
Return Period of 100 Years	36,355	389.97
Jakarta Flood	40,204	420.76

The flood discharge was calculated for the main model and the flood return period. It was measured with RRI hydro calculation at Water Gate Manggarai on the Ciliwung River (106°12′27.48″ S and 106°50′54.55″ E). The maximum discharge of each model is presented in Table 4. The discharge of the main model was compared with floods with 2-, 5-, 10-, 25-, 50- and 100-year return periods (Figure 9). The maximum discharge of the main model exceeded the 100-year return period. Therefore, the main model corresponded better with the rain-gauge data, which was more than the 100-year return-period rain data.

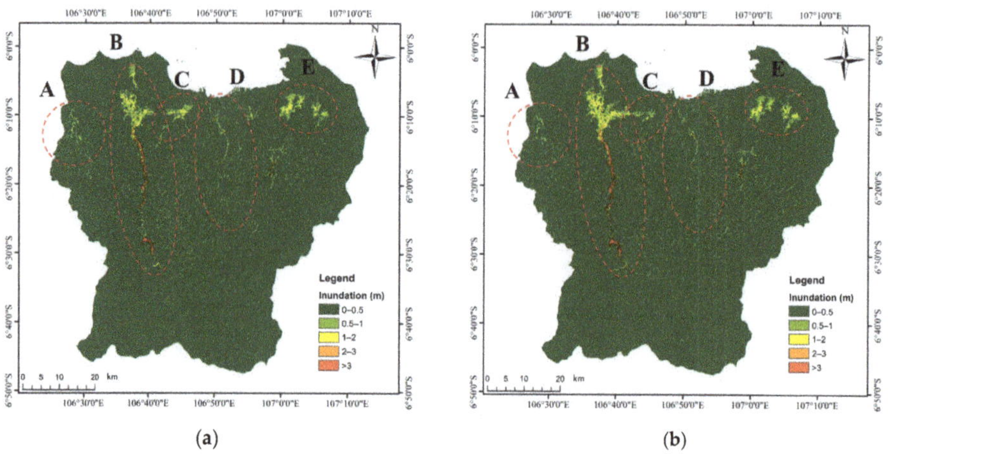

Figure 8. The comparison between (**a**) the 100-year flood return period and (**b**) the flood on 1 January 2020.

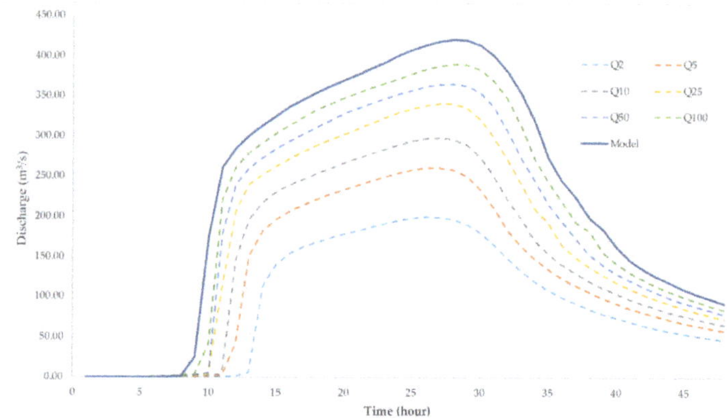

Figure 9. Discharge at Water Gate Manggarai.

4. Conclusions

This study attempted to conduct a rapid assessment of the flood on 1 January 2020 in Jakarta and several areas of the Ciliwung–Cisadane Watershed. This study was conducted to determine the return period of the flood that occurred. The rainfall–runoff–inundation (RRI) model is a distributed model that is able to simulate rainfall–runoff and flooding simultaneously. The model based on the inundation map from remote-sensing satellites was calibrated with ground-station rainfall data. The resolution of the Digital Elevation Model (DEM), which was too high, hindered the simulation process; it made it necessary for the resolution of the DEM to be decreased. The DEM was upscaled; then, the grid size was changed from 30 m to 100 m. After the model was calibrated, a return-period flood model was performed with a return-period rainfall input. There were six return periods of rainfall being simulated: 2, 5, 10, 25, 50 and 100 years.

The discharge and inundation areas of all return periods were compared with the simulation of the flood on 1 January. The comparison showed that the discharge extent on 1 January exceeded the discharge for a return period of 100 years. This result corresponds with the rain recorded at several rainfall stations, which surpassed the 100-year rainfall return period. Thus, there are several conclusions. First, the flood that occurred on 1

January 2020 was superior to a 100-year return period of flooding. Second, the rainfall, which was the most in history, was the main effect of this flood because it exceeded the 100-year return period of rainfall.

In this study, the DEM implemented in this model had a low resolution because of the limitation of data sources. In addition, the model needed to be compared with the existing discharge during the flood. These conditions need to be considered in future research.

Author Contributions: Conceptualization, M.F., N.F.J. and I.R.M.; methodology, M.F., N.F.J. and I.R.M.; software, Y.I.S. and N.F.J.; validation, Y.I.S. and A.R.; formal analysis, Y.I.S. and A.R; investigation, Y.I.S. and A.R.; resources, M.F. and I.R.M.; data curation, Y.I.S. and A.R.; writing—original draft preparation Y.I.S., A.R. and N.F.J.; writing—review and editing, M.F. and I.R.M.; visualization, Y.I.S. and A.R.; supervision, M.F. and I.R.M.; project administration, M.F. and I.R.M.; funding acquisition, M.F. and I.R.M. All authors have read and agreed to the published version of the manuscript.

Funding: This research was funded by the Institute for Research and Community Services of Institut Teknologi Bandung (LPPM ITB) through ITB Excellent Research 2023 and by the Indonesia Endowment Funds for Education (LPDP) through grant scheme PRJ-102/LPDP/2021.

Institutional Review Board Statement: Not applicable.

Informed Consent Statement: Not Applicable.

Data Availability Statement: Data is unavailable.

Conflicts of Interest: The authors declare no conflict of interest.

References

1. Chow, V.T.; Maidment, D.R.; Mays, L.W. *Applied Hydrology*; McGraw-Hill Book Company: New York, NY, USA, 1988. Available online: https://ponce.sdsu.edu/Applied_Hydrology_Chow_1988.pdf (accessed on 1 February 2023).
2. Penning-Rowsell, E.C.; Ward, R. Floods: A Geographical Perspective. *Geogr. J.* **1980**, *146*, 139. [CrossRef]
3. Webster, T.L.; Forbes, D.L.; Dickie, S.; Shreenan, R. Using topographic lidar to map flood risk from storm-surge events for Charlottetown, Prince Edward Island, Canada. *Can. J. Remote Sens.* **2004**, *30*, 64–76. [CrossRef]
4. Merz, B.; Hall, A.J.; Disse, M.; Schumann, A.H. Fluvial flood risk management in a changing world. *Nat. Hazards Earth Syst. Sci.* **2010**, *10*, 509–527. [CrossRef]
5. Zhou, Q.; Mikkelsen, P.S.; Halsnæs, K.; Arnbjerg-Nielsen, K. Framework for economic pluvial flood risk assessment considering climate change effects and adaptation benefits. *J. Hydrol.* **2012**, *414–415*, 539–549. [CrossRef]
6. Ganguli, P.; Merz, B. Extreme Coastal Water Levels Exacerbate Fluvial Flood Hazards in Northwestern Europe. *Sci. Rep.* **2019**, *9*, 13165. [CrossRef]
7. Ferreira, S.; Ghimire, R. Forest cover, socioeconomics, and reported flood frequency in developing countries. *Water Resour. Res.* **2012**, *48*, W08529. [CrossRef]
8. Li, F.; Wang, L.; Zhao, Y. Evolvement rules of basin flood risk under low-carbon mode. Part I: Response of soil organic carbon to land use change and its influence on land use planning in the Haihe basin. *Environ. Monit. Assess.* **2017**, *189*, 377. [CrossRef]
9. Maes, J.; Molombe, J.M.; Mertens, K.; Parra, C.; Poesen, J.; Che, V.B.; Kervyn, M. Socio-political drivers and consequences of landslide and flood risk zonation: A case study of Limbe city, Cameroon. *Environ. Plan. C Politics Space* **2019**, *37*, 707–731. [CrossRef]
10. Mahmoud, S.H.; Gan, T.Y. Urbanization and climate change implications in flood risk management: Developing an efficient decision support system for flood susceptibility mapping. *Sci. Total Environ.* **2018**, *636*, 152–167. [CrossRef]
11. Blume, T.; Zehe, E.; Bronstert, A. Rainfall—Runoff response, event-based runoff coefficients and hydrograph separation. *Hydrol. Sci. J.* **2007**, *52*, 843–862. [CrossRef]
12. Baker, V.R.; Pickup, G. Flood geomorphology of the Katherine Gorge, Northern Territory, Australia. *Geol. Soc. Am. Bull.* **1987**, *98*, 635. [CrossRef]
13. Bauer, B.O.; Baker, V.R.; Kochel, R.C.; Patton, P.C. Flood Geomorphology. *Geogr. Rev.* **1990**, *80*, 341. [CrossRef]
14. Texier, P. Floods in Jakarta: When the extreme reveals daily structural constraints and mismanagement. *Disaster Prev. Manag.* **2008**, *17*, 358–372. [CrossRef]
15. Lee, M.-H.; Bae, D.-H. Climate Change Impact Assessment on Green and Blue Water over Asian Monsoon Region. *Water Resour. Manag.* **2015**, *29*, 2407–2427. [CrossRef]
16. Baker, V.R. Stream-Channel Response to Floods, with Examples from Central Texas. *Geol. Soc. Am. Bull.* **1977**, *88*, 1057–1071. [CrossRef]
17. Dasanto, B.D.; Pramudya, B.; Boer, R.; Suharnoto, Y. Effects of Forest Cover Change on Flood Characteristics in the Upper Citarum Watershed. *J. Manaj. Hutan Trop. (J. Trop. For. Manag.)* **2014**, *20*, 141–149. [CrossRef]

18. Farid, M.; Mano, A.; Udo, K. Modeling flood runoff response to land cover change with rainfall spatial distribution in urbanized catchment. *J. Jpn. Soc. Civ. Eng. Ser. B1 (Hydraul. Eng.)* **2011**, *67*, I_19–I_24. [CrossRef]
19. Rosso, R.; Rulli, M.C. An integrated simulation method for flash-flood risk assessment: 2. Effects of changes in land-use under a historical perspective. *Hydrol. Earth Syst. Sci.* **2002**, *6*, 285–294. [CrossRef]
20. Lane, S.N.; Reid, S.C.; Tayefi, V.; Yu, D.; Hardy, R.J. Reconceptualising coarse sediment delivery problems in rivers as catchment-scale and diffuse. *Geomorphology* **2008**, *98*, 227–249. [CrossRef]
21. Rizaldi, A.; Moe, I.R.; Farid, M.; Aribawa, T.M.; Bayuadji, G.; Sugiharto, T. Study of flood characteristic in Cikalumpang River by using 2D flood model. *MATEC Web Conf.* **2019**, *270*, 04010. [CrossRef]
22. Zaghloul, M.S.; Ghaderpour, E.; Dastour, H.; Farjad, B.; Gupta, A.; Eum, H.; Achari, G.; Hassan, Q.K. Long Term Trend Analysis of River Flow and Climate in Northern Canada. *Hydrology* **2022**, *9*, 197. [CrossRef]
23. Karim, M.F.; Mimura, N. Impacts of climate change and sea-level rise on cyclonic storm surge floods in Bangladesh. *Glob. Environ. Change* **2008**, *18*, 490–500. [CrossRef]
24. Hallegatte, S.; Ranger, N.; Mestre, O.; Dumas, P.; Corfee-Morlot, J.; Herweijer, C.; Wood, R.M. Assessing climate change impacts, sea level rise and storm surge risk in port cities: A case study on Copenhagen. *Clim. Change* **2010**, *104*, 113–137. [CrossRef]
25. Darsan, J.; Asmath, H.; Jehu, A. Flood-risk mapping for storm surge and tsunami at Cocos Bay (Manzanilla), Trinidad. *J. Coast. Conserv.* **2013**, *17*, 679–689. [CrossRef]
26. Buchanan, M.K.; Kopp, R.E.; Oppenheimer, M.; Tebaldi, C. Allowances for evolving coastal flood risk under uncertain local sea-level rise. *Clim. Change* **2016**, *137*, 347–362. [CrossRef]
27. Du, J.; Qian, L.; Rui, H.; Zuo, T.; Zheng, D.; Xu, Y.; Xu, C.-Y. Assessing the effects of urbanization on annual runoff and flood events using an integrated hydrological modeling system for Qinhuai River basin, China. *J. Hydrol.* **2012**, *464–465*, 127–139. [CrossRef]
28. Aich, V.; Liersch, S.; Vetter, T.; Fournet, S.; Andersson, J.C.; Calmanti, S.; van Weert, F.H.; Hattermann, F.F.; Paton, E.N. Flood projections within the Niger River Basin under future land use and climate change. *Sci. Total Environ.* **2016**, *562*, 666–677. [CrossRef]
29. BNPB Editorial. Hujan Ekstrem Penyebab Banjir Jakarta. Available online: https://bnpb.go.id/berita/hujan-ekstrem-penyebab-banjir-jakarta (accessed on 16 May 2020).
30. Biro Komunikasi Publik Kementrian PUPR. Menteri Basuki Siapkan Langkah Penanganan Banjir di Jakarta dan Sekitarnya. Available online: https://pu.go.id/berita/view/17794/menteri-basuki-siapkan-langkah-penanganan-banjir-di-jakarta-dan-sekitarnya (accessed on 16 May 2020).
31. Caljouw, M.; Nas, P.J.; Pratiwo, M. Flooding in Jakarta: Towards a blue city with improved water management. *J. Humanit. Soc. Sci. Southeast Asia* **2009**, *161*, 454–484. [CrossRef]
32. Kurniawan, B.E.; Corps, M. Adapting cities to climate variability and change: Balance between community engagement and supporting facilitation roles of the local government to reduce the impact of climate change. In Proceedings of the Human(e) Settlements: The Urban Challenge Conference, Johannesburg, South Africa, 17–21 September 2012.
33. Octavianti, T.; Charles, K. The evolution of Jakarta's flood policy over the past 400 years: The lock-in of infrastructural solutions. *Environ. Plan. C Politics Space* **2019**, *37*, 1102–1125. [CrossRef]
34. Miceli, R.; Sotgiu, I.; Settanni, M. Disaster preparedness and perception of flood risk: A study in an alpine valley in Italy. *J. Environ. Psychol.* **2008**, *28*, 164–173. [CrossRef]
35. Petry, B. Keynote Lecture: Coping with Floods: Complementarity of Structural and Non-Structural Measures. In *Flood Defence*; Science Press: New York, NY, USA, 2002; pp. 60–70.
36. Hadihardaja, I.K.; Kuntoro, A.A.; Farid, M. Flood Resilience for Risk Management: Case Study of River Basin in Indonesia. *Glob. Asp.* **2013**, *3*, 16–19.
37. Budiyono, Y.; Aerts, J.C.J.H.; Tollenaar, D.; Ward, P.J. River flood risk in Jakarta under scenarios of future change. *Nat. Hazards Earth Syst. Sci.* **2016**, *16*, 757–774. [CrossRef]
38. Mishra, B.K.; Emam, A.R.; Masago, Y.; Kumar, P.; Regmi, R.K.; Fukushi, K. Assessment of future flood inundations under climate and land use change scenarios in the Ciliwung River Basin, Jakarta. *J. Flood Risk Manag.* **2017**, *11*, S1105–S1115. [CrossRef]
39. Moe, I.R.; Kure, S.; Januriyadi, N.F.; Farid, M.; Udo, K.; Kazama, S.; Koshimura, S. Effect of land subsidence on flood inundation in Jakarta, Indonesia. *J. Jpn. Soc. Civ. Eng. Ser. G (Environ. Res.)* **2016**, *72*, I_283–I_289. [CrossRef]
40. Moe, I.R.; Kure, S.; Januriyadi, N.F.; Farid, M.; Udo, K.; Kazama, S.; Koshimura, S. Future projection of flood inundation considering land-use changes and land subsidence in Jakarta, Indonesia. *Hydrol. Res. Lett.* **2017**, *11*, 99–105. [CrossRef]
41. Emam, A.R.; Mishra, B.K.; Kumar, P.; Masago, Y.; Fukushi, K. Impact Assessment of Climate and Land-Use Changes on Flooding Behavior in the Upper Ciliwung River, Jakarta, Indonesia. *Water* **2016**, *8*, 559. [CrossRef]
42. Januriyadi, N.F.; Kazama, S.; Moe, I.R.; Kure, S. Evaluation of future flood risk in Asian megacities: A case study of Jakarta. *Hydrol. Res. Lett.* **2018**, *12*, 14–22. [CrossRef]
43. Moe, I.R.; Rizaldi, A.; Farid, M.; Moerwanto, A.S.; Kuntoro, A.A. The use of rapid assessment for flood hazard map development in upper citarum river basin. *MATEC Web Conf.* **2018**, *229*, 04011. [CrossRef]
44. Bronowicka-Mielniczuk, U.; Mielniczuk, J.; Obroślak, R.; Przystupa, W. A Comparison of Some Interpolation Techniques for Determining Spatial Distribution of Nitrogen Compounds in Groundwater. *Int. J. Environ. Res.* **2019**, *13*, 679–687. [CrossRef]

45. Liu, J.; Doan, C.D.; Liong, S.-Y.; Sanders, R.; Dao, A.T.; Fewtrell, T. Regional frequency analysis of extreme rainfall events in Jakarta. *Nat. Hazards* **2015**, *75*, 1075–1104. [CrossRef]
46. Jarvis, A.; Rubiano, J.; Nelson, A.; Farrow, A.; Mulligan, M. *Practical Use of SRTM Data in the Tropics—Comparisons with Digital Elevation Models Generated from Cartographic Data;* Centro Internacional de Agricultura Tropical (CIAT): Cali, Colombia, 2004.
47. Zhang, H.; Li, Z.; Saifullah, M.; Li, Q.; Li, X. Impact of DEM Resolution and Spatial Scale: Analysis of Influence Factors and Parameters on Physically Based Distributed Model. *Adv. Meteorol.* **2016**, *2016*, 8582041. [CrossRef]
48. Sayama, T.; Ozawa, G.; Kawakami, T.; Nabesaka, S.; Fukami, K. Rainfall–runoff–inundation analysis of the 2010 Pakistan flood in the Kabul River basin. *Hydrol. Sci. J.* **2012**, *57*, 298–312. [CrossRef]
49. Chini, M.; Pelich, R.; Pulvirenti, L.; Pierdicca, N.; Hostache, R.; Matgen, P. Sentinel-1 InSAR Coherence to Detect Floodwater in Urban Areas: Houston and Hurricane Harvey as a Test Case. *Remote Sens.* **2019**, *11*, 107. [CrossRef]
50. Ogden, F.L.; Sharif, H.O.; Senarath, S.U.S.; Smith, J.A.; Baeck, M.L.; Richardson, J.R. Hydrologic analysis of the Fort Collins, Colorado, flash flood of 1997. *J. Hydrol.* **2000**, *228*, 82–100. [CrossRef]
51. Arnaud, P.; Bouvier, C.; Cisneros, L.; Dominguez, R. Influence of rainfall spatial variability on flood prediction. *J. Hydrol.* **2002**, *260*, 216–230. [CrossRef]

environmental sciences proceedings

MDPI

Proceeding Paper

Estimation of the Crop Water Requirements and Crop Coefficients of Multiple Crops in a Semi-Arid Region by Using Lysimeters †

Fazal Hussain [1,2,*], Muhammad Adnan Shahid [1,2,*], Muhammad Danish Majeed [1,2,*], Sikandar Ali [1,2] and Muhammad Shahid Ibni Zamir [3]

1 Agricultural Remote Sensing Lab (ARSL), National Center of GIS and Space Applications (NCGSA), University of Agriculture, Faisalabad 38000, Pakistan; sikandarali@uaf.edu.pk
2 Department of Irrigation and Drainage, Faculty of Agricultural Engineering and Technology, University of Agriculture, Faisalabad 38000, Pakistan
3 Department of Agronomy, Faculty of Agriculture, University of Agriculture, Faisalabad 38000, Pakistan; ibnizamir@uaf.edu.pk
* Correspondence: 2016ag6994@uaf.edu.pk (F.H.); muhammad.shahid@uaf.edu.pk (M.A.S.); danish.majeed@uaf.edu.pk (M.D.M.)
† Presented at the 7th International Electronic Conference on Water Sciences, 15–30 March 2023; Available online: https://ecws-7.sciforum.net.

Abstract: Pakistan is primarily an agricultural nation. Agriculture is the backbone of Pakistan economy and accounts for around 19.3% of its total gross domestic product. Pakistan ranks among the world's top 10 producers of wheat, maize, rice, cotton, and sugarcane, according to the United Nations Food and Agriculture Organization (FAO). It also cultivates edible oil seed crops. Water is the most important element for the cultivation of various crops. Pakistan is a country where irrigation water is extremely valuable. The irrigation schedule affects a crop's agronomic and economic viability. For the sake of water conservation as well as enhanced crop yields, a correct irrigation schedule is essential. Based on monitoring the soil water levels and crop water requirements, irrigation scheduling can be used to apply water for cultivation in accordance with predefined programs. How much water to apply to a particular crop and when depends heavily on the kind of soil and on the weather circumstances and is an important practical component of irrigation. Crop water requirements and crop coefficients are the most significant parameters that must be determined precisely locally to determine how much water is required for crop growth at various stages. There are many approaches for determining a crop water requirements, but lysimeters are the most effective devices. Lysimeters are devices that are routinely used to determine agricultural water requirements all around the world. The water required by different crops such as wheat, oat, carrot, and maize at different growth stages was determined using non-weighing-type lysimeters in this study. To compute the crop coefficient, the Penman–Monteith equation was used, which takes into account the daily mean temperature, wind speed, humidity, and solar radiation as inputs to the equation. The reference values for evapotranspiration during this period were obtained from a metrology station. The water requirement of the wheat varieties Fakhre Bhakhar and Anaj-17 were 361.8 mm and 379.5 mm, and their crop coefficient (Kc) values were between 0.79 and 1.19 and 0.27 and 1.27, respectively. The water required by carrot was 94.42 mm, and its crop coefficient was between 0.82 and 1.16. The water requirement of a maize hybrid variety was found to be 403.07 mm, and its crop coefficient was between 0.62 and 1.07. The water required by oat throughout its growth season including three harvests was 331.89 mm, and its crop coefficient was between 0.66 and 1.13. The water requirements of each crop and the crop coefficients calculated in this study are close to those recommend by FAO.

Keywords: lysimeters; crop water requirements; crop coefficients; climate change; water resources

check for updates

Citation: Hussain, F.; Shahid, M.A.; Majeed, M.D.; Ali, S.; Zamir, M.S.I. Estimation of the Crop Water Requirements and Crop Coefficients of Multiple Crops in a Semi-Arid Region by Using Lysimeters. *Environ. Sci. Proc.* **2023**, *25*, 101. https://doi.org/10.3390/ECWS-7-14226

Academic Editor: Athanasios Loukas

Published: 16 March 2023

1. Introduction

Evapotranspiration and moisture standards for crops are two important factors limiting the development of agricultural practices. The roots of crops that are submerged or filled with groundwater will be depleted, and the crop will not grow. This will make the crop wither. Lysimeters are used to determine how much water plants lose through evaporation. They record the amount of precipitation and water that is delivered to the crop through irrigation as well as the amount of water that is lost through the soil due to percolation. This information can be used to determine how much water and how much evapotranspiration the crop needs. Lysimeters are very important instruments, consisting of tanks or vessels that can set limits on how much water can stay in the soil and measure the soil water balance, how much water moves vertically, or how much water is present in the soil.

In agriculture, a crop water requirement is defined as the depth of water required to meet the water loss through evapotranspiration of a disease-free crop growing in large fields under intolerable soil conditions, including soil water and fertility status, and achieving its full grain production potential under the given environmental conditions (including soil water and fertility status). Identifying water requirements and crop coefficients is critical for irrigation scheduling and agricultural water management in field management. The assessment of crop evapotranspiration (ETc), which is dependent on crop features, management techniques, crop development stage, as well as climate and environmental circumstances, is directly related to the crop water requirement. There are two ways for calculating crop evapotranspiration (ETc), direct and indirect. In the direct technique, lysimeters are used to estimate ETc, whereas in the indirect method, reference evapotranspiration (ET) and crop coefficients (Kc) are used indicating the water requirements of the crop [1].

In terms of crop variation and output, water is the most significant limiting factor. Most of the water resources are devoted to agricultural irrigation. The proper use of irrigation water is becoming increasingly crucial as a means of dealing with the impending water crisis. An irrigation system ensures that plants receive the proper amount of water at the appropriate time for growth. Subsidized irrigation is the most important aspect of irrigation because it is used to augment natural rainfall in locations where rain falls irregularly or infrequently. Irrigation is used in arid places when the soil is extremely fertile and can support crop productivity, but there is little or no rainfall. Total irrigation is the name given to this method of irrigation. Due to a lack of irrigation, crops suffer from water stress, which lowers their yield. Plant nutrients can leach, and runoff and soil erosion can cause pollution of water sources if excessive irrigation is applied. In most cases, factors such as soil texture, rooting pattern, topography, evapotranspiration (ET) rate, rainfall, and crop type and/or development stage are likely to influence the frequency of irrigation. Carrot production can benefit from an efficient use of irrigation water. Irrigation scheduling has the potential to increase the irrigation efficiency, decrease the irrigation costs, and reduce the water opportunity costs. Similarly, a study found that boosting carrot production in a sustainable way relies on appropriate irrigation design, management, and scheduling. Thus, scheduling irrigation in accordance with a crop's water needs reduces the risk of under- or overwatering. Crop failure and fertilizers leaching beyond the root zone are also reduced, and growers see an increase in their profit when considering well-established crop water need references. The goal of this review study was to extend the knowledge on carrot water requirements, irrigation schedules, and water use efficiency by combining material from various studies on the subject [2].

The evaluation of the water requirements of a crop begins with the collection of fundamental data, which is represented by the estimation of evapotranspiration (ET). One of the most common approaches to calculating ET is based on what is known as the "crop coefficient", which is defined as the ratio of total evapotranspiration to reference evapotranspiration, ETo This approach is very popular. The recent standard methods of the FAO have shown that the value of the crop coefficient is connected to canopy factors reflecting the crop growth stage. These variables include canopy height and fractional vegetation cover. This relationship holds true under given climatic conditions [3].

Soil–plant–atmosphere water cycling includes evapotranspiration (ET), which is a major component. Water-related studies and applications, such as irrigation system design, irrigation scheduling, water resource planning and management, water allocation, water balance computation, crop yield forecast, and so on, rely heavily on its trustworthy determination. Although a variety of methods exist for measuring ET, they are time-consuming, expensive, and labor-intensive. As a result, it is typically calculated as the product of reference crop evapotranspiration (ETo) and crop coefficient in a well-known two-step approach [4,5].

2. Materials and Methods

2.1. Study Area

In the semi-arid climate of Faisalabad, experimental research was carried out to determine the water requirements and crop coefficients of wheat, oat, carrot, and maize. For this study, experimental trials were cultivated in the experimental site of the faculty of Agricultural Engineering and Technology of the University of Agriculture, Faisalabad. This study was carried out at an experimental site with 12 lysimeters. The First 4 Lysimeters are in a rectangular geometry as shown in Figure 1 and the other Eight Lysimeters are in a circular geometry as shown in Figure 2.

Figure 1. Experimental site with four lysimeters (University of Agriculture, Faisalabad, Pakistan).

Figure 2. Experimental site with eight lysimeters (University of Agriculture, Faisalabad, Pakistan).

2.2. *Data Collection and Analysis*

Irrigation Scheduling

Sandy loam was the soil in each site examined with the lysimeters. Sandy loam soil has a field capacity of 22% and a permanent wilting point of 8%. At 50% depletion of soil moisture, each crop is irrigated. In order to re-establish the field capacity, irrigation was applied when the soil moisture levels fell below 15%. Soil moisture sensors detected when irrigation should be applied based on the amount of moisture present in the soil. As a result, irrigation was applied anytime the soil moisture percentage in each lysimeter approached 15%. For irrigation purposes, the depth of water needed to re-establish the field capacity is measured using the following formula:

$$\text{Depth} = d = (FC - M.C)/100 \times R.Z \quad (1)$$

in which

d = depth of water required for irrigation
FC = field capacity
M.C = moisture content
R.Z = root zone depth of the crop

Flood irrigation has a 30 to 40% efficiency rate, but our field was levelled and tiny; therefore, we supposed that water was applied at a 70 percent efficiency rate. As a result, the actual depth required to reach field capacity was once again an important factor:

$$D = d/0.7 \quad (2)$$

In order to determine how long it will take to irrigate a certain area, one simply needs to know the discharge area of a lysimeter and the depth of that area:

$$T = (A \times d)/Q \quad (3)$$

T = time required for irrigation
A = area of the lysimeter
d = depth required
Q = discharge of the pump

As a result, the amount of time it takes to re-establish a soil's field capacity (FC) is obtained with the above equation.

ETo is critical to determine crop coefficients and water requirements. The ETo value for each month or each day might vary based on a variety of factors, including weather, time of the day, and wind velocity. It was from the UAF meteorological observatory that evapotranspiration data by month were collected from 2013 to 2021. As a result, an average of 8 years' worth of ETo data were utilized to estimate the crop water needs and crop coefficients. The amount of water a crop needs from seeding to harvest is known as the crop water requirement. This water is required by crops at several stages of growth, including planting, maturation, and flowering. It is possible to estimate the amount of water a crop will require by summing all the amounts of irrigation water used to reach the soil capacity and the actual rainfall:

$$CWR = TAW + Re \quad (4)$$

CWR = crop water requirement
TAW = total applied water
Re = effective rainfall

Thus, by the above relation, we calculated the water requirement of each crop sown in the experiment, i.e., wheat, oat, carrot and, maize. The crop coefficient Kc can be calculated by a simple relation between ETc and ETo. The calculation of the crop coefficient is based on the Penman–Monteith Equation:

$$\text{ETc} = \text{ETo} \times \text{Kc} \quad (5)$$

$$\text{Thus, Kc} = \text{ETc}/\text{ETo} \quad (6)$$

By using the above relation between ETc and ETo, the Kc for wheat, oat, carrot, and maize were calculated.

3. Results and Discussion

The crop water requirements and crop coefficients for the years 2021–2022 of wheat, carrot, oat, and maize growing in Faisalabad's semi-arid climate were analyzed using Lysimeters.

3.1. Soil Characterictics

The movement of soil along the water's surface is determined by the soil's physical properties. Infiltration rate, field capacity, permanent wilting point, bulk density, and water holding capacity of the soil were considered.

3.2. Soil Bulk Density

In terms of physical properties, bulk density refers to the soil's composition, the number of pores, and how much organic matter is present. The bulk density of the treated plots was computed using ASAE Norm S269.4, and the results were analyzed. We found that the bulk density ranged between 1.52 and 1.60 g/cm^3.

3.3. Soil Infiltration Rate

Before the analysis, the soil penetration intensity was determined at the testing site in September 2021. If runoff is to be avoided, the irrigation rate must be less than or equal to the infiltration rate. The infiltration rates ranged from 0.70 to 0.83 cm/h. Soil infiltration has a direct effect on the movement of water in the soil. A decrease in soil porosity was seen as a result of clay particle dispersion and growth in irrigation water containing high concentrations of Na ions, both of which were caused by irrigation water of low quality.

3.4. Effect of the Climate on ETo

The temperature is increasing throughout the study region, especially in the northern parts of the country. Pakistan has an agrarian economy, and climate change and the consequent threat of global warming have become a great challenge. The principal weather parameters affecting evapotranspiration are radiation, air temperature, humidity, and wind speed. In a warming climate, increased evapotranspiration may shift the fraction of precipitation that runs off as surface water or infiltrates the subsurface as recharge.

3.5. Determination of Effective Rainfall

The amount of rainfall water that is accessible for use by plants is referred to as "effective rainfall." Results regarding the effective rainfall have been presented in Table 1. The formulas provided by FAO were used to determine this value:

$$\text{Pe} = 0.8\text{P} - 25 \qquad \text{if P} > 75 \text{ mm} \quad (7)$$

$$\text{Pe} = 0.6\text{P} - 10 \qquad \text{if P} < 75 \text{ mm} \quad (8)$$

Table 1. Effective rainfall.

Month	P (mm/month)	Formula	Pe (mm/month)
December	20	Pe = 0.6P − 10	2
January	48.4	Pe = 0.6P − 10	2
February	2.5	Pe = 0.6P − 10	0
March	75	Pe = 0.8P − 25	35
April	57	Pe = 0.6P − 10	24.2
May	54	Pe = 0.6P − 10	22.4

3.6. Water Requirement and Kc of the Wheat Variety Fakhre Bhakkar

Wheat was sown on 1 December 2021 and harvested on 28 April 2022. The water required by wheat during this season was 361.8 mm as the same results achieved by [6]. The Kc value for the wheat variety Fakhre Bhakkar varied from 0.79 to 1.19 during the whole crop season. Results regarding the crop coefficients of the wheat variety Fakhare Bhakhar is shown in Figure 3.

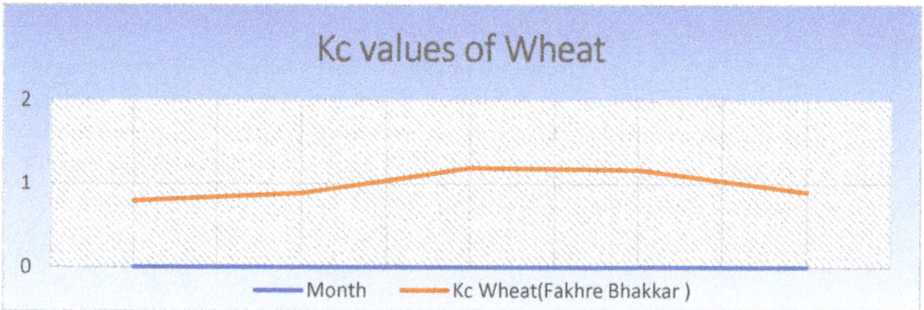

Figure 3. Crop coefficients of the wheat variety Fakhare Bhakhar.

3.7. Water Requirement and Kc of the Wheat Variety Anaj-17

Wheat was sown on 1 December 2021 and harvested on 28 April 2022. The water required by wheat during this season was 379.5 mm, and its Kc value was between 0.87 and 1.27 as the results match to this research [7,8]. Results regarding the crop coefficients of the wheat variety Anaj-17 is shown in Figure 4.

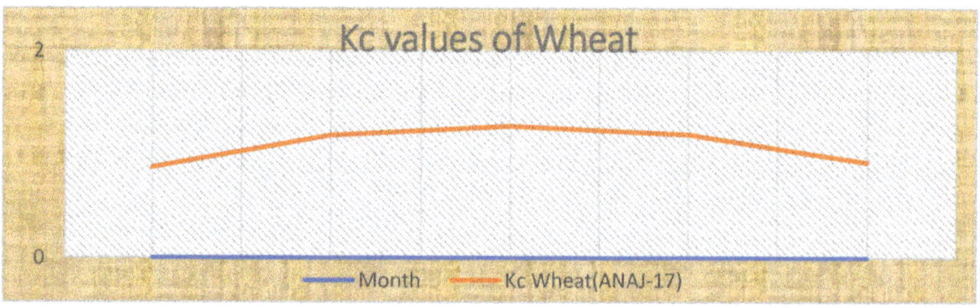

Figure 4. Crop coefficients of the wheat variety Anaj-17.

3.8. Water Requirement and Kc of Carrot

Carrot was sown on 1 December 2021 and harvested on 20 February 2022. The water required by carrot during this season was 94.41 mm, and the Kc value was between 0.79 and 1.16. Results regarding the crop coefficients of carrot is shown in Figure 5.

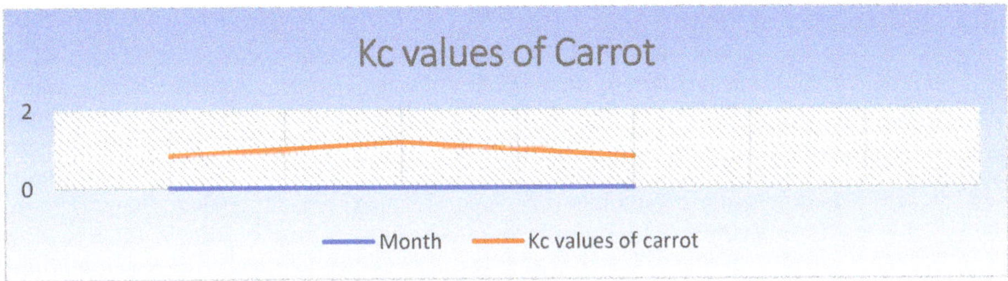

Figure 5. Crop coefficients of carrot.

3.9. Water Requirement and Kc of Maize

Maize was sown on 1 March 2022 and harvested on 15 July 2022. The water required by maize during this season was 403.07 mm, and the Kc value was between 0.62 and 1.09. as the conclusion and results were helped out form this research [9–11]. Results regarding the crop coefficients of Maize is shown in Figure 6.

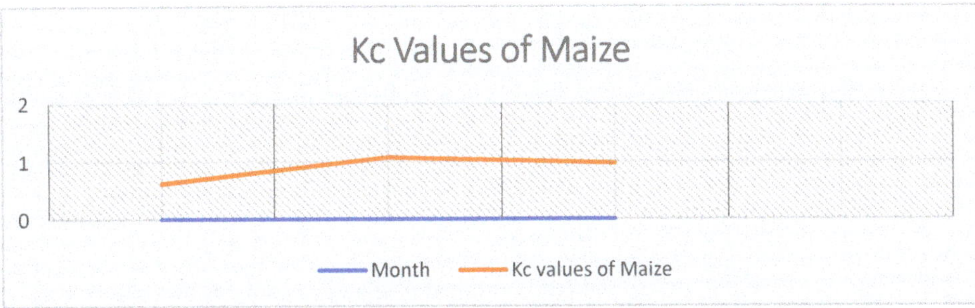

Figure 6. Crop coefficients of maize.

3.10. Water Requirement and Kc of Oat

Oat was sown on 1 December 2021 and harvested three times, as it germinated again after each harvest. The last harvest was carried out on 28 April 2022. The water required by oat during this season was 331.89 mm, and the value of Kc was between 0.66 and 1.13. As the results are same from this level of research my results are most relative to that research [12–14]. Results regarding the crop coefficients of Oat is shown in Figure 7.

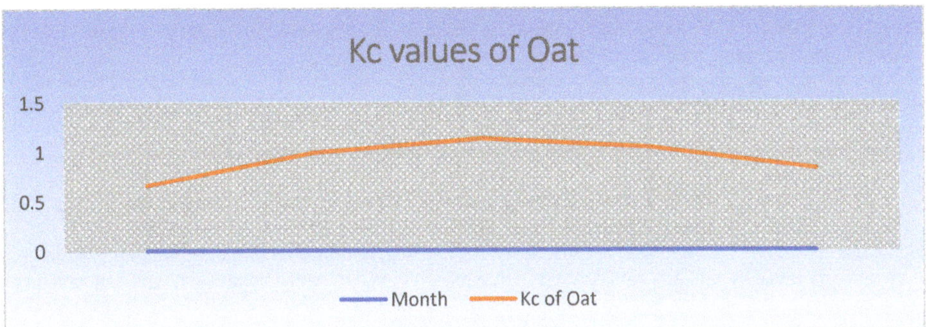

Figure 7. Crop coefficients of oat.

4. Conclusions

In this study, we determined that the water needs of the wheat varieties Fakhre Bhakhar and Anaj-17 were 361.81 mm and 379.5 mm, respectively. The crop coefficients (Kc) of the Fakhre Bhakhar variety were between 0.79 and 1.19. The crop coefficient (Kc) of the Anaj-17 variety was between 0.87 and 1.27. These values are in the range of Kc values (0.2–1.5) suggested by FAO. The yields from these two experiments were different, nevertheless. The Anaj-17 output was larger than that of Fakhre Bhakhar, using similar farming techniques. The Anaj-17 yield was 491 g per m^2 and was greater than the Fakhre Bhakhar yield, which was 491 g per m^2. The water required by carrot was 94.41 mm, and its crop coefficient resulted to be between 0.79 and 1.16. The water demand of the maize hybrid variety examined was found to be 403.07 mm, and its crop coefficient was between 0.62 and 1.07. The water required by oat throughout its growing season was 331.89 mm, and its crop coefficient was between 0.66 and 1.13, values that are in the range of Kc values (0.9–0.85) recommended by FAO.

Author Contributions: Conceptualization, F.H. and M.A.S.; Introduction, S.A., M.S.I.Z. and M.A.S.; materials and methods, F.H. and M.D.M.; data collection; F.H.; results and discussions, F.H., M.A.S. and M.D.M.; conclusions and recommendations, S.A. and M.A.S.; writing original draft preparation, F.H.; editing, F.H. and M.D.M. All authors have read and agreed to the published version of the manuscript.

Funding: This research received funding from Agricultural Remote sensing Lab University of Agriculture Faisalabad, 38000.

Institutional Review Board Statement: Not applicable.

Informed Consent Statement: Not applicable.

Data Availability Statement: Not applicable.

Conflicts of Interest: The authors declare no conflict of interests.

References

1. Shahrokhnia, M.H.; Sepaskhah, A.R. Single and dual crop coefficients and crop evapotranspiration for wheat and maize in a semi-arid region. *Theor. Appl. Climatol.* **2013**, *114*, 495–510. [CrossRef]
2. Beshir, S. Review on estimation of crop water requirement, irrigation frequency and water use efficiency of cabbage production. *J. Geosci. Environ. Prot.* **2017**, *5*, 59. [CrossRef]
3. D'Urso, G.; Calera Belmonte, A. Operative approaches to determine crop water requirements from Earth Observation data: Methodologies and applications. *AIP Conf. Proc.* **2006**, *14*, 852.
4. Allen, R.G.; Pereira, L.S.; Raes, D.; Smith, M. *Crop Evapotranspiration Guidelines for Computing Crop Water Requirements*; Food and Agriculture Organization of the United Nations: Rome, Italy, 1998.
5. Liu, X.; Xu, C.; Zhong, X.; Li, Y.; Yuan, X.; Cao, J. Comparison of 16 models for reference crop evapotranspiration against weighing lysimeter measurement. *Agric. Water Manag.* **2017**, *184*, 145–155. [CrossRef]
6. Mahmoud, M.; El-Bably, A. Crop water requirements and irrigation efficiencies in Egypt. In *Conventional Water Resources and Agriculture in Egypt*; Springer: Cham, Switzerland, 2017; pp. 471–487.
7. Abid, S. Forecasting wheat production using time series models in Pakistan. *Asian J. Agric. Rural. Dev.* **2019**, *8*, 172–177.
8. Mehta, R.; Pandey, V. Crop water requirement (ETc) of different crops of middle Gujarat. *J. Agrometeorol.* **2016**, *18*, 83–87. [CrossRef]
9. Abideen, Z.U. Comparison of Crop Water Requirements of Maize Varieties Under Irrigated Condition in Semi-Arid Environment. *J. Environ. Earth Sci.* **2014**, *4*, 2224–2226.
10. Anapalli, S.S.; Ahuja, L.R.; Gowda, P.H.; Ma, L.; Marek, G.; Evett, S.R.; Howell, T.A. Simulation of crop evapotranspiration and crop coefficients with data in weighing lysimeters. *Agric. Water Manag.* **2016**, *177*, 274–283. [CrossRef]
11. Sun, H.; Zhang, X.; Liu, X.; Liu, X.; Shao, L.; Chen, S.; Wang, J.; Dong, X. Impact of different cropping systems and irrigation schedules on evapotranspiration, grain yield and groundwater level in the North China Plain. *Agric. Water Manag.* **2019**, *211*, 202–209. [CrossRef]
12. Ali, M.H. *Fundamentals of Irrigation and On-Farm Water Management*; Springer: Heidelberg, Germany, 2010; Volume 1, pp. 453–487.

13. Pereira, L.; Paredes, P.; Jovanovic, N. Soil water balance models for determining crop water and irrigation requirements and irrigation scheduling focusing on the FAO56 method and the dual Kc approach. *Agric. Water Manag.* **2020**, *241*, 106357. [CrossRef]
14. da Silva, G.H. Biodegradable mulch of recycled paper reduces water consumption and crop coefficient of pak choi. *Sci. Hortic.* **2020**, *267*, 109315. [CrossRef]

environmental sciences proceedings

MDPI

Proceeding Paper

Assessment of Flood Frequency Pattern in a Complex Mountainous Terrain Using the SWAT Model Simulation †

Nada Joumar [1,*], Amal Markhi [2], Jamal Eddine Stitou El Messari [1] and Lahcen Benaabidate [3]

1 Laboratory of Applied and Marine Geosciences, Geotechnics and Geohazards (LR3G), Abdelmalek Essaadi University, Tetouan 93000, Morocco; jstitouelmessari@uae.ac.ma
2 GEOHYD Laboratory, Faculty of Sciences Semlalia, Cadi Ayyad University, Marrakech 40000, Morocco; amal.markhi32@gmail.com
3 Laboratory of Functional Ecology and Environment Engineering, Faculty of Sciences and Techniques, University of Sidi Mohamed Ben Abdellah, Fez 30000, Morocco; lahcen.benaabidate@usmba.ac.ma
* Correspondence: nada.joumar@etu.uae.ac.ma

† Presented at the 7th International Electronic Conference on Water Sciences, 15–30 March 2023; Available online: https://ecws-7.sciforum.net/.

Abstract: Understanding the relationship between rainfall and runoff is one of the requirements and necessities in flood modeling, predicting, and recording annual runoff contributions. This study aimed to evaluate the use of hydrological modeling and flood frequency analysis (FFA) in studying the extent and occurrence of floods in complex mountain basins and the impact of dams on downstream flooding. The N'fis subbasin, the study area, is located in the High Atlas Mountains of Morocco; it drains a total area of 1700 km^2 and is characterized by an arid to semi-arid climate in the plains and a subhumid climate in the mountains. Flood modeling in this catchment is very difficult due to the lack of sufficient spatial and temporal flood data available for FFA. Therefore, the SWAT (Soil and Water Assessment Tool), a physics-based continuous model, was used to simulate and reproduce the hydrological behavior upstream of N'fis. The model's parameters were calibrated and validated using data collected from 2000 to 2016, and the model performed well using Nash–Sutcliffe statistics with a calibration period of 0.52 and a validation of 0.69. Finally, using daily flood data (1982–2016), we performed FFA using the L-moments method (Gumbel, normal, and log-Pearson III). Furthermore, a comparison of the goodness of fit of the Gumbel, GEV, and LP3 distributions to the flood frequency analysis in the N'fis basin highlighted that the GEV distribution gave good results and appears to be the more appropriate distribution. This research will enable better assessment of floods and help water managers and decision makers to better plan and manage flood mitigation.

Keywords: flood frequency analysis; SWAT; flood forecast; High Atlas; Morocco

Citation: Joumar, N.; Markhi, A.; El Messari, J.E.S.; Benaabidate, L. Assessment of Flood Frequency Pattern in a Complex Mountainous Terrain Using the SWAT Model Simulation. *Environ. Sci. Proc.* **2023**, *25*, 102. https://doi.org/10.3390/ECWS-7-14195

Academic Editor: Silvia Kohnova

Published: 14 March 2023

1. Introduction

Morocco's rainfall distribution is varied in both spatial and temporal scales. It can reach more than 800 mm at high altitudes, but hardly more than 300 mm on the plains [1]. The Moroccan High Atlas has seen some of the most destructive flood occurrences in history, such as the 1995 Ourika flood, which killed 732 people and cost MAD 80 million in economic damage [2]. The use of flood frequency analysis in dry and semi-arid environments, particularly in developing nations, can be tremendously beneficial for better assessment of, and planning for, flood risk and reducing the disastrous effects of this phenomena [3]. The N'fis watershed was studied because it is located in the High Atlas area, which is prone to flooding [4]. This area is marked in particular by substantial spatiotemporal variability of precipitation and relative irregularity of surface runoff [5].

Many studies have been carried out in the N'fis Wadi watershed. For instance, [6] investigated the relative performance of the Snowmelt Runoff Model (SRM) to simulate streamflow in five sub-catchments of the High Atlas Mountain range; this study found that snowmelt contribution to surface runoff in the N′fis watershed was lower compared to neighboring mountainous catchments, and the hydrological model tended to underestimate peak flows due to their absence in the input weather data [7]. On the other hand, the first to develop a water erosion risk map of the watershed via digital processing of various satellite products, this study indicated that more than three quarters of the catchment area located south-east and west of the sub-basin is subject to moderate, high, and very high risk of water erosion. This work was followed by [8], who applied the RUSLE and SEDD models over two periods to analyze the impact of land use change on potential erosion, as well as the suspended sediment yield (SSY); this study revealed that, in general, there was a decrease in the spatial annual average erosion rate (and subsequently the sediment delivery ratio) over the years due to significant changes in land use in the area. Similarly, but using a different approach, [9] used the SWAT model (Soil and Water Assessment Tool) to estimate potential soil erosion and sediment yield in the semi-arid N′fis basin; this study determined soil loss within each hydrological response unit in the watershed and indicated a high yield rate (123 t Ha^{-1} for an average annual rainfall of 315 mm yr^{-1}).

Additionally, a previous study carried out regarding the Ourika watershed by [3] compared 12 frequency models using the maximum likelihood method and the criteria to guide the selection of the most appropriate model; it showed that the most suitable method was the GEV law. Another study tackled regional flood frequency analysis in the catchments (N′Fis, Rheraya, Ourika, Zat, and R′dat) [10], using the generalized extreme value (GEV) model. However, no studies have compared statistical probability distributions (such us generalized extreme value (GEV), log-Pearson III (LP3), and Gumbel (EV1)) in the N′fis watershed; these constitute an important aspect of flood modeling and forecasting. These statistical probability distributions are generally used to estimate the magnitude and probability of the occurrence of extreme events and to obtain accurate results [11]. The data series required to proceed with FFA generally exceeds 50 years [12]; due to data scarcity, hydrological modeling is required to acquire long-term flood time series. The SWAT model (Soil and Water Assessment Tool) was used, which is a physically based, continuous model. Major model components include weather, hydrology, soil temperature and properties, land use, etc. [13]. This model was chosen for its ability to take into account the different important factors that impact floods, especially in a rural area (which is the case for the N′fis Wadi watershed), and for its proven good performance in different watersheds in Morocco [9,14–16]. The overarching goals of this study were to first, generate a long time series of flow using the SWAT model, and second, apply flood frequency analysis to these data series, compare the three common statistical distributions used in FFA, and investigate the goodness of fit of the three distributions methods selected.

2. Materials and Methods

2.1. Study Area

The Tensift watershed is located in west-central Morocco; it covers a total area of 24,000 km^2 and includes the seven prefectures and provinces, including Marrakech, Al Haouz, and Al Youssoufia (Figure 1). The study area, N′fis, is the largest subbasin of the Tensift watershed, which drains a total area of 1700 km. This complex-terrain watershed is characterized by an elevation that ranges from 641 m to 4080 m, with a mean elevation of 1860 m. The watershed is 96% rural land and only 6% urban [8]. The land use consists mostly of forested areas (44%) and the dominant soil type is zonal brown soils on shale socle, which explains the impermeable nature of the basin (79.3%) [17]. The basin′s climate is semi-arid, characterized by hot (max: 46°C) and dry summers and then cold (min: −7.4 °C) humid winters [18]. The annual rainfall ranges from 254.1 mm in low altitude to 796.9 mm in high altitudes.

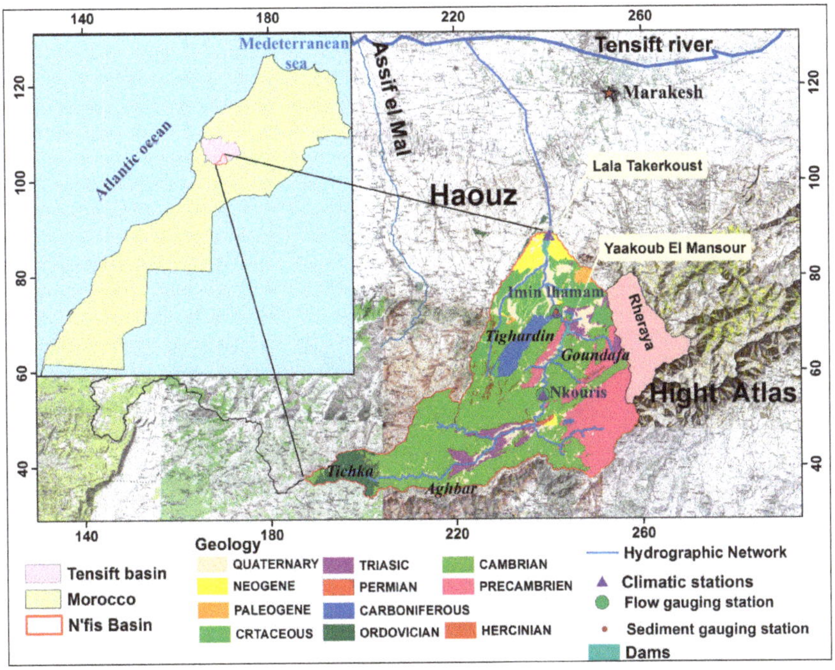

Figure 1. Location of the N'fis watershed and the hydrometeorological stations.

2.2. SWAT Model

The Soil and Water Assessment Tool (SWAT) is a widely used hydrological model that simulates the hydrological processes of a watershed [13]. It is a comprehensive model that integrates various sub-models to simulate the water balance, erosion, sediment transport, nutrient cycling, and crop growth within a watershed. The SWAT uses a semi-distributed approach to simulate the complex interactions between land use, climate, topography, soil properties, and vegetation in a watershed [19].

The mathematical equation of the SWAT is based on the water balance equation [20], which is expressed as:

$$P = Q + S + E + G + A \tag{1}$$

where P is precipitation, Q is runoff, S is soil water storage, E is evapotranspiration, G is groundwater flow, and A is lateral flow. The SWAT model uses this equation to represent the water balance of the watershed and to simulate various hydrological processes.

The SWAT has been used in a variety of applications, including water resources management, land use planning, and climate change impact assessments [21]. Its flexibility, adaptability, and comprehensiveness make it a valuable tool for studying the interactions between land use, climate, and water resources in watersheds. The dataset used in the SWAT set up are presented in Table 1. The weather data inputs used in this study were daily precipitation and minimum and maximum air temperatures. The SWAT can simulate wind, solar, and humidity data. The SWAT was set up for a period of 16 years, including 2 years warm-up (2000 and 2001) to ensure that the model was initialized, calibration (2002–2012), and 4 years validation (2013–2016).

Table 1. Geo-spatial and hydrometeorological datasets.

Data	Source	Spatial Resolution	Temporal Resolution
DEM	STRM-United States Geological Survey (USGS) https://earthexplorer.usgs.gov/ (accessed on 16 June 2020)	30 m	-
Land Use Map	(MODIS) Land Cover Type (MCD12Q1)	500 m	Yearly
Soil Map	Tensift Basin Hydraulic Agency (TBHA)	ArcInfo Format (scale 1:100,000)	-
Soil Data	Field Work [9]	ArcInfo Format (scale 1:100,000)	-
Observed Hydrometeorology	Tensift Basin Hydraulic Agency (TBHA)	-	Daily

2.3. Model Calibration and Validation

Among various methods used to perform calibration and uncertainty analysis is the widely used Sequential Uncertainty Fitting 2 (SUFI-2) approach with the SWAT Calibration Uncertainty Procedure (SWAT-CUP) [22,23]. The SUFI-2 is a semi-automated approach that is used to perform parameterization, sensitivity analysis, uncertainty analysis, calibration, and validation of hydrologic parameters [24]. Sensitivity analysis, in particular, is necessary to understand which particular input parameter has a significant impact on the model outflow [25]. Thus, the SUFI-2 algorithm was used to analyze 16 input parameters and further calibrate and validate the SWAT model. The model's performance was evaluated using Nash–Sutcliffe efficiency (NSE) [26,27], percent of bias (PBIAS) [28], and the root-mean-square error (RMSE) [29].

2.4. Flood Frequency Analysis

Using the calibrated model output, 50 years of annual discharge (m3/s) (1966 to 1960) was simulated and used to derive the flood frequency curve by applying commonly used probability distribution functions (the Gumbel distribution (EV1), log-Pearson III, and generalized extreme value distribution (GEV)). These distributions are widely used to estimate extreme values of available datasets [30]. Two methods were used to estimate the parameters of the distributions: first, the method of moments (MOM) technique, which is most used in Canada to estimate parameters for EV1 (Gumbel) [31] and the Pearson logarithm type III (log-Pearson III) (which is widely used by U.S. federal agencies for flood frequency analysis) [32]; second, the probability-weighted moments (PWM) method was used to calculate the generalized extreme value distribution (GEV), which is beginning to be accepted [31]. GEV and LP3 distributions include three parameters (location, scale, and shape), whereas Gumbel and normal distributions include two parameters (location and scale).

The probability density functions of the three distributions are presented in Table 2. The mathematical details regarding the above-mentioned probability distributions can be found in the reference book [33]. Kolmogorov–Smirnov (K–S) and Anderson–Darling (A–D) tests were used to assess the performance of each distribution.

Table 2. Probability density function equations.

Name	Equation	Symbols
Log-Pearson III	$f(x) = \frac{1}{\alpha * x * \Gamma(\beta)} * \left[\frac{ln(x) - \gamma}{\alpha}\right]^{\beta-1} * e^{-\left\{\frac{ln(x)-\gamma}{\alpha}\right\}}$	α = shape parameter (α > 0) β = scale parameter (β ≠ 0) γ = location parameter Γ(β) = Gamma distribution function for the parameter β.
Gumbel (EV1)	$f(x) = \frac{1}{\sigma} * exp\left[-\left(\frac{x-\mu}{\sigma}\right) - e^{-\left\{\frac{x-\mu}{\sigma}\right\}}\right]$	μ = shape parameter (−∞ < α < ∞) σ = scale parameter (β > 0)
Generalized Extreme Value Distribution (GEV)	$f(x) = \frac{1}{\sigma} * \left[1 - k * \frac{x-\mu}{\sigma}\right]^{1/k-1} * e^{-\left[1-k*\frac{x-\mu}{\sigma}\right]^{1/k}}$	σ = scale parameter (σ > 0) k = shape parameter μ = location parameter

3. Results and Discussion

3.1. SWAT Model Performance

The calibration of the model was performed using data from the Imin Lhamam and Iguir Nkouris gauging stations for the period of 2002-2012. Subsequently, the model was validated using data from the period of 2013-2016. The effect of melting snow was taken into account during the calibration stage.

Figure 2a,b show the results of the calibration and verification of the model as the hydrographs of measurements and modeling for the calibration and verification periods; Table 3 shows the respective results of the statistical indicators. The results of NSE were >0.5 and were considered satisfactory for both the calibration and validation periods (Table 3). The indicator PBIAS and RMSE for both stations (Iguir Nkouris and Imin Lhamam) presented good performance. Observing the hydrographs, the surface runoff was overestimated for a number of years (e.g., 2009); it could not be reduced because doing so would have affected the base flow. However, the model generally underestimated the runoff. A study by [34] explained that this underestimation was due to the limited number of meteorological stations or an inadequate description of the rainfall input.

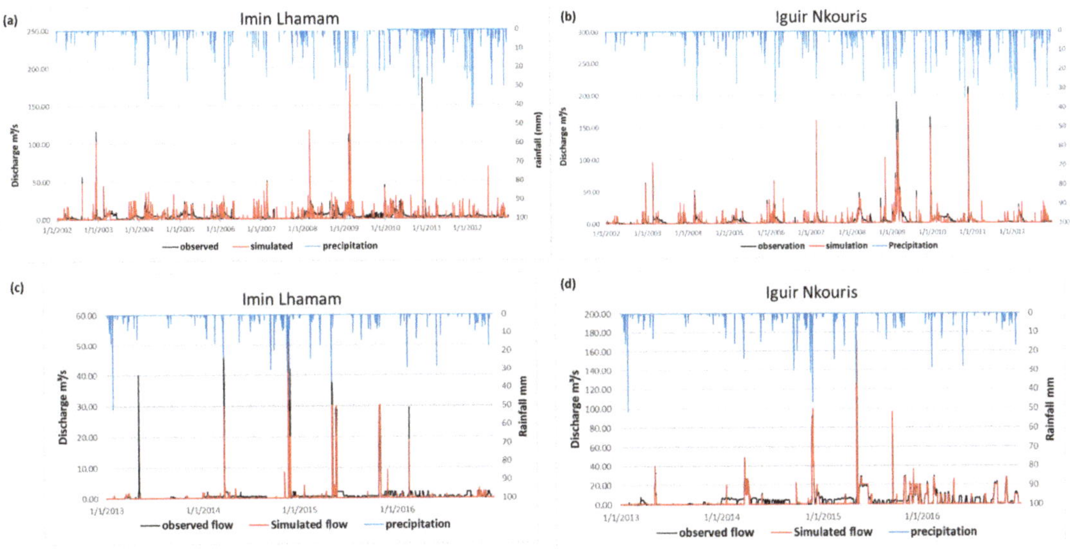

Figure 2. (**a**) Daily simulated and observed discharge (m^3/s) for the 2002–2012 period at Imin Lhamam station. (**b**) Daily simulated and observed discharge (m^3/s) for the 2002–2012 period at Iguir Nkouris station. (**c**) Daily validated discharge (m^3/s) for the 2002–2012 period at Imin Lhamam station. (**d**) Daily validated discharge (m^3/s) for the 2002–2012 period at Iguir Nkouris station.

Table 3. Results of statistical evaluation for calibration.

Parameter	Imin Lhamam		Iguir Nkouris		Condition
	Calibration	Validation	Calibration	Validation	
NSE	0.51	0.56	0.54	0.62	satisfactory > 0.5
PBIAS	16.4	15.3	22.9	21.03	satisfactory 25%
RMSE	4.00	3.25	3.72	3.12	

3.2. Flood Frequency Analysis

The return period LP3, Gumbel, and GEV results are given in Table 4. The distributions were able to provide maximum discharge for return periods between 2 and 100 years. It appears that the GEV was the most appropriate to estimate the occurrence probability of N'fis' floods. The value of the decennial return period was 303.09 m3/s, and the 100-year flood reached 949.5 m^3/s. These quantiles are very high compared to the mean discharge (5.3 m^3/s).

Table 4. Results of Return Period Analysis for Log-Pearson III, Gumbel, and GEV Models.

T	Log-Pearson III	Gumbel	GEV
2	82.3	114.8	78.5
4	187.5	231.4	163.2
8	304.4	333.2	264.9
10	343.4	364.6	303.0
20	466.0	460.1	442.7
40	588.3	553.7	623.6
80	707.5	646.5	859.2
100	745.0	676.2	949.5

The GEV distribution better approximated the relationship between the return period and discharge (Figure 3). In both tests, GEV demonstrated favorable outcomes and was found to be highly comparable to the log-Pearson III results, as evidenced by Table 5. The Gumbel distribution deviated from the norm (based on the Anderson–Darling test) and at the same time provided lower flow values for long return periods; therefore, it lagged behind the other two distributions and was considered unequable. A notable observation is that the 95% confidence interval for longer return periods was expanded for the log-Pearson III and GEV log distributions, while remaining unchanged for the Gumbel distribution.

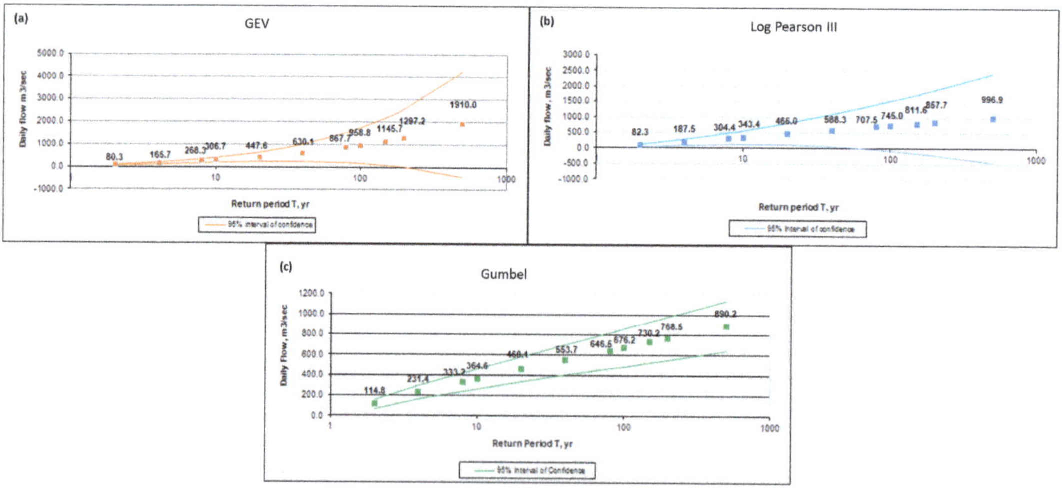

Figure 3. Flood Frequency Analysis of the N'fis Watershed (1966–2016) Based on Imin Lhamam Data (**a**) GEV (**b**) Log-Pearson III (**c**) Gumbel (EV1).

Table 5. Goodness-of-fit test results for Imin Lhamam.

T	Log-Pearson III	Gumbel	GEV
Kolmogorov–Smirnov	0.978	0.179	0.141
Anderson–Darling	−8.009	4.524	2.763

4. Conclusions

This study's main finding is that the SWAT (Soil and Water Assessment Tool) model managed to adequately simulate the ungauged N'fis watershed. The predicted values showed quite good agreement with the observed data, based on statistical criteria. The calibration process requires a long time to provide as many cases as possible to reach the best scenario. Another problem is that the SWAT model is not able to simulate single events and reach high peaks of flow during dry periods. These high peaks affect the Nash–Sutcliffe and RSR values.

Comparing the goodness-of-fit of log-normal, Gumbel, GEV, and LP3 distributions for flood frequency analysis in the N'fis watershed, the GEV distribution showed good results and appears to be the most suitable.

Author Contributions: Conceptualization, N.J. and A.M.; methodology, N.J.; software calibration/validation, N.J.; data curation, A.M.; supervision, L.B. and J.E.S.E.M.; writing—review and editing, all authors. All authors have read and agreed to the published version of the manuscript.

Funding: This research received no external funding.

Institutional Review Board Statement: Not applicable.

Informed Consent Statement: Not applicable.

Data Availability Statement: Not applicable.

Conflicts of Interest: The authors declare no conflict of interest.

References

1. Zribi, M.; Nativel, S.; Le Page, M. Analysis of Agronomic Drought in a Highly Anthropogenic Context Based on Satellite Monitoring of Vegetation and Soil Moisture. *Remote Sens.* **2021**, *13*, 2698. [CrossRef]
2. Moumen, Z.; Nabih, S.; Elhassnaoui, I.; Lahrach, A. Hydrologic Modeling Using SWAT: Test the Capacity of SWAT Model to Simulate the Hydrological Behavior of Watershed in Semi-Arid Climate. In *Decision Support Methods for Assessing Flood Risk and Vulnerability*; IGI Global: Hershey, PA, USA, 2020; pp. 162–198.
3. El Alaoui El Fels, A.; Alaa, N.; Bachnou, A.; Rachidi, S. Flood Frequency Analysis and Generation of Flood Hazard Indicator Maps in a Semi-Arid Environment, Case of Ourika Watershed (Western High Atlas, Morocco). *J. African Earth Sci.* **2018**, *141*, 94–106. [CrossRef]
4. Karmaoui, A.; Balica, S.F.; Messouli, M. Analysis of Applicability of Flood Vulnerability Index in Pre-Saharan Region, a Pilot Study to Assess Flood in Southern Morocco. *Nat. Hazards Earth Syst. Sci. Discuss.* **2016**, *2*, 1–24. [CrossRef]
5. Riad, S.; Mania, J.; Bouchaou, L. Variabilité Hydroclimatique Dans Les Bassins-Versants Du Haut Atlas de Marrakech (Maroc). *Sécheresse* **2006**, *17*, 443–446. [CrossRef]
6. Boudhar, A.; Hanich, L.; Boulet, G.; Duchemin, B.; Berjamy, B.; Chehbouni, A. Evaluation of the Snowmelt Runoff Model in the Moroccan High Atlas Mountains Using Two Snow-Cover Estimates. *Hydrol. Sci. J.* **2009**, *54*, 1094–1113. [CrossRef]
7. Amaya, A.; Algouti, A.; Algouti, A. The Use of Remote Sensing and GIS to Identify Water Erosion Risks Areas in the Moroccan High Atlas. The Case Study of the N'Fis Wadi Watershed. *Int. J. Sci.* **2014**, *2*, 43–51.
8. Gourfi, A.; Daoudi, L. Effects of Land Use Changes on Soil Erosion and Sedimentation of Dams in Semi-Arid Regions: Example of N'fis Watershed in Western High Atlas, Morocco. *J. Earth Sci. Clim. Chang.* **2019**, *10*, 513. [CrossRef]
9. Markhi, A.; Laftouhi, N.; Grusson, Y.; Soulaimani, A. Assessment of Potential Soil Erosion and Sediment Yield in the Semi-Arid N'fis Basin (High Atlas, Morocco) Using the SWAT Model. *Acta Geophys.* **2019**, *67*, 263–272. [CrossRef]
10. Zkhiri, W.; Tramblay, Y.; Hanich, L.; Berjamy, B. Regional Flood Frequency Analysis in the High Atlas Mountainous Catchments of Morocco. *Nat. Hazards* **2017**, *86*, 953–967. [CrossRef]
11. Kidson, R.; Richards, K.S. Flood Frequency Analysis: Assumptions and Alternatives. *Prog. Phys. Geogr.* **2005**, *29*, 392–410. [CrossRef]
12. Engeland, K.; Wilson, D.; Borsányi, P.; Roald, L.; Holmqvist, E. Use of Historical Data in Flood Frequency Analysis: A Case Study for Four Catchments in Norway. *Hydrol. Res.* **2018**, *49*, 466–486. [CrossRef]

13. Arnold, J.G.; Srinivasan, R.; Muttiah, R.S.; Williams, J.R. Large Area Hydrologic Modeling and Assessment Part I: Model Development. *J. Am. Water Resour. Assoc.* **1998**, *34*, 73–89. [CrossRef]
14. Aqnouy, M.; El Messari, J.E.S.; Ismail, H.; Bouadila, A.; Moreno Navarro, J.G.; Loubna, B.; Mansour, M.R.A. Assessment of the SWAT Model and the Parameters Affecting the Flow Simulation in the Watershed of Oued Laou (Northern Morocco). *J. Ecol. Eng.* **2019**, *20*, 104–113.
15. Briak, H.; Moussadek, R.; Aboumaria, K.; Mrabet, R. Assessing Sediment Yield in Kalaya Gauged Watershed (Northern Morocco) Using GIS and SWAT Model. *Int. Soil Water Conserv. Res.* **2016**, *4*, 177–185. [CrossRef]
16. Brouziyne, Y.; Abouabdillah, A.; Bouabid, R.; Benaabidate, L.; Oueslati, O. SWAT Manual Calibration and Parameters Sensitivity Analysis in a Semi-Arid Watershed in North-Western Morocco. *Arab. J. Geosci.* **2017**, *10*, 1–13. [CrossRef]
17. Sirtou, M. Etude Hydro-Climatologique Des Bassins Du N' Fis, Du Rheraya, de l' Ourika et Du Zat (Maroc). Ph.D. Thesis, Université de Metz, Metz, France, 1995.
18. Abourida, A.; Razoki, B.; Errouane, S.; Leduc, C.; Prost, J. Impact de l'irrigation Sur La Piezometrie Du Secteur N'fis Au Haouz Central de Marrakech (Maroc). *Iahs Publ.* **2003**, *278*, 389–395.
19. Neitsch, S.L.; Arnold, J.G.; Kiniry, J.R.; William, J.R. *Soil and Water Assessment Tool—Theoretical Documentation, Version 2005;* Agricultural Research Service: Temple, TX, USA; Texas Agricultural Experiment Station: Temple, TX, USA, 2005; p. 476.
20. Neitsch, S.L.; Arnold, J.G.; Kiniry, J.R.; Williams, J. *Soil and Water Assessment Tool Theoretical Documentation, Version 2009;* Texas A&M University: College Station, TX, USA, 2009.
21. Arnold, J.G.; Moriasi, D.N.; Gassman, P.W.; Abbaspour, K.C.; White, M.J.; Srinivasan, R.; Santhi, C.; Harmel, R.D.; Van Griensven, A.; Van Liew, M.W.; et al. SWAT: Model Use, Calibration, and Validation. Trans. *ASABE* **2012**, *55*, 1491–1508. [CrossRef]
22. Abbaspour, K.C.; Vaghefi, S.A.; Srinivasan, R. A Guideline for Successful Calibration and Uncertainty Analysis for Soil and Water Assessment: A Review of Papers from the 2016 International SWAT Conference. *Water* **2018**, *10*, 6. [CrossRef]
23. Mehan, S.; Neupane, R.P.; Kumar, S. Coupling of SUFI 2 and SWAT for Improving the Simulation of Streamflow in an Agricultural Watershed of South Dakota. *Hydrol. Curr. Res.* **2017**, *8*, 280. [CrossRef]
24. Hui, J.; Wu, Y.; Zhao, F.; Lei, X.; Sun, P.; Singh, S.K.; Liao, W.; Qiu, L.; Li, J. Parameter Optimization for Uncertainty Reduction and Simulation Improvement of Hydrological Modeling. *Remote Sens.* **2020**, *12*, 4069. [CrossRef]
25. Feyereisen, G.W.; Strickland, T.C.; Bosch, D.D.; Sullivan, D.G. Evaluation of SWAT Manual Calibration and Input Parameter Sensitivity in the Little River Watershed. Trans. *ASABE* **2007**, *50*, 843–855. [CrossRef]
26. Nash, J.E.; Sutcliffe, J. V River Flow Forecasting Through Conceptual Models Part I-a Discussion of Principles. *J. Hydrol.* **1970**, *10*, 282–290. [CrossRef]
27. Krause, P.; Boyle, D.P.; Bäse, F. Comparison of Different Efficiency Criteria for Hydrological Model Assessment. *Adv. Geosci.* **2005**, *5*, 89–97. [CrossRef]
28. Gupta, H.V.; Sorooshian, S.; Yapo, P.O. Status of Automatic Calibration for Hydrologic Models: Comparison with Multilevel Expert Calibration. *J. Hydrol. Eng.* **1999**, *4*, 135–143. [CrossRef]
29. Legates, D.R.; McCabe Jr, G.J. Evaluating the Use of "Goodness-of-fit" Measures in Hydrologic and Hydroclimatic Model Validation. *Water Resour. Res.* **1999**, *35*, 233–241. [CrossRef]
30. Farooq, M.; Shafique, M.; Khattak, M.S. Flood Frequency Analysis of River Swat Using Log Pearson Type 3, Generalized Extreme Value, Normal, and Gumbel Max Distribution Methods. *Arab. J. Geosci.* **2018**, *11*, 216. [CrossRef]
31. Millington, N.; Das Samiram, S.S.P. *An Integrated System Dynamics Model for Analyzing Behaviour of the Social-Energy-Economic-Climatic System: Model Description*; Water Resources Research Report; The University of Western Ontario: London, ON, Canada, 2011; p. 233.
32. Griffs, V.W.; Stedinger, J.R. Log-Pearson Type 3 Distribution and Its Application in Flood Frequency Analysis. I: Distribution Characteristics. *J. Hydrol. Eng.* **2007**, *12*, 482–491. [CrossRef]
33. Rao, A.R.; Hamed, K.H. *World Meteorological Organization Statistical Distributions for Flood Frequency Analysis;* World Meteorological Organization: Geneva, Switzerland, 2000.
34. Stehr, A.; Aguayo, M.; Link, O.; Parra, O.; Romero, F.; Alcayaga, H. Modelling the Hydrologic Response of a Mesoscale Andean Watershed to Changes in Land Use Patterns for Environmental Planning. *Hydrol. Earth Syst. Sci.* **2010**, *14*, 1963–1977. [CrossRef]

MDPI
St. Alban-Anlage 66
4052 Basel
Switzerland
Tel. +41 61 683 77 34
Fax +41 61 302 89 18
www.mdpi.com

Environmental Sciences Proceedings Editorial Office
E-mail: environsciproc@mdpi.com
www.mdpi.com/journal/environsciproc

www.ingramcontent.com/pod-product-compliance
Lightning Source LLC
Chambersburg PA
CBHW061445220926
51850CB00004B/15